M. G. Nichols

PIPING SYSTEMS: Drafting and Design

Louis Gary Lamit

Department of Technology and Occupational Education
Northern Kentucky University

Prentice-Hall, Inc., Englewood Cliffs, N.J. 07632

Library of Congress Cataloging in Publication Data

LAMIT, LOUIS GARY.
Piping systems, drafting and design.

Bibliography: p.
Includes index.
1. Pipe lines–Drawings. 2. Pipe lines –Design and construction. I. Title.
TJ930.L35 621.8'672 80-19666
ISBN 0-13-676445-2

DEDICATION

To my daughter Corina
and the enlightenment of all beings.

Editorial production/interior design: Margaret McAbee
Cover design: Judith Winthrop
Manufacturing buyer: Anthony Caruso

Printed in the United States of America

10 9 8 7 6 5

Prentice-Hall International, Inc., *London*
Prentice-Hall of Australia Pty. Limited, *Sydney*
Prentice-Hall of Canada, Ltd., *Toronto*
Prentice-Hall of India Private Limited, *New Delhi*
Prentice-Hall of Japan, Inc., *Tokyo*
Prentice-Hall of Southeast Asia Pte. Ltd., *Singapore*
Whitehall Books Limited, *Wellington, New Zealand*

CONTENTS

PART III
APPLICATION OF PRINCIPLES

14 SOLAR POWER PIPING 516

15 MODEL BUILDING 537

APPENDIX A: Standards 557

APPENDIX B: Tables, Charts, and Components 560

ABBREVIATIONS 588

GLOSSARY 592

INDEX 609

PREFACE

Piping Systems, Drafting and Design is based on the author's experience teaching piping drafting at the college level and working in industry for a number of years. Basically, it is a four-quarter course that enables students, upon completion, to enter various areas of industry that employ piping drafters and designers.

The need for piping drafters and designers seems to increase daily. Worldwide needs for multiple unit housing, basic services, and industrial products are making piping drafting and design major fields. Within the fields of power generation, oil refining, chemicals, food processing and others, 35% to 40% of a project's total cost can be attributed directly to piping costs, which include the drafting and design, construction, and materials.

Because piping has become one of the fastest growing fields open to a designer or drafter, salary ranges are considerably higher than in the parallel fields of architectural, mechanical, or structural drafting and design.

Rapid increase in the number and size of power generation systems and continuing growth in the production of petroleum products indicate the need for more engineers, technicians, and drafters skilled in these areas. Among the most important support groups for engineering services are the drafters, piping designers and modelers. Working with the design engineers to translate their analytical calculations and basic design into a project that civil and construction engineers can build requires many piping and instrumentation drafters. After the chemical or nuclear engineers have developed a production process, the mechanical engineer must specify or design equipment to mass-produce the product based upon required temperatures, fluid velocities, volume of flow, physical distances, materials that must be piped, and the like. Then the piping and instrumentation drafters must display the system in drawings of many kinds. All of these drawings are needed so that contractors can make accurate bids and so that equipment and supplies can be ordered according to an efficient schedule.

This text covers many new techniques that have been developed to aid the efficiency and accuracy of the piping drafter and designer. These new techniques increase the productivity of the people involved in the field and produce higher quality work with more accuracy and less waste than the old method.

Among these innovations are scale models, which are becoming widely used throughout the piping field as design and checking tools. In some cases a piping drafter is put to work directly on a model.

Among the many improvements covered are the simplification of graphic symbols and legends, standardization of aspects that are repetitive within the piping field, and the use of computers for material control as well as for actual graphic representation of piping systems.

The one slant in my approach which I hope will somehow bring readers beyond the scope of usual school learning, is my firm belief that what one learns should be applicable to getting and holding a job. I am concerned not only with the presentation of material for a course in piping drafting and design, but also with giving each student a feel for the industrial skills that will ease the transition to a work situation. Years of experience and observation have impressed upon me the need for a wide range of knowledge and techniques. Self-confidence is needed to put these skills into practice in the real world.

The drafting prerequisite for this course should be one or two terms of high-school drafting or one term of mechanical or structural drafting at a college or technical trade institute. The basics of drawing or drafting procedures, lettering, line work, and orthographic projection are only reviewed in this text, except where those procedures in the piping subfield differ from established practice for the mechanical drafter.

An understanding of basic algebra, trigonometry and descriptive geometry would be useful in many of the chapters.

High schools should find this book easy to use because it does not assume that the reader knows anything about piping drafting. A basic understanding of mechanical drafting skills would help, because the text only touches on the use of drafting equipment and lettering, basic line-work, and projections.

Vocational-technical, and trade [community college] courses in piping drafting, process drafting, process design, and power generation should find that the material in this text fits into their present programs or may enable them to expand into new areas and courses.

College students majoring in petrochemical engineering, industrial engineering, model building, and drafting and design will find this text an excellent reference.

Industry level drafters and modelers and those wishing to move from structural, mechanical, architectural, or electrical drafting into piping design drafting should find this text invaluable for easy transition into this ever-expanding specialty. Traditionally, a majority of pipe drafters and designers have come from these parallel fields because piping has not been offered in a wide variety of educational institutions until recently.

What follows is a general description of this text. Use of the parts and the possible selection sequence of chapters to be included in courses at various levels is discussed in the introduction.

The first part covers background and reference information. The second part covers the basic areas of piping drawing, symbols and fittings, flow diagrams and instrumentation, isometric drawing, pipe supports, structural uses of piping, and the essentials of piping design as they apply to all fields within the piping industry. These six chapters cover common elements that are similar for all piping systems.

The third part of the text introduces prospective piping drafters to the particular subfields in which they may find employment. These chapters represent piping drafting in practice in specific applications. Process petrochemical piping and design are covered in Chapter 11. Chapter 12 covers conventional power generation. Chapters 13 and 14 cover nuclear power generation and solar energy piping, respectively. The final chapter covers model building.

Throughout the text numerous examples, from simple instructional drawings to advanced drawings, are provided. Plans for power plants, both conventional and nuclear; process refineries; structural piping; pipe support designs; solar heating systems; and models are included to familiarize readers with the vast area included in the piping field. The many photographs should be a rich source of knowledge and provide readers with a feel for the various piping subfields.

Graphic guidelines that have been standardized by industry, and comprehensive coverage of design features and theories are included to instruct readers in various piping rules and procedures as to the why and how of piping systems. Based on the industrial uses of piping, the text is not only a guide and introduction to piping drawing, but also describes much valuable first-hand experience in what might be encountered "out in the field." Chapters are based primarily on information received from industrial sources.

Another important value of this text derives from the author's care in graduating the depth of the material presented, so that readers are gradually introduced to increasing depth and higher level work. The final chapters individually apply that accumulated knowledge to special areas of industry where one might find employment.

Each chapter opens with a short introduction and closes with a list of problems and assignments. The review quiz provided at the end of each chapter tests the retention of the subject matter. All problems and quiz questions are drawn from the text.

A large appendix with a glossary and catalog reprints aids readers in selecting and ordering component parts and in constructing a materials list. It is advisable for students to familiarize themselves with the particular catalogs and available equipment options of piping systems.

I have tried to include within this text those areas and uses of piping drafting that, to my knowledge, are not in print at this time. Chapter 7, on isometrics, includes traditional isometric drawings and drawing aids but also goes deeper into the area of piping-in-the-round. This type of isometric is unique to the power generation field and is used extensively in nuclear plants.

The pipe support chapter, Chapter 8, covers standard pipe supports plus design problems encountered in industry and critical service areas in petrochemicals and power generation, where pipe support and design drafting are fast becoming a separate area in and of themselves. Industry's attempts to deal with stringent and necessary codes regarding seismic and thermal considerations for pipe supports are presented in this chapter.

The book closes with a section on the budding field of model making. Students and instructors should build many of the drawing and design projects in the book. Modei making gives students a much deeper understanding of the two-dimensional world of drawing by providing the visual and tactile scope encountered in the third dimension. Models are used throughout industry to design piping and to locate equipment within and around plant installations. I strongly recommend that any piping drafting and design course include the building of a model of a piping system that has been drawn by the students. See *Industrial Model Building* by L. G. Lamit, a Prentice-Hall text.

Competent drafters need to have high quality printing and line work. A student's style and competence in printing and line work should clearly exhibit improvement during the course of the class. When inquiring about my students' qualifications, employers have told me that 50% to 60% of their hiring criteria is based upon the prospective employee's lettering on the application. Although this seems harsh, I want to emphasize that all the knowledge obtained through this and other books will be of little avail unless the student's basic drafting skills are of high quality.

Throughout engineering schools and technical trade schools, I have found a low level of social responsibility among students, instructors, and official staff. It is my hope that educational institutions become more aware of the need for a rounded education and offer short courses in environmental education, history, psychology, sociology, and other fields which give a prospective employee in industry a better understanding of the needs of society and the effects of one's work upon that society. The impact of one's life work and the ramifications of the projects with which one is engaged should be understood clearly, such as in nuclear and petrochemical industries. The nuclear field has experienced some extremely difficult times, and the petrochemical field also touches on some very marginal areas of acceptability to the environment and to the social fabric of the society.

This text represents a culmination of my years in industry and teaching. I hope you will benefit as much from its use as I have from assembling the material and writing the text. Without the cooperation of industry I could not have entertained the possibility of such a task. I sincerely thank my students, my assistants, Steven Richard Heimans and John Hitchcock, and my wife Margie. I apologize to my daughter Corina, who competed for time that I had to spend on the manuscript. Lastly I would like to thank my father, who secured my first drafting job for me at age 16, and who influenced my decision to remain in engineering after college.

I sincerely hope that those who use this book will make available to me criticisms, corrections, or comments on the text, including suggestions for future publications or revisions. Most people do not realize the amount of influence they can have on a writer.

Louis Gary Lamit

ACKNOWLEDGEMENTS

The author is extremely grateful for the help and assistance received from industrial sources. An appropriate credit line identifies all of these contributors to the text. Much of the book, including 40% of the photographs and drawings, has been derived from other large petrochemical and construction companies who, for one reason or another, have chosen to remain anonymous.

Special appreciation must be extended to Ken McIntire of Grove Valve and Regulator Company, who not only supplied a great many of the photographs throughout the book, but also took the time to edit Chapter 4 on Valves; Ron Billard of Pacific Pipe Company for help in assembling Chapter 9 on Pipe Fabrications; Ron Roberts of Automated Drafting Corporation and Brian Newick for contributions to the CAD section of Chapter 5; Steven Shelly for research material on solar piping; The American Engineering Model Society, which supplied much of the text for Chapters 10, 11, and 15, and special thanks to Charles Bliss and the Engineering Model Associates, Inc. (EMA) staff, who were so kind as to give complete use of their material, photographs, and books.

The author is also indebted to those who worked directly with the manuscript: John Hitchcock, who did the technical content editing for the whole manuscript and also contributed heavily to Chapter 13 on Nuclear Piping and Chapter 14 on Solar piping; and Steve Heimans who did much of the footwork for the first eight chapters, including the xeroxing, typing, and illustrations.

A majority of the illustrations were drawn by Pat Scheetz, John Higgins, and Steve Heimans. Sincere appreciation is conveyed to them here and to Laura Bottarini and Sharon Bolls who typed the manuscript and put up with so much for so long and did an excellent job.

Lastly, the three people who most influenced the creation

and completion of this text: my wife, Margie, who unselfishly weathered this project; Jere Kinnan of Prentice-Hall, who originally planted and nurtured the idea of doing and finishing this book; and my editor at Prentice-Hall, Hank Kennedy, who returned the best possible contract in an unbelievably short time.

Louis Gary Lamit

PART I

Basic Background Information

Chemical plant installation: two gear-operated ball valves installed on a steel pipeline by means of welded flanges and fittings. (Courtesy of Grove Valve and Regulator Company)

1 PIPING AND INSULATION

Most people do not recognize the many refinements of modern life that depend on the existence of pipes and their related components. Without pipes and piping, there would be no plumbing. There would be no gasoline to drive cars, to heat homes, or to generate electricity (Fig. 1-1). These are the obvious uses, but in modern society almost every product is processed with the use of piping: chemicals, foods, beverages, drugs, energy itself. The elimination of wastes in these industries, and in the community at large, deserves special mention with regard to piping.

Industrial growth in any modern or emerging society is made necessary by increasing population and the corresponding need for jobs, services, and housing. Along with this growth there is a per capita increase in the use of energy and industrial products. The use of piping in new industrial complexes and manufacturing centers is more important than ever with as much as 40 percent of the total cost of the project going into piping. This use applies not only to refineries and chemical plants but also includes power generation, food processing, office buildings, and housing.

Piping, as we can see, is an indispensable component in modern society, yet most men and women know little about pipe sizes, schedules, fittings, valves, and the like. A former student, after taking piping drafting for one quarter, began to realize the vastness of piping applications. He noted that his 700-unit apartment building was a maze of piping, valves, drains, and tanks in the underground garage and concluded that "you can't have four walls without piping." Piping can be viewed as a building's bloodstream and digestive system. It is just as essential as electricity to the functioning of modern society.

It is important for students of piping to familiarize themselves with all the forms of piping systems—not just to read and complete piping drawings, but to start noticing all the

FIG. 1-1 Model of fossil fuel steam generation power plant.

various uses and applications of piping that influence our daily existence. Take time to explore the many piping installations in your home, in your apartment building, at school, on the job. Noting the infinite variety of piping configurations, valves, and equipment placements and how the systems are supported will give you firsthand knowledge that is far more valuable than book learning. Get a good look at the uses of piping: how it is installed, what fittings are used (welded, screwed, flanged), how the flow of pipe contents is controlled, what valves and meters are used. This firsthand understanding of piping will become invaluable in your work as a drafter or piping designer.

Because of vast changes in society and the corresponding needs of industry, the piping designer and drafter have increasingly important roles to play in the total construction effort. Many new and efficient methods of piping design and drafting are coming into common use—computer ordering and computer graphics, international standardization, specialized symbols and models, as shown in Fig. 1-2, among others. The person employed as a designer/drafter in the piping field should know the basics of piping throughout industry, even if they are not used in his or her immediate job, for precise symbols must be employed to distinguish components from those used in other areas.

HISTORY

Archaeologists have discovered the use of pipe in civilizations ranging back thousands of years. In ancient times, pipe materials were limited to natural substances and local forms. Thus bamboo, wood, and stone were used early on. Eventually, with the ability to smelt metals, society turned to iron, brass, copper, and lead. Today aluminum, magnesium, titanium, and highly specialized metal alloys have been added to the available piping materials. Piping also comes in concrete, clay, glass, plastic, rubber, or a combination of materials. Some piping is insulated, lined, or reinforced, depending upon its intended use.

FIG. 1-2 Model of a petrochemical installation.

Clay piping that dates back to about 4000 B.C.E. was found in the ruins of Babylon, where the king had a bathroom constructed with tile drainpipes. Both the Greeks and the Romans experimented with the use of wood and lead, although wood tended to deterioriate quickly and lead tended to poison certain substances that were carried within it. A lead pipe system was uncovered in the ruins of Pompeii, complete with bronze plug valves just as they were in 87 B.C.E. when the city was covered with volcanic ash. Wooden pipes made from logs and staves have been found in many parts of the world and are still used in the western United States. Both the Romans and the Carthaginians transported their water supply through piping systems as early as 800 B.C.E.

The invention of the cast iron cannon brought about the eventual adaptation to cast iron pipe for water conduit. Evidence shows that cast iron pipe was used in Germany as early as 1400 C.E. The French utilized a cast iron pipe system to supply water for their fountains at the Palace of Versailles in 1660 C.E. Before 1800, the United States imported all its metal piping from England. The first U.S. foundry was established in 1834 in Millville, New Jersey. The first use of piping systems in America was for the carrying of water, but as homes were built closer and closer together, the need for adequate plumbing and sewer systems became quite evident. Although iron pipes were used throughout Europe, iron was scarce in America, so the first pipelines laid in Boston, Philadelphia, and New York used bored logs in their construction as early as 1752. Many of these lines were still in use in the late 1800s. In certain applications, wooden pipelines are still used to carry substances for which other materials are inadequate.

The first strong but economical pipe was made in 1825 when a method of fabricating pipe from long strips of hot metal was devised. The discovery of the Bessemer process in 1855 and the open hearth process in 1861 were important steps in the sophistication of piping manufacture—now steel became widely available. With the invention of the steam engine, the need for strong high-pressure and high-temperature piping materials became acute.

Many of the pipes laid in the eighteenth century have given long and trouble-free service. In San Francisco, as early as 1863, a 5-mile-long water supply line was installed and parts of it are still in use today. In 1870 riveted lines were installed and later, in 1890, lines of welded steel pipe. Records show that many of these pipes have been in service for more than 50 years. These records attest to the tremendous durability of steel, especially since steel pipes at that time were mostly laid without the protective coatings that are available today.

In post-World War I America, the power and process industries were beginning to expand greatly with the eventual need for high-pressure and high-temperature materials. Seamless pipe formed from solid billets of steel then became common. Steel pipe began to see use in the late eighteenth century and by the early twentieth century had become the predominant type of pipe, though the valves and fittings were manufactured from cast iron. Piping joints and methods of connecting pipe to valves and other fittings conformed to the established practices of the seventeenth and eighteenth centuries. Bolted flange and threaded connections were still used throughout the industry.

Some of the worst problems encountered with the construction of high-pressure and high-temperature systems involved the use of unreliable joining methods. Now welded joints have become the norm for extreme service (see Chapter 2). The first portable welding method used oxygen-acetylene torches, which are still used in some shops today though the method is primarily used for cutting pipe. Many welding processes, especially electric arc types, have been invented within the last thirty years, enabling pipe and joints to be fabricated for safer high-pressure and high-temperature service.

More efficient power plants, both nuclear and fossil fuel, and the expansion of the petrochemical industry throughout this century, have required new materials and designs for higher and higher temperatures and pressures. In 1935 the American Standards Association (ASA) code for pressure piping was established to standardize piping valves and fittings throughout the industry. Many different codes have been added to the original ASA code to cover the many fields that have sprung up since 1935. Throughout the last 30 years, the piping industry has slowly moved toward standardization so that pipes, valves, fittings, and other components are readily interchangeable even though manufactured by different companies. Eventually there will be international standardization of all piping materials and components enabling worldwide distribution and use.

A hundred years ago, water and sewage were still the main fluids conveyed by piping from one point to another. Today almost every conceivable fluid and many slurries are carried through piping. With the advent of atomic energy plants as shown in Fig. 1-3, and rocket power, the use of piping has expanded to the transportation of liquid metals and extremely cold liquefied gases in addition to common fluids such as gases, oils, acids, drugs, and beverages. Thin-walled piping or tubing is also used for hydraulic and pneumatically operated control mechanisms.

Through the use of catalogs and descriptive material from pipe and component manufacturers, the drafter and designer are able to select the proper materials and parts for the design at hand. Wall thickness, size, shape, and material all depend on the application. The drafter and the designer should be familiar with all aspects of piping, proper methods of representation, and how to order pipes, fittings, valves, and other components. Manufacturer's catalogs, standards, and listings, are essential to the piping drafter.

FIG. 1-3 Nuclear power plant under construction.

BASIC METHODS OF MANUFACTURING PIPE

Most pipe is still made of cast iron—either centrifugally cast in metal molds or poured in sand-lined molds. In centrifugally cast iron pipe, molten metal is spread through the center of a rapidly spinning assembly to produce a homogeneous pipe of cylindrical form and a constant wall thickness manufactured in normal lengths of 18 ft (5.51 m); this type is created from metal molds. In the sand-lined molds, molten metal is poured in the end of a rapidly spinning horizontally positioned flask. The opening of one end holds a baked oil-sand core which extends into the mold just far enough to shape the socket end of the pipe. This process permits casting of units of 16½ and 18 ft (5.0 and 5.48 m) lengths, the most commonly manufactured.

Various methods are used to produce welded pipe (Fig. 1-4). With electric weld pipe, a coil of metal (thin sheet metal) is fed into an uncoiling unit and then into a forming section which begins the transformation of the strip of steel into a round pipe section. After the flat steel has been formed into a U shape, it passes through a section of forming rolls which finishes the rounding process and also contours the edges of the strip for the eventual welding process. An automatic, high-speed, continuous welding operation takes place at 2600°F (1400°C) at the fusion point. Pressure is also exerted on the heated edges to help form the fusion weld. After this process, the weld is inspected and goes to a seam normalizer where the weld area is given postweld heat treatment to remove the stresses created by welding and to produce a normal grain structure in the piping wall. The pipe is cooled by air and water before it is sized. In the sizing mill, the pipe passes through a series of rollers which produce the proper outside diameter and straighten the pipe at the same time. The pipe is then cut into sections of a designated length.

In an alternative continuous welding method—standard weld pipe—the coil of flat metal is fed into rollers which flatten the metal and trim the edges square. The coil then travels to a furnace where it is heated to approximately 2500°F (1370°C) and then passes through a series of forming rolls that bend it into an oval with the opening downward. It proceeds from there through the welding area where both heat and pressure are exerted by the rolls to form a fusion weld. Then the pipe passes through a stretch reduction mill where the outside diameter (OD) is reduced and wall thickness is standardized. The pipes are then cut to predetermined rough lengths and travel to the outside diameter sizing mill. After this, the pipe is inspected, cut to finished lengths, and straightened.

To produce double submerged arc weld standard pipe, a flat piece of metal, rectangular steel plate, is rounded through a series of presses and then put through the welding process to form a fusion weld. This process, described in greater detail in Chapter 2, is used predominantly for the fabrication of large-diameter pipe (Fig. 1-5).

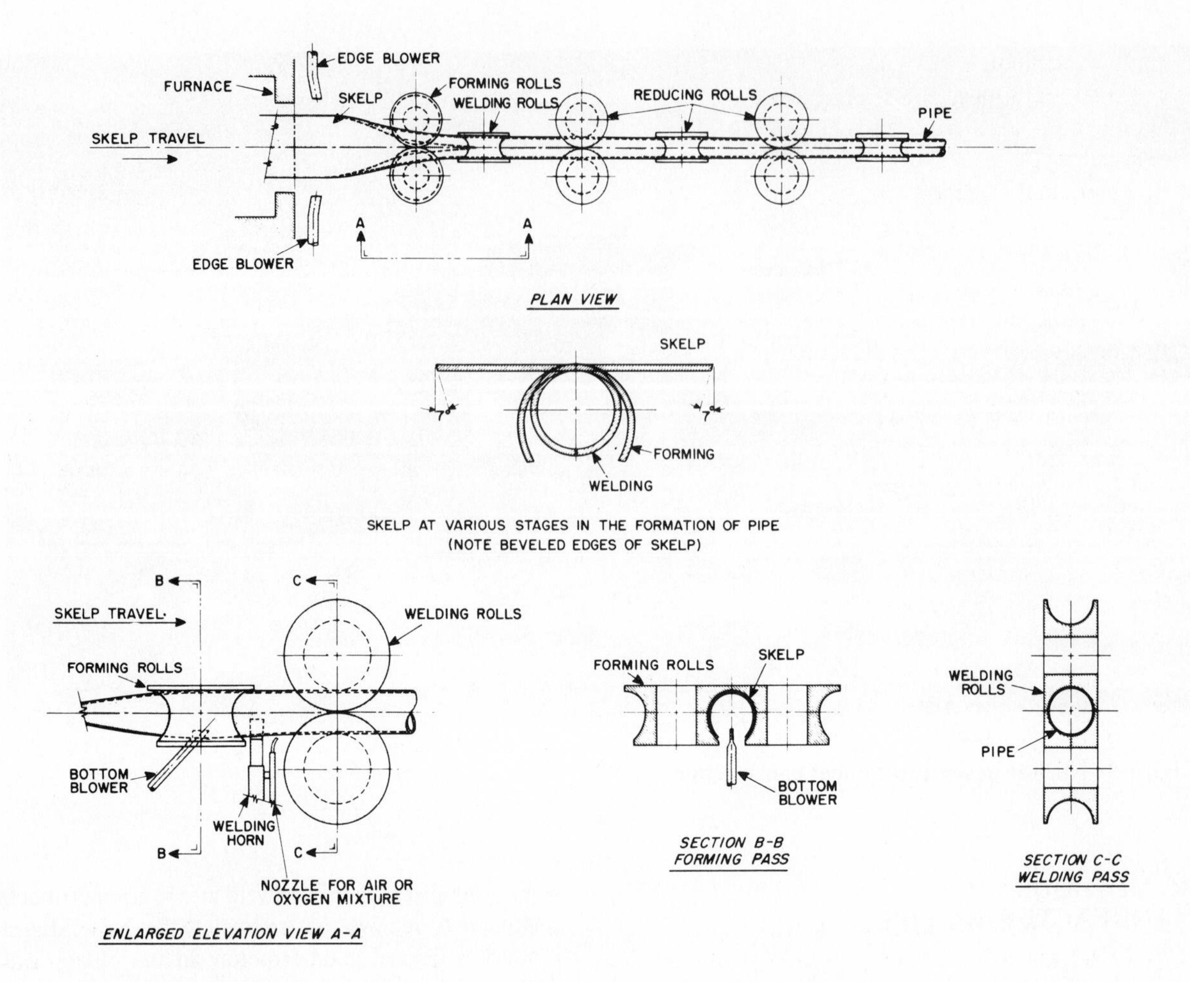

FIG. 1-4 Operations performed in a continuous forming and welding mill. (Courtesy of United States Steel Corporation)

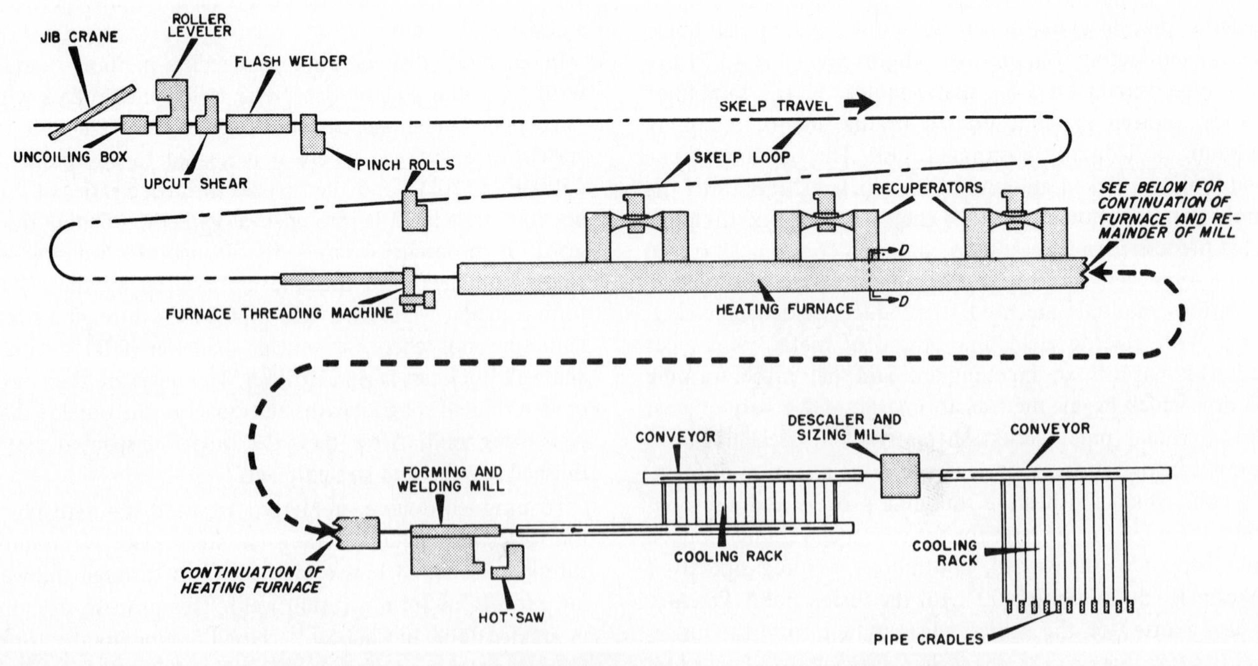

FIG. 1-5 Layout of a typical continuous butt-weld pipe mill. (Courtesy of United States Steel Corporation)

Another common process for manufactured pipe is the seamless standard pipe process (Fig. 1-6). In this type, solid metal billets are conditioned by a machine similar to a lathe and then brought to a proper temperature for piercing. In the piercing mill, the billet is gripped by rolls that rotate and advance it over the piercing point, forming a hole through the length of the billet. Large sizes must go through more than once. From the piercing mill, the pierced billet is put through a plug-rolling mill where the billet is rolled over a mandrel to reduce the diameter and wall thickness and to increase the length of the pipe. A rotary mill (shown in Fig. 1-7) is used for 16 in. (40 cm) diameter and over, the diameter is enlarged, and the wall thickness is reduced to approximate finished dimensions. From there the pipe passes through the sizing mill on a series of rolls that create a true round shape and the diameter required. The pipe then passes through a straightener and on to the hydrostatic testing and inspection stations.

Figure 1-8 shows another method of manufacturing pipe: the reeling machine. Although there are many other processes for manufacturing metal piping, the ones described here are the most common.

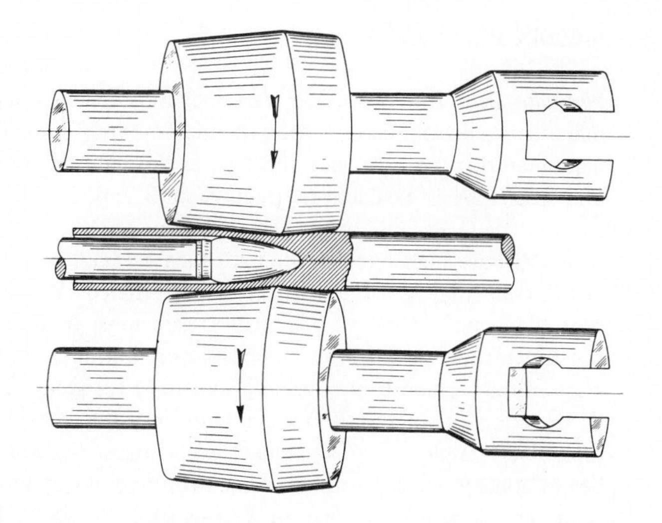

FIG. 1-6 Action of different parts of Mannesmann piercer in the piercing of a solid billet. (Courtesy of United States Steel Corporation)

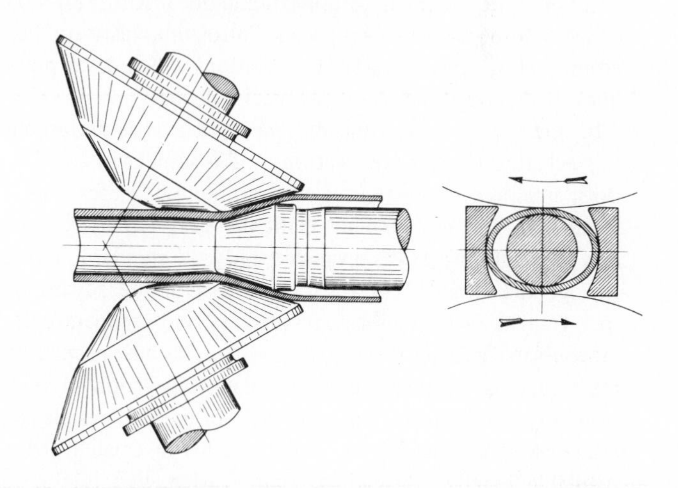

FIG. 1-7 Rotary rolling mill. (Courtesy of United States Steel Corporation)

PIPING MATERIALS

The service for which a pipe is to be used determines the material and the manufacturing and assembling operation. Pipe can be made from a variety of materials, including glass, rubber, plastic, wood, aluminum, clay, concrete, combinations of metals, copper, lead, brass, wrought iron, and steel. Special pipes are also manufactured with exotic materials such as titanium, with aluminum and steel alloys such as stainless steel, and with plastic and space age materials. These materials are used for special situations, and their physical properties and temperature and pressure limitations must be taken into account along with their cost. Because service conditions vary so widely regarding heat, strain, pressure, and chemical reactivity, the selection of pipe becomes an important factor in the design of the total system. In general, cast iron, wrought iron, and steel are the most common materials still used today although there are about 40 metallurgical materials on the market.

The chief variables that must be taken into consideration in the selection of pipe are corrosion, cost, pressure, temperature, and safety. With corrosion, the problem cannot be solved by manufacturer's recommendations alone because the conditions under which the system will be operating can affect its corrosion properties even if the same fluid is being transported. It is usually up to the engineers and designers to choose materials for the project. Though cost is always a consideration, it must never take precedence over the safety conditions necessary for high-temperature and high-pressure service.

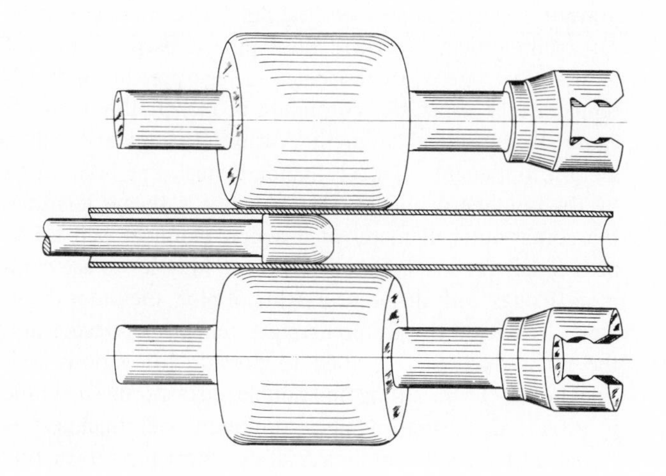

FIG. 1-8 Reeling machine. (Courtesy of United States Steel Corporation)

Metallic Pipe and Tubing

Wrought Iron and Steel The vast majority of pipes are manufactured from carbon steel; chemical composition and manufacturing methods vary throughout the industry. Carbon steel pipe is available in grade A or B and, in a few cases, grade C. Because of the higher carbon content, grades B and C have higher tensile strength than grade A but are less ductile—that is, more brittle. Although the cost usually runs the same for the various grades, care must be taken not to substitute grade A for applications requiring grade B steel (though grade B may usually substitute for grade A). Other common compositions are wrought iron, chrome, moly steel, stainless steel, and nickel steel along with a number of nonferrous metals that are manufactured in iron pipe sizes. The American Society of Testing Materials (ASTM) publishes most of the specifications for piping materials. A great majority of grades recognized in ASTM specifications are approved also by the American Standards Association (ASA) and the American National Standards Institute (ANSI) codes. Moreover, the American Petroleum Institute has established its own material specifications which cover pipelines for the oil and gas and chemical industries.

In highly corrosive situations, wrought iron is preferable to steel although it is not as strong and is somewhat costlier. Both steel and wrought iron are commonly used for water, steam, oil, and gas services. Until the early twentieth century wrought iron was available only as standard, extra strong, and double extra strong, but with the advent of steam services of increasing pressures and temperatures, more diversified wall thicknesses and sizes became necessary. Steel pipe is obtained in either black pipe form or, less commonly, the galvanized type. Both of these materials are available in lengths up to 40 ft (12.2 m) in smaller sizes. Available lengths decrease with increasing wall thickness and size. A majority of pipe comes in random (±20 ft) or double random (±40 ft) lengths.

Steel pipe which has been treated with molten zinc to prevent rust is called *galvanized pipe.* This pipe is suitable for drinking water services. Steel pipe is used throughout industry because of its resistance to high pressure and high temperatures, especially chromium or nickel alloy pipe because of the added strength. Various fittings and valves are also obtainable in steel. Standard steel pipe is specified by the nominal diameter. This diameter is always less than the actual outer diameter of the pipe. Nominal wall thickness equals the average wall thickness. In order to use common fittings with different weights of pipe, the outer diameter of each of the different pipes remains the same and the inner diameter is varied to provide for various wall thicknesses. Extra strong and double extra strong steel pipe is produced in this manner. Minimum wall thickness is found by taking the nominal wall thickness and subtracting the mill tolerance—12½% for seamless and 0.01% for pipe made from plate stock.

Cast Iron Cast iron pipe is used mainly for the transportation of water or gas or, in some cases, for soil pipe—applications used extensively in underground piping systems. For these applications the push-on joint is used, which makes installation in a trench easier than screwed, flanged, or welded pipe. Generally, water or gas pipes are connected with bell and spigot joints or flanged joints. Soil pipe is available from 2 to 15 in. (5.0 to 38.1 cm) in diameter, and in lengths of 5 ft (1.5 m), and in various weights. Soil pipe is almost exclusively manufactured with bell and spigot joints, though threaded joints are sometimes available. Special external loading conditions for buried installations (including ground settlement), along with internal pressure, must be considered by the designer when using cast iron pipe because such loads can fracture the pipeline if it is not sufficiently flexible.

Cast iron has been shown to have an excellent resistance to destruction by external loads when compared to wrought iron or steel, although it cannot be used under conditions of high temperature, high pressure, or vibration. For cast iron, steam temperatures should not exceed 450°F (230°C) and oil not over 300°F (150°C). Although the first years of installation show progressive corrosion caused by rust, the rate of penetration decreases considerably after a number of years of operation.

The nominal size of cast iron pipe, unlike steel pipe, always indicates the inside diameter (ID) regardless of size. When ordering cast iron pipe, one must indicate wall thickness and outside diameter along with nominal size because terms such as strong and extra strong are not used for cast iron pipe. Cast iron piping is available in a variety of standards and weights when manufactured with bell and spigot joints and in sizes 1¼ to 12 in. (3.17 to 30.48 cm) with threaded ends.

Brass and Copper Brass and copper pipes are manufactured in two weights: extra strong and regular. They are available from sizes 1/8 in. to 12 in. (0.31 to 30.4 cm) and are similar in wall thickness and inside diameter to steel pipe classified as strong and extra strong and standard. The outside diameters of brass and copper pipe are exactly the same as the outside diameters of the corresponding nominal sizes of steel pipe. Because of their cost, brass and copper pipes are used only for services where longer life expectancy will compensate for higher price.

Pipe made from these compositions resists many of the chemical solutions in the pulp and paper and chemical industries. Red brass pipe has a composition of 85 percent copper and 15 percent zinc alloy and has a greater allowable stress value than copper at temperatures up to 300°F (150°C). At 400°F (204°C) these stress values become equal at 3000 psi. Standard piping codes do not permit the use of brass or copper pipes at temperatures over 400°F. Brass and copper pipes are available in straight lengths of up to 12 ft (3.6 m) and should be joined with fittings of copper-base alloys. Both copper ASTM B42 and brass ASTM

B43 pipe are available in only two grades: regular and extra strong. They are very similar to schedule 40 and 80 steel pipes, respectively. This material, when used in piping runs, must be supported at more frequent intervals than other piping (which is usually stronger). When soft copper tubing is arranged in parallel runs, a trough construction is used to ensure continuous support of the piping.

Copper Water Tubes Copper water tubes are used for waste and vent lines, as well as for heating services and general plumbing. They are available in three weights only: type K, type L, and type M. One should consult the manufacturer's catalog for weights and sizes. Type K and L are available as hard or soft tubes; type M is available as a hard tube only. Hard-tempered tubing is much stiffer than soft tubing and is used where extra rigidity is necessary. Soft tubing is used in situations that require frequent bending without the use of fittings. Copper tubing is available in lengths up to 20 ft (6.09 m) or, in soft tubing, in coils up to 60 ft (18.28 m) long. The ASTM B88 and B251 specifications give the dimensions for copper tubing. The actual diameter is consistently 0.1258 in. (0.3195 cm) greater than the normal size in copper water tubing, which is available from 1/8 in. to 12 in. (0.31 to 30.4 cm) in diameter. The wall thickness is considerably less than that of other piping materials. When specifying the size of tubing, one must specify wall thickness and outside diameter.

Flexible or soft tubing is used in many situations where equipment vibration is present. Tubing whose thickness corresponds to standard steel pipe is referred to as *pipe* instead of tubing and should be avoided in systems that experience severe vibration, high temperatures, and excessive stress.

Seamless brass pipe is used for water condenser tubes and heating lines, but it is expensive. Copper tubing is used in plumbing and heating where vibration and misalignment are factors and in automotive, hydraulic, and pneumatic design. Copper and brass are also excellent for handling liquids containing salts, though because of the extra expense they are used only in special situations. If a copper water system is to be connected to a steel system, special insulating fittings must be used; otherwise, electrolytic corrosion will quickly destroy the fittings.

Lead Lead pipe or lead-lined pipe resists the chemical action of many acids. Therefore this type of pipe is found in chemical work and in systems that transport acids.

Aluminum Aluminum, with a weight one-third that of steel, is used throughout the piping field for industrial applications. Since temperature affects its strength, this material's design capabilities must be thoroughly examined before it is used in high-pressure and high-temperature situations. Many alloys of aluminum are used in piping and for pipe fittings. The manufacturer's catalog of specifications, weights, and alloy composition should be consulted if aluminum piping is considered for use—there are vast differences in manufacturing details throughout the industry.

Alloys In highly corrosive service involving high fluid velocity, shock (thermal and mechanical), and high temperatures and pressures, the use of alloy piping has increased in recent years. Certain alloys are able to resist corrosion even at high temperatures, which generally increase a fluid's chemical activity. It is always important to consider the solution, its chemical concentration, and its velocity when deciding on piping material. In many cases the alloy's resistance to chemical compositions stems from the formation of a surface film which, when the velocity of the fluid is high, may prove unsatisfactory because the fluid will scour the protective film. Thermal shock and rapid fluctuation of temperature also affect this protective coat.

Stainless Steel A common alloy used throughout the nuclear field and in other special situations is stainless steel—a class of alloy highly resistant to corrosion primarily because of the 10 percent or more of chromium added to the steel. This pipe material is extremely resistant to oxidizing solutions. The ability of stainless steel to resist corrosion is directly proportional to the chromium content. The American Iron and Steel Institute codes divide stainless steels into chromium nickel types (the AISI 300 series) and straight chromium types (the AISI 400 series).

Titanium One of the more recent exotic materials used throughout the piping field in special services is titanium, which is found to be of great help in the chemical and pulp industries. There are a number of alloys containing aluminum and also chromium magnesium which, when added to titanium, change the material's properties for specific applications.

Nonmetallic Pipe and Tubing

Nonmetallic pipe and tubing are used predominantly for low-pressure and low-temperature services. The major types include plastic, glass, clay, and wood pipe.

Plastic Plastic pipe is one of the most commonly used nonmetallic materials. Plastics offer considerable savings in initial expense and cost of support systems and have proved to be resistant to many corrosive chemicals. It is important to consult the manufacturer's catalog when utilizing plastic piping because of the varying strengths and resistances of the many different types on the market today. Fluorocarbon plastics are considered the most resistant to attack by acids, alkalies, and organic compounds. Many organic compounds, such as petroleum products and chlorinated hydrocarbons, are readily handled by some, but not all, of the plastic materials. Plastics can also be used as liners in high-pressure and low-temperature service in steel pipe. Synthetic rubbers are also used to line steel pipe. Major drawbacks of plastic piping are that it deteriorates in direct sunlight and cannot be used for high-temperature service. It is important to weigh the initial cost and corrosion resistance of a piping system against the eventual cost of replacement. Because of the industry's new exotic materials, many plastics and other molten products provide excellent strength and rigidity compared with earlier materials. Polyvinyl chloride (PVC) piping was first produced in Germany and since then has become one of the most widely used forms of plastic piping. It is used predominantly for intermediate-strength and high-flexibility applications, for water services, or for waste piping where temperatures and pressures are not extreme. This type of piping is considered very tough and exceptionally resistant to chemical attack. PVC pipe is manufactured by extrusion; the fittings and flanges are manufactured by injection molding. Plastic pipe, in general, is available in iron pipe sizes. The manufacturer's catalog should be consulted when ordering.

Glass Glass resists many acidic materials and can be found throughout the chemical industry, though it has a 400°F temperature limitation in most instances and is vulnerable to pressure and vibration. This type of piping is used predominantly in the chemical, beverage, pharmaceutical, and food industries where corrosive contamination of the line contents is undesirable. Glass piping can also be found in manufacturing plants such as paper mills and textile plants.

Clay Clay piping has been used for sewage, industrial waste, and water storage in diameters from 4 to 36 in. (10.1 to 91.4 cm). Clay pipe is manufactured predominantly from clays and shales in combination. High-temperature firing of the piping clay produces a strong, chemical-proof composition that has been used with exceptional success throughout the industrial waste and sewage field. It can carry away every known chemical waste with the exception of hydrochloric acid, which is not discharged through normal sewage channels.

Wood Wood pipe has been used throughout history for the transportation of water. It is manufactured by boring holes through solid logs and is used predominantly for transportation of municipal water supplies, sewers, mining, irrigation, and hydroelectric power developments. It is found mainly in the western states where timber is plentiful. Douglas fir and redwood are the most common types of wood pipe, but excessive logging has made redwood extremely rare and expensive.

Continuous-stave wood piping is made in diameters as large as 20 ft (6.09 m) and is shipped unassembled and then erected at the site. The wood is chemically prepared with creosote to ensure against chemical attack and fungus. Continuous-stave wood piping is generally made of Douglas fir bound by steel or iron bands. Certainly wood piping has its positive aspects: It acts as its own insulation; it is not affected by corrosion; and it is affected only slightly by vibration. Wire-wound wood pipe can be purchased in sizes up to 24 in. (60.9 cm). This factory-made product is wrapped with wire instead of steel bands to retain the staves.

Wood pipe is often used in underground installations after it has been treated to withstand rotting. It is also possible to obtain wood-lined steel pipe for service where high pressures make wood pipe unsatisfactory but chemical resistance is needed. This type of pipe is used throughout the paper and pulp mill industry.

AMERICAN STANDARDS ASSOCIATION

Standards and Codes

The American Standards Association (ASA) is a conglomeration of five major engineering societies that joined together in 1918. Within it are more than 100 specialized technical organizations that legally provide services to industry—including the standardization of many different aspects of piping and design. The federal government also issues many standards and works closely with the ASA to promote standards for new products.

Standardization enables companies in an industry to use interchangeable parts without worrying about each other's peculiar manufacturing details. The American Society for Testing Materials (ASTM), the American Society of Mechanical Engineers (ASME), and the American Waterworks Association (AWWA) have established most of the standards throughout the piping field. Eventually, the adoption of the metric system will allow worldwide standardization. (See the appendix for ASA standards.) One of the frequently used codes for pressure piping is ASA B31.1. This code gives minimum design requirements which enable the drafter/designer to make decisions that are compatible with other areas in the piping industry. There are many standard organizations throughout Europe and Canada. For piping that is meant for foreign application, consult these other standard organizations for their piping codes.

All pipelines in a drawing are usually identified by a line and specification number. Chapter 6 discusses methods for listing pipelines. The pipeline schedule (or *list callouts*) specifies the line size, the schedule number, the temperature, pressure, and insulation requirements, the painting or color code, and special specification numbers along with the line's termination point, drawing sheet, and material takeoff procedure.

Pipe Sizes, Dimensions, and Schedule Numbers

Any tubular product that has sizes governed by the American Petroleum Institute Standards (APIS) is referred to as *pipe*. The pipe's dimensions are set by the American National Standards Institute. Wall thickness varies with schedule number, but the outside diameter remains constant for

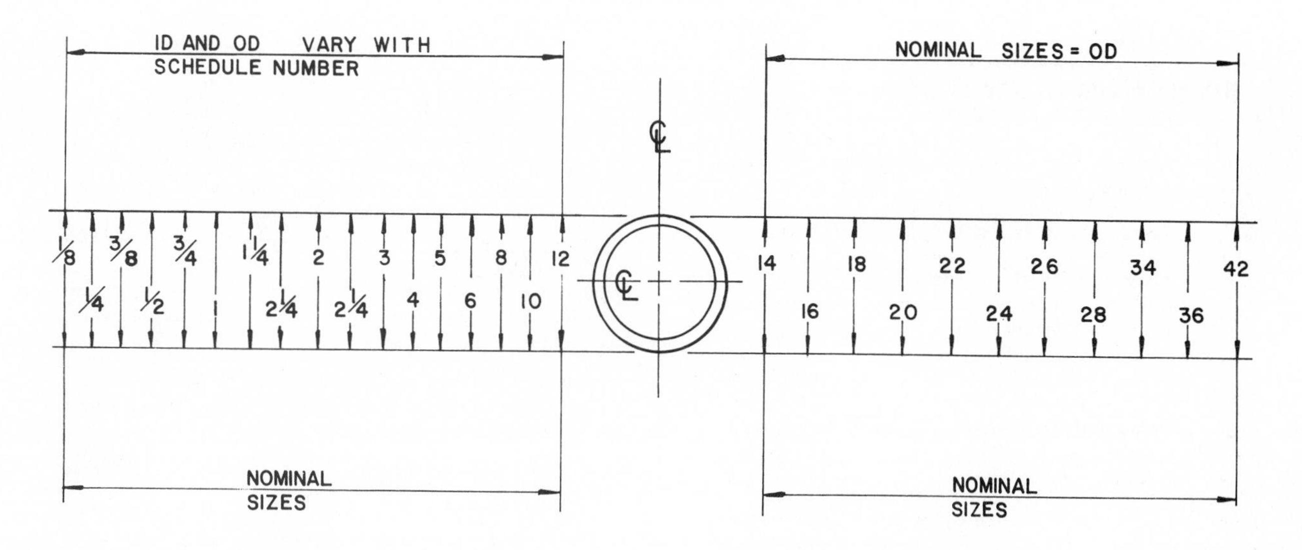

FIG. 1-9 Relationship of nominal size to inside diameter, outside diameter, and schedule number.

14 in. (35.5 cm) diameter and over (Fig. 1-9). As the thickness changes, the inside diameter is altered. To call out a pipe, it is necessary to give both the schedule number and the weight/strength designation for both pipes and fittings. All pipes under 14 in. (35.5 cm) in diameter are designated by the nominal inside diameter and schedule number. Those over 14 in. (35.5 cm) are designated by the actual outside diameter and thickness of walls. In Fig. 1-9, pipe sizes from 1/8 in. to 12 in. (0.31 to 30.4 cm) are shown by the nominal inside diameter (which is different from the actual inside diameter). The inside diameter and the outside diameter will vary with the schedule number from 1/8 in. to 12 in. (0.31 to 30.4 cm) diameters and the nominal size will equal the outside diameter from 14 to 42 in. (35.5 to 106.6 cm).

During the early years of pipe manufacturing, the walls of the smaller sizes were too thick and manufacturers, in correcting this error, took the excess from inside diameter, thus avoiding changes in fittings. To distinguish pipe sizes from actual measured diameters, the terms IPS (iron pipe size) or NPS (nominal pipe size) are usually used. The outside diameter will remain constant or relatively constant for all pipe sizes. Variations in wall thickness usually affect only the inside diameter. To distinguish weights for pipe, three traditional designations are used—standard wall, extra strong wall, and double extra strong wall (SW, XS, XXS). In some cases extra heavy wall (XH) and double extra heavy wall (XXH) are used instead of extra strong wall and double extra strong wall, respectively.

The American National Standards Institute, sponsored by the ASTM and the ASME, published ANSI B36.10 to standardize pipe dimensions throughout the industry. This publication also contains the broadened range of wall thicknesses available. Schedule numbers from 10 through 160 were adopted for steel pipe. These schedule numbers are indicative of approximate values of 1000 times the pressure/stress ratio. Stainless steel schedule numbers from schedule 5S through 80S are published in ANSI B36.19 for sizes up to 12 in. (30.4 cm). (The additional S pertains to stainless steel.) Refer to Table 1-1 for schedule number and pipe sizes for steel and stainless steel pipes.

The ASTM tolerances on regular pipe products specify that wall thickness should not vary more than 12½ percent under the nominal wall thickness that is specified for regular mill-rolled pipe. Pipe is usually supplied at random length from 16 to 22 ft (4.8 to 6.7 m), though it is possible to get lengths cut to order at a higher price. When specifying pipes and schedule numbers, the drafter should follow closely the recommendations of the design engineer. Besides piping, tubing is also manufactured and controlled by certain specifications. Any round tubular products that are not in standard pipe sizes are called *tubes* or *tubing.* Tubing is designated by an outside diameter for the basic size when a variety of inside diameters is offered. It is important to specify the outside diameter, weight per foot, and wall thickness when ordering tubing.

Calculation of Schedule Numbers

The ASA standard schedule numbers which specify wall thickness can be calculated by using the formula $1000 \times p/s$. Here p is equal to the internal pressure of the pipe (pressure in pounds per square inch) and s is the allowable fiber stress (in pounds per square inch). (Consult manufacturer's table or Table 1-2 for the s value.) As pressure increases, so does the pipe thickness requirement. The temperature of the line medium, besides putting thermal stress on the system, will also affect the pipe thickness. The s value takes into account temperature, pressure, and material.

TABLE 1-1 SCHEDULE NUMBERS AND PIPE SIZES FOR STEEL AND STAINLESS STEEL PIPE
(Reprinted courtesy ITT-Grinnell)

properties of steel pipe

The following formulas are used in the computation of the values shown in the table:

† weight of pipe per foot (pounds) $= 10.6802t(D-t)$
weight of water per foot (pounds) $= 0.3405d^2$
square feet outside surface per foot $= 0.2618D$
square feet inside surface per foot $= 0.2618d$
inside area (square inches) $= 0.785d^2$
area of metal (square inches) $= 0.785(D^2-d^2)$
moment of inertia (inches[4]) $= 0.0491(D^4-d^4) = A_m R_g^2$
section modulus (inches[3]) $= \frac{0.0982(D^4-d^4)}{D}$
radius of gyration (inches) $= 0.25\sqrt{D^2+d^2}$

A_m = area of metal (square inches)
d = inside diameter (inches)
D = outside diameter (inches)
R_g = radius of gyration (inches)
t = pipe wall thickness (inches)

† The ferritic steels may be about 5% less, and the austenitic stainless steels about 2% greater than the values shown in this table which are based on weights for carbon steel.

*** schedule numbers**

Standard weight pipe and schedule 40 are the same in all sizes through 10-inch; from 12-inch through 24-inch, standard weight pipe has a wall thickness of ⅜-inch.

Extra strong weight pipe and schedule 80 are the same in all sizes through 8-inch; from 8-inch through 24-inch, extra strong weight pipe has a wall thickness of ½-inch.

Double extra strong weight pipe has no corresponding schedule number.

a: ANSI B36.10 steel pipe schedule numbers

b: ANSI B36.10 steel pipe nominal wall thickness designation

c: ANSI B36.19 stainless steel pipe schedule numbers

nominal pipe size *outside diameter,* in.	schedule number*			wall thickness, in.	inside diameter, in.	inside area, sq. in.	metal area, sq. in.	sq ft outside surface, per ft	sq ft inside surface, per ft	weight per ft, lb†	weight of water per ft, lb	moment of inertia, in.[4]	section modulus, in.[3]	radius gyration, in.
	a	b	c											
⅛ *0.405*			10S	0.049	0.307	0.0740	0.0548	0.106	0.0804	0.186	0.0321	0.00088	0.00437	0.1271
	40	Std	40S	0.068	0.269	0.0568	0.0720	0.106	0.0705	0.245	0.0246	0.00106	0.00525	0.1215
	80	XS	80S	0.095	0.215	0.0364	0.0925	0.106	0.0563	0.315	0.0157	0.00122	0.00600	0.1146
¼ *0.540*			10S	0.065	0.410	0.1320	0.0970	0.141	0.1073	0.330	0.0572	0.00279	0.01032	0.1694
	40	Std	40S	0.088	0.364	0.1041	0.1250	0.141	0.0955	0.425	0.0451	0.00331	0.01230	0.1628
	80	XS	80S	0.119	0.302	0.0716	0.1574	0.141	0.0794	0.535	0.0310	0.00378	0.01395	0.1547
⅜ *0.675*			SS	0.065	0.710	0.396	0.1582	0.220	0.1859	0.538	0.1716	0.01197	0.0285	0.2750
			10S	0.065	0.545	0.2333	0.1246	0.177	0.1427	0.423	0.1011	0.00586	0.01737	0.2169
	40	Std	40S	0.091	0.493	0.1910	0.1670	0.177	0.1295	0.568	0.0827	0.00730	0.02160	0.2090
	80	XS	80S	0.126	0.423	0.1405	0.2173	0.177	0.1106	0.739	0.0609	0.00862	0.02554	0.1991
½ *0.840*			5S	0.065	0.710	0.3959	0.1583	0.220	0.1859	0.538	0.171	0.0120	0.0285	0.2750
			10S	0.083	0.674	0.357	0.1974	0.220	0.1765	0.671	0.1547	0.01431	0.0341	0.2692
	40	Std	40S	0.109	0.622	0.304	0.2503	0.220	0.1628	0.851	0.1316	0.01710	0.0407	0.2613
	80	XS	80S	0.147	0.546	0.2340	0.320	0.220	0.1433	1.088	0.1013	0.02010	0.0478	0.2505
	160			0.187	0.466	0.1706	0.383	0.220	0.1220	1.304	0.0740	0.02213	0.0527	0.2402
		XXS		0.294	0.252	0.0499	0.504	0.220	0.0660	1.714	0.0216	0.02425	0.0577	0.2192
¾ *1.050*			5S	0.065	0.920	0.665	0.2011	0.275	0.2409	0.684	0.2882	0.02451	0.0467	0.349
			10S	0.083	0.884	0.614	0.2521	0.275	0.2314	0.857	0.2661	0.02970	0.0566	0.343
	40	Std	40S	0.113	0.824	0.533	0.333	0.275	0.2157	1.131	0.2301	0.0370	0.0706	0.334
	80	XS	80S	0.154	0.742	0.432	0.435	0.275	0.1943	1.474	0.1875	0.0448	0.0853	0.321
	160			0.218	0.614	0.2961	0.570	0.275	0.1607	1.937	0.1284	0.0527	0.1004	0.304
		XXS		0.308	0.434	0.1479	0.718	0.275	0.1137	2.441	0.0641	0.0579	0.1104	0.2840
1 *1.315*			5S	0.065	1.185	1.103	0.2553	0.344	0.310	0.868	0.478	0.0500	0.0760	0.443
			10S	0.109	1.097	0.945	0.413	0.344	0.2872	1.404	0.409	0.0757	0.1151	0.428
	40	Std	40S	0.133	1.049	0.864	0.494	0.344	0.2746	1.679	0.374	0.0874	0.1329	0.421
	80	XS	80S	0.179	0.957	0.719	0.639	0.344	0.2520	2.172	0.311	0.1056	0.1606	0.407
	160			0.250	0.815	0.522	0.836	0.344	0.2134	2.844	0.2261	0.1252	0.1903	0.387
		XXS		0.358	0.599	0.2818	1.076	0.344	0.1570	3.659	0.1221	0.1405	0.2137	0.361
1¼ *1.660*			5S	0.065	1.530	1.839	0.326	0.434	0.401	1.107	0.797	0.1038	0.1250	0.564
			10S	0.109	1.442	1.633	0.531	0.434	0.378	1.805	0.707	0.1605	0.1934	0.550
	40	Std	40S	0.140	1.380	1.496	0.669	0.434	0.361	2.273	0.648	0.1948	0.2346	0.540
	80	XS	80S	0.191	1.278	1.283	0.881	0.434	0.335	2.997	0.555	0.2418	0.2913	0.524
	160			0.250	1.160	1.057	1.107	0.434	0.304	3.765	0.458	0.2839	0.342	0.506
		XXS		0.382	0.896	0.631	1.534	0.434	0.2346	5.214	0.2732	0.341	0.411	0.472
1½ *1.900*			5S	0.065	1.770	2.461	0.375	0.497	0.463	1.274	1.067	0.1580	0.1663	0.649
			10S	0.109	1.682	2.222	0.613	0.497	0.440	2.085	0.962	0.2469	0.2599	0.634

nominal pipe size *outside diameter*, in.	schedule number* a	b	c	wall thick-ness, in.	inside diam-eter, in.	inside area, sq. in.	metal area, sq. in.	sq ft outside surface, per ft	sq ft inside surface, per ft	weight per ft, lb†	weight of water per ft, lb	moment of inertia, in.[4]	section modu-lus, in.[3]	radius gyra-tion, in.
1½ *1.900*	40	Std	40S	0.145	1.610	2.036	0.799	0.497	0.421	2.718	0.882	0.310	0.326	0.623
	80	XS	80S	0.200	1.500	1.767	1.068	0.497	0.393	3.631	0.765	0.391	0.412	0.605
	160	......	......	0.281	1.338	1.406	1.429	0.497	0.350	4.859	0.608	0.483	0.508	0.581
	......	XXS	......	0.400	1.100	0.950	1.885	0.497	0.288	6.408	0.412	0.568	0.598	0.549
	......	......	......	0.525	0.850	0.567	2.267	0.497	0.223	7.710	0.246	0.6140	0.6470	0.5200
	......	......	......	0.650	0.600	0.283	2.551	0.497	0.157	8.678	0.123	0.6340	0.6670	0.4980
2 *2.375*	......	......	5S	0.065	2.245	3.96	0.472	0.622	0.588	1.604	1.716	0.315	0.2652	0.817
	......	......	10S	0.109	2.157	3.65	0.776	0.622	0.565	2.638	1.582	0.499	0.420	0.802
	40	Std	40S	0.154	2.067	3.36	1.075	0.622	0.541	3.653	1.455	0.666	0.561	0.787
	80	XS	80S	0.218	1.939	2.953	1.477	0.622	0.508	5.022	1.280	0.868	0.731	0.766
	160	......	......	0.343	1.689	2.240	2.190	0.622	0.442	7.444	0.971	1.163	0.979	0.729
	......	XXS	......	0.436	1.503	1.774	2.656	0.622	0.393	9.029	0.769	1.312	1.104	0.703
	......	......	......	0.562	1.251	1.229	3.199	0.622	0.328	10.882	0.533	1.442	1.2140	0.6710
	......	......	......	0.687	1.001	0.787	3.641	0.622	0.262	12.385	0.341	1.5130	1.2740	0.6440
2½ *2.875*	......	......	5S	0.083	2.709	5.76	0.728	0.753	0.709	2.475	2.499	0.710	0.494	0.988
	......	......	10S	0.120	2.635	5.45	1.039	0.753	0.690	3.531	2.361	0.988	0.687	0.975
	40	Std	40S	0.203	2.469	4.79	1.704	0.753	0.646	5.793	2.076	1.530	1.064	0.947
	80	XS	80S	0.276	2.323	4.24	2.254	0.753	0.608	7.661	1.837	1.925	1.339	0.924
	160	......	......	0.375	2.125	3.55	2.945	0.753	0.556	10.01	1.535	2.353	1.637	0.894
	......	XXS	......	0.552	1.771	2.464	4.03	0.753	0.464	13.70	1.067	2.872	1.998	0.844
	......	......	......	0.675	1.525	1.826	4.663	0.753	0.399	15.860	0.792	3.0890	2.1490	0.8140
	......	......	......	0.800	1.275	1.276	5.212	0.753	0.334	17.729	0.554	3.2250	2.2430	0.7860
3 *3.500*	......	......	5S	0.083	3.334	8.73	0.891	0.916	0.873	3.03	3.78	1.301	0.744	1.208
	......	......	10S	0.120	3.260	8.35	1.274	0.916	0.853	4.33	3.61	1.822	1.041	1.196
	40	Std	40S	0.216	3.068	7.39	2.228	0.916	0.803	7.58	3.20	3.02	1.724	1.164
	80	XS	80S	0.300	2.900	6.61	3.02	0.916	0.759	10.25	2.864	3.90	2.226	1.136
	160	......	......	0.437	2.626	5.42	4.21	0.916	0.687	14.32	2.348	5.03	2.876	1.094
	......	XXS	......	0.600	2.300	4.15	5.47	0.916	0.602	18.58	1.801	5.99	3.43	1.047
	......	......	......	0.725	2.050	3.299	6.317	0.916	0.537	21.487	1.431	6.5010	3.7150	1.0140
	......	......	......	0.850	1.800	2.543	7.073	0.916	0.471	24.057	1.103	6.8530	3.9160	0.9840
3½ *4.000*	......	......	5S	0.083	3.834	11.55	1.021	1.047	1.004	3.47	5.01	1.960	0.980	1.385
	......	......	10S	0.120	3.760	11.10	1.463	1.047	0.984	4.97	4.81	2.756	1.378	1.372
	40	Std	40S	0.226	3.548	9.89	2.680	1.047	0.929	9.11	4.28	4.79	2.394	1.337
	80	XS	80S	0.318	3.364	8.89	3.68	1.047	0.881	12.51	3.85	6.28	3.14	1.307
	......	XXS	......	0.636	2.728	5.845	6.721	1.047	0.716	22.850	2.530	9.8480	4.9240	1.2100
4 *4.500*	......	......	5S	0.083	4.334	14.75	1.152	1.178	1.135	3.92	6.40	2.811	1.249	1.562
	......	......	10S	0.120	4.260	14.25	1.651	1.178	1.115	5.61	6.17	3.96	1.762	1.549
	......	......	......	0.188	4.124	13.357	2.547	1.178	1.082	8.560	5.800	5.8500	2.6000	1.5250
	40	Std	40S	0.237	4.026	12.73	3.17	1.178	1.054	10.79	5.51	7.23	3.21	1.510
	80	XS	80S	0.337	3.826	11.50	4.41	1.178	1.002	14.98	4.98	9.61	4.27	1.477
	120	......	......	0.437	3.626	10.33	5.58	1.178	0.949	18.96	4.48	11.65	5.18	1.445
	......	......	......	0.500	3.500	9.621	6.283	1.178	0.916	21.360	4.160	12.7710	5.6760	1.4250
	160	......	......	0.531	3.438	9.28	6.62	1.178	0.900	22.51	4.02	13.27	5.90	1.416
	......	XXS	......	0.674	3.152	7.80	8.10	1.178	0.825	27.54	3.38	15.29	6.79	1.374
	......	......	......	0.800	2.900	6.602	9.294	1.178	0.759	31.613	2.864	16.6610	7.4050	1.3380
	......	......	......	0.925	2.650	5.513	10.384	1.178	0.694	35.318	2.391	17.7130	7.8720	1.3060
5 *5.563*	......	......	5S	0.109	5.345	22.44	1.868	1.456	1.399	6.35	9.73	6.95	2.498	1.929
	......	......	10S	0.134	5.295	22.02	2.285	1.456	1.386	7.77	9.53	8.43	3.03	1.920
	40	Std	40S	0.258	5.047	20.01	4.30	1.456	1.321	14.62	8.66	15.17	5.45	1.878
	80	XS	80S	0.375	4.813	18.19	6.11	1.456	1.260	20.78	7.89	20.68	7.43	1.839
	120	......	......	0.500	4.563	16.35	7.95	1.456	1.195	27.04	7.09	25.74	9.25	1.799
	160	......	......	0.625	4.313	14.61	9.70	1.456	1.129	32.96	6.33	30.0	10.80	1.760
	......	XXS	......	0.750	4.063	12.97	11.34	1.456	1.064	38.55	5.62	33.6	12.10	1.722
	......	......	......	0.875	3.813	11.413	12.880	1.456	0.998	43.810	4.951	36.6450	13.1750	1.6860
	......	......	......	1.000	3.563	9.966	14.328	1.456	0.933	47.734	4.232	39.1110	14.0610	1.6520

TABLE 1-1 (cont'd.)

properties of steel pipe

nominal pipe size *outside diameter,* in.	schedule number* a	b	c	wall thick-ness, in.	inside diam-eter, in.	inside area, sq. in.	metal area, sq. in.	sq ft outside surface, per ft	sq ft inside surface, per ft	weight per ft, lb†	weight of water per ft, lb	moment of inertia, in.⁴	section modu-lus, in.³	radius gyra-tion, in.
	...	...	5S	0.109	6.407	32.2	2.231	1.734	1.677	5.37	13.98	11.85	3.58	2.304
	...	...	10S	0.134	6.357	31.7	2.733	1.734	1.664	9.29	13.74	14.40	4.35	2.295
	...	...	...	0.219	6.187	30.100	4.410	1.734	1.620	15.020	13.100	22.6600	6.8400	2.2700
	40	Std	40S	0.280	6.065	28.89	5.58	1.734	1.588	18.97	12.51	28.14	8.50	2.245
6 *6.625*	80	XS	80S	0.432	5.761	26.07	8.40	1.734	1.508	28.57	11.29	40.5	12.23	2.195
	120	...	...	0.562	5.501	23.77	10.70	1.734	1.440	36.39	10.30	49.6	14.98	2.153
	160	...	...	0.718	5.189	21.15	13.33	1.734	1.358	45.30	9.16	59.0	17.81	2.104
	...	XXS	...	0.864	4.897	18.83	15.64	1.734	1.282	53.16	8.17	66.3	20.03	2.060
	...	...	...	1.000	4.625	16.792	17.662	1.734	1.211	60.076	7.284	72.1190	21.7720	2.0200
	...	...	...	1.125	4.375	15.025	19.429	1.734	1.145	66.084	6.517	76.5970	23.1240	1.9850
	...	...	5S	0.109	8.407	55.5	2.916	2.258	2.201	9.91	24.07	26.45	6.13	3.01
	...	...	10S	0.148	8.329	54.5	3.94	2.258	2.180	13.40	23.59	35.4	8.21	3.00
	...	...	...	0.219	8.187	52.630	5.800	2.258	2.150	19.640	22.900	51.3200	11.9000	2.9700
8 *8.625*	20	...	...	0.250	8.125	51.8	6.58	2.258	2.127	22.36	22.48	57.7	13.39	2.962
	30	...	...	0.277	8.071	51.2	7.26	2.258	2.113	24.70	22.18	63.4	14.69	2.953
	40	Std	40S	0.322	7.981	50.0	8.40	2.258	2.089	28.55	21.69	72.5	16.81	2.938
	60	...	...	0.406	7.813	47.9	10.48	2.258	2.045	35.64	20.79	88.8	20.58	2.909
	80	XS	80S	0.500	7.625	45.7	12.76	2.258	1.996	43.39	19.80	105.7	24.52	2.878
	100	...	...	0.593	7.439	43.5	14.96	2.258	1.948	50.87	18.84	121.4	28.14	2.847
	120	...	...	0.718	7.189	40.6	17.84	2.258	1.882	60.63	17.60	140.6	32.6	2.807
8 *8.625*	140	...	...	0.812	7.001	38.5	19.93	2.258	1.833	67.76	16.69	153.8	35.7	2.777
	160	...	...	0.906	6.813	36.5	21.97	2.258	1.784	74.69	15.80	165.9	38.5	2.748
	...	...	...	1.000	6.625	34.454	23.942	2.258	1.734	81.437	14.945	177.1320	41.0740	2.7190
	...	...	...	1.125	6.375	31.903	26.494	2.258	1.669	90.114	13.838	190.6210	44.2020	2.6810
	...	...	5S	0.134	10.482	86.3	4.52	2.815	2.744	15.15	37.4	63.7	11.85	3.75
	...	...	10S	0.165	10.420	85.3	5.49	2.815	2.728	18.70	36.9	76.9	14.30	3.74
	...	...	...	0.219	10.312	83.52	7.24	2.815	2.70	24.63	36.2	100.46	18.69	3.72
	20	...	...	0.250	10.250	82.5	8.26	2.815	2.683	28.04	35.8	113.7	21.16	3.71
	30	...	...	0.307	10.136	80.7	10.07	2.815	2.654	34.24	35.0	137.5	25.57	3.69
	40	Std	40S	0.365	10.020	78.9	11.91	2.815	2.623	40.48	34.1	160.8	29.90	3.67
10 *10.750*	60	XS	80S	0.500	9.750	74.7	16.10	2.815	2.553	54.74	32.3	212.0	39.4	3.63
	80	...	...	0.593	9.564	71.8	18.92	2.815	2.504	64.33	31.1	244.9	45.6	3.60
	100	...	...	0.718	9.314	68.1	22.63	2.815	2.438	76.93	29.5	286.2	53.2	3.56
	120	...	...	0.843	9.064	64.5	26.24	2.815	2.373	89.20	28.0	324	60.3	3.52
	...	...	...	0.875	9.000	63.62	27.14	2.815	2.36	92.28	27.6	333.46	62.04	3.50
	140	...	...	1.000	8.750	60.1	30.6	2.815	2.291	104.13	26.1	368	68.4	3.47
	160	...	...	1.125	8.500	56.7	34.0	2.815	2.225	115.65	24.6	399	74.3	3.43
	...	...	...	1.250	8.250	53.45	37.31	2.815	2.16	126.82	23.2	428.17	79.66	3.39
	...	...	...	1.500	7.750	47.15	43.57	2.815	2.03	148.19	20.5	478.59	89.04	3.31
	...	...	5S	0.156	12.438	121.4	6.17	3.34	3.26	20.99	52.7	122.2	19.20	4.45
	...	...	10S	0.180	12.390	120.6	7.11	3.34	3.24	24.20	52.2	140.5	22.03	4.44
	20	...	...	0.250	12.250	117.9	9.84	3.34	3.21	33.38	51.1	191.9	30.1	4.42
	30	...	...	0.330	12.090	114.8	12.88	3.34	3.17	43.77	49.7	248.5	39.0	4.39
	...	Std	40S	0.375	12.000	113.1	14.58	3.34	3.14	49.56	49.0	279.3	43.8	4.38
	40	...	...	0.406	11.938	111.9	15.74	3.34	3.13	53.53	48.5	300	47.1	4.37
	...	XS	80S	0.500	11.750	108.4	19.24	3.34	3.08	65.42	47.0	362	56.7	4.33
12 *12.750*	60	...	...	0.562	11.626	106.2	21.52	3.34	3.04	73.16	46.0	401	62.8	4.31
	80	...	...	0.687	11.376	101.6	26.04	3.34	2.978	88.51	44.0	475	74.5	4.27
	...	...	...	0.750	11.250	99.40	28.27	3.34	2.94	96.2	43.1	510.7	80.1	4.25
	100	...	...	0.843	11.064	96.1	31.5	3.34	2.897	107.20	41.6	562	88.1	4.22
	...	...	...	0.875	11.000	95.00	32.64	3.34	2.88	110.9	41.1	578.5	90.7	4.21
	120	...	...	1.000	10.750	90.8	36.9	3.34	2.814	125.49	39.3	642	100.7	4.17
	140	...	...	1.125	10.500	86.6	41.1	3.34	2.749	139.68	37.5	701	109.9	4.13
	...	...	...	1.250	10.250	82.50	45.16	3.34	2.68	153.6	35.8	755.5	118.5	4.09
	160	...	...	1.312	10.126	80.5	47.1	3.34	2.651	160.27	34.9	781	122.6	4.07

TABLE 1-1 (cont'd.)

properties of steel pipe

nominal pipe size *outside diameter,* in.	schedule number* a	b	c	wall thick-ness, in.	inside diam-eter, in.	inside area, sq. in.	metal area, sq. in.	sq ft outside surface, per ft	sq ft inside surface, per ft,	weight per ft, lb†	weight of water per ft, lb	moment of inertia, in.[4]	section modu-lus, in.[3]	radius gyra-tion, in.
	......	......	5S	0.156	13.688	147.20	6.78	3.67	3.58	23.0	63.7	162.6	23.2	4.90
	......	......	10S	0.188	13.624	145.80	8.16	3.67	3.57	27.7	63.1	194.6	27.8	4.88
	......	......	......	0.210	13.580	144.80	9.10	3.67	3.55	30.9	62.8	216.2	30.9	4.87
	......	......	......	0.219	13.562	144.50	9.48	3.67	3.55	32.2	62.6	225.1	32.2	4.87
	10	......	......	0.250	13.500	143.1	10.80	3.67	3.53	36.71	62.1	255.4	36.5	4.86
	......	......	......	0.281	13.438	141.80	12.11	3.67	3.52	41.2	61.5	285.2	40.7	4.85
	20	......	......	0.312	13.376	140.5	13.42	3.67	3.50	45.68	60.9	314	44.9	4.84
	......	......	......	0.344	13.312	139.20	14.76	3.67	3.48	50.2	60.3	344.3	49.2	4.83
14	30	Std	......	0.375	13.250	137.9	16.05	3.67	3.47	54.57	59.7	373	53.3	4.82
14.000	40	......	......	0.437	13.126	135.3	18.62	3.67	3.44	63.37	58.7	429	61.2	4.80
	......	......	......	0.469	13.062	134.00	19.94	3.67	3.42	67.8	58.0	456.8	65.3	4.79
	......	XS	......	0.500	13.000	132.7	21.21	3.67	3.40	72.09	57.5	484	69.1	4.78
	60	......	......	0.593	12.814	129.0	24.98	3.67	3.35	84.91	55.9	562	80.3	4.74
	......	......	......	0.625	12.750	127.7	26.26	3.67	3.34	89.28	55.3	589	84.1	4.73
	80	......	......	0.750	12.500	122.7	31.2	3.67	3.27	106.13	53.2	687	98.2	4.69
	100	......	......	0.937	12.126	115.5	38.5	3.67	3.17	130.73	50.0	825	117.8	4.63
	120	......	......	1.093	11.814	109.6	44.3	3.67	3.09	150.67	47.5	930	132.8	4.58
	140	......	......	1.250	11.500	103.9	50.1	3.67	3.01	170.22	45.0	1127	146.8	4.53
	160	......	......	1.406	11.188	98.3	55.6	3.67	2.929	189.12	42.6	1017	159.6	4.48
	......	......	5S	0.165	15.670	192.90	8.21	4.19	4.10	28	83.5	257	32.2	5.60
	......	......	10S	0.188	15.624	191.70	9.34	4.19	4.09	32	83.0	292	36.5	5.59
	10	......	......	0.250	15.500	188.7	12.37	4.19	4.06	42.05	81.8	384	48.0	5.57
	20	......	......	0.312	15.376	185.7	15.38	4.19	4.03	52.36	80.5	473	59.2	5.55
	30	Std	......	0.375	15.250	182.6	18.41	4.19	3.99	62.58	79.1	562	70.3	5.53
16	40	XS	......	0.500	15.000	176.7	24.35	4.19	3.93	82.77	76.5	732	91.5	5.48
16.000	60	......	......	0.656	14.688	169.4	31.6	4.19	3.85	107.50	73.4	933	116.6	5.43
	80	......	......	0.843	14.314	160.9	40.1	4.19	3.75	136.46	69.7	1157	144.6	5.37
	100	......	......	1.031	13.938	152.6	48.5	4.19	3.65	164.83	66.1	1365	170.6	5.30
	120	......	......	1.218	13.564	144.5	56.6	4.19	3.55	192.29	62.6	1556	194.5	5.24
	140	......	......	1.437	13.126	135.3	65.7	4.19	3.44	223.64	58.6	1760	220.0	5.17
	160	......	......	1.593	12.814	129.0	72.1	4.19	3.35	245.11	55.9	1894	236.7	5.12
	......	......	5S	0.165	17.670	245.20	9.24	4.71	4.63	31	106.2	368	40.8	6.31
	......	......	10S	0.188	17.624	243.90	10.52	4.71	4.61	36	105.7	417	46.4	6.30
	10	......	......	0.250	17.500	240.5	13.94	4.71	4.58	47.39	104.3	549	61.0	6.28
	20	......	......	0.312	17.376	237.1	17.34	4.71	4.55	59.03	102.8	678	75.5	6.25
	......	Std	......	0.375	17.250	233.7	20.76	4.71	4.52	70.59	101.2	807	89.6	6.23
18	30	......	......	0.437	17.126	230.4	24.11	4.71	4.48	82.06	99.9	931	103.4	6.21
	......	XS	......	0.500	17.00	227.0	27.49	4.71	4.45	93.45	98.4	1053	117.0	6.19
18.000	40	......	......	0.562	16.876	223.7	30.8	4.71	4.42	104.75	97.0	1172	130.2	6.17
	60	......	......	0.750	16.500	213.8	40.6	4.71	4.32	138.17	92.7	1515	168.3	6.10
	80	......	......	0.937	16.126	204.2	50.2	4.71	4.22	170.75	88.5	1834	203.8	6.04
	100	......	......	1.156	15.688	193.3	61.2	4.71	4.11	207.96	83.7	2180	242.2	5.97
	120	......	......	1.375	15.250	182.6	71.8	4.71	3.99	244.14	79.2	2499	277.6	5.90
	140	......	......	1.562	14.876	173.8	80.7	4.71	3.89	274.23	75.3	2750	306	5.84
	160	......	......	1.781	14.438	163.7	90.7	4.71	3.78	308.51	71.0	3020	336	5.77
	......	......	5S	0.188	19.634	302.40	11.70	5.24	5.14	40	131.0	574	57.4	7.00
	......	......	10S	0.218	19.564	300.60	13.55	5.24	5.12	46	130.2	663	66.3	6.99
	10	......	......	0.250	19.500	298.6	15.51	5.24	5.11	52.73	129.5	757	75.7	6.98
	20	Std	......	0.375	19.250	291.0	23.12	5.24	5.04	78.60	126.0	1114	111.4	6.94
20	30	XS	......	0.500	19.000	283.5	30.6	5.24	4.97	104.13	122.8	1457	145.7	6.90
20.000	40	......	......	0.593	18.814	278.0	36.2	5.24	4.93	122.91	120.4	1704	170.4	6.86
	60	......	......	0.812	18.376	265.2	48.9	5.24	4.81	166.40	115.0	2257	225.7	6.79
	......	......	......	0.875	18.250	261.6	52.6	5.24	4.78	178.73	113.4	2409	240.9	6.77
	80	......	......	1.031	17.938	252.7	61.4	5.24	4.70	208.87	109.4	2772	277.2	6.72
	100	......	......	1.281	17.438	238.8	75.3	5.24	4.57	256.10	103.4	3320	332	6.63

properties of steel pipe

nominal pipe size *outside diameter,* in.	schedule number*			wall thick-ness, in.	inside diam-eter, in.	inside area, sq in.	metal area, sq in.	sq ft outside surface, per ft	sq ft inside surface, per ft	weight per ft, lb†	weight of water per ft, lb	moment of inertia, in.[4]	section modu-lus, in.[3]	radius gyra-tion, in.
	a	b	c											
20 *20.000*	120			1.500	17.000	227.0	87.2	5.24	4.45	296.37	98.3	3760	376	6.56
	140			1.750	16.500	213.8	100.3	5.24	4.32	341.10	92.6	4220	422	6.48
	160			1.968	16.064	202.7	111.5	5.24	4.21	379.01	87.9	4590	459	6.41
22 *22.000*			5S	0.188	21.624	367.3	12.88	5.76	5.66	44	159.1	766	69.7	7.71
			10S	0.218	21.564	365.2	14.92	5.76	5.65	51	158.2	885	80.4	7.70
	10			0.250	21.500	363.1	17.18	5.76	5.63	58	157.4	1010	91.8	7.69
	20	Std		0.375	21.250	354.7	25.48	5.76	5.56	87	153.7	1490	135.4	7.65
	30	XS		0.500	21.000	346.4	33.77	5.76	5.50	115	150.2	1953	177.5	7.61
				0.625	20.750	338.2	41.97	5.76	5.43	143	146.6	2400	218.2	7.56
				0.750	20.500	330.1	50.07	5.76	5.37	170	143.1	2829	257.2	7.52
	60			0.875	20.250	322.1	58.07	5.76	5.30	197	139.6	3245	295.0	7.47
	80			1.125	19.750	306.4	73.78	5.76	5.17	251	132.8	4029	366.3	7.39
	100			1.375	19.250	291.0	89.09	5.76	5.04	303	126.2	4758	432.6	7.31
	120			1.625	18.750	276.1	104.02	5.76	4.91	354	119.6	5432	493.8	7.23
	140			1.875	18.250	261.6	118.55	5.76	4.78	403	113.3	6054	550.3	7.15
	160			2.125	17.750	247.4	132.68	5.76	4.65	451	107.2	6626	602.4	7.07
24 *24.000*	10			0.250	23.500	434	18.65	6.28	6.15	63.41	188.0	1316	109.6	8.40
	20	Std		0.375	23.250	425	27.83	6.28	6.09	94.62	183.8	1943	161.9	8.35
		XS		0.500	23.000	415	36.9	6.28	6.02	125.49	180.1	2550	212.5	8.31
	30			0.562	22.876	411	41.4	6.28	5.99	140.80	178.1	2840	237.0	8.29
				0.625	22.750	406	45.9	6.28	5.96	156.03	176.2	3140	261.4	8.27
	40			0.687	22.626	402	50.3	6.28	5.92	171.17	174.3	3420	285.2	8.25
				0.750	22.500	398	54.8	6.28	5.89	186.24	172.4	3710	309	8.22
			5S	0.218	23.564	436.1	16.29	6.28	6.17	55	188.9	1152	96.0	8.41
				0.875	22.250	388.6	63.54	6.28	5.83	216	168.6	4256	354.7	8.18
	60			0.968	22.064	382	70.0	6.28	5.78	238.11	165.8	4650	388	8.15
	80			1.218	21.564	365	87.2	6.28	5.65	296.36	158.3	5670	473	8.07
	100			1.531	20.938	344	108.1	6.28	5.48	367.40	149.3	6850	571	7.96
	120			1.812	20.376	326	126.3	6.28	5.33	429.39	141.4	7830	652	7.87
	140			2.062	19.876	310	142.1	6.28	5.20	483.13	134.5	8630	719	7.79
	160			2.343	19.314	293	159.4	6.28	5.06	541.94	127.0	9460	788	7.70
26 *26.000*				0.250	25.500	510.7	19.85	6.81	6.68	67	221.4	1646	126.6	9.10
	10			0.312	25.376	505.8	25.18	6.81	6.64	86	219.2	2076	159.7	9.08
		Std		0.375	25.250	500.7	30.19	6.81	6.61	103	217.1	2478	190.6	9.06
	20	XS		0.500	25.000	490.9	40.06	6.81	6.54	136	212.8	3259	250.7	9.02
				0.625	24.750	481.1	49.82	6.81	6.48	169	208.6	4013	308.7	8.98
				0.750	24.500	471.4	59.49	6.81	6.41	202	204.4	4744	364.9	8.93
				0.875	24.250	461.9	69.07	6.81	6.35	235	200.2	5458	419.9	8.89
				1.000	24.000	452.4	78.54	6.81	6.28	267	196.1	6149	473.0	8.85
				1.125	23.750	443.0	87.91	6.81	6.22	299	192.1	6813	524.1	8.80
28 *28.000*				0.250	27.500	594.0	21.80	7.33	7.20	74	257.3	2098	149.8	9.81
	10			0.312	27.376	588.6	27.14	7.33	7.17	92	255.0	2601	185.8	9.79
		Std		0.375	27.250	583.2	32.54	7.33	7.13	111	252.6	3105	221.8	9.77
	20	XS		0.500	27.000	572.6	43.20	7.33	7.07	147	248.0	4085	291.8	9.72
	30			0.625	26.750	562.0	53.75	7.33	7.00	183	243.4	5038	359.8	9.68
				0.750	26.500	551.6	64.21	7.33	6.94	218	238.9	5964	426.0	9.64
				0.875	26.250	541.2	74.56	7.33	6.87	253	234.4	6865	490.3	9.60
				1.000	26.000	530.9	84.82	7.33	6.81	288	230.0	7740	552.8	9.55
				1.125	25.750	520.8	94.98	7.33	6.74	323	225.6	8590	613.6	9.51
30 *30.000*			5S	0.250	29.500	683.4	23.37	7.85	7.72	79	296.3	2585	172.3	10.52
	10		10S	0.312	29.376	677.8	29.19	7.85	7.69	99	293.7	3201	213.4	10.50
		Std		0.375	29.250	672.0	34.90	7.85	7.66	119	291.2	3823	254.8	10.48
	20	XS		0.500	29.000	660.5	46.34	7.85	7.59	158	286.2	5033	335.5	10.43
	30			0.625	28.750	649.2	57.68	7.85	7.53	196	281.3	6213	414.2	10.39

TABLE 1-1 (cont'd.)

properties of steel pipe

nominal pipe size *outside diameter,* in.	schedule number* a	b	c	wall thick-ness, in.	inside diam-eter, in.	inside area, sq. in.	metal area, sq. in.	sq ft outside surface, per ft	sq ft inside surface, per ft	weight per ft, lb†	weight of water per ft lb	moment of inertia, in.[4]	section modu-lus, in.[3]	radius gyra-tion, in.
	40			0.750	28.500	637.9	68.92	7.85	7.46	234	276.6	7371	491.4	10.34
30				0.875	28.250	620.7	80.06	7.85	7.39	272	271.8	8494	566.2	10.30
30.000				1.000	28.000	615.7	91.11	7.85	7.33	310	267.0	9591	639.4	10.26
				1.125	27.750	604.7	102.05	7.85	7.26	347	262.2	10653	710.2	10.22
				0.250	31.500	779.2	24.93	8.38	8.25	85	337.8	3141	196.3	11.22
	10			0.312	31.376	773.2	31.02	8.38	8.21	106	335.2	3891	243.2	11.20
		Std		0.375	31.250	766.9	37.25	8.38	8.18	127	332.5	4656	291.0	11.18
	20	XS		0.500	31.000	754.7	49.48	8.38	8.11	168	327.2	6140	383.8	11.14
32	30			0.625	30.750	742.5	61.59	8.38	8.05	209	321.9	7578	473.6	11.09
32.000	40			0.688	30.624	736.6	67.68	8.38	8.02	230	319.0	8298	518.6	11.07
				0.750	30.500	730.5	73.63	8.38	7.98	250	316.7	8990	561.9	11.05
				0.875	30.250	718.3	85.52	8.38	7.92	291	311.6	10372	648.2	11.01
				1.000	30.000	706.8	97.38	8.38	7.85	331	306.4	11680	730.0	10.95
				1.125	29.750	694.7	109.0	8.38	7.79	371	301.3	13023	814.0	10.92
				0.250	33.500	881.2	26.50	8.90	8.77	90	382.0	3773	221.9	11.93
	10			0.312	33.376	874.9	32.99	8.90	8.74	112	379.3	4680	275.3	11.91
		Std		0.375	33.250	867.8	39.61	8.90	8.70	135	376.2	5597	329.2	11.89
	20	XS		0.500	33.000	855.3	52.62	8.90	8.64	179	370.8	7385	434.4	11.85
34	30			0.625	32.750	841.9	65.53	8.90	8.57	223	365.0	9124	536.7	11.80
34.000	40			0.688	32.624	835.9	72.00	8.90	8.54	245	362.1	9992	587.8	11.78
				0.750	32.500	829.3	78.34	8.90	8.51	266	359.5	10829	637.0	11.76
				0.875	32.250	816.4	91.01	8.90	8.44	310	354.1	12501	735.4	11.72
				1.000	32.000	804.2	103.67	8.90	8.38	353	348.6	14114	830.2	11.67
				1.125	31.750	791.3	116.13	8.90	8.31	395	343.2	15719	924.7	11.63
				0.250	35.500	989.7	28.11	9.42	9.29	96	429.1	4491	249.5	12.64
	10			0.312	35.376	982.9	34.95	9.42	9.26	119	426.1	5565	309.1	12.62
		Std		0.375	35.250	975.8	42.01	9.42	9.23	143	423.1	6664	370.2	12.59
	20	XS		0.500	35.000	962.1	55.76	9.42	9.16	190	417.1	8785	488.1	12.55
36	30			0.625	34.750	948.3	69.50	9.42	9.10	236	411.1	10872	604.0	12.51
36.000	40			0.750	34.500	934.7	83.01	9.42	9.03	282	405.3	12898	716.5	12.46
				0.875	34.250	920.6	96.50	9.42	8.97	328	399.4	14903	827.9	12.42
				1.000	34.000	907.9	109.96	9.42	8.90	374	393.6	16851	936.2	12.38
				1.125	33.750	894.2	123.19	9.42	8.89	419	387.9	18763	1042.4	12.34
				0.250	41.500	1352.6	32.82	10.99	10.86	112	586.4	7126	339.3	14.73
		Std		0.375	41.250	1336.3	49.08	10.99	10.80	167	579.3	10627	506.1	14.71
	20	XS		0.500	41.000	1320.2	65.18	10.99	10.73	222	572.3	14037	668.4	14.67
42	30			0.625	40.750	1304.1	81.28	10.99	10.67	276	565.4	17373	827.3	14.62
42.000	40			0.750	40.500	1288.2	97.23	10.99	10.60	330	558.4	20689	985.2	14.59
				1.000	40.000	1256.6	128.81	10.99	10.47	438	544.8	27080	1289.5	14.50
				1.250	39.500	1225.3	160.03	10.99	10.34	544	531.2	33233	1582.5	14.41
				1.500	39.000	1194.5	190.85	10.99	10.21	649	517.9	39181	1865.7	14.33

TABLE 1-2 ALLOWABLE *s* VALUES FOR SEAMLESS ALLOY, CARBON STEEL, AND WROUGHT IRON PIPE
(in 1000 psi units)

	SEAMLESS ALLOY PIPE				CARBON STEEL & WROUGHT IRON PIPE									
PIPE MATERIAL	FERRITIC			AUSTENITIC	WROUGHT IRON		CARBON STEEL		CARBON STEEL					
SEAMLESS/ WELDED	S	S	S	S	BUTT WELD	LAP WELD	ERW (RESISTANCE)	ERW (RESISTANCE)	ELECTRIC FUSION WELDED					
ASTM Spec.	A335 A369	A335 A369	A335 A369	A312 A376	A72	A72	A135	A135	A155	A155	A155	A139	A139	A139
GRADE / TYPE	TYPE P5 & FP5	TYPE P2 FP2	TYPE P1 FP1	TYPE TP 321			B	A	C55	C50	C45	B	A	A283D
MIN. ULT. STRENGTH	60	55	55	75	40	40	60.0	60.0	55.0	50.0	45.0	60.0	48.0	60.0
-20 to 100	15	13.75	13.75	18.75	6.0	8.0	12.75	10.02	12.4	11.25	10.1	12.0	9.6	10.1
MAXIMUM TEMPERATURES														
200	15			18.75										
300	15			17.0										
400	15			15.8										
450	14.7			15.4										
500	14.5			15.2										
600	14			14.4										
650	13.7			14.85	6.0	8.0	12.75	10.02	12.4	11.25	10.1	12.0	9.6	10.1
700	13.4			14.8			12.12	9.9	11.9	10.9	9.8	11.35	9.25	
750	13.1	13.75	13.75	14.7			11.0	9.1	10.85	9.9	8.9	9.95	8.3	
800	12.8	13.45	13.45	14.55					9.2	8.45	7.5			
850	12.4	13.15	13.15	14.3					7.0	6.55	5.95			
900	11.5	12.5	12.5	14.1										
950	10	10		13.85										
1000	7.3	6.25		13.5										
1050	5.2			13.1										
1100	3.3			10.3										
1150	2.2			7.6										
1200	1.5			5										

CARBON STEEL PIPE

CARBON STEEL						CARBON STEEL								BLACK OR GALVANIZED CARBON STEEL		
ELECTRIC FUSION WELDED						SEAMLESS		SEAMLESS		ELECTRIC RESISTANCE WELDED		BUTT WELD	LAP WELD	S	BUTT WELD	LAP WELD
A134						A106		A53		A53		A53	A 53	A120	A120	A120
A283C	A283B	A283A	A245C	A245B	A245A	B	A	B	A	B	A					
55.0	50.0	45.0	55.0	52.0	48.0	60.0	48.0	60.0	48.0	60.0	48.0	45.0	45.0			
10.1	9.2	8.3	10.1	9.6	8.8	15.0	12.0	15.0	12.0	12.75	10.2	6.75	9.0	10.8	6.5	8.8
▲	▲	▲	▲	▲	▲	▲	▲	▲	▲	▲	▲	▲	▲	10.6	6.35	8.6
														10.2	6.1	8.2
														9.8	5.85	7.8
														9.6	5.7	7.6
▼	▼	▼	▼	▼	▼	▼	▼	▼	▼	▼	▼	▼	▼			
10.1	9.2	8.3	10.1	9.6	8.8	15.0	12.0	15.0	12.0	12.75	10.2	6.75	9.0			
						14.33	11.65	14.35	11.65	12.2	9.9					
						12.95	10.7	12.95	10.7	11.0	9.1					
						10.9	9.0	10.9	9.0							
						7.8	7.1	7.8	7.1							
						5.0	5.0	5.0	5.0							

At any given point along a pipe, the pressure tends to expand the pipe's cross-section. This expansion also puts a lengthwise or longitudinal tension in the pipe. This stress can be visualized easily as two taut cords strung between fixed walls. If the two cords are wedged apart by means of a stick, the tension in the cords is increased. Here the cords are like the diametrically opposite walls of a pipe, which are wedged apart by the pressure in the pipe. This tension occurs at every point along the length of the pipe (Fig. 1-10). A small piece of pipe wall is thus being stretched in all directions, but the pressure tends to produce a longitudinal burst in a pipe.

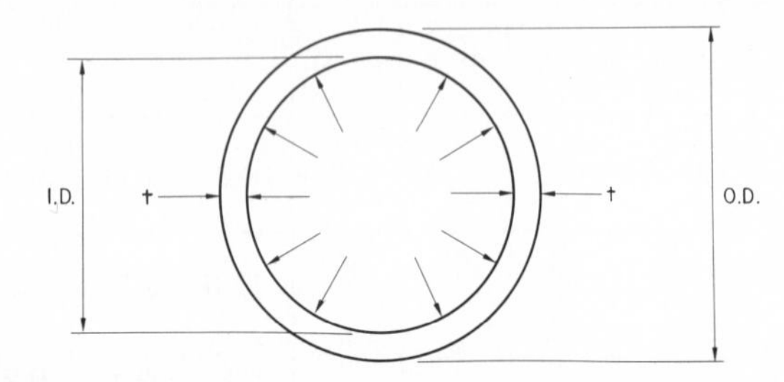

FIG. 1-10 Inside diameter (ID), outside diameter (OD), and thickness (*t*) along with the direction of pressure exerted on the cross-sectional plane.

A single formula, with several algebraic variations, can be used to find the maximum pressure for a given pressure, wall thickness, and pipe diameter or to find the required wall thickness if pressure demand is known along with the material and pipe diameter. Usually the diameter is determined first, from flow requirements, but the formula could also be used to select the pipe material if all other factors were determined and the only quantity to vary was the tensile strength of the material. Here are the calculations:

$$t = \frac{p \times D}{2 \times s} = \frac{pD}{2s}$$

$$p = \frac{2 \times s \times t}{D} = \frac{2st}{D}$$

$$\text{Schedule number} = 1000 \times \frac{p}{s} = 1000\frac{p}{s}$$

where p = pressure in pipe (psi)

D = inside diameter of pipe (in.)

s = unit tensile stress (allowable fiber stress in psi) (from manufacturer's table or Table 1-2)

t = thickness of pipe wall

To find the thickness of the wall we take the pressure per square inch p times the diameter (ID) divided by 2 times the unit's tensile stress (s). Use the next greater t and schedule number after calculating. This formula enables us to find the required thickness of pipe for the design and therefore the schedule number for ordering.

To find the safe internal unit pressure (or psi) in a pipe, given the schedule number or thickness, we can use p (psi) equals 2 times s (unit tensile stress) times t (thickness of pipe) divided by D (internal diameter of pipe).

The following simplified calculations and sample problems are for practice only. They are by no means a substitute for a manufacturer's listed recommendations for materials and products or for the engineer's calculations.

Sample Problems

1. How thick must a 16-in. butt-welded wrought iron pipe be if it is to hold 600 psi steam pressure at 450°F? Table 1-2 shows the s value for wrought iron pipe at 450°F to be 6000. What is the schedule number of the pipe?

   $$t = \frac{p \times D}{2 \times s} = \frac{600 \times 16}{2 \times 6000} = t = 0.8 \text{ in.}$$

   See Table 1-1 for schedule number.

   0.8 = 0.843 standard wall = schedule 80

   Use the next highest standard wall thickness.
   Answer: Wall thickness = 1.031; schedule = 100.

2. What is the maximum psi in a 12-in.-diameter stainless steel pipe with a wall thickness of ½ in. and an allowable unit tensile stress of 5000 psi?

   $$p = \frac{2 \times st}{D} = \frac{2 \times 5000 \times 0.5}{12} = 416 \text{ psi}$$

   Also find the schedule number from Table 1-1.
   Answer: 416 psi, schedule XS.

3. Find the schedule number for a seamless ¾-in. A106 carbon steel pipe with a unit tensile stress (s) of 12,000. The pipe is subject to 950 psi at 600°F. Table 1-1 gives an s value of 1200 for A106 pipe at 600°F.

   $$\text{Schedule} = \frac{p}{s} \times 1000$$

   $$= \frac{950}{12{,}000} \times 1000$$

   $$= 79$$

   Answer: 79 = schedule 80.

Threaded Joints

Threaded fittings must be clean-cut and uniform to provide strength, tightness, and durability in service. The American Standard pipe tapered thread is tapered 1/16 in. (0.16 cm) per inch (3/4″ per ft) to ensure a tight joint at the fitting. The crest of the thread is flattened, and the root is filled so that the depth of the thread is equal to 0.80P (80 percent of the pitch). The number of threads per inch for a given nominal diameter can be obtained from Table 1-3. The American Standard tapered thread or pipe thread is very similar to the American Standard thread or straight thread.

TABLE 1-3 AMERICAN NATIONAL STANDARD TAPER PIPE THREADS (NPT)
(Reprinted courtesy ITT-Grinnell)

AMERICAN NATIONAL STANDARD TAPER PIPE THREADS (NPT)

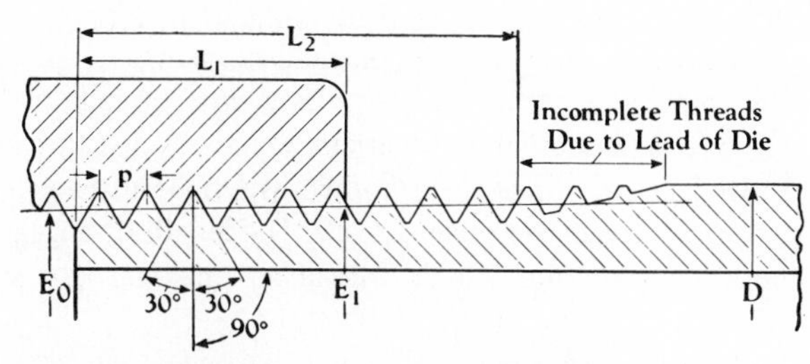

L1

Flush by Hand

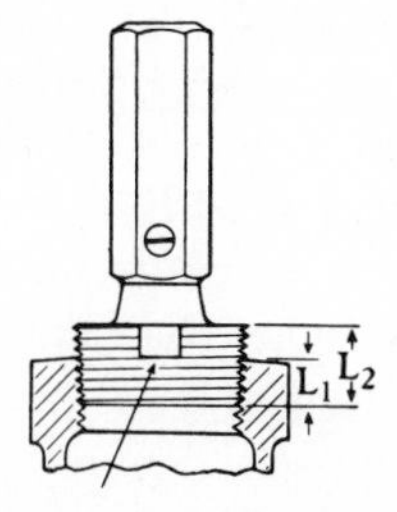

$E_0 = D - (0.050D + 1.1)p$
$^{*}E_1 = E_0 + 0.0625\, L_1$
$L_2 = (0.80D + 6.8)p$

p = Pitch
Depth of thread = $0.80p$
Total Taper ¾-inch per Foot

Tolerance on Product
One turn large or small from notch on plug gauge or face of ring gauge.

Notch flush with face of fitting. If chamfered, notch flush with bottom of chamfer

Dimensions in Inches

Nominal pipe size	D Outside diameter of pipe	Number of threads per inch	p Pitch of thread	E_0 Pitch diameter at end of external thread	E_1† Pitch diameter at end of internal thread	L_1¶ Normal engagement by hand between external and internal threads	L_2§ Length of effective external thread	Height of thread
1/16	0.3125	27	0.03704	0.27118	0.28118	0.160	0.2611	0.02963
⅛	0.405	27	0.03704	0.36351	0.37360	0.1615	0.2639	0.02963
¼	0.540	18	0.05556	0.47739	0.49163	0.2278	0.4018	0.04444
⅜	0.675	18	0.05556	0.61201	0.62701	0.240	0.4078	0.04444
½	0.840	14	0.07143	0.75843	0.77843	0.320	0.5337	0.05714
¾	1.050	14	0.07143	0.96768	0.98887	0.339	0.5457	0.05714
1	1.315	11.5	0.08696	1.21363	1.23863	0.400	0.6828	0.06957
1¼	1.660	11.5	0.08696	1.55713	1.58338	0.420	0.7068	0.06957
1½	1.900	11.5	0.08696	1.79609	1.82234	0.420	0.7235	0.06957
2	2.375	11.5	0.08696	2.26902	2.29627	0.436	0.7565	0.06957
2½	2.875	8	0.12500	2.71953	2.76216	0.682	1.1375	0.10000
3	3.500	8	0.12500	3.34062	3.38850	0.766	1.2000	0.10000
3½	4.000	8	0.12500	3.83750	3.88881	0.821	1.2500	0.10000
4	4.500	8	0.12500	4.33438	4.38712	0.844	1.3000	0.10000
5	5.563	8	0.12500	5.39073	5.44929	0.937	1.4063	0.10000
6	6.625	8	0.12500	6.44609	6.50597	0.958	1.5125	0.10000
8	8.625	8	0.12500	8.43359	8.50003	1.063	1.7125	0.10000
10	10.750	8	0.12500	10.54531	10.62094	1.210	1.9250	0.10000
12	12.750	8	0.12500	12.53281	12.61781	1.360	2.1250	0.10000
14 O.D.	14.000	8	0.12500	13.77500	13.87262	1.562	2.2500	0.10000
16 O.D.	16.000	8	0.12500	15.76250	15.87575	1.812	2.4500	0.10000
18 O.D.	18.000	8	0.12500	17.75000	17.87500	2.000	2.6500	0.10000
20 O.D.	20.000	8	0.12500	19.73750	19.87031	2.125	2.8500	0.10000
24 O.D.	24.000	8	0.12500	23.71250	23.86094	2.375	3.2500	0.10000

† Also pitch diameter at gauging notch.

§ Also length of plug gauge.

¶ Also length of ring gauge, and length from gauging notch to small end of plug gauge.

* For the ⅛-27 and ¼-18 sizes . . . E_1 approx. = $D - (0.05\,D + 0.827)\,p$

Above information extracted from American National Standard for Pipe Threads, ANSI B2.1.

In general, threaded couplings and pipes are used for piping 2½ in. (6.3 cm) or less in diameter except for high temperature and pressure situations. It is also common practice to use external straight threads for male ends and taper threads for female ends since piping materials are usually sufficiently ductile that threads can adjust themselves and create a tight joint. Taper threads are recommended by the ASA for most threaded joints except for pressure-tight joints for couplings, pressure-tight joints for raised-cup fuel or oil filter fittings, loose-fitting mechanical joints for hose couplings, and free-fitting mechanical joints for fixtures. Other than these four situations, the normal American Standard pipe thread is to be used. Figure 1-11 shows the normal engagement for tight joints for the length of pipe entering the fitting. The normal engagement specified for American Standard pipe thread is based on parts being threaded to the American Standard pipe thread or the API standard for pipeline threads. ANSI B2.1, which covers pipe threads for various purposes, was one of the first standards established and most commonly used. The API standards 5A, 6A, and 5L cover oil field tubular material such as line pipe and casing threads. Line pipe threads have the same form and taper as American Standard pipe threads.

All internal threads are known as female threads and all external threads are male threads. The length of engagement between male and female threads to make a tight joint is

normal engagement for tight joints *(does not apply to companion flanges)*

length of pipe entering fitting

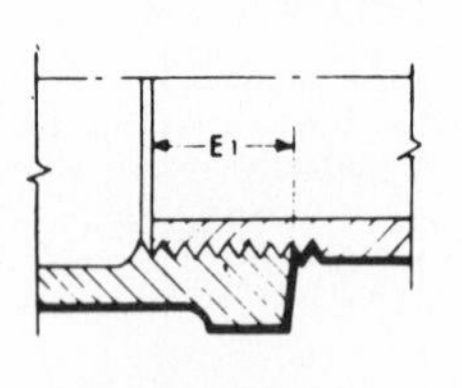

American Standard and API line pipe threads

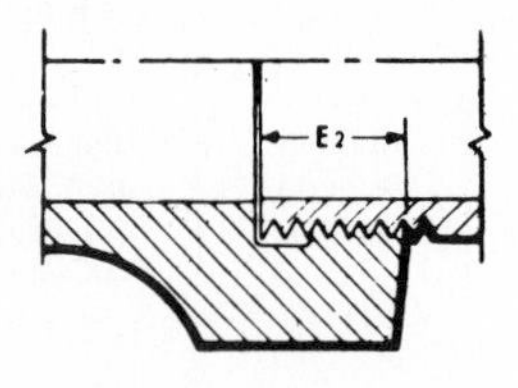

shoulder type drainage fitting threads

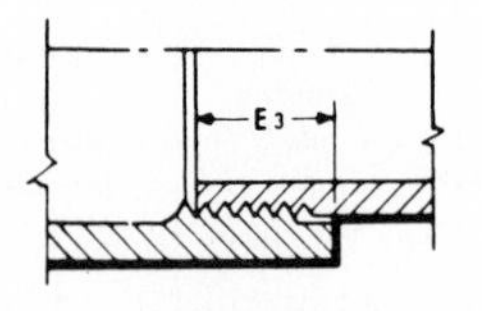

railing fitting thread assembly

dimensions, in.

size	1/8	1/4	3/8	1/2	3/4	1	1 1/4	1 1/2	2	2 1/2	3	3 1/2	4	5	6	8	10	12
E1, E2	1/4	3/8	3/8	1/2	9/16	11/16	11/16	11/16	3/4	15/16	1	1 1/16	1 1/8	1 1/4	1 5/16	1 7/16	1 5/8	1 3/4
E3	......	......	......	1/2	1/2	5/8	11/16	3/4	3/4	......	......	......	......	......	......	......	......	......

FIG. 1-11 Normal engagement for tight joints. (Courtesy of ITT Grinnell Corporation. WARNINGS are published in each catalog for proper installation and performance of products, beyond which the company does not warrantee performance.)

The symbols recommended for use in designating the various types of pipe threads are as follows:

NPT: American standard taper threads
NPSC: American standard straight pipe threads in pipe couplings
NPSF: American standard straight pipe threads for pressure-tight joints for use without lubricant
NPSI: American standard internal straight pipe or sealer* (Dryseal)
NPSM: American standard straight pipe threads for mechanical joints
NPSL: American standard straight pipe threads for locknuts and locknut pipe threads.
NPSH: American standard straight pipe threads for hose couplings and nipples
NPTR: American standard taper pipe threads for railing fittings

*Lubricant may be used in making up these joints when desired.

based on the thread being machined to the American Standard for pipe threads or the API standard for line pipe threads which have been established under practical working conditions. For designating various pipe thread types, it is important to call out one of the various symbols:

NPT American Standard pipe threads

NPTR American Standard taper pipe threads for rail fittings

NPSH American Standard straight pipe threads for hose couplings and nipples

NPSL American Standard pipe threads for locknuts and locknut pipe threads

NPSI American Standard internal straight pipe or sealer

NPSM American Standard straight pipe threads for mechanical joints

NPSC American Standard straight pipe threads and pipe couplings

NPSF American Standard straight pipe threads for pressure-tight joints for use without lubricant

The following are examples of typical pipe thread callouts and explanations:

6–8 NPT	*Explanation*	1/8–27 NPT
6 =	nominal pipe diameter	= 1/8
8 =	number of threads per inch	= 27
N =	American Standard thread	= N
P =	pipe	= P
T =	taper	= T

1/4–18 NPSC	*Explanation*	1½–11½ NPSM
1/4 =	nominal pipe diameter	= 1½
18 =	number of threads per inch	= 11½
N =	American Standard thread	= N
P =	pipe	= P
S =	straight	= S
C =	couplings	
	mechanical joints	= M

3/4–14 NPSL	*Explanation*	12–8 NPTR
3/4 =	nominal pipe diameter	= 12
14 =	number of threads per inch	= 8
N =	American Standard thread	= N
P =	pipe	= P
S =	straight	
	taper	= T
L =	locknuts & locknut pipe threads	
	rail fittings	= R

Table 1-4 shows the American Standard straight pipe threads. An American Standard straight pipe thread having the same number of threads per inch as the tapered thread is used for pressure-tight joints or couplings, for pressure-tight joints for grease and oil fittings, for hose couplings and nipples, and for free-fitting mechanical joints. As we have seen, a tapered external thread is often used with a straight internal thread to provide a tighter fit. Dimensions for American National Standard pipe threads are given in ASAB 2. Table 1-4 on American Standard straight pipe threads gives the nominal pipe size, the number of threads per inch and pitch of the thread.

Figure 1-12 shows the various drafting procedures—the regular method, the simplified method, and the pictorial method—for representing tapered pipe thread and straight pipe thread for male and female parts. It is important to realize that the normal thread engagement for tapered thread varies with the size of the pipe. The normal thread engagement is taken into consideration when drawing pipe assemblies that are to meet certain specifications. The taper on a pipe thread is so slight that it does not show up on drawings unless it is exaggerated. In general it is drawn to 1/8 in. (0.32 cm) taper per inch. Pipe shown in Fig. 1-12 is represented by the same conventional symbols used for the American Standard thread. Note that Fig. 1-12 exaggerates the pipe OD on the female end shown. For each method, OD remains the same, regardless of male or female pipe

MALE
FEMALE
REGULAR METHOD
SIMPLIFIED METHOD
PICTORIAL METHOD
TAPERED PIPE THREAD

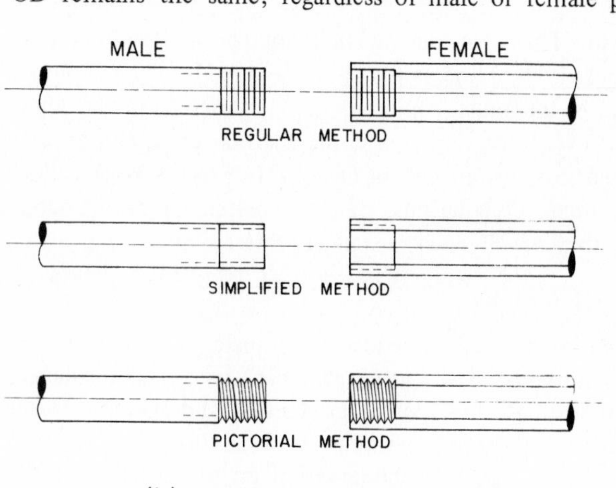

FIG. 1-12 Drafting representations for male and female pipe threads.

TABLE 1-4 AMERICAN STANDARD STRAIGHT PIPE THREADS (Reprinted courtesy Crane Company).

American Standard straight threads

Information abstracted from American Standard Pipe Threads ANSI B2.1.

■ Straight thread gages are used to gage mechanical joint straight pipe threads.

The actual pitch diameters of the tapped hole will be slightly smaller than the values given.

♦ American Standard taper pipe thread plug gages are used to gage straight pipe threads in couplings with the gaging notch coming flush with the edge of the thread or with the bottom of chamfer, if chamfered, allowing a tolerance of one and one half turns large or small.

basic dimensions (inches)

			pitch diameter									
			straight pipe threads♦ in pipe couplings (pressure tight joints) internal		straight pipe threads for mechanical joints■ (free fitting)				straight pipe threads for locknut connections (loose fitting mechanical joints)			
					external		internal		external		internal	
nominal pipe size	threads per inch	pitch of thread	max	min	max	min	max	min	max	min	max	min
1/8	27	0.0370	0.3782	0.3713	0.3748	0.3713	0.3782	0.3748	0.3840	0.3805	0.3898	0.3863
1/4	18	0.0556	0.4951	0.4847	0.4899	0.4847	0.4951	0.4899	0.5038	0.4986	0.5125	0.5073
3/8	18	0.0556	0.6322	0.6218	0.6270	0.6218	0.6322	0.6270	0.6409	0.6357	0.6496	0.6444
1/2	14	0.0714	0.7851	0.7717	0.7784	0.7717	0.7851	0.7784	0.7963	0.7896	0.8075	0.8008
3/4	14	0.0714	0.9956	0.9822	0.9889	0.9822	0.9956	0.9889	1.0067	1.0000	1.0179	1.0112
1	11 1/2	0.0870	1.2468	1.2305	1.2386	1.2305	1.2468	1.2386	1.2064	1.2523	1.2739	1.2658
1 1/4	11 1/2	0.0870	1.5915	1.5752	1.5834	1.5752	1.5915	1.5834	1.6051	1.5970	1.6187	1.6106
1 1/2	11 1/2	0.0870	1.8305	1.8142	1.8223	1.8142	1.8305	1.8223	1.8441	1.8360	1.8576	1.8495
2	11 1/2	0.0870	2.3044	2.2881	2.2963	2.2881	2.3044	2.2963	2.3180	2.3099	2.3315	2.3234
2 1/2	8	0.1250	2.7739	2.7505	2.7622	2.7505	2.7739	2.7622	2.7934	2.7817	2.8129	2.8012
3	8	0.1250	3.4002	3.3768	3.3885	3.3768	3.4002	3.3885	3.4198	3.4081	3.4393	3.4276
3 1/2	8	0.1250	3.9005	3.8771	3.8888	3.8771	3.9005	3.8888	3.9201	3.9084	3.9396	3.9279
4	8	0.1250	4.3988	4.3754	4.3871	4.3754	4.3988	4.3871	4.4184	4.4067	4.4379	4.4262
5	8	0.1250			5.4493	5.4376	5.4610	5.4493	5.4805	5.4688	5.5001	5.4884
6	8	0.1250			6.5060	6.4943	6.5177	6.5060	6.5372	6.5255	6.5567	6.5450
8	8	0.1250							8.5313	8.5196	8.5508	8.5391
10	8	0.1250							10.6522	10.6405	10.6717	10.6600
12	8	0.1250							12.6491	12.6374	12.6686	12.6569

thread. The ASA recommendation for representing pipe threads is the same as for all other threads. The simplified form is the most common, the pictorial form the least. Whichever form is chosen, the method of representation should remain constant throughout the project. When calling out thread designations, state the nominal pipe diameter, threads per inch, and the standard letter symbol.

INSULATION

Insulation is used to reduce the transfer of heat from the pipe contents to the atmosphere or vice versa. Insulation helps decrease the cost of production and also provides a degree of safety for operating personnel. Insulation for cold lines helps eliminate condensation on the exterior of the pipe which may cause damage by corrosion to surrounding equipment. The thickness of the insulation varies with the temperature difference between the line contents and the surroundings. Overinsulating and the material costs entailed is as poor a practice as underinsulating. Various types of insulation are needed for different services.

It is usually the job of the design engineer to provide the necessary information for the types and thickness of insulation. Heat conservation insulation is installed to conserve the heat of the line contents. Cold insulation is usually provided for line temperatures of 35°F (2°C) and below.

Insulation is often required for the personal safety of the operators even if heat conservation is not important. In areas around walkways and platforms where safety is essential, insulation is added to pipes where contact might be made. In steam service, hot insulation is applied to pipes, fittings, valves, and flanges, as in Figure 1-13. These should be insulated with the same material as the pipe itself. In petrochemical plants, valves are usually not insulated because heat loss at a valve is not considered significant; in

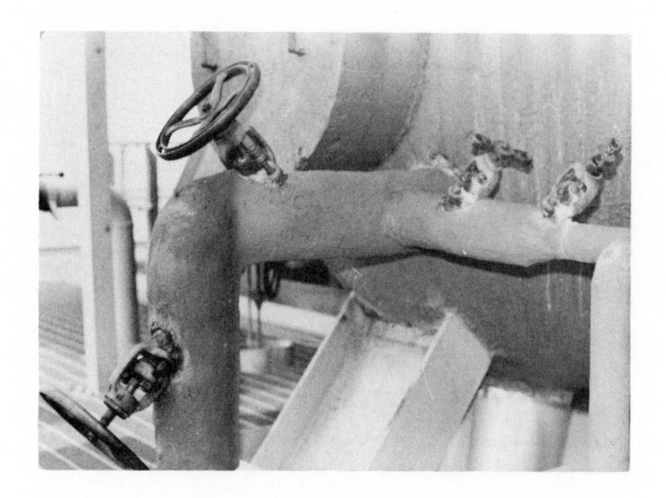

FIG. 1-13 Insulated pipelines and four Y valves.

nuclear power plants where lines are insulated, however, it is important to calculate heat loss even through pipe hangers. Insulation is often covered with aluminum shielding to protect it from weather and to provide rigidity (see Fig. 1-14). Hot insulation lines 3 in. (7.62 cm) and larger in diameter are supported by insulation shoes. (See Chapter 8 for information on shoes.) The shoes have to be at least ½ in. (1.27 cm) higher than the insulation thickness. Figure 1-15 shows an example of the use of pipe shoes. Check the manufacturer's insulation thicknesses before calling out the information on the drawing.

Insulation comes in a wide variety of materials. For pipes up to 600°F (315°C), 85 percent magnesia insulation is used. This molded insulation is quite durable and has excellent waterproof and fireproof characteristics. The recommended thickness for 85 percent magnesia insulation varies according to pipe size and temperature. Check the manufacturer's recommendations for adequate sizing of insulation.

FIG. 1-14 Fireproofed, insulated piping lines with aluminum shielding at a petrochemical installation. View shows a series of motor-operated pumps.

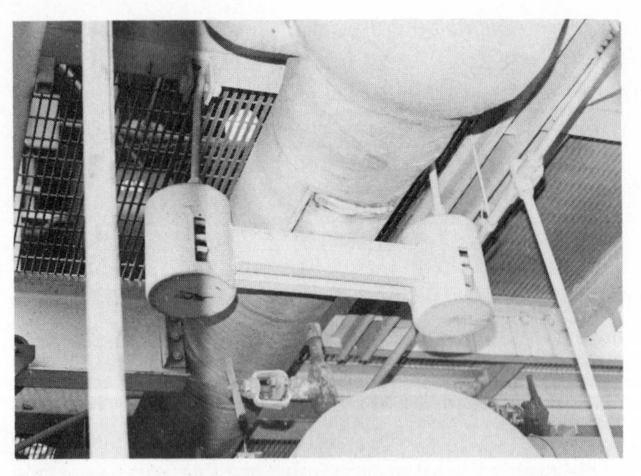

FIG. 1-15 Insulated pipeline suspended by a double-spring hanger. Note the use of a reinforcing shoe between pipe insulation and hanger.

FIG. 1-16 Insulated steam valve with bypass.

Fiberglass is another type of insulation used at up to 600°F (315°C). This material has excellent structural strength because of its rigidity. Asbestos sponge felted insulation can be used up to 700°F (371°C). This insulation has great mechanical strength and high resistance to moisture and is usually applied in layers. Diatomaceous silica and asbestos fiber insulation are used for temperatures of 600 to 1900°F (315 to 1037°C). Wool felt is commonly used for hot and cold water lines and is made up of layers of wool felt with a felt liner.

In piping systems that must have steam tracers, as in Fig. 1-16 a small steam line, usually 3/4 in. (1.9 cm) in diameter, is run directly beneath or around the line that must remain hot. Both lines are enclosed in a cover of insulation. (See Fig. 1-17 for an isometric drawing showing steam tracing; Chapter 7 and Chapter 10 discuss steam trace lines.)

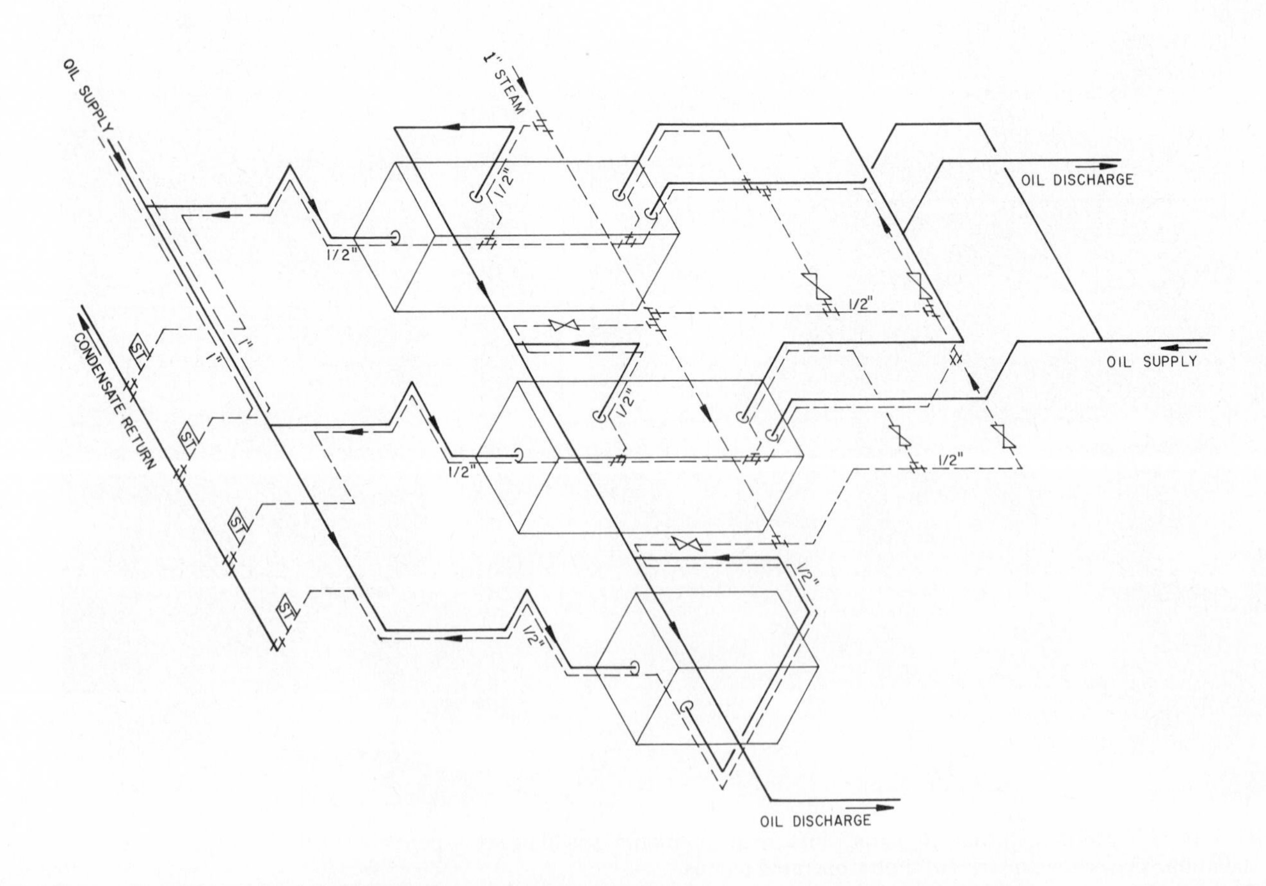

FIG. 1-17 Isometric drawing of a piping system utilizing steam tracing.

Antisweat pipe insulation is used to prevent sweating of cold water lines and can be seen on regular plumbing installations. When insulating pipes to prevent freezing, the most important factor is the velocity of the line material, especially in conditions that range far below zero. To prevent refrigeration losses, cold lines are insulated with cork, which has very low thermal conductivity and high moisture resistance. Cork comes in three basic thicknesses: *ice water thickness* is used for refrigerated cold lines 35°F (1.6°C) and above; *brine thickness* is used for carbon dioxide, ammonia, brine, and other fluids at 0 to 35°F (-18 to 1.6°C); *specialty brine thickness* is used on cold lines to -25°F (-32°C) or where surrounding temperatures are extremely high in comparison to the cold line.

This chapter is but an introduction to the many types of pipes and piping materials and insulation (Fig. 1-18). In general, piping drafters will not be called upon to know all this material upon entrance into the profession, but they should become familiar with it as they receive on-the-job training and experience. In all cases that concern piping materials, schedule numbers, and insulation, manufacturer's standard catalogs should be consulted continually to ensure that the manufacturer's design requirements are not disregarded.

FIG. 1-18 Underground cross-country oil pipeline being installed. The section being lowered into the trench has a large reinforced gate valve already attached to it. (Courtesy of Grove Valve and Regulator Company)

QUIZ

1. Name five common pipe materials.
2. What substance was mainly carried in pipes before 1900?
3. Explain how one type of welded pipe is manufactured.
4. What is a metal billet and what is it used for?
5. What is wood piping used for?
6. Why would pipe be lined with wood?
7. What is a schedule number?
8. What does ASTM stand for?
9. Cast iron pipes are used mainly for what services?
10. Why is steel pipe useful?
11. What is stainless steel?
12. What does *std.* mean when referring to pipe?
13. What is the composition of brass?
14. Name three uses of soft copper tubing.
15. What services are lead pipes recommended for?
16. Why does galvanized pipe resist corrosion?
17. Name two steel alloy piping materials.
18. What does PVC mean?
19. Name two services of glass pipe.
20. What does *continuous-stave pipe* mean?
21. Define ASA.
22. Define 3-8 NPSL.
23. What is a female thread?
24. Name three reasons for insulating a pipe.
25. Name two insulation materials for temperatures below 600°F (315°C).
26. What is fiberglass insulation and how is it used?
27. Describe cork insulation and its uses.
28. Why insulate cold lines?
29. What is rock cork?
30. Define *extra strong.*

PROBLEMS

Find the schedule number and calculate the thickness (t) for the following steel steam pipes. (You must calculate pipe inside diameter.)

	Pipe Dia. (in.)	Pressure p (psi)	Unit Tensile Stress (s)
1.	15	450	4000
2.	1-1/2	300	8000
3.	3/4	200	3000
4.	6	800	5000
5.	32	900	8500

	Pipe Dia. (in.)	Pressure p (psi)	Material	Temp.
6.	16	925	Galvanized carbon STL A120	450°F
7.	12	1500	Carbon STL A53 LapW	650°F
8.	4	475	Carbon STL A135-B	750°F
9.	1-3/4	250	Seamless alloy A335 P2	200°F
10.	3/4	150	Wrought iron A72	300°F

Find the maximum internal pressure p (psi) for the following pipes.

	Pipe Dia. (in.)	Schedule No.	Safe Unit Tensile Stress (s)
11.	3/4	160	5500
12.	1-1/4	80	4500
13.	4	40	3500
14.	5	120	3000
15.	20	30	4000
16.	18	100	8000
17.	42	20	6000
18.	34	40	10,000
19.	30	10	5000
20.	26	20	11,000

Find the total pressure on a diametral plane for the following stainless steel water pipes. (You must use inside pipe diameter.)

	psi	Pipe Length (L)	Nominal Pipe Diameter
21.	500	18 ft	3/4 in.
22.	2000	12 ft	32 in.
23.	600	15 in.	16 in.
24.	350	32 ft	12 in.
25.	800	45 in.	4 in.

Find the total resisting stress (s) for the following steel pipes.

	Pipe Length (L)	Unit Tensile Stress (s)	Schedule No.
26.	10 ft	3000	20
27.	16 in.	5500	120
28.	32 ft	6000	40
29.	18 ft	5000	XS
30.	12 ft	4500	Std

Find the schedule numbers for the following piping situations. Use this formula: schedule number = $p/s \times 1000$.

	Unit Tensile Stress (s)	Pipe Dia. (in.)	psi
31.	1500	2	350
32.	10,700	16	550
33.	12,000	10	875
34.	5000	8	970
35.	6000	4	150

Using the simplified and regular thread pipes, draw the following:

36. A 3–8 NPT external pipe and thread.
37. A ½–14 NPT external pipe and thread.
38. A 6–8 NPSC internal pipe and thread.

Electric arc welding of gate valve hub. (Courtesy of Grove Valve and Regulator Company)

2 WELDING

Knowledge of welding is important for the piping drafter for many reasons. In the first place, welds are used in manufacturing pipe and in attaching components of a piping system together. Moreover, they are used to attach the system to the supporting structure or building. In the latter case, various types and directions of loads are encountered which call for knowledge of a variety of means of support including pipe racks and pipe hangers—all taken in relation to the nature of the building that houses the system. Welding is also used in the prefabrication of pipe configurations.

The drafter will need to learn, through experience, the essentials necessary to let the welder know the type, dimensions, and position of weld to be used. In practice, the welder will, through experience in the shop, know better than the designer and drafter what to do, but the drawings provide an essential guide for the welding process and a safety check for the system.

In designing and drafting piping systems, it is important to keep the welds simple to eliminate misunderstanding and save time. Intricate welds take time and are therefore expensive; moreover, time is often lost just in interpreting the symbols.

At first the drafter will be familiar only with welds and their symbols. But, through experience, drafters will develop the knowledge of welding theory necessary for designing a system by determining adequate weld sizes and placement and welding procedure. The many fields include manufacturing of pipe, the construction of piping lines, the fabrication of special pipe configurations, the design of structural aspects of the system, and the general engineering design of pipe supports, a major field in itself.

It will be the responsibility of the designer/drafter to determine when subsystems of the piping system should be welded fabrications. Costs and material options will determine whether welding or bending will be used for directional changes in the system. Among the many considerations in the design of a system is the use of welded, flanged, or other connections. Welded fittings are used often in severe service, such as high temperature and pressure situations in petrochemical facilities or power plants.

WELDED PIPE SYSTEMS

Welded piping systems are possible for almost all services today. This type of system is a closed container with fittings, valves, and equipment forming a leakproof, maintenance-free, safe piping system. Where internal pressures are high, the welded unit offers a great safety margin. Welded fittings form smooth joints that are easy to insulate. The most common weld is the butt-weld, where both ends are machine beveled to form a groove between the mating parts. This forms an efficient joint when fitted together and fused by welding. Socket-welded fittings are also used, but in less critical situations.

Welding is preferred for systems that involve infrequent dismantling and require strong leakproof connections. A drafter must take care to design a system which will allow adequate room for on-the-job welding construction and also for convenience—especially when welded valves and fittings or other equipment must be taken from the line for servicing and replacement. (See Chapter 4 on welds and welded ends for valves.)

The drafter will eventually learn to avoid sharp bends in the system to reduce turbulence and air pockets. These bends may be designed out of the system by the use of fabricated bends or special welded connections and fittings. (See Chapter 5 for symbols used in the drawing of weld-connected pipe systems.) Welding provides for cheaper, lighter, stronger, safer joints and connections. Other types of connections that are easier to assemble and dismantle must also be taken into account, however. Flanged connections for valves, instruments, and the like, utilize welding to connect the pipeline to the flange. Welded lines require more working space in fabrication.

Welding plays an important part in the fabrication of the pipe and the piping system itself. The drafter should be familiar with all the welds and welding processes and know when to use them.

WELDING METHODS

Welding is the procedure by which two pieces of metal are fused together along a line or a surface between them or at a certain point. Welding can be classified by process. *Non-pressure welding* (fusion and brazing) predominates in the piping field. No mechanical pressure is applied. The pieces of metal are welded at the point of contact by heat which is created by an electric arc or by a gas or oxyacetylene flame. *Pressure welding* (forging) or resistance welding is used to form a joint by the passage of electrical current through the area of the joint as mechanical pressure is applied. This type of welding is usually done by machines only and is not used very often in the construction of piping systems. It is usually more convenient, however, to classify welds into three separate categories: *resistance welding, gas welding,* and *arc welding.*

Resistance Welding

Resistance welding is the process by which heat and pressure are applied at the same time, usually by a machine. Two or more parts can be welded together by passing an electrical current through the work as pressure is applied. This process is not as common as the others in the piping field. The main types of resistance welds are spot, seam, projection, flash, and upset.

Gas Welding

In the form of fusion welding known as gas welding, heat is created by the combustion of a gas and air or pure oxygen. A welding filler is sometimes used as a flux. In *oxyacetylene welding* (sometimes called autogenous welding or gas welding), a flame is produced by the combustion of oxygen and acetylene gases (Fig. 2-1). This type of welding is used less

FIG. 2-1 Welder completing weld on elbow. (Courtesy of Babcock & Wilcox, Tubular Products Division.)

frequently than the others and is manually applied. Besides guiding the torch, the welder also introduces the welding rod as the extra material. Gas welding is usually used only for small (2 or 3-in.-diameter) pipes; arc welding techniques are used for larger piping. In addition to its use for welding, this type of heat production is used often in the fabrication of piping—as a cutting torch, for creating different piping shapes, and for removing fittings and valves from the line after they have been welded.

Arc Welding

The third type of welding, which is the most commonly used at present, especially in the piping field, is arc welding. This section examines several arc welding processes: submerged arc welding, shielded arc welding, and gas-metal arc welding.

Submerged Arc Welding In this welding process, coalescence is produced by the heating of an electric arc between the electrodes and the work or between the electrode and the base metal. The work is shielded by a blanket of granular, fusible material (*flux*). In most cases, no pressure is used, and some filler material is created by the melting flux which also forms a slag shield that coats the molten metal. The slag must be removed at the end of the process.

The filler material can also be obtained from a supplementary welding rod or from the electrode itself. In this process, loose flux (also called melt or welding composition) is placed over the joint to be welded. The welding zone is completely covered by a blanket of this material. After the arc is established, the flux melts to form a shield which coats the molten metal (Fig. 2-2). A bare wire electrode is used in this process instead of a coated electrode, and the flux is supplied separately. In the manufacture of pipes, especially circumferential pipe joints, this type of welding is often automated, as shown in Figure 2-3.

FIG. 2-3 Machine welding using the submerged arc process: As the assembly rotates under the welding machine, the weld rod is fed automatically at a predetermined rate and powered flux is continuously applied just ahead of the arc, effectively burying it. The fused flux is curling away from the area already welded. Depending on wall thickness, multiple passes must be made. Note that the welder does not need the usual welding helmet. (Courtesy of Grove Valve and Regulator Company.)

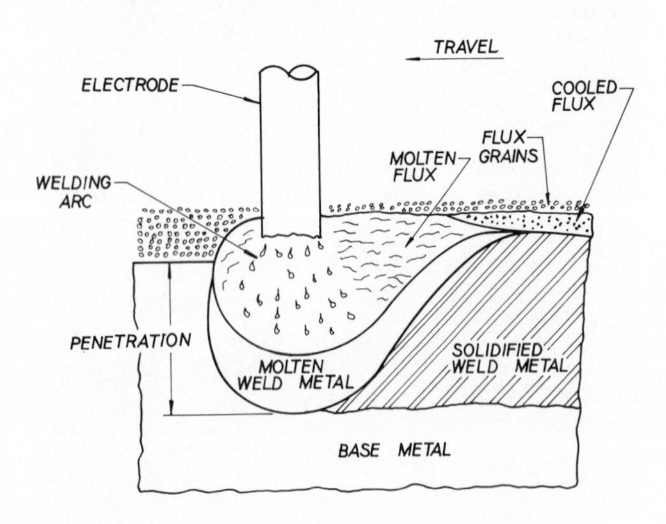

FIG. 2-2 Submerged arc welding.

Shielded Arc Welding In shielded arc welding, the metal arc is produced by the contact of a coated metal electrode and the material to be welded. Fusion takes place by heating the metal in the electrode to a temperature that causes them to melt together. Figure 2-4 shows how the manual shielded arc weld proceeds. Note how slag is formed on top of the base metal or on top of the solidified weld metal. This slag must be removed after the welding process is completed. Figure 2-2, by way of contrast, shows how the electrode in submerged arc welding extends into the work itself and how the flux and base materials are fused together in molten weld metal; note that the penetration is much deeper than in shielded arc welding.

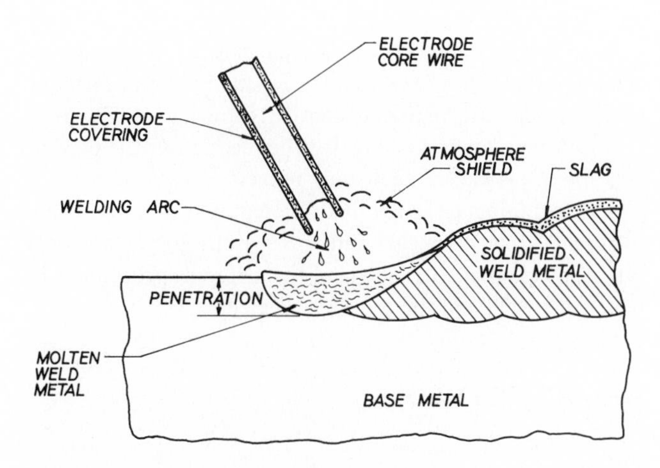

FIG. 2-4 Manual shielded metal-arc welding.

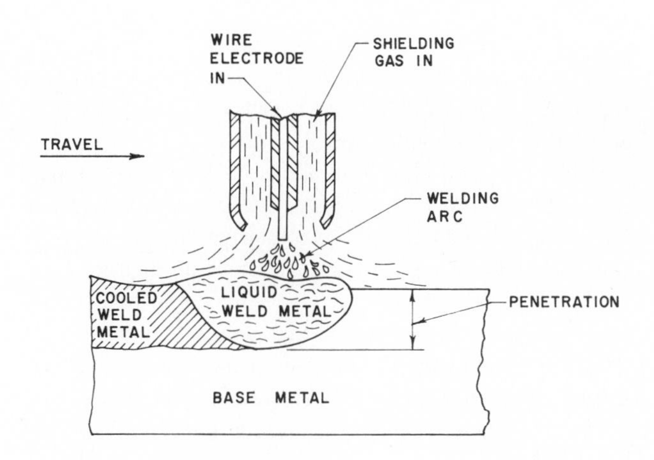

FIG. 2-5 Gas-metal arc welding.

Shielded metal arc welding is almost always accomplished manually by a trained welder rather than an automated machine. It is used quite often in the structural fabrication of piping systems because of its ease of application in the field. It is also used for *tack welding*–holding parts in position prior to welding up a series of deposits in the joint manually. Tack welds are used on almost all pipes to be joined. After the pipes are aligned, they are joined by four or more tack welds before undergoing the complete welding process.

Gas-Metal Arc Welding This process is used for fabricating pipe supports and connecting them to structural members in a piping system. For gas arc welding, heat is created electrically as in the process just described, but the shielding is accomplished by a blanket of gas. Pressure may or may not be used; in the piping field, welding is generally pressureless. A filler metal may or may not be used; in most cases it is added during the process itself. The filler metal may also be added to the welding zone prior to welding. In this process, inert gases are released onto the welding area to form a blanket. In the construction of pipes or piping joints, gas also is applied to the inside of the pipe until it is sealed off after the first pass. This welding procedure is used for exotic metals such as magnesium, aluminum, alloyed steels, and mild steel pipes, especially where pipes are to be used for high-pressure service.

Gas-metal arc welding is also referred to as *gas-tungsten arc welding* (GTAW). Since the electrode in this case is usually of tungsten and not a filler metal, it does not melt. The GTAW process produces root beads of high quality and is seldom used for the entire weld unless there are stringent standards to be met. Characteristics of this process include high-quality welds of nearly all metals, no filler metal across arc stream (consequently no splatter), little postweld cleaning, multiposition welding, and no slag. The base metal is supplemented by a filler metal except where unnecessary (as in the welding of thin-walled pipe).

Figure 2-5 shows gas-metal arc welding, a process in which the electrode wire is inserted through to the arc area. This type of welding does not create a deep penetration. The figure shows how the shielding gas–an inert gas such as helium or argon–is used to shield the work. Filler metal is sometimes added.

WELDING MATERIALS

Weldability is the capacity of a metal to be welded in relation to its suitability to the design and service requirements. Metals that become fused during the welding process undergo changes similar to those which occur during manufacture. Chemical, thermal, physical, and metallurgical changes make it essential for the engineer to understand the nature of the materials to be fused. The metallurgy of welds is the metallurgy of the material.

Since the weldability of different materials varies greatly, so does the process by which the weld is completed. When welding cast iron to steel, for instance, cast iron rods are used as the welding material and the steel must be preheated before an adequate weld can be made. When welding steel castings, there is no easy rule for the process because carbon content can vary drastically between types of steel. Steel welding rods usually produce an adequate weld.

In the piping field, brass is usually brazed instead of welded because of the high temperature produced by the welding process and the low melting point of brass in comparison to other metals. In welding copper, that metal must be used as a filler and care must be taken not to produce oxidation.

Carbon steel welding is usually completed by the use of shielded metal arc welding with a covered electrode. Rod iron has characteristics similar to those of mild steel, and a similar process is used in its welding. Carbon molybdenum steel is becoming less popular because of its unfavorable reactions to high temperatures.

In aluminum and aluminum alloys, most of the commercial welding and brazing processes can be used, though the most common are the inert gas tungsten arc and inert gas consumable metal arc processes. Various problems are encountered when using the acetylene process to weld aluminum because of an oxide film that prevents metal flow at welding temperatures. Aluminum is characterized by its low melting point and high thermal conductivity. Great care must therefore be taken when welding this material. In the nickel and nickel alloy group, adequate welding can be accomplished by welding similar materials to themselves.

When dissimilar metals are welded, the weld deposit can be of a different chemical composition than the parts to be welded or the pieces themselves can be of different metallurgical composition. In addition to differentiating between the weld deposit and the joined pieces or between one piece and another, it must also be determined whether the metals themselves have dissimilar major constituents and dissimilar alloying metals. Three different metals can result in a joint where the weld metal becomes a composite of the filler metal and the base metals.

Dissimilar metals are usually welded by fusion welding, resistance welding, soldering, brazing, or pressure gas welding. The major concern is the melting temperature of the metals to be welded. If the melting temperatures are similar, it is usually possible to use gas or arc welding with the addition of a filler metal of the same composition as one of the two pieces to be joined. If the melting points are quite different, then soldering, braze welding, and brazing are used.

With the introduction of special plastics in piping, the welding process has been adapted for the fusion of different parts of a plastic piping system to form an airtight weld. Care must be taken to allow for the reaction of the particular type of plastic to the heating process.

The beginning drafter will not be required to know all the different kinds of welds and the reasons for using them, but it is essential to understand the different procedures by which the weld and the joint are created.

WELDING SYMBOLS

Welding symbols are the shorthand by which the drafter can communicate to the fabricator the total process of joining two pieces of metal. All welds can be identified by their profile or cross-sectional view. In piping drafting the fillet and groove welds are probably the most common.

The welding drawing shows the parts or units that are to be made up by welding. The symbols define and locate the specific welds to be used. Each joint in the welding process must be described fully. The *weld symbol* (Fig. 2-6) denotes the desired type of weld: fillet, square, beveled, J, U, V, flare bevel, flare V, backed, weld, arc seam, spot, plug, slot. The *complete weld symbol* takes into account all welding information that might be needed: weld type, size, length, location, place of construction (field or shop). Figure 2-7 shows the standard weld symbol and the location of its various elements. This type of symbol, with the exception of the field weld flag which is used in only a few companies, is standard throughout the industry. Figure 2-8 shows the components of the complete welding symbol and their location.

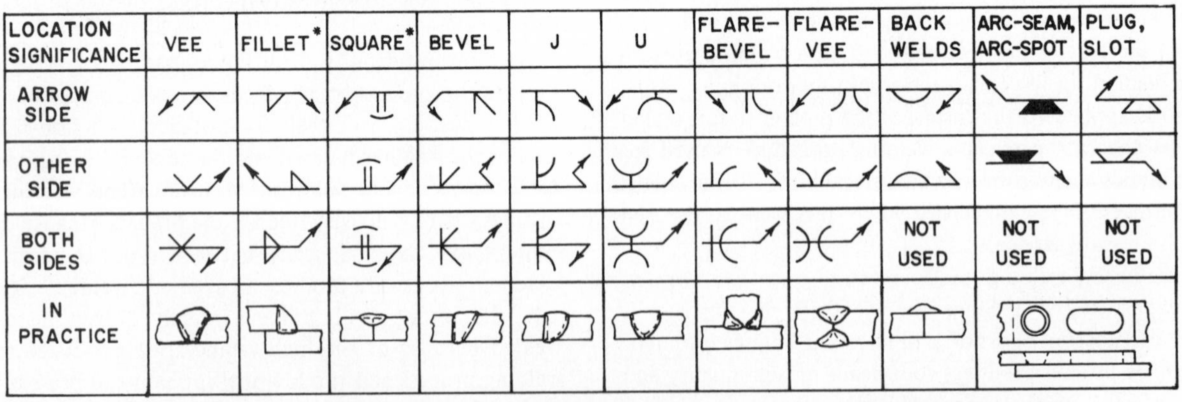

LOCATION SIGNIFICANCE	VEE	FILLET*	SQUARE*	BEVEL	J	U	FLARE-BEVEL	FLARE-VEE	BACK WELDS	ARC-SEAM, ARC-SPOT	PLUG, SLOT
ARROW SIDE											
OTHER SIDE											
BOTH SIDES									NOT USED	NOT USED	NOT USED
IN PRACTICE											

* NOTE: THERE IS NO JOINT PREPARATION ON FILLET & SQUARE WELDS. THE SYMBOLS FOR THESE REPRESENT THE WELDS; OTHERS REPRESENT THE PREPARATION FOR THE WELD.

FIG. 2-6 Weld symbols.

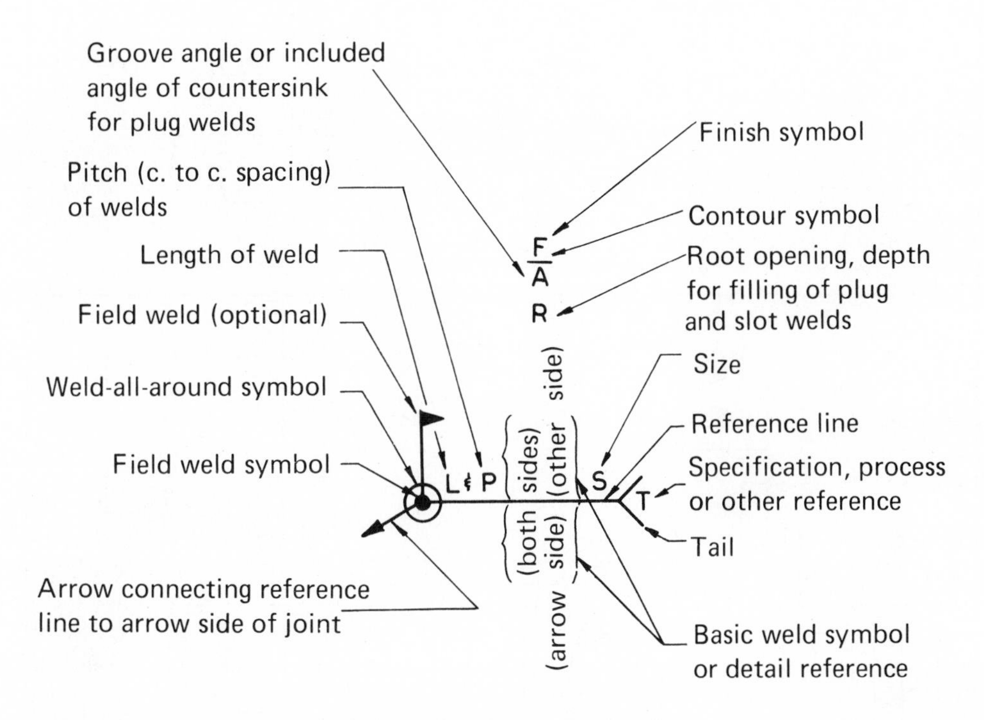

FIG. 2-7 Location of elements for standard welding symbols.

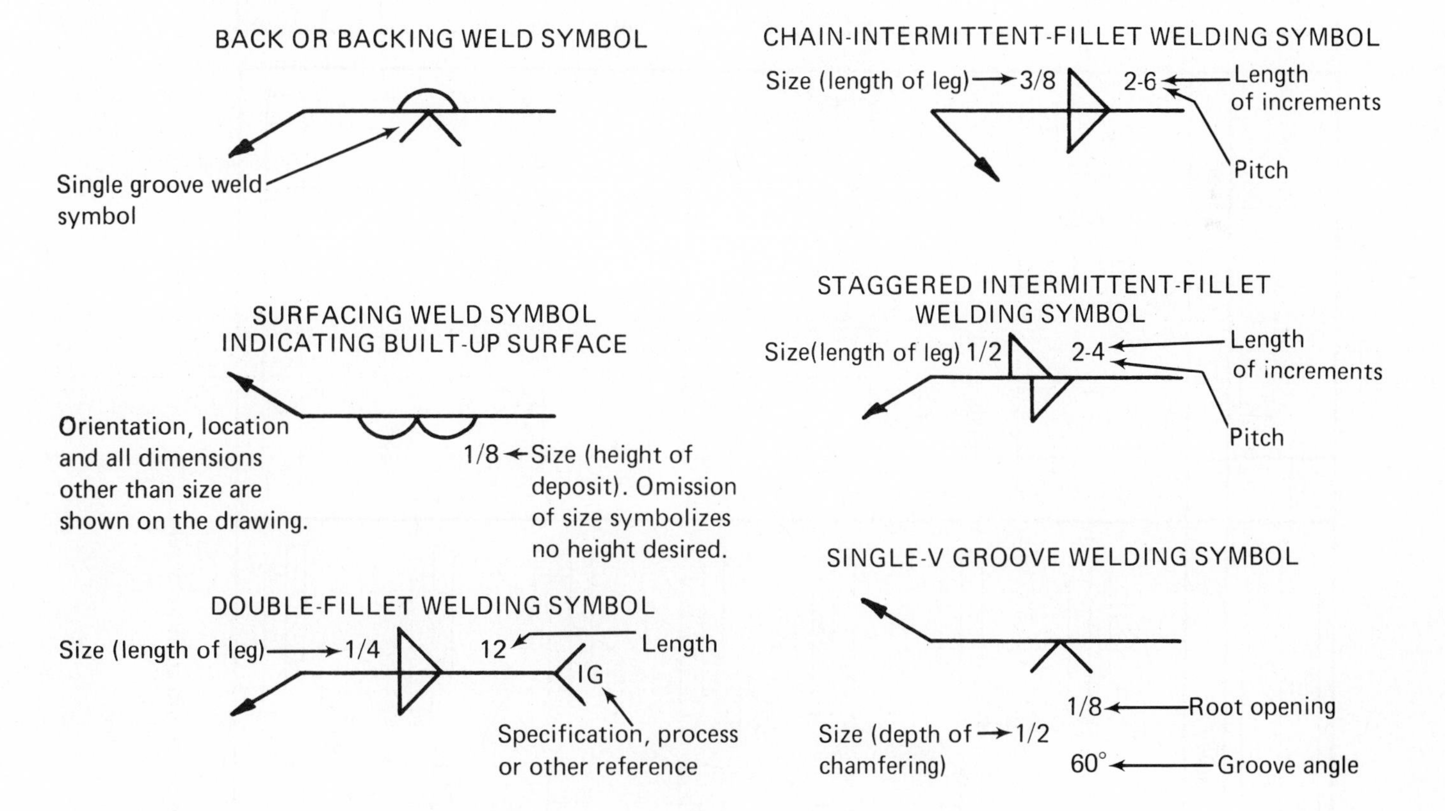

FIG. 2-8 Typical application for welding symbols.

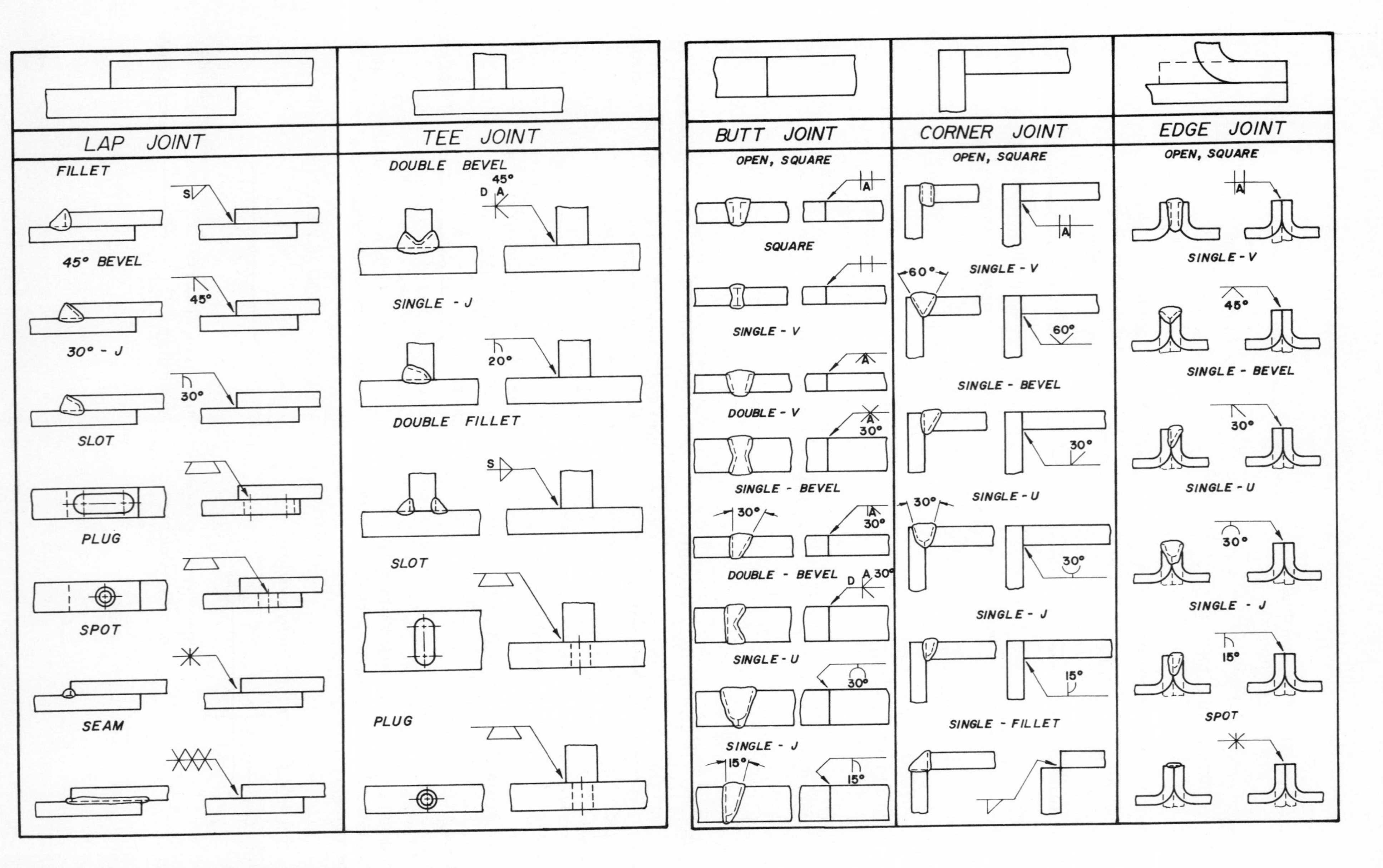

FIG. 2-9 Major joints and applicable welds.

Figure 2-9 shows five different joints that may be encountered by a drafter in the construction of welded connections. This figure offers a sample of various welds as applied to the particular joint.

Welding symbols are composed of three basic parts:

1. An arrow which points to the joint
2. A reference line upon which all the dimensions and other data are placed
3. The weld symbol itself, which indicates the type of weld required.

Usually the same welding process is used throughout a drawing. If this is not the case–say the drawing contains submerged arc welding by machine and manual welding–each process must be noted on the symbol when pointing to the joint to be completed. The welding process should be placed at the tail of the welding symbol. (The tail is omitted when references are not needed to supplement the symbol.) Drafters may use templates for drawing welding symbols but should not become dependent on them because drawings require various sizes of symbols, depending on whether or not it will be reduced onto microfilm.

The welding symbol should be of an adequate size to be readily visible to the fabricator. At the junction of the arrow and the reference line within the welding symbol is usually placed a flag or a solid field weld designation for any work that is to be performed on the job site.

The edge preparation for such welds is completed at the fabricating plant, seldom at the job site itself, so shop drawings contain all the edge preparation designations and only field drawings contain field weld symbols. The use and placement of welding symbols discussed here are adapted from ANSI Y32.3.

The arrow in a welding symbol connects the reference line to the joint. The side that the arrow is on is called the *arrow side* of the joint. The side opposite the arrow is the *opposite* of the joint, except for plug, slot, seam, and other projection welds where the arrow connects the symbol to the surface to be acted upon. The arrow side of the joint is always considered the near side. Welds on the arrow side of the joint are shown by placing the weld symbol on the side nearest to the reader. To show welds on both sides, the weld symbol is placed on both sides of the reference line.

In many situations one welding symbol is shown on the drawing and a note is included that specifies the type of weld to be used on the whole drawing–such as all welds to be 1/4 in. unless otherwise noted. On welds that are to be on all sides of a particular joint, the weld all around symbol is used at the junction of the arrow line and the reference line, which may also contain a field weld symbol at the same joint. When welds must be finished or contoured, this requirement must be shown on the symbol as in Fig. 2-8. It is also possible to combine different weld symbols in one welding symbol. The weld symbols must be placed according to the side of the joint they are to be made on.

TYPES OF WELDS

The drafter will be most concerned with symbols for the various welds and joint preparations. Figure 2-10 shows how weld symbols and types of welds are classified by process. Resistance weld symbols are grouped under flash or upset, projection, seam, and spot with supplementary descriptions such as contour weld and field weld. The supplementary types are used in both arc and gas welding and in resistance welding.

Arc and gas weld symbols are divided into groove types (bevel, square, J, U, V), bead, fillet, plug, or slot welds. For bead and fillet welds, there are no special preparations for the metal. The essential difference between the groove welds is the end preparation of the material to be welded, whether it is to remain square, beveled, or machined in a V, U, or J shape. Although these welds can be combined, an effort should be made to keep welding symbols of similar joints both similar and simple.

Groove Welds

Figure 2-10 defines the five basic groove welds (bevel, square, J, U, and V) though they also have various combinations: single V, single bevel, single J, single U, double V, double bevel, double J, and double U.

For small to medium-diameter pipes, single V-groove joints are used to join pipe sections. In double-bevel groove joints, applicable to 24 in. (60.96 cm) and larger, including vessels and tanks, the welding operation is usually completed both inside and outside. This size of pipe allows for better inspection and inside finishing. Double-bevel joint makeup is the common procedure for this operation. These welds are made in a groove between adjacent ends, edges, or surfaces to be joined on various joints.

The edges in a groove weld are usually prepared by a flame cutting torch. In welded joints, a flame cuts V bevels for groove welds. Whether single or double, these are probably the easiest, most economical welded joints used to join two ends, especially when dealing with pipe. With this type of joint preparation, the most common faulty connection, whether using submerged arc, shielded arc, or another process, is the lack of penetration which sometimes occurs at the root of the weld. When pipe is welded on one side only, as is often the case because the inside of the pipe is not accessible for back welding or other inside preparation, lack of penetration or incomplete fusion can be a major factor in the failure of pipes whose service demands strong, safe, accurate welds under high temperature, high pressure, high thermal or mechanical fatigue, stress, corrosion, or subzero temperatures. These welds are classified according to whether they entail complete or partial penetration.

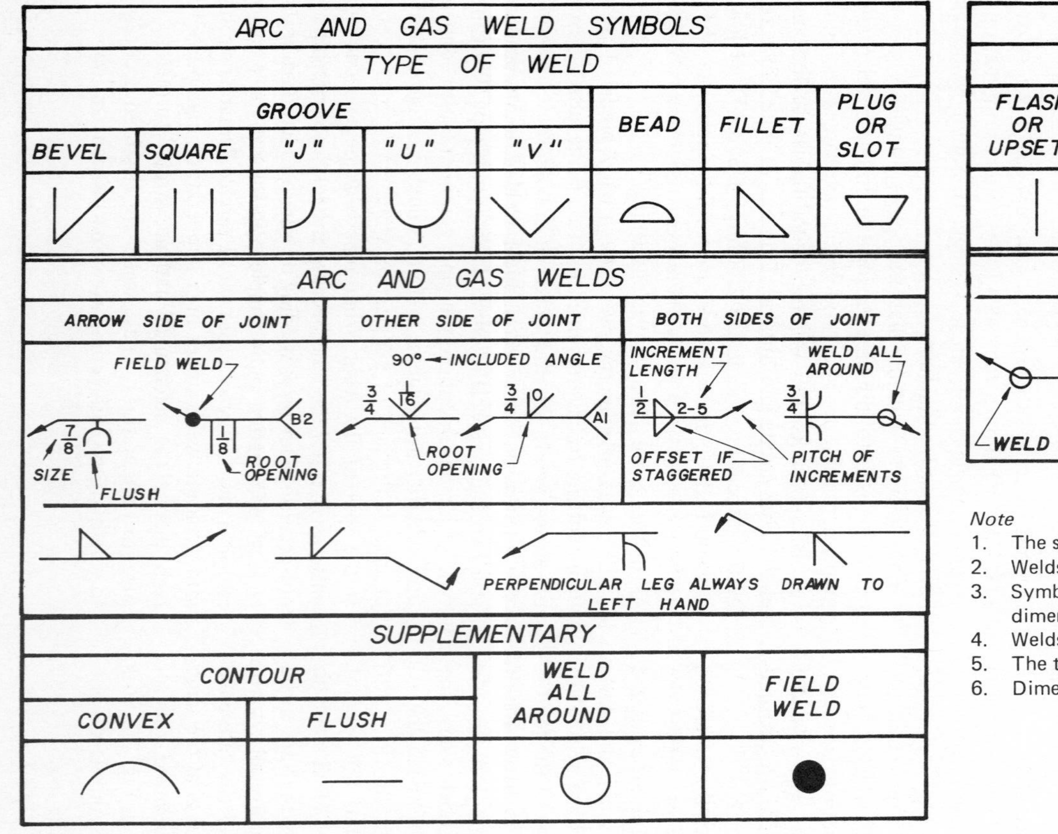

Note

1. The side of the joint to which the arrow points is the arrow side.
2. Welds on both sides of a joint of the same type are of the same size.
3. Symbols apply between abrupt changes in direction of joint or as dimensioned unless the all-around symbol is used.
4. Welds are continuous and of user's standard proportions.
5. The tail of an arrow may be used for specification reference.
6. Dimensions of weld size, increment length, and spacing are in inches.

FIG. 2-10 Gas and arc weld symbols.

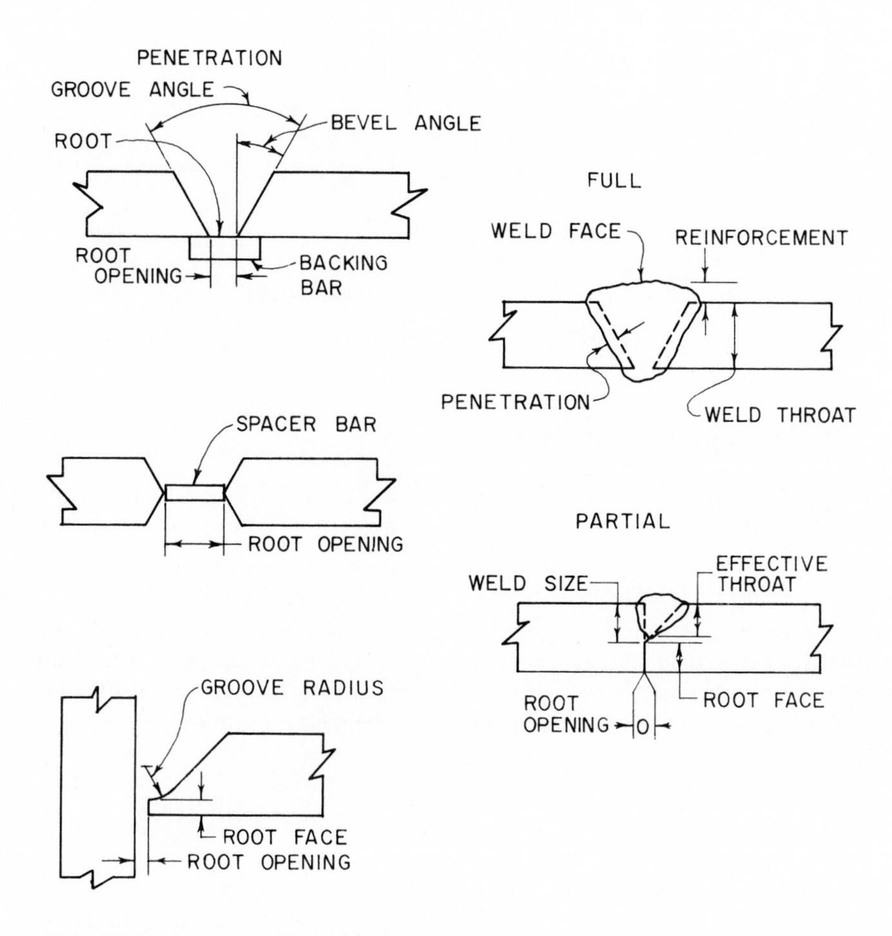

FIG. 2-11 Groove weld terminology.

When a joint is prepared for welding, the thickness of the material must be taken into account. In small thicknesses such as 1/16 to 1/8 in. (0.16 to 0.32 cm), welding can be successful when the edges are square. When the edges are thicker than this, bevels must be made to create an adequate joint. Otherwise the flame will not be hot enough to produce adequate fusion. A general rule to follow when material is thicker than 1/8 in. (0.32 cm) is that the edges must be prepared for a groove channel and during the welding process a filler material must be added.

In most cases steel and iron should be beveled at an angle of 45 degrees. The included angle of the bevel should be approximately 90 degrees and V-shaped. Many groove welds employ a backing bar or backing weld. Figure 2-11 shows cross-sectional views of various groove welds and the end preparation necessary for that particular joint including the use of backing bars and spacer bars.

The ASME boiler code and other regulating codes require full penetration welds, especially for pressure vessels and piping services that are considered critical. Backup rings are therefore employed in the construction of vessels and pipe joints where full penetration is required. Backing rings reinforce the root of the joint and vary in thickness and shape depending on the joint requirements. Figure 2-12 shows the groove weld joint breakdown and the terms for parts of a groove weld joint. It also shows the difference between partial and full penetration. In this figure the penetration is partial. In full penetration, the root opening may be called out or not be used at all.

Root spacing (Fig. 2-13) is used to minimize lack-of-penetration notches caused by insufficient spacing or tight butting. Root spacing is just one of the considerations that determine the quality of the joint. Alignment is also important, although slight imperfections are tolerable.

Open roots are used to give total joint penetration in butt welds. They also create a secure attachment to the backing rings. The drawback of using open roots is that they may create weld metal penetration beyond the root face or improper root fusion. In cases where this result is important, a single or double U-butt joint with closed roots is employed (at increased cost).

In many cases, groove weld symbols can be combined on

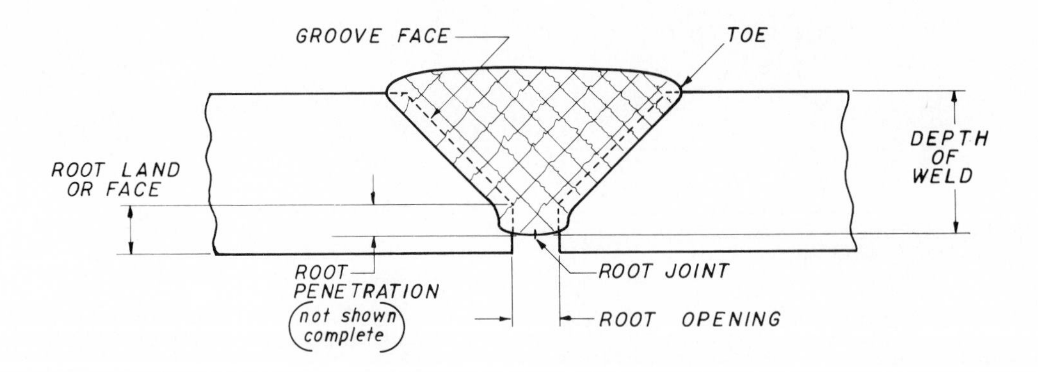

FIG. 2-12 Groove weld joint breakdown.

the same weld. Here again, it must be mentioned that symbols, though they may be correct, should not be overly complicated but should reflect a simplified accurate method of joining two ends without confusing the welder or fabricator. Often a drafter will be called upon to detail different cross-sectional views of the joint geometry including backing, space, or extension bars and to show whether the weld is a full or partial penetration and its various angles and dimensions. *Joint geometry* is the basic cross-sectional shape of the joint prior to welding. Figure 2-9 shows the five most

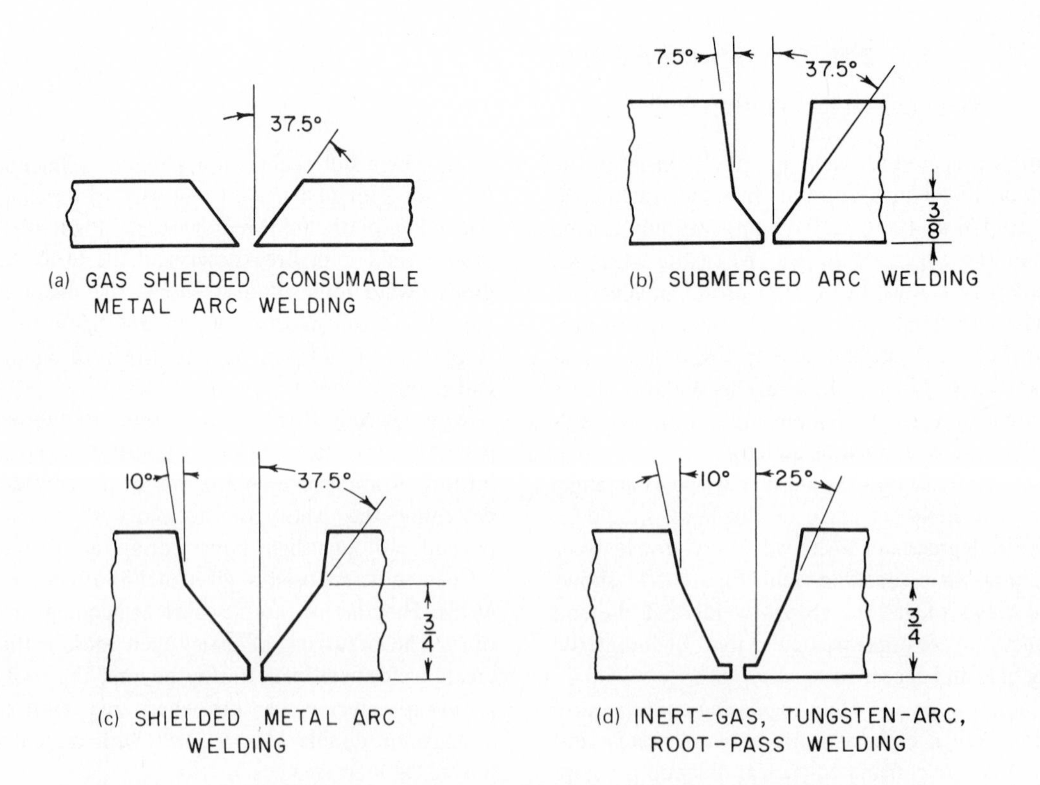

FIG. 2-13 Recommended groove joint preparation for various welding processes.

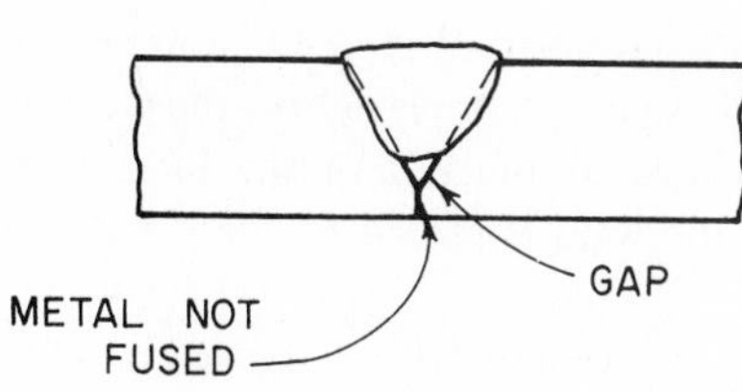

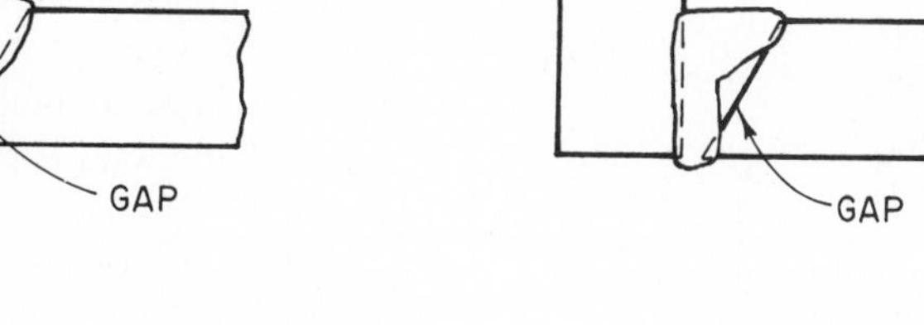

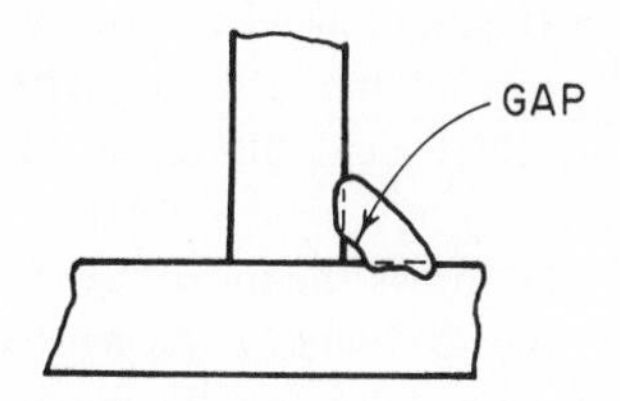

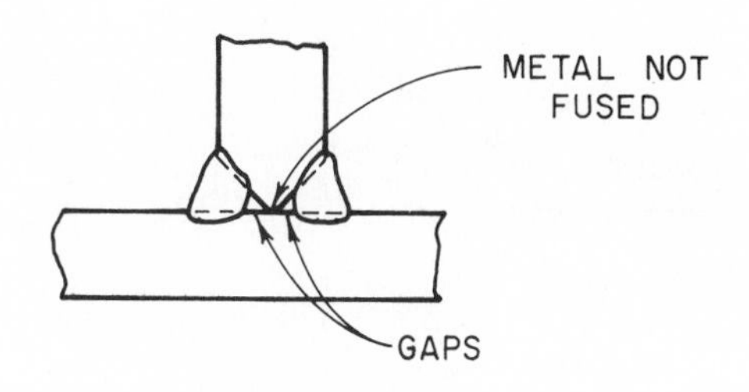

FIG. 2-14 Incomplete fusion and penetration.

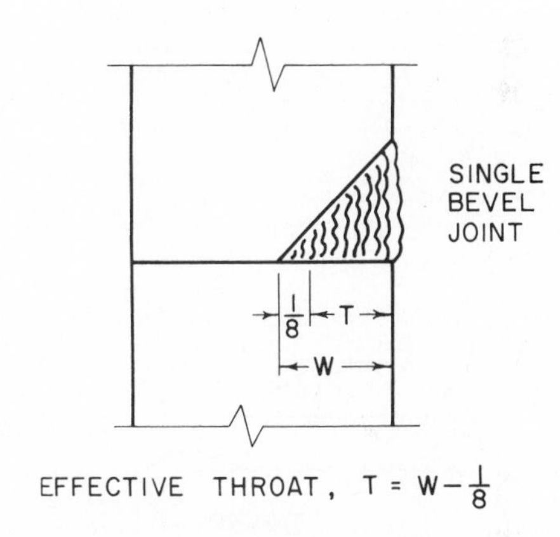

FIG. 2-15 Single-bevel joint (effective throat).

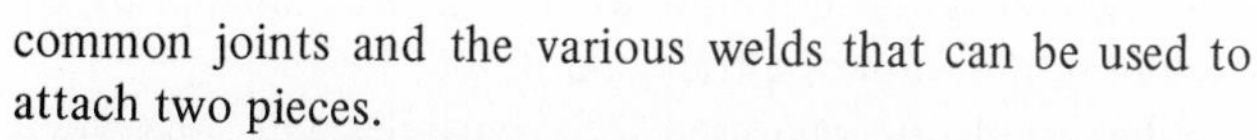

common joints and the various welds that can be used to attach two pieces.

Complete penetration is where the weld and the base metal are fused through the entire depth of the joint, which may or may not entail backing bars or bead welds. Partial penetration groove welds are used when full penetration is not necessary because of stress. The only difference between the two penetrations is in the depth of the end preparation or edge preparation: Partial penetration does not cover the full thickness of the two materials to be joined. Incomplete fusions, as shown in Figure 2-14, can cause failures in the welded joint.

The flush and convex supplementary weld symbols are also applied to groove welds, as when the outer contour of the weld must be altered by grinding or machining. All dimensions of groove welds are shown on the side that the weld symbol is on. Figure 2-15 shows the effective throat of a single-bevel joint used in groove welding. Figure 2-16 shows both the weld sizes and how they are specified in both a groove and a fillet weld.

Generally, the length of a groove weld is not placed on the symbol because the groove weld is for the entire length of the piece unless otherwise noted (as in circumferential pipe welding). There are variations regarding end preparation (V, bevel, J, U, groove), root opening, weld size, and angle. Most of these variations are provided for in the book of standards used by most companies.

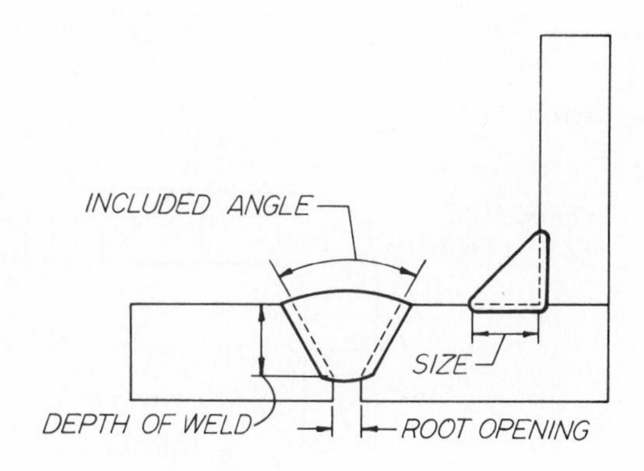

FIG. 2-16 Weld sizes.

Fillet Welds

Figure 2-17 describes fillet weld terminology. The dimensions of fillet welds are placed on the welding symbol. The weld size goes to the left of the fillet weld symbol; the length of the weld is placed to the right of the basic weld symbol.

Fillet welds usually have a triangular cross-section and join two or more surfaces at right angles—such as lap, tee, and corner joints. They are often used in combination with groove welds for corner joints. The length dimensions are omitted in the fillet weld symbol when the weld extends the full length of the material. The all-around symbol is used in the fillet weld the same way as in a groove weld symbol.

Figure 2-18 shows the maximum size of an effective fillet weld and how to calculate the necessary dimensions. Weld size is determined by the thicker of the two parts to be joined. In most cases weld size need not exceed the thickness of the thinner part unless the larger size is required by calculated stress.

Dimensions for fillet welds, as for groove welds, are shown on the same side as the weld symbol. The various parts of a fillet weld include the leg size and the throat of the weld. The cross-sectional view of a fillet weld shows the throat area, which is the effective weld bead area created by the welder in the form of a triangular fillet.

Fillet welds are also used in larger holes and slots where

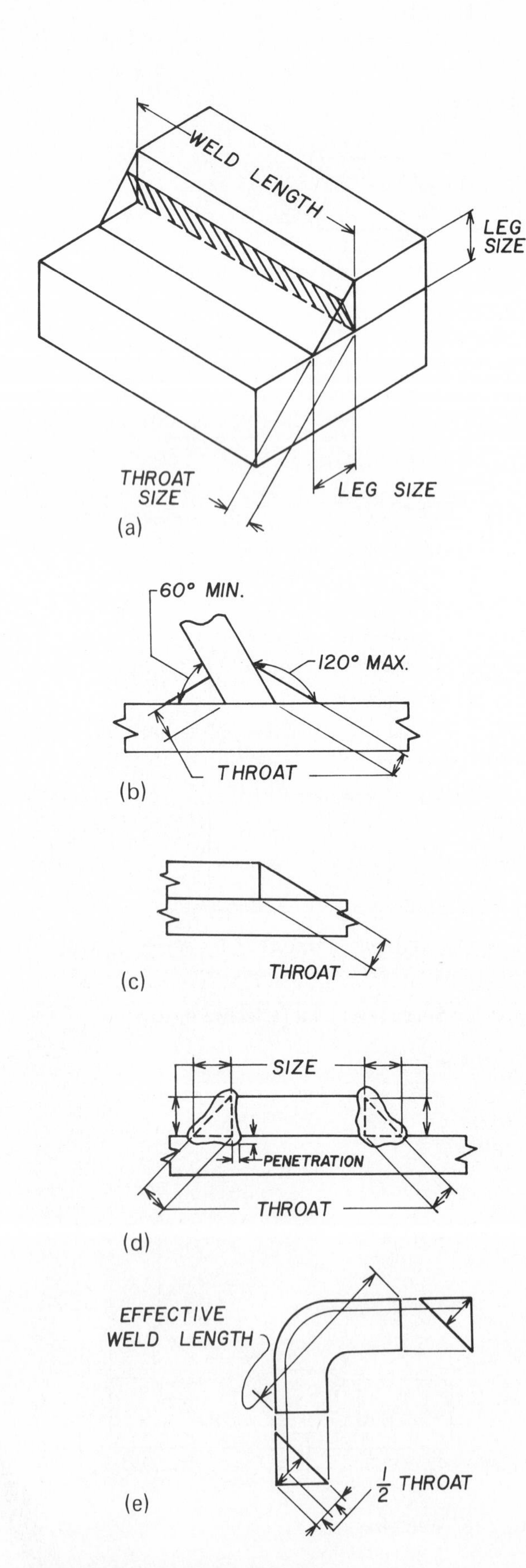

FIG. 2-17 Fillet weld terminology.

MAXIMUM SIZE OF EFFECTIVE FILLET WELDS

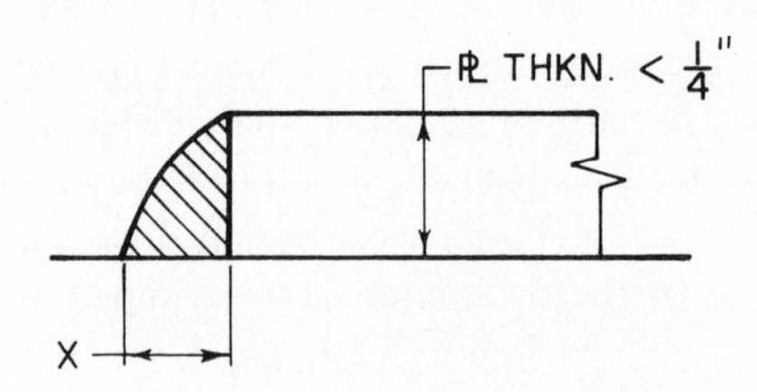

Maximum effective leg size of fillet welds for 1/4" thick material or less shall be equal to the plate thickness (PL).

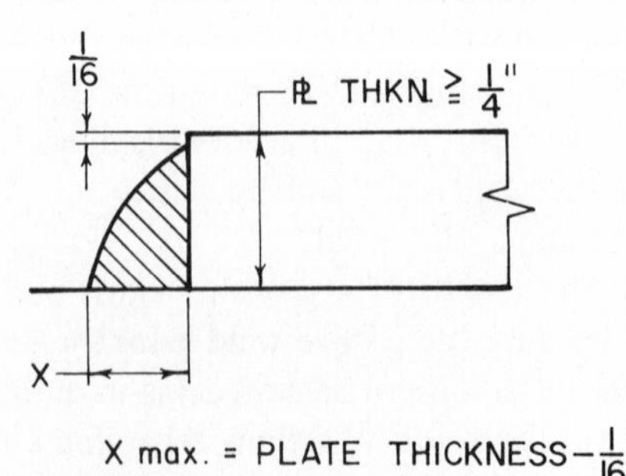

Maximum effective leg size of fillet weld for 1/4" thick material or more shall be equal to the plate thickness (PL) less 1/16".

FIG. 2-18 Fillet weld calculation.

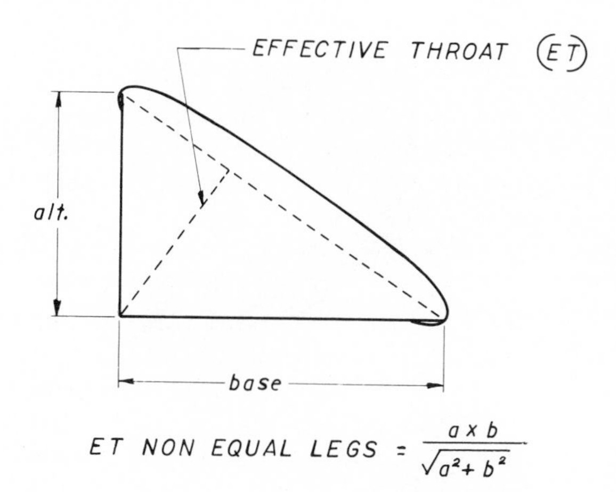

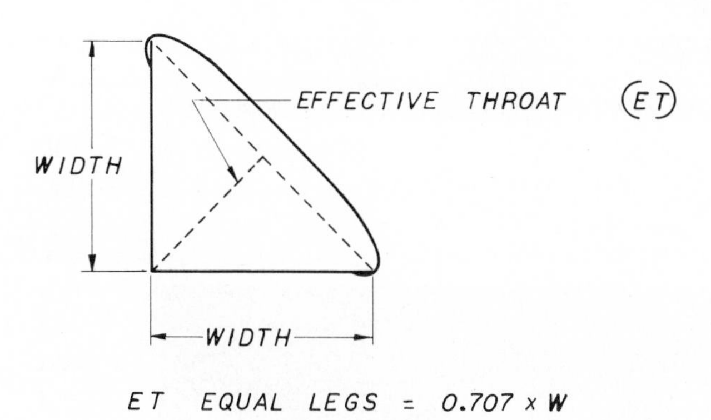

FIG. 2-19 Fillet weld effective throat.

plug or slot welds are inappropriate. Generally, fillet welds are not finished unless a specific finishing process is called out on the symbol (which is rare in piping and pipe support welding). Figure 2-19 provides the formulas for calculating the effective throat of the fillet weld. There are two basic types of fillet welds: those with equal legs and those with unequal legs.

Plug and Slot Welds

The trapezoidal basic weld symbol is used for designating plug and slot welds. All the rules for drawing symbols and their locations apply for this type of weld as well. Plug and slot welds are often used in joint and lap joints for reinforcement, but they are not common in piping welding. When the slot or hole is too large to make plug or slot welds effective or economical, fillet welds are used.

Other welding procedures are also available, such as seam, flash, upset, and resistance welds, but, generally, they are not used as often. A drafter should, when necessary, consult one of the many welding standard publications to deepen his or her knowledge in this field.

In Fig. 2-20, a valve body has been made out of thick rolled steel with a submerged arc weld along its seam. Figure 2-21 shows a welder putting the finishing touches on an end connection for a large pipeline ball valve. Figure 2-22 shows three separate shop-fabricated pipes with their corresponding elbows and tees. The shop will receive drawings similar to Fig. 2-23 which give the exact dimensions from which the piece may be fabricated. All joints in this case, as in Fig. 2-22, are welded connections which employ a bevel joint for a groove weld.

FIG. 2-20 Finished gate valve body using all welded construction.

FIG. 2-21 Welder completing the end connections for a large pipeline ball valve.

FIG. 2-22 Prefabricated welded pipes and fittings. (Courtesy of Flowline Corporation)

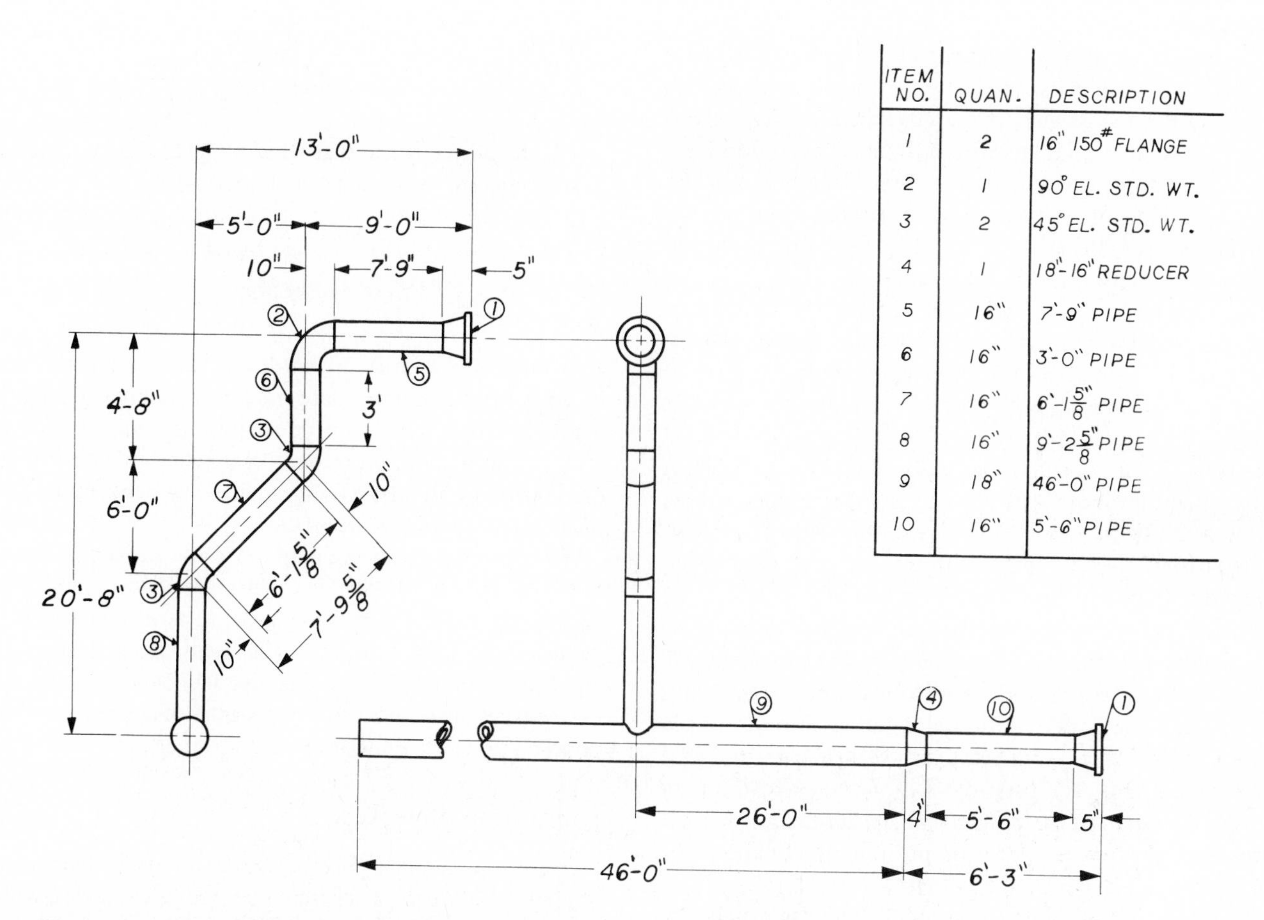

ITEM NO.	QUAN.	DESCRIPTION
1	2	16" 150# FLANGE
2	1	90° EL. STD. WT.
3	2	45° EL. STD. WT.
4	1	18"-16" REDUCER
5	16"	7'-9" PIPE
6	16"	3'-0" PIPE
7	16"	6'-1 5/8" PIPE
8	16"	9'-2 5/8" PIPE
9	18"	46'-0" PIPE
10	16"	5'-6" PIPE

FIG. 2-23 Orthographic spool (all-welded assembly).

PIPE END PREPARATION

Pipe end preparations are usually performed in the shop prior to shipping, so the end preparation instructions, symbols for welding, and any other callouts must be on the shop drawing. With each welding process, there are joint variations that may not work well in all cases. When one is designing specific joints to be welded in pressure vessels and piping, it is essential to take into account the welding procedure and, with the welder in mind, to create an effective weld—not one that is difficult to fabricate even if it is theoretically correct. This practical concern is the designer's main responsibility to the welder who must put the design into practice and therefore produce a safe and accurate welded assembly. The welding configuration must be readily accessible, not inherently prone to defective conditions, and readily weldable.

The cost of the welding operation is another important factor in the designing of welded fabrications and their corresponding joints, though cost never take precedence over safety and quality. Cost must be weighed against the possibility of reducing weld defects and poor notch conditions.

JOINT PREPARATION

Proper weld joint preparation is essential to the smooth operation of the system. Inspection methods do not always find improper weld penetration or flaws, so it is important to specify joint preparation according to welding operation and the materials to be welded. An inadequately designed weld joint is likely to fail in a piping system subject to radical thermal and mechanical shock aggravated by excessive rigidity. In this situation it becomes essential to avoid a design with sharp notches. Figure 2-13 gives examples of pipe and weld end preparations that are recommended for various welding processes.

Joint preparation for shielded arc welding of thick-walled pipe requires a 75-degree included angle (37.5-degree bevel angle) on the pipe ends. Care must be taken in designing and preparing the ends to be joined. This process can be used in almost all types of joints, the most common being butt, lap, corner, edge, and tee joints performed both manually and by machine. This type of welding is excellent for thin metals of 1/8 in. (0.32 cm) or less because its extreme heat ensures rapid fusion and fast operation. For thicker metals, it becomes important to use multiple passes. For joints made up without filler metal, it is essential to have properly designed ends, especially for square, butt, and lap joints. For the most elaborate joints where a filler metal is used, the joint fitup is less critical. In many cases backing is used to improve the quality of a joint. Penetration in this process is one of its most positive aspects. Even when a backing is used, the penetration will be smooth, clean, and accurate with little cleanup necessary.

Joint preparation for submerged arc welding is similar to end preparation in other fusion welding processes, but because of the deep penetration created by the submerged arc weld it is possible to weld plates of considerable thickness without bevel preparation. It is a common practice to bevel the plate or pipe ends only partially so that the root face is larger than in other types of joints and yet becomes fused in the weld. Because of greater penetration with this process it is possible to use a much narrower included angle when preparing the joint for a single V-groove butt joint.

Submerged arc weld properties are similar to those of the base metal and, as usual, the quality of the base metal will affect both. This process is used to make butt welds, plug welds, and fillet welds. Butt welds can be made on thicknesses of up to 3 in. (7.62 cm) with a single pass, though it is better to limit each pass to 3/4 in. (1.90 cm). With fillet welds, the fillet can be considerably smaller than with other processes because of excellent penetration. Limiting the size does not make the joint weaker.

In the design of joints for welding dissimilar metals, the procedure is the same as that followed for welding like metals. Standard fillet and groove joints are used in all cases. The composition does not influence joint preparation.

Thermal expansion or contraction is usually a factor, especially in a piping system where high temperatures and pressures produce stresses and strains in the welds. Thermal expansion must always be taken into consideration in these situations. As a general rule, 2 percent of the length of weld is allowed for expansion and contraction. This is most often the case when the weld is long.

SERVICE FAILURES

The responsibility for weld failure can generally be assigned to one of five classifications:

1. Base metal defects (occurring during manufacturing and shaping of components)
2. Design (notches, joint location, structure, welding end configuration)
3. Fabrication (cleaning of pressure vessels, fabrication, heat treatment, welding, field erection)
4. Materials (selection and handling)
5. Service (severe conditions)

One should recognize that when a piping system or vessel is subjected to mechanical and thermal fatigue, cracking may start at the surface or at internal notches. Design notching or sharp corners should therefore be avoided.

Weld Defects

Pipe failures can result in shutdowns which are costly and in some cases extremely dangerous to personnel and equipment. Improper pipe joint preparation is a major cause of failure. Joints must be made up according to specified requirements by the designer or engineer who must take adequate safety standards into account.

Distortion caused by welding can cause the piping installation to be out of alignment. In cross-country pipelines, clamps are used to align the pipe until a sufficient weld is deposited to ensure accurate alignment. Backup supports are sometimes used for welding pipe fabrications, and welding fixtures may have to be designed and drawn by the drafter.

Fabrication Defects

Pressure vessels and piping systems are normally fabricated, welded, and inspected according to established codes. When failures occur, the resulting losses in production frequently cost far more than the pressure vessel or piping component that failed. A one-day shutdown of a boiler unit in a modern steam power plant may involve a loss exceeding $20,000. Failures in chemical and refinery plants can be even more costly. Sound fabrication, based on knowledge of the many different types of failure resulting from poor procedures, is of great importance.

Stress and Fatigue

Even when pressure vessels, tanks, and piping have been properly designed, assembled, fabricated, and inspected, a service or environmental condition that has not been anticipated can cause a system to fail. Some of the most common causes of failures are

1. External loading
2. Excessive stresses
3. Hydrogen blistering
4. Hydrogen embrittlement
5. Mechanical fatigue and shock
6. Overheating
7. Overpressure
8. Thermal fatigue and shock

FIG. 2-24 All welded construction pipe fabrications: Burst test photo showing that elbows are stronger than the corresponding pipe. Tee in center is stronger than pipe because of reinforcement in the throat area of the tee. The other tees failed because of poor welds and lack of reinforcement in the throat area. (Courtesy of Flowline Corporation)

Improper assembly or support can result in excessive stresses leading to service failures, although a good design would normally avoid excessive stresses. In some chemical process equipment, failures caused by overpressure result in a severe pressure rise. Normally, equipment is protected by safety valves, but on occasion these do not function properly, leading to rupture of the equipment. Figure 2-24 is an example of pressure testing pipe welds to ensure proper design and fabrication techniques. The freezing of water or other liquids in piping systems is a fairly common cause of failure also.

External Loading Supports, legs, brackets, hangers, and lugs may cause improper external loading of vessels, tanks, and piping. Walkways, platforms, and ladders may also cause external loading, the latter because of thermal expansion. Wind and ice may become external loading factors as well.

Mechanical Fatigue and Shock Fatigue is the condition leading to a fracture under repeated or fluctuating stresses having a maximum value less than the tensile strength of the material. The fractures are progressive, starting as minute cracks or notches. Causes of mechanical fatigue in pressure vessels and piping are

1. External factors
2. Pressure cycling
3. System changes
4. Variations in flow

Low vibration levels will not usually cause cracking or crack propagation in vessels and piping. Severe mechanical shock can cause failure and, because of its suddenness, can be disastrous.

Thermal Fatigue and Shock Thermal fatigue is the effect on material life of fluctuating temperatures or alternating exposures at high and low temperatures. When the temperature gradient is severe, the failure should be ascribed to thermal shock. Thermal shock may involve sudden heating or sudden cooling (heat shock, cold shock). Pressure vessels and piping materials in normal service conditions in steam power plants, refineries, and chemical plants are not usually subject to thermal shock. Sudden localized heating, cooling, and quenching will immediately produce stresses and strains which could cause cracking and failure.

Wall Thickness Because of sharp variations in wall thickness, numerous failures have occurred in cast, machined, and welded components involving both thick and thin sections where cracking has started from undesirable design notches. Design should provide for gradual changes in section thickness and for rounded rather than sharp corners to minimize the possibility of these failures. Service that involves mechanical or thermal fatigue between slip-on flanges and pipes is particularly susceptible to failure.

WELDED PIPE FABRICATION

Many advances have been made in the fabrication of automatically welded steel pipe. This pipe has excellent qualities: good flow, durability, strength, economy, and great carrying capacity. The capacity is due to the large diameter obtainable by this procedure—up to 240 in. (609.6 cm) total diameter. Submerged arc welding or double submerged arc welding is used to weld the length of the seam created by the rolled plate. This kind of piping must go through a series of manufacturing processes: crimping the edges of the steel plate; bending the plate into a U-shaped cross-section; creating a tubular pipe that has not been welded. The outside of the pipe is then welded and inside welding is done. Thereafter the piping may be hydrostatically expanded (in some cases) and X-ray inspected for flaws in the weld.

In piping fabrications, electric arc welding is most often used. An arc between two electrodes or between electrode and the piece being welded creates heat for fusing the metals. The goal in welded pipe fabrication is to create a joint as strong as any part of the total fabricated portion.

Inert-gas tungsten arc welding is used on most ferrous and nonferrous piping materials. Hardened steel piping can also be welded in the root of butt joints with this welding process. Maximum thickness for this type of welding is limited by equipment and the material to be welded. Hand welding can be accomplished down to approximately 1/32 in. (0.08 cm) in thickness. For pipe wall thicknesses from 1/4 to 3/8 in. (0.64 to 0.95 cm), this process is commonly used to complete only the root pass; then other welding processes are used to complete the pipe butt weld. Gas shielded metal arc welding can also be adapted for steel piping, though it is used more extensively on aluminum piping (with argon as the shielding gas).

Submerged arc welding, whether automatic or semiautomatic, is used extensively in the shop welding of chromium alloyed steel, carbon steel, and stainless steel. Oxyacetylene welding is generally confined to small-size steel piping. Although its use has declined as arc welding has increased over the past 20 years, at one time this process was used extensively for ferrous and nonferrous piping systems. For small noncritical piping systems, brazing, braze welding, and soldering are quite common. Soldering is used with pipe of brass or bronze or copper alloy.

In general, the circumferential butt weld is used in the fabrication of welded pipe systems of all sizes. Figure 2-25 shows the bevel weld preparation for mating pipes and fittings. A double angle preparation is used in thicknesses over 7/8 in. (2.22 cm). For smaller thicknesses a single bevel with a 1/16-in. (0.16 cm) root face is used.

Figure 2-26 shows the welding procedure for pipes of unequal thickness. The length of the taper is usually four times the offset between the pipe or the vessel thickness. This figure shows the two preferred methods of alignment when different thicknesses of piping are joined. Backup rings are also employed when it is permissible to obstruct the flow (as in some steam plant piping systems). In general, backup rings are not used for chemical and refinery piping systems. In most cases, the wall thickness of a valve or fitting is greater than that of the pipe to which it is to be joined. Figure 2-27 is a good example of the end preparation necessary for joining the fitting and the pipe or the valve and the pipe.

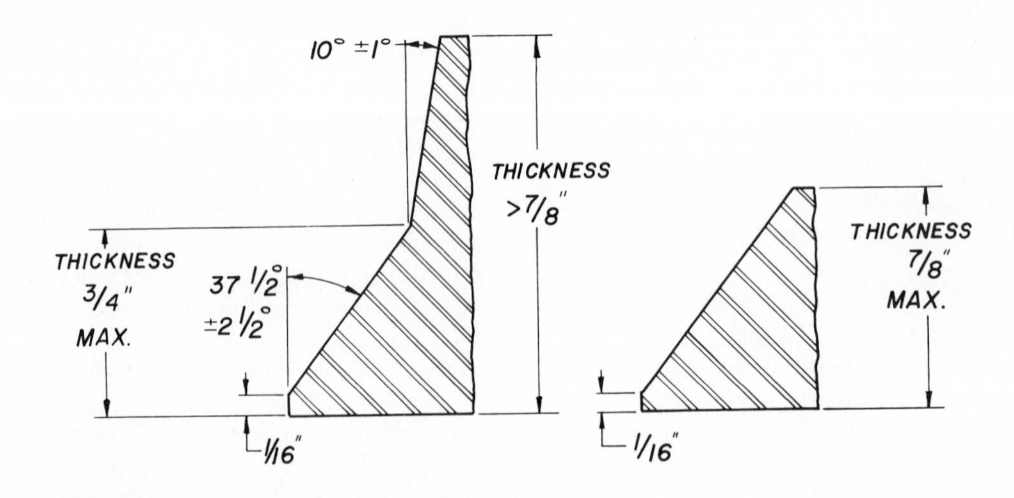

FIG. 2-25 Bevel weld preparation.

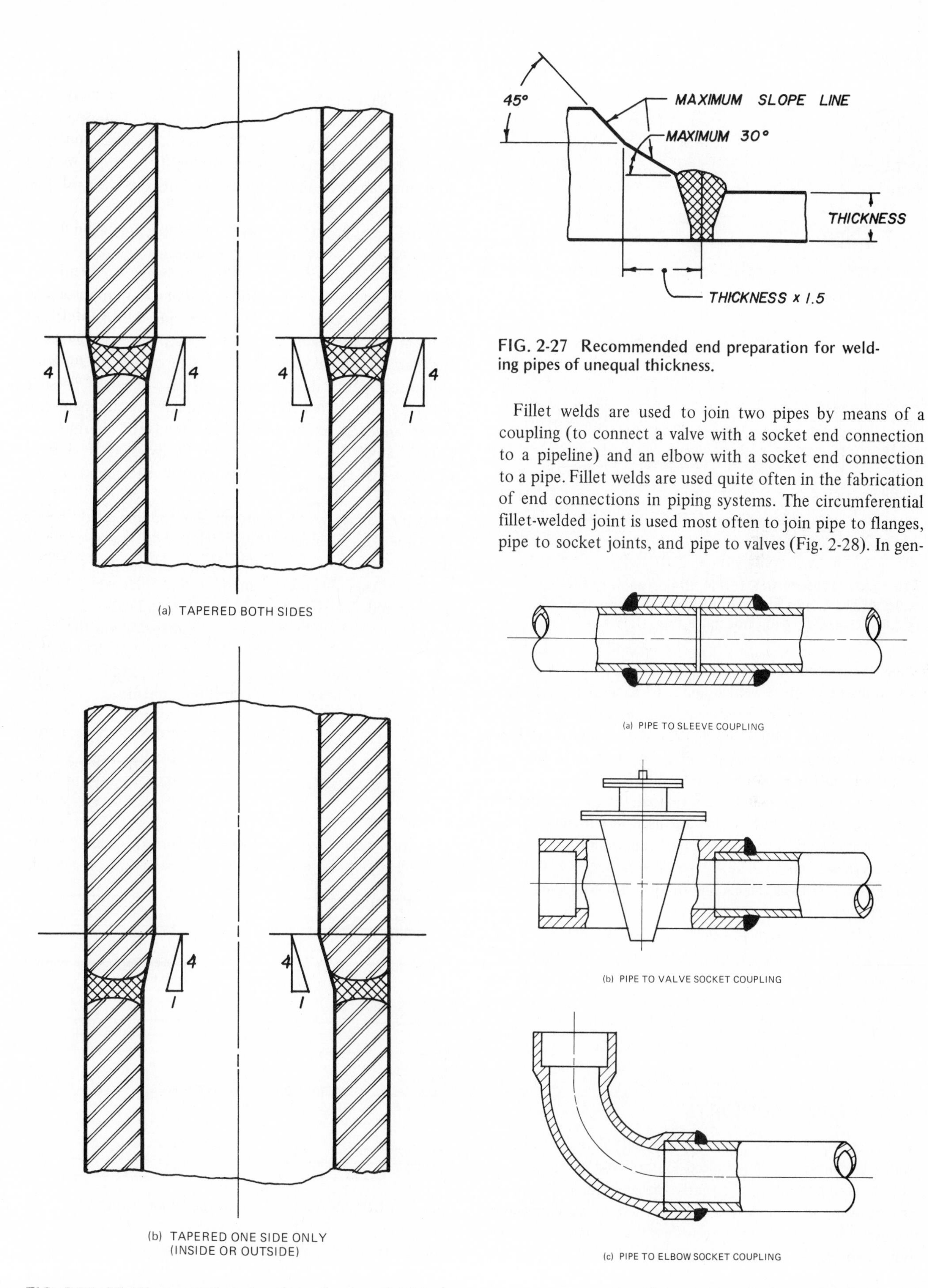

FIG. 2-26 Welding procedure for pipes of unequal thickness.

FIG. 2-27 Recommended end preparation for welding pipes of unequal thickness.

Fillet welds are used to join two pipes by means of a coupling (to connect a valve with a socket end connection to a pipeline) and an elbow with a socket end connection to a pipe. Fillet welds are used quite often in the fabrication of end connections in piping systems. The circumferential fillet-welded joint is used most often to join pipe to flanges, pipe to socket joints, and pipe to valves (Fig. 2-28). In gen-

FIG. 2-28 Fillet weld joints for pipe joints.

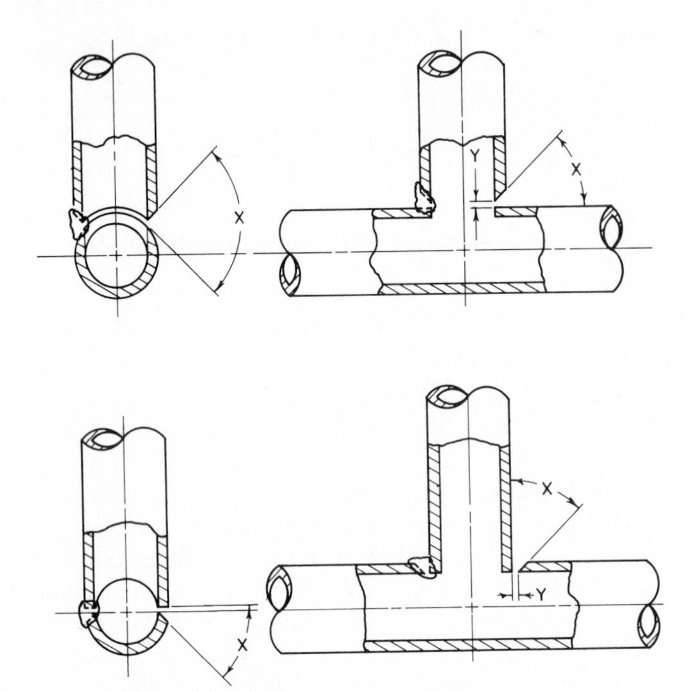

FIG. 2-29 Preparation for a 90-degree intersection weld. Angle X: not less than 15 degrees. Dimension Y: not less than 0.062 in., not more than 0.250 in.

eral, this procedure is used in pipe sizes 2 in. (5.08 cm) or less in diameter. Butt-welded joints are used in critical situations, such as nuclear or radioactive services, because socket connections are subject to inadequate penetration. Butt welds, properly applied, can yield complete weld penetration. Intersections between tees, laterals, Y's, and vessel openings are somewhat harder to join than other connections. Nonetheless butt welds are used extensively in this type of connection when complete fusion and penetration are necessary.

Figure 2-29 shows the general rule for the 90-degree intersection weld penetration between two pipes. By using an angle of not less than 45 degrees, the intersection of the two pipes can create a sound and adequate penetration of the weld deposit. Especially in severe conditions, there is a necessity for complete weld penetration and fusion of the two pieces being welded. In general, though, intersection welds should be avoided where pressure exceeds 125 psi.

WELDED PIPE SUPPORTS

In the construction of pipe supports and related structures, welds are preferred over bolted joints. Weldments are usually lighter and more efficiently designed than bolted or riveted supports. Sometimes the weight of an equivalent welded fabrication is 25 percent less than an equally strong bolted support—and thus the cost is far less. Welded joints offer a higher quality than any other joining method. Welding can also make use of a greater variety of modern materials, whereas in bolted fabrication it is not always possible to match bolts and materials for strength. Lastly, a welded structure can be designed more efficiently and design variation options are greater.

Welds for supports on vessels and piping are called for by the designer. The designer should minimize the use of welding directly to pipes or vessels because heavy welds for supports and hangers connected to pipes and equipment may cause severe mechanical strain (increasing the likelihood of thermal and mechanical shock) or may result in cracks in the wall. In critical services, the design of welded supports and attachments should limit notches whenever possible and take into account the susceptibility of equipment corrosion and erosion. In nuclear power plants, thermal and seismic shock are a major concern. Designers must provide efficient placement of supports and other attachments (see Chapter 8).

Figure 2-30 shows the common pipe support welds, which must be kept simple. These eight basic joints are used throughout the construction of piping supports. Figure 2-31(a) shows the use of an all-around fillet weld. In Fig. 2-31(a) and 2-31(b), the weld length is indicated by an all-around symbol or by placing the dimension on the reference line of the weld symbol. In Fig. 2-31(c), the length is not specified and it is assumed that the 1/4-in. (0.64-cm) fillet weld continues the length of the I beam.

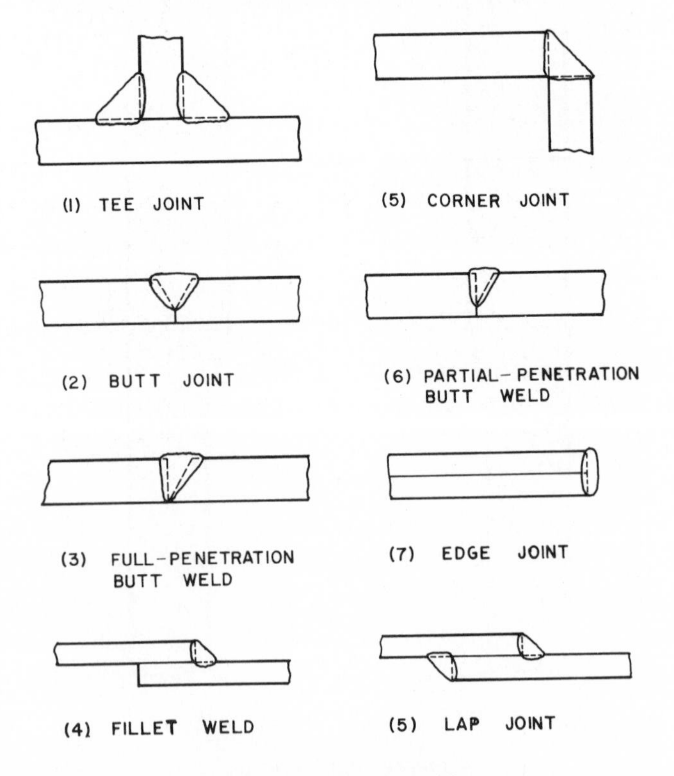

FIG. 2-30 Common pipe support welds.

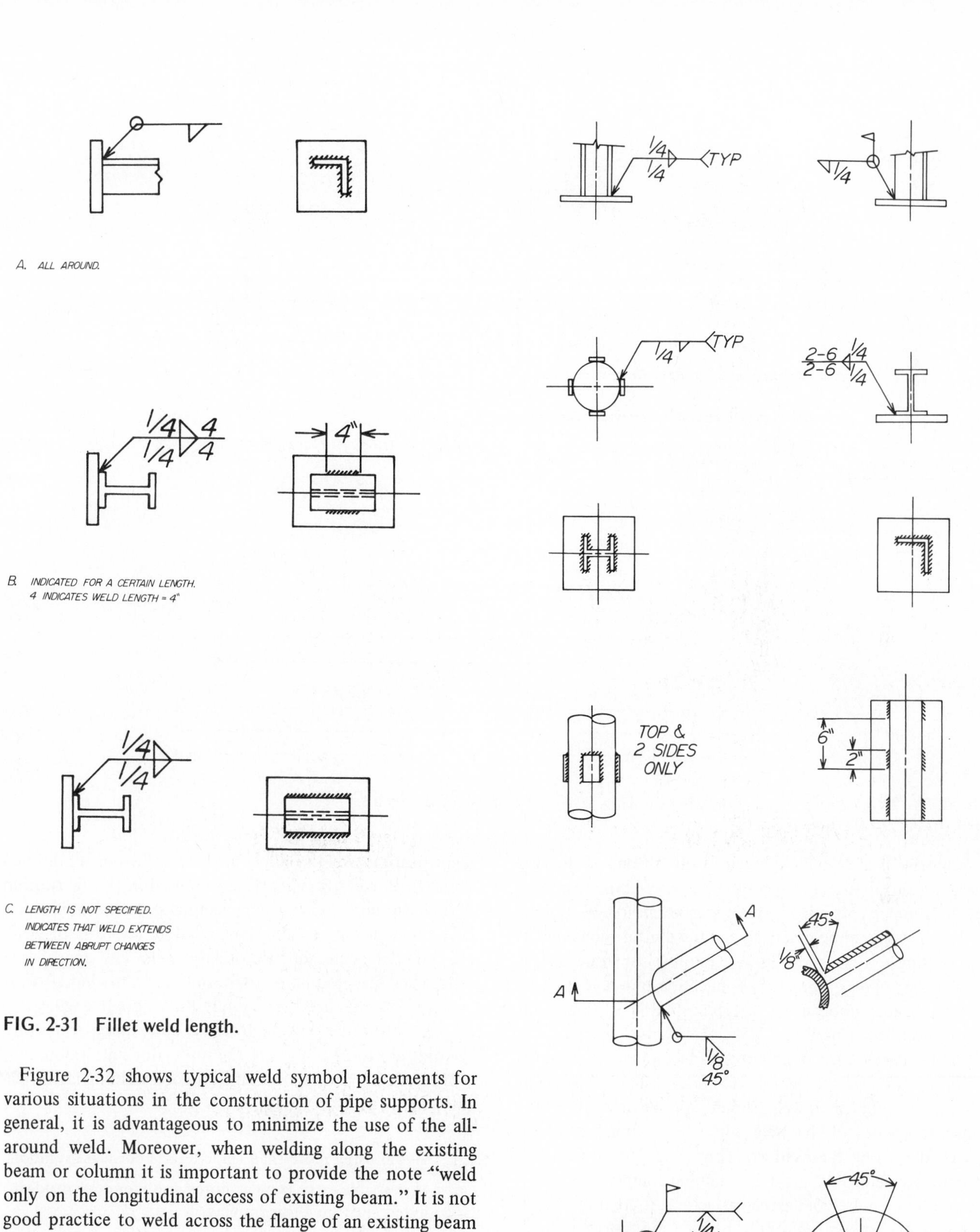

FIG. 2-31 Fillet weld length.

Figure 2-32 shows typical weld symbol placements for various situations in the construction of pipe supports. In general, it is advantageous to minimize the use of the all-around weld. Moreover, when welding along the existing beam or column it is important to provide the note "weld only on the longitudinal access of existing beam." It is not good practice to weld across the flange of an existing beam or column.

In pipe support design, the fillet weld is most commonly used for structural connections. The faces of the weld which are in contact with the parts joined are called the *legs*. The *throat* is the shortest distance from the root of the weld to the hypotenuse (see Figs. 2-11, 2-15, 2-17 and 2-19). A fillet weld is designated by its size and length.

FIG. 2-32 Typical weld symbols for pipe supports.

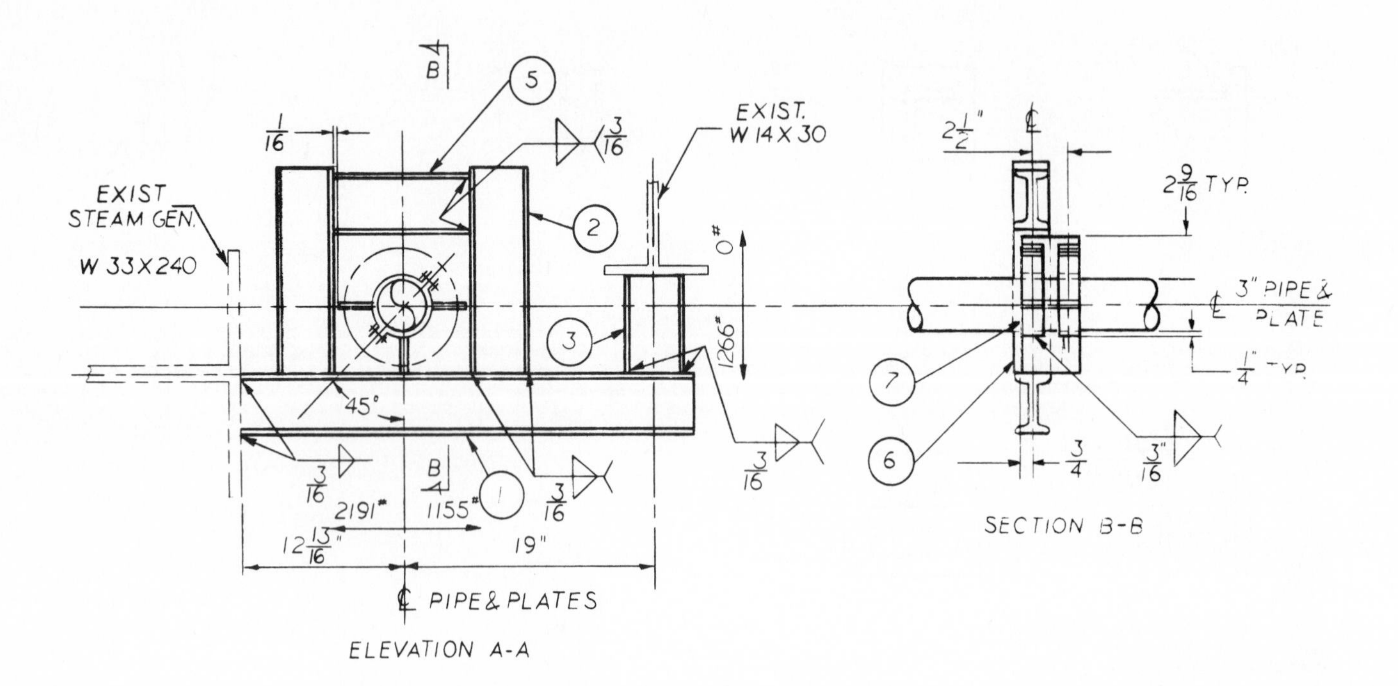

PARTS					
PART	QUAN.		PART	QUAN.	
1	1	S4X7.7, 2'-11" LG.	5	1	S4X7.7, 8 3/4" LG
2	2	S4X7.7, 13 1 4 LG.	6	3	1/2"X2-9/16", C.S., 4" LG.
3	1	S4X 7.7, 6 13 16 LG.	7	2	8-P PART 298, 3" PIPE

FIG. 2-33 Typical pipe support drawing showing 3/16-in. (0.48-cm) fillet welds.

Weld size is always determined by the thicker of the two parts joined, though the size need not exceed the thickness of the thinner part joined unless it is necessary. Weld sizes and shapes should be practical, and the drafter should avoid designating too many different sizes of welds in one design. Figure 2-33 shows a practical application of a pipe support and its weld symbols using a 3/16-in. (0.48-cm) fillet weld throughout the structure.

On field welds for beams and girders under load, or for columns, the effective length of a fillet weld or segment of an intermittent fillet weld should not be less than four times the width of the weld within a minimum of 1½ in. (3.81 cm), along the weld axis (Fig. 2-34). No attachment should be welded to pressure-retaining components unless conventional bolted or damped attachments are impractical.

Figure 2-35 shows the parallel loading and transverse loading shear stress that a fillet weld will sustain the length of its axis. In this application, it is necessary to analyze the weld as though it existed without a cross-sectional area.

There are two general rules to observe when employing weld symbols for pipe supports. First, always use the standard weld symbols of the American Welding Society for identification. Weld symbols must be shown on all finished sketches to identify all welds to be used in the fabrication of that support. Second, engineering sketches should provide the following welding information: location of weld, type of weld required, field or shop weld, and appropriate weld size. The drafter is responsible for showing all weld symbols on the finished support drawings. If engineering sketches are not adequate to show all welding situations, the drafter should bring it to the supervisor's attention.

The following rules are useful in good welding identification:

1. Avoid using more than three letters per weld symbol.
2. If possible, use typical notes for identical welds to avoid cluttering the drawing.
3. The tail in the weld symbol should be omitted unless a reference is necessary.
4. The arrowhead should always touch the area or surface to be welded.
5. Avoid using any verbal description on welds. A symbol is all that is necessary.

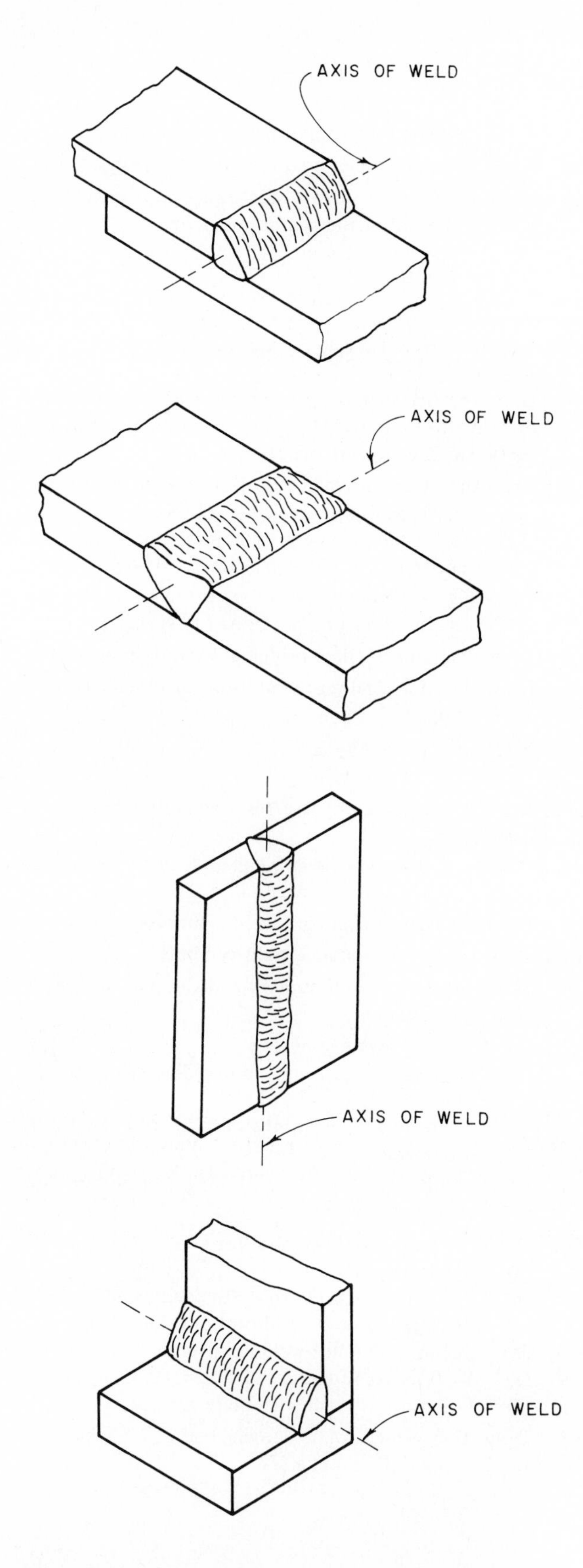

FIG. 2-34 Groove and fillet weld axis. This is a theoretical line through the length of a weld at its center of gravity, perpendicular to its cross-section.

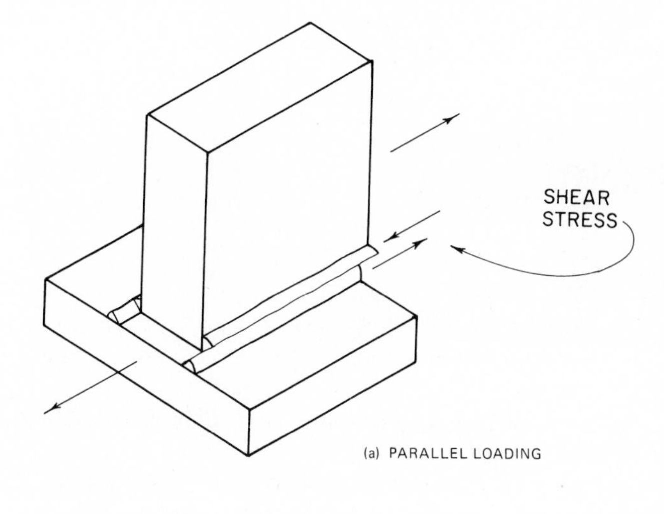

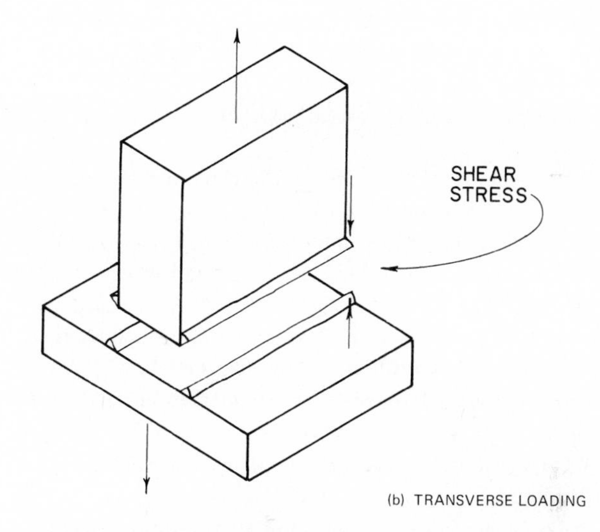

FIG. 2-35 Shear and stress for transverse and parallel loadings.

To conclude, then, high-quality welds are extremely important throughout the piping industry—whether they are on supports, on the pipes themselves, or on special pipes and fittings. Because of severe stresses and strains encountered in certain services, a failed pipe weld will not only lead to loss of service but also increase the likelihood of injury to operators of the system.

The drafter will be called on to prepare drawings of piping systems, to prepare drawings of pipe supports, and to prepare the essential drawings that will show all welding necessary for the system. Welders know how to prepare the end connections for all the basic joints they have to construct, but drafters must know the variations and cross-sectional drawings that may be necessary for the preparation of end connections, machining, and beveling, in addition to the appropriate weld symbols and sizes.

QUIZ

1. What welding process uses pressure in addition to heat?
   (a) Arc
   (b) gas
   (c) resistance
   (d) machine
2. What are the three types of arc welding?
3. Deeper weld penetration results from __________ arc welding.
4. What welding process is used most often in pipe fabrication?
5. The main difference between V, bevel, U, and J welds is in the __________ of the parts to be joined.
6. What are the two methods of indicating a field weld?
7. Define *partial penetration.*
8. What is the weld throat?
9. The size of the fillet weld is shown to the __________ of the __________ symbol.
10. Plug and slot welds are used primarily in __________ joints.
11. What type of weld is used to join pipes and fittings in sizes 2 in. (5.08 cm) and less?
12. The tail of the weld symbol should be omitted when __________ or __________ are not used.
13. Describe the welding process that may be used to weld two pieces of steel together without melting the base metals and without adding filler metal.
14. Explain the difference between shielded arc and unshielded arc welding processes.
15. What is the difference between bare and shielded arc welding?
16. What is the name for the granulated welding material used in submerged arc welding?
17. What is the difference between the joint preparation for submerged arc and shielded metal arc welding?
18. In submerged arc welding, what types of backing can be used?
19. What is inert-gas shielded welding?
20. Define *resistance welding.*
21. Define *weldability.*
22. What factors can affect the weldability of metals?
23. Which process is most commonly used to weld dissimilar materials?
24. Is a bolted or riveted structure lighter or heavier than a welded structure of comparable strength?
25. Is welding the best process for joining materials?
26. Why are open roots sometimes used?
27. What is a tack weld and what are some of its uses?
28. Why is it important to bevel plate metals before welding?
29. How do you measure a fillet weld and a groove weld?
30. What are the advantages of using weld symbols?
31. Name the elements of a complete weld symbol.
32. What is meant by arrow side and opposite side and what is the tail of the symbol used for?

PROBLEMS

1. Draw the five major joints and a cross-sectional view of two different welds applied to each. Use Fig. 2-8 (draw two times book scale).
2. Draw the three figures in Fig. 2-28 (the pipe-to-sleeve, pipe-to-valve, and pipe-to-elbow couplings) using 2-in. (5.08-cm) diameter pipe and fittings. Use fitting catalogs or appendix to find dimensions.
3. Draw Fig. 2-29 using 1.250 dia. as the small pipe and 1.500 for the joining pipe. (Draw full scale.)
4. Draw diagrams of three principal welding processes.
5. Draw a schematic representation of shielded metal arc welding.
6. Sketch five basic joints.
7. Sketch three basic welds.
8. Draw a cross-section of a fillet weld and call out its basic parts.
9. Draw a double V-groove weld and name its various parts.
10. Draw the basic arc and gas weld symbols.
11. Draw the basic resistance weld symbols.
12. Sketch the all-around symbol and the two versions of field weld symbols.
13. Create the following symbols:
    (a) 1/4 in. (0.64 cm) continuous fillet weld on arrow side.
    (b) 3/8 in. (0.95 cm) intermittent fillet weld on other side

    Each weld is 3 in. (7.62 cm) long spaced on 5-in. (12.7-cm) centers.
14. Create an appropriate symbol for a butt joint:
    (a) Single V-groove weld.
    (b) 3/8-in (0.95-cm) weld on opposite side.
    (c) 60-degree groove weld on both sides.
    (d) 1/2-in. (1.27-cm) weld on arrow side.
    (e) Workpieces to be placed 1/8 in. (0.32 cm) apart.
    (f) Grind welds flush with plate surface.
15. Draw the welded joints shown in Fig. 2-36, showing weld configuration and sizes.

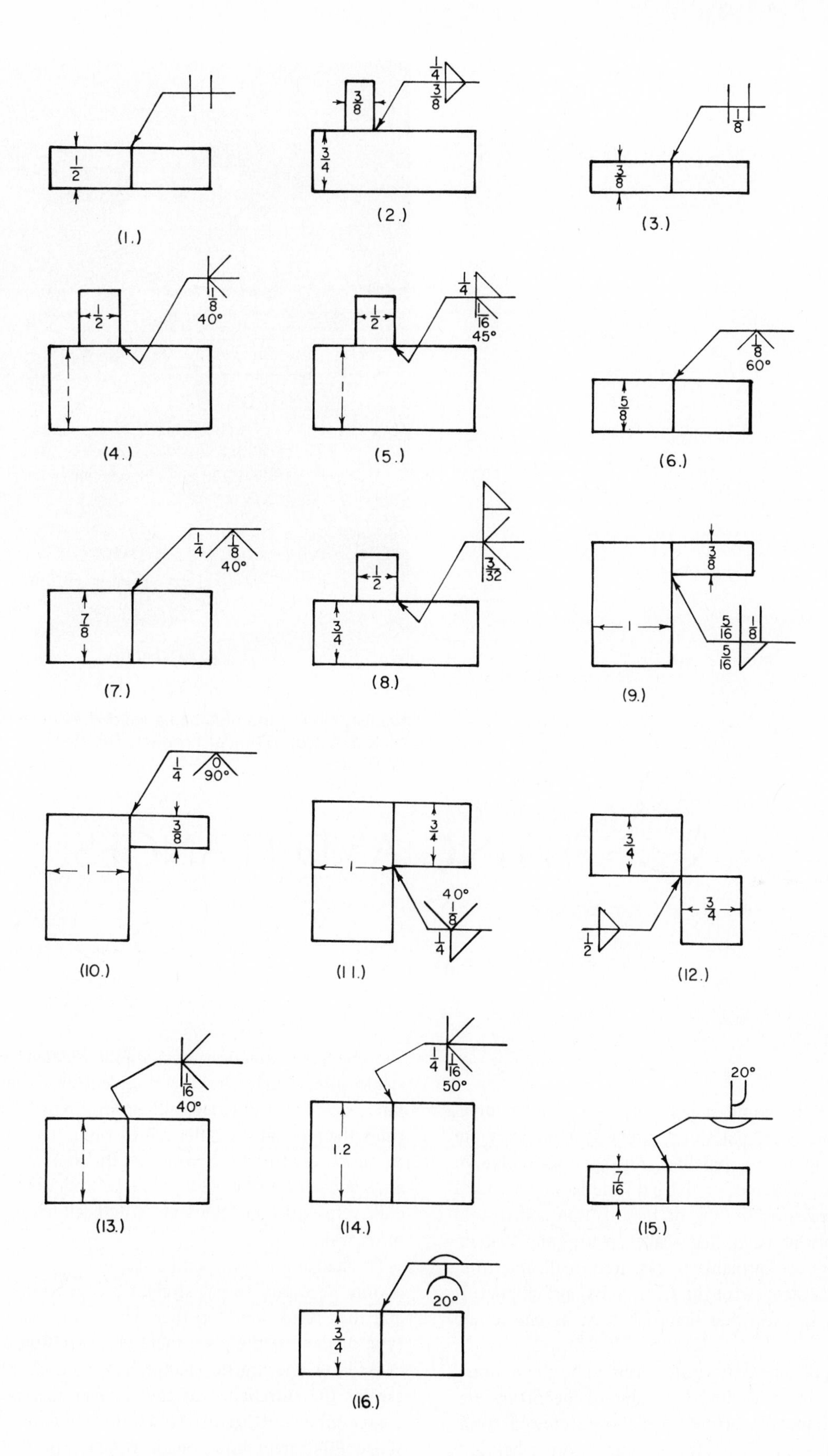

FIG. 2-36 Weld configuration problem.

Flange, elbow, and pipe being welded. (Courtesy Babcock & Wilcox, Tubular Products Division)

3 FITTING AND FLANGES

FITTINGS

Pipe fittings are the components which tie together pipelines, valves, and other parts of a piping system. They are used in "making up" a pipeline. Fittings may come in screwed, welded, soldered, or flanged varieties and are used to change the size of the line or its direction and to join together the various parts that make up a piping system. Standard fittings are available in cast iron, malleable iron, forged steel, cast steel, wrought steel, brass, and copper. It is also possible to use other materials such as plastic and glass.

The majority of pipe fittings are specified by the nominal pipe size, type, material, and the name of the fitting. Besides the end connections mentioned above (screwed, welded, soldered, flanged), it is also possible to order bell and spigot fittings, which are usually cast iron and used for low-pressure service.

One of the major functions of fittings in a piping system is to change direction. Figure 3-1 shows three procedures for changing direction in a piping system: bending, the single-miter bend, and the double-miter bend. Bending a pipe is common practice, but in some cases it stretches and thins the pipe on the outer radius while the inner wall tends to bunch or thicken. Because of this tendency, flow resistance and erosion increase at the bend. Moreover, the pipe's ability to withstand pressure is reduced because of the thin outer wall.

To change direction with a single- or double-miter joint it is only necessary to cut straight pieces to the proper angle and then weld them together. The main drawback with this type of bend is the great increase in friction and resistance caused by the drastic change in direction, which also increases the turbulence of the flow at the turn. A full 90-degree elbow fitting offers six times less resistance than a single 90-degree miter bend. Because of these considerations, fittings that change direction are used more often than bending or mitered joints.

In general, then, a fitting is any component in a piping system that changes its direction, alters its function, or

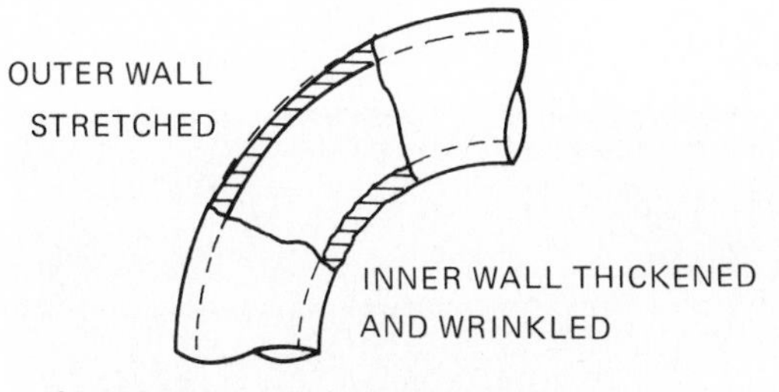

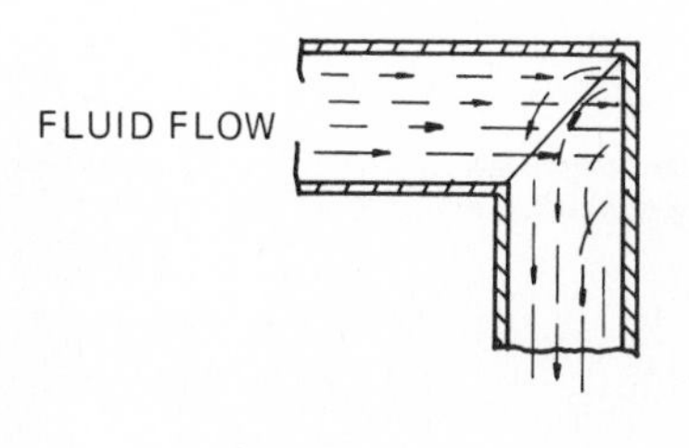

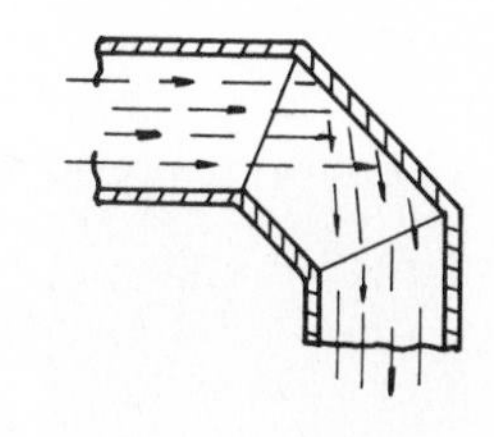

FIG. 3-1 Bends and mitered pipe joints. (a) Bending the pipe is the easiest way to change direction, but is not very satisfactory. (b) Mitered joints maintain cross-section and wall thickness, but sacrifice efficiency because of friction and turbulence.

simply makes end connections. A fitting is joined to the system by bolting, welding, or screwing, depending on many variables in the system. Fittings, therefore, are the most economical and efficient way of redirecting or constructing a piping system, especially where flow rate, pressure, and other considerations are important.

One of the most commonly used means of joining fittings to pipe is the butt weld, which employs fittings made just for that purpose. In this type of fitting, the ends are machine-grooved to form a bevel which fits together with a piece of pipe which is also beveled at the end (Fig. 3-2). They are fused together by welding. Welded piping systems

FIG. 3-2 A series of welded tees and elbows. (Courtesy Babcock & Wilcox, Tubular Products Division)

FIG. 3-3 Piping manifold using 12-in. (30.48-cm) flanged gate valves and fittings. (Courtesy Grove Valve and Regulator Company)

have many advantages over the screwed or flanged variety. They require less maintenance and provide a permanent leak-proof bond. Moreover, because of their high-pressure/high-temperature safety record, they cut down on the total maintenance cost of the system. Welded construction also weighs much less than the flanged type and does not require as much clearance, since flanged or screwed fittings are larger and heavier. Therefore welded construction requires a minimum of space, which enables the designer to place the system closer to ceilings, walls, or other equipment. Another advantage of welded construction is that it is easier to insulate.

Flanged fittings are bolted to the piping system by means of welded or screwed flanges (Fig. 3-3). This type of connection is used in petrochemical work at places where it is necessary to dismantle lines frequently. Flanges are also used where welding would be impossible because of fire hazard. One drawback of an all flanged system is that a flanged fitting may weigh three to four times as much as its welded or screwed counterpart, making the whole system heavier and more difficult to support.

Threaded or screwed fittings are usually limited to smaller-sized piping. They come machined and threaded with standard size pipe threads. Forged steel screwed fittings are used primarily for high-temperature/high-pressure service.

Cast iron screwed fittings are more common for heating systems and low-pressure/low-temperature requirements. One disadvantage of the threaded system is the loss of pipe thickness—cutting the threads on a pipe can remove up to 50 percent of the pipe wall thickness. Exposure of threads can also lead to faster corrosion. All fittings are manufactured to standard dimensions. Threaded fittings are manufactured to national pipe size standards.

Figure 3-4 shows the four most common methods of joining pipes and fittings. Figure 3-5 exhibits the seven major pipe joints. The following section covers the diverse types of fittings and their connections in more depth.

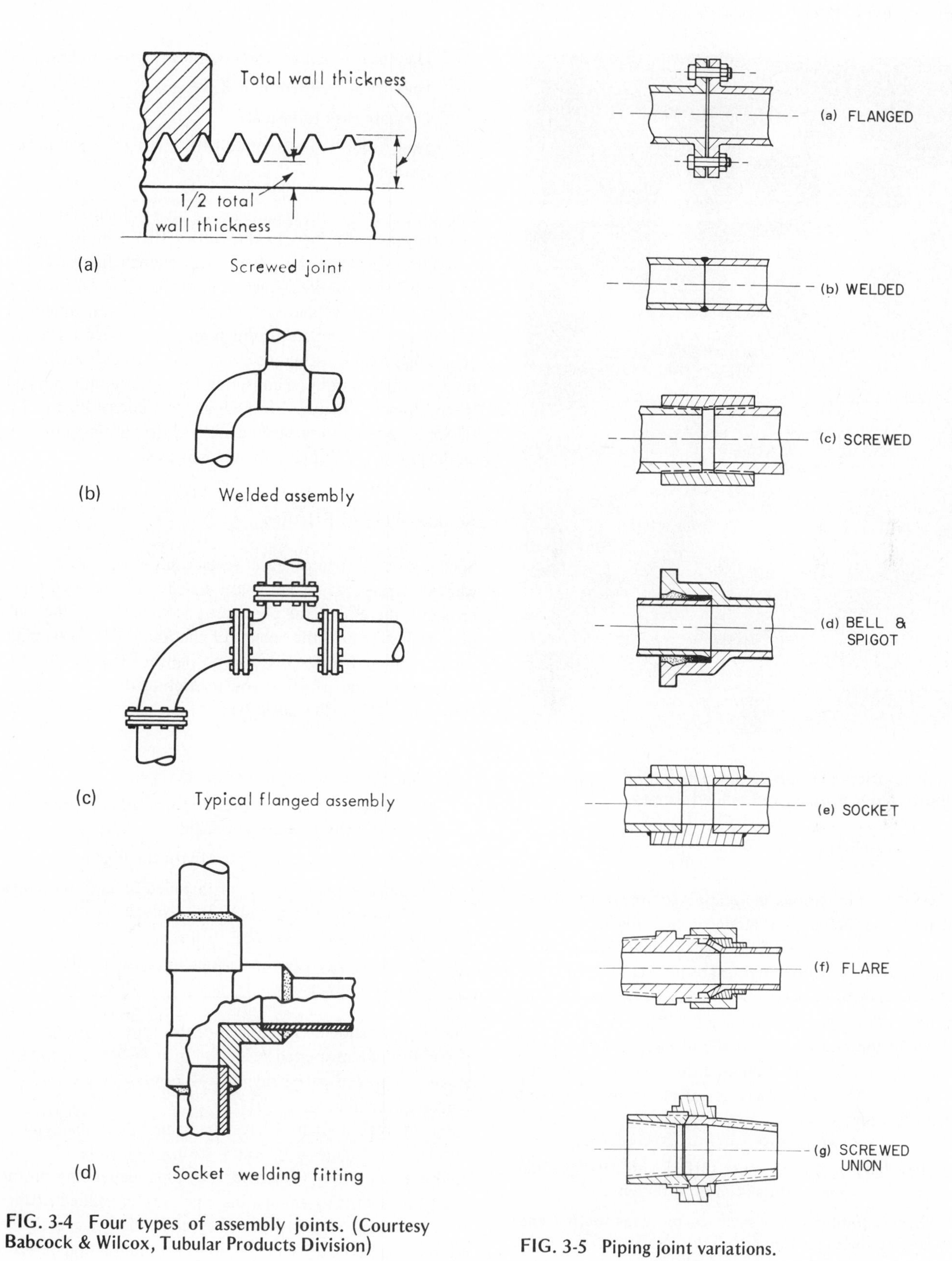

FIG. 3-4 Four types of assembly joints. (Courtesy Babcock & Wilcox, Tubular Products Division)

FIG. 3-5 Piping joint variations.

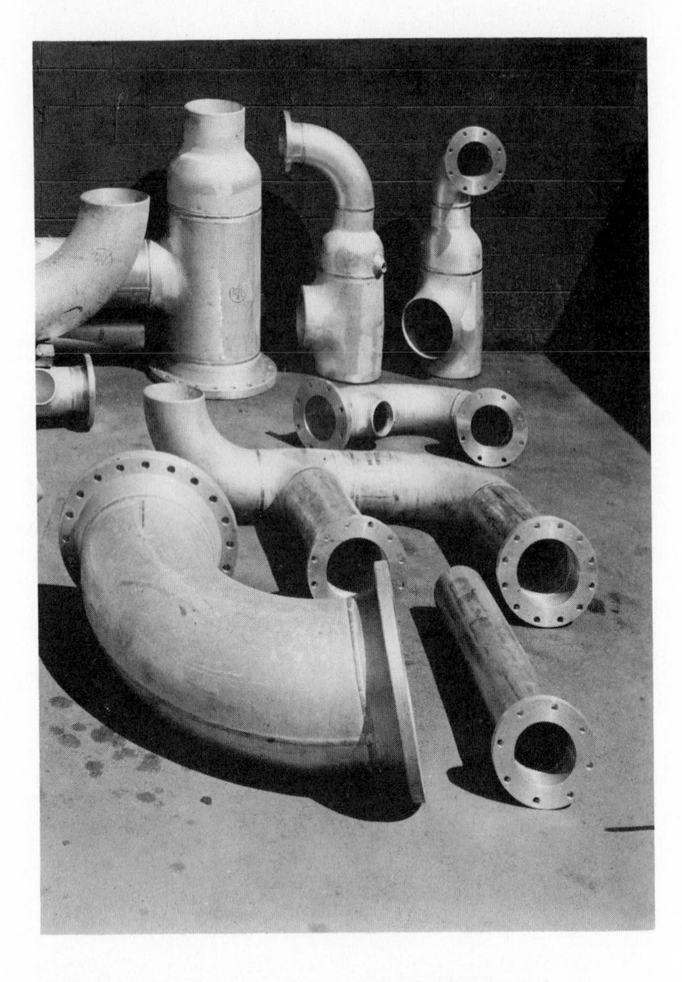

FIG. 3-6 A variety of welded piping fabrications showing elbows, reducers, tees, and welded flanges.

Butt-Welded Fittings

Because of simplifications in welding equipment it is now possible to use welding on piping systems that in the past could only accept screwed or flanged fittings. Today, even in small jobs such as buildings, small industrial plants, and schools, a completely welded system is within reach, thus providing a permanent, self-contained system.

Welded fittings are used primarily in systems meant to be permanent. They have the same wall thickness as the mating pipe. Among the many advantages of butt-welded systems are the following:

1. They have a smooth inner surface and offer gradual direction change with minimum turbulence.
2. They require much less space for constructing and hanging the pipe system.
3. They form leakproof connections.
4. They are almost maintenance free.
5. They have a higher temperature and pressure limit.
6. They form a self-contained system.
7. They are easy to insulate.
8. They offer a uniform wall thickness throughout the system.

One of the major disadvantages of butt-welded systems is that they are not easy to dismantle. Therefore it is often advisable to provide the system with enough flanged joints so that it can be broken down at intervals. (One of the main uses of the butt-welded system is for steam lines, which are usually in high-temperature/high-pressure service.) The appendix has a table listing the dimensions of steel butt-welded fittings which should be consulted for practice ordering and when drawing fittings throughout the book. Figure 3-6 shows a variety of fabricated welded fittings and pipe, including reducers, flanges, elbows, and tees.

Socket-Welded Fittings

Socket-welded fittings have certain advantages over butt-welded fittings. They are easier to use on small-size pipelines and the end of the pipes need not be beveled since the pipe end slips into the socket of the joint (Fig. 3-5). With socket-welded fittings there is no danger of the weld protruding into the pipeline and restricting flow or creating turbulence. Thus the advantages of the socket-welded system are:

1. The pipe does not need to be beveled.
2. No tack welding is necessary for alignment since the joint and the pipe are self-aligning.
3. Weld material cannot extend into the pipeline.
4. It can be used in place of threaded fittings, therefore reducing the likelihood of leaks which usually accompany the use of threaded fittings.
5. It is less expensive and easier to construct than other welded systems.

One of the major disadvantages of this type of fitting is the possibility of a mismatch inside the fitting where improperly aligned or mated parts may create a recess where erosion or corrosion could start.

Socket-welded fittings have the same inside diameter as standard (schedule 40), extra strong (schedule 80), and double extra strong (schedule 160) pipe, depending on the weight of the fitting and mating pipe. Socket-welded fittings are covered in ASA B16.11. They are drilled to match the internal diameter of schedule 40 or schedule 80 pipe. Typical dimensions of commercial standard forged steel socket-welded fittings are provided in the appendix.

Flanged Fittings

Flanged connections are found on piping systems throughout the petrochemical and power generation fields on pipelines that are a minimum of 2 in. (5.08 cm) in diameter. The majority of flanged fittings are made of cast steel or cast iron. The appendix provides standard dimensions and callouts for both of these types.

Flanged steel fittings are used in place of cast iron where the system is subject to shock or in high-temperature/high-pressure situations where the danger of fire is prevalent, because cast iron has a tendency to crack or rupture under certain stresses. A flange may be cast or forged onto the ends of the fitting or valve and bolted to a connecting flange which is screwed or welded onto the pipeline, thereby providing a tight joint. An assortment of facings, ring joint grooves, and connections are available in flange variations.

One advantage of flanged systems is that they are easily dismantled and assembled. One of the disadvantages is that they are considerably heavier than an equally rated butt-welded system, because of the large amounts of metal that go into making up the joints and flanges. Moreover, flanged fittings occupy far more space than the butt-welded or screwed equivalents. Because of this higher weight load, a flanged system becomes far more expensive to support or hang from the existing structure.

Cast steel and forged steel flanged fittings are available in ASA ratings from 150 to 2500 lb. The cast steel flange fittings comply with the specifications of ASA B16.5. Companion flanges are covered in more detail at the end of this chapter.

Screwed or Threaded Fittings

Screwed fittings are used primarily on low-temperature/low-pressure installations and can be found throughout the typical house or apartment building (in water systems, for example). They are used for small-size pipes usually 2½ in. (6.35 cm) or less in diameter. The threads on a screwed fitting correspond to the American Standard pipe threads covered in Chapter 1.

Fittings can come with a tapered thread, which is more common, or with a straight thread for special applications. In practice the pipe fitter will often use a fitting with a straight thread and pipe with a tapered thread, thereby providing a tighter joint. To form a tight joint with a threaded fitting, dope or pipe compounds are employed. Threaded or screwed connections are considered the least leakproof joint possible in a piping system. In general, pipes 2 in. (5.08 cm) or larger are not fitted with screwed connections. Because of the advantages of all-welded systems, they are being employed in place of screwed fittings because of ease in makeup and leak resistance. Threaded joints should be used with caution in systems that have a high-temperature/high-pressure requirement, since cutting the threads reduces the wall thickness of the affected pipe and thereby increases the chance of failure.

Screwed fittings are available in steel, malleable iron, bronze, brass, and cast iron. Cast brass screwed fittings are covered by ASA B16.15 and used in systems with a maximum working steam pressure of 250 psig. Cast iron screwed fittings are covered by ASA B16.4 and rated at a maximum saturated steam pressure of 250 psig. Malleable iron screwed fittings are covered under ASA B16.3 and are used at a maximum of 300 psig. All screwed fittings have been standardized by the ASA.

Soldered Fittings and Flared Fittings

Soldered joint fittings are used primarily for brass or copper water tubes. These fittings are covered by ASA H23.1 and manufactured in accordance with ASA B16.22 and B16.18. Soldered joints are used primarily for thin-wall tubing sizes 1/4 to 4 in. (0.64 to 10.16 cm) OD. This type of joint has a low-temperature/low-pressure rating and therefore will not be encountered often in petrochemical or power generation facilities. Soldered joint fittings come in wrought metal, cast brass, or bronze and are used mainly for fittings on water systems for domestic use.

Flared fittings or joints are used on soft copper tubing for low-pressure/low-temperature situations.

Bell and Spigot Fittings

The bell and spigot fitting is used for cast iron, low-pressure/low-temperature situations and is not found very often in petrochemical, power generation, or heating systems. It is used primarily in sewage, water, and gas lines for underground service. There are also other methods of connecting cast iron pipe including welding, sleeve coupling, universal joints, and ball and socket types. For steel and wrought iron pipes the welded, threaded, or flanged fittings are the most common.

Common Fittings

Figure 3-7 shows 20 commonly available pipe fittings for butt-welded, screwed, flanged, bell and spigot, and socket-welded types.

Fitting 1, a bushing, is used on screwed piping systems to reduce the size of an opening in a fitting when a reducer would be called for but the bushing could save money and makeup time.

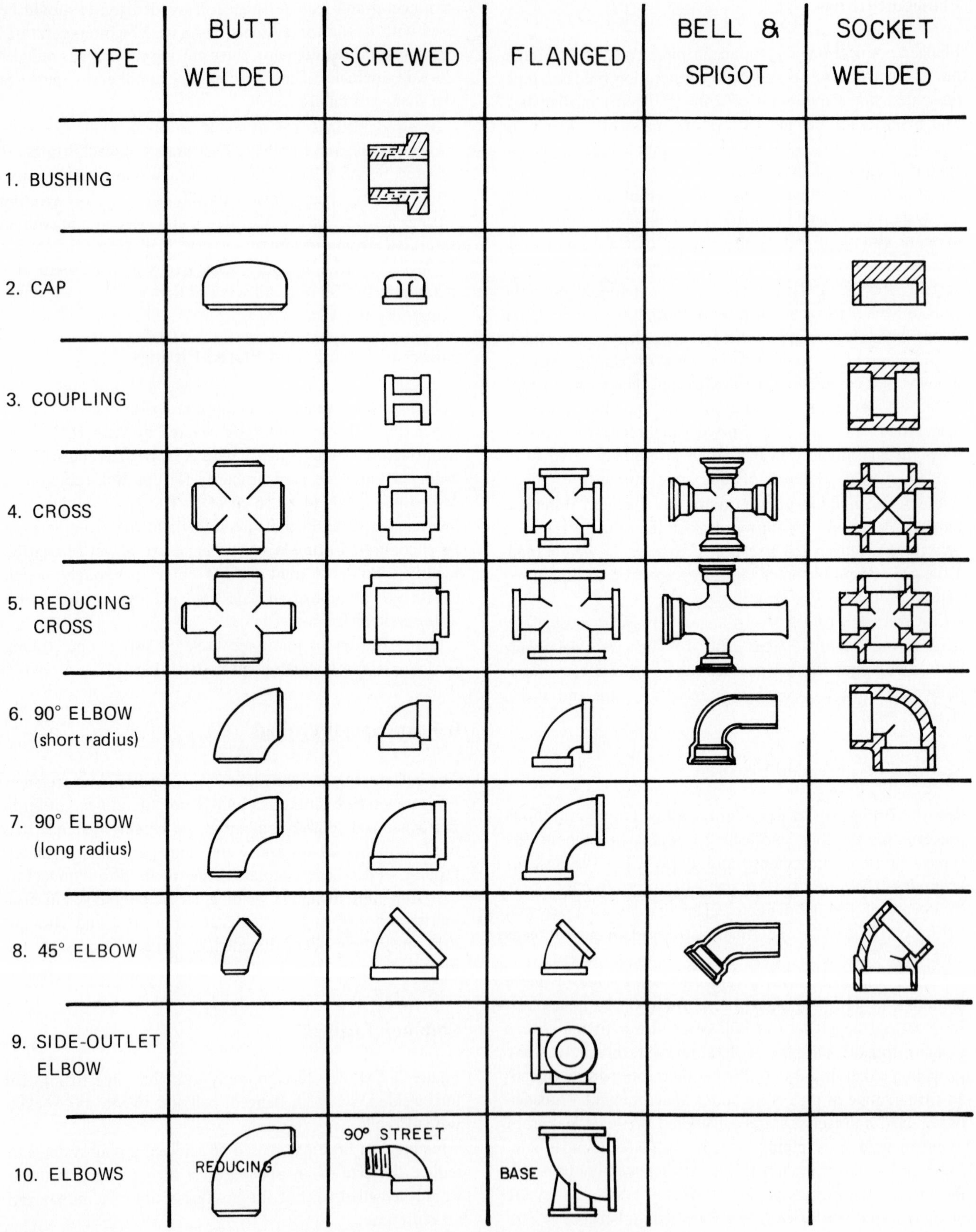

TYPE	BUTT WELDED	SCREWED	FLANGED	BELL & SPIGOT	SOCKET WELDED
1. BUSHING					
2. CAP					
3. COUPLING					
4. CROSS					
5. REDUCING CROSS					
6. 90° ELBOW (short radius)					
7. 90° ELBOW (long radius)					
8. 45° ELBOW					
9. SIDE-OUTLET ELBOW					
10. ELBOWS	REDUCING	90° STREET	BASE		

FIG. 3-7 Butt-welded, screwed, flanged, bell and spigot, and socket-welded fittings.

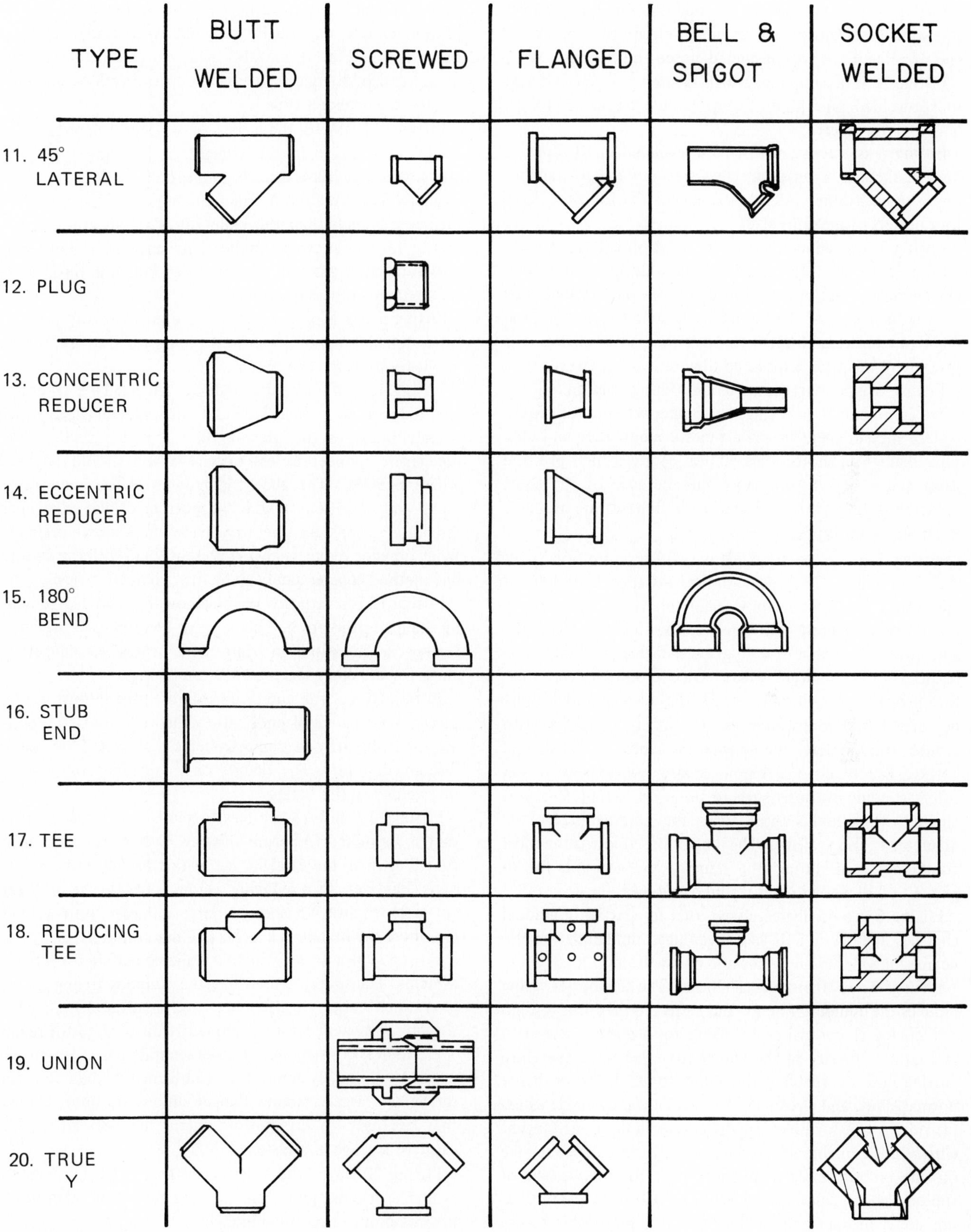

TYPE	BUTT WELDED	SCREWED	FLANGED	BELL & SPIGOT	SOCKET WELDED
11. 45° LATERAL					
12. PLUG					
13. CONCENTRIC REDUCER					
14. ECCENTRIC REDUCER					
15. 180° BEND					
16. STUB END					
17. TEE					
18. REDUCING TEE					
19. UNION					
20. TRUE Y					

FIG. 3-7 Butt-welded, screwed, flanged, bell and spigot, and socket-welded fittings. (cont'd)

Fitting 2 is a cap, which is used on the butt-welded, screwed, or socket-welded system to close off or block the end of a line. Caps are considered closed-end fittings.

Fitting 3 is a coupling, which is used for screwed and socket-welded pipe joints to enable two pieces of pipe to be joined together.

Fitting 4 is a cross. Crosses are available in all types of joint makeups. A cross has two side outlets or branches opposite each other. All crosses, except the following have four outlets of the same size.

Fitting 5 is a reducing cross. Its function is to enable the outlets to be of different sizes to take different size pipe. On a straight cross, as in Fitting 4, all four outlets are equal; on a reducing cross, the branch outlets are smaller than the main run outlet and inlet. Through the use of crosses it is possible to maintain a specified flow rate and pressure.

Fitting 6 is a short-radius elbow. Elbows are among the more commonly used fittings in a piping system. The center-to-face measurement in a short-radius elbow is equal to the pipe diameter. Short-radius elbows create a high pressure drop and a higher turbulence rate because of the short distance through the 90-degree turn. Elbows are available in all pipe joint types.

Fitting 7 is a long-radius elbow, which is preferred over the short radius. The center-to-face distance is one and a half times the nominal pipe diameter. It is important to take into consideration the cost, space, flow requirements, and flow rate when selecting a 90-degree elbow. Short-radius elbows are used where space is limited and flow turbulence is not important. If the line material must be carried over long distances or it might be necessary to reduce friction loss, the long-radius elbow is considered better. The long-radius elbow is also less expensive and offers a minimum resistance to flow. An extra-long-radius elbow with a radius three times the nominal size of its diameter is also available. The long-radius elbow is the most commonly used throughout industry and accounts for 90 percent of all elbows manufactured and sold.

Fitting 8 is a 45-degree elbow used for partial or gradual changes in direction. It has advantages similar to the 90-degree long-radius elbow because of low friction loss.

Fittings 9 and 10 are special elbows. The side-outlet elbow reduces the number of bends and turns required in a system. A reducing elbow makes it possible to turn the pipe system and reduce the size of the line at the same time, therefore cutting cost by enabling the elbow to do the work of the normal elbow and a reducer. A base elbow is available which provides a pad for pipe support connections. Long-tangent elbows are available which have a straight mounting service on one or both ends to provide space for the welding of slip-on flanges, since the conventional elbow has a continuous arc from one end to the other it is impossible to use slip-on flanges and maintain the 90-degree angle. Long-tangent elbows are usually provided with straight sections on one end. Elbows are also available in 22½-degree segments and odd angle segments for special orders, but only for the butt-welded type.

Fitting 11 shows the types of 45-degree laterals. The lateral is available in the straight or reducing type. The straight lateral has a branch connection which is the same diameter as the main run. The reducing lateral has a smaller-diameter branch than the main run. Laterals are used primarily for low-pressure applications since they have little structural strength and do not meet pressure vessel code requirements unless reinforced.

Fitting 12 is a plug which is used to close off a tee or other female end of a screwed piping run and has similar applications as the cap fitting.

Fitting 13 is a concentric reducer. Fitting 14 is an eccentric reducer. A reducer reduces the pipeline size, which usually increases the pressure in the piping system. The concentric reducer is less expensive and should be used where possible. The eccentric reducer is used on piping racks and situations where the bottom of the pipe must remain at a constant level to provide for adequate drainage. When drawing reducers always present them with the double-line method to adequately show their characteristics.

Fitting 15 is a variation of the elbow in a 180-degree bend or change in direction. This type of bend is used primarily for special situations in heater coils or heat exchangers to allow for the close spacing of return piping.

Fitting 16 is a stub end. Stub ends are used in butt-welded piping systems. Stub ends, also called lap joint stub ends, are used primarily in piping systems that must have quick-disconnect joints. The flanges on a lap joint stub end are not welded to the fitting.

Fittings 17 and 18 are tees. A tee is a reinforced fitting that is branched to permit flow at 90 degrees to the main run. Shown is the straight tee, which has all branches the same diameter. In a reducing tee, branches can be of different sizes. Figure 3-8 shows a large-diameter, butt-welded, tee where the branch is a larger size than the run.

Fitting 19 is a union, which is available mainly for screwed systems. The union is used like the coupling fitting though it is much easier to disconnect and is advantageous when the piping system must be frequently dismantled for repair. The union is constructed of three separate parts. Unions are used in piping systems 2 in. (5.08 cm) in diameter and smaller; in larger systems, flanged unions are used. Though they are easy to disconnect, unions are also a frequent cause of leakage in a system.

Fitting 20 shows the Y fitting or true Y fitting, which is available in most joint makeups and allows the main run to be broken into two equal branches.

FIG. 3-8 A 16-in. (40.64-cm) schedule 10S stainless steel welding elbow. (Courtesy Flowline Corporation)

Figure 3-9 exhibits a stub-in and weld saddle. Because a welded tee is expensive and hard to install, the piping system using butt-welded fittings often employs a reinforcing saddle or reinforcing pad and stub-in the connecting pipe to make up a 90-degree branch to a main line. Most of the prevalent fittings and their dimensions are listed in the appendix, which should be consulted regularly.

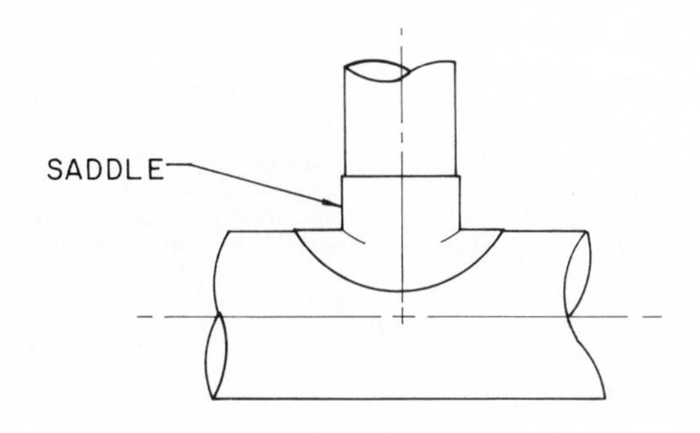

FIG. 3-9 Stub-in and weld saddle.

Designating Sizes

When designating pipe fittings, give the dimension of the run before that of the branches. In all cases the dimension of the larger opening should precede the dimension of the smaller.

Figure 3-10 demonstrates three typical reducing fittings—the reducing tee, cross, and 45-degree lateral—and an example of how they would be called out for ordering. When specifying a fitting, give the nominal inside diameter of the pipe for which the opening is to be threaded if it is a screwed fitting and also the type of fitting and material. When specifying a reducing fitting, give the largest opening of the through run and then the opposite end followed by the outlet or branches. In Fig. 3-10 the reducing tee has a 4-in. (10.16-cm) main run and a 2-in. (5.08-cm) branch and is therefore called out as a 4 x 4 x 2 inch tee.

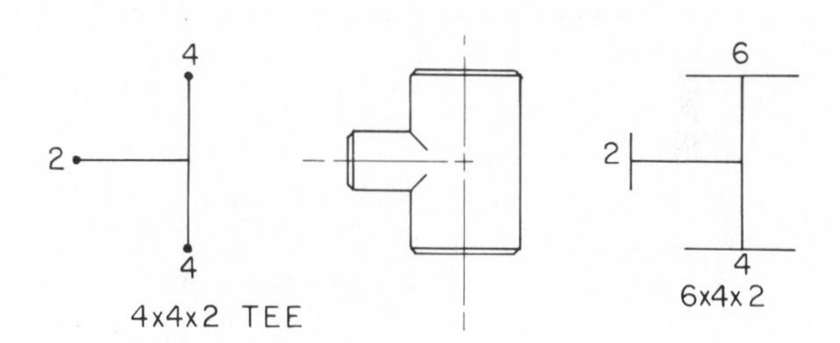

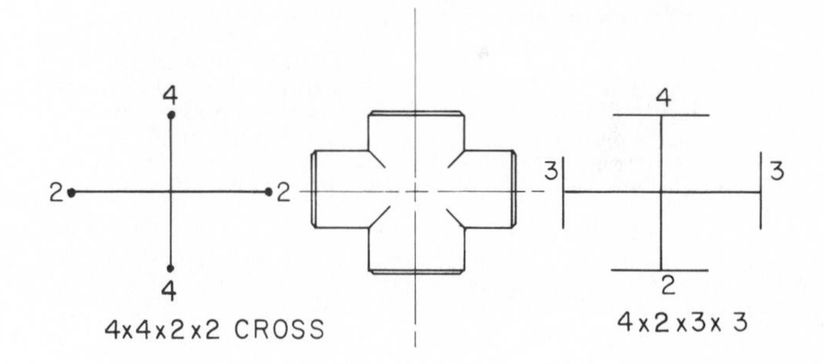

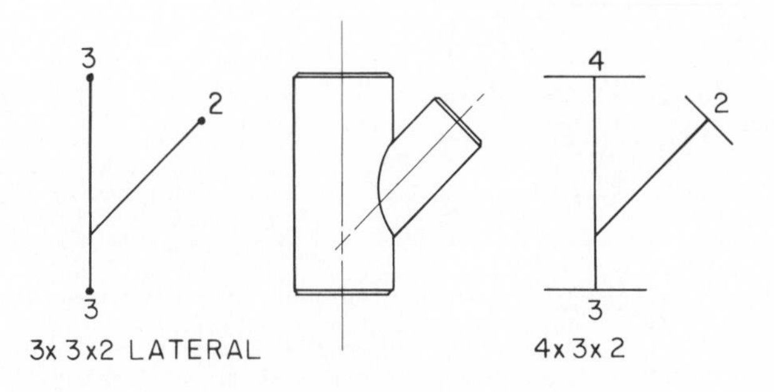

FIG. 3-10 Reducing tee, cross, and 45-degree lateral showing proper callout and specification sequence.

Selection and Application of Forged Steel Companion Flanges

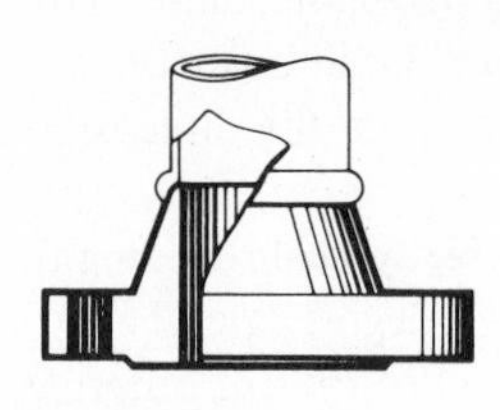

Welding Neck Flanges are distinguished from other types by their long tapered hub and gentle transition of thickness in the region of the butt weld joining them to the pipe. The long tapered hub provides an important reinforcement of the flange proper from the standpoint of strength and resistance to dishing. The smooth transition from flange thickness to pipe wall thickness effected by the taper is extremely beneficial under conditions of repeated bending, caused by line expansion or other variable forces, and produces an endurance strength of welding neck flanged assemblies(1) equivalent to that of a butt welded joint between pipes, which, in practice, is the same as that of unwelded pipe. Thus this type of flange is preferred for every severe service condition, whether this results from high pressure or from sub-zero or elevated temperature, and whether loading conditions are substantially constant or fluctuate between wide limits; welding neck flanges are particularly recommended for handling explosive, flammable or costly liquids, where loss of tightness or local failure may be accompanied by disastrous consequences.

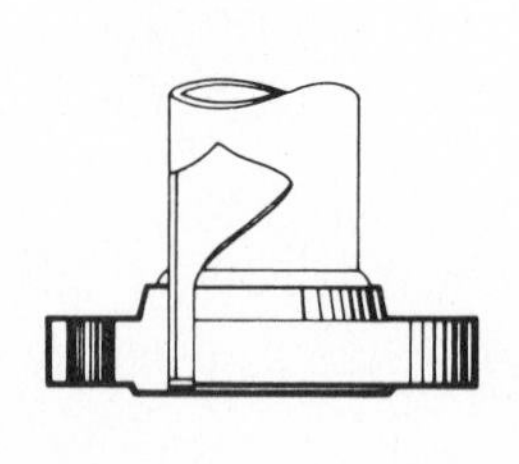

Slip-On Flanges continue to be preferred to welding neck flanges by many users on account of their initially lower cost, the reduced accuracy required in cutting the pipe to length, and the somewhat greater ease by alignment of the assembly; however, their final installed cost is probably not much, if any, less than that of welding neck flanges. Their calculated strength under internal pressure is of the order of two-thirds that of welding neck flanges, and their life under fatigue is about one-third that of the latter. For these reasons, slip-on flanges are limited to sizes NPS ½ to 2½ in the 1500 standard.

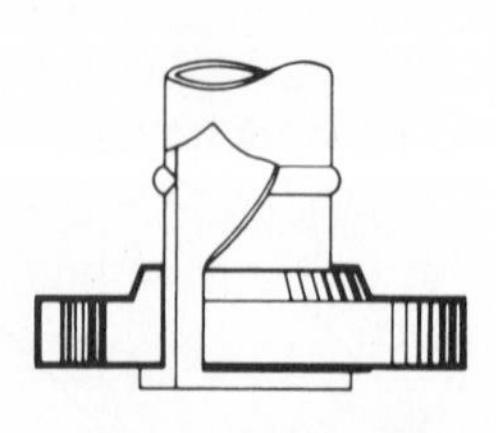

Lap Joint Flanges are primarily employed with lap joint stubs, the combined initial cost of the two items being approximately one-third higher than that of comparable welding neck flanges. Their pressure-holding ability is little, if any, better than that of slip-on flanges and the fatigue life of the assembly is only one-tenth that of welding neck flanges. The chief use of lap joint flanges in carbon or low alloy steel piping systems is in services necessitating frequent dismantling for inspection and cleaning and where the ability to swivel flanges and to align bolt holes materially simplifies the erection of large diameter or unusually stiff piping Their use at points where severe bending stress occurs should be avoided.

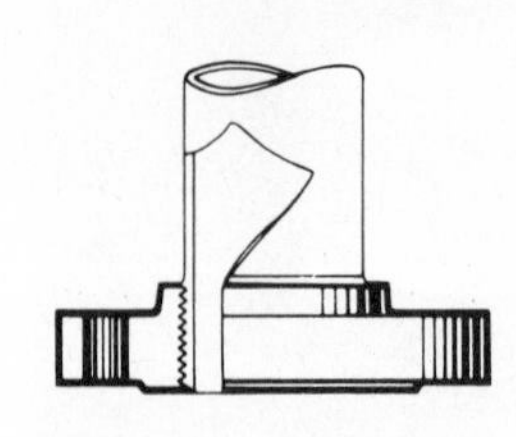

Screwed Flanges made of steel are confined to special applications. Their chief merit lies in the fact that they can be assembled without welding; this explains their use in extremely high pressure services, particularly at or near atmospheric temperature, where alloy steel is essential for strength and where the necessary post-weld heat treatment is impractical. Screwed flanges are unsuited for conditions involving temperature or bending stresses of any magnitude, particularly under cyclic conditions, where leakage through the threads may occur in relatively few cycles of heating or stress; seal welding is sometimes employed to overcome this, but cannot be considered as entirely satisfactory.

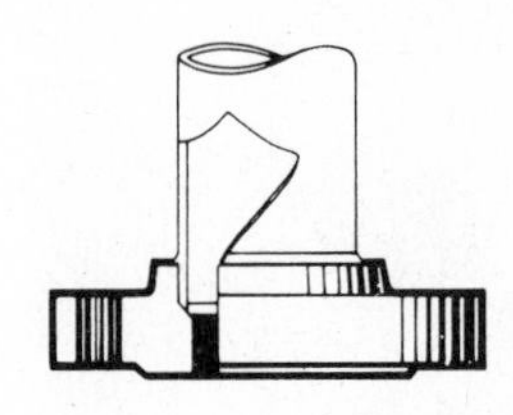

Socket Welding Flanges were initially developed for use on small-size high pressure piping. Their initial cost is about 10% greater than that of slip-on flanges; when provided with an internal weld as illustrated, their static strength is equal to, but their fatigue strength 50% greater than double-welded slip-on flanges. Smooth, pocketless bore conditions can readily be attained (by grinding the internal weld) without having to bevel the flange face and, after welding, to reface the flange as would be required with slip-on flanges. The internally welded socket type flange is becoming increasingly popular in chemical process piping for this reason.

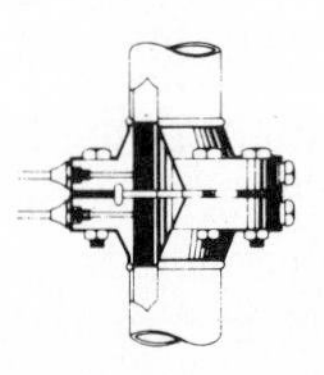

Orifice Flanges are widely used in conjunction with orifice meters for measuring the rate of flow of liquids and gases. They are basically the same as standard welding neck, slip-on and screwed flanges except for the provision of radial, tapped holes in the flange ring for meter connections and additional bolts to act as jack screws to facilitate separating the flanges for inspection or replacement of the orifice plate. In choosing the type of orifice flange, the considerations affecting the choice of welding neck, slip-on and screwed standard flanges apply with equal force.

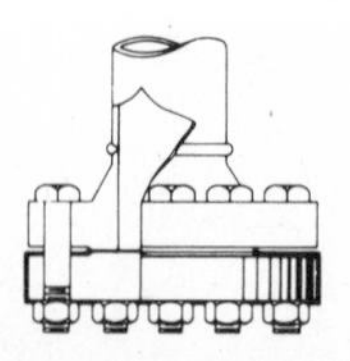

Blind Flanges are used to blank off the ends of piping, valves and pressure vessel openings. From the standpoint of internal pressure and bolt loading, blind flanges, particularly in the larger sizes, are the most highly stressed of all American Standard flange types. However, since the maximum stresses in a blind flange are bending stresses at the center, they can safely be permitted to be higher than in other types of flanges. Where temperature is a service factor, or repeated severe water hammer, consideration should be given to closures made of welding neck flanges and caps.

(1) In tests of all types of flanged assemblies, fatigue failure invariably occurred in the pipe or in an unusually weak weld, never in the flange proper. The type of flange, however, and particularly the method of attachment, greatly influence the number of cycles required to cause fracture.

FIG. 3-11 Selection and application of forged steel companion flanges. (Courtesy ITT Grinnell Corp. WARNINGS are published in each catalog for proper installation and performance of products, beyond which the company does not warrantee performance.)

FLANGES

One of the most common methods of joining pipe components and equipment is the flange. Flanged piping can be dismantled to repair valves, meters, and other equipment without disturbing the rest of the system. All flanges and bolting patterns are standardized. Flanges are found almost exclusively on piping systems 2 in. (5.08 cm) and larger in diameter where screwed connections would be inconvenient or impossible. Flanges are also used on piping subassemblies to pressure vessels and for meters and valves and the like. The appendix lists the standard dimensions for steel flanges with ring joints and other flanges that are available. These dimensions should be consulted when ordering for the problems at the end of each chapter.

Flanges are basically forged or cast steel rings designed to join two sections of pipe together. They are attached to the pipe by threading, welding, or lap joints. Flange joints are made up with two flanges that are bolted together with a gasket between the flange faces. Flanges come in an assortment of face types which include flat face, raised face, male and female, tongue and groove, and with ring joints. In most cases the thickness and outside diameter of the flange increases along with the flange's pressure rating.

Figure 3-11 exhibits the most common forged steel companion flanges on the market. They include the weld-neck flange, which is available in a variety of pressure ratings and sizes. This type of flange is butt-welded to the end of the pipe and used in high-pressure/high-temperature situations. A variation of the weld-neck flange is the long-neck flange shown in Fig. 3-12. This type of flange is also referred to as a nozzle and is used as an outlet on vessels. Because of its long, heavy-wall straight hub, it has a greater pressure rating. The long-neck flange comes with a square-cut end. The extra thickness on this type of flange hub provides for greater resistance to piping strains caused by connecting piping. Moreover, this flange resists corrosion and erosion better than the other types. The thick wall also makes it ideal for use in pressure vessels and tanks for nozzles. In many cases the weld-neck flange and long-neck flange are available in services up to 2500 lb.

Threaded flanges are similar to slip-on flanges except that their inside diameter is threaded. Threaded flanges are used in low-temperature/low-pressure situations and where welded flanges cannot be used because of fire hazard. The threaded flange is ideal for small-diameter pipes 1½ in. (6.35 cm) and under and is often seal-welded to provide leakproof service.

The blind flange is available in several sizes and pressure ratings. It is used to close off the end of a pipeline that may be temporarily out of service or is needed for future expansion. A blind flange can be altered to be bored, tapped, or drilled for high reduction in pipe inside diameter to reduce the flow in the system or to serve as an instrument signal source.

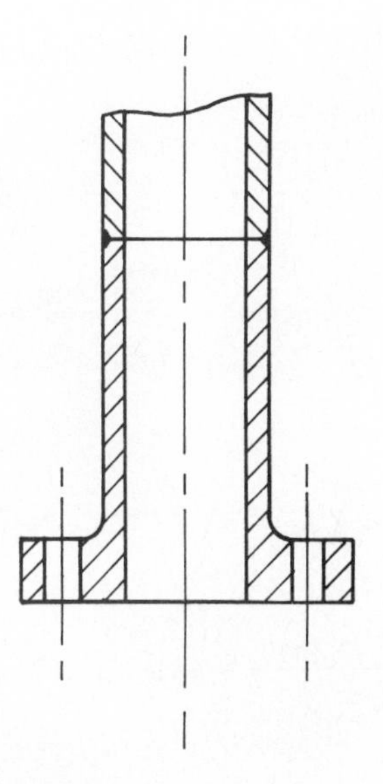

FIG. 3-12 Long-neck flange.

The lap joint or Van Stone flange comes equipped with a lap joint. This type is available in many pressures and sizes. The flange slips over the pipe and is usually not welded or fastened to the pipe itself. Bolting pressure and gasket maintain an adequate joint. The lap joint allows fast makeup of the pipeline because of the ease in aligning the holes of the two flange pieces. In this type of joint, line material does not come in contact with the flange so the flanges are reusable, even after long service. In Fig. 3-13, lap joint flanges are shown on the elbow of the left-hand pipe fabrication.

The socket-welded flange is used in small-diameter piping systems where a screwed fitting is inadequate because of leakage. The socket-welded flange is counterbored larger than the outside diameter of the matching pipe so the pipe can be inserted into the flange. The inside diameter on both the flange and the pipe are the same. Socket-welded flanges, in other words, have a recess in the back of the flange that receives the pipe end. These flanges are usually confined to 4 in. (10.10 cm) and smaller diameter pipes because of their low pressure and temperature ratings.

The orifice flange draws its name from the fact that an orifice plate is clamped between a pair of flanges. An orifice flange is commonly used for measuring or regulating fluid flow in piping systems. The flange is equipped with two radially drilled and tapped holes for metering the flow rate. This type comes available in the weld-neck, slip-on, and threaded varieties.

FIG. 3-13 Three welded pipe fabrications. Piece at left shows an elbow welded to a pipe. The elbow has a lap joint flanged on one end and a butt-welded end on the other. Middle fabrication shows an elbow to a reducing tee. (Courtesy Flowline Corporation)

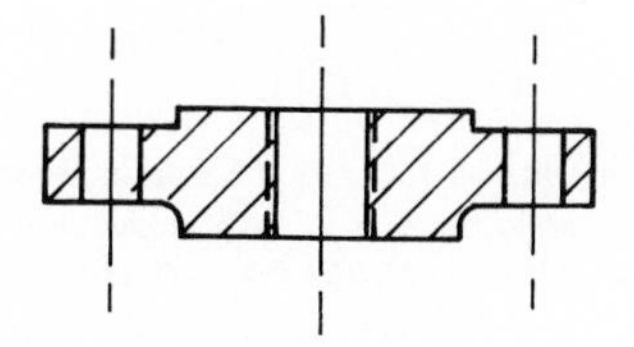

(a) **THREADED**

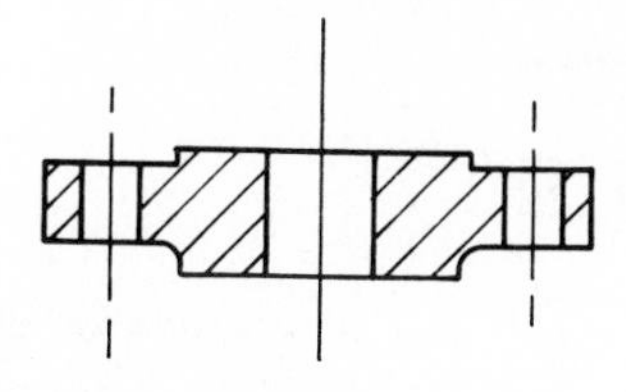

(b) **SLIP-ON**

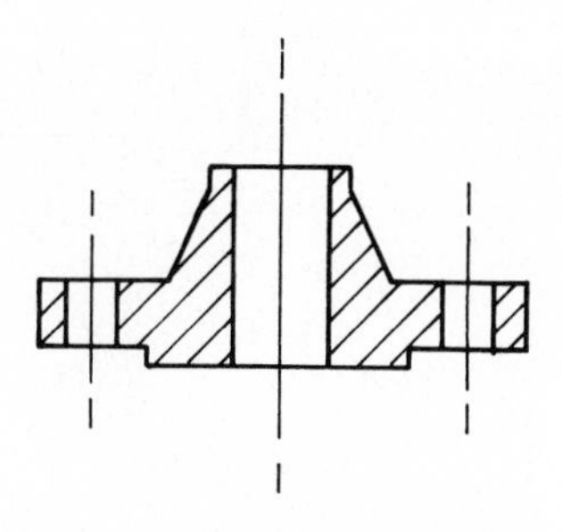

(c) **WELDING NECK**

FIG. 3-14 Three types of reducing flanges.

Reducing flanges, shown in Fig. 3-14, are used to reduce the size of the line. They are used in similar ways to concentric reducers. Figure 3-11 provides the essential information needed to select companion steel flanges.

Flange Facings

One of the most important aspects of a flange is its facing. Facings come in raised, ring type, male and female, tongue and groove, and flat face types. Each facing has a variety of finishes. The raised facing is commonly used in piping systems. The faces are raised 1/16 in. (0.16 cm) for 150# and 300# class flanges and 1/4 in. (0.64 cm) for 400# and above on typical flanges.

The ring-type facing is used in high-temperature/high-pressure situations and is more expensive. Male and female facings are for special service. Tongue and groove types are used for services that require a retained gasket and where contact between line material and flange must be kept to a minimum. The flat face is similar to the raised face, but there is no raised portion. This type is used for low-pressure cast iron systems.

Figure 3-15 displays typical facings and dimensions that are common to all flanges. Figure 3-16 shows the various ring joint facings and their dimensions for the 150#-class.

Gaskets

Gaskets are used for all flanges and are available in metallic or nonmetallic materials, depending on intended service. Table 3-1 is a general reference chart for gasket materials and their temperature and service recommendations. Be sure to choose gasket materials that will not be affected chemically by the line material.

Metal ring-joint gasket materials are covered by ASA B16.20; nonmetallic gaskets are covered by ASA B16.21. Figure 3-17 presents the various types of facings and gaskets. Figure 3-18 shows a RTJ (ring-type joint) flange.

Bolting

All flange joints require bolting; the requirements are standardized by ANSI. Bolt patterns come in groups of four (4, 8, 12, 16, and so on) and are equally spaced on the bolt circle. All temperature and pressure ratings on flanges and fittings are standardized by ANSI. In high-pressure/high-temperature situations flanges require considerably more bolt holes. The higher the pressure, the more bolts will be needed. Figure 3-19 shows the 4-, 8-, and 16-bolt pattern. Notice that the holes are never on the vertical or horizontal axis, but are evenly straddled.

To find the number of degrees between bolt holes, divide 360 by the number of bolts required. The pressure/temperature rating for the system will determine the flange rating; the allowable pressure will decrease as the temperature rises. For services up to 700°F (371.1°C) carbon and low alloy steel bolts are commonly provided. Consult the manufacturer's recommendations for flanges, bolting, gaskets, and the like for specific service applications.
flanges, bolting, gaskets, and the like for specific service applications.

American Standard facings

Facing dimensions shown are in accordance with ANSI B16.5. Gaskets for all flanges are available in accordance with ANSI B16.5.

■ Tolerance of plus or minus 1/64 inch is allowed on the inside and outside diameters of all facings other than ring joint.

Ӿ Regular facing for Classes 150 and 300 steel flanged fittings and companion flange standards is a 1/16 inch raised face included in the minimum flange thickness. A 1/16 inch raised face may be supplied also on the Classes 400, 600, 900, 1500 and 2500 flanges, but it must be added to the minimum thickness.

◈ Regular facing for Classes 400, 600, 900, 1500 and 2500 flange standards is a 1/4 inch raised face not included in minimum flange thickness dimensions.

facings applicable to all types of flanges

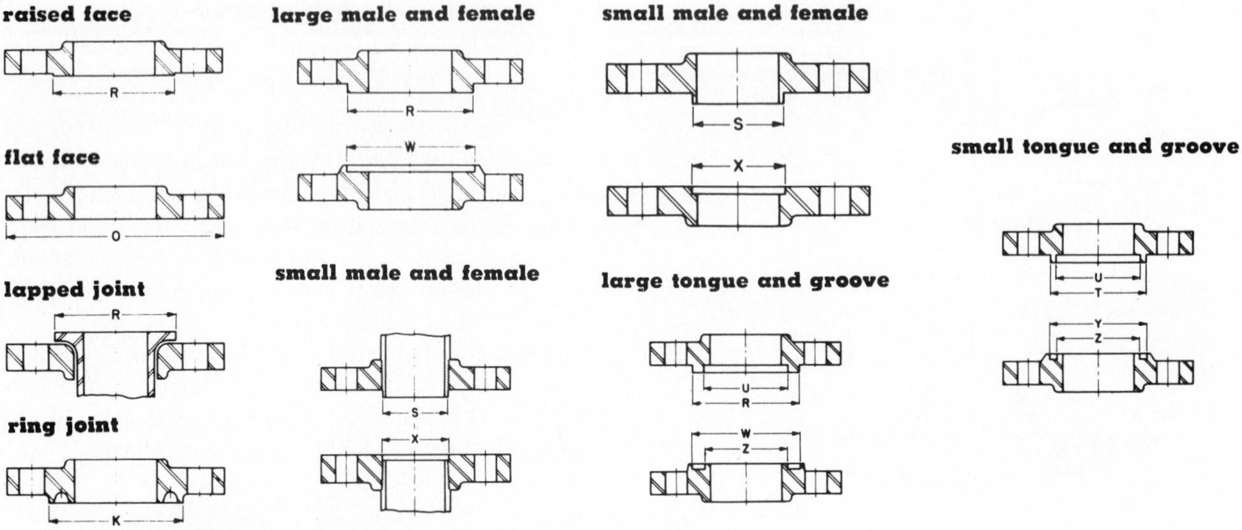

raised face

flat face

lapped joint

ring joint

large male and female

small male and female

small male and female

large tongue and groove

small tongue and groove

facing dimensions *(except for ring joint)*

nominal pipe size	outside diameter■ raised face, lap joint, large male and large tongue R	small male S	small tongue T	I. D. of large and small tongue■ U	outside diameter■ large female and large groove W	small female X	small groove Y	I. D. of large and small groove■ Z	height raised face 150 & 300 Ӿ	raised face, large and small male and tongue 400-2500 ◈	depth of groove or female
1/2	1 3/8	23/32	1 3/8	1	1 7/16	25/32	1 7/16	15/16	1/16	1/4	3/16
3/4	1 11/16	15/16	1 11/16	1 5/16	1 3/4	1	1 3/4	1 1/4	1/16	1/4	3/16
1	2	1 3/16	1 7/8	1 1/2	2 1/16	1 1/4	1 15/16	1 7/16	1/16	1/4	3/16
1 1/4	2 1/2	1 1/2	2 1/4	1 7/8	2 9/16	1 9/16	2 5/16	1 13/16	1/16	1/4	3/16
1 1/2	2 7/8	1 3/4	2 1/2	2 1/8	2 15/16	1 13/16	2 9/16	2 1/16	1/16	1/4	3/16
2	3 5/8	2 1/4	3 1/4	2 7/8	3 11/16	2 5/16	3 5/16	2 13/16	1/16	1/4	3/16
2 1/2	4 1/8	2 11/16	3 3/4	3 3/8	4 3/16	2 3/4	3 13/16	3 5/16	1/16	1/4	3/16
3	5	3 5/16	4 5/8	4 1/4	5 1/16	3 3/8	4 11/16	4 3/16	1/16	1/4	3/16
3 1/2	5 1/2	3 13/16	5 1/8	4 3/4	5 9/16	3 7/8	5 3/16	4 11/16	1/16	1/4	3/16
4	6 3/16	4 5/16	5 11/16	5 3/16	6 1/4	4 3/8	5 3/4	5 1/8	1/16	1/4	3/16
5	7 5/16	5 3/8	6 13/16	6 5/16	7 3/8	5 7/16	6 7/8	6 1/4	1/16	1/4	3/16
6	8 1/2	6 3/8	8	7 1/2	8 9/16	6 7/16	8 1/16	7 7/16	1/16	1/4	3/16
8	10 5/8	8 3/8	1C	9 3/8	10 11/16	8 7/16	10 1/16	9 5/16	1/16	1/4	3/16
10	12 3/4	10 1/2	12	11 1/4	12 13/16	10 9/16	12 1/16	11 3/16	1/16	1/4	3/16
12	15	12 1/2	14 1/4	13 1/2	15 1/16	12 9/16	14 5/16	13 7/16	1/16	1/4	3/16
14	16 1/4	13 3/4	15 1/2	14 3/4	16 5/16	13 13/16	15 9/16	14 11/16	1/16	1/4	3/16
16	18 1/2	15 3/4	17 5/8	16 3/4	18 9/16	15 13/16	17 11/16	16 11/16	1/16	1/4	3/16
18	21	17 3/4	20 1/8	19 1/4	21 1/16	17 13/16	20 3/16	19 3/16	1/16	1/4	3/16
20	23	19 3/4	22	21	23 1/16	19 13/16	22 1/16	20 15/16	1/16	1/4	3/16
24	27 1/4	23 3/4	26 1/4	25 1/4	27 5/16	23 13/16	26 5/16	25 3/16	1/16	1/4	3/16

FIG. 3-15 Standard American facings for flanges. (Courtesy ITT Grinnell Corp. WARNINGS are published in each catalog for proper installation and performance of products beyond which the company does not warrantee performance.)

ring joint facing

rings

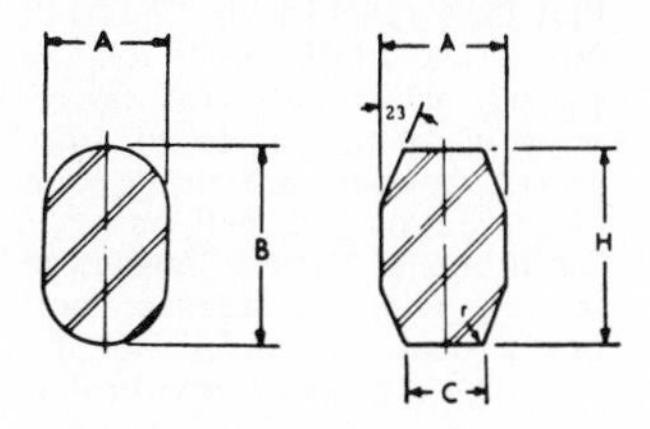

groove △

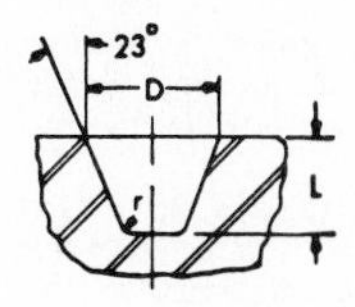

Ring joint facings are in accordance with ANSI B16.5, except above NPS 24 are not covered by ANSI B16.5.

Dimension "Q" is given on pages showing flanges and does not include facing height "L."

The dimensions of the rings and grooves of ring joint flanges rated at Classes 300, 400, and 600 are identical.

✦ The depth of groove is added to the minimum flange thickness.

△ Octagonal groove (per ANSI B16.5) furnished unless otherwise ordered; suitable for either octagonal or oval ring.

♦ For calculating the "laying length" of fittings with ring joints, the space dimensions given in these tables must be added.

For NPS 3 lapped flanges in Classes 300, 400 and 600, the pitch diameter of the ring should be 4 5/8 inch instead of 4 7/8 inch and the ring number should be R30 instead of R31.

For dimensional tolerances, see page wff-87.

Ring joint gasket dimensions shall conform to ANSI B16.20.
All dimensions in inches.

threaded **slip-on** **welding neck**

Q L P K

nominal pipe size	pitch diameter of ring and groove P	ring: width A	ring: height oval B	ring: height octagonal H	ring: width of flat on octagonal rings C	groove: width D	groove: depth✦ L	radius r	diameter of raised face or lap K	distance (approx) between flanges when ring is compressed♦	ring and groove number
class 150											
1	1 7/8	5/16	9/16	1/2	.206	11/32	1/4	1/32	2 1/2	5/32	B15
1 1/4	2 1/4	5/16	9/16	1/2	.206	11/32	1/4	1/32	2 7/8	5/32	R17
1 1/2	2 9/16	5/16	9/16	1/2	.206	11/32	1/4	1/32	3 1/4	5/32	R19
2	3 1/4	5/16	9/16	1/2	.206	11/32	1/4	1/32	4	5/32	R22
2 1/2	4	5/16	9/16	1/2	.206	11/32	1/4	1/32	4 3/4	5/32	R25
3	4 1/2	5/16	9/16	1/2	.206	11/32	1/4	1/32	5 1/4	5/32	R29
3 1/2	5 3/16	5/16	9/16	1/2	.206	11/32	1/4	1/32	6 1/16	5/32	R33
4	5 7/8	5/16	9/16	1/2	.206	11/32	1/4	1/32	6 3/4	5/32	R36
5	6 3/4	5/16	9/16	1/2	.206	11/32	1/4	1/32	7 5/8	5/32	R40
6	7 5/8	5/16	9/16	1/2	.206	11/32	1/4	1/32	8 5/8	5/32	R43
8	9 3/4	5/16	9/16	1/2	.206	11/32	1/4	1/32	10 3/4	5/32	R48
10	12	5/16	9/16	1/2	.206	11/32	1/4	1/32	13	5/32	R52
12	15	5/16	9/16	1/2	.206	11/32	1/4	1/32	16	5/32	R56
14	15 5/8	5/16	9/16	1/2	.206	11/32	1/4	1/32	16 3/4	1/8	R59
16	17 7/8	5/16	9/16	1/2	.206	11/32	1/4	1/32	19	1/8	R64
18	20 3/8	5/16	9/16	1/2	.206	11/32	1/4	1/32	21 1/2	1/8	R68
20	22	5/16	9/16	1/2	.206	11/32	1/4	1/32	23 1/2	1/8	R72
22	24 1/4	5/16	....	1/2	.206	11/32	1/4	1/32	25 1/2	1/8	R80
24	26 1/2	5/16	9/16	1/2	.206	11/32	1/4	1/32	28	1/8	R76
26	28 5/8	3/8	....	9/16	.250	13/32	5/16	1/32	30 3/8	1/8	....
30	32 7/8	3/8	....	9/16	.250	13/32	5/16	1/32	34 5/8	1/8	....
34	37	1/2	....	11/16	.341	17/32	3/8	1/16	38 7/8	3/16	....
36	39 1/4	1/2	....	11/16	.341	17/32	3/8	1/16	41 1/8	3/16	....
42	46	1/2	....	11/16	.341	17/32	3/8	1/16	47 7/8	3/16	....

FIG. 3-16 Ring joint facings for RTJ flanges. (Courtesy ITT Grinnell Corp. WARNINGS are published in each catalog for proper installation and performance of products beyond which the company does not warrantee performance.)

facings & gaskets

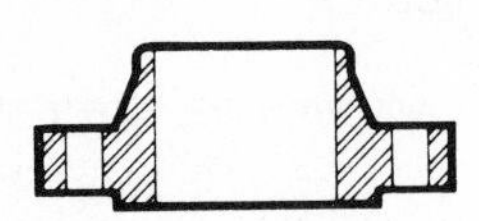

The Raised Face is the most common facing employed with steel flanges; it is 1/16" high for Class 150 and Class 300 flanges and 1/4" high for all other pressure classes. The facing is machine-tool finished with spiral or concentric grooves (approximately 1/64" deep on approximately 1/32" centers) to bite into and hold the gasket. Because both flanges of a pair are identical, no stocking or assembly problems are involved in its use. Raised face flanges generally are installed with soft flat ring composition gaskets. The width of the gasket is usually less than the width of the raised face. Faces for use with metal gaskets preferably are smooth finished.

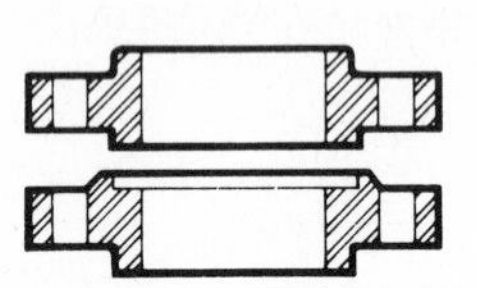

Male-and-Female Facings are standardized in both large and small types. The female face is 3/16" deep and the male face 1/4" high and both are usually smooth finished since the outer diameter of the female face acts to locate and retain the gasket. The width of the large male and female gasket contact surface, like the raised face, is excessive for use with metal gaskets. The small male and female overcomes this but provides too narrow a gasket surface for screwed flanges assembled with standard weight pipe.

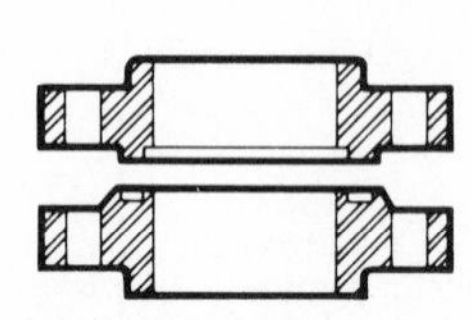

Tongue-and-Groove Facings are also standardized in both large and small types. They differ from male-and-female in that the inside diameters of tongue and groove do not extend to the flange bore, thus retaining the gasket on both its inner and outer diameter; this removes the gasket from corrosive or erosive contact with the line fluid. The small tongue-and-groove construction provides the minimum area of flat gasket it is advisable to use, thus resulting in the minimum bolting load for compressing the gasket and the highest joint efficiency possible with flat gaskets.

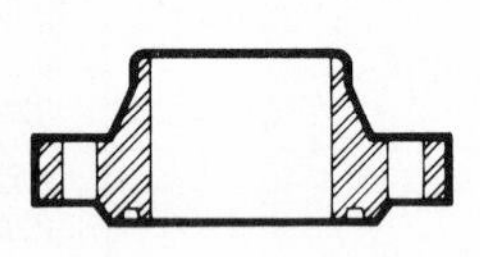

Ring Joint Facing is the most expensive standard facing but also the most efficient, partly because the internal pressure acts on the ring to increase the sealing force. Both flanges of a pair are alike, thus reducing the stocking and assembling problem found with both made-and-female and tongue-and-groove joints. Because the surfaces the gasket contacts are below the flange face, the ring joint facing is least likely of all facings to be damaged in handling or erecting. The flat bottom groove is standard.

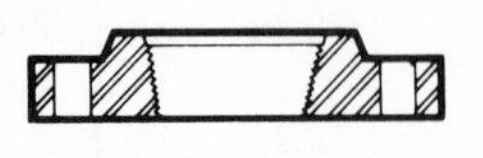

Flat Faces are a variant of raised faces, sometimes formed by machining off the 1/64" raised face of Class 150 and Class 300 flanges. Their chief use is for mating with Class 125 and Class 250 cast iron valves and fittings. A flat-faced steel flange permits employing a gasket whose outer diameter equals that of the flange or is tangent to the bolt holes. In this manner the danger of cracking the cast iron flange when the bolts are tightened is avoided.

Flat Ring Gaskets are made of numerous materials such as paper, cloth, rubber, compressed asbestos, ingot iron, nickel, copper, aluminum and other metals, as well as combinations of metals and non-metals. Their thicknesses normally range from 1/64" to 1/8". Widths range from 1/4" upwards; narrow gaskets are preferable since they require lower bolt loads for joint tightness but they must not be too narrow lest the bolt load crush them, if non-metallic, or indent them into the flange face, if metallic. Paper, cloth and rubber gaskets should not be used for temperatures over 250F. Asbestos may be employed up to 650F or somewhat higher, while ferrous or nickel-base metal gaskets are generally satisfactory for the maximum temperature the flanges themselves will withstand.

Serrated Gaskets are flat, metal gaskets having concentric grooves machined into their face. With the contact area reduced to a few concentric lines, the required bolt load is greatly reduced as compared with an unserrated gasket and hence an efficient joint is obtained for applications where soft gaskets are unsuited; serrated gaskets are used with smooth-finished flange faces.

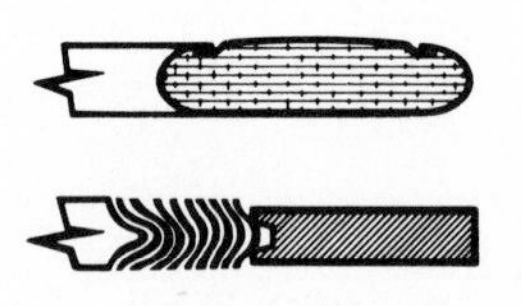

Laminated Gaskets made of metal and soft filler (the filler usually being soft asbestos sheet) are also used; in some gaskets (jacketed type) the laminations are in the plane of the flange face; in others (spiral wound type) the laminations are formed in an axial or edgewise direction. Laminated gaskets with asbestos filler are considered suitable for about 100F higher temperatures than plain asbestos gaskets and require less bolt load to compress them than solid metal gaskets, hence tend to make high-pressure, high-temperature joints more efficient than those using flat, solid metal gaskets.

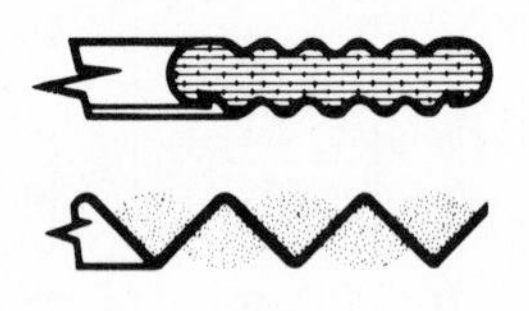

Corrugated Gaskets are a less commonly used type intermediate in stiffness between flat non-metallic and metallic gaskets. The ridges of the corrugations again tend to concentrate the gasket loading along concentric rings. This type is available plain, but to prevent crushing of the corrugations it is preferable to use the asbestos filled or asbestos inserted varieties.

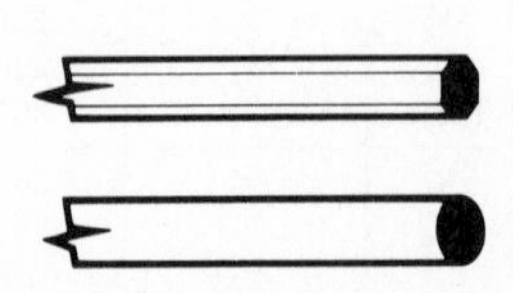

Ring Joint Gaskets are available in two types, octagonal and oval cross-section, both of which are standardized, but the former is considered the superior type. Either may be employed with flat bottom grooves which are now standard. Such rings are almost always made of metal, usually of the softest carbon steel or iron obtainable. In very high temperature or severe corrosion service, they may be made of alloy steel in which case they should be heat-treated to make them as much softer than the flanges proper as possible. For relatively low temperatures, rings made of plastic may be employed to resist corrosion or insulate the joint from electric currents.

FIG. 3-17 Facings and gaskets for pipe flanges. (Courtesy ITT Grinnell Corp. WARNINGS are published in each catalog for proper installation and performance of products beyond which the company does not warrantee performance.)

TABLE 3-1 RECOMMENDATIONS FOR GASKET MATERIALS

Material	*Maximum Temperature (°F)*	*Recommended Service*
Aluminum	800	
Asbestos	Cold	Water
Composition	750	Steam, oil, air, gas
Spiral wound	750	Steam
Metal	600	Steam
Woven	600	Gas
Composition blue	Hot-cold (varies)	Acids
Woven blue	Hot-cold (varies)	Acids
Metallic	1000	Ammonia, gas
Composition		Ammonia
Spiral-wound composition	1000	Steam, air
Graphitic	1000	Gasoline, kerosene
Brass, admiralty	500	
Brass, high	500	
Chrome, moly steel	1200 (varies)	Acids
Chrome, steel	1300 (varies)	Acids
Copper	600	Steam
Cork fiber	212	Cold oil
Everdur	600	
Gold	1200	
Hastelloy	2000	
Inconel	2000	
Iron, ingot	1000	Steam, oil
Lead	212	
Lead sheet	Cold	Ammonia
Magnesium	400	
Monel	1500	Steam
Nickel	1400	
Platinum	2300	
Rubber		
Red	Hot	Water
Red	220	Steam, air, gas
Black	Cold	Water, ammonia
Soft	Cold	Water
Brown	Hot	Water
Silver	1200	
Steel		
Corrugated	1000	Steam
Low carbon	1000	
Sheet alloy		Acids
Stainless	1000	Steam
Stainless (304, 316)	800	
Stainless (347)	1700	
Tantalum	3000	
Tin	212	
Zinc	212	

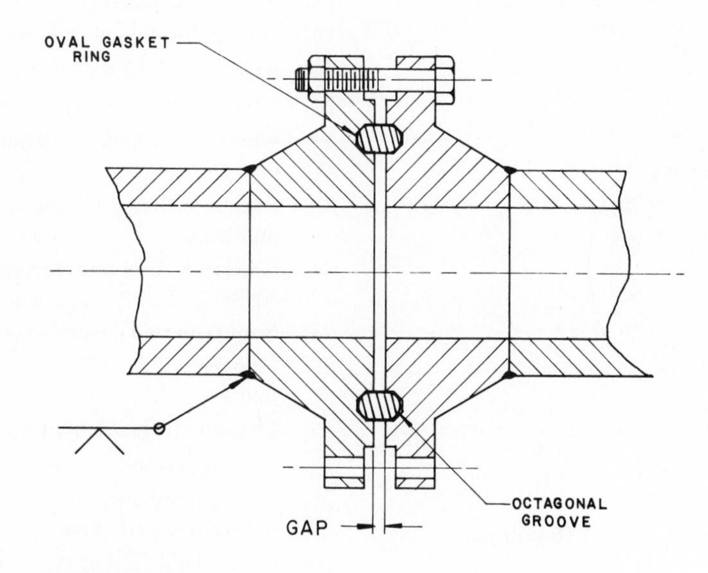

FIG. 3-18 Weld-neck ring type (RTJ) flanges.

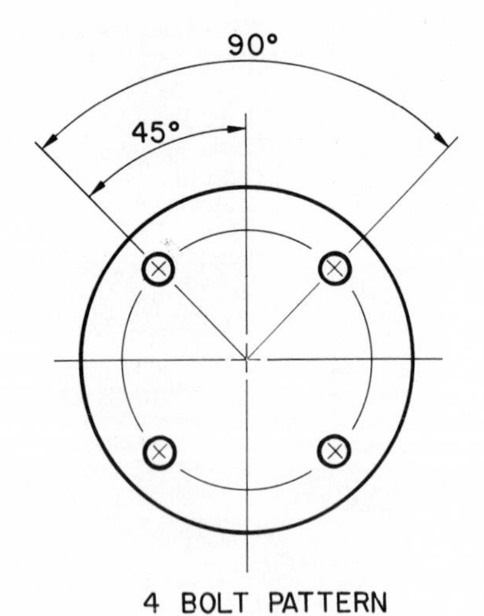

45°
45°
$22\frac{1}{2}°$
8 BOLT PATTERN

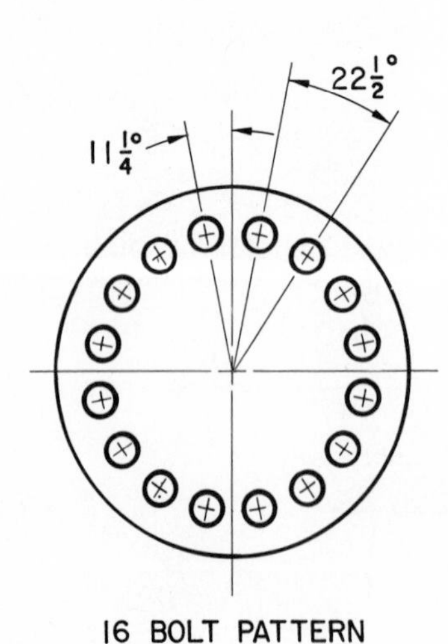

FIG. 3-19 Flange bolting for 4-, 8-, and 16-bolt patterns.

QUIZ

1. What is the major function of pipe fittings?
2. Pipe fittings are specified by nominal pipe size, name of fitting, and ____________.
3. Name three characteristics of welded fittings that make them one of the most desirable types of connection.
4. The pipe size and fitting should always be drawn to ____________.
5. Describe the difference between a short radius, long radius, long tangent, and extra long welded elbow.
6. What service situations call for the use of screwed fittings?
7. What is the difference between a concentric and an eccentric reducer?
8. How can the use of a welded tee be avoided?
9. Describe the service recommendations for socket-welded fittings.
10. Flanged fittings are used in systems that require frequent dismantling. True or false?
11. What are some of the chief drawbacks of flanged fittings?
12. Screwed connections are found on piping systems that are usually __________ inches (cm) and __________ inches (cm) in diameter.
13. In piping systems of brass and copper, ____________ or ____________ fittings are the most common.
14. Copper tubing and pipe have an upper temperature limit of ____________ °F.
15. What is a pipe joint?
16. Name four common pipe joints and their major characteristics.
17. What is a Van Stone flange?
18. What is the function of a blind flange?
19. Orifice flanges are used to ____________________.
20. What does RTJ mean?
21. What is meant by flange facing?
22. Why are gaskets sometimes used to make up pipe joints?
23. Flange bolting patterns come in 4, 8, ____________ numbers.
24. American Standard facings come in accordance with ANSI ____________ standards.
25. Name the two general types of flanges.
26. Flanges are ____________ together to form a tight joint.
27. Flared fittings are used on what type of joint?
28. What is a saddle?
29. For high-pressure/high-temperature situations, ______ fittings are preferred.
30. Name the different types of elbows.

PROBLEMS

1. Draw one-half scale a 4 x 3 x 2 x 2 in. reducing welded cross.
2. Draw one-half scale a pictorial view of a flanged base elbow for a 3-in. (7.62-cm) diameter pipe.
3. Draw one-quarter scale a flanged 90-degree elbow connected to a 4-in. (10.16-cm) pipe by slip-on flanges and an eight-bolt pattern.
4. Draw one-quarter scale a 6-in. (15.24-cm) pipe connected to a flanged gate valve with a weld-neck flange.
5. Draw one-quarter scale a 5-in. (12.7-cm) diameter pipe closed off by a blind flange with a 16-bolt pattern.
6. Draw one-half scale a 2-in. (5.08-cm) pipe joint using a socket-weld connection coupling (section view).

6″ to 12″ series G-4 Valve, handwheel operated. (Courtesy Grove Valve and Regulator Company)

4 VALVES

Valves are one of the most important parts of a piping system. Without them the liquids and gases in the line could not be used. A valve is a mechanical device used to start, stop, or regulate the flow in piping. It is used in connection with various sources of pressure which establish movement of the material in the piping line. We can usually divide this source of pressure into four basic origins: a boiler for steam; an air tank and compressor for air; head pressure from a pump or a tank for liquids as shown in Fig. 4-1; and pressure derived from natural sources such as oil, gas, and geothermal wells. It is essential to know the medium–air, steam, liquid, or earth resource–and its possible effects on the system, especially chemical effects.

Careful consideration must be given to valve selection in order to maximize pipeline effectiveness. Accurate selection is determined by a multitude of considerations–most of them outside the drafter's province–but the competent drafter will understand the relationship of all parts of the system on which he or she is working. Generally valves are selected by the project engineer, designers, and in some cases senior drafters.

All piping systems encountered by the drafter will have as their chief function the transfer of line material (liquid and gases) in a controlled direction and rate. These substances will be under a variety of pressures and temperatures which affect the rate of flow. When drawing pipe systems, the drafter uses symbols to represent various components. Besides a knowledge of symbols, understanding the uses of valves and conditions of operation will aid in their placement and support. The following list gives the standard abbreviations used throughout the valve industry. It is helpful to the would-be professional to memorize this list.

FIG. 4-1 Vessel.

1. Valve descriptions

AI	All iron
Al	Aluminum
BR	Bronze
CI	Cast iron
Cr	Chromium
Cr 13	Type 410 stainless steel
CS	Cast steel
DI	Ductile iron
FS	Forged steel
HF	Stellite face (hard face)
IBBM	Iron body bronze mounted
M	Monel metal
MI	Malleable iron
Mo	Molybdenum
N	Nickel
NI	Nickel iron
NICU	Nickel copper alloy
PVC	Polyvinyl chloride
SA	Sludge acid
SA	Sulfuric acid
SS	Stainless steel
Tef	Teflon
18-8 Mo	Type 316 stainless steel

2. Ratings

CWP	Cold working pressure
S	Steam pressure
SP	Steam pressure
WOG	Water, oil, gas pressure
WP	Working pressure
WSP	Working steam pressure

3. Operating mechanism

NRS	Nonrising stem
OS&Y	Outside screw and yoke
RS	Rising stem

4. Facings, disks, and joints

DD	Double disk
FF	Flat face
RF	Raised face
RTJ	Ring-type joint

5. End connections

BW	Butt-welded ends
FE	Flanged ends
FFD	Flanged, faced, and drilled
Flg	Flanged ends
Scr	Screwed ends
SE	Screwed ends
SJ	Soldered ends
SW	Socket-welded ends

6. Societies

ANSI	American National Standards Institute
API	American Petroleum Institute
ASME	American Society of Mechanical Engineers
ASTM	American Society of Testing Materials
MSS	Manufacturers Standardization Society of the Valve and Fittings Industry
SAE	Society of Automotive Engineers

7. Measurements

cfm	cubic feet per minute
gpm	gallons per minute
psi	pounds per square inch
rpm	revolutions per minute

SELECTING THE RIGHT VALVE

More than half this chapter is devoted to the description of major types of valves, each with possible variations in construction for specific operations. Naturally, the type of valve is the prime consideration in selection for a given use, but before we discuss valve types let us consider certain aspects that apply to all valves.

Service Conditions

Choosing the right valve and its design options (bonnet, stem, disk) is determined mainly by the conditions under which it will be in operation. The nature of the service, critical or noncritical, is of primary importance. Noncritical services usually include low-pressure/moderate-temperature plumbing and heating systems. These uses are probably more familiar to the reader than the critical services—which entail high pressures, high temperatures, and fire hazards in the power generation and petrochemical industries. Critical services demand close evaluation of valve types and their suitability to the job. Whenever there is danger to people or property and valve operation is critical, extreme caution is taken in selecting the valve, its connections, bonnet construction, and safety features.

Every liquid and gas has its own characteristics and subsequent effect on piping materials and equipment. Erosion and corrosion of pipes and valves, crystallization, vaporization of line contents—all are aspects which need to be considered when selecting the valve type. Air, gas, oil, water, and solids suspended in liquid all have a specific effect on the system.

A valve's pressure classification is always given in pounds per square inch (psi). Each item is rated for steam pressure (SP) first, then water, oil, and gas (WOG). Sometimes the WOG rating is replaced by a cold working pressure (CWP) rating. The temperature rating will sometimes follow the SP rating (for example, 150 SP 450°F-250 WOG). Many valves have only the SP rating cast on their bodies (Fig. 4-2). For example, a 150 SP-300 WOG bronze globe valve might have 150 SP on its side. Each product is tested and rated by design and allowable stress at a certain temperature. Temperature restrictions are listed in catalogs provided by the valve company. Valves may be used at or below the SP rating, but the temperature must not exceed the rating established for the valve material. This rating takes into account the hazard of controlling high-temperature fluids and has an ample safety factor.

On pipeline valves, it is traditional to mark only the ANSI rating (no SP rating). Thus class 150 is understood to have a CWP (WOG) of 275 psi; while class 300 is understood to

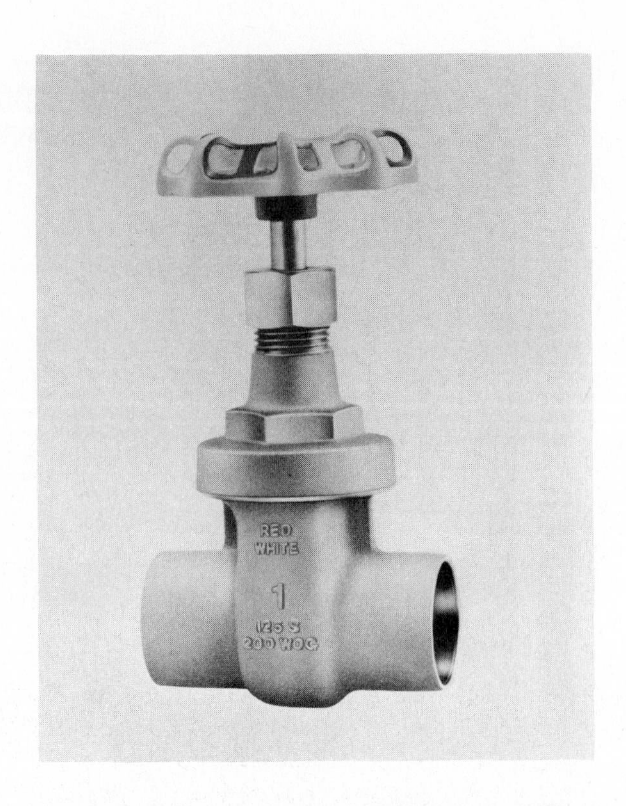

FIG. 4-2 Bronze gate valve, nonrising stem, soldered ends. (Courtesy Red and White Valve Co.)

have a CWP (WOG) of 720 psi; and class 600 is understood to have a CWP (WOG) of 1440 psi.

The WOG rating designates the maximum nonshock pressure (at atmospheric temperatures) at which the product may be used. This rating is usually cast on the product along with the SP rating if space permits. Exclusive WOG valves have only a WOG rating. One should become familiar with these ratings because a drafter's final job on most drawings is to construct a bill of materials and callout list.

Shock is a critical consideration in the design and placement of valves in a system. Physical shock from *external* sources becomes a concern when the valve is connected to moving or vibrating equipment such as pumps, motors, and compressors (Fig. 4-3). Physical shock due to frequency of operation, maintenance, and the surrounding environment must also be taken into account. A prime consideration in many areas of the world where the service is critical is seismic movement—the impact caused by earthquakes. When referring to *internal* shock one must consider all movements and stress produced by overpressure in the pipeline. Overpressure is caused by the sudden arresting of flow, which when superimposed on static pressure may overload the valve and also damage fittings and pipes. In these cases, it is advisable to select higher-rated valves for the service or provide a surge relief system.

FIG. 4-3 Alkylation plant hydrofluoric acid makeup pump.

Thermal shock is caused by rapidly rising and falling temperatures in the line which cause expansion and contraction of materials. This type of shock is of extreme seriousness in the nuclear power field.

Placement

Adequate attention must be given to the placement of valves. Ease of access, frequency of operation, accessibility for maintenance, hanger placement, possible temperature losses through the valve—all are considerations. Sometimes the placement will even dictate the type and style of valve selected for the job.

Environmental considerations include the following questions: Is the valve suitable for outside service where it may be exposed to extreme heat, cold, or moisture? Will it be buried in ground which freezes (Fig. 4-4)? Can the valve materials withstand the conditions under which it must operate effectively? Will insulating the valve affect its placement and its ability to be operated and maintained? These and other questions are essential to the proper selection and placement of valves.

FIG. 4-4 Cylinder-operated gate valve being installed in a pipeline. (Courtesy Grove Valve and Regulator Company)

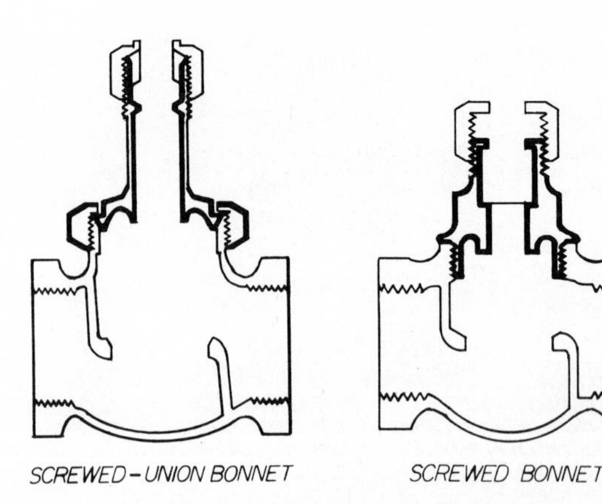

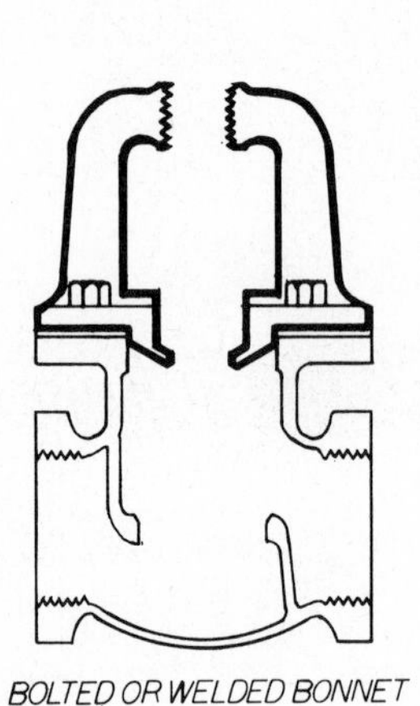

Design Options

Bonnet Assemblies (Fig. 4-5) The bonnet is the part of a valve which covers the internal moving parts and through which the stem threads must pass. *Screwed bonnets* are the least expensive type and are used for low-pressure plumbing. They provide easy access to the valve's inner parts but are not advised for frequent dismantling because of wear and tear on the threads and on the valve itself. These bonnets come on smaller valves which are not intended for critical assignments (Fig. 4-6).

Bolted bonnets (Fig. 4-7) are easily dismantled, come in larger sizes, and are recommended for critical service. The steel type is preferred in the power and process areas. A variation of the bolted bonnet is the *flanged type with ring joint,* which is used for high temperatures and high pressures. *Flanged with gasket bonnets* are common for 900°F (482.2°C) and below. The *pressure-seal* type is self-sealing and requires periodic adjustment. This bonnet design is good for frequent dismantling.

Welded bonnets for high temperatures and high pressures are used in critical situations that do not involve frequent

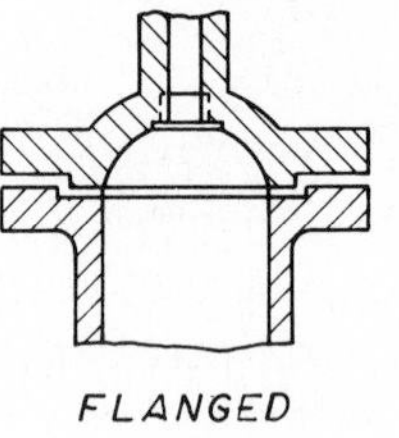

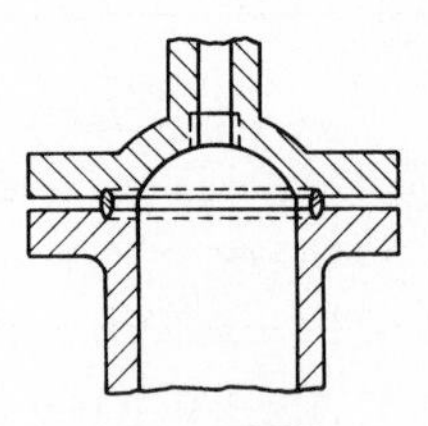

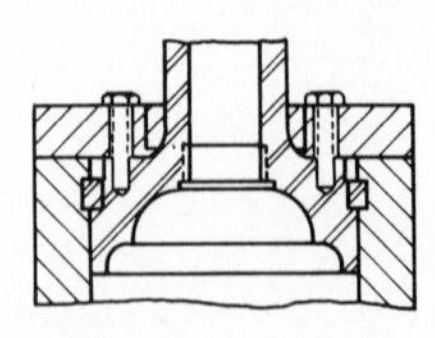

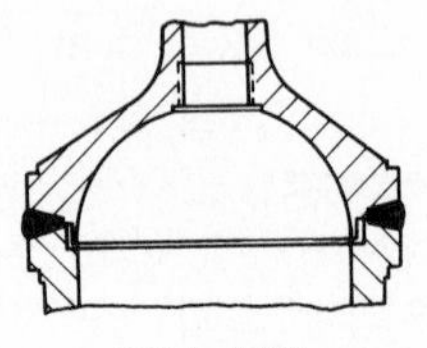

FIG. 4-5 Bonnet variations.

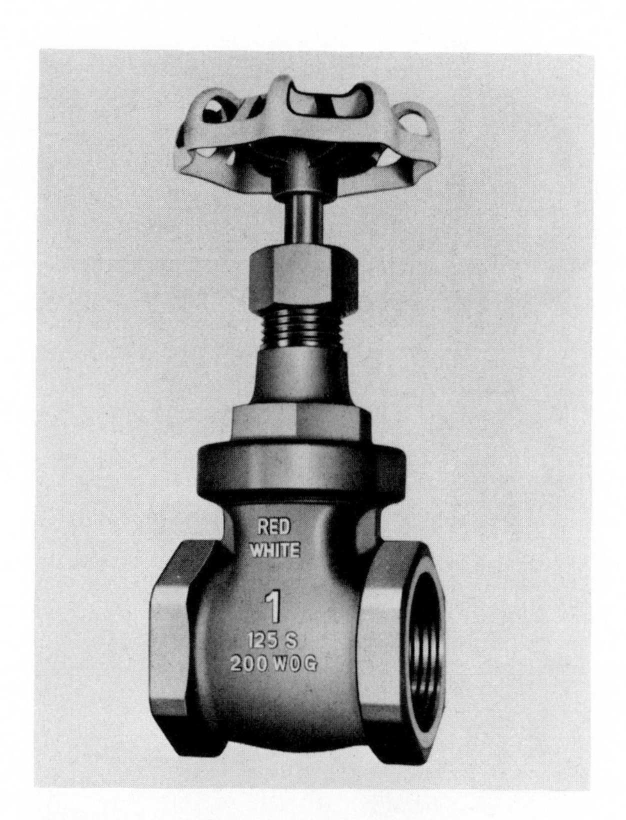

FIG. 4-6 A 125 S200 WOG gate valve. (Courtesy Red and White Valve Co.)

dismantling. This design is excellent for corrosive materials and provides a tight leakproof joint.

Screwed union bonnets, as shown in Fig. 4-8, are used for small valves with high temperature and high pressure ratings because, unlike screwed bonnets, there is no danger of their being screwed off in normal operation. This type is widely used in petrochemical plants.

Stems Stem selection is important because of headroom clearance. Stem options are available in several designs from which to choose. Frequency of operation and ease in reading disk position are just two considerations involved in choosing an option (Fig. 4-9).

Outside screw and yoke (OS&Y) is the best construction for high temperatures and pressures. Stem threads are not subject to severe thermal shock and are easy to lubricate. Corrosion possibilities are minimal. The need for adequate headroom for operation and maintenance is a major drawback, however. A variation of this style has both a rising stem and stationary handwheel. The stem position indicates disk position. Stem and threads must be protected from external damage. In the second option–*rising stem and stationary handwheel*, the only differences are that disk position is less easily noticed by stem position and the danger of external damage to threads is greater unless they are covered (Fig. 4-10).

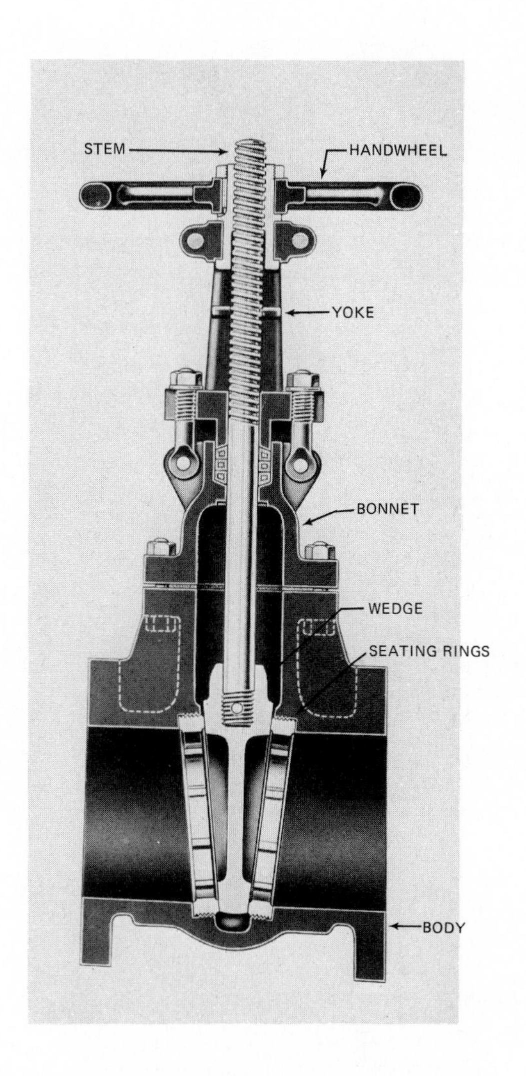

FIG. 4-7 OS&Y solid-wedge gate valve. (Courtesy The Lunkenheimer Co., A Division of Conval Corp.)

Inside screw options feature a *rising stem and inside screw* (Fig. 4-6), which is most common in small bronze valves that are 2 in. (5.08 cm) and less. Such valves are used in situations where threads are not damaged by line contents (water, hydrocarbons, steam). Stem position indicates the disk position, and adequate headroom is necessary. The second variation is the *nonrising stem and inside screw* (Fig. 4-11). This option is mainly for valves 4 in. (10.16 cm) and below and is not recommended for critical services. Since line fluids are in constant contact with threads, high-temperature use is undesirable. Two positive aspects of this option are low wear on packing (because only the stem revolves) and less headroom is needed. This valve must be equipped with a stem indicator to designate disk position and is usually used as a plumbing valve.

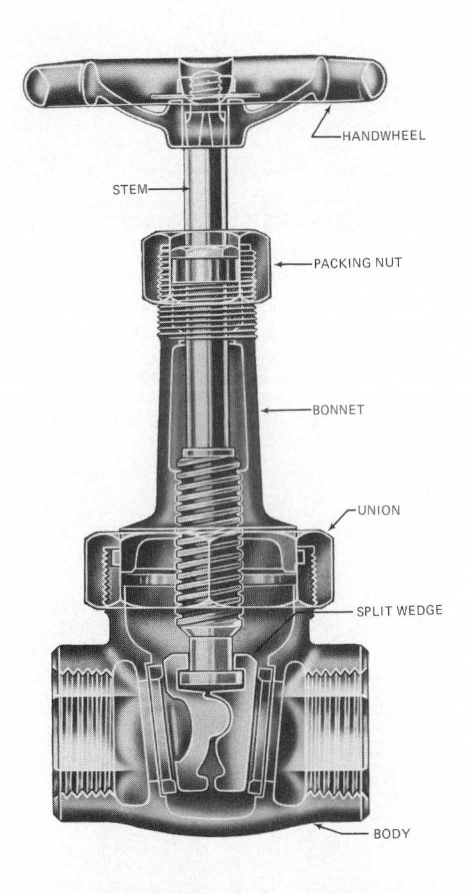

FIG. 4-8 Split-wedge gate valve (rising stem, screwed ends). (Courtesy The Lunkenheimer Co., A Division of Conval Corp.)

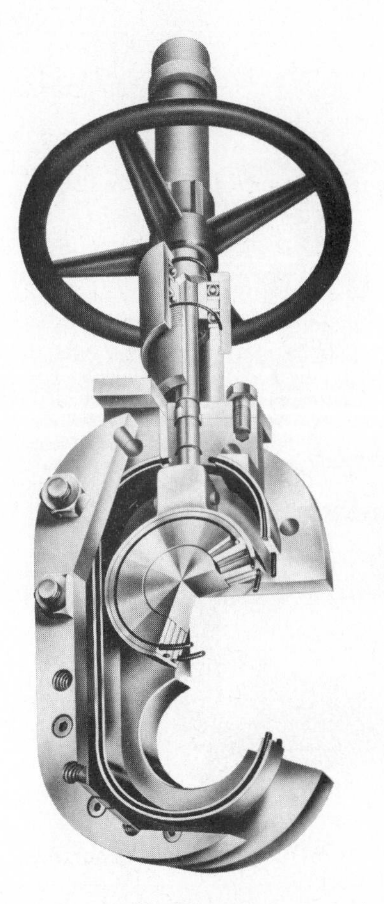

FIG. 4-10 Parallel-disk gate valve. (Courtesy Grove Valve and Regulator Company)

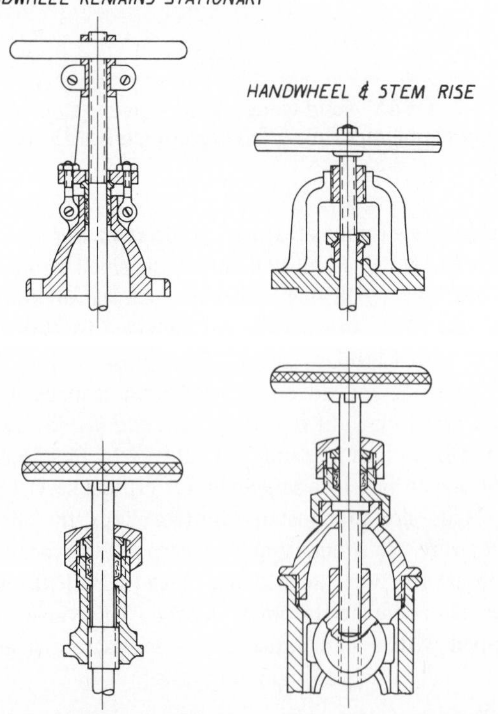

FIG. 4-9 Valve stem variations.

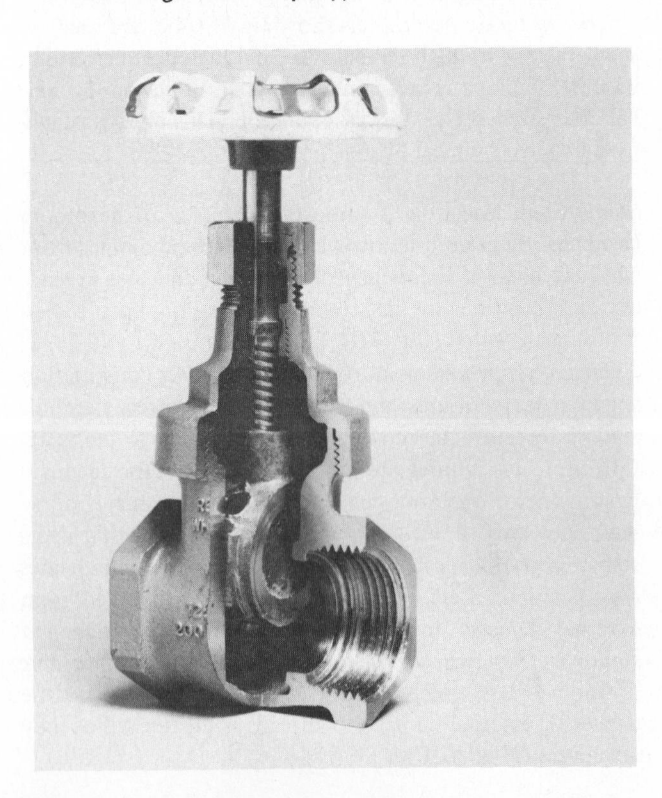

FIG. 4-11 Wedge gate valve (nonrising stem, screwed bonnet, screwed ends). (Courtesy Red and White Valve Co.)

CONSTRUCTION MATERIALS

Valve bodies come in several types of metal and plastic. Bronze, brass, iron, and steel alloys are most common. Selection of the body material depends on the valve's function (shutoff or throttling), the pressure and temperature ratings, line contents and their corrosive effects, line stresses, and fire hazards. One should always check the manufacturer's guide to selection and recommendations before ordering. Valve bodies are often classified by material (Fig. 4-12).

Brass is a copper-zinc alloy used for noncritical service at temperatures below 400°F (204.4°C) and low pressures. Both brass and bronze are copper-based alloys. Check with the manufacturer's data as to their true composition. *Bronze,* a copper-tin alloy, is stronger than brass and used at 500°F (259.9°C) maximum and 300 psi and below. This material is usually found in valves 3 in. and smaller.

Iron comes in a variety of types and compositions. Since manufacturers have different grades, it is important to check the specific valve type one wants and its specifications. *Cast iron* valve bodies come in most sizes. The SP ratings of 250 and a temperature of 450°F (232.2°C) are maximum. This material is an alloy of iron, carbon, silicon, and manganese and comes in a wide variety of grades and compositions. *Malleable iron* valves are excellent for service subject to thermal shock. Their tightness, stiffness, and toughness are desirable characteristics.

Steel valves come either forged or cast, or they can be fabricated from plate stock. *Cast steel* is used for large valves where service is critical and other materials are inadequate. *Forged steel* valves are drilled and machined from solid stock and hence are not practical for anything but small sizes. They are used for highly corrosive conditions. *Fabricated steel* is used for extremely large valves where cast steel would be impractical because of weight considerations.

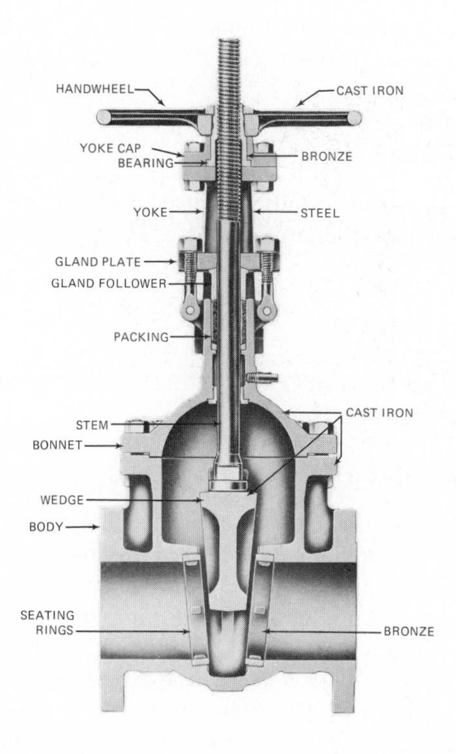

FIG. 4-12 Typical materials on a cast iron gate valve. (Courtesy Jenkins Bros.)

TRIM MATERIALS

Valve trim consists of the seat, ring, disk, facing, and stem, all of which come in an assortment of steel and bronze alloys. The trim is as important as the body itself, for without proper trim the valve cannot perform satisfactorily. When selecting the trim, consideration must be given to operating temperature, chemical stability of the line fluid, and tensile properties (hardness and toughness). (See Table 4-1.) Seat-dish facings, their seizing possibilities, and the trim's corrosive resistance must also be taken into account. These data can usually be found in the valve company's catalog.

The drafter must be familiar with valve designs and options, know how to specify and order valves, and be able to draw them and their corresponding symbols. As a practical exercise in callouts, specification of parts, and parts lists, the problems encountered in this chapter and other parts of the book can be used for practice ordering from manufacturer's catalog pages or the appendix.

When specifying valves, one may order in the following way:

1. SP rating
2. Temperature rating (where applicable)
3. WOG rating (CWP)
4. Body material
5. Type (gate, globe) and size
6. Trim (ends, stem variations, bonnet, disk)
7. Catalog figure and catalog page

For example: 150 SP 300°F 250 WOG; 2-in. bronze globe valve, screwed ends, inside screw, nonrising stem, composition disk, screwed bonnet; Fig. 20, p. 150A, Lunkenheimer Valve Company. On the job a drafter need only call out SP and WOG, size, type, company, and figure or order number (sometimes only order number), but the practice described here will help familiarize one with the parts and functions of valves.

End connections are also important in valve selection because not all valve types and sizes are adaptable to all means of connection. The drafter must be familiar with the possible end connection variations, which will also influence the symbols for valves—there are different symbols for most of

TABLE 4-1 VALVE TRIM SELECTION GUIDE
(Courtesy Stockham Valves and Fittings)

Range	*Temperature (°F)*	*Material*
Very high	2000	Refractory metals, ceramics
High	1200 to 1600	High-temperature alloy steels
Intermediate	1000	Carbon steel
	650	Ductile iron
	550	Bronze
	450	Cast iron
	150	PVC plastic
Low	–250	Low-alloy steels, bronze
Very low (cryogenic)	–450	Bronze, austenitic ductile iron, austenitic stainless steels

the end connections (see Chapter 5). In many instances, industries that deal exclusively in one type of joint will establish their own simplified method of drawing end connections for fittings, valves, and other equipment. If only flanged fittings are used, for example, no flanges are drawn or symbolized.

But not all companies do things the same way. A drafter's greatest asset will be an ability to adjust to the job at hand. Most companies have compiled a book of standards they have found useful throughout many years of experience. Employers do not always conform to all the rules and regulations found in drafting books or taught in schools. These are only meant to be guidelines, not laws.

END CONNECTIONS

Welded Ends

Socket-welded connections are used on small pipelines, usually 2 in. (5.08 cm) and below. They are good for severe service because they can eliminate leakage. These connections are self-aligning and make valve installation simple.

Butt-welded connections are available only in steel and are used for high-temperature/high-pressure service where they require infrequent dismantling. Good for process and power generation situations because of their excellent safety record, these connections provide tight, leakproof joints.

Screwed or Threaded Ends

Screwed or *threaded* ends are limited to smaller pipes. Light and inexpensive because of less finishing, they are used in common plumbing at ordinary pressures and are easily fitted in original equipment installations. They are available in a variety of materials: bronze, iron, steel, brass alloys. Valve inlets and outlets are tapped with ANSI standard female pipe threads (Fig. 4-11).

Flanged Ends and Hub Ends

Flanged ends are available in valves of 3 in. (7.62 cm) and larger. They are excellent for critical situations in petrochemical installations, where the transfer of heavy viscous material is essential. Connections of this type require more finishing and, being composed of a greater amount of material, are heavy. Thus they are more expensive. Flanged ends should be used in applications which entail frequent dismantling. They provide strong leakproof connections and are available in a variety of materials. When selecting them, it is essential to know the conditions of operation and match the connecting flanges according to strength, bolting, and mating found elsewhere in the system.

Hub ends are used mainly for underground low-pressure gas and water and sewage systems.

Socket and Soldered Ends

Soldered (Fig. 4-2) and socket-brazed ends are used in nonflammable services on copper and red brass piping. Special packing is an important consideration because the standard packing available in valves with this connection style may not be suitable. Socket-brazed connections withstand high temperatures, but one should consult the manufacturer's service recommendations if temperature is critical.

Bell and Spigot Ends

Bell and *spigot* connections are for low-pressure/low-temperature services such as water, gas, and sewage pipelines and are usually made of cast iron, which has a high resistance to corrosion. The connections are caulked as in other bell and spigot fittings.

Flared and Compressed Ends

Flared and *compressed* ends are predominantly used for instrumentation on thin-walled tubing. They can be easily dismantled.

TYPES OF VALVES

Valves are classified here by their normal use, though many can be utilized for a variety of services.

On-Off, Start-Stop Service

These valves suffer a minimum of pressure loss when open, are usually operated in a fully opened or closed position, and offer minimum resistance to flow. Where a tight shutoff is desirable and the valve is operated infrequently, these types are preferred.

Gate valves, ball valves, and *plug valves,* as shown in Figures 4-7, 4-13, and 4-14, fall almost exclusively in the on-off category. *Butterfly valves* are also used for this type of

FIG. 4-13 Motor-operated 30-in. (76.2-cm) ball valve. (Courtesy Grove Valve and Regulator Company)

FIG. 4-14 Flow characteristics of a plug valve. (Copyright 1976 by Stockham Valves and Fittings)

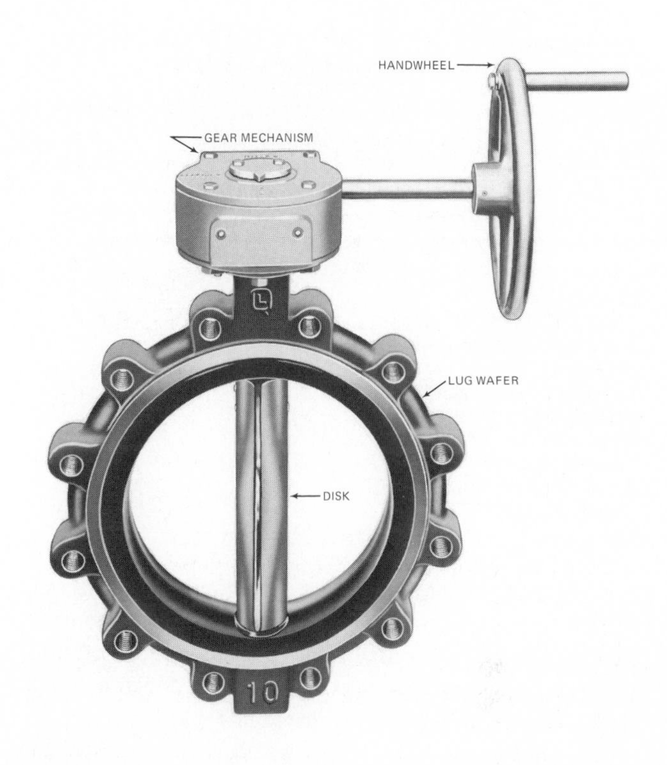

FIG. 4-15 Gear-operated butterfly valve (lug type). (Courtesy The Lunkenheimer Co., A Division of Conval Corp.)

operation because of their low pressure drop and the tight shutoff made possible by the development of plastic seals. Still, the butterfly valve is best suited for regulation of flow (Fig. 4-15). *Diaphragm valves* can also be used for start-stop operation, especially in highly corrosive services and where contamination of line substances is a problem. Because of the diaphragm valve's wide range of uses, it has been classified in a separate category.

Gate Valves Gate valves (Figs. 4-7 and 4-8) are probably the most frequently used valve. Designed primarily for infrequent operation and free-flow service, they have tight shutoff characteristics. This valve is not good for throttling, because accurate regulation of flow is not possible. Moreover, throttling causes uneven wear, and vibration can damage the gate. Because pressure always forces the disk down against the seat, the gate wears on one side only. Erosion of the gate disk downstream is a major problem with high-velocity regulation, but it can be reduced with gate guides (Fig. 4-16).

Gate valves are available in variations of body, end connections, bonnets, and stems. Types of gates include an assortment of disk shapes: solid wedge (Fig. 4-7), split wedge (Fig. 4-8), and parallel wedge (Fig. 4-10). Seating for the disks is available in the *screwed* variety for moderate serv-

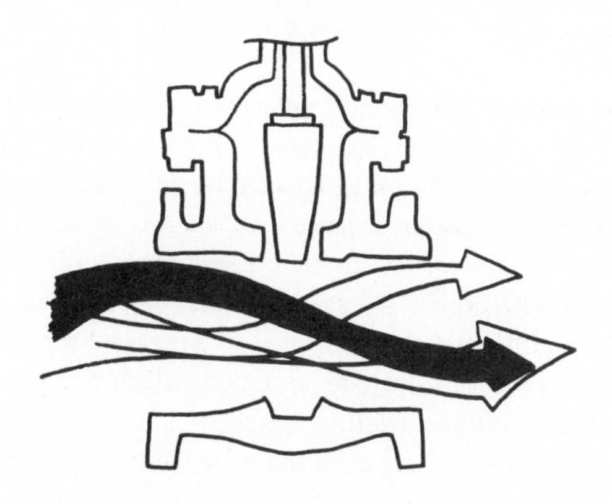

FIG. 4-16 Flow characteristics of a gate valve. (Copyright 1976 by Stockham Valves and Fittings)

TABLE 4-2 SERVICE CONDITIONS FOR GATE VALVES
(Courtesy Stockham Valves and Fittings)

	Disk Options		*Stem Options*	
Service Conditions	*Solid*	*Split Wedge, Double, or Flexible*	*Rising Stem*	*Nonrising Stem*
Viscous or gummy fluids	•		•	
Limited operating space	•			•
Varying temperatures (expansion/contraction)		•	•	•
Visual position indication of disk	•	•	•	
Steam service (and general oil, gas, and water services)	•		•	
Gases, volatile fluids, light liquids		•	•	•
Water mains and distribution		•		•

ices, *welded* or *rolled* seat rings for critical situations, and *integral* seat for uses not requiring replacement. One drawback of the gate valve is that the pocket at the bottom of the body between the seats can fill with foreign material. Disk and stem options are compared in Table 4-2.

The *solid wedge* (Fig. 4-11), which is the simplest and most common design, is of one-piece construction and can function in any position in the line. The tapered gate disk wedges itself on both sides of the seat for positive shutoff. There is no chance of line material getting trapped in the bonnet with this type. Many solid-wedge gate valves have accurately machined disk wings which guide the gate disk to a tight seating. These valves are excellent for high flow rates with low line turbulence. In steam service, solid-wedge gates are preferred over the split or double-disk gate because there is less rattle.

The split wedge (Fig. 4-8) has a freely adjusting disk-to-seating surface, even when one side is out of alignment due to foreign material. It is used for moderate temperatures and is available in iron and bronze bodies. Because of its tapered seat, a split wedge does not require a spreader for tight shutoff.

The *parallel-seat, double-disk gate valve* has a spreader in some models to force the disks against the seats; others seat by pressure on one side only. These valves are excellent for use with motor actuators because they are less likely to jam or seize during operation. In steam service, parallel disks are employed because they are less likely to stick or fail due to thermal shock.

Lubricant-seal gate valves are equipped with lubricant seal systems which reduce friction between disk and seating surface. Injected from a port in the valve body into grooves in the seat rings, the lubricant fills valve seats which are scored or pitted. These valves are available in cast iron or cast steel bodies.

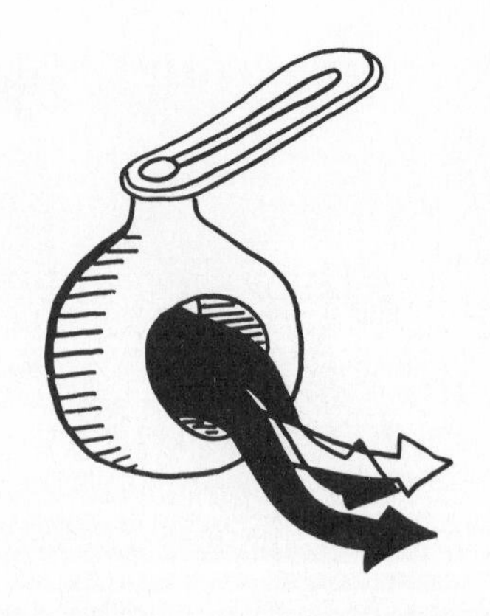

FIG. 4-17 Flow characteristics of a ball valve. (Copyright 1976 by Stockham Valves and Fittings)

Ball Valves Ball valves (Figs. 4-17 and 4-18) fall within the on-off, nonthrottling category, though some types are used for regulation of flow. Because of their quick, one-quarter-turn operation they are readily adapted to compact mechanical or hydraulic actuators and remote control.

Ball valves offer a low profile, are nonsticking and compact, have a low torque requirement, and are self-sealing. Their smooth flow characteristics make them ideal for viscous fluids and slurries, though on many models the temperature is restricted because the seating materials are usually plastic or derivatives of rubber. Seals are easily replaced and the valve is usually repaired without much difficulty unless large valves—for example, 48 in. (121.92 cm)—must be cut from the line to repair or replace the ball. A tight shutoff is achieved by the use of plastomer or elastomer seat rings. Elastomeric seat seal rings are used in large valves up to 54 in. (137.16 cm) and at temperatures to about 350°F (176.7°C). Services with temperatures as high as 1200°F (648.9°C), as with the transfer of crude oil over long distances, have proved the reliability of ball valves. For fire-safe service, metal backup seats are used.

Several types of ball valves are available. *Top-entry* types are easily maintained and repaired in the line and offer minimum flow resistance. *End-entry* types (Fig. 4-13) can be operated in any position. *Union-end* types (Fig. 4-19) are exclusively seen on smaller sizes (they are easily installed) and are excellent for pneumatic and electric operation.

Various ball port shapes are common. The *venturi port*, *full port* (Fig. 4-18), and *reduced port* all have low pressure drop and are available with a lubricant seal system.

Plug Valves Plug valves are sometimes called *cock valves* (Fig. 4-20). These, like ball valves, shut off with a quarter turn of the stem (90 degrees) and are quick-acting flow controls. They have a high pressure drop through the port when open. Positive characteristics of plug valves include minimum installation space, limited throttling ability, and

FIG. 4-18 Section view of a ball valve. (Courtesy Grove Valve and Regulator Company)

FIG. 4-19 Ball valve with union end.

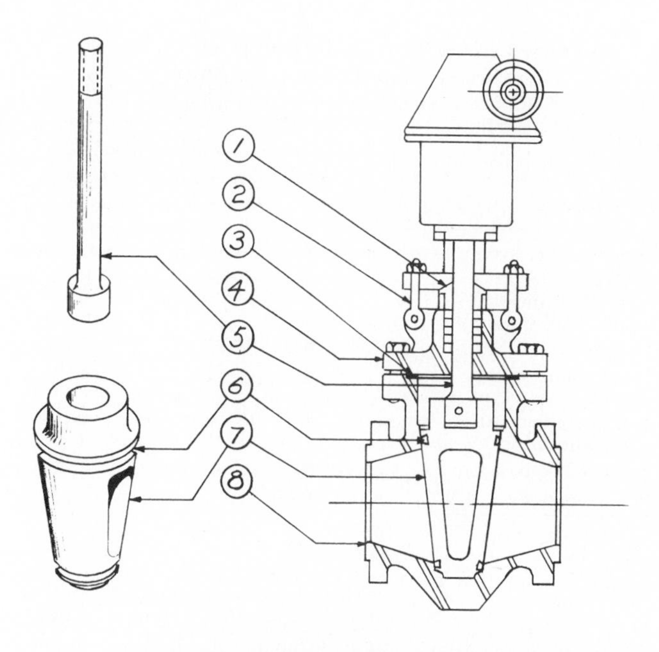

1	GLAND FOLLOWER
2	PACKING GLAND EYEBOLT
3	GASKET
4	BONNET
5	STEM
6	O-RING
7	PLUG
8	BODY

FIG. 4-20 Gear-operated plug valve.

more complete shutoff than gate valves. Moreover, the internal contours give maximum flow efficiency and the unexposed seats eliminate corrosion and erosion due to the action of line contents. This type can be used for services similar to those of gate valves; it is ideal for slurry service since line fluids cannot enter the body cavity.

Plug valves come in either lubricated or nonlubricated types (Table 4-3). In the lubricated type a pressure gun exerts a jacking action on the plug and lifts it for an instant, providing ease in turning and overall operation. Lubricant pressure is greater than line pressure so there is no chance of solids lodging between plug and body. These seating surfaces are protected from wear caused by corrosion or erosion and offer a positive seal against both external and internal leakage. Lubricant seal valves have a positive rotary action, are compact, and provide simplified service in many applications. They are excellent for critical services where repacking is done under pressure. Negative characteristics of lubricated plug valves include a maximum temperature of 1000°F (537.8°C), and usage limited to services where con-

TABLE 4-3 SELECTION GUIDE FOR LUBRICATED vs. NONLUBRICATED PLUG VALVES
(Courtesy Stockham Valves and Fittings)

		Nonlubricated		
Feature	*Lubricated*	*Lift Type*	*Nonmetallic Sleeve Type*	*Eccentric Plug Type*
Temperatures above 450°F		●		
Some throttling (with minimum abrasion)	●			●
Slurries	●	●		
Minimum operating torques		●		●
Low operating cost (no lubrication)		●	●	●
Protected body seats		●		
Coking service		●		

tamination of line fluid is not a problem. The lubrication, of course, requires extra maintenance time. Nonlubricated plug valves are compact, low-maintenance mechanisms which are adaptable to high temperatures and have low plug-to-seat friction. A cam and crank apparatus lifts and revolves the plug. These valves have a less positive seal, and packing is not replaceable under pressure (Fig. 4-20).

Port variations for plug valves include a somewhat rare and expensive *full-bore circular port* where valve ID equals pipe ID, pressure drop is minimal, and efficiency level is high. *Multiport* variations are also available (Fig. 4-21). They are excellent for convenience, economy, and simplification of piping. The *regular pattern* has a tapered form of port which equals 70 to 80 percent of the pipe cross-sectional area, and face-to-face dimensions are greater than the standard gate valve. The *short pattern* has face-to-face dimensions similar to gate valves. The *venturi port* comes in a flanged body design in 6 in. (15.24 cm) and larger diameters. This type has a reduced port size as small as 35 percent of the pipe's cross-sectional area.

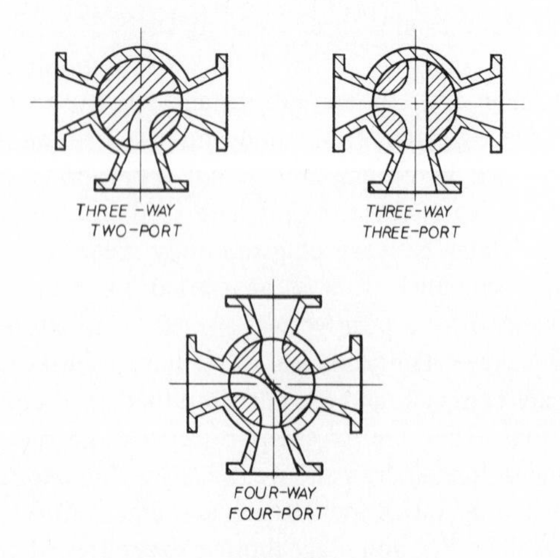

FIG. 4-21 Multiple-port valves.

Regulation of Flow

Valves designed to regulate flow include the following types: globe, Y pattern, angle, needle, butterfly, ball, and sometimes diaphragm. All of these types offer close control over the quantity of fluid passing through the valve. Resistance to flow is governed by varying the effective area of the valve bore, by change of direction, or by some other obstruction, enabling the flow to be closely adjusted. These valves are excellent for throttling or suppressing the rate and amount of flow. The closest regulation is possible with the globe variations. All the devices in this section have a high pressure drop through the port.

Globe Valves Globe valves (Figs. 4-22 and 4-23) are excellent for flow control, throttling, and frequent operation. Since the seat is parallel to the pipe axis, the flow must change direction twice in an S pattern. This increases flow resistance and pressure drop.

Most common in sizes 3 in. (7.62 cm) and under, they are available up to 12 in. (30.48 cm), though cost and throttling efficiency decline rapidly above 6 in. (15.24 cm). In most uses, an inside screw and rising stem are standard but OS&Y variations exist in the larger sizes. Usually seats and disks are repairable and replaceable while the valve is in the line. Seat erosion is minimal because seat and disk break contact

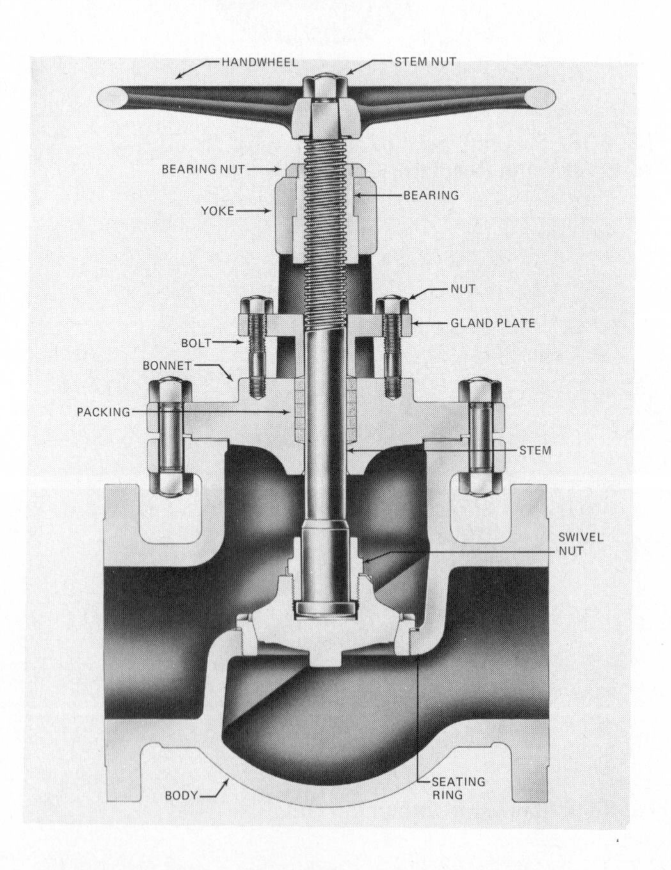

FIG. 4-22 OS&Y flanged-end, bolted-bonnet globe valve. (Courtesy Jenkins Bros.)

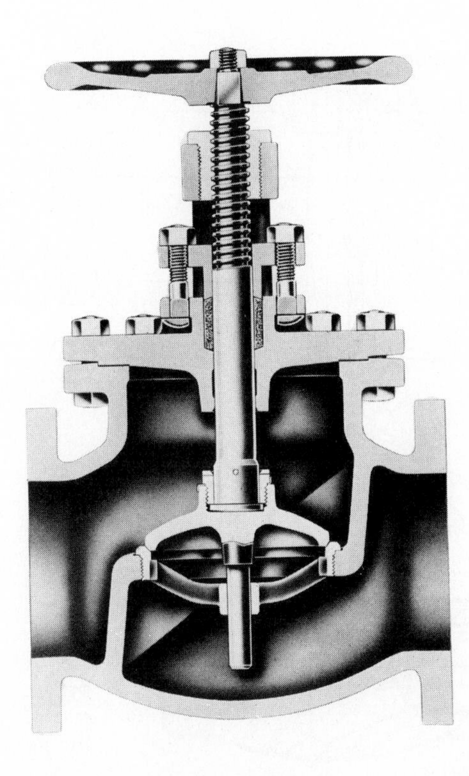

FIG. 4-23 Full-port globe valve. (Courtesy Jenkins Bros.)

as flow begins. When a globe valve operates in an extreme throttled position, though, disk very close to seat, the high fluid velocity accelerates erosion of the seating surface and that of the disk or poppet—this is called *wire-drawing.* Positive shutoff is also a major advantage in globe valves, as is the short disk travel needed for quick manual operation. The major drawback of this valve type, besides high pressure drop and size limitations, is that complete drainage of the system is not readily accomplished because of body pockets.

Globe valves have similar flow characteristics to lift check valves and are used in conjunction with them (Fig. 4-24).

FIG. 4-24 Flow characteristics of a globe valve. (Copyright 1976 by Stockham Valves and Fittings)

Large globe valves are often equipped with a bypass in applications where line pressure or thermal conditions can cause seizing of the disk to the seat.

The major option in globe valves is the disk type (Fig. 4-25). *Plug-type disks* offer wide seating and bearing surfaces which provide a high resistance to erosion by line dirt or foreign matter. *Composition* or *nonmetallic disk* (NMD)

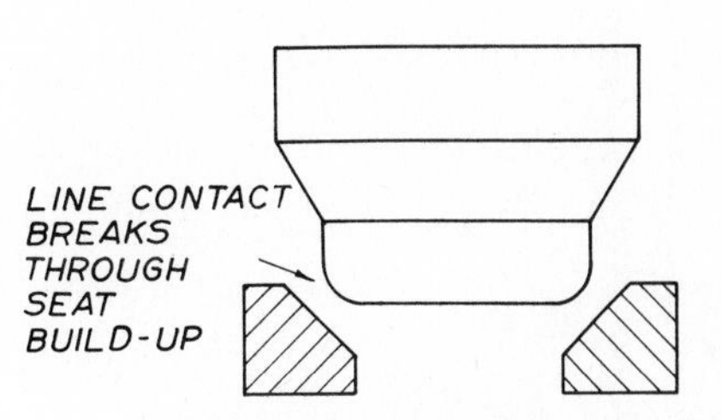

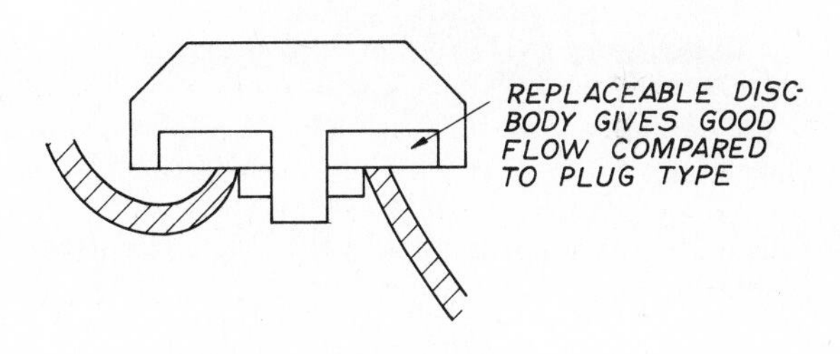

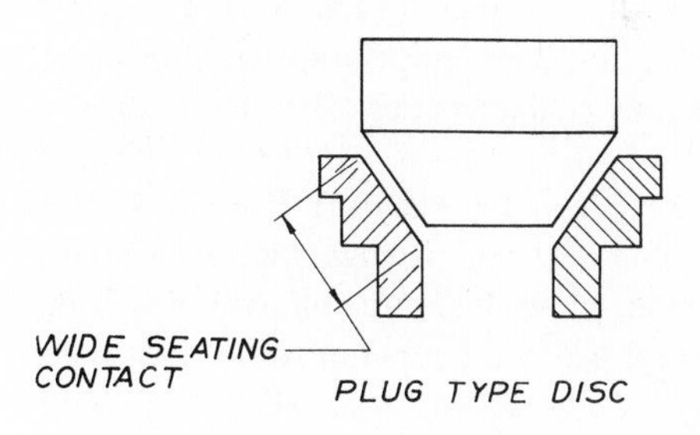

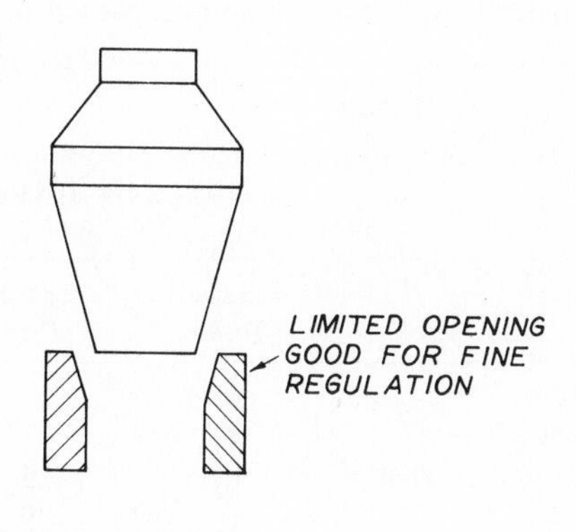

FIG. 4-25 Globe disk variations. (Copyright 1976 by Stockham Valves and Fittings)

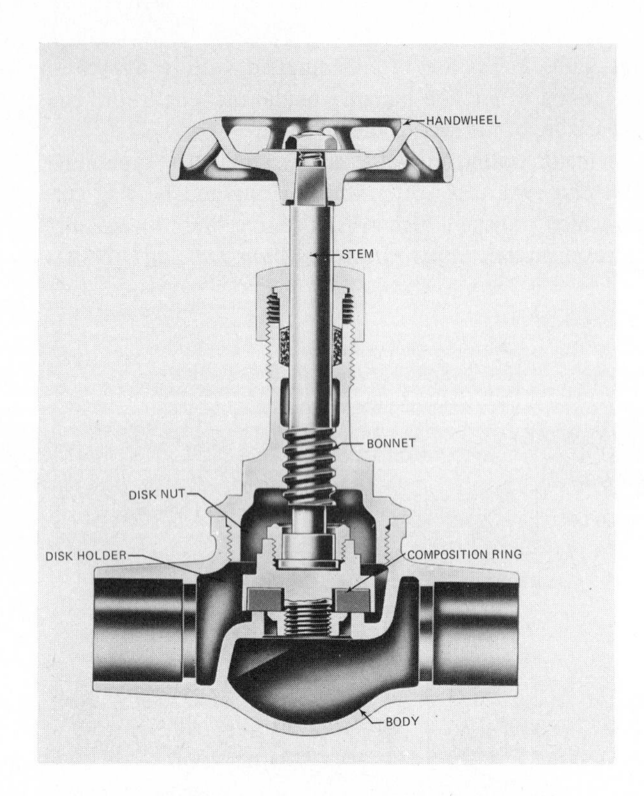

FIG. 4-26 Composition disk globe valve. (Courtesy Jenkins Bros.)

valves, shown in Fig. 4-26, have a readily replaceable disk body and offer great adaptability because disks can be changed according to line contents (Table 4-4). The NMD valve is good only for moderate temperatures and pressures and has poor throttling characteristics. The disk is usually made of laminated asbestos or resins that have a flat face pressed against the seat for a tight shutoff. Even in services that have considerable dirt and foreign matter content, the disk can become embedded but still will not leak. Flow characteristics for NMDs are superior to the plug type. *Conventional, spherical,* or *full-way disks* (Fig. 4-25) have a narrow line of seat contact which breaks through any

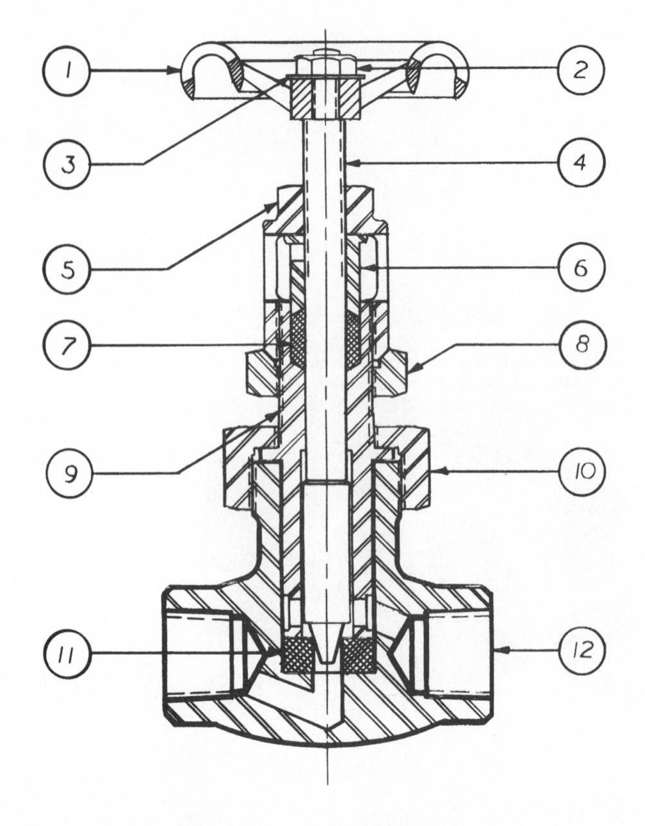

PARTS LIST			
1	HANDWHEEL	7	PACKING
2	HANDWHEEL NUT	8	PACKING NUT LOCK
3	IDENTIFICATION PLT.	9	BONNET
4	STEM	10	UNION RING
5	PACKING NUT	11	SEAT
6	GLAND FOLLOWER	12	BODY

FIG. 4-27 Needle valve.

buildup on the seat, thus assuring tight shutoff. The *needle disk,* shown in Fig. 4-27, is limited to fine adjustment and is not recommended for steam service (see needle valves).

TABLE 4-4 SELECTION GUIDE FOR DISK VARIATIONS FOR GLOBE VALVES
(Courtesy Stockham Valves and Fittings)

Disk Type	*Renewable Disks*	*Regrinding Disks*	*Renewable Seats*	*Hardened Seats to Resist Erosion*	*Limited Sealing Contact*	*Close Regulation*	*Tight Shutoff of Gases*
TFE	●						●
Composition	●						●
Spherical or full-way		●	●†		●		
Plug		●	●†	●†			
Needle					●	●	

†In some cases.

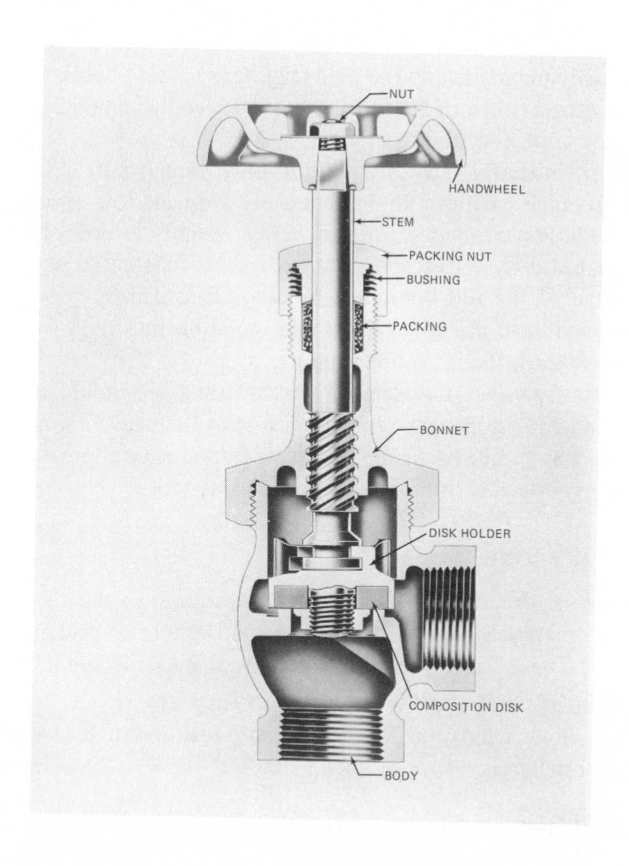

FIG. 4-28 Composition disk angle valve. (Courtesy Jenkins Bros.)

Angle Valves Angle valves (Fig. 4-28) have the same stem, disk, and seating design as a globe valve. The only difference is that the fluid makes a 90-degree turn (Fig. 4-29). In noncritical situations, using an angle valve will replace a costlier globe and elbow combination with a smaller pressure drop and less resistance to flow. Reducing the number of joints in a line also saves money and installation time. In critical services an angle valve should not replace a globe and elbow because the valve may not be designed for abnormal stress. Designed primarily as a flow regulation valve, the angle valve is limited in size. The body also traps line material.

FIG. 4-29 Flow characteristics of an angle valve. (Copyright 1976 by Stockham Valves and Fittings)

FIG. 4-30 Flow characteristics of a Y valve. (Copyright 1976 by Stockham Valves and Fittings)

Needle Valves Needle valves (Fig. 4-27) are used for throttling to obtain exact flow regulation in applications such as instrumentation, gauges, meters, and other services that require close control. They are not recommended for steam service. These valves are small and compact with extremely fine threads for close, accurate regulating. Used for high temperatures and pressures, they are available in regular and angle patterns. The disk is tapered and usually made of bar stock, which is very rugged. Positive shutoff is not always possible, however, because of seat ring scoring caused by shutting the valve too tightly. Needle valves are available from 200 SP 400 WOG to a maximum of 1200 WOG for bronze and up to 5000 WOG in steel for critical services.

Y-Pattern Valves Y-pattern valves (Fig. 4-30) are a useful variation that offers less obstruction to fluid flow and lower pressure drop. They are quick-operating and excellent for throttling (depending on disk type) (Fig. 4-31).

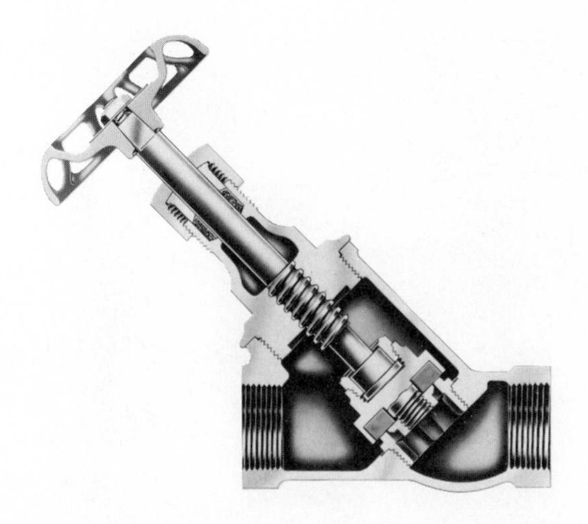

FIG. 4-31 Composition disk Y-globe valve (screwed bonnet). (Courtesy Jenkins Bros.)

FIG. 4-32 Hand-operated butterfly valve (groove ends, rubber lined). (Courtesy The Lunkenheimer Co., A Division of Conval Corp.)

Butterfly Valves Butterfly valves (Figs. 4-15 and 4-32) are good for on-off service as well as for throttling and flow regulation. They have a low initial cost and extremely simple design, are lightweight, need little installation space, and are easily maintained in the field (Fig. 4-33). Being operated by a quarter turn of the disk, this type is easily adapted to remote control actuators.

Some butterfly valves have a full-flow-through port (Fig. 4-32) which creates a minimal pressure drop and low velocity head loss compared to other valves when fully open. In some butterfly valves, the valve occupies a substantial percentage of the full bore area. Newer models usually come equipped with elastomer liners and metal backup rings for positive shutoff and seating support.

Butterfly valves are excellent for throttling and should be considered first in many services because of their advantages. Their major drawbacks are the temperature limitations of liner material and unsuitability for steam service.

FIG. 4-33 Butterfly valve.

Slurry Valves

Slurry valves are for coarse, viscous, and solid-suspended-in-liquid materials like red mud, alumina, liquor, or coal in liquid. These valves come in a variety of patterns and are sometimes lined for rugged service. They are similar to other flow regulation valves but are manufactured for coarser material.

Backflow Prevention

Check valves, as shown in Fig. 4-34, are kept open by gas or liquid movement and are closed by gravity or flow reversal. This device can make quick automatic reactions to flow change and limit flow to one direction. Most check valves are installed in horizontal position, but some are made for vertical line service. They are available in several materials for both body and disk. The two major categories of check valves are the swing and the lift types (Figs. 4-35 and 4-36).

The *swing check valve,* shown in Fig. 4-37, can be used in both horizontal and vertical service. It offers minimum resistance to flow and has similar flow characteristics to gate valves, with which it is usually used in conjunction. Swing check valves are used mainly with liquids, especially at low velocities. They depend on disk weight, gravity, and flow reversal for closing. The disk must travel through a 90-degree arc to seat. This type is not recommended for steam, pulsation flow, frequent reversals, or high pressure because it has a tendency to slam and clatter, being quite noisy in many services. However, many can be equipped with coil spring or cylinder-type slam retarders to anticipate flow reversals.

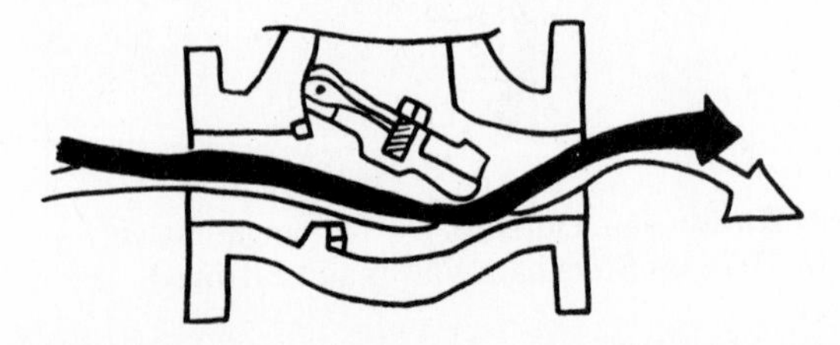

FIG. 4-34 Flow characteristics of a swing check valve. (Copyright 1976 by Stockham Valves and Fittings)

FIG. 4-35 A 48-in. check valve for 1200 WOG oil pipeline. (Courtesy Grove Valve and Regulator Company)

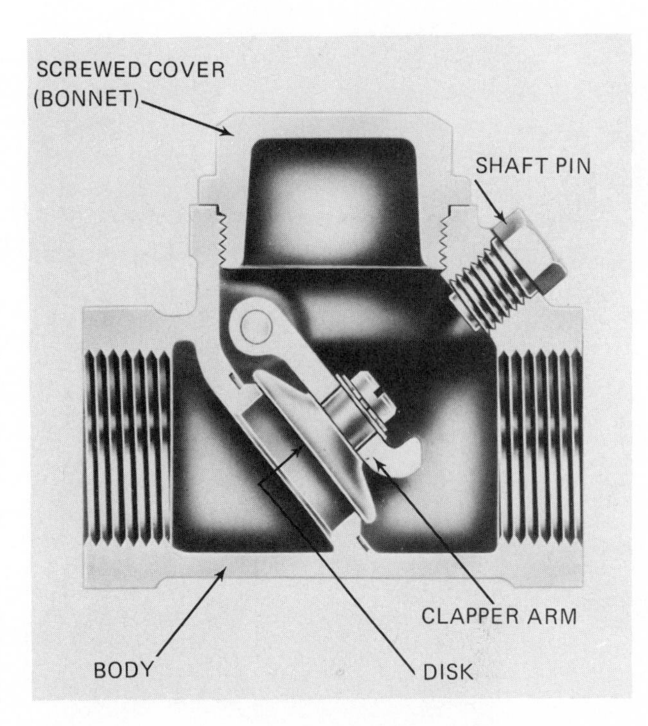

FIG. 4-37 Swing check valve (screwed bonnet, screwed ends). (Courtesy Jenkins Bros.)

The *flapper-type swing valve* (NMD), shown in Fig. 4-38, has a rubber disk, seats at 45 degrees with less travel of the disk, and, consequently, has less flow resistance, is faster acting, and shuts off quicker. The effect of line material and temperature on the disk must be considered.

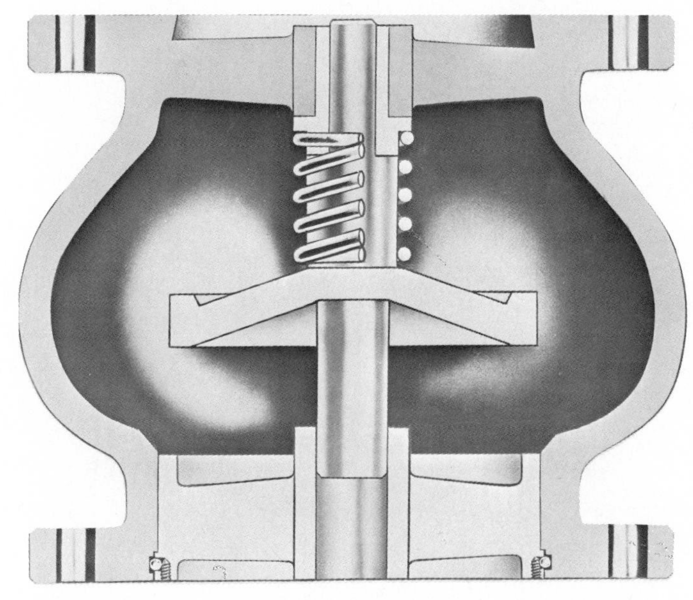

FIG. 4-36 Globe-style "lift" check valve. (Courtesy APCO/Valve and Primer Corp., Schaumberg, Ill. 60196)

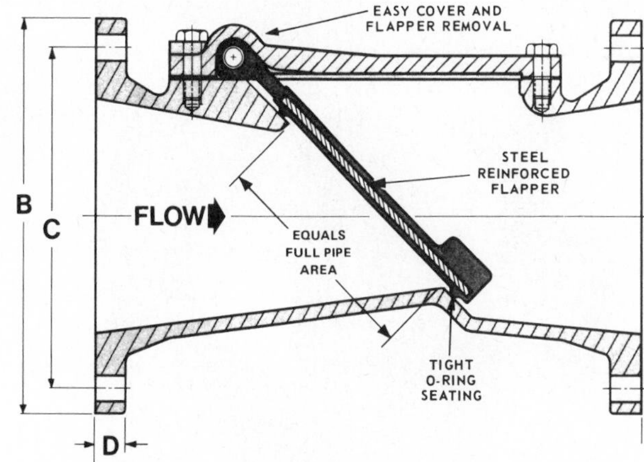

FIG. 4-38 Flapper-style check valve. (Courtesy APCO/Valve and Primer Corp., Schaumberg, Ill. 60196)

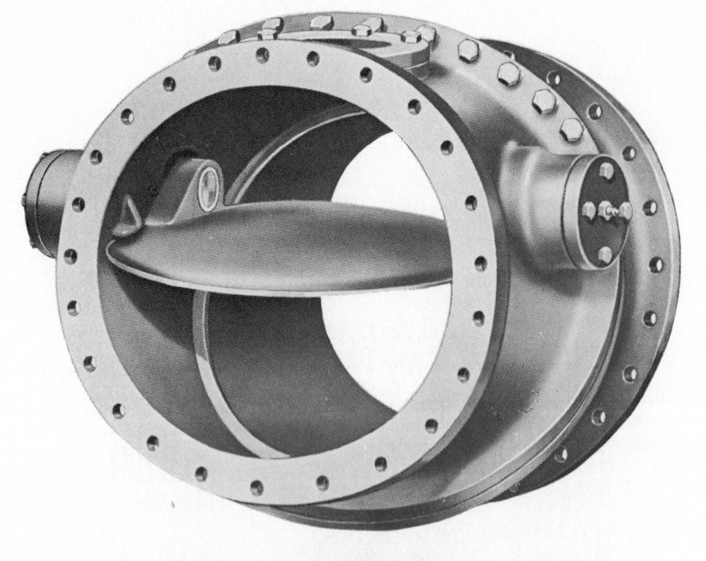

The *slanting* or *pivot-disk valve,* shown in Fig. 4-39, pivots off center with 30 percent flow above pivot and 70 percent below. The shorter disk travel and the amount of flow above the disk help to eliminate slamming. The *double-disk check valve,* shown in Fig. 4-40, has a split disk which allows quicker opening and low resistance to flow. Disk travel is lessened and slamming is reduced.

Lift check valves, as shown in Figs. 4-41 and 4-42, have flow characteristics similar to the globe valve and are, therefore, usually found in service with it. Some manufacturers use the same body for both their lift check and their globe valves. Being gravity-actuated, lift check valves come in both horizontal and vertical models. These valves are excellent for air, steam, and gas service, especially where an increase

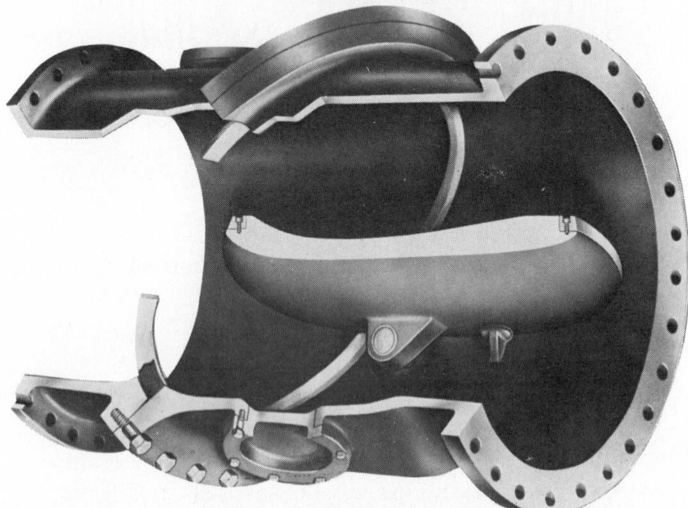

FIG. 4-39 Pivot-disk check valve. (Courtesy APCO/Valve and Primer Corp., Schaumberg, Ill. 60196)

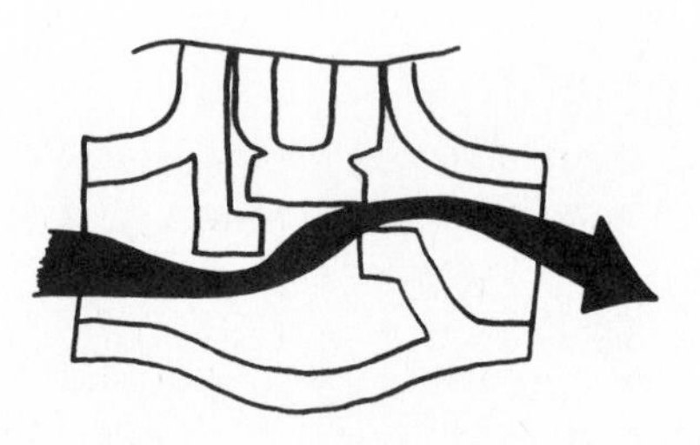

FIG. 4-41 Flow characteristics of a lift check valve. (Copyright 1976 by Stockham Valves and Fittings)

FIG. 4-40 Split-disk check valve. (Courtesy APCO/Valve and Primer Corp., Schaumberg, Ill. 60196)

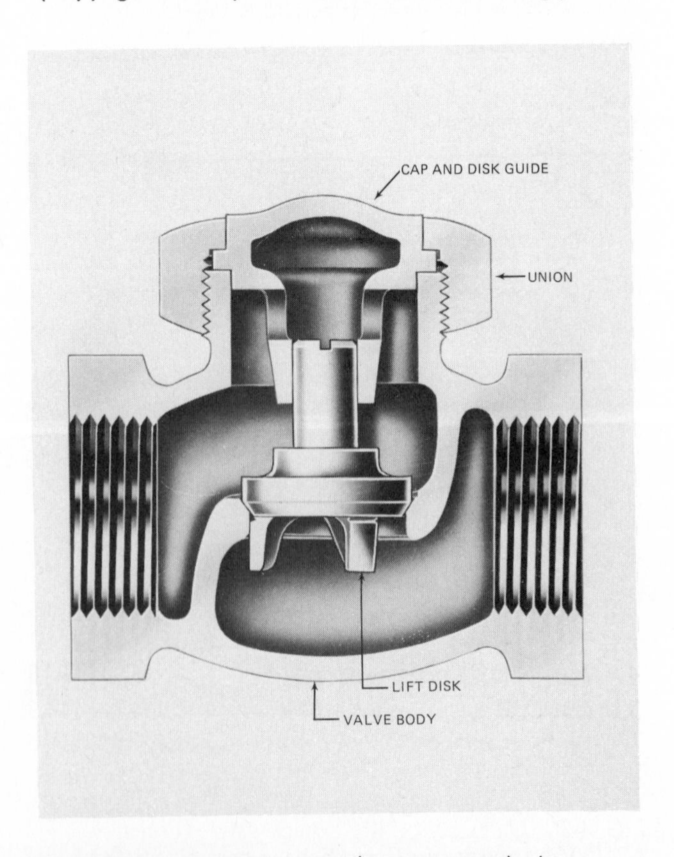

FIG. 4-42 Lift check valve (screwed ends). (Courtesy Jenkins Bros.)

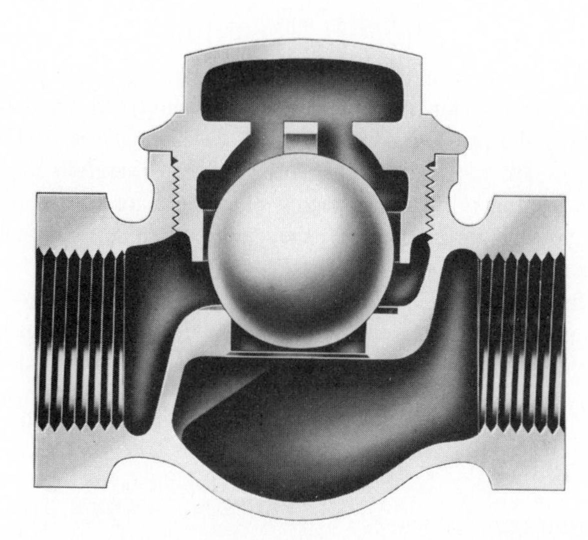

FIG. 4-43 Ball valve (screwed bonnet, screwed ends). (Courtesy Jenkins Bros.)

in flow resistance is not objectionable. A full port opening is available. The disk (globe, wafer, piston, poppet) is usually center-guided and is recommended for frequent operation. The major drawback is size limitation to 6 in. (15.24 cm) and under.

Ball check valves, shown in Fig. 4-43, are unique lift check valves because there is no mechanical method of seating the ball disk to the seat. For the valve to be repaired, the ball must be replaced or the seats must be refinished with a lathe.

Pressure Safety Devices

Pop safety valves, as shown in Fig. 4-44, are exclusively steam, air, or gas mechanisms *never* to be used for liquids. These devices are used for protecting equipment and workers from sudden and dangerous excess pressures. Pop safety valves are spring-loaded and open automatically to expel excess pressure. In operation, this safety valve pops wide open at a set pressure and then snaps back for total shutoff when lower pressure is achieved. This valve is susceptible to corrosion and must be periodically inspected. An adjustable huddling chamber regulates popping action and the amount of blowdown or blowback. (Blowdown is the difference between relieving the closing line pressure.) Lifting levers, as depicted in Fig. 4-45, and drainage holes, are required on the discharge side by ASME codes, except when releasing noxious substances would create a hazard. In such cases, pop safety valves discharge into a blowdown holding tank.

Relief valves, shown in Fig. 4-46, are used for liquids only. They have no huddling chamber like pop safety valves. A fixed lip on the disk gives increased lift area at a predetermined pressure. After opening, the valve closes slowly as

FIG. 4-45 Safety valve installed at a steam power generation plant.

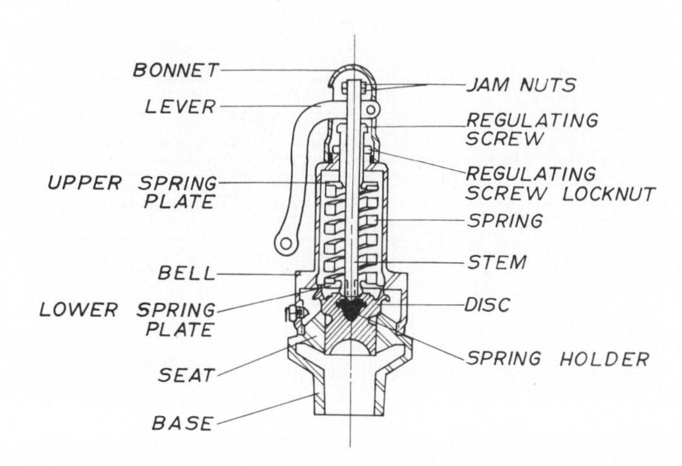

FIG. 4-44 Pop safety valve.

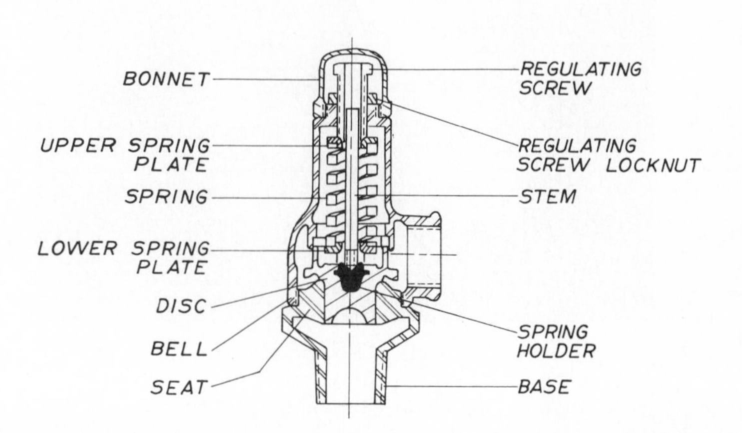

FIG. 4-46 Relief valve.

the pressure drops to a predetermined level. This is not a backpressure regulating valve, but a safety device. It prevents dangerous pressure buildup. With liquids, the discharge is channeled through the side opening into a blowdown tank or outflow area, not expelled to the outside as with some pop safety valves.

The *rupture disk* is another safety device. In this case, the disk ruptures at a predetermined pressure. Since there is no reseat capability, this device is used only in places where emergencies are rare. Rupture disks provide quick relief, but the disk must be replaced after each use. A gate or plug valve (normally open) is usually installed ahead of the rupture disk, though this is a dangerous practice—if the operator forgets to open the valve after a new rupture disk is installed, there is no protection.

Pressure-Reducing Valves

The pressure-reducing valve maintains a constant pressure in the downstream line from a high-pressure inlet source. Size is determined by flow rate and pressure conditions. A gate valve is usually installed on either side of the regulator with a globe bypass line around the reducing valve and a relief/safety valve on the low-pressure side.

FIG. 4-47 Diaphragm control valve station. (Copyright 1976 by Stockham Valves and Fittings)

The *pilot type* is for severe service (more than 20:1 pressure reduction ratio) and is controlled by a pilot valve that reduces the pressure which operates the main valves. These devices are more expensive and harder to maintain than direct-acting rubber diaphragms. Since they are very sensitive to changes in pressure, they are used in services that demand a high degree of accuracy.

The *diaphragm (rubber) type* valve, shown in Fig. 4-47, depends on line pressure acting on the diaphragm to regulate the opening. Reduced pressure from a port in the valve body actuates the diaphragm to open the valve when pressure decreases; a preset spring closes the valve when line pressure increases.

Diaphragm Valves

Diaphragm valves, which are shown in Fig. 4-48, have a great range of uses and operate effectively in services where

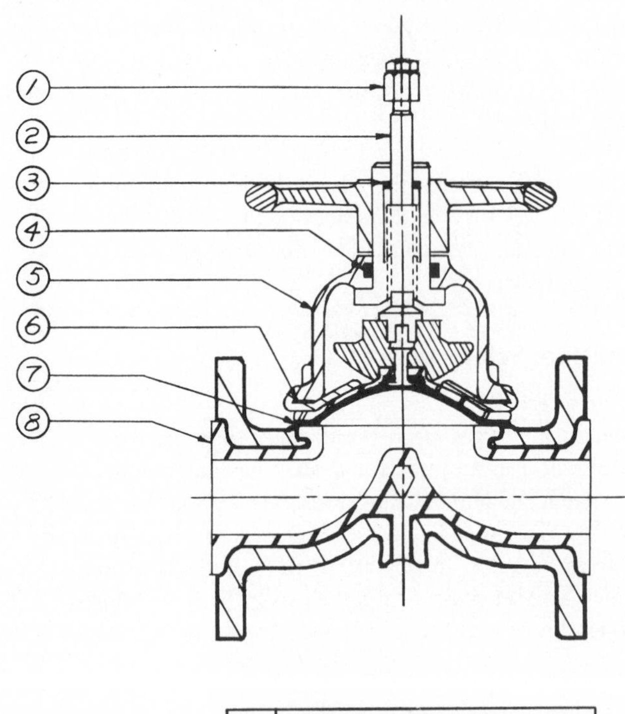

1	TRAVEL STOP
2	INDICATING STEM
3	O-RING
4	O-RING
5	SEALED BONNET
6	GASKET
7	DIAPHRAM
8	PLASTIC LINING

FIG. 4-48 Hand-operated diaphragm valve (weir type).

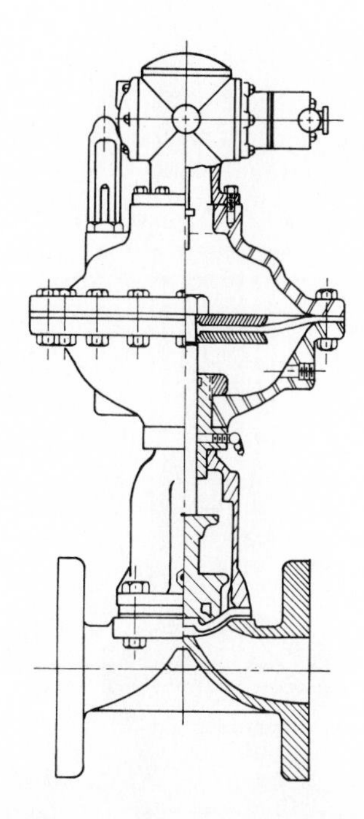

FIG. 4-49 Motor-operated diaphragm valve (weir type).

other valves are inadequate. These valves are simple to operate and maintain. The bonnet isolation eliminates contamination of the line fluid, provides a leakproof closure, and separates the working parts from the line medium. This type of valve has no packing glands, is not subject to corrosion from liquid or gas services, and has no seats, disk holders, or metal-to-metal moving contact. These valves are used for both on-off service and regulation of flow. They are excellent for corrosive, viscous materials, slurries, sludge, and the like. Only the diaphragm is subject to wear.

Diaphragm valves are widely used in the food, beverage, and drug industries because they are easily cleaned and do not contaminate the line substance. Most diaphragm valves are restricted to 400°F (204.4°C) and also have problems at very cold temperatures. They are available in the conventional weir, full-bore straightway, and weir full-bore variations. Diaphragm valves are excellent for adaption to remote control actuators (Fig. 4-49).

Control Valves
(See Chapter 10)

Control valves are used for sequence of operation, pressure reducing, and directional control. For directional control the disk is usually a plug, poppet, or piston type with a four-way opening, but two, three, five, and six-way valves are available (Fig. 4-21). All these variations start or stop or direct flow by a movable disk which opens, shuts, or obstructs one or more ports or makes interchangeable connections between various inlets and outlets. These mechanisms are excellent for moving large masses of material with a minimum of operation effort and also for time interval spacing. The pressure control type is used on certain hydraulic control circuits which require reduced pressure for operation. For sequence control the valve channels the flow through secondary circuits.

Motor and Cylinder-Actuated Valves

Actuators, depicted in Fig. 4-50, are available on many valve types. They are excellent for remote, inaccessible locations (Fig. 4-51) or in emergency services for fast and effective operation. *Cylinder-operated actuators,* which are shown in Fig. 4-52, are operated by a double-acting cylinder using water, air, or oil as the medium. They have low operation and maintenance cost.

Electric motor-actuated valves, shown in Figs. 4-53 and 4-54, also provide laborsaving service, safety in emergencies, and extreme flexibility in piping systems. These actuators are excellent for rapid operation of single valves or simultaneous operation of valve groups. Service features on electric motor-operated valves include automatic operation through the use of pressure switches and other electrical devices. There is also a reduction of effort for accurate valve control. Valves close and open rapidly and completely, though not as rapidly as cylinder-operated valves.

Air may be used to operate motors provided it is free of foreign materials which may affect the motor. *Air-* or *gas-operated motors* will operate under water and have an advantage over electric motors since there are no wires that can cause short circuits. Air/gas actuators are used in many refineries where lines convey oil and gas and the fire hazard is critical.

FIG. 4-50 Motor-actuated valves. (Copyright 1976 by Stockham Valves and Fittings)

FIG. 4-51 A 96-in. (243.8-cm) motor-operated gate valve for a wind tunnel. (Courtesy Grove Valve and Regulator Company)

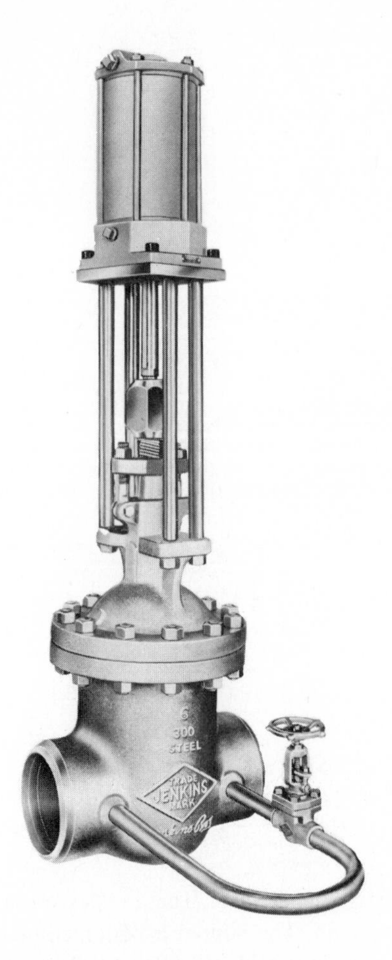

FIG. 4-52 Cylinder-operated wedge gate valve (butt-weld ends) with bypass. (Courtesy Jenkins Bros.)

FIG. 4-53 Motor-operated iron gate valve (125 WSP 200 OWG). (Courtesy Jenkins Bros.)

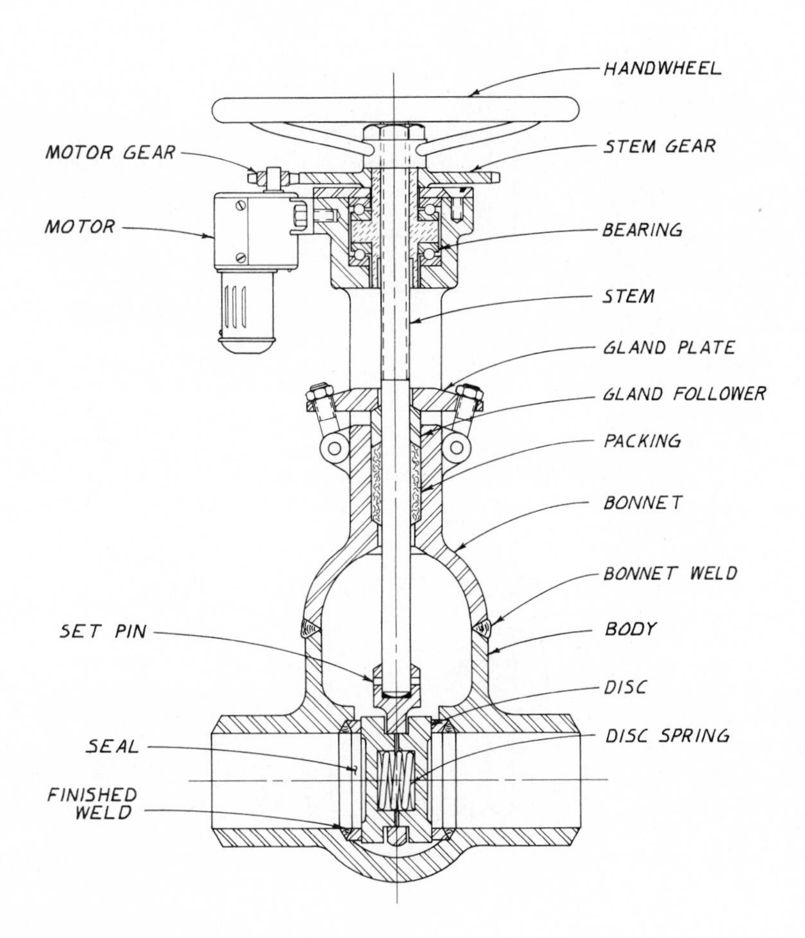

FIG. 4-54 Motor-operated parallel-disk gate valve (welded bonnet and ends).

Steam Traps

Steam traps are used in lines where condensate must be discharged as it accumulates, while still holding steam. The *float trap,* shown in Fig. 4-55, has a hollow float that rises with the water level; levers discharge the water more continuously than the *bucket trap,* shown in Fig. 4-56, where the water spills over the edge and intermittently opens a valve to discharge the condensate. *Inverted open float steam traps,* shown in Fig. 4-57, can return condensate to a lower pressure area in the system. The *impulse trap* uses hot condensate to force a piston to discharge water. Steam traps are essential for all steam piping, separators, and steam-actuated equipment. They are selected on the basis of their discharge capacity, not on pipe size.

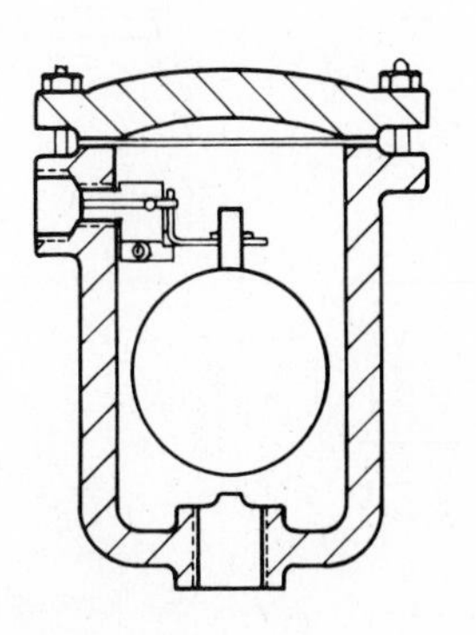

CROSS-SECTION SHOWING SIMPLIFIED FLOAT AND BALL-TYPE SEATING MECHANISM. RANGE OF WORKING PRESSURES:
1 TO 150 LBS AS AIR VENT VALVES
1 TO 125 LBS AS WATER DRAIN VALVES.

FIG. 4-55 Automatic air vent or water drain. (Courtesy of Crane Co.)

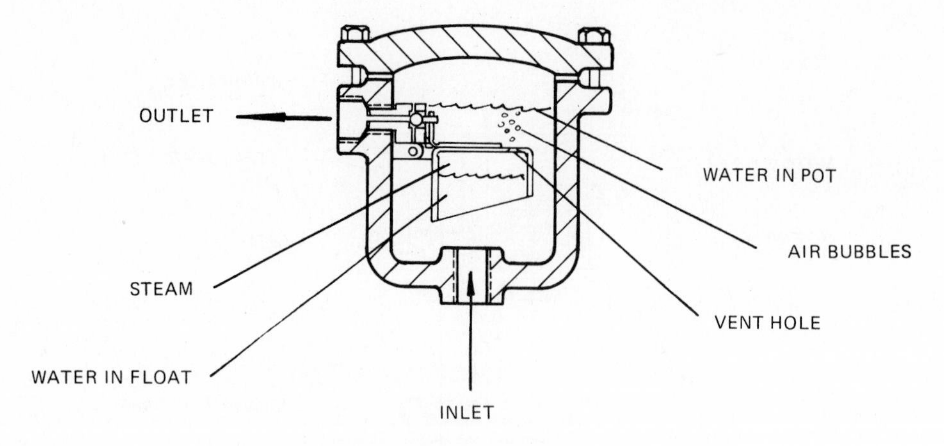

(a) TRAP CLOSED: INCOMING STEAM UNDER THE FLOAT BUOYS THE FLOAT UP, CLOSING THE VALVE.

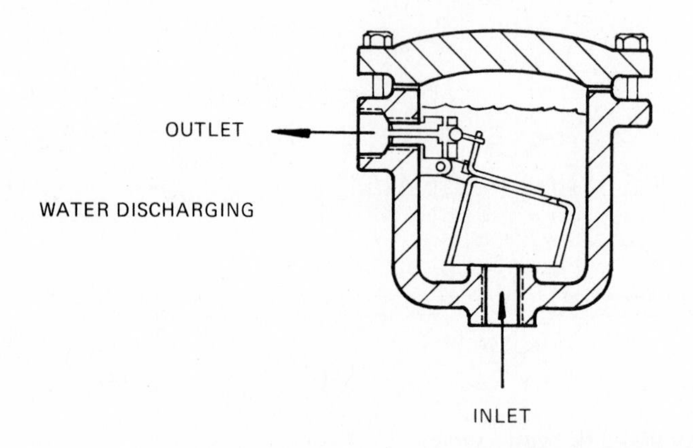

(b) TRAP OPEN: INCOMING CONDENSATE FILLS THE FLOAT, SINKS IT, AND OPENS THE VALVE.

FIG. 4-56 Steam trap. (Courtesy of Crane Co.)

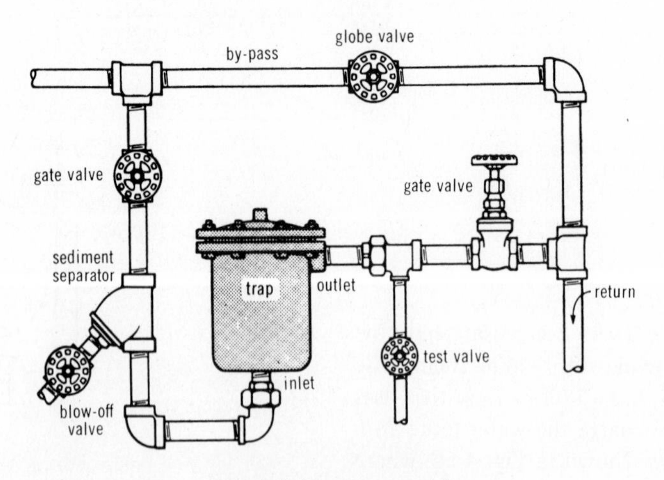

FIG. 4-57 Installation of steam trap. (Courtesy of Crane Co.)

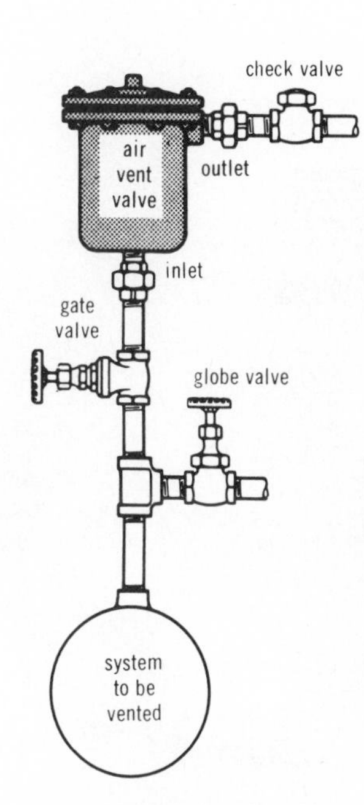

As air vent valves:

Install the valve in an upright position at the highest point in the system. Always pitch the inlet line upward to the valve. Normally, valve is filled with the liquid in the line. As air accumulates, it displaces this liquid, causing the float to drop, thus opening the valve to release the air. As the air is discharged, line liquid re-enters the valve and lifts the float, thereby closing the valve. Operation is fully automatic and keeps repeating with the accumulation of air in the system.

As water drain valves:

Install the valve in an inverted position at the lowest point in the system. Always pitch inlet line downward to the valve. In air lines, the valve is normally filled with air. As the liquid (water or oil) accumulates, it displaces the air, causing the float to rise and thereby opening the valve. Liquid in the valve is discharged by line pressure. As air re-enters the valve, the float drops and closes the valve to the escape of air. This cycle is automatic and continuous.

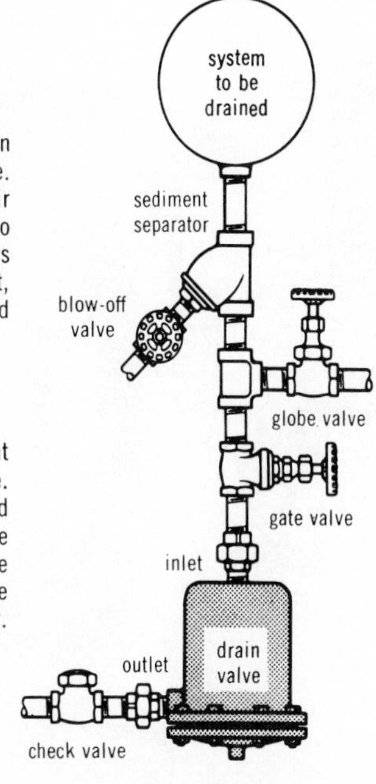

FIG. 4-58 Recommended installation of traps. (Courtesy of Crane Co.)

Air Vents or Water Drain Valves

This type of valve frees trapped air in oil and water systems when used as an air vent valve (Fig. 4-58). They are also used in compressed air systems to discharge accumulated water and oil.

Flow Rate Regulators

Regulators, as shown in Fig. 4-59, maintain accurate, constant, and uniform flow in the line. They are automatic and come in a wide range of variations depending on the service (Fig. 4-60). Regulators should be selected by capacity to provide the desired rate of flow in the line (Figs. 4-61 and 4-62). Many regulators are specifically designed to prevent surging, chattering, or cycling. An oversized regulator can cause extreme wear on valve poppet and seat areas. An undersized regulator may clatter, especially at low pressures, and will not be able to hold the reduced pressure at the desired level.

This chapter has presented only an overview of valves, their types, services, and peculiarities. Nevertheless, the information presented here should lead to a basic understanding of valves and develop the drafter's ability to apply that knowledge in various on-the-job situations. Only many years of experience will enable the reader to understand their full range of possibilities and limitations.

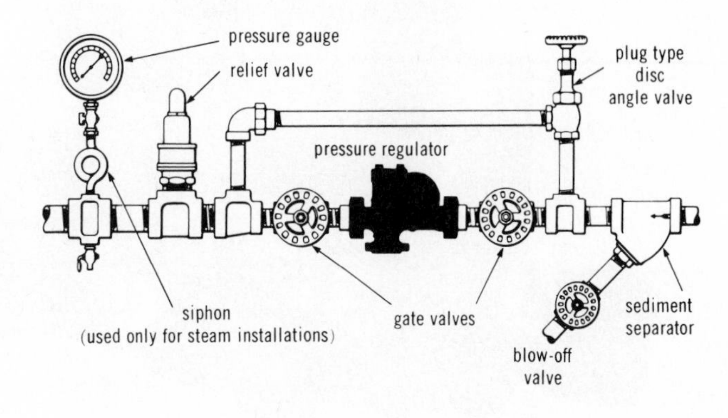

FIG. 4-59 Recommended pressure regulator arrangement. (Courtesy of Crane Co.)

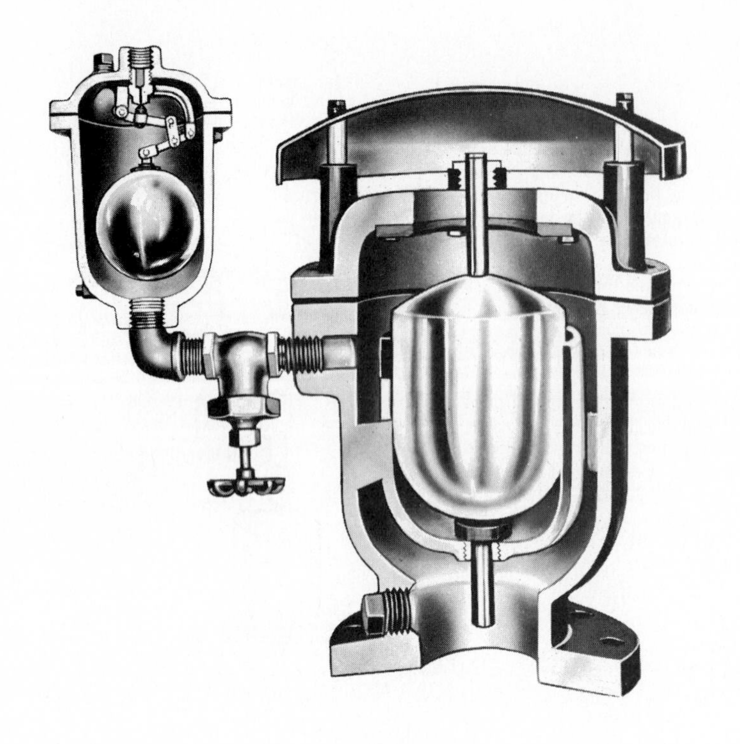

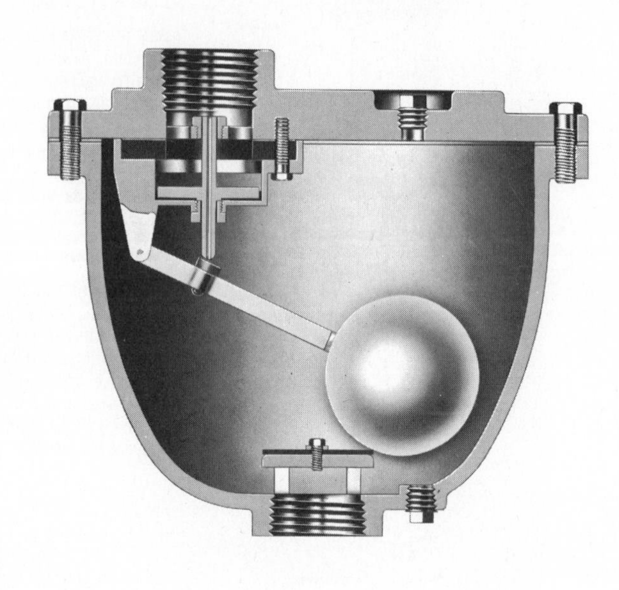

CUSTOM STANDARD

FIG. 4-60 Air release valve. (Courtesy APCO/Valve and Primer Corp., Schaumberg, Ill. 60196)

FIG. 4-61 Dual regulator manifold. (Courtesy Grove Valve and Regulator Company)

FIG. 4-62 A 10,000-psi gate valve and air dome "powreactor" regulator. (Courtesy Grove Valve and Regulator Company)

BIBLIOGRAPHY

Choosing the Right Valve. New York: Crane Company.

Piping Pointers–Application and Maintenance of Valves and Piping Equipment. New York: Crane Company.

Product Selection. Birmingham, Ala.: Stockham Valves and Fittings.

Product Selection. Oakland, Calif.: Grove Valve and Regulator Co.

Stockham Makes Your Job Easier. Birmingham, Ala.: Stockham Valves and Fittings.

The One Great Name in Valves. Cincinnati: Lunkenheimer.

What Is a Valve? Cincinnati: Lunkenheimer.

Which Valve Should I Use? Schauburg, Ill.: Valve and Primer Company.

QUIZ

1. What safety pressure devices are used for steam, air, or gas lines exclusively?
   (a) Globe
   (b) Gate
   (c) Safety
   (d) Relief
   (e) Pilot
2. What valve material should be used in an area experiencing thermal shock?
   (a) Steel
   (b) Malleable iron
   (c) Copper
   (d) Brass
3. What material can be used at 400°F (204.4°C) and lower pressures in noncritical service?
   (a) Steel
   (b) Brass
   (c) Zinc
   (d) Aluminum

4. Welded bonnets are ideal for what conditions?
   (a) High temperatures and pressures and infrequent dismantling
   (b) Low pressures
   (c) Where there is a fire hazard
   (d) Temperatures of 900°F (482.1°C) and higher
5. Traps serve what purpose?
   (a) Keep lines clean
   (b) Discharge condensate while still holding steam
   (c) Discharge liquids continuously
   (d) Act as filters
6. What type of globe disk offers wide seating and bearing surfaces?
   (a) Nonmetallic
   (b) Plug
   (c) Conventional
   (d) Needle
7. What type of check valve should be used with a gate valve?
8. Define OS&Y.
9. In what valve stem option do the line contents have constant contact with the threads?
   (a) Rising stem and handwheel
   (b) Nonrising stem, inside threads
   (c) Rising stem, inside threads
   (d) Rising stem, stationary handwheel
10. What is the function of a valve bypass?

PROBLEMS

1. Sketch a cross-sectional view of a body and disk area of the following valves and identify the various parts.
   (a) Plug or globe valve
   (b) Swing check valve
   (c) Solid-wedge gate valve
   (d) Weir diaphragm valve
2. Draw the plan and elevation views of a pressure-reducing valve line with accompanying bypass and support valves. Use the single-line method of representing lines and valve symbols.
3. Choose the trim, end connections, bonnet, and so forth for a gate valve to be installed in a 300 WOG, 600°F (315°C) oil line. Order the valve from the appendix or catalogs.
4. Draw an outside view of a gate valve (top, front, and side) with a bypass line.
5. Draw a cross-sectional view of a weir diaphragm valve and show its flow characteristics.

PART II Principles in Practice

Student using a large drafting board equipped with a parallel bar.

5 PIPING DRAFTING

Piping drafting is a specialized area that uses a language of its own to transmit the necessary information for the construction of a project. Symbols, dimensioning systems, special notation, and various lines are used to convey to the fabricator or construction crew a set of calculations and design situations peculiar to the piping field (Fig. 5-1).

The piping drafter has to provide convenient, accurate, and detailed drawings of a system. Thus the drafter in the field of piping will have to learn the various symbols and peculiarities of pipe drafting in order to display on drawings or models sufficient information. All drafting involves lines, symbols, dimensions, and notations to represent the engineer's design and calculations in a working form that can be easily read and translated into a complete project.

Mathematics is used throughout the piping field as it is in other drafting areas and constitutes an intricate part of the drafter's background and knowledge. Mathematics is used to calculate rate of flow, cold-spring, pipe temperature, structural needs, and various other requirements.

Throughout history, drawings have proved to be the most reliable way of conveying visual information that can be easily and economically reproduced and distributed for the construction of a project. Drawings are used throughout a company in various capacities: sales, procurement, and mechanical, electrical, and structural design. The information given on the drawings must be concise and readily discernible.

Drawings are usually completed on paper, mylar (a plastic paper), or linen (a treated cloth) in pencil or ink. All these mediums of paper, plastic, and cloth enable the drawing to be reproduced with a blueprint machine.

The piping drafter must use symbols to convey the necessary information and, to avoid confusion, keep to a bare minimum the amount of detail that is shown. Pictorial

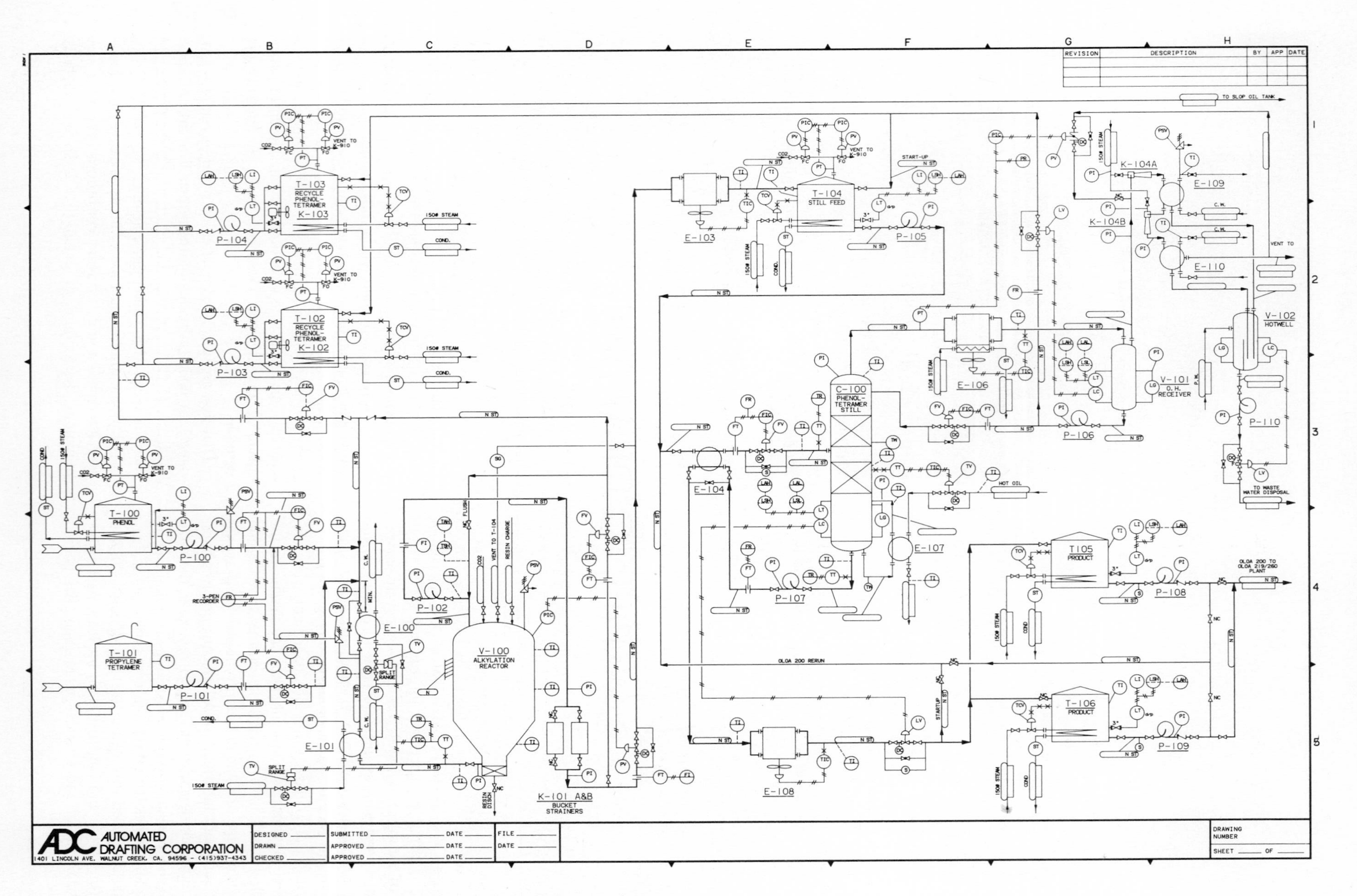

FIG. 5-1 P&ID drawn by a CAD system. (Courtesy Automated Drafting Corp.)

representation is used on piping drawings only where absolutely necessary to show clearances and relationships of equipment size to the surrounding situation.

Besides being able to draw and represent a system, the piping drafter must be familiar with the components that make up a piping system. Chapters 1 through 4 provide essential background information on pipes, valves, welding, and fittings. The drafter must also be familiar with the use of standards and catalog descriptions of material used on piping drawings. All catalogs and standards must be readily available if the piping drafter is to display, dimension, and order parts for a project.

Mechanical drafting ability is the most important aspect of every type of drafting. Thus the piping drafter's success in this field will depend on his or her skill in mechanical drafting and lettering. Certainly the piping drafter will come in contact with a vast amount of equipment and machinery that will require adequate detailing. *Architectural drafting* is also important to the piping drafter because many piping problems involve the use of architectural plans for building complexes. The piping drafter and designer uses architectural, structural, piping, and mechanical drafting, along with model making, and the mathematical calculations needed for pipe sizing, valve selection, fluid flow, and stress evaluation.

Most piping drafters used to come from allied fields because formal instruction in piping drafting was not available and had not yet become a separate discipline. The knowledge demanded of the piping drafter was so vast that no standard skills could be determined apart from the structural, architectural, and mechanical fields that have been mentioned. Another roadblock in the solidification of the piping drafting field was the fact that many companies used their own standards and requirements in their drawings. Only recently has standardization spread throughout the piping industry.

This chapter covers the basics of drawing and lettering—the calling card of every drafter no matter what the field. The main job of the piping drafter is to represent flow diagrams and piping installations and the details of a pipe system as designed by the head drafter or designer. Problems such as calculations for stress, temperature evaluations, and selection of component parts are usually up to a senior drafter, designer, or the piping engineer who, after designing the project, will give this information to the drafter. As one moves up in the field of piping drafting one must develop the ability to do these calculations and eventually be able to design the system itself, but at the beginning stages the piping drafter is just required to know the basics of drafting and the background needed for piping—including the background presented in the first four chapters of this book.

Most companies still expect the designer to provide most of the information to the piping drafter who will in turn lay out the plant in orthographic views with a sufficient number of elevations, plan, and section views so that the system is represented adequately. This method requires a highly qualified drafter who can read architectural, structural, civil, and equipment layout drawings and apply the design information to the layout of the system. This time-consuming procedure is one of the main reasons for the increasing use of models, though not all companies have adopted them yet. Various support plans and drawings are required before the system is put down on paper: electrical, structural, architectural, civil, site, mechanical, and HVAC drawings, vendor drawings, the original design flow diagram, and the equipment placement drawing.

Electrical drawings cover the electrical systems to be installed in the building, including separate systems for power communications and service. It is important to know the location and the electrical design requirements in the beginning stages of a piping job because of the need to run cable trays and electrical conduit, and to provide adequate space for service and installation while avoiding interference between the piping and electrical systems. The electrical drawings will include light locations, control panels, conduit runs, junction boxes, and other electrical needs.

The structural information needed for the drawing is an extremely important aspect of any piping system. The piping system must be attached to, hung from, and supported by the structural system. *Structural drawings* provide the boundaries in which the piping system is to be maintained. Structural drawings will include steel versus concrete construction, size of the plant, pads, and floors, building elevations, floor to ceiling heights, wall bracing, platforms, column placement, and structural design standards. These are just a few of the aspects that must be considered before starting the physical plant design. The architectural plans include elevations of the buildings involved in the piping system—including a sufficient number of cross-sectional views showing foundations, floor penetrations for stairwells and elevators, wall locations, floor elevations, windows, ceiling heights, ducts, doors, and interior and exterior sectional views. All these allow the piping drafter to work within the limits that have been prescribed whether the pipe system is to fit in a building, a complex, or a processing plant.

Site plans provided by the civil engineering department show the location of the plant in regard to elevation, grade, and physical (geographic) surroundings. The site plan locates the boundaries and perimeters on which the building, plant, or power station must be laid out. This drawing is usually executed at a very small scale and includes all natural and artificial boundaries, including ponds, roads, railroad lines, power lines, and waste disposal units.

Plot plans are drawings that depict the design situation of the proposed project, including north-south orientation and all necessary coordinates. (See Chapter 10.) Much of the structural and architectural design information will show on the plot plan—building outlines (existing and proposed), equipment centerlines, and any physical information that is essential to the construction and design of the piping system.

Equipment placement drawings are prepared by the piping design group; in some cases simple models are used to position equipment options. Simplified graphic shapes are used to represent types of equipment, such as vessels, towers, pumps, storage tanks, and turbines.

HVAC drawings as referred to here involve the heating, ventilation, and air conditioning systems. These systems are as important as electrical systems because of the amount of ducting and equipment needed for adequate replacement of air circulation, air conditioning, and heating to meet federal and state standards. All this requires a considerable amount of space for installation. These requirements must be known well in advance so they do not interfer with piping and electrical runs.

Flow diagrams are a separate category which indicates the basic equipment, flow, temperature, pipe size, valve, and instrumentation needs for the system. Chapter 6 describes flow diagrams and their role at the start of a job.

Vendor's drawings include all essential data pertaining to manufactured equipment, valves, and instrumentation. Anything that is ordered from a catalog or is not fabricated must have dimensions, sizes, flange types, ratings, and so forth supplied before the system is designed (see appendix).

Besides drawing the plan, section, and elevation views, the drafter will be called upon to make small detailed drawings showing standard parts and special arrangements—including subdetails of vessels and other fabricated items—that cannot be seen on the large plan, section, and elevation views. All these drawings require the drafter to understand the equipment, valves, fittings, pipes, and other components. This information is available through the catalogs from which the drafter must order the constituent parts of the piping system.

The piping designer and engineer coordinate the beginning stages of a project—including the special studies and calculations needed to produce the desired results (production of energy, for example, or manufacture of a commodity).

Although all flow diagrams are prepared *before* the drafting and laying out of pipe in plan, section, elevation, detail, and isometric views, this chapter introduces the student to the actual drafting practices involved in construction of simple orthographic piping drawings utilizing lettering, line work, and other essential knowledge which can be applied also to the flow diagrams. The flow diagram is the most important document in the initial stages of a project. Only after this drawing is completed can the piping designer and drafter start the layout of the plant or piping system. The proper drawing scale for a project will vary, though 3/8 in. = 1 ft is common in the petrochemical industry. The power generation field has found this size to be too large and in general uses 1/4 in. = 1 ft where possible.

DRAFTING METHODS

There are basically three procedures used throughout the piping industry for the drafting design stages of a project requiring extensive piping: the elevation, plan, and section method, the isometric method, and the model method.

Elevation, Plan, and Section Method

The first method is the elevation, plan, and section method. Conventional orthographic views are drawn of the system in its totality using a multitude of section, elevation, and plan views. The method is used in all industries, especially in the power generation field where the systems are complicated and large. This drawing procedure shows most of the equipment, pipe runs, pipe support location, and other components that compose the project, with the exception of extremely small pipelines or complicated areas that must be shown on blowups or details. This method requires a vast quantity of drawings, including a large number of sectional views. Vertical dimensions are called out by elevation designations and horizontal dimensions are shown in the plan view. Dimensions which cannot be seen on the major views are provided in smaller cutaway or sectional views in order to dimension the system completely.

This drafting method demands extremely well qualified personnel. This is one reason why many companies are adopting design methods which require less experienced personnel. The elevation, plan, and section method has other drawbacks as well. Above all, piping systems do not always lend themselves to orthographic projection because lines and components cannot be shown in relation to objects that lie in front, on top, or behind them.

Often the head drafter completes the major plan and elevation views and the less experienced drafters cut sections and do small detailing of the project. This practice may lead to overdetailing of certain areas, since duplications can be detected only after a multitude of sections and details have been made. Orthographic projection used in the elevation, plan, and section method makes it extremely difficult for the construction group and fabricators to follow the operational sequence needed to construct the project. With this method, isometric drawings are made of all individual lines on a project to show the entire course of a pipeline, including its components. This enables the drafter to dimension one line of the system at a time completely (which aids the fabricator). Chapter 7 explains the various shop, and system isometric drawings and their uses.

Isometric Method

The second drafting method is the isometric method. For heating and plumbing in architectural areas of buildings, schools, and houses, this method has proved quite successful and less cumbersome than the orthographic method (see Chapter 7). It is not used for large projects, however, because of the complexity of the isometric. Large isometric drawings are made of whole piping systems; certain areas of plants and projects may also be drawn in isometric. This three-dimensional procedure enables the plan and elevation to be combined in one drawing, thereby reducing the amount of drawing needed to represent the system.

As in the elevation, plan, and section method, isometric spool sheets are sometimes made of individual lines in order to do material takeoff, dimensioning, and pipe support placement (unless the system isometric is sufficient to show these aspects clearly). The isometric method can be rendered by the less experienced drafter more easily than the elevation, plan, and section method. Isometric systems are discussed in Chapter 7 along with isometric fabrication spool sheets.

When making isometric drawings, the drafter sketches each pipeline from the original system drawing and then prepares a detailed, completely dimensioned, finished drawing from the sketch—showing piping penetration in walls, floors, and other essential data including connections to equipment and clearances between vessels and pipelines.

Revisions to isometric system drawings are not so hard to make as with the plan, section, and elevation method. Nevertheless, any changes in the design must be transferred to every drawing affected—a long and expensive process.

Material takeoff is the same in this method as in the plan, section, and elevation method. Checking the three-dimensional isometric method of pipe design is easier and less time consuming than with orthographic drawings because clearances and relationships of lines can be seen more readily. It must be emphasized, though, that this method of pipe drafting cannot be used for complicated systems.

Model Method

The third method, and one that is gaining more acceptance in the power generation field, is modeling. The petrochemical industries have done the essential groundwork for this type of design. The model method is covered in detail in Chapter 15.

The model method still requires many of the original support drawings, including architectural, civil, electrical, site, mechanical, and flow diagrams, at the onset of the project. Moreover, the model shop is usually staffed with people who understand only the basics of piping design. Therefore many industries use model makers only for large construction items such as bases, walls, elevations, steel, equipment, hanger placement, and other items not within the design realm.

The piping designer is usually called upon to run pipe and position components in the system. Along with the support drawings and data that are required, the piping designer and drafter requisition the model parts, bases, and equipment necessary for construction of the model. Model design often takes place in the drafting area of the company.

Scale models of equipment are used on the larger model. The equipment is never completed in great detail but is only boxed in or shaped to represent the part for the job (such as a turbine which comes as a standard item ordered from a company). It is not necessary for the model shop to show great detail on most pieces of equipment.

The model shop usually provides all support construction for the building of the model by the pipe designers and drafters. This method of pipe design is extremely useful. Having physical objects to manipulate, the designer can see clearances or interferences between objects on the model which might not be apparent on drawings. Equipment layout drawings and design sketches may still be used with this method.

One use of a model is for checking the piping system after it has been designed and drawn. However, it is advantageous to use the model in the design stage, thus significantly lessening design time. All the electrical equipment, heating, air conditioning, and ventilation requirements, structural and architectural requirements, piping, and equipment can then be readily viewed, placed, and designed on the model. A model usually constitutes only one-half of 1 percent of the project's total cost. The model's favorable aspects will always outweigh its cost.

In petrochemical design where sections of plants or plants themselves are designed with models, the piping drafter may be put to work on the model running pipe, constructing vessels, placing component parts, and so forth under the supervision of a head designer. The head designer can oversee the project much more easily than in the power generation field, where systems are so complicated that whole design groups work on one portion of a large-scale model. The petrochemical industry uses a scale of 3/8 in. = 1 ft wherever possible in the model and on drawings. The power generation industry uses smaller-scale drawings and usually larger-scale models with 1/4 in. = 1 ft for the drawings and 1/2 in. = 1 ft for the model to show the information needed for project construction.

After equipment requirements have been decided, the placement and construction of equipment on the base of the model are determined. This practice enables designers to check for sufficient design capabilities and allows them to change the layout without revising a lot of drawings. Plan views and some sectional views are still used for this method, though to a lesser extent than for the plan, elevation, and section method.

The one type of drawing used in the model design sequence is the sketched isometric view of each line on the model. These sketches are then converted into isometric drawings or fed into a computer graphic system that draws the isometric line and does a material takeoff.

Checking the model, listing materials, and making revisions all require less manpower and experience than do other methods of design. The checkers of the model must review all flow diagrams, equipment placement, isometrics, and any other drawings that have been made to produce the model, but not to such as extent as required by the other methods. The model itself will represent all the other drawings that have been made, making it much easier for the checker to see problems that may be encountered during the fabrication, installation, and operational stages of the project.

The main drawback of the model building method is the human element. Resistance within the company and industry comes from seasoned veterans in the piping field who look upon the model method as a waste of money. Many utility companies and other power generation groups have found it hard to institute model programs because of the built-in resistance to change among piping designers and drafters. The petrochemical field, because of its competitive nature, has adapted much more rapidly to this advantageous system of piping design.

The model also presents problems in reproduction. Since photography of models has come into play, design groups can photograph stages or completed views of the model and ship them to the job site along with the model. Still, photographs do not offer the convenience that blueprinting and xeroxing have brought to the drawing method of design. Chapter 15 presents a number of photographs of models and explains how they can be used advantageously in the construction stage of a project.

Completed models are shipped to the job site. There the construction leaders monitor the job through the steps that have been laid out for construction and use the model much more effectively than a cumbersome set of drawings. The model remains at the job site possibly 10 to 20 years after construction of the plant and becomes part of the training program for new personnel. Thus the company can train personnel, coordinate services, and maintain the facility without a large staff. Revisions or additions of new equipment can be designed on the model before they are tested in the actual plant. (See "Industrial Model Building" by L. G. Lamit, Prentice-Hall, Inc., 1981)

A Comparison

When comparing the three methods it is possible to find deficiencies in each case. Industry seems determined, therefore, to use portions of all three methods of piping design. Modeling, with the use of flow diagrams, limited plan and sectional views, and isometric takeoffs, will be the way of the future for pipe design and drafting. The three-dimensional or isometric method will continue to be used for small-system pipe design for buildings and plumbing systems where models are not an efficient use of time or money. The plan, elevation, and section method will be used throughout the power generation field longer than in the petrochemical industry.

Another method of advanced piping design is starting to make its way into the field: the use of computers for graphics and material control. Though this procedure is costly, it will eventually take over some of the design or drawing stages of piping projects. The computer-aided drafting (CAD) technique is used to create piping drawings. From the input of piping designers who make sketches and feed calculations into the computer, the computer produces, by the use of a plotter, an ink drawing as in Fig. 5-2. The computer can also do material takeoff and pipeline specifications. More will be said on this aspect of piping design at the end of the chapter.

Material Take-off When completing assignments for this chapter and those elsewhere in the text, the student should attempt to list valves and fittings as shown in Chapters 3 and 4. For all valves, list the class, size, and name.

EXAMPLE: Class 150 lb, 4" flanged globe valve.

The appendix can be used along with manufacturer's catalogs for practice ordering. The ordering list may also include the companies' name.

EXAMPLE: Class 300 lb, 6" flanged gate valve, Crane Co., part number.

Fittings should be called out by listing name, class, size, and order number, along with company name if appropriate.

EXAMPLE: 3-1/2" welded straight cross, ex strong, or 10" weld neck flange, Class 300 lb. ITT Grinnell.

When constructing a material ordering list, provide a logical sequence of categories.

EXAMPLE: Size, class, description (name of part), company name, part number.

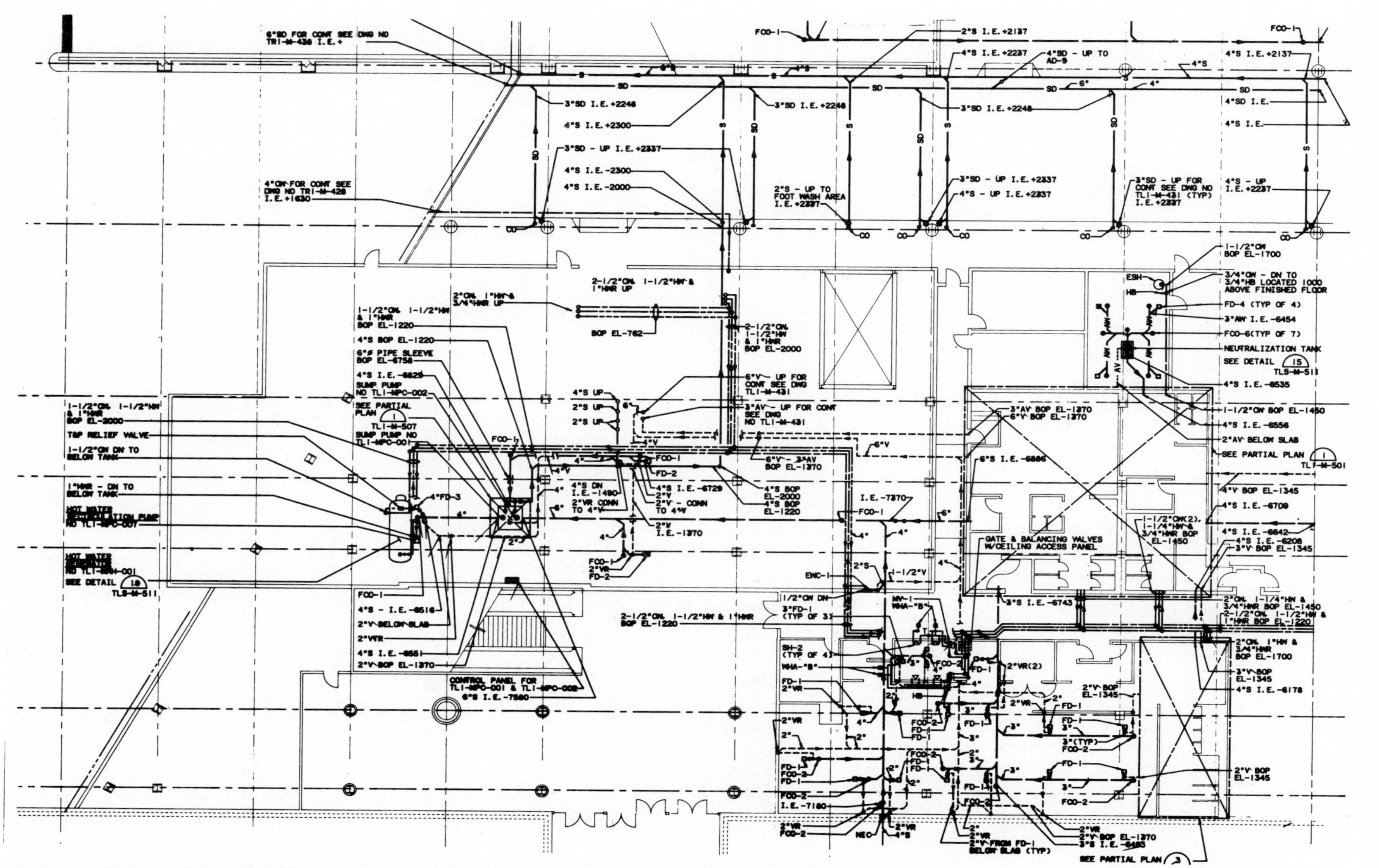

FIG. 5-2 Plumbing piping layout drawn by a CAD system. (Courtesy Automated Drafting Corp.)

Fittings are available in wall thicknesses of standard, extra strong, schedule 160, and double extra strong. When ordering fittings for pipelines that do not correspond to these four weight classes, the next higher commonly stocked weight class should be specified.

EXAMPLE: a 4" schedule 120 pipe (0.438 wall thickness) would require a schedule 160 fitting (0.531 wall thickness).

DRAFTING EQUIPMENT

In all forms of drafting it is necessary to have adequate equipment to ensure high-quality line work and lettering. Though it is possible to do without many of the variations in drafting equipment, special templates and drawing instruments can make a world of difference in the quality and ease of drawing. Tools and supplies are essential for the proper representation of drafting symbols and line work. Students may be familiar already with much of the drafting equipment listed here:

1. *Drafting tape:* Drafting tape is used throughout the profession. Staples and tacks have been used, but they tend to mar the drawing surface and can also have adverse effects on the drafting board and equipment.
2. *Ames lettering guide:* This item is used to draw light guidelines so that the drafter can keep all lettering within the prescribed boundaries.
3. *RapiDesign lettering aid:* This item can be extremely useful to a person with average printing ability. The template allows the drafter to print within prescribed boundaries established by slots. Printing remains freehand and individualized, though the aid ensures that the height and line spacing of the lettering remain constant.
4. *Lead holders:* It is advisable to have more than one lead holder so that it is not necessary to change lead for different types of drawings.
5. *Lead:* Hardness 2H and H are the only types of lead the drafter needs. After proficiency is reached in line work and lettering, the drafter should be able to adjust to the use of H lead, which provides sufficient darkness and proficient line work. 2H can be used for laying out lines that may need to be changed and doing preliminary pipe drafting. HB, a soft, dark lead, is preferred by drafters who are proficient in keeping their drawing clean, but most people smear their work when using HB lead.
6. *Fine-line pencils:* These pencils are coming into vogue in the drafting profession. They require no sharpening to maintain a particular thickness of line.
7. *Pencil pointer:* This instrument is extremely useful and produces a better and cleaner point than the sandpaper block provides. Many drafters now have electric pencil pointers which produce cleaner, neater, more accurate points. Their use generally results in better line work.
8. *Sandpaper pad:* This pad is needed for sharpening compass lead points and in some cases pencils.
9. *30-degree triangle:* It is advisable to have more than one size of triangle though the 12-in. size is a useful length.
10. *45-degree triangle:* This triangle comes in various sizes, and it is useful to have small and large ones, depending on the work one is doing.
11. *Adjustable triangles:* These instruments are extremely helpful when one is using the parallel bar or T square, though modern drafting rooms may have drafting machines which make adjustable triangles obsolete.
12. *Brush:* The brush is needed to keep the drawing clean. Frequent brushing will improve the finished drawing.
13. *Erasers and erasing shields:* These items are essential to the drafter. Erasers of different types and hardness are available, and the choice depends on the paper surface and line-work procedure (ink or pencil).
14. *Scales:* Both mechanical and architectural scales are used throughout the profession. With the increasing use of metrics, it has become advantageous to have a metric scale also. Architectural scales are available in different degrees or scales including 1 in. = 1 ft, 1/4 in. = 1 ft, 1/8 in. = 1 ft, 1/16 in. = 1 ft. The architectural scale is used on most of the drawings the piping drafter will prepare; the mechanical scale is useful for the drawing of details.
15. *Color pencils:* Color pencils may be used to represent different pieces of equipment that are not part of the project but whose relationship needs to be shown on the drawings. They are also used for checking and showing revisions on blueprints.
16. *Drawing set or compass and dividers:* The use of drawing sets has been somewhat controversial throughout the history of drafting. A majority of schools require drafting students to purchase fancy chrome-plated sets that do not provide a long-lasting quality instrument (and also contain unusable or inaccurate extras which only drive up the price of the set). Essential are the following: a bow compass of sufficient size to draw medium to large diameters, a pair of dividers with which to lay off dimensions, and, as optional equipment, a beam compass to draw extremely large diameters and a set of proportional dividers.
17. *Paper:* Each student will be told the paper size and type required by the instructor. The most common paper sizes are standardized by ANSI Y14.1 and include 8½ × 11 in., 11 × 17 in., 17 × 22 in., 22 × 36 in., and 34 × 44 in. For large projects like those in Chapters 12, 13, and 14, 15- to 50-ft-long rolls of vellum paper 3 ft wide are essential.
18. *Templates:* Circle templates, ellipse templates, piping templates, instrument templates, isometric templates, valve templates, and fitting templates are all extremely useful to the piping drafter. They provide accurate representation for items that would normally call for time-consuming rendering.

19. *Curves:* Since irregular, French, and sweep curves are used throughout engineering drawings, it is necessary to have one or more of the diverse sizes and shapes.

Besides these essential items, there is a great deal of optional equipment of use to the piping drafter. Inking, though not often used in piping drafting because of its time-consuming and permanent qualities, is used for civil drawings, equipment location drawings, and sometimes even flow diagrams. *Inking pens* provide an accurate, clean, and easy way to draw permanent lines. Inking pens have rendered obsolete the ink holders that are available in drafting sets. *Leroy lettering sets* are not used in the piping field except for special circumstances, for advertising, or for artistic renderings of a project. The use of both inking pens and Leroy sets should be taught in every drafting course.

Drafting boards, T squares, drafting machines, parallel bars, and straightedges are supplied by the employer. Many drafters prefer to have equipment of their own at home for use off the job. The *T square* consists of a blade and a head which allows the drafter to draw horizontal lines or to set up the triangle to draw vertical lines. It is rapidly becoming obsolete.

The *parallel bar* or *parallel edge* can be found throughout the drafting profession, especially on drafting boards that are extremely large for projects requiring long horizontal lines. The parallel bar consists of a straightedge and a series of cables, wires, and pulleys which keep the edge of the bar parallel to the drafting board or at a preset angle.

The *drafting machine* has almost totally replaced both the T square and the parallel bar in many areas of the drafting profession, including the piping field. It takes the place of the T square, parallel edge, and triangle and can establish all angles on the 360-degree circle. Though it is extremely expensive, it compensates in accuracy and time. The only place where the drafting machine is not preferable is for large drawings of considerable length; in such cases the parallel bar is handier for drawing long horizontal lines accurately. Figure 5-3 shows a close-up of both a template and a drafting machine. Figures 5-4, 5-5, and 5-6 all show piping templates which are extremely useful to the drafter.

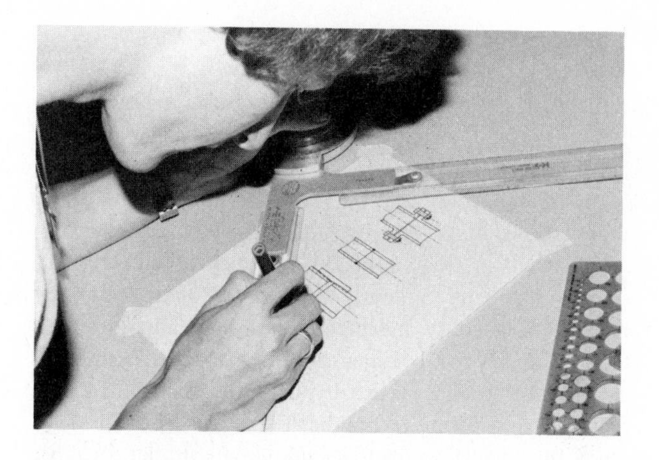

FIG. 5-3 Student using a drafting machine.

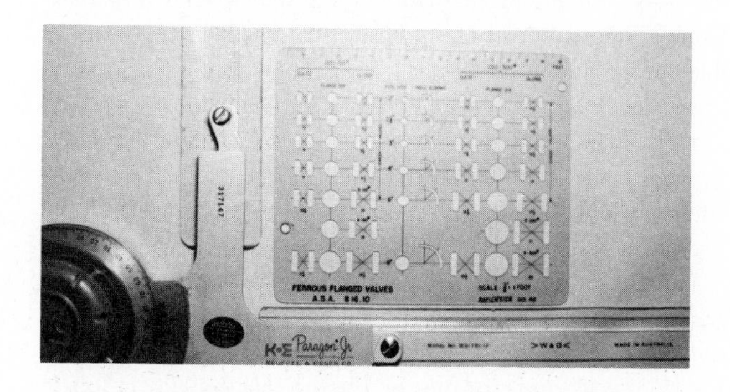

FIG. 5-4 Drafting machine and a ferrous flanged valve template showing ASA B15.10 standard sizes.

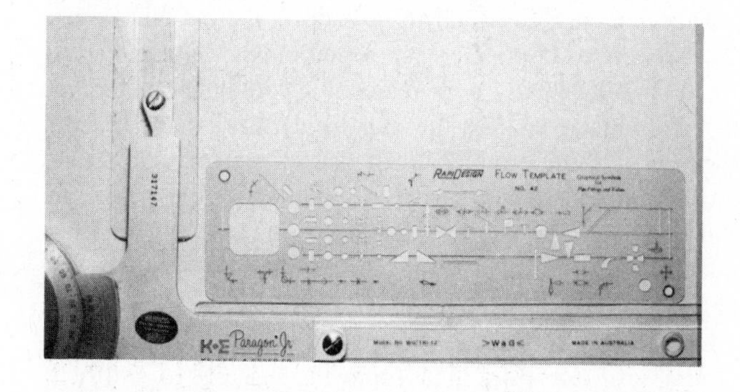

FIG. 5-5 RapiDesign template for flow diagrams.

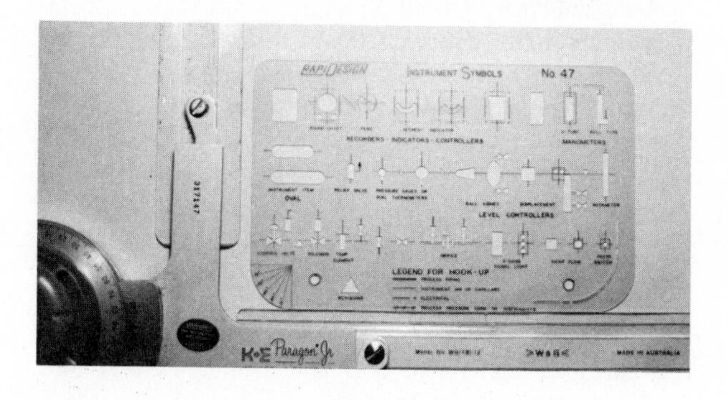

FIG. 5-6 RapiDesign template for instrument symbols.

BASICS OF DRAFTING

Piping drafting, like all forms of engineering, is an applied science. It is concerned with communication through a special graphic language that is used to convey on paper, accurately and efficiently, information for a project. Two of the most important aspects of all piping drafting and engineering are accuracy and efficiency. Piping engineering and drafting have a symbolic language, theories, rules, and conventions that express ideas that are unique to the field.

The person who aspires to be a piping drafter and eventually a designer must learn that symbolic language. Not all of the practices set forth in this book are followed in every company, but they do represent a wide range of standard procedures for drawing and designing piping projects. The drafter can develop drafting skills and understand these standard procedures only by completing the drafting projects provided in this book. This experience will be worth far more than any amount of textual explanation. Only through serious study and practice can the student hope to attain the level of skill that is necessary for future employment as a drafter and, eventually, a piping designer.

Elementary drafting techniques are covered here, but, in general, mechanical drafting techniques are similar for most aspects of piping drafting insofar as quality of line work and lettering and standard orthographic and isometric projections are concerned. The practices set forth in this chapter deal with the areas of piping in orthographic projection as applied to plan, section, and elevation views of projects including the detailing of smaller components such as spool drawings and pipe supports. Like all drawings they serve their full purpose only if they are completely legible to the user—who may not be a drafter, designer, or engineer.

The drawings must provide legible blueprints and in some cases xeroxed prints which are then reduced to microfilm copies. Because of this reduction in size it is important that the line work and printing on the original drawing be made extremely accurate and legible: A loss of sharpness always occurs when the original drawing is run through a blueprint machine, xeroxed, or reduced to microfilm.

This chapter is designed to develop skills that will enable the student to complete engineering piping drawings accurately and efficiently with reasonable speed. A neat, clean, precise drawing enables the user to identify all the aspects essential to the production of the item that is drawn—nothing should be left to the imagination of the worker in shop or field. The drafter's line work, the lettering, and the quality of the finished drawing will reflect his or her ability and character as a drafter.

One of the drafter's main concerns will be the beauty and accuracy of line work. The following suggestions will help maintain the quality that is necessary on all engineering drawings:

1. *Lead selection:* On all piping drawings that are completed for this course, attempt to use a minimum of different pencil or lead weights on a single drawing. Two weights will usually be sufficient for conveying precise lines. In general, 2H and H leads are the only ones necessary. Some drafters prefer HB lead, which smears easily but, with extreme care, produces a dark high-quality line. The 2H lead is used primarily for laying out the original drawing lightly; H lead can be used to darken all types of lines and also for printing. Be careful to keep all lines sharp and clear. Avoid using lead that is too soft and creates a fuzzy dull gray line which tends to smear. Lines which are not precise will not reproduce accurately.
2. *Use of erasers:* In general, erasers of the pink or soft white art gum type are preferred because they cause less smearing of the pencil line. The use of electric erasers, especially in inking drawings, can produce excellent results if great care is taken not to burn or erode the paper. Avoid gritty erasers that can damage the face of the paper. Erasing is hard to avoid no matter how excellent the drafter, so every drafter should have an erasing shield, two or three different types of erasers, and if possible an erasing machine. An erasing shield will help protect lines near the area being erased. Do not use razor blades or attempt to scratch the paper surface to remove the line—this only destroys the paper surface and ruins the drawing's legibility. Electric erasers and highly abrasive erasers should be avoided when one is using plastic paper (mylar). They destroy the finish of the paper, making it impossible to redraw on the area that has been erased.
3. *Procedures for accurate line work:* Rotating the pencil and keeping a *semisharp* pencil point will enable the drafter to control line thickness and maintain a sharp clear line. Many different methods have been set down to help the beginning drafter prepare drawings. Students should use various techniques until they are satisfied with the line quality throughout a complete drawing. Since each person's physical and mental abilities differ, not all procedures will be of value.

   Try to maintain a constant angle between the pencil and the sheet of paper by leaning the top of the pencil in the direction of travel. This practice will result in a consistent manner of line work which in turn produces a consistent line quality throughout the drawings.

   To keep line work and paper clean, use a dusting pad or eraser powder. Brush frequently after each line is drawn and apply the erasing powder or pad when necessary. Another hint for keeping the drawing clean is to avoid drawing over areas that have been frequently erased. Remember to keep the pencil semisharp at all times to avoid fuzzy and dirty lines. Sharp pencils tend to break frequently.

Line Types

Lines in piping drafting vary from the standard in mechanical drafting, since different line thicknesses are used to represent various aspects of piping lines. The one thing that all lines have in common is their darkness. In all cases, lines should be black—whether or not they are thick or very thin, the color remains constant. This enables the blueprint or xerox machine to reproduce the drawing accurately no matter what size or type of lines are used. Only the *width* will change in the different types of lines, not the density or blackness. Contrast between lines is maintained in the variation of thickness, not color. Figure 5-7 displays the line thicknesses and what they represent.

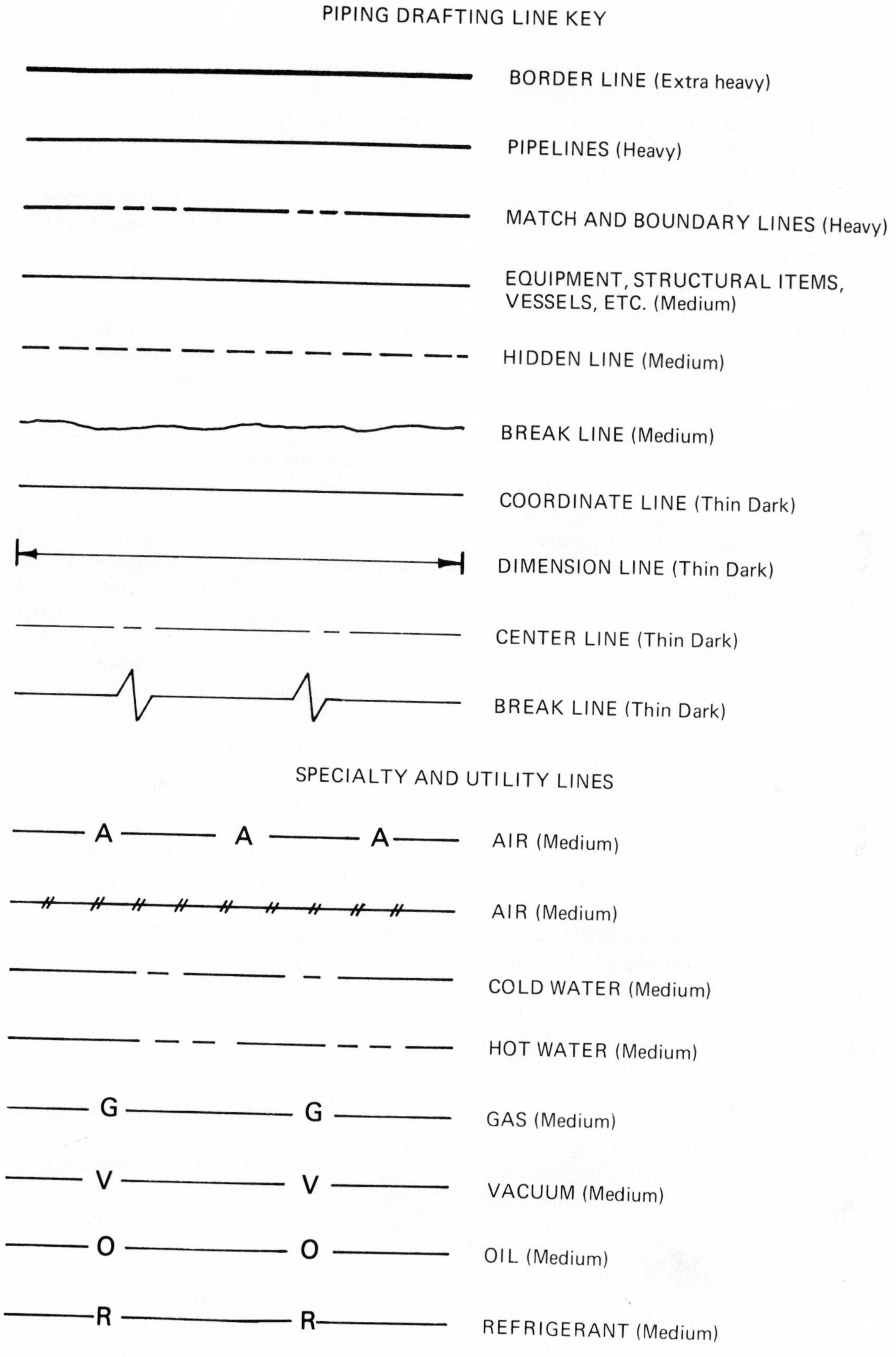

PIPING DRAFTING LINE KEY

BORDER LINE (Extra heavy)

PIPELINES (Heavy)

MATCH AND BOUNDARY LINES (Heavy)

EQUIPMENT, STRUCTURAL ITEMS, VESSELS, ETC. (Medium)

HIDDEN LINE (Medium)

BREAK LINE (Medium)

COORDINATE LINE (Thin Dark)

DIMENSION LINE (Thin Dark)

CENTER LINE (Thin Dark)

BREAK LINE (Thin Dark)

SPECIALTY AND UTILITY LINES

A A A AIR (Medium)

AIR (Medium)

COLD WATER (Medium)

HOT WATER (Medium)

G G GAS (Medium)

V V VACUUM (Medium)

O O OIL (Medium)

R R REFRIGERANT (Medium)

FIG. 5-7 Piping drafting line key with specialty and utility lines.

The thickest line, the *extra-heavy full line,* is used for borders and drainage pipe and, in some cases, for cutting plane lines. The *heavy line* or heavy full line is used for major pipelines mainly on single-line piping drawings. Variations of this heavy line are used for match and boundary lines. The *medium* and *medium light lines* are used for hidden lines, for vessels, for equipment, and for structural lines. Break lines and all fixture, equipment, specialty, and utility lines are drawn with the medium thickness line, including air, cold water, hot water, gas, vacuum, oil, and refrigerant. Different employers may have slightly different procedures than those set forth in this drafting line key, but a majority of companies follow these practices.

The *thin line* is used for coordinate lines, dimension lines, centerlines, and break lines. It is also used for building outlines and backgrounds and where a secondary or future portion of a system is represented lightly to show its relationship to the present structure. Some companies use light lines to represent existing equipment.

Centerlines are also drawn thin and dark. They are used to indicate centerlines of vessels, pipe, equipment, or other portions of the drawing and are used on *double line* and *pictorial drawings* to represent the centerline of the pipe.

Inking

Inking is not used very often in piping drafting work. In some cases, border lines, permanent match lines, boundary lines, and other coordinate lines may be put in with ink, so it is useful to be proficient with inking pens.

Lettering

Notations and comments on lettering are given in Fig. 5-8. The student learning to letter in an accepted form should be careful not to develop bad habits in the beginning. In all drawing it is important to keep lettering clear, neat, and legible—do not cram together the various letters and numbers. Only through long practice can lettering skills be developed. Though each person's lettering technique will differ, all must have in common certain qualities: readability, precision, neatness, clarity, and complete intelligibility along with excellent darkness and spacing so that the drawing can be blueprinted, xeroxed, or reduced in size without loss of clarity or legibility.

In general, throughout this book either slant or vertical lettering can be used. Since each company and industry varies in its established practice, the drafter should be able to switch from vertical to slant lettering and back when necessary. Most piping drawings use slanted lettering. In isometric drawings the printing must run according to the line on which it is drawn. When completing changes on existing drawings, the drafter must be sufficiently adaptable to match the size and lettering style that is already used throughout the drawing so that revisions are consistent with the original.

In regard to lettering size throughout this text, a maximum of 1/8 in. (0.32 cm) lettering is used except in titles and for special callouts and headings which can be 3/16 to 1/2 in. (0.48 to 1.27 cm) high. The 1/8 in. (0.32 cm) rule also applies to each character of a fraction with the use of a diagonal division line. As a rule, larger printing is sloppy. Lettering should be oriented consistently to be read from the bottom and right-hand side of the sheet (in some cases, the left-hand side of the sheet).

Besides the notations and comments provided on the formation of numbers and letters, Fig. 5-8 consists of a practice lettering sheet to be completed by the student. Set aside a short period of time each day for the practice of lettering. Anyone can learn to letter proficiently, but it takes a great deal of effort for real improvement to show. Whether to use slanted or vertical lines for characters is up to the drafter, though both styles should be mastered to some degree.

Lettering is one of the primary aspects of drafting. Inaccurate or sloppy lettering can be extremely costly if misunderstandings or errors are conveyed to the shop or construction site. Lettering and line work are the two credentials of the professional drafter. They can make or break the drafter and also the drawing. Only through repetition and serious effort can a person develop sufficient lettering skills.

Two main requirements for good lettering are accurate spacing and uniformity throughout the drawing. Always use guidelines when lettering on a drawing to help prevent irregular printing. In some cases, a lettering guide or aid (not a template) will greatly improve the printing, keeping it legible and uniform. These lettering aids are available at drafting supply shops. A student or drafter should not use a lettering template unless it is for a title block, special heading, or callout that is acceptable to the instructor or employer. (Many drawings in this text have been done with a template or Leroy set, but only to ensure clarity for reproduction.)

Uniformity of height is extremely important in keeping the lettering visually pleasing, as is the spacing between letters, which should appear equal in most cases. Not all characters are of equal thickness. Careful attention must be given to the actual width of each printed number or letter along with accurate spacing so that no uneven gaps are left between characters.

Number or Letter		Notations and Comments	Wrong	Possible Mistakes
A	*A*	Make upper part larger than bottom.	*A*	
B	*B*	Lower part slightly larger than upper part.	*B*	*8*
C	*C*	Full open area, elliptical letter body.	*C*	*O*
D	*D*	Horizontal bars and straight line back.	*D*	*O*
E	*E*	Short line bar slightly above center.	*E E*	*L*
F	*F*	Short bar slightly above centerline.	*F F*	*T E*
G	*G*	Based on true ellipse, short horizontal line above centerline.	*G G G*	*C O 6*
H	*H*	Bar slightly above centerline.	*H H*	
I	*/*	No serifs, except when next to Number 1.	*I 1*	
J	*J*	Wide full hook with no serifs.	*J J*	
K	*K*	Extend lower branch from upper branch.	*K K*	*R*
L	*L*	Make both lines straight.	*L*	
M	*M*	Not as wide as W: center part extends to bottom of letter.	*M M*	
N	*N*	Do not cram lines together.	*N N*	*V U*
O	*O*	Full true ellipse.	*O O*	*Q*
P	*P*	Middle bar intersects at letter's middle.	*P P*	*K T D*
Q	*Q*	Based on true wide ellipse.	*Q*	*O*
R	*R*	Make upper portion large.	*R R*	*K*
S	*S*	Based on number 8; keep ends open.	*S*	*8*

FIG. 5-8 Lettering practice sheets.

Number or Letter		Notations and Comments	Wrong	Possible Mistakes
T	*T*	Draw full width of letter E.	*T T*	*7*
U	*U*	Lower portion elliptical, vertical bars parallel.	*U*	*V*
V	*V*	Bring bottom to points.	*U*	*U*
W	*W*	Widest letter; center extends to top of letter.	*W W*	*N*
X	*X*	Cross lines above centerlines.	*X*	
Y	*Y*	Upper part meets below center.	*Y Y*	*V T*
Z	*Z*	Horizontal lines parallel.	*Z*	*2*
0	*0*	Same as letter O.	*0*	*Q*
1	*/*	Same as letter I.	*1 1*	*7*
2	*2*	Based on number 8; open hook.	*2 2*	*Z*
3	*3*	Based on number 8; upper part smaller than lower.	*3 3*	*8 5*
4	*4*	Horizontal bar below center of figure.	*4 4 4*	*7 9 H*
5	*5*	Based on ellipse; keep wide.	*5 5*	*6 3 S*
6	*6*	Based on ellipse; open.	*6*	*8*
7	*7*	Keep as wide as letter E.	*7 7 7*	*/*
8	*8*	Based on ellipse; keep wide.	*8 8 8*	*B*
9	*9*	Composed of two ellipses; keep full.	*9*	*8*

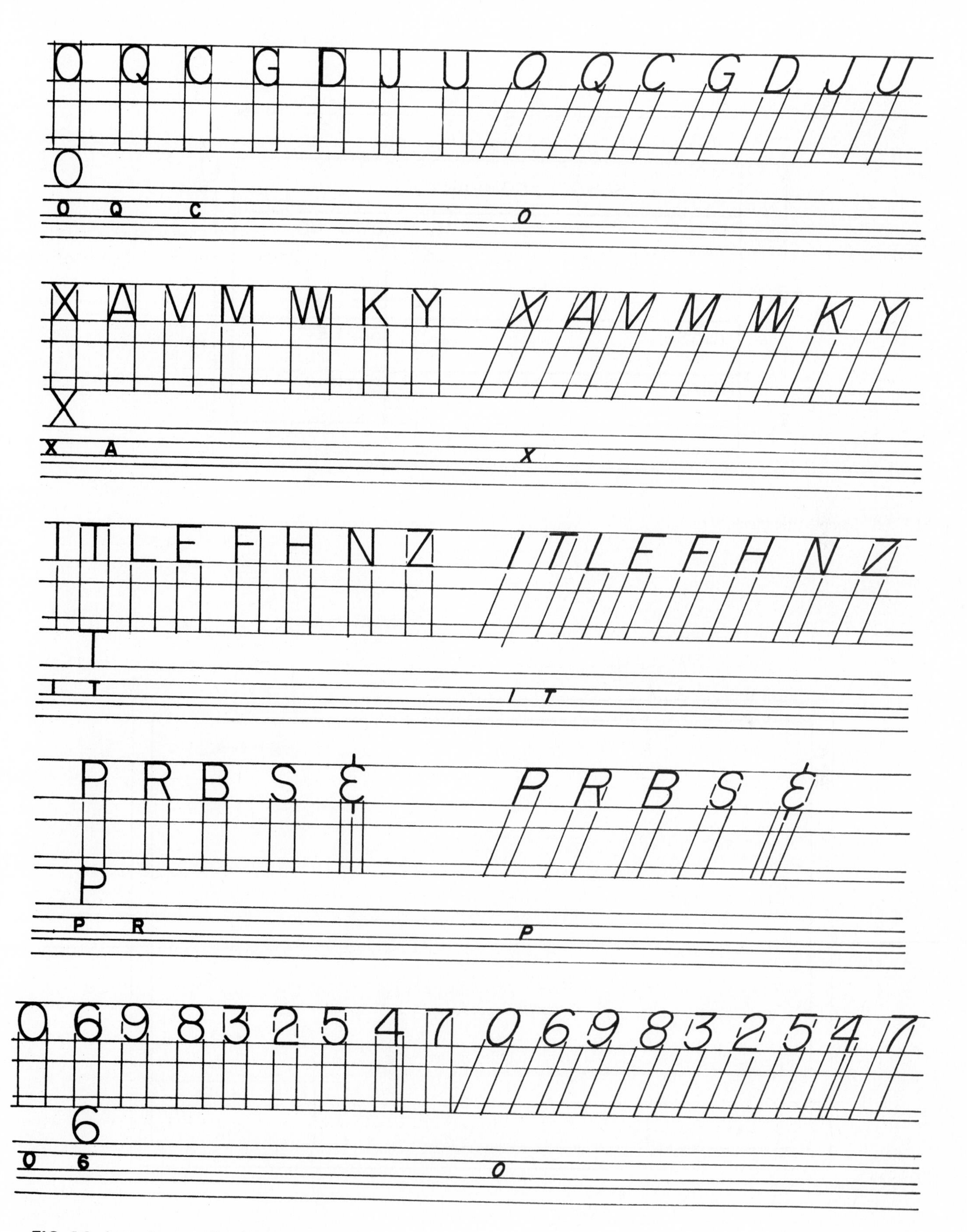

FIG. 5-8 Lettering practice sheets.

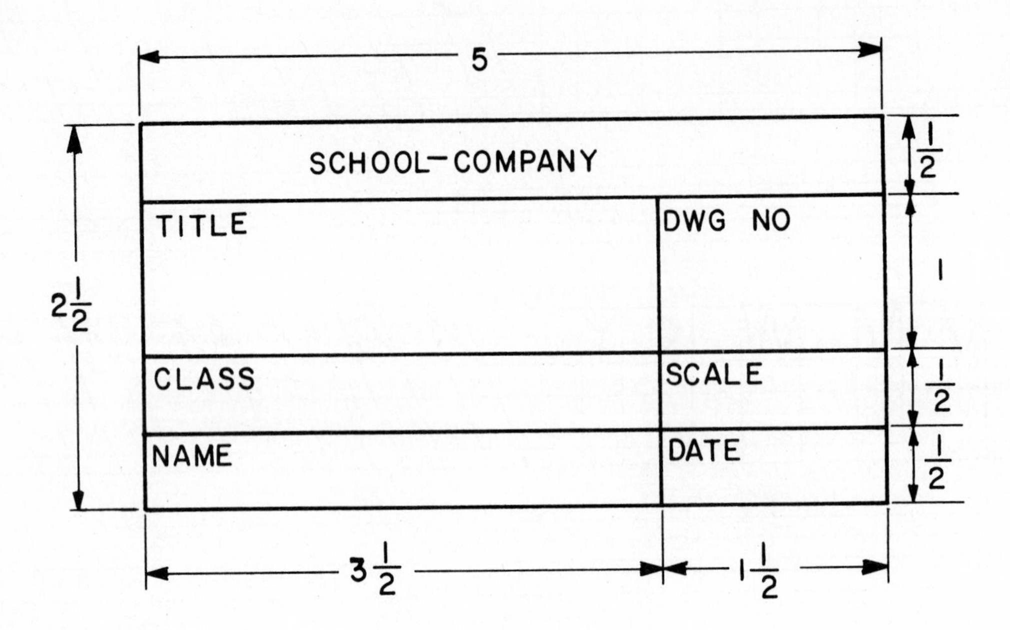

FIG. 5-9 Sample title block.

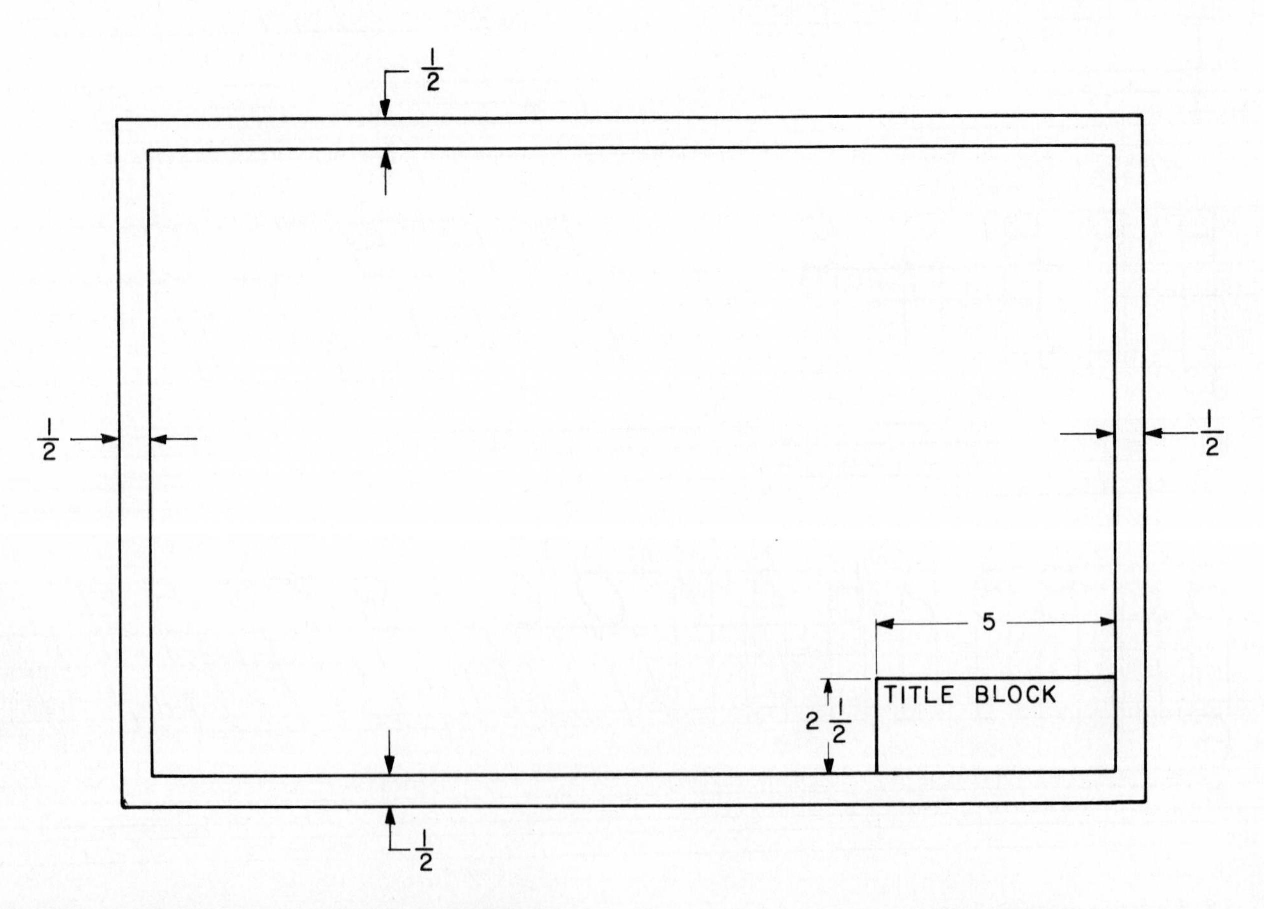

FIG. 5-10 Borderline setup and title block placement.

Title Blocks

Figure 5-9 is an example of a title block that can be used throughout this book unless otherwise specified by the instructor. Title blocks are drawn or stamped in the lower right-hand corner of each drawing. Some types of paper come with preprinted title blocks. The title block suggested here includes a 2½ x 5 in. (6.35 x 12.7 cm) area divided to provide room for the school or company name, title of the drawing, class number, drafter's name, drawing number, scale, and the date of the drawing. Along with these details, room for the date of checking, the checker or teacher's grade, sheet number, drawing description block, and other information may be provided. The area directly above the title block on all drawings should be kept empty for the use of notes and stock lists. All information within the title block should be printed freehand except for major titles. Besides stock lists, revisions, and notes, the area above the title block should be used for special callouts.

Lettering in the title block should remain 1/8 in. (0.32 cm) high, except for major headings and titles which should be 3/16 to 1/4 in. (0.48 to 0.64 cm) in height. Figure 5-10 shows how to set up the border line for each drawing no matter what size the finished project is to be. Standard size sheets of paper are available at any drafting supply store—including large rolls 15 to 50 ft (4.57 to 15.24 m) long by 3 ft (0.91 m) in width. A border line and title block should be provided for all drawings in a project, no matter what their finished size.

PIPING SYMBOLS

The piping symbols drawn in Fig. 5-11 represent various flanges, fittings, and other components of piping systems. The components are shown in single-line flanged, screwed, bell and spigot, welded, and soldered variations; in some cases, one variation of double-line and pictorial style are shown. This figure, along with the symbols provided in the isometric, pipe support, and flow diagram chapters, enables the student to learn all the standard variations of drawing piping symbols. Figure 5-11 corresponds to the recommended configuration of piping symbols from the American Standards Association and ANSI Z32.23 along with variations and extra symbols that are prevalent throughout the piping industry. Many companies still have their own specialized symbols, but Fig. 5-11 provides a sufficient variety to acquaint the student with what may be encountered in the field. Only slight modification of symbols will be found throughout the piping industry. The student should be able to adapt his or her knowledge to the accepted practice within all areas of the profession.

Because of the repetition of so many elements, pipe drafting, like electrical and electronic drafting, uses symbols. Thus drawings can be completed accurately and consistently without undue pictorial line work which proves costly and time consuming. Small drawings of piping systems or pipelines should be sketched first and the symbols located on the pipeline. This step will help determine the sheet size and area requirements for the drawing in question. This procedure is especially useful when setting up spool drawings either in orthographic or isometric projection but cannot be used when laying out a complete plan, elevation, and section drawing. Orthographic and isometric piping symbols are to be drawn using medium to heavy line weight consistently; symbols should be well formed and accurately placed on the drawing.

Symbols must be made clearly and concisely so they are not confused with similar symbols. As can be seen in Fig. 5-11, many of the symbols have only slight differences between them, making it extremely important for the drafter to know, and be able to draw, each and every symbol accurately. Sometimes no symbol will exist for a particular component or special piece of equipment. In this case, the drafter may use a simplified pictorial representation or a new symbol outlined or drawn to resemble the component. When creating new symbols the drafter must call out or flag the item to call attention to its appearance and function.

Symbols must be laid out on the centerlines which stand for all piping runs and lines. Centerlines are the first lines in the drawing. Symbols should be put in only after centerlines of pipe and equipment and coordinate lines have been laid out on the drawing. Symbols are located from ends of pipe, coordinate lines, or equipment by means of construction lines that cross the pipe's centerline. End to end, face to face of a fitting, valve, or piece of equipment are not always drawn to scale but they are dimensioned accurately, as shown in Fig. 5-12. It is necessary to refer to the manufacturer's or standard dimensions for these items when completing a piping drawing (see appendix).

Understanding symbols and their uses is essential to the training of a piping drafter. Each symbol represents a definite object. There are three basic types of symbols used in orthographic piping drawings: single-line symbols, double-line symbols, and pictorial symbols.

Single-line symbols are the easiest, simplest, and quickest to render. Used in a majority of cases for all piping drawings, they are the most common symbols found in the piping industry. All symbols shown in Fig. 5-11 have their isometric equivalents (see Chapter 7). An attempt has been made to give a fairly comprehensive list of piping symbols in Fig. 5-11, though there are many variations of piping equipment and components available. The single-line method of drawing symbols is also used in Chapter 6, which covers flow diagrams.

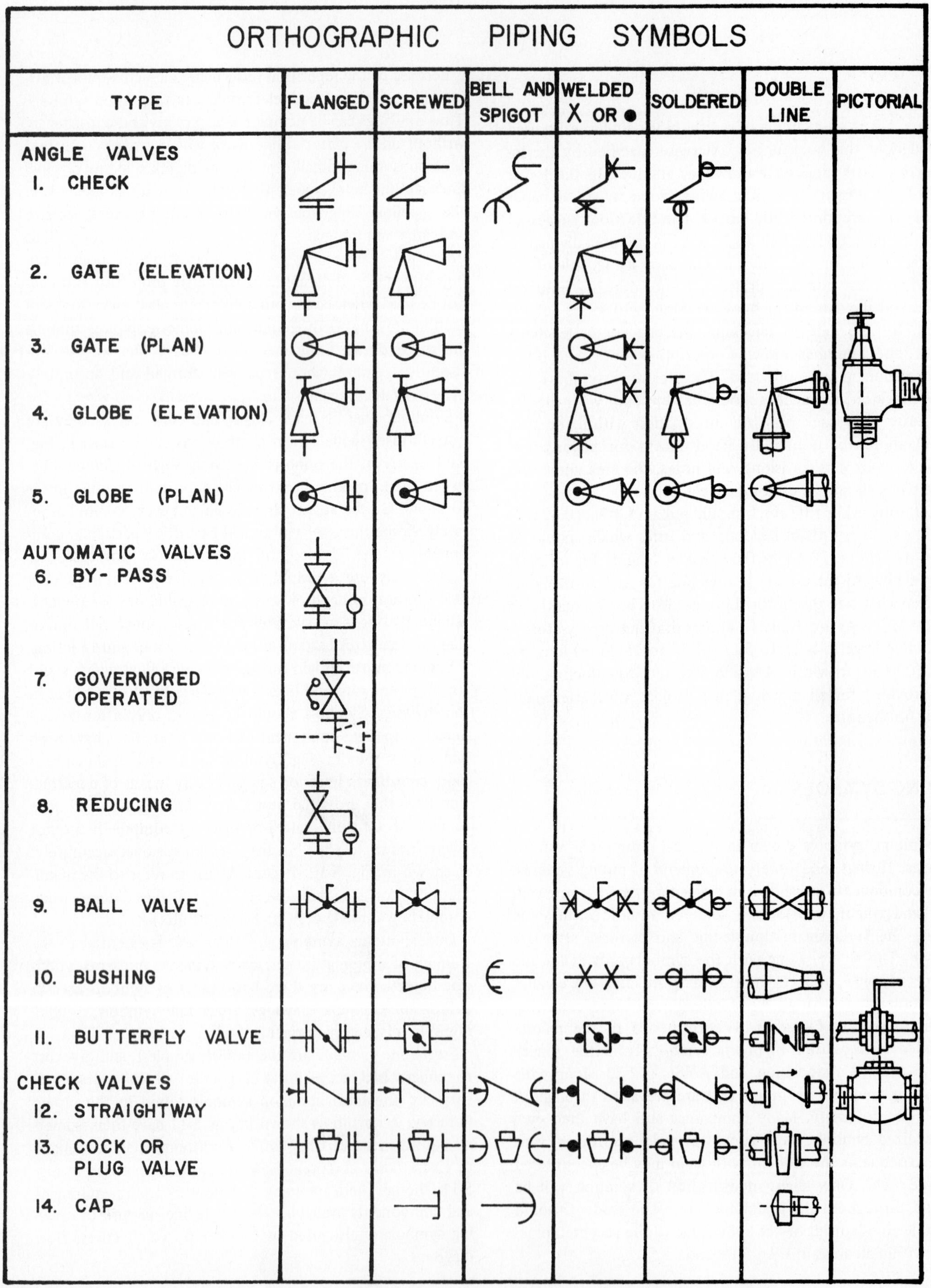

ORTHOGRAPHIC PIPING SYMBOLS							
TYPE	FLANGED	SCREWED	BELL AND SPIGOT	WELDED X OR ●	SOLDERED	DOUBLE LINE	PICTORIAL
ANGLE VALVES							
1. CHECK							
2. GATE (ELEVATION)							
3. GATE (PLAN)							
4. GLOBE (ELEVATION)							
5. GLOBE (PLAN)							
AUTOMATIC VALVES							
6. BY-PASS							
7. GOVERNORED OPERATED							
8. REDUCING							
9. BALL VALVE							
10. BUSHING							
11. BUTTERFLY VALVE							
CHECK VALVES							
12. STRAIGHTWAY							
13. COCK OR PLUG VALVE							
14. CAP							

FIG. 5-11 Orthographic drafting symbols.

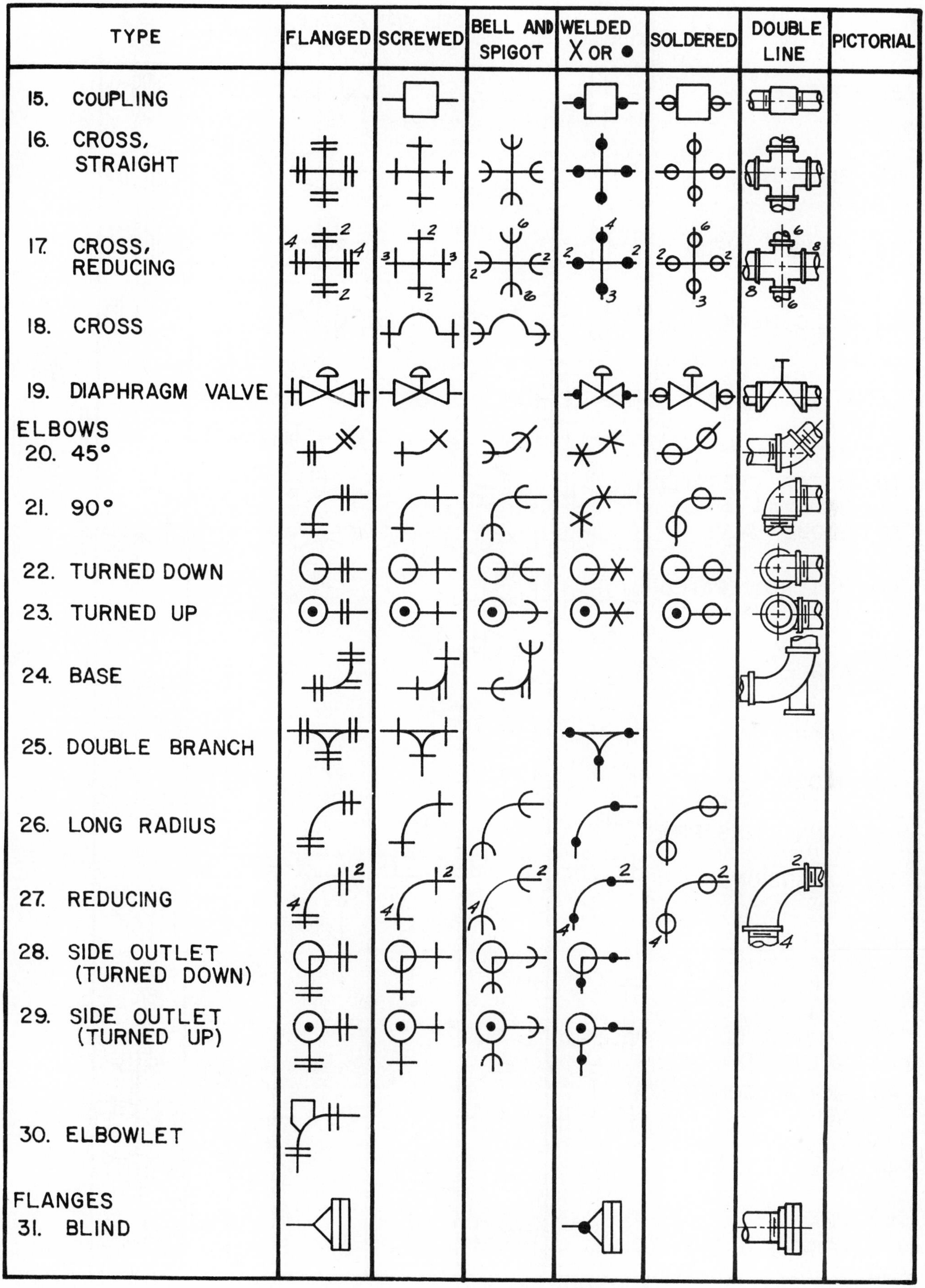

TYPE	FLANGED	SCREWED	BELL AND SPIGOT	WELDED X OR ●	SOLDERED	DOUBLE LINE	PICTORIAL
15. COUPLING							
16. CROSS, STRAIGHT							
17. CROSS, REDUCING							
18. CROSS							
19. DIAPHRAGM VALVE							
ELBOWS							
20. 45°							
21. 90°							
22. TURNED DOWN							
23. TURNED UP							
24. BASE							
25. DOUBLE BRANCH							
26. LONG RADIUS							
27. REDUCING							
28. SIDE OUTLET (TURNED DOWN)							
29. SIDE OUTLET (TURNED UP)							
30. ELBOWLET							
FLANGES							
31. BLIND							

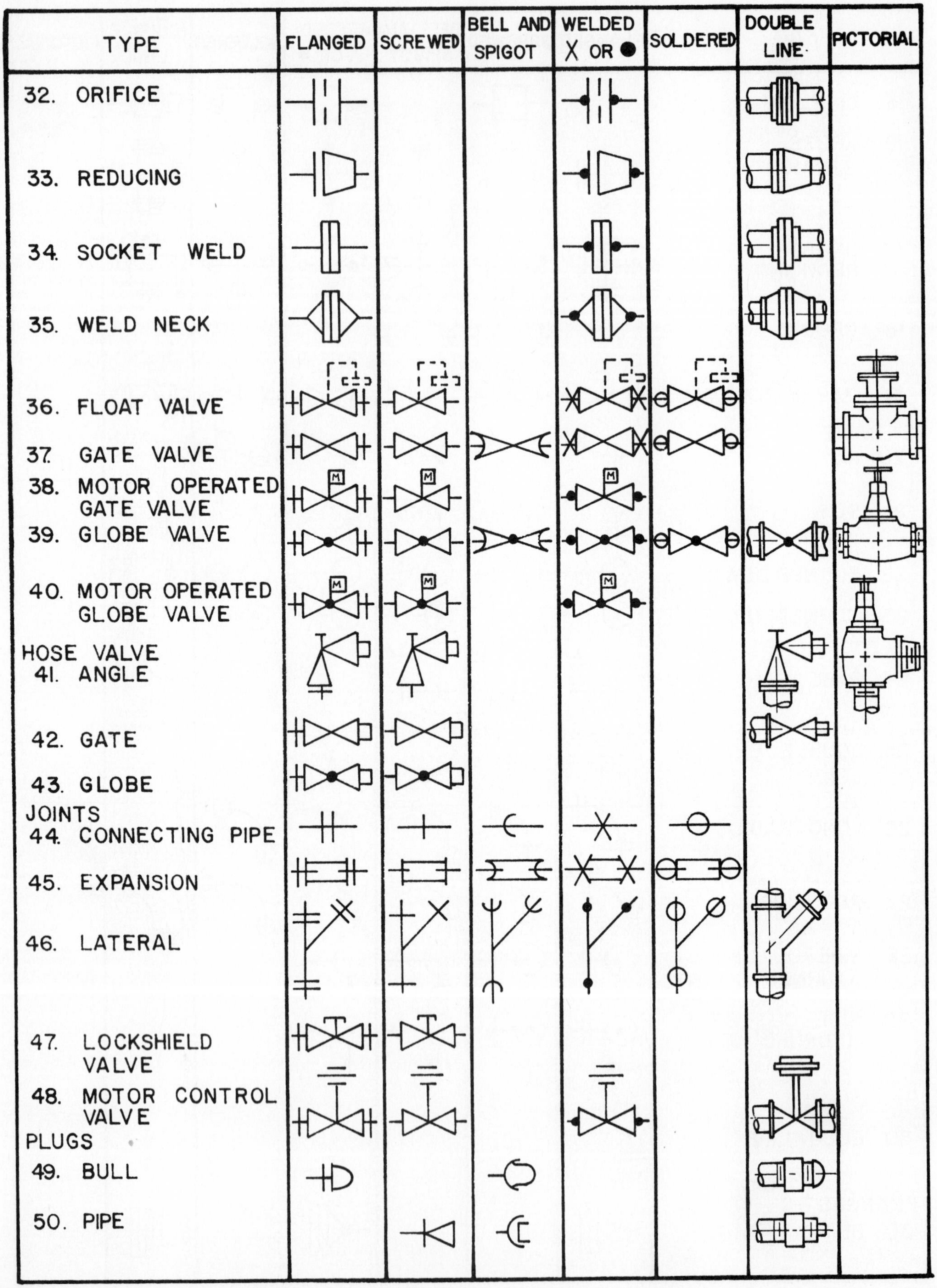

TYPE	FLANGED	SCREWED	BELL AND SPIGOT	WELDED X OR ●	SOLDERED	DOUBLE LINE	PICTORIAL
32. ORIFICE							
33. REDUCING							
34. SOCKET WELD							
35. WELD NECK							
36. FLOAT VALVE							
37. GATE VALVE							
38. MOTOR OPERATED GATE VALVE	M	M		M			
39. GLOBE VALVE							
40. MOTOR OPERATED GLOBE VALVE	M	M		M			
HOSE VALVE							
41. ANGLE							
42. GATE							
43. GLOBE							
JOINTS							
44. CONNECTING PIPE							
45. EXPANSION							
46. LATERAL							
47. LOCKSHIELD VALVE							
48. MOTOR CONTROL VALVE							
PLUGS							
49. BULL							
50. PIPE							

TYPE	FLANGED	SCREWED	BELL AND SPIGOT	WELDED X OR ●	SOLDERED	DOUBLE LINE	PICTORIAL
51. QUICK OPENING							
REDUCERS							
52. CONCENTRIC							
53. ECCENTRIC							
54. SOLENOID VALVE							
55. RELIEF VALVE							
56. SAFETY VALVE							
57. SLEEVE							
58. STRAINER							
TEES							
59. STRAIGHT SIZE							
60. OUTLET UP							
61. OUTLET DOWN							
62. DOUBLE SWEEP							
63. REDUCING							
64. SINGLE SWEEP							
65. SIDE OUTLET DOWN							
66. SIDE OUTLET UP							
67. UNION							
68. Y-VALVE							

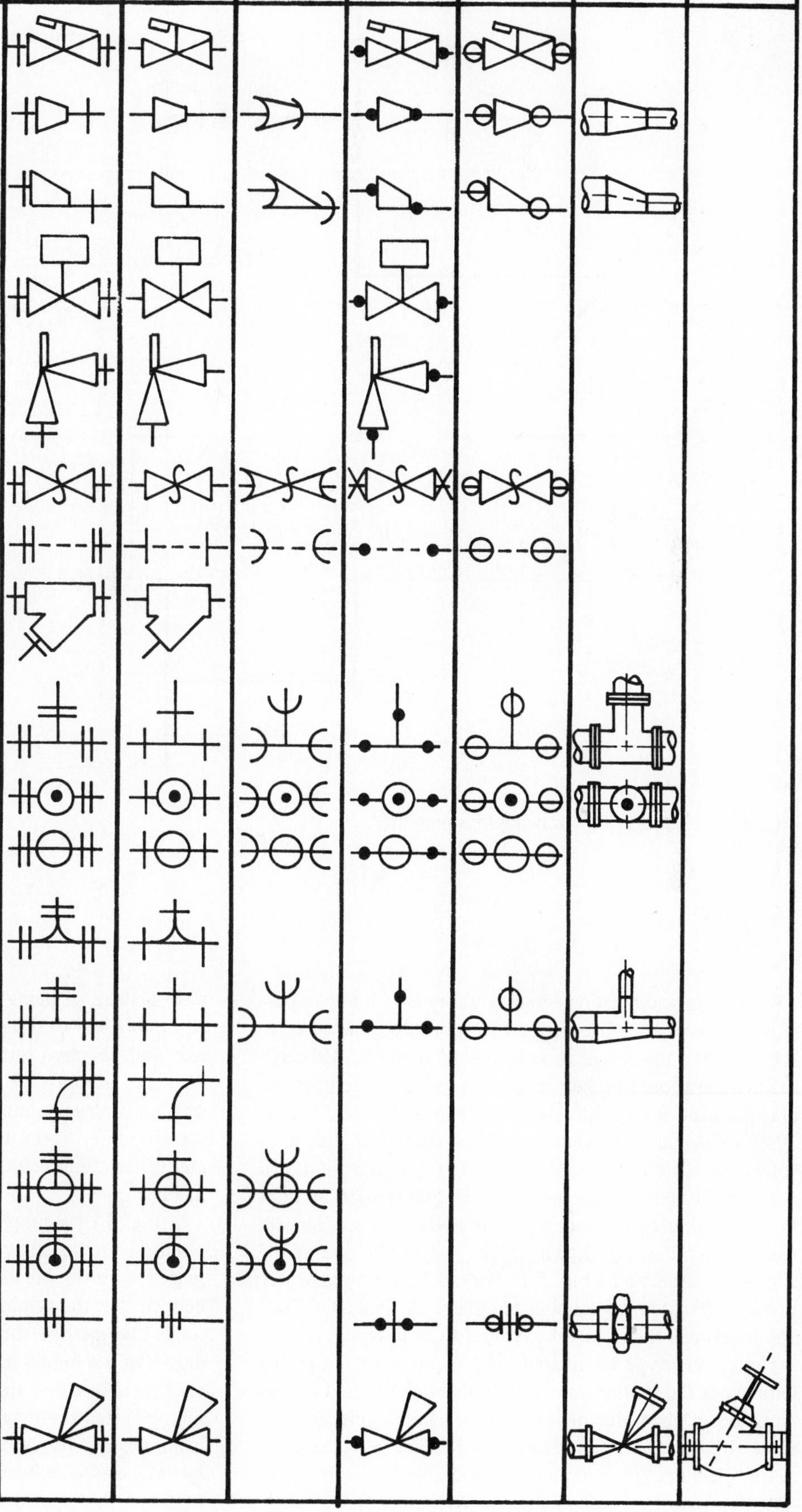

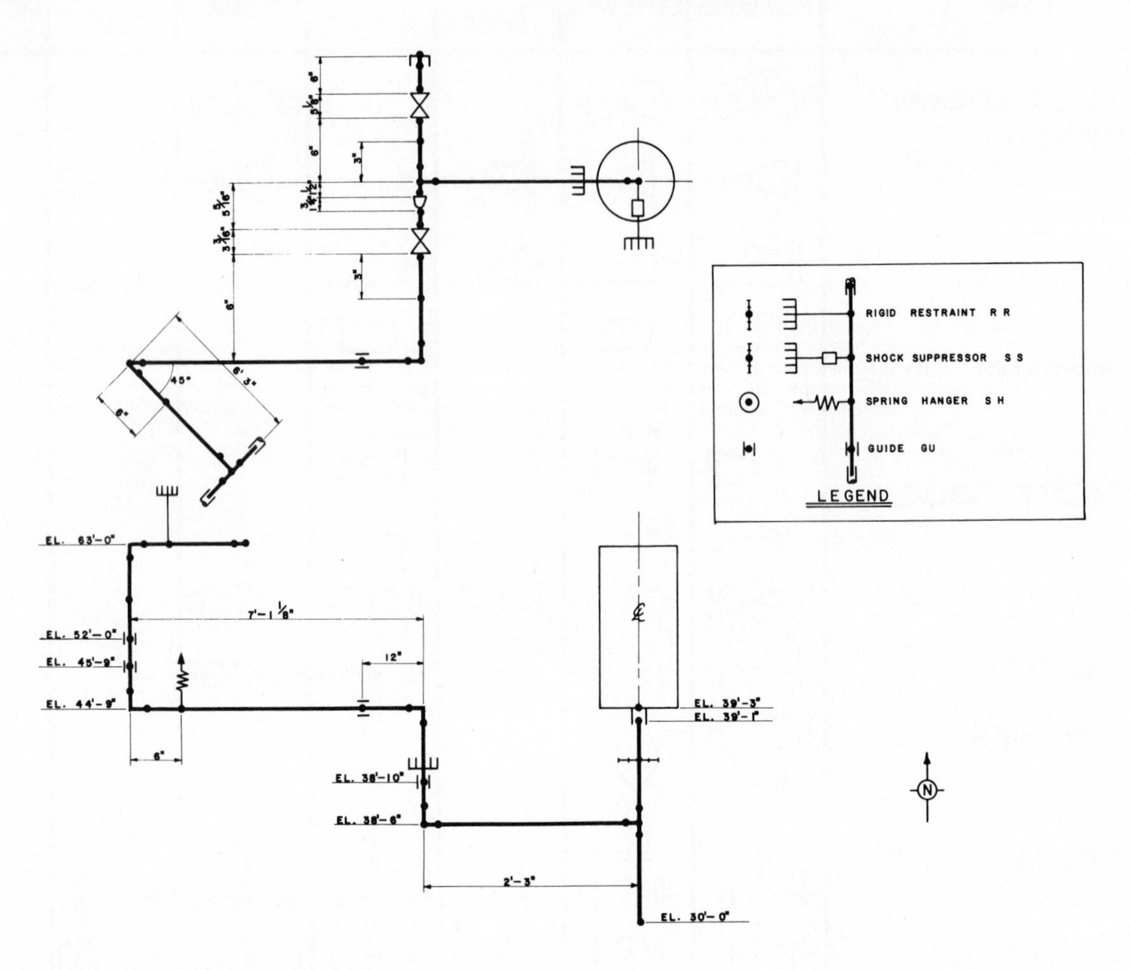

FIG. 5-12 Orthographic drawing of a pipeline.

When constructing single-line symbols it is more important to show the relative size and relationship between components than the actual to-scale size of symbols. Although the symbol should be kept roughly to scale, sometimes the drawing scale is so small—such as 1/8 in. = 1 ft (0.32 cm = 30.48 cm)—that the symbol would disappear if drawn to scale. Therefore throughout the industry symbols are drawn considerably larger than their real-life counterparts and not in relationship to the scale of the drawing. For problems in this chapter that are drawn 3/8 in. = 1 ft, the size of the symbols presented in Fig. 5-11 should be followed where possible. Since piping symbol templates are available, these will determine the size of the symbol to be drawn.

The second type of symbol used throughout the piping industry is the *double-line.* The double-line method is used more often than the pictorial type. Double-line drawing shows interferences and clearances between pipes and between pipelines and equipment because the line is drawn to scale instead of being represented by a single centerline. On all double-line drawings the pipeline must include a centerline and the two outer lines which represent the outside diameter of the pipe. In this form of representing orthographic piping systems, the symbols for valves are still in the simplified form, not the pictorial type, though fittings and other components are usually drawn in a semipictorial manner.

In Fig. 5-13 a simple piping orthographic drawing is represented in single-line symbols. This same system is represented with double-line symbols in Fig. 5-14, where it can be seen that the symbols for the valves remain similar to the single-line method though the elbows, tees, and unions are drawn in a semipictorial method. Double-line drawings take considerably more time than single-line representations of piping systems and symbols. *Pipelines 12 in. (30.48 cm) in diameter and larger are always represented in the double-line manner to show adequate clearances and pipe size.*

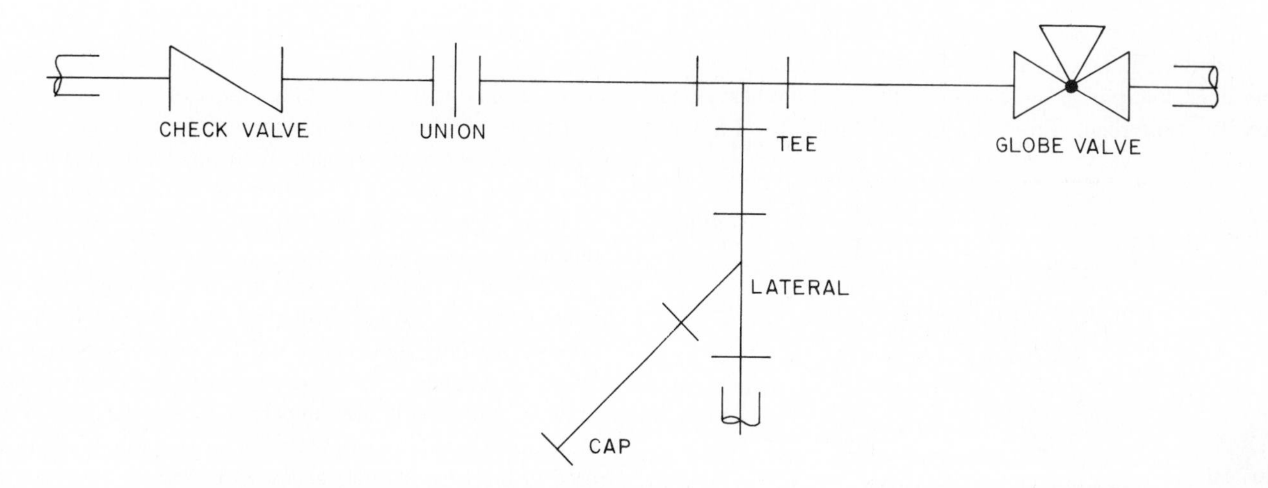

FIG. 5-13 Single-line piping drawing using screwed connections.

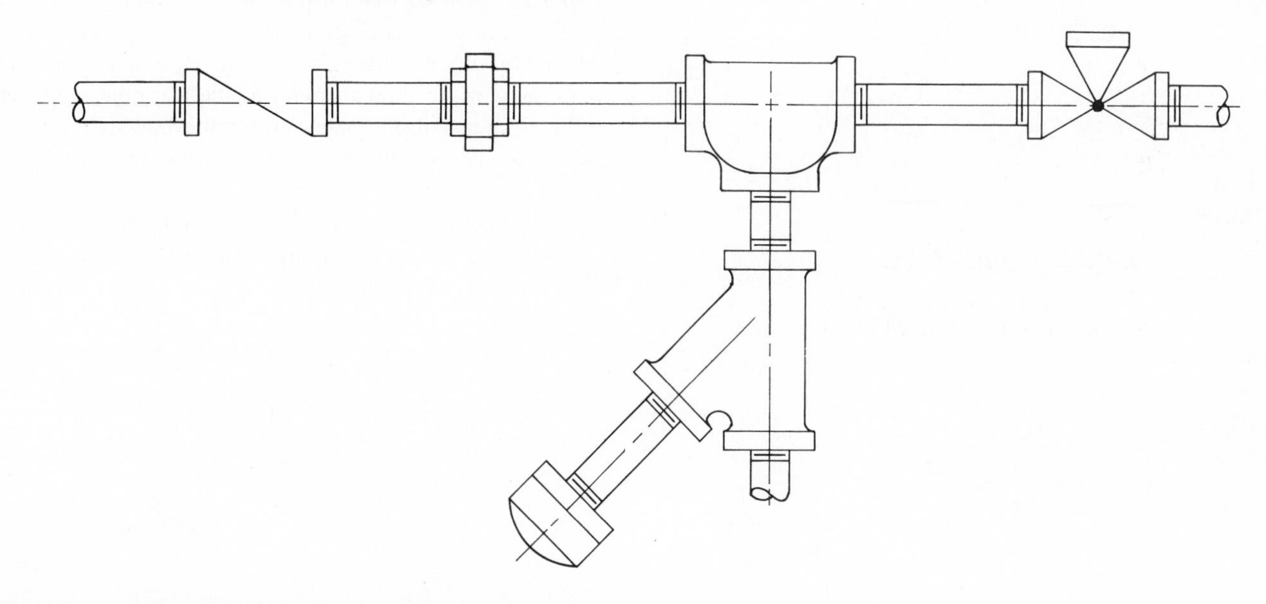

FIG. 5-14 Double-line piping drawing of Fig. 5-13.

The double-line symbol is used predominantly on large-scale drawings that need to show the exact relationship between lines and equipment. *The outside diameter of the pipe is always drawn to scale.* The only difference in valve symbols in double-line drawings is shown in Fig. 5-14, where the valve ends are drawn differently. In many cases, the face-to-face and end-to-end dimensions are drawn to scale according to standard sizes provided in the manufacturer's catalog for the particular component or valve (see appendix). Most drawings used throughout the petrochemical and power generation fields have flanged and welded fittings and valves.

Threaded and soldered pipelines are almost always drawn single line with no regard to face-to-face and end-to-end dimensions for fittings, valves, and the like. On threaded and soldered piping systems, the field or shop fabricator determines most of the dimensions unless they are extremely important. For drafting double-line drawings to scale or finding product dimensions, the appendix of this book includes reprints for a variety of fittings and valves for flanged, soldered, welded, and screwed connections. These dimensions are standardized in a majority of cases.

Double-line symbols are used for drawings as small as 1/8 in. = 1 ft (0.32 cm = 30.48 cm), though it is good practice to keep the scale of the drawing larger when using the double-line system to avoid confusing the lines. Chapters 3 and 4, which cover fittings, flanges, and valves, provide the essential information necessary for drawing and understanding double-line components for piping systems along with the theory and background necessary for all types of piping components. (See also Chapter 9 on double-line spool drawings.)

Figure 5-15 shows the single-line and double-line pipe representation. For 1½ in. (3.81 cm) diameter and smaller the single-line pipe is represented with a medium heavy line for its centerline. From 10 in. (25.4 cm) to 1-5/8 in. (4.12 cm) nominal size, the pipeline is usually represented with the single-line method, drawing the outside diameter to scale only at the ends of the pipe. For 12 in. (30.48 cm) and larger, double-line pipe representation is used, drawing the outside diameter to scale. With the double-line method it is always important to include the centerline of pipe for accurate location, dimensions, and placement of pipelines on the drawing.

The third symbol variation is the *pictorial* type. This type is used only in rare cases when no symbol exists for a component or when the drawing will be used for books, magazines, or publications where an accurate "photographic" drawing will enable the layperson to see the system and understand what it represents. Pictorial symbols are rarely used in industry because of the expense and time they entail. This symbol is used where clarity and beauty are more important than cost and speed. A few examples of pictorial symbols are given in Fig. 5-11. Special valves and components that do not have assigned symbols must be drawn pictorially or be shown similar to the real object. In these cases they must also be called out on the drawing to identify the component for those who will later use the drawing.

The drafter must be able to understand, draw, and complete single-line, double-line, and, when necessary, pictorial symbols that convey the necessary information for the construction of the piping system. Figure 5-16 is a pictorial

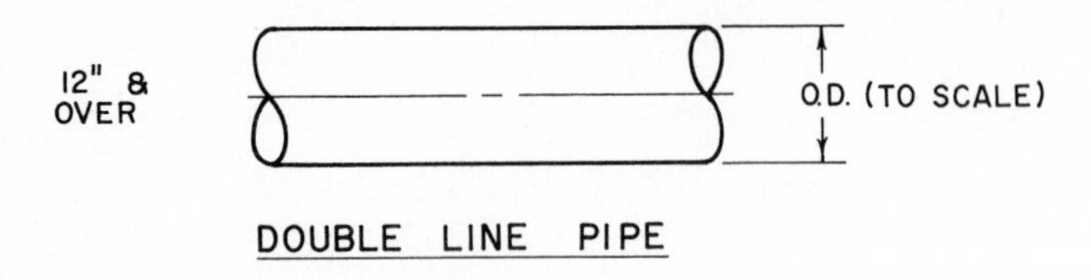

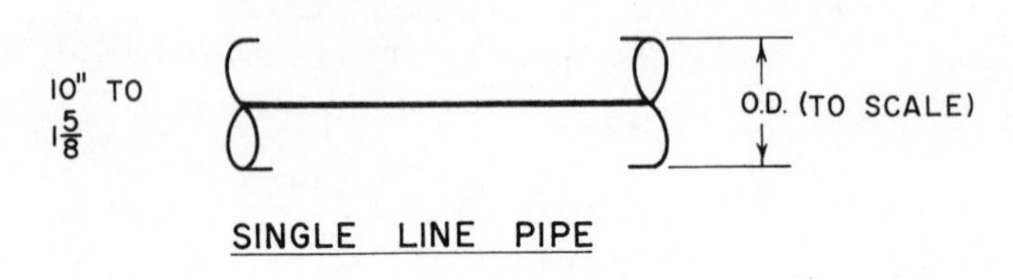

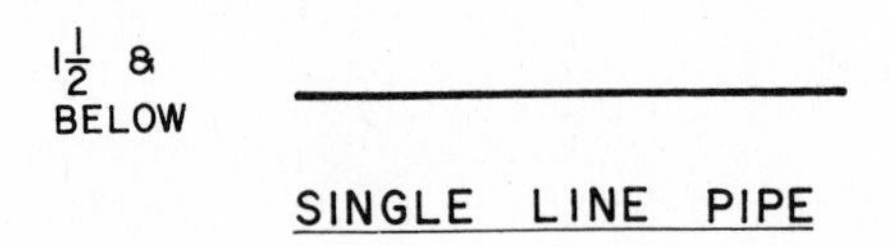

FIG. 5-15 Pipeline identification of different pipe sizes.

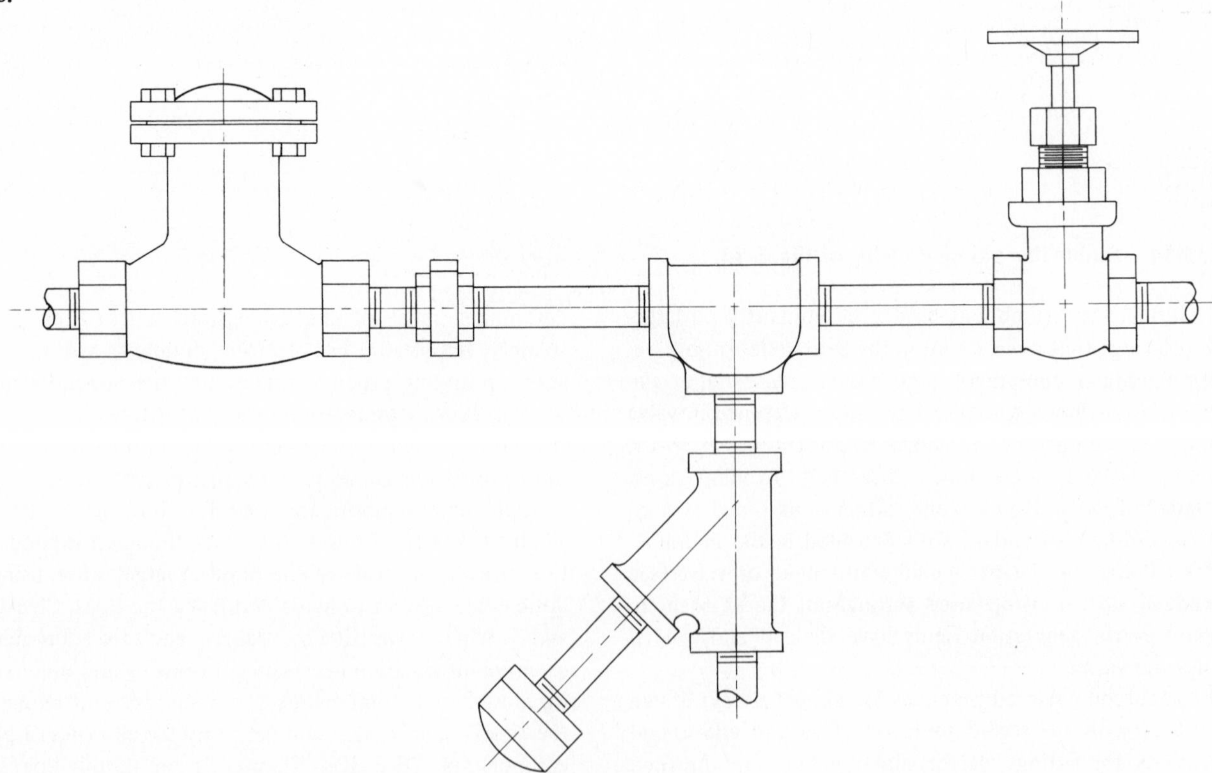

FIG. 5-16 Pictorial drawing of pipeline and components of Fig. 5-13.

drawing of the system represented in single-line in Fig. 5-13 and in double-line in Fig. 5-14. These three figures should give sufficient information for the student to see the difference between the three variations of piping symbols.

In Fig. 5-11 certain symbols are shown in both elevation and plan views and are referred to throughout as being in elevation, turned down, or turned up—in other words, the pipe is coming toward the reader or going away. Elevation means the way a fitting, valve, or symbol looks when viewed in the vertical position from its side or front.

DRAWINGS AND SCALES

Unlike mechanical drawing, piping drafting is rarely done in full scale. Piping drawings are projected as in architectural and structural drafting: A reduced scale is used throughout. Although 3/8 in. = 1 ft (0.96 cm = 30.48 cm) is the most commonly found scale in the piping field, scales as small as 1/8 in. = 1 ft (0.32 cm = 30.48 cm) and 1/4 in. = 1 ft (0.64 cm = 30.48 cm) are used when the project is too large to accommodate a larger scale. Each project should use a given scale consistently throughout all drawings where possible. A scale shows the relative size and configuration of the project, including equipment location, piping, and components. In many cases, as in isometric spool drawings and orthographically projected spool drawings, an actual scale is not adhered to and the face-to-face dimensions and lengths of pipe are either foreshortened or lengthened in order to accommodate dimensions which appear on the drawing.

A graphic scale is often represented in the corner of the drawing—such as 3/8 in. = 1 ft (0.96 cm = 30.48 cm). This scale is actually drawn on the paper, making it possible to scale the drawing after it has been reduced, as in Fig. 12-6. The drafting scale is always noted in the title block of the drawing.

In many cases modern methods of piping drafting and design have eliminated the necessity for extremely accurate drawings. Freehand sketches of isometric and orthographic spool drawings can eliminate the need for the drafter to create a picture-perfect drawing, thereby saving cost and time. At present the use of drawings is the least expensive and most logical way of communicating a designer's and engineer's set of calculations and instructions to the fabricator and manufacturer. Drawings provide an economical way for distributing ideas and designs. Pencil and paper are the most commonly used medium, though in some cases linen or mylar along with ink is used. Certainly pencil and paper are the least expensive medium.

The working drawings of a drafter in the piping field include a full set of plan, elevation, and section views, detail drawings of pipe supports, equipment details for towers and vessels, structural drawings, civil drawings, floor plans, equipment placement drawings, flow diagrams, and isometric spool sheets. Chapter 6 covers engineering flow diagrams and Chapter 7 discusses isometric projection. The present chapter is mainly concerned with the use of orthographic projection for piping projects.

Orthographic Projection

Orthographic projection for piping is exactly the same as that used for mechanical drafting. The student using this book should thus have sufficient training to grasp the procedures of orthographic projection for piping systems. Orthographic projection consists of drawing top, front, and side or section views. In piping these are usually referred to as plan, elevation, and section views. When a model is constructed from drawings, it is sometimes only necessary to draw the plan view and one or two elevation views instead of a full set of drawings or to work from the flow diagram to design the system directly on the model.

Orthographic projection is used in the representation of large and complicated piping systems and smaller simplified piping spools. Thus the drafter can show relationships between structural, electrical, heating, air conditioning, and mechanical equipment. Details can be drawn by using orthographic projection to represent parts similar to those found in mechanical drafting as these lend themselves well to dimensioning in this type of projection. Moreover, orthographic projection is used to represent piping systems for buildings, enabling the drafter to show the relationship of the building in its architectural aspects to the piping system. Almost all large orthographic drawings in the piping area use the single-line for pipe representation and symbol drawing. Double-line symbols and pipes are used for pipelines 12 in. (30.48 cm) and larger in diameter.

Orthographic projection creates problems for the piping drafter, though. Complicated areas of plans cannot be shown in sufficient detail to eliminate all inconsistencies, interferences, and clearance problems. In most cases, orthographic projections need to be accompanied by isometric drawings to show exact relationships between equipment and piping. Many section and auxiliary views are necessary for complicated piping systems in order to show hidden features of the project.

Orthographic projections, though still in common use throughout the piping industry, have become less important in recent years because of the introduction of models and isometrics. Orthographically projected piping systems, because of their complicated views and the need to detail and cut accurate sections in large quantities, are time consuming and expensive. Among the drawbacks of orthographic views of piping projects are these:

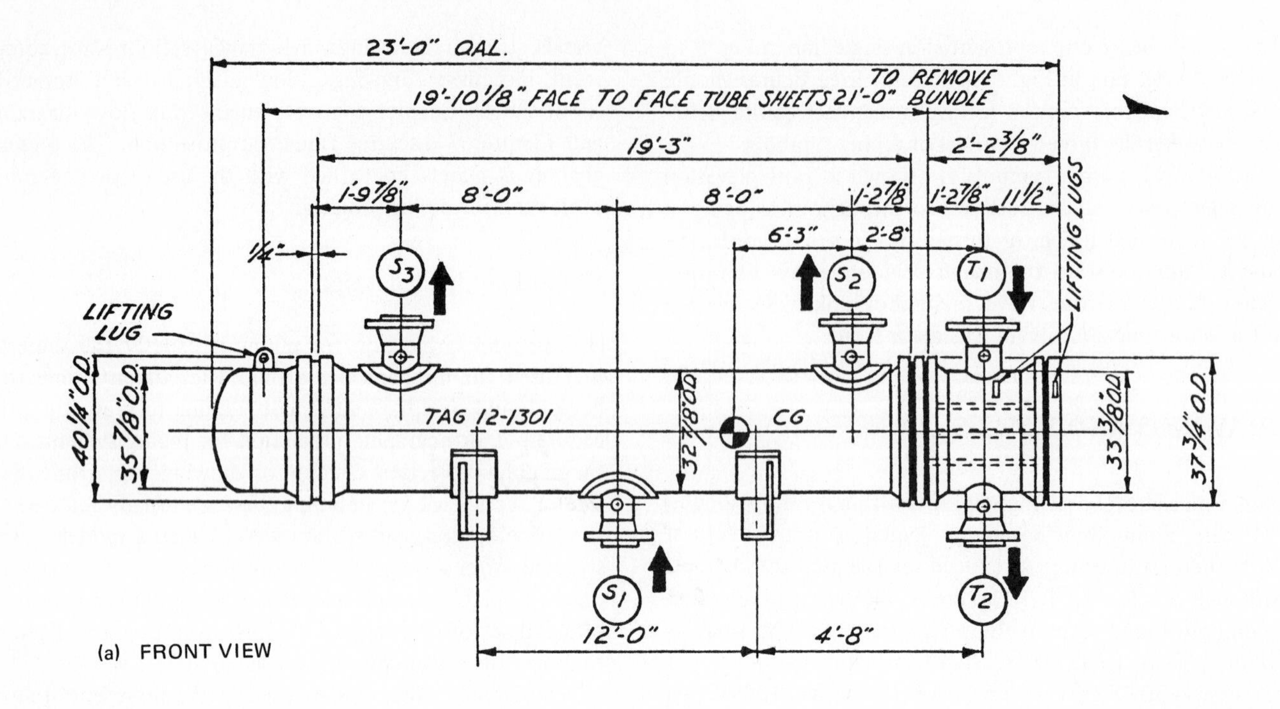

(a) FRONT VIEW

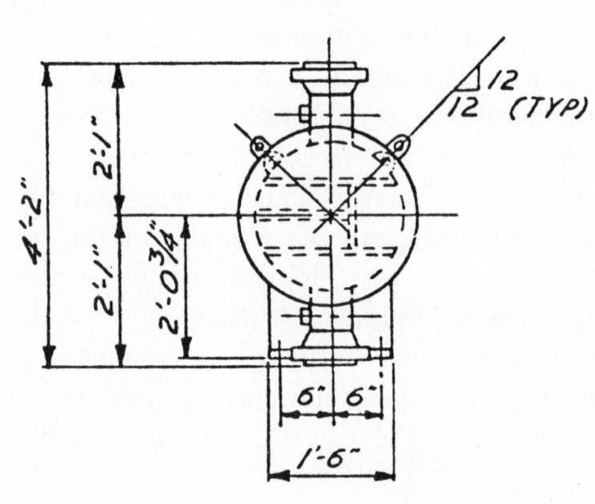

(b) RIGHT SIDE VIEW

FIG. 5-17 Stabilizer reboiler.

1. Large complicated systems become cluttered and hard to read.
2. Intricate piping systems require numerous sections and auxiliary views, increasing the cost and time required for the design and production of the project.
3. In many cases, orthographic drawings need to be supplemented by models and isometric spool sheets.
4. Pipes, fittings, and components that lie behind portions of other lines, etc., disappear because only the equipment that is viewed first in that view can be represented on paper. Therefore clearances and component positions cannot always be readily ascertained by the fabricator or construction crew, including the checker of the drawing.

Figure 5-17 is a two-view, orthographically projected vessel fabrication drawing. Figure 5-18 shows three single view details of level gauges and level controllers using the single-line method for pipe and fitting representation.

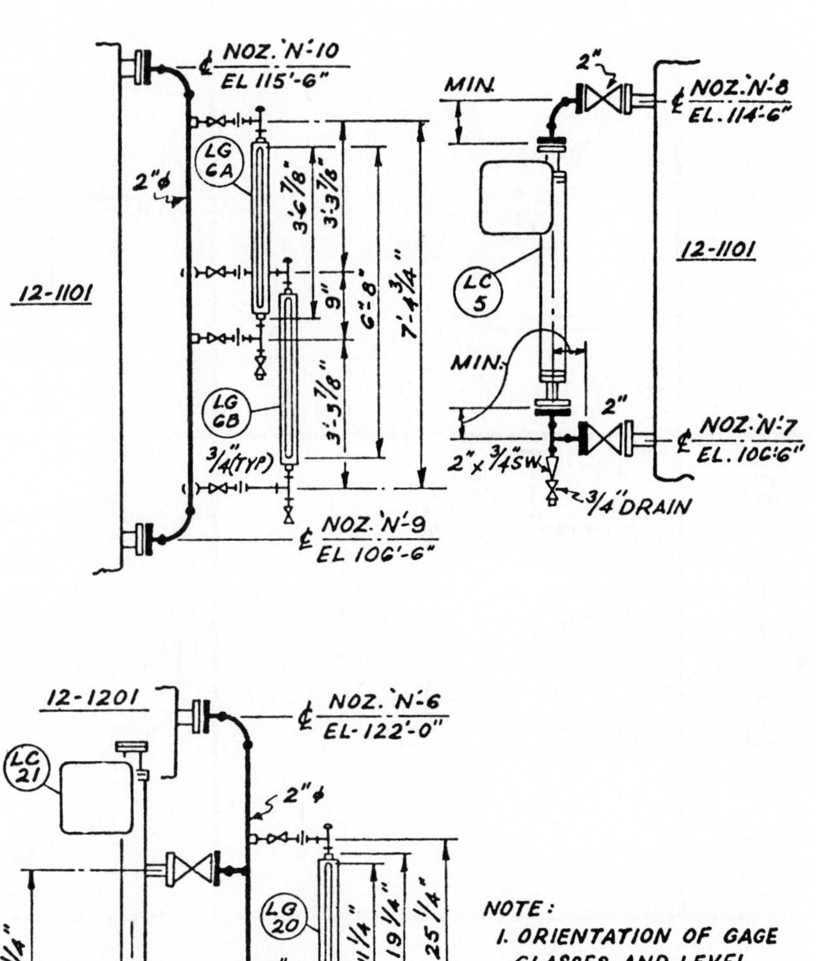

FIG. 5-18 Single-line views of level glass and level controller details.

Projection Methods

Whether the piping system being represented is drawn with the single-line, double-line, or pictorial method, the drafter must master all of the aspects of projection. Figures 5-19 and 5-20 show portions of piping systems and projected views. The student is to complete the missing views in both figures—being careful always to locate the major components and fittings first on the piping line before completing the rest of the view. When the view is not taken at 90 degrees to the pipeline (as in Fig. 5-20), the top view is projected as an orthographic view but the fittings and components are shown elliptically because the plane of projection is not 90 degrees to the pipeline. The left-side view must be completed by the student.

Pipelines that are coming toward the reader are represented either by a dot at the end of the pipe or by a half-darkened

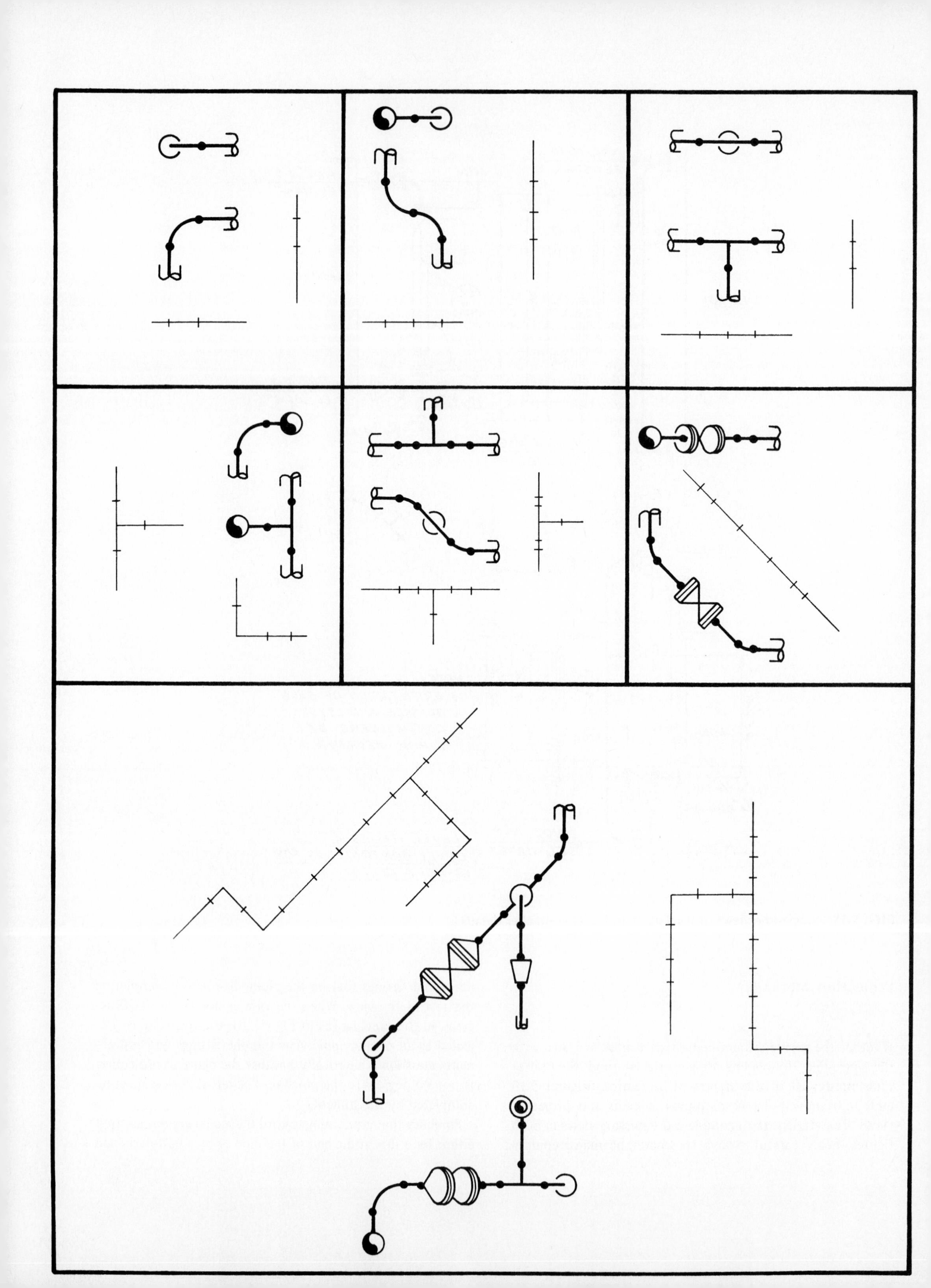

FIG. 5-19 Projection exercise sheet.

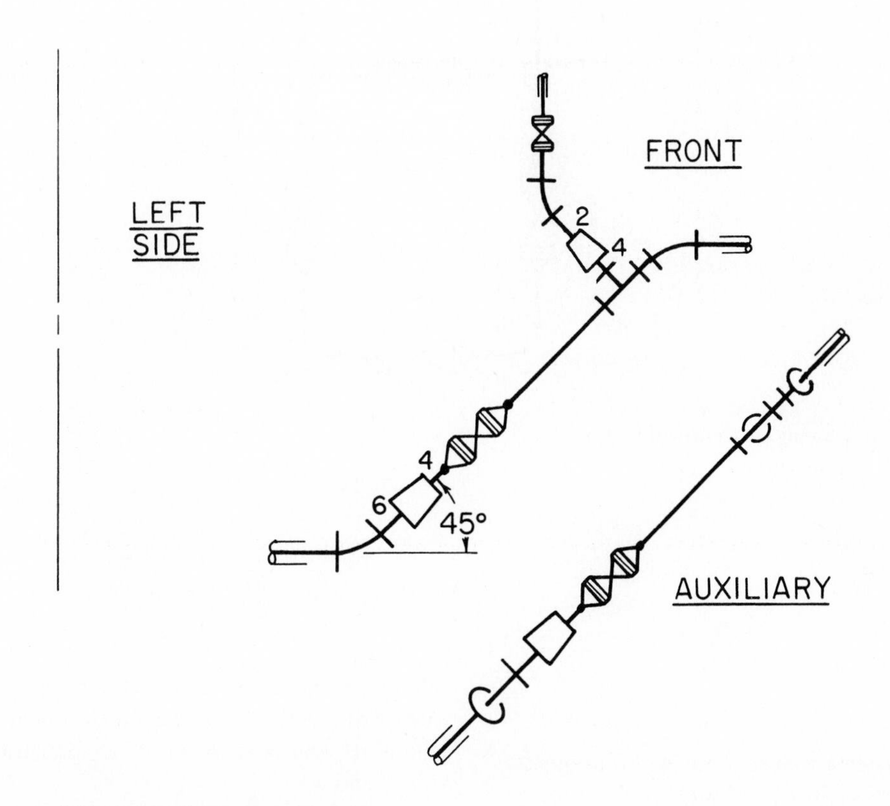

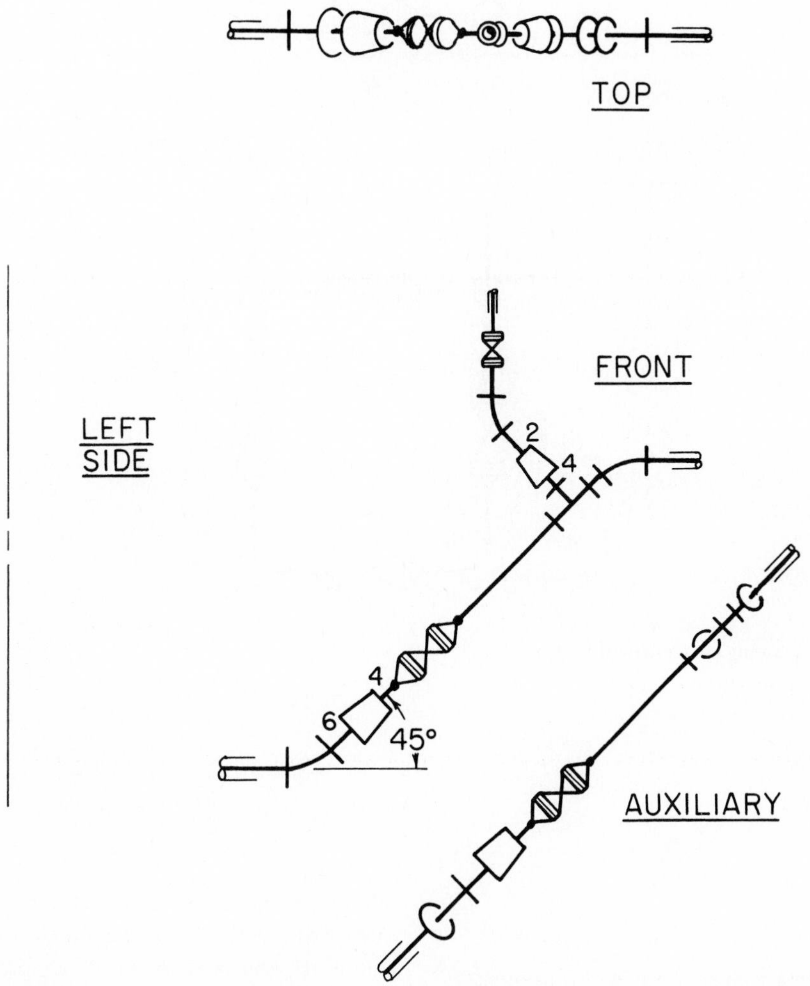

FIG. 5-20 Projection exercise.

pipe end as in Fig. 5-19. When the lines move away from the reader, the pipe appears as a disk or circle with the pipeline drawn to the center. When laying out symbols and components on a pipeline, always locate the end-to-end or face-to-face dimensions and the centerline of the fitting first so that they are placed accurately. If the drawing is drawn to scale, these dimensions must be followed exactly in order to represent the system correctly. All components, including valves and fittings, should be drawn in lightly; templates should be used where the scale permits. Figure 5-21 represents a valve control station showing a single pipeline in plan and elevation views. This system has all-welded fittings.

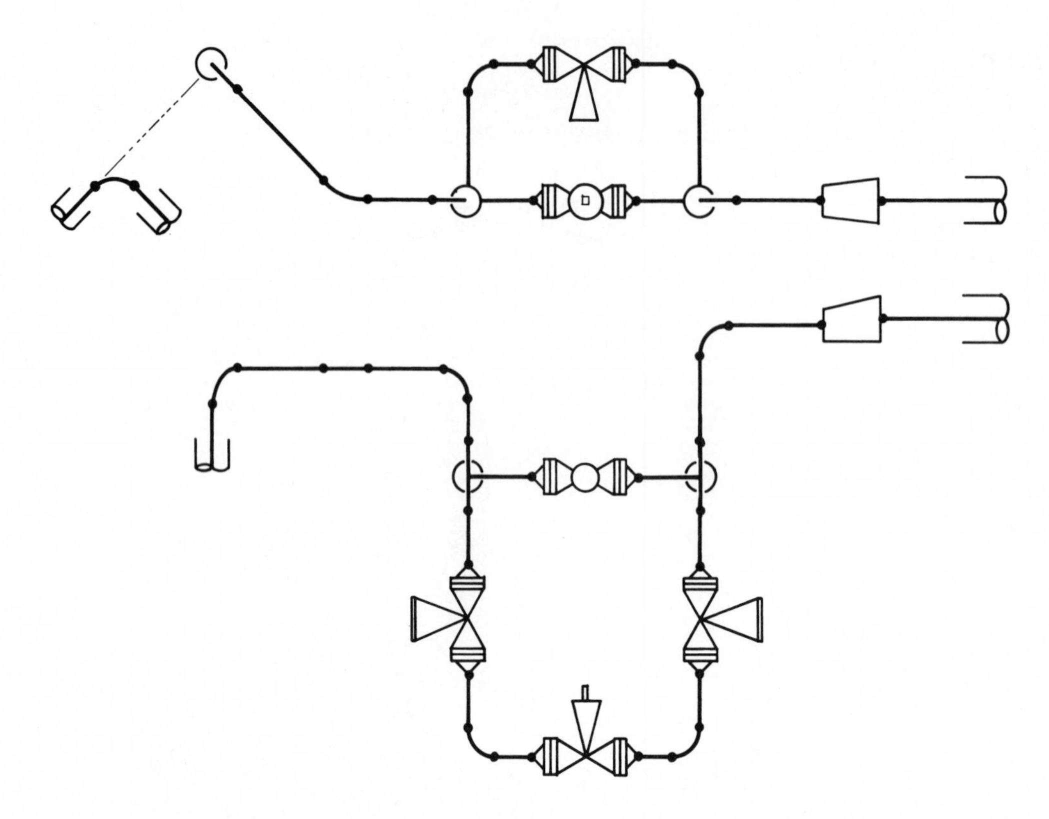

FIG. 5-21 Orthographic viewing of a control station.

Developed Views

In simplified piping systems such as those used for domestic heating, developed views allow the drafter to draw the entire system in one plane. Figure 5-22 shows an isometric view of a pipeline and the same line rotated so that all lines lie within the plane of the paper (flattened on one plane). This method of projection enables the drafter to represent a simplified piping system quickly, using the drawing for material takeoff, callouts, and listing parts. Developed views are not used for complicated or fabrication drawings that must show the system's actual configuration.

Laying Out A System Orthographically

Figure 5-23 is a simple piping project including vessels, a tower, and major piping along with battery limits, and elevation callouts. Note that this drawing deviates from the accepted practice of calling out coordinate dimensions from the north and west. In Fig. 5-23 dimensions are given from south and west.

Before starting construction of a piping project, it is necessary to have the engineer's flow diagram; a complete set of drawings indicating plot plans; civil engineering considerations; plant architectural plans; building designs; heating, air conditioning, and ventilation requirements; and the electrical and structural requirements for the project. The first step in the construction of a piping project is to establish plant datum lines, match lines, or battery limits of the overall refinery or power plant (Fig. 5-25).

Geographic details such as rivers, lakes, hills, roads, and sewage disposal must be established on the site plan if their relationship to the project is important (Fig. 5-24). Use the established plant datum line system, match lines or existing battery limits, throughout the project in a consistent manner. The following is a *specification list* for setting up the plot plan.

1. Locate pumps to the face of the pump foundation and to the longitudinal centerline of a foundation (or centerline of discharge and inlet).
2. Locate multiple pumps to the face of the pump foundation and the longitudinal centerline of the first and last pump and row. Locate intermediate pumps by centerline dimension.
3. Locate all vertical equipment to main centerlines.
4. Locate horizontal equipment to centerlines of concrete piers and the longitudinal center of equipment.
5. Locate all structures to centerlines or columns. Give the width and length of all building construction.
6. Locate boundary lines, pipeways, and pipe access areas.
7. Orient all drawings with the north arrow in the same direction as the plot plan.

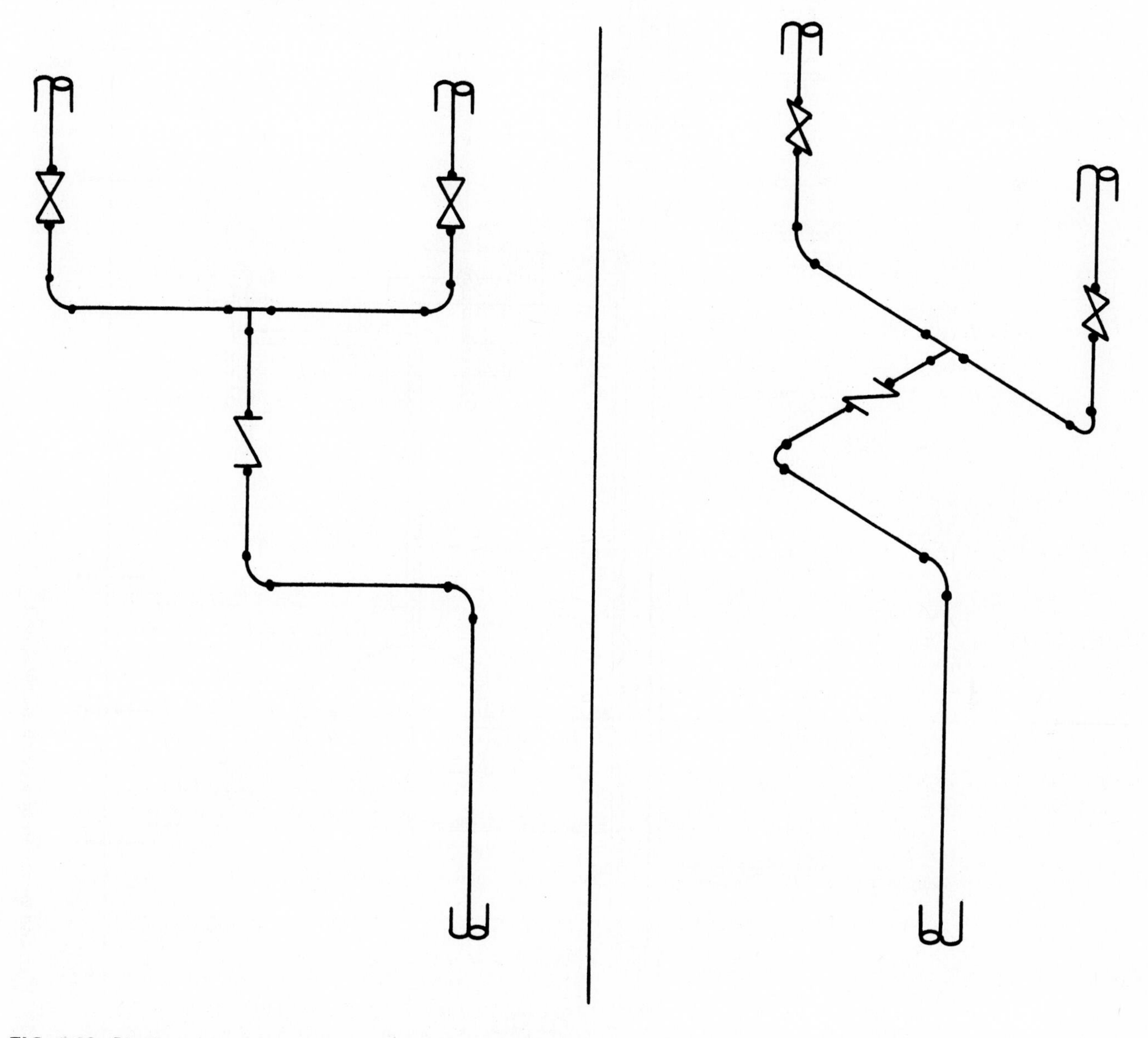

FIG. 5-22 Pipeline shown in isometric and in a developed view.

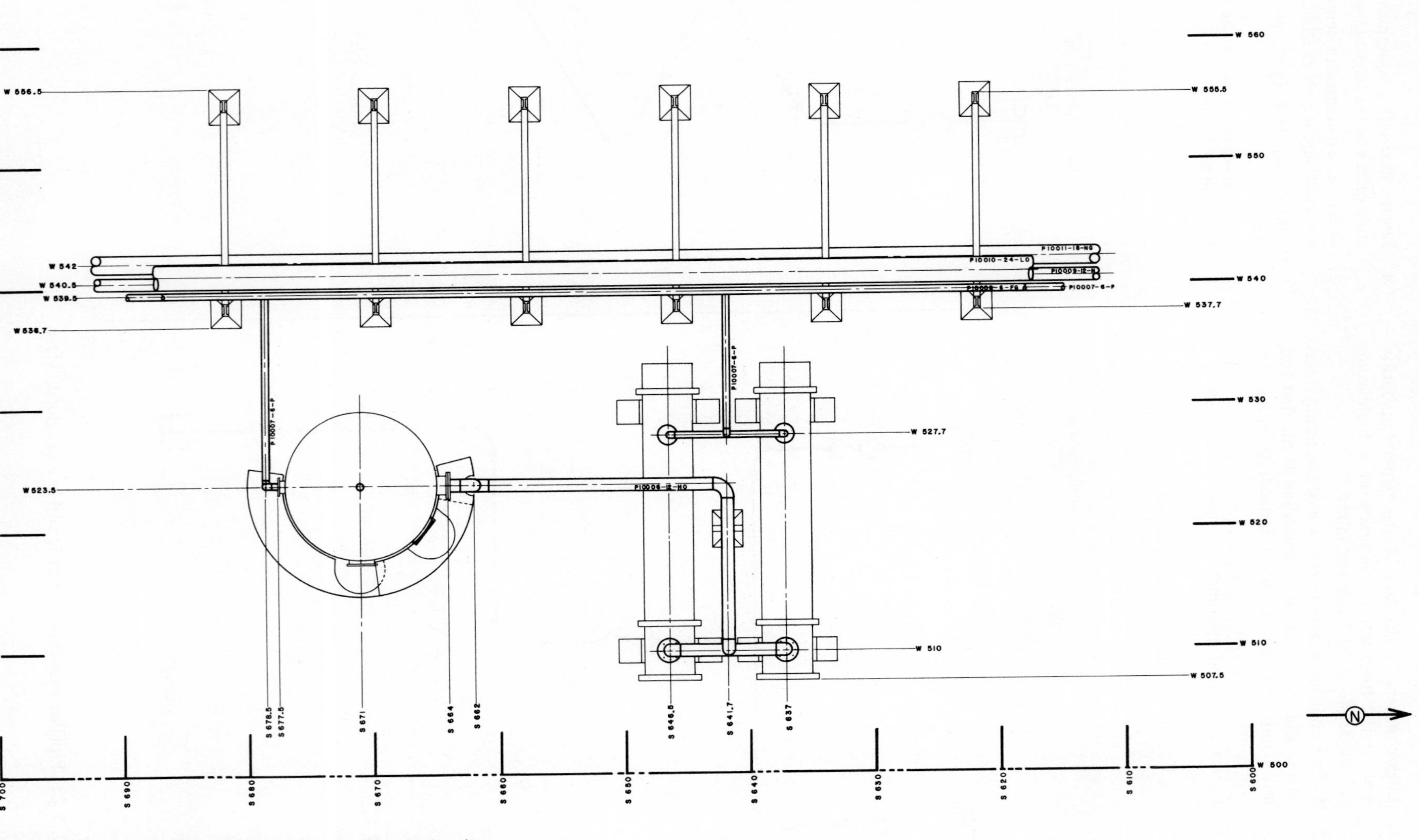

FIG. 5-23 Plan and elevation of a petrochemical unit.

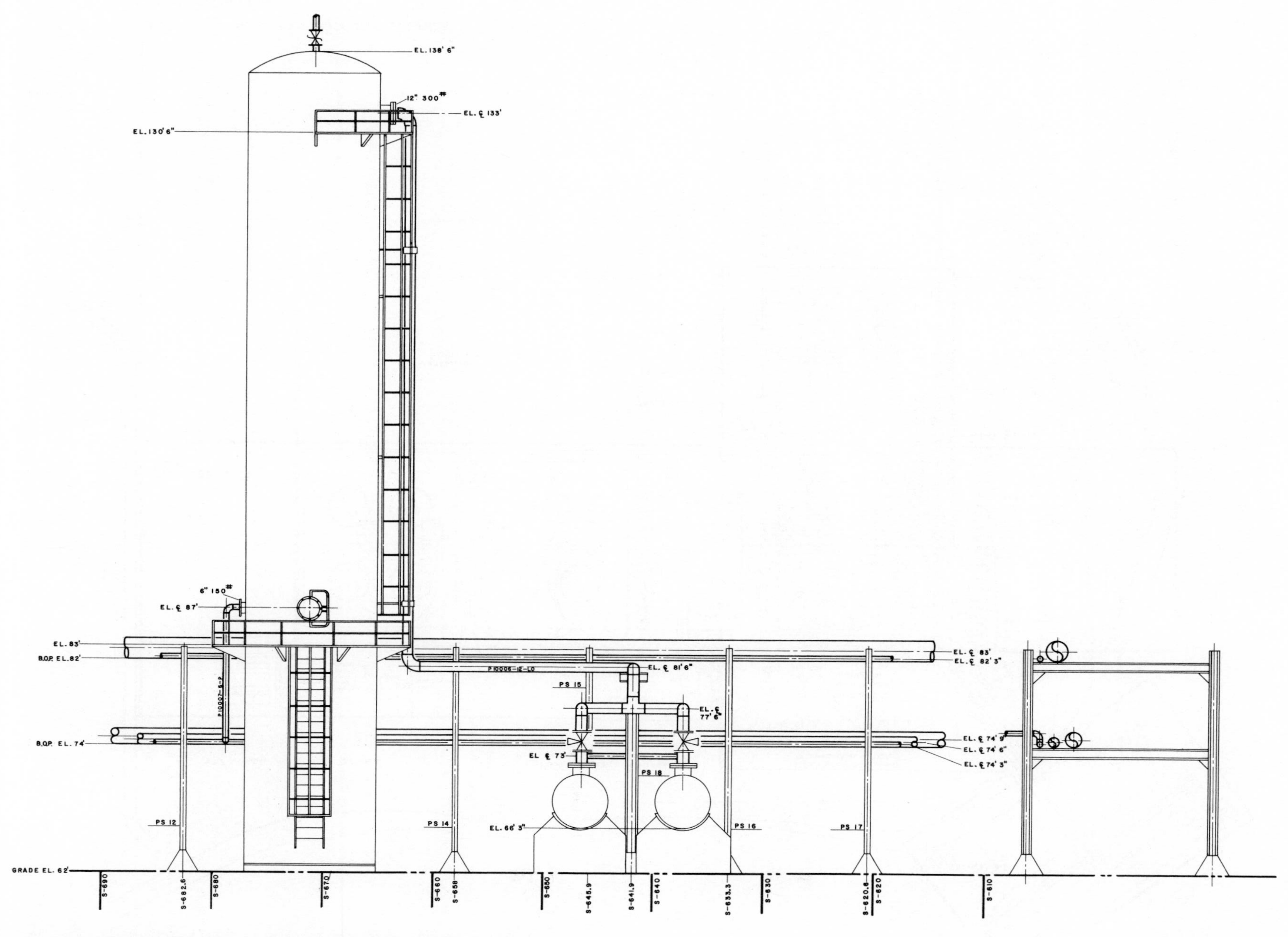

FIG. 5-23 Plan and elevation of a petrochemical unit. (cont'd)

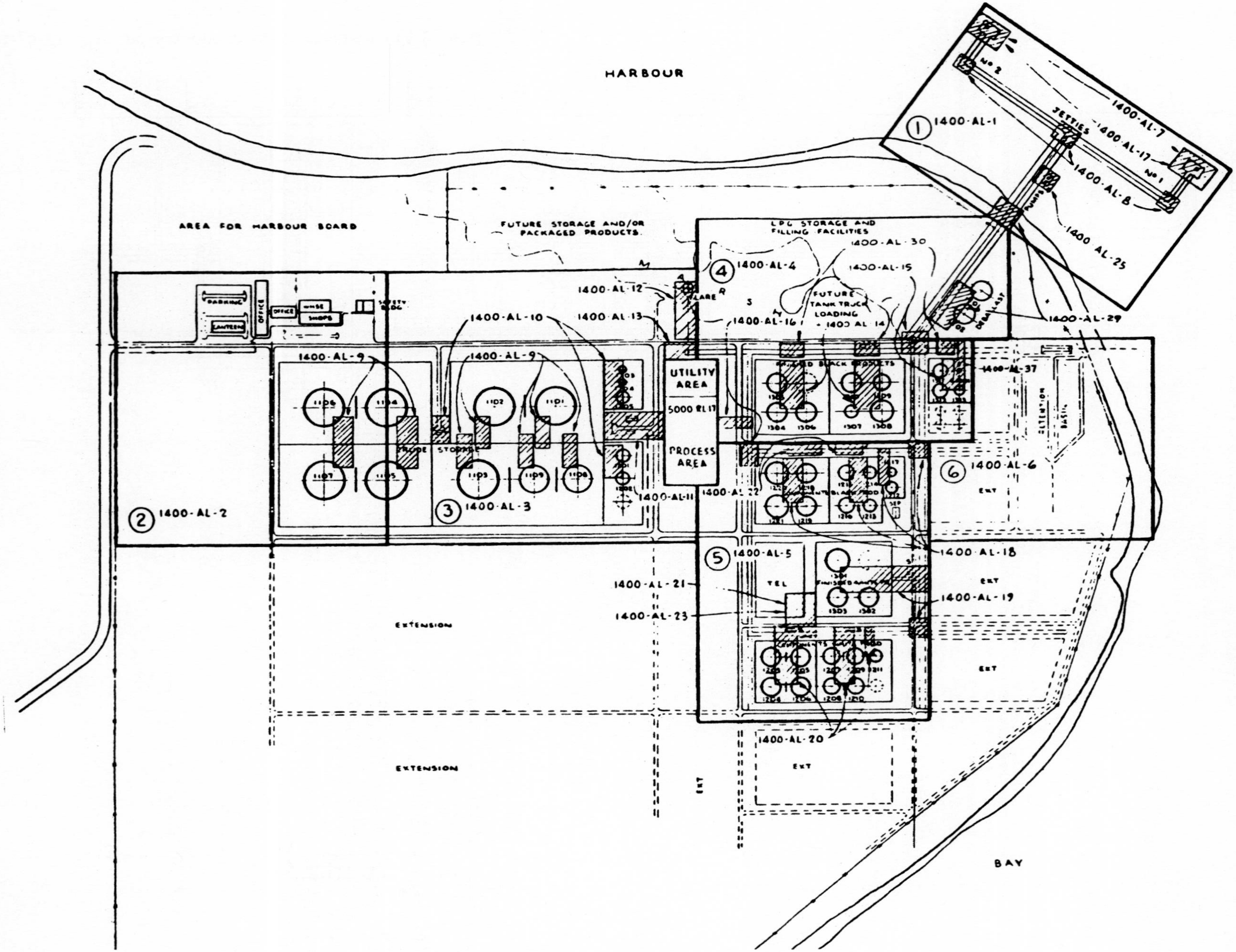

FIG. 5-24 Off-plot site plan.

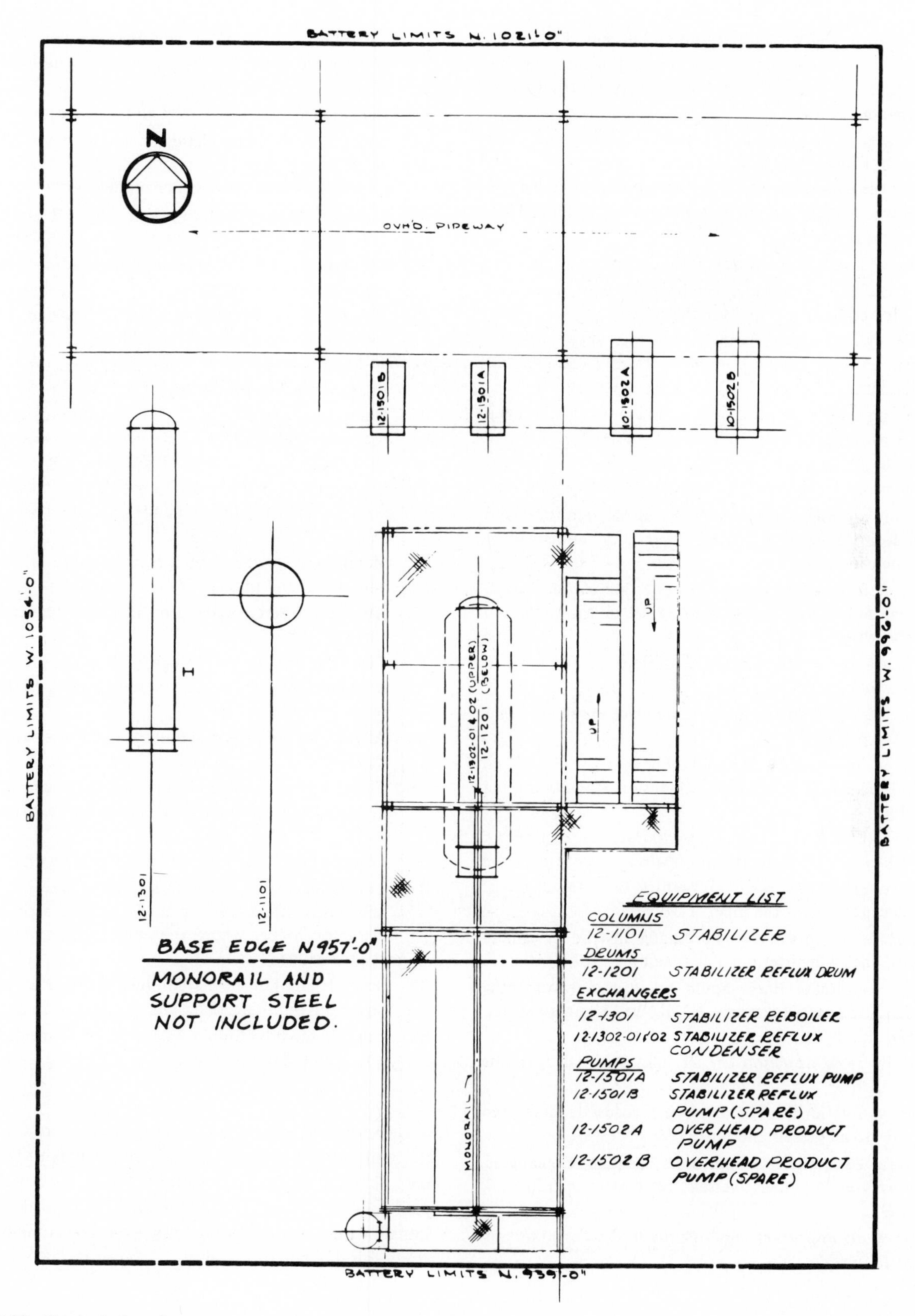

FIG. 5-25 Typical plot plan.

The following list presents commonly accepted practices and specifications throughout the piping industry for the construction and drawing of piping projects in orthographically projected views:

1. Locate all drawings in relationship to an established north, south, east, west arrow.
2. Use single-line method unless clearances must be shown. Use the double-line method for 12-in. (30.48-cm) pipes and larger.
3. Show all reducing fittings in double line.
4. Follow the established symbols for equipment and piping components and use the suggested line weights.
5. Draw all fittings and valves single line where possible. Use symbols that are provided by the company or standards.
6. Use an appropriate scale on all drawings. Try to keep a consistent scale throughout the whole project.
7. Show all structural steel and architectural aspects of the project single-line; use thin, dark line weights.
8. Avoid double dimensioning by using elevation and coordinate designations wherever possible instead of dimension lines.
9. Show face-to-face dimensions and fabrication dimensions only on isometric spool sheets unless instructed otherwise.
10. Locate valve handwheel orientations in all views to show clearances.
11. Locate north arrows in the same place on all drawings in the project.
12. Show dimensions in feet and inches for 1 ft (30.48 cm) and over and use metric dimensions for all foreign jobs.
13. Do not use pictorial representation for equipment such as compressors, pumps, heaters, and exchangers.
14. Draw single-line pipes with thick lines to represent the centerline of the pipe.
15. Locate equipment by coordinates instead of dimension lines whenever possible (example, N576').
16. All coordinate lines should be located toward the north and west (example, N376', W511') unless otherwise noted.
17. Show flow direction on all lines and call out line number.
18. Show relationship of platforms, access ladders, and cages to all vessels and pipelines.
19. Draw equipment (compressors, heaters, exchangers, conveyors, turbines, vessels, towers) with thin dark lines.
20. Draw all important equipment such as concrete supports to scale.
21. Show clearance between pipeways and equipment.
22. Draw valve stems in open position to indicate clearance.
23. Indicate clearances between all major equipment.
24. Show pipe supports, hangers, and guides in order to avoid interference. In many cases these will be shown on the isometric drawings or model.
25. Start elevation designations at 100 ft (30.48 m) in order to keep the basement at a positive elevation.
26. Maintain 12 ft (3.66 m) minimum headroom beneath main process pipeways.
27. Maintain a minimum clearance of 7 ft 6 in. (2.29 m) for pipelines located inside buildings.
28. Maintain a clearance of 7 ft 6 in. (2.29 m) for all pipelines over aisles and platforms.
29. Maintain a clearance of 17 ft 6 in. (5.33 m) over main roadways and 15 ft (4.57 m) clearance over secondary roadways.
30. Space pipe rack supports 25 ft (7.62 m) maximum or refer to standard recommended spacing (see Chapter 10).
31. Maintain a clearance of 2 ft 6 in. (0.76 m) between equipment and piping for maintenance.
32. Attempt to keep main pipe racks on one level (see Chapter 10).
33. Keep the number of fittings used on the pipeline to a minimum by avoiding unnecessary bends, elbows, and tees.
34. Always follow piping standards and specifications for a project.
35. Equip all operating valves 7 ft 6 in. above grade or above platforms with chain operators or other automatic controls.
36. Consult pipe standards before starting a project.
37. Locate all instrumentation so that it is readily visible and easily operated—especially in the case of level glass controllers where the operator must be able to see the gauge glass.
38. Support all hot insulated lines 3 in. (7.62 cm) and larger on 3-in.-high (7.62 cm) insulation shoes.
39. Consult manufacturer's information on insulation requirements for a project.

Specification lists provided in relevant chapters will enable the reader to follow accepted practice throughout the industry. Figure 5-26 is an elevation view of a petrochemical plant with notations and dimensions for typical specifications for the design of piping complexes. The student should study this drawing thoroughly.

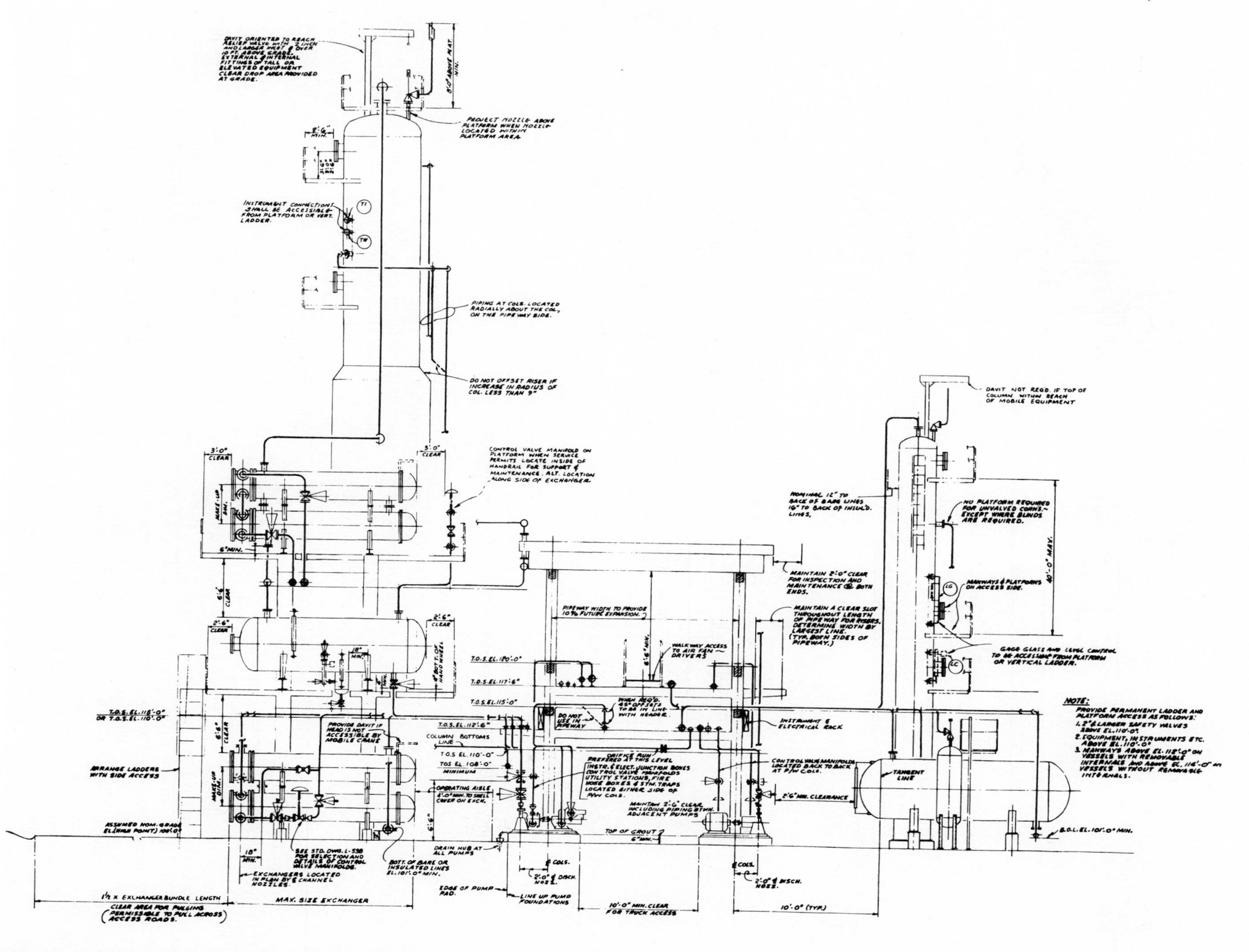

FIG. 5-26 Specifications for the layout and design of petrochemical projects.

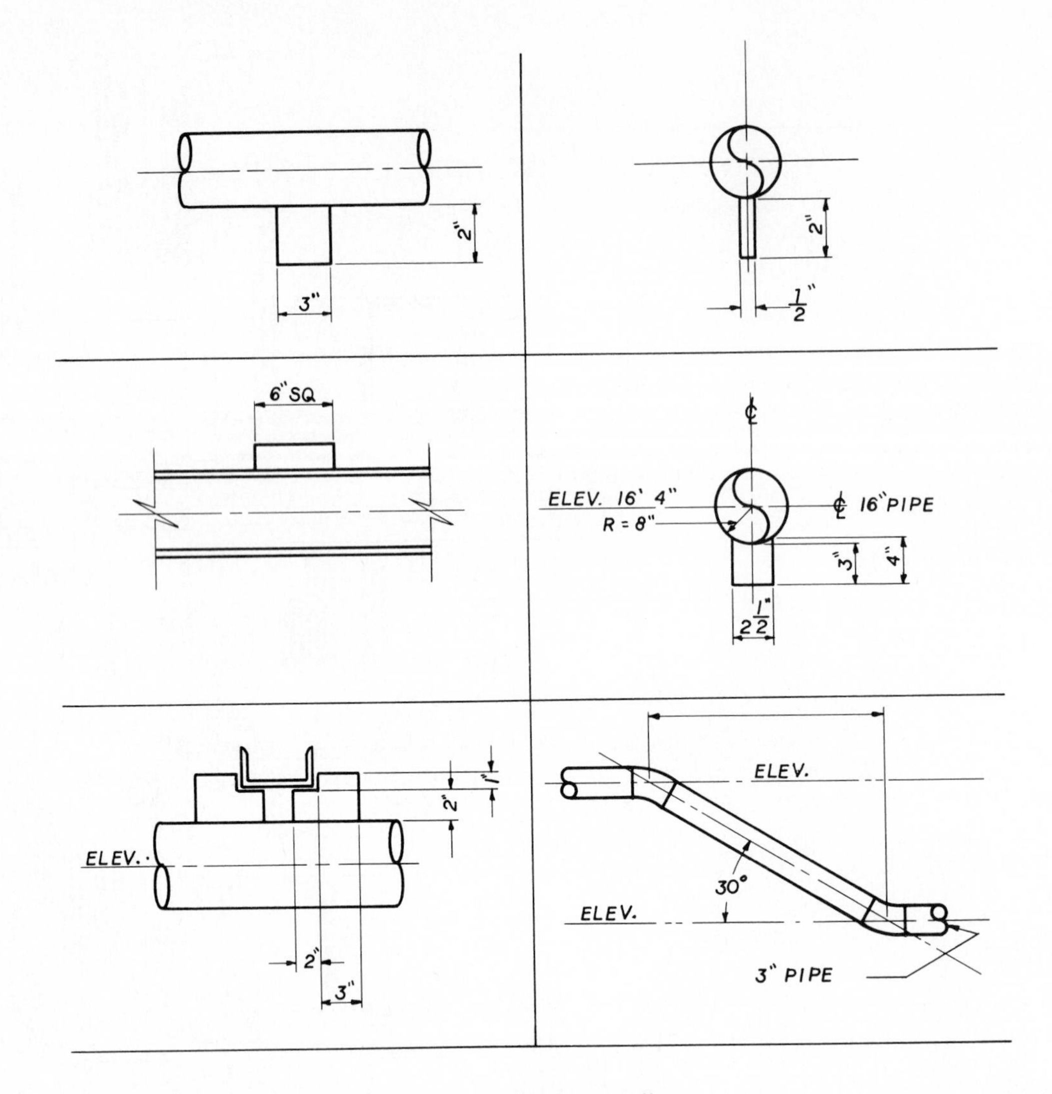

FIG. 5-27 Typical dimensioning practices for pipe details.

DIMENSIONING

Dimensioning practices for piping drawings are similar to those for mechanical drafting. Views that give the shape of the object are organized and laid out to provide room for dimensions and for arrows and leader lines for special callouts. Different piping drawings require different dimensioning procedures. Fabrication drawings, isometric drawings, shop drawings, field drawings, plan, elevation, and section drawings, pipe support drawings—all have their own dimensioning peculiarities. (See Chapter 9 for dimensioning of fabrication spools.)

Flow diagrams are never dimensioned (see Chapter 6). Chapter 7 gives the dimensioning practices for isometric drawings, which are similar to the dimensioning practices for mechanical objects in isometric. Intricate dimensions are often called out in isometric because the shape of the pipeline or system is easier to ascertain since all three dimensions (length, width, height) can be shown on one view. All drawings need adequate dimensions to convey sufficient information to the construction crew or fabricator. Dimensions are also used for engineering sketches of piping isometrics or detailed drawings. The engineer and designer must give sufficient information so the drafter can translate the sketch into an accurate drawing.

In the plan, elevation, and section method of drawing piping systems, elevations are given (instead of using dimension lines) for most vertical dimensions, except for those that are intrinsic to individual pieces of equipment. The elevation at grade is the basis for all other elevations to ensure coordinated drawings. Figure 5-23 shows a simplified plan, and elevation, drawing of a petrochemical installation. In this drawing, elevations are given for all the important levels in the project. Figure 5-23 shows the use of coordinates to establish dimensions in the plan view, therefore eliminating the use of dimension lines. Elevation designations are used in all views to establish vertical dimensions. (Some projects are completed with only plan view and elevation callouts.)

Sometimes the plan, elevation, and section method involves dimensions and dimension lines besides the use of

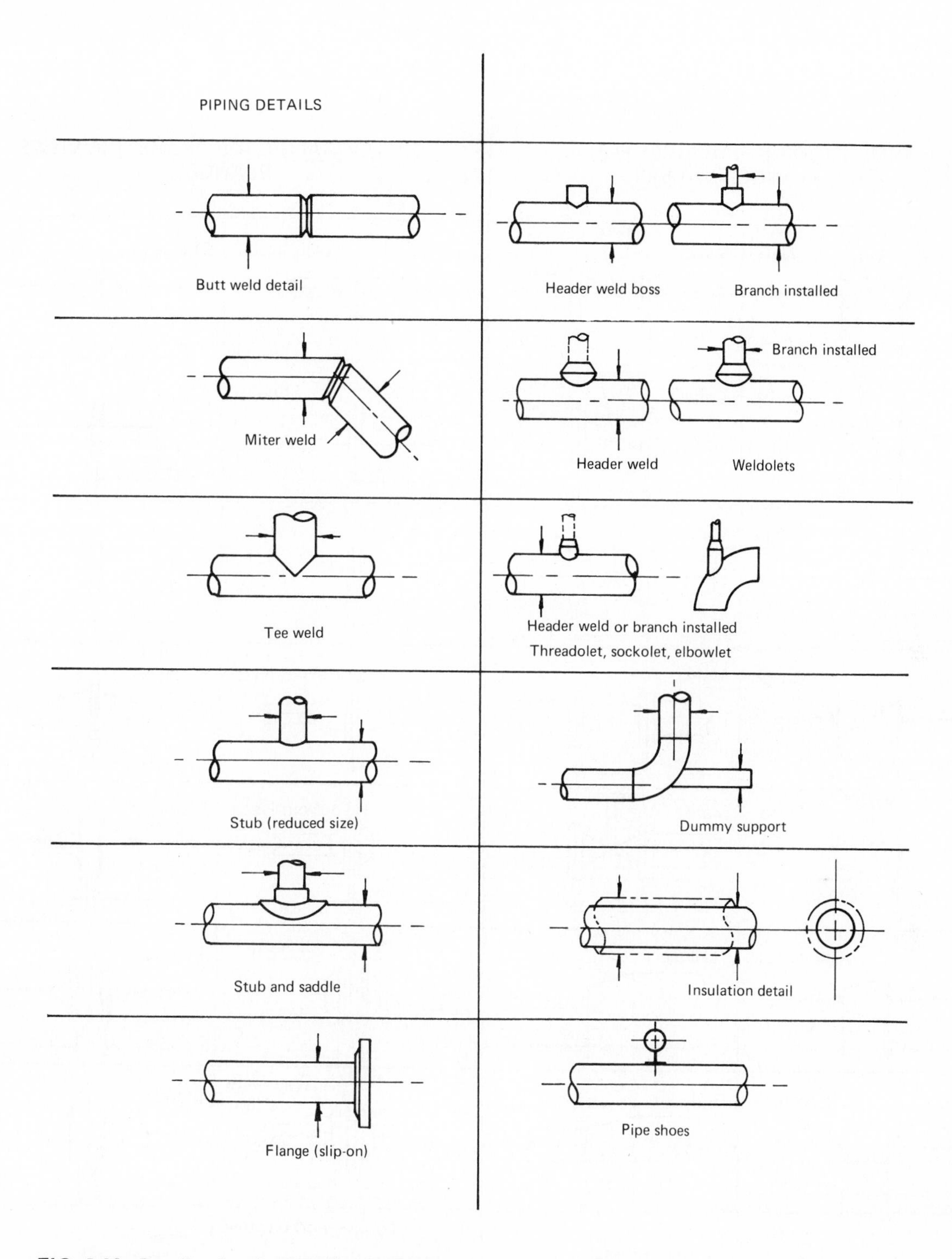

FIG. 5-28 Details of various piping situations.

coordinate systems. In these cases, keep most horizontal dimensions in the plan view, making sure that dimensioned aspects of the project are not called out more than once. (Avoid double dimensioning.) In dimensioning piping projects, whether the plan, elevation, section, isometric, or other method is used, the dimension line is never broken as in mechanical drafting. The drafter can complete the dimensioning more efficiently by keeping the dimension line unbroken and placing the dimension above the dimension line. Either arrowheads or slashes are used to terminate the dimension line.

Figures 5-27 and 5-28 show typical dimensioning practices for a variety of small cutaway sections and details that might occur on a piping detail.

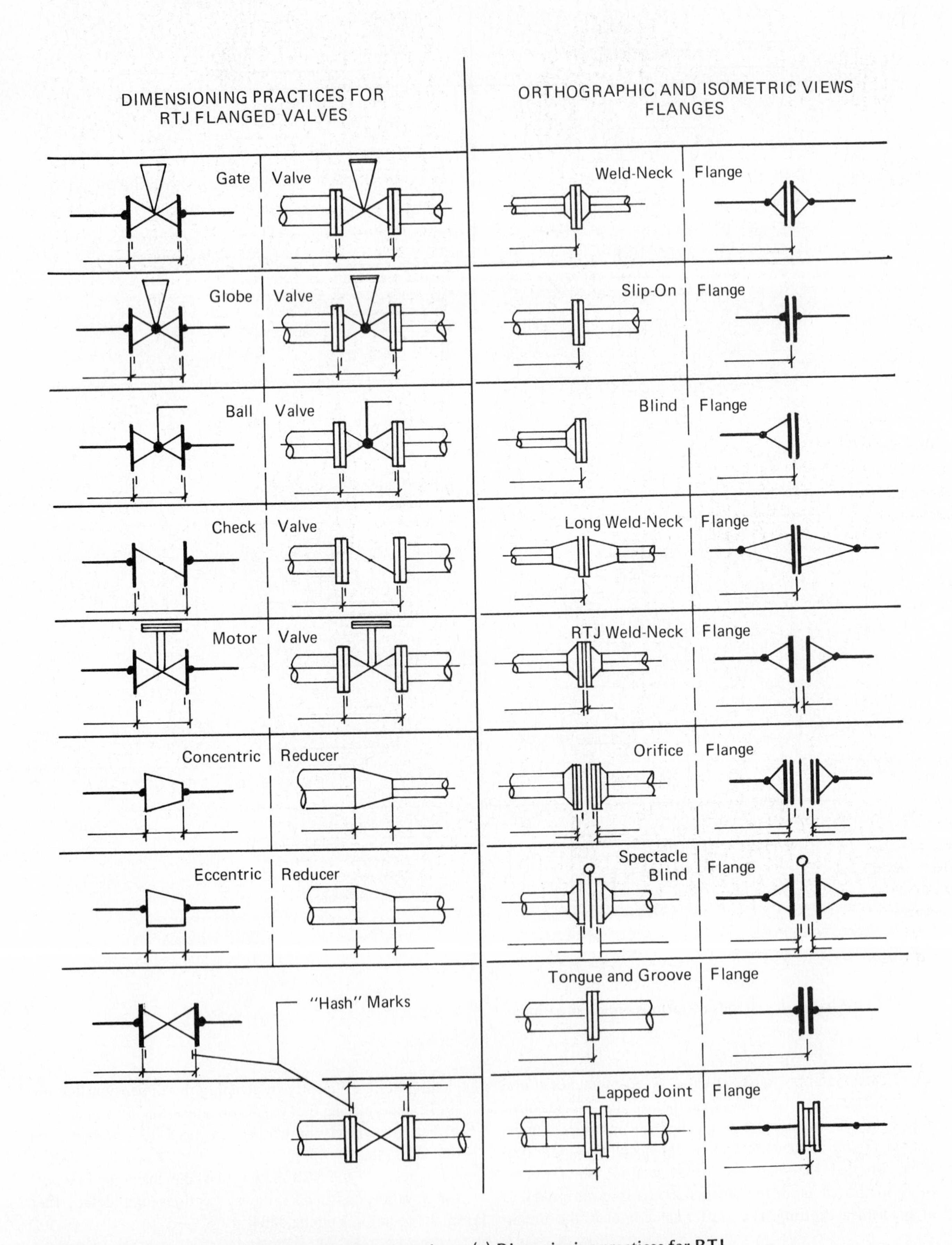

FIG. 5-29 Flange, valve, and reducer detailing procedures. (a) Dimensioning practices for RTJ flanged valves (hash marks inside dimension lines mean face-to-face dimension includes gasket). (b) Orthographic and isometric views of flanges.

Fittings are usually called out on the spool drawing (as in the isometric drawings of Chapter 7). A designation number for a component is used instead of a callout if the item is standard. When preparing orthographic drawings, the drafter does not usually show overall dimensions and face-to-face dimensions of valves, fittings, and other components—these are left mainly to fabrication drawings. Be sure to locate to the center of each component—for example, the center of a valve and the distance between the centerlines of pipes and components.

On detail drawings, isometrics, and other spool drawings, face-to-face dimensions are given in most cases. Figure 5-29 provides dimensioning practices for both orthographic and isometric views of spool drawings. This figure also depicts dimension lines for end-to-end dimensions and face-to-face dimensions, for valves, flanges, and fittings (including ring-type joints). Remember when detailing piping projects to give a sufficient number of location dimensions so that all piping runs and equipment can be located in relationship to match lines or battery limits. The use of a baseline or assumed elevation from grade should also be established in all elevation views in order to eliminate dimension lines. Establish grade at 100 ft elevation.

When using dimension lines, take care to space them evenly (where possible) and to avoid dimensioning on or through objects (equipment, valves, and so on). A neat, uncluttered drawing is the main objective for all drafters.

Basic Rules for Dimensioning Piping Systems

The basic elements are the dimension line itself and the extension line, including either a slash or an arrowhead to show the extent of the dimension. Some of the following rules also apply to mechanical, structural, and architectural drawing:

1. Dimension figures, numbers, and letters are placed above the dimension line for all types of piping system drawings. Do not break the dimension line as in mechanical drafting.
2. Dimension figures are to be read from the bottom of the page or the right side of the drawing unless another practice is used for a particular company.
3. Do not give double dimensions or duplicate dimensions.
4. Always place long dimension lines outside shorter ones to avoid crossing lines. In most cases the dimensions for piping drawings are strung together (stacked).
5. Show dimensions from plot limits (match and battery limits). Centerlines of vessels or structural columns can also be used to locate components and equipment.
6. Dimensional notes such as those that appear on the right-hand column above the title block are to be read from the bottom of the sheet.
7. When there are several parallel dimension lines, stagger the dimensions and numbers to make them easier to read.
8. Use feet and inch marks on all dimension figures.
9. On detail drawings it is not necessary to call out zero feet when expressing a dimension that is in inches only, though this should be done on all plan, elevation, and section drawings.
10. All dimensions should be shown on the drawing instead of using notes, unless a descriptive note will enable the drafter to eliminate repetitive dimensions and data—for example, by using a note "typical."
11. Whenever possible, horizontal dimensions should be shown on plan views and vertical dimensions on vertical section and elevation views.
12. Never place dimensions inside objects or pieces of equipment.
13. If a dimension is indicated by the engineer as being calculated or scaled, place the scale or calculation in parentheses under the dimension line.
14. All dimensions should be given precisely and clearly so they cannot be misinterpreted. Form the characters so they will not run together if the drawing is reduced.
15. Dimensions should be given between piping or surfaces which have a relationship to each other or which control the location of mating parts such as centerlines of equipment or edges of equipment, between equipment, or between structural elements.
16. Place dimensions in views where the objects are detailed in true size wherever possible. This cannot always be done—in spool drawings, for example, lines and components are seldom drawn to scale.
17. When objects are to be centered on another object, call out the centerline of both objects and omit the locating dimension from the centerline.
18. When dimensioning from centerlines, call out the name of the object on which the centerline is drawn.
19. Remember that dimensions are as important as the views and objects. Be scrupulous in following correct dimensioning practices.
20. Always consult the company's list of standard practices for dimensioning.

Though general practices will vary between drafters regarding the positioning and construction of dimensions, including the style of printing and line work, all information given must follow accepted standards and company practices. These procedures ensure that the drawing will exhibit, clearly and concisely, sufficient dimensions for the construction of the system. Since many people will be using the drawings produced in the drafting department, it is the drafter's responsibility to create a legible drawing. Dimensions are determined by the drawing's function. Thus fabrication drawings (isometric and orthographic spool sheets) may follow dimensioning practices slightly different from those for plan, elevation, and section views.

Most dimensioning practices are similar for pipe supports, orthographic projections, and so forth. The differences are called out in the relevant chapters: Chapter 7 (isometric dimensioning), Chapter 8 (pipe support dimensioning), and Chapter 9 (pipe fabrications).

Components of Dimensioning Systems

Dimensional Figures and Units The following rules apply to dimensioning figures and units:

1. Follow the lettering procedures for the formation of numbers, fractions, and other characters as explained in the beginning of this chapter.
2. Letter dimensions so that they read from the bottom or right-hand side of the page. When dimensions must be given in a direction other than horizontal or vertical, follow the company's guidelines.
3. Show all property and grid coordinates for dimensional units in feet and hundredths. For dimensions on plot plans, equipment location drawings, and structural and architectural drawings, use dimensions up to but not including 1 ft 0 in. (30.48 cm) to the nearest 1/16 in. (0.61 cm) unless they are noncritical. Round off end dimensions that are not critical (in feet and inches).
4. For all piping detail drawings, column and vessel drawings, and electrical drawings, show dimensional units up to and including 24 in. (60.96 cm).
5. On machine drawings and sheet metal drawings, show dimensions up to and including 72 in. (182.88 cm).
6. For pipe support drawings use dimensions up to but not including 24 in. (60.96 cm). Dimensions 24 in. and larger are expressed in feet and inches except when specifying pipe size.

Arrowheads The following practices should be observed:

1. Arrowheads should be approximately 3/16 in. (0.47 cm) in length and about one-third the length in width. They should be black and completely darkened. Open arrowheads are not used unless it is company practice.
2. Arrowheads should touch the extension line of the object to be dimensioned.
3. Avoid carelessly drawn arrowheads which will smudge easily and reproduce poorly.
4. Do not use slashes and dots in place of arrowheads unless that is company practice.

Figure 5-26 shows the use of arrowheads in dimensioning; Figure 5-29 shows the use of slashes.

Dimension and Extension Lines The following rules apply to the use of dimension lines (including centerlines and leaders):

1. Keep dimension lines thin and sharp by using 2H or H lead. (See line types in Fig. 5-7.)
2. Always keep dimension and extension lines thinner than object lines. Object lines for equipment and piping should stand out clearly from dimension lines.
3. Centerlines should be of the same line weight as dimension and extension lines.
4. Always draw dimension and extension lines continuously. Do not break them for the insertion of characters as in mechanical drafting.
5. Use the aligned method when drawing dimension and extension lines and group them for consistency.
6. Never pass dimension or extension lines through lettering or notes.
7. Spacing between dimension lines should be consistently 1/4 in. (0.64 cm) or 3/8 in. (0.96 cm) apart and in some cases 1/2 in. (1.27 cm), depending on the size and the available area. Great care must be taken not to space dimension lines randomly.
8. Extension lines should extend about 1/16 in. (0.16 cm) to 1/8 in. (0.32 cm) past the dimension line. Leave a gap of about 1/16 in. (0.16 cm) from the object to be dimensioned (unless, of course, the centerline is used as the extension line).
9. Try not to cross dimension lines, though extension lines and centerlines will in many cases cross.
10. Place smaller dimensions inside larger dimensions so the dimensions do not cross.
11. Leaders should be drawn from the first letter or number or last letter or number on the first line of a note, whichever is closer to the object.
12. Leaders are to be drawn with the same line weight as extension lines.
13. Avoid dimensioning to hidden lines whenever possible.
14. A leader line can be straight or curved, depending on company practice, and should not be broken when

crossing dimension and extension lines or when crossing an object line unless this is accepted practice.

Elevations The following guidelines should be observed:

1. Elevations should be placed above extension lines or centerlines whenever possible. Arrowheads pointing to the respective elevation level may be used when space is not available.
2. The following abbreviations for elevation descriptions are used throughout the profession:
   (a) Floor = FL
   (b) Centerline = CL
   (c) Elevation = EL
   (d) Top or bottom of concrete = TOC or BOC
   (e) Top or bottom of steel = TOS or BOS (or T/S or B/S)
3. In section and elevation views, calling out an object's elevation should replace the use of vertical dimensioning wherever possible—unless the dimension is for a piece of equipment (such as a tower or vessel) which needs to be dimensioned in relation to itself and on detail drawings of equipment.
4. Call out pipe centerlines or bottom of pipe elevations on all drawings.
5. Express elevation dimensioning in feet and inches.
6. Elevations for all sloped pipe should be given in plus or minus dimensional units.
7. Give elevations in feet and hundredths on grade plans, and sewer drainage plans.
8. Give elevations in feet and inches for foundations. (Give top of grout or leveling plate and bottom of foundation.) For all structural drawings, give dimensions to top of framing numbers, tops of concrete beams, and finished concrete floors.

Dimensioning to Equipment

The following practices are recommended for dimensioning to valves, nozzles, equipment, flanges, and fittings:

1. Valves
   (a) Consult ANSI standards for all face-to-face dimensions. If these standards are not available, use manufacturer's catalog dimensions. (Also see appendix.)
   (b) Use face-to-face dimensions on all shop fabrication spools and for flanged valves.
   (c) On plan, elevation, and section drawings, field fabrication spools, and system isometrics, dimension to the centerline or stem of valves if pipe lengths are left to the field crew.
   (d) Dimension from the face of the valve to the end of the line (flange face or fitting face for shop fabrication or field fabrication drawings).
   (e) For screwed, bell and spigot, welded, or socket ends, dimension only to the valve centerline or stem.
2. Joints (flanges, pipe ends, fitting ends)
   (a) Terminate dimensions at major break points—nozzles, valves, flanges.
   (b) Keep noncritical dimensions to a minimum unless they are needed to place unions, couplings, or welded joints accurately (which does not occur often).
   (c) For all ring-type joints dimension to the flange face and then give gasket space dimension by showing face-to-face dimensions of the assembled joint (Fig. 5-29).
   (d) Refer to standard gasket measurements when determining face-to-face dimensions on all RTJ valves and flanged joints. (See Appendix and Chapter 3.)
3. Vessels and equipment
   (a) Dimension to a vessel nozzles face from vessel centerline.
   (b) Dimension from vessel nozzle to valve face, flange face, and so on.
   (c) Call out all vessel nozzle elevations on isometrics, and in all vertical (elevation) views.
   (d) Dimension from centerlines of equipment (pumps, compressors) to major connecting joints.

Figure 5-27 shows typical dimensioning practices for details such as pipe supports and sloping pipes. Figure 5-30 shows a standard orthographic projection of a piping system in the round (spool drawing) for a nuclear reactor pipeline. This drawing shows all the elevations and centerline-to-containment center dimensions for pipe supports and components and all degrees between each segment of piping corresponding to the 360-degree circle. Many times drawings such as Fig. 5-30 are shown in isometric as in Fig. 5-31.

Figure 5-32 shows an orthographic projection that expresses how the drawing can become confused when one is using this method. The isometric drawing in Fig. 5-33, by contrast, gives a three-dimensional view of the whole line, thereby eliminating any misunderstanding regarding its configuration.

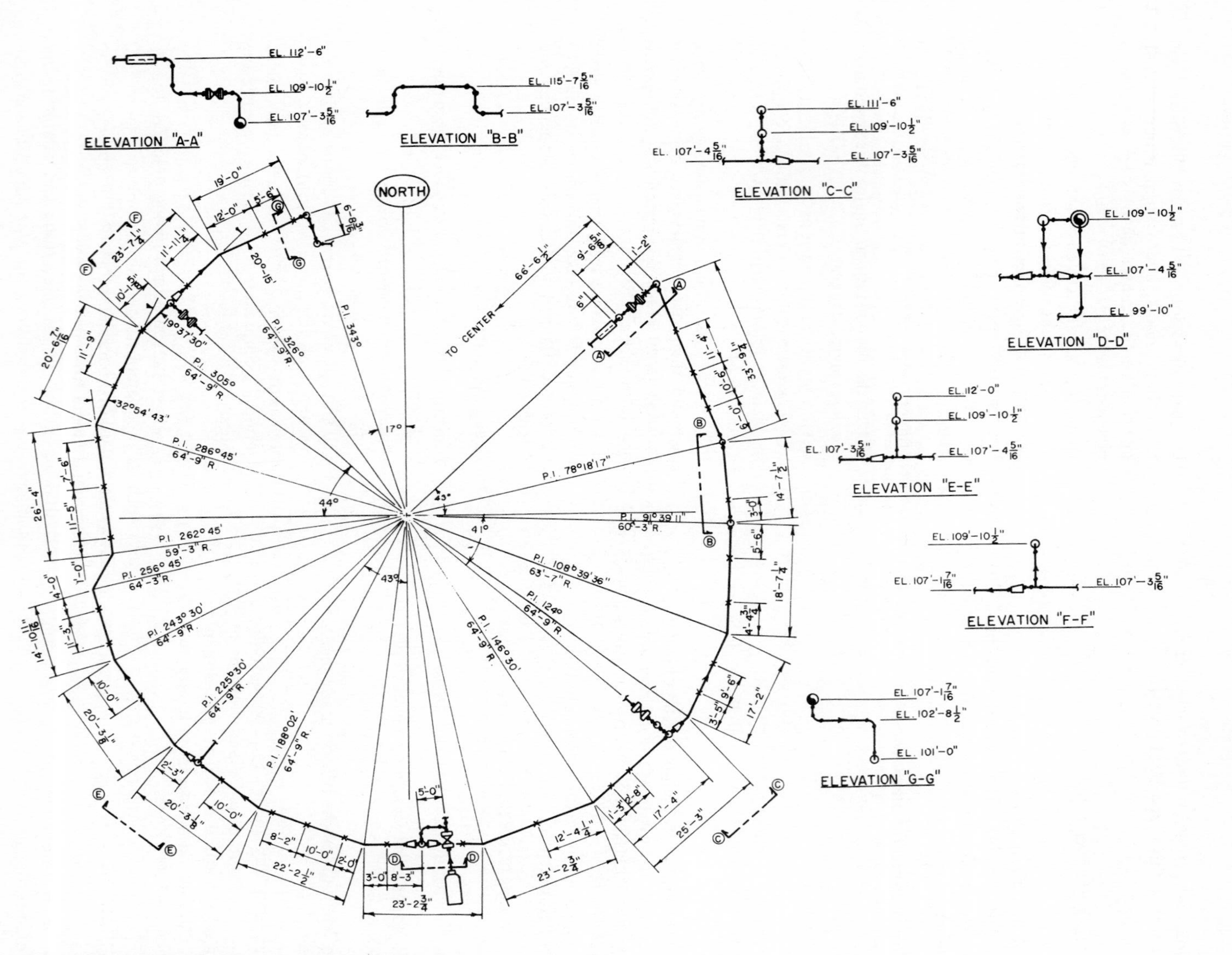

FIG. 5-30 Nuclear piping: orthographic projection.

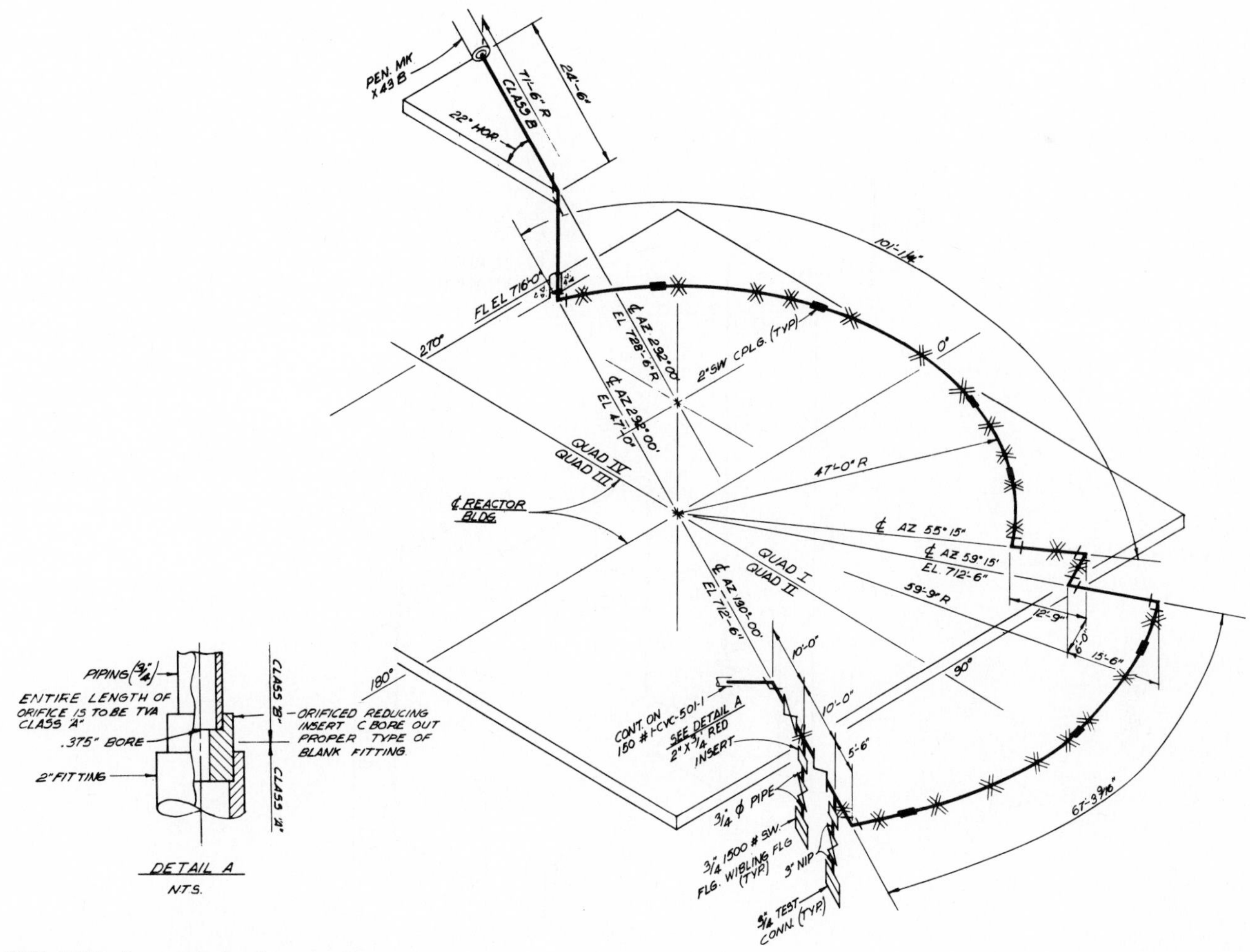

FIG. 5-31 Isometric in the round.

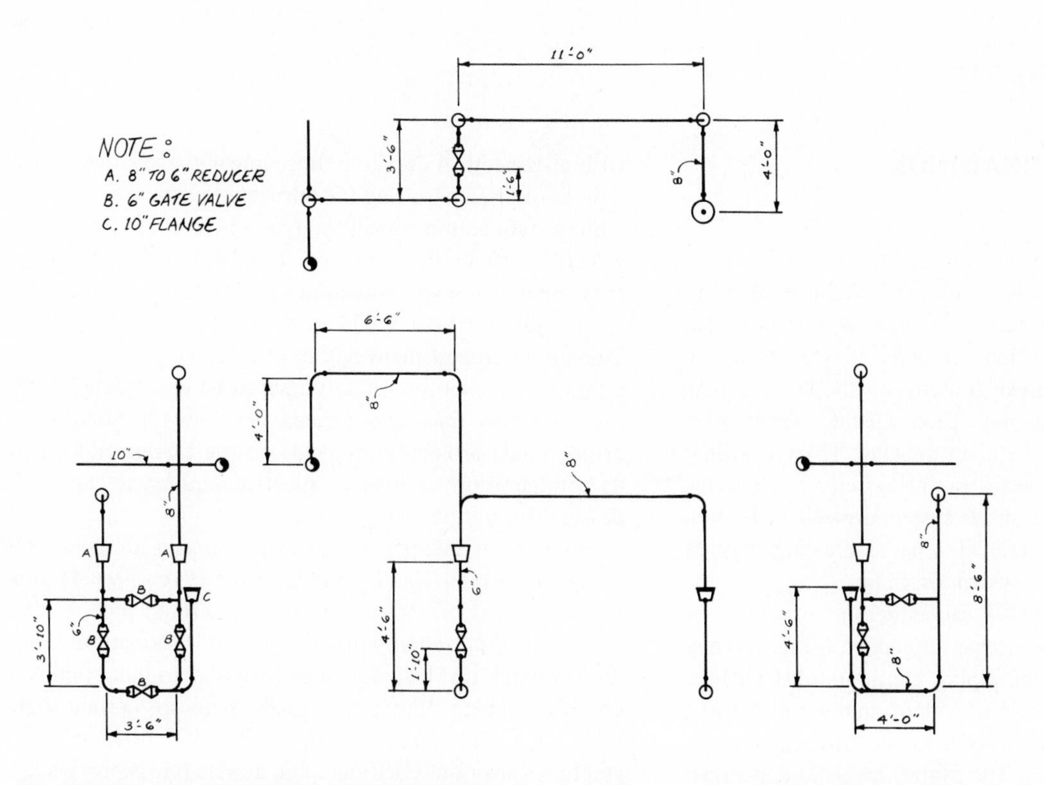

FIG. 5-32 Orthographic drawing of Fig. 5-33.

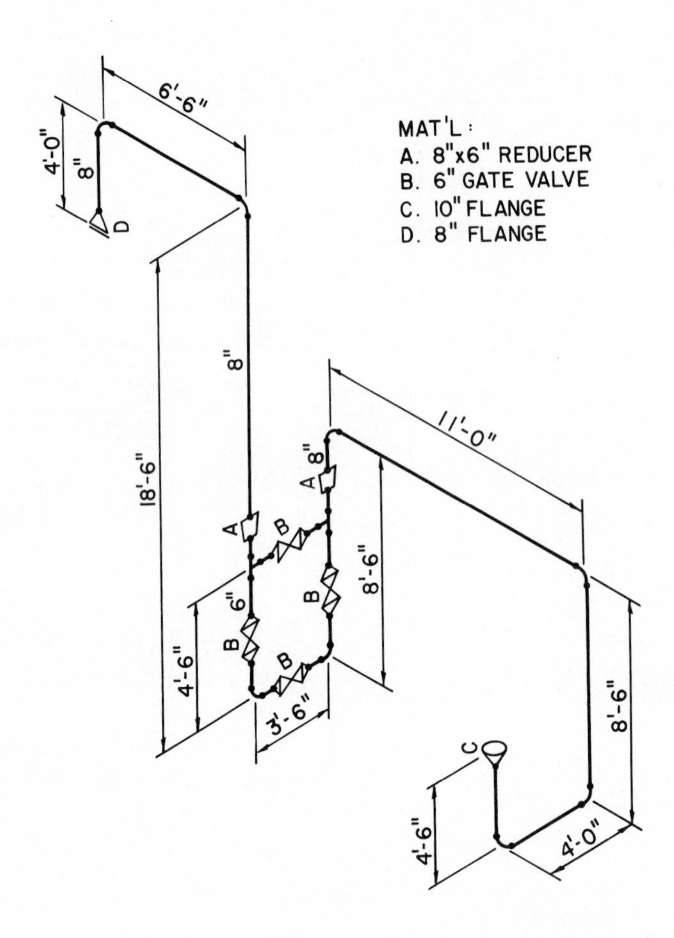

FIG. 5-33 Isometric of Fig. 5-32.

DETAIL AND SHOP DRAWINGS

Detail drawings such as those represented throughout the book are used for a variety of purposes and are taken directly from piping plan and elevation views and other piping drawings which need portions clarified in greater detail. Sometimes details are shown directly on the larger piping system drawing, though in most cases separate drawings are used and referenced on the main drawing. These drawings may include whole portions of piping systems or small detailed objects that appear on the system drawing or isometric. In some cases, the scale of a detail drawing may be larger than the scale of the system drawing.

Shop drawings (Chapter 9) can either be isometric or orthographically projected views of piping systems. They identify various aspects of piping configurations such as special bend radii, informational callouts and identification, and all the dimensions a shop needs to construct the portion of the system in question. The drafter must be concerned with all aspects of detailing shop drawings and other special aspects of piping drafting for fabricated parts.

Shop fabrication drawings (*spools*) are used for most welded configurations of 3-in. (7.62-cm) pipe and larger. Most small-diameter fabricating–2 in. (5.08 cm) and under –is usually left to the field crew to make up at the job site. The more frequent the welds and the larger the pipe size for a fabrication, the more likely it is to be constructed in the shop. Whether these *spool sheets* are drawn in isometric or orthographic projection depends on the fabricated section, its configuration, and the perspective required to convey all design information clearly.

Spool sheets identify the line number and location, pipe size, welded joints (and sometimes the type of weld), pipe materials, heat treatment, painting, coatings, pipe lengths, fittings, flanges, and any other pertinent information. Detail dimensioning on shop drawings is similar to that employed on other piping drawings, though there are certain variations when the system is put in either isometric or orthographic projection. (Orthographic double-line is preferred.)

It is usually on the detail drawings or spool drawings that center-to-center and face-to-face dimensions for flanges and fittings are given. These dimensions are usually omitted on large system piping drawings. Figure 5-29 shows the basic dimensioning procedures for detailing face-to-face and end-to-end dimensions of fittings and flanges. Note that hash marks are used inside dimension lines for face-to-face dimensions that include the gasket dimensions for valves and flanges. The manufacturer's gasket and face-to-face dimension should be taken into account when detailing these components.

Callouts are another important aspect of piping drawings. Reducing fittings, tees and special items such as reducing elbows must be called out; all valves and other components are also referred to on the detail drawings. (See Chapter 7 for isometric use of dimensioning and spool callouts.) Great care must be taken always to detail adequately and show callouts on the piping system. Leave nothing to the imagination of the construction crew, unless the drawing is for field construction, in which case dimensions will be worked out in the field (see Chapter 9).

REPRODUCTION METHODS

Blueprinting is used throughout the drafting profession, as are xeroxing and microfilm reduction of drawings. The drawings themselves are completed on transparent paper—plastic, mylar, linen, bond, or vellum. Choice of paper depends on the cost and length of time the drawing will be used or stored. Blueprints are still the least expensive way of reproducing drawings for all types of drafting, especially when it is necessary to produce a large quantity of similar prints. Brownlines are used so that more prints can later be run off the brownline instead of the original drawing. This practice saves wear and tear on the original drawing and allows users in the field to have reproductions made without the original.

Microphotography, xeroxing, and microfilming are becoming widely accepted practices in the drafting field—especially in piping, where there is a need for permanent records stored in a small space. Drawings that must be reduced need well-spaced lettering and line work with all dimensions a minimum of 1/16 in. (0.16 cm) from dimension lines so they do not blend when reduced.

PHOTO DRAFTING

Photo drafting has been widely used in the piping field in the last few years. Photo drawings are vellum transparencies made by projecting a picture of a facility through a halftone screen on sensitized vellum paper. The resulting photo drawing can then be drawn upon with standard drafting equipment to show alterations that may be required for a project. Photo drawings can then be reproduced in the same way with the xerox or blueprinting technique. Photo drawings have become useful for piping additions and alterations, for connections at pumps, exchangers, tanks, and pipeways, for instrument installations, for electrical installations and alterations such as starter racks, for conduit work, and for other details such as structural supports.

COMPUTER-AIDED DRAFTING

Drawing with a computer is becoming cost-effective in many of the drafting disciplines, including the piping, architectural, civil, mapping, and electrical fields. Computer programs on the market today have proved the effectiveness of *computer-aided drafting*—they allow for extremely fast, high-quality, error-free line work. With a few weeks of training in the system being used, a skilled drafter can become an operator capable of finishing drawings in a fraction of the time required for conventional hand-drafting procedures. Employers in the computerized drafting profession have found it much easier to train a drafter in computer techniques than to give a computer technician the necessary drafting background.

The main features of an automated drafting system are the computer, an input station, and a plotter. The *input station* translates pictures, lines, and notes into the computer's binary language of electronic *bits.* The *plotter* reads this encoded information in order to make finished drawings.

This section discusses the general procedure for using the computer-aided drafting (CAD) system. In most cases, the designer or engineer will sketch the required project, including calculations and dimensions for component placement (as in the case of isometric spool sheets). Then the project is constructed at an input station (Fig. 5-34). The drafter is

FIG. 5-34 CAD input station. (Courtesy Automated Drafting Corp.)

seated in front of a visual display unit called a cathode-ray tube (CRT) where the drawing in progress is projected.

With the use of a *digitizer,* sketches, drawings, and other details are converted into information that can be stored in the computer bank by means of coordinate designations. The digitizer allows the designer or drafter to record engineering information in the computer graphically. The digitizer has the ability to produce *X*-axis and *Y*-axis line drawings. The digitizer is also used in conjunction with the CRT. The digitizer will automatically change the information programmed into the computer and display it graphically on the CRT. Thus the drafter is able to draw by means of a keyboard and an electronic cursor.

All coded information is visually displayed on the CRT screen (Fig. 5-35), where it is possible to make extensive rearrangements and changes in design. Thus, by pointing with the cursor (Fig. 5-36), the drawing on the table is transferred to the CRT. If the drafter is working from sketches, or engaging the computer in direct calculations, he or she can use the CRT console and render precise drawings by using simple commands (Fig. 5-37). Instant symbol change, mirror reversals, quick change in size, and reduction of drawing (including scale changes) are all available. Elaborate symbols can be reduced and recalled for future use (Fig. 5-38).

Computer drafting also provides a fast and accurate method of checking the drawing and making design changes at any time with a minimum of time and expense. The computer system can perform a variety of shortcuts in computations, design analysis, material takeoff, reductions, and creative design. Figure 5-39 shows a closeup of the table. For simple drawing procedures, there are crosshairs (inside the circle) and four buttons on the console. The drafter positions the cursor and pushes the first button: Begin a

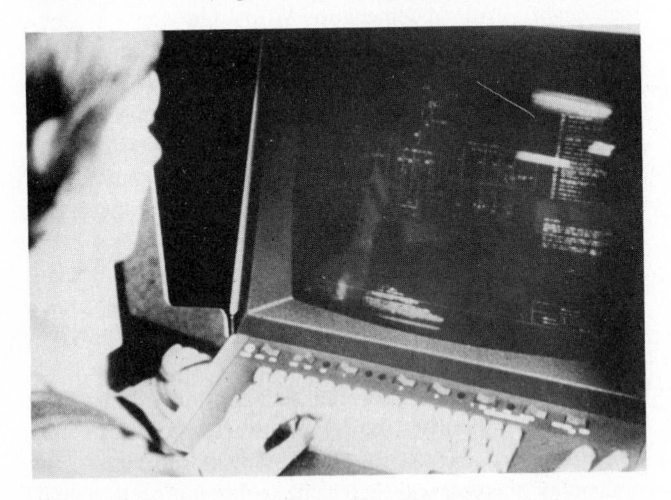

FIG. 5-35 CRT and input keys. (Courtesy Automated Drafting Corp.)

FIG. 5-37 Dual input stations. (Courtesy Automated Drafting Corp.)

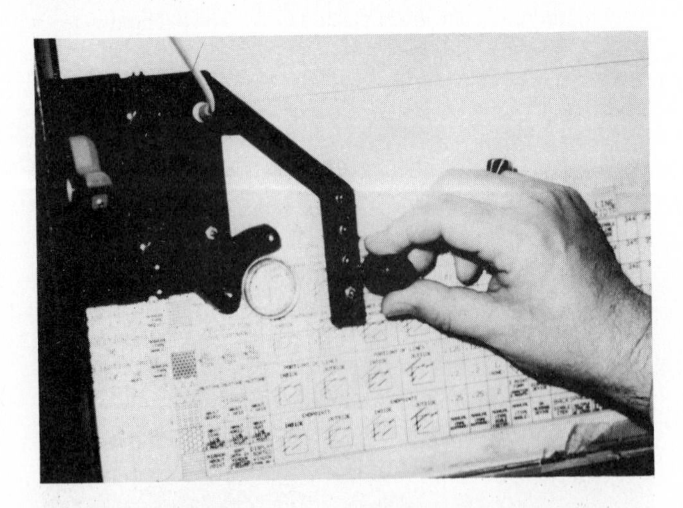

FIG. 5-36 Symbols are selected by positioning the cross hairs directly over the symbol on the existing menu and pressing the input button. The symbols will then appear on the CRT as positioned. (Courtesy Automated Drafting Corp.)

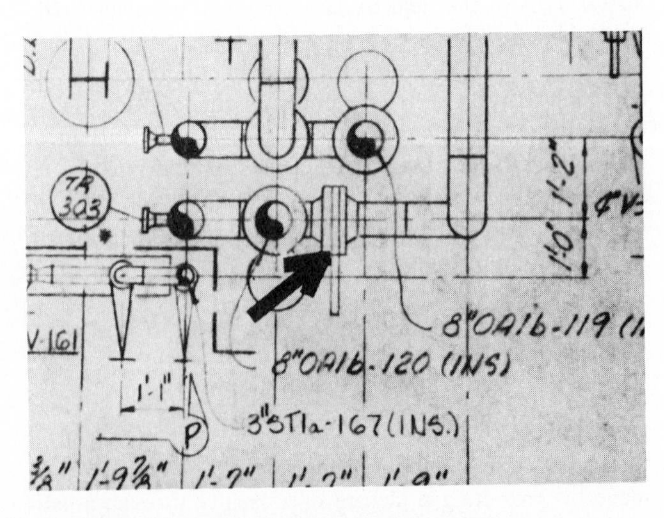

FIG. 5-38 Drawing showing the positioned symbol. (Courtesy Automated Drafting Corp.)

FIG. 5-39 The drafter is taking input directly from an existing drawing, which will then appear on the CRT. (Courtesy Automated Drafting Corp.)

FIG. 5-41 Computer systems for a CAD system. (Courtesy Automated Drafting Corp.)

Line. Then, moving the endpoint of the line, the drafter pushes the second button: Solid Line. The line is encoded and displayed instantly on the CRT. The third button is Dashed Lines; the fourth is Insert a Figure. As shown in Fig. 5-36, symbols previously drawn and stored in the computer's memory can be selected by the drafter and inserted anywhere on the drawing. Figure 5-40 shows a *menu* of piping drafting symbols. By placing the crosshairs on the reference point in the chosen symbol, the drafter can immediately project it into the drawing with the Insert a Figure button. Another method consists of a menu made of clicker buttons—the cursor stays in position on the drawing while the clicker signals the insertion. The drafter has the option of working directly on the CRT screen. The keyboard is used to instruct the computer (Fig. 5-35).

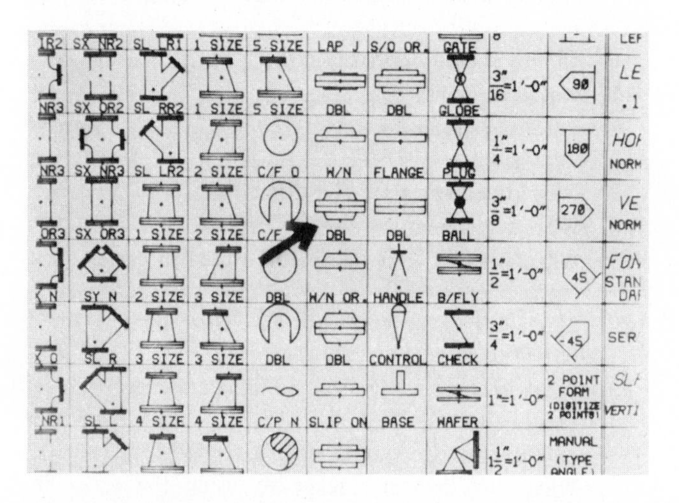

FIG. 5-40 Close-up showing the symbol menu. The chosen symbol will then be positioned on the CRT. (Courtesy Automated Drafting Corp.)

Using the cathode-ray tube and a light pen, the drafter can sketch the original design directly on the tube, recall various symbols, and position them with the light pen. With this method the design work is completed directly on the tube, making it possible to make extremely fast and accurate changes of symbols and equipment.

After the drawing is completed and projected on the cathode-ray tube, it is possible to get a printout or xerox copy of the displayed format in order to verify the information and see the finished product. After a drawing has been approved, the computer encodes all drawings and design information necessary for the project on magnetic tape, disks, drums, or punch cards in its own binary language (Fig. 5-41). This permanent record can be retrieved at any time and displayed on the CRT. Extensive design or drafting changes can then be made using the same technology that created the original drawing. The operator, by recalling stored information, can create a perfect drawing and reproduce the drawing in the desired quantity (by means of an ink or electrostatic copier). Some plotters stand vertically, as in Fig. 5-42. In this version, a drawing head moves across the top of the plotter while rollers move the paper vertically. Complicated as this arrangement may seem, the plotter is capable of drawing very fast (10 in. per second). Figure 5-43 shows a flatbed plotter (which is somewhat slower) used for larger drawings and more accurate line work.

Because of the variety of procedures and programming options available, the CAD system is widely used in civil, mapping, and architectural drafting. It has also been suc-

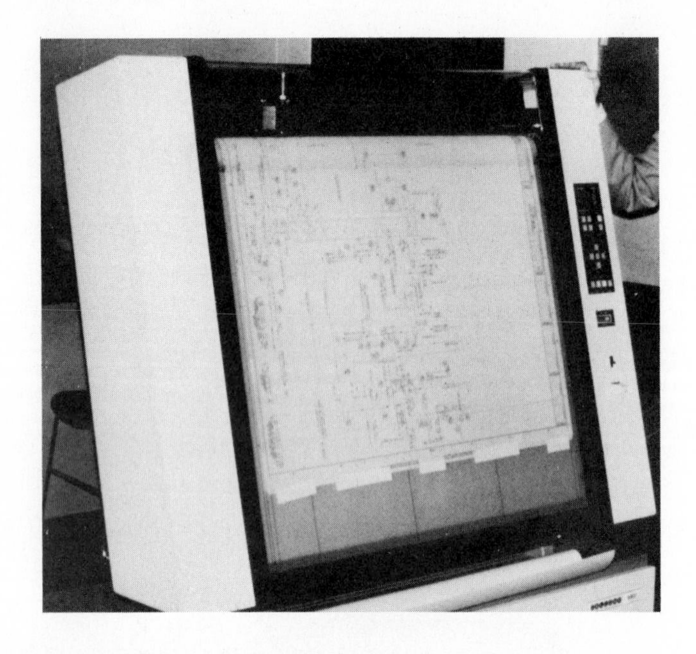

FIG. 5-42 Vertical plotter. (Courtesy Automated Drafting Corp.)

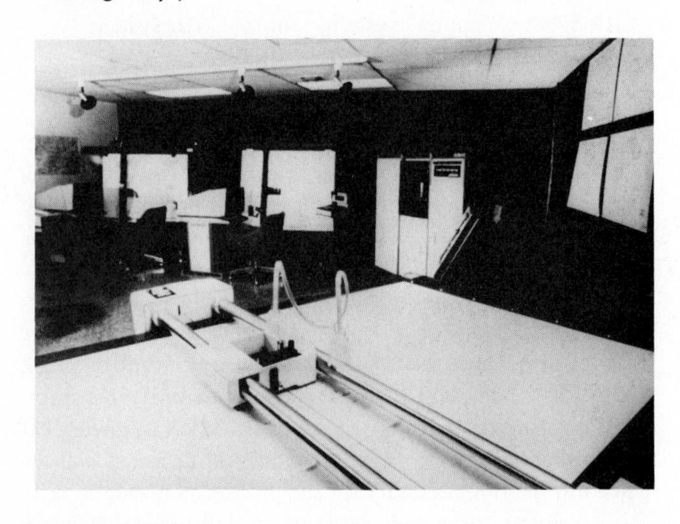

FIG. 5-43 Flatbed pen plotter. (Courtesy Automated Drafting Corp.)

cessfully adapted to electronics and electrical drafting and lately has entered the field of mechanical and piping drafting and design. The CAD system is used in the piping field for drawing flow diagrams (which are usually presketched) and isometric spool drawings. The system does not, however, lend itself to large plan, section, and elevation drawings with present-day techniques.

Designing with the CAD system has been gaining acceptance, though computer time is so expensive that designing directly through the system is feasible only in certain situations. The CAD system offers a somewhat simplified method of designing because of the computer's ability to store a large bulk of piping symbols, details, and equipment variations, all precoded and available for immediate recall. Symbols such as those for valves, vessels, pumps, and fittings are immediately projected on a screen, thereby eliminating the repetitious drafting procedures usually associated with piping.

QUIZ

1. A piping drafter should be well versed in standard ____________ drafting practices and ____________.
2. Knowledge of piping handbooks, ____________, and company standards is essential to the piping drafter.
3. Orthographic views are used in piping drafting for ____________ and ____________.
4. Name three types of support drawings that a pipe drafter must have before he or she can start to lay out a system.
5. How is the isometric drawing used in relation to the total piping system design?
6. What are the advantages to using a model?
7. What scale is usually used on a model?
8. Name three drawbacks to using models for piping systems.
9. What is a pipeline schedule?
10. How are computers used in piping system design and drafting?
11. What is a drafting machine and what equipment does it replace?
12. What type of pencil lead should be used for the finished drawing?
13. How can a constant line thickness be maintained?
14. Heavy drafting lines are used for what representations?
15. One of the main requirements for lettering is ____________.
16. Letters should be spaced ____________.
17. It is preferable to use ____________ lettering.
18. Lettering should be consistently ____________ inches in height.
19. Name the major uses of a title block.
20. What determines the size of symbols?
21. When is the double-line method used in piping drafting?
22. Single-line drawings always have their symbols represented to scale. True or false?
23. What scale is usually used in piping drawings?
24. Dimensions for symbols and fittings are derived from manufacturer's catalogs and correspond to ____________.
25. What are the advantages to using the single-line method?

26. In the single-line method the pipe's ____________ is shown to scale at the ends of the pipe.
27. What fittings are always rendered with double lines?
28. When is the pictorial method used?
29. What is a developed view?
30. How are drawings reproduced?
31. All piping lines are represented double-line for ______ ____________ and over.
32. Dimensions are put ____________ of the dimension line.
33. What are the uses of coordinate lines?
34. What scale is preferred on most pipe drawings?
35. Arrowheads should touch the ____________ and be about ____________ inches in length.
36. Spacing between dimension lines should be about ____________ inches.
37. Define T/S, B/S, El.
38. Double dimensions should ____________ be used.
39. When are gaskets dimensioned?
40. Vertical dimensions should be established from ______ ____________.

PROBLEMS

Use the appendix to find the size and configuration of valves and fittings and to construct the bill of materials for the appropriate problems. Valve and fitting catalogs may be used also. Use Fig. 5-9 and Fig. 5-10 to set up paper.

1. Prepare lettering exercise sheets from Fig. 5-8.
2. Complete missing projections in Figs. 5-19 and 5-20.
3. Complete a single-line drawing of Fig. 5-44. Place all symbols and show the proper configuration of line. No scale is used for this project. Note that problem 3, 4, and 5 may have aspects that need to be changed, or valves and fittings that either do not exist or are not placed correctly. There is no flow direction for these three projects, so they would not "work" if

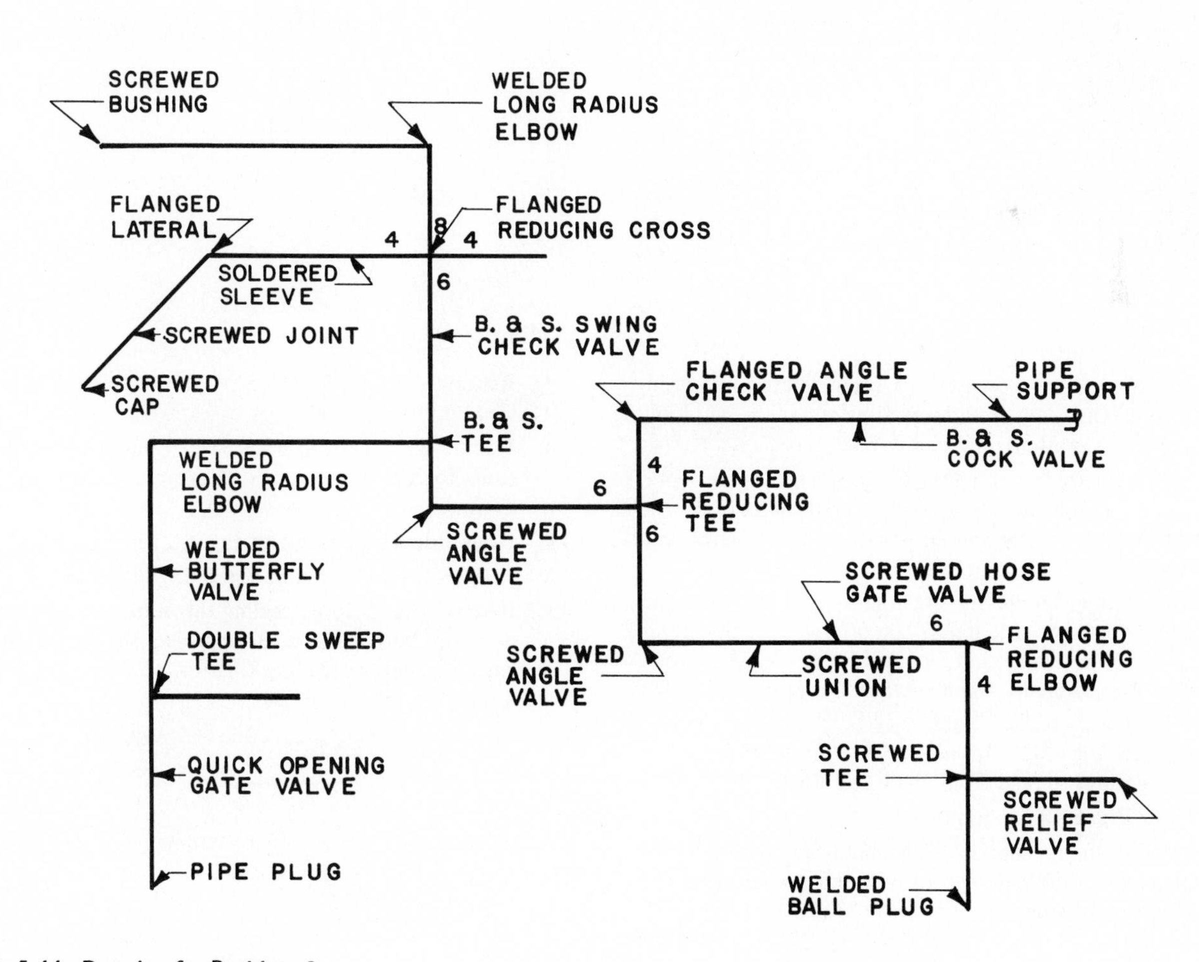

FIG. 5-44 Drawing for Problem 3.

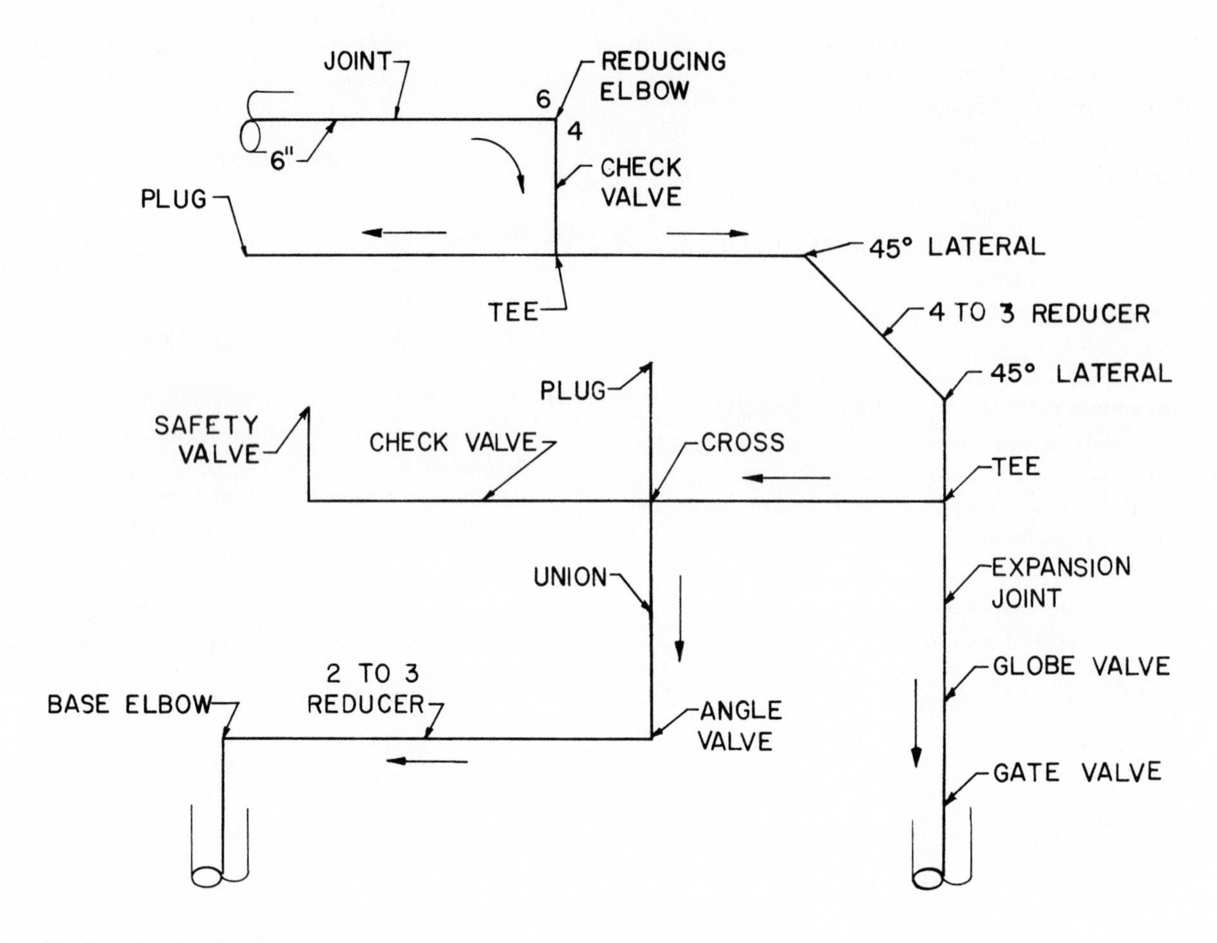

FIG. 5-45 Drawing for Problem 4.

constructed as shown. These projects are only to introduce the single-line, double-line and pictorial piping drawings, symbols, and component parts.

4. Using the pictorial method, complete Figure 5-45. Use the appendix to do a material takeoff of the valves and fittings and to construct the general configuration of the components.
5. Complete a double-line drawing of Fig. 5-46. Use the appendix to do a material takeoff for as many items as possible. Valve and fitting catalogs may be used also for problems 4 and 5 when the appendix is not sufficient.
6. Redraw Fig. 5-23, using 3/8″ = 1′ scale. Change the coordinates to read from north and west.
7. Draw Fig. 5-30 3× book size.
8. Choose an appropriate size and draw Fig. 5-31 in plan and sufficient elevation views to describe the piping configuration.
9. Redraw Fig. 5-17 to 3/8″ = 1′ scale. Project a top view for this vessel.
10. Draw Fig. 5-18 2× book size. Follow all specifications for the drawing of orthographic, single-line pipe drawings.
11. Draw a developed view of Fig. 5-21 or Fig. 5-30.
12. Redraw the plot plan in Fig. 5-25.
13. Redraw Fig. 5-26 by scaling the book and making the drawing 2× book size. This project should be studied carefully before reading Chapter 10 on piping system design.

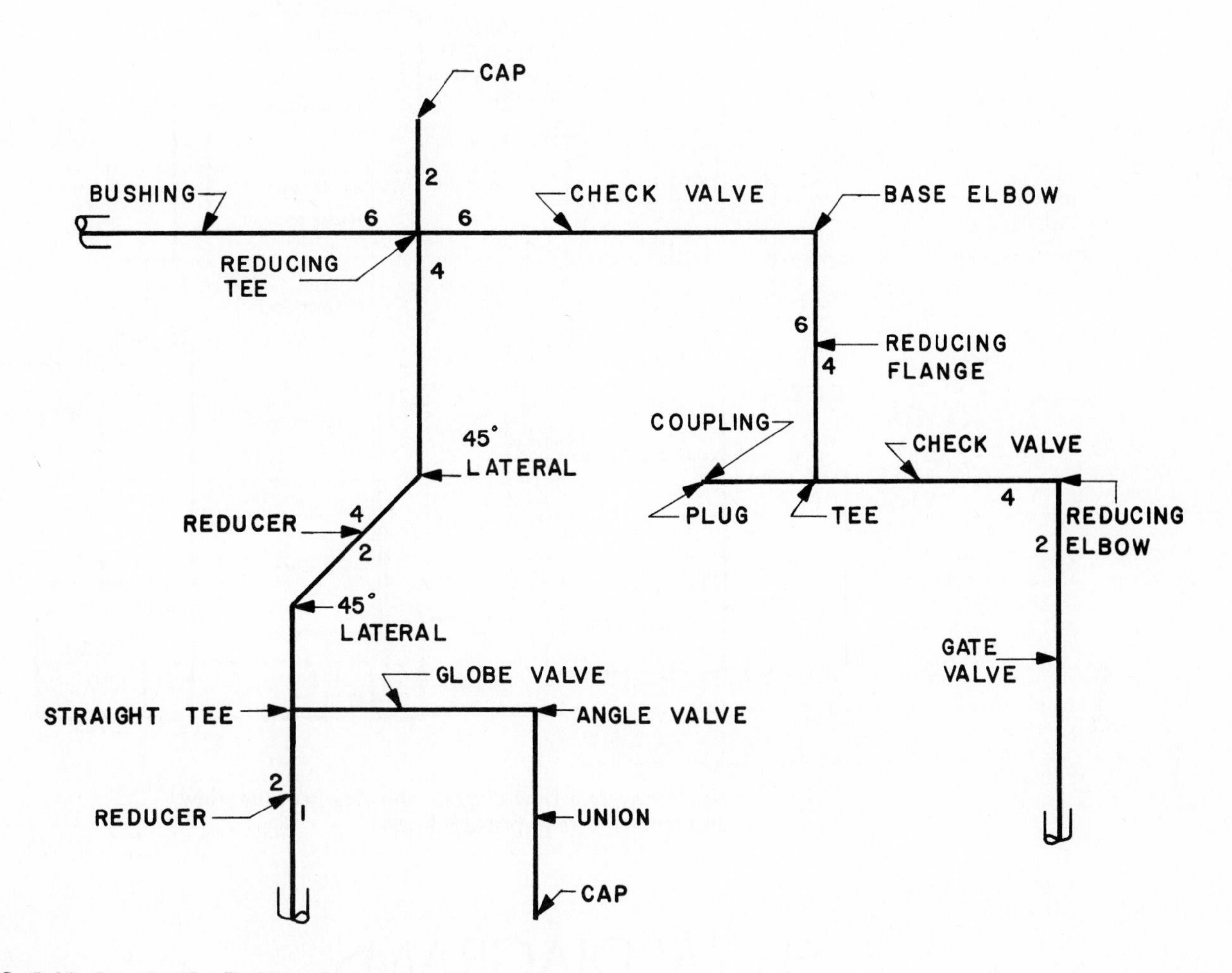

FIG. 5-46 Drawing for Problem 5.

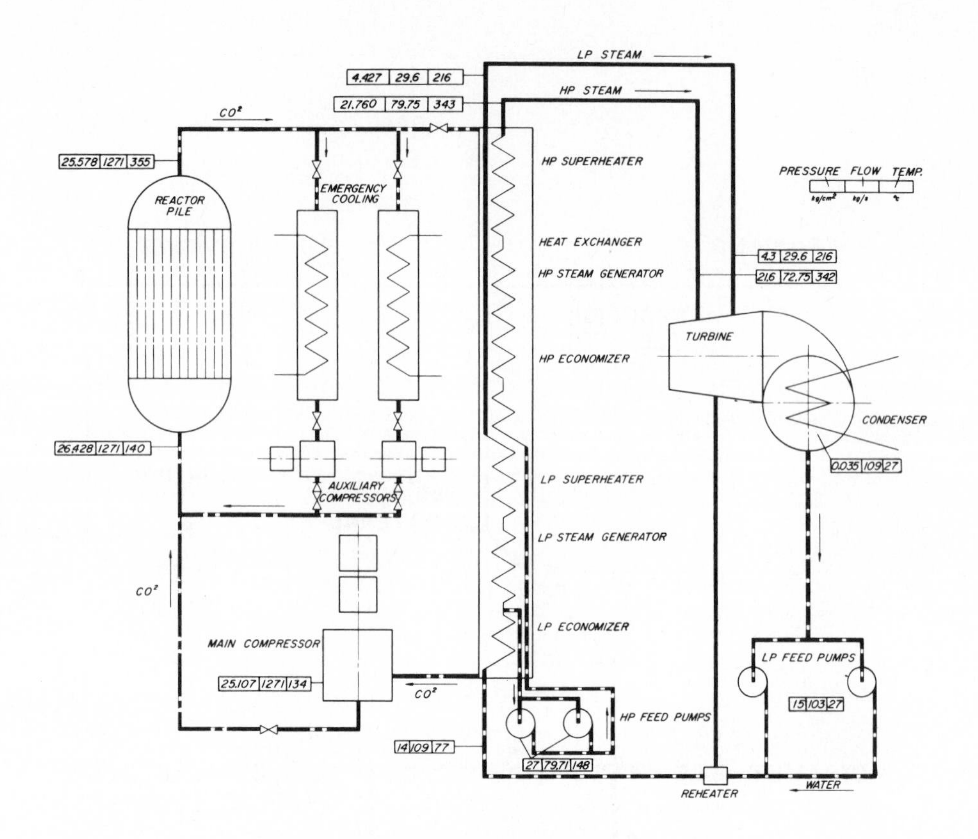

Nuclear system flow diagram showing pressure, flow, and temperature of primary lines.

6 FLOW DIAGRAMS AND INSTRUMENTATION

Flow diagrams are the basic blueprint of the process, function, and equipment for a job. They provide all the data necessary to begin the design and layout of the system. In most cases, flow diagrams can be regarded as pictorial statements of purpose for the whole project or part of it, forming the link between process calculations and the designed functional system. Schematic flow diagrams are essential in the early phases of development and process study. After the preliminary calculations and feasibility studies are complete, the more complicated stage of creating a process design with the aid of a flow diagram is instituted—and, from this, the mechanical flow diagram (piping and instrumentation diagram) (Fig. 6-1).

The flow diagram is the first drawing to be completed and used by most departments in a project involving piping systems. Starting from a sketch, the drafter lays out the flow diagram according to standard procedures utilizing symbols and piping line representations. The flow diagrams are the drafter's first responsibility in the sequence of steps toward a completed set of design prints or model which will be shipped to the construction site.

A drafter without much experience will not readily grasp all the engineering and background information of the system. He or she must be able to draw the flow diagram accurately, usually from a rough sketched form, with only the essential information. This sketch is designed and completed by the project or instrument engineer. The drafter, through experience with many types of flow diagrams, will eventually come to understand the function and operating principles behind each project as represented by the flow diagrams.

Flow diagrams are used for all applications of piping systems. The drafter must be able to read the designer's sketch and convert it into a symbolic flow diagram of the system—or, if it is an instrumentation sketch, into a *piping and instrumentation diagram* (P&ID). From this symbolic draw-

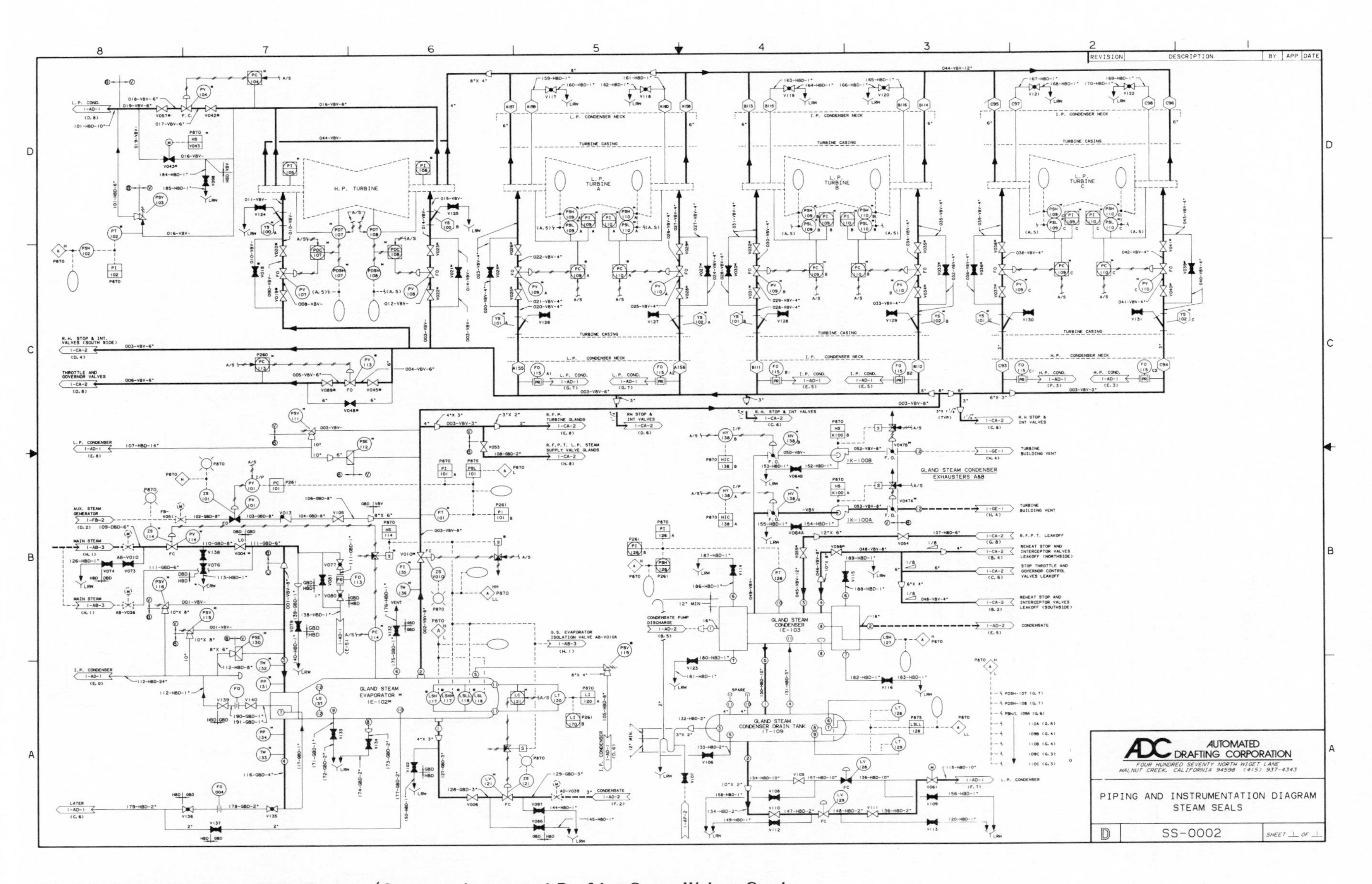

FIG. 6-1 Computer-drawn flow diagram. (Courtesy Automated Drafting Corp., Walnut Creek, CA)

ing, the drafter and designer will lay out the complete system in orthographic drawings or on the model without the aid of drawings.

In many companies involved in petrochemical process work, this procedure of designing, placing equipment, and running the pipelines directly on the model (without the use of orthographic views) from the P&ID has increased the importance of understanding, drawing, and designing the flow diagram efficiently and accurately. This knowledge and practice is essential for the engineer, designer, and drafter in the modern piping industry.

If the drafter is to comprehend the basic elements of a flow diagram, the *uses* of the diagram must be understood. In industry, the engineer prepares the preliminary flow diagram, which gives the desired process for a project along with the material and balance sheet. From this information, a complete P&ID is designed and sent to all major departments in the company's engineering and design section (Fig. 6-1). From the planning group to the system design department, all the way to the marketing and advertising department, flow diagrams are used by many groups. The preliminary flow diagram is the central source of information for all the engineering departments on a project, including the structural, electrical, piping, mechanical, and instrumentation design groups. To be intelligible to all these divergent groups, the symbols, notations, and abbreviations must be coordinated.

SCHEMATIC FLOW DIAGRAMS

There are several guidelines to follow when drawing schematic diagrams. Because each diagram represents a completely new system, there is no way to set up narrow limits for layout and drawing, but some general rules apply to all. Moreover, since they do represent various types of systems they have their own characteristics. To emphasize the difference between flow diagrams it is sometimes necessary to draw or construct them in a manner that is somewhat unique each time they are produced.

An essential consideration in the drawing of a flow diagram is the general physical arrangement, which must be well planned so the user of the drawing can readily understand and interpret the system (Fig. 6-2). High-quality line work, lettering, and symbol drawing are, of course, indispensable.

It is helpful to convey some idea of the future plot plan. The general arrangement of the equipment should also be understood before the diagram is drawn. This practice helps the design and planning groups to construct and design the plant because the flow diagram conveys an impression of the total system. Though there is no overall scale used in the drawing of flow diagrams, short pipelines should not be represented on the flow diagram as extremely long lines and vice versa. Pieces of equipment should be represented in proportion to each other and to other parts of the system.

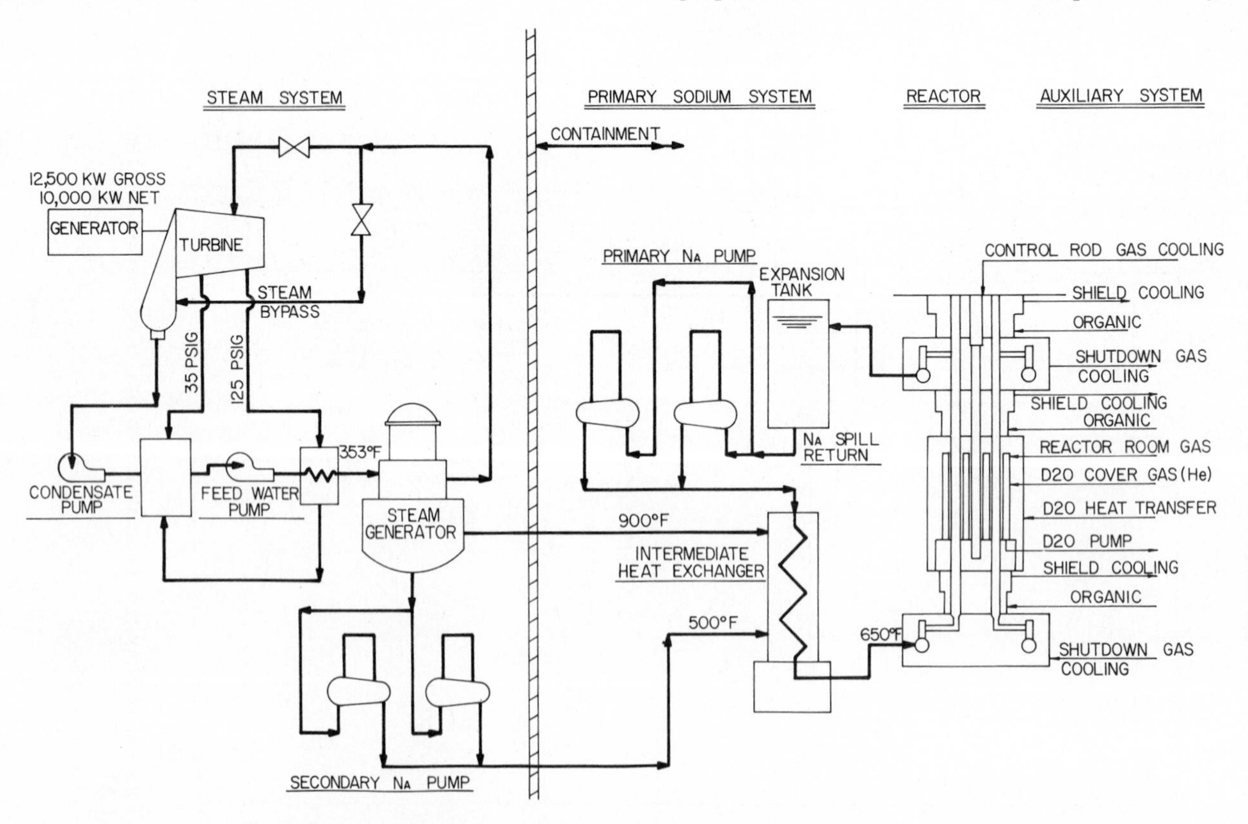

FIG. 6-2 Secondary sodium system flow diagram.

Most of the diagrams shown in this chapter are from a plan view perspective; one nuclear diagram is drawn pictorially in isometric. In general, the flow diagram is drawn in the plan view perspective except in rare cases where a process is primarily vertical or where more clarity is necessary. (In this case, an elevation diagram is prepared.) See the nuclear reactor P&ID in Chapter 13 for an example of a vertical diagram.

By establishing the sequence of flow by the sequential placement of equipment and conveying the plot plan arrangement, the flow diagram can prevent misconceptions at the early stages of the design. The piping lead drafter and the project engineer must discuss the diagram's requirements and review diagrams that have been done in the past. The lead drafter then determines the number and size of diagrams that will be required for the job. A variety of criteria are involved:

1. Drawing package systems (refrigeration units, compressor units, and so on, on separate diagrams
2. Using the size of paper (sheet size) that will accommodate the diagram without waste
3. Keeping utility and process systems distinct on the diagram
4. Preparing a line list arranged by line identification number

When starting to construct the flow diagram, the drafter can use these general instructions to set up the problem project:

1. Study similar equipment locations on other diagrams from former installations.
2. Use standard P&ID templates or follow the practice of drawing main valves 3/8 to 1/4 in. (0.95 to 0.64 cm).
3. Show consistency in line work.
   (a) All major pipelines which are part of the process sequence of flow are drawn with thick, heavy lines.
   (b) Utility lines, bypass piping, and instrument lines are shown with medium lines.
   (c) Equipment: pumps, vessels, heaters, towers, exchangers, etc., are drawn with thin lines.
   (d) All lines are black.
4. Label the description of equipment as well as its number. Do not describe the function of pumps serving equipment already labeled.
5. Refer to project instructions for equipment and line numbers. Obtain instrument numbers from the instrument and electrical group.
6. Place pumps on the drawing below other equipment whenever possible.
7. Maintain a constant elevation when lines on a flow diagram must pass onto another sheet so that the continuation of the line on the next sheet can be easily recognized and located.
8. Describe the origin and destination of process lines entering or leaving the diagram.
9. Arrange all equipment above a common baseline, with flow sequence to equipment going from left to right across the drawing.
10. Indicate the plot limit for lines entering or leaving the physical limits of the facility.
11. Break minor lines where they intersect major lines. (Some companies break all vertical lines where they intersect horizontal lines.)
12. Plan the line routing to minimize the number of changes of direction.
13. Make lines long enough to place the line code, balloon box, or number above the line.
14. Show valve and nozzle sizes when they differ from the size in the line identification code.
15. Do not overcrowd the flow diagram. If necessary, break the diagram into a number of separate drawings.
16. Add symbols to the legend to identify special items and codes which are not standardized.
17. List all drawings associated with P&ID in the reference drawing column.
18. Show utility lines on one drawing and process lines on another where possible.
19. Do not overcrowd when combining utility and process lines on the same drawing.

Symbols and Abbreviations

All flow diagrams use symbols and abbreviations. Many industries have standardized the symbols that are peculiar to the specific branches of piping. Many companies use their own symbols, and often a flow or instrumentation legend appears on the drawing as in Fig. 6-3. Drafters must know all the standard symbols and be adaptable enough to utilize any new symbols with which they come in contact.

Figures 6-4 and 6-5 give a fairly comprehensive list of piping flow diagram and instrumentation symbols. Abbreviations are used throughout diagrams to simplify callouts and descriptions of the process. (For a comprehensive list of abbreviations, see the appendix at the back of this book.) Among the common abbreviations for line materials and line identification letters are these:

Ac	Acid	HO	Hot oil
AI	Instrument air	LO	Lube oil
Am	Ammonia	N	Nitrogen
Au	Instrument air	NG	Natural gas
BFW	Boiler feedwater	P	Process lines
BO	Blowdown	PW	Process water
Ca	Caustic	R	Refrigeration
CW	Cooling water	RW	Refrigerated water
DW	Drinking water	S	Steam
FG	Fuel gas	Sc	Steam condensate
FL	Fire water line	SW	Salt water
FO	Fuel oil	V	Vent
G	Inert gas		

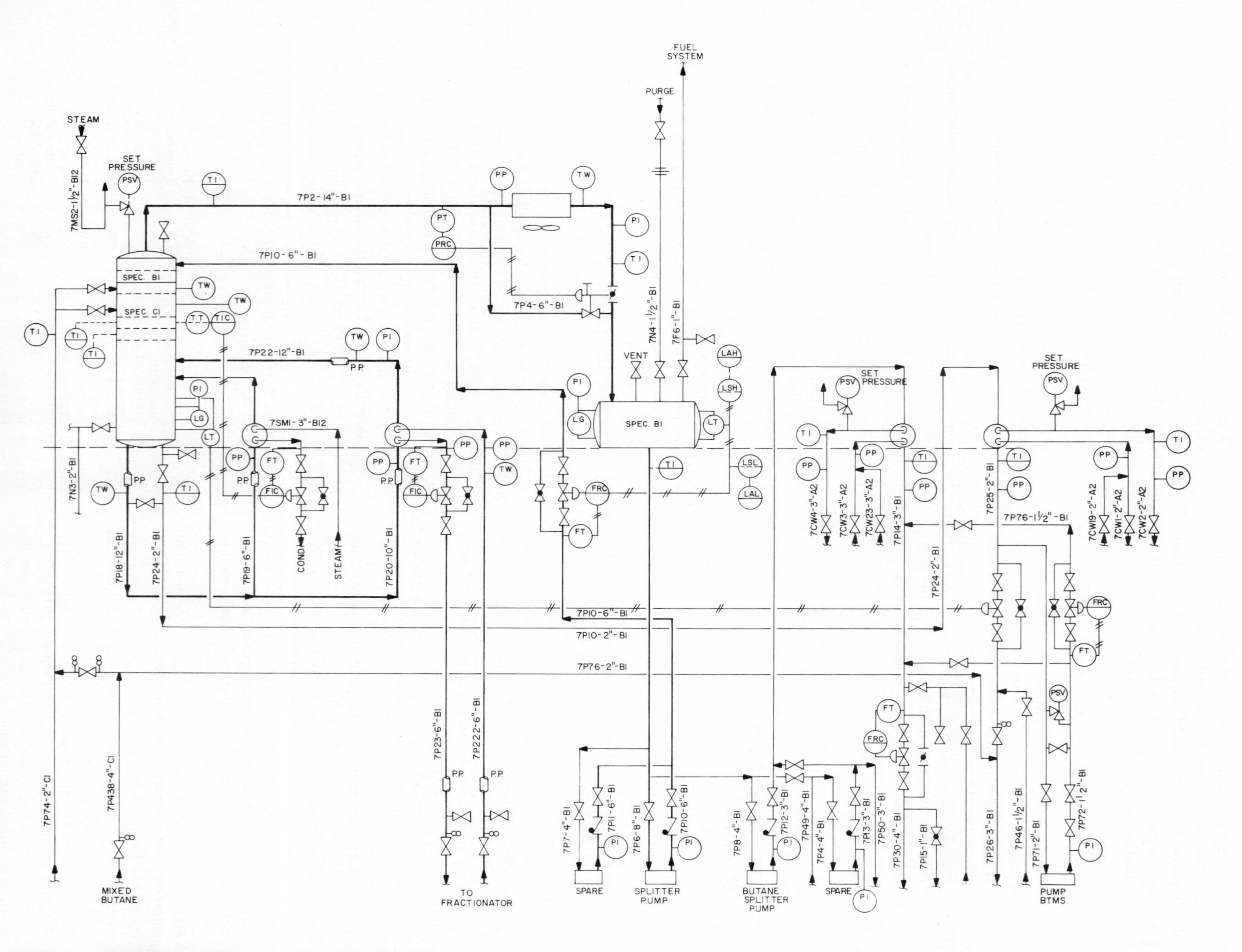

FIG. 6-3 P&ID diagram for a petrochemical installation.

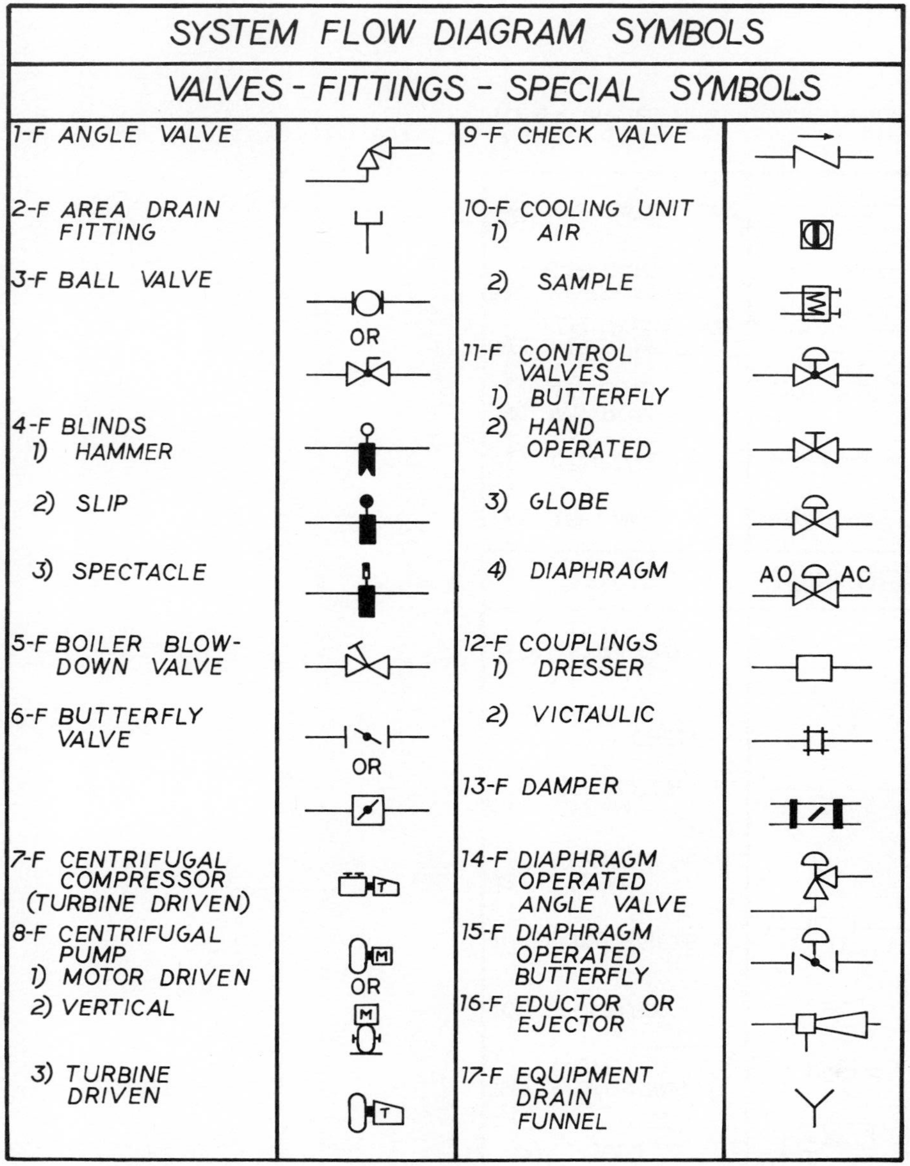

SYSTEM FLOW DIAGRAM SYMBOLS

VALVES - FITTINGS - SPECIAL SYMBOLS

- 1-F ANGLE VALVE
- 2-F AREA DRAIN FITTING
- 3-F BALL VALVE
- 4-F BLINDS
  1) HAMMER
  2) SLIP
  3) SPECTACLE
- 5-F BOILER BLOW-DOWN VALVE
- 6-F BUTTERFLY VALVE
- 7-F CENTRIFUGAL COMPRESSOR (TURBINE DRIVEN)
- 8-F CENTRIFUGAL PUMP
  1) MOTOR DRIVEN
  2) VERTICAL
  3) TURBINE DRIVEN
- 9-F CHECK VALVE
- 10-F COOLING UNIT
  1) AIR
  2) SAMPLE
- 11-F CONTROL VALVES
  1) BUTTERFLY
  2) HAND OPERATED
  3) GLOBE
  4) DIAPHRAGM
- 12-F COUPLINGS
  1) DRESSER
  2) VICTAULIC
- 13-F DAMPER
- 14-F DIAPHRAGM OPERATED ANGLE VALVE
- 15-F DIAPHRAGM OPERATED BUTTERFLY
- 16-F EDUCTOR OR EJECTOR
- 17-F EQUIPMENT DRAIN FUNNEL

FIG. 6-4 System flow diagram symbols (valves, **fittings**, and special symbols).

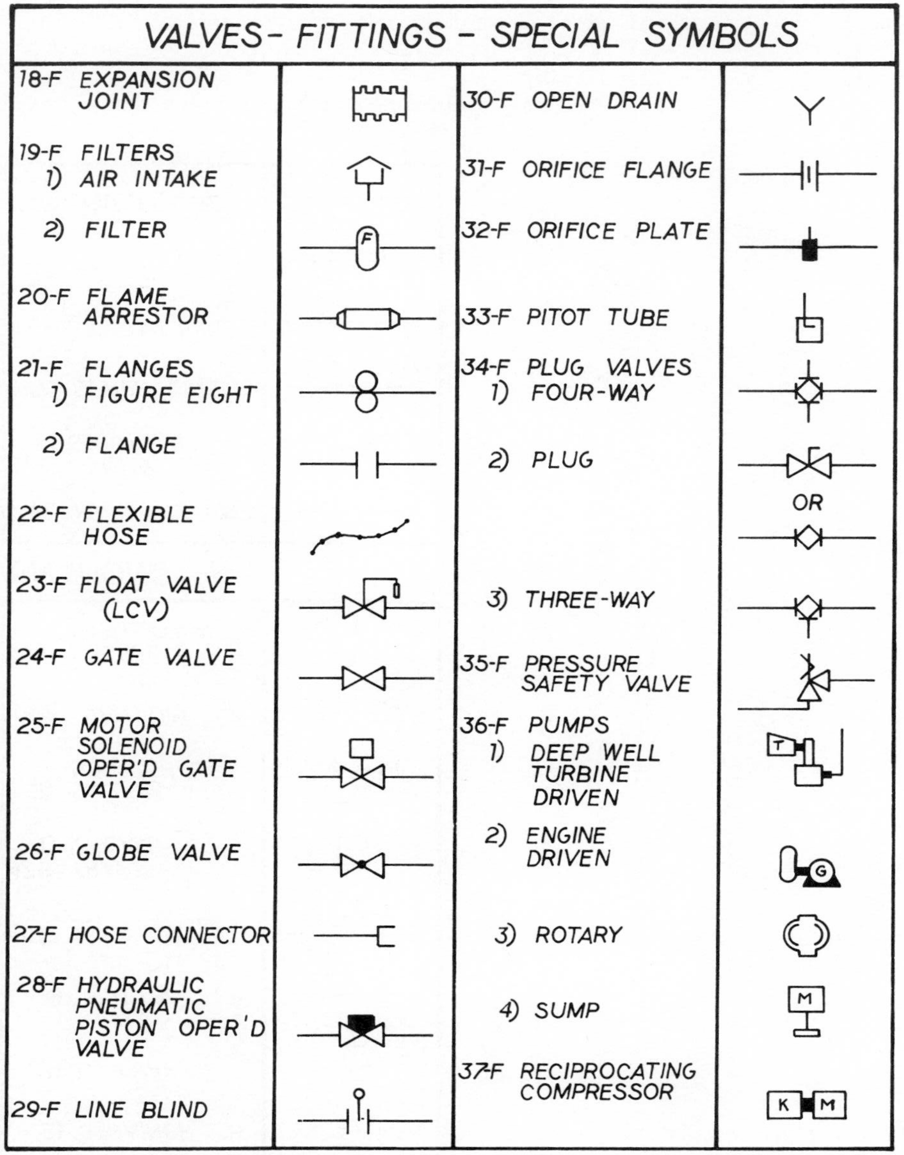

VALVES - FITTINGS - SPECIAL SYMBOLS

- 18-F EXPANSION JOINT
- 19-F FILTERS
  1) AIR INTAKE
  2) FILTER
- 20-F FLAME ARRESTOR
- 21-F FLANGES
  1) FIGURE EIGHT
  2) FLANGE
- 22-F FLEXIBLE HOSE
- 23-F FLOAT VALVE (LCV)
- 24-F GATE VALVE
- 25-F MOTOR SOLENOID OPER'D GATE VALVE
- 26-F GLOBE VALVE
- 27-F HOSE CONNECTOR
- 28-F HYDRAULIC PNEUMATIC PISTON OPER'D VALVE
- 29-F LINE BLIND
- 30-F OPEN DRAIN
- 31-F ORIFICE FLANGE
- 32-F ORIFICE PLATE
- 33-F PITOT TUBE
- 34-F PLUG VALVES
  1) FOUR-WAY
  2) PLUG
  3) THREE-WAY
- 35-F PRESSURE SAFETY VALVE
- 36-F PUMPS
  1) DEEP WELL TURBINE DRIVEN
  2) ENGINE DRIVEN
  3) ROTARY
  4) SUMP
- 37-F RECIPROCATING COMPRESSOR

FIG. 6-4 System flow diagram symbols (valves, fittings, and special symbols). (cont'd)

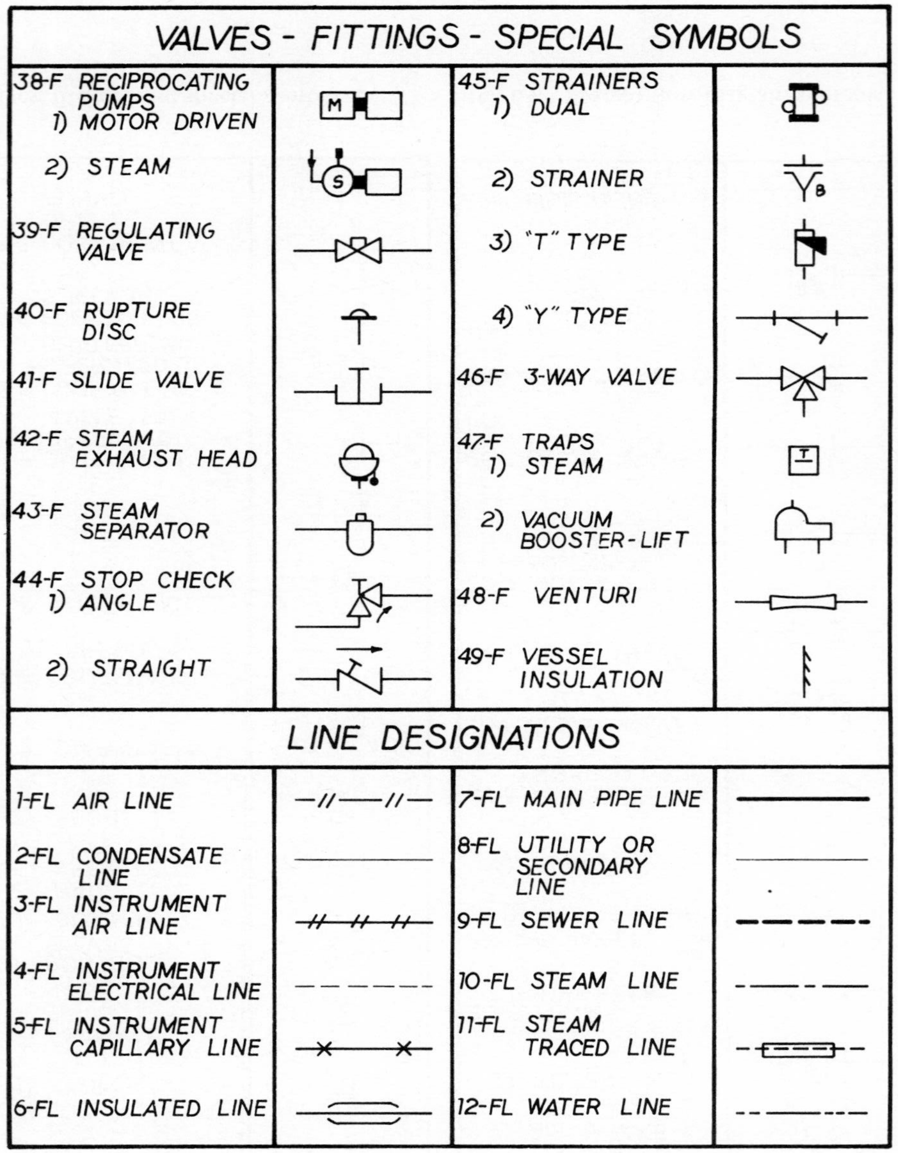

VALVES - FITTINGS - SPECIAL SYMBOLS			
38-F RECIPROCATING PUMPS 1) MOTOR DRIVEN		45-F STRAINERS 1) DUAL	
2) STEAM		2) STRAINER	
39-F REGULATING VALVE		3) "T" TYPE	
40-F RUPTURE DISC		4) "Y" TYPE	
41-F SLIDE VALVE		46-F 3-WAY VALVE	
42-F STEAM EXHAUST HEAD		47-F TRAPS 1) STEAM	
43-F STEAM SEPARATOR		2) VACUUM BOOSTER-LIFT	
44-F STOP CHECK 1) ANGLE		48-F VENTURI	
2) STRAIGHT		49-F VESSEL INSULATION	

LINE DESIGNATIONS			
1-FL AIR LINE		7-FL MAIN PIPE LINE	
2-FL CONDENSATE LINE		8-FL UTILITY OR SECONDARY LINE	
3-FL INSTRUMENT AIR LINE		9-FL SEWER LINE	
4-FL INSTRUMENT ELECTRICAL LINE		10-FL STEAM LINE	
5-FL INSTRUMENT CAPILLARY LINE		11-FL STEAM TRACED LINE	
6-FL INSULATED LINE		12-FL WATER LINE	

FIG. 6-4 System flow diagram symbols (valves, fittings, and special symbols). (cont'd)

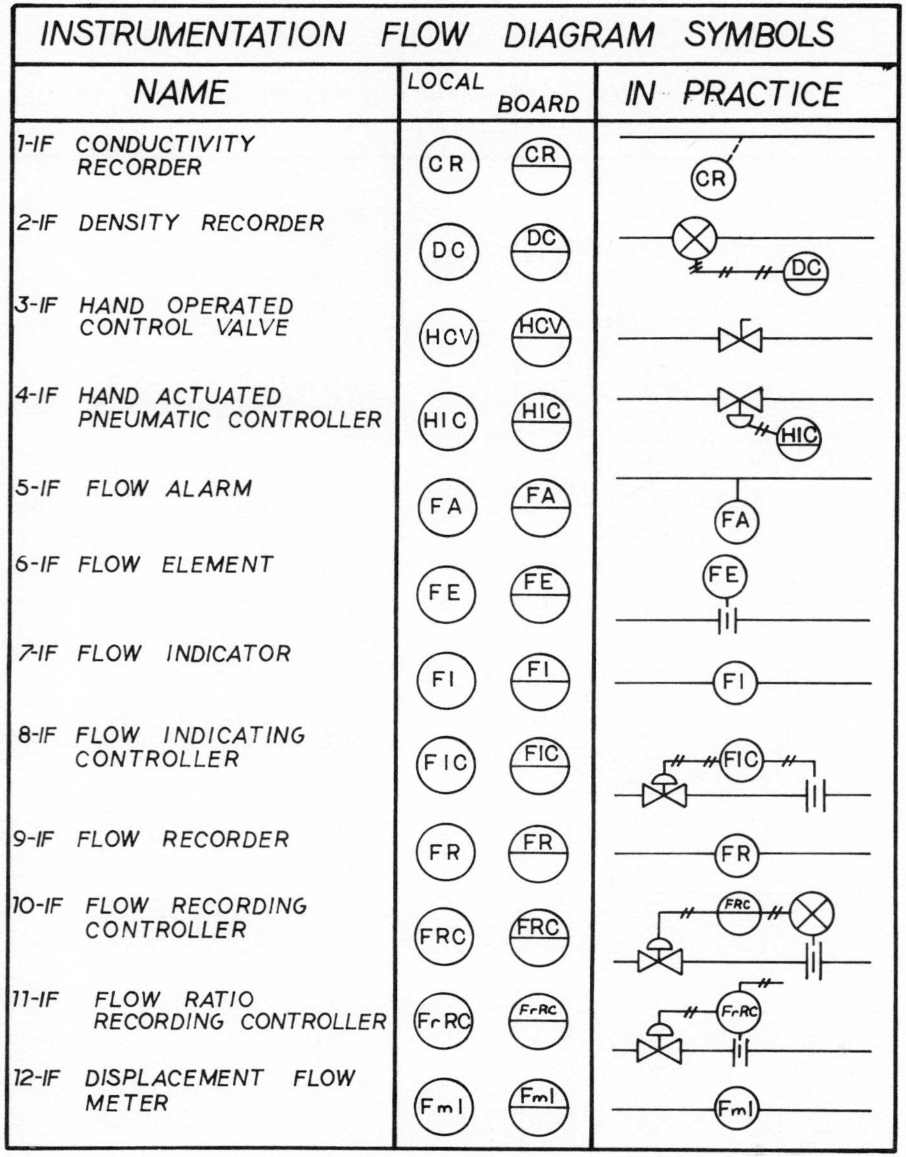

INSTRUMENTATION FLOW DIAGRAM SYMBOLS

NAME	LOCAL	BOARD	IN PRACTICE
1-IF CONDUCTIVITY RECORDER	CR	CR	CR
2-IF DENSITY RECORDER	DC	DC	DC
3-IF HAND OPERATED CONTROL VALVE	HCV	HCV	
4-IF HAND ACTUATED PNEUMATIC CONTROLLER	HIC	HIC	HIC
5-IF FLOW ALARM	FA	FA	FA
6-IF FLOW ELEMENT	FE	FE	FE
7-IF FLOW INDICATOR	FI	FI	FI
8-IF FLOW INDICATING CONTROLLER	FIC	FIC	FIC
9-IF FLOW RECORDER	FR	FR	FR
10-IF FLOW RECORDING CONTROLLER	FRC	FRC	FRC
11-IF FLOW RATIO RECORDING CONTROLLER	FrRC	FrRC	FrRC
12-IF DISPLACEMENT FLOW METER	Fml	Fml	Fml

FIG. 6-5 Instrumentation flow diagram symbols.

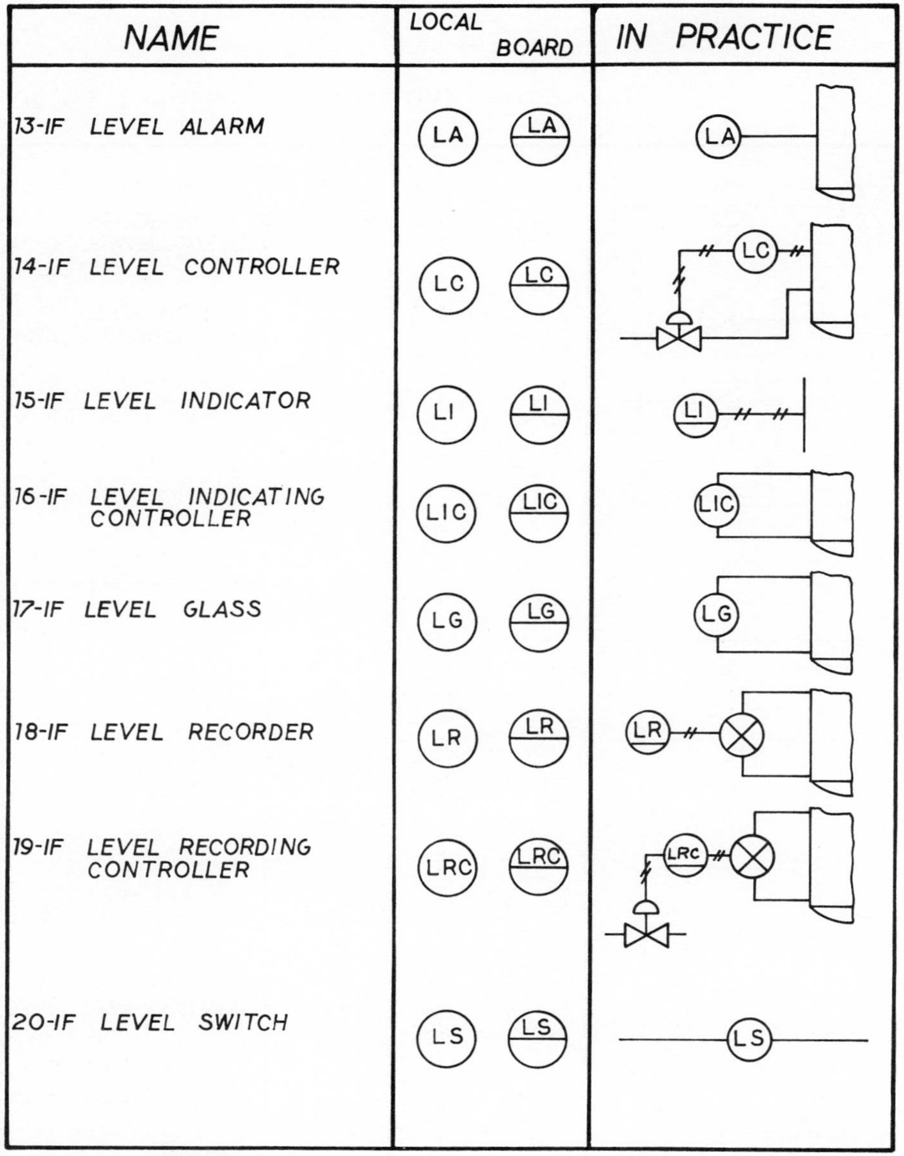

NAME	LOCAL	BOARD	IN PRACTICE
13-IF LEVEL ALARM	LA	LA	LA
14-IF LEVEL CONTROLLER	LC	LC	LC
15-IF LEVEL INDICATOR	LI	LI	LI
16-IF LEVEL INDICATING CONTROLLER	LIC	LIC	LIC
17-IF LEVEL GLASS	LG	LG	LG
18-IF LEVEL RECORDER	LR	LR	LR
19-IF LEVEL RECORDING CONTROLLER	LRC	LRC	LRC
20-IF LEVEL SWITCH	LS	LS	LS

FIG. 6-5 Instrumentation flow diagram symbols. (cont'd)

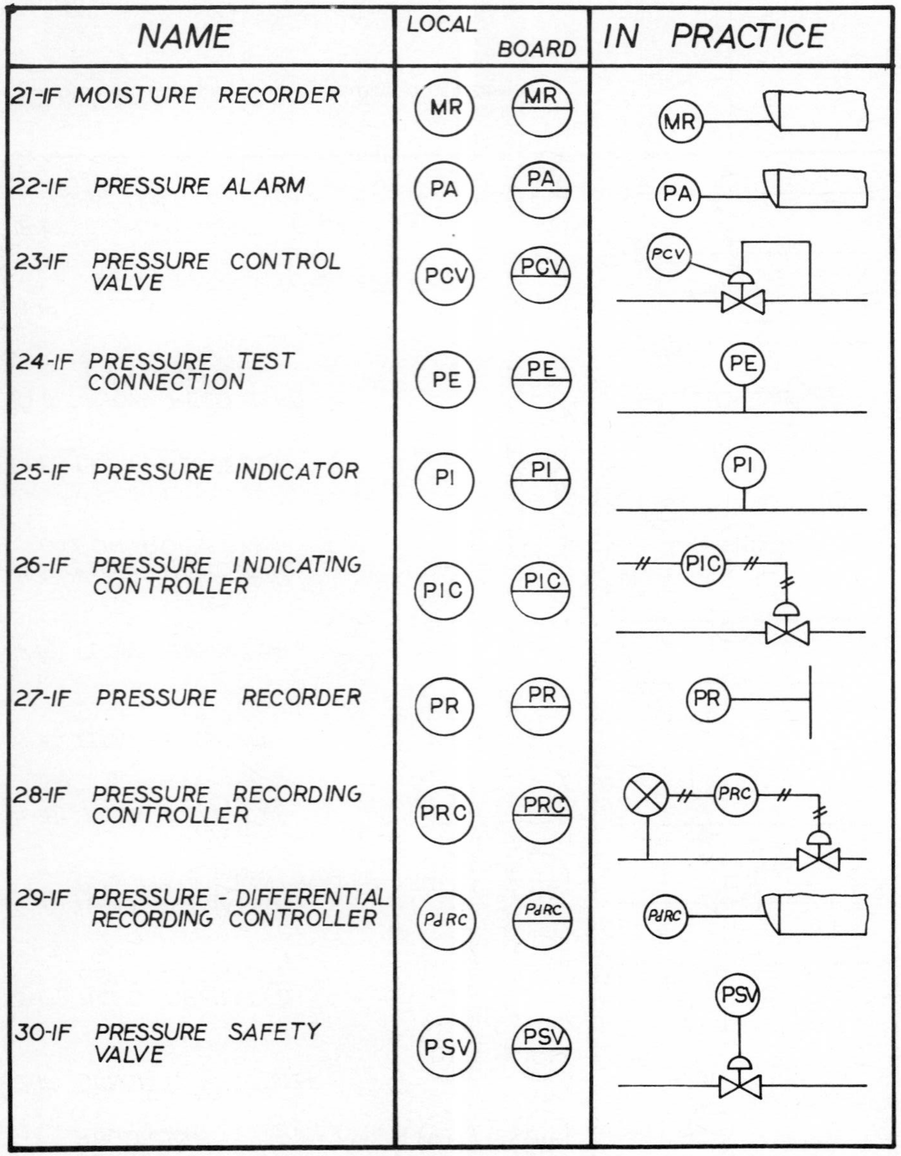

NAME	LOCAL	BOARD	IN PRACTICE
21-IF MOISTURE RECORDER	MR	MR	MR
22-IF PRESSURE ALARM	PA	PA	PA
23-IF PRESSURE CONTROL VALVE	PCV	PCV	PCV
24-IF PRESSURE TEST CONNECTION	PE	PE	PE
25-IF PRESSURE INDICATOR	PI	PI	PI
26-IF PRESSURE INDICATING CONTROLLER	PIC	PIC	PIC
27-IF PRESSURE RECORDER	PR	PR	PR
28-IF PRESSURE RECORDING CONTROLLER	PRC	PRC	PRC
29-IF PRESSURE - DIFFERENTIAL RECORDING CONTROLLER	PdRC	PdRC	PdRC
30-IF PRESSURE SAFETY VALVE	PSV	PSV	PSV

FIG. 6-5 Instrumentation flow diagram symbols. (cont'd)

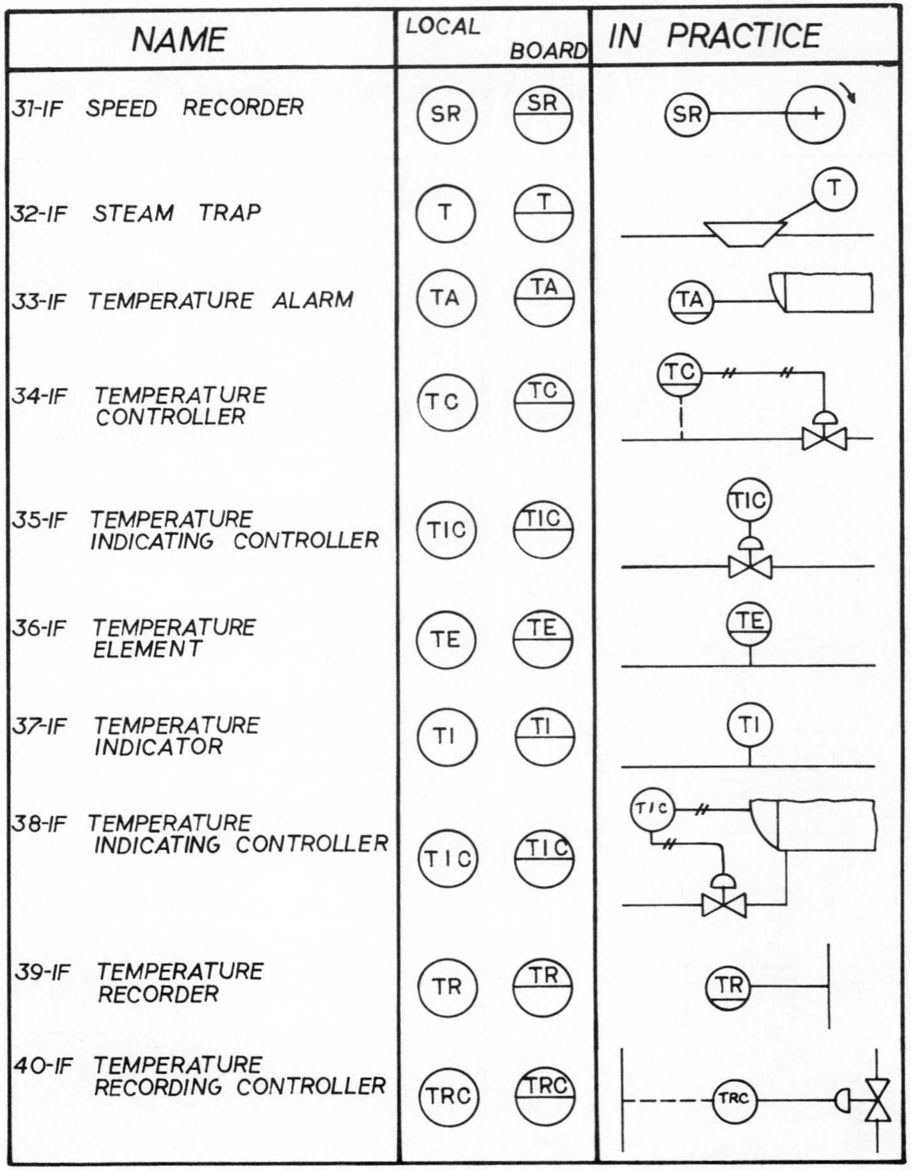

NAME	LOCAL	BOARD	IN PRACTICE
31-IF SPEED RECORDER	SR	SR	SR
32-IF STEAM TRAP	T	T	T
33-IF TEMPERATURE ALARM	TA	TA	TA
34-IF TEMPERATURE CONTROLLER	TC	TC	TC
35-IF TEMPERATURE INDICATING CONTROLLER	TIC	TIC	TIC
36-IF TEMPERATURE ELEMENT	TE	TE	TE
37-IF TEMPERATURE INDICATOR	TI	TI	TI
38-IF TEMPERATURE INDICATING CONTROLLER	TIC	TIC	TIC
39-IF TEMPERATURE RECORDER	TR	TR	TR
40-IF TEMPERATURE RECORDING CONTROLLER	TRC	TRC	TRC

FIG. 6-5 Instrumentation flow diagram symbols. (cont'd)

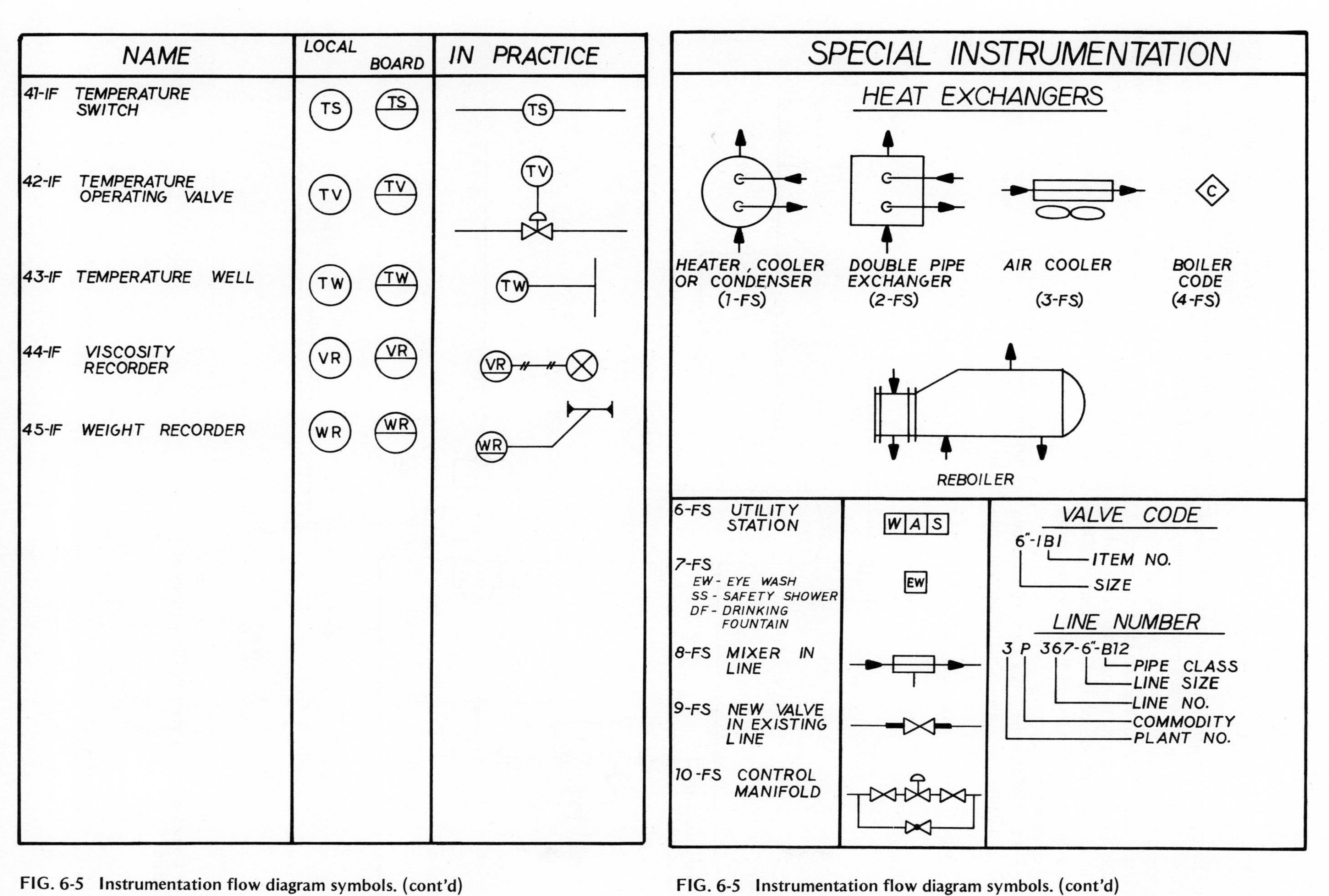

FIG. 6-5 Instrumentation flow diagram symbols. (cont'd)

FIG. 6-5 Instrumentation flow diagram symbols. (cont'd)

Notations

When constructing any diagram, provide sufficient codes, notations, or callouts next to the equipment or directly on the diagrammatic symbol that is used to represent the piece of equipment. Each company has a standard set of equipment designations, developed through experience, which provide for an efficient and easily read drawing.* When equipment is not standard, it is sometimes necessary to call out special operating characteristics.

Figure 6-6 is a set of drawings giving a general indication of the use of notations and callouts, including item numbers on the pieces of equipment. Most companies have their own item codes or item numbering procedures. A flow diagram may be divided into sections and numbered to provide the layout drafter with information to start designing the system. Section and area limits are indicated on many flow diagrams, especially when the system being represented involves a large and intricate set of equipment and pipes.

*On many of the diagrams in this chapter, designations, line numbers, and other descriptive callouts are missing because these diagrams have been donated by firms who requested that their identifying information be excluded.

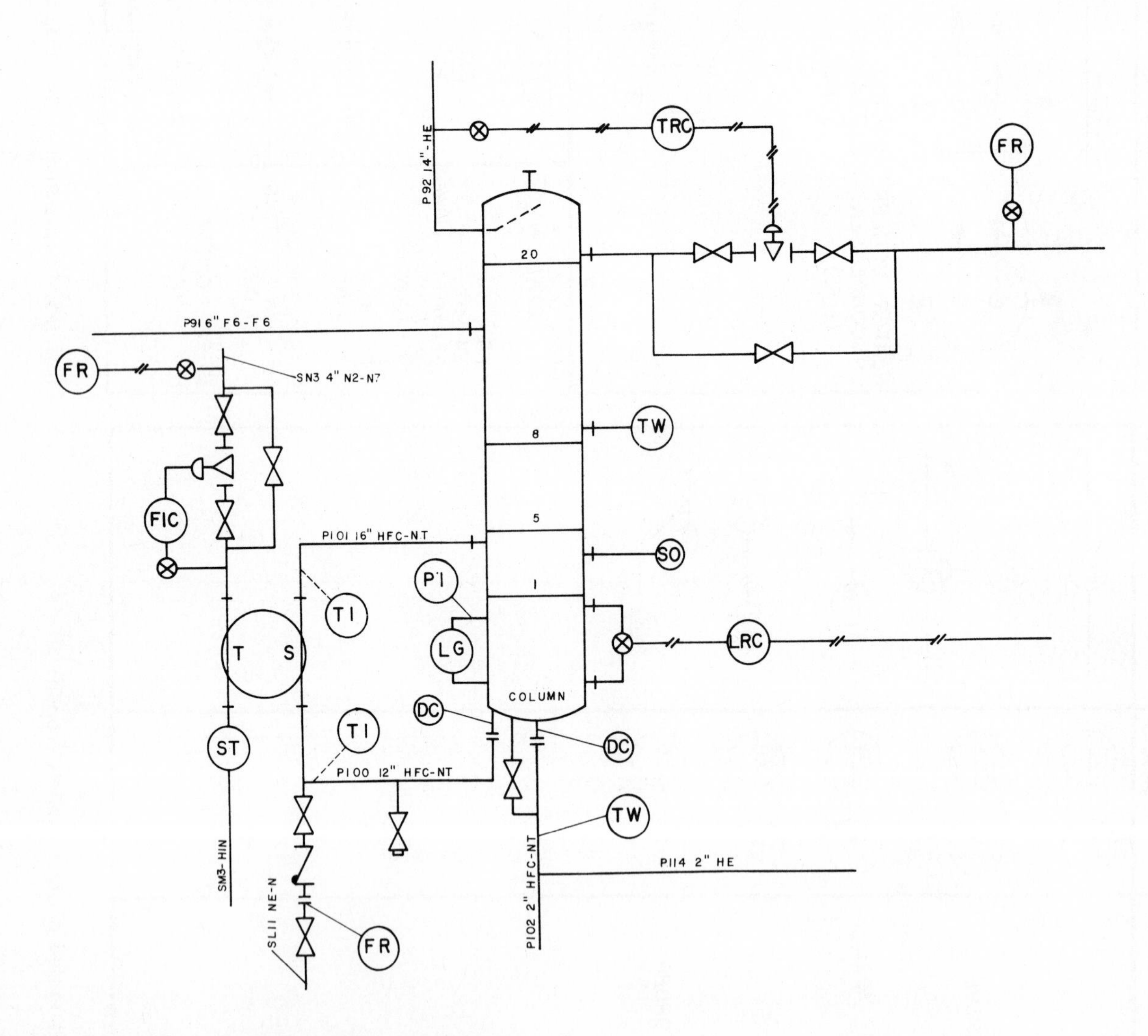

FIG. 6-6 Common process plant P&ID: sections of flow diagram.

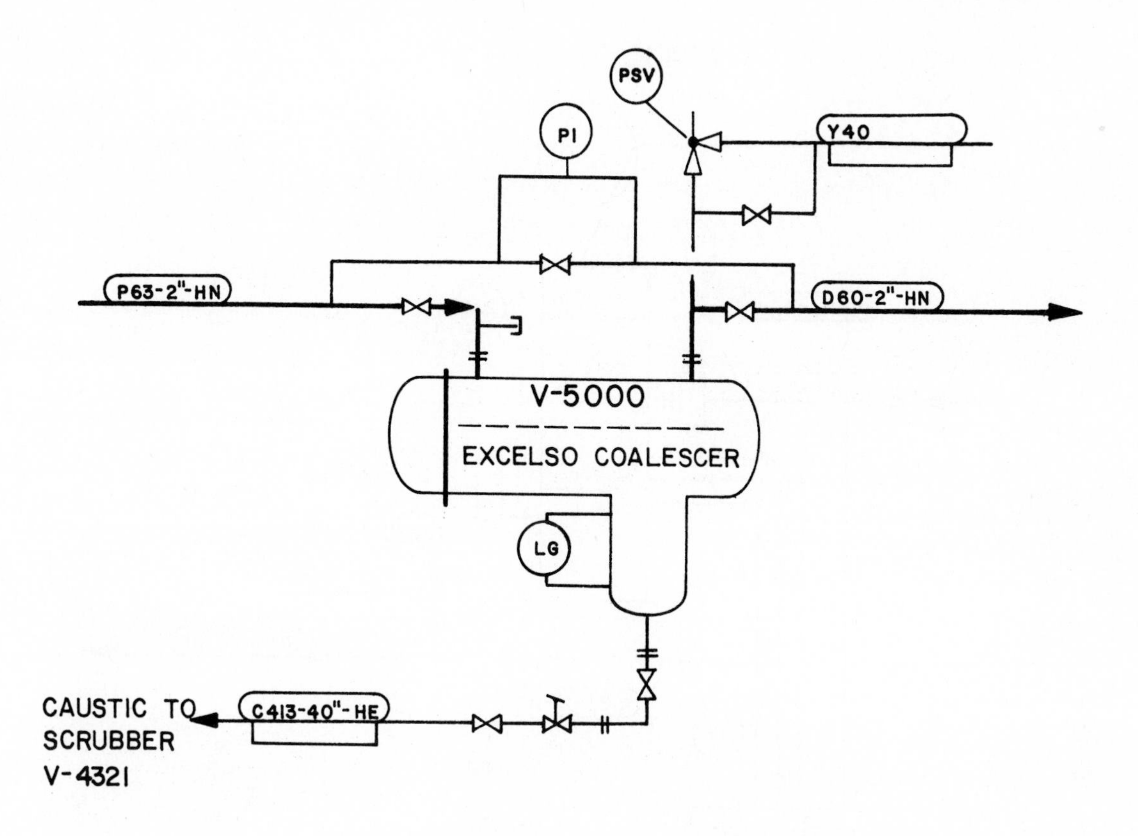

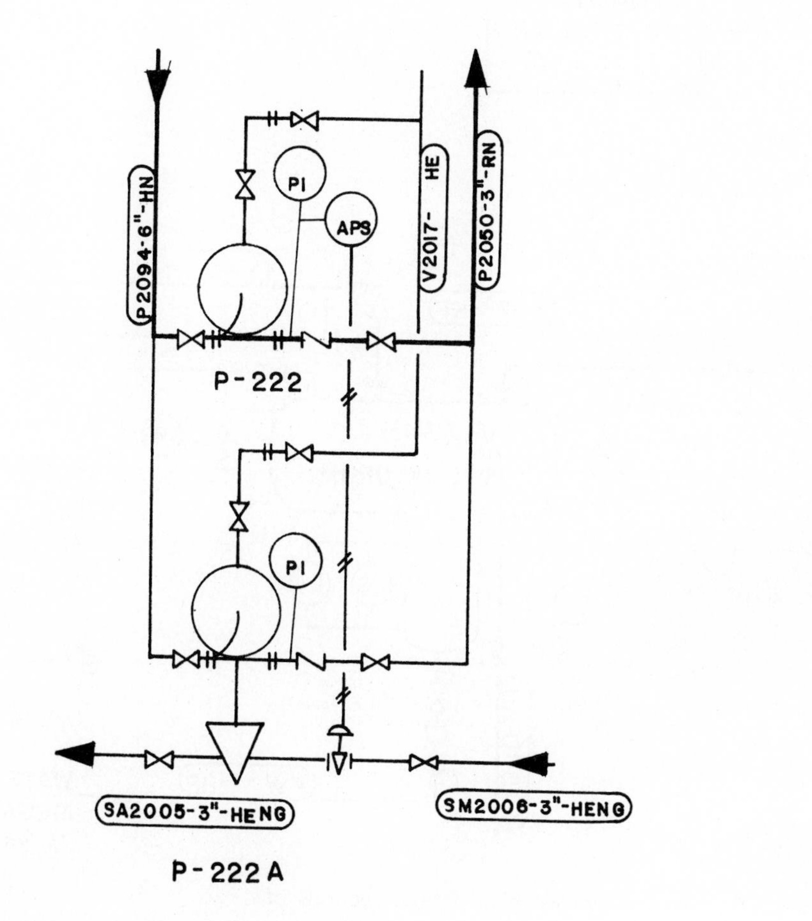

FIG. 6-6 Common process plant P&ID: sections of flow diagram. (cont'd)

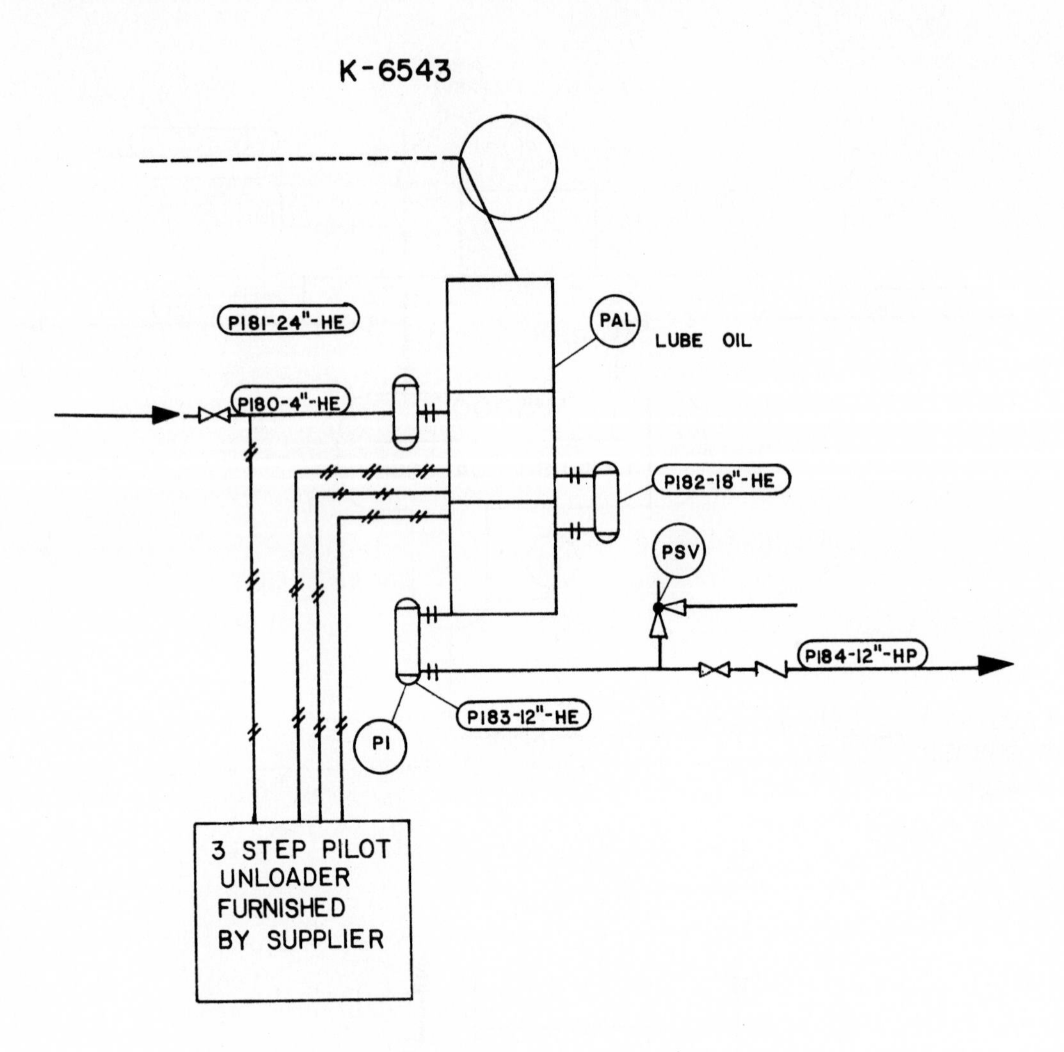

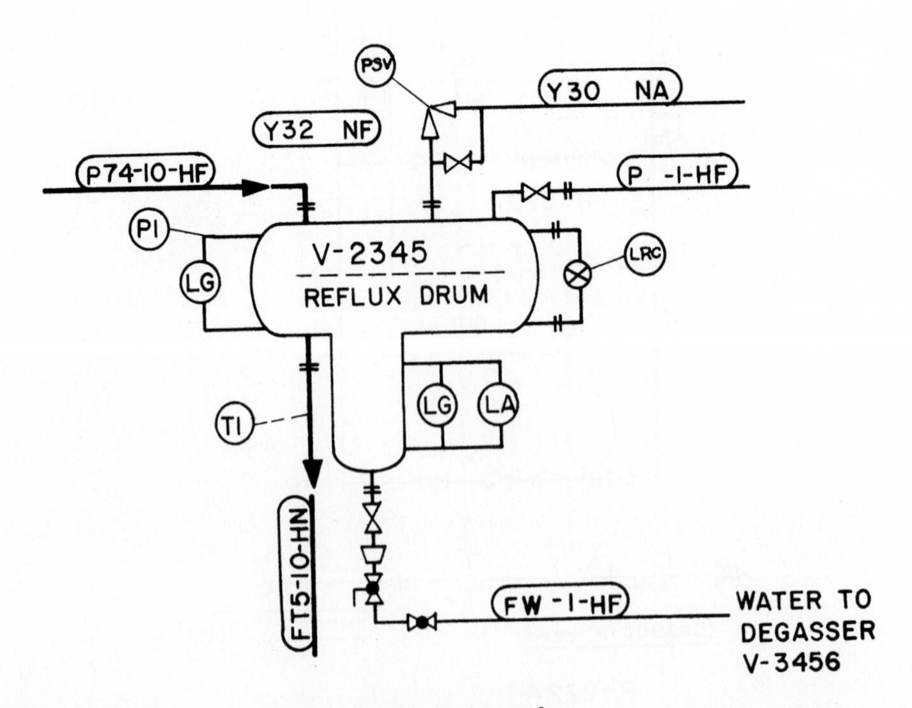

FIG. 6-6 Common process plant P&ID: sections of flow diagram. (cont'd)

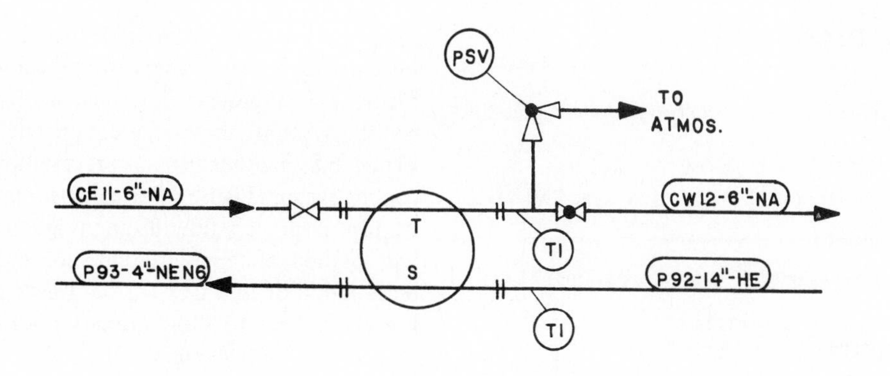

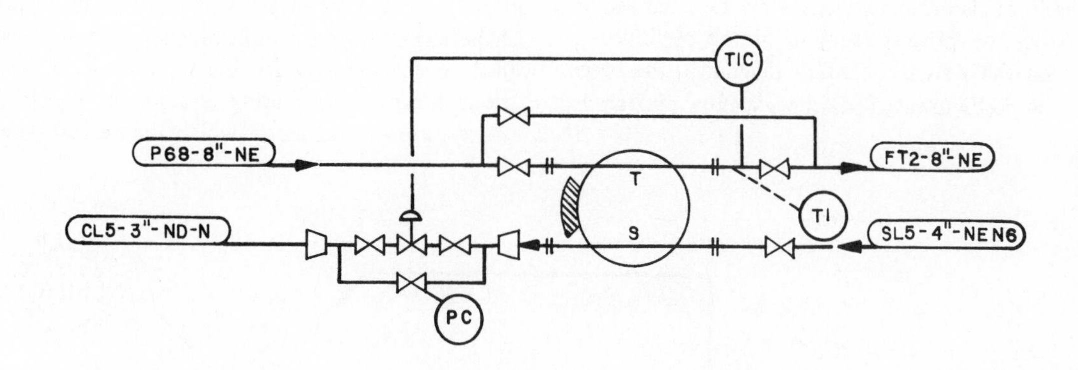

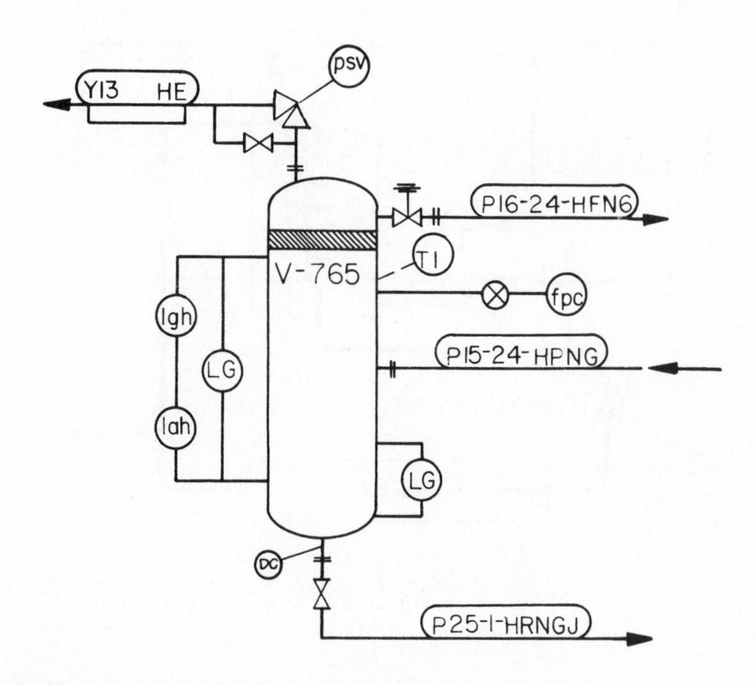

FIG. 6-6 Common process plant P&ID: sections of flow diagram. (cont'd)

TYPES OF FLOW DIAGRAMS

Flow diagrams are divided into four major types:

1. Process flow diagrams
2. Mechanical flow diagrams (also referred to as P&ID)
3. Utility flow diagrams
4. Block or geometric diagrams (general diagrams)

Process Flow Diagrams

Process flow diagrams are schematic drawings used as a guide to laying out the total piping system. The process flow diagram illustrates function, major equipment, and primary lines. It is also frequently used by the engineer or designer making the process study or doing preliminary design work on the system. This flow diagram is the basis from which the P&ID drawing (mechanical flow diagram is constructed.

The process flow diagram is used in the early stages of system design. Many engineers will come in contact with it. Figure 6-7, a process flow diagram for solar heating and cooling systems, shows the equipment and flow direction. Figure 6-8, another general process flow diagram, shows a cooling system. Figure 6-9 is a nuclear process flow diagram. Almost all process flow diagrams are drawn with the single-line method of representation, depending on the contents or function of that line on the drawing. It is important to use heavy lines to show primary lines throughout the system. Figure 6-10 shows the basic flow diagram for a steam generation plant. In this drawing, three separate types of lines are used to represent different functions; steam, condensate, and water systems are included. Arrows are used to show the direction of flow throughout the system.

The process flow diagram represents equipment in simple shapes, as do all the sample diagrams in this chapter. Sometimes a shape resembles the equipment it represents, in proportion to surrounding equipment. Figure 6-11 is an isometric flow diagram in which the nuclear core is pictori-

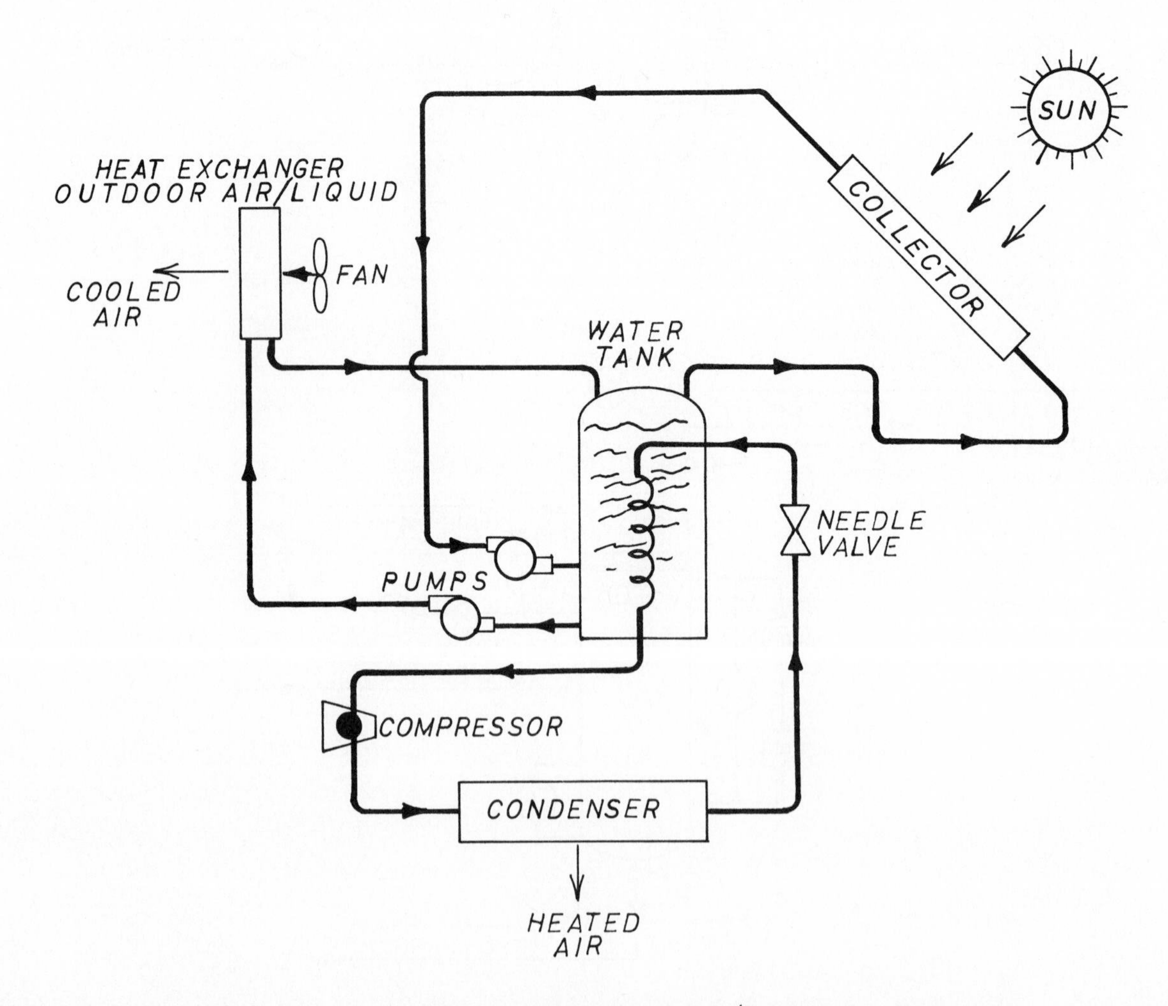

FIG. 6-7 Process flow diagram for solar heating and cooling of buildings (liquid-to-air pump cycle).

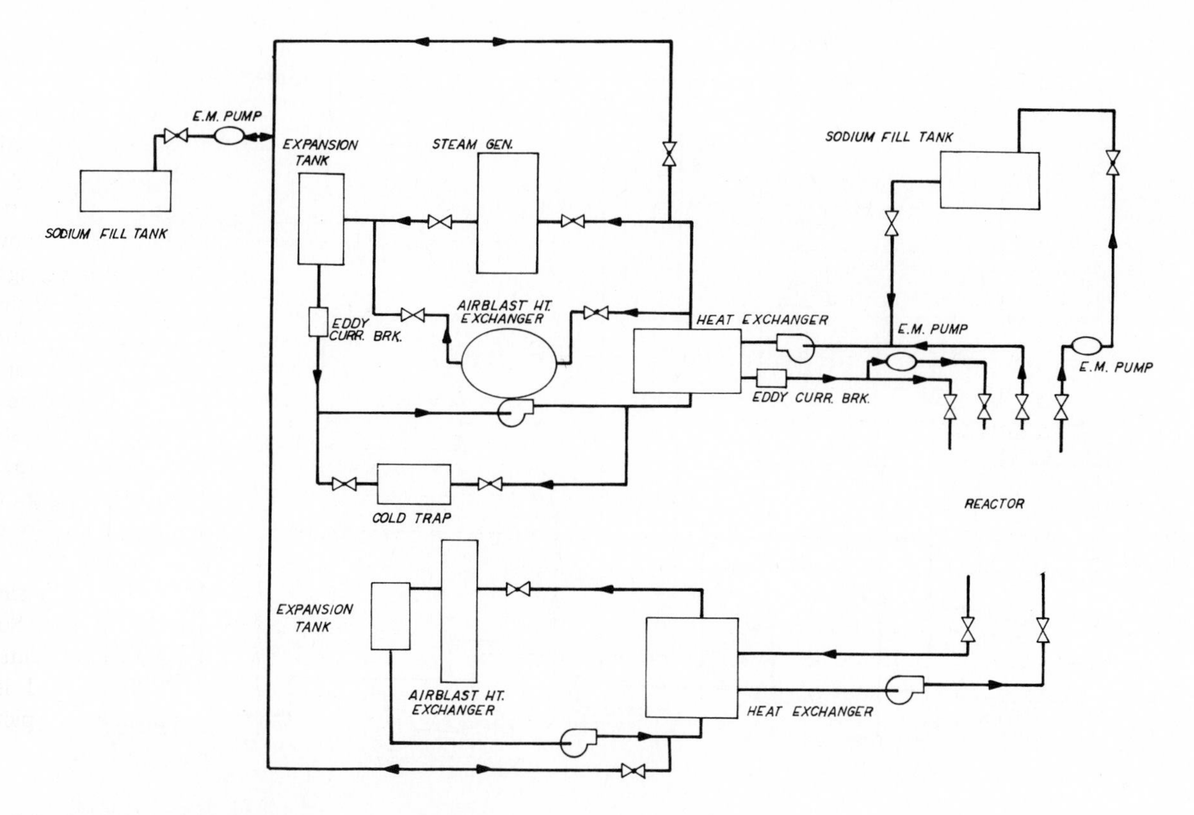

FIG. 6-8 Cooling system diagram for a nuclear plant.

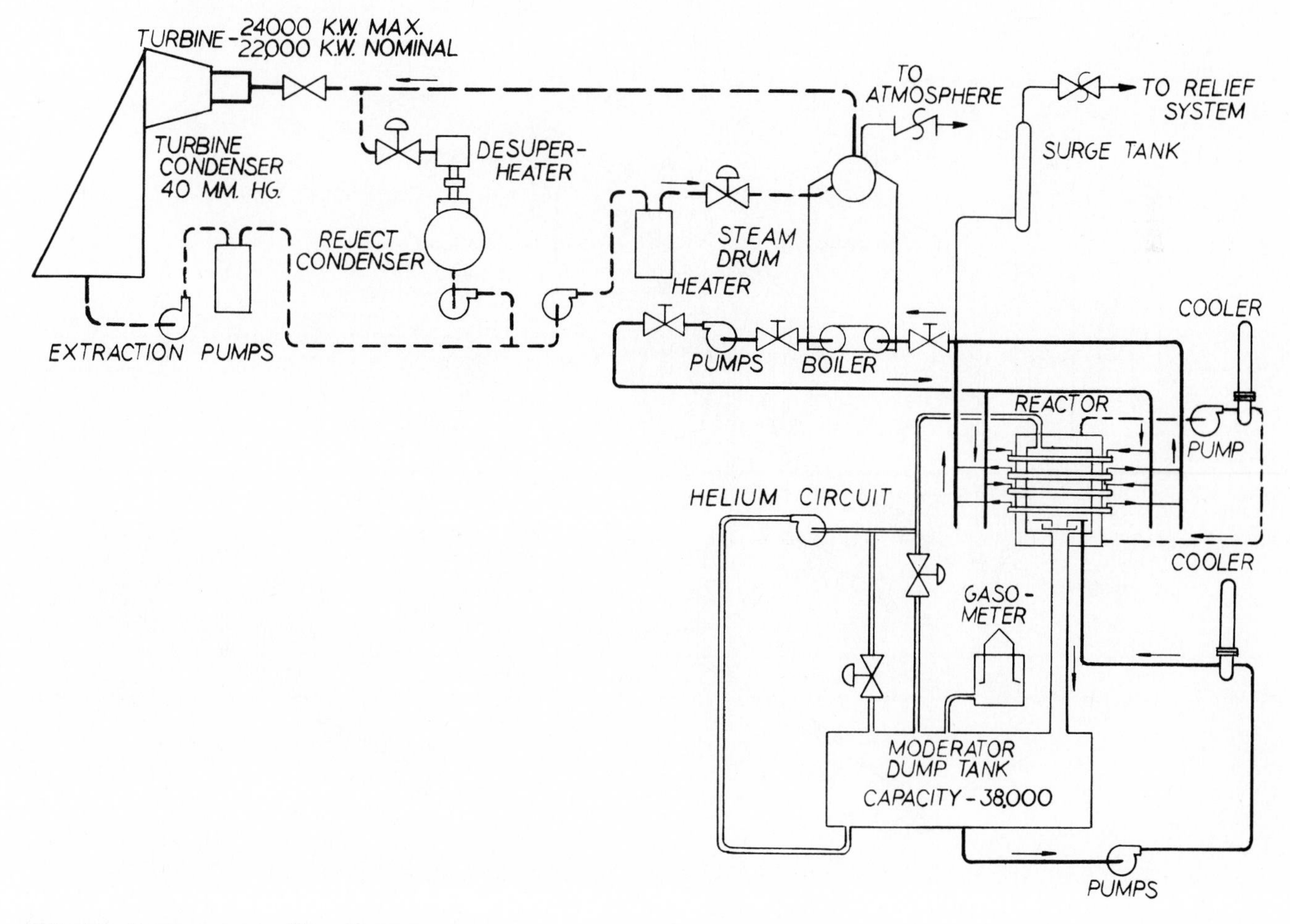

FIG. 6-9 Nuclear process flow diagram.

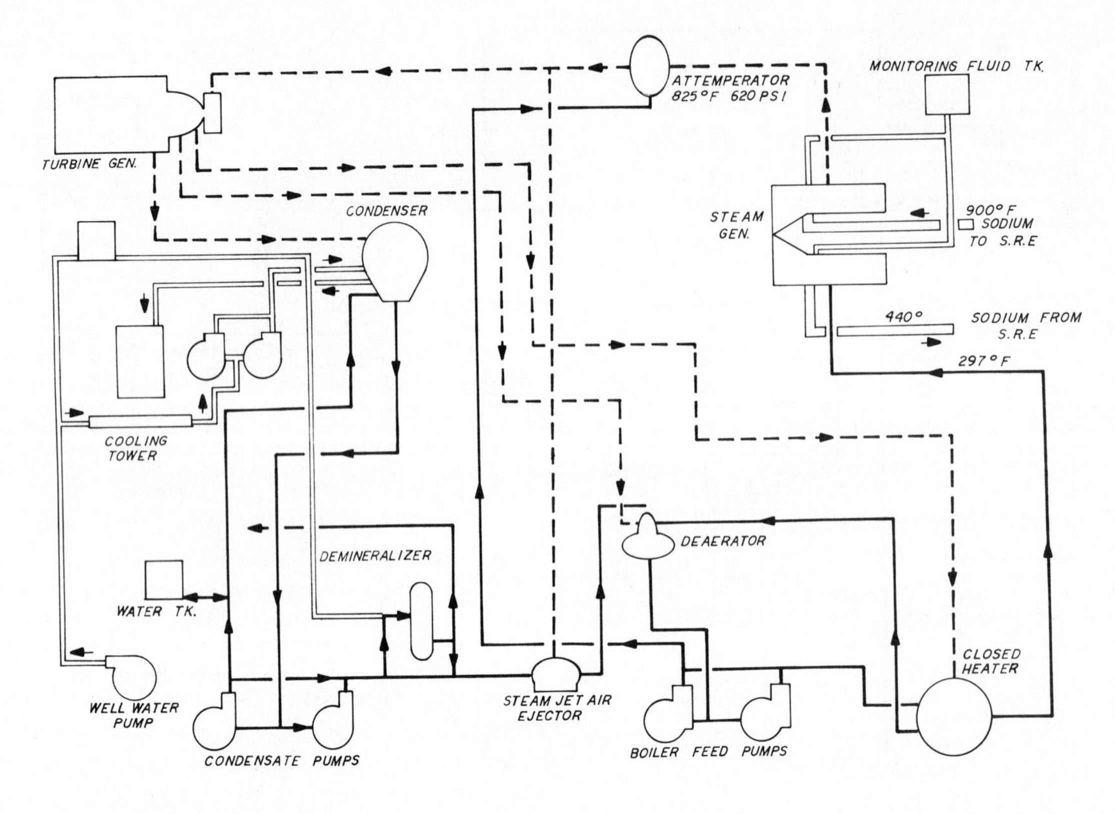

FIG. 6-10 Basic flow diagram for a steam electrical generation plant.

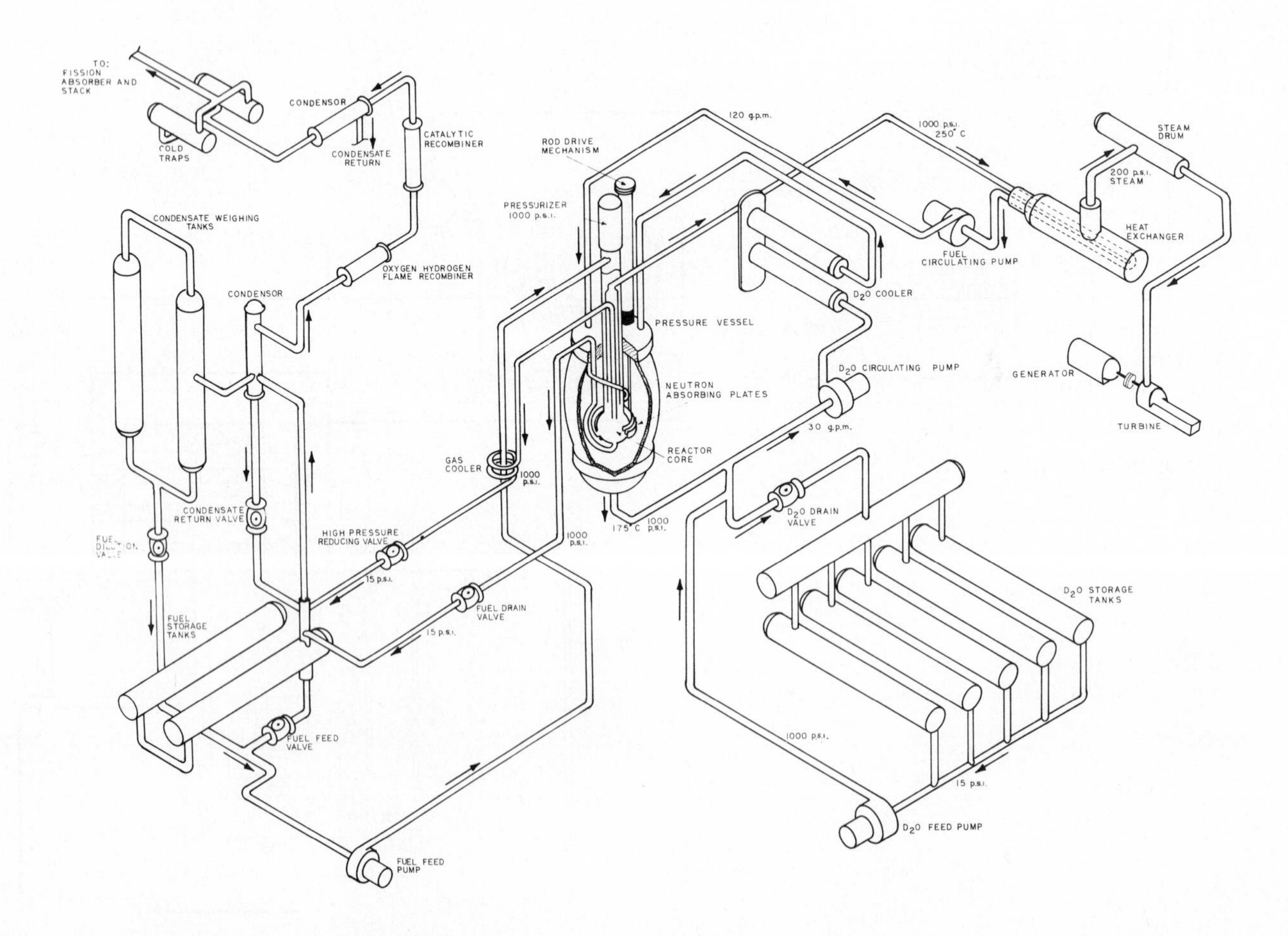

FIG. 6-11 Isometric flow diagram.

ally represented. In Fig. 6-9 the turbine shape resembles the actual turbine. No scale is used in a process flow diagram or any other flow diagram. This type of drawing is laid out neatly and with some sense of proportion for the system as a whole so it can be easily interpreted by the piping layout drafter.

The process flow diagram depicts only essential information—major pipes, equipment, direction of flow, pressures, temperatures, flow quantities, all of which are called out on the lines at important points along with other essential information peculiar to the system. Eliminate details that are not essential for the beginning stages of the job. Other design information—such as elevations of equipment, vessel design, codes, boiler callouts, economizer information—can also be placed on the process flow diagram. Most companies have a set of standard symbols they use to show frequently used equipment such as pumps, generators, heat exchangers, traps, and tanks.

The process flow diagram does not show utility lines or callouts for spare equipment or instrumentation unless they are essential to the process. Valves are shown on the diagram only for special situations; in Fig. 6-9, the nuclear process flow diagram shows major valves necessary for the control of the process.

After the piping designer receives the diagrams, specifications, plot plans, and related information, he or she can begin to design the process diagram. All vendor subcontractor prints, vessel requirements, and other piping data essential to the job are used to draw the process flow diagram. In the planning stages for process diagrams, the drafter should use the following set of instructions:

1. Space equipment in anticipation of additions to the piping in that area.
2. Always allow space for positioning spare pumps and additional units.
3. Place pumps in either vertical or horizontal lines whichever suits the space requirement.
4. In petrochemical process work, try to position equipment that handles light fractions toward the top of the sheet and equipment handling heavy fractions toward the bottom (see Chapter 11).
5. Direct main flow across the sheet from left to right whenever possible.
6. Locate the process lines that enter connecting diagrams at approximately the same distance from the bottom of the sheet.

When drawing process diagrams, the drafter should take into account the following suggestions:

1. Show only the size of relief, bypass, and control valves.
2. Call out block and bypass valve sizes only when they differ from the control valve size or line size.
3. Show pump block and check valve sizes when they differ from line size.
4. Show pump nozzle sizes.
5. Code other equipment nozzle sizes only when they differ from line size.
6. Locate control valves as close to their transmitters as practical in order to keep instrument leads short (when instrumentation is called for).
7. Indicate the tube and shell side of heat exchangers.
8. Extend the relief system piping beyond the relief valves and bypass valves.
9. Show all steam lines and steam traps appearing on process diagrams, but do not show trap discharge lines. (Draw traps 3/8 to 1/2 in. square.)

The process flow diagram becomes the central focus of the entire array of design groups. It expresses the engineer's conception of the purpose, function, and operating sequence of the system, which can be anything from a refinery to a nuclear power station. The complete diagram bears little resemblance to the finished system, but the system and its function must be represented with clarity. Size and scale are not critical on the flow diagram unless relative sizes of equipment are significant.

Mechanical Flow Diagrams

In the *mechanical flow diagram* (P&ID) or engineering flow diagram, the system is shown in greater detail. Each pipeline has a number code as in Fig. 6-12. All equipment and valves appear on this diagram (Fig. 6-13). Lines are usually assigned a code number and a location designation. Valves, fittings, and the like are also represented by symbols on the mechanical flow diagram.

The P&ID shows instrumentation and all control devices. Pipe sizes are called out, including line numbers to indicate location, size, specification, and line material (Fig. 6-14). Operating temperatures and pressures may also be included. On Fig. 6-15 all valves are shown along with the instrumentation needed for this system. Each line is given a designation for flow direction.

On the P&ID all process lines must be shown, including essential utility lines serving the process equipment. As in all flow diagrams, the primary process lines are shown with a heavier than usual line. Flow arrows should be provided on all lines. Flanges and fittings are usually not shown on the P&ID drawing unless they are connected to equipment. Equipment nozzles are shown for connections to vessels as in Fig. 6-16.

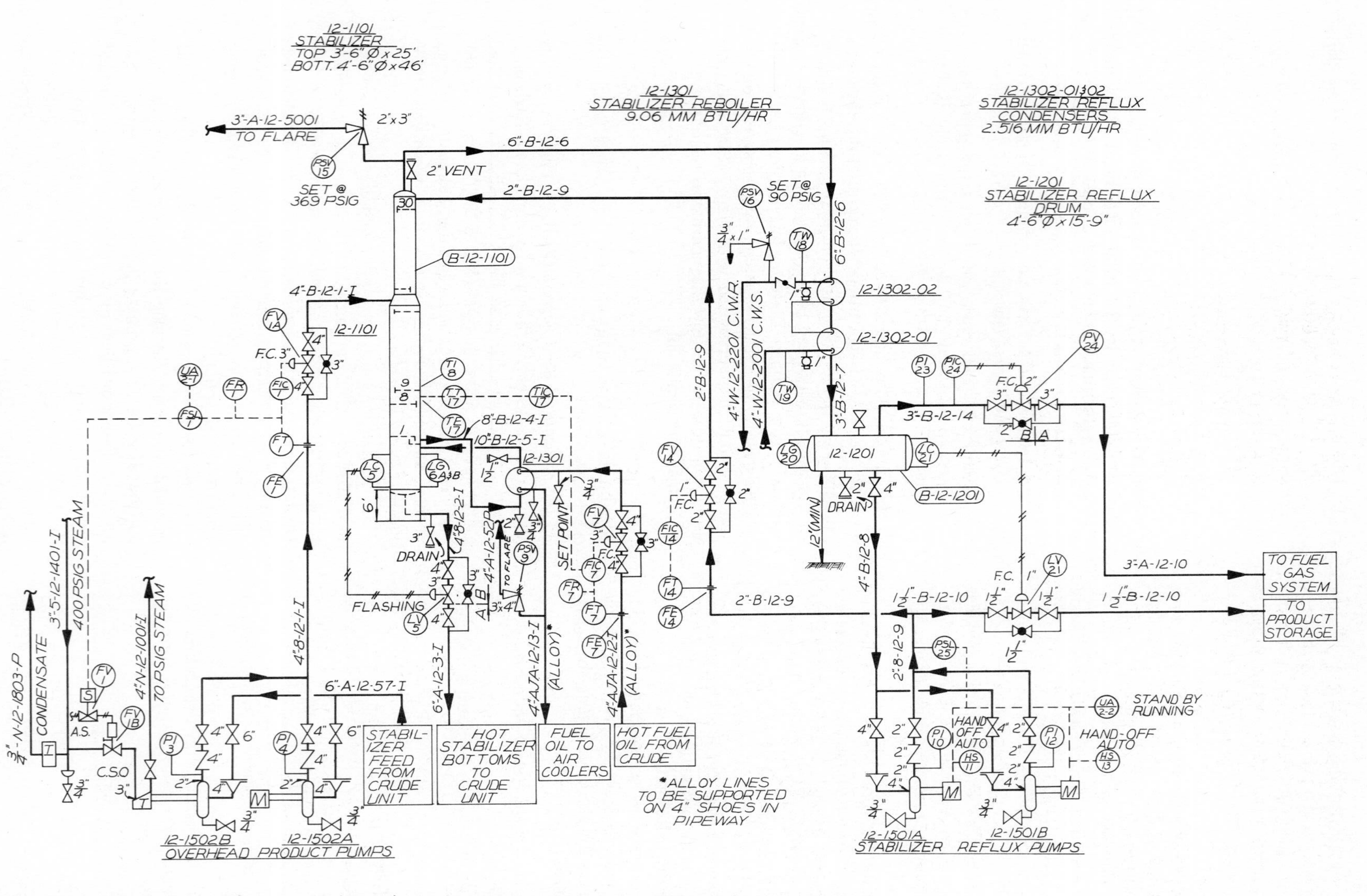

FIG. 6-12 Mechanical flow diagram (P&ID) showing instrumentation and valves for a process plant.

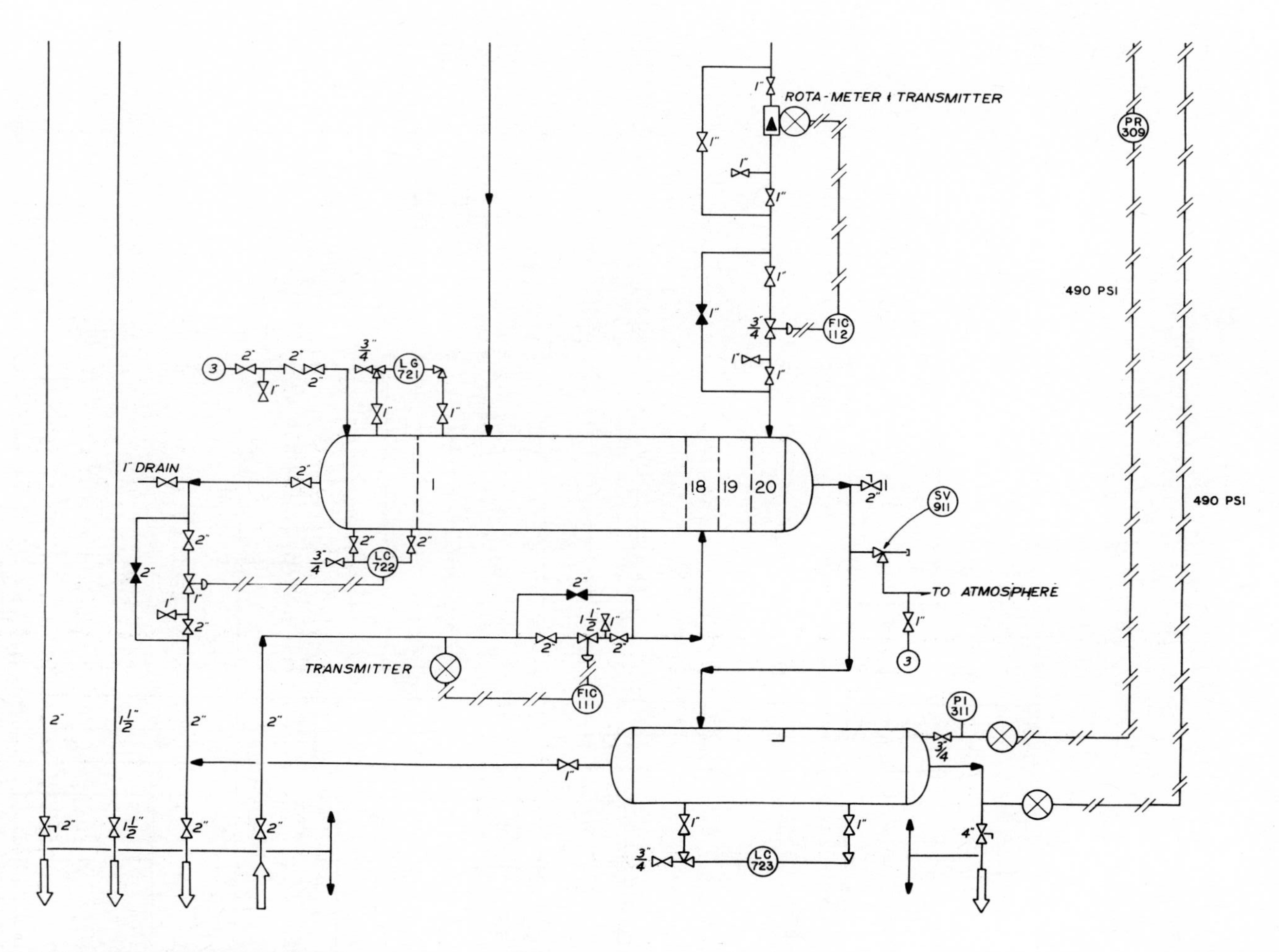

FIG. 6-13 Simplified P&ID.

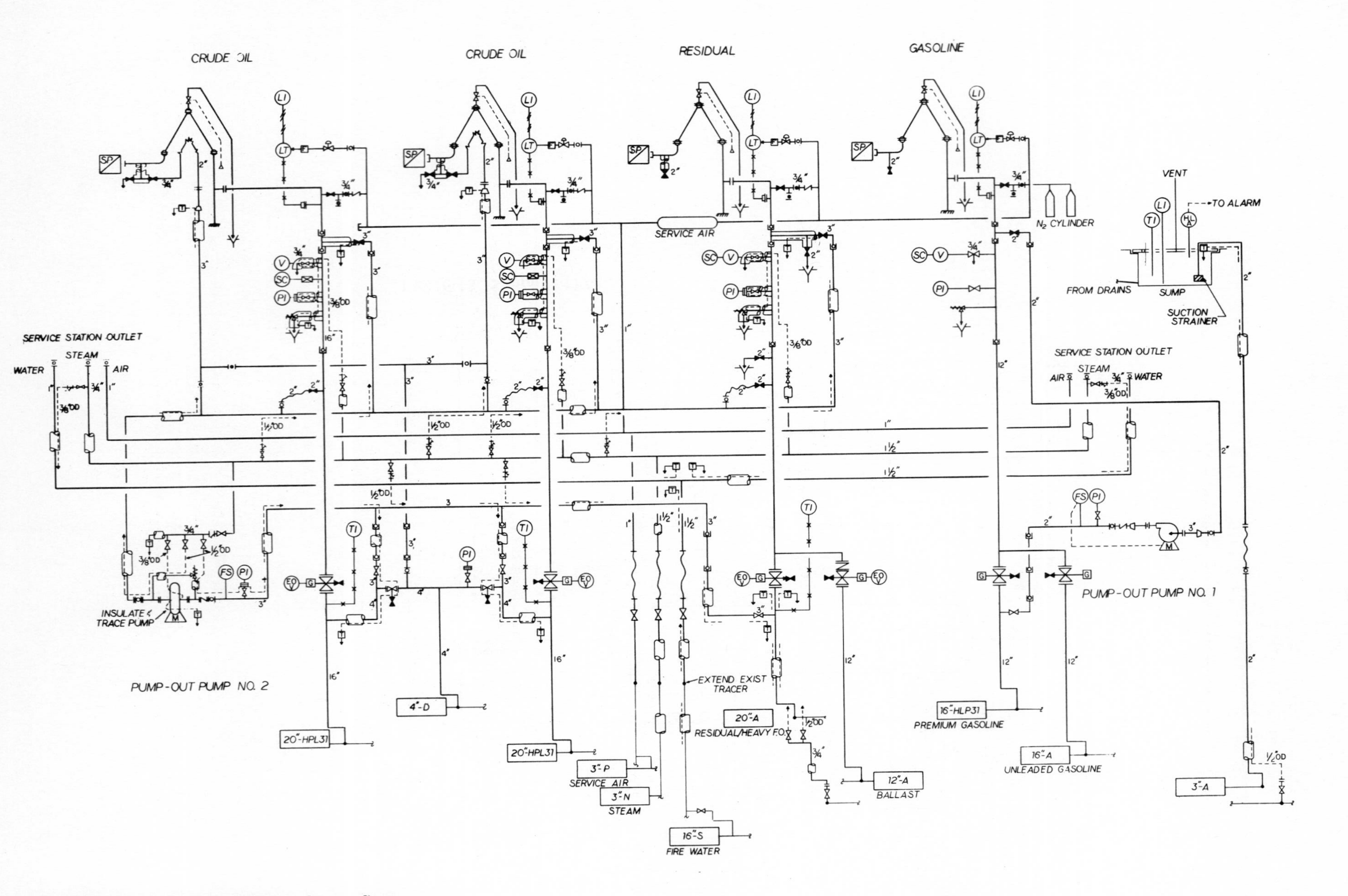

FIG. 6-14 Mechanical flow diagram for a refinery.

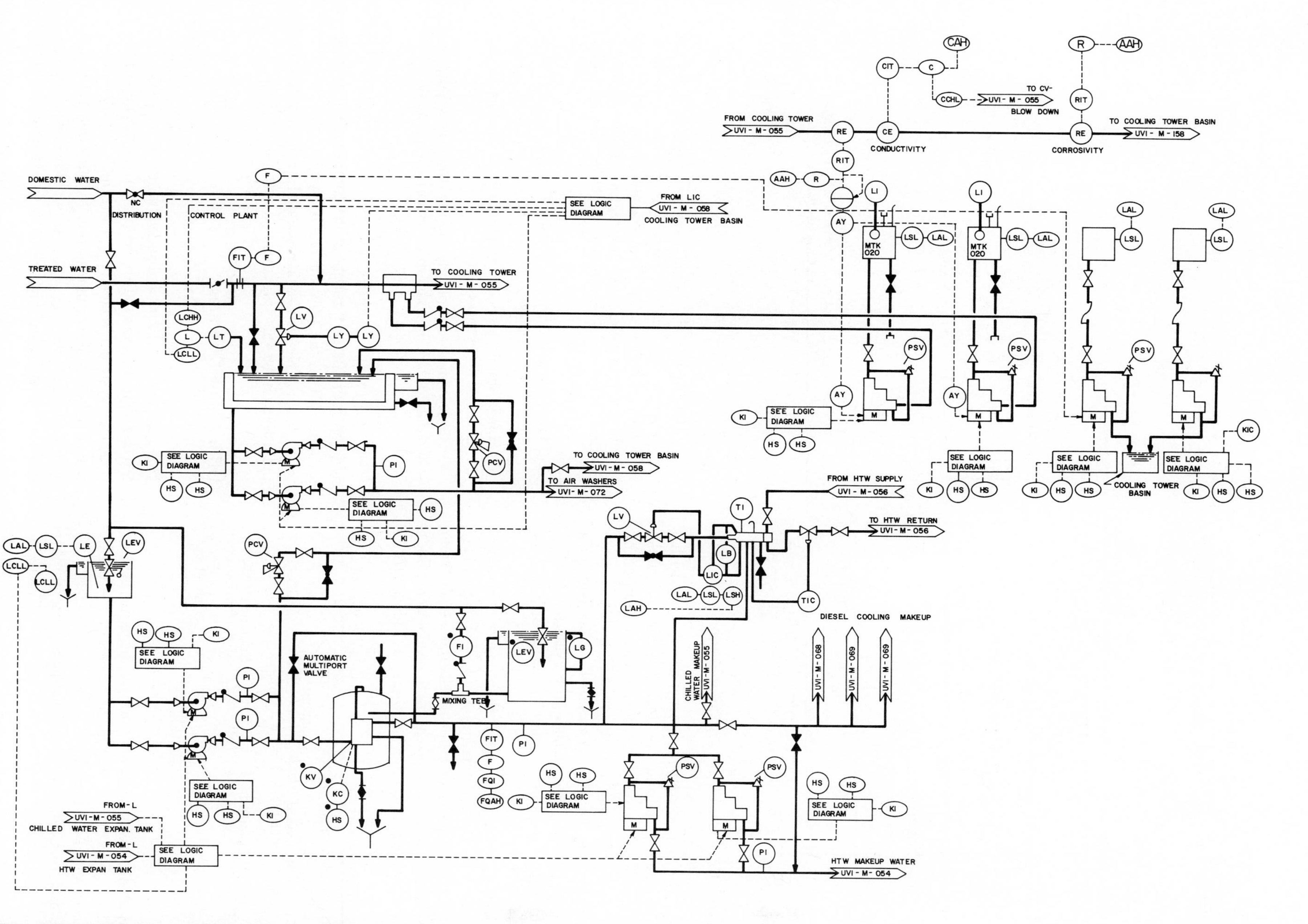

FIG. 6-15 P&ID of a water treatment plant showing the electrical and logic aspects of the system.

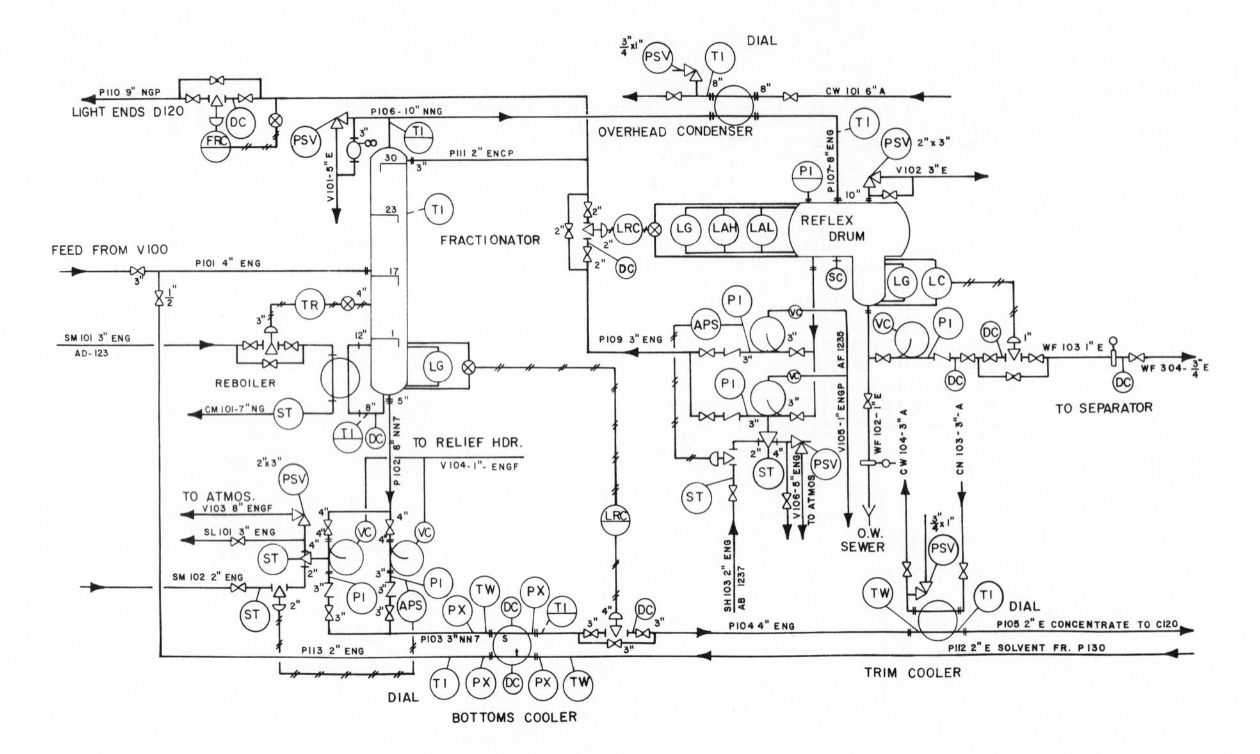

FIG. 6-16 P&ID of the distillation section of a refinery.

The P&ID provides the piping designer and engineer with a variety of essential information including pipe schedule, class of flanges, valve sizes and types, material carried by the line, line sizes, equipment types, and instrumentation locations. All lines on the mechanical flow diagram must be given an identification number to provide the designer with information for an easy transition to the orthographic views or model.

Line identification codes are used throughout the industry, though they are not standardized (see Fig. 6-5). Many of the diagrams and figures in this chapter use different systems because they are taken from several companies. One way of constructing a line identification code is this:

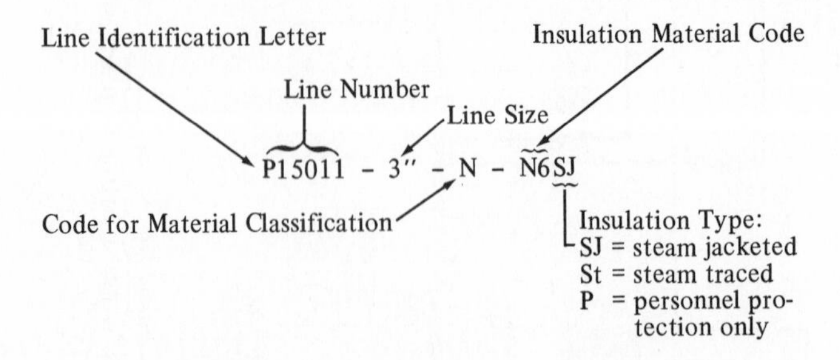

The line identification code must provide the piping designer with all the information that might be required in the design and layout of the piping.

Some useful procedures for designating pipeline numbers are:

1. Continuous lines should have the same line number whether they change direction or come in contact with valves or minor components—except where the line size changes or the line comes into contact with a major piece of equipment (column, vessel, pump, compressor).
2. In most cases branch lines will need separate line numbers even if the line size remains constant.
3. Lines should be coded sequentially according to line medium and process.
4. Codes must completely define the line's purpose, content, size, and any other essential information.

The use of line numbers, line codes, and designations must be coordinated throughout the company. Valves must be identified on the flow diagram by symbols and, in many cases, codes. These codes are essential in the construction and listing on the valve procurement list.

Utility Flow Diagrams

The *utility flow diagram* displays utilities for a system—services that are essential to the proper functioning of the complete system. Diagrams must exhibit all the utility piping, valves, and instrumentation for such items as water, steam, sewage, gas, fuel, electricity, blowdown, air, condensate, oil, and flare. Figure 6-17 is an example of a utility

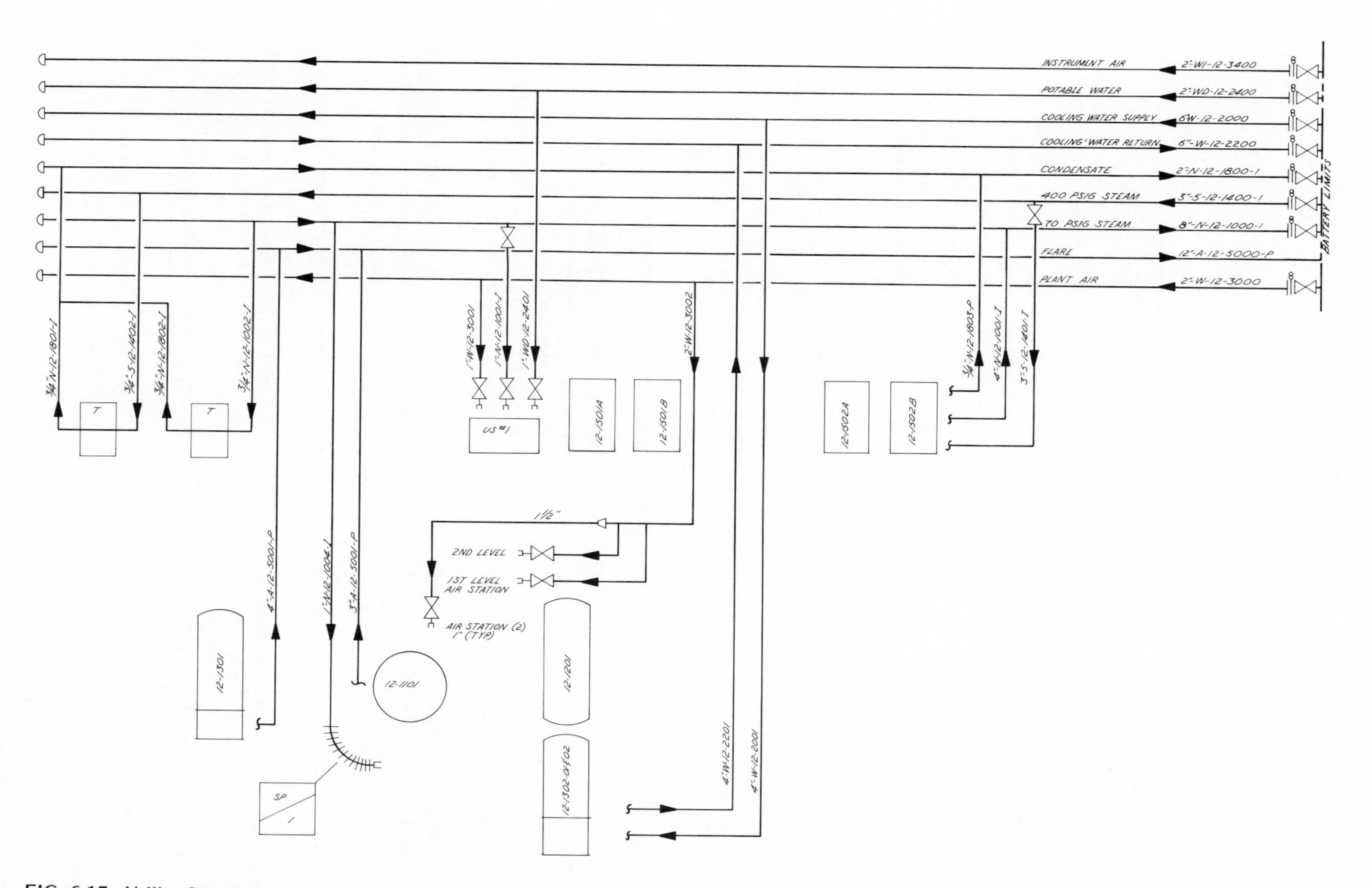

FIG. 6-17 Utility flow diagram.

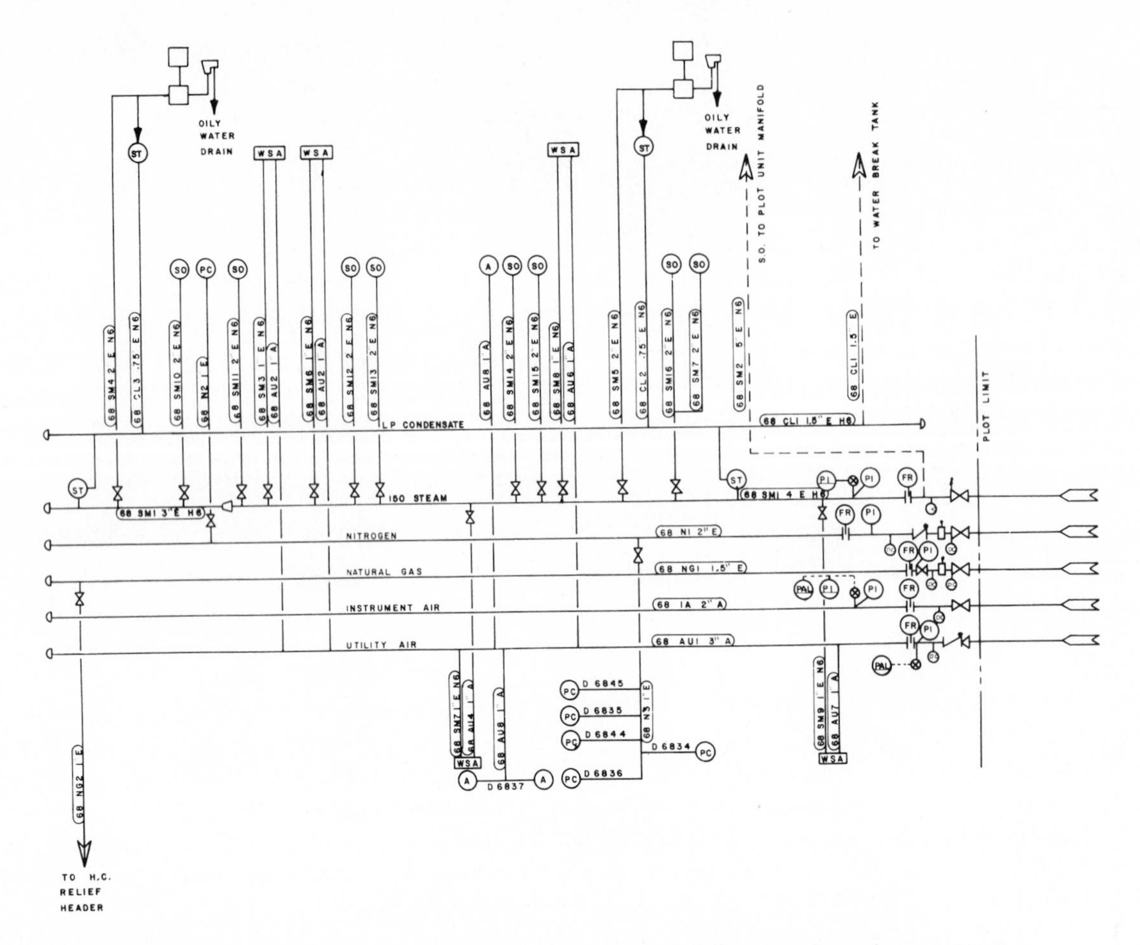

FIG. 6-18 Flow, piping, and instrument diagram for utilities—steam, condensate, air, and gas systems.

cooling system. Figure 6-18 shows the flow, piping, and instrumentation for utilities including the steam, condensate, air, and gas system for a distillate treating facility.

The difference between the utility flow diagram, the process flow diagram, and the mechanical flow diagram is that the utility flow diagram lies plot-plan-wise as in the actual plant or system being represented. Though some utilities appear on the other types of flow diagram, industry procedure usually requires that all the utilities be represented on the complete utility flow diagram.

Utilities represented on these diagrams include air, water, and fuel. Utility air is used for cleaning or for air-driven services like power tools. Water covers services such as cleaning, cooling, fire protection, and sanitation. Fuel is an important utility that includes liquid fuels, gaseous fuels, fluidized solid fuels, and waste product fuels. Steam is included on the utility flow diagram when used for such operations as heating and cleaning. Steam may be divided into separate high-pressure, low-pressure, condensate, and exhaust systems.

Waste requires another service. It includes drains for a variety of substances that are solid, liquid, or gaseous, as well as complete piping systems that transfer these substances to other areas of the plant for further breakdown or disposal. Blowdown is the service that redirects or disposes of the substances automatically or manually discharged from relieving mechanisms and valves. Another service to be represented on the utility flow diagram is the flare stack, found exclusively on petrochemical plants and used for removing burnable waste gases.

The piping drafter, before constructing utility flow diagrams, must take into account the following in the planning stage:

1. Decide which systems can be grouped on one diagram. The following combinations should be considered:
   (a) Water (hydrocarbon free), air, steam and condensate.
   (b) Natural gas, nitrogen, inert gas
   (c) Fuel oil, fuel gas
2. Plan to combine on a single sheet auxiliary systems serving compressors and blowers (seal oil, lube oil, jacket water).
3. Avoid adding complex diagram systems to simple diagram systems.

When drawing the utility diagram, the drafter should take into account the following suggestions:

1. Do not show steam tracing or jacketing traps. (These are shown on the P&ID.)
2. Show reducers on headers and subheaders.
3. Show sizes of branches on subheaders.
4. Describe the service on each header.

5. Show the control valve only, not the complete manifold.
6. Terminate lines serving utility stations with a code symbol and note the location.
7. Represent utility control valves detailed on the process diagram only where necessary.
8. Draw process equipment in simplified form showing only connections concerned with utilities.
9. Allow 5/8 to 3/4 in. (1.59 to 1.90 cm) minimum spacing between lines representing pipeway headers.

Block or Geometric Diagrams

The *block* (or geometric) *diagram* is the simplest and easiest to read, though it is usually used only in books or catalogs for illustration purposes. It is used in all industries to represent economic, psychological, sociological, and historical data and studies for a particular presentation. The shapes that represent the aspects of the block diagram are kept as simple as possible. This diagram allows more artistic latitude than most piping drawings.

INSTRUMENTATION

The P&ID indicates all instrumentation in the four major categories of *instruments: temperature, pressure, flow,* and *level.* Symbols are used to designate all instruments on the diagram. The symbols given in Fig. 6-5 are used when drawing a diagram. Instruments, an essential aspect of all piping systems, provide accurate and safe operation with a minimum of personnel attention. They are used separately or in combination to monitor the various aspects of a piping system. *Pressure, flow, temperature,* and *level* are the primary functions that instruments must *sense, monitor, record, control, indicate,* and *transmit.*

Instruments that are locally operated and read are indicated by a balloon with an identification abbreviation.

(PI) Means "locally read pressure indicator."

Instruments mounted or operated on a board in a control room are shown with a line through the center of the balloon.

(LR) Means "board-mounted level recorder."

The lower part of the balloon is for line identification.

The process engineer indicates all essential instrumentation on the flow diagram sketch. The drafter should be familiar with the function and various uses of instruments.

Figure 6-12 is a petrochemical plant P&ID drawing which shows all equipment, temperature indicators, recorders, level glasses, and other instruments. Figures 6-12 and 6-16 are examples of instrumentation drawings and the use of symbols. Most symbols have been standardized by the Instrument Society of America, though many companies still use their own variations.

Instrumentation piping must take into account the service conditions of temperature, pressure, strength, and operation frequency of the instruments. Air and hydraulically operated control valves and other devices are indicated on instrumentation drawings. Figure 6-19 shows a nuclear power

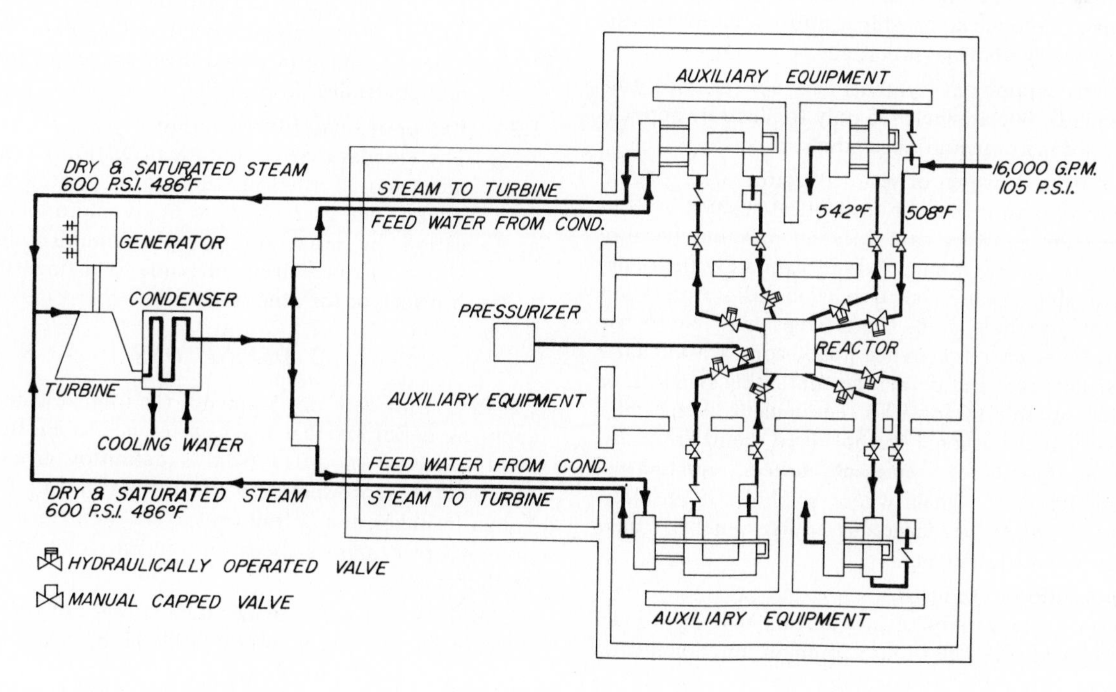

FIG. 6-19 Control valve flow diagram.

plant flow diagram used exclusively for the placement of hydraulically and manually operated control valves from the reactor to the auxiliary equipment.

This chapter is designed to acquaint novice drafters with the many different diagrams they will encounter, so that they will learn, through experience, the processes and rational behind the design of the flow diagram. This chapter should be consulted frequently as the student progresses through subsequent chapters of the book. Chapters 10, 11, 12, and 13 also cover instruments and utilities.

INSTRUMENTATION SPECIFICATIONS

This section is an essential element in the piping drafter's education and should be studied thoroughly. The following information has been extracted from the Instrument Society of America's 1975 standard on instrumentation symbols and identification (ISA-S5.1) and corresponds to ANSI Y32.20-1975. It will give the reader an in-depth description of instruments and symbol representation.

1. SCOPE

1.1 General

1.1.1 The purpose of this Standard is to establish a uniform means of designating instruments and instrumentation systems used for measurement and control. To this end, a designation system is presented that includes symbols and an identification code.

1.1.2 The differing established procedural needs of various organizations are recognized, where not inconsistent with the objectives of the Standard, by providing alternative symbolism methods. A number of options are provided for adding information or simplifying the symbolism, if desired.

1.1.3 Process equipment symbols are not part of this Standard, but are included only to illustrate applications of instrumentation symbols.

1.1.4 If a given drawing, or set of drawings, uses graphic symbols that are similar or identical in shape or configuration and that have different meanings because they are taken from different standards, then adequate steps shall be taken to avoid misinterpretation of the symbols used. These steps may be to use caution notes or reference notes, comparison charts that illustrate and define the conflicting symbols, or other suitable means. This requirement is especially critical if the graphic symbols used, being from different disciplines, represent devices, conductors, flow lines, or signals whose symbols, if misinterpreted, might be dangerous to personnel or cause damage to equipment.

1.2 Application to Industries

1.2.1 Despite the variety of instruments that have been developed, they all fit into common functional categories. The Standard is suitable for use in the chemical, petroleum, power generation, air conditioning, metal refining, and numerous other industries.

1.2.2 Certain fields, such as astronomy, navigation, and medicine, use very specialized instruments that are different from the conventional industrial process instruments. No specific effort was made to have the Standard meet the requirements of those fields. However, it is expected that the Standard will be flexible enough to meet many of the needs of special fields.

1.3 Application to Work Activities

1.3.1 The Standard is suitable for use whenever any reference to an instrument is required. Such references may be required for the following uses as well as others:

Flow diagrams, process and mechanical
Instrumentation system diagrams
Specifications, purchase orders, manifests, and other lists
Construction drawings
Technical papers, literature, and discussions
Tagging of instruments
Installation, operating, and maintenance instructions, drawings, and records

1.3.2 The Standard is intended to provide sufficient information to enable anyone reading a flow diagram and having a reasonable amount of plant knowledge to understand the means of measurement and control of the process without having to go into the details of instrumentation that require the knowledge of an instrument specialist.

1.4 Extent of Functional Identification

The Standard provides for the identification and symbolization of the key functions of an instrument. The full details of the instrument are left to be described in a suitable specification, data sheet, or other document intended for those people interested in such details.

1.5 Extent of Loop Identification

The Standard covers the identification of an instrument and all other instruments associated with it in a loop. The user is free to apply additional identification—by serial number, plant number, or otherwise—as he may deem advisable to distinguish among projects, or for other purposes.

2. DEFINITIONS

For the purposes of this Standard, the following definitions apply. A definition that makes reference to another document has been modified from a definition given in that document to correspond to the format of this Standard. Terms in italics in a definition are also defined in this section. Where examples are given, the list is not intended to be all-inclusive.

ALARM—A device that signals the existence of an abnormal condition by means of an audible or visible discrete change, or both, intended to attract attention.

BALLOON—The circular symbol used to denote an instrument or instrument tagging.

BEHIND THE BOARD—A term applied to a location that (1) is within an area that contains the instrument board, (2) is within or in back of the board or is otherwise not accessible to the operator for his normal use, and (3) is not designated as *local.*

BOARD—A structure that has a group of instruments mounted on it and is chosen to have an individual designation. The board may consist of one or more component panels, cubicles, desks, or racks.

BOARD-MOUNTED—A term applied to an instrument that is mounted on the board and is accessible to the operator for his normal use.

COMPUTING RELAY—A relay that performs one or more calculations or logical functions or both and sends out one or more resultant output signals.

CONTROLLER—A device that has an output that can be varied to maintain a controlled variable at a specified value or within specified limits or to alter the variable in a specified manner. An automatic controller varies its output automatically in response to a direct or indirect input of a measured process variable. A manual controller is a manual loading station, and its output is not dependent on a measured process variable but can be varied only by manual adjustment. A controller may be integral with other functional elements of a control loop. [See Table 6-1, note 13.]

CONTROL STATION—A manual loading station that also provides switching between manual and automatic control modes of a control loop. It is also known as an auto-manual station and auto-selector station.

CONTROL VALVE—A device, other than a common hand-actuated on-off valve, that directly manipulates the flow of one or more fluid process streams. In some applications, it is commonly known as a damper or louver. [See Table 6-1, note 13 and Fig. 6-21.] It is expected that use of the designation hand control valve will be limited to hand-actuated valves that (1) are used for process throttling or (2) are special valves for control purposes and that are to be specified by an instrumentation group or instrument engineer.

CONVERTER—A device that receives information in the form of an instrument signal, alters the form of the information, and sends out a resultant output signal. A converter is a special form of relay. A converter is also referred to as a transducer, although transducer is a completely general term and its use specifically for signal conversion is not recommended.

FINAL CONTROL ELEMENT—The device that directly changes the value of the manipulated variable of a control loop.

FUNCTION—The purpose of or action performed by a device.

INSTRUMENT—A device used directly or indirectly to measure or control variables or both. The term includes control valves, relief valves, and electrical devices such as annunciators and pushbuttons. The term does not apply to parts, e.g., a receiver bellows or a resistor, that are internal components of an instrument.

INSTRUMENTATION—The application of instruments.

LOCAL—The location of an instrument that is neither on nor behind a board. Local instruments are commonly in the vicinity of a primary element or a final control element.

LOCAL BOARD—A board that is not a central or main board. Local boards are commonly in the vicinity of plant subsystems or subareas.

LOOP—A combination of one or more interconnected instruments arranged to measure or control a process variable, or both.

MANUAL LOADING STATION—A device having a manually adjustable output that is used to actuate one or more remote devices. Although the remote devices may be controller elements, the station does not provide switching between manual and automatic control modes of a control loop. (See Controller and Control Station.) The station may have integral gauges, lights, or other features. It is also known as a manual station or a remote manual loader.

TABLE 6-1 SPECIAL ABBREVIATIONS

Abbreviation	*Meaning*
A	Analog signal
ADAPT	Adaptive control mode
AS	Air supply
AVG.	Average
C	Patchboard or matric board connection
D	Derivative control mode Digital signal
DIFF.	Subtract
DIR.	Direct-acting
E	Voltage signal
ES	Electric supply
FC	Fail closed
FI	Fail indeterminate
FL	Fail locked
FO	Fail open
GS	Gas supply
H	Hydraulic signal
I	Current (electrical) signal Interlock
M	Motor actuator
MAX.	Maximizing control mode
MIN.	Minimizing control mode
NS	Nitrogen supply
O	Electromagnetic or sonic signal
OPT	Optimizing control mode
P	Pneumatic signal Proportional control mode Purge or flushing device
R	Automatic-reset control mode Reset of fail-locked device Resistance (signal)
REV.	Reverse-acting
RTD	Resistance (-type) temperature detector
S	Solenoid actuator
S.P.	Set point
SQ.RT.	Square root
SS	Steam supply
T	Trap
WS	Water supply
X	Multiply Unclassified actuator

MEASUREMENT—The determination of the existence or magnitude of a variable. Measuring instruments include all devices used directly or indirectly for this purpose.

PILOT LIGHT—A light that indicates which of a number of normal conditions of a system or device exists. It is unlike an alarm light, which indicates an abnormal condition. The pilot light is also known as a monitor light.

PRIMARY ELEMENT—That part of a loop or of an instrument that first senses the value of a process variable and assumes a corresponding predetermined and intelligible state or output. The primary element may be separate from or integral with another functional element of a loop. The primary element is also known as a detector or sensor.

PROCESS—Any operation or sequence of operations involving a change of energy state, of composition, of dimension, or of other property that may be defined with respect to a datum. The term process is used in the Standard to apply to all variables other than instrument signals.

PROCESS VARIABLE—Any variable property of a process.

RELAY—A device that receives information in the form of one or more instrument signals; modifies the information or its form, or both, if required; sends out one or more resultant output signals; and is not designated as a controller, a switch, or otherwise. (Also see Computing Relay.) The term relay is also specifically applied to an electric switch that is remotely actuated by an electric signal. However, for the purposes of the Standard, the term is not so restricted. [See Fig. 6-21, note 13.] The term is also applied to the functions performed by relays.

SCAN—To sample each of a number of inputs intermittently. A scanning device may provide additional functions such as record or alarm.

SWITCH—A device that connects, disconnects, or transfers one or more circuits and is not designated as a controller, a relay, or a control valve. [See Fig. 6-21, note 13.] The term is also applied to the functions performed by switches.

TELEMETRY—The practice of transmitting and receiving the measurement of a variable for readout or other uses. The term is most commonly applied to electric signal systems.

TEST POINT—A process connection to which no instrument is permanently connected, but which is intended for temporary, intermittent, or future connection of an instrument.

TRANSDUCER—A general term for a device that receives information in the form of one or more physical quantities; modifies the information or its form or both, if required; and sends out a resultant output signal. Depending on the application, the transducer can be a primary element, a transmitter, a relay, a converter, or other device.

TRANSMITTER—A device that senses a process variable through the medium of a primary element and has an output whose steady-state value varies only as a predetermined function of the process variable. The primary element may or may not be integral with the transmitter.

3. OUTLINE OF THE IDENTIFICATION SYSTEM

3.1 General

3.1.1 Each instrument shall be identified first by a system of letters used to classify it functionally. To establish a loop identity for the instrument a number shall be appended to the letters. This number will, in general, be common to other instruments of the loop of which this instrument is a part. A suffix is sometimes added to complete the loop identification. A typical tag number for a temperature recording controller is shown below. [See Figs. 6-20 and 6-21.]

3.1.2 The instrument tag number may include coded information such as plant area designation. (See Section 3.3.2.)

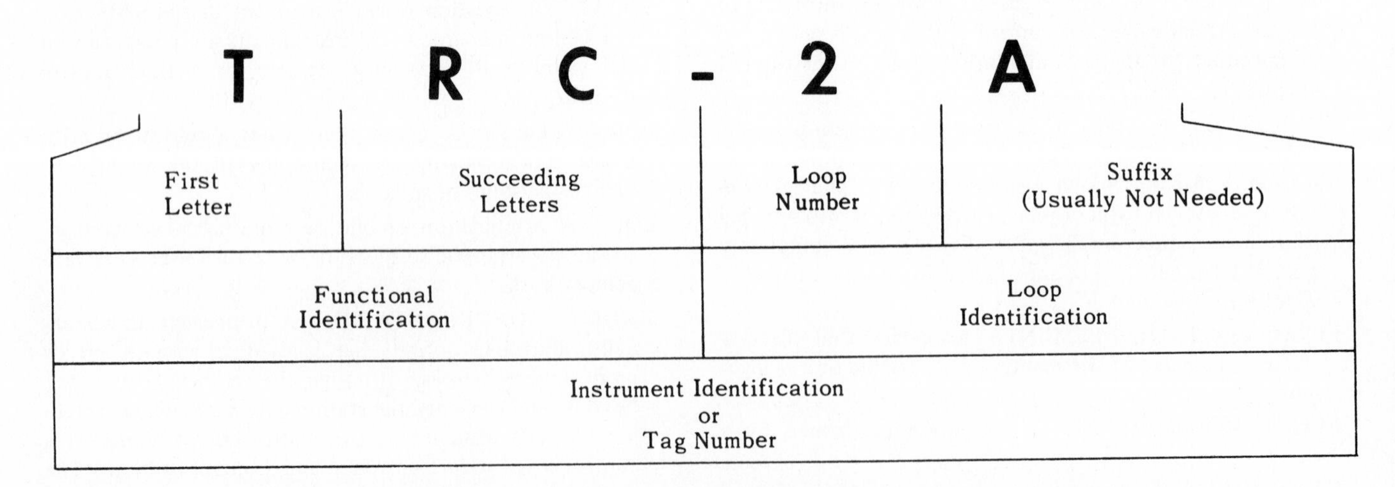

FIG. 6-20 Instrumentation identification or tag number. (Reprints used with permission from ISA S5.1 Instrumentation and Identification, © Instrument Society of America, 1975.)

MEANINGS OF IDENTIFICATION LETTERS

This table applies only to the functional identification of instruments. Numbers in table refer to notes following.

	FIRST LETTER		SUCCEEDING LETTERS (3)		
	MEASURED OR INITIATING VARIABLE (4)	MODIFIER	READOUT OR PASSIVE FUNCTION	OUTPUT FUNCTION	MODIFIER
A	Analysis (5)		Alarm		
B	Burner Flame		User's Choice(1)	User's Choice(1)	User's Choice(1)
C	Conductivity (Electrical)			Control (13)	
D	Density (Mass) or Specific Gravity	Differential (4)			
E	Voltage (EMF)		Primary Element		
F	Flow Rate	Ratio (Fraction) (4)			
G	Gaging (Dimensional)		Glass (9)		
H	Hand (Manually Initiated)				High (7, 15, 16)
I	Current (Electrical)		Indicate (10)		
J	Power	Scan (7)			
K	Time or Time-Schedule			Control Station	
L	Level		Light (Pilot) (11)		Low (7, 15, 16)
M	Moisture or Humidity				Middle or Intermediate (7, 15)
N (1)	User's Choice		User's Choice	User's Choice	User's Choice
O	User's Choice(1)		Orifice (Restriction)		
P	Pressure or Vacuum		Point (Test Connection)		
Q	Quantity or Event	Integrate or Totalize (4)			
R	Radioactivity		Record or Print		
S	Speed or Frequency	Safety (8)		Switch (13)	
T	Temperature			Transmit	
U	Multivariable (6)		Multifunction (12)	Multifunction (12)	Multifunction (12)
V	Viscosity			Valve, Damper, or Louver (13)	
W	Weight or Force		Well		
X (2)	Unclassified		Unclassified	Unclassified	Unclassified
Y	User's Choice(1)			Relay or Compute (13, 14)	
Z	Position			Drive, Actuate or Unclassified Final Control Element	

FIG. 6-21 Meanings of identification letters used in Figure 6-20.

MEANINGS OF IDENTIFICATION LETTERS
FIGURE 6-21

1. A user's choice letter is intended to cover unlisted meanings that will be repeated in a particular project. If used, the letter may have one meaning as a first letter and another meaning as a succeeding letter. The meanings need be defined only once in a legend, or otherwise, for that project. For example, the letter N may be defined as "modulus of elasticity" as a first letter but "oscilloscope" as a succeeding letter.

2. The unclassified letter, X, is intended to cover unlisted meanings that are used only once or to a limited extent. If used, the letter may have any number of meanings as a first letter and any number of meanings as a succeeding letter. Except for its use with distinctive symbols, it is expected that the meanings will be defined outside a tagging balloon on a flow diagram. For example, XR-2 may be a stress recorder, XR-3 may be a vibration recorder, and XX-4 may be a stress oscilloscope.

3. The grammatical form of the succeeding-letter meanings may be modified as required. For example, "indicate" may be applied as indicator or indicating, "transmit" as transmitter or transmitting, and so forth.

4. Any first letter, if used in combination with modifying letters D (differential), F (ratio), or Q (integrate or totalize), or any combination of them, shall be construed to represent a new and separate measured variable, and the combination shall be treated as a first-letter entity. Thus, instruments TDI and TI measure two different variables, namely, differential-temperature and temperature. These modifying letters shall be used when applicable.

5. First letter A for analysis covers all analyses that are not listed in Fig. 6-21 and are not covered by a user's choice letter. It is expected that the type of analysis in each instance will be defined outside a tagging balloon on a flow diagram. Readily recognized self-defining symbols, such as pH, Q2, and CO, have been used optionally in the past in place of first letter A. This practice may cause confusion and misunderstanding, particularly when designations are printed by machines that have only uppercase letters.

6. Use of first letter U for a multivariable in lieu of a combination of first letters is optional.

7. The use of modifying terms "high," "low," "middle" or "intermediate," and "scan" is preferred, but optional.

8. The term "safety" shall apply only to emergency protective primary elements and emergency protective final control elements. Thus, a self-actuated valve that prevents operation of a fluid system at a higher-than-desired pressure by bleeding fluid from the system shall be a back-pressure-type PCV, even if the valve were not intended to be used normally. However, this valve shall be a PSV if it were intended to protect against emergency conditions that are not expected to arise normally. The designation PSV applies to all valves intended to protect against emergency pressure conditions regardless of whether the valve construction and mode of operation place them in the category of the safety valve, relief valve, or safety relief valve.

9. "Passive function glass" applies to instruments that provide an uncalibrated direct view of the process.

10. The term "indicate" applies only to the readout of an actual measurement. It does not apply to a scale for manual adjustment of a variable if there is no measurement input to the scale.

11. A pilot light that is part of an instrument loop shall be designated by a first letter followed by succeeding letter L. For example, a pilot light that indicates an expired time period may be tagged KL. However, if it is desired to tag a pilot light that is not part of a formal instrument loop, the pilot light may be designated in the same way or alternatively by the single letter L. For example, a running light for an electric motor may be tagged either EL, assuming that voltage is the appropriate measured variable, or XL, assuming that the light is actuated by auxiliary electric contacts of the motor starter, or simply L. The action of a pilot light may be accompanied by an audible signal.

12. Use of a succeeding letter U for multifunction instead of a combination of other functional letters is optional.

13. A device that connects, disconnects, or transfers one or more circuits may be either a switch, a relay, an ON-OFF controller, or a control valve, depending on the application. If the device manipulates a fluid process stream and is not a hand-actuated ON-OFF block valve, it shall be designated as a control valve. For all applications other than fluid process streams, the device shall be designated as follows:

A "switch," if it is actuated by hand.

A "switch" or an "ON-OFF controller" if it is automatic and is the first such device in a loop. The term switch is generally used if the device is used for alarm, pilot light, selection, interlock, or safety. The term controller is generally used if the device is used for normal operating control.

A "relay," if it is automatic and is not the first such device in a loop, i.e., it is actuated by a switch or an ON-OFF controller.

14. It is expected that functions associated with the use of a succeeding letter Y will be defined outside a balloon on a flow diagram when it is convenient to do so. This need not be done when the function is self-evident, as for a solenoid valve in a fluid signal line.

15. Use of modifying terms "high," "low," and "middle" or "intermediate" shall correspond to values of the measured variable, not of the signal, unless otherwise noted. For example, a high-level alarm derived from a reverse-acting level transmitter signal shall be an LAH even though the alarm is actuated when the signal falls to a low value. The terms may be used in combinations as appropriate. (See Section 5.9A.)

16. The terms "high" and "low," when applied to positions of valves and other open-close devices, are defined as follows: "High" denotes that the valve is in or approaching the fully open position, and "low" denotes in or approaching the fully closed position.

3.1.3 Each instrument may be represented on diagrams by a symbol. The symbol may be accompanied by an identification.

3.2 Functional Identification

3.2.1 The functional identification of an instrument shall consist of letters from [Table 6-1] and shall include one first letter, covering the measured or initiating variable, and one or more succeeding letters covering the functions of the individual instrument. An exception to this rule is the use of the single letter "L" to denote a pilot light that is not part of an instrument loop. [See Fig. 6-21, note 11.]

3.2.2 The functional identification of an instrument shall be made according to the function and not according to the construction. Thus, a differential-pressure recorder used for flow measurement shall be identified as an FR; a pressure indicator and a pressure switch connected to the output of a pneumatic level transmitter shall be identified as LI and LS, respectively.

3.2.3 In an instrument loop, the first letter of the functional identification shall be selected according to the measured or initiating variable and not according to the manipulated variable. Thus, a control valve varying flow according to the dictates of a level controller is an LV, not an FV.

3.2.4 The succeeding letters of the functional identification designate one or more readout or passive functions, or output functions, or both. A modifying letter may be used, if required, in addition to one or more other succeeding letters. Modifying letters may modify either a first letter or other succeeding letters, as applicable.

3.2.5 The sequence of identification letters shall begin with a first letter. (See exception in Section 3.2.1.) Readout or passive functional letters shall follow in any sequence, and output functional letters shall follow these in any sequence, except that output letter C (control) shall precede output letter V (valve); e.g., HCV, a hand-actuated control valve. However, modifying letters, if used, shall be interposed so that they are placed immediately following the letters they modify.

3.2.6 An instrument tagging designation on a flow diagram may be drawn with as many tagging balloons as there are measured variables or outputs. Thus, a flow-ratio recording transmitter with a flow-ratio switch may be identified on a flow diagram by two tangent circles, one inscribed FFRT-3 and the other FFS-3. The instrument would be designated FFRT/FFS-3 for all uses in writing and reference. If desired, however, the abbreviated FFRT-3 may serve for general identification or purchasing while FFS-3 may be used for electric circuit diagrams.

3.2.7 The number of functional letters grouped for one instrument should be kept to a minimum according to the judgment of the user. The total number of letters within one group should not exceed four. The number within a group may be kept to a minimum by these means:

(1) Arrange the functional letters into subgroups. This practice is described in Section 3.2.6 for instruments having more than one measured variable or output, but it may also be done for other instruments.

(2) If an instrument both indicates and records the same measured variable, then the I (indicate) may be omitted.

3.2.8 All letters of the functional identification shall be uppercase.

3.3 Loop Identification*

3.3.1 The loop identification of an instrument shall generally use a number assigned to the loop of which the instrument is a part. Each instrument loop shall have a unique number. An instrument common to two or more loops may have a separate loop number, if desired.

3.3.2 A single sequence of loop numbers shall be used for all instrument loops of a project or sections of a project regardless of the first letter of the functional identification of the loops.† A loop numbering sequence may begin with a number 1 or with any other convenient number, such as 301 or 1201, that may incorporate coded information such as plant area designation.‡

3.3.3 If a given loop has more than one instrument with the same functional identification, then, preferably, a suffix shall be appended to the loop number; e.g., FV-2A, FV-2B, FV-2C, etc., or TE-25-1, TE-25-2, TE-25-3, etc. However, it may be more convenient or logical in a given instance to designate a pair of flow transmitters, for example, as FT-2 and FT-3 instead of FT-2A and FT-2B. The suffixes may be applied according to the following guidelines:

(1) Suffix letters, which shall be uppercase, should be used, i.e., A, B, C, etc.

(2) For an instrument such as a multipoint temperature recorder that prints numbers for point identification, the primary elements may be numbered TE-25-1, TE-25-2, TE-25-3, etc. The primary element suffix numbers should correspond to the point numbers of the recorder. Optionally, they may not correspond.

(3) Further subdivisions of a loop may be designated by alternating suffix letters and numbers. (See Section 5.9R (3).)

3.3.4 An instrument that performs two or more functions may be designated by all of its functions. For example, a flow recorder FR-2 with pressure pen PR-4 is preferably designated FR-2/PR-4; alternatively, it may be designated UR-7 (see Section 5.8U and 5.9U); a two-pen pressure recorder may be PR-7/8; and a common annunciator window for high- and low-temperature alarm may be TAH/L-9.

3.3.5 Instrument accessories, such as purge rotameters, air sets, and seal pots that are not explicitly shown on a flow diagram but need a tagging designation for other purposes should be tagged individually according to their function and shall use the same loop number as that of the instrument they directly serve. Application of such a designation does not imply that the accessory must be shown on the flow diagram. Alternatively, the accessories may use the identical tag number as that of their associated instrument, but with clarifying words added, if re-

***The rules for loop identification need not be applied to those instruments or accessories, e.g., steam traps, pressure indicators, temperature wells, that are purchased in bulk quantities, if such is the user's practice.**

†Although not recommended, parallel numbering is a method that has been used to designate loops. In this method, a new numbering sequence is begun for each first letter.

‡An optional method that has been used for designating plant areas is to use coded numbers to prefix the functional identification of an instrument. (See Section 5.2(4).)

quired. Thus, an orifice flange union associated with orifice plate FE-7 should be tagged FX-7, but may be tagged FE-7 flanges. A purge rotameter-regulator associated with pressure gauge PI-8 should be tagged FICV-8 but may be tagged PI-8 purge. A thermowell used with thermometer TI-9 should be tagged TW-9, but may be tagged TI-9 thermowell.

3.4 Symbols

3.4.1 [Figures 6-22 and 6-23] illustrate the symbols that are intended to depict instrumentation of flow diagrams and other drawings, and cover their application to a variety of processes. The applications shown were chosen to illustrate principles of the methods of symbolization and identification. Additional applications that adhere to these principles may be devised as required. The examples show numbering that is typical for the pictured instrument interrelationships, but the numbering may be varied to suit the situation. The symbols indicating the various locations of instruments have been applied in typical ways in the illustrations; this does not imply, however, that the applications or the designations of the instruments are therefore restricted in any way. No inference should be drawn that the choice of any of the schemes for illustration constitutes a recommendation for the illustrated methods of measurement or control. Where alternative symbols are shown without a statement of preference, the relative sequence of the symbols does not imply a preference.

3.4.2 The circular balloon may be used to tag distinctive symbols, such as that for a control valve, when such tagging is desired. (In such instances, the line connecting the balloon to the instrument symbol shall be drawn close to but not touching the symbol.) In other instances, the balloon serves to represent the instrument, proper.

3.4.3 A distinctive symbol whose relationship to the remainder of the loop is easily apparent from a diagram need not be individually tagged on the diagram.* For example, it is expected that an orifice plate or a control valve that is part of a larger system will not usually be shown with a tag number on a diagram. Also, where there is an electrical primary element connected to another instrument on a diagram, use of a symbol to represent the primary element on the diagram is optional. (See Section 5.8C.)

3.4.6 Aside from the general drafting requirement for neatness and legibility, all symbols may be drawn with any orientation. Likewise, signal lines may be drawn on a diagram entering or leaving the appropriate part of a symbol at any angle. Directional arrowheads shall be added to signal lines when needed to clarify the direction of flow of intelligence.

3.4.7 The electric, pneumatic, or other power supply to an instrument is not expected to be shown unless it is essential to an understanding of the operation of the instrument or the loop. (See Sections 5.1(1) and 5.10(9), (12), and (26).)

3.4.8 In general, one signal line will suffice to represent the interconnections between two instruments on flow diagrams even though they may be connected physically by more than one line.

3.4.9 The sequence in which the instruments of a loop are connected on a flow diagram shall reflect the functional logic; this arrangement will not necessarily correspond to the signal connection sequence. Thus, a loop using analog voltage signals requires parallel wiring while a loop using analog current signals requires series wiring, but the diagram in both instances shall be drawn as though all the wiring were parallel. This will show the functional interrelationships clearly while keeping their aspect of the flow diagram independent of the type of instrument system installed. The literal and correct wiring interconnections are expected to be shown on a suitable electric wiring diagram.

3.4.10 For process flow diagrams or other applications where it may be desired to depict only those instrumentation end functions that are needed for the operation of the process, proper, the intermediate instrumentation and other derails may be omitted, provided that this is done consistently for a given type of drawing throughout a project. Minor instruments and loop components, e.g., pressure gauges, thermometers, transmitters, converters, may thus be eliminated from the diagrams.

3.4.11 It is common practice for mechanical flow diagrams to omit the symbols of interlock-hardware components that are actually necessary for a working system, particularly when symbolizing electric interlock systems. For example, a level switch may be shown as tripping a pump, or separate flow and pressure switches may be shown as actuating a solenoid valve or other interlock device. In both instances, auxiliary electrical relays and other components may also be required, but these additional components may be considered details to be shown elsewhere. By the same token, the current transformer shown under Section 5.8I is omitted sometimes and its receiver shown connected directly to the process–in this case the electric motor.

*In some instances, the illustrations tend to apply more tagging balloons than are usually required. This has been done for clarity of illustrations.

5. DRAWINGS

5.1 INSTRUMENT LINE SYMBOLS

All lines shall be fine in relation to process piping lines.

(1) Connection to process, or mechanical link, or instrument supply* ————————

(2) Pneumatic signal †, or undefined signal for process flow diagrams —//——//——//—

(3) Electric signal - - - - - - -

(4) Capillary tubing (filled system) —×——×——×—

(5) Hydraulic signal —L——L——L—

(6) Electromagnetic § or sonic signal (without wiring or tubing) ∿∿∿

FIG. 6-22 Instrument line symbols used in Figure 6-23.

The following abbreviations denote the types of power supply and may also be applied for purge fluid supplies:

AS	Air Supply
ES	Electric Supply
GS	Gas Supply
HS	Hydraulic Supply
NS	Nitrogen Supply
SS	Steam Supply
WS	Water Supply

The power supply level may be added to the instrument supply line—for example AS (100-psig air supply) and ES 24DC (24-volt direct current supply).

The pneumatic signal symbol applies to a signal using gas as the signal medium. If a gas other than air is used, the gas shall be identified by a note on the signal symbol or otherwise.

Electromagnetic phenomena include heat, radio waves, nuclear radiation, and light.

5.2 GENERAL INSTRUMENT SYMBOLS – BALLOONS

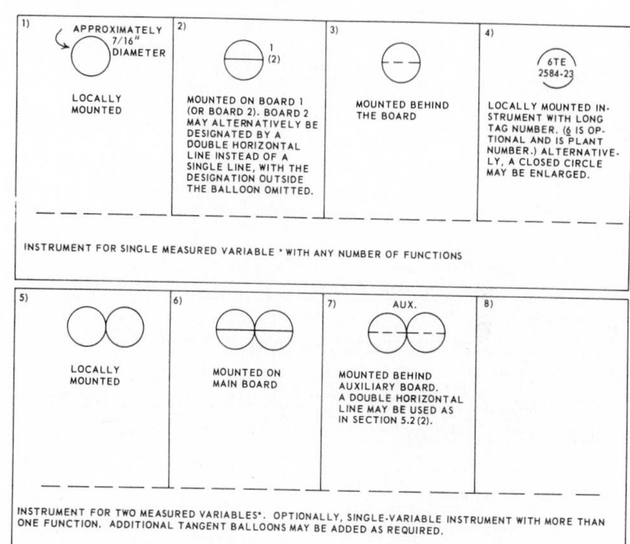

*Certain measured variables may have more than one input (see Table 1, note 4). An instrument that only indicates differential-pressure, for example, shall use only one balloon, tagged PDI, even though it has two inputs.

5.3 CONTROL VALVE BODY SYMBOLS*

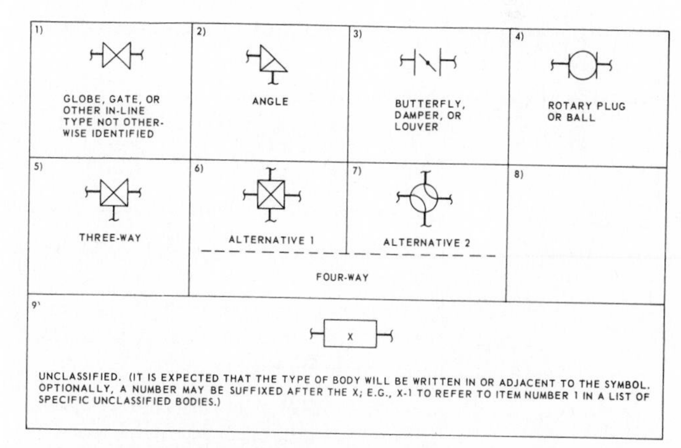

* If special valve details are required for supplementary detailed instrument drawings (other than flow diagrams), then symbols may be taken from References 4 and 5. These symbols are not recommended for use on flow diagrams, in accordance with the statement of scope of this Standard. (See Section 1.3.2.)

FIG. 6-23 Instruments and identification. (Reprints used with permission from ISA S5.1 Instrumentation and Identification, © Instrument Society of America, 1975.)

5.4 ACTUATOR SYMBOLS* †

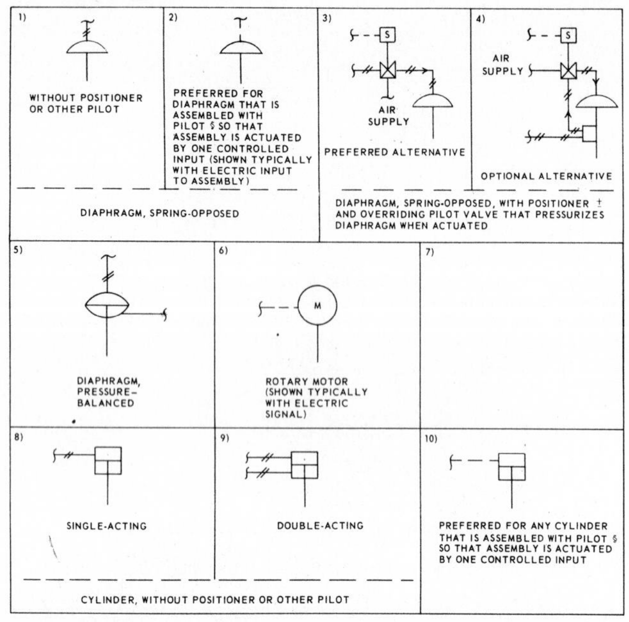

* See footnote, page 19.

† Normally, modes of control valve action will not be designated on a flow diagram. However, an on-off valve mode may be designated, if desired, by placing the symbol 1-0 or ON-OFF near the valve symbol.

§ Pilot may be positioner, solenoid valve, signal converter, etc.

‡ The positioner shall preferably not be shown unless an intermediate device is on its output. The positioner tagging, ZC, shall preferably not be used even if the positioner is shown. The positioner symbol, a box drawn on the actuator shaft, is the same for all types of actuators. When the symbol is used, the type of instrument signal, i.e., pneumatic, electric, etc., shall be drawn as appropriate. If the positioner symbol is used and there is no intermediate device on its output, then the positioner output signal need not be shown.

5.4 ACTUATOR SYMBOLS (Contd.)

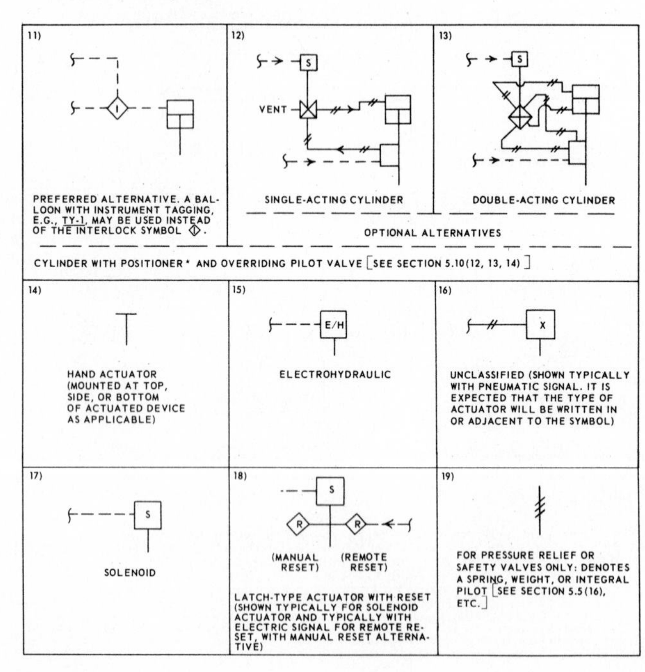

*See footnote § on page 20.

5.5 SYMBOLS FOR SELF-ACTUATED REGULATORS, VALVES, AND OTHER DEVICES

FLOW

1) FICV 5 — AUTOMATIC REGULATOR WITH INTEGRAL FLOW INDICATION. TAG REGULATOR FCV-5 IF IT DOES NOT HAVE INTEGRAL FLOW INDICATION.

2) (UPSTREAM ALTERNATIVE) (DOWNSTREAM ALTERNATIVE) FI 6 — INDICATING ROTAMETER WITH INTEGRAL MANUAL THROTTLE VALVE

3)

HAND

4) HCV 9 — HAND CONTROL VALVE IN PROCESS LINE

5) HS 10 — HAND-ACTUATED ON-OFF SWITCHING VALVE IN PNEUMATIC SIGNAL LINE

6) HO 11 — MANUALLY ADJUSTABLE RESTRICTION ORIFICE IN SIGNAL LINE [SEE SECTION 5.8F (17)]

LEVEL [SEE SECTION 5.5 (29)]

7) TANK, LCV 14 — LEVEL REGULATOR WITH MECHANICAL LINKAGE

8)

9)

PRESSURE

10) PCV 17 — PRESSURE-REDUCING REGULATOR, SELF-CONTAINED

11) PCV 18 — PRESSURE-REDUCING REGULATOR WITH EXTERNAL PRESSURE TAP

12) PDCV 19 — DIFFERENTIAL-PRESSURE-REDUCING REGULATOR WITH INTERNAL AND EXTERNAL PRESSURE TAPS

5.5 SYMBOLS FOR SELF-ACTUATED REGULATORS, VALVES, AND OTHER DEVICES (Contd.)

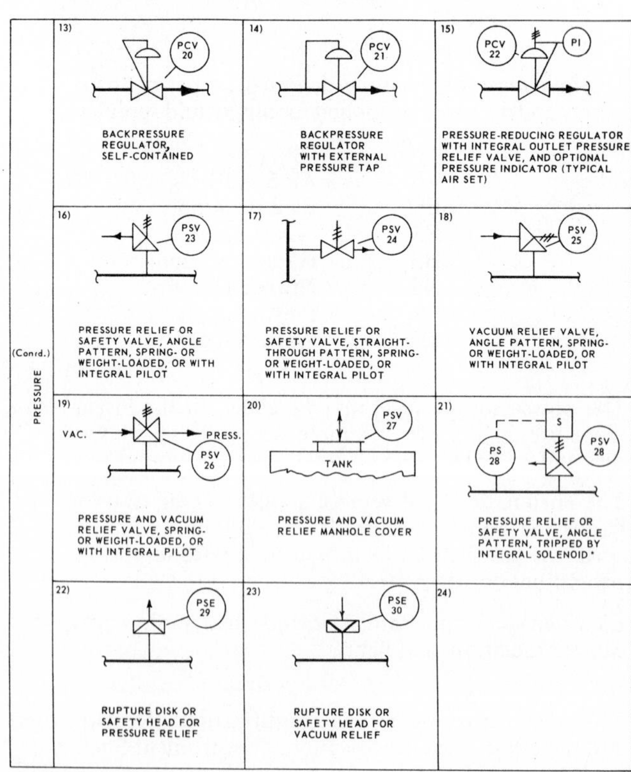

*The solenoid-tripped pressure relief valve is one of the class of power-actuated relief valves and is grouped with the other types of relief valves even though it is not entirely a self-actuated device.

- 23 -

FIG. 6-23 Instruments and identification. (cont'd)

5.5 SYMBOLS FOR SELF-ACTUATED REGULATORS, VALVES, AND OTHER DEVICES (Contd.)

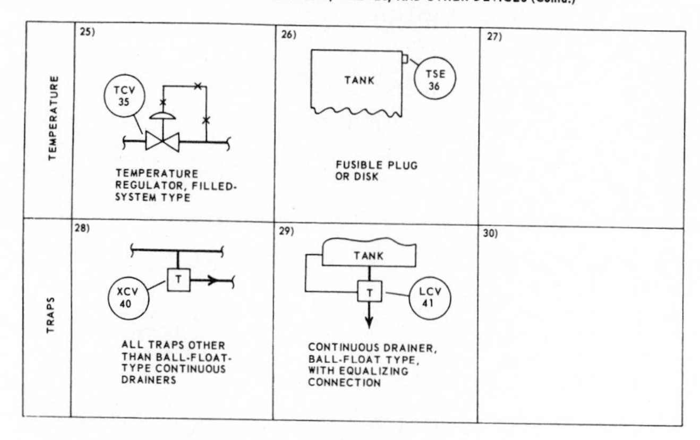

5.6 SYMBOLS FOR ACTUATOR ACTION IN EVENT OF ACTUATOR POWER FAILURE. (SHOWN TYPICALLY FOR DIAPHRAGM-ACTUATED CONTROL VALVE)

1) FO TWO-WAY VALVE, FAIL OPEN	2) FC TWO-WAY VALVE, FAIL CLOSED	3) A B FO C THREE-WAY VALVE, FAIL OPEN TO PATH A-C
4) FO A B C D FO FOUR-WAY VALVE, FAIL OPEN TO PATHS A-C AND D-B	5) FL ANY VALVE, FAIL LOCKED (POSITION DOES NOT CHANGE)	6) FI ANY VALVE, FAIL INDETERMINATE

5.7 MISCELLANEOUS SYMBOLS

1) APPROXIMATELY 7/16" DIAMETER PILOT LIGHT	2) APPROXIMATELY 7/16" SQUARE C 12 BOARD-MOUNTED PATCHBOARD OR MATRIX BOARD CONNECTION, NUMBER 12	3) APPROXIMATELY 1/4" SQUARE P PURGE OR FLUSHING DEVICE (MEANS OF REGULATING PURGE MAY BE SHOWN IN PLACE OF SYMBOL)
4) R RESET FOR LATCH-TYPE ACTUATOR	5) CHEMICAL SEAL	6)
7) I GENERALIZED – FOR UNDEFINED OR COMPLEX INTERLOCK LOGIC	8) AND INTERLOCK IS EFFECTIVE ONLY IF ALL INPUTS EXIST	9) OR INTERLOCK IS EFFECTIVE IF ANY ONE OR MORE INPUTS EXIST

INTERLOCK

5.8 PRIMARY ELEMENT SYMBOLS

	1)	2)	3)
A ANALYSIS	O_2 COMBUSTIBLES AIT 45 AIT 46 RECEIVER RECEIVER DUAL ANALYSIS INDICATING TRANSMITTER FOR OXYGEN AND COMBUSTIBLES CONCENTRATIONS		
B BURNER FLAME	FIRED FURNACE BE 51 BI 51 ONE BURNER FLAME DETECTOR CONNECTED TO ANALOG-TYPE FLAME INTENSITY INDICATOR	FIRED FURNACE BE 52A BE 52B BS 52 ALARM TWO BURNER FLAME SENSORS CONNECTED TO COMMON SWITCH	FIRED FURNACE BE 53 BX 53 TV RECVR. TELEVISION CAMERA AND RECEIVER TO VIEW BURNER FLAME
C CONDUCTIVITY	CE 56-5 CJR 56-5 CONDUCTIVITY CELL CONNECTED TO POINT 5 OF MULTIPOINT SCANNING CONDUCTIVITY RECORDER		
D DENSITY OR SPECIFIC GRAVITY	TANK DT 59 RECVR. DENSITY TRANSMITTER, DIFFERENTIAL-PRESSURE TYPE, EXTERNALLY CONNECTED	RADIO-ACTIVE SOURCE DX 60 CONVEYOR DE 60 DR 60 RADIOACTIVE-TYPE DENSITY ELEMENT CONNECTED TO DENSITY RECORDER ON BOARD	DT 61 RECEIVER SPECIFIC GRAVITY TRANSMITTER, THROUGH-FLOW TYPE

5.8 PRIMARY ELEMENT SYMBOLS (Contd.)

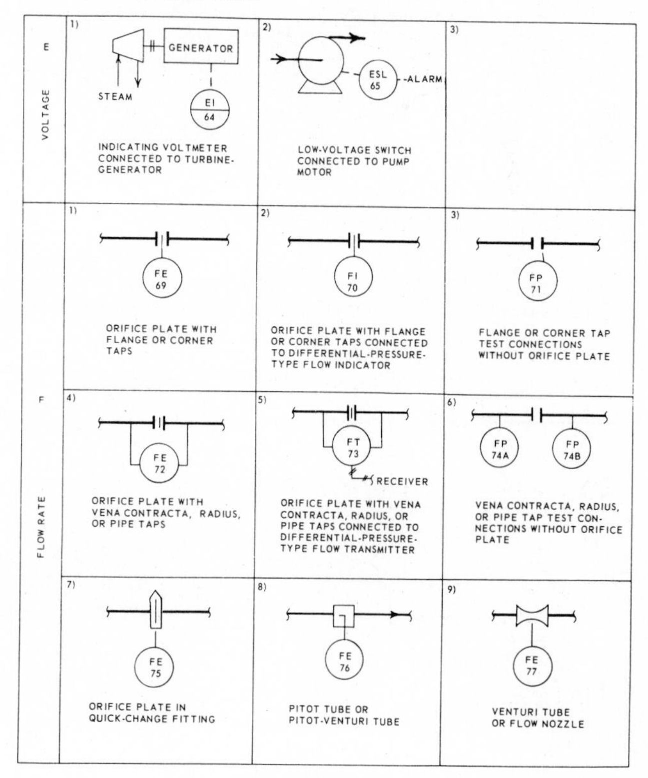

FIG. 6-23 Instruments and identification. (cont'd)

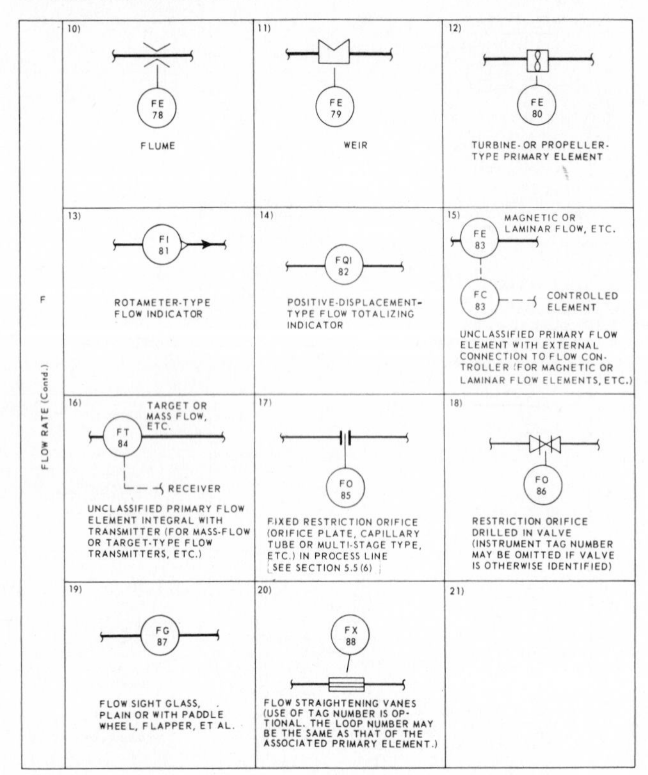

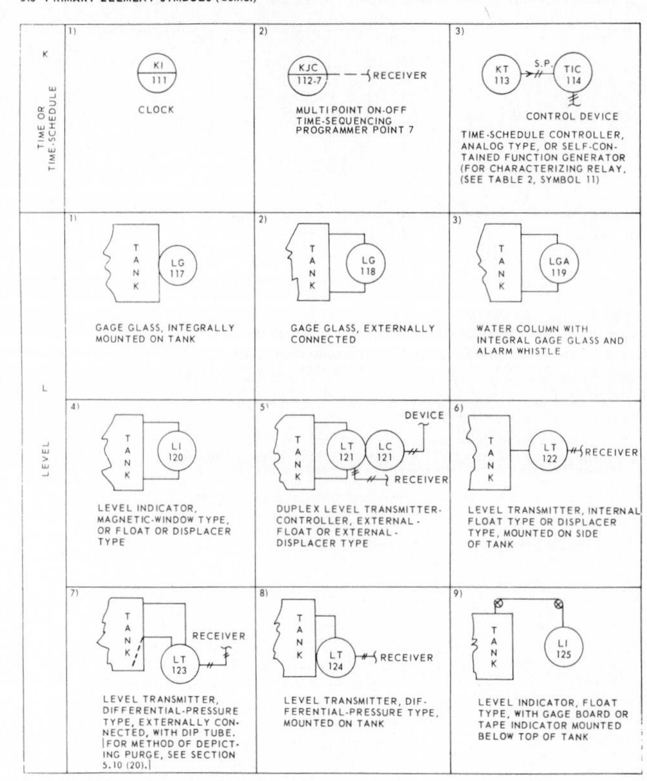

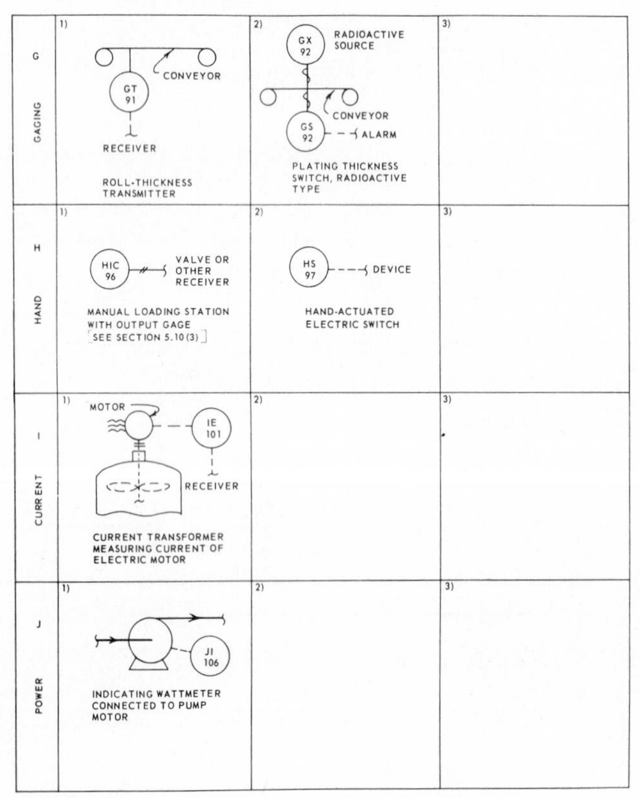

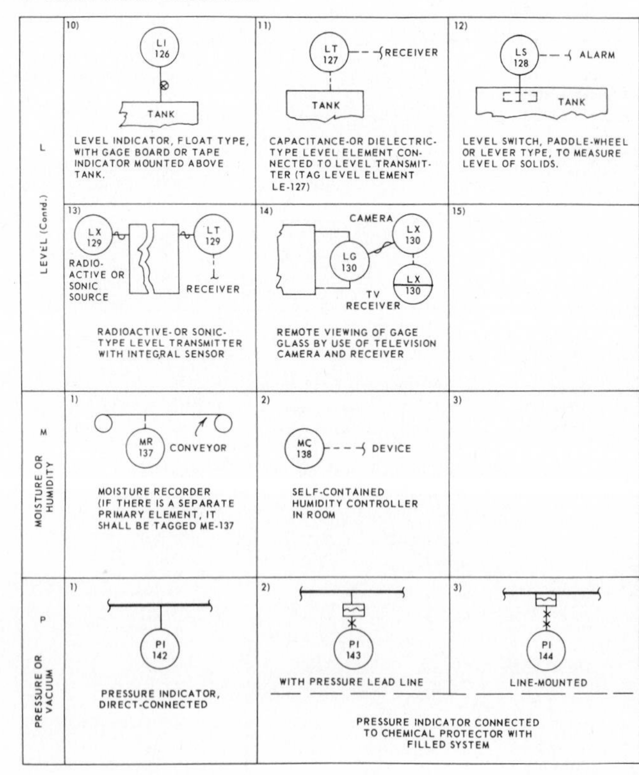

FIG. 6-23 Instruments and identification. (cont'd)

5.8 PRIMARY ELEMENT SYMBOLS (Contd.)

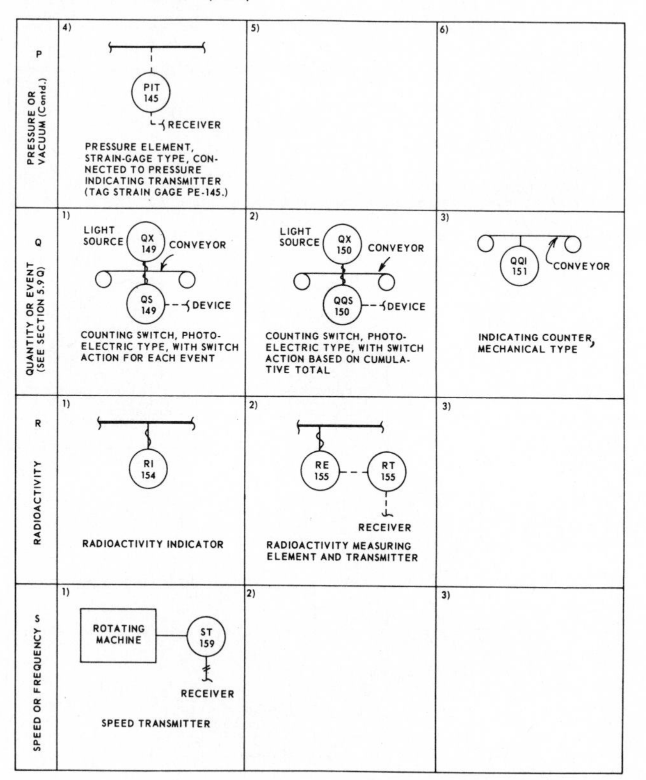

5.8 PRIMARY ELEMENT SYMBOLS (Contd.)

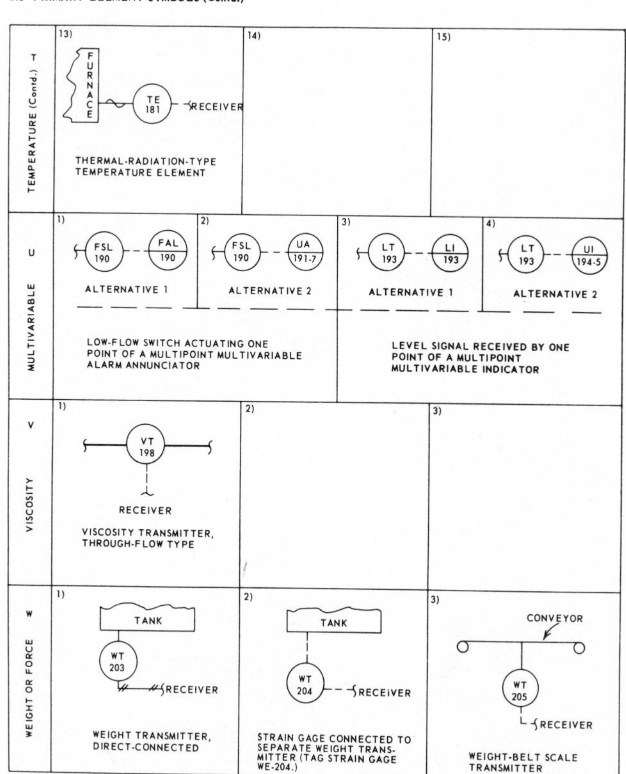

5.8 PRIMARY ELEMENT SYMBOLS (Contd.)

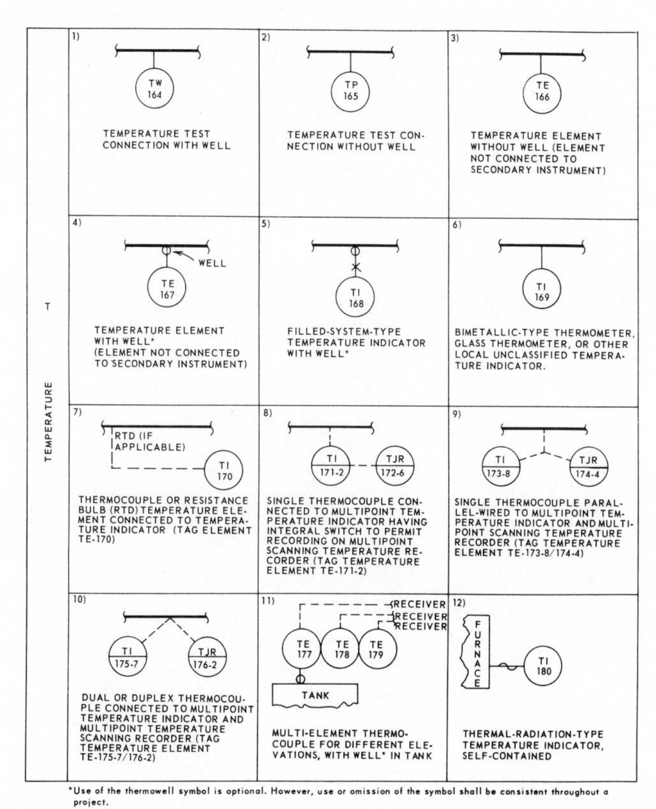

*Use of the thermowell symbol is optional. However, use or omission of the symbol shall be consistent throughout a project.

5.8 PRIMARY ELEMENT SYMBOLS (Contd.)

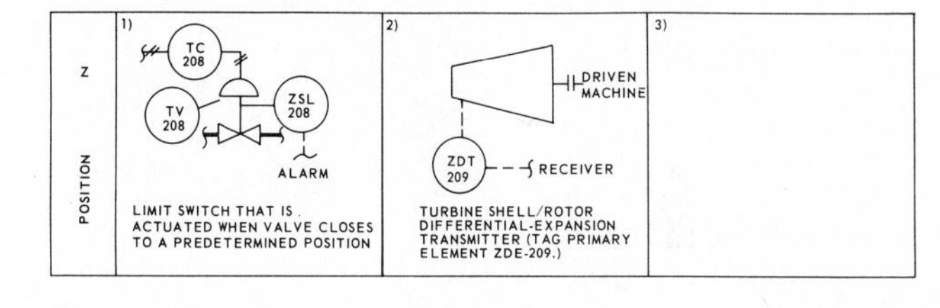

FIG. 6-23 Instruments and identification. (cont'd)

5.9 FUNCTION SYMBOLS

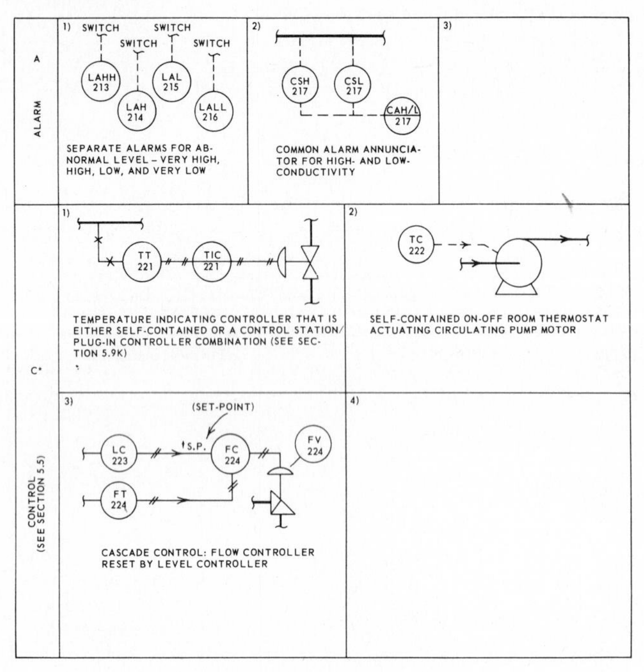

* It is expected that control modes will not be designated on a diagram. However, the following designations may be used outside the controller symbol, if desired, in combinations such as %, ∫, D.

CONTROL MODE	DESIGNATION	CONTROL MODE	DESIGNATION
ON-OFF	1-0 OR ON-OFF	INVERSE DERIVATIVE	1/D
DIFFERENTIAL-GAP, TWO-POSITION	△1-0 OR △ON-OFF	OPTIMIZING	OPT. OR MAX. OR MIN. (as applicable)
PROPORTIONAL	% OR P	ADAPTIVE	ADAPT.
AUTOMATIC RESET, FLOATING, OR INTEGRAL	∫ OR I	UNCLASSIFIED	AS REQUIRED
DERIVATIVE OR RATE	D OR d/dt	DIRECT ACTING	DIR.
		REVERSE ACTING	REV.

† A controller is understood to have integral manual set-point adjustment unless means of remote adjustment is indicated.

5.9 FUNCTION SYMBOLS (Contd.)

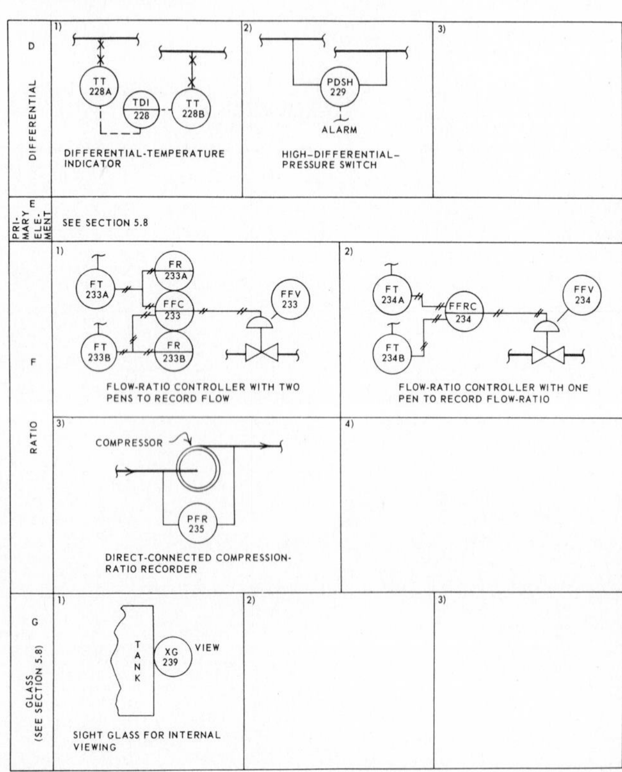

5.9 FUNCTION SYMBOLS (Contd.)

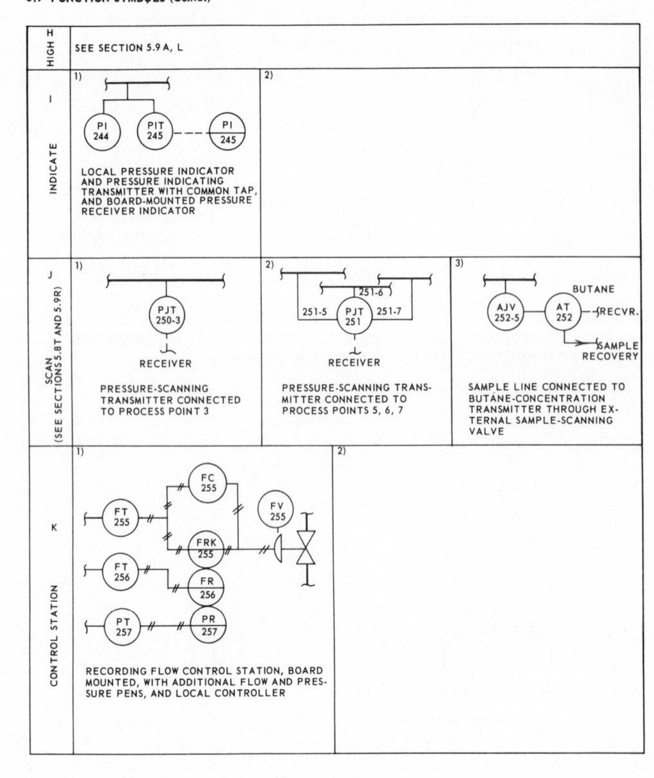

5.9 FUNCTION SYMBOLS (Contd.)

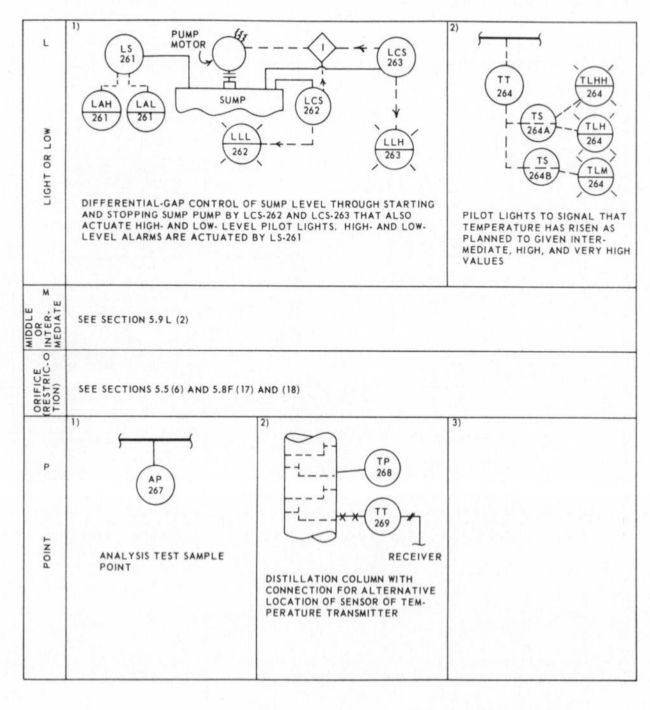

FIG. 6-23 Instruments and identification. (cont'd)

5.9 FUNCTION SYMBOLS (Contd.)

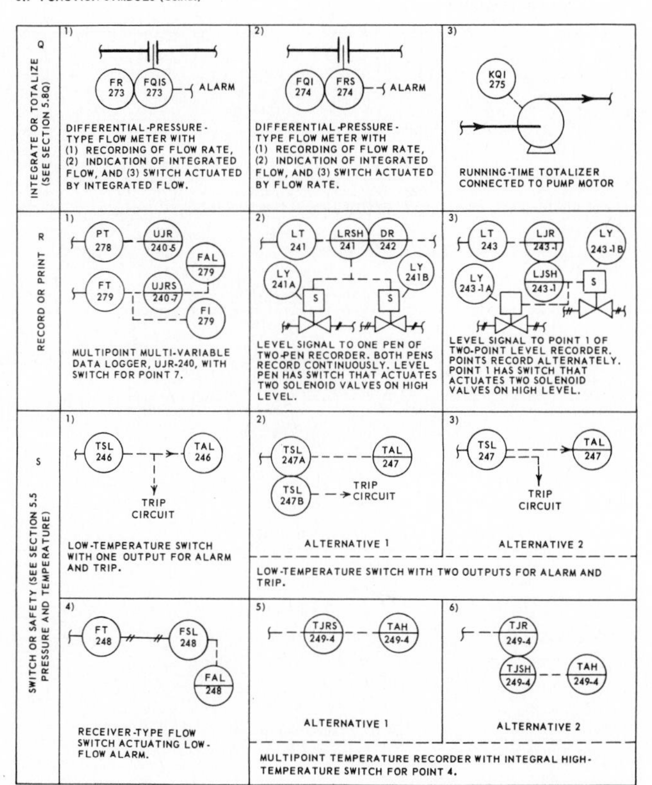

5.9 FUNCTION SYMBOLS (Contd.)

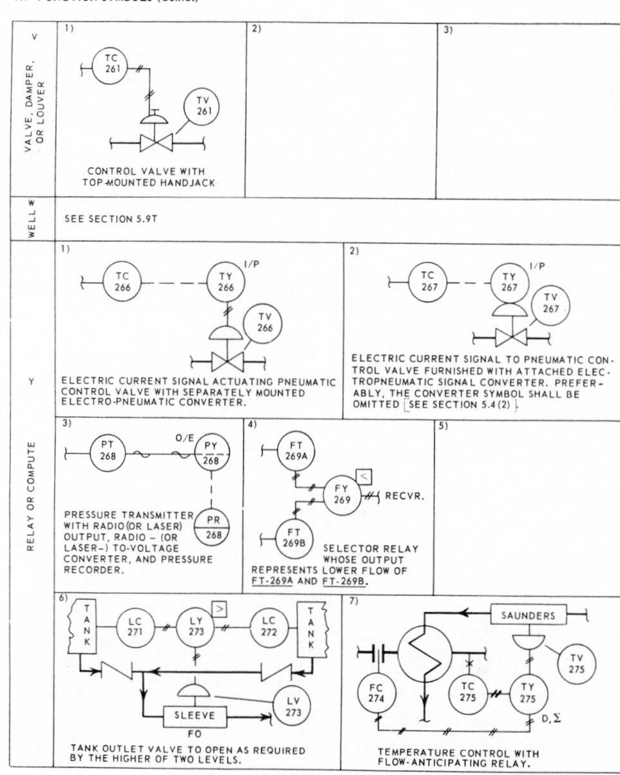

5.9 FUNCTION SYMBOLS (Contd.)

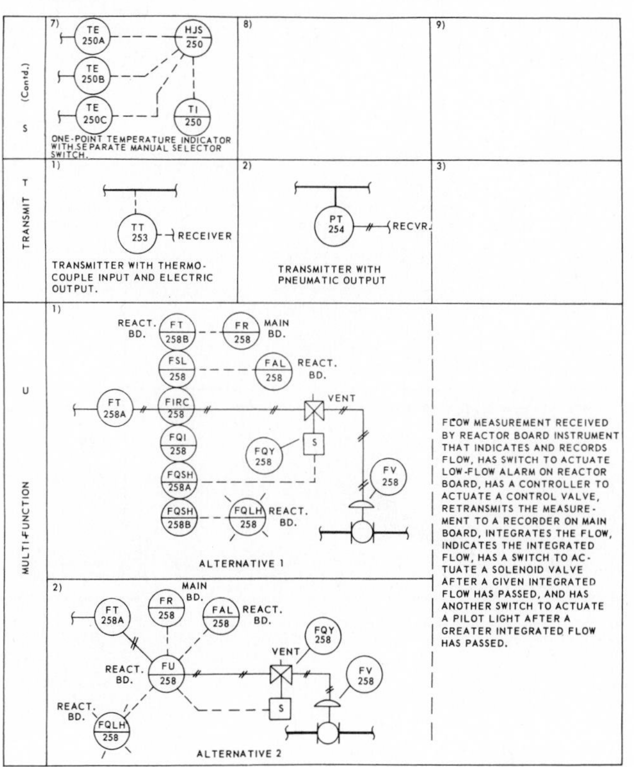

5.9 FUNCTION SYMBOLS (Contd.)

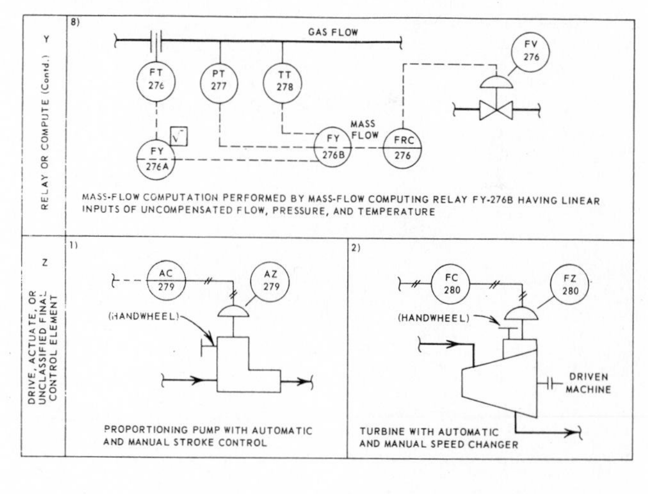

FIG. 6-23 Instruments and identification. (cont'd)

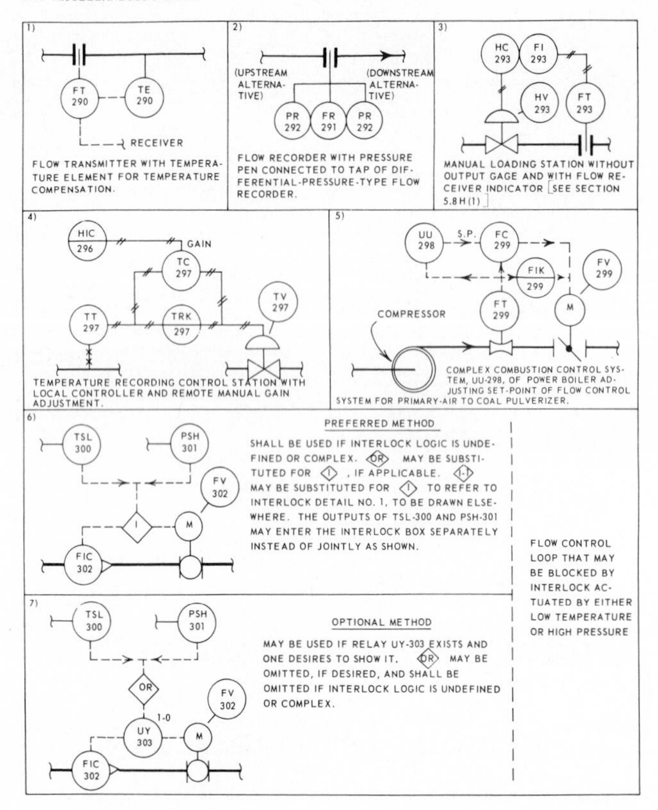

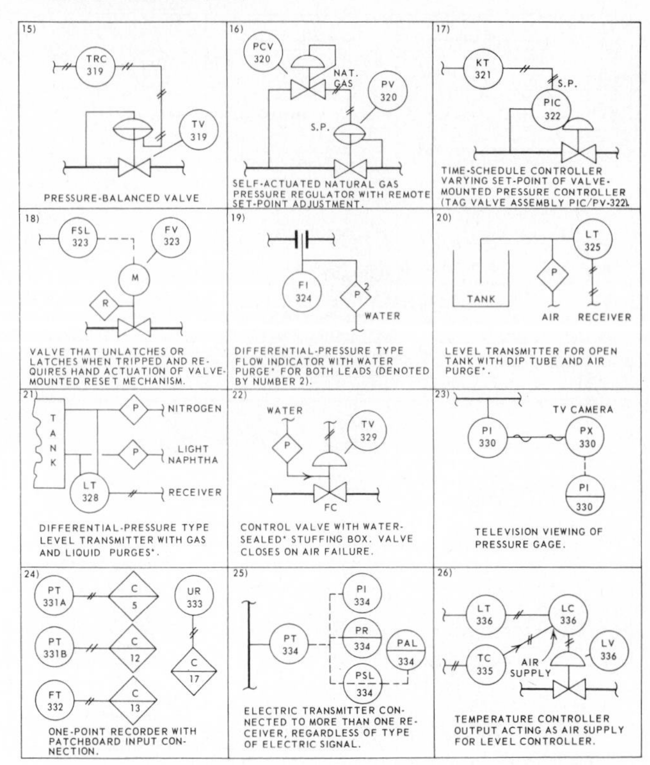

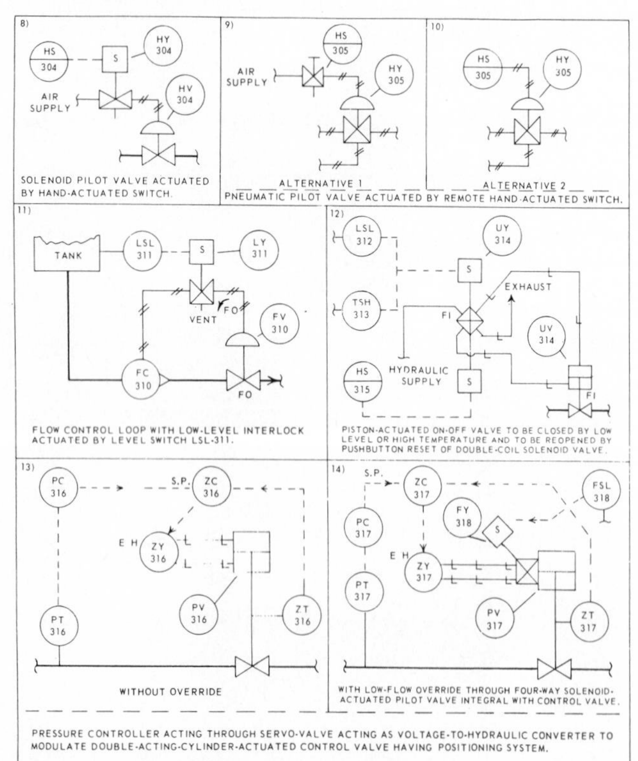

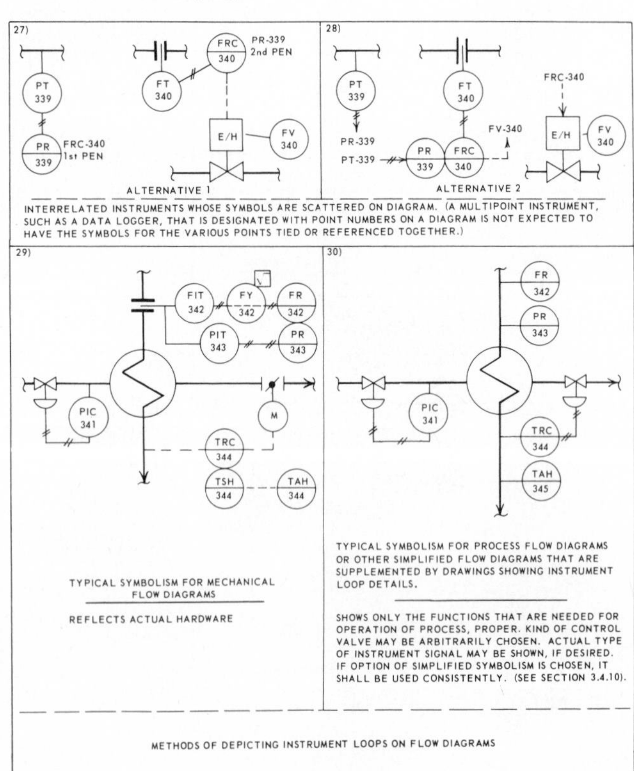

FIG. 6-23 Instruments and identification. (cont'd)

QUIZ

1. Define the four major types of flow diagrams (process, utilities, mechanical, block).
2. At what stage is the flow diagram completed?
3. Pipe sizes are called out on ____________ flow diagrams.
4. All equipment is shown on the ____________ flow diagram.
5. The ____________ flow diagram shows the pipeline plot plan-wise.
6. What are the four major categories of piping instrumentation?
7. Describe the types of gas utilities.
8. Drains are shown on the ____________ flow diagram.
9. Equipment is drawn to scale on all flow diagrams. True or false?
10. Flare systems are part of:
    (a) Utility diagrams
    (b) Process diagrams
    (c) Waste diagrams
    (d) Instrumentation diagrams
11. Define PR.
12. Level glasses are mounted and read ____________.
13. What is the function of an FR?
14. What is a TIC.
15. How is an instrument with two functions represented?

PROBLEMS

1. Redraw any of the flow diagrams in this chapter or elsewhere in the book on the appropriate size of paper. Give special attention to spacing, lettering, and line weight.
2. Draw an isometric flow diagram of Fig. 6-12.
3. Draw a simplified process flow diagram from the water treatment P&ID (Fig. 6-15).
4. In a larger scale, draw the P&ID sections in Fig. 6-6 using standard symbols.
5. Redraw Fig. 6-16. On a separate sheet of paper, explain the line code for each pipeline.
6. Redraw Fig. 6-17.

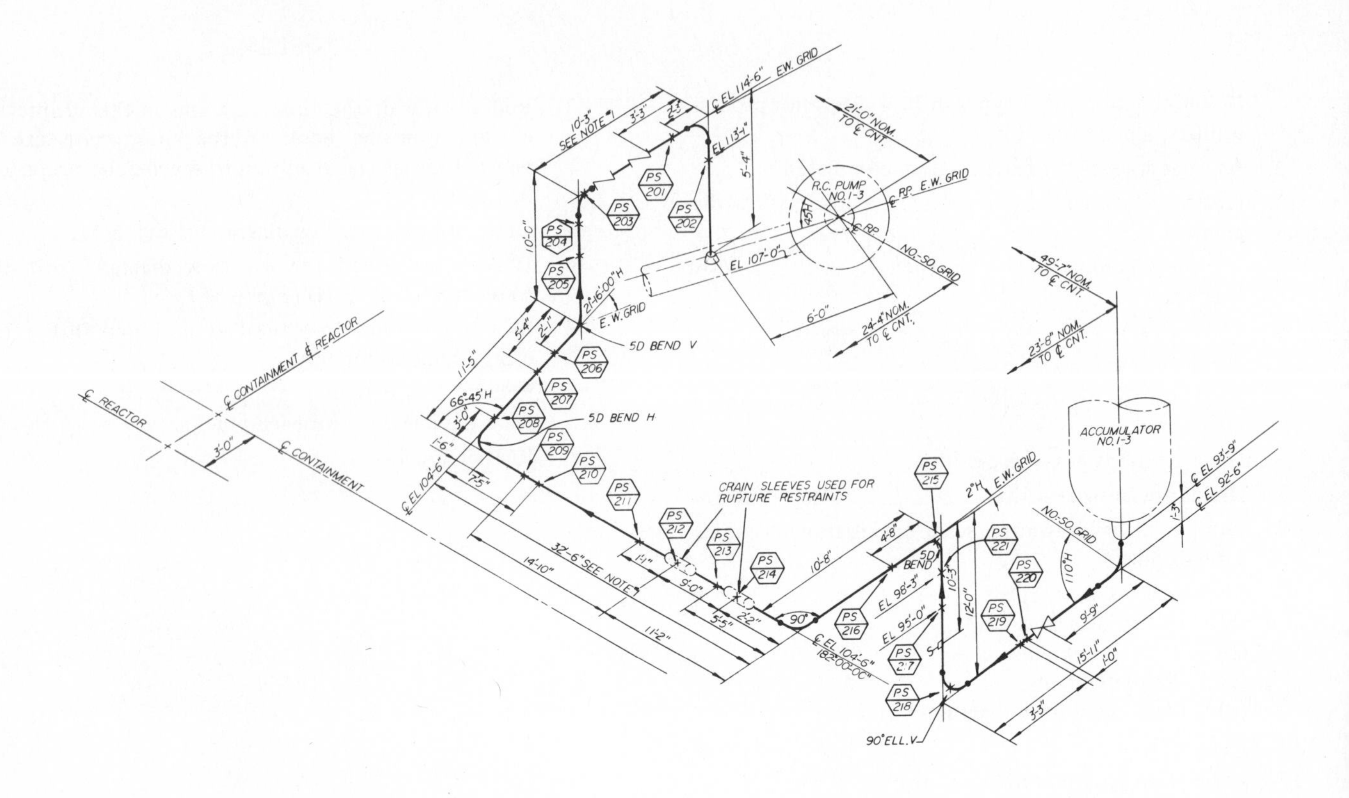

7 ISOMETRICS

Isometric drawings are a fast, accurate method of representing piping systems. In industries that employ piping drafters, many people work entirely on isometrics doing pipe support isometrics, fabrication "isos", conducting pipe support and stress studies, and in some cases using computers for calculations and graphic representations. Power companies, petrochemical firms, and large construction companies employ many drafters to do isometric drawing.

Isometric projections are among the most commonly used methods of representing piping. Isometric drawings provide accurate three-dimensional views of a pipeline and components, enabling the fabricator and construction crew to visualize complete systems or lines.

In an isometric drawing the elevation, plan, and side views are presented to the viewer simultaneously. All horizontal and vertical dimensions are shown in one view instead of the normal three views presented in orthographic projection. An isometric drawing is usually taken of each pipeline in a project to show the details and dimensions required to make up that line (spool).

In drafting it is helpful to possess a sense of visualization. When beginning an isometric drawing, the first step in constructing the line should be to visualize it in space. The next step is to sketch out the full length of the pipeline from origin to termination—along with all component parts such as vessel connections and other essential equipment attached to the line, including valves and fittings. On this sketch all components should be located in proportion to one another.

Isometric drawings can be used for detailing and dimensioning a pipeline (Fig. 7-1), for ordering and specifying component parts, for flagging special callouts and piping fabrication notes, and for positioning pipe supports. In critical industries such as the nuclear field, piping isometrics are also used for seismic and thermal movement studies. By programming the isometric drawing into a computer with

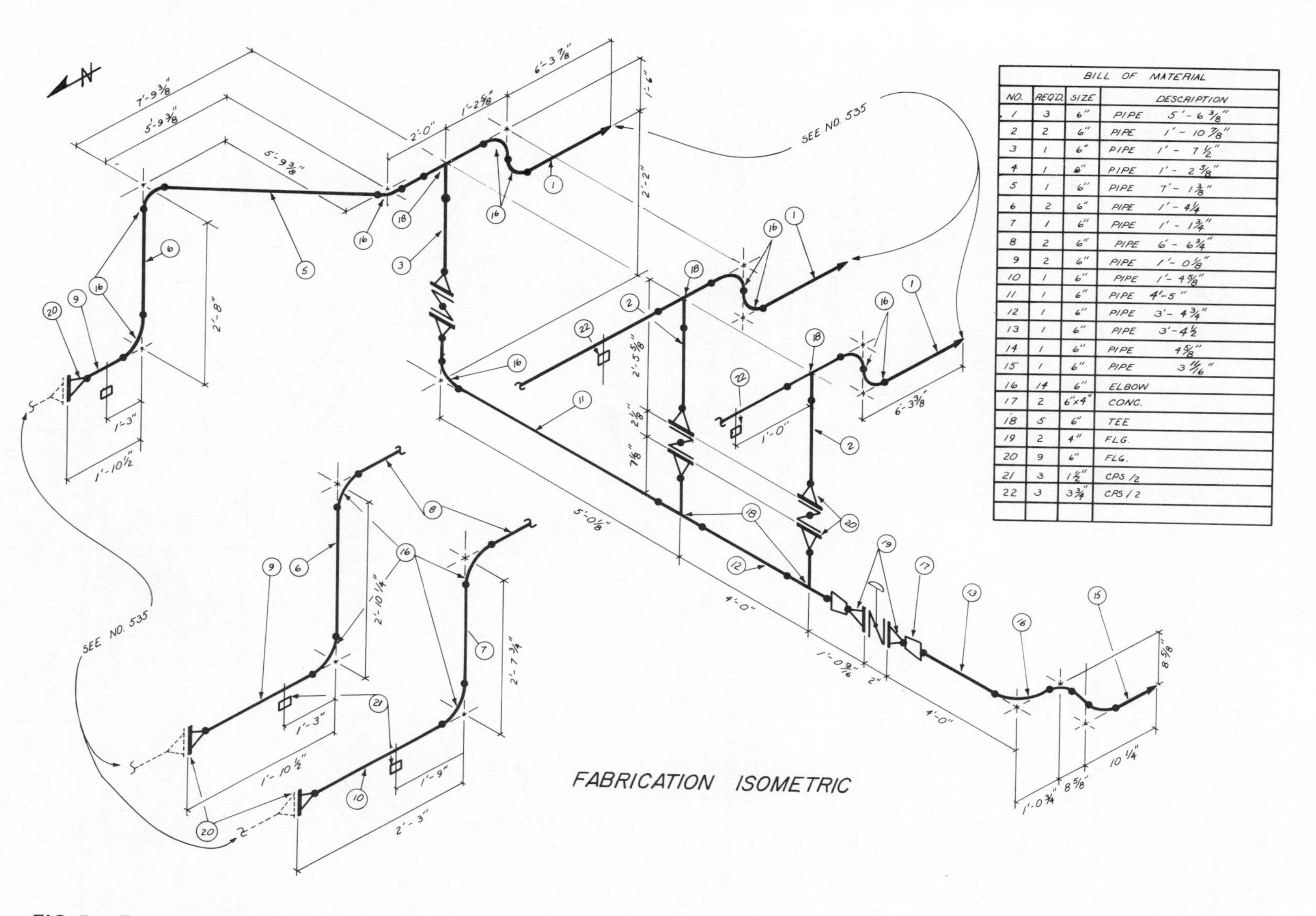

BILL OF MATERIAL			
NO.	REQ'D.	SIZE	DESCRIPTION
1	3	6"	PIPE 5'-6 3/8"
2	2	6"	PIPE 1'-10 7/8"
3	1	6"	PIPE 1'-7 1/2"
4	1	6"	PIPE 1'-2 5/8"
5	1	6"	PIPE 7'-1 3/8"
6	2	6"	PIPE 1'-4 1/4
7	1	6"	PIPE 1'-1 3/4"
8	2	6"	PIPE 6'-6 3/4"
9	2	6"	PIPE 1'-0 1/8"
10	1	6"	PIPE 1'-4 5/8"
11	1	6"	PIPE 4'-5"
12	1	6"	PIPE 3'-4 3/4"
13	1	6"	PIPE 3'-4 1/2
14	1	6"	PIPE 4 5/8"
15	1	6"	PIPE 3 11/16"
16	14	6"	ELBOW
17	2	6"x4"	CONC.
18	5	6"	TEE
19	2	4"	FLG.
20	9	6"	FLG.
21	3	1 1/2"	CPS/2
22	3	3 3/4"	CPS/2

FIG. 7-1 Fabrication isometric.

the preliminary placement of supports and hangers and then analyzing the system's reaction to seismic and thermal disturbances before installation, the designer can accurately place special and standard pipe supports, hangers, and shock suppressors (snubbers).

Isometric drawings can be divided into four basic types:

1. *System isometrics:* This type of isometric shows a total or partial system. It is used less often than the other types because it is more difficult to draw, exhibiting a complete system instead of a single piping line. System isometrics are also used for pictorially representing a piping system for sales or advertising purposes. The piping system isometric drawing is usually limited to small-diameter pipelines, especially where threaded connections are used. In this case, many dimensions are left to the fabricator for field placement. System isometrics are usually limited to relatively simple piping systems.
2. *Field fabrication isometrics:* This type of drawing locates only the critical aspects of the system, such as major equipment. Dimensions are given only for overall pipeline sizes; many dimensions are calculated in the field by the construction crew. Note that in Fig. 7-2 only the dimensions (D) to components and between major lines are given. Face-to-face dimensions for fittings, flanges, and valves are not shown on this type of drawing; many of the dimensions are provided and verified at the job site.
3. *Shop fabrication isometrics:* An example of this type of isometric is given in Fig. 7-3. All face-to-face dimensions for valves, flanges, and fittings are given, including dimensions for the placement and fabrication of the pipeline. Figure 7-4 is an example of a shop fabrication drawing which is also used for the placement of pipe supports. Piping fabrication drawings must be complete because lines on the same project are often fabricated at different shops—thus all necessary information must be included. In general, the shop fabrication piping isometric is used for pipelines of 2 in. (5.08 cm) OD and greater. (See the section in Chapter 9 on orthographic spools.)
4. *Detail isometrics:* This type shows particular details and special items. Figure 7-5, for example, details steam tracing on a piping isometric line. Special trim and other essential information are shown on the detail isometric drawing. This type of isometric sometimes shows a small portion of a pipeline or a piece of equipment in order to express the dimensions for its fabrication.

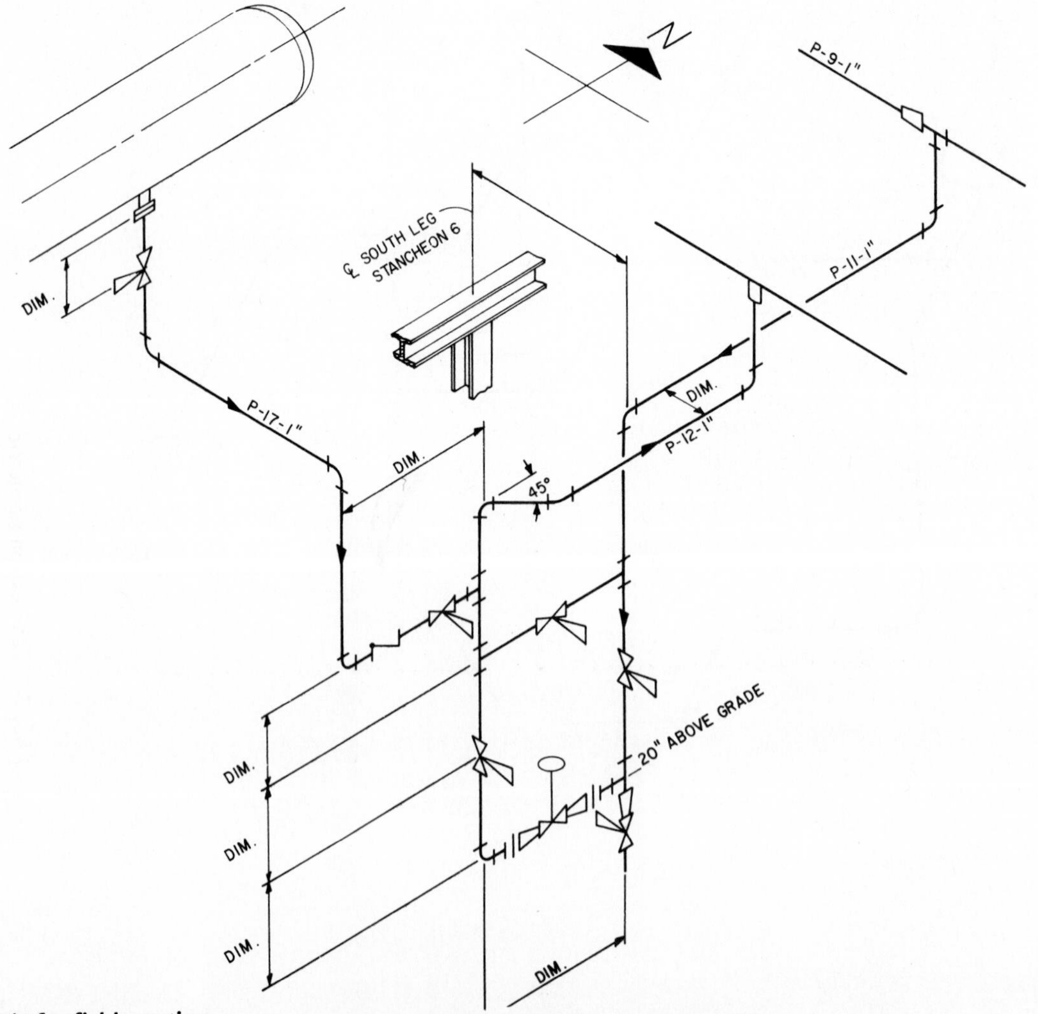

FIG. 7-2 Isometric for field erection.

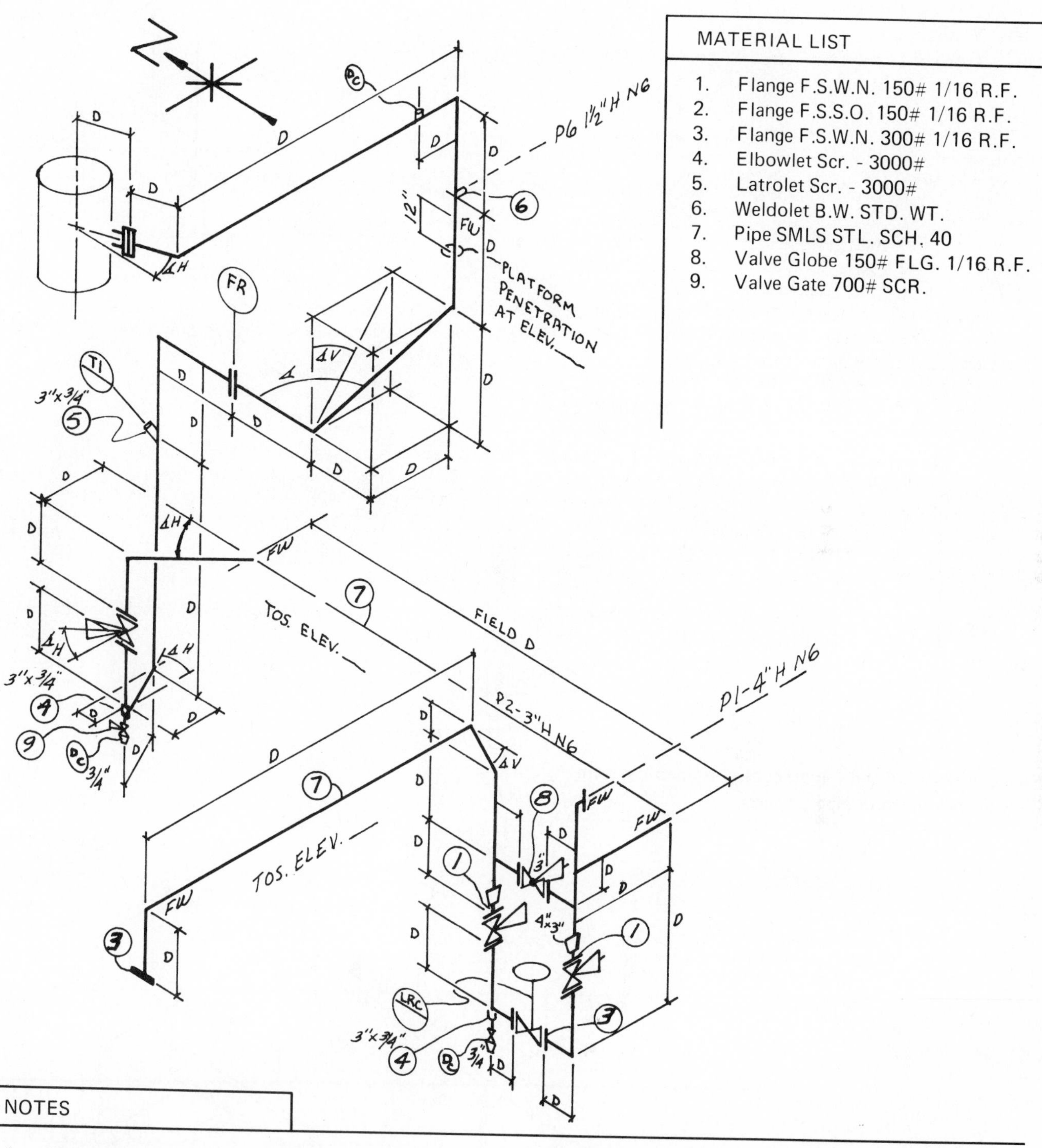

NOTES

A. D = Dimension. B. FW = Field weld. C. ∡ = Angle to be given.
D. ∡H = Angle from horizontal E. ∡V = Angle from vertical
F. TOS = Top of steel G. Do not dimension face-to-face on valves.
H. North arrow in top lefthand corner I. Dimension lines not broken.
J. Show all field welds on shop isometric.
K. Show handwheel orientation and check for minimum 4″ clearance to nearest line or equipment.

FIG. 7-3 Isometric for shop fabrication.

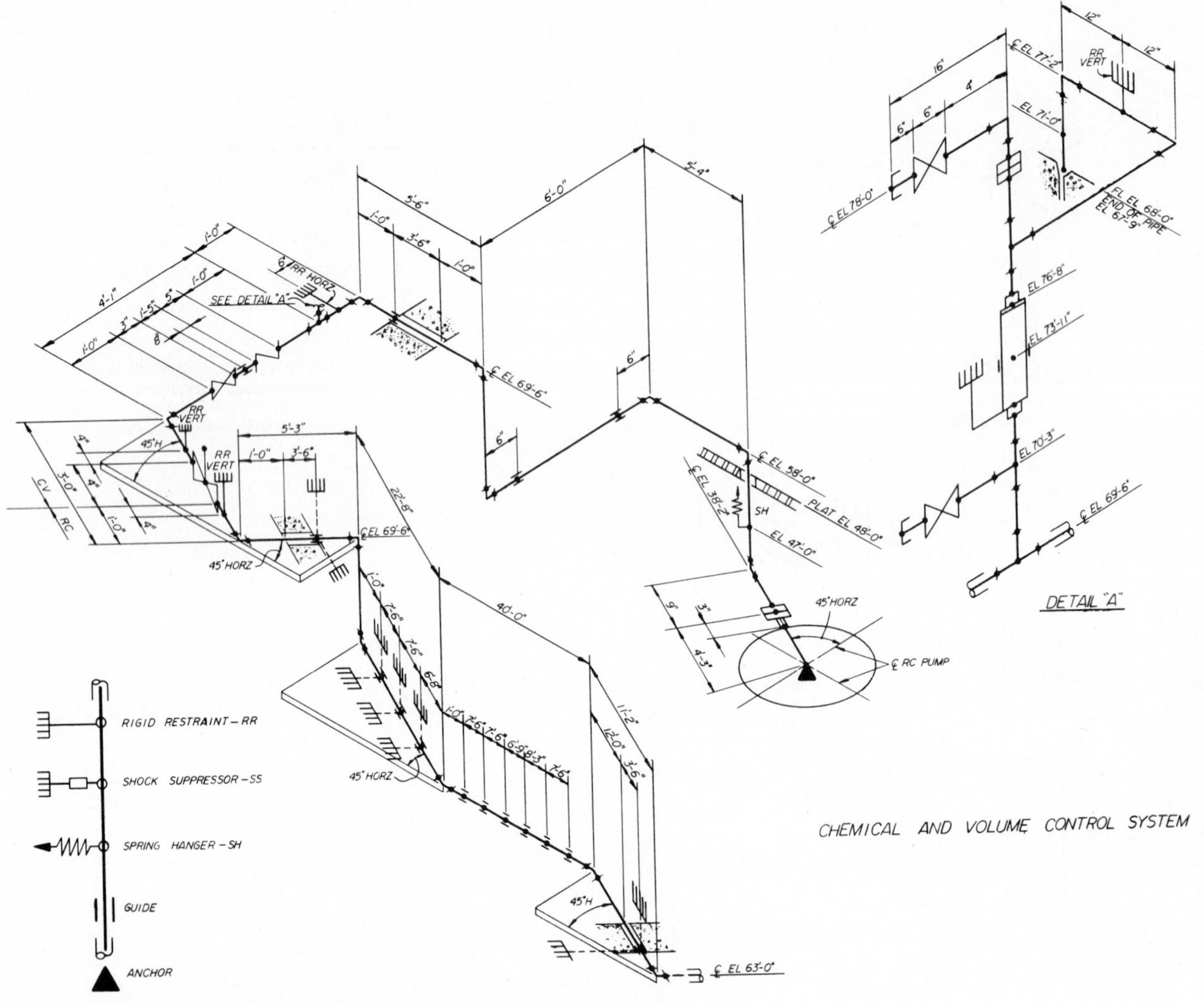

FIG. 7-4 Chemical and volume control isometric system showing placement of pipe supports.

FIG. 7-5 Isometric for steam tracing.

For drawing a piping isometric a 3/8 in. = 1 ft scale is accepted practice throughout the piping industry. However, *in many cases it is necessary to disregard scale altogether and only draw the isometric in proportion to itself and to existing equipment and other lines in the piping system.* If the system is immense (as in power plants), a much smaller scale must be used. Because of paper size restrictions, the drafter must shorten the lines to fit within prescribed boundaries. This practice has been followed in many of the figures in this chapter.

MODELS AND ISOMETRICS

Models (see Chapter 15) are employed extensively throughout the piping field. In many cases the piping system is designed directly on the model without the aid of the plan or sectional views, taking the design information from only the P&ID. In these cases the drafter works from dimensions, coordinates and elevations tagged on the model, constructing an isometric drawing of each line to be fabricated. When using this method, the drafter draws or sketches from each line on the model an isometric view of the complete pipeline from beginning to end, including all valves, fittings, and other equipment related to the line. This type of drawing is called a *spool drawing* (see Chapter 9). The sketched piping line isometric shows all the dimensions and components required to make up the pipeline. A spool drawing can also be made from the plan and elevation drawing when that design method is used.

COMPUTERS AND ISOMETRICS

In some companies the use of computer graphics for drawing isometrics has become standard practice. After the isometric has been sketched from the model, the dimensions, coordinates, equipment, and calculations (flow, temperatures, and so on) are fed into a computer. With this data input the computer is able to produce an isometric drawing and, in some cases, complete a bill of materials, thus bypassing the need for a perfectly drawn isometric prepared by a drafter. (See the discussion of computer-aided drafting in Chapter 5.)

Some computer-aided drafting (CAD) systems can do all computations and construct the drawing. Computer graphics are used in the piping field primarily for the construction of isometrics and flow diagrams. This use of a computer is relatively new to the industry. Although the present cost and scarcity of the process will no doubt change drastically within the next few years, computers and models are not likely to totally replace the need for hand-drafted orthographic and isometric drawings.

SYMBOLS AND ISOMETRICS

Figure 7-6 shows the most commonly used symbols to represent piping components in isometric; many slight variations of these symbols are employed throughout the piping field. This book attempts to show the symbols the drafter may encounter in the field. The examples in this chapter are drawn with one of the four basic methods of symbol representation. In a majority of cases the single-line symbol is used. These drawings are taken from industrial sources and vary somewhat in style.

All the symbols represented in Chapter 5 have isometric variations. In this chapter Fig. 7-6 shows only a few of the many possible symbols. Isometric templates are available at drafting supply stores. These templates enable the drafter to draw symbols and portions of piping systems efficiently, accurately, and more rapidly than would be possible if each symbol were individually constructed.

The *single-line isometric symbol* is the simplest and quickest method and the one commonly found throughout industry (Fig. 7-7). When constructing valves and fittings on isometric lines, the drafter draws cross lines for flange fittings so that they lie in the plane of the bend of the pipe nearest the valve or fitting.

The second type, the *single-line elliptical isometric symbol,* is seldom used because of the time involved in its construction. Nevertheless, all symbols can be drawn in this manner. Figure 7-8 is an example of this type of symbol.

The third type of symbol, the *double-line symbol,* is used more often than the elliptical symbol. The double-line isometric drawing allows the system to show interference and clearance between pipes and between pipelines and equipment because the line is drawn to an actual scale instead of being represented by a single centerline. Figure 7-9 employs double-line symbols and lines. In this form of representing isometric piping systems, the symbols for valves are still the simplified form, not the pictorial type. Note that the double-line symbols shown in Figs. 7-6 and 7-9 use an elliptical variation for valves and fittings instead of the more common method of double-line symbols in Chapter 5. (Pipelines 12 in. (30.48 cm) and larger in diameter are sometimes represented in the double-line isometric method.)

The fourth type of symbol, the *pictorial symbol,* is used mainly in books, magazines, and other publications where a photographic representation enables the viewer to understand the system with a minimum of technical knowledge. This method is extremely time consuming, but it is beneficial where clarity and beauty take precedence over cost and speed. It is an excellent procedure for exploded views and display drawings as in Fig. 7-10. Pictorial isometrics are rarely used in the piping industry for fabrication or design drawings.

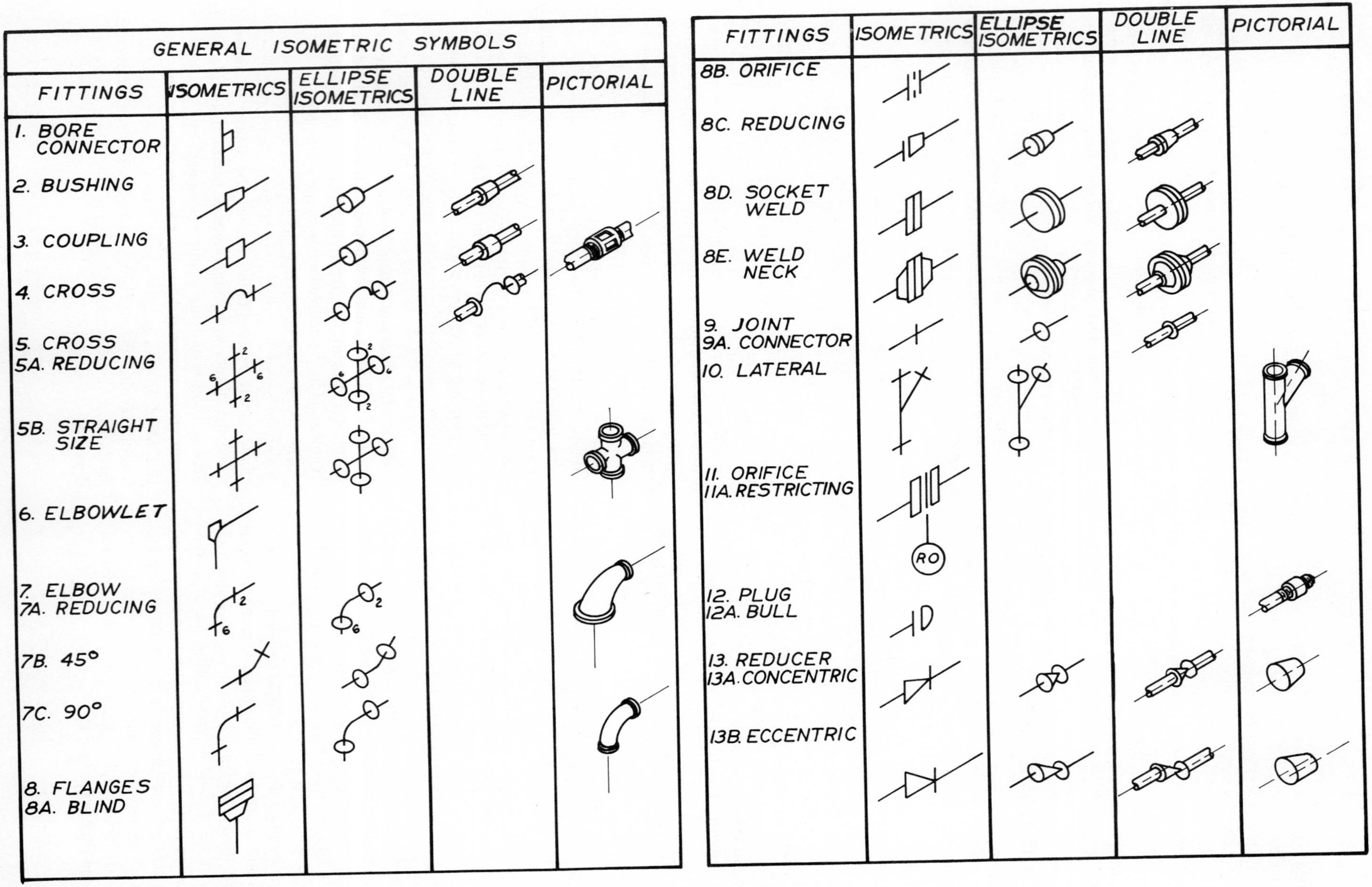

GENERAL ISOMETRIC SYMBOLS

FITTINGS	ISOMETRICS	ELLIPSE ISOMETRICS	DOUBLE LINE	PICTORIAL
1. BORE CONNECTOR				
2. BUSHING				
3. COUPLING				
4. CROSS				
5. CROSS 5A. REDUCING	2, 6, 6, 2	2, 6, 6, 2		
5B. STRAIGHT SIZE				
6. ELBOWLET				
7. ELBOW 7A. REDUCING	2, 6	2, 6		
7B. 45°				
7C. 90°				
8. FLANGES 8A. BLIND				

FITTINGS	ISOMETRICS	ELLIPSE ISOMETRICS	DOUBLE LINE	PICTORIAL
8B. ORIFICE				
8C. REDUCING				
8D. SOCKET WELD				
8E. WELD NECK				
9. JOINT 9A. CONNECTOR				
10. LATERAL				
11. ORIFICE 11A. RESTRICTING	RO			
12. PLUG 12A. BULL				
13. REDUCER 13A. CONCENTRIC				
13B. ECCENTRIC				

FIG. 7-6 Isometric symbol sheets.

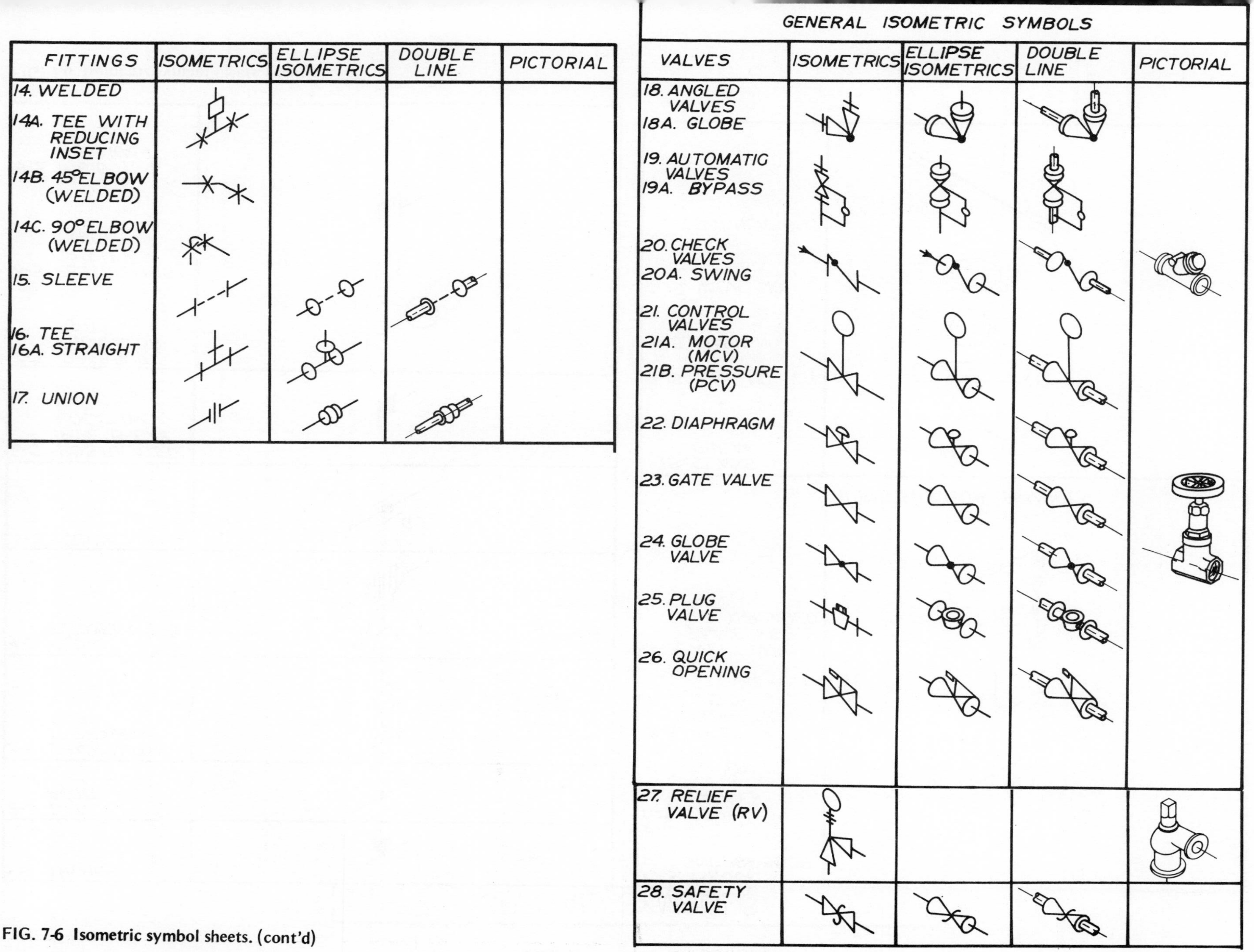

GENERAL ISOMETRIC SYMBOLS

FITTINGS	ISOMETRICS	ELLIPSE ISOMETRICS	DOUBLE LINE	PICTORIAL
14. WELDED				
14A. TEE WITH REDUCING INSET				
14B. 45° ELBOW (WELDED)				
14C. 90° ELBOW (WELDED)				
15. SLEEVE				
16. TEE 16A. STRAIGHT				
17. UNION				

VALVES	ISOMETRICS	ELLIPSE ISOMETRICS	DOUBLE LINE	PICTORIAL
18. ANGLED VALVES 18A. GLOBE				
19. AUTOMATIC VALVES 19A. BYPASS				
20. CHECK VALVES 20A. SWING				
21. CONTROL VALVES 21A. MOTOR (MCV) 21B. PRESSURE (PCV)				
22. DIAPHRAGM				
23. GATE VALVE				
24. GLOBE VALVE				
25. PLUG VALVE				
26. QUICK OPENING				
27. RELIEF VALVE (RV)				
28. SAFETY VALVE				

FIG. 7-6 Isometric symbol sheets. (cont'd)

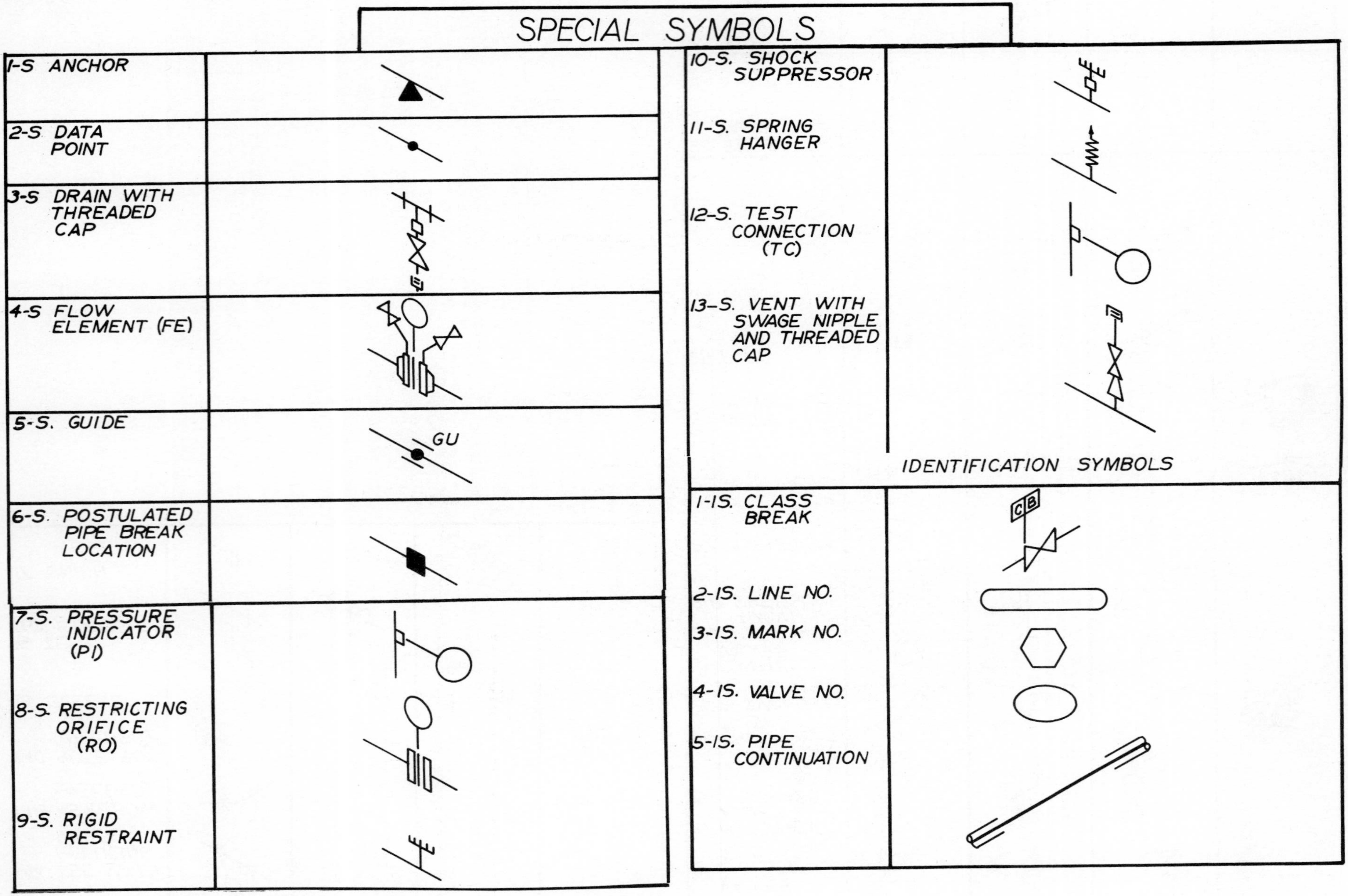

SPECIAL SYMBOLS

1-S ANCHOR

2-S DATA POINT

3-S DRAIN WITH THREADED CAP

4-S FLOW ELEMENT (FE)

5-S. GUIDE

GU

6-S. POSTULATED PIPE BREAK LOCATION

7-S. PRESSURE INDICATOR (PI)

8-S. RESTRICTING ORIFICE (RO)

9-S. RIGID RESTRAINT

10-S. SHOCK SUPPRESSOR

11-S. SPRING HANGER

12-S. TEST CONNECTION (TC)

13-S. VENT WITH SWAGE NIPPLE AND THREADED CAP

IDENTIFICATION SYMBOLS

1-IS. CLASS BREAK

CB

2-IS. LINE NO.

3-IS. MARK NO.

4-IS. VALVE NO.

5-IS. PIPE CONTINUATION

FIG. 7-6 Isometric symbol sheets. (cont'd)

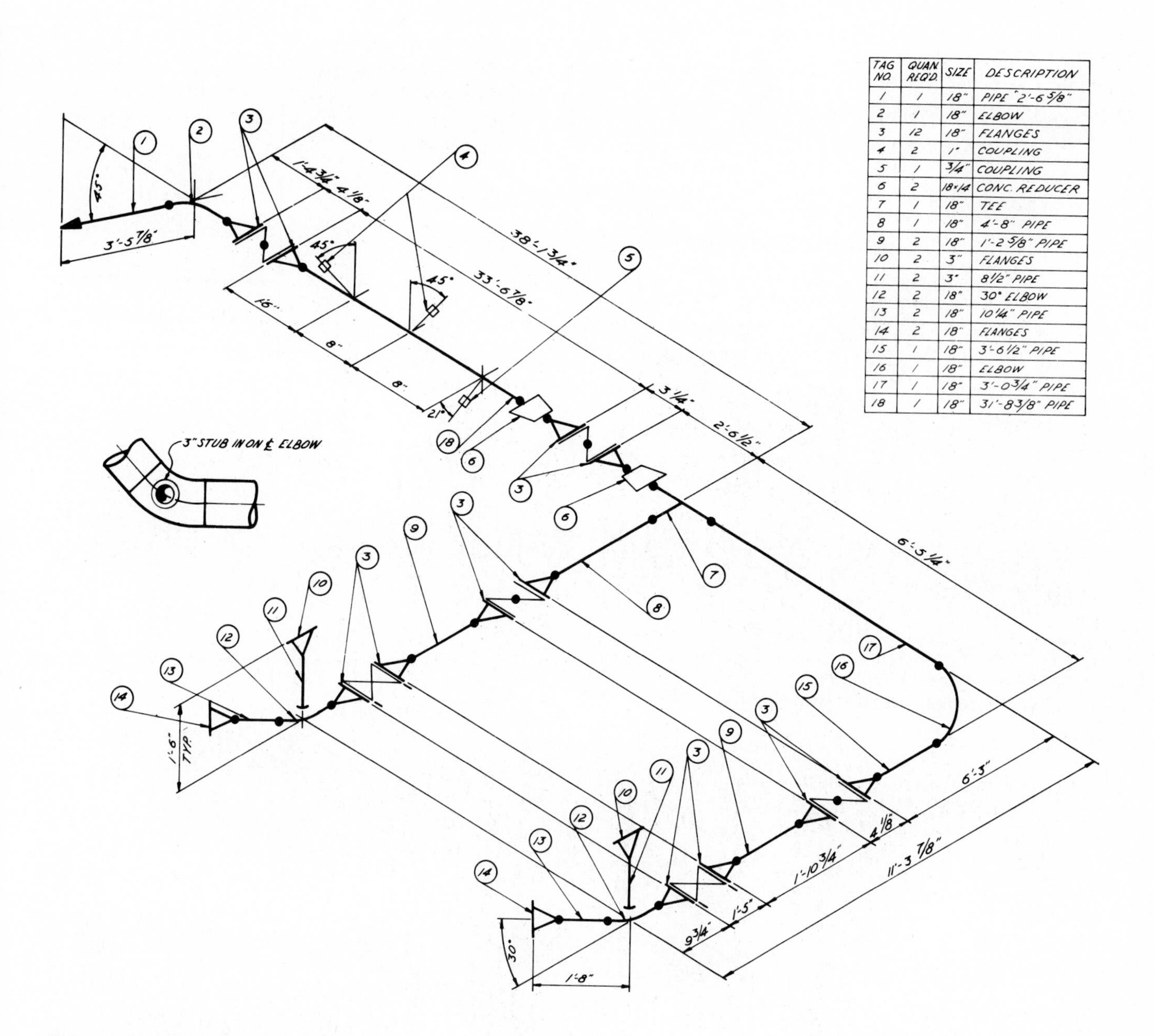

TAG NO.	QUAN REQ'D.	SIZE	DESCRIPTION
1	1	18"	PIPE 2'-6 5/8"
2	1	18"	ELBOW
3	12	18"	FLANGES
4	2	1"	COUPLING
5	1	3/4"	COUPLING
6	2	18×14	CONC. REDUCER
7	1	18"	TEE
8	1	18"	4'-8" PIPE
9	2	18"	1'-2 5/8" PIPE
10	2	3"	FLANGES
11	2	3"	8 1/2" PIPE
12	2	18"	30° ELBOW
13	2	18"	10 1/4" PIPE
14	2	18"	FLANGES
15	1	18"	3'-6 1/2" PIPE
16	1	18"	ELBOW
17	1	18"	3'-0 3/4" PIPE
18	1	18"	31'-8 3/8" PIPE

FIG. 7-7 Pipe fabrication isometric.

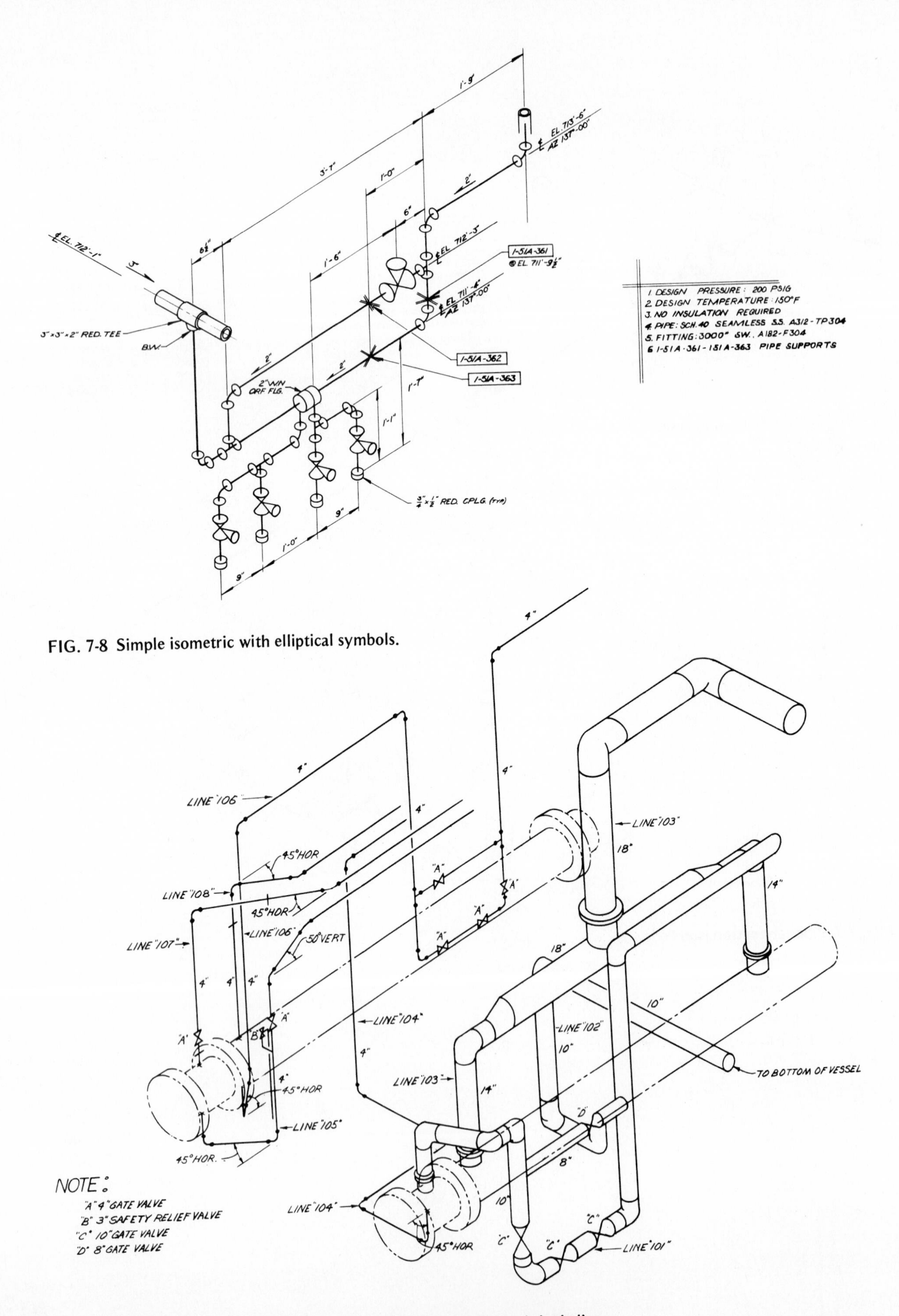

FIG. 7-8 Simple isometric with elliptical symbols.

FIG. 7-9 Isometric drawing showing welded connections in double and single lines.

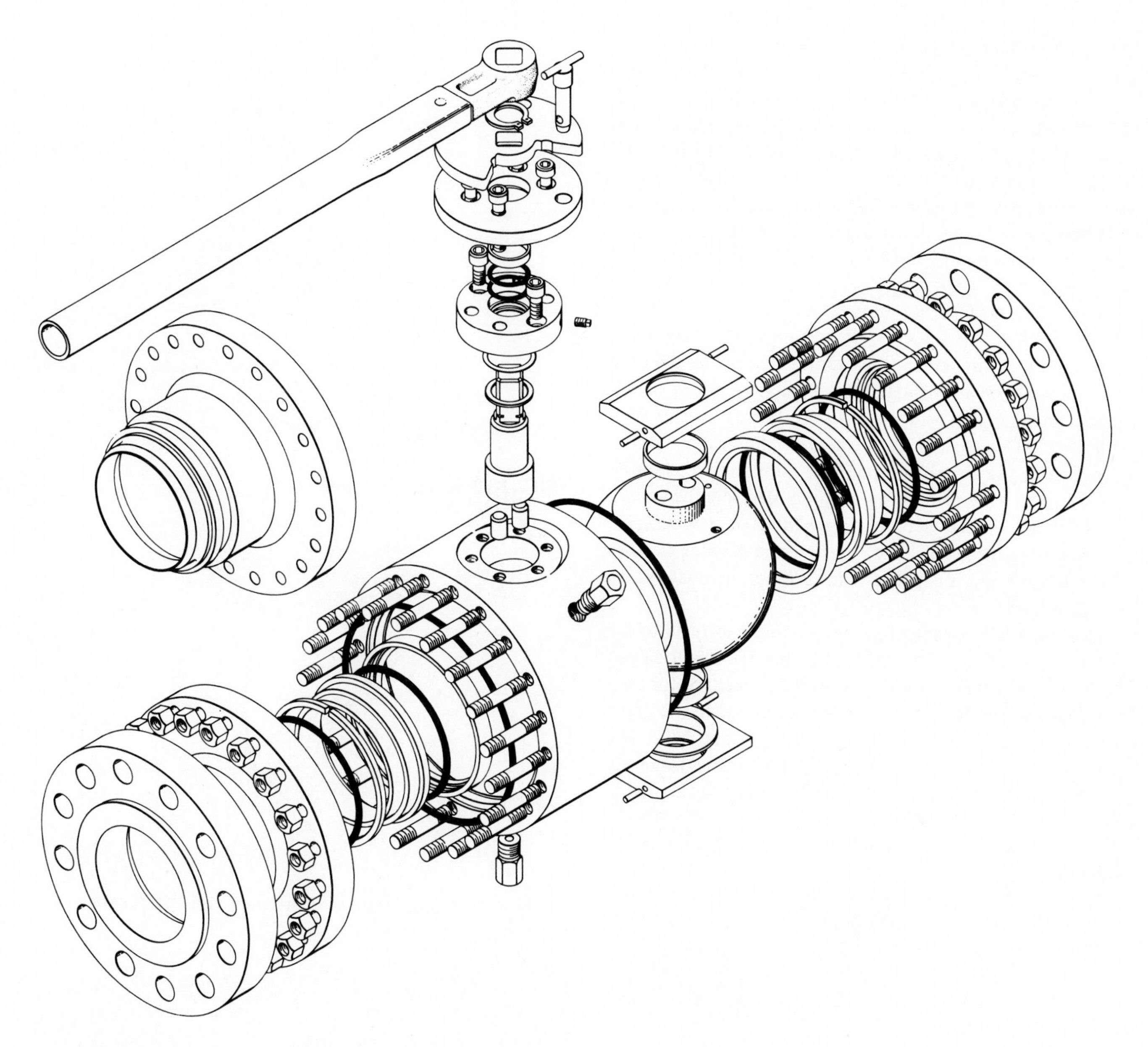

FIG. 7-10 Isometric pictorial exploded view of a ball valve. (Courtesy Grove Valve and Regulator Company)

Special fittings and valves, or any equipment that does not have an assigned symbol, may be drawn pictorially or somehow represented and then be labeled on the drawing as a *spool callout* to identify the component for those who will later use the drawing. Special symbols that are unique to certain industries (guides, snubbers, pipe supports) are provided in Fig. 7-6. The symbols shown are used throughout the power generation field. (See also the pipe support symbols in Chapter 8.) The drafter must be able to use all the variations of symbols and construct special symbols where necessary using excellent line work and lettering.

ISOMETRIC DRAWINGS

Most companies still use plan and elevation assembly drawings which are completed after the flow diagram. Then the drafter proceeds to make the isometric views for spool fabrication, though many large companies now use models. The isometrics are then taken directly from the model.

Isometrics are usually drawn to a certain proportion instead of to an accurate scale, as shown in the isometrics in the round of nuclear piping in this chapter. The diameters from the center of the reactor are so large that they (and many other lines on the drawing) must be shortened to fit on the sheet. Fittings and valves that are too small in relation to other parts in the system are enlarged in scale to accommodate their dimensions. When the isometric is not true to scale, the drawing is represented to facilitate the efficient placement of hangers, components, dimensions, and ease of reading for the fabricator. In general, the drafter employs conventional drafting practices when preparing an isometric drawing. (See Chapter 5 for line weights and lettering.)

Isometric drawings are used extensively to dimension portions of the pipe system by clearly showing, in one view, both the horizontal and vertical dimensions rather than the three views used in orthographic projection. In an orthographic view, piping symbols in front of other lines or symbols hide all that is behind them. Only the line or symbol seen first in the view is drawn. Thus the dimensions on an orthographic view or drawing can only describe the system partially. (See Chapter 5 for an example of orthographic drawing.) On the other hand, an isometric drawing exposes the total pipeline, or system, and provides adequate space for dimensions.

All isometric horizontal lines recede to the left and to the right at 30-degree angles. In Fig. 7-11 the placement of two 30-degree triangles exhibits the simple procedure for constructing an isometric box. The axes in the isometric drawing are at 120 degrees to each other as in Fig. 7-12. The dimensions are not foreshortened as in a true isometric projection but are drawn at true length along the isometric lines. Therefore dimensions can be taken directly from the orthographic view or model and used to construct the piping isometric.

Isometric drawings are always oriented to the north arrow (Figs. 7-13 and 7-14), which in most drawings points along

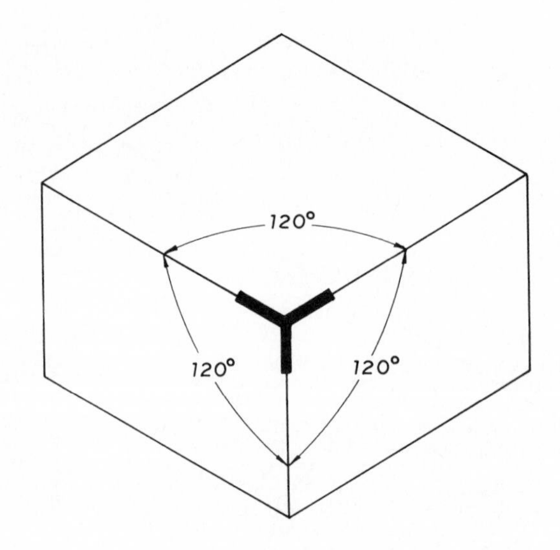

FIG. 7-12 Isometric cube.

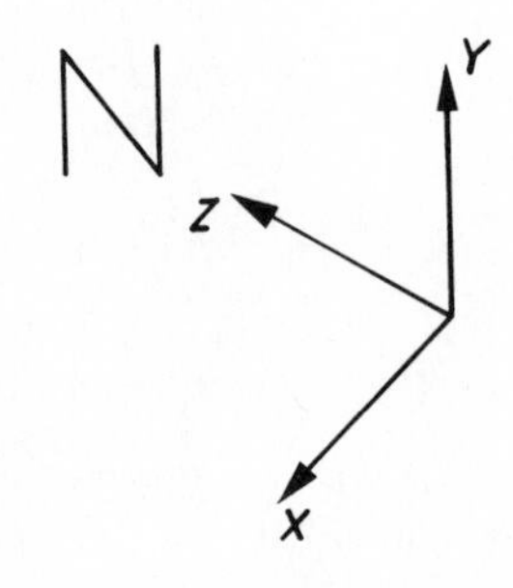

FIG. 7-13 North arrow orientation and *XYZ* coordinates.

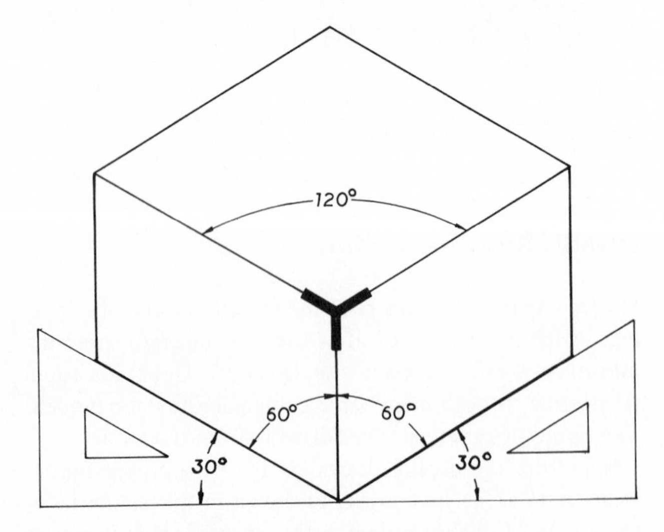

FIG. 7-11 Setting up an isometric drawing.

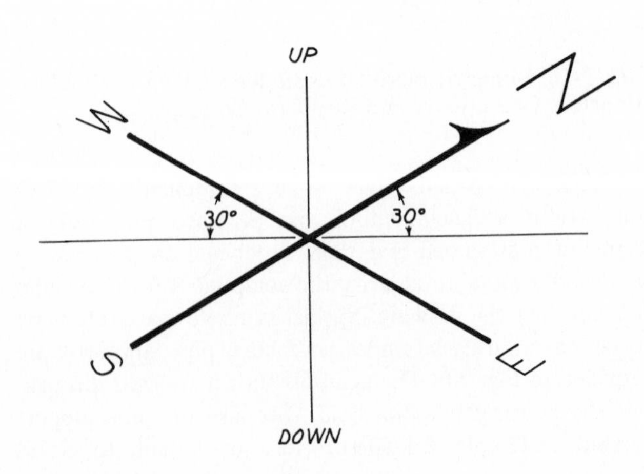

FIG. 7-14 North arrow orientation.

one of the horizontal axes toward the upper right or left of the drawing. The north arrow orientation is determined by the best view of the system or pipeline. When drawing an isometric, students should orient their drawing to the north arrow from the major point of intersection of the drawing, which is usually a major line crossing. For system isometrics, large equipment such as vessels and towers should be located first, followed by pipelines and components, with pipe supports, valves, and fittings drawn last.

Isometric Lines

Isometric lines are all lines that fall along axis lines are parallel to the axes or are vertical lines. Their dimensions can be laid off at true length. Figure 7-15 shows a half-scale orthographic drawing in the upper left-hand corner: When put into its isometric form, dimension *A* can be laid off at true length because it is parallel to the axis line (or, in other words, is an isometric line). Dimension *D* (like dimension *B*) is an isometric line and can be laid off at true length at 30 degrees to the horizontal.

Since line 1-3 is not parallel to the axes or a vertical line and therefore will not show up as a true-length line in the isometric, it is referred to as a nonisometric line. *In isometric drawing only isometric lines appear true length.* All dimensions can be taken directly from the model or from the orthographic views and laid off at true length in isometric if the line is parallel to the axes, falls on the axes, or is a vertical line. Figures 7-16 and 7-17 give examples of lines and angles in isometric.

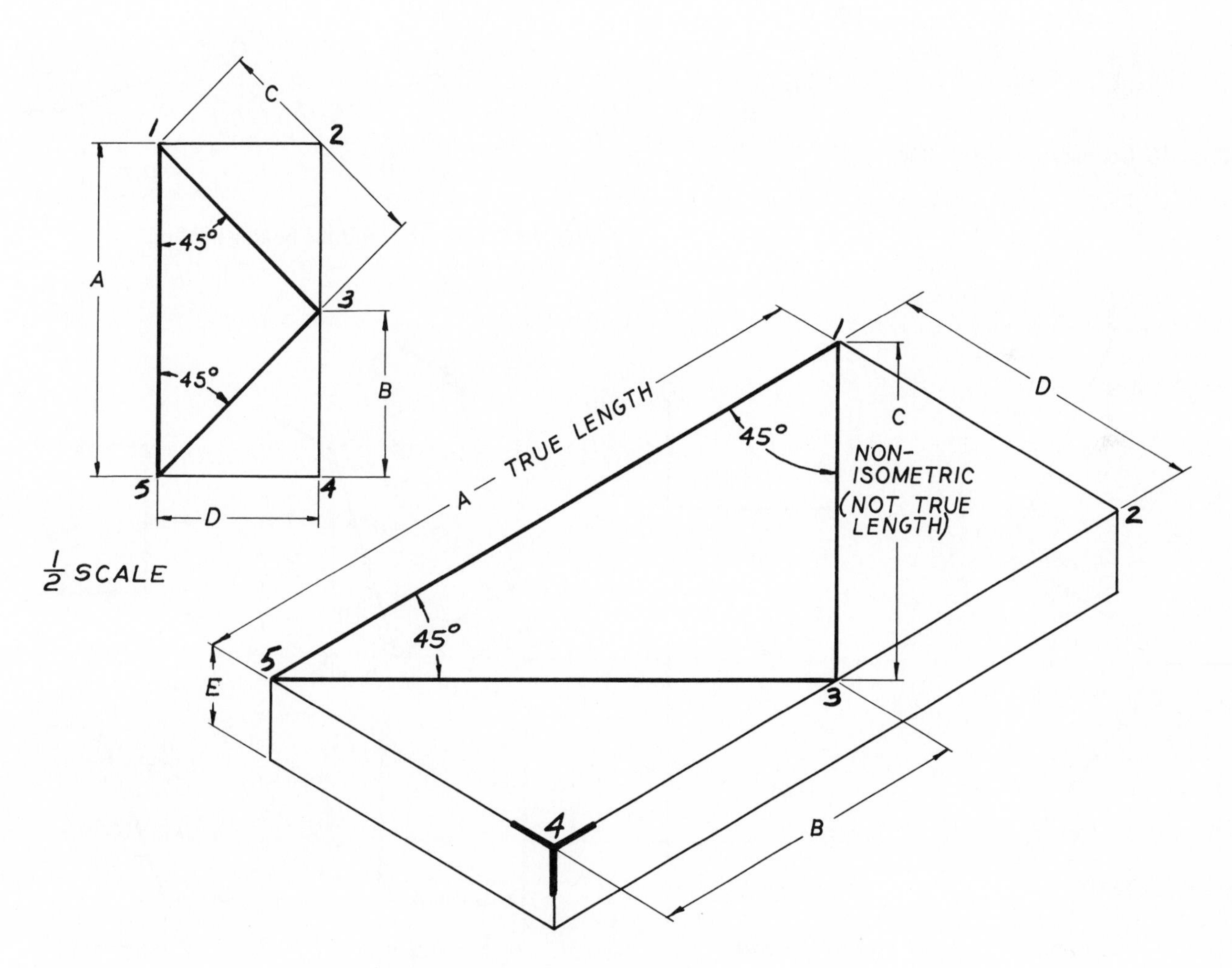

FIG. 7-15 True length, isometric lines, nonisometric lines, and angles in isometric.

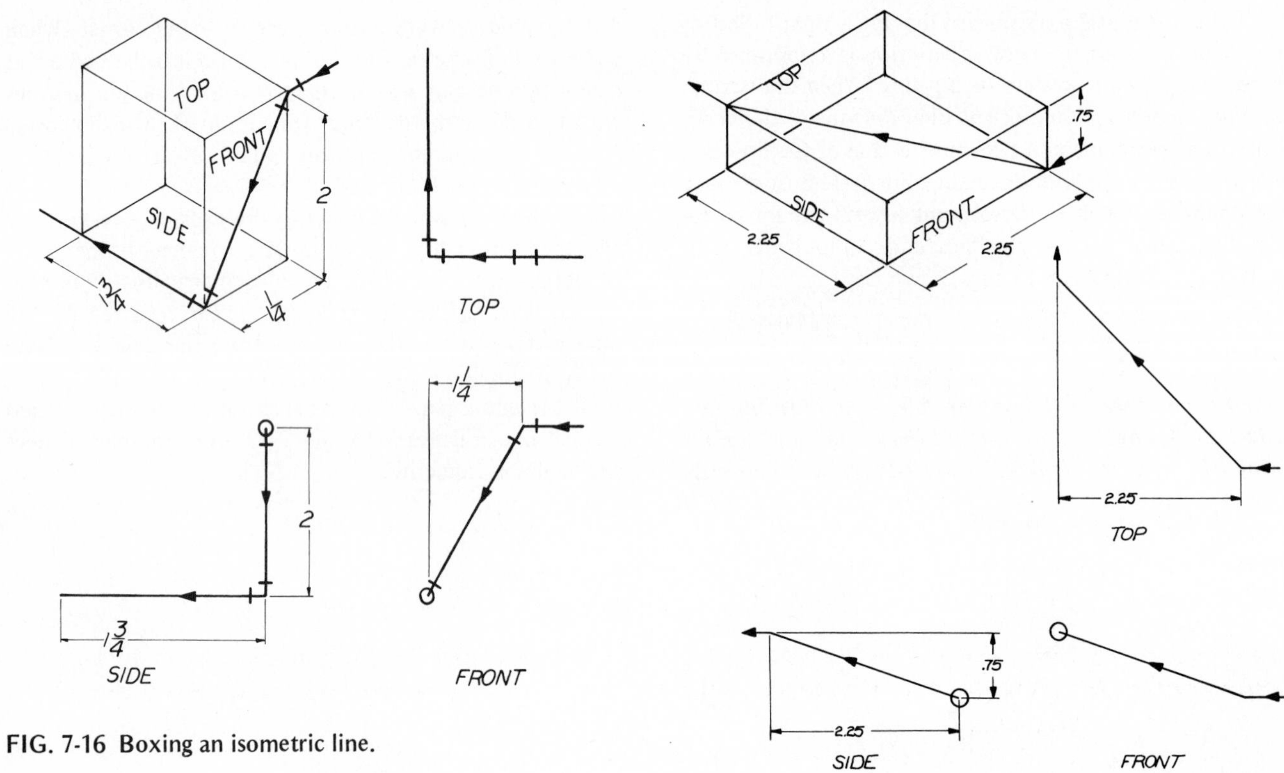

FIG. 7-16 Boxing an isometric line.

FIG. 7-17 Isometrics and the boxing method.

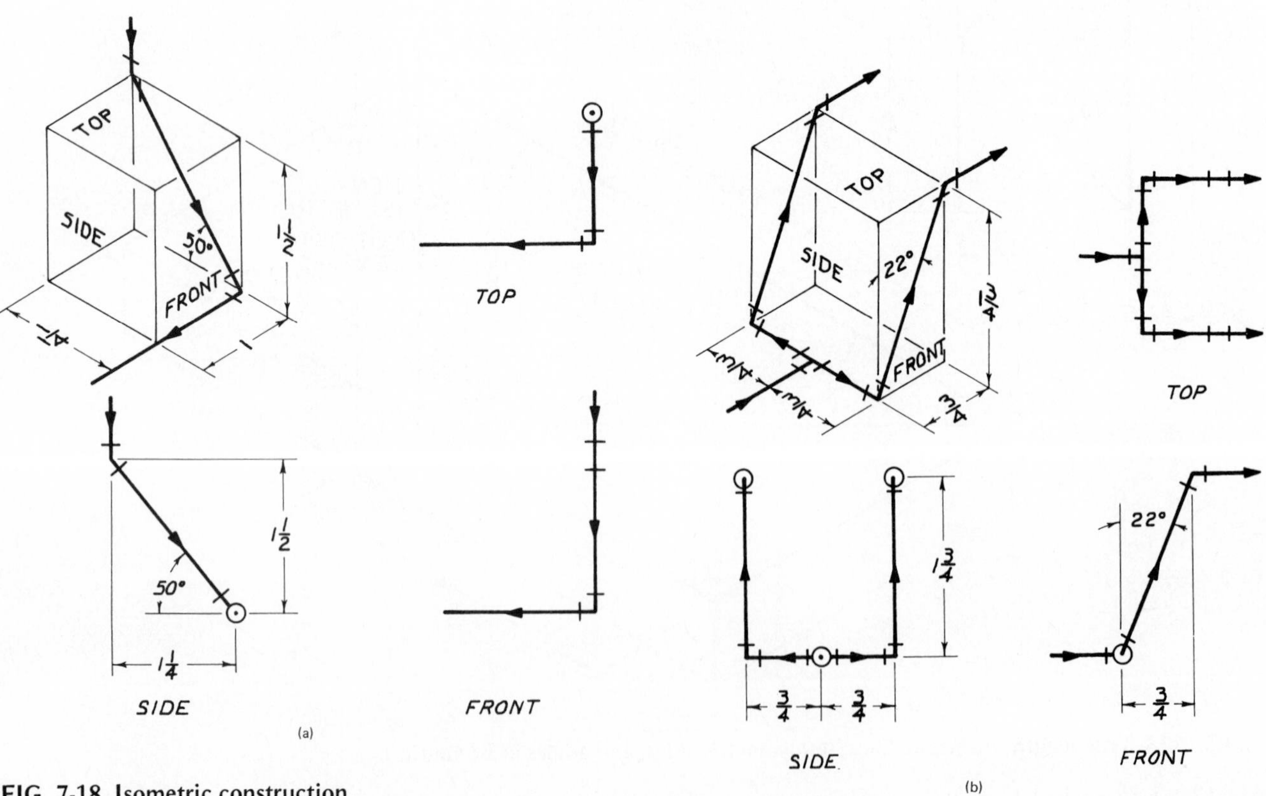

FIG. 7-18 Isometric construction.

FIG. 7-18 Isometric construction. (cont'd)

Nonisometric Lines

Nonisometric lines are oblique lines, inclined lines, or any lines that are not parallel to the axes or vertical. A nonisometric line will not appear true length; the measurement cannot be laid off directly on the isometric view. Nonisometric lines are constructed by means of offset dimensions which are laid off on isometric (true-length) lines.

In Fig. 7-15 dimension *C* (line 1-3) cannot be laid off on an isometric true-length line or measured directly from the orthographic view. By using dividers to lay off dimensions *A, B,* and *D,* it is possible to construct lines 1-3 and 3-5. Neither of these lines will be true length in the isometric: Line 1-3 will scale shorter than true length and line 3-5 will scale longer than true length.

The boxing method is one useful way of constructing an isometric drawing. The height, length, and depth of a box are laid off as isometric lines on the major axes. It is convenient to build an isometric box and drawing from the lower or upper front corner of the figure. In many cases, however, piping drawings are too large for the boxing method, though small portions of the drawing can be shown in this method.

Figures 7-18 and 7-19 are examples of the boxing method for the construction of pipe angles and bends. When the boxing method cannot be applied for piping drawings, offset dimensions must be taken along isometric lines in order to construct oblique, inclined, and all other nonisometric lines.

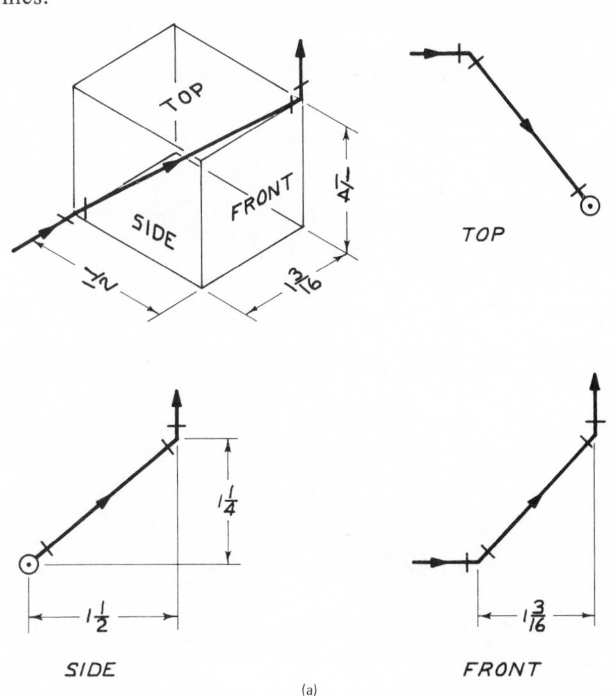

FIG. 7-19 Isometric construction.

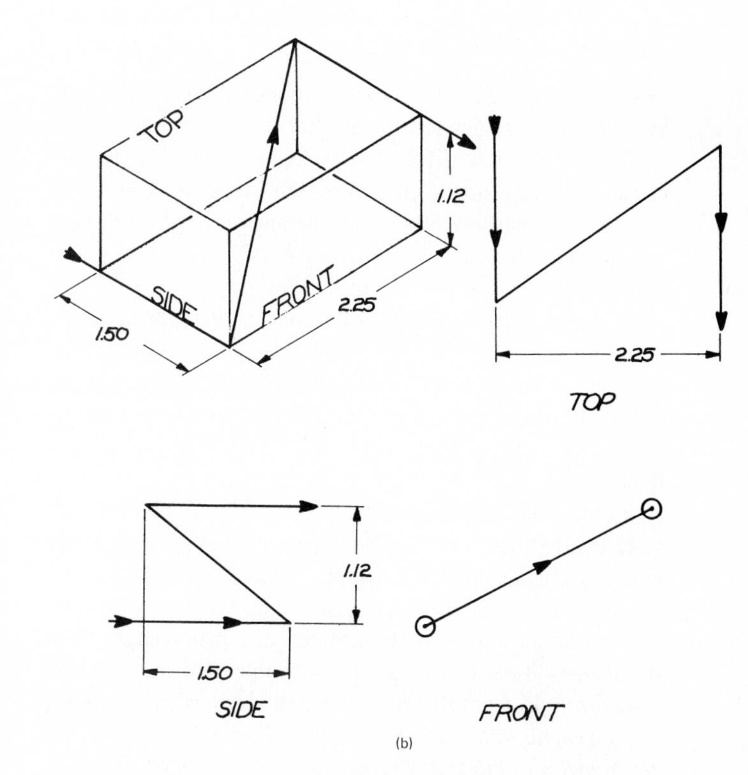

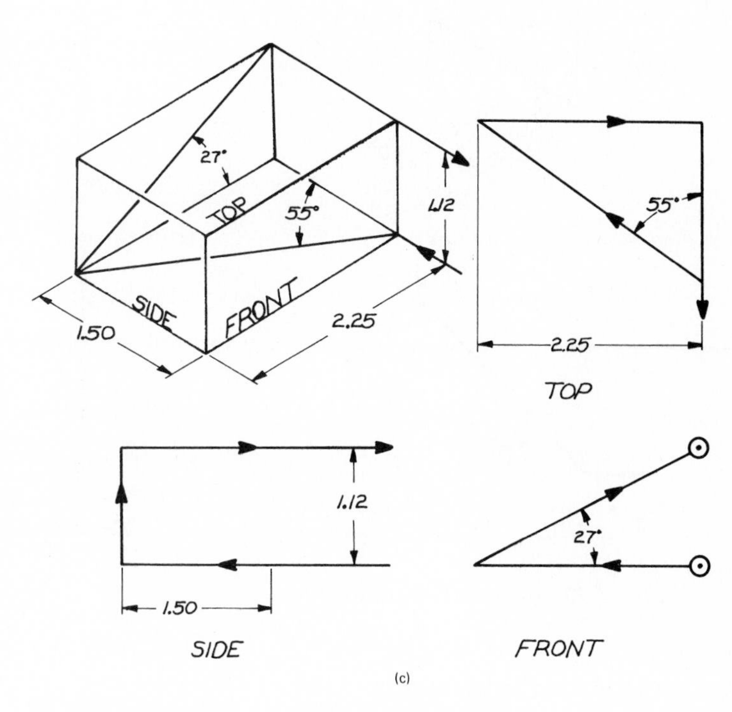

FIG. 7-19 Isometric construction. (cont'd)

Isometric Angles

The major axes along the isometric cube in Fig. 7-12 appear at 120-degree angles to one another in the isometric view. In reality, all lines within the cube are at 90 degrees. Because of this distortion created by the isometric view of the cube, few angles will appear as true angles. *Angles will project at true size only when the angle itself is parallel to the plane of projection, which is rarely the case.*

All angles must be laid off by means of offset dimensions as in Fig. 7-15. Angles will appear larger or smaller than true size in the isometric depending on their position in the view. The protractor cannot be used to set up an angle in isometric—measurements *must* be taken along isometric lines.

In Fig. 7-20 the orthographically projected drawing shows a 45-degree-angle. In the isometric view, the same angle appears smaller than 45 degrees. Figure 7-15 shows two identical angles in the plan view. When put into isometric one becomes less than 45 degrees, the other larger than 45 degrees. To set up this type of problem all construction must proceed from dimensions taken along *isometric lines. As a general rule, angles in isometric must be converted from angles to linear measurements along isometric lines.*

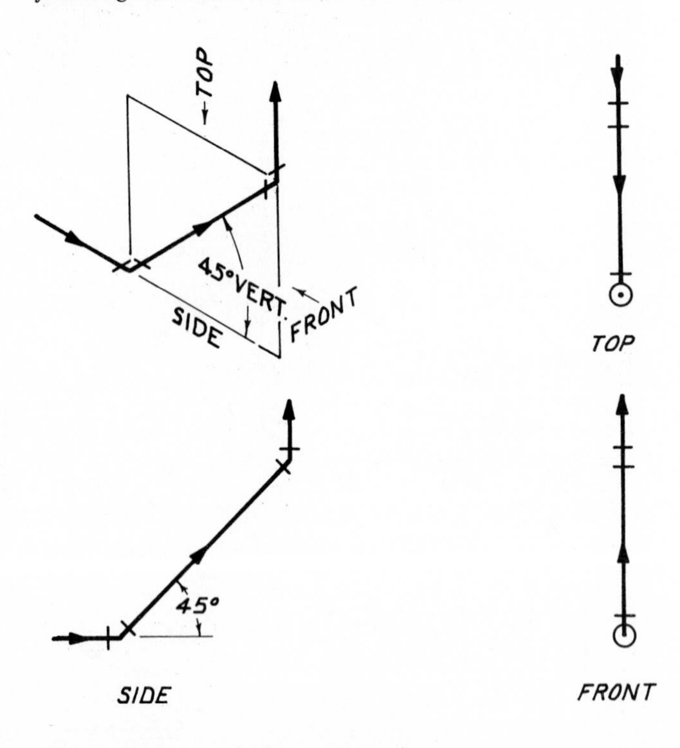

FIG. 7-20 Isometric line construction.

Circles and Isometrics

All circles in an isometric view appear as ellipses. Therefore the use of an ellipse template is necessary when drawing an isometric view of a piping system. Only when absolutely necessary should an ellipse be set off by measurements, and even then it is sufficient to draw the ellipse by means of the four-center method. Though the four-center ellipse is somewhat shorter and fatter than the true ellipse, it is the quickest and easiest method and quite adequate in almost all situations.

Figure 7-21 shows the step-by-step construction of an ellipse. First lay off lines *DA* and *DC* at 30 degrees along

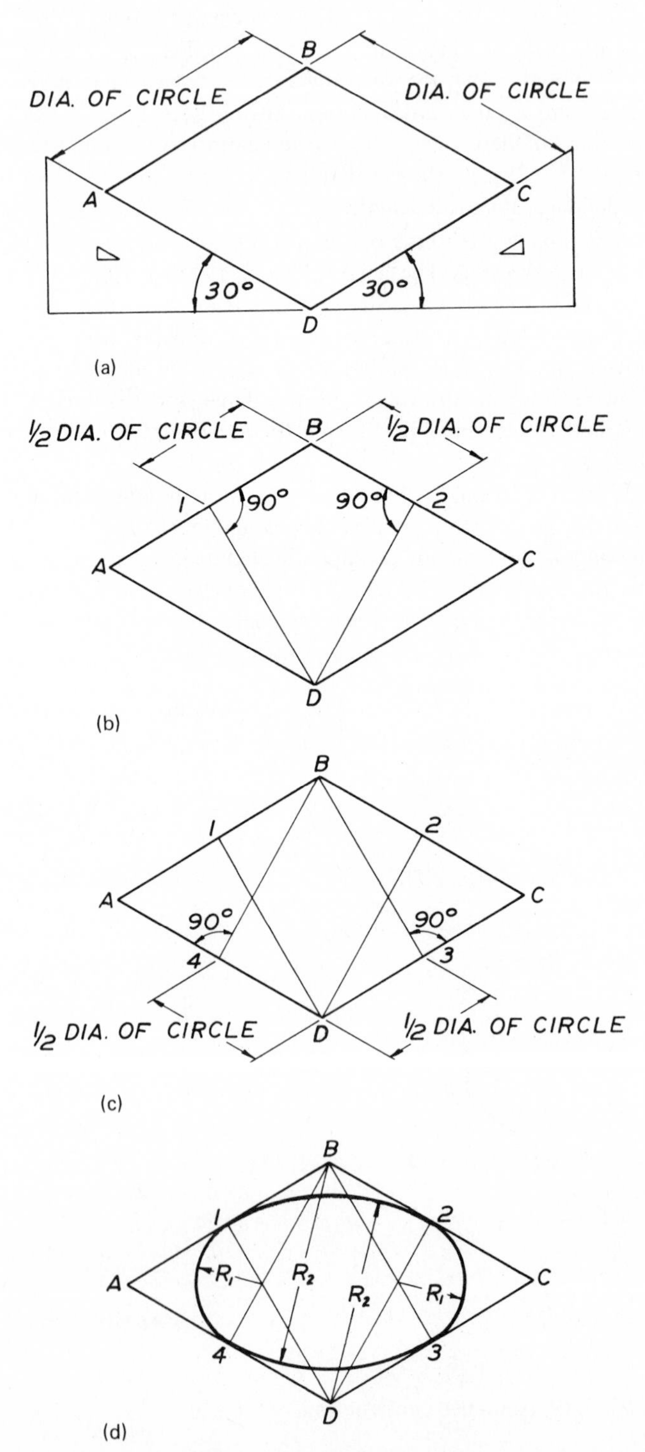

FIG. 7-21 Ellipse: construction of a circle in isometric.

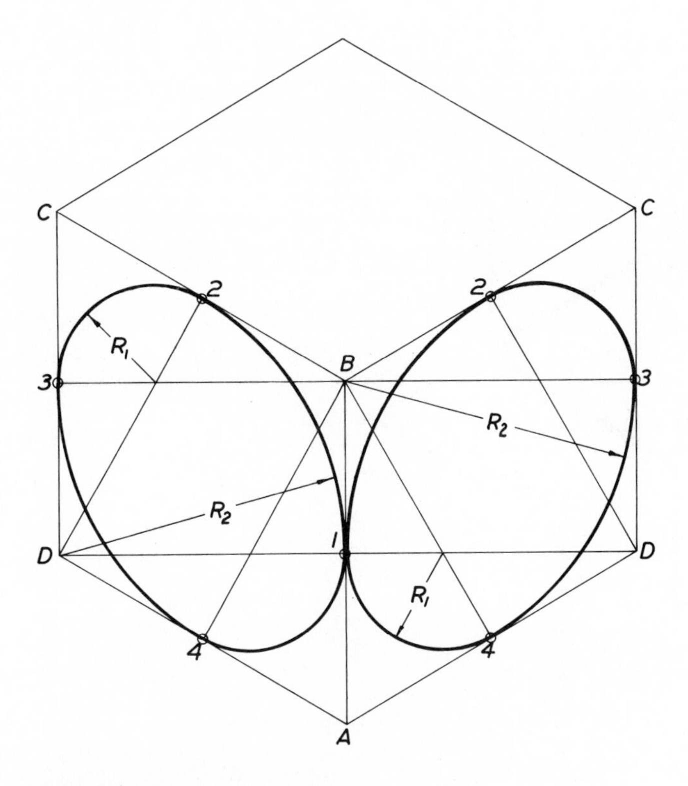

FIG. 7-22 Four-center method for the construction of ellipses for vessels in the horizontal plane.

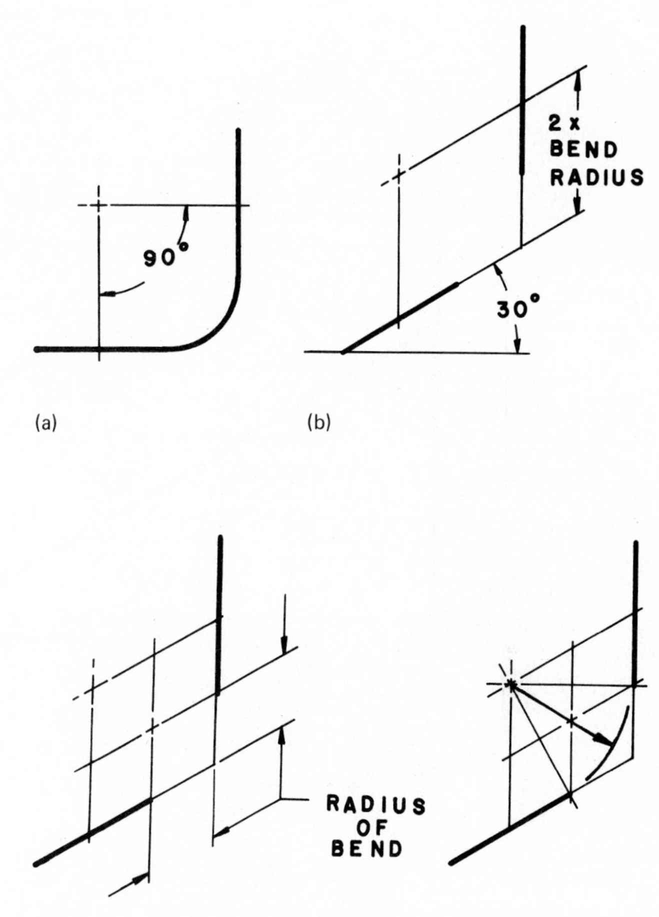

the isometric axes using the diameter of a given circle as the isometric line measurement (DC); then find the midpoint between the lines that corresponds to half the diameter of the circle. Point 1 falls halfway between points *A* and *B*. Points 2, 3, and 4 also fall midpoint on the corresponding lines. By drawing a line from point *D* to points 1 and 2, and from point *B* to points 3 and 4, the newly formed lines 1*D* and 4*B* cross, as do 2*D* and 3*B*. At these intersections radius *R*1 can be laid off by swinging an arc from point 1 to point 4 and point 2 to point 3. Radius *R*2 is set off by taking the distance from point *B* to point 4 and swinging arc 4-3 then from point *D* by swinging arc 1-2.

Figure 7-22 demonstrates the construction of circles in isometric when they appear in the vertical plane. The basic method is the same as in the horizontal plane. When constructing piping bends and elbows in isometric, the ellipse method can be used to swing the necessary arcs (Fig. 7-23). The square method for showing piping elbows is exhibited in part of Fig. 7-24 and 7-4. This method is much quicker and is becoming standard practice for many companies throughout the piping field.

Figure 7-25 shows construction setups for vessels, tanks, and towers, all of which frequently appear in isometric piping drawings. The four-center method can be used for the ellipse constructions, though a template would normally be used. Figure 7-26 shows a photograph of a typical vessel.

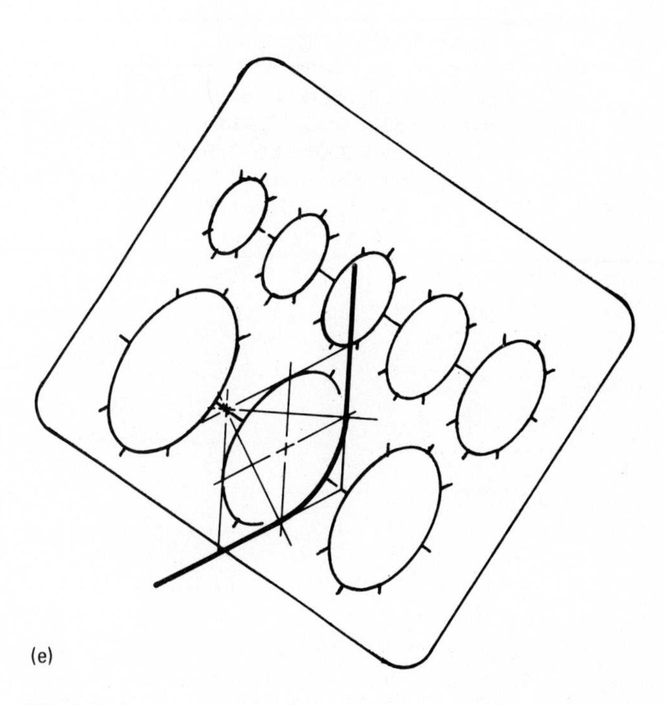

FIG. 7-23 Using an isometric template to draw a pipe bend.

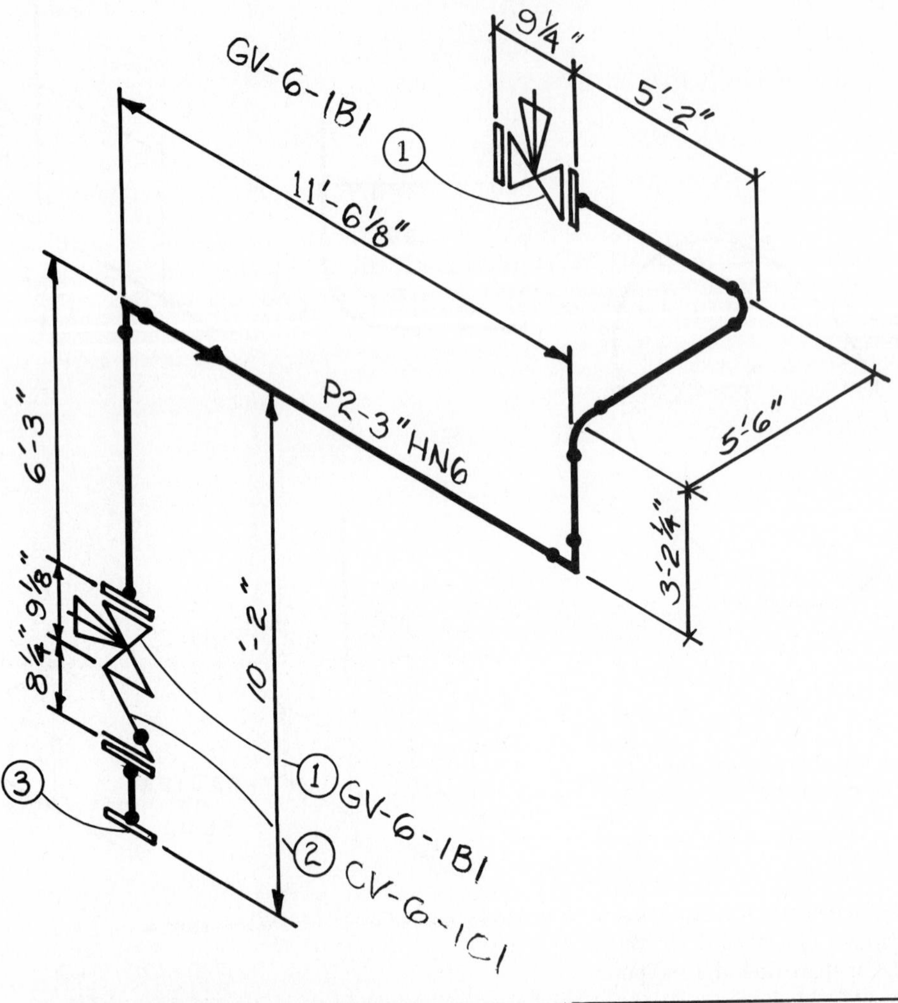

MATERIAL DESCRIPTION LIST	ISOMETRIC DIMENSIONING PRACTICES
Pipe—SMLS STL. SCH. 40 1. Valve Gate 300# FLG. 1/16 RF. 2. Valve SW. check 300# FLG. 1/16 RF. 3. FLG. F.S.W.N. 300# 1/16 RF.	1. Elbows square or round corners. 2. Show spool and line number. 3. Components numbered. 4. Arrowheads or slashes for dimensions. 5. Show handwheel orientation, and check for clearance between it and the nearest line or equipment. 6. Flow arrow shown on line. 7. Dimension lines not broken. 8. Dimensions in feet and inches. 9. Dimensions not necessarily drawn to scale. 10. Valves dimensioned face-to-face. 11. Lettering is slanted. 12. Show valve identification number.

FIG. 7-24 Isometric showing various dimensioning practices.

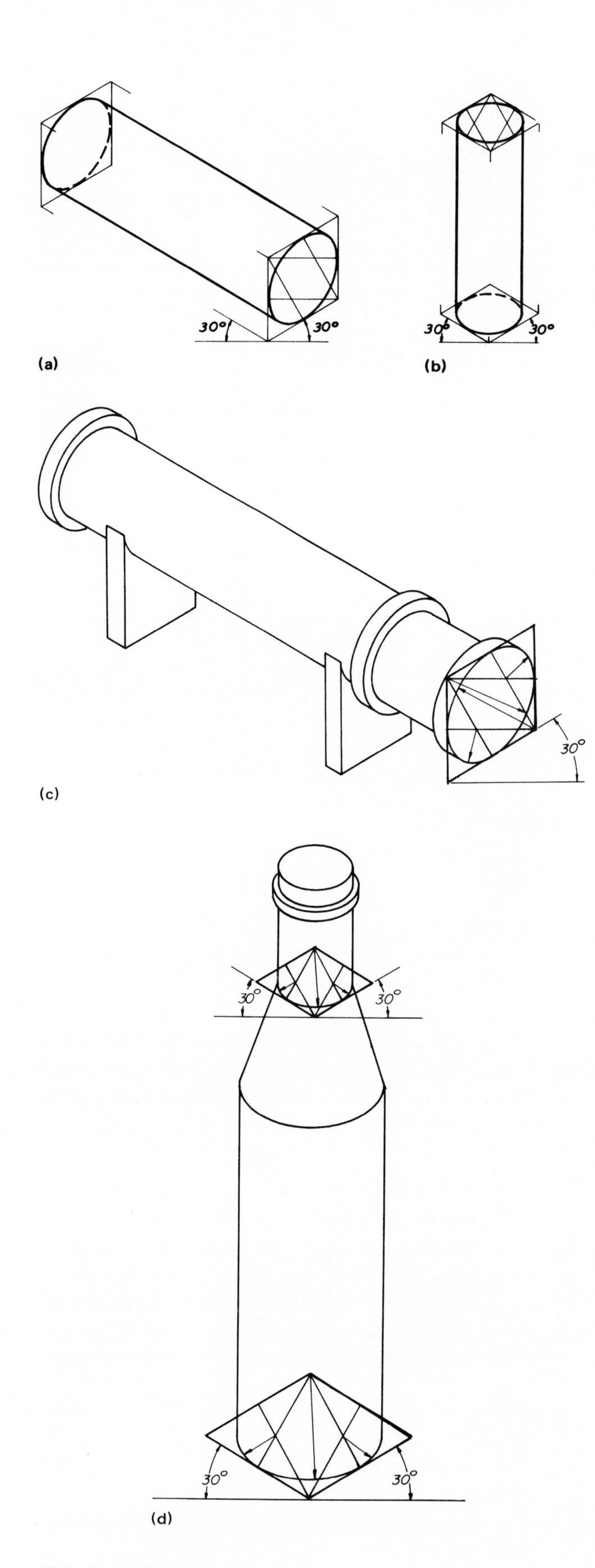

FIG. 7-25 Vessels in isometric.

FIG. 7-26 Vessel, piping, and instruments.

Dimensioning

The dimensioning of isometrics is similar to dimensioning on orthographic drawings. The aligned method of printing is preferred when dimensioning isometric views. Dimension lines are seldom broken as in mechanical drafting, though in some cases this method may be used. See Fig. 7-24 for an example of the proper way to setup a piping drawing in isometric. Figures 7-1 and 7-4 show typical dimensioning procedures for isometric drawings.

All dimension lines must be parallel to the pipeline to be dimensioned. Placement of dimensions must be clear and concise and thus readily available to the fabricator and others using the drawing. Most dimensions are taken from centerlines or from the end of the pipe to the center of

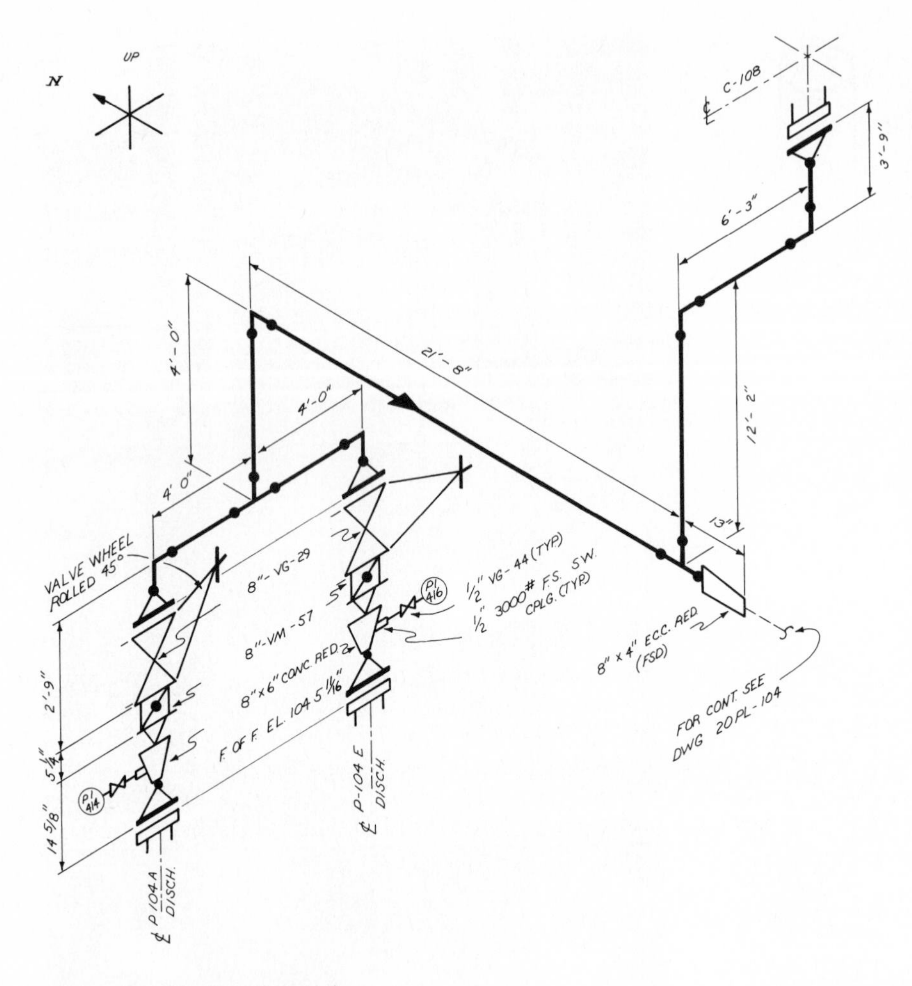

BILL OF MATERIAL		
REQ.	UNIT	DESCRIPTION
1	LF	½" SML'S S/30 PIPE A-106-B
35	LF	8" SML'S S/40 PIPE A-106-B
5	EA	8" 90° LR BW ELBOW S/40 A-234
2	EA	8" BW TEE S/40 A-234
1	EA	8" x 4" ECC. REDUCER S0/40 A-234
2	EA	8"x6" CONC REDUCER S0/40 A-234
3	EA	½" 3000 # F.S. SW CPLG
		ERECTION MAT'L
2	EA	6" 300# FLEXLTALLIC TYPE OG. GASK.
7	EA	8" 300" 〃
24	EA	¾" Ø x 5" LG. B-7 STUD BOLTS W/NUTS
24	EA	⅞" Ø x 5 ¾" LG 〃
24	EA	⅞" Ø x 10¼" LG. 〃
2	EA	½" VG-44 GATE VA
2	EA	8" VG-29 GATE VA
2	EA	8" VM-57 CHECK VA
5	EA	8" 300# RFWN FLANGE S/40 BORE

FIG. 7-27 Pipeline isometric.

valves, fitting ends, pipe supports, and other components, depending on the use of the isometric drawing. The face-to-face dimensions between flanges, valves, and fittings are also important. Most of the drawings provided throughout this chapter are taken from industrial sources and offer a variety of correct dimensioning practices (Fig. 7-27).

Angles dimensioned in isometric are located in the horizontal or vertical plane from isometric lines. It is essential to locate the angle from isometric lines and planes only. Many angles in isometric are dimensioned as in Fig. 7-20, where the 45-degree angle is called out as being in the vertical plane. Figure 7-4 shows the same method of locating and dimensioning the angle.

In isometrics in the round (Fig. 7-29), all angles based from the reactor centerline are dimensioned from the 0 degree axis line.

Centerlines are drawn when needed to define symmetry and to establish vessels, towers, elevation designations, and distances between lines in a system. Only when they confuse the drawing should they be omitted. Centerlines are always used in double-line drawings.

The *English system* of measurement using fractions, inches, and feet remains the dominant practice throughout American industry. The metric system is employed whenever the drawing is meant for foreign construction.

Procedures and Specifications

When drawing special or pictorially represented equipment that does not have a symbol, the drafter should call out all components that are not standard and specify them on the isometric drawing. The following list presents common procedures used throughout the piping industry for the construction, drawing, and dimensioning of isometric drawings:

1. Define areas within the project that must be shown in isometric.
2. Determine the maximum allowable size of the spool sheets.
3. Locate and designate logical field and shop material break points.
4. Prepare isometric drawings as soon as the study drawings are adequate or the model is at a sufficient stage.

5. Draw all fittings and symbols (except reducing fittings) with the single-line method of representation.
6. Show pipelines 12 in. (30.48 cm) OD and larger in double line when this is established procedure.
7. Draw isometrics to 3/8 in. = 1 ft, if drawn to scale.
8. Show all piping isometric drawings dimensioned in feet and inches. Use inches for dimensions under 2 ft (24 in.; 60.96 cm).
9. Use the metric system for all foreign jobs.
10. Draw isometrics so that they are oriented from the same direction throughout all the drawings of a project (toward the same direction).
11. Locate the north arrow symbol in the upper left-hand corner of the drawing.
12. Point the north arrow to the upper right or upper left on each isometric spool sheet.
13. Draw or sketch isometric views freehand before constructing an accurate isometric drawing.
14. Use standard or company symbols to represent all piping components.
15. Show rotation of valve stems and handwheel orientation.
16. Dimension and orient bosses and weldolets clearly on all fabrication isometric drawings.
17. Show dimensions to valves, between fittings, and all face-to-face dimensions on isometric fabrication drawings.
18. Draw all reducing fittings double line except outlet tees.
19. Draw all equipment in phantom or light lines in relation to major and auxiliary pipelines.
20. Always give the dimension for top of support or bottom of pipeline for pipeways.
21. Show work points on centerlines for sloping lines at support points (Fig. 7-3).
22. End all branch lines that connect to the isometric at a clearly defined fitting or component (nozzles).
23. Code all items that need to be clarified throughout the fabrication drawing.
24. Identify each line throughout the piping isometric by showing line number, line code specification, and insulation requirements.
25. Locate a common match point when a line appears on more than one sheet.
26. Indicate all field welds on lines that extend through platforms approximately 12 in. (30.48 cm) above the platform level. Show all platform locations (Fig. 7-3).
27. Show all pipeline elevations and details when a line extends through a containment area or floor.
28. Avoid double dimensioning (duplication of dimensions) on all isometric drawings.
29. Do not draw equipment such as pumps and compressors pictorially.
30. List straight runs of piping that are connected to isometric subassemblies.
31. Record all insulation shoes for an isometric line.
32. Locate and size all shop-fabricated shoes for insulation.
33. Do not dimension small piping on isometric drawings unless necessary for fabrication.
34. Show flow arrows on all drawings.
35. Use heavy dashed lines on steam tracing isometrics as in Fig. 7-5.
36. Locate all tracer traps for steam tracing isometric drawings (Fig. 7-5).
37. Locate and show the condensate return line for steam tracing lines (Fig. 7-5).

These suggestions are meant only as guidelines. In all cases, isometric drawings must contain sufficient data to enable the fabricator to position and construct the piping configuration accurately.

Isometrics in the Round

Isometrics in the round are used to represent circular piping systems—such as the containment area of a nuclear reactor used as an example here. These drawings must be based on the isometric ellipse, using the horizontal axes as centerlines for the reactor and establishing quadrants.

The *X*, *Y*, and *Z* axes are sometimes used in these drawings to originate or orient the system. All angles for isometrics in the round are taken from the horizontal axes formed by the centerlines of the ellipse. These centerlines establish the reactor quadrants. Figure 7-28(a) is an example of a nuclear reactor containment pipeline in the plan view. All angles are taken from quadrant 1, 0 degrees, and all radii are measured from the centerline of the reactor building. When laying out the system in isometric, the quadrant axes are drawn at 30 degrees receding to the right and to the left as in Fig. 7-28(b).

When drawing an isometric in the round, use the four-center ellipse method of construction because isometric templates of adequate size are not available (Fig. 7-28c). Figure 7-28(d) shows the construction of radius 11 ft 6 in. (3.5 m) at elevation 731 ft 0 in. (217 m) from the pipeline at 316 degrees to the bend at 56 degrees. Figure 7-28(e) shows the formation of a second isometric construction plane at elevation 715 ft 0 in. (217.9 m) for the drawing of radius 13 ft 6 in. (4.1 m) and the continuation of the line at 715 ft 0 in. (217.9 m) in elevation. This line runs from 56 degrees to 116 degrees and then drops down to the original elevation at 713 ft 0 in. (217 m). Figure 7-28(f) displays the isometric construction plane for radius 15 ft 6 in. (4.8 m) at 713 ft (217 m) elevation extending from 116 degrees to 140 degrees. The last drawing in this series, Fig. 7-28(g), shows the completed isometric in the round, including all elevations, radii, and angular dimensions.

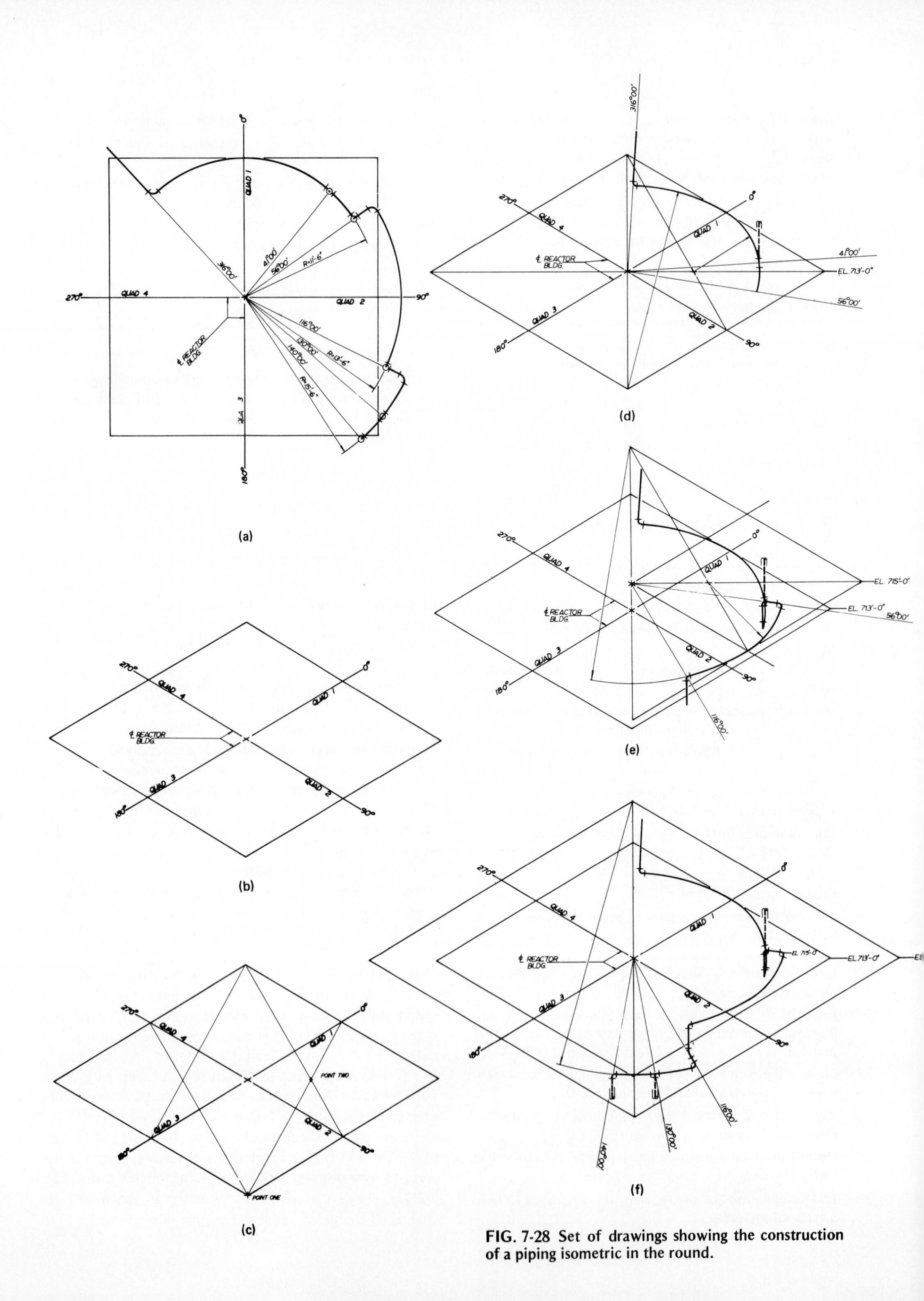

FIG. 7-28 Set of drawings showing the construction of a piping isometric in the round.

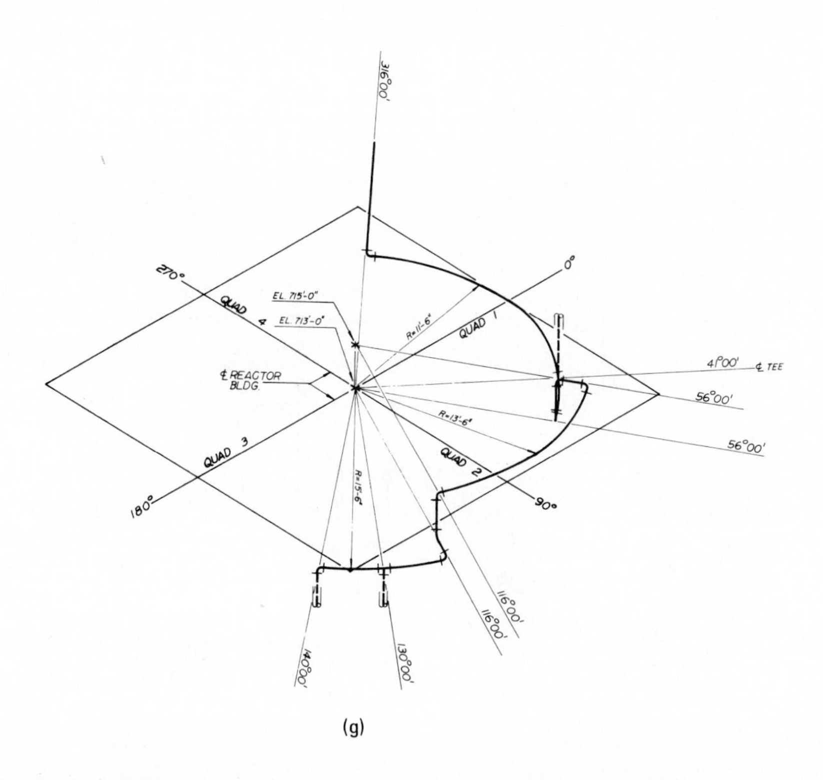

FIG. 7-28 Set of drawings showing the construction of a piping isometric in the round. (cont'd)

Isometrics and Pipe Supports

One of the main uses for isometric drawings lies in the placement of hangers, supports, guides, and shock suppressors (snubbers) (Figs. 7-29, 7-30, and 7-31). The isometric coordinates are fed into a computer along with all support

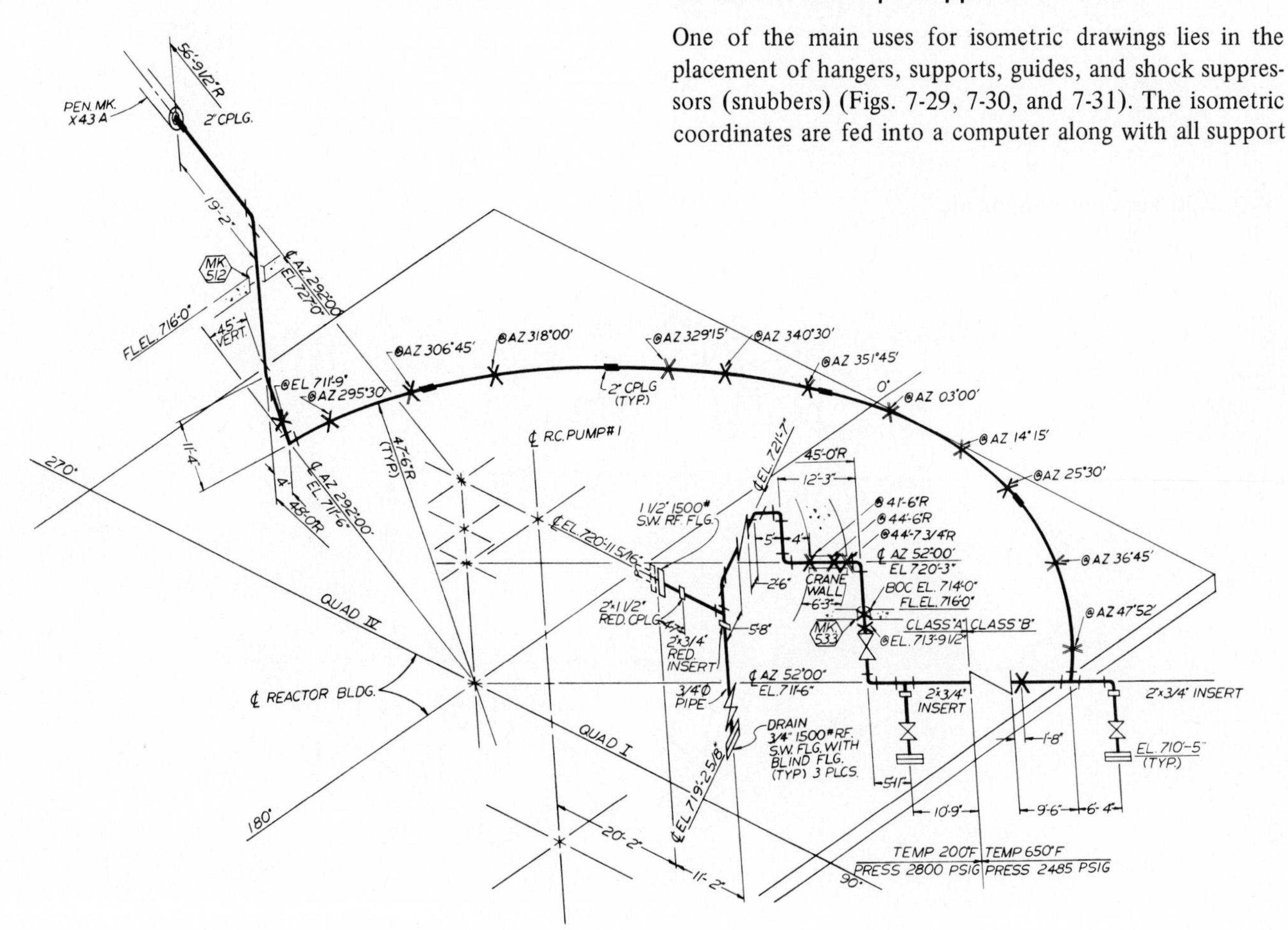

FIG. 7-29 Isometric in the round.

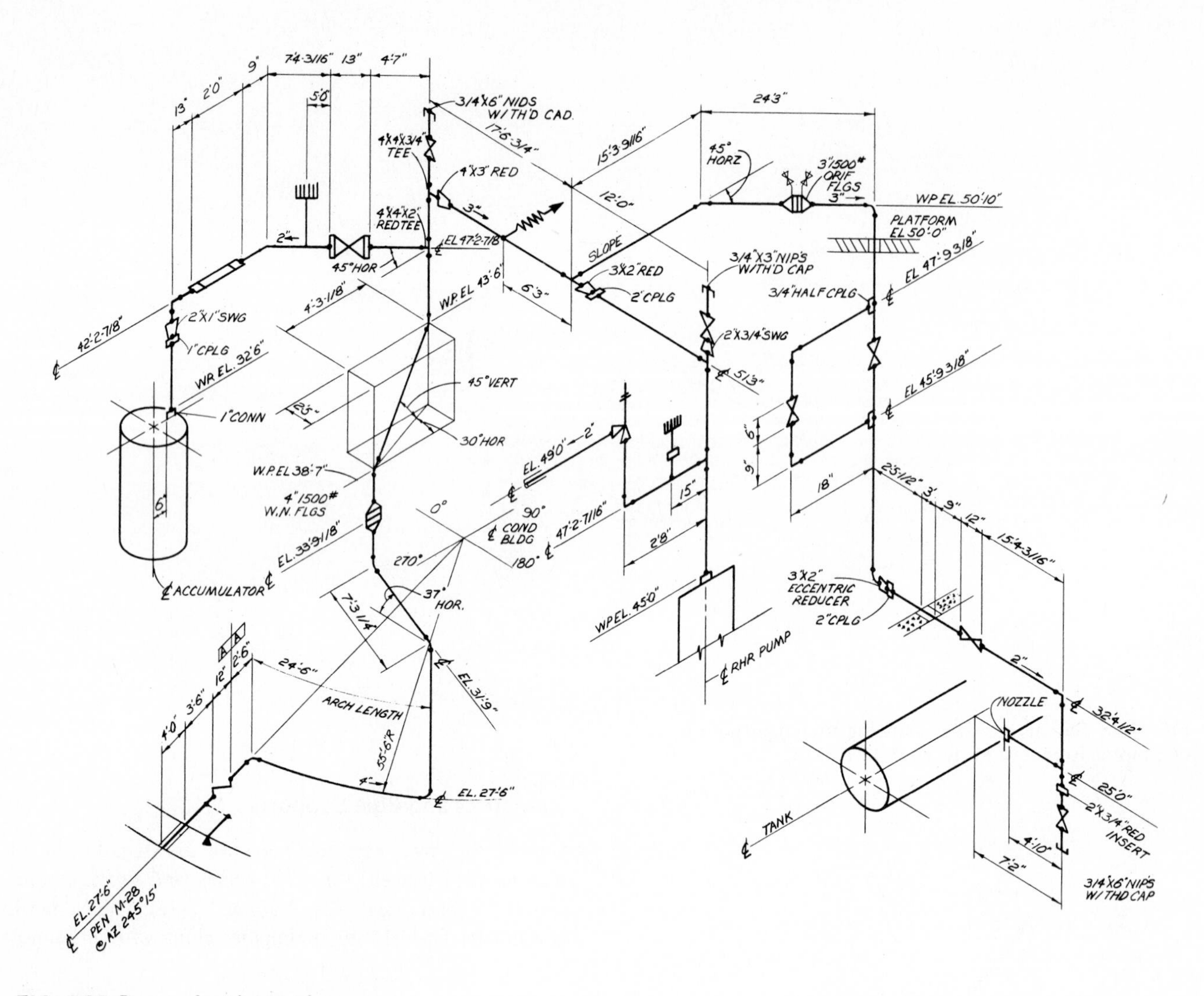

FIG. 7-30 Power plant isometric.

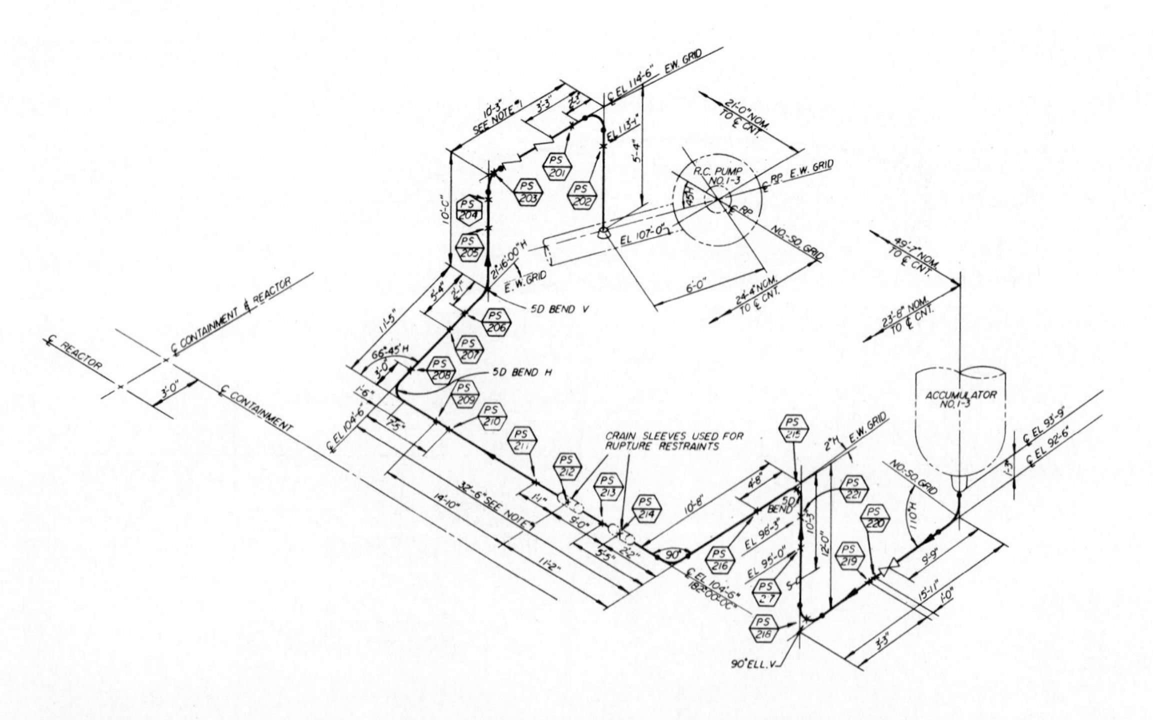

FIG. 7-31 Nuclear isometric.

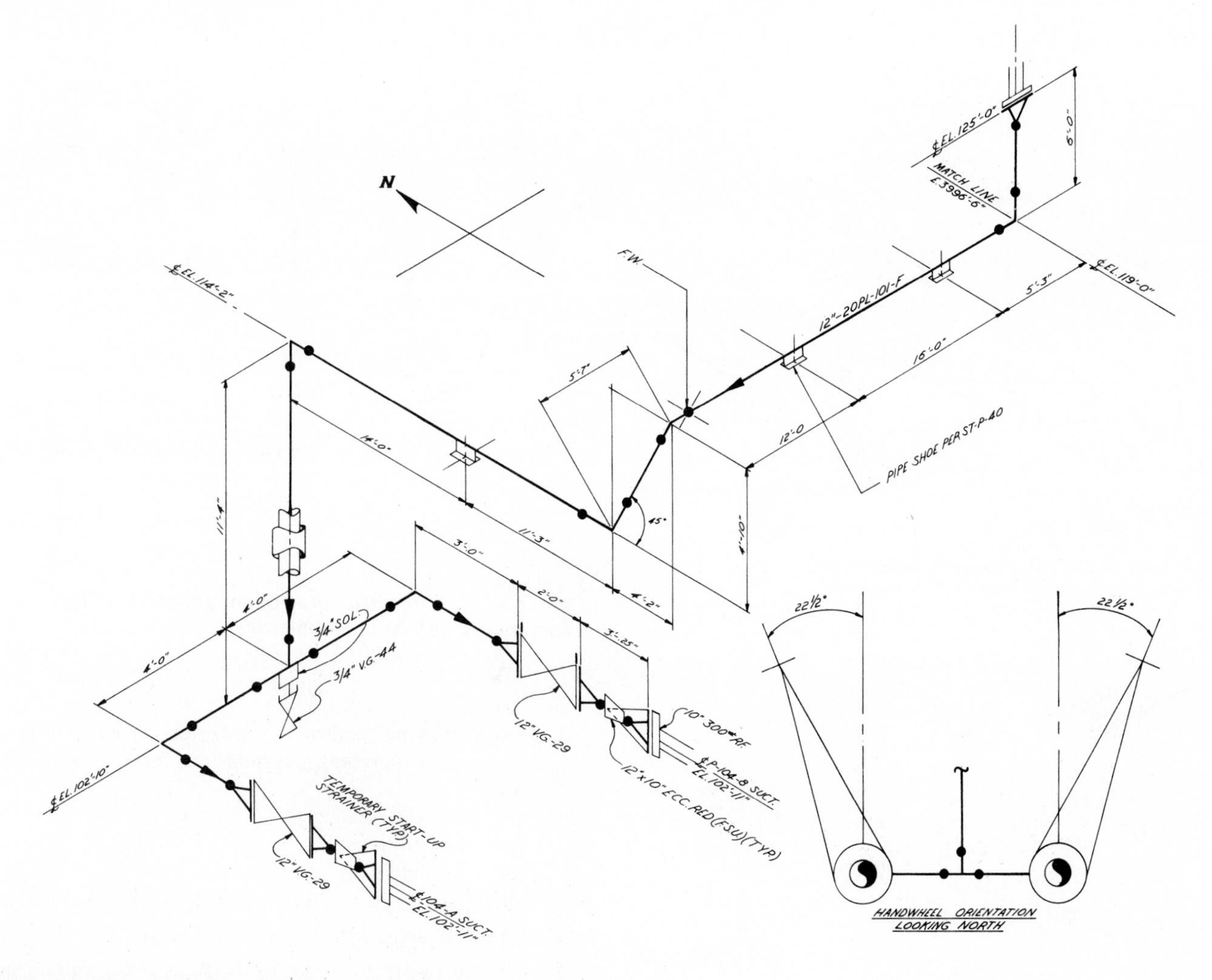

FIG. 7-32 Pump suction line isometric spool.

FIG. 7-33 Model of a large chemical plant.

locations, seismic movement studies, thermal movement data, and other essential information. The computer then does a complete study of the possible safety hazards, location problems, and movement effects on the system. Because of the dangers inherent in a nuclear power plant, strict codes have been enacted to supposedly ensure the safety of the system under normal and abnormal stresses. drawing is an essential aspect of this work.

Figure 7-32 is an isometric from the model shown in Fig. 7-33.

Figures 7-34 and 7-35 are typical isometric details of containment penetration and elevation designations. Isometrics for nuclear power plants often use symbols that are unique to this area of piping. The isometric drafter will sometimes be used for positioning pipe supports on the model, as well as constructing and drawing the isometric view.

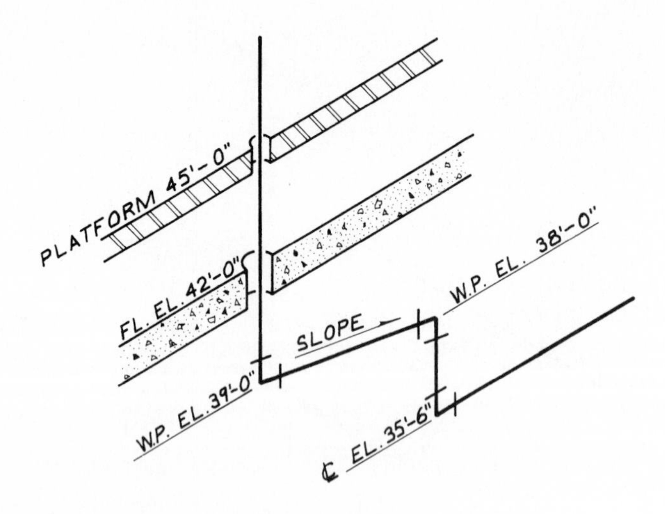

FIG. 7-34 Isometric elevation designations.

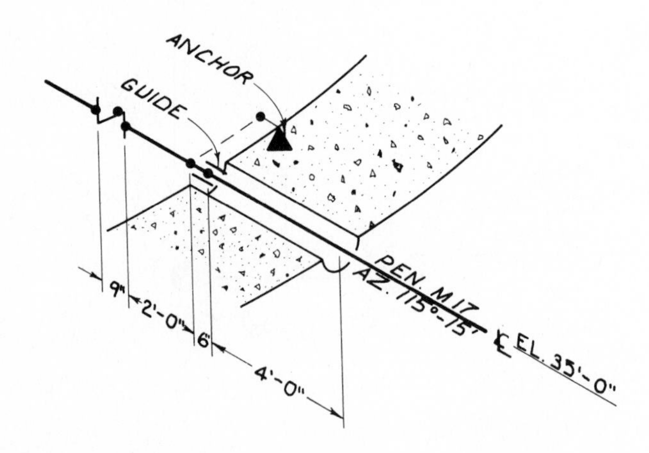

FIG. 7-35 Containment penetration detail in isometric.

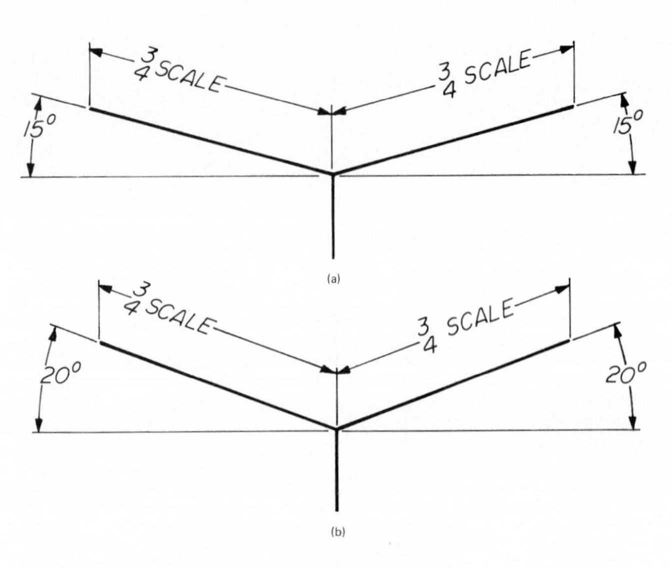

FIG. 7-36 Axonometric projection angles: (a) 15° axonometric, (b) 20° axonometric.

Axonometric Projection

The axonometric projection is very similar to the isometric projection. The difference lies in the use of a recession angle other than 30 degrees. Figure 7-36 shows two common angles used in axonometric projections. Horizontal dimensions are also shortened to three-quarter scale along the horizontal axes. Using different angles of projection allows the drafter to show more detail and construct a more realistic view of the piping system.

With the information presented in this chapter, the student should be able to apply the various construction used in industry for the three-dimensional views of isometric and axonometric projection.

QUIZ

1. What drawing method is used most often in piping drafting to represent a system in a three-dimensional form?
   (a) Dimetric
   (b) Axonometric
   (c) Isometric
2. What scale is preferred in piping drawing?
3. Name three uses of an isometric drawing.
4. The north arrow on an isometric drawing should be oriented to the:
   (a) Upper right
   (b) Upper left
   (c) Lower left
   (d) Top of paper
5. Isometric lines will appear:
   (a) As angles in the isometric
   (b) True length
   (c) Out of scale
   (d) Three-quarters their own size
6. Nonisometric lines are parallel to the axes. True or false?
7. Isometric angles that are not parallel to the plane of projection are ____________.
8. Vertical vessels or towers have an elliptical shape that is parallel to the horizontal. True or false?
9. The four-center method is ____________.
10. To dimension the pipe drawing in English measurements use:
    (a) Decimals
    (b) Centimeters
    (c) Fractions
    (d) Yards

11. Angles should be located from the ______________ and ______________ plane.
12. When should centerlines be used in an isometric?
13. Elevation dimensions eliminate the need for ________ in some cases.
14. The usual axonometric drawing uses an angle of:
    (a) 45 degrees
    (b) 35 degrees
    (c) 30 degrees
    (d) 15 degrees
15. Isometrics should be oriented to be viewed from ______________ direction throughout a project.
16. When should an isometric be drawn freehand?
17. How are angles dimensioned in isometric?
18. How do models pertain to isometric drawings?

PROBLEMS

1. Draw isometric views of Figs. 7-1 and 7-4.
2. Prepare an isometric spool drawing of Fig. 7-7.
3. Draw a pictorial isometric of Fig. 7-7.
4. Construct an axonometric drawing of Fig. 7-8.
5. Draw the appropriate orthographic views of the Fig. 7-29.
6. Complete an isometric spool drawing of Fig. 7-30 (250 psi, 1000°F, ASTM A376, seamless alloy pipe). Do a complete material takeoff for pipe, valves, and fittings.
7. Draw the plan and elevation views for Fig. 7-27.
8. Do an isometric spool drawing of Fig. 7-30 or Fig. 7-32.
9. Redraw Fig. 7-4 in orthographic projection.
10. Redraw Fig. 7-1 as an isometric 3× book scale.
11. Redraw Fig. 7-4 using an appropriate scale.
12. Redraw Fig. 7-31 3× book scale.
13. Complete an isometric spool drawing from an existing model.
14. Draw a pictorial isometric from Fig. 7-37 or Fig. 7-32.
15. Complete a double-line drawing of Fig. 7-27 or Fig. 7-32, using pictorial representation of the pumps.
16. Complete an isometric spool sheet for one assigned line from the projects in Chapters 11, 12, 13, and 14. (Use Tables 1-1 and 1-2 for pipe size and schedule calculations.)

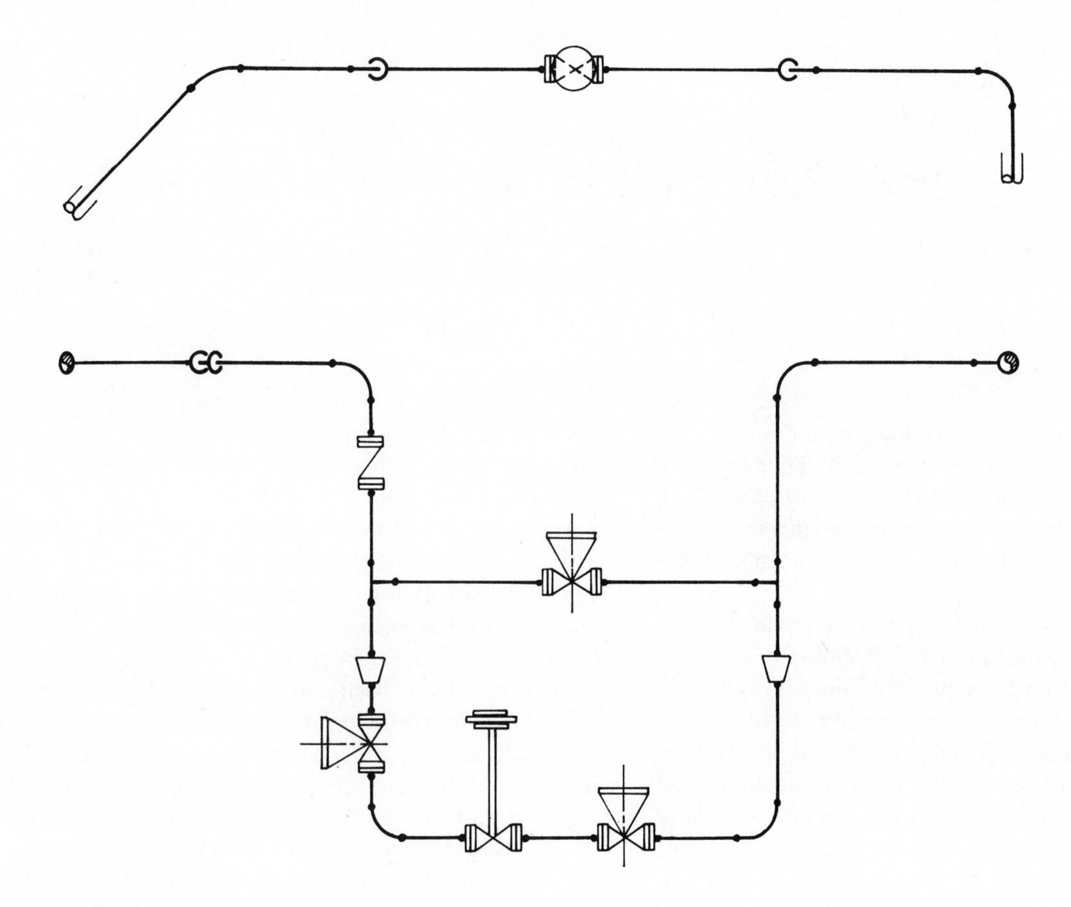

FIG. 7-37 Orthographic view of control station.

Double-spring trapeze constant-load hanger for a large steam line.

PIPE SUPPORTS

This chapter covers pipe supports, hangers, and other accessories needed to support piping systems. All piping systems and pipe lines require supports and hangers to carry, hold, and sustain the pipeline when under stresses due to thermal expansion, mechanical vibration, and the weight of the piping system.

In modern terminology, *pipe support* refers to any support, hanger, or mechanism which will allow the system to be suspended from above or supported from below. Specifically, though, *pipe supports* are mechanisms which allow the piping to be supported from below whereas *pipe hangers* are mechanisms which enable the piping to be hung from above. This distinction is observed throughout the chapter when referring to specific support or hanger mechanisms, but when referring to general categories (such as pipe support design group) the term *pipe support* is used loosely to designate all mechanisms which enable the piping system to be maintained by compensating for various types of mechanical, thermal, or shock movement and the physical weight of the system.

The early part of this chapter deals primarily with standard hangers (or catalog items as they are sometimes called); the latter part deals with critical hangers for nuclear power, conventional power, and petrochemical piping installations.

The work of pipe support detail and design groups is becoming a field in itself, because of the large number of critical piping situations that occur when designing petrochemical, nuclear, and conventional power plants. Many drafters are employed directly and exclusively in this field.

For noncritical uses, the simplest pipe hangers, consisting of split rings, clamps, brackets, clips, pipe rolls, pipe stands, and simple hangers, are used by themselves or in combination with other pipe support accessories. These simplified catalog items are used primarily for services such as domestic plumbing and building heating systems where stresses and strains due to thermal and mechanical vibration are minimal and the total weight of the piping system is small. In the noncritical area, rods and straps and other simple pipe support mechanisms are usually adequate for small-OD pipe when suspending the system in horizontal runs. Dummy supports, bottom spring supports, and special pipe support configurations are used for supporting vertical runs which usually involve greater weights.

Numerous pipe support companies are represented in this chapter, and the drafter and designer must carefully consider the design capabilities of all the components available through standard catalogs. The manufacturer's design load recommendations and suitable design applications should be adhered to strictly. It must be stressed here that pipe supports are not placed indiscriminately or left for field installation except in noncritical plumbing situations.

Whether the pipeline is insulated or not, it is important to consider all the variables necessary for the adequate suspension and support of the total piping system. In a majority of cases, whether noncritical or critical, the supports are suspended or attached to the steelwork, or concrete embedments are used to support or hang the piping system, thereby tying the pipe system into the structural framework of the facility.

Steel, wrought iron, and special alloys are the primary materials used for a majority of pipe support components. For the support and suspension of small pipelines and tubing (depending on the length of the run) it is usually sufficient to support the system by means of rods and straps connecting the pipe to various structural members that are close to the pipeline—such as ceilings, walls, structural steel members, columns, posts, and concrete flooring.

Standard catalog items are available for most pipe supports for noncritical pipe installations. ANSI B31.1, the code for pressure piping, covers all piping support situations and systems and gives the codes for braces, guides, shock suppressors, and supports. It is usually unnecessary to fabricate pipe supports because of the vast number of standard components available, which can either be used separately or in combination to ensure adequate suspension of the piping system and regulate its movement due to mechanical or thermal causes.

STANDARD COMPONENT SUPPORTS

Standard component supports usually involve small assemblies consisting of an assortment of catalog items. The proper use of catalog items in conjunction with each other simplifies the design and drafting process. Thus the system can be designed and produced more cheaply and quickly than if all the various pipe support mechanisms had to be fabricated individually. One of the most important prerequisites for the pipe support drafter and designer is familiarization with catalog items.

Catalog items are accompanied by a variety of information for ordering and designing the pipe support system—including the item's name, materials, suggested applications, part numbers, major dimensions, code ratings, and design information. There are many kinds of pipe supports available and a considerable number of individual companies producing similar (though not always comparable) items.

PIPE SUPPORT SERVICE

Pipe support mechanisms are used for a variety of situations. Besides controlling both the thermal and mechanical vibration and supporting the piping system's deadweight (gravitational weight), other considerations are relevant to adequate placement and design of pipe supports:

1. The weight of the system includes not only the pipeline weight filled with the medium it is carrying, but also the insulation and other loading factors. Climatic conditions may influence the total weight of the pipeline by means of external loading, such as the buildup of snow or ice or high wind conditions if the system is outdoors. Seismic shock, which is dealt with later in the chapter, is another important consideration.
2. Pipe supports such as sway braces, sway struts, and snubbers are used to dampen or redirect different types of vibration that is caused by mechanical or seismic activity.
3. Pipe support systems must be able to handle shifting loads caused by thermal expansion and contraction of the pipeline under normal and abnormal operating situations. Many hangers and supporting equipment are designed to permit the pipe to have free movement due to thermal expansion, which is independent of pipe schedule and size. Never immobilize the pipe in a manner which may cause the adverse transfer of loads from one support to another or from the supports to the piping. For instance, a 3-in. (7.62-cm)-diameter steel pipe with 1/2-in. (1.27-cm) wall thickness, if held immobile, generates a tension or compression of 50 tons force if its temperature is lowered or raised by 50°F. Larger pipes are capable of generating much larger thermal forces. Thermal movement of all types must be compensated or allowed for by the use of the pipe support materials, guides, rollers, and spring supports. The thermal expansion and contraction of the system are determined by the pipe material and by the operating temperatures.
4. Pipe systems must be designed with sloping lines to permit adequate line drainage. This is especially critical for steam lines because of the buildup of condensate which must be drained from the system. The

slope of a pipeline is always in the direction of the flow of steam, and at appropriate intervals traps, separators, valves, and other devices are used to expel liquid (condensate) from the system. Also, pipe support systems must be sloped to allow for drainage of nonsteam lines and eliminate pocketing of the medium. Areas of the line may sag—for example, heavily loaded sections of a pipeline where valves or other components increase the weight. By locating supports close to heavy valves and equipment, it is possible to reduce the gravitational load which causes pocketing and overloading of the pipe section.

5. Stresses and strains caused by twisting or torsion must be dealt with so that no undue stress is placed on the pipeline.
6. The pipe system should be designed with a generous safety factor—usually a 3:1 ratio between designed capability and actual load. In the critical services such as nuclear power piping systems, which undergo high thermal and mechanical shock, this safety factor is higher. Many nuclear projects have been delayed until the pipe support system has been completely modified to satisfy a higher safety level.

PIPE SUPPORT CLASSIFICATIONS

The most commonly used pipe supports are the simple catalog items, which include rods, hangers, springs, guides, and other restraints and anchors, struts, shock suppressors, snubbers, and sliding supports.

For nuclear application, hangers are usually classified into types 1, 2, or 3, depending on importance and safety considerations. The class 1 hanger is used for class 1 piping and indicates an application or system vital to plant safety. Classes 2 and 3 denote supports on systems with lower safety requirements. For nonnuclear applications, hangers and supports are categorized in accordance with the complication or criticality of the piping they support. In most cases, the classification is indicated on drawings showing the detail of the pipe support mechanism.

Chapters 11, 12, and 13 cover petrochemical process piping, fossil fuel power generation piping, and nuclear piping. In all these situations, equipment and machines are joined together by pipes forming complete systems for turning electrical generators or producing a product such as gasoline.

The power generation field requires many types of piping systems—for main steam, feedwater, condensate extraction, cooling, safety, and backup systems. Auxiliary and backup systems are needed in nuclear power plants where high levels of radiation and high temperatures are generated. Among these auxiliary and backup systems are those designed for radiation waste removal, residual heat removal, cooling ponds, and reactor emergency core cooling. The total amount of piping needed in a nuclear power plant—or for that matter even in a conventional power plant—may be anywhere from 75,000 to 200,000 ft per generating unit, creating an enormous demand for pipe supports. In fact, pipe supports may number between 10,000 and 30,000 for one installation—not even including small-OD piping 2 in. in diameter and smaller which is field-erected and supported. The placement of the piping systems and equipment along with all electrical, HVAC, and other systems must be coordinated throughout all phases of design because of the obvious need to prevent congested areas and interference between the different systems, including interference caused by pipe supports.

Since safety considerations must be coordinated with all other design factors, federal and state codes and specifications have been instituted to ensure adequate safety and design. In the power generation field, critical and noncritical piping situations have made pipe support design and drafting one of the most important areas. In petrochemical and other process piping installations a variety of high temperatures and pressures and caustic material transfer through pipelines create hazards throughout the plant. Since the possibility of loss of life or harm to the environment is extremely high, all pipe support systems are designed with safety in mind.

Simple Pipe Hangers

Figures 8-1 and 8-2 show a few types of simple pipe hanger and support variations. Besides the clips, rollers, and split and unsplit rings that are shown, the pipe support assembly may use a variety of small items such as brackets, clamps, bolts, and turnbuckles, which can be seen in the pipe support details provided throughout the chapter. For instance, Fig. 8-3 demonstrates a simple pipe support using welded-beam attachments, eye nuts, hex nuts, rods, turnbuckles, and bolt pipe clamps. In this case, a pipe clamp or ring is connected to a threaded rod which in turn is connected to the structural beam by attachments and bolted to concrete anchors embedded in the concrete wall.

The items in Fig. 8-1 are used throughout industry for noncritical support or in conjunction with other standard items in critical situations. Note the variety of pipe clamp arrangements using steel channel, which is commonly used throughout piping systems to attach or suspend pipes. The majority of the devices in Fig. 8-2 are for noncritical pipe support. With the structural channel shown in Fig. 8-2, the piping system can be adjusted and the support is not permanently attached to a structural member.

In Fig. 8-3, the turnbuckle (item 5) enables the system to be adjusted after erection. Trial operation of the system

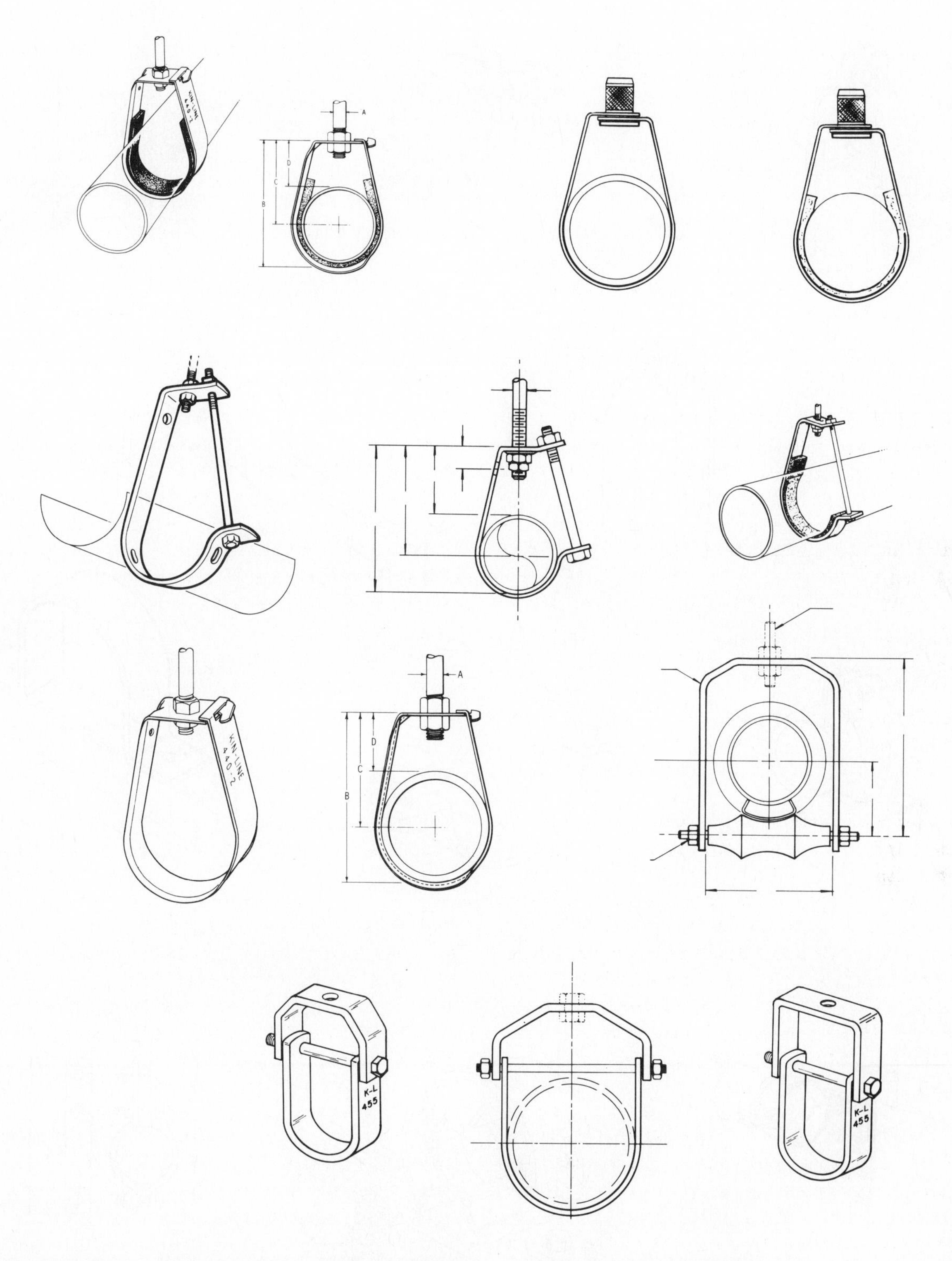

FIG. 8-1 Pipe hanger variations (not for nuclear services). (Courtesy Kin-Line, Inc., Emeryville, CA)

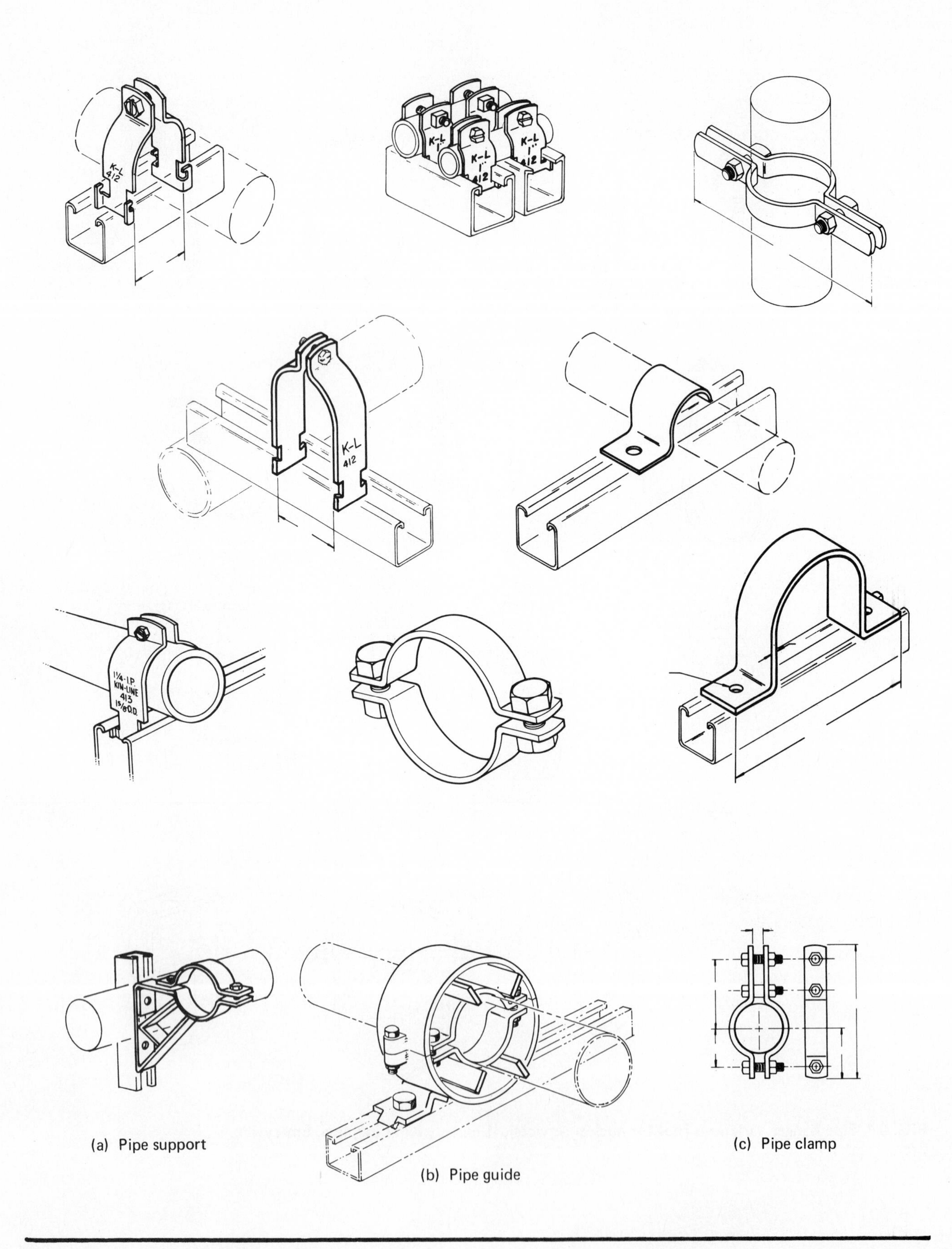

FIG. 8-2 Pipe clamp, and pipe guide variations. (Courtesy Kin-Line, Inc., Emeryville, CA)

ITEM NO.	NO. REQ'D	PART NO.	SIZE	DESCRIPTION
1	1	113	#3	WELDED BEAM ATTACHMENT W/LOCKNUT
2	2	274	#5	WELDLESS EYE NUT
3	3	–	5/8"	HEX NUTS
4	1	133	5/8" X 3'-0"	ROD W/6" TBE
5	1	132	5/8"	TURNBUCKLE
6	1	133-A	5/8" X 1'-4"	ROD W/6" RHT & 6" LHT
7	1	304	4"	THREE-BOLT PIPE CLAMP W/LOCKNUT
8	1	–	W4 X 13	BEAM X 5'-8" LG.
9	1	–	3/8" X 11" X 11"	C. S. PLATE
10	4	512	5/8"	CONCRETE ANCHOR
11	4	–	5/8" X 1-1/4"	BOLT

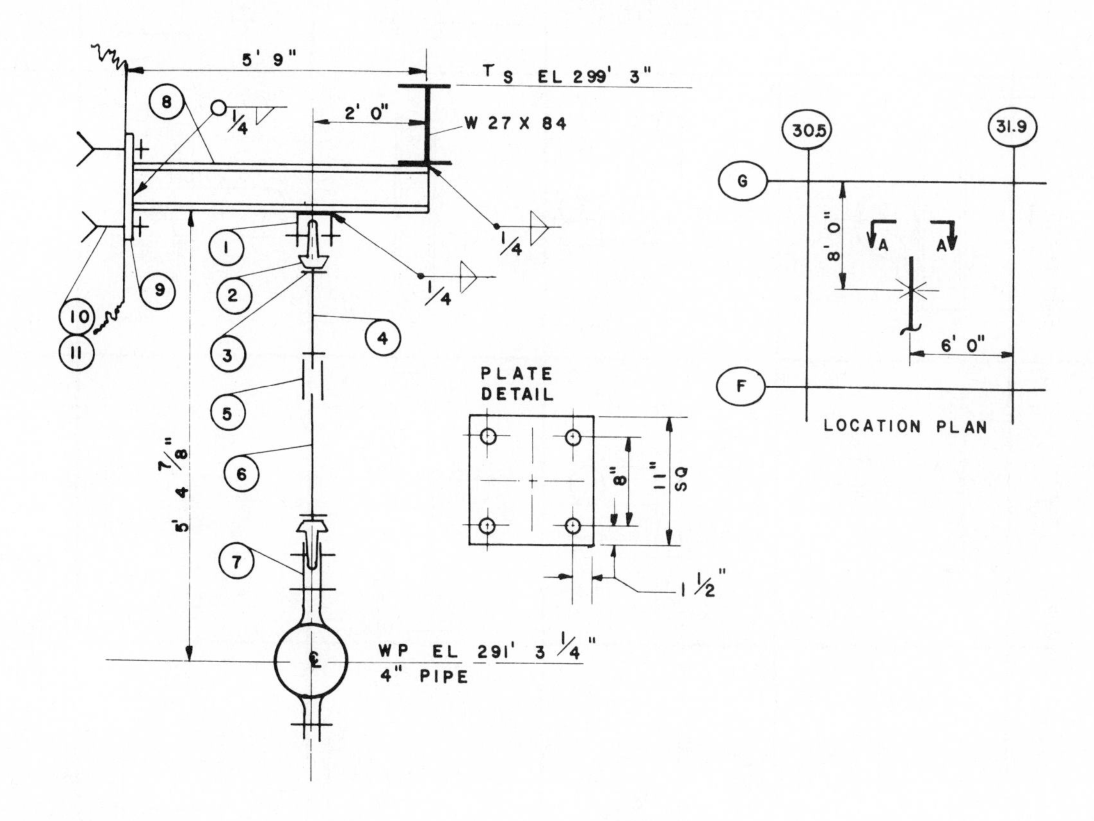

FIG. 8-3 Simple pipe support detail using rod, turnbuckle, and pipe clamp.

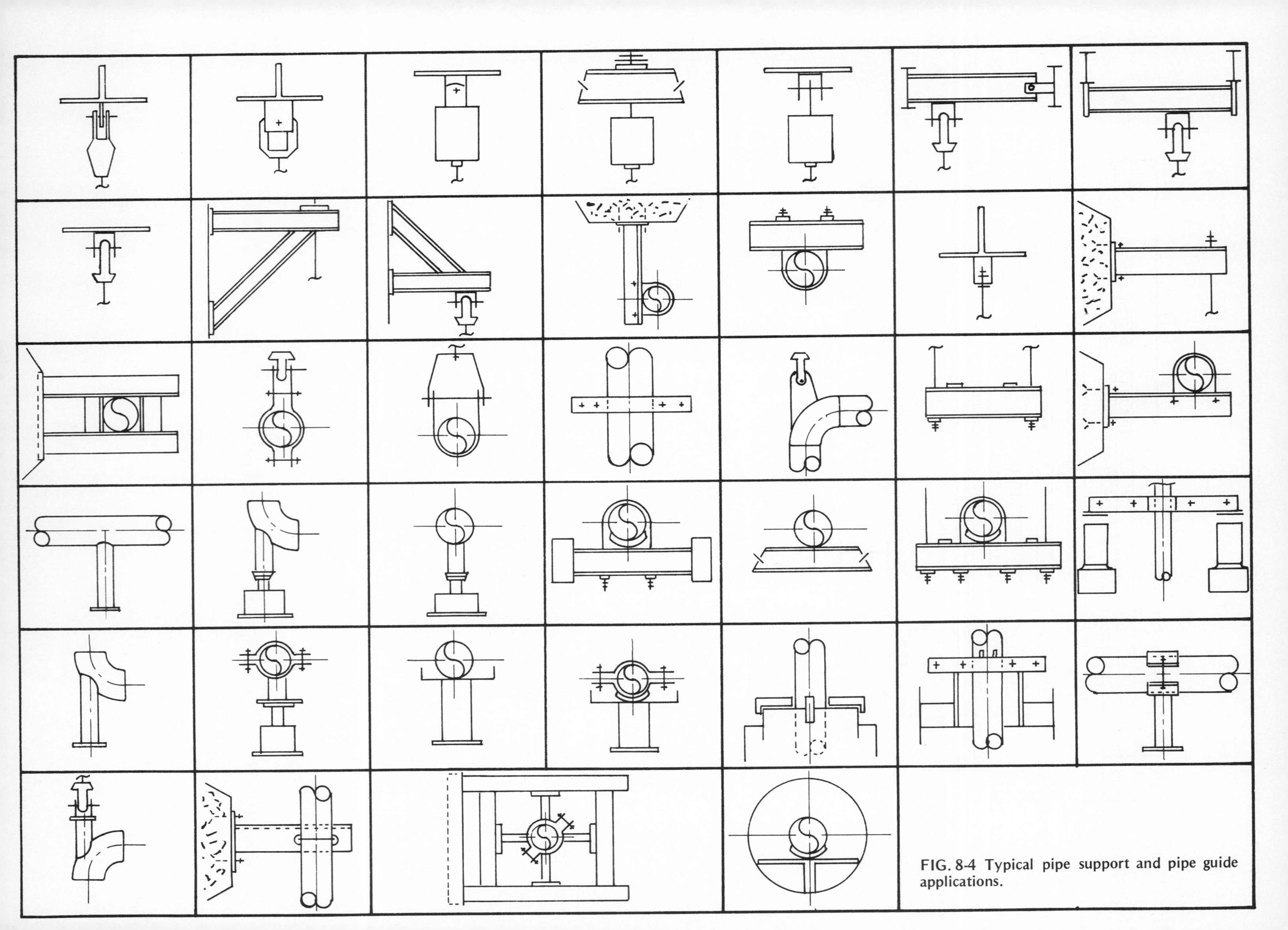

FIG. 8-4 Typical pipe support and pipe guide applications.

provides the data used to adjust the support to allow for thermal, mechanical, and vibrational movement. Figure 8-3 would be considered a typical rod hanger detail. Rod hangers allow for free movement of the pipe in horizontal directions within specified limits. Rigid rod hanger assemblies can accommodate a certain amount of vertical deflection (movement) if a proper combination of offsetting, rod length, and axial thermal expansion is taken into account by the designer. They are one of the commonly used pipe supports. In some cases, two rod hangers are tied together in a trapeze-type mechanism. In some of these cases, spring hangers are used in conjunction with rods and trapeze assemblies to provide for a variety of movements. Sometimes a trapeze assembly will support a riser for the suspension of vertical pipe.

Figure 8-4 displays common pipe supports and variations in their application. All rod hangers have as their main component the hanger rod itself (items 4 and 6 in Fig. 8-3). A number of other components, such as springs, turnbuckles, structural embedments, eye nuts, and special beam attachments, are used in conjunction with the hanger rod to form an assortment of hinge joints that provide for the different movements a system may undergo. Turnbuckles are used in the majority of rod hanger applications. Their use depends on the total rod length and the need for adjustability. When the required length is 3 ft (91 cm) or less, turnbuckles are not used; adjustability can normally be provided through the threaded ends of the rod and the eye units.

When pipe risers are used, riser clamps are utilized and lugs are welded to the riser. The drafter or designer must consult the established design specifications and calculations provided in the catalog by the pipe support manufacturer. The minimum rod size is based on pipe size and load requirement. Hanger rods that are less than 1/2 in. (1.27 cm) in diameter are not used for pipe supports except in noncritical situations or with spring hangers; in the latter case, rod hanger size is determined by the size of the spring hanger. When using a rod for a rigid riser, it is important to design the system to withstand the total design load, including stresses.

Hanger rods are usually supplied with right-hand thread at both ends unless two rods are joined by a turnbuckle. In this case, the left-hand thread should be provided at one end of the lower rod so it can engage with the turnbuckle. Thread length is also important for rods because of its effect on adjustability. In all situations, the maximum limit for pipe rod length should be 20 ft (6 m), and in the majority of cases it is good design practice to keep the rod length 10 ft (3 m) and under. When drawing pipe support details, locate the turnbuckle for the pipe rod in the lower third of the total rod length for ease of installation.

Pipe Rollers

Pipe rollers provide for the free horizontal movement (along the length of the pipe) caused by thermal or other mechanical movement of the pipeline parallel to the pipe. They should be designed to support the full weight (load) of the pipe and can accommodate both insulated and noninsulated lines (Fig. 8-5).

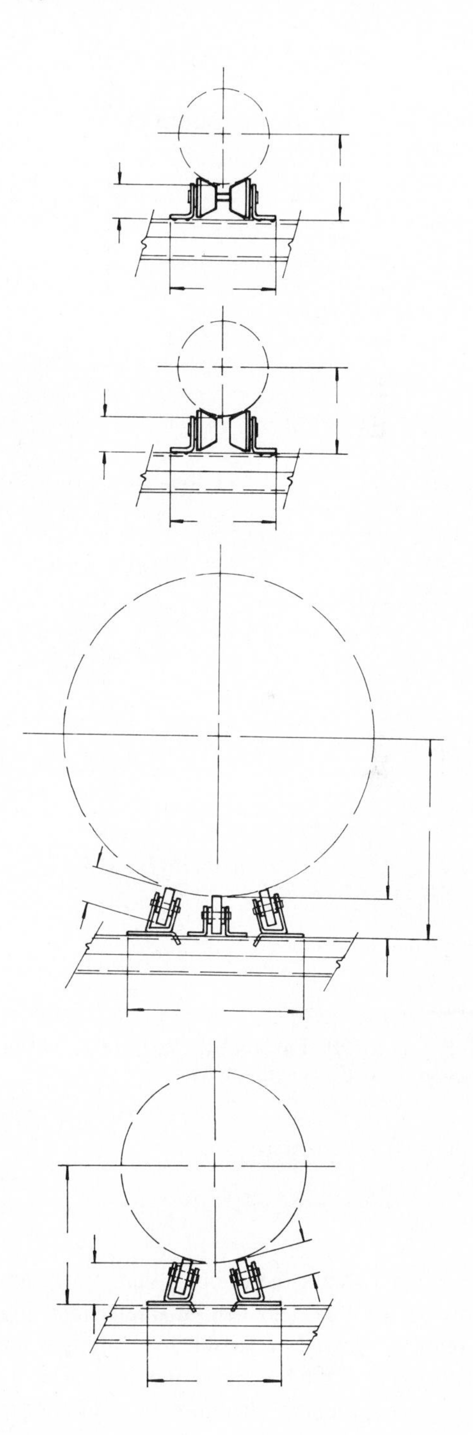

FIG. 8-5 Pipe roller variations. (Courtesy Kin-Line, Inc., Emeryville, CA)

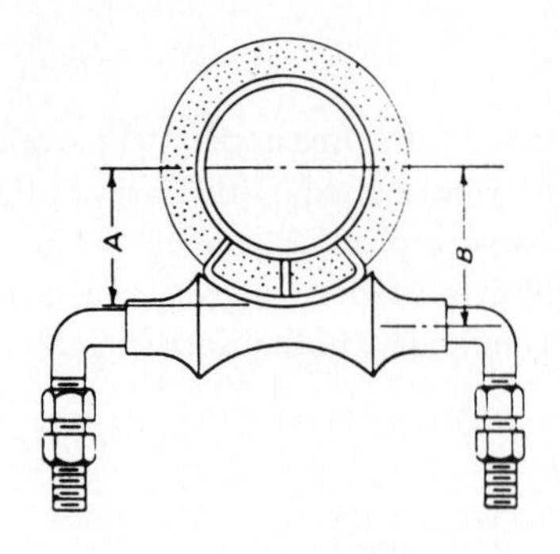

(a) Roller Support

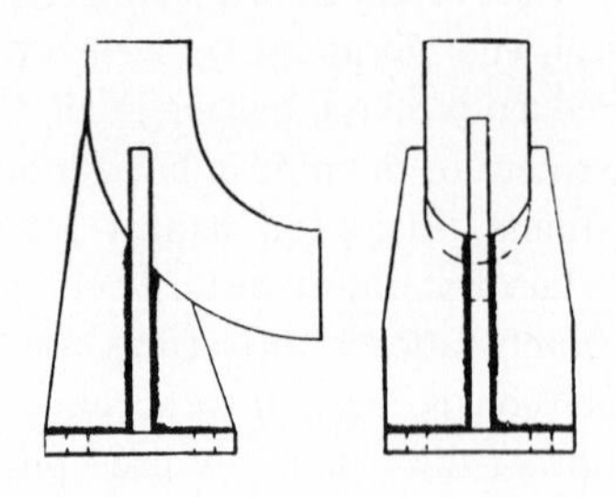

(b) Welded-base Anchor Support

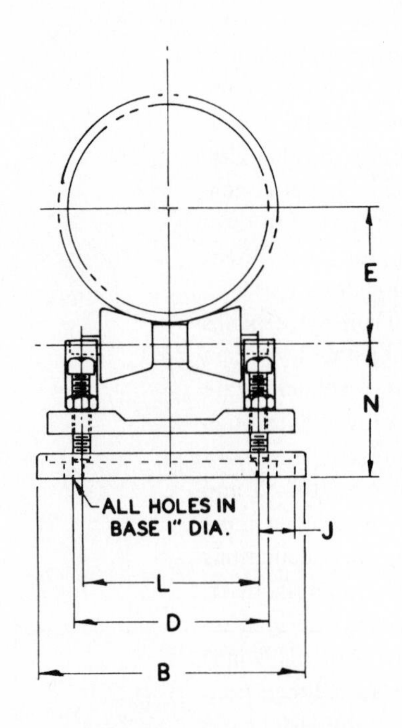

(c) Adjustable Chair and Roll

FIG. 8-5 (cont'd) Pipe roller variations. (Courtesy Kin-Line, Inc., Emeryville, CA)

Figure 8-6 shows the use of a dummy leg welded to the elbow of the 8-in. (20.3-cm)-diameter pipe and supported on item 2, which is a variable-load spring support. In Fig. 8-6, the plate at the bottom of the dummy leg is not welded to the spring support, therefore allowing for a minimum of horizontal movement of a piping system. Dummy legs are used for both critical and noncritical pipe support. To avoid undue load and stress buildup, consult the manufacturer's suggested allowable stress on the elbow and pipeline to which the dummy leg will be attached.

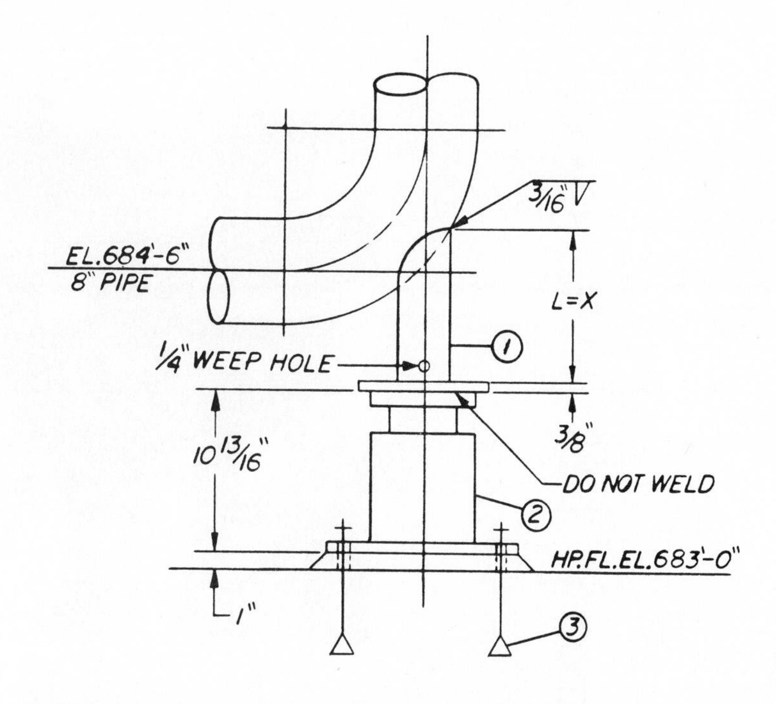

ITEM NO.	NO. REQ'D	FIG NO.	SIZE DESCRIPTION	TW NO.	MAT'L
1	1	HS63	3" STD. WT. PIPE STANCHION, TYPE A	8	PIPE
			FOR 8" PIPE, D= , E=		
			G=3/8" X 5" X 5" (OR SA515GR65)		SA36
2	1	82	7 VARIABLE SUPPORT TYPE F,		SHELL
			HL= 477, CL= 491		
			TRAVEL STOPS & LOAD FLANGE		
3	4	—	3/4" X 7" STUD HILTI KWIK (OR EQUAL)		

FIG. 8-6 Pipe support detail.

Spring Supports

Spring supports are divided into two separate kinds: the *variable-load* spring support and the *constant-load* spring support. The choice between the two depends on a number of variables such as the amount of movement caused by thermal or seismic disturbances, the cost of the hanger, and other design situations which differ for each pipe support installation. All spring supports consist of a spring enclosed in a metal casing, usually fabricated from steel pipe.

Figure 8-7 shows typical applications of variable-load supports. Figure 8-8 shows a constant-load spring support at a conventional power plant. A cheaper alternative to the spring support is the use of a counterweight and pulley mechanism. Spring supports are preengineered and available through commercial catalogs as standard items. They are used in combination with other pipe support devices, such as rod hangers, structural attachments, and trapeze assemblies.

The two categories of variable-load and constant-load spring supports also have subcategories of three basic series for each hanger type. Both variable-load and constant-load spring supports come in short-spring, medium-spring, and double-spring supports. Short-spring supports are used in few situations compared to the other two. They have a heavier spring and approximately 50 percent of the deflection for a given load range. The short-spring type is used where thermal movement is slight. The medium-spring support is the most frequently used type in both variable-load and constant-load supports. The double-spring support closely resembles the medium-spring type except that the mechanism is composed of a large spring with a smaller spring inside it. It allows about 200 percent of the total deflection of a medium spring for the same load range. This spring is used in situations involving considerable movement caused by thermal expansion.

Each manufactured spring support must be used in relation to the amount of movement either required or per-

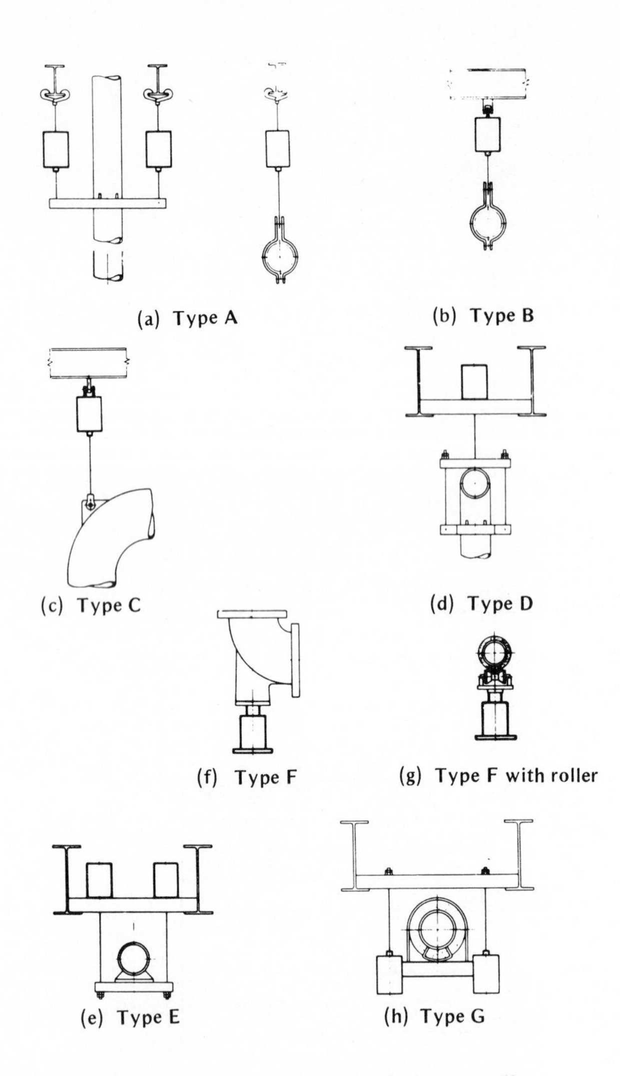

FIG. 8-7 Typical variable-spring pipe hanger applications. (Courtesy ITT Grinnell Corp., Pipe Hanger Div., Emeryville, CA)

FIG. 8-8 Constant-spring pipe hanger installed at a power generation plant.

missible in the design situation, and the series type will depend on the manufacturer's suggested design limits. Figure 8-7 shows the seven basic applications of variable spring hangers available in all three series. Selection depends on whether the pipe run is vertical or horizontal, the amount of thermal and seismic movement encountered, available headroom, and the structure from which the hanger will be hung or supported.

Type A in Fig. 8-7 is used for hanging both vertical and horizontal pipe runs from above. Type-A spring supports are designed for attachment to supporting members by screwing a rod into a tapped hole in the top cap of a hanger to the full depth of the cap. Hanger load is adjusted by turning the coupling or turnbuckle on the lower hanger rod until the hanger picks up the load and the load indicator points to the desired position.

Type B is used with a single lug plate on top for attachment to a structure. The single lug permits the use of a clevis for attachment. This type is used predominantly where headroom is limited.

Type C is usually furnished with two lug plates at the top for attachment to a building structure. The two lugs permit the use of eye rods or a single plate for attachment. This type is also used where headroom is limited.

Type D permits adjustment of the hanger from the top. In this situation, the hanger has a piece of tubing which passes through a hole in the top of the cap. The hanger is set above supporting beams and the pipe is suspended below.

Type E is designed to permit adjustment from either above or below the hanger. It is installed on top of the supporting member and the pipe is suspended below. In most cases, a large turnbuckle is furnished.

Type F is used under base elbows, dummy supports, or pipe that is supported directly from the floor. This type of spring support has a base flange for fastening to the floor or to a structural beam and can be adapted to pipe rollers.

Type E is a complete trapeze assembly similar to type G. Type G consists of two standard spring units in any of the three series plus a pair of back-to-back channels or steel attachments creating an enclosed hanger casing and trapeze arrangement.

Constant-load spring supports are divided into types similar to variable-load supports in their application. The only difference lies in the fact that the spring support is of a constant-load type. Selection of spring supports, constant load and variable load, is determined by the type of suspension problems encountered, the amount of headroom, whether a pipe is supported from above or below the spring, amount of thermal movement, amount of seismic movement, and so forth. Besides these standard applications, it is also possible to fabricate special hangers for unusual pipe support situations, using spring supports.

ITEM NO.	NO. REQ'D	PART NO.	SIZE	DESCRIPTION
1	1	113	#6	WELDED BEAM ATTACHMENT
2	1	VS 1 B 133	#11	VARIABLE SUPPORT, TYPE B, HL–#1301 CL–#1279 W/TRAVEL STOPS
3	2	–	3/4″	HEX NUT
4	1		3/4″ X 5′-0″	ROD
5	1	274	#6	WELDLESS EYE NUT
6	1	304	12″	THREE BOLT PIPE CLAMP
7	1	–	4W = 13	BEAM
8	1	–	7/8″ X 12″ X 12″	C.S. PLATE
9	4	–	3/4″	THREADED ROD
10	4	–	3/4″	HEX NUT

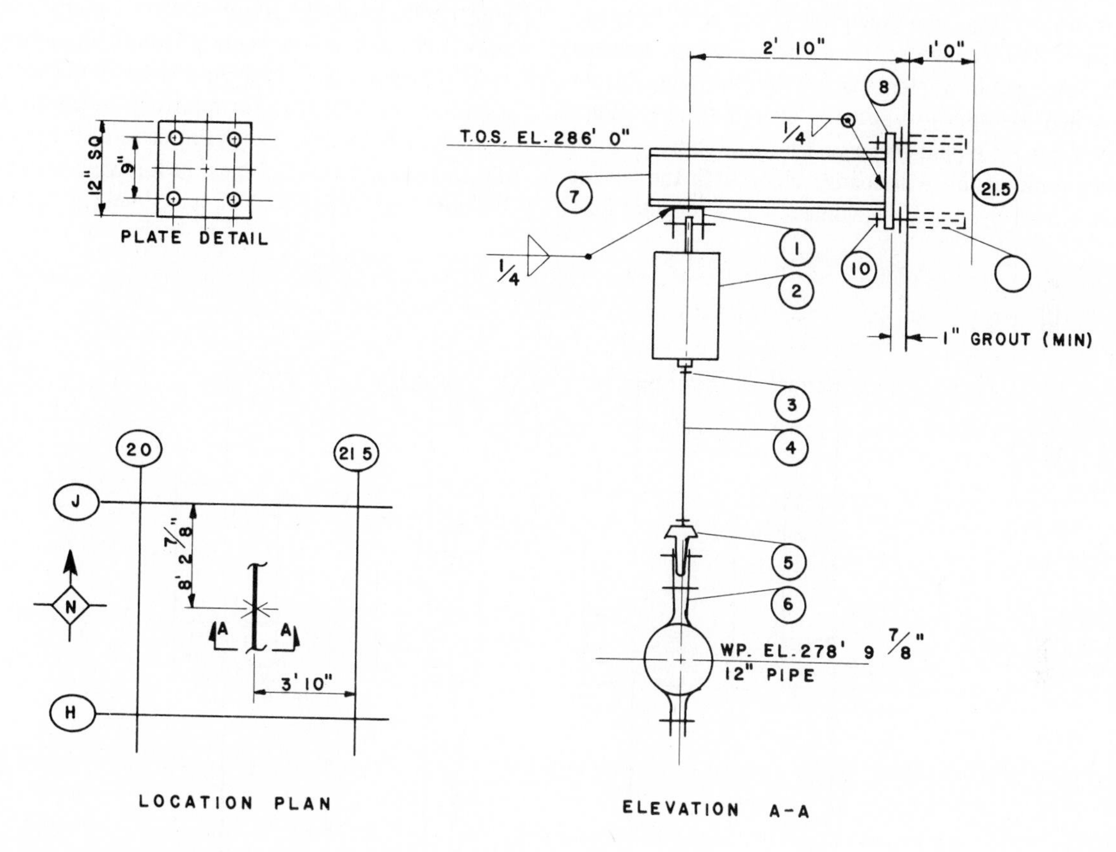

FIG. 8-9 Type B variable pipe support spring used to hang a 12-in.-diameter pipe.

Variable-Load Hangers Variable-load spring hangers are used when it is necessary to carry only the working dead-weight or load of a piping system while allowing the pipeline to move in both horizontal and vertical directions. With variable-load spring supports, deflection (variation of the pipe's vertical position) means there is a variation in the supporting force, which is shown on the spring scale.

This hanger is used for special situations–for example, to avoid overloading vessel nozzles. It is also used in many critical applications where it is referred to as a *flexible support* in contrast to *rigid supports,* such as pipe guides and rod hangers. As has been stated, variable-spring hangers are divided into three basic series, depending on the size of the spring, which determines the amount of travel distance each hanger allows. These three series of hangers meet different suspension requirements for varying situations.

In a majority of cases, variable-load spring hangers are combined with the rod hanger assembly to form a complete spring hanger mechanism. Care must be taken when situating pipe supports, especially spring hangers, so that they are accessible for installation and adjustment.

Figure 8-9 shows a variable support (type B) used with rod hanger and pipe clamp assembly. Selection for variable-spring hangers is determined by the load. The load deflection chart in most pipe support catalogs should be used to determine the spring size. Select a spring size with hot load and cold load inside the working range.

Hot Load and Cold Load All spring hangers work like the spring-type weighing scale. Figure 8-10(a) shows an installed hanger in an unloaded position with the pipe not yet hung from the spring. The deadweight that is carried by the spring after installation is called *cold load* and the system is said to be in *cold-loaded condition* (Fig. 8-10c). When the system is in operation, the pipe moves because of temperature change. The change in spring compression depends upon the thermal movement caused by the temperature change. When the spring support is altered from its cold-loaded position, it is said to be in *hot load;* consequently, the spring compression will change along with the weight and load carried by the spring (unless it is a constant-load spring support). Though it must be understood that the calculated gravitational weight of the system is the same in both the hot and cold position, the load shifts from one support to another but does not change as a whole. Figure 8-10(b) shows the loaded pipe hanger in the hot position. Figure 8-10(c) shows the piping as it moves down from cold to hot position in operation. Figure 8-10(d) shows the cold-loaded piping as it moves up from cold to hot position during operation.

Hot load is greater than cold if the pipe moves downward; cold load is greater than hot if the pipe moves upward. The pipe stress analysis group usually determines the total amount of thermal movement from cold to hot load. One task of the designer in a pipe support division will be to calculate the cold load for resetting the spring for installation and adjustment. The spring scale of the hanger and the deflection of the piping must be known in order to establish the cold load setting for a variable-load spring hanger.

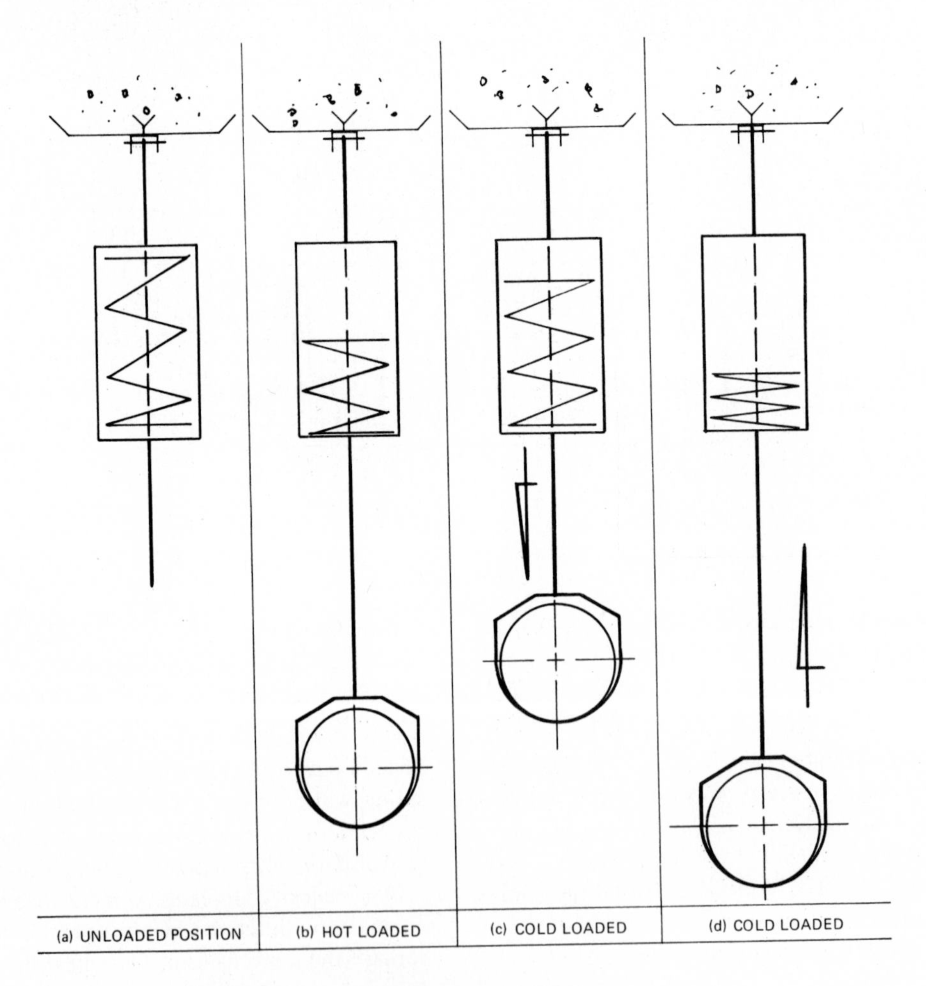

(a) INSTALLED HANGER IN THE UNLOADED POSITION
(b) LOADED PIPE HANGER IN THE HOT POSITION
(c) PIPE IN COLD-LOADED POSITION MOVES DOWN FROM COLD TO HOT POSITION
(d) PIPE IN COLD-LOADED CONDITION MOVES UP FROM COLD TO HOT POSITION

FIG. 8-10 Spring hanger in four different load conditions.

Cold load can be determined by using the following formula. If the pipe moves down from cold to hot, cold load will equal hot load minus the spring constant in pounds per inch times the absolute value of thermal movement of the pipe in inches. If the pipe moves up from cold to hot, cold load equals hot load plus the spring constant in pounds per inch times the absolute value of thermal movement of the pipe in inches:

If pipe moves down from cold to hot:

$$CL = HL - K|\Delta|$$

If pipe moves up from cold to hot:

$$CL = HL + K|\Delta|$$

where CL = cold load in pounds
HL = hot load in pounds
K = spring constant in pounds per inch
$|\Delta|$ = absolute value of thermal movement of pipe in inches

Hot load always equals the deadweight of the pipe as calculated by pipe stress analysis.

With variable-load spring hangers, the load changes as the thermal movement acts on the piping system. In most cases, control over the change of load is necessary to keep the piping system within certain movement limitations. If this is not done, the preexisting load on the spring in the cold position may be too high or too low. When the system is in cold position (not in operation), the variation in support resulting from deflection will be less serious if the hanger being installed is adjusted to support the weight in the hot position first. Check the variability of the spring by using the following formula: Variability equals cold load minus hot load divided by hot load times 100 percent (or variability equals required travel times spring constant divided by hot load times 100 percent:

$$V = \frac{CL - HL}{HL} \times 100\%$$

or

$$\frac{\text{required travel} \times \text{spring constant}}{HL} \times 100\%$$

For nuclear systems, the maximum variability generally ranges from 25 to 10 percent; the variability for pipelines is usually 15 percent and for nozzles 10 percent. Consult the designer's allowable maximum variability for the system in question. In the design section of this chapter, a sample problem is provided for calculating the variability and hot and cold load in a typical pipe support arrangement.

Constant-Load Hangers Both constant-load and variable-load spring hangers have essentially the same function, but the load of the constant-spring hanger remains the same throughout its travel range up to a certain point. Constant-load spring hangers are designed with a special spring and lever mechanism which counterbalances the force of movement and fluctuation of the spring. The constant-load spring hanger is used where additional stress due to load fluctuation is unacceptable or where vertical movement of the pipeline is so great that the maximum allowable variability cannot be met with a variable-load spring hanger. The supporting force is the same in the hot and cold positions when constant-load hangers are used. The constant-load spring hanger costs considerably more than the variable-load spring, and it is necessary to provide for a larger installation and operation space.

The constant-load spring hanger is ideally suited for high-temperature piping systems where thermal movement is considerably greater than in other installations. When using a constant-load spring hanger, minimize the vertical distance between the pipe and the structure supporting the pipe. Because a larger space is needed for the constant-load hanger, an attempt should be made to reduce the length of rods, turnbuckles, and the like.

Constant-load hangers are divided into two types—vertical and horizontal—and vary mainly in their relationship to the building structure member, the pipe, and the method of securing the hanger. In all cases, the travel allowed for a spring hanger should not exceed the designed capabilities of the support. When drawing pipe support details, the hot to cold and cold to hot position is usually indicated on the detailed drawing.

Constant-load hangers enable the system to move according to thermal necessities without transferring loads from one area of the pipeline or its system to another, which may cause undue stress on the system. Constant-load spring supports are designed to withstand a calculated fluctuation in load and spring tension in a variety of pipeline positions determined by the amount of seismic or thermal movement.

Figure 8-11 shows constant-load supports and typical arrangements. These arrangements, which correspond closely to those of Fig. 8-7, are limited to types A, A1, B1, B2, C1, D1, and G. Study both these figures carefully to see the possibilities for placement of standard, constant-load, and variable-load spring support arrangements. These standard applications allow for unlimited design variations, depending on the pipe support situation.

Figure 8-12 is a typical pipe support detail drawing where a horizontal constant-load support is used with a welded eye rod and three-bolt pipe clamp to form the pipe support assembly. Take note of the dimensioning and layout applications on all the pipe support details provided in this chapter. This detail defines both the hot and cold elevations.

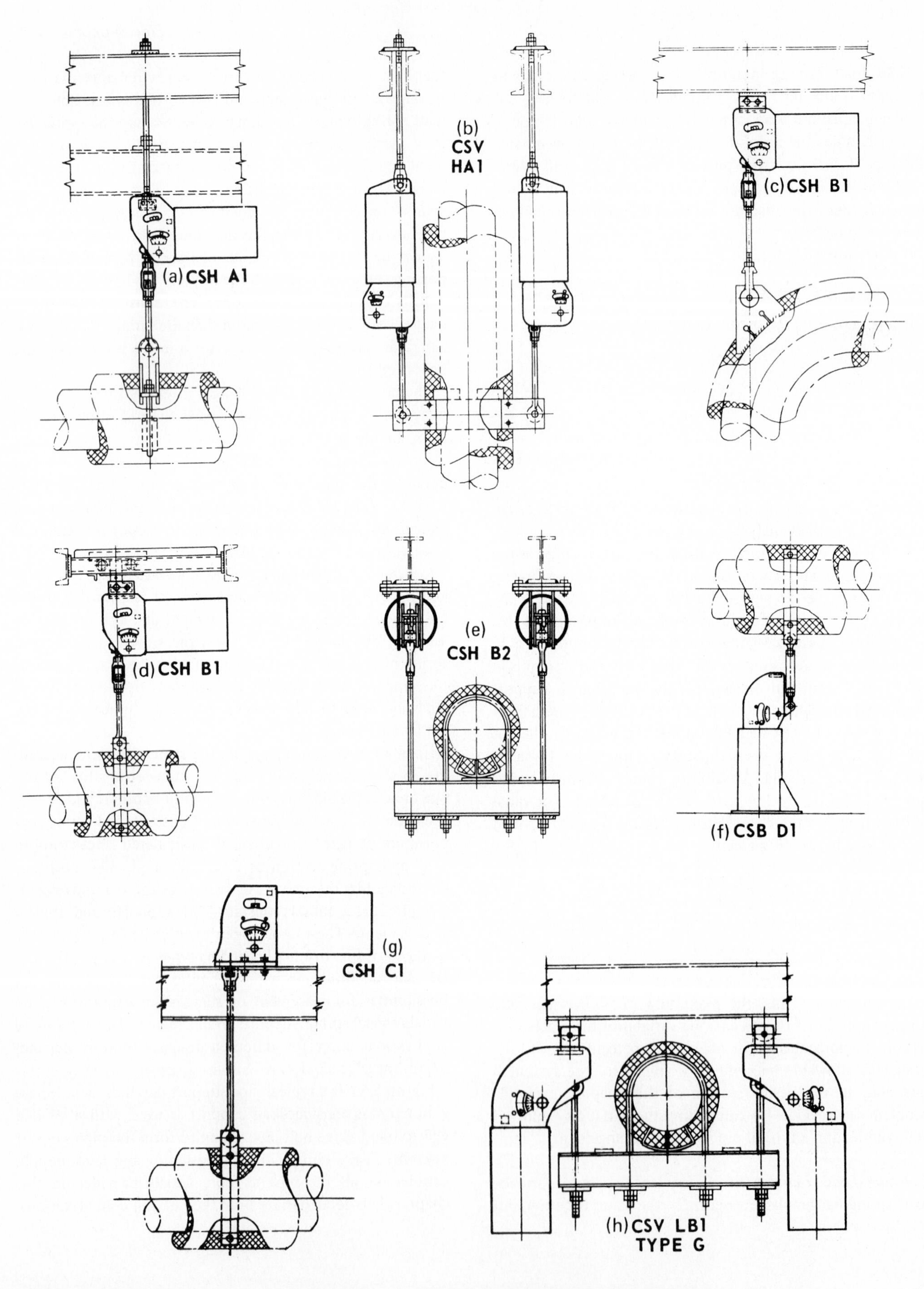

FIG. 8-11 Constant supports—typical arrangements. (Courtesy Bergen-Patterson Corp., A Massachusetts Corp.)

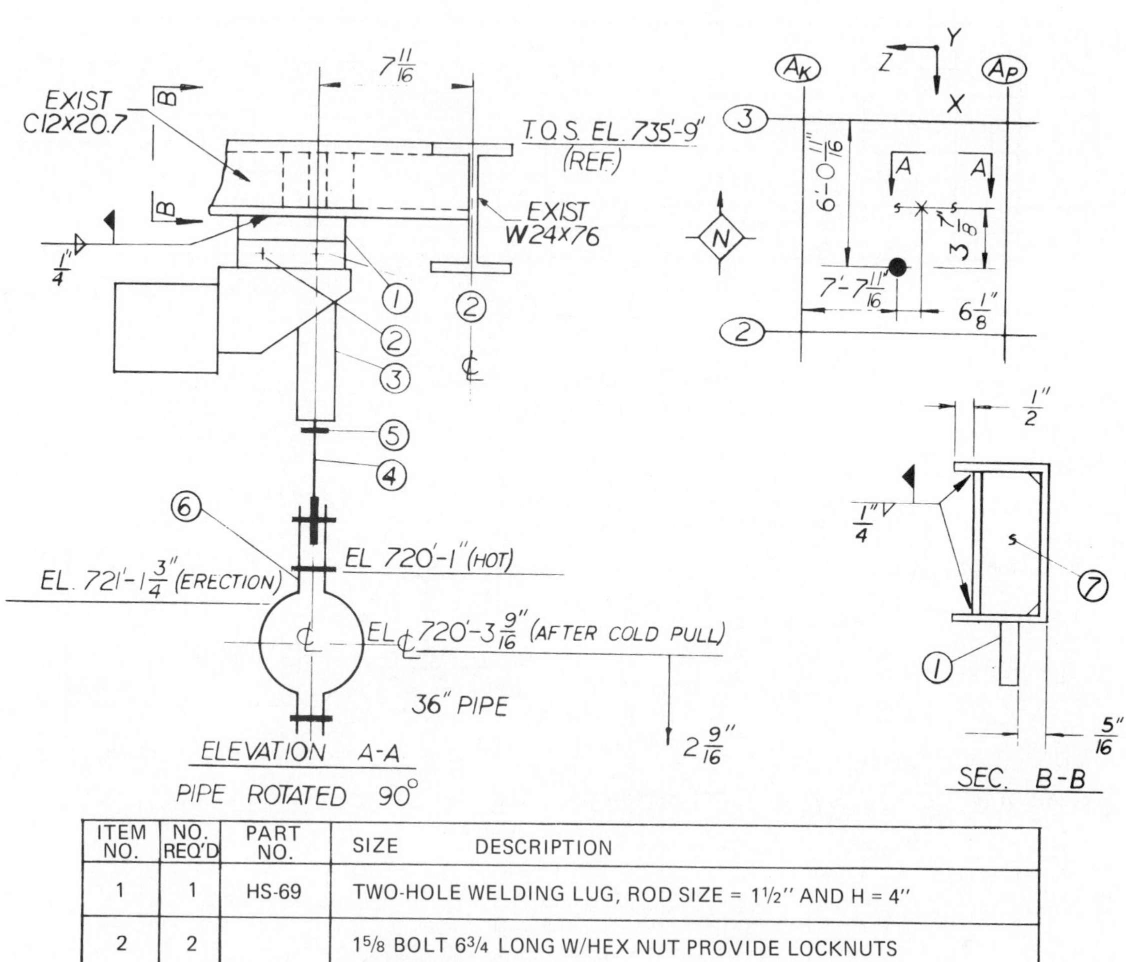

ITEM NO.	NO. REQ'D	PART NO.	SIZE DESCRIPTION
1	1	HS-69	TWO-HOLE WELDING LUG, ROD SIZE = 1½" AND H = 4"
2	2		1⅝ BOLT 6¾ LONG W/HEX NUT PROVIDE LOCKNUTS
3	1	81-H	51 HORIZONTAL TYPE CONSTANT SUPPORT W/18 TURN B
			SUPPORTED LOAD = 8200#
			COLD TO HOT TRAVEL = 2 9/16
			DIRECTION OF TRAVEL = DOWN
			TOTAL TRAVEL RANGE = 4
4	1	278	1½ WELDED EYE ROD W/20 RHT × 8-5 LG.
5	1		1½ HEX NUT
6	1	HS41	36" THREE-BOLT PIPE CLAMP DL/HC = 79¾" PROVIDE LOCKNUTS
7	1		WT 6 × 25 STRUCT'L. TEE × 10⅞" LG. (CUT STEM TO SUIT)

FIG. 8-12 Pipe support detail using a constant-spring support.

Struts and Braces

Struts are rigid rods of adjustable length. They are similar to rod hangers except that they are attached to the pipe and to the structure by means of lubricated guides which allow motion in a single plane. *Braces* are similar to struts except that they are spring-loaded devices and therefore of variable length during operation, depending on the applied load due to temperature variations. The compression in the spring increases if the pipe moves to either side of a selected control position. Spring-loaded devices have much greater vibration control than rigid devices.

Swing struts (*sway struts*) or *sway braces* are used to dampen vibration in the piping system and counteract pipe movement which may adversely affect the piping system. The swing strut is different from a rod hanger in that it resists not only tension but also compression. It carries loads only in the axial direction while providing a limited amount of restriction to the pipe in any direction perpendicular to the axis within certain limits.

ITEM NO.	NO. REQ'D	FIG. NO.	SIZE	DESCRIPTION
1	1	211	#2	SWAY STRUT ASSY., PIPE O.D. = 4-1/2"
				W = 0'-10-7/8", LOAD = 10,000#
2	1	—	P 3/4" X 12" X 12"	
3	1	—	W 8 X 31 X 3'-0" LG.	
4	1	—	W 8 X 31 X 3'-4" LG.	
5	2	—	P 1/2" X 3-7/8" X 7-1/16"	
6	4	—	∠ 3" X 3" X 3/8" X 8-13/16"	

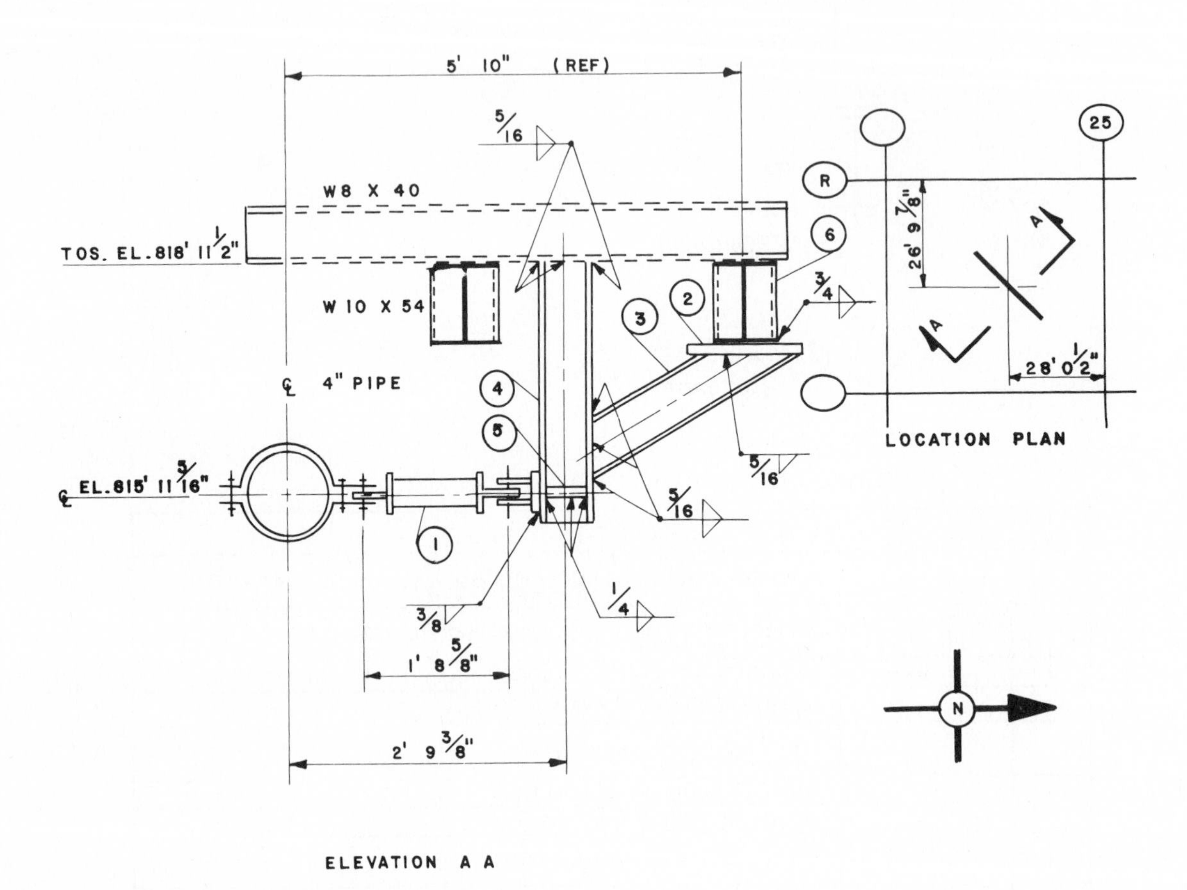

FIG. 8-13 Sway strut assembly.

Figure 8-13 shows a sway strut assembly. In general, the strut assembly consists of brackets, rods, and pipe clamps. Swing struts can be used for vertical, horizontal, or inclined pipe supports. Moreover, two struts may be used together to form a support to restrain lateral pipe movement. In selecting the proper swing strut, consider the following design requirements:

1. Be sure that the strut diameter is large enough to be well within the design load range. The distance between the structural steel or embedment to the centerline of the pipe must be determined and then subtracted from the takeout distance for the pipe clamp to be installed and the rear bracket for the selected size. This will give the required center-to-center dimension between the rear bracket centerline and the pipe clamp centerline. The required center-to-center dimension is the catalog dimension of the strut or brace and must be checked against the allowable maximum and minimum for the swing strut selected.
2. Check the swing angle for pipe movement.
3. Determine the required length of the extension piece by subtracting two times the distance from the end of the extension piece to the centerline of the pipe clamp clevis.

There are numerous differences in structural attachment, and modification must be made for them depending upon the swing struts available on the market.

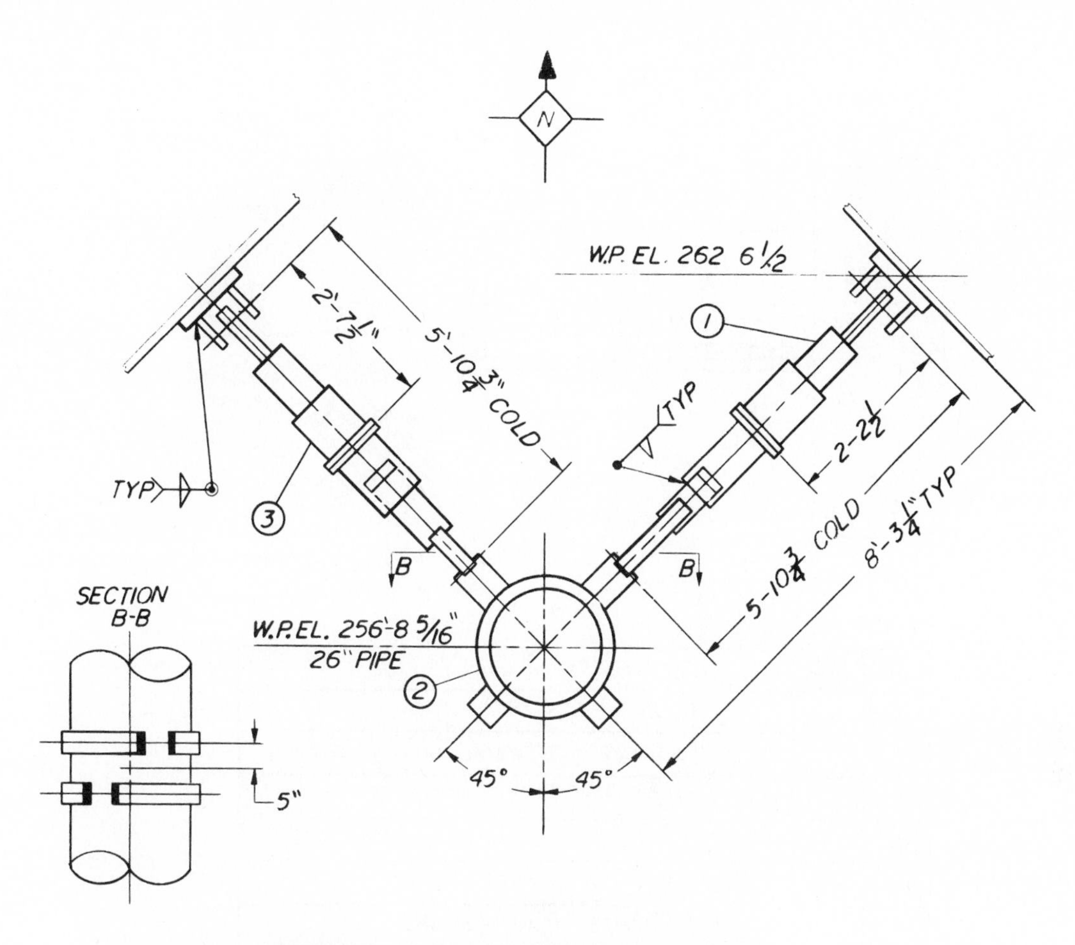

VERTICAL EAST-WEST SNUBBER

ITEM NO.	NO. REQD	PART NO.	SIZE, DESCRIPTION
1	1	2540-50	MECHANICAL SHOCK ARRESTOR
			STROKE = 10" SET = 2½"
			THERMAL MTV (T) = 4 15/16"
			L=2-2½" PIN TO PIN = 5-10¾
2	2	2640-50	26" PIPE ATTACHMENT (MAT'L. C.S.)
			FOR 2540-50
3	1	2540-50	MECHANICAL SHOCK ARRESTOR
			STROKE = 10 THERMAL MTV (C) = 4 15/16
			L=2-7½" PIN TO PIN 5-10¾" SET 7½

FIG. 8-14 Pipe support detail showing the use of two mechanical shock arrestors (snubbers).

Snubbers

There are two kinds of snubbers: the *hydraulic snubber* and the *mechanical snubber.* Both are used for protection of piping equipment that is subject to shock loading, swaying, or vibration. All shock suppressors or snubbers are used to resist shock forces which may cause damaging motion in piping and related systems, including pipe supports.

Shock may come about through a variety of means—water-hammer impact, seismic events, sudden temperature changes (thermal shock), or violent movements of the pipe system. Snubbers are designed to permit the slow movement of the pipeline as in normal thermal expansion and contraction. In this way, the piping system is protected from overstressing in normal operation. Shock suppressors also enable the system to resist wind damage in outdoor piping installations.

Snubbers resist axial or lateral shock movements. One or more snubbers may be used at a support point. When two snubbers are used, as in Fig. 8-14, it is important to select them with matched performance ratings. Hydraulic snubbers are velocity-sensitive—that is, the snubber reacts as a rigid restraint when the pipe's movement exceeds a certain velocity. Mechanical snubbers are acceleration-sensitive—they restrain the pipe when its acceleration exceeds 2 percent of gravitational acceleration. Both types are used to transfer sharp displacement forces imposed on the piping or equipment directly to the structure to which it is attached, at the

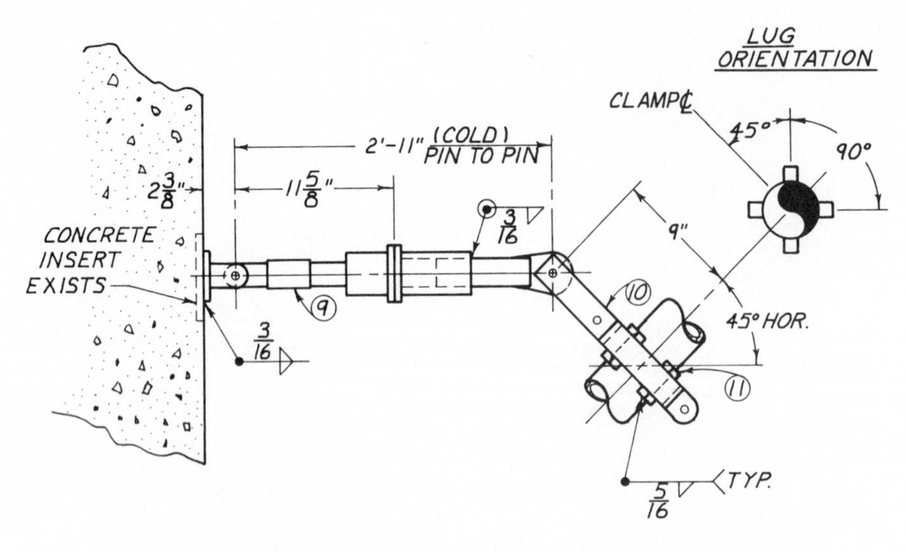

ITEM NO.	NO. REQ'D	DESCRIPTION
9	1	SHOCK ARRESTOR
		5" STROKE, THERMAL MVT. $\frac{1}{8}$" (C)
		L = 11$\frac{5}{8}$", PIN TO PIN = 2'-11"
10	1	20" PIPE ATTACHMENT (MAT'L. C.R.S.)
		WITH LOCK NUTS
11		PIPE LUGS 1" X 1" X 3" C.R.S.

FIG. 8-15 Pipe detail of snubber assembly.

instant of shock, while at all other times the piping or equipment is free to move in its normal operating range.*

The minimum length and the range of length must be calculated to determine a complete snubber assembly. The minimum-length setup generally uses a rear bracket welded to the structural member, followed by the snubber and the forward bracket, then a pipe clamp or another bracket for attachment to the pipe. Figure 8-15 shows the typical mechanical shock arrestor application using a concrete-insert shock arrestor, pipe attachments, and nuts. The pin-to-pin dimension is given in this arrangement (see page 280 for a discussion of snubber selection).

Anchors

Anchors provide rigid support and restraint, thereby preventing any thermal and mechanical movement either translational or rotational. They are used wherever pipeline movement may adversely affect surrounding equipment such as nozzle connections to vessels.

Restraints are different from anchors in that they prevent the pipe from moving in one way, two ways, or three ways, depending on the type of restraint. Restraints are often referred to as north-south, east-west, vertical, or combination. In the restrained directions, the restraint will be subjected to forces from the pipe load. In order to allow the pipe to move in a nonrestrained direction, clearance of 1/16 in. or more is provided between the pipe and restraining member, as in Figs. 8-16 and 8-17.

Pipe Shoes, Slides, and Guides

Figure 8-18 demonstrates example situations in which pipe shoes are used to connect the pipeline to the shoe base. In general, shoes are not used on uninsulated pipes. In many cases, the shoe is welded directly to the pipeline; sometimes the shoe must be bolted or strapped, however. Figure 8-18 shows only welded shoes. Welding directly to the pipeline may cause a problem with the pipe stress allowance. Shoes are sometimes used for sloping runs where the lines must be adequately drained, such as steam pipe systems or pipe racks.

Type 1 in Fig. 8-18 shows a pipe shoe welded to a base plate with a Teflon base enabling the pipe to move if thermal or mechanical movement is necessary. Types 2, 3, 5, and 6 show the use of pipe slides and guides. Type 7 is a more elaborate mechanism showing the pipe shoe welded to a

***The preceding information on snubbers was adapted from a Bergen Patterson Corporation publication.**

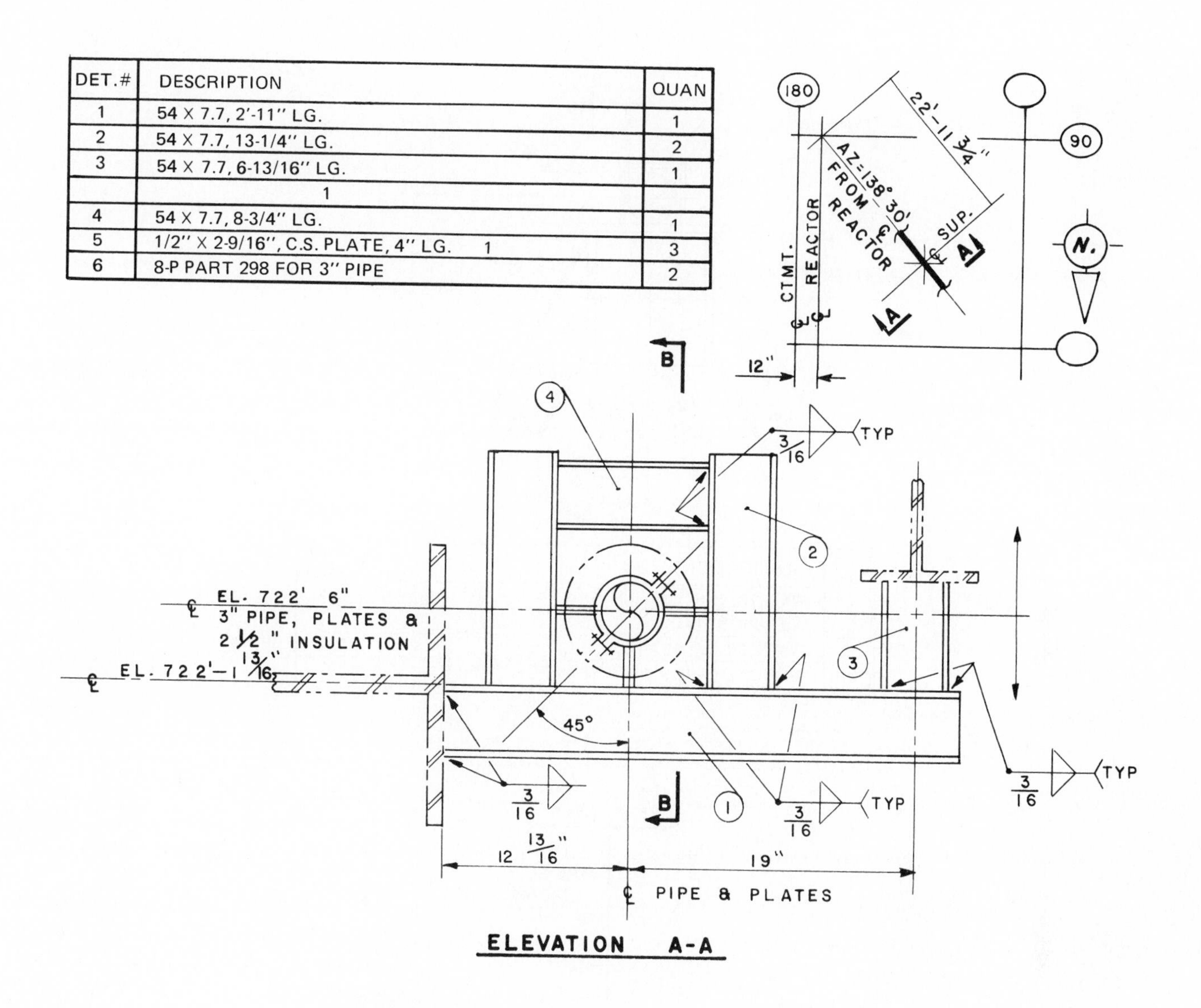

DET.#	DESCRIPTION	QUAN
1	54 X 7.7, 2'-11" LG.	1
2	54 X 7.7, 13-1/4" LG.	2
3	54 X 7.7, 6-13/16" LG.	1
	1	
4	54 X 7.7, 8-3/4" LG.	1
5	1/2" X 2-9/16", C.S. PLATE, 4" LG. 1	3
6	8-P PART 298 FOR 3" PIPE	2

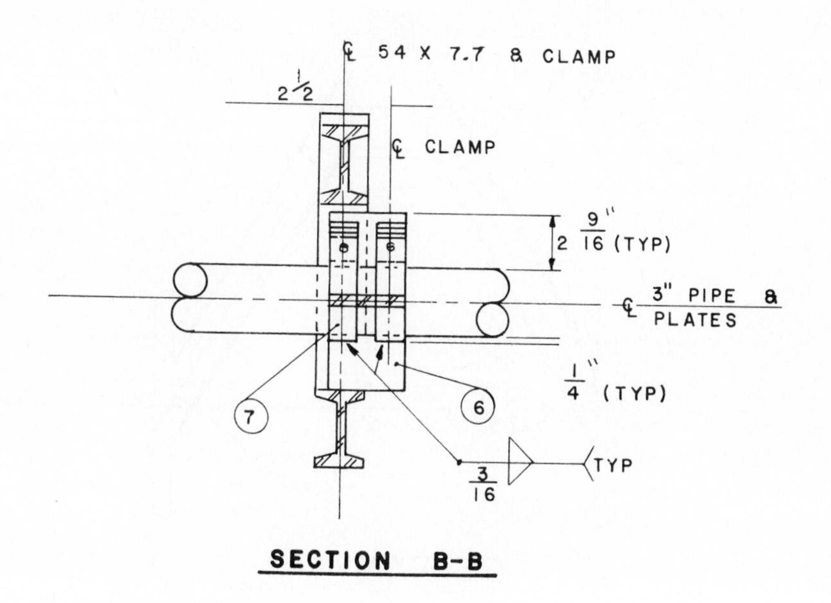

FIG. 8-16 Fabricated pipe guide for nuclear project.

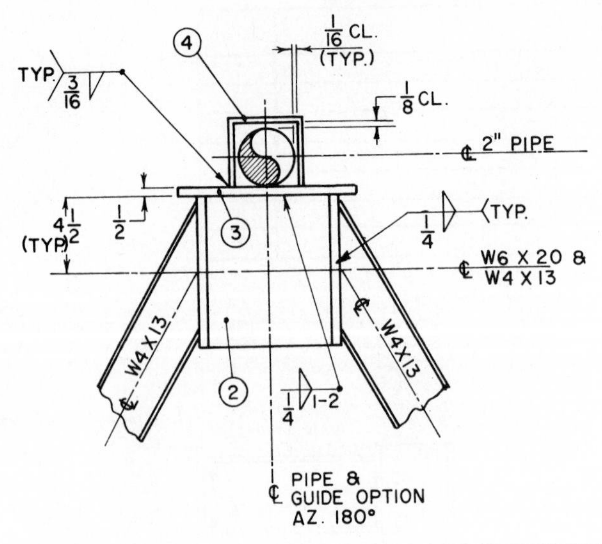

ITEM NO.	DESCRIPTION	NO. REQ'D
1	W4 X 13, 7'-6" LG.	2
2	W6 X 20, 9" LG.	1
3	1/2" X 8" C.S. PLATE, 8" LG.	1
4	TYPICAL GUIDE OPTION #1 FOR 2" PIPE	1

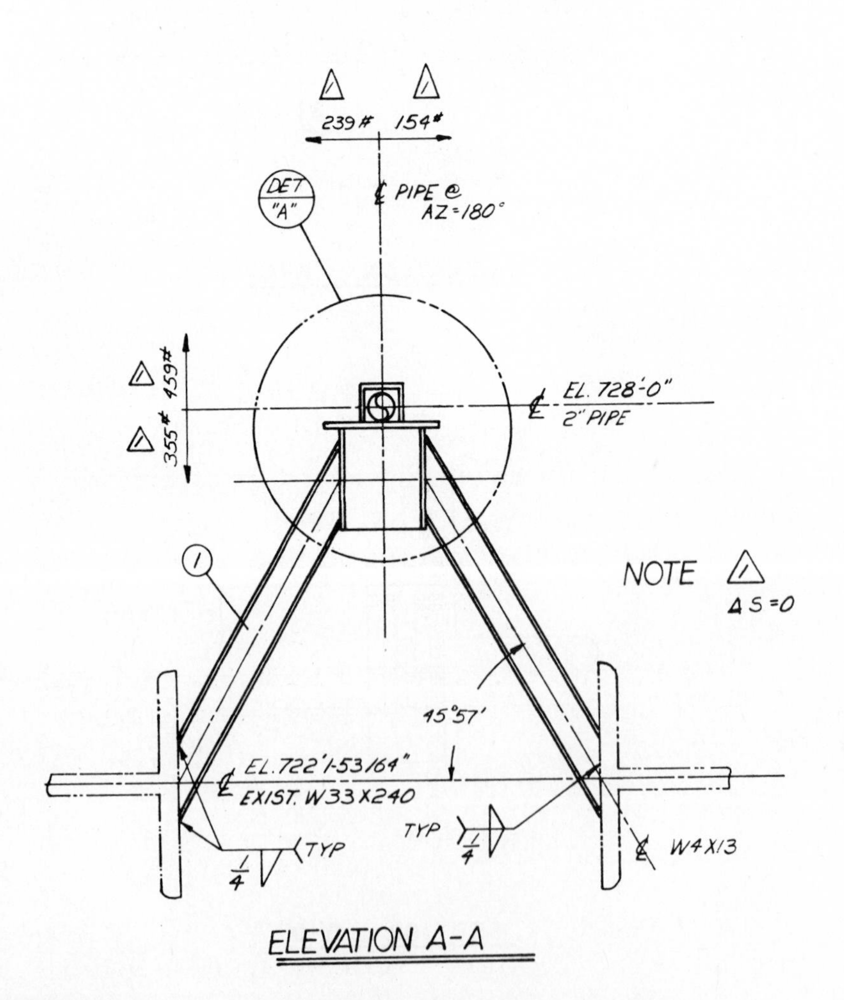

FIG. 8-17 Pipe support detail.

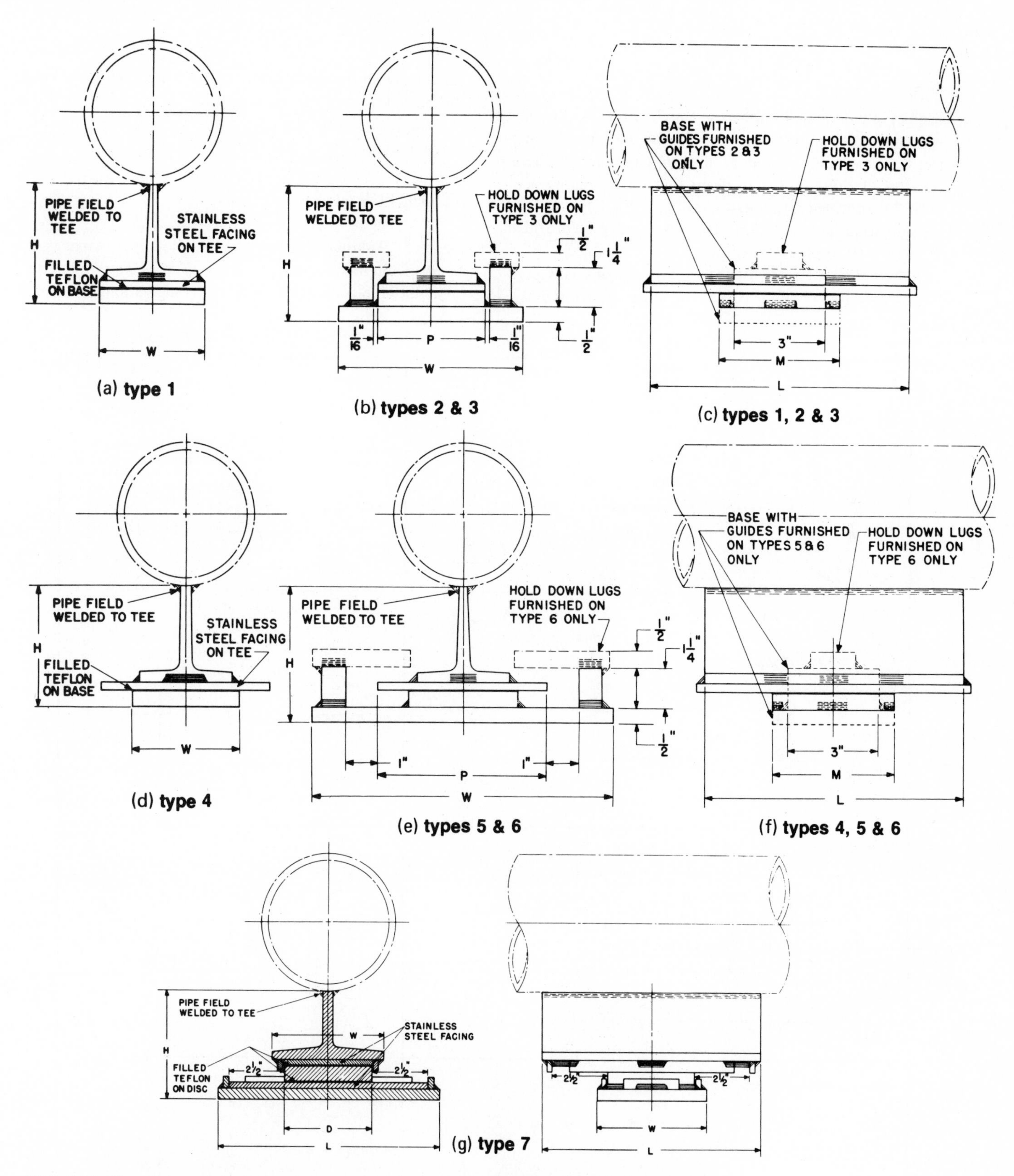

FIG. 8-18 Pipe slide variations using shoes, guides, and slide assemblies. (Courtesy Super Strut, Inc., Oakland, CA)

pipe slide device enabling the movement of pipe in only one direction.

Pipe guides are used to limit or direct thermal movement in power plants and other piping systems. In many critical situations, nuclear pipe supports (Fig. 8-16) involve the use of pipe guides and restraints.

Saddles

Saddles are used for supporting insulated pipes on pipe racks or pipe supports. Insulated pipelines are usually subject to movements caused by thermal expansion and contraction. On long runs, this means that the pipe movement must be provided for longitudinally without the vertical considerations involved in shorter runs. Saddles are provided in these cases to protect the insulation and pipe itself.

Figure 8-4 shows a few of the saddle applications. Figure 8-7, type G, shows how the variable-spring trapeze assembly, supporting an insulated pipeline, has a pipe saddle to keep the insulation from being crushed. Figure 8-11 (CSV LB1 and CSH B2) shows the use of pipe saddles also. The pipe saddle is usually welded directly to the pipeline.

STRUCTURAL SHAPES

Figure 8-19 shows a variety of structural steel shapes used in pipe support design. Steel is the most common material used for pipe supports because of its high strength and its

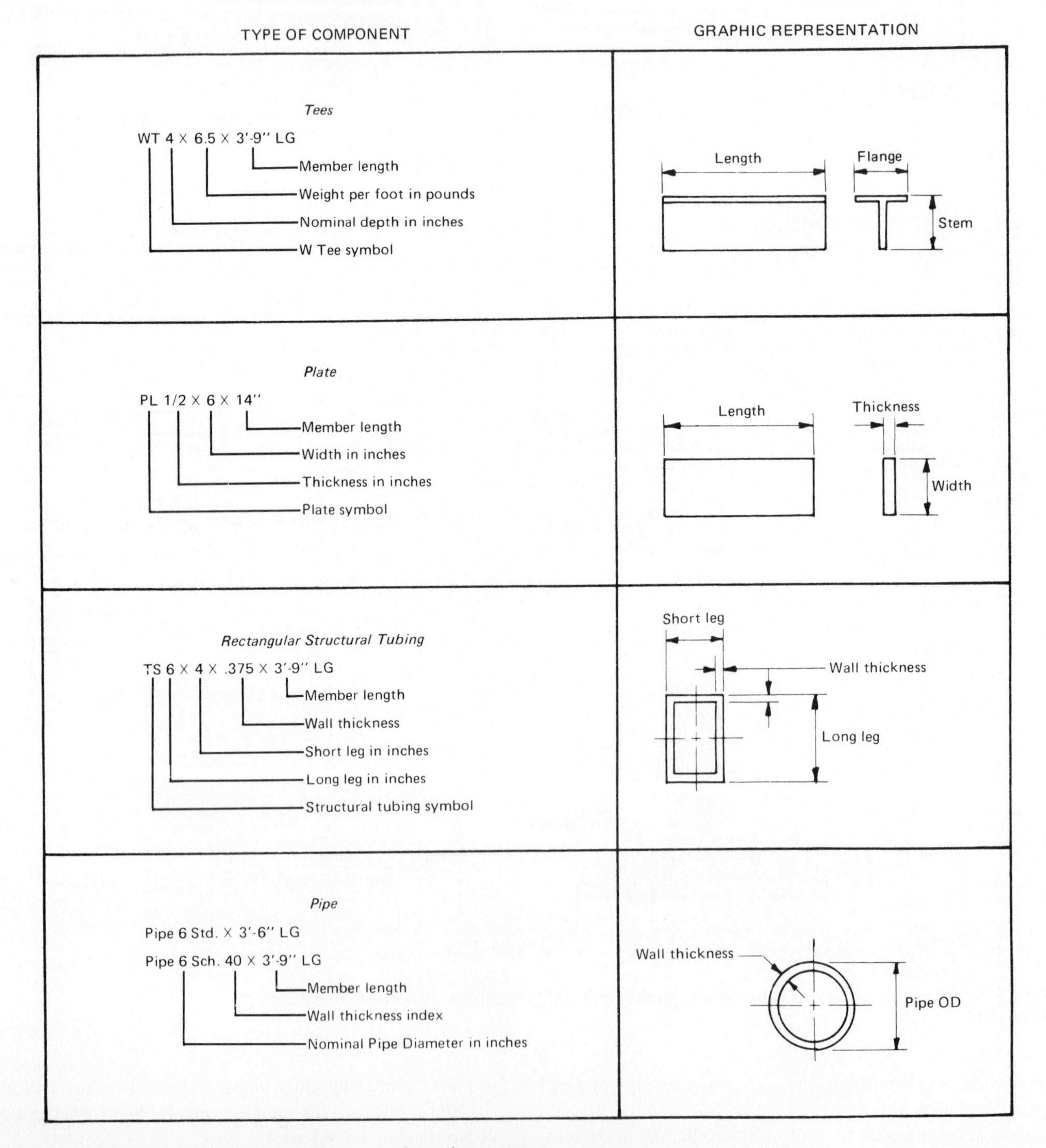

FIG. 8-19 Common structural steel shapes used on pipe supports.

ability to maintain its load-carrying capacity at high temperatures. It is relatively long lasting and is more economical than any other material. All structural shapes are prefabricated by steel manufacturing companies and though the shapes are limited, they come in many sizes.

On pipe support details and in all piping system design and layout, the structural shapes should not be drawn sectioned, though some companies use the method of darkening the whole end section and showing the actual contour, including the fillet in the corner. In general, to speed up the drafting procedure, they should be shown as square corners. Since the majority of shapes, such as tees, angles, and channels, have slight taper angles in the flat stock portions, they should be drawn without tapers to save time. In many cases, dimension size and unit weight of the structural shape will be required in the design and detailing of pipe supports. Pipes are also used for structural shapes (see Chapter 9).

Figure 8-19 shows the common structural steel shapes.

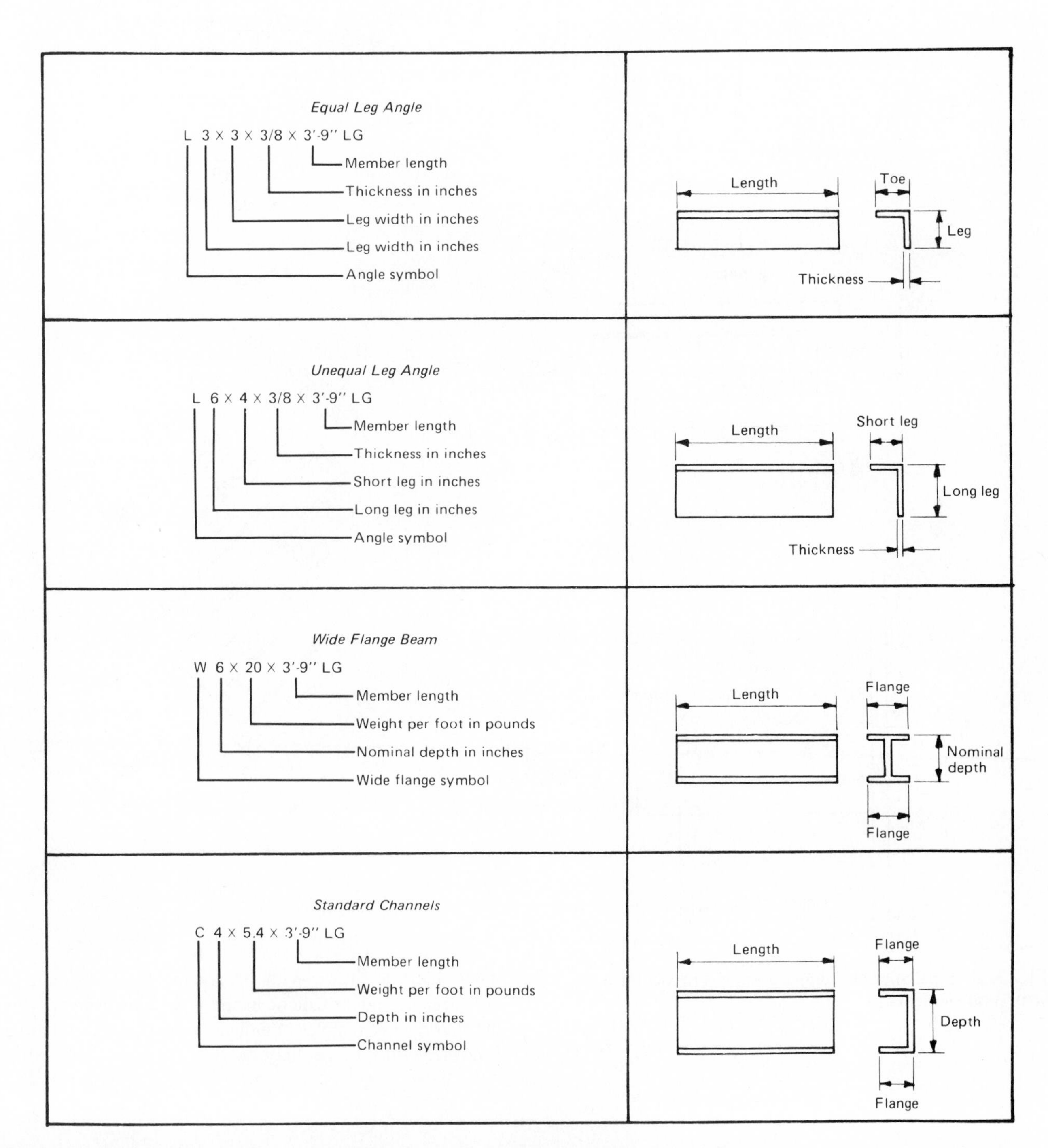

FIG. 8-19 Common structural steel shapes used on pipe supports. (cont'd)

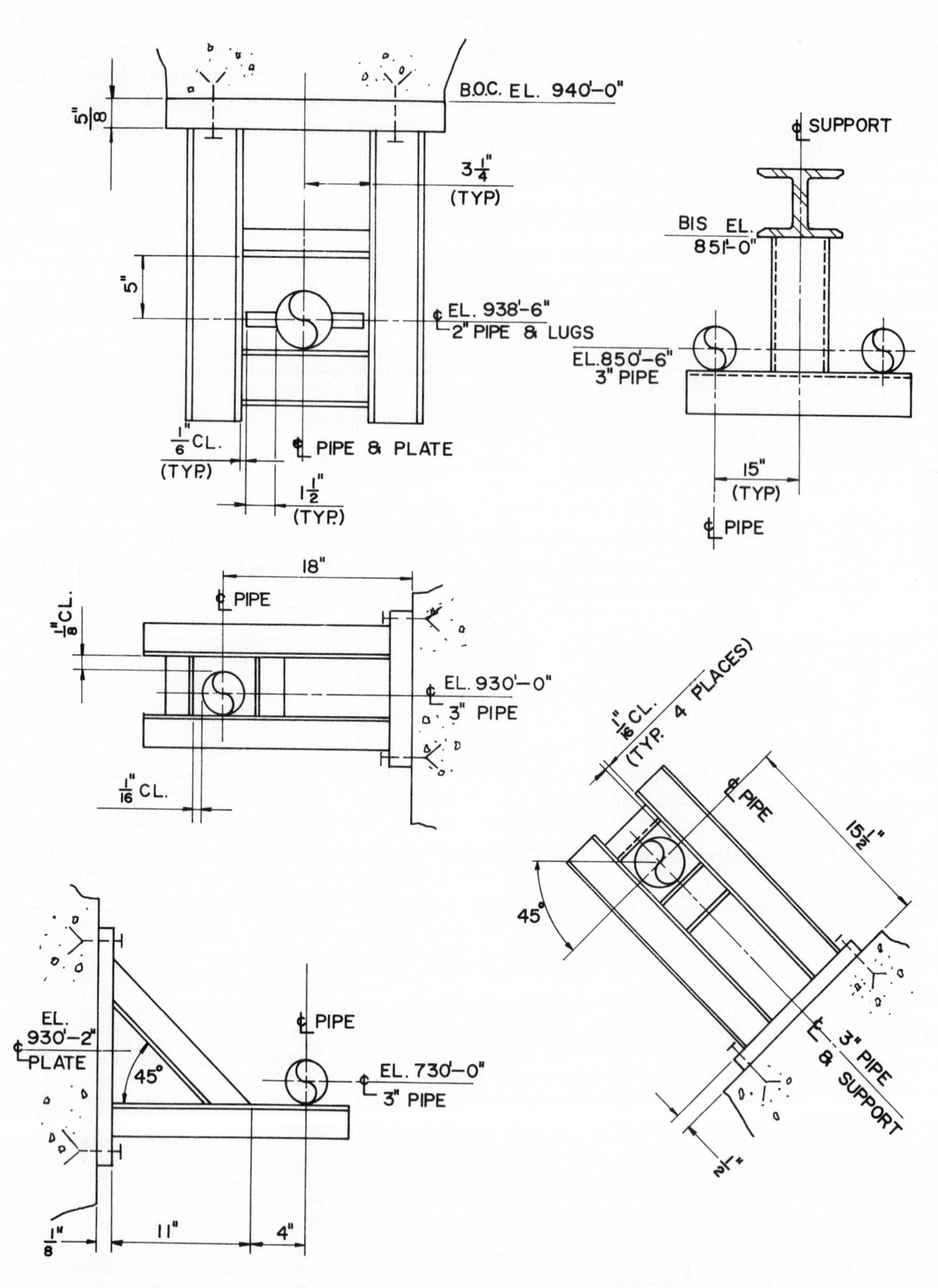

FIG. 8-20 Pipe support dimensioning practices for structural elements.

Each shape is shown with its graphic representation, its designation, and its callout in the pipe support detail. Figure 8-20 gives examples of dimensioning practices for structural elements on pipe supports.

THE PIPE SUPPORT GROUP

In the majority of large engineering companies, the pipe support group is a special section that works closely with the stress analysis department and in conjunction with the piping design department. This separation into a subfield of its own has come about because of the importance of pipe support location and design. In less critical areas, such as residential plumbing, pipe support placement and design is still usually handled by the piping design and layout group.

The pipe support group is responsible for the design and location of pipe supports needed to meet standard specifications and safety requirements. Each project in the pipe support group has its own leader who coordinates the design and detail drawing of the pipe supports. The pipe support group works closely with the stress analysis group, which determines the thermal, mechanical, and vibration problems which might occur in all portions of the projected pipe system. In determining the piping system's movement and potential problems, the stress analysis group uses a computer for many of its calculations.

The pipe support group uses data supplied by other groups, including electrical, HVAC piping, stress analysis, civil, structural, and piping layout groups. The usual design procedure follows a general plan. First the pipe support designer, using the calculations provided by the stress group, determines the logical placement and location of pipe supports throughout the piping system with reference to the structural configuration of the plant, electrical runs, HVAC ducting, and other considerations. The designer then shows these locations on the isometrics of individual pipelines. Standard specifications for the project will determine the overall placement and span requirements between pipe supports for straight runs of pipe. After the locations have been determined and shown on the isometrics, they are transferred to all the drawings that are necessary throughout the drafting groups—such as the piping layout drawings, if these are being used in the design procedure. The placement of supports depends on interference with surrounding equipment and other plant piping runs. An attempt should be made to align the supports with existing framing, walls, columns, and structural members.

The designer then makes a rough sketch of the developed pipe support, providing all the information needed for finished detailing by the pipe support drafter. Remember to keep the design simple and functional: Eliminate any elaborate or expensive variations, and use standard supports and equipment whenever possible.

Within the pipe support group, the checker verifies all the different aspects of the design—interferences, welding procedures, structural member size, adequate load allowances, and so forth. The pipe support drafter then receives the rough sketch showing the required pipe support design and from it will be able to complete a final drawing and bill of materials for ordering.

The many pipe support details in this chapter are examples of the pipe support drafter's finished drawing. A variety of information is shown—such as elevations, rod length, beam length, hot and cold positions, and spring load setting. Some of this information must be calculated by the drafter. The checker then reviews the completeness, accuracy, and clarity of the drawing thoroughly.

BASIC CONSIDERATIONS

The following sections discuss thermal movement, mechanical movement, and spacing of pipe supports—all of which are important considerations in the construction of pipe supports. Pipe support drawings and details are covered later in depth; as are pipe support specifications and design. The drafter need not learn all the design considerations involved in pipe support work since most of the relevant design procedures will be learned through direct experience on the job with people who have had years of experience.

Thermal and Mechanical Movement

Thermal and mechanical movement are major considerations for the adequate placement of pipe supports. The amount of thermal movement in piping systems is calculated with computers by the stress analysis group. These calculations must take into account the total load of the piping system the weight of the pipe and its contents along with all component parts, such as valves, fittings, traps, and insulation. Besides the load of the pipeline and its components, any external loading or stress caused by weather (ice build-up, snow, winds) must also be considered.

The total load that is supported by the various pipe supports and hangers is referred to as the hot load of the piping system. Thermal expansion will in many cases transfer the load from a pipe support to another part of the piping system. The maximum load for pipe supports is given in the catalogs of standard items. The tables provided by the manufacturer must be strictly followed to avoid overloading a pipe support.

There are many ways of securing the pipeline and ensuring that its movement is in the desired direction. Spring hangers provide for both vertical and horizontal movement, as do rod hangers and other such supports. Anchors and guides are used to secure portions of the pipeline, enabling the line to move in a certain direction, guiding it, or anchoring it at certain points. Anchors eliminate movement near nozzles or other connections that may be damaged by thermal movement of the pipeline. The designer must make sure

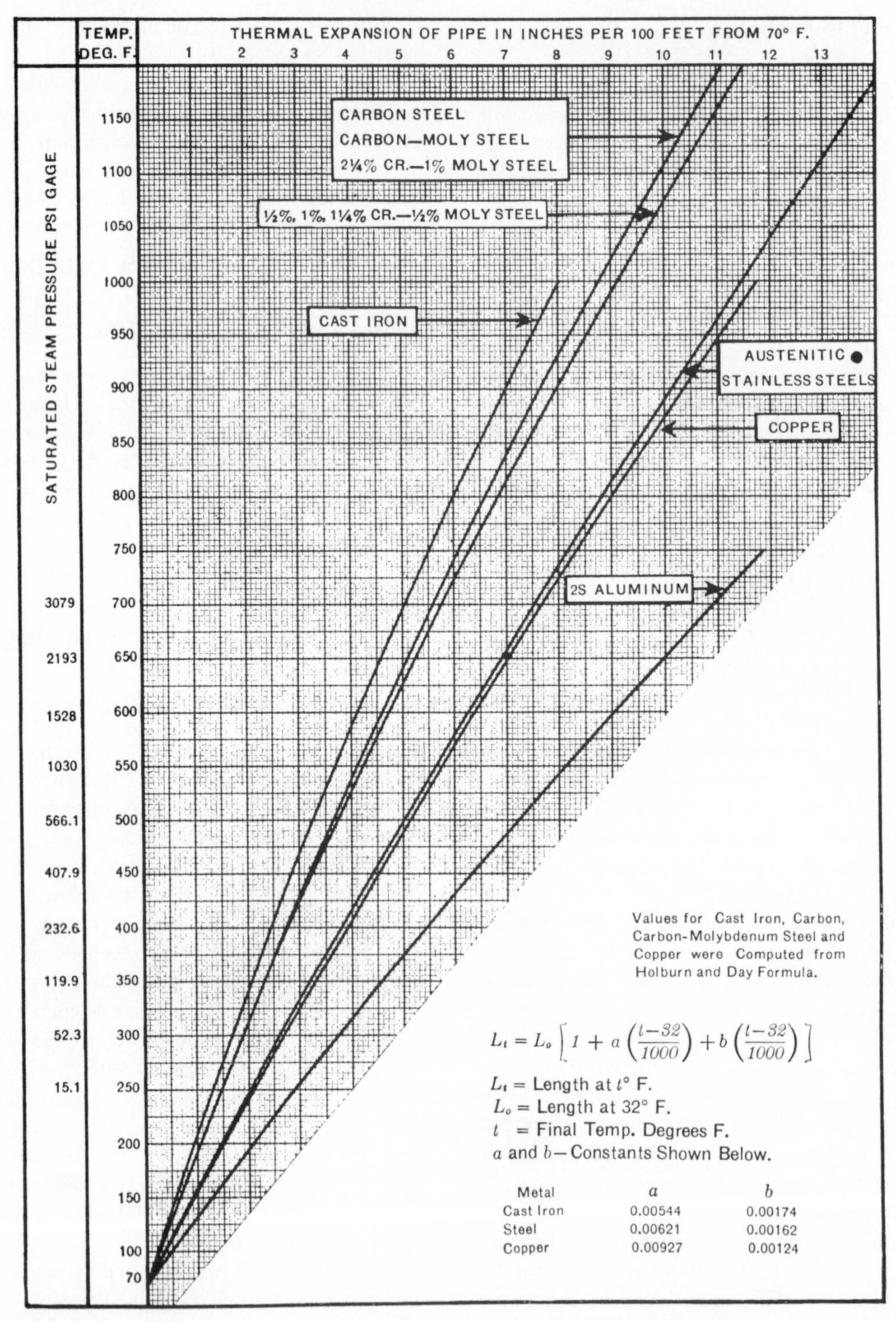

FIG. 8-21 Thermal expansion of pipe. (Courtesy Bergen-Patterson Pipe Support Corp., A Massachusetts Corp.)

that the placement of anchors and guides does not severely restrict the movement of the pipeline and cause undue stress in other areas of the system. In many systems that are subject to drastic thermal, seismic, or mechanical movement, expansion joints are used to transfer stress to the joint instead of to the pipeline and its component parts. All piping systems that are subject to thermal movement must be designed to permit the free movement of piping within specified limits. The placement and design of hanger supports, guides, and restraints must take into consideration movements encountered in that system. The amount of movement possible for each pipe support is usually given in the design information on the standard manufactured item.

Figure 8-21 provides a chart for calculating the thermal expansion of pipe in inches per 100 ft from 70°F. Mechanical movement of piping systems can be caused by equipment vibration, external loading, and seismic activity. Figures 8-22, 8-23, and 8-24 show a variety of seismic bracing possibilities that can be considered when designing piping supports in areas that undergo geological disturbances. All the seismic applications shown in these three figures involve standard pipe support items along with steel channel. These bracing arrangements show both transverse and longitudinal applications for the distribution of seismic movement.

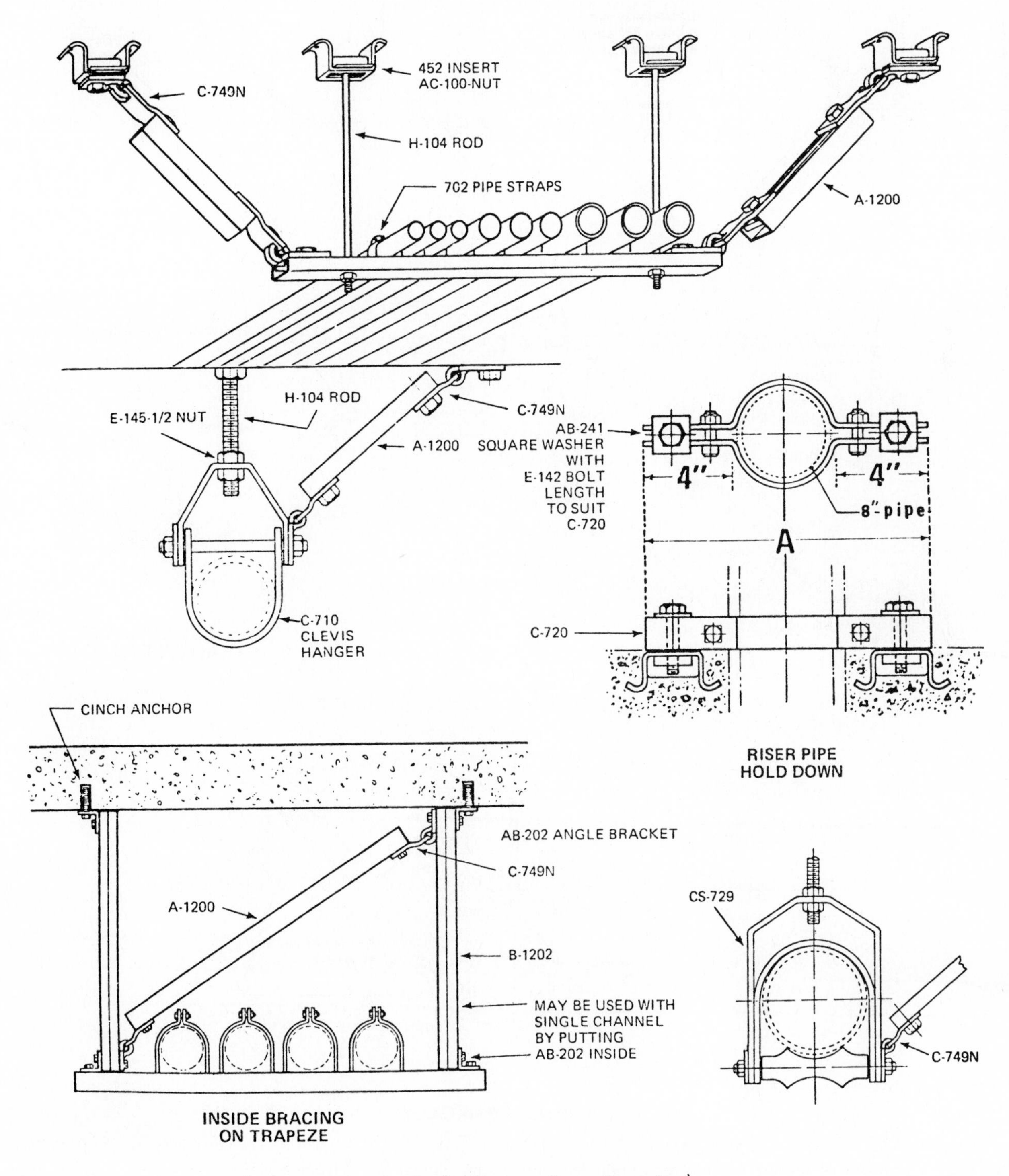

FIG. 8-22 Seismic bracing (not for nuclear services). (Courtesy Super Strut, Inc.)

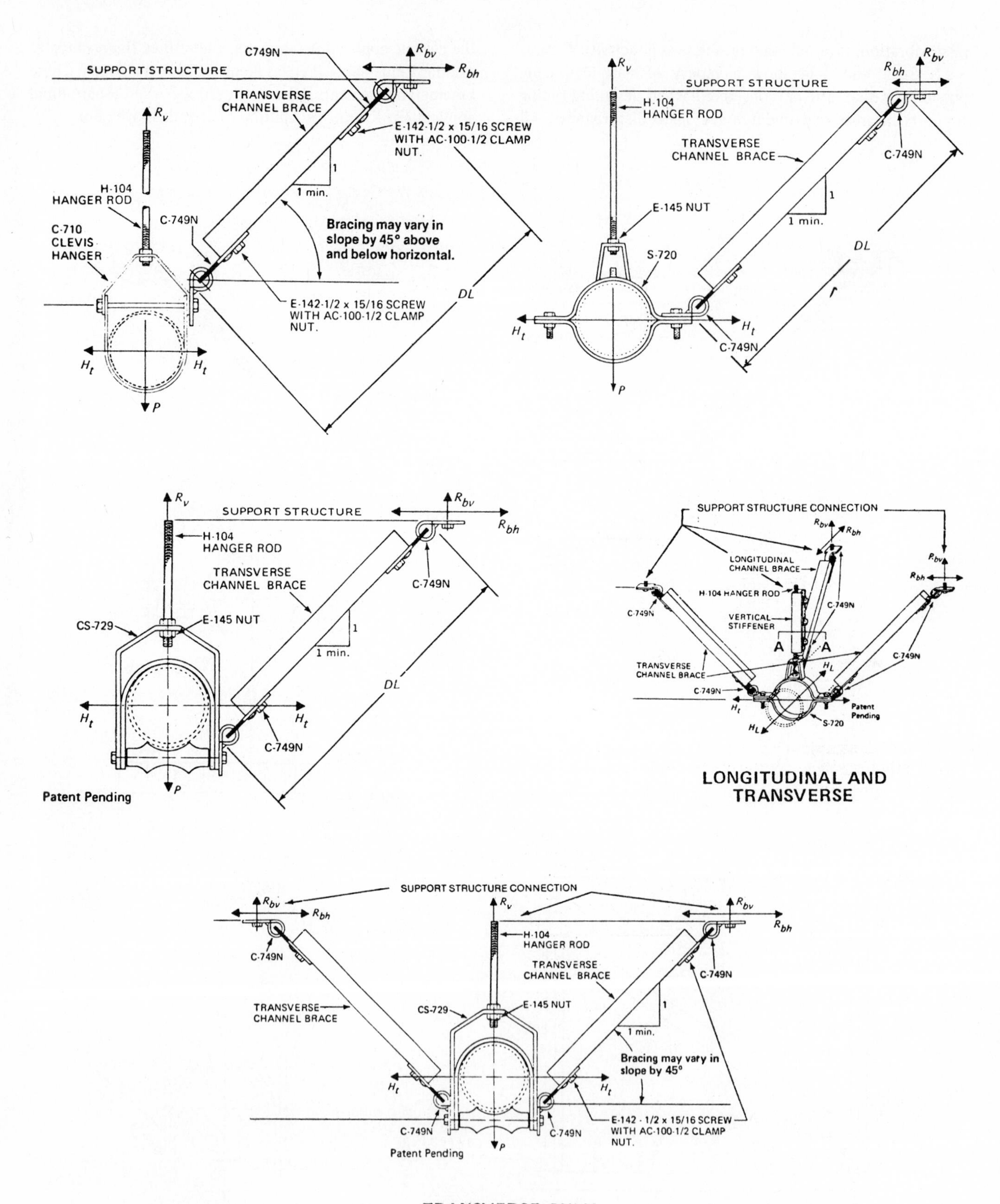

FIG. 8-23 Variations of seismic bracing for nonnuclear pipe supports: (a) 0.5 G force bracing, (b) 1.0 G force bracing. (Courtesy Super Strut, Inc., Oakland, CA)

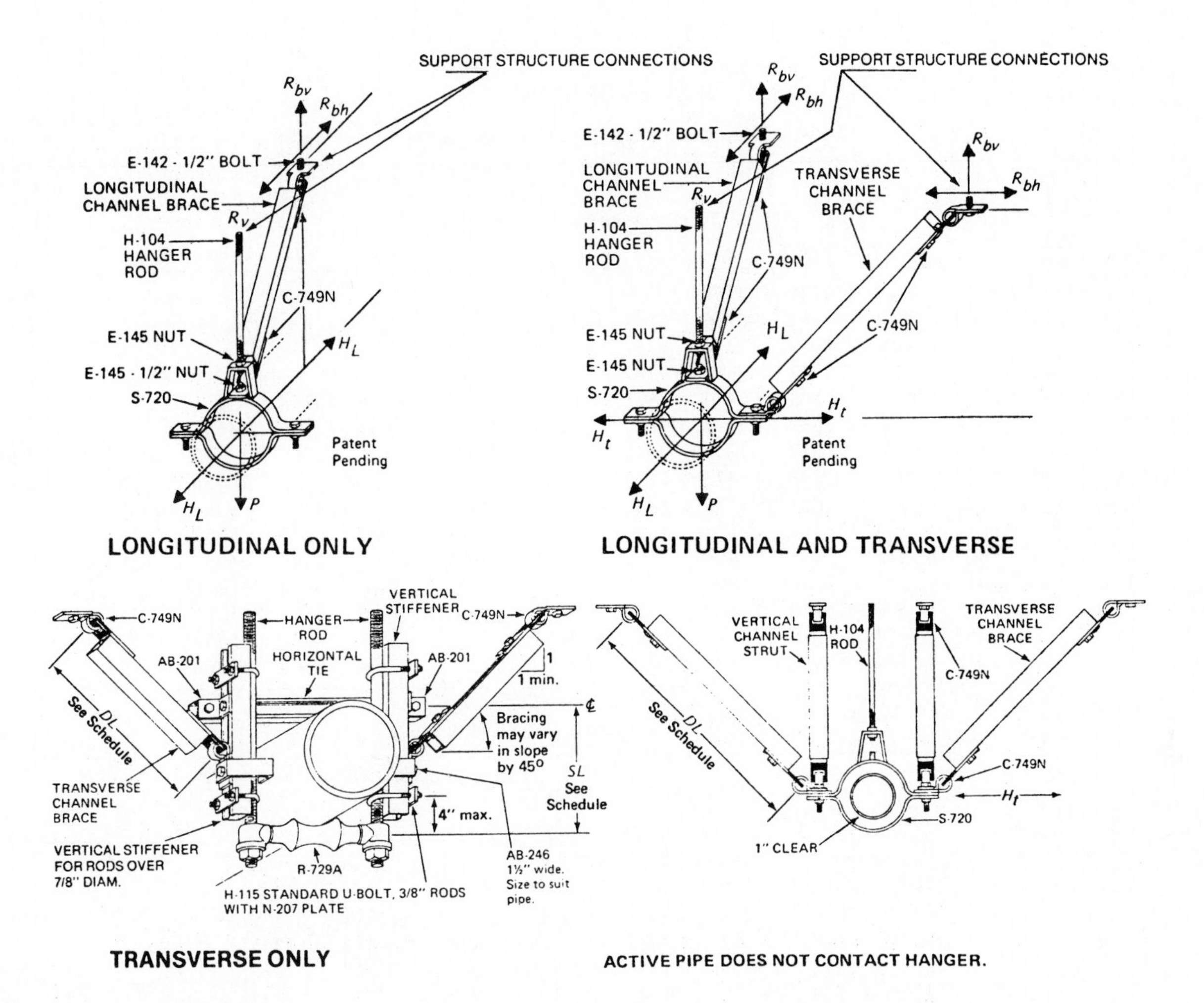

FIG. 8-24 Longitudinal and transverse bracing applications for seismic movement. 0.5 G forces for assemblies: (a) longitudinal only, (b) longitudinal and transverse, (c) transverse only. (Courtesy Super Strut, Inc., Oakland, CA)

Spacing and Spans

All horizontal piping runs and lines are subject to sagging unless the pipeline is continuously supported (as it must be in the case of thin-walled tubing). The total amount of sag in the pipeline depends on the medium to be carried, insulation, and components such as valves and fittings. All these must be taken into account when determining the span between pipe supports. In most cases, pipelines are designed to provide for a slight downward pitch so that areas between supports will not create pockets for the accumulation of line medium or condensation as in steam lines. It is important to create an adequate downward pitch so that the outlet of each span is at a lower elevation than the preceding sag that may be encountered for the length of pipe.

The designer should consult the manufacturer's suggested spacing for pipeline size and contents. A chart for the deflection of horizontal pipeline is usually supplied. Table 8-1 provides charts for the spacing of pipe supports. They are calculated for stresses of standard weight pipes. Table 8-1(a) shows the stress due to sag with a pipe filled with water; Table 8-1(b) shows the stress due to sag with the pipe empty.

Pipe racks are used where a number of piping runs lie parallel to each other and can thus be supported by the same supports. Pipe racks, which are covered in detail in Chapter 10, are used primarily for outdoor piping runs where there are no surrounding buildings or structural elements that might be used to support the pipeline. In buildings such as conventional power plants, piping is routed with consideration of all the equipment and other systems, such as electrical runs and HVAC. In all cases, piping within a structure must be connected to the existing columns, beams, walls, ceilings, and floors for primary pipe support. From these primary connections, subconnections of structural steel and other items can be conveniently added to provide adequate support. Small lines 2 in. (5.08 cm) and less in diameter are usually left for the construction crew to locate and support in the field unless the piping is critical.

When placing pipe supports and determining the span between each pipe support location, a variety of items must

TABLE 8-1 SPACING OF PIPE SUPPORTS
STRESSES CALCULATED FOR STANDARD WEIGHT PIPE

(a) Stress Due to Sag—Pipe Filled with Water

Pipe Size (in.)	*Span Between Supports (ft)*													*Pounds of Water per Lineal Foot*
	10	*12*	*14*	*16*	*18*	*20*	*24*	*30*	*36*	*42*	*48*	*54*	*60*	
½	2387	3438	4680	6113	7736	9551								0.13164
1	1531	2205	3002	3921	4962	6126	8822	13785						0.37345
1½	1342	1932	2630	3436	4349	5369	7731	12081						0.88260
2	903	1301	1771	2313	2927	3614	5205	8133	11711	15941				1.4541
3	620	892	1215	1587	2008	2480	3571	5580	8035	10936	14284			3.0032
4	503	725	986	1288	1631	2013	2900	4531	6525	8881	11600	14681		5.5172
5	424	610	831	1085	1373	1696	2442	3816	5495	7480	9769	12365	15265	8.6666
6	368	530	721	942	1192	1472	2120	3313	4771	6495	8483	10736	13255	12.530
8	319	459	625	816	1033	1276	1837	2871	4135	5628	7351	9304	11487	22.206
10	283	407	554	724	917	1132	1630	2547	3668	4993	6522	8254	10191	35.454
12	238	343	467	610	772	954	1374	2146	3091	4207	5496	6955	8587	49.760
14 OD	213	308	419	547	701	855	1232	1925	2772	3774	4929	6239	7702	60.000
16 OD	197	284	387	506	640	790	1138	1778	2561	3486	4554	5763	7115	79.187
18 OD	181	261	355	464	588	726	1045	1633	2353	3201	4182	5292	6534	100.48
20 OD	173	250	340	444	562	694	1000	1563	2251	3064	4003	5066	6254	125.30

(b) Stress Due to Sag—Pipe Empty

Pipe Size (in.)	*Span Between Supports (ft)*												
	10	*12*	*14*	*16*	*18*	*20*	*24*	*30*	*36*	*42*	*48*	*54*	*60*
½	2063	2971	4044	5282	6685	8254	11886						
1	1249	1799	2449	3199	4049	4998	7198	11247					
1½	1010	1454	1979	2586	3273	4040	5818	9091	13092				
2	644	977	1262	1649	2087	2577	3711	5798	8350	11365	14844		
3	434	625	851	1111	1406	1736	2500	3907	5627	7659	10003	12661	15631
4	331	477	650	849	1075	1327	1911	2986	4390	5853	7644	9675	11945
5	265	381	519	678	858	1060	1526	2385	3435	4675	6106	7728	9541
6	220	317	432	656	667	823	1186	1853	2669	3893	5085	6436	7945
8	167	240	327	427	541	668	962	1503	2165	2998	3915	4956	6118
10	131	189	258	337	426	526	758	1185	1706	2323	3034	3840	4741
12	110	159	217	283	359	443	638	997	1436	1955	2554	3232	3990
14 OD	101	145	198	259	332	405	583	911	1312	1786	2333	2953	3646
16 OD	88	127	172	225	285	352	508	793	1143	1555	2032	2571	3175
18 OD	78	112	153	200	253	313	450	704	1014	1380	1802	2281	2817
20 OD	70	100	137	179	226	230	403	630	907	1235	1613	2041	2520

be taken into consideration, such as the center of gravity of each section of pipe between supports. These locations must be determined with regard to component parts for the pipeline unless the pipe run has no valves, traps, and the like. Any extra item that is attached to the pipeline—heavy valves, headers, flanges, meters—will change the loading of the system and this must be taken into account when placing pipe supports. It is also important to reduce the torsional forces that may be encountered in the pipeline.

Sloped Pipelines

Pipelines are sloped to provide the line with sufficient drainage and to counteract the effects of sagging between pipe supports. For steam lines, pipes are usually sloped to provide for drainage of condensate that may form in the pipeline, though this procedure is not always used because of the problems it entails. It is possible to slope piping lines by using mechanical devices such as pipe shoes and brackets; in cases where the pipe is hung, turnbuckles provided on the pipe support rod enable the installer to adjust the pipeline accordingly. Pipelines that need to be sloped on pipe racks often have shoes or other spacers enabling the construction crew to adjust the elevation of the pipe rather than the rack. Pipelines that are supported within the confines of a plant or building are usually sloped by means of adjustable rods, brackets, and turnbuckles. Figure 8-3 shows a typical pipe support detail utilizing the turnbuckle and rod assembly.

PIPE SUPPORT DRAFTING

Much of the following discussion applies to the construction and placement of pipe supports in the critical areas of power generation and in some cases petrochemical and process piping. It is not meant to indicate the procedures for drawing and detailing noncritical areas for plumbing and heating.

Pipe support detail drawings are fabrication instructions. With them the designer can convey the necessary information on the pipe support type, design, and materials to the shop or field fabricator. The pipe support detail drawing is used throughout many engineering groups in the company, since it contains instructions for many disciplines. Besides showing the design and construction of the pipe support, the detail also contains a bill of materials for standard components and other items to be purchased, such as structural shapes, attachments, and standard manufactured items. The drawing indicates what portion of the support assembly will be fabricated. All pipe supports go through this sequence and are eventually installed, tested, inspected, and documented—especially where safety considerations are essential, as in nuclear power plants. The detail of the pipe support thus becomes an extremely important piece of information for the total construction effort. The pipe support detailer should consider the following questions:

1. What information is necessary?
2. What information is redundant?
3. Is there an adequate number of views?
4. Are all the components and structural shapes called out on the bill of materials?
5. What information is needed for fabricating and ordering the pipe supports?

It is the responsibility of the pipe support drafter to develop detail drawings which are consistent with the drafting practices discussed in Chapter 5. Many of the drafting procedures and detail drawings are similar to those found in the mechanical drafting field.

The pipe support drafter must prepare a total bill of materials and calculate rod length, beam length, and dummy leg lengths—to name just a few responsibilities. All the pipe support drawings must have the correct weld symbols adequately placed (see Chapter 2). The pipe support drafter must also detail all complicated structural framing and calculate offsets that may be necessary for placement of the pipe support.

To be able to draw the pipe support detail, the drafter must be familiar with a variety of standard components, including predrawn pipe support standards and decals, available from pipe support companies. The background information must include knowledge of standard items such as spring hangers, sway struts, and vibration shock absorbers and snubbers, along with many other standard attachments.

Detail Information

The following list suggests the basic information to be conveyed by the pipe support detail drawing. There are many possible variations of this list, and individual companies may have procedures and information requirements which differ from those set forth here:

1. Pipe support drawings must contain location and area plans. Figure 8-25 (upper right-hand corner) shows a typical location plan of a pipe support detail. Since each support is designed for a particular pipe at a definite point, this point must be indicated on a location plan with reference to major building column lines in two perpendicular directions. The elevation of the pipeline at the centerline of the pipe where the pipe support connection is to be made should also be given. Without this location plan and subsequent piping information, the pipe support would be impossible to install correctly. Building, plant, and area number should also be indicated on the plan whenever necessary.
2. When showing the pipe support assembly, draw it in its installed position as it would be oriented in the surrounding building, plant, or other piping situation. All details are drawn in orthographic views with the use of simplified drawing methods and, where possible, symbols are used for standard catalog items. All structural shapes and steel that are not part of the building structure must be shown adequately with the individual connections (such as welding or bolting) used for the pipe support assembly.
3. Pipe supports must be identified by individual numbers. The procedure differs from company to company, but in general the pipe support number is shown on isometrics as in Fig. 8-26 or 8-27. On the pipe support isometric drawings and the plan, elevation, and section drawings, pipe supports should be indicated by symbols or represented by another method to show the location and configuration of the support in relation to its surrounding structure.
4. On all pipe support details it is essential to show how the support will be attached to the surrounding building or plant structure. The structure it will be attached to must be clearly shown—concrete walls, floors, ceilings, structural beams, steel members, columns, or embedded steel plates. All this, including all structural or concrete embedments, must be indicated graphically on the pipe support detail.
5. All dimensions required for fabrication of the pipe support must be accurately placed to provide the fabricator with all necessary information.
6. Each pipeline that is being supported must be called out and numbered. In some cases, the working medium of the line called out should also be shown on the detail drawing. Line designations are sometimes included as part of the support number. Other information necessary for callout of the pipe should be given, such as nominal pipe sizes and reference information.

7. Bills of materials must be made for all pipe support details. Item number, part number, size and description, type of material, and number of required items must be specified.
8. Any special notes or references necessary for the construction, fabrication, or placement of the pipe support must be shown on the detail drawing. Code, piping class, size, categories, and hanger classifications are important for design, fabrication, inspection, documentation, and ordering and must therefore be noted on the drawing. Any orders for the fabrication and installation of the pipe support—such as field welds, design fabrication, or installation instructions—must be noted on the detail drawing also.

ITEM NO.	NO. REQ'D	PART NO.	SIZE	DESCRIPTION
1	2	113	#7	WELDED BEAM ATTACHMENT
2	4	274	#7	WELDLESS EYE NUT
3	2	133	7/8" X 2'-5"	ROD W/6" TBE
4	2	132	7/8"	TURNBUCKLE
5	2	133A	7/8" X 2'-3"	ROD W/6" RHT & 6" LHT
6	6	–	7/8"	HEX NUT
7	1	370	8" (4-1/2" X 5/8")	RISER CLAMP, TYPE-1; PIPE O.D. 8-5/8", OPERATING
				LOAD 1370#, TEMP = 70° F; C-C = 1'-8"; LOAD BOLT, 3/4"
8	4	–	1-1/2" X 1" X 3"	LUG
9	1	–	W4 X 13	BEAM X 6'-2" LG

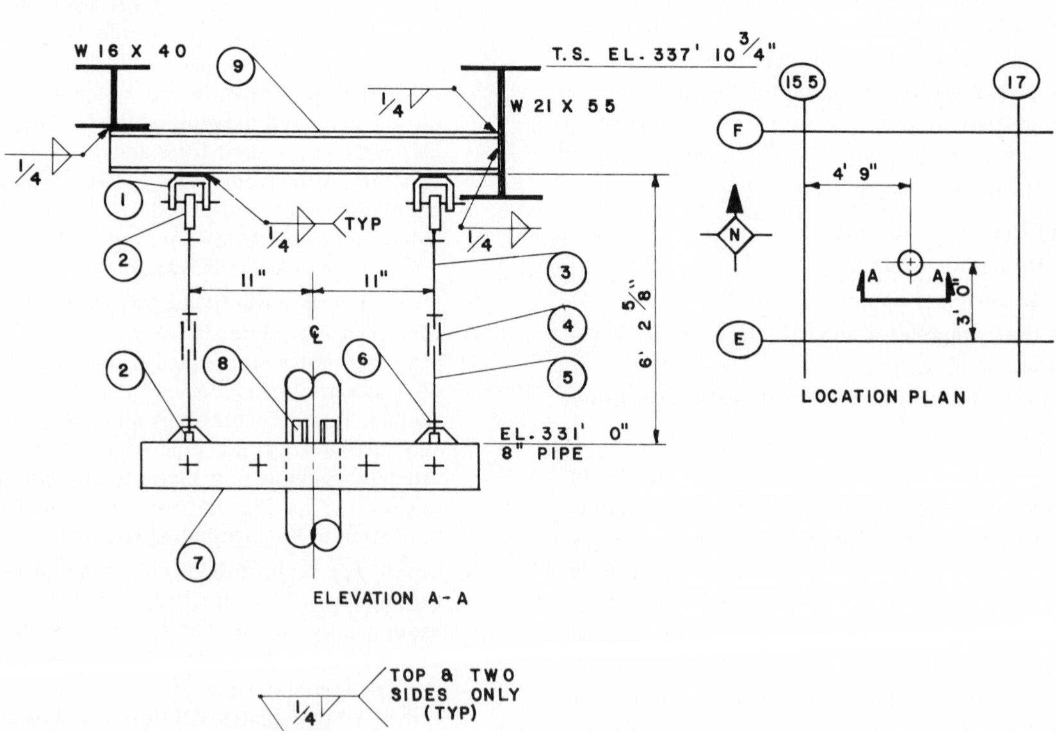

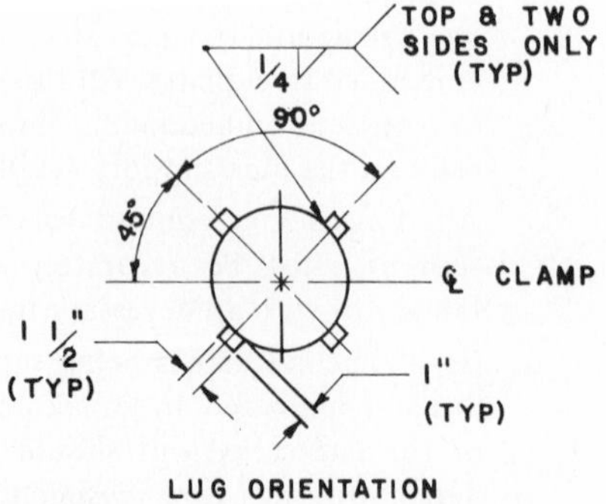

FIG. 8-25 Pipe support detail.

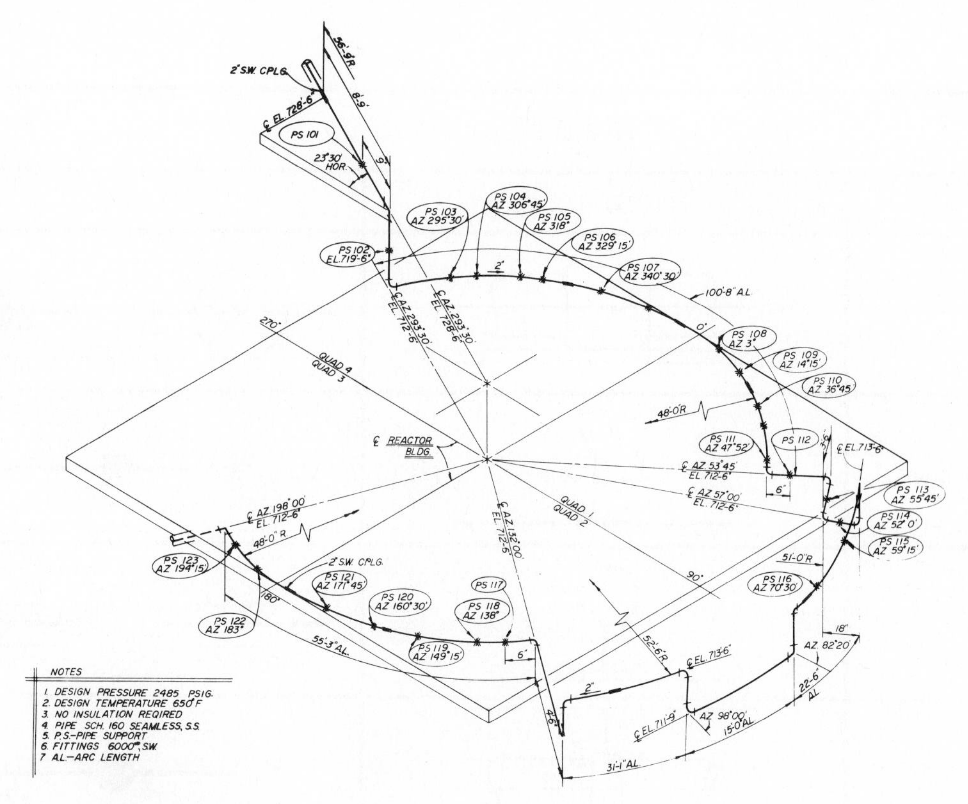

FIG. 8-26 Pipe support isometric drawing.

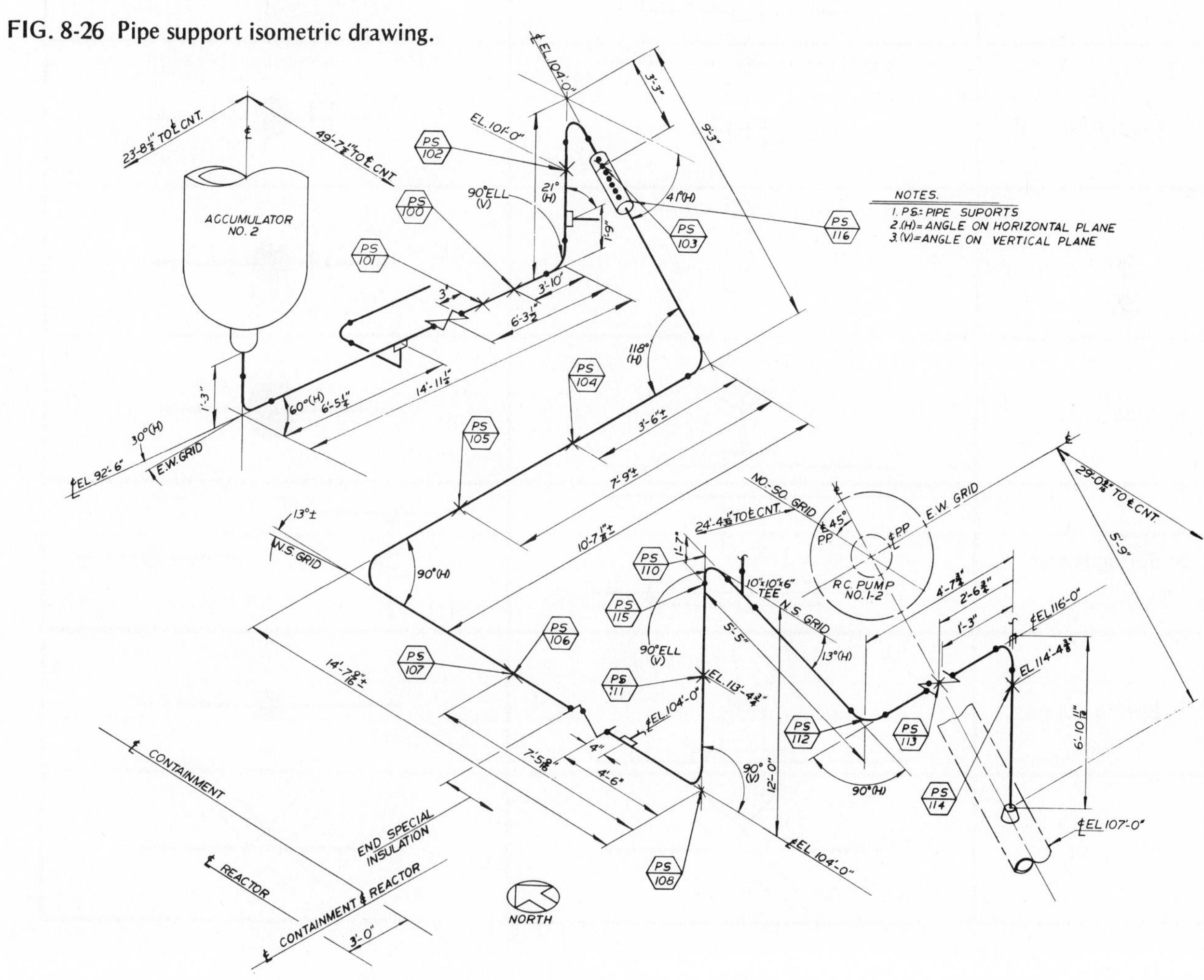

FIG. 8-27 Nuclear pipe support placement isometric.

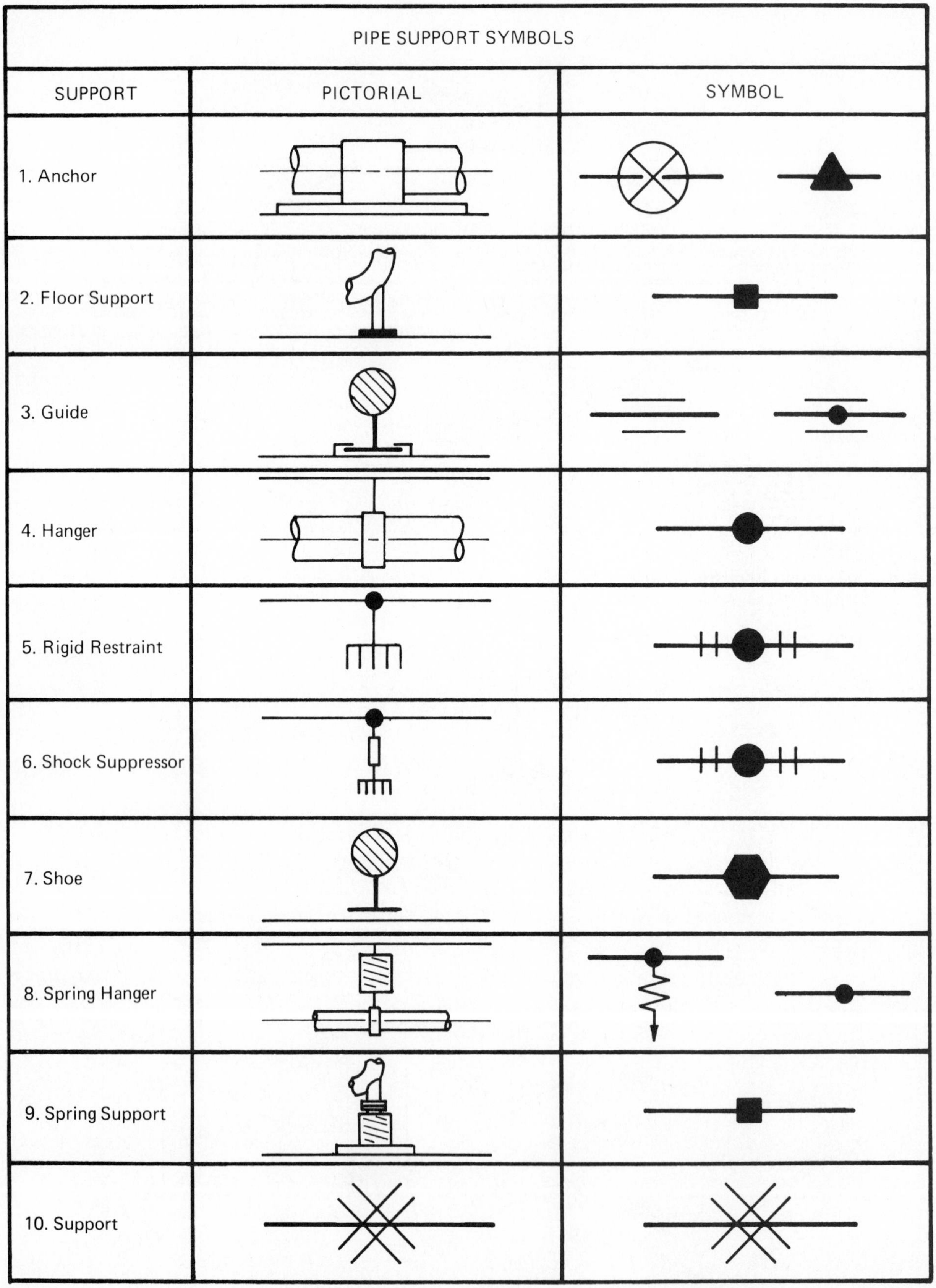

PIPE SUPPORT SYMBOLS		
SUPPORT	PICTORIAL	SYMBOL
1. Anchor		
2. Floor Support		
3. Guide		
4. Hanger		
5. Rigid Restraint		
6. Shock Suppressor		
7. Shoe		
8. Spring Hanger		
9. Spring Support		
10. Support		

FIG. 8-28 Pipe support symbols.

Pipe Support Symbols

Figure 8-28 shows a variety of pipe support symbols. Though not standard throughout the piping field, these are the commonly accepted methods of showing pipe supports and designations. These symbols and drawing methods are used primarily for the location and representation of pipe supports on piping drawings such as isometrics, plan elevations, sections, and other orthographic drawings—but not on pipe detail sheets. Figures 8-26 and 8-27 show the use of these symbols. As is the case with all symbols, it is important to place and dimension their location accurately and provide information for location, design, and fabrication.

Detail Views

All pipe support drawings require an adequate number of views to represent the pipe support. In some cases, a single view is sufficient (as in Fig. 8-9); when the pipe support is complicated, however, several views are necessary (as in Fig. 8-17). The pipe support drafter should be able to construct a pipe support detail accurately, clearly, and concisely—that is, without overdetailing. The selection of views is determined by several factors:

1. Choose the combination of perspectives that will describe the shape and configuration of the support completely, economically, and with regard to its *best* view.
2. In general, as is the case throughout this chapter, the overall elevation is the principal view used on the pipe support detail. This elevation view is determined with the aid of the location drawing plan. The main view of the pipe support detail must be consistently oriented with regard to its actual placement. The elevation view should be oriented on paper in the location plan so the plan view can be eliminated in a majority of cases.
3. The plan view should only be used when the elevation and sectional views do not provide sufficient information for detailing the object.
4. Provide sufficient details and sectional views to describe the arrangement of the pipe support detail—especially in cases such as Fig. 8-17, where a blowup view is necessary to detail the guide portion of the support.
5. In general, an 8½ × 11 in. sheet of paper should be sufficient for each pipe support. For elaborate configurations, provide larger sheets of paper and more detailed views of the prospective pipe support.

Dimensioning Practices

All dimensioning practices outlined in Chapter 5 should be strictly followed when constructing pipe support detail drawings. A short overview is given here along with dimensioning practices that are used exclusively for pipe support details.

Pipe support detail drawings are divided into two categories: fabrication drawings and engineering sketches. Engineering sketches are provided by the pipe support designer and include sufficient dimensional and detail information so that all missing dimensions can be calculated later by the pipe support drafter, who will complete the fabrication detail. The details provided in this chapter are fabrication details.

One of the main aspects of dimensioning a pipe support detail is the positioning and placement of dimensions. Practice varies from person to person, but certain guidelines should be followed. Dimensions should be clearly placed to avoid cluttering the drawing. Dimensions must be clearly printed above dimension lines; ballooning and detailing must be concise. Dimensions that run parallel to each other should be staggered so they will be easier to read.

Feet and inch marks are used throughout the detail drawings. On support details, it is not usually necessary to use zero feet. For instance, if the dimension is 3 in., do not use 0 ft 3 in.; 3 in. is sufficient. Figure 8-25 shows a cleanly drawn, well-spaced detail providing the necessary dimensions. The following dimensioning specifications should be followed when completing pipe support details. They are similar to those given in Chapter 5 but are given here in simplified form:

1. Show horizontal dimensions on plan views; show vertical dimensions on sections and elevations.
2. Do not place dimensions inside objects.
3. Show all dimensions on details. Do not use notes unless the note will eliminate the drawing of an extra view.
4. Use "typical" as a callout to eliminate redundant dimensioning.
5. Never double dimension.
6. Note in parentheses, next to or underneath the dimension line, any special dimensions that are indicated by the designer as being calculated or scaled.
7. Print all dimensions clearly.
8. Dimension between objects that have a relationship to each other or to centerlines of mating parts.
9. Try to place dimensions on views where the object appears true size.
10. Call out the object's name when using the centerline of an object.
11. Follow the lettering procedures suggested in Chapter 5.
12. Express dimensions over 24 in. in feet and inches (except pipe sizes).
13. Letter all dimensions so that they read from the bottom right-hand side of the page.
14. Follow all line thickness and size guidelines set forth in Chapter 5.
15. Do not break dimension lines for the insertion of the dimension.
16. Do not pass dimension or extension lines through lettering.
17. Try to align all dimension and extension lines.
18. Space dimension lines a minimum of 1/4 in. apart.
19. Place long dimensions outside small dimensions.
20. Do not cross dimension lines unless absolutely necessary.
21. Use straight leaders where possible and curved leaders only when necessary.
22. Dimension to hidden lines only if absolutely necessary.
23. Place elevation dimensions above extension lines and centerlines. Point to the level of elevation with an extension line with leader and arrowhead only where necessary.
24. Always call out elevations for pipe centerlines.
25. Use elevation designations for all major objects, thereby eliminating vertical dimensions unless they are needed to calculate component lengths.
26. *Do not draw pipe supports to scale unless absolutely necessary.*
27. *Draw pipe support objects in proportion to each other instead of to scale.*
28. Review the total size of components before attempting to proportion them on a pipe support detail.

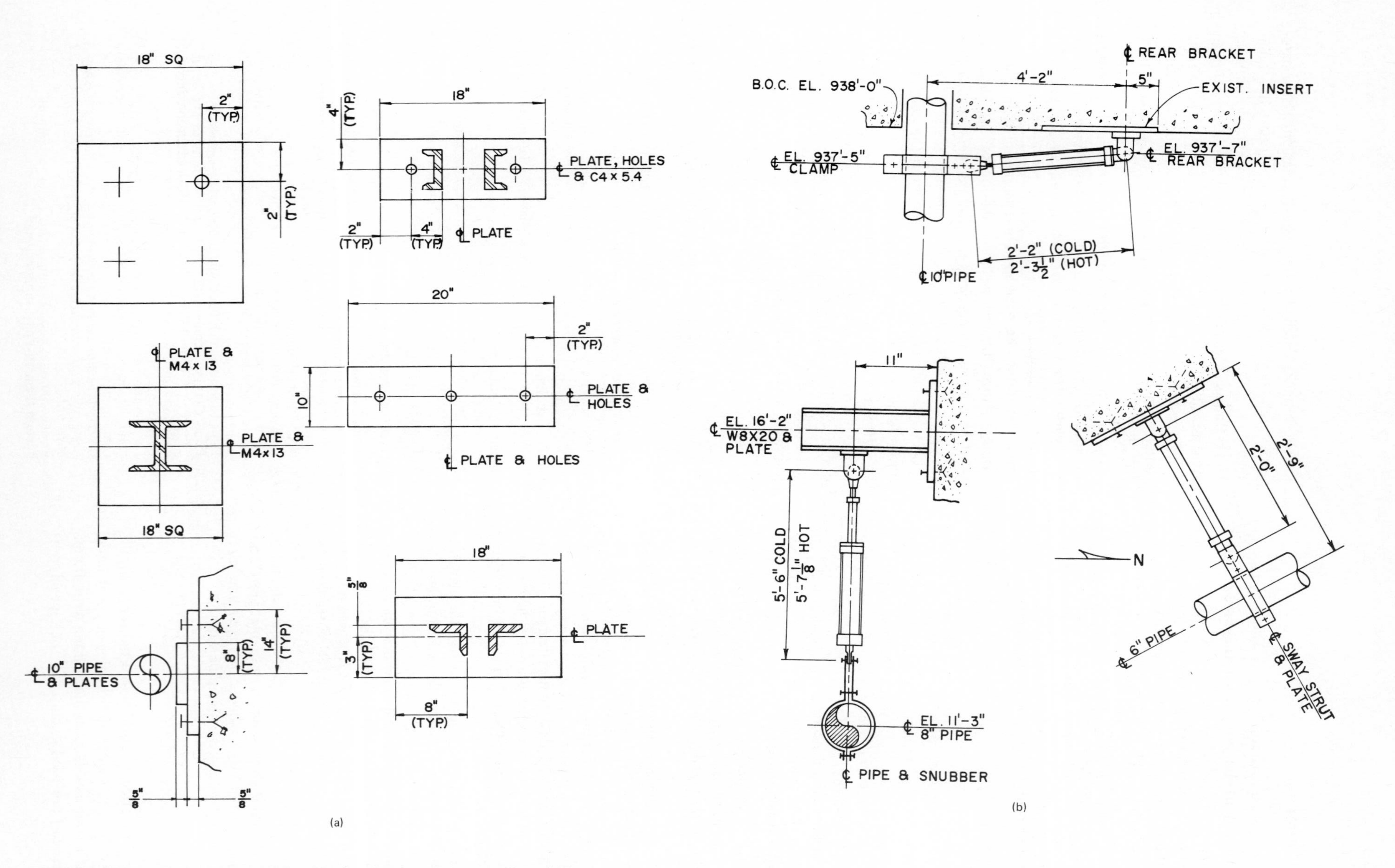

FIG. 8-29 Typical dimensioning procedures for base plates and snubbers.

Figure 8-29 shows six examples of dimensioning pipe support face plates and snubbers. These are not an exhaustive representation of all the dimensioning practices that are possible, but they do indicate typical procedures. When dimensioning face plates and other steel plates for pipe supports, adhere to the following specifications:

1. Use the callout "SQ" for all square plates. Dimension only one side, as in Fig. 8-29, thus eliminating the necessity of showing two dimensions.
2. For all plates and lugs, provide two overall dimensions unless you are using the square callout.
3. Dimension from the centerline to the edge of the plate. The use of typical dimensions, as in Fig. 8-29, is acceptable.
4. It is acceptable to stack dimensions in a chain and eliminate the overall dimension so long as a complete chain of dimensions is used.
5. All plates that are not square must be completely dimensioned, including cutoff angles.
6. Always use the centerline callout for the centerline of plates where steel is attached to the center.
7. All holes should be dimensioned from the edge of the plate. Use typical callouts whenever possible.
8. Use the plan view of the plate for dimensioning unless a side view is absolutely necessary.

For the dimensioning of structural items such as steel channels, angles, beams, and various types of tubing, including pipe, the following specifications are provided as guidelines:

1. It is unnecessary to dimension the width and length of the structural shape.
2. All distances showing the clearance between the structural member and the piping must be shown.
3. Locate dimensions of steel from established reference points.
4. Dimension from the edge of the steel when dimensioning angles and beams.
5. Dimension from the centerline of the structural shape if it is tubing or pipe.
6. All reference angles for cutting of structural shapes should be shown where necessary.
7. Use the centerline callout in place of dimensions whenever possible.
8. Provide one angle and one length dimension for diagonal braces.
9. Clearly identify all structural shapes on the detail through callouts or, preferably, through the bill of materials by means of a detail or item number.

Spring hangers, sway struts, sway braces, and shock absorbers involve a variety of dimensional practices that are shown on details throughout the chapter. When dimensioning spring supports or spring hangers in elevation or section views, use elevation callouts rather than vertical dimensioning. Swing angles and offsets should be given on the detail sheet, which must be calculated by the pipe support designer or drafter. Always call out the offset in the view that most clearly describes the offset procedure.

Dimension spring hangers, snubbers, sway braces, sway struts, and the like in their cold-loaded position—though in a majority of cases both the hot and the cold dimension are called out, as in Fig. 8-29, which shows typical snubber applications and dimensioning practices. As can be seen in Fig. 8-29, the overall pin-to-pin dimensions are the principal dimensions given for the snubber detail. Above and below the pin-to-pin dimension line, the hot and cold dimensions can also be indicated. If the rear bracket and clamp elevations differ, both elevations must be shown on the detail.

Figures 8-17 and 8-20 show typical dimensioning practices for structural elements used in pipe supports and guides.

Welding Symbols

Welding symbols and welding practices used for pipe supports are similar to those used for all welding. Chapter 2 discusses the major welding practices that will be encountered in all areas of piping drafting. Standard symbols of the American Welding Society are always used for welding identification. The drafter must not invent new or complicated welding symbols. Pipe support detail sketches provided by the support designer or engineer must show all of the necessary welding information—weld sizes and weld locations—and indicate whether the welding should be done in the field or at the shop. If the drafter notices insufficient welding information, he or she should consult the supervisor. The following specifications apply to welding symbols and pipe support drawings:

1. Use "typical" whenever possible for welding that is similar or identical.
2. Eliminate the welding symbol tail if no reference notes are indicated.
3. Be sure the welding arrowhead meets the surface to be welded.
4. Use a maximum of three and preferably two leader lines from the welding symbol to avoid confusing the detail.
5. Keep the size of the weld continuous throughout each pipe support detail whenever possible, as in Fig. 8-30, which uses typical 1/4-in. welds.

Piping Detail

Since pipe support details have as their main objective the support of a pipeline, all support drawings must show the orientation of the pipe support to an existing pipeline. Detail only the portion of the pipeline or section of the pipeline to be supported. The accompanying fittings (valves, elbows) of the pipelines should be shown on the location plan only, unless they are required for pipe support placement or for clarity. In some cases, as with the use of dummy legs, the pipe support is attached directly to a fitting. A fitting may be shown when it is necessary to dimension from that component to the pipe support for field location.

ITEM NO.	NO. REQ'D	PART NO.	SIZE	DESCRIPTION
1	2	113	#5	WELDED BEAM ATTACHMENT
2	2	274	#5	WELDLESS EYE NUT
3	2	133	5/8" X 5'-7"	ROD W/6" TBE
4	2	—	C6 X 8.2	CHANNEL 5'-11" LONG
5	6	—	5/8"	HEX NUT
6	1	283	6"	U-BOLT TANGENT 11-1/2"
7	2	—	3/8" X 2" X 4-1/4"	C. S. PLATE
8	4	260	#5	WASHER PLATES
9	1	—	W6 X 15.5	BEAM 10'-4" LONG

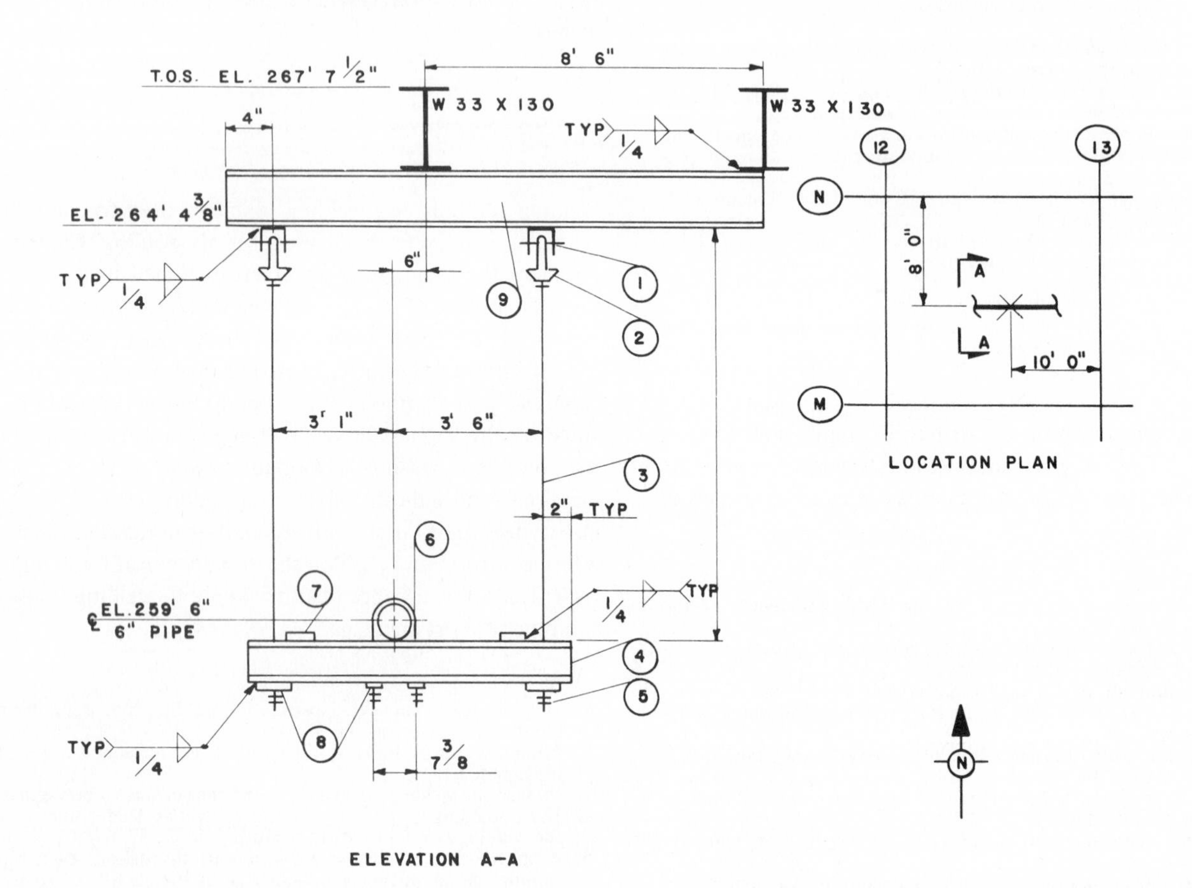

FIG. 8-30 Pipe support trapeze assembly.

The following specifications are used for the location and callout of piping on pipe support details:

1. Pipe size, pipe number, and centerline elevation must be given on all pipe support details.
2. The pipeline is to be shown in double-line symbol representation, never single line, especially in plan, elevation, and section views.
3. When using double-line representation, the centerline of the pipe must always be shown.
4. When an existing pipeline is detailed for reference only, its elevation code and size must be called out.
5. On the location plan, the pipeline is to be shown in single-line representation only.
6. Existing piping should be shown in phantom or hidden lines.
7. Tables for pipeline thermal movement or for total load, hot and cold, and so on are all provided on the pipe support detail—as in Fig. 8-31, where cold load, travel, and thermal movement are called out in the material list. In some cases, a separate table is provided to show the movement and forces. A table should be completed on the first sheet of each pipe support detail, showing positive and negative loads in *X, Y,* and *Z* directions using plus or minus. Forces should be given in pounds; movement in *X, Y,* and *Z* should be given in inches; movements should be given in foot-pounds. This information should be supplied by the pipe support designer and engineer and calculated by the stress analysis group. The information necessary for the construction and implementation of pipe support details should be shown on the detail drawing. Forces that act on the pipe support, as well as movements of the pipe support, are to be shown on the plan or section views where necessary or in the elevation views as in Fig. 8-31, which shows the hot and cold pin-to-pin dimension.

When movements, forces, and displacements are detailed on pipe support drawings, they should be dimensioned from the centerlines of pipe to corresponding components. For axial movements and forces, these should be shown from the center of support. The abbreviation for pound (lb) should be placed at the right side of all load values. Where possible, use the main view to show all forces and movement of the pipeline. The abbreviations *t* (thermal displacement) and *s* (seismic displacement) should be used wherever necessary. Thermal movement and seismic movement could also be called out on the materials list pertaining to the part which undergoes the movement, as in the snubber application of Fig. 8-31. Be sure to show an adequate number of views to express all the movement undergone by a pipe.

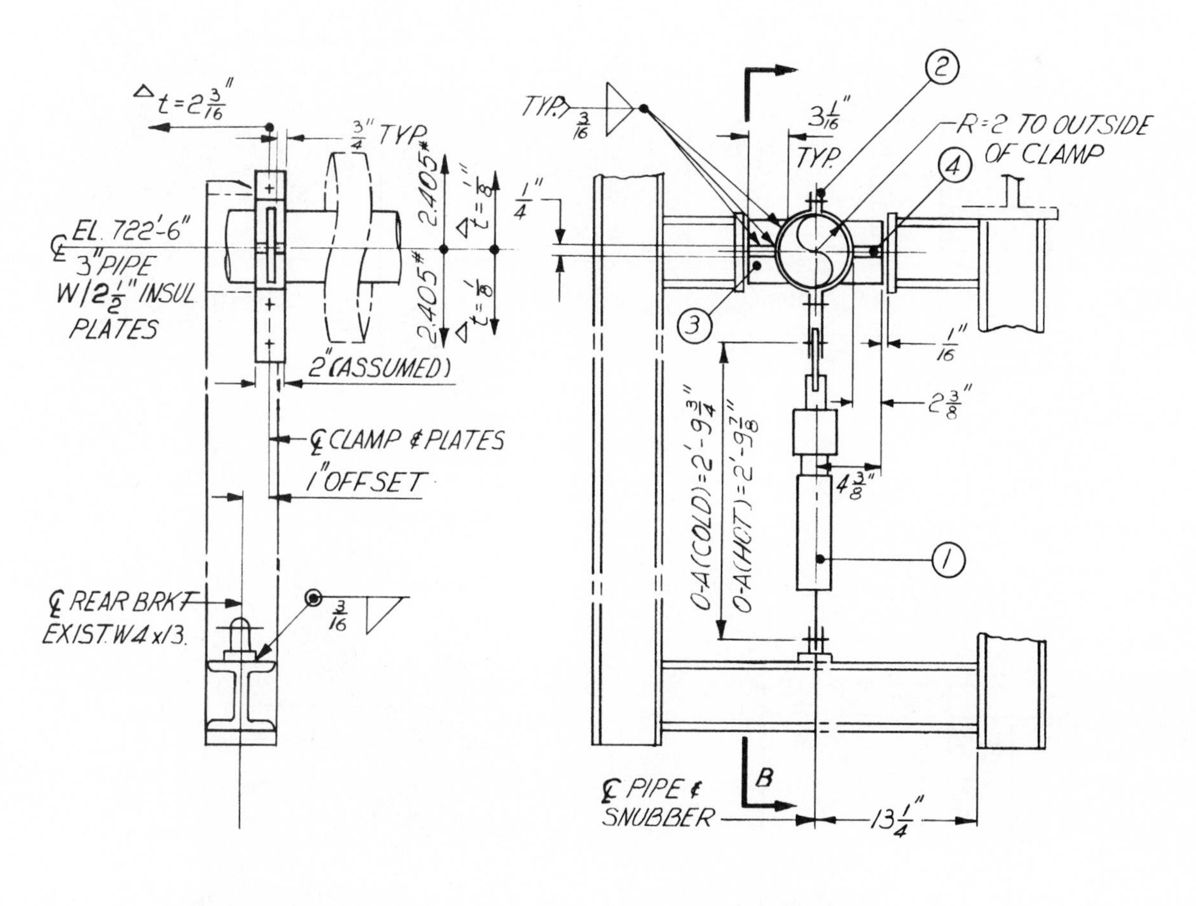

MATERIALS		
NO	DESCRIPTION	REQ
1	PART 2540 PSA-3 W/5 TRAVEL	1
	OA (COLD) = 2'-9$\frac{3}{4}$" L (COLD) = 17"	
	E = 10$\frac{1}{4}$", $\frac{1}{8}$" THERMAL MVMT. (T), &	
	$\frac{1}{8}$" THERMAL MVMT (C) GOV. LOAD =	
	2405# (FAULTED)	
2	PART 2640 SIZE 6 FOP 3" PIPE	1
	MAT. L.C.S.	
3	$\frac{1}{2}$" x 3" C.S. PLATE. 3$\frac{1}{16}$" LG.	2
	(CUT AS SHOWN)	
4	$\frac{1}{4}$" x $\frac{3}{4}$" C.S. PLATE. 2$\frac{3}{8}$" LG	4

FIG. 8-31 Pipe guide and snubber assembly detail.

Location Plans

Figure 8-32 shows six typical location plans and should be studied carefully. The location plan allows the pipe support to be located in regard to established points and coordinates. Location plans are extremely important for engineers, field direction crews, and fabricators. Do not clutter the pipe support location plan: Give only the information necessary for the placement and construction of the pipe support.

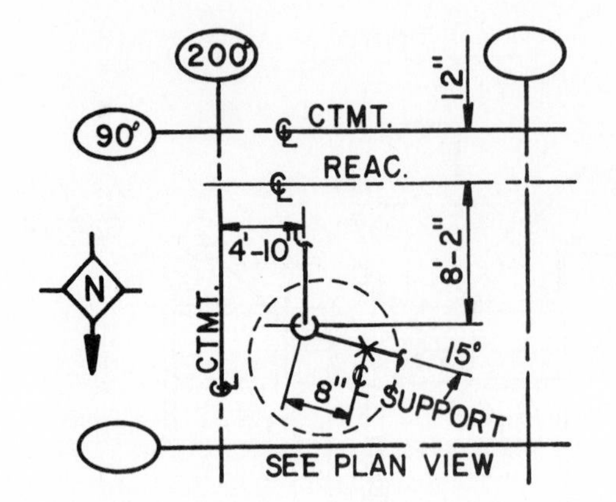

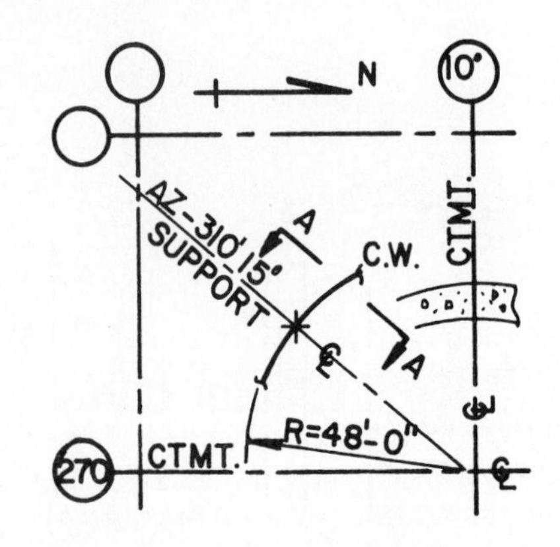

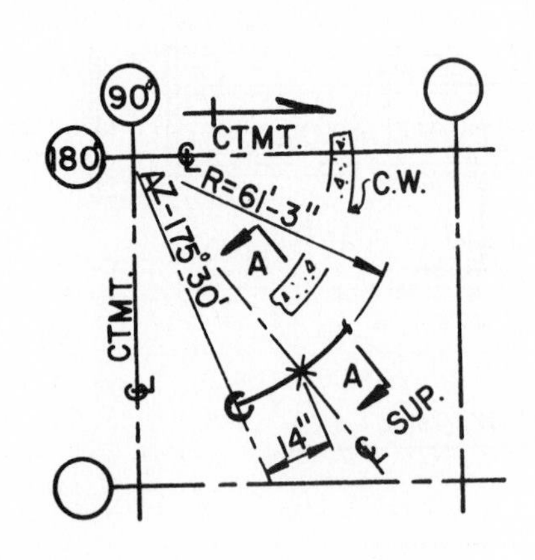

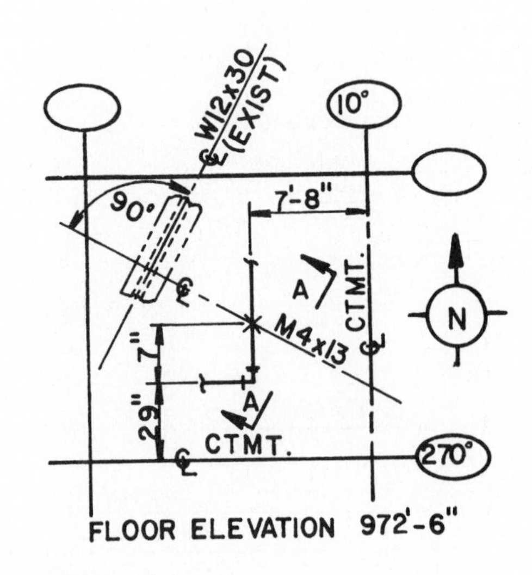

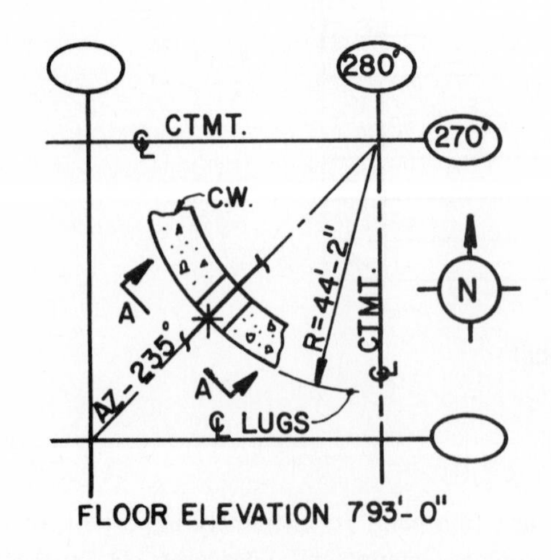

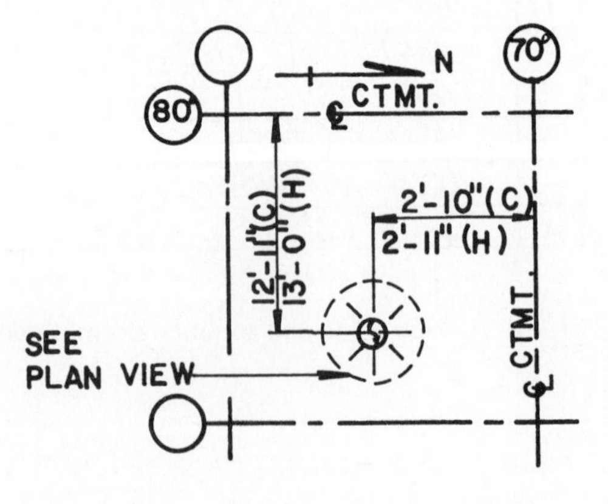

FIG. 8-32 Area location drawings.

The location plan is the only method by which a pipe support detail can be located in a piping system. The following specifications will help in the setup and drawing of pipe support location plans:

1. Show the location plan on the first sheet of each pipe support detail. (Many location plans have been omitted from the pipe support details in this chapter.)
2. Show a north arrow orientation on location plans as in Fig. 8-32.
3. The location of the pipe support can be clearly defined through the following procedures:
   (a) Locate the pipe support with regard to the nearest change-of-direction fitting; then locate the support along the piping run from that fitting.
   (b) If there is no change-of-direction fitting to dimension from, the support should be located in two directions from surrounding columns or structural members, from the centerline of a piece of equipment close to the pipe support, or, in nuclear containment areas, from a centerline of the containment, using an azimuth, radius and callout.
   (c) Always show dimensions for cold positions of pipe; show the hot position only when necessary.
   (d) Pipe supports in sleeves or walls (such as containment walls) should be located with reference to the wall–as in the pipe support location provided in Fig. 8-32 in the lower left-hand corner, showing the floor elevation 793 ft 0 in. The containment wall is detailed and the pipe support is located outside the containment wall. An azimuth of 235 degrees is given to locate the pipe support along with the radius 44 ft 2 in.
   (e) Always specify centerlines of columns and large structural elements that are dimension or location reference points.
4. The callouts for center of support, center of steel, center of clamp, and so forth must be given on the pipe support location plan to avoid confusion in placement of the support.
5. Use the plan view to orient the location plan. If it is not possible to use the plan view, section notations must be given in the location plan.
6. All floor elevations should be noted on the location plan if possible.
7. Use the cross or double cross symbol for showing the location of the pipe support on the location plan, as is done throughout Fig. 8-32.

Bill of Materials

The bill of materials is necessary for all pipe support details, though some have been eliminated in this chapter. Bills of materials provide concise information for all the components necessary to fabricate a pipe support. A bill of materials is always presented in chart form with separate columns for detail or item number, quantity, part, component, or figure number, size, description, weights and forces, unit cost, and in some cases total cost. In general, the drafter is concerned only with item number, quantity, part number, size, and description. The location of the bill of materials depends on the views necessary for the neat arrangement of the pipe support detail itself and may be at the top or bottom of the page. Whenever possible, the pipe support materials should be provided on the first sheet of the pipe support detail; in no case should the bill of materials be split between two sheets–when the list is long, a separate sheet should be used.

On the bill of materials, the drafter must use excellent printing and lettering. The bill of materials is the last item the pipe support detailer will complete, and great care should be taken to see that every component or portion of the pipe support is called out. The following specifications or guidelines can be used to construct a bill of materials:

1. Major items such as snubbers and springs should be called out first.
2. Use abbreviations where possible: avoid long descriptions.
3. Attempt to group similar components–all beams, all nuts, all bolts, all structural members–in sequential order.
4. Structural steel items and shapes should be described by using the name or item before the length and size.
5. For bar stock in steel plates or metal plates, the thickness, width, material type, item description, length, and material specification should be provided in that order.
6. When specifying nuts, bolts, and anchors, give the size, name, length, and vendor in that order.
7. Always follow the catalog ordering procedure for the company that will provide the component. Many companies use different abbreviations, and these must be adhered to. Each item on the pipe support detail should be ballooned with the item number. This number is used to list the materials for the construction of the pipe support. When ballooning the pipe support detail, adhere to the following instructions:
   (a) Use a 1/2-in.-diameter or 1/4-in.-diameter balloon, depending on the size of the pipe support detail.
   (b) Always point directly to and touch the item that the balloon is calling out.
   (c) Ensure that the balloon numbers and the bill of materials numbers correspond.
   (d) Point to and balloon items that are shown in solid lines; use hidden views only when absolutely necessary.
   (e) Balloon an item only once on each detail drawing unless it is necessary to show identical pieces in different views.
   (f) Balloon details on the first page of a pipe support detail in the main view wherever possible.

After the bill of materials is completed, the material take-off list is created from all the pipe support details for the job. This is not usually the assignment of the pipe support drafter.

MODELS AND PIPE SUPPORTS

Figure 8-33 shows a typical model in the piping field. Figure 8-34 shows some structural guides and supports. The importance and necessity of models in the piping field has already been discussed. Designing the pipe support on the model enables the department to reduce drawing time and facilitates accurate spotting of the pipe support locations with regard to other equipment. In most cases, the model totally eliminates the interference problem in the placement of pipe supports and improves the overall planning for placement of pipe supports in combination with each other. By using a model, the pipe support group can greatly reduce problems between different groups in the company–between the HVAC group, for example, and the pipe support location group. The model should be used for evaluating all structural embedments and structural framing methods.

FIG. 8-33 Typical model requiring pipe support details.

FIG. 8-34 Model assembly showing structural bracing and supports.

Before locating pipe supports on the model, identify the surroundings, important equipment, and objects that may influence placement of the pipe support–surrounding piping runs and components (such as valves and fittings), the building structural elements involved, electrical runs and trays, all HVAC ducts and equipment, all mechanical equipment and vessels. By establishing dimensions with respect to elevations and coordinates for the location of pipe supports, one can eliminate interference problems.

In many cases, the pipe support drafter is called upon to fabricate the pipe support for the model and position it. When spotting pipe supports on a model, it is important to take into account the following information:

1. Consider the access and installation space necessary for the construction of pipe supports in an area.

2. Note common supports which may be used to support parallel or adjacent pipe runs.
3. Beware of any situation where an incomplete structural or concrete area is not shown on the model.
4. Be careful to allow sufficient pulling space for equipment when placing pipe supports close to surrounding vessels and other components. Do not locate pipe supports too close to traps or pipe vents and drains.

Overall, a model greatly facilitates the design and placement of pipe supports on a project and eliminates a great number of difficulties that can be encountered when only the drawing method is used.

REFERENCE DRAWINGS

The pipe support drafter will come in contact with a variety of other drawings and, in some cases, will be called upon to move from department to department or group to group. The pipe support designer will need to consider a vast array of reference drawings: piping drawings, steel structural drawings, embedments, sleeve drawings, hanger drawings, electrical drawings, civil and concrete drawings, HVAC drawings, fabrication isometric drawings, system isometrics, stress isometrics, piping layout drawings, electrical tray layout drawings, ducting layout drawings, piping and instrumentation diagram drawings, and all penetration drawings for the building. These are supplied by subgroups in the company and, because of interference problems, are extremely important when designing pipe supports.

Models have greatly reduced the problem of interference, and in some cases the pipe support designer need not review the reference drawings unless checking for problems. Reference drawings that may be needed when placing, fabricating, and installing a pipe support should be noted on a pipe support detail sheet. It is the duty of the drafter to list the reference drawings on the pipe support detail sheet when necessary. The following information is given on the reference drawings:

1. *Fabrication isometrics* give the line designation, pipe size and elevation, piping configuration, elbow-to-elbow distances, field weld numbers and locations, reference dimensions to columns and floor elevation, and total pipe footage. Also given are the types of valves necessary for the pipeline, pipe support locations and markups, equipment number and location with regard to the pipeline, and any other fabrication or construction information needed to implement the system (see Chapter 7).
2. *Stress isometrics* include the following information: line designations, pipe size and elevation, piping configuration, elbow-to-elbow distances, reference columns, pipe support locations and number, equipment numbers and location, pipe support loads and type, pipe support symbols, data points, continuation numbers, and a consistent numbering system. Figure 8-35 is an example of a stress isometric.
3. *System isometric drawings* (Chapter 7) are large three-dimensional piping configurations that include all the necessary piping for a single process system. This isometric is usually complicated but gives an overall picture of a piping system that enables the designer to locate anchors, restraints, guides, and supports accurately. The system isometric drawing is used less often than other isometrics. It shows line designations, pipe size elevations, piping configuration, elbow-to-elbow distances, field weld location, reference columns, pipe footage, pipe support numbers and locations and symbols, attachment details, continuation numbers, and an accurate numbering system.
4. *The P&ID drawing* (see Chapter 6) provides the necessary information for the direction of flow, line numbers, pipe sizes, pipe classification, all equipment designations, component parts, instrumentation connections, and the processes in the various systems.
5. *Heating, ventilation, and air conditioning (HVAC) drawings* provide interference checks for pipe supports. They show the general configuration of the ducting, whether horizontal or vertical, all sizes and locations of ducting, duct supports, openings and vents, and the bottom of duct elevations.
6. *Electrical tray layout and electrical drawings* show the electrical lines, callout size of cable trays, electrical tray locations, conduit and junction boxes, stack trays and their elevations, vertical trays, and the like.
7. *Penetration drawings* show penetrations of pipes through containments and floors. They give the sleeve schedules, type of sleeve, sleeve size, and line designation.
8. *Civil drawings* (including structural steel drawings) show column size, top of steel elevations, beam sizes, spacing, plate girders and schedules, platform drawing, grading, and concrete enclosing steel.
9. *Concrete drawings* show wall thickness, slab thickness, column location, whether north, south, east, or west, concrete or block walls, penetration, floor and wall, rebar locations, openings, drains, embedded plates, channels, unistruts, and poured-in-place anchor bolts.
10. *Piping layout drawings* enable the drafter to locate pipe supports in the structure. They include floor elevation area, pipe routing, plan view, sections and elevations, proximity of all other pipes and equipment, insulation, and all equipment necessary for the completion of the project. Layout drawings are the most important drawings for reference in the pipe support group.

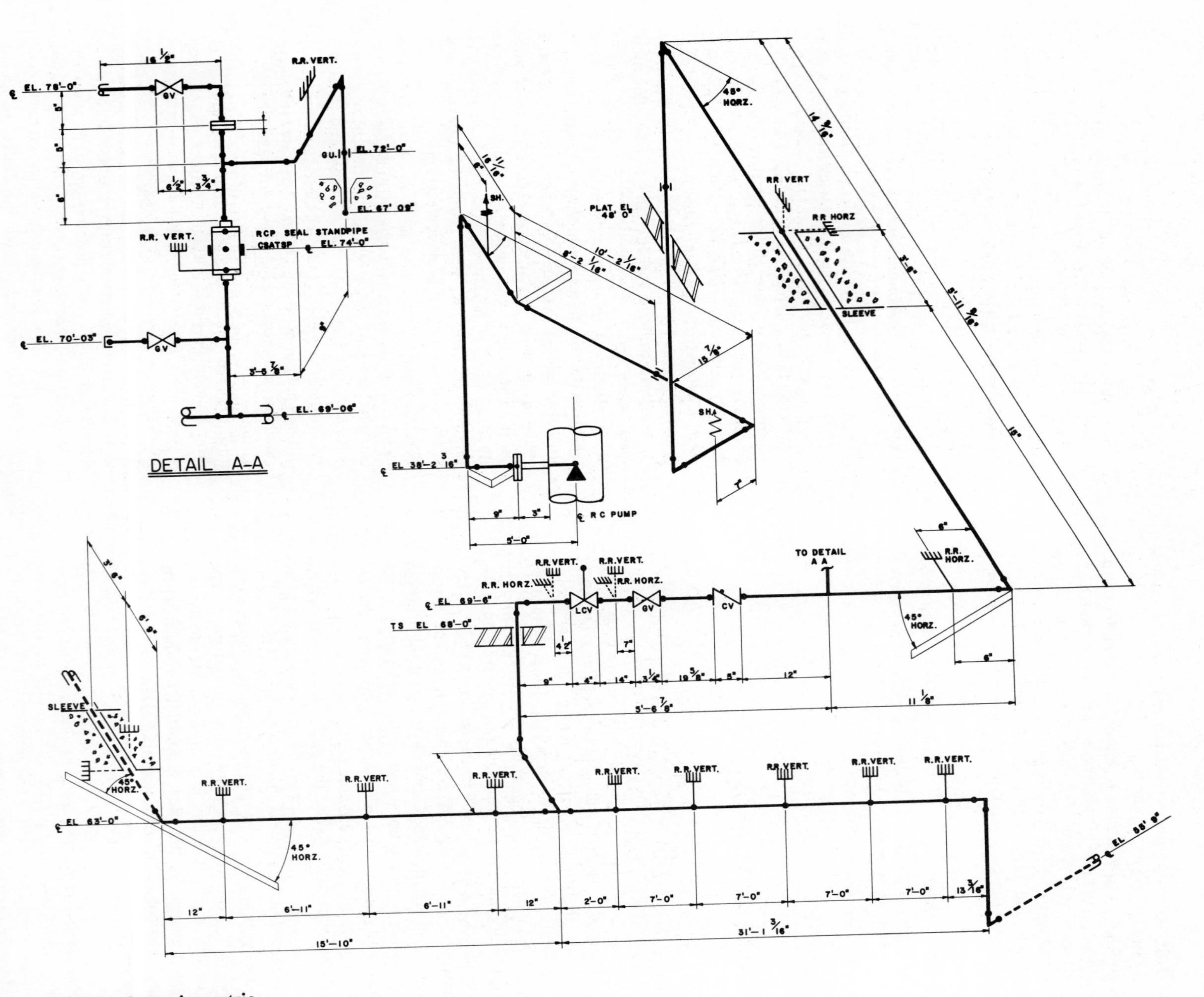

FIG. 8-35 Stress isometric.

CHECKING FOR INTERFERENCE

Checking for pipe support interference is extremely necessary and constitutes one of the major aspects of pipe support design. References for checking pipe support locations include piping layout drawings, hanger location drawings, electrical tray layout drawings, tray support drawings, HVAC drawings, duct support, structural steel drawings, concrete floor plan drawings, and the model. In checking for pipe support interference, the following list is useful:

1. Check the surrounding area for pipe runs that are located near the pipeline to be supported.
2. Always check the distance between pipelines in order to provide sufficient room for structural steel members.
3. Check hanger location drawings for other parts of the job to avoid conflicting with other hangers in the general area.
4. Always determine whether an insulated pipe has enough clearance around it for the pipe support.
5. Always check the tolerances designated for pipelines and equipment to avoid underestimating the area available for construction and location of the support.
6. Check all interferences that may be caused by seismic or thermal movement. Check whether or not the hangers would clear other objects in the area.
7. Check whether or not any part of the support could hit the surrounding equipment or structural elements, such as trays, ducts, steel members, ceiling, equipment, vessels, or other pipelines.
8. When support framing is connected to structural beams, the support must clear the bottom of the concrete.
9. Be sure to allow sufficient room for moving equipment.
10. Provide sufficient room for the installation, operation, and servicing of components on a pipeline such as valves and traps.
11. Keep a minimum of 7 ft (2.1 m) headroom clearance for all pipe supports.
12. Always refer to the electrical tray and ducting layout drawings whenever interferences might occur between pipe supports, electrical trays, and HVAC systems. It may be necessary to cut sections through various areas of the reference drawings to determine whether or not the pipe support will have sufficient clearance. Interferences must be checked on all the reference drawings, especially when the surrounding equipment is close.

PIPE SUPPORT CALCULATIONS

This section discusses a variety of methods, based on elementary trigonometry, selecting snubbers and variable springs, checking swing angles, and determining offsets, stanchion length, rod length, and the elevation of a sloping pipe at a particular point.

Basic Calculations

Figure 8-36 shows the calculations necessary for determining rod swing angle acceptability. The swing angle of a hanger rod is the rotation angle of the rod due to pipe movement from cold to hot position. In most cases, the specifications allow for a maximum swing angle of 4 degrees, depending on the service. This specification should be checked for each job. In cases where the swing angle is 2 degrees or more, or the total movement is in excess of 2 in., the hanger itself should be offset two-thirds of the thermal movement in cold position. Figure 8-36 shows dimension *A*, the vertical pin distance; dimension *B*, the horizontal linear movement; dimension *C*, the rod length; and angle *B*, the rod swing angle. The calculations are set up so that the tangent of angle *B* is equal to B/A. The maximum allowable swing angle without an offset is 4 degrees, so we need a tangent of 4 degrees, which equals 0.0699. This is the maximum tangent for B/A. When calculating the conditions for swing angle acceptability, dimensions for *A*, *B*, and *C* may be plugged into this calculation, thereby determining the angle at *B*. At no time should the maximum tangent for B/A equal more than the tangent of 4 degrees or 0.0699. Problem 14 at the end of the chapter tests the student's ability to determine the swing angle acceptability.

The offset should be 66 percent (two-thirds) of the thermal movement in cold position. In other words, if the total movement exceeds 2 degrees (or 2 in.) the hanger should be moved so that the total rod swing angle in the hot position becomes 2 degrees or less with a total movement of less than 2 in. This is done by moving the rod attachment in the direction of thermal movement (66 percent or two-thirds of the total thermal movement), which is the *B* dimension.

Figure 8-37 gives the calculations for determining the length of a structural brace. This is a simple trigonometric calculation, and for each problem dimension *B*, *A*1, and *A*2 should be available. When determining the calculations for the structural brace, the *A*1 and *B* dimensions will be known and *A*2 will be derived by calculating one-half the height of the structural member for the *B* dimension that covers the horizontal structural components. (See Problem 9 at the end of this chapter.)

Figure 8-38 provides the calculations for determining stanchion length. In this case, the radius of the pipe will be known, as will the height dimension or dimension from the centerline of the pipe to the bottom of the stanchion. The important calculation will be solving for *X*, which will equal

$$R^2 - \frac{(\text{diameter of stanchion})^2}{2}$$

After *X* is determined, the length of the stanchion can be found by taking distance *H* and subtracting the calculated distance *X*. (See Problem 15 at the end of this chapter.)

Figure 8-39 provides the calculations for determining the elevation at a given point on a sloping pipeline. In this case, the elevation above and below the unknown point to be

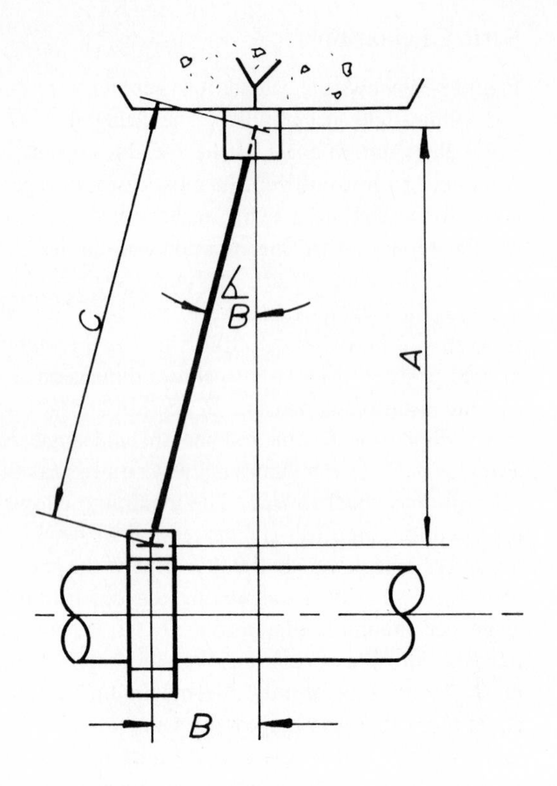

CALCULATIONS FOR DETERMINING
ROD SWING ANGLE ACCEPTABILITY

A = Vertical pin distance
B = Horizontal movement
C = Rod length
angle B = Rod swing angle

$$\tan \text{angle B} = \frac{B}{A}$$

Maximum allowable rod swing angle without offset is 4 degrees.

$$\tan 4 \text{ degrees} = 0.0699$$

$$0.0699 = \text{maximum tan for } \frac{B}{A}$$

FIG. 8-36 Calculations for determining acceptability of rod swing angle.

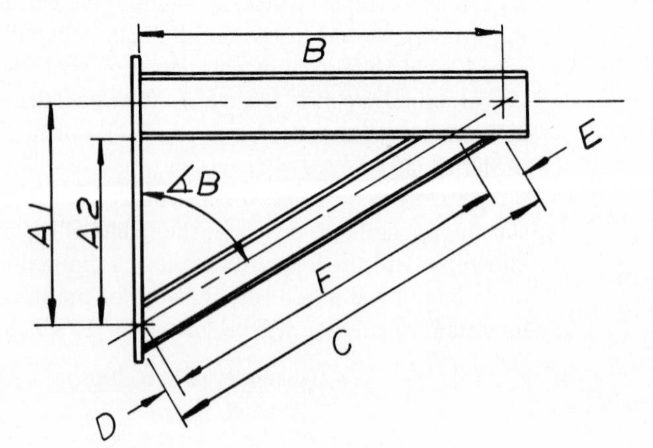

CALCULATIONS FOR DETERMINING
LENGTH OF STRUCTURAL BRACE

$$\tan \text{angle B} = \frac{B}{A1}$$

then find cos angle B

A2 = A1 − 1/2 thickness of upper brace

$$F = \frac{A2}{\cos B}$$

E = 2 × tan B
D = 2/tan B
C = D + F + E = brace length

FIG. 8-37 Calculations for determining length of structural brace.

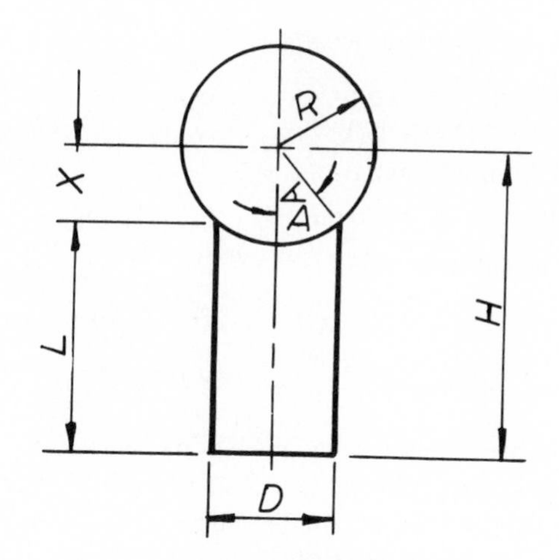

CALCULATIONS FOR DETERMINING STANCHION LENGTH

D = diameter of stanchion
H = distance from pipe centerline to base of stanchion
R = radius of pipe
$X = R^2 - (D/2)^2$
$L = H - X$ = length of stanchion

FIG. 8-38 Calculations for determining stanchion length.

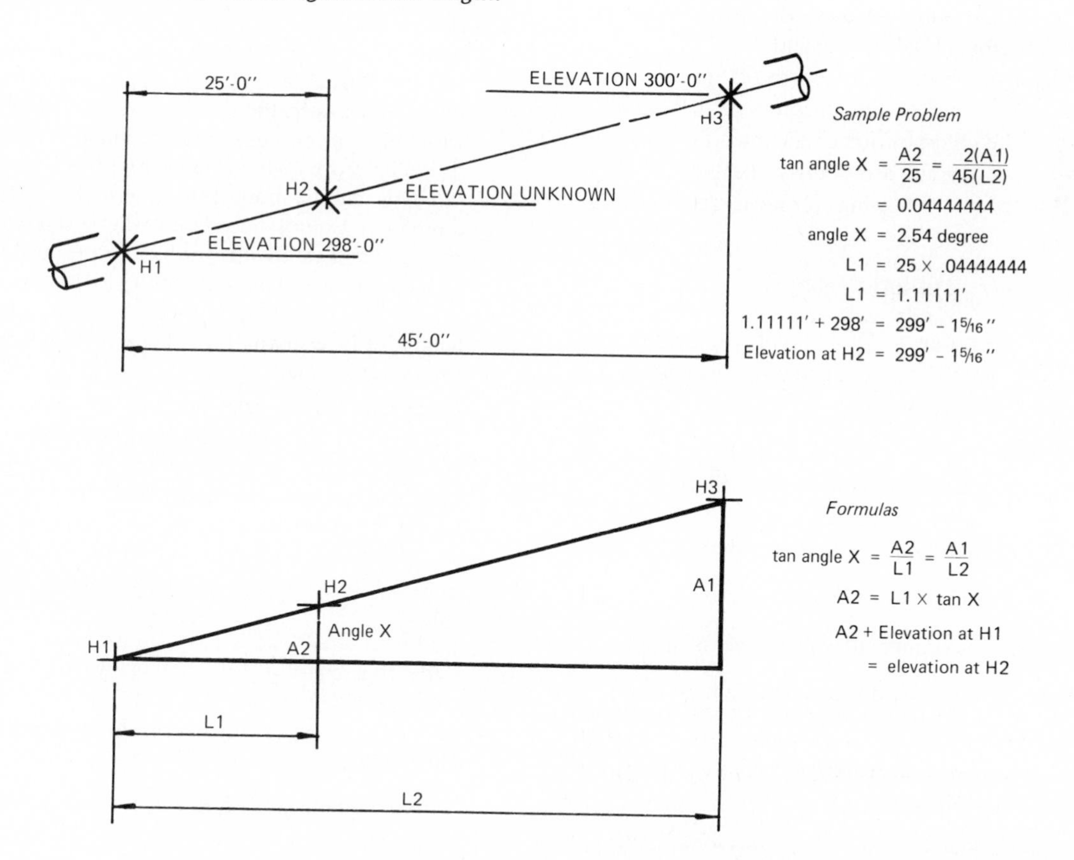

Determining the elevation at a given point on a sloping pipeline.

H1 = Hanger location 1
H2 = Hanger location 2
H3 = Hanger location 3
L1 = Horizontal distance between H1 and H2
L2 = Horizontal distance between H1 and H3
A1 = Vertical distance between H1 and H3
A2 = Vertical distance between H1 and H2

FIG. 8-39 Determining the elevation at a given point on a sloping pipeline.

determined must be available along with the horizontal distance between the unknown point and one of the other elevation points. These calculations require a knowledge of simple trigonometry.

Snubber Selection

There are different sizes of snubbers with allowable design load ranges from 350 to 120,000 lb. For each size there are minimum and maximum overall lengths. If the actual space is less than the minimum, the assembly cannot be installed. If the actual space is larger than the maximum, the assembly will have a load capacity less than the rated value. A snubber lets the pipe move slowly. The maximum movement a snubber can cope with is defined as the *stroke.* Some sizes have two different allowable strokes. The stroke must be larger than the anticipated pipe movement.

When ordering a snubber, specify length L^{**} (see Bergen-Patterson data at the end of this chapter) of the snubber corresponding to the cold position so the manufacturer can set the snubber scale reading accordingly. Length L^{**} may be computed with the following formulas (all units in inches):

$$L^{**} = L_{min} + \text{set}$$

Set = (stroke - MVMT)/2 if thermal movement from cold to hot will cause snubber to extend

= (stroke + MVMT)/2 if thermal movement from cold to hot will cause snubber to contract

where L_{min} = minimum length of snubber

set = scale reading in cold position

stroke = maximum allowable travel of snubber

MVMT = anticipated thermal movement of pipe

In designing a snubber assembly, the swing angle is limited to 6 degrees maximum as suggested by the manufacturer. Here is the procedure for selecting a snubber assembly:

1. Determine snubber size and clamp size.* Suppose the design load is 7283 lb, so select a size which has an allowable load of 15,000 lb for strut length up to about 7 ft. The anticipated movement is 1¼ in., so a stroke of 6 in. is sufficient. The pipe clamp is for 18-in. pipe.
2. Determine the L^{**} dimension. The movement is southward from cold to hot.

   Set = (6 - 1.25)/2 = 2.375 = 2 3/8 in.

   L^{**} = 17 1/8 + 2 3/8 = 19 1/2 = 1 ft 7 1/2 in.

3. Determine strut length:

Distance from pipe to supporting steel	4 ft 4 3/4 in.
Clamp takeout (E)	- 1 ft 4 in.
Rear bracket takeout (G)	- 3 1/4 in.
Actual strut length (pin to pin)	2 ft 9 1/2 in.

The actual strut length is less than 7 ft 0 in.—acceptable.

Variable-Spring Selection

The problem: Select a variable spring for a 12-in. pipe in a cooling system. The require load is 1345 lb. The thermal movement is 0.05 in. up. The distance between the elevation of the pipe centerline and top of the assembly is 6 ft 10 1/16 in. Here is the procedure:

1. Determine the series number and type of spring. The series number is determined by the required travel and the allowable variability. The main difference between series is in the spring constant. Spring hangers with high spring constants are small and less expensive. Therefore a spring hanger with a low spring constant should be used only where the variability requirement cannot be met by the spring with higher spring constant. In this problem, the travel is 0.05 in. up and the maximum variability is 15 percent. Both type A and type B spring may be used, since the rod length will be about 6 ft. Type B is selected as it is somewhat simpler.
2. Determine spring size. The size of the spring is determined by the load. The load provided in the stress guidance should be combined with the weight of the members between the pipe and the bottom of the spring to arrive at the hot load. A load deflection chart should be used to determine spring size. Cold and hot load should never be outside the working range.

   The required load is 1345 lb. By inspecting the load deflection chart in the catalog, we pick size 11 for trial. Based on this spring size, the rod size is found to be 3/4 in. After calculating the rod length, we may check the spring size as follows:

3/4-in.-diameter rod (5 ft 0 in.)	1.5 × 5 = 7.5 lb
Part	1.6
Part	+23.9
Subtotal	33.0 lb
Required load	+1345.0
Hot load	1378.0 lb

   Cold load = 1378 × 546 × 1/16 = 1378 + 34 = 1412 lb

   The hot load and the cold load are within the working range of the spring—use size 11.

3. Check variability and swing angle:

$$\text{Variability} = \frac{34 \times 100}{1412}\,\% = 2.4\% < 15\%\ \text{(acceptable)}$$

*Generally, snubber size is based on design load. In some projects, the selection is based on both design load and pipe size.

Rod Length Determination

The basic way to determine rod length is to find the distance between the centerline of the pipe to the bottom of the steel member to which the rod hanger assembly will be attached. Then subtract the lengths occupied by other components of the assembly. The lengths to be subtracted are usually referred to as TO (takeout) dimensions. Here is a typical calculation:

TOS (existing) El.		199 ft 3 in.
Height of W27 x 84 (26.69 in.)		- 2 ft 2 3/4 in.
TO W27 x 84 El.		197 ft 0 1/4 in.
Height of W4 x 13 (4.16 in.)		- 4 1/8 in.
TO W4 x 13 El.		196 ft 8 1/8 in.
TO of pipe El.		- 191 ft 3 1/4 in.
Distance between centerline of pipe to TO W4 x 13		5 ft 4 7/8 in.
Total takeout		
Welded beam attachment	2 in.	
Two weldless eye nuts	2 in.	
Turnbuckle (about 2/3 from top elevation)	3 in.	
Pipe clamp	+6 1/2 in.	
	1 ft 1 1/2 in.	
Total rod length	5 ft 4 7/8 in.	
	- 1 ft 1 1/2 in. (total take out)	
	4 ft 3 3/8 in. (total rod length)	
Number rod length (upper)		3 ft 0 in.
Number rod length (lower)		1 ft 3 in.

SPECIFICATIONS

The following specifications have been adapted from ITT-Grinnell Corporation and Bergen-Patterson Corporation. Pipe support designers and drafters should have at their disposal the standards provided by ANSI and MSS and any other pipe support information or codes that may be required for a job.

*PIPE SUPPORT SPECIFICATION**

A. *SCOPE*

This specification shall apply for the design and fabrication of all hangers, anchors and guides. Where piping design is such that exceptions to this specification are necessary, the particular system will be identified and the exceptions clearly listed through an addendum which will be made part of the specification.

B. *DESIGN*

1. All supports and parts shall conform to the latest requirements of the USA Standard Code for Pressure Piping B31.1.0 and MSS Standard Practice SP-58, except as supplemented or modified by the requirements of this specification.
2. Designs generally accepted as exemplifying good engineering practice, using stock or production parts, shall be utilized wherever possible.
3. Accurate weight balance calculations shall be made to determine the required supporting force at each spring hanger location and the pipe weight load at each equipment connection.
4. Pipe hangers shall be capable of supporting the pipe in all conditions of operation. They shall allow free expansion and contraction of the piping, and prevent excessive stress resulting from transferred weight being induced into the pipe or connected equipment.
5. Wherever possible, pipe attachments for horizontal piping shall be pipe clamps.
6. Horizontal or vertical pipes should be supported preferably at locations of least vertical movement.
7. For critical, high-temperature piping, at hanger locations where the vertical movement of the piping is 3/4" or more, or where it is necessary to avoid the transfer of load to adjacent hangers or connected equipment, pipe hangers shall be of an approved constant support design, as Grinnell Fig. 80-V and Fig. 80-H, or equal.

   An exception may be made in the instance where the piping movement occurs at a hanger supporting a portion of a piping riser on which a rigid support is also located. In this case, variable spring hangers may be used for any amount of expansion up to the full recommended working range of the spring, provided the change in supporting effect of the variable spring is added to the design load of the rigid support assembly.

   Where transfer of load to adjacent hangers or equipment is not critical, and where the vertical movement of the piping is less than 3/4", variable spring hangers may be used, provided the variation in supporting effect does not exceed 25% of the calculated piping load through its total vertical travel.
8. The total travel for constant support hangers will be equal to actual travel plus 20%. In no case will the difference between actual and total travel be less than 1".
9. Variable springs shall be furnished with travel stops. The travel stops are to be factory installed so that the piston cap is set at the "cold" position. The travel stop shall be of such design as to be easily removable, but yet during erection will act as a rigid hanger.
10. Constant supports shall be furnished with travel stops. The travel stops will be factory installed so that the hanger level is at the "cold" position. The travel stops will be of such design as to permit future reengagement, even in the event the lever is at a position other than "cold," without having to make hanger rod adjustments.
11. For low temperature, non-critical systems, where vertical movements up to 2" are anticipated, an approved precompressed variable spring design equal to Grinnell Fig. B-268 may be used. Where the vertical movement is greater than 2", a variable spring hanger equal to Grinnell Fig. 98 may be used. Where movements are of a small magnitude, spring hangers equal to Grinnell Fig. 82 may be used.
12. All rigid hangers shall provide a means of vertical adjustment after erection.
13. Where the piping system is subject to shock loads, such as seismic disturbances or thrusts imposed by the actuation of safety valves, hanger design shall include provision of shock absorbing devices of approved design.
14. Selection of vibration control devices shall not be part of the hanger contractor's work. If vibration is encountered after the piping system is in operation, appropriate vibration control equipment will be installed at the direction of the engineers.
15. Hanger rods shall be subjected to tensile loading only. At hanger locations where lateral or axial movement is anticipated, suitable linkage shall be provided to permit swing.
16. Where horizontal piping movements are such that the hanger rod angularity from the vertical is greater than 4 degrees from the cold to hot position of the pipe, the hanger pipe and structural attachments shall be offset in such a manner that the rod is vertical in the hot position.
17. Hangers shall be designed so that they cannot become disengaged by movements of the supported pipe.
18. For piping 350° and over, hangers shall be spaced in accordance with the latest requirements of the USAS Code for Pressure Piping B31.1.0.

***This specification is reprinted by permission of ITT-Grinnell Corporation.**

Spacings of hangers for other piping may be greater than listed by the above code, provided that:

(a) The sag of the pipe between supports will permit drainage of the system.

(b) No excessive bending or shear stresses result from the pipe weight, or concentrated loads, between supports.

19. Where practical, riser pipe shall be supported independently of the connected horizontal piping.

Pipe support attachments to the riser piping shall be riser clamps of a design equal to Grinnell Hanger Standard 40.

The design loads for rigid riser supports shall be as follows:

STEAM LINES—Stock sizes for riser clamps shall be selected using allowable stresses listed in MSS SP-58 and USAS B31.1.0 Section 121.1.2 for the maximum operating load or hydrostatic test load, whichever is greater.

WATER-FILLED LINES—Design shall be based on the maximum operating load. Selection of riser clamp stock size shall be based on MSS SP-58 and USAS B31.1.0 Section 121.1.2.

20. Supports, guides, and anchors shall be so designed that excessive heat will not be transmitted to the building steel. The temperature of supporting parts shall be based on a temperature variation factor of 100°F per inch distance from the outside surface of the pipe.

*PIPE SUPPORT SPECIFICATIONS FOR PLUMBING AND HEATING**

A. GENERAL

All support components shall conform to Manufacturers' Standardization Society Specification SP-58, 1967 Edition.

B. INSTALLATION–HORIZONTAL PIPING

1. Hangers shall adequately support the piping system. They shall be located near or at changes in piping direction and concentrated loads. They shall provide vertical adjustment to maintain pitch required for proper drainage. They shall allow for expansion and contraction of the piping.
2. Hangers shall be fastened to existing building steel members wherever practicable.
3. Horizontal *steel* piping shall be supported as below:

Pipe Size	*Rod Diameter*	*Maximum Spacing*
Up to 1-1/4"	3/8"	8 Ft.
1-1/2" and 2"	3/8"	10 Ft.
2-1/2" and 3-1/2"	1/2"	12 Ft.
4" and 5"	5/8"	15 Ft.
6"	3/4"	17 Ft.
8" to 12"	7/8"	22 Ft.

4. Horizontal lines of *copper tubing* shall be supported as below:

Nom. Tube Size	*Rod Diameter*	*Maximum Spacing*
Up to 1"	3/8"	6 Ft.
1-1/4" and 1-1/2"	3/8"	8 Ft.
2"	3/8"	9 Ft.
2-1/2"	1/2"	9 Ft.
3" and 4"	1/2"	10 Ft.

5. Support horizontal *cast iron* soil pipe with one hanger for each pipe length. Hangers shall be sized for actual O.D. of cast iron pipe and located close to hub.
6. Horizontal PVC piping shall be supported by a continuous member which itself may be supported at spacings compatible with the location of building structural members.

***These specifications are reprinted by permission of ITT-Grinnell Corporation.**

RECOMMENDED SPECIFICATIONS FOR PIPE SUPPORTS

SCOPE

This specification covers the design, fabrication and furnishing of all hangers, supports, anchors, guides, supplementary steel and special appurtenances required for the proper support of all piping included under this contract.

The extent of design and fabrication of hangers for the various classifications of piping systems shall be as delineated herein.

DESIGN

(a) Hangers and supports shall be designed to meet all static and operational conditions to which the piping and equipment may be subjected. The supporting systems shall provide for and control, subject to the requirements of the piping configuration, the free or intended movement due to the thermal expansion and contraction of the piping and connected equipment.

(b) All supports and hanger components shall conform to the latest requirements of the ASA Code for Pressure Piping B-31.1 and MSS Standard Practice SP-58.

(c) The design should take full advantage of commercially available load rated and tested hanger components to the greatest extent possible. An effort should be made to maintain uniformity and simplicity in design. All hangers shall be functional, provide means for piping elevation adjustment, and be readily installable using normal field labor and equipment.

(d) Support points shall be selected on the basis of proper location and spacing for optimum load distribution and weight balance, taking into consideration available building structure from which hangers can be suspended. Spacing shall be limited so as to prevent excessive sag, bending and shear stresses in the piping, and to keep within the allowable building loading limitations. Maximum spans between hangers where not otherwise limited shall be in accordance with the recommendations of ASA B-31.1.

(e) The supporting force required and movement at each hanger location shall be determined by approved methods of calculation. Effects of shock forces, testing and cleaning fluids on the overall loading pattern, shall also be determined by formal calculation and provision made to accomodate same.

(f) At points of support subject to horizontal movement, suspended type hangers shall provide for the movement by the swing of long hanger rods. Normally accepted practice limits rod swing to a maximum of 4 degrees from the vertical. Where the angular rotation exceeds this maximum some form of roller or trolley should be incorporated to accomodate for the movement. For piping supported from below, an approved form of slide must be incorporated or in the case of assured longitudinal movement, a pipe roll may be used.

(g) All piping shall be carried on adjustable hangers or properly leveled supports.

(h) All screw or equivalent adjustments shall be provided with suitable locking features.

(i) All parts shall be fabricated and assembled so that they will not be disengaged by movement of the supported piping.

(j) Pipe attachments for horizontal piping shall be pipe clamps wherever possible.

(k) Pipe attachments for vertical risers shall be approved riser clamps with hold-down lugs, or fabricated welded lugs. Welded attachments shall be of compatible material to that of the pipe, designed and welded in accordance with governing codes.

(l) Covering on insulated piping shall be protected from damage at all hanger locations. Saddles, bases or suitable metal shields, properly constructed and secured to the covered pipe, shall be used at points of roller, base and trapeze support. Pipe clamps for insulated pipe shall have the load bolt located outside of the insulation.

(m) Auxiliary supporting steel shall be designed on the basis of the allowable stresses as per the AISC Steel Handbook or applicable local building code.

(n) Attachment to building steel structure shall be in the form of standard recognized welded lug components. All welding to the underside flange shall be parallel to the beam web.

(o) Embedded anchor bolts or inserts are preferred whenever possible for attachment to building concrete structure. Where required, approved concrete fasteners may be used. Hanger rods shall not be threaded directly into a fastener; attachment by conventional means to an intermediate plate or member held in place by the concrete fastener is preferred.

(p) At points of support, subject to vertical movement, springs of suitable design to prevent excessive variation in supporting effect shall be used to provide for the movement. The amount of variation that can be tolerated shall be based on such considerations as bending effect, control of piping elevation, allowable terminal loadings, etc. In general, the deviation in supporting effect shall be limited to plus or minus 6% for critical systems such as main steam, hot and cold reheat, extraction lines over 750°F, and the portions of the boiler feed discharge system in the vicinity of the pumps and boiler terminal connection. On non-critical systems, the variation in supporting effect shall be limited to 25%. For all systems a greater allowance in percent load change is permissible at points of support where the variation in supporting effect is transferred directly to a rigid support or terminal connection specifically designed for the resulting loading condition.

(q) Constant supports shall be of rugged construction, with spring and all moving parts suitably covered or protected. Each unit shall be individually shop calibrated to support the operating load with provision for possible field load adjustment equal to a minimum of plus or minus 10% of the set load. Load deviation shall not exceed 6% thru-out the total travel range. It is recommended that the total travel of the unit selected be the calculated travel plus 20% (minimum ½ inch), considered good practice to take care of any possible discrepancy between the calculated and the actual thermal movement. Constant supports shall have travel stops, distant visible travel indicators and scale, hot and cold setting buttons, and shall have means for shop pre-setting to required installing position and provision to accomodate hydrostatic test, as in Bergen CSH, CSV, CSB, CSm, or equal.

(r) Variable supports shall have an affixed load and travel scale with hot and cold position clearly indicated. All units shall be factory pre-set to the required installing position and shall have positive means for locking the piston plate to facilitate erection and accommodate hydrostatic test, as in Bergen VS1, VS2, VS4, or equal.

(s) Piping installations located in seismic areas, outdoor lines subject to wind loadings, points in piping systems known to have possible shock loadings such as generated by quick closing valves, water hammer and possible relief valve reaction forces shall be provided with shock absorbing devices of approved design. The specific means provided shall not interfere with the normal or intended thermal movement of the piping system.

(t) Provision for vibration, at other than commonly known points susceptible to vibration, shall be considered as being beyond the scope of this contract. Should vibration be experienced after initial operation, any additional design or vibration control equipment required shall be furnished as an addition to the contract.

HANGER DESIGN SCOPE AND PRESENTATION

Purchasing specification should clearly define the scope of the piping support contract and designate which piping systems or pipe sizes shall be supplied as fully engineered, fabricated and shop assembled hangers. Recommended practice calls for design of rigid hangers for piping 6" and larger and spring hangers for piping 3" and larger.

Engineered hangers shall be completely shop fabricated and individually assembled insofar as is practical, tagged and bundled prior to shipment. A location plan showing specific location of each hanger with respect to building structure, movement offsets, and any required supplemental framing shall be furnished for erection purposes.

Rigid hangers for piping 5" down to and including 1" shall be furnished in the form of random material for field fabrication. A set of piping drawings shall be marked with approximate hanger locations and a standard hanger sheet showing typical random support arrangements will be furnished for field guidance.

Springs for piping 2½" and smaller shall be sized by approximate methods, identified and located by mark number and shipped separately for incorporation in the field fabricated hanger.

It is recommended that hangers for piping smaller than 1" be fabricated in the field using a stock of hanger components, field purchased under a separate contract.

*PIPE SUPPORT SPECIFICATIONS**

*SELECTING PIPE SUPPORTS**

10 BASIC STEPS

1. Make isometric piping sketch.
2. Spot preliminary location of hangers on sketch.
3. Study building steel, and adjust location of hangers to suit.
4. Check for interference.
5. Calculate distribution of weight of piping.
6. Summarize hanger loadings.
7. Calculate distribution of vertical expansion to hangers.
8. Calculate distribution of equipment vertical movement to hangers.
9. Summarize hanger movements.
10. Choose hangers for loadings and movements.

As in all fields of engineering, the most effective working tool a man can be given is a simple, clear-cut procedure to follow in arriving at a sound, economic solution. A well thought out procedure will be applicable to all jobs; only the extent to which each individual step is carried out will vary depending upon the nature of the particular problem at hand. The procedure for the design and selection of pipe supports has been broken down into the ten basic steps shown above. With the use of a typical pipe support problem, we will endeavor to point out some of the considerations and methods involved in performing the various steps.

Before starting with a typical problem, it would be well to discuss hangers generally and also to review some of the basic data peculiar to the pipe support study.

The two main factors governing selection of pipe hangers are:

(1) Changes resulting from thermal expansion which causes movement of pipe due to increase of length of legs, and displacement of equipment connections.

(2) Weight to be supported, which depends on pipe used, flowing medium, insulation chosen, and number and type of fittings in line supported.

A formal solution for movement and loads would approach that of pipe stress calculations in complexity. In general, these costly, time consuming calculations can be replaced by simple practical solutions that give results well within the pipe tolerances that can be expected in a normal installation. Most plant piping can be hung or supported correctly by using common sense and a minimum of theory and calculation.

To hang or support pipe properly, certain basic data must be on hand. Facts assembled here have been helpful in the solution of practical hanger and support problems. Let us examine these data before seeing how they are used.

Pipes expand with temperature increase. Movement accompanying this expansion is one of the two most important factors that must be taken into consideration when selecting hangers or supports. Table I gives expansion in inches per foot for various carbon and alloy steels based on an installation temperature of 70°F. As an example, assume a line with a 70°F installation temperature and 900°F operating temperature; expansion for low chrome pipe is 7.81 inches per 100 feet or .0781 inches per foot. For a run of 50 feet of straight pipe under these conditions, the total expansion at operating temperature is 50 (.0781) = 3.905 inches.

Hanger spacing must be close enough to prevent excessive sagging that overstresses pipe or interferes with drainage. Table II, page 86, lists maximum recommended spacing between supports for various sizes of pipe. For estimating purposes on piping 3" and larger, a good rule of thumb for determining hanger spacing is pipe size in inches plus 10 to obtain allowable span in feet.

Now that we know what data are used when selecting pipe supports, we can proceed with the sample problem. For fast accurate hanging of pipes, follow these ten steps.

1. Draw a freehand isometric sketch of the piping system. A sketch similar to Fig. 1 is good. It is not necessary to draw it to exact scale; if proportions are approximate, results will be satisfactory.

2. Spot in hangers tentatively. First put in end hangers as near terminal connections as possible. Keeping hangers close prevents overloading connections on equipment. Locate hangers at or near any concentrated loads such as heavy valves, risers, etc. Pick up all horizontal bends to prevent any excessive overhang. Next, refer to Table II and space intermediate hangers so that recommended spacing is not exceeded for size pipe being hung.

3. Examine building framing in vicinity of hanger locations and adjust locations on sketch to minimize need for additional hanger supporting steel.

4. Check for possible interference with other piping and equipment.

5. Calculate distribution of weight of pipe. Use the simple and practical methods shown for distributing weights of vertical legs, bends, and for obtaining zero load at equipment flange.

6. Summarize hanger loadings in tabular form as shown. This will also provide a weight check of the piping system. It is also desirable to show the hanger loading on the isometric sketch.

7. Calculate distribution of pipe expansion to each hanger. Select the method for this from those shown on page 85 based on the number of vertical legs. Follow calculation procedure shown. Establish movement of top and bottom of vertical legs first. Next, distribute this movement to hangers.

8. Calculate distribution of equipment vertical movement to each hanger. In some cases, the movement of the flange is given by the manufacturer. If information is not available, and unit operates hot, establish the amount of movement at unit flange, to which the pipe connects. Do this by taking the vertical distance of center of flange face from the point of no movement of the unit. This is the point where the unit is fastened to the cold, structural steel or concrete. Multiply this distance by the expansion per foot at the operating temperature of the unit.

9. Summarize the hanger movements in tabular form, combining the movements due to pipe expansion with those due to equipment movement as shown. If required, the same principles may be used to determine horizontal movements. It is desirable to mark the movement of each hanger on the isometric sketch.

10. Now we have gathered sufficient data to select the proper hangers. Type hanger selected is based on position of support point with relation to supporting structure, interferences that must be cleared and amount of piping movement the hanger must accommodate. For use of constant and variable support spring hangers please refer to applicable sections of this catalog.

*Reprinted by permission of Bergen-Patterson Pipe Support Corporation.

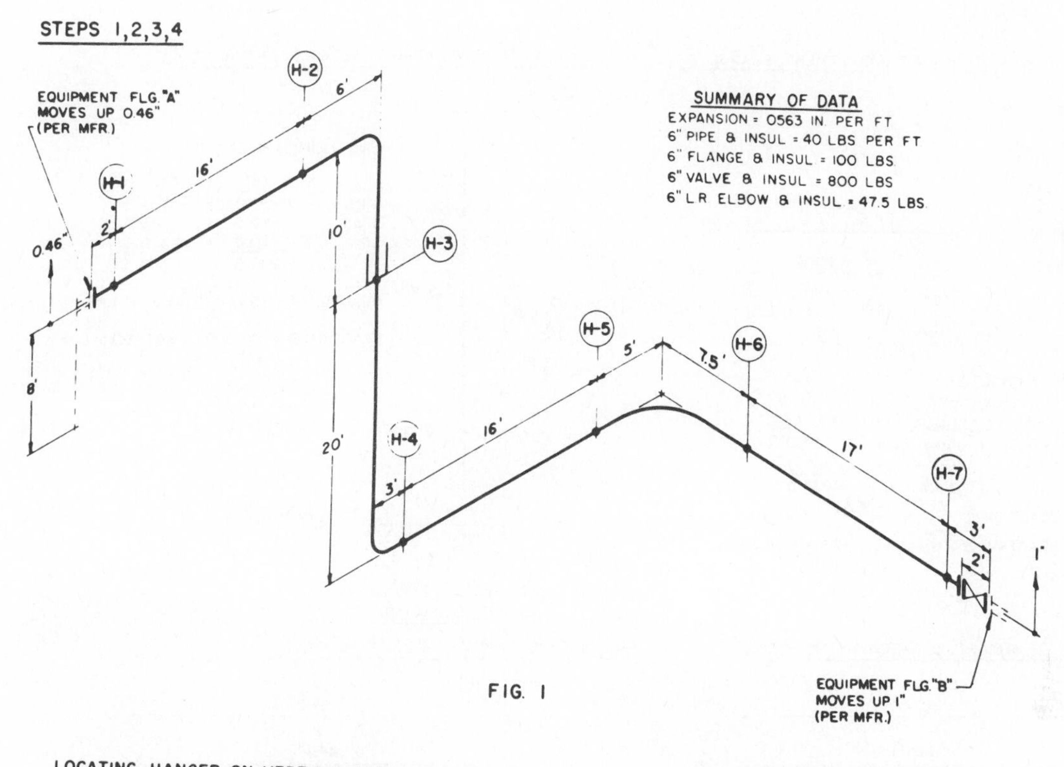

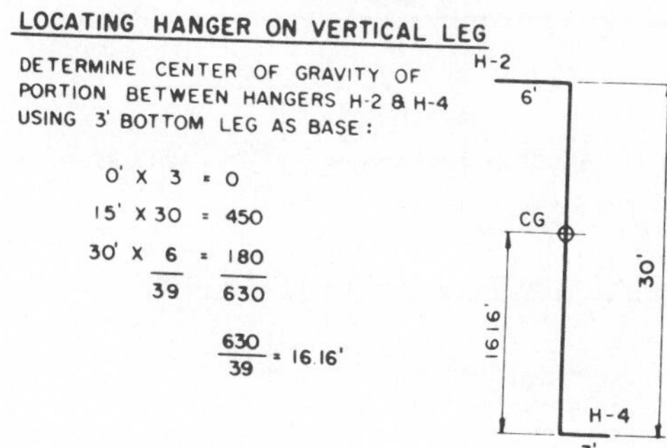

FIG. 8-40 Selection and placement of pipe supports. (Courtesy Bergen-Patterson Corp., A Massachusetts Corp.)

STEP 5

DISTRIBUTION OF WEIGHT BETWEEN EQUIPMENT FLANGE "A" & H-1

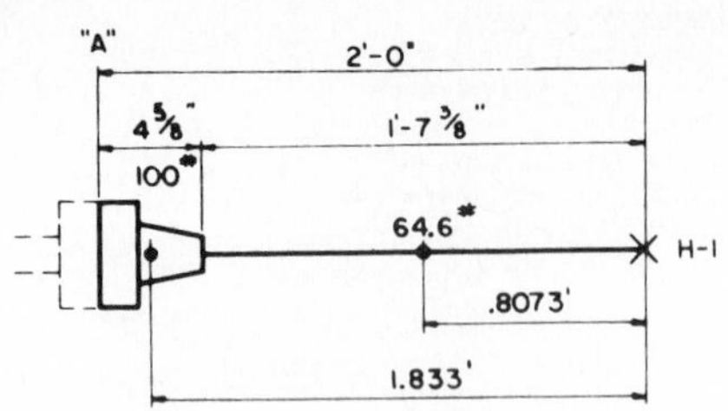

TAKING MOMENTS ABOUT H-1

FT.	X	LBS.	=	FT.-LBS.
.8073		64.6	=	52.2
1.833		100.0	=	183.3
		164.6		235.5

REACTION @ FLG."A" = $\frac{235.5}{2.0}$ = 117.8 LBS.

REACTION @ H-1 = 164.6 − 117.8 = 46.8 LBS.

DISTRIBUTION OF WEIGHT BETWEEN H-3 & H-4.

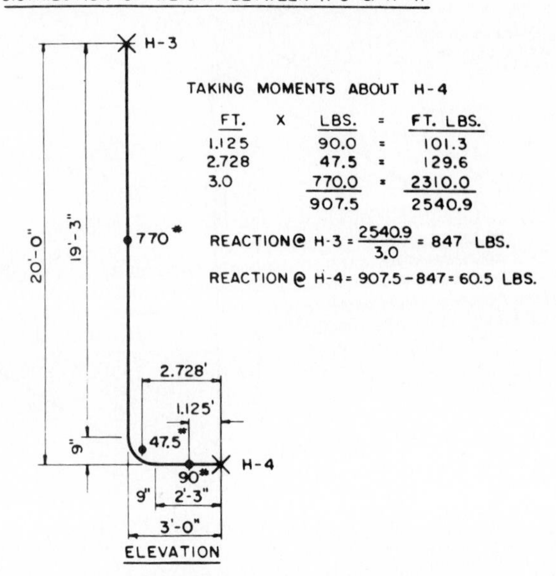

TAKING MOMENTS ABOUT H-4

FT.	X	LBS.	=	FT. LBS.
1.125		90.0	=	101.3
2.728		47.5	=	129.6
3.0		770.0	=	2310.0
		907.5		2540.9

REACTION@ H-3 = $\frac{2540.9}{3.0}$ = 847 LBS.

REACTION @ H-4 = 907.5 − 847 = 60.5 LBS.

DISTRIBUTION OF WEIGHT BETWEEN H-1 & H-2

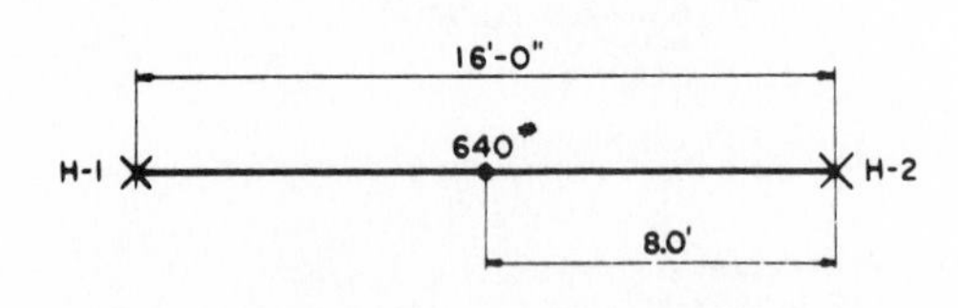

REACTIONS H-1 & H-2 = $\frac{640}{2}$ = 320 LBS.

DISTRIBUTION OF WEIGHT BETWEEN H-4 & H-5

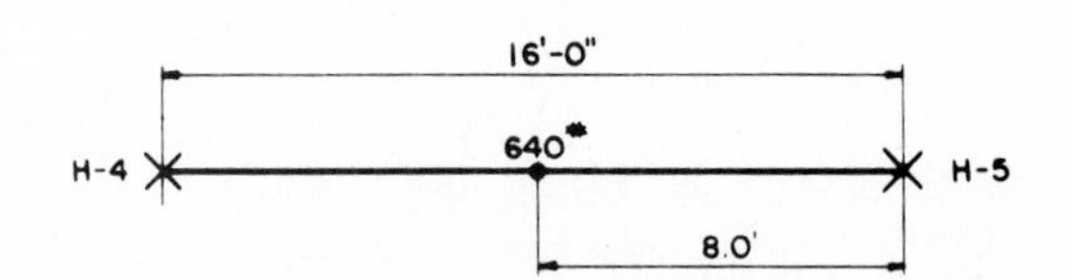

REACTION@ H-4 & H-5 = $\frac{640}{2}$ = 320 LBS.

DISTRIBUTION OF WEIGHT BETWEEN H-2 & H-3

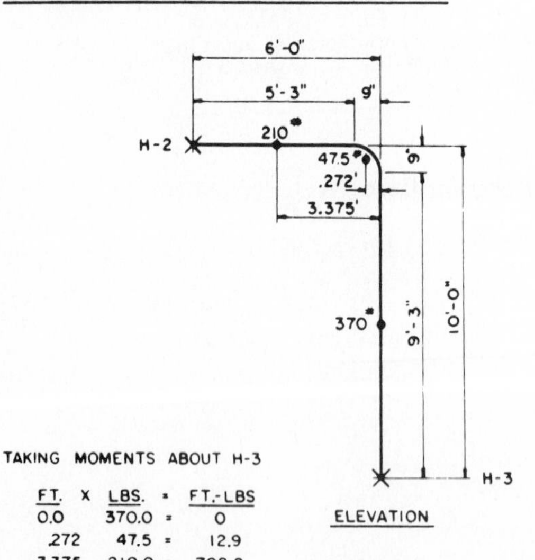

TAKING MOMENTS ABOUT H-3

FT.	X	LBS.	=	FT.-LBS
0.0		370.0	=	0
.272		47.5	=	12.9
3.375		210.0	=	708.8
		627.5		721.7

REACTION@ H-2 = $\frac{721.7}{6.0}$ = 120.3 LBS.

REACTION @ H-3 = 627.5 − 120.3 = 507.2 LBS.

DISTRIBUTION OF WEIGHT BETWEEN H-5 & H-6

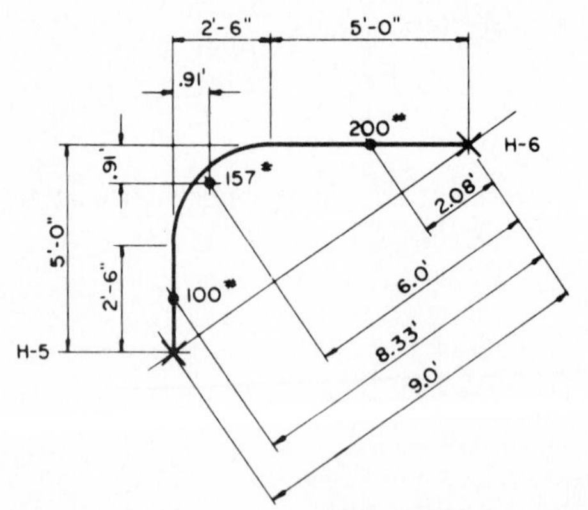

TAKING MOMENTS ABOUT H-6

FT.	X	LBS	=	FT.-LBS.
2.08		200	=	416.0
6.00		157	=	942.0
8.33		100	=	833.0
		457	=	2191.0

REACTION @ H-5 = $\frac{2191}{90}$ = 243.4 LBS.

REACTION @ H-6 = 457 − 243.4 = 213.6 LBS.

FIG. 8-40 Selection and placement of pipe supports (cont'd). (Courtesy of Bergen-Patterson Corp., A Massachusetts Corp.)

STEP 5 (CON'T.)

DISTRIBUTION OF WEIGHT BETWEEN H-6 & H-7 TO MAINTAIN ZERO REACTION ON FLANGE "B"

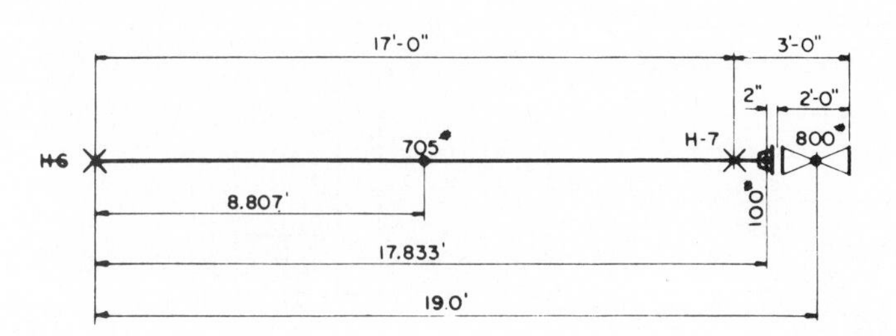

TAKING MOMENTS ABOUT H-6

FT.	X	LBS.	=	FT.-LBS.
8.807		705	=	6210
17.833		100	=	1783
19.0		800	=	15200
		1605		23193

REACTION @ H-7 = $\frac{23193}{17.0}$ = 1364 LBS.

REACTION @ H-6 = 1605 − 1364 = 241 LBS.

STEP 6

SUMMARY OF HANGER LOADINGS

HANGER MARK	REACTIONS							HANGER LOAD
	A TO H-1	H-1 TO H-2	H-2 TO H-3	H-3 TO H-4	H-4 TO H-5	H-5 TO H-6	H-6 TO H-7	
FLANGE "A"	117.8							117.8
H-1	46.8	320.0						366.8
H-2		320.0	120.3					440.3
H-3			507.2	847.0				1354.2
H-4				60.5	320.0			380.5
H-5					320.0	243.4		563.4
H-6						213.6	241.0	454.6
H-7							1364.0	1364.0
FLANGE "B"							0.0	0.0
							TOTAL	5041.6

STEP 7

DISTRIBUTION OF VERTICAL EXPANSION TO HANGERS

FLATTEN OUT PIPE SHAPE INTO ONE PLANE AND ESTABLISH MOVEMENT AT TOP AND BOTTOM OF VERTICAL LEG.
USE METHOD FOR ONE VERTICAL LEG SHOWN ON PAGE 85.

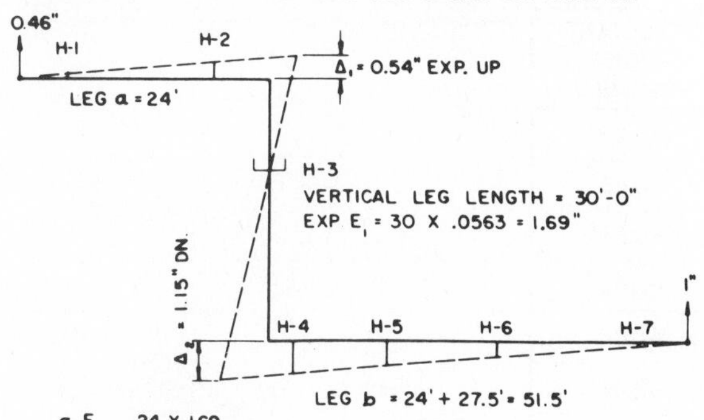

$\Delta_1 = \frac{a E_1}{a+b} = \frac{24 \times 1.69}{24 + 51.5} = 0.54"$ UP

$\Delta_2 = E_1 - \Delta_1 = 1.69 - 0.54 = 1.15"$ DN.

USING CASE 2 FORMULA ON PAGE 85, DETERMINE DISTRIBUTION OF MOVEMENT AT TOP OF VERTICAL LEG H-1 & H-2.

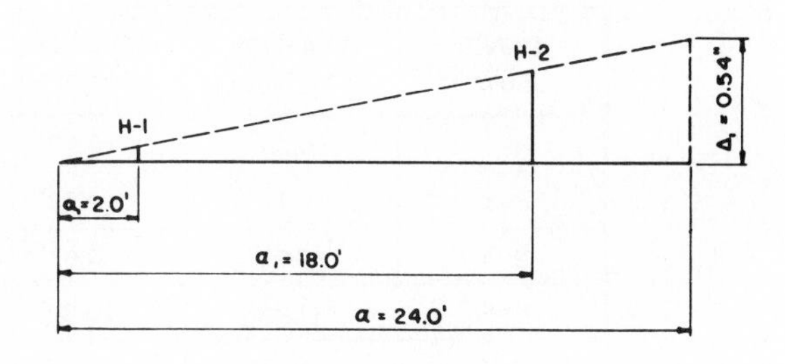

$\Delta_x = \frac{a_1 \Delta_1}{a}$

Δ_x @ H-1 = $\frac{2.0 \times 0.54}{24.0}$ = .045"

Δ_x @ H-2 = $\frac{18.0 \times 0.54}{24.0}$ = 0.405"

FIG. 8-40 Selection and placement of pipe supports (cont'd). (Courtesy of Bergen-Patterson Corp., A Massachusetts Corp.)

STEP 7 (CON'T)

MOVEMENT AT HANGER H-3

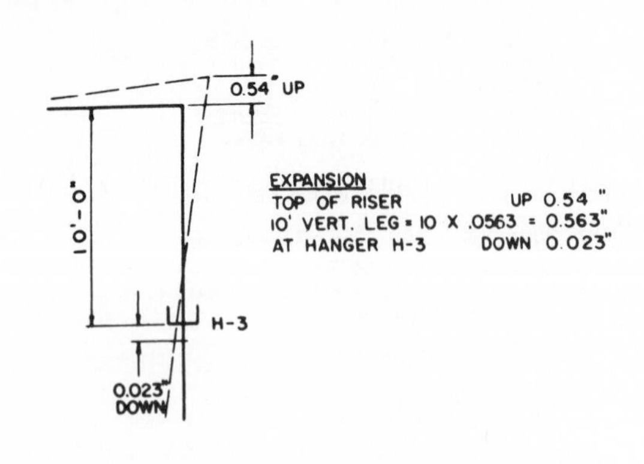

USING CASE 2 FORMULA ON PAGE 85 DETERMINE DISTRIBUTION OF MOVEMENT AT BOTTOM OF VERTICAL LEG TO HANGERS H-4, H-5, H-6 & H-7.

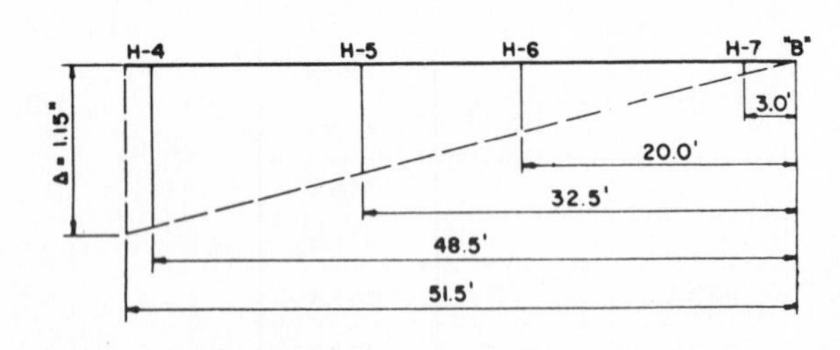

Δ_x@ H-4 = 48.5 X 1.15 ÷ 51.5 = 1.082" DOWN
Δ_x@ H-5 = 32.5 X 1.15 ÷ 51.5 = 0.725" DOWN
Δ_x@ H-6 = 20.0 X 1.15 ÷ 51.5 = 0.447" DOWN
Δ_x@ H-7 = 3.0 X 1.15 ÷ 51.5 = 0.067" DOWN

STEP 8

USING CASE 3 FORMULA ON PAGE 85 DETERMINE DISTRIBUTION OF EQUIPMENT VERTICAL MOVEMENT TO HANGERS.

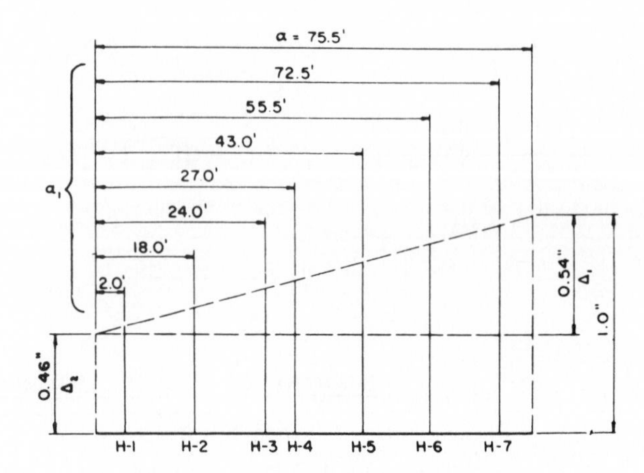

$$\Delta_x = \frac{a_1\,\Delta_1}{a} + \Delta_2$$

Δ_x@ H-1 = $\frac{2 \times 0.54}{75.5}$ + 0.46 = 0.014 + 0.46 = 0.47" UP

Δ_x@ H-2 $\frac{18 \times 0.54}{75.5}$ + 0.46 = 0.129 + 0.46 = 0.58" UP

Δ_x@ H-3 = $\frac{24 \times 0.54}{75.5}$ + 0.46 = 0.172 + 0.46 = 0.63" UP

Δ_x@ H-4 = $\frac{27 \times 0.54}{75.5}$ + 0.46 = 0.193 + 0.46 = 0.65" UP

Δ_x@ H-5 = $\frac{43 \times 0.54}{75.5}$ + 0.46 = 0.308 + 0.46 = 0.77" UP

Δ_x@ H-6 = $\frac{55.5 \times 0.54}{75.5}$ + 0.46 = 0.397 + 0.46 = 0.86" UP

Δ_x@ H-7 = $\frac{72.5 \times 0.54}{75.5}$ + 0.46 = 0.518 + 0.46 = 0.98" UP

STEP 9

SUMMARY OF HANGER MOVEMENTS
(FROM STEP 7 AND STEP 8)

HANGER MARK	EXPANSION MOVEMENT	EQUIPMENT MOVEMENT	RESULTANT HANGER MOVEMENT
H-1	+0.05"	+0.47"	+0.52"
H-2	+0.41"	+0.58"	+0.99"
H-3	-0.02"	+0.63"	+0.61"
H-4	-1.08"	+0.65"	-0.43"
H-5	-0.73"	+0.77"	+0.04"
H-6	-0.45"	+0.86"	+0.41"
H-7	-0.07"	+0.98"	+0.91" *

(+) DENOTES UP MOVEMENT; (-) DENOTES DOWN MOVEMENT

FIG. 8-40 Selection and placement of pipe supports (cont'd). (Courtesy of Bergen-Patterson Corp., A Massachusetts Corp.)

SAMPLE WEIGHT BALANCE LOAD CALCULATION

THE FOLLOWING EXAMPLE IS USED TO ILLUSTRATE THE CENTER OF GRAVITY METHOD FOR DETERMINING HANGER LOADINGS. THIS METHOD IS VERY USEFUL FOR LOCATING HANGERS AND DETERMINING HANGER LOADS WHERE LOADINGS ON EQUIPMENT FLANGES MUST EQUAL ZERO OR NOT EXCEED A SPECIFIED LIMIT.

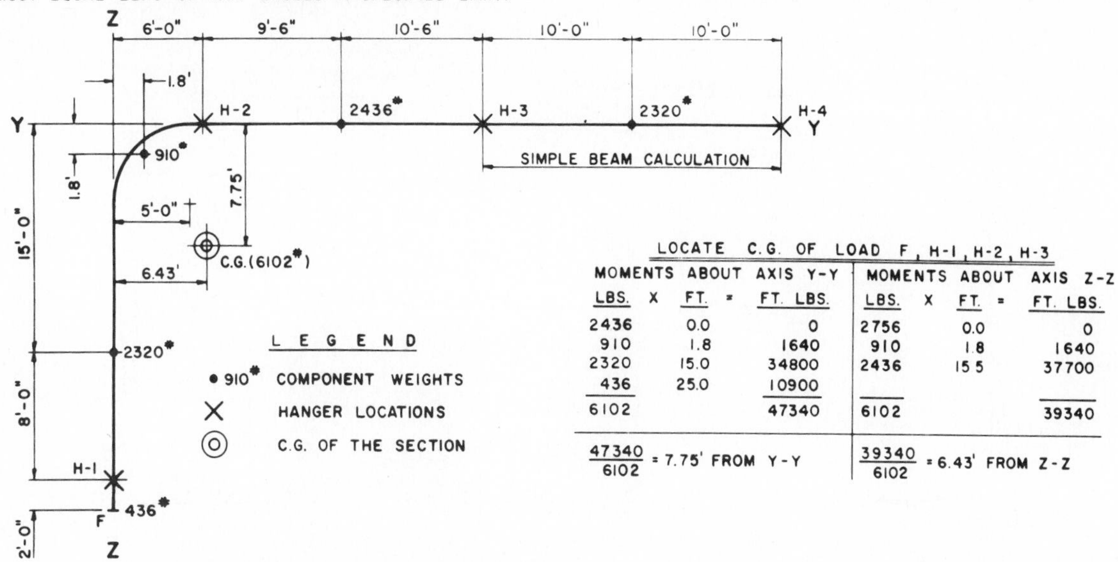

LOCATE C.G. OF LOAD F, H-1, H-2, H-3

MOMENTS ABOUT AXIS Y-Y				MOMENTS ABOUT AXIS Z-Z			
LBS.	X	FT. =	FT. LBS.	LBS.	X	FT. =	FT. LBS.
2436		0.0	0	2756		0.0	0
910		1.8	1640	910		1.8	1640
2320		15.0	34800	2436		15.5	37700
436		25.0	10900				
6102			47340	6102			39340

$\frac{47340}{6102}$ = 7.75' FROM Y-Y

$\frac{39340}{6102}$ = 6.43' FROM Z-Z

CASE 1 – THREE HANGERS WITH ZERO REACTION AT TERMINAL FLANGE "F".

CONSTRUCT THE TRIANGLE H-1, H-2, H-3 SUPERIMPOSING LOCATION OF C.G. AND CALCULATE REACTIONS AT H-1, H-2 & H-3.

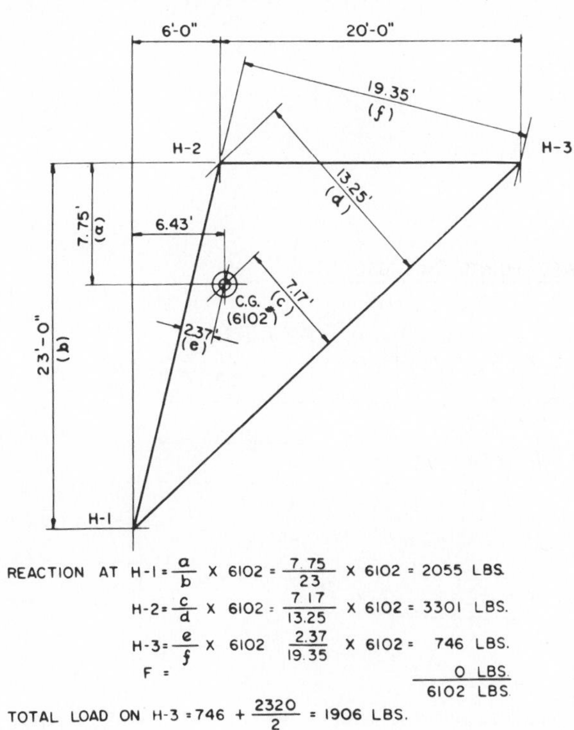

REACTION AT H-1 = $\frac{a}{b}$ X 6102 = $\frac{7.75}{23}$ X 6102 = 2055 LBS.

H-2 = $\frac{c}{d}$ X 6102 = $\frac{7.17}{13.25}$ X 6102 = 3301 LBS.

H-3 = $\frac{e}{f}$ X 6102 $\frac{2.37}{19.35}$ X 6102 = 746 LBS.

F = 0 LBS.

6102 LBS.

TOTAL LOAD ON H-3 = 746 + $\frac{2320}{2}$ = 1906 LBS.

CASE 2 – TWO HANGERS WITH ZERO REACTION AT TERMINAL FLANGE "F".

H-1-ELIMINATED
H-2-RELOCATED
H-3-REMAINING IN SAME POSITION

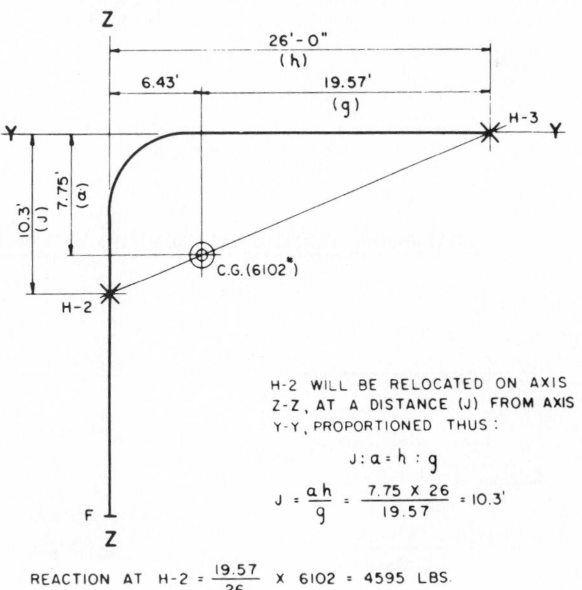

H-2 WILL BE RELOCATED ON AXIS Z-Z, AT A DISTANCE (J) FROM AXIS Y-Y, PROPORTIONED THUS:

$$J : a = h : g$$

$$J = \frac{ah}{g} = \frac{7.75 \times 26}{19.57} = 10.3'$$

REACTION AT H-2 = $\frac{19.57}{26}$ X 6102 = 4595 LBS.

H-3 = $\frac{6.43}{26}$ X 6102 = 1507 LBS.

6102 LBS.

TOTAL LOAD ON H-3 = 1507 + $\frac{2320}{2}$ = 2667 LBS.

FIG. 8-40 Selection and placement of pipe supports (cont'd). (Courtesy of Bergen-Patterson Corp., A Massachusetts Corp.)

SIMPLIFIED METHOD FOR APPROXIMATING PIPING MOVEMENTS

ESTABLISHING MOVEMENTS AT TOP AND BOTTOM OF VERTICAL LEGS

FOR COMPLEX PIPING SYSTEMS FLATTEN OUT PIPE SHAPE INTO ONE PLANE AND COMBINE ADJACENT HORIZONTAL LEGS.

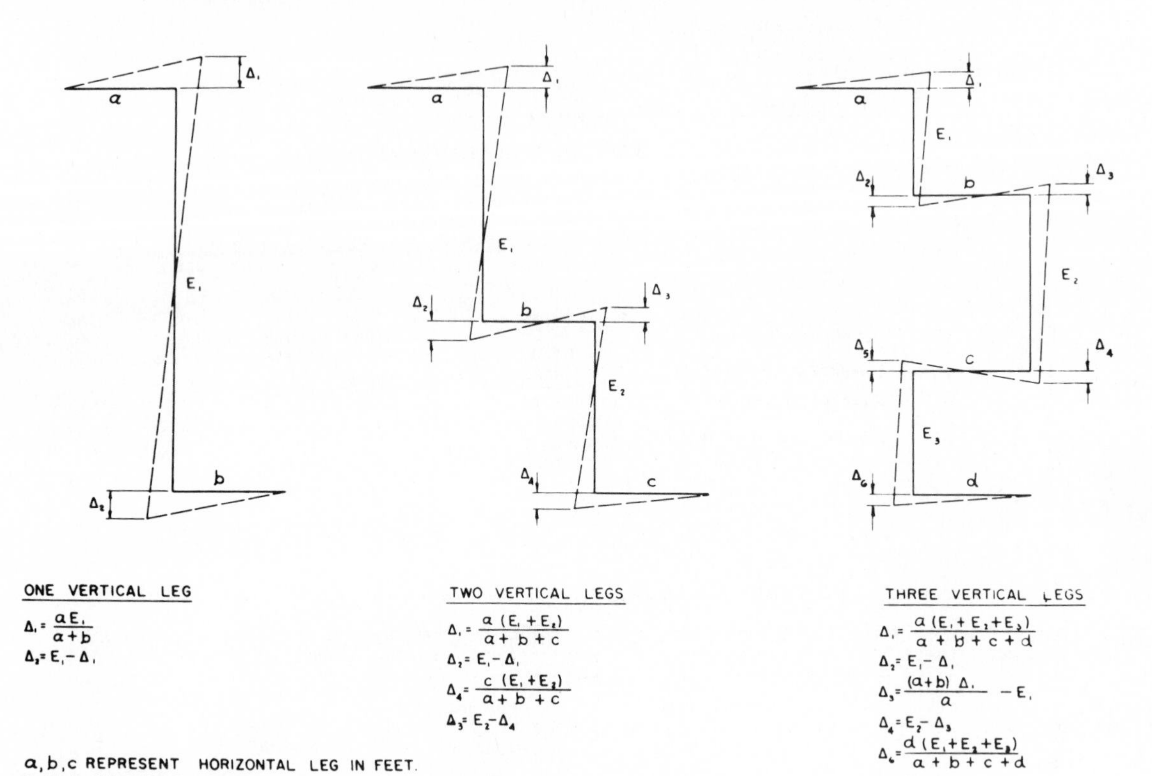

DISTRIBUTING VERTICAL MOVEMENT TO INTERMEDIATE POINTS ON HORIZONTAL LEGS

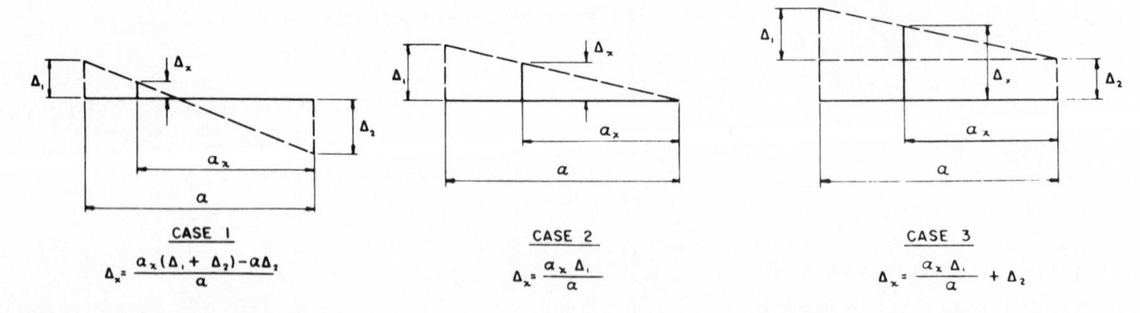

NOTE: IF Δ_x IS PLUS (+) MOTION IS UP
IF Δ_x IS MINUS (−) MOTION IS DOWN

ALL CASES MAY BE SOLVED GRAPHICALLY BY DRAWING DEFLECTIONS TO FULL SIZE AND THE LENGTH OF HORIZONTAL LEGS AND LOCATION OF HANGERS TO SOME CONVENIENT SCALE.

FIG. 8-40 Selection and placement of pipe supports (cont'd). (Courtesy of Bergen-Patterson Corp., A Massachusetts Corp.)

TABLE I

THERMAL EXPANSION OF PIPE MATERIALS

INCHES PER FOOT – FROM 70°

SATURATION GAGE PRESSURE		Temp. °F.	Carbon Carbon-Moly Low Chrome-Moly (Thru 3% Cr)	Intermediate Alloy Steels (5 thru 9 CrMO)	Austenitic Stainless Steels	Straight Chromium Stainless Steels 12Cr, 17Cr, 27Cr
Vacuum in. of Mercury	Press. psi					
29.18	–	70	0.00	0.00	0.00	0.00
27.99	–	100	.0022	.0021	.0033	.0019
22.35	–	150	.0060	.0057	.0089	.0052
6.46	–	200	.0099	.0093	.0146	.0085
	15.1	250	.0140	.0131	.0203	.0120
	52.3	300	.0182	.0170	.0261	.0156
	119.9	350	.0225	.0210	.0321	.0192
	232.6	400	.0270	.0250	.0380	.0230
	407.9	450	.0316	.0292	.0440	.0269
	666.1	500	.0362	.0335	.0501	.0308
	1030	550	.0411	.0379	.0562	.0349
	1528	600	.0460	.0424	.0624	.0390
	2193	650	.0511	.0469	.0687	.0431
	3079	700	.0563	.0514	.0750	.0473
		750	.0616	.0562	.0815	.0516
		800	.0670	.0610	.0880	.0560
		850	.0725	.0658	.0946	.0604
		900	.0781	.0707	.1012	.0649
		950	.0835	.0756	.1080	.0694
		1000	.0889	.0806	.1148	.0740
		1050	.0946	.0855	.1216	.0785
		1100	.1004	.0905	.1284	.0831
		1150	.1057	.0952	.1362	.0875
		1200	.1110	.1000	.1420	.0920
		1250	.1166	.1053	.1488	.0965
		1300	.1222	.1106	.1556	.1011
		1350	.1278	.1155	.1624	.1056
		1400	.1334	.1205	.1692	.1101

TABLE II

MAXIMUM RECOMMENDED SPACING BETWEEN PIPE SUPPORTS

Nominal Pipe Size Inches	Maximum Span Feet
1	7
1½	9
2	10
2½	11
3	12
3½	13
4	14
5	16
6	17
8	19
10	22
12	23
14	25
16	27
18	28
20	30
24	32

FIG. 8-40 Selection and placement of pipe supports (cont'd). (Courtesy of Bergen-Patterson Corp., A Massachusetts Corp.)

QUIZ

1. Describe the major functions of a pipe support.
2. What are the different elements in the design and placement of pipe supports when comparing non-critical versus critical situations?
3. What are the seismic categories of pipe supports?
4. What is the reason for sloping a piping system?
5. What is the difference between a pipe hanger and a pipe clamp?
6. What are the main components of a rod hanger?
7. What is the swing angle of a pipe support?
8. Define adjustability with respect to a pipe support.
9. What is the difference between a constant-load and variable-load spring hanger?
10. For preengineered hangers (spring), the ____________ type is the most common.
11. What is a double-spring hanger and what is its function?
12. Describe a trapeze assembly and two applications.
13. Variable-spring hangers are divided into three series based on ____________ allowed.
14. Why is thermal movement an important factor in pipe support design?
15. Define hot load versus cold load.
16. Give a definition for calculating variability.
17. What does a constant-load spring hanger do?
18. Describe the difference between a swing strut and a snubber.
19. What are the available types of shock suppressors and their differences?
20. When using two snubbers at one location on a pipe, they should be of ____________ performance.
21. Snubbers have a ____________ degree maximum swing angle.
22. What is the difference between a restraint and an anchor?
23. What are three uses of pipe slides?
24. Pipe shoes are not used on ____________ pipes except for ____________ of the pipe system.
25. Pipe guides are used for thermal movement to ____________.
26. What are the functions and uses of pipe saddles?
27. Define structural attachments and explain their uses in pipe supporting.
28. Name the eight major structural shapes for pipe supports.

29. What are the major functions of a pipe support group?
30. What is the working load of a pipe system?
31. Why are pipe saddles used on insulated pipes in some cases?
32. Define seismic bracing.
33. Piping systems must not be supported from or by equipment because ___________.
34. Why is pipe support spacing important?
35. Steam lines need to be ___________ because of condensation buildup.
36. What are pipe support details used for?
37. What is a pipe support location drawing used for?
38. What are the five major callouts needed for ordering pipe support equipment?
39. Describe the procedure for completing a pipe support detail.
40. How are pipe support detail views determined?
41. List three typical pipe support specifications.
42. Are pipe supports drawn to scale?
43. What are elevation designations used for on pipe support details?
44. Name five different types of information needed on a pipe support detail.
45. Name ten different things to look for when checking for interferences.
46. How are models used in pipe support design?
47. When determining pipe support locations on models, what are some of the important factors?
48. List the reference drawings needed before starting a pipe support design.
49. What is a stress isometric and what are its uses?
50. What do structural steel drawings show?
51. What is a HVAC drawing and how is it used?
52. What are pipe support specifications and how are they used?

PROBLEMS

1. Redraw the pipe support location isometric (Fig. 8-27) two times book size.
2. Sketch three pipe clamp and pipe hanger variations for 4-in.-diameter pipe. (Use an appropriate scale.)
3. Redraw the pipe support detail in Fig. 8-3, inserting a plate detail between items 1 and 8. Also draw a right-side view showing the complete assembly.
4. Using Fig. 8-6 as an example, draw a pipe support detail for a 10-in. pipe bend (short bend). Pipe is at elevation 321 ft 10 in.; floor elevation is 319 ft 6 in. For item 1, use 6-in.-diameter pipe. Draw front elevation and top view.
5. For this variable-spring selection and calculation problem, use Fig. 8-9 as the support situation. Check the variability and swing of the hanger. Required load is 1301 lb. Thermal movement is 3/4 in. down. The distance between pipe centerline elevation and top of hanger assembly is 6 ft 9.5 in. The maximum variability is 15 percent. Hot load is 1301 lb total; cold load is 1279 lb.
6. Check the angle for Fig. 8-3. Draw a side elevation view to complete the problem. Direction of travel is forward. Pin-to-pin dimension is 4 ft 8¼ in. Horizontal movement is 2 5/16 in. (B). Maximum swing angle is to be 4 degrees.
7. Redraw Fig. 8-13, 8-14, or 8-15.
8. Draw a top view of Fig. 8-15.
9. For this angle brace problem, use Fig. 8-37. The B dimension has been determined to be 4 ft 3 in. and $A1$ is 3 ft 7 in. Determine the length of the brace (C); also determine C if B = 3 ft 1 in. and $A1$ = 2 ft 0 in.
10. Redraw Fig. 8-16, putting the detail views in proper orthographic projection.
11. Draw Fig. 8-17, 8-25, or 8-30.
12. Draw Fig. 8-26 (two times book scale), 8-27, or 8-35.
13. Redraw Fig. 8-31 or 8-41 in an appropriate scale.
14. This is a swing angle problem. If the maximum allowable rod swing angle is 4 degrees and the pin-to-pin dimension is 3 ft 5 in., what is the horizontal movement:
    (a) If L is 2 ft 3 in. and the swing angle is 3 degrees?
    (b) If L is 5 ft 3 in. and the swing angle is 4 degrees?
15. Calculate the stanchion length for the following situations:
    (a) 14-in. pipe; stanchion diameter = 10 in.; H = 22 in.
    (b) 8-in. pipe; stanchion diameter = 6 in.; H = 12 in.
    (c) 20-in. pipe; stanchion diameter = 14 in.; H = 17.5 in.
16. Calculate the location for the following hanger ($H2$) of an inclined pipe. Suppose $H1$ is at an elevation of 435 ft 2 in. and $H3$ is at 435 ft 11 in.; the distance between $H1$ and $H3$ is 39 ft and the distance between $H3$ and $H2$ is 21 ft.
17. *Advanced Problem:* Refer to the Bergen-Patterson specification on Selecting Pipe Supports. Following each step and doing the calculations, draw each part of Fig. 8-40. Draw each figure on a separate 8½ × 11 in. sheet of paper and letter all calculations that belong to the figure.
18. Using Fig. 8-40, redo cases 1 and 2 in the same way as Problem 17.

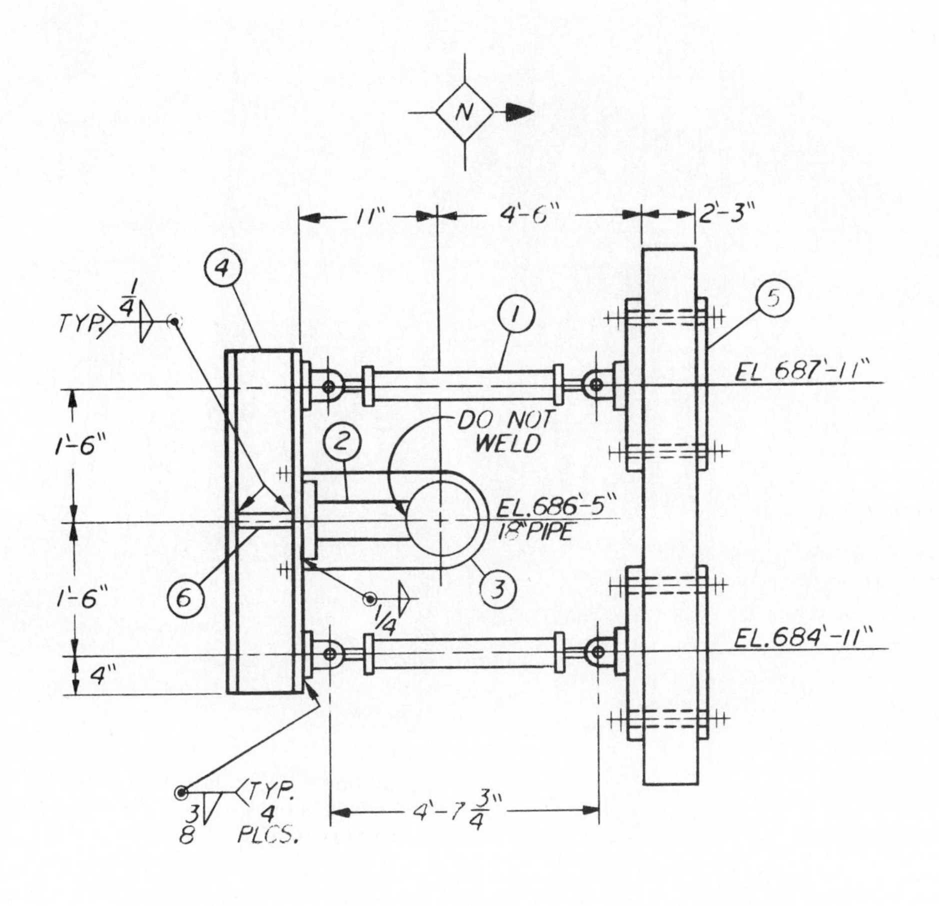

ITEM NO.	NO. REQD	FIG. NO.	SIZE, DESCRIPTION	TW. NO.	MAT'L
1	2	211	NO.5 SWAY STRUT ASSY., W=3'-5 3/4" LG.,		
			LOAD=26487 NO.PH74, ROD=SA-36		
			REAR BRKT. PIN=A-193 GR. B7,		
			PL=SA-36 (OR SA-515 GR.65)		
1	2	306	NO.35 MECHANICAL SHOCK ARRESTOR		
			(OPTION-1) 6" STROKE C 2'-3"		
2	1	H563	8" STD. WT. PIPE STANCHION TYPE C	28	SA-36
			FOR 18" PIPE D=11", E=2 3/8, G=3/4 X 10 X 10		
3	2	137	1" X 18" STD U BOLT	27	
4	1		W10 X 49 X 3'-8" LG	180	SA-36
5	4		PL 1 1/4 X 15 X 15 (SA-515 GR65)		
6	2		PL 1/2 X 8 7/8" X 4 3/4" (OR SA 515 GR65)	10	SA-36

FIG. 8-41 Pipe support detail.

Tug pulling a large fabrication constructed entirely of pipe, which is to be sunk in the North Sea and used as the base for an offshore oil rig.

9 PIPING FABRICATION

Pipe fabrication is the construction of pipe configurations that are joined or altered by an assortment of procedures, which include shaping, machining, welding, flanging, belling, and extruding. All these different procedures for the construction of unit fabrications are used throughout the pipe industry. Pipe fabrications are necessary for all pipe systems, whether the system is for a petrochemical installation, nuclear power plant, conventional power plant, or building, heating, and plumbing unit. Pipe fabrications are sections of the total piping system.

The pipe drafter and designer must be quite knowledgeable in this area in order to design and draw a variety of configurations, including headers, exchangers, coils, and pipe sections with nonstandard bends. The different pipe fabrication procedures convert standard straight pipe into various configurations in conjunction with components, such as valves, fittings, and flanges, which are usually welded or flanged together to form units that can be shipped to the construction site in one piece and assembled along with other sections of the piping system. In the case of pipe fabrications in the field, components are welded and assembled at the job site.

The drafter is responsible for drawing the isometric or orthographic spool sheets which show the portions of the piping system that need to be fabricated, complete with specifications for fabrication. Isometric and orthographic drawings are discussed in Chapters 5 and 7. Chapter 2 explains the processes associated with welding piping systems.

Shop fabrication is always preferred to **field fabrication.** Since the general purpose of pipe fabrication is to maximize the amount of work done in the shop and minimize that done in the field, special configurations and assemblies of pipe and fittings should be fabricated in the shop whenever possible. Many of the fabrication procedures discussed here require special equipment which is more easily operated in the shop than in the field. In some cases, however, it pays to move the equipment to the job site. New techniques and equipment have greatly increased the field capabilities of pipe fabricators. Whether the pipe fabrication is prefabricated in the shop in small sections or totally created at the job site, field construction is always the last stage in the completion of a pipe system project.

Shop fabrication allows for a variety of welding processes that are not always available with field fabrication. Smaller welded assemblies are ideal for shop welding. Large fabrications like the offshore drilling rig shown on page 294 must be field welded. A typical small pipe fabrication might consist of a short run of pipe, with a variety of welded fittings, and flanges, with branch lines or headers connected to the main section (see Fig. 9-1). Whether the fabrication will be done in the field or the shop depends on the design, size, number of fabrications to be made, labor requirements, and union agreements.

FIG. 9-1 A variety of pipe fabrications—using elbows, tees, reducers, flanges—and joined by all-welding. Each spool or fabrication unit has an identification number painted on its side, and all pipe ends have been sealed to keep out dirt and foreign substances. (Courtesy Flow Line Corporation, New Castle, Penn.)

FIG. 9-2 Overhead view of a pipe fabrication shop. (Courtesy Pacific Pipe Company, Oakland, CA)

SHOP FABRICATION (Fig. 9-2)

For large diameter pipes shop fabrication has many advantages. Figure 9-3 shows a large steel elbow, 54 in. in diameter, which has been shop fabricated. In the case of piping and fittings joined by flanged or welded joints, it is generally agreed that units 3 in. in diameter and larger are to be shop fabricated if the fabrication can be created in sections that meet shipping requirements.

An organized fabricating shop has many advantages over field construction techniques:

1. *Accuracy:* All piping systems are predetermined configurations that must result in acceptable service and follow the specified guidelines of design laid down by the engineer or designer. These requirements can be fulfilled more easily and economically in a shop because of the use of experienced workers and excellent materials along with shop techniques that are not available in the field. Shop techniques also offer increased accuracy and high-quality workmanship because the shop permits greater concentration on the job at hand. Figure 9-4 shows a large-diameter pipeline being bent to a required radius bend within extremely high tolerances. This degree of accuracy can be attained only through shop techniques.
2. *Proficiency:* Fabrication shops employ experienced workers and mechanics who have as their sole responsibility the production of pipe fabrications. In a shop with a high volume of work, each worker specializes in a particular operation. Thus the proficiency level is much higher for shop fabrication—especially in welding and other complex procedures (see Figs. 9-5 and 9-6).
3. *Safety:* Especially for high-pressure and high-temperature systems, safety becomes a primary concern for

every piping fabrication and installation. Shop fabrication can ensure that all safety precautions are followed. All welds can be tested to ensure that gas pockets or faulty penetrations will not cause pipe failure. Threaded connections for flanges or other joints can be inspected to ensure that the fabricated unit and assembled system do not fail in service.

4. *Working conditions:* Shop fabrication offers working conditions far superior to those in the field, especially for the manufacture of bulky, heavy, or large pieces of equipment and components. Figure 9-3 shows a large, bulky pipe elbow that has been shop fabricated. Notice the use of rollers and large dolly to support the pipe section. The metal rings around the elbow are used to provide guideways for the rollers. This section is slowly turned during the submerged arc welding procedure. Shop techniques allow the fabricator to work in a safe and comfortable position protected from the weather.
5. *Simplified assembly:* It should not be necessary to call upon field-directed crews to assemble piping units that a modern fabrication shop can produce. The less work done on the job site, the better that work will be. This is especially important for overall operation economics. Less field work leads to fewer failures in operation.
6. *Testing:* Shop fabrication ensures that the product will be tested and proved totally dependable before it is shipped to the field. Piping fabrications are usually tested for defects before they are shipped for field assembly. Electronic testing devices and hydrotesting machines, which are available to shop fabricators, can ensure that the quality of the pipe fabrication is adequate under the given service conditions. The necessity for removal, replacement, or repair of portions of the erected system could prove economically disastrous for many projects and cause considerable delay in the implementation of the system.
7. *Shop techniques:* A variety of shop fabrication techniques are available because of special equipment, welding procedures, clamps, appliances, jigs, dies, and the ability to extrude pipe. All these techniques are much more conveniently done in the shop. Stability and availability of special equipment ensures accurate alignment, squaring up of flanges or pipe ends, finishing to accurate dimensional design specifications, and holding angles within close limits. Techniques such as stress relieving, heat treatment, annealing, and hot bending are readily performed in the shop but almost impossible to accomplish in the field.
8. *Labor:* Shop fabrication is done by a permanent skilled labor force at the shop, making it unnecessary to recruit skilled mechanics and laborers in the field. Since workers recruited in the field sometimes lack the experience necessary for the fabrication techniques involved, the contractor may be obliged to hire temporary skilled help at great expense.
9. *Presence of a drafter:* Shop fabrication also allows the drafter to be present at the construction stage of the project, so design changes or alternative drawings can be made upon request. All fabrication shops employ drafters for a variety of design and drawing needs—from the detailing of isometrics and orthographic spool sheets to the design and drawing of special machines such as jigs and fixtures, dies and extruding devices, and many kinds of bending machines. The shop fabrication drafter must be well versed not only in piping and drafting procedures but in mechanical and machine drawing. Most of the drawings in this chapter have been provided by the **Pacific Pipe Company**. They are typical examples of the drawings a drafter will encounter in a pipe fabrication shop.

In sum, shop fabrication allows for precise construction with a diversity of techniques which enable the fabricator to produce safe, high-quality welded or flanged piping components and unit fabrications. Modern assemblies such as nuclear components, pressure vessels, headers, coils, and heat exchangers are more easily, and more economically, prefabricated, tested, and constructed in the shop than in the field. Moreover, the vast variety of forming techniques—bending, extruding, belling—along with the automatic welding procedures explained in Chapter 2 make shop fabrication the most advantageous procedure in the construction of pipe system components and units. Shop fabrication tech-

FIG. 9-3 This 54-in.-diameter steel elbow is to be used for a refinery fitting. Worker is overseeing the submerged arc welding process. (Courtesy Pacific Pipe Company, Oakland, CA)

niques also allow a degree of standardization which surpasses that available in field construction. The Pipe Fabrication Institute (PFI) has a list of codes and specifications that should be available to drafters working for a pipe fabrication shop.

Figure 9-1 shows several configurations of welded pipe fabricated sections which have been cleaned and sealed for shipping to the job site for assembly. Notice that each unit is clearly identified with a number painted on its side. The majority of the drawings and photographs in this chapter show shop-fabricated items. For more photographs of pipe fabrications, see Chapters 2 and 3.

FIG. 9-4 Bending machine cold-bending a large-daimeter pipe. (Courtesy Pacific Pipe Company, Oakland, CA)

FIG. 9-5 Worker completing the welding for the fabrication of a mitered elbow. (Courtesy Pacific Pipe Company, Oakland, CA)

FIG. 9-6 Heat exchange unit. (Courtesy Pacific Pipe Company, Oakland, CA)

FIELD FABRICATION

In situations where field fabrication is extensive and necessary, such as the construction of oil refineries, nuclear reactors, and offshore drilling rigs, which are located in less populated areas, the best course is usually to construct a fabrication shop at the job site so that critical fabrication techniques can be handled locally. The fabrication shop can become a maintenance shop after the plant is built.

Field construction techniques are determined by pipe size and the operation necessary to fabricate the system. In almost all cases, screwed piping and pipes 3 in. (7.62 cm) in diameter and smaller are fabricated in the field. Some pipe systems that have few subcomponents or units, such as those shown in isometric in-the-round drawings in Chapter 7, must be field fabricated because of their size. Structural applications of pipe, such as those for building reinforcements and pipe racks, may also require field fabrication.

Almost all threaded (screwed) piping is fabricated and fitted at the job site. The majority of tubing using flared or socket joints is also field fitted. In these situations, the drawings show general dimensions for the fabricator to follow within tolerance limits established by the designer. Thus the field fabrication often has some leeway in the fitup procedure—especially for plumbing, heating, and water supply systems in buildings. Only when close tolerances must be maintained is it necessary for small-diameter tubing or piping to be precisely detailed and dimensioned extensively on the drawings.

Field fabrication includes the precise fitup of the total piping system, which is accomplished with the use of a variety of construction techniques for the hoisting and alignment of shop fabrications along with the longer runs of pipe that may be required. Since pipe fabrications must be clean and absolutely free of foreign matter, the ends of the pipe section are sealed before being shipped to the field (Fig. 9-1). Field fabrication must allow for calculation and repositioning within certain tolerance limits in the field. Screwed fittings, valves, and pipes, for instance, require adequate assembly space. The engagement length for screwed piping (Chapter 1) must be taken into account when ordering pipe length, since the thread length differs for various pipe sizes and types.

CODES AND STANDARDS

High-temperature/high-pressure piping systems include many types of pressure vessels, heat exchangers, and the like that must be carefully fabricated and inspected to ensure that, under working conditions, they will meet the safety codes. Even with the institution of stringent codes by standards organizations, the chance of failure and resulting loss in life and property is still a major concern throughout the piping industry. Shutdowns in modern steam power plants, nuclear plants, and process petrochemical plants can cost thousands of dollars. Even more important is the possibility of a failure which could be fatal to plant personnel or, in the case of nuclear power plant failure, contaminate the surrounding environment and cause long-term injury to the community. Because of these safety and operating considerations, the knowledge of sound fabrication is extremely important to the designer and drafter. Chapter 2 discusses the fundamentals of accurate welding design in piping systems, whether shop or field erected; Chapter 13 discusses nuclear applications.

Codes and standards have been established by a variety of institutes to govern the fabrication of welded piping systems. These codes state the minimum requirements for the safe operation of a piping system. For severe service, it is important to design far more conservatively than even the codes require. Full penetration welds and pipe-forming procedures, such as bending and extruding, require modern safety checks including the use of radiographic examination and hydrostatic testing. In most cases, the pipe fabrication methods are standardized by the American Standard Code for Pressure Piping; the PFI has a series of standards that supplement this code. The Boiler and Pressure Vessel Code of the American Standard Society of Mechanical Engineers covers the power boilers, nuclear vessels, unfired pressure vessels, and welding standards for pressure vessels and boilers. The American Standards Association has prepared codes for pressure piping, which includes power piping, industrial gas, air piping, petroleum refinery piping, oil transportation piping, refrigeration piping, chemical industrial process piping, nuclear piping, and gas transmission and distribution systems. The American Petroleum Institute has a variety of pressure pipe specifications in addition to those available from ASA. Another standards organization is the American Waterworks Association. The American Welding Society (Chapter 2) has prepared a set of recommended practices that apply to the welding of pipe.

Pipe designers must have at their disposal all these codes along with the PFI codes which cover design, fabrication, and inspection throughout the pipe fabrication industry. Government agencies and other commercial codes are also available and should be strictly adhered to whenever possible. Up-to-date versions of all piping standards must be kept for the piping industry to which the code applies, since standards are constantly being changed. The vigilant attitude taken by the standards associations enables the codes to be updated and kept in compliance with modern design variations. Standardization of fabrication methods allows for dissemination of pipe design methods and procedures throughout industry.

PIPE FABRICATION PROCESSES

This section is devoted to the three basic pipe fabrication processes: *welding, bending, and forming.*

Welding

Pipe fabrication welding procedures are covered primarily in Chapter 2, which should be consulted throughout. Fabricated piping comprises an almost infinite assortment of shop-welded products. Welding techniques have been continually modernized and refined throughout the industry. Welding pipe components, units, and sections is the primary work of pipe fabrication shops, and most of the photographs and drawings in this chapter show welded pipe fabrications.

Bending

The art of pipe bending (Figs. 9-4, 9-7, and 9-8) has been developed to a high degree over the last 75 years. Today pipe bends of many sizes and shapes are available. Because pipe installations are never alike, piping bends must be adapted to space requirements and special configurations. The majority of pipe bending is done in shop facilities, which allows for an almost infinite variety of bends, coils, spirals, and other shapes. Though a wide range of standard elbows and bends are available (see Chapter 3), the use of bending for tubing and piping is still one of the most economical procedures for changing the direction of a pipe run or creating a special fabrication. Bending pipe offers numerous advantages over standard pipe fittings:

1. Reduces fabrication cost
2. Reduces flow turbulence and resistance
3. Allows the pipe medium to travel at a high velocity through the bend
4. Offers a infinite variety of piping bend radii
5. Reduces erosion

But there are drawbacks in using pipe bending:

1. There is a tendency to stretch the outside bend of the pipe beyond the acceptable wall thickness limits.
2. The bendability of the pipe depends on the composition and ductility of the pipe material, which must be taken into consideration when comparing the use of standard elements to pipe bending.
3. Bending may create compression, crinkling, or wrinkling of the inside radius of the pipe.
4. Improper bends can result in high turbulence.
5. Friction loss and erosion at the pipe bend can be a problem if the inside surface is wrinkled.
6. The pipe ID and OD shape may be distorted beyond acceptable limits.
7. Distortion, wall stretching, and wrinkling could cause pipe failure.

FIG. 9-7 View of a large-diameter pipe bend. (Courtesy Pacific Pipe Company, Oakland, CA)

FIG. 9-8 View of a pipe fabrication shop showing a bent section of pipe. (Courtesy Pacific Pipe Company, Oakland, CA)

All these situations must be taken into account before determining whether bending is a suitable operation for the piping system in question. Under any circumstances, the metal or alloy should have a combination of high ductility and high ultimate strength.

In many cases, pipe sections are bent to a radius five times the pipe diameter. A radius of three times the pipe diameter can be used in some situations, depending on material and pipe size. Any variation of five times the pipe diameter or greater can easily be produced by the pipe fabricator (Figs. 9-4 and 9-7).

Pipe can be bent by using cold or hot techniques. When the pipe wall thickness is thin and is of 2½ in. (6.35 cm) OD and smaller, cold bending is done exclusively. In the past, the larger the diameter and the thicker the pipe wall, the greater the preference for hot bending. Today, however, with the use of modern equipment and quality control of standards, cold bending can compete with hot bending in a majority of circumstances up to 42 in. (106.68 cm) in diameter and 3/4 in. (1.91 cm) wall thickness (see Fig. 9-4). The choice of whether to use cold or hot bending for pipes 2½ in. (6.35 cm) to 40 in. (101.6 cm) in diameter depends on wall thickness, cost, the number of similar bends to be made, the radius of the bend, the metallurgical and chemical composition of the pipe, and whether or not the fabrication can take place in the shop or must be done in the field. For both hot and cold bending, it is important to monitor the bending procedure carefully to maintain an accurate degree of tolerance control.

Cold Bending Cold bending used to be restricted to small-diameter pipe and tubing with thin walls. Modern machinery and bending techniques now enable shop fabrication equipment to cold-bend pipe as large as 42 in. (106.68 cm). Tubing with 5 in. (12.7 cm) OD and smaller can also be bent readily. When cold bending large-diameter pipe, it is essential to use the proper equipment and dies to avoid wrinkling, thinning, and other pipe distortions. A variety of design and engineering data indicate the bendability of certain metals and alloys whose thermal properties must be known prior to the bending operation. Whenever possible, cold bending should be used because of its economy. Ram-type bending machines have been used to cold-bend pipe, especially for thick walls. This machine flattens the bend area and may not be suitable for certain situations. The rotary bending machine and the bending machine shown in Fig. 9-4 are examples of modern pipe-bending devices.

Hot Bending Hot bending is used for making bends in pipes 3 in. (7.62 cm) and larger. The larger the pipe, the more applicable the hot bending process. In general, hot bending requires that the pipe be filled with sand or, in some cases, with high-pressure argon or nitrogen gas. The use of gases eliminates or at least reduces the cleaning operations that accompany the formation of pipe scale on the pipe wall. The use of sand helps to maintain the pipe at a hot bending temperature, giving the fabricator more time to finish the pipe-forming operation. Besides maintaining elevated temperatures, sand packing reduces crinkling and distortion of the pipe cross section. Temperatures between 1500°F (815.5°C) and 2000°F (1093.3°C) are normal for bending carbon, alloy, and stainless steels. In cases where the pipe to be bent has a heavy wall thickness, the gas or sand filling may be eliminated.

Pipes to be hot-bent are heated in furnaces that maintain even temperatures throughout, thus avoiding hot spots and excess oxidation of the pipe material. After the pipe has been heated to the necessary temperature, it is bent to the specified radius by bending tables or machines. The pipe is then cooled and the sand or gas removed; the cleaning operation takes place at this time, including blasting, scrubbing, or turbining the pipe's inner surface. Hot bending of pipe allows for a great range of short bending radii, especially when the forming process is closely supervised to prevent buckling and other distortions. Many problems can be eliminated by prolonging the bending time (for both cold and hot bending).

Hot bending can cause crackling or other internal defects, which are usually a result of the pipe material. Safety inspection must accompany any pipe bends that will be used in severe service, because visual detection of cracks may not be possible, especially on the inner surfaces. Another drawback of hot bending procedures is that wall thicknesses are liable to change more readily and more drastically than with cold bending. When using hot bending, it is usually standard procedure to allow for thinning in the outer arc of the bend by increasing the minimum wall thickness requirements, so that the design thickness is still met after the bending.

A bend radius of five pipe diameters is commonly used for hot pipe bending, though smaller radii can be achieved by this procedure more readily than by using cold bending techniques. Crease bends allow the bending radii to be much smaller (as low as two times the diameter of the pipe), though they can only be used when a wrinkled inner surface does not produce undesirable flow characteristics or erosion.

Before bending aluminum or aluminum alloys, the chemical and metallurgical composition of the pipe must be understood. In general, the same methods described above can be used for bending aluminum pipe. When aluminum is hot-bent, the temperature is much lower than for other metal pipes. One of the most common applications of bending is for copper or copper alloy tubing (Figs. 9-9 and 9-10), which can be bent to extremely small radii. This material readily lends itself to the production of coils and spirals for heat exchangers.

Standard Pipe Bends A variety of standard bends are available for the fabrication of piping units. When specifying a standard pipe bend to be ordered from a fabrication shop, give the recommended radius, the minimum radius, and the maximum radius, along with the minimum length of tangent. In cases where welded flanges are used, pipe bends with plain or beveled ends are furnished without tangents. In Fig. 9-11, the *T* dimension denotes the length of the tangent for the standard bends.

In general, pipe ends are finished with tangents to allow for convenience in erection, and it is recommended that pipe bends with beveled ends have tangents equal to the minimum length necessary for the use of jigs, clamps, and

FIG. 9-9 Double-coil heat exchange unit of small-diameter pipe. (Courtesy Pacific Pipe Company, Oakland, CA)

other alignment equipment that may be needed for field fabrication and assembly. For pipe bends that require slip-on welded flanges, the minimum length of tangent furnished on the pipe end is equal to the length through the hub of the slip-on welding flange to be used for that pipe size (see Chapters 2 and 3). For pipe bends fitted with weld-neck flanges, the bends can be fitted without tangents since these flanges are usually butt-welded directly at the end of the arc of the bend.

Examples of bends and welded flanges can be seen in Chapters 2 and 3. Figure 9-11 shows the standard pipe bends available. The bends for weld applications are shop fabricated, usually with hot bending procedures involving sand packing. The bends are heated before beveling, threading, flanging, or other machining. Compound and unusual bends can be fabricated by techniques available in the modern factory.

When designing or ordering pipe bends, take into account the maximum allowable lengths for welded or seamless

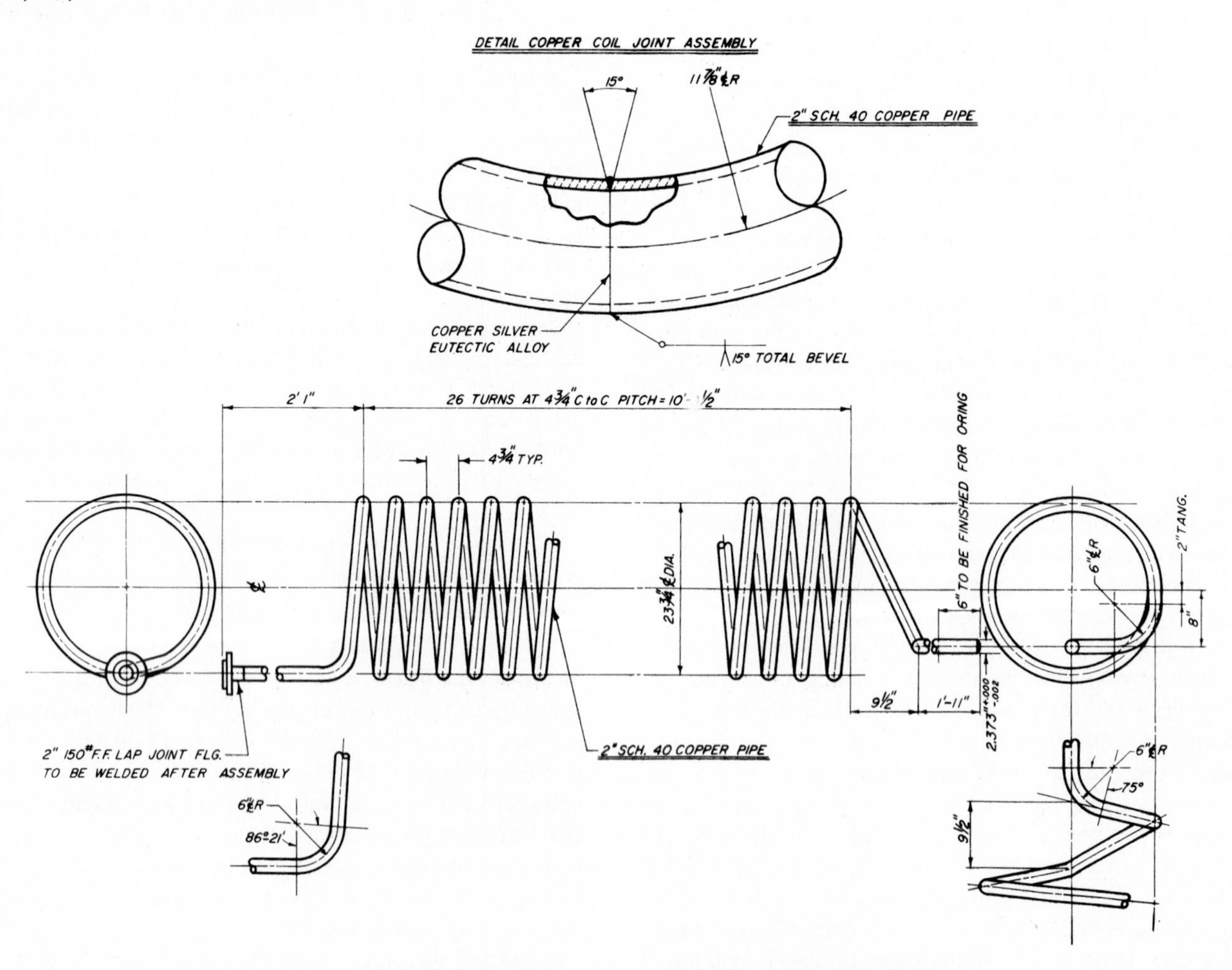

FIG. 9-10 A 2-in. schedule 40 copper pipe coil used for heat exchange in an acid cooler reaction plant (see Fig. 9-17). (Courtesy Pacific Pipe Company, Oakland, CA)

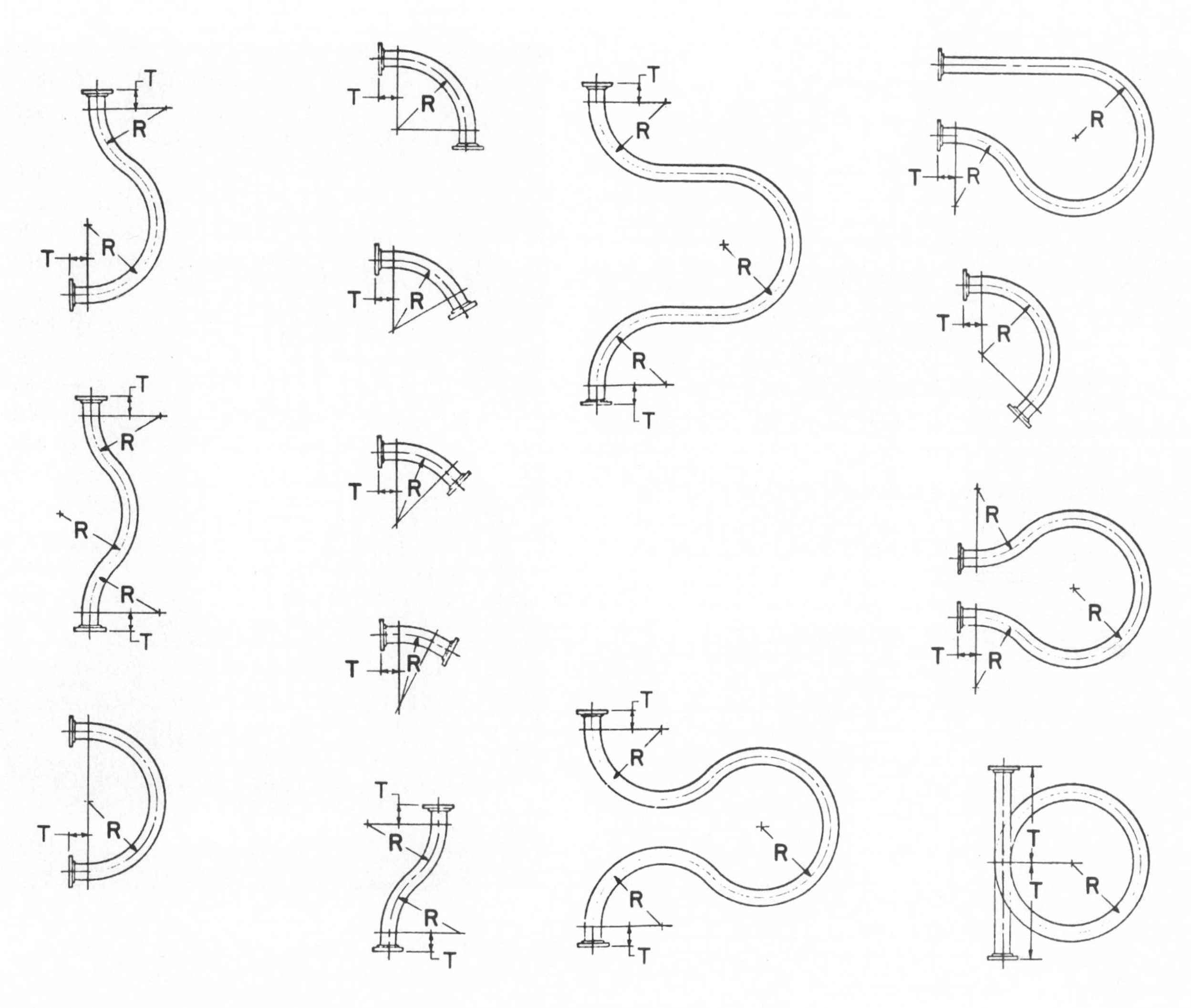

FIG. 9-11 A variety of standard pipe bends. (Courtesy Crane Co.)

pipe. Steel pipe is usually available in stock, extra long, or double random lengths, which may be necessary for the fabrication of pipe bends or coils of a certain size. Figure 9-10 shows a detail of a copper coil joint assembly; because of the coil's length, it is necessary to bevel-weld the pipe. When it is necessary to weld pipes together before bending, the location of the intermediate joint should be indicated on the fabrication drawing.

Standard pipe bends come furnished with plain, beveled, threaded, or flanged ends, depending on the design requirements. Pipe bends are available in many materials, including steel alloys, aluminum, brass, nickel, bronze, stainless steel, and rubber-lined pipe. When considering commercial pipe bends, remember that they are made to approximate dimensions only. In most cases, the bends are furnished at lesser expense for noncritical service.

Pipe Coils Figure 9-12 exhibits three standard coil configurations; Figs. 9-6, 9-13, and 9-14 show a variation of the coiling process. Most of these coils are used for heat exchange units. Figure 9-9 shows a double coil. Coils and their variations—including flat, oval, and double cylindrical —are all available through shop fabrication. Figures 9-10, 9-15, and 9-16 display typical pipe coils and spiral drawings that are necessary for shop fabrication. Coils provide efficient and economical heat transfer from one fluid to another—the coil in Fig. 9-10 will be used in the heat exchange vessel shown in Fig. 9-17. Pipe coils are made of an assortment of materials, including carbon steel for seamless or welded pipe, and copper (Fig. 9-18). Copper and copper alloy tubing is an excellent material that can be used for pipe coils, though the temperature range of service is limited in comparison to other materials.

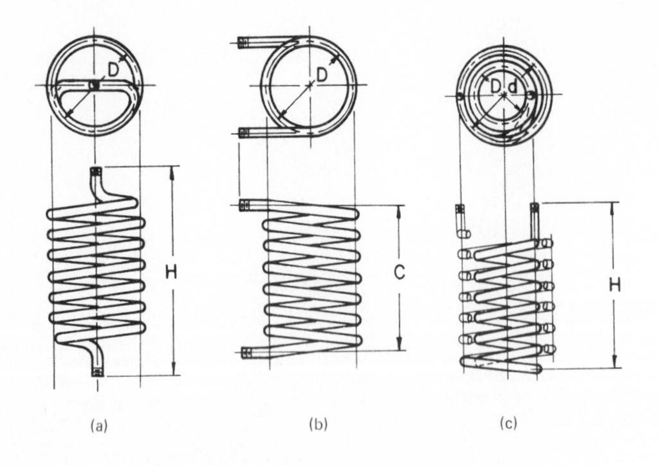

FIG. 9-12 Three typical pipe coil configurations.

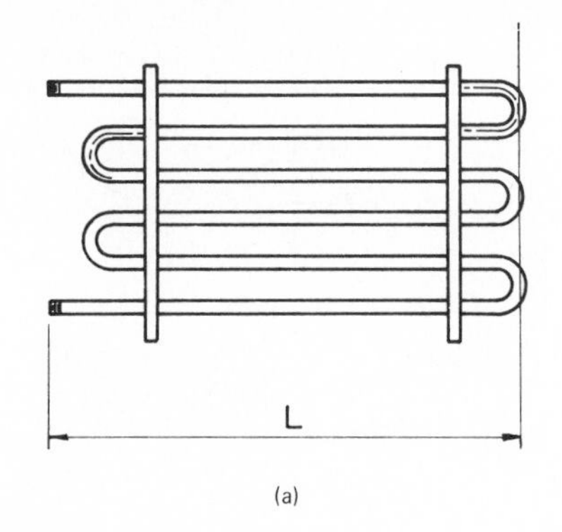

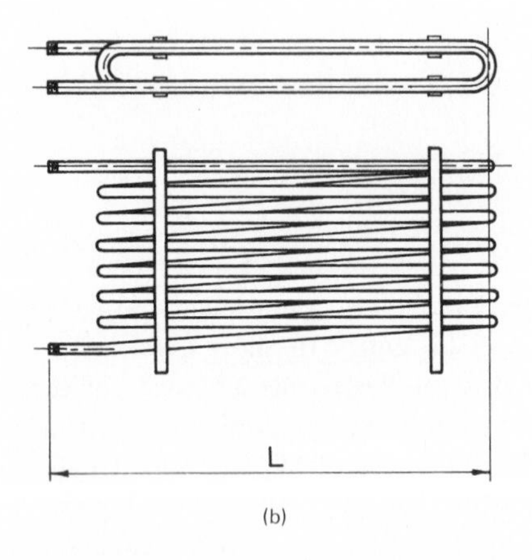

FIG. 9-13 Two common pipe fabrications.

FIG. 9-14 Piping fabrication for heat exchanger unit. (Courtesy Pacific Pipe Company, Oakland, CA)

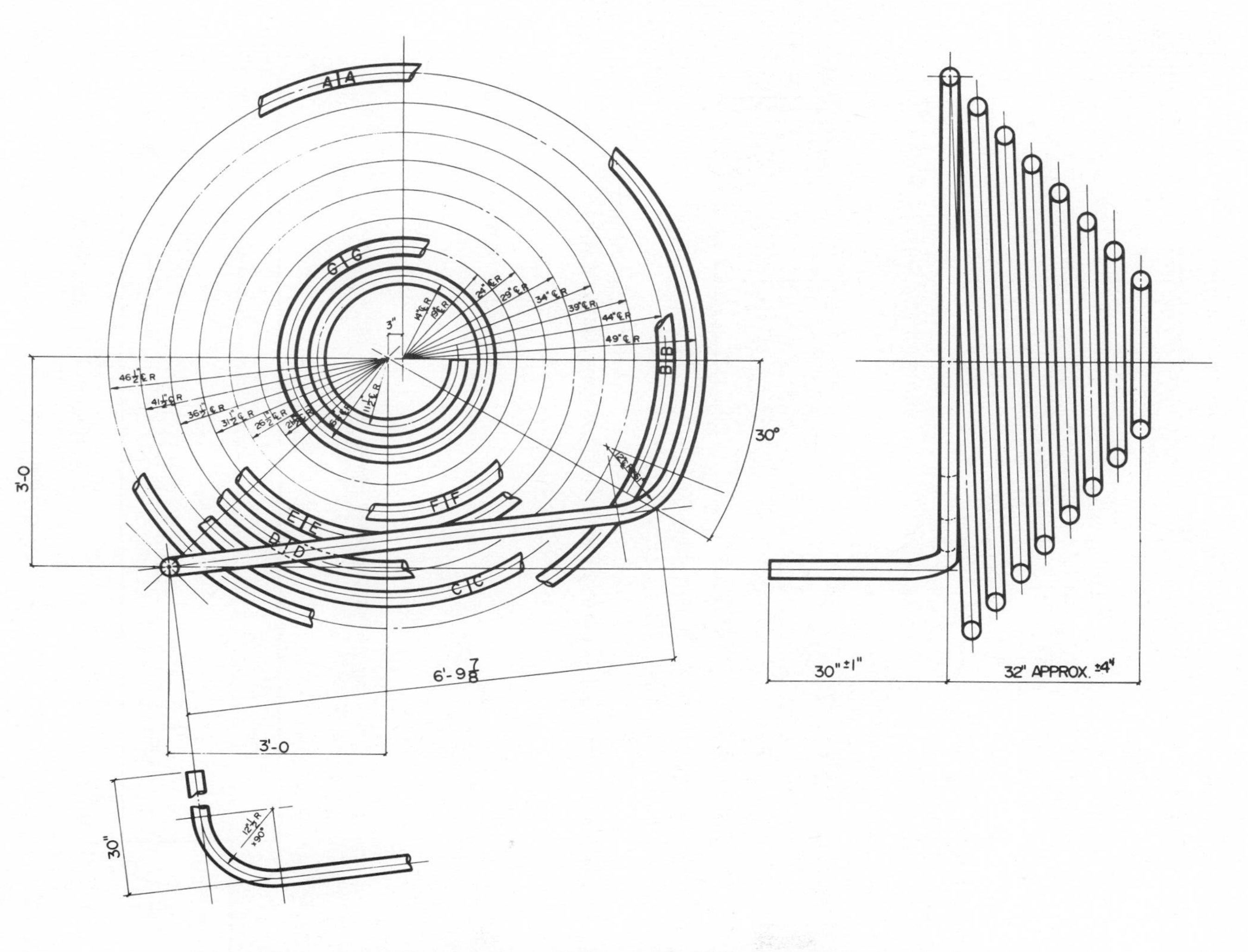

FIG. 9-15 A 3-in.-OD tubing coil. (Courtesy Pacific Pipe Company, Oakland, CA)

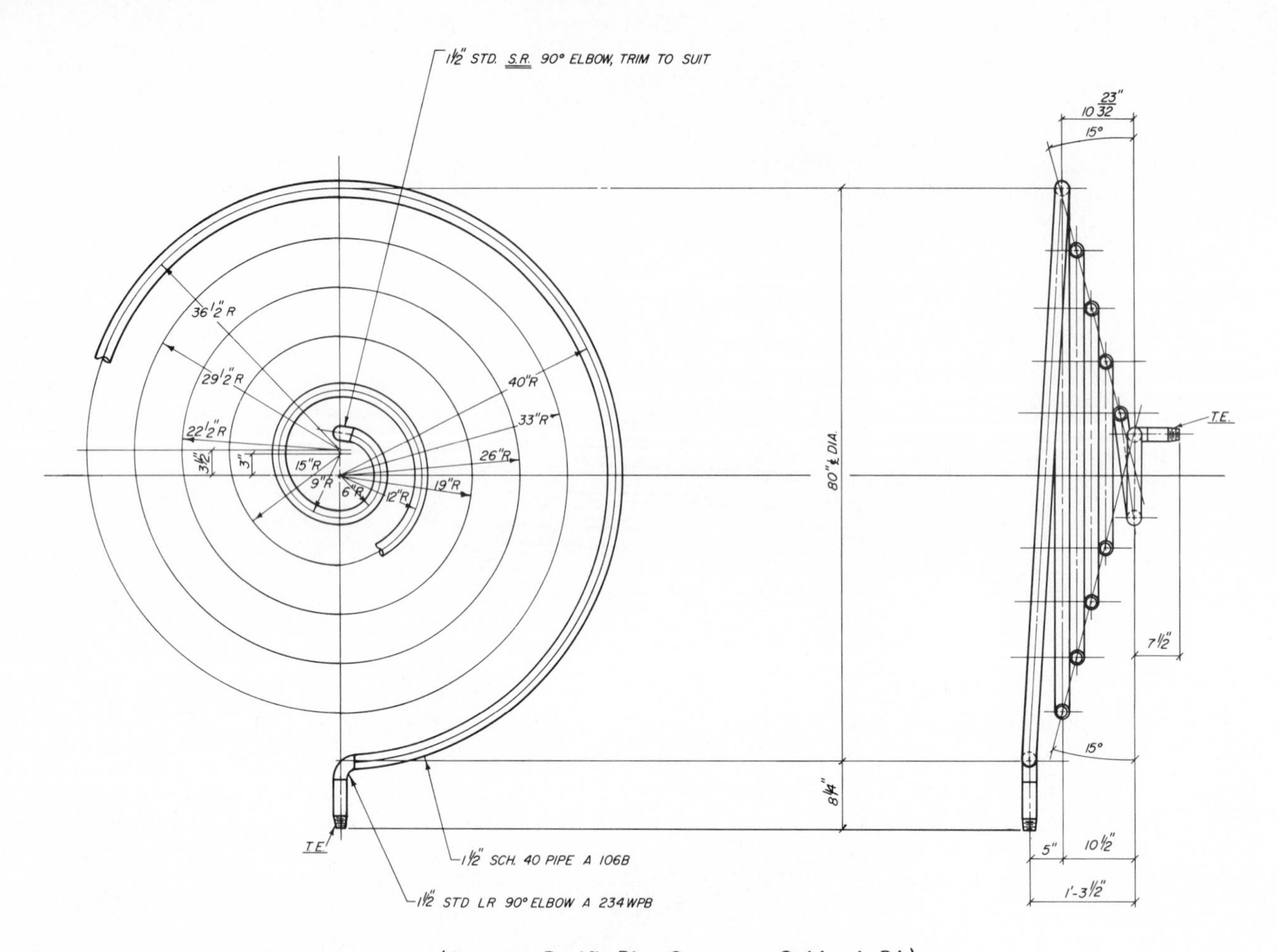

FIG. 9-16 1½-in. schedule 40 pipe spiral. (Courtesy Pacific Pipe Company, Oakland, CA)

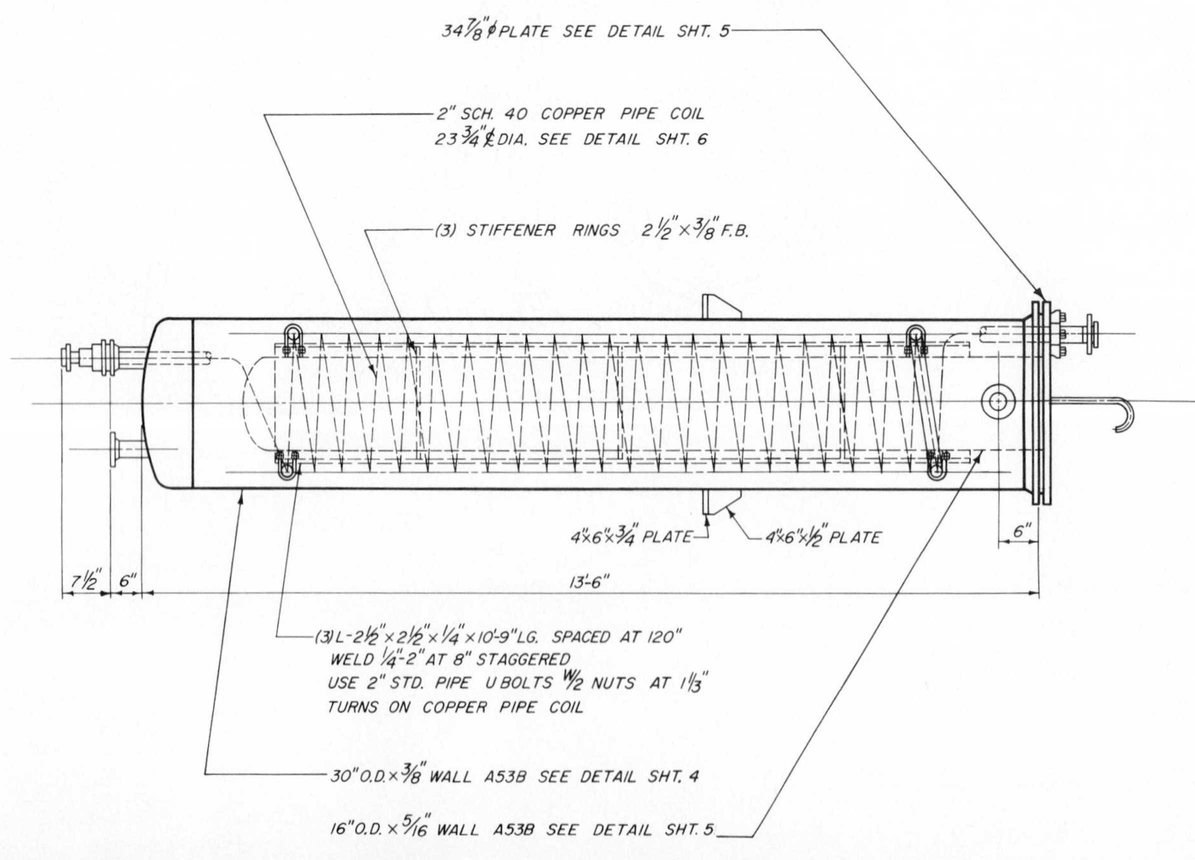

FIG. 9-17 Assembly drawing of heat exchange unit for acid cooler reaction plant. Vessel is 30 in. OD. (Courtesy Pacific Pipe Company, Oakland, CA)

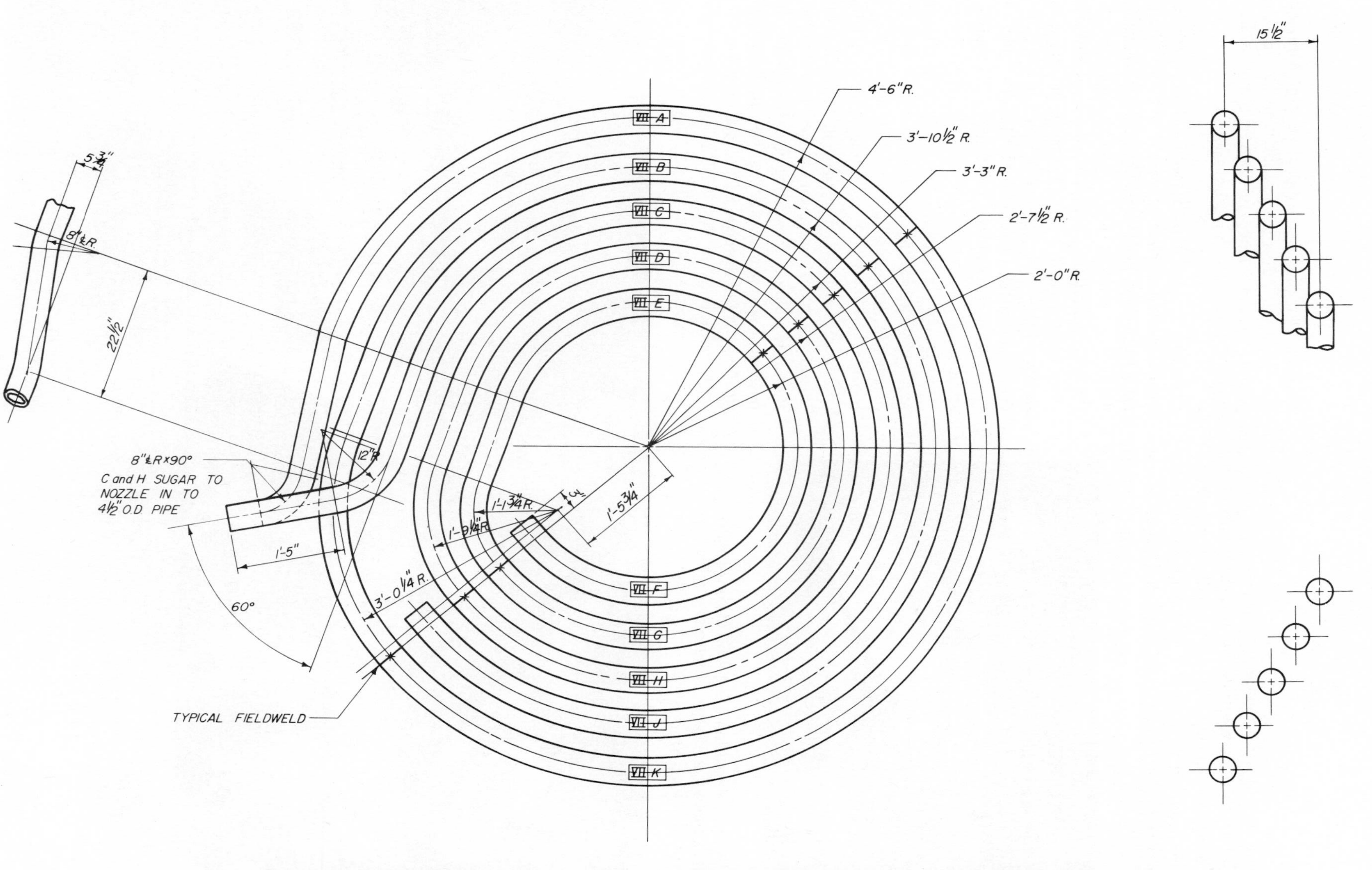

FIG. 9-18 4½-in.-OD copper 12 ga. tubing spiral. (Courtesy Pacific Pipe Company, Oakland, CA)

Forming

Forming involves all the manipulations of pipe and components necessary to achieve a desired design result, including flanging, swagging, lapping, belling, extruding, and their combinations. Special equipment is necessary for all these operations and, in general, is available only in the shop.

FIG. 9-19 Pipe fabrication using welded construction for flanges, elbows, and pipe. (Courtesy Pacific Pipe Company, Oakland, CA)

Flanging Flanging is an aspect of pipe fabrication. (See Figs. 9-19 and 9-20.) Shop facilities enable the fabricator to produce a wide range of flanged joints, either in their standard form or with modifications.

Extruding *Extrusion* is the forcing of metal spheres through smaller outlets (holes) in pipe, therefore eliminating the need for welded headers or nozzles. Extrusions are usually limited to materials such as stainless or nickel alloy steels, though aluminum and copper alloys are also extruded in some situations. In Fig. 9-21, the large gate valve body is shown with extruded inlet and outlet openings. Figure 9-22 depicts a multiple-outlet extruded header. These two figures are excellent examples of the use of extruding for pipe sections.

Extrusions are usually made by heating the pipe to the required temperature and forcing a steel ball through an opening in the side of the object, in this case the pipe. This provides for a smoother contour than would be possible with welded headers or nozzles. In Fig. 9-22, note that the extruded nozzles have been finished at the ends to provide for butt welding to the mating pipe. Extrusions can be made on almost any size of pipe and vary in diameter from very small to the full diameter of the pipe extruded.

The valve body in Fig. 9-21 has extruded inlet and outlet ports in excess of 48 in. (121.92 cm) in diameter. The pipe

FIG. 9-20 Large pipe header to be used for an underground line at an oil refinery. (Courtesy Pacific Pipe Company, Oakland, CA)

FIG. 9-21 Fabricated gate valve.

used for the body of this valve is 5 ft (1.5 m) in diameter and 3 in. (7.62 cm) thick and has been rolled and fabricated in the shop.

Headers Welded headers (Fig. 9-20) and extruded headers (Fig. 9-22) are both examples of shop-fabricated units. Figures 9-20, 9-23, and 9-24 all show header assemblies that have been manufactured in the shop. Notice the drawing and dimensioning practices on the fabrication drawings in this chapter.

FIG. 9-23 Close-up view of pipe header. Small-diameter pipe is welded onto a large pipeline. (Courtesy Pacific Pipe Company, Oakland, CA)

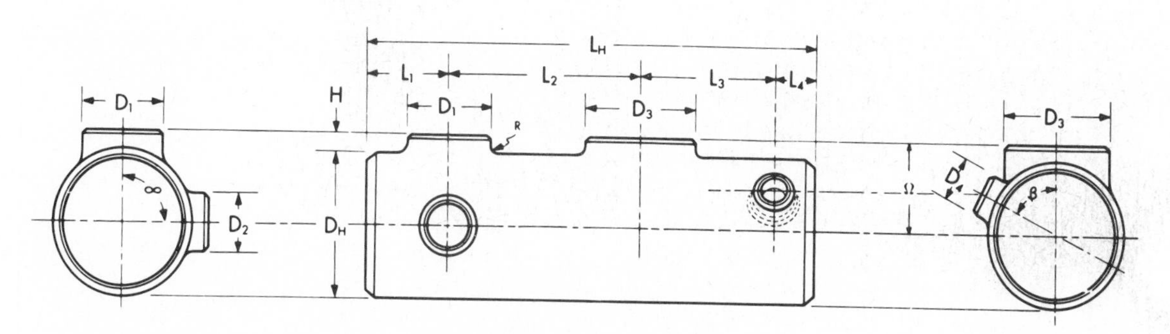

FIG. 9-22 Multiple-outlet extruded header. (Courtesy Crane Co.)

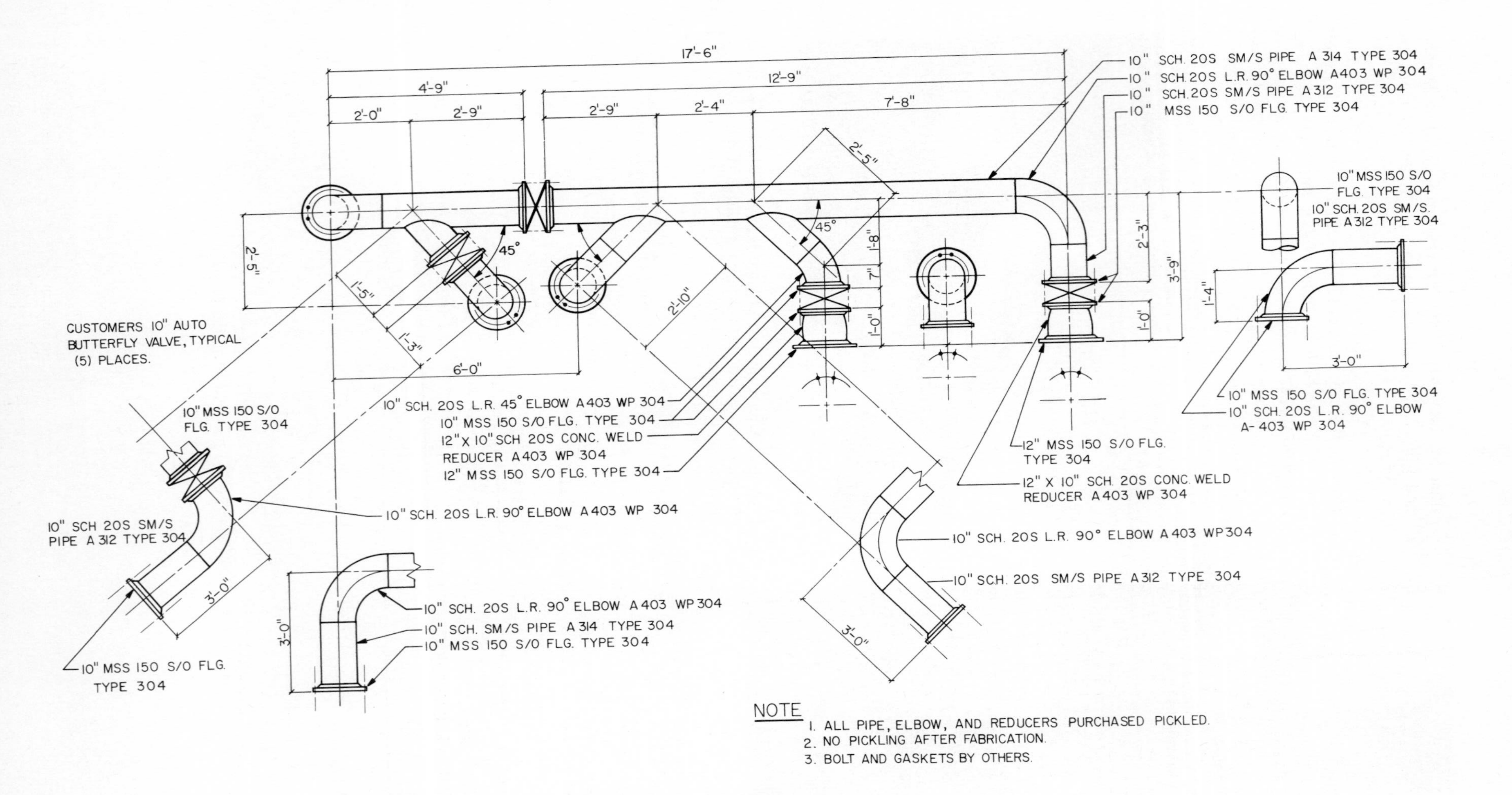

FIG. 9-24 Shop pipe fabrication drawing of a header assembly. (Courtesy Pacific Pipe Company, Oakland, CA)

DRAWINGS FOR PIPE FABRICATION

The drafter will be called upon to do many types of pipe fabrication drawings. All fabrication drawings for specific pipelines are identified with line number and spool designations as in Fig. 9-1, which shows identification numbers painted on the side of the welded fabrication.

Pipe fabrication shops employ drafters and designers to do drawings for bending, templates, spirals, coils, intersections and developments, clusters, headers, fabrication machines, bending devices, dies, jigs, extrusions, and fixtures. Thus the shop fabrication drafter must be well versed in both mechanical and pipe drafting techniques and procedures.

Individual fabrication drawings are usually referred to as *spool drawings* and can be done in isometric or orthographic projection. Pipe fabrication shops prefer to use orthographic projection in a majority of cases. Figures 9-24 and 9-25 are excellent examples of orthographically projected piping fabrications. Figure 9-26 shows the detail of a compound bend, which if shown in isometric would totally confuse both the fabricator and the drawer. In Chapters 5 and 7, a wide selection of single-line, double-line, isometric, and orthographic fabrication drawings are provided.

Fabrication drawings must supply all dimensions necessary to locate component parts (flanges, fittings, valves), size of pipe bends, and any other dimensions for the construction of the pipeline configuration. In many cases, the engineering company provides isometric projections of spools and the pipe fabrication shop converts them into orthographically projected drawings.

Drawing Spools

The determination of whether to use double-line or single-line drawing procedures depends on the piping configuration and the complexity of the unit. Double-line drawings are much preferred over single-line drawings for shop fabrication drawings. Figure 2-37, showing the pipe fabrication detail on all-welded assemblies, is drawn in double-line. This unit will have sections fabricated in the shop and then assembled at the job site because of its size. The simplified bill of materials that appears on this spool drawing will enable the fabricator or ordering department to procure the items for the fabrication. In some cases, such as Fig. 9-24, items that are necessary for the spool are flagged on the drawing and the ordering department or drafter will later do a material takeoff. Care must be taken to flag and list every item that is necessary for fabrication of the unit. Missing components can mean extremely costly delays in construction—especially if the items are not missed before the construction stage.

Dimensioning

Spool drawings, whether orthographically or isometrically projected, double line or single line, must give the graphic and dimensional description of a section or unit of piping. Dimensions that are necessary for fabrication drawings depend on the fabrication process and where it is to take place. In general, lengths of welded pipe, face-to-face dimensions, and location of dimensions for valves, fittings, and flanges must all be given along with the necessary angles for the piping configuration and bending sizes. For small-diameter pipe, especially for screwed and socket connections, dimensions are not as stringent as those for welded assemblies. See Chapters 5 and 7 on the dimensioning of pipe fabrications for a detailed explanation.

The following list presents the basic rules of dimensioning for pipe fabrication:

1. Figures, numbers, and letters of the dimension are placed above the dimension line on all fabrication drawings. Do not break the dimension line as in mechanical drafting.
2. Dimension figures are to be read from the bottom of the page or the right side of the drawing.
3. Avoid double-dimensioning fabrication sections.
4. Place larger dimension lines outside shorter ones.
5. Where there are several parallel dimension lines, stagger the dimensions and numbers to make them easier to read.
6. Use feet and inch marks for dimensional figures.
7. Show all dimensions for the pipe fabrication unit. Use notes only where necessary.
8. Avoid placing dimensions inside objects, pieces, or components.
9. Dimensions should be given between piping or surfaces which have a relationship to each other, such as end-to-end dimensions of flanges or valves.
10. Great care must be taken to provide all the views that are necessary to show the total configuration of the piping unit.
11. Use ANSI standards for face-to-face dimensions for all valves and fittings.
12. Show a dimension from the face of the valve to the end of line or to flange face or fitting face for shop fabrication and field fabrication drawings.
13. When dimensioning screwed, bell and spigot, or socket ends, dimension only to the valve centerline or stem.
14. When ring-type joints are used, dimension to the flange face and then give the gasket space dimension by showing the face-to-face dimensions of the assembled joint. (See Chapter 5.)
15. Show rotation of valve stems and handwheel orientation.
16. Dimension and orient bosses and weldolets clearly on all fabrication drawings.
17. Show all dimensions between fittings and major components on the fabrication drawings.
18. In general, use double-line techniques for representing the pipe and pipe components.
19. Code all items throughout the fabrication drawing that need to be clarified.
20. Identify each line throughout the unit with a line number or code specification.
21. Locate and size all shop-fabricated shoes for insulation that are required as part of the fabricated unit.
22. Show centerline-to-centerline dimensions between pipelines that are part of the same fabricated detail.

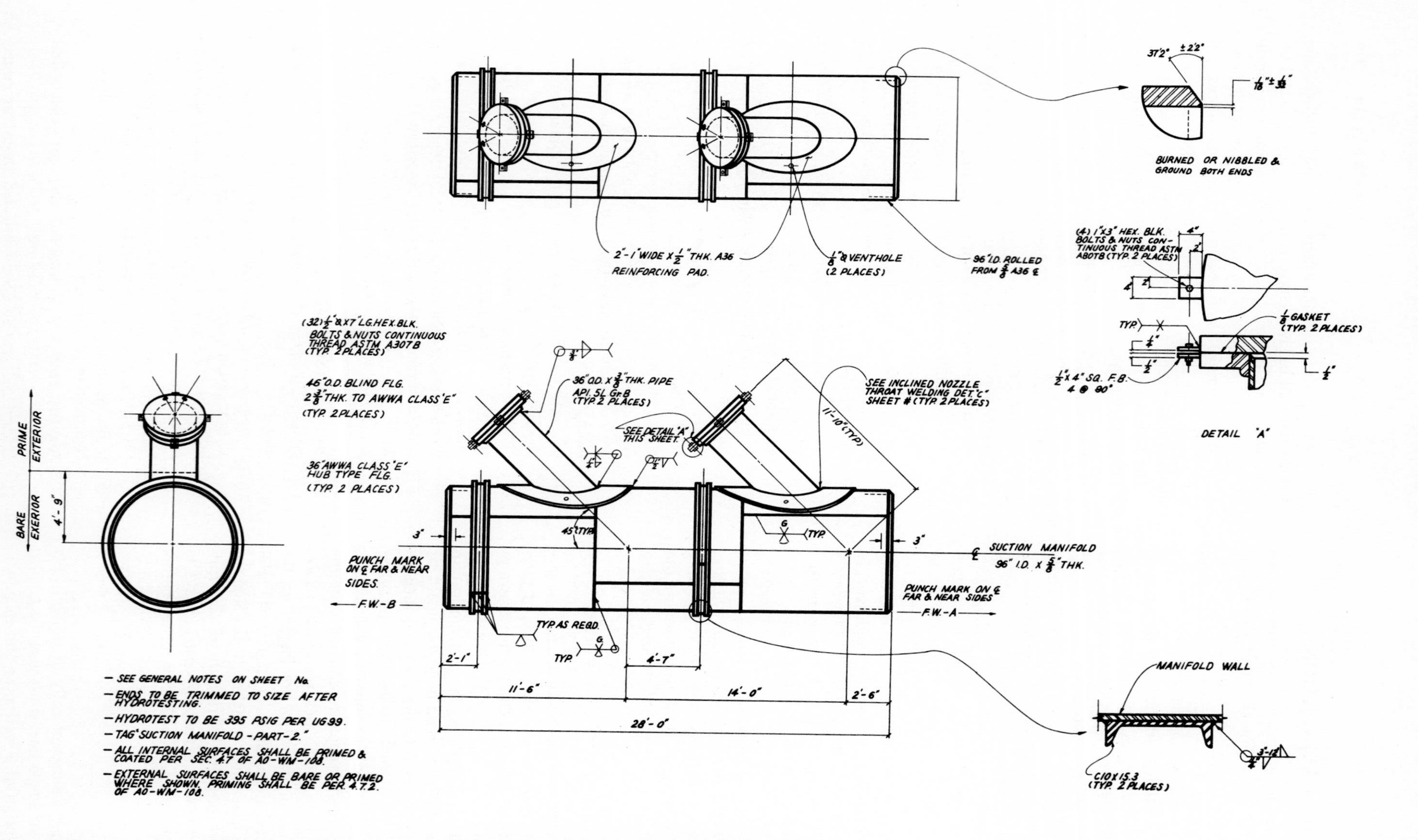

FIG. 9-25 Suction manifold. Note the dimensioning practices used for this spool drawing. (Courtesy Pacific Pipe Company, Oakland, CA)

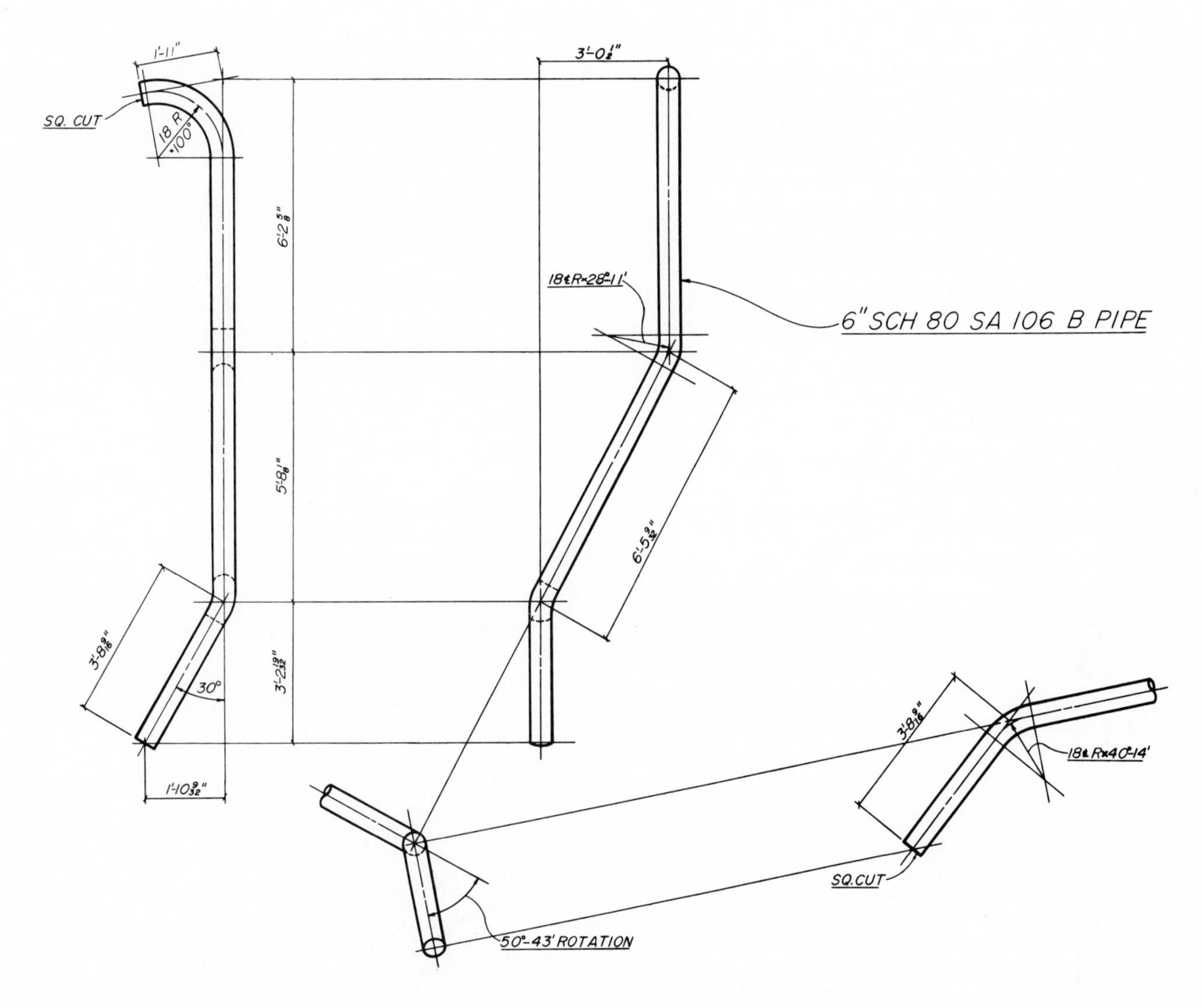

FIG. 9-26 Spool sheet showing a variety of compound bends. (Courtesy Pacific Pipe Company, Oakland, CA)

Study the drawings in Chapters 5 and 7 to see the dimensional techniques that are necessary for the adequate drawing and description of a unit fabrication, whether it is field fabricated or shop fabricated.

PIPING INTERSECTIONS

Figures 9-27 and 9-28 display two variations of pipe intersections. In Fig. 9-27, two pipes of identical diameter are shown as they intersect each other at an angle of 45 degrees. This drawing shows the steps that determine the intersection line to be cut on the pipe wall for the pipes to be fitted together at the specified angle. This drawing also shows the projection of the two piping templates (developments) which, after being drawn and cut out, can be wrapped around the pipe and scribed onto the pipe surface, enabling the fabricator to cut that portion of the pipe with a cutting torch. Headers and piping intersections require this type of procedure, and in many cases the drafter will be called on to draw the piping intersections, developments, and templates.

In Fig. 9-28, a large-diameter pipe is being intersected by a smaller pipe at 45 degrees; the projection techniques for the drawing of the intersection, developments, and templates are shown. The template length always corresponds to the circumference of the pipe. The end of the pipe diameter is shown divided into equal segments. By projecting that portion of the end view back into the side view of the pipe, the intersection of the two mating pieces can be determined. Figure 9-29 shows the intersection of a 5-in. (12.7-cm) standard pipe into a 14-in. (35.56-cm) pipe run at 90 degrees along with its corresponding template. Figure 9-30 demonstrates another example of this procedure.

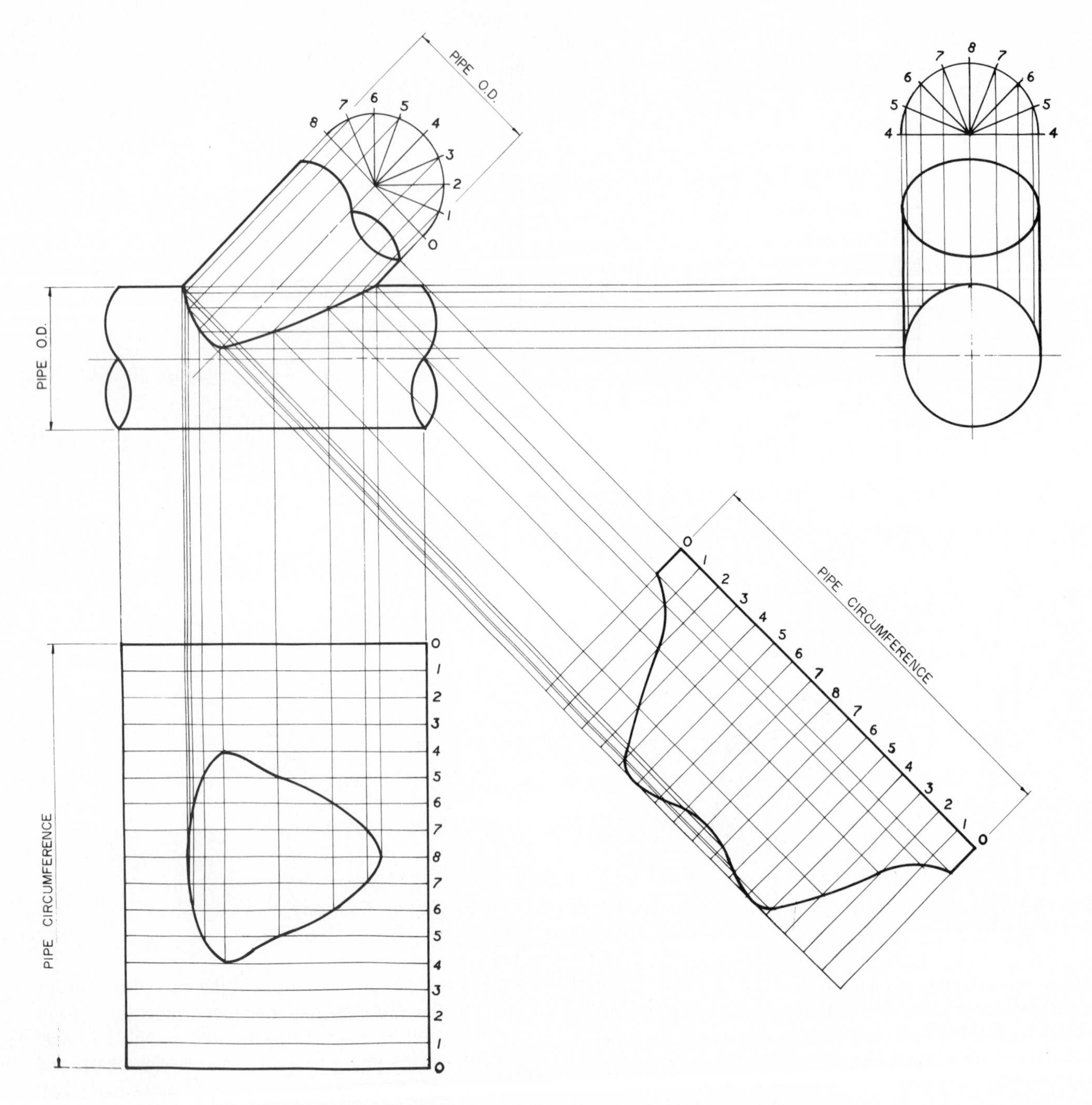

FIG. 9-27 Intersection, development, and template construction for 45-degree pipe intersection of similar diameter pipes. Outer diameter of pipe is divided evenly into sections along the circumference. These points are projected back into the intersecting pipe, thereby establishing points of intersection between two cylinders. The total length of the template corresponds to the pipe circumference.

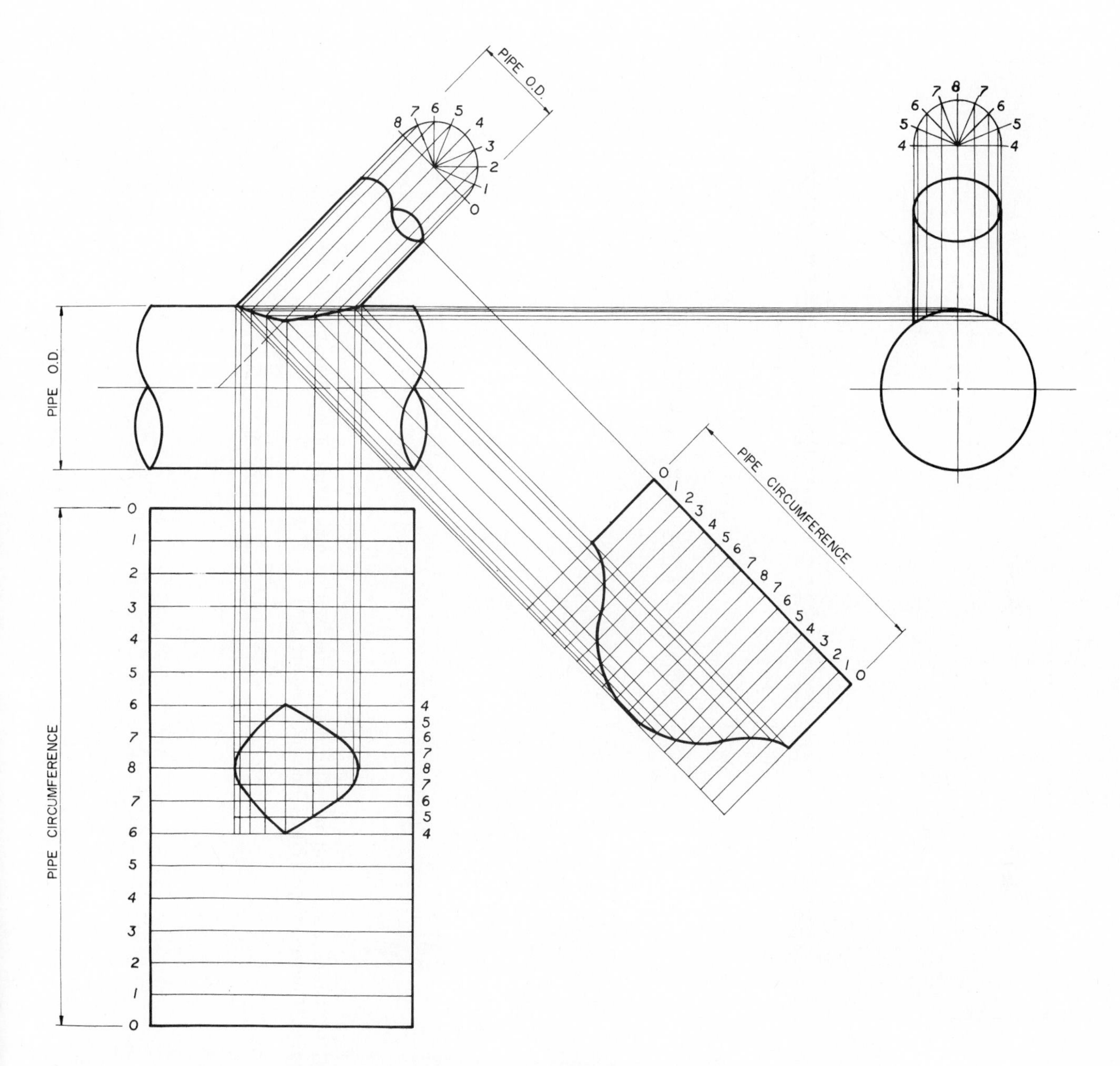

FIG. 9-28 A 45-degree intersection of a small-diameter pipe into a larger-diameter pipe. The intersection, development and template is constructed as in Fig. 9-27.

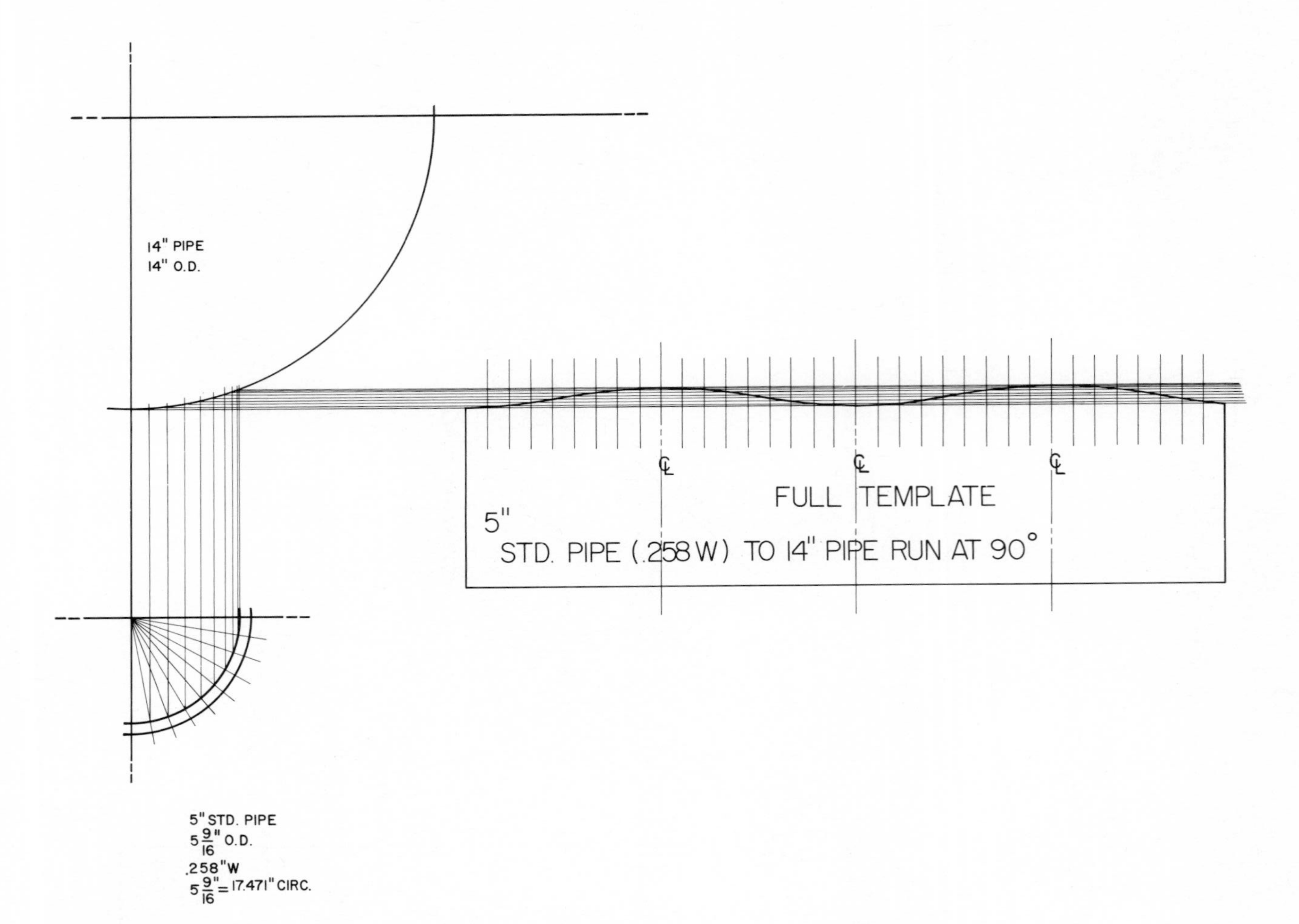

FIG. 9-29 Intersection and template drawing for 5-in. standard pipe into 14-in. pipe at 90 degrees. (Courtesy Pacific Pipe Company, Oakland, CA)

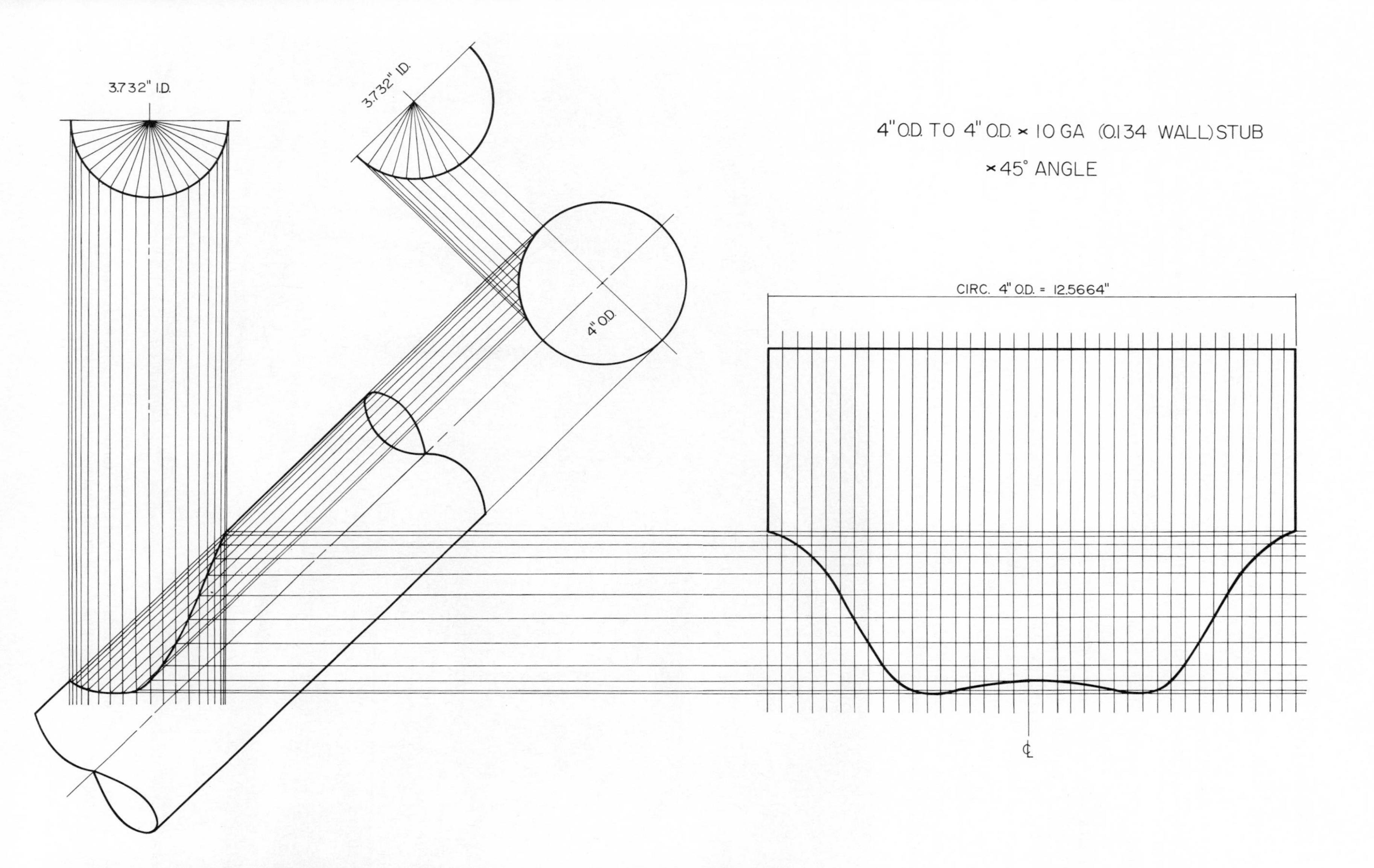

FIG. 9-30 A 4-in. OD to 4-in. OD stub-in at a 45-degree angle showing intersection and template development. (Courtesy Pacific Pipe Company, Oakland, CA)

STRUCTURAL PIPING FABRICATIONS

The opening page of this chapter shows a tugboat pulling the framework of an offshore drilling rig—a good example of pipe being used for structural purposes. Most of the structural elements on this oil rig are large-diameter pipe. The construction industry has been steadily increasing the use of pipe for structural elements: Oil rigs, structural elements of buildings, and pipe racks all use piping for structural purposes. Figures 9-31, 9-32, and 9-33 depict the fabrication of pipe racks. These photographs show the different piping racks that can be created by using piping instead of other structural shapes. The strength and stress capability of the piping material must be known before it is used for structural fabrications.

Figure 9-34 is a drawing of a pipe cluster. This cluster is not meant for conveying liquids or gases; it is a structural fitting in the lower stand of a sports arena. This complicated drawing shows intersections of six 10-in. (25.4-cm) OD, 1/2-in. (1.27-cm) wall pipe and four 8-in. (20.32-cm) OD, 1/2-in. (1.27-cm) wall thickness pipes. All the pipes are belled to 1 in. (2.54 cm) larger ID than the original pipe section. All the total lengths from the centerline of the cluster to the face of the pipe are 22 in. (55.88 cm); the belled length is 4 in. (10.16 cm). This drawing is an excellent example of how descriptive geometry, intersection and development, and template drawings are used for structural piping. The instructor can assign simplified versions of this cluster (such as Fig. 9-35), but it is recommended that this drawing be completed by all who use this book. The result should make a handsome addition to the student's portfolio of drawings.

Figure 9-36 exhibits the descriptive geometry procedure for finding the true angle between two pipes in the cluster.

FIG. 9-31 View of a refinery section of a chemical plant and close-up view of a pipe rack constructed completely of pipe for structural purposes. All cross-bracing and columns are made from pipe.

Descriptive geometry problems are necessary because the true angles do not show on any orthographic projections in Fig. 9-34. Figure 9-37, showing a truss and frame fabrication, is a typical example of structural piping used for architectural purposes. This project will make an excellent assignment also.

This chapter on pipe fabrications should aid students in understanding and drawing the many variations of pipe components, unit fabrications, and pipe sections they may encounter throughout their industrial experience. This material supplements that provided in Chapter 2 (welding and pipes), Chapter 3 (flanges and fittings), Chapter 5 (piping drawings and dimensioning), and Chapter 7 (isometric drawings)—all of which include fabrication and spool drawings in their variations.

FIG. 9-32 At right, a view of a pipe rack constructed solely of pipe.

FIG. 9-33 Pipe rack showing the use of pipe for structural purposes.

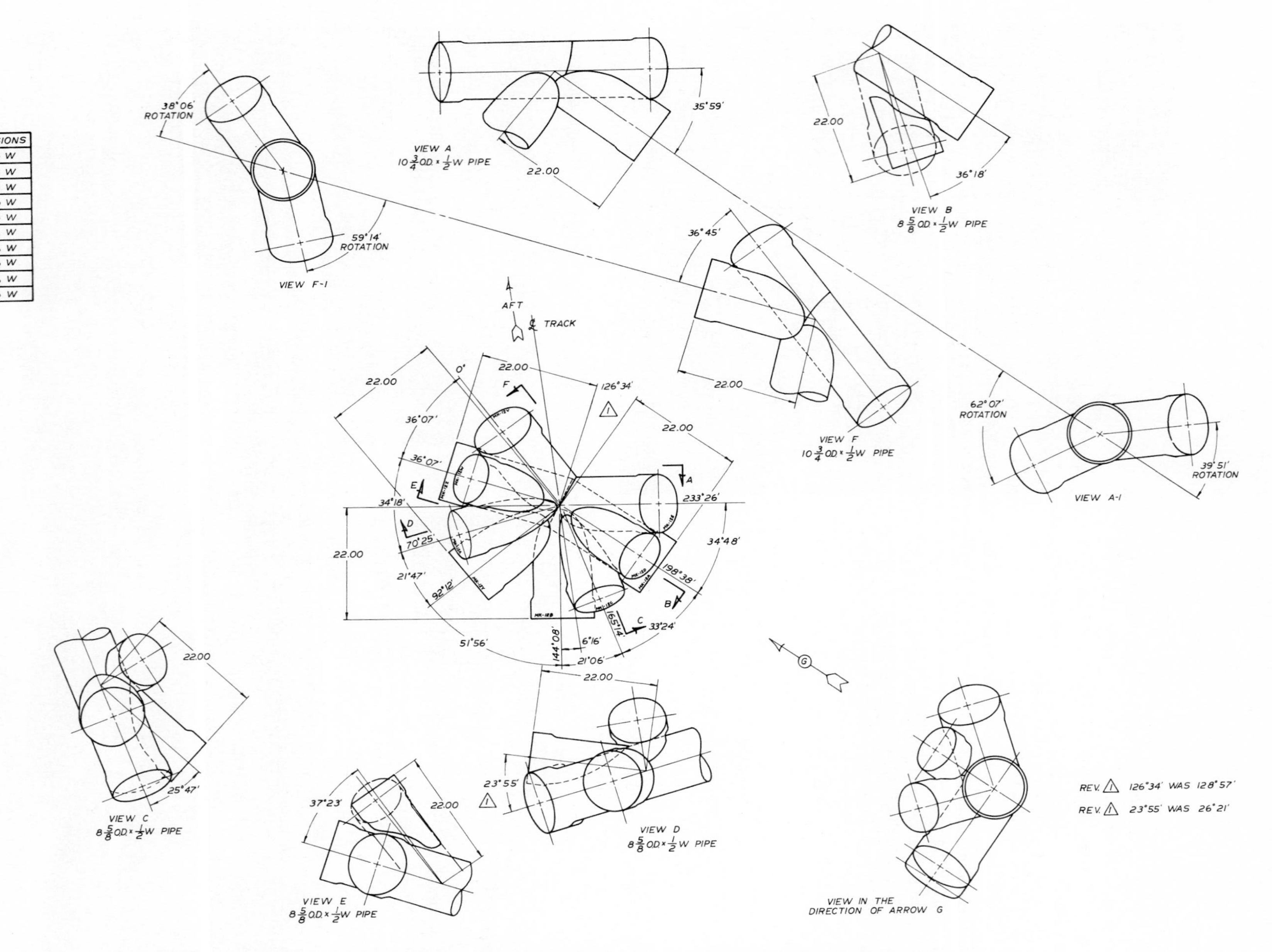

PIPE #	PIPE DIMENSIONS
MK-12 A	$10\frac{3}{4}$ O.D. × $\frac{1}{2}$ W
MK-12 B	$10\frac{3}{4}$ O.D. × $\frac{1}{2}$ W
MK-12 C	$8\frac{5}{8}$ O.D. × $\frac{1}{2}$ W
MK-12 D	$8\frac{5}{8}$ O.D. × $\frac{1}{2}$ W
MK-12 E	$10\frac{3}{4}$ O.D. × $\frac{1}{2}$ W
MK-12 V	$10\frac{3}{4}$ O.D. × $\frac{1}{2}$ W
MK-12 W	$8\frac{5}{8}$ O.D. × $\frac{1}{2}$ W
MK-12 X	$8\frac{5}{8}$ O.D. × $\frac{1}{2}$ W
MK-12 Y	$10\frac{3}{4}$ O.D. × $\frac{1}{2}$ W
MK-12 Z	$10\frac{3}{4}$ O.D. × $\frac{1}{2}$ W

FIG. 9-34 Intersection and multiview drawing of a complex assembly of 8-in. and 10-in.-diameter pipes used for structural connections on rollout bleachers in a modern sports arena. Each belled end of this fabrication will receive a pipe. This is excellent for practicing intersection, development, and template drawings. (Courtesy Pacific Pipe Company, Oakland, CA)

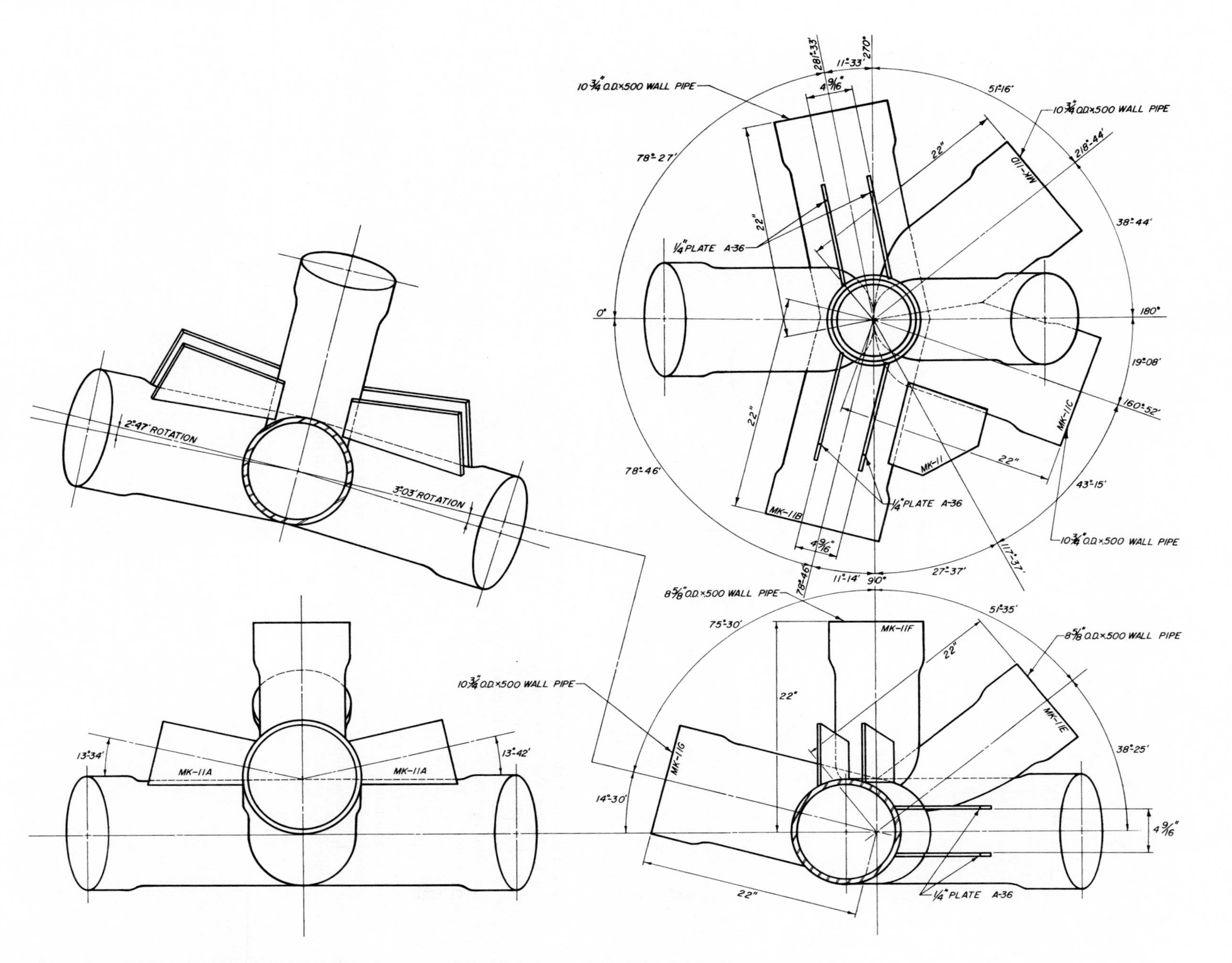

FIG. 9-35 Simple version of Fig. 9-34 with only six 8 and 10-in.-diameter pipes intersecting. (Courtesy Pacific Pipe Company, Oakland, CA)

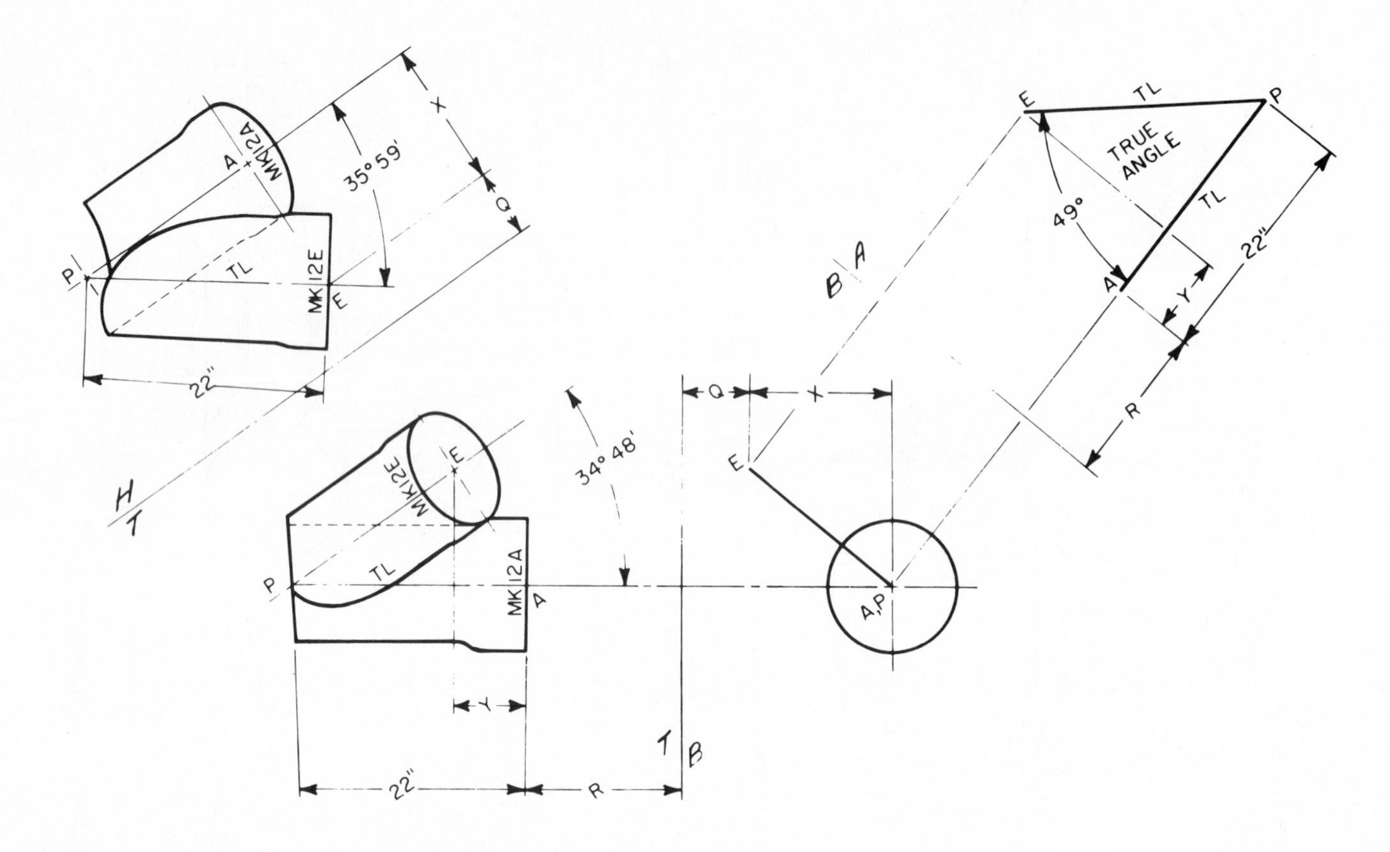

FIG. 9-36 Descriptive geometry worksheet used to find true angles between mk12A and mk12E pipes in Fig. 9-34. In Fig. 9-34, the orthographically projected view does not show true angles between various sections of 8- and 10-in. pipe. It is necessary to use descriptive geometry to find the true view and true angle and hence establish the correct intersection.

The plan view in Fig. 9-34 shows mk12E and mk12A at a 34° 48′ angle; in this view mk12A is true length at 22 in. View *A* in Fig. 9-34 shows mk12E true length at 22 in. and at 35° 59′ from mk12A. It can be seen, therefore, that the true angle between mk12A and mk12E has not been determined.

Both of these views are repeated in this figure. The plan view is now labeled *T* and view *A* is now labeled *H*. The intersecting point of mk12E and mk12A is labeled *P*. By projecting view *B* to attain a point view of line *A-P* and bring point *E* into this view, *AP-E* forms a line. It is now possible to take a true view of *APE* by projecting view *A* at 90 degrees to line *AP-E*. View *A* establishes angle *APE*. Therefore the true angle between mk12A and mk12E is 49 degrees.

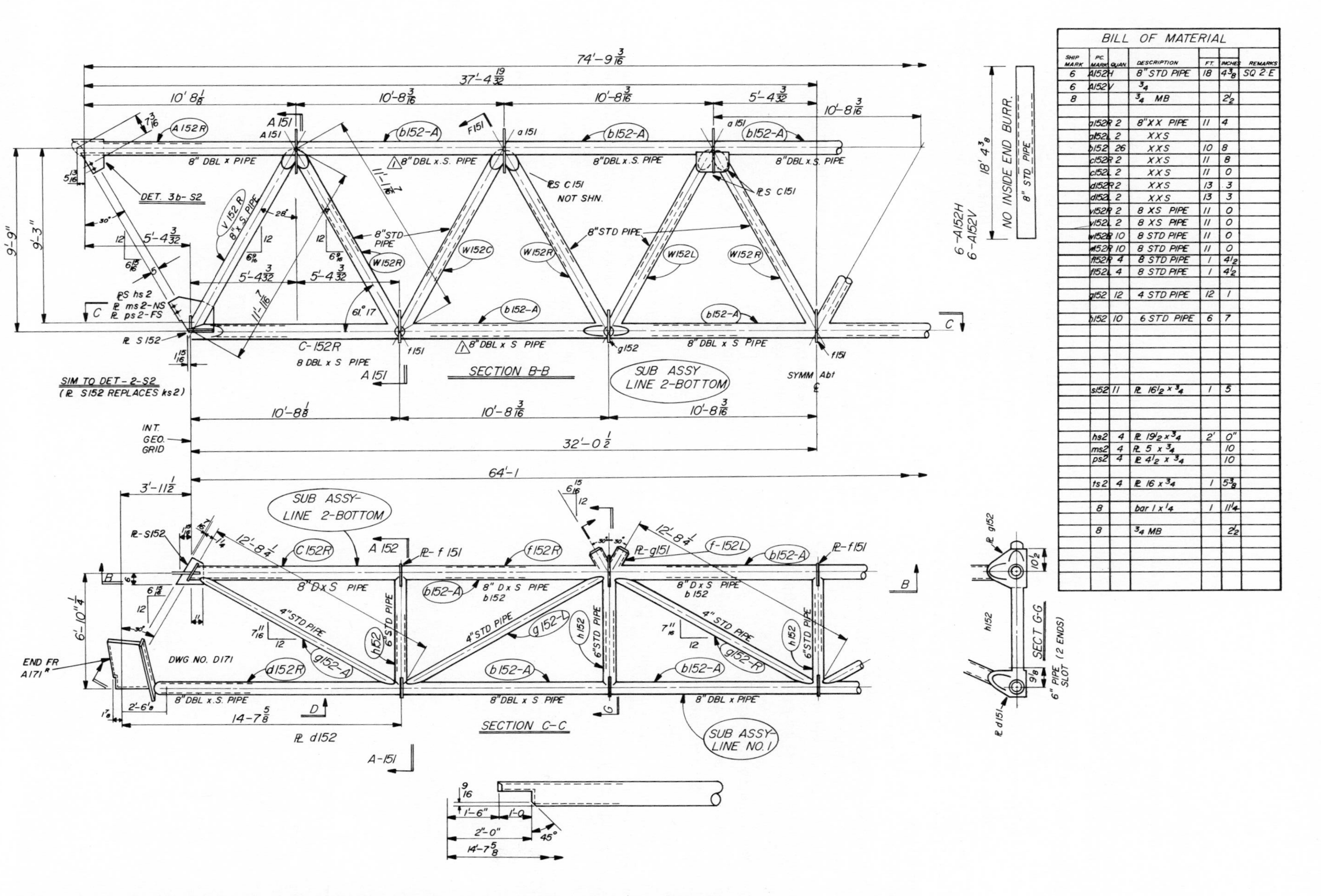

FIG. 9-37 A truss and frame detail utilizing all-welded pipe construction (Courtesy Pacific Pipe Company, Oakland, CA)

QUIZ

1. Define the term *pipe fabrication.*
2. Name five processes that are involved in pipe fabrication.
3. Which is preferable: shop fabrication or field fabrication? Why?
4. Pipe fabrication is usually performed on pipes ______ ____________ in diameter and above.
5. Name three processes performed only in shops.
6. When is it absolutely necessary to fabricate piping in the field?
7. Screwed piping is usually____________ fabricated.
8. What purpose do codes serve in pipe fabrication?
9. Name three uses of pipe for structural fabrication.
10. Small-diameter piping is usually bent ____________ .
11. Name three uses for pipe bends.
12. What is the difference between hot and cold bending?
13. At what temperature range is hot bending of pipe usually done?
14. Name five standard pipe bends.
15. What are five processes involved in pipe forming?
16. What is a pipe extrusion and how is it produced?
17. Name three uses for pipe coils.
18. Pipe fabrication drawings are usually called ________ drawings.
19. What are pipe templates used for?

PROBLEMS

1. Prepare a complete set of drawings showing the intersection of a 2½-in.-diameter pipe at 90 degrees to a 6½-in.-diameter pipe. Show the intersection and development used and draw the templates needed to fabricate the assembly. Use an appropriate scale. Use the intersection and development template drawings provided in this chapter as an example of how to set up the problem. (Instructors may assign a great number of variations of this problem.)
2. Redo Problem 1, but this time make the branch line a 5-in.-diameter pipe entering a 10-in.-diameter pipe at 45 degrees. (See Fig. 9-30)
3. Redraw the pipe fabrication in Fig. 9-24, or Fig. 9-25.
4. Redraw the compound bend in Fig. 9-26.
5. Draw a pipe coil using a 2-in.-diameter pipe coiled to an outside diameter of 30 in. with four coils per foot and a total of 6 ft long. Start and finish the coil with straight piping.
6. Redraw spiral in Fig. 9-15.
7. Redraw truss fabrication in Fig. 9-37.
8. Redraw the pipe spiral in Fig. 9-18.
9. Draw the vessel assembly in Fig. 9-17.
10. Redraw the pipe coil in Fig. 9-10.
11. Redraw the pipe spiral in Fig. 9-16.
12. Redraw the cluster in Fig. 9-34 or 9-35, completing all the necessary views. Do two descriptive geometry calculation drawings similar to the one provided in Fig. 9-36 to find the true angles between pipes assigned by the instructor.
13. Find the true angles for MK-12B and MK-12C, or MK-12B and MK-12W, on Fig. 9-34.

PART IIII APPLICATION OF PRINCIPLES

Petrochemical installation.

10 PIPING SYSTEM DESIGN AND LAYOUT

Much of the material in this chapter has been extracted from the American Engineering Model Society Manual.* It will give the prospective designer, drafter, and modeler of industrial piping systems some idea of the background data, calculations, and design variables that may be encountered on the job.

PIPING

General Piping Information

The objective of this section is to acquaint the "modeler, piping designer" with the basic factors which must be considered in the development of any piping system. To be able to apply all the factors involved may seem a formidable task; however, in most piping systems, only a few of the factors will apply to any one pipe. Considerable experience and the application of sound judgment are necessary for a designer to reach a high degree of proficiency.

Before piping design is begun, process engineers develop process flow diagrams and piping and instrument diagrams (P&ID). This information serves as a basis for the design of the complete system.

The piping designer must have a knowledge of the process equipment and its arrangement before he can begin his work. . . . A good equipment arrangement is essential for an efficient overall design. The piping designer is usually permitted to recommend minor revisions of the arrangement to facilitate the piping design. In some extreme cases, major revisions may be necessary.

The variables that affect individual pipelines will be examined. Although the variables will be treated individually, the reader must recognize that they must be applied collectively to the line being designed. Usually it will be found that many of the variables have conflicting requirements and compromises become necessary. Therefore, the designer must use judgment and recognize that absolute perfection in most cases is impossible.

*** Reprinted courtesy AEMS.**

Design Considerations

The function of piping is to transfer fluids from one point to another. Therefore, the variables are divided into two groups: first, the group that has a direct bearing on flow, and second, the group that does not directly affect flow but may be equally important.

Group I Direct Bearing on Flow
- Positive Pressure Head
- Friction Loss
- Viscosity
- Hydraulic Equilibrium
- Slurries
- Split Flow
- Flashing
- Vapor Lock
- Condensables

Group II No Direct Bearing on Flow
- Economics
- Hazardous Fluids
- Operating Conditions
- Instruments
- Maintenance
- Vibration

Group I

Positive Pressure Head

The driving force used to transfer fluids through pipes is known as pressure head and is usually stated in feet of water or pounds per square inch. In order to successfully transfer the fluid, the pressure head must have a positive value at the point of discharge. The methods for creating the needed pressure head are pumps, gravity, displacement, and suction.

Friction Loss

Friction loss is the loss of pressure head due to the friction between the external surfaces of the fluid and the pipe wall. The friction loss varies approximately with the square of the velocity of the fluid. Tables are available which list the friction losses for various sizes of pipe at given flows. Charts are also available to convert fittings and valves to equivalent lengths of straight pipe.

Thus, the friction loss at a given flow may be calculated for a given pipe size. If the friction loss is greater than the pressure head available, the fluid cannot be transferred in the given pipe size. The flow must be reduced to a point where the velocity of the fluid is such that the available pressure head is greater than the friction losses or the pipe size must be increased.

In order to keep friction losses to a minimum, the piping designer must keep the run of pipes as short as possible and use as few fittings as possible consistent with good design.

Viscosity

Viscosity is the internal friction within the fluid itself. A major concern of the piping designer is a fluid whose viscosity will be subject to wide variation over the temperatures which may be encountered during plant operation. When such fluids are encountered, the piping must be designed in such a manner as to guard against flow failure. The piping must have maximum slope and a minimum of fittings. External heat sources may be necessary. In extreme cases, wash out facilities are required. Valving should be located such that there is a minimum quantity of fluid holdup in the line when it is shut down. Flush bottom valves are often used for this service.

Hydraulic Equilibrium

Extreme care must be given to vents and vent systems when designing piping. It is quite common to speak of an empty bottle in everyday life. Of course, this is not true. The liquid in the bottle has been displaced with a volume of air. To refill the bottle with liquid, the air must be free to escape. In the process plant, the vents provide the path of escape for air or vapors which must be displaced as the vessels are filled. Many times the displaced vapors are valuable and/or harmful and therefore the vapors must be collected and condensed and returned to the process. Air and other noncondensables are discharged to the atmosphere after suitable treatment.

The vessel from which the liquid is being drawn is subject to similar conditions as the vessel receiving the liquid. In order to remove liquid from the vessel, something must take its place. The common practice is to permit air or an inert gas to be drawn in through a vent.

Thus far, the displacement of air by liquid and vice versa has been discussed but the positive pressure head must also be considered. Take the case of the receiving vessel first; if there is a positive pressure head of 25 psig at the point of discharge and if the tank is unvented, liquid will flow into the tank until the air above the liquid is compressed to a pressure of 25 psig. When this pressure is reached, the flow of liquid into the tank will cease. At this time, the positive pressure at the point of discharge will be balanced by the pressure in the receiving vessel.

In the case of an unvented vessel from which the liquid is being discharged, the liquid will flow out of the vessel until the pressure due to the height of liquid equals the atmospheric pressure. For example: the unvented vessel is 40 feet high and filled with water. The end of the discharge pipe is placed below the liquid surface in an open tank of water (this provides a seal so the vessel will remain in an unvented condition). As the valve in the discharge pipe is opened, the water will flow out of the tank until the liquid in the vessel is approximately 30 feet above the surface of the water in the open tank. Under this condition, the pressure above the water in the vessel is almost zero while the pressure on the water in the open tank is approximately 15 psig. The barometric condenser uses this principle to maintain vacuum for vacuum distillation.

Process requirements may require the discharging and receiving vessels to be under pressure. In this situation, it is obvious that the vessels may not be vented to the atmosphere. However, the air must be removed from the receiving vessel. This is accomplished by connecting a pipe from the top of the receiving vessel to a point above the surface of the liquid on the discharging vessel. The pressure above the surfaces of the liquid in both vessels will now be equal. The positive pressure head will equal the gravity head from the surface of the liquid in the discharging vessel to the point of discharge in the receiving vessel. Only the gravity head is considered in such cases as the pressure of the system will not create additional positive pressure head.

Slurries

Quite often solids are transferred from one point to another by suspending solid particles in a liquid. These fluids are known as slurries. When piping slurries, the solid particles must not be allowed to settle. The turbulence which keeps the particles in suspension is due to the velocity of the fluid. Therefore, adequate velocity must be maintained at all times.

Changes of direction should be kept to a minimum. If a change occurs, a five diameter pipe bend should be used, where possible, to maintain proper flow requirements. Dead spots such as might be caused by bypass valves must be minimized by placing the bypass outlet on the top or side of the normal flow pipe and placing the bypass valve as close as possible to the outlet of the tee.

Split Flow

In many processes it is necessary to split the flow of a fluid and direct the partial flows to separate destinations. The most common device used to accomplish this is the automatic control valve. In order to obtain the desired control results, the control valve must have a high pressure drop. There must be a positive pressure head upstream of the control valve which is greater than the pressure drop across the valve plus the static and friction head downstream of the valve. The pressure upstream of the valve is not necessarily equal to the pressure upstream of the split. The pressure upstream of the valve depends on the back pressure developed in the uncontrolled line. If the back pressure is found to be inadequate, additional back pressure must be added. This can be done in several ways. First a restricting orifice can be added in the uncontrolled line. Second, a vertical loop of pipe may be inserted in the pipe run to increase the static head. However, the loop must be vented or the static head is filled with liquid. Decreasing the pipe size is usually an unsatisfactory method of increasing the back pressure. The amount of reduction of size for minimum flow is so great that the friction losses are greater than the available pressure head at maximum flow.

Flashing

Flashing is a sudden change from the liquid state to the vapor state until a state of equilibrium exists between the liquid phase and the vapor phase. The conditions of this equilibrium depend on pressure and maximum heat capacity of the liquid. As the pressure is reduced, the maximum heat capacity is also reduced. Therefore, if the pressure in a pipe transferring a liquid at maximum heat capacity is reduced, there is more heat available than the maximum heat capacity of the liquid at the lower pressure. This heat must be disposed of in order to maintain the equilibrium. The excess heat transforms a part of the liquid to the vapor state to maintain the balance.

Flashing usually occurs at points in the pipe where large pressure drops are encountered. If there is a suitable path of escape for the vapor that is formed the problem is solved.

Flashing may also be due to the reduction of pressure caused by pump suction. This may result in serious damage to the pump if these vapor pockets are drawn into the pump. As the impeller repressurizes the fluid, the vapor pocket will collapse, causing severe shock. This condition is known as cavitation. It may be avoided by locating the pump at an elevation which will ensure a positive suction head.

Vapor Lock

The discussion of flashing shows how vapors are formed in piping. Originally, the term "vapor lock" was associated with vapors formed by external heat causing highly volatile liquids such as gasoline to vaporize in the piping. The same term is often applied to process piping even though the external heat source does not exist. It is also applied when noncondensables such as air are trapped in the pipe.

Vapor lock results from blocking the path of escape for the vapors or noncondensables. It occurs when there is a point in the path of the pipe which is higher than either of the end points. This can be avoided by sloping toward the receiving vessel which permits the vapor bubbles to travel upward against the flow and escape into the discharging vessel. If it is impossible to avoid high points, these high points must be vented.

Vapor lock only occurs in piping where the velocity of the liquid is low, such as in gravity systems.

Condensables

When vapors are being piped, the design problem is different. In this case, liquids must not be allowed to block the flow of the vapor. The designer must avoid pockets if possible. Suitable drains are necessary when pockets are unavoidable.

Drains must not be confused with seal legs, because in some cases, it is desirable to block the flow of vapor while permitting the flow of liquid. The seal leg is a pocket purposely placed in the pipe to block the flow of vapor. When seal legs are required, they must be located so that there is a minimum vertical drop on the downstream side of the leg. Since venting would defeat the purpose of the seal, the seal liquid must be prevented from siphoning.

There are cases where steam is brought in direct contact with process liquids. The designer must consider the result of closing the steam valve in such cases. As the steam condenses, a partial vacuum is created and the process liquids will be drawn into the steam line. The steam pipe and valve must be constructed of a material suitable to withstand the chemical attack of the process liquid. If the process fluid will solidify, the problem is more serious, especially if it cannot be readily remelted. In such cases, the steam valve should be located as close as possible to the process vessel.

Group II

This group of variables must be carefully considered as they have a direct bearing on each pipe while the problems previously discussed may have a direct bearing on as little as 10% of the piping system.

Economics

Economics is the major reason for the existence of piping designers. The designer justifies his existence by reducing costs to a minimum consistent with good design. In order to assign the proper economic values, the designer must have some idea of the items involved in piping systems and their costs. The installed cost of these items is the important figure.

Many of the materials used in process piping may be ten to twenty times the cost of carbon steel. When dealing with alloy or other "exotic" materials runs, the distance between two pieces of equipment should be kept to a minimum, thereby assuring that unnecessary fittings and pipe footage are not used. To prevent this a line study should be made. Also, preference must be given to the more expensive material when clearance problems exist between pipes of different materials. In some cases, 6" carbon steel pipes may have additional elbows installed in order to avoid additional elbows in a 2" nickel pipeline.

As better construction methods are developed, the installed cost of various items will be altered. Therefore, the economically preferred installation several years ago may not be that preferred today. For example, when better pipe bending techniques were developed and welding costs increased, bends were preferred over elbow fittings.

Finally, the designer must recognize the limits to which the economic studies are justified. There is a point at which further savings are more expensive to attain than the saving itself.

Hazardous Fluids

The piping designer must use extreme care if the fluid being transferred is of hazardous nature. Flanged connections should not be placed at eye level. Bottom unloading of tanks must be avoided and suitable safety valves must be used to protect the system from excessive pressures.

Operating Conditions

With the constant increase of material and labor costs, efficient plant operation is growing more important every day.

The operating group judges a proposed design from a different point of view than the designer. Most of their requests for design changes are based on practical field experience and should be given careful consideration. In contrast, a surprisingly high percentage of design personnel have had no operating experience whatsoever and only a few have had extensive experience. An alert designer can learn a great deal from operating personnel if their suggestions are carefully evaluated. In addition, start-up costs can be reduced by eliminating some of the minor, but costly, rearrangements that are usually necessary. However, it is the designer's responsibility to report any additional costs resulting from such requests.

Instruments

Instruments which may be installed in a pipe by welding are a minor problem to the designer. Usually, a suitable location for reading and maintenance is not difficult to find.

An instrument that must be inserted into a pipe may be installed directly through the wall of the pipe if the pipe is of sufficient diameter to accept it. If the diameter is insufficient, the device must be inserted parallel to the flow at an elbow. Therefore, the elbow must be located at a spot suitable for maintenance and display of the reading.

When orifice plates are used for flow measurement, there must be a straight run of pipe ahead of and behind the orifice. This is necessary to obtain accurate readings. The length of straight run will depend upon the location of the orifice taps and the degree of accuracy required. Usually 15 to 20 pipe diameters upstream and 5 pipe diameters downstream are adequate. If it is impossible to route the pipe in such a manner as to obtain a suitable straight length, a straightening vane can be used. The additional expense of the straightening vane should be avoided wherever possible. Flow measuring devices must not be installed in a section of pipe carrying both liquid and vapor.

Maintenance

Consideration of the maintenance department's function becomes more important as costs increase. More plants are operating on continuous processes and maintenance functions must be performed with minimum down time. A good design must recognize the maintenance problems.

The maintenance requirements are:

1. Adequate space to perform work.
2. Minimum disturbance of items not requiring maintenance.
3. Ease of access for adjustment, assembly, and disassembly.
4. Availability of services such as steam, air, and water.
5. Special lifting and rigging devices if needed.

The designer must be able to visualize the space required for maintenance as in most cases these spaces are not shown on drawings. Internals such as exchanger tube bundles, pump impellers, and dip tubes must have space clear of piping or other interferences. When it is impractical to keep the piping clear of the maintenance spaces, break-out flanges should be installed in the pipe.

When it is necessary to remove part of the piping for maintenance, a 90° elbow should be installed in the line to permit assembly and disassembly.

Devices requiring packing, lubricating, and adjustment must be readily accessible.

Vibration

The two most common causes of vibration in piping are water hammer and reciprocating pumps and compressors. Another cause is excessive turbulence caused by high velocities.

Water hammer can be eliminated by proper removal of condensate from vapor piping. See Vapor Lock for further discussion. Water hammer may also be caused by sudden stoppage of a flowing liquid by the rapid closing of a valve. This type of water hammer can be avoided by installing an air chamber to buffer the shock.

Reciprocating equipment causes intermittent flow or pulsations. The pulsation may be eliminated for the most part by a suitable surge tank.

PIPING SYSTEM DESIGN FACTORS

This section covers the essential factors in the design of most piping systems:

1. Thermal movement (flexibility and expansion)
2. Temperature control (insulation, jacketing, tracing)
3. Condensate and air removal (drains, traps, vents)
4. Valves (including control valves)
5. Instruments

Thermal Movement

All piping systems must take into account thermal expansion and contraction. For each degree of temperature change, a corresponding expansion or contraction of the pipeline takes place. The extent of the elongation or shrinkage is related to both the temperature change and the type of material. Different materials have different changes in size when exposed to different temperatures. The designer/drafter should have charts showing the amount of expansion and contraction that can be expected during temperature change in the material. Thermal movement caused by temperature change introduces thermal stress in different parts of the piping system, especially where the pipe has been anchored, such as at nozzles and connections to vessels and other equipment. These stresses must be absorbed by other parts of the system which can be designed for flexibility.

In general, it has been common practice to provide loops in pipelines which undergo large temperature changes. Piping loops made from elbows and lengths of straight pipes provide excellent flexibility for certain pipe stress problems. Figure 10-1 shows a variety of pipe loops, as do many of the photographs in this chapter and the next. Another method used to increase the flexibility of a pipe system is to offset the pipes instead of connecting directly to a piece of equipment. It is possible to offset the pipe run to provide for an extra length of pipe and corresponding elbows. This method of offsetting allows a degree of flexibility which cannot be attained through the direct connection of pipe to equipment.

In many cases, it is impossible to avoid anchoring pipelines to certain equipment, but by using offsets and the maximum number of elbows and extra pipe lengths to produce loops, it is possible to take up much of the thermal stress on the pipe system (at extra piping costs).

Many pipe hangers and supports are designed to permit the slow movement which accompanies thermal change in a piping system while still protecting the pipeline against excessively quick movements such as those caused by seismic disturbances or other types of mechanical loading. (See Chapter 8 for a detailed explanation.) Another source of stress to the piping system is caused by the settlement of foundations, especially for large, heavy equipment such as storage tanks and vessels. Piping connections to such equipment must be designed to allow for this settling stress.

Flexibility In normal operation of a petrochemical installation or a power generation unit, piping increases in temperature several hundred degrees in comparison to its cold, nonoperational temperature. Therefore, great flexibility must be built into the system to reduce the amount of stress associated with this thermal change. Expansion joints are one way of eliminating excessive thermal stress. Though they are not usually used for petrochemical installations, they are frequently found in power generation units.

For petrochemical installations and plants in which the use of expansion joints should be limited, loops, offsets, and change-of-direction design should be substituted (Fig. 10-1). In general, the larger the offset or loop size, the greater the flexibility of the piping system. Piping systems are under extreme stress during operation and it is generally assumed that some plasticity helps to eliminate built-up stress, especially in carbon steel pipe.

In general, the piping drafter should attempt to maximize the use of expansion loops, offsets, and change-of-direction fittings. This will provide the flexibility necessary for the piping system. The L bend, C bend, and U bend are common configurations found in piping systems which must undergo high-temperature/high-pressure operation. Figure 10-1 shows various change-of-direction fittings, offsets, loops, and corner arrangements that can be designed into piping systems to relieve thermal stress. The creation of flexibility through layout is of primary significance in the design process. For piping systems which must undergo great temperature changes during operation, it is important to eliminate, in the beginning stages of design, straight runs of pipe which are connected directly to fixed points or nozzles, equipment, and so forth. It is during the layout and design stages that these ideas must be taken into account, not after the pipe runs have been determined. The simple corner bend or L turn provides a degree of flexibility which the straight connection does not. The Z bend provides extra flexibility, though not as much as when using a pipe loop or U bend.

Where possible, it is good practice to introduce flexibility in the layout of the pipe run in combination with expansion joints. For pipe connections to equipment which are anchored at nozzles with a straight run of pipe in between, an expansion joint can provide the necessary flexibility for pipeline undergoing thermal expansion and contraction. For a petrochemical installation, expansion joints are usually limited to service and utility piping. Many standard catalog-ordered expansion joints are available on the market, and the design specifications and suggested applications should be consulted.

Figure 10-1 shows how to group a series of expansion loops on pipe racks. It is general practice to run the larger pipelines on the outside of the rack and also to group loops so that the hotter lines are on the outside and extend over the other pipelines on the rack. When providing for these group loops, the hotter, larger-diameter pipeline will be on the outside of the loop with cooler and smaller pipelines on the inside. This practice ensures that the largest and hottest pipeline has the largest available pipe loop while the smaller-diameter and lower-temperature lines have smaller loops. Piping loops on pipe racks should always extend over the top of the rack and never protrude from the side, which may cause interference and support problems. These loops must be supported by the piperack bent or by other structural steel.

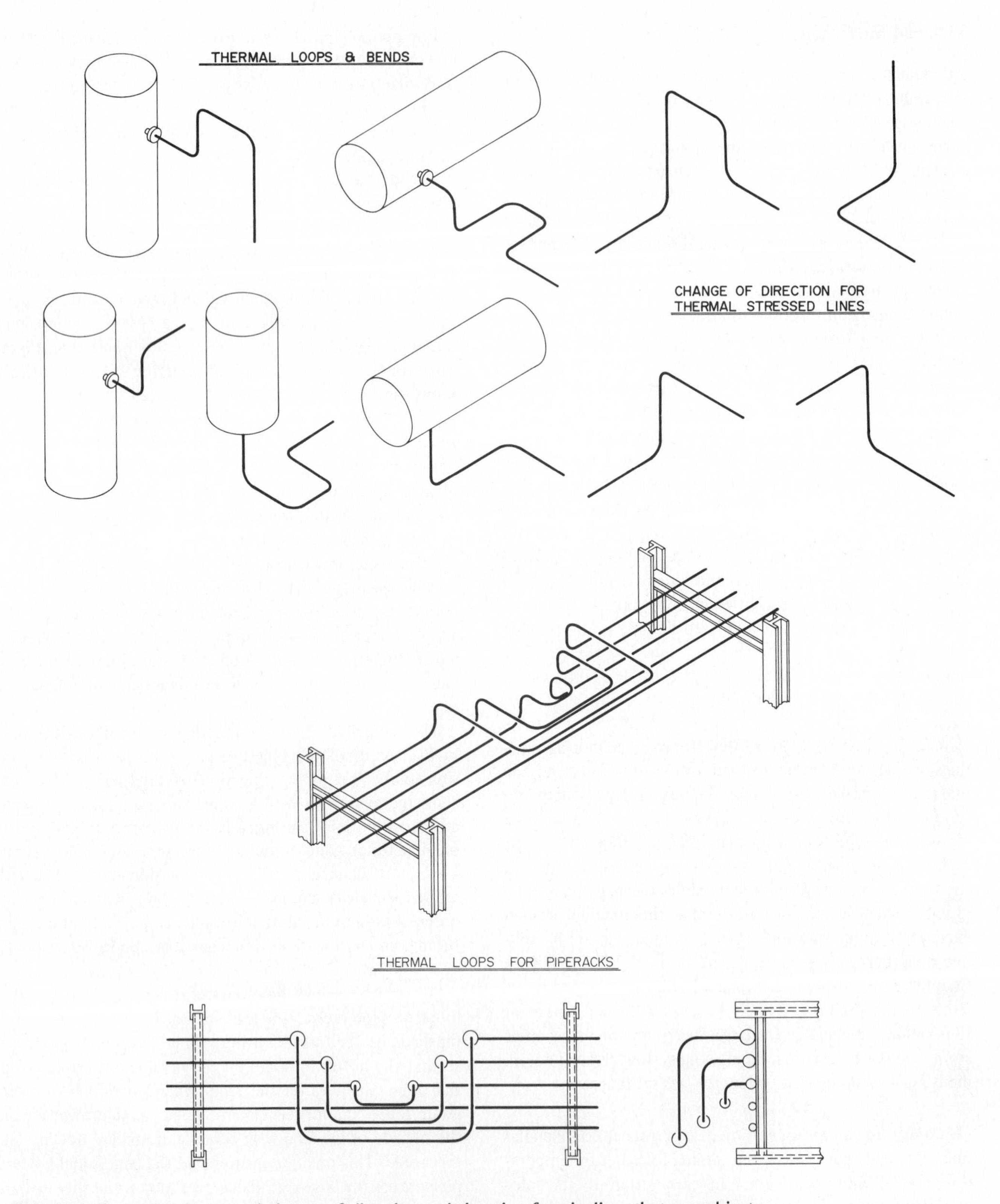

FIG. 10-1 Expansion loops and change of direction and elevation for pipelines that are subject to thermal movement. Thermal loops on pipe racks should be arranged so the larger pipeline is outside of the pipe rack and the expansion loop extends over the top of smaller lines. The coldest line should be near the middle of the rack. All loops need to be supported.

To reduce tension created by thermal movement, offset and change the elevation of the piping when turning corners. This method allows for greater flexibility and unhampered movement of the pipeline. It is also possible to provide a loop at a change of direction and elevation.

Cold Springing One method of preventing temperature stress from building up is the use of *cold springing.* Cold springing is a process by which the distance between hot and cold positions is calculated and divided in half; the pipeline is then cut somewhat smaller than necessary and sprung (stretched) into position at 50 percent of the total thermal movement expected. Therefore the pipe will contract or expand 50 percent in either direction when moving to both the cold or hot position instead of moving 100 percent of a thermal movement length from the cold to hot position. By springing the pipe into place, it is possible to provide flexibility and lessen the total amount of thermal stress on a pipeline. In general, cold springing means splitting the amount of thermal movement that will take place during operation of a system halfway between the cold and the hot location. Cold springing provides a significant reduction in stress and may eliminate the need for an expansion joint.

Another procedure is to offset pipe supports halfway toward the hot position. Therefore, during operation, the pipeline will move in relationship to the pipe support only 50 percent of the total movement. Nevertheless, cold springing should be avoided whenever possible because of the problems involved in installation and because the pipe is always under a certain amount of stress.

Temperature Control

Insulation Insulation plays an important part for both hot and cold lines, depending on the service. Chapter 1 covers many aspects of insulation, including materials. Insulation is found in general use in petrochemical, power generation, and many other piping installations and is used to reduce heat loss in line material. By providing insulation, it is also possible to eliminate pipeline freezing, which can be prevalent in the colder climates, especially in steam lines where condensate buildup is a common occurrence. Insulation is normally thought of as a material used around pipelines and equipment to eliminate heat transfer from the material and pipeline to the surrounding areas. When used to protect personnel from hot lines, it is referred to as protective insulation. Figure 10-2 shows a refrigeration unit which is heavily insulated.

FIG. 10-2 Chilled water pumps and piping installation. Notice the use of cement pads for the pump installation, the various drains and instruments provided on the equipment, and the heavily insulated pipelines, fittings, and valves. Aluminum shielding has been provided in certain areas. (Courtesy Stockham Valves and Fittings)

In most cases, pipeline insulation consists of insulating material, which depends on design requirements, along with some sort of shielding material. The shielding may be aluminum and is usually provided with attachment banding or another arrangement for fastening to the pipeline. As a general rule, the maximum thickness of insulation is usually 2 in. because thicker insulation does not yield significantly greater heat retention for the added expense. The designer and drafter should consult the thermal conductivity tables provided by insulation manufacturers when selecting the type of insulation for a particular job, along with the necessary thickness requirements. Besides insulating straight runs of pipe, it is also possible to insulate valving and fittings (Fig. 10-2). Standard formed insulation sections are available for many standard valves and fittings, including elbows, flanges, and tees. The pipeline code and number include a specification for insulation where needed. It usually designates the type of insulation and reason, such as for personnel protection or heat retention.

Jacketing In some cases, it is important to provide pipe jackets, which are usually shop-fabricated or standard items. These form a double-walled pipe to circulate two fluids simultaneously between the pipe walls without mixing the two streams. In general, jacketing is used only in special circumstances because of its cost and complicated design and installation. Jacketing can be used for heating one stream or for cooling it. Jacketing, though expensive, is probably still the best method for the retention or transfer of heat or cooling to the process stream. The engineer will decide on the type of jacketing for the system in question.

Heat Tracing Tracing is thin tubing or electrical lines attached to the pipeline to provide a constant source of heat. Heat tracing is usually divided into two types: electrical tracing and steam tracing. In some cases, tracing lines are spiraled around important pieces of equipment or pipelines. Petrochemical installations use it for many design situations, especially where the charge or feedstock has a high viscosity and must be kept at an elevated temperature to flow from the storage tank to the process unit. Tracing is also used in colder climates to prevent freezing, especially in lines which are used infrequently or where condensate buildup is a factor.

In general, electrical tracing is more expensive, though it can be regulated at a more constant level than steam tracing and provides for a much greater range in temperature requirements. The pipeline and the tracing tube or cable are enclosed in insulation and covering to maximize the effectiveness of tracing on piping systems.

Steam tracing is the most prevalent type of tracing used for power generation and petrochemical installations. Steam is provided from utility steam lines, which may be completely separate from other steam services in the plant. In many cases, it is important to trace not only the pipeline but also the equipment to which the pipeline connects, such as pumps and valves. It is general procedure to run the tracing element under the pipe to provide maximum heat distribution. A series of traps and drains must be provided for all steam tracing lines. Where it is necessary to provide a great deal of heating for a pipeline, the use of tandem or multiple tracers should be considered. Other methods include spiraling the tracer around the pipeline to be heated or attaching the tracer to the pipeline by welding or cementing.

Drawings of pipelines that need to be traced should show the location of steam headers, traps, and drains and the configuration of tracing, such as spiraling or positioning along the bottom of the pipeline. Thermal movement in the tracer line must be provided for. This can be taken up by the use of spirals. It is important to provide room for the tracer to expand inside the insulation, especially where the tracer line enters or leaves the pipeline. The drawing should also call out the type and thickness of insulation to be used and also the type of tracing element. Steam tracers should be provided with drains at low points and traps at the end of the line, using a trap for each tracer. Do not connect multiple tracers to one trap.

Condensate and Air Removal

Drains, Traps, and Vents. Drains, traps, and vents are of primary importance, especially for steam piping and steam tracing lines, where condensate may build up. Figure 10-3 shows a multiple drain used for pipelines where it is necessary to remove condensate or unwanted liquid from the line. Note the small level glass, which indicates the level of liquid buildup in the drains.

On steam lines where unwanted air buildup is a problem, it is important to vent the system. Chapter 4 shows a variety of APCO air vents that remove unwanted air from the pipeline. Air vents are usually temperature-sensitive devices. It is essential to remove air from steam lines, especially where the steam will push the air in advance when the system is started up, thereby causing steam hammer which may damage the pipeline or components. In pipelines that have a problem with condensate buildup during startup, condensate pushed through the pipeline will cause water hammer, which might seriously damage the piping system and components. Therefore unwanted air, liquid, or condensate buildup must be removed or reduced to safe levels.

As has been stated throughout, one way of eliminating condensate buildup is to slope the pipeline, which reduces

FIG. 10-3 Multiport drainer. Notice the use of a miniature level glass which allows the operator to check the level in the drainer. (Courtesy The Crane Co., Cochrane Div.)

pocketing—especially in areas where the pipeline tends to sag, such as at valve installations. Drip legs and steam separators, along with trapping and drains, help to remove condensate buildup from steam lines. Drip legs are small diameter of pipes connected by weldolets or other small connections to the pipeline where condensate buildup is a problem. The small-diameter pipeline has a valve that can be used to remove the condensate buildup in that portion of the system.

Steam Traps Steam traps are provided on all steam lines and also on steam tracing lines in order to remove condensate and air. Various traps are shown in Chapter 4. Traps are designed to collect the condensate and allow its removal from the system without affecting the steam. They are also used to remove excess gas in the system, especially during startup operation. Many types of traps are available on the market, and their design recommendations should be strictly followed when specifying types for a piping system. Many have strainers for the removal of foreign matter. In general, the following specifications should be adhered to when designing piping systems with steam traps:

1. Determine whether the steam trap should react to temperature, pressure, or density changes.
2. Arrange traps in groups whenever possible, especially for steam tracing lines, to facilitate maintenance and operation.
3. Provide steam traps for all separate lines, never using the same trap for different pipelines.
4. Install traps so that they are at a lower elevation than the surrounding equipment they are servicing.
5. Install drip legs at low points where pocketing or condensate buildup might be a problem, especially on such equipment as economizers and heat exchangers.
6. Use temperature-sensitive steam traps and impulse traps for installations which may encounter freezing temperatures.
7. Consider insulation and tracing for all cold-weather steam pipelines, especially where there is a problem of condensate buildup.

Valving

Valves are covered in detail in Chapter 4, which should be consulted throughout. The following specifications will help in the location, use, and orientation of valves for piping systems. Valves are used throughout piping systems for controlling the flow of material, whether it be for startup, emergency shutdown, movement of materials in a sequence for process operations, discharging, draining, bypassing of equipment, or other service situations. Chapters 11, 12, and 13 also cover valves for specific uses in industry.

Location and Orientation One of the primary concerns of all piping designers is the location and orientation of valves.

These specifications are only general guidelines that should be considered during the design process:

1. Locate all valves for easy accessibility, especially those for processes which involve frequent operation and maintenance.
2. Situate control valve stations and accompanying valving so that they can be easily operated from platforms or aisleways adjacent to pipe racks whenever possible.
3. Place all valves with maximum clearance around the body of the valve stem for easy operation.
4. Locate valves whenever possible to be operable from platforms at grade; equip them with chain operators if necessary.
5. Locate emergency valving well away from hazardous areas so that they are operable during emergencies.
6. Equip valves with chain operators or extension stems when located between 7 ft 3 in. (2.2 m) and 18 ft (5.4 m) in elevation unless platforms can be provided.
7. Consider the weight of the valve during the location process and be sure that sufficient support is provided.
8. Orient valves so that they fall on a common centerline and at a similar elevation to surrounding valves. Whenever possible, install valves on horizontal lines from headers, not on vertical takeoff lines, which can cause pocketing.
9. Consider maintenance and operation frequency when locating valves in relation to surrounding equipment.
10. Eliminate the use of valves on lines which are on pipe racks whenever possible. Pipe racks should be constructed without valves.
11. Group sets of valves for equipment at similar elevations to eliminate unnecessary platforming.
12. Locate valves controlling hazardous services or major process lines so that they can be operated from grade whenever possible.
13. Use hydraulic or pneumatic valves rather than motor-operated valves in and around process equipment or hazardous areas.
14. Take into account the access needed for installation, especially for large, heavy valves.
15. Provide easy access around the valve body for all valves which need frequent servicing, such as plug valves.
16. Consider the use of a fire shield, or isolating the valves well away from the process area, for valving which controls hazardous or flammable piping materials.
17. Do not use chain operators for screwed valves.
18. Show the orientation of the valve stem on all lines to ensure adequate clearance around the valve.
19. Orient the valve stem so that it is easily accessible and operable and does not protrude into aisleways, stairwells, or other areas where it may interfere with personnel and movable equipment.
20. Always consult the manufacturer's recommendation for the installation orientation of the valve. In general, gate valves, plug valves, and check valves are for horizontal piping and should not be installed upside down with their stems pointed downward.
21. Remember that the primary concerns for valve location are easy access for maintenance, installation, and operation.

Control Valve Stations and Manifolds Control valves are used to regulate the flow of a pipeline, either remotely by an actuator or locally by a handwheel. Remote control is accomplished by an electric or pneumatic signal. *Control valves monitor pressure and flow of the line material.* The control valve station includes the control valve and the isolating valves on either side of it which shut down the pipe loop to service or remove the control valve. Bypasses are usually provided so that the system can still operate during this period. In Chapter 11 and elsewhere in this book, there are views of control valve stations. They are in general situated at piping loops, where a control valve and associated valves and piping provide for the reduction and regulation of the stream (Figs. 10-4, 10-5, and 10-6).

The specifications set down in this section are general recommendations for the selection and design of control valve manifolds. Figures 10-4 and 10-5 show some of the common configurations used throughout industry. Control valves must be designed to fit the needs of the piping. Since control valve stations are expensive, they should be designed as economically and efficiently as possible. In general, they should be oriented so that they fall on a line parallel to the pipe rack and are accessible from major aisleways. They should be supported so that the control valve can be serviced and removed without disturbing the surrounding piping and valving for the station. Another important consideration that must be taken into account in the design stage is the effect of thermal stress on the control manifold station.

A side- or top-mounted handwheel in combination with the automatic actuator may be part of the control valve mechanism. When locating control valves, consideration should be given to areas around equipment to which the control valve directly relates. For stations which must be reached from platforms, these should be designed to facilitate installation, maintenance, and operation. The control should be an independent station, and the manifold should be independently supported and not attached to the surrounding equipment. Whenever possible, the control valve station should be located at grade—especially for equipment at low elevation and for stations located under or next to pipe racks. The following specifications apply to Fig. 10-4, which shows control valve manifold configurations. Specifications which are ballooned on Fig. 10-4 are general specifications used on all control valve manifolds.

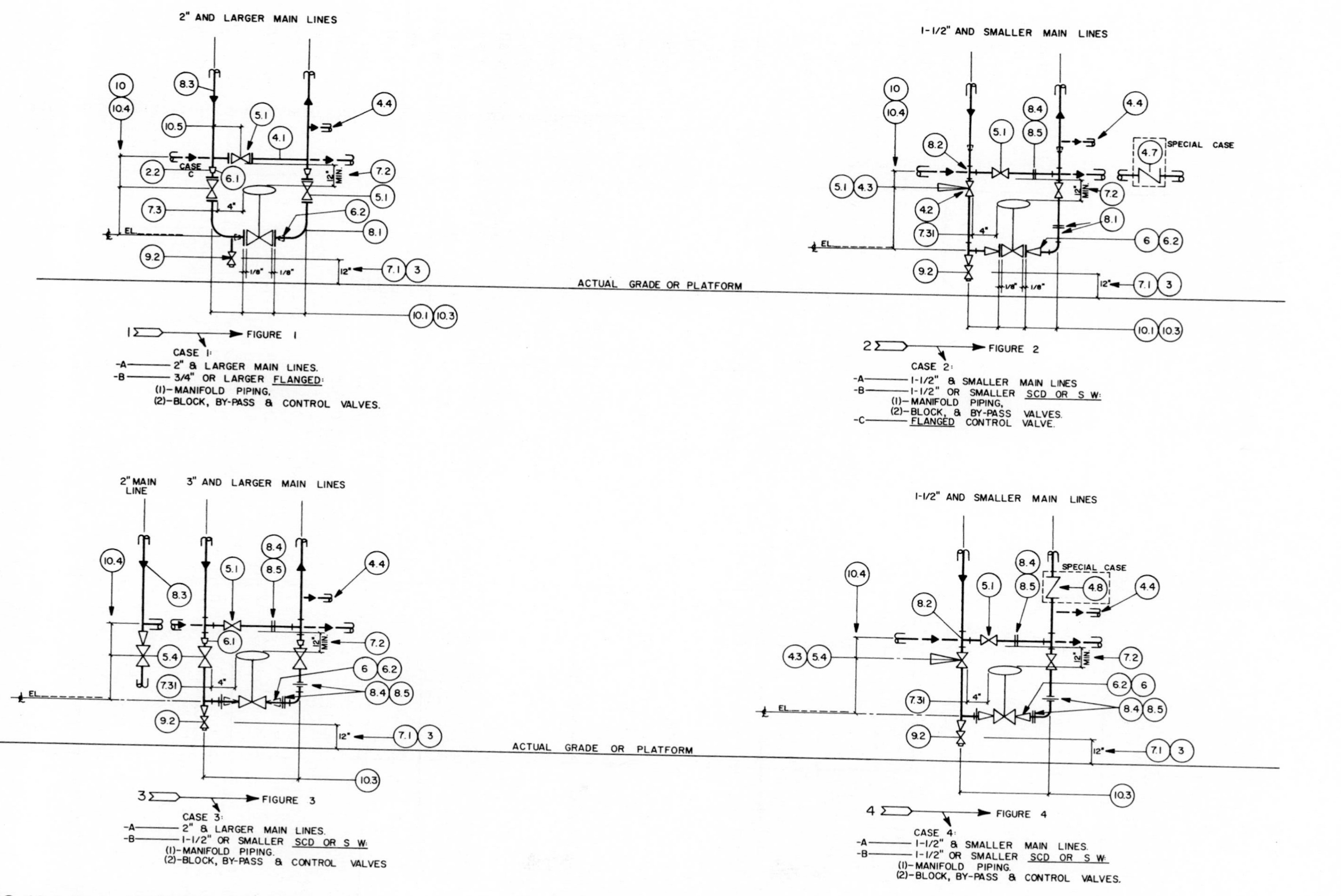

FIG. 10-4 Control valve manifold specifications.

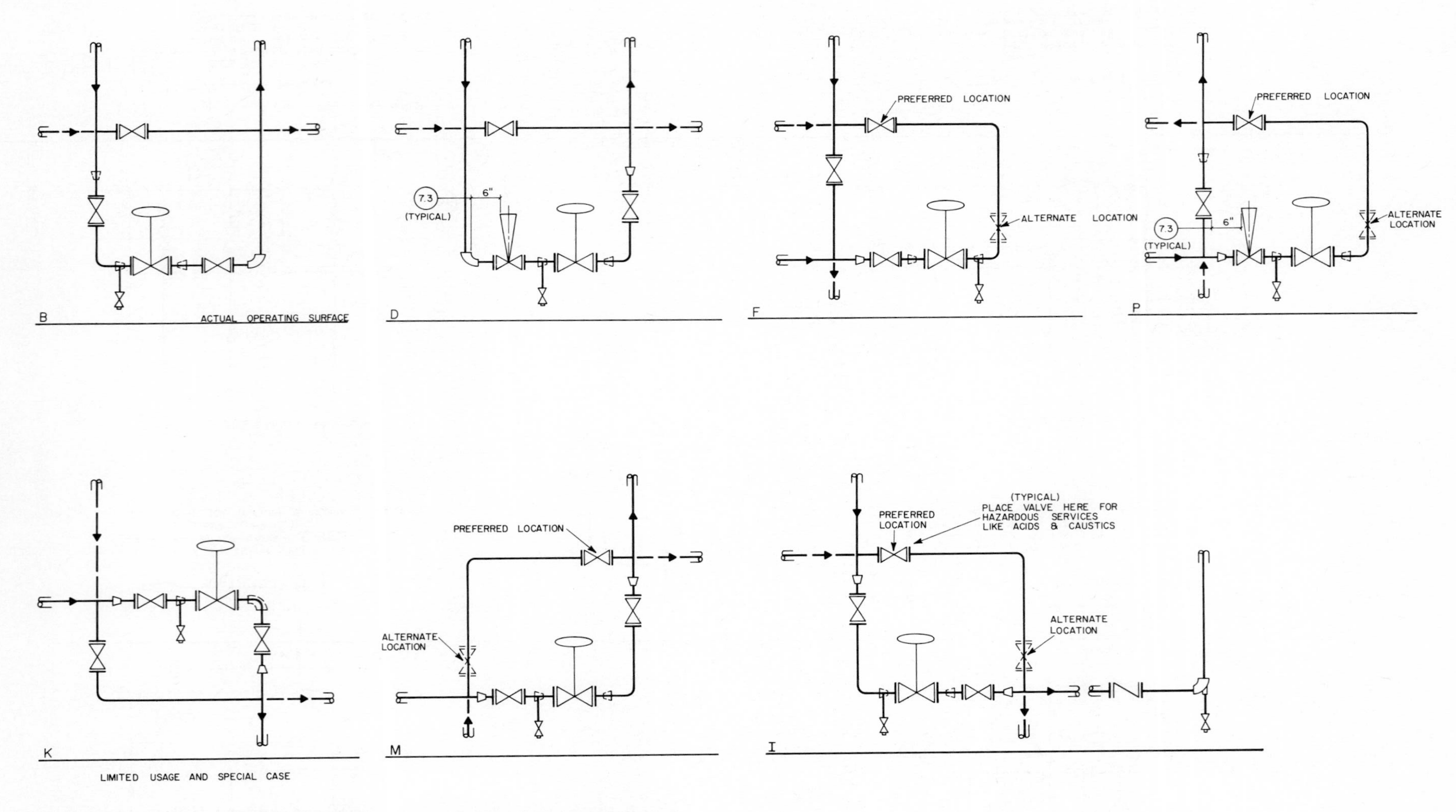

FIG. 10-5 Typical control valve manifold configurations (see specifications).

FIG. 10-6 Control valve installation. Many types of controls, instrumentation, and insulation are shown on this petrochemical control valve station. Various diaphragm control valves can be seen with appropriate instrumentation. Notice the drain from the bottom of the control valve loop. Note also the different pipe supports and fittings.

1. Control Valve Manifolds

   1.1 This list of specifications covers recommendations for the selection and design of control valve manifolds.

2. Manifold Sizes

   2.1 The following instructions should be used as a guide in sizing the manifold's components, except final sizing shall be shown on the P&ID.

   2.2 Basic sizes usually are:
   Control Valves—one size smaller than the main line
   Bypass Valves—the same capacity and size as control valve
   Bypass Piping—the same size as the bypass valve
   Block valves and related piping as follows:
   (a) For 1½" and smaller mainline, control valve should be smaller than line size. For block size and related piping size, it should be equal to the mainline size. The reduction location point should be at the control valve.
   (b) For 2" and larger mainlines, the control valve line should be 1½" and smaller. For block valve size and related piping, 1½" is the minimum size.
   (c) For 3" and larger mainlines, control valve should be 2" and larger. For the block valve size and related piping, it may be reduced, usually to one size smaller than the mainline or to control valve size. For the reduction location, the outside of the block valves, and in addition at the control valve, is a usual location.

3. Locations of manifolds shall be as follows:

   3.1 An accessible location, preferably on grade or on permanent platforms or walkways.

   3.2 Within view of control or associated instruments (or a transmitted air or receiver gauge shall be used and made visible from the manifold's bypass valve).

   3.3 At the downstream end of the piping run in services where vaporization is likely to occur.

4. Locations of miscellaneous items as noted shall be as follows:

   4.1 Bypass at or above the elevation of the control valve.

   4.2 Valve handwheels at operational orientations and heights consistent with specifications.

   4.3 Stems of small valves in the main plane of the manifold.

   4.4 Sampling connections or lines on or above the centerline of the bypass line.

4.5 Pipe supports located so manifold is self-supporting with control valve removed.
4.6 Large swing or lift check valve in low horizontal lines accessible from grade or platform.
4.7 Small swing or lift check valve in horizontal lines at accessible locations.
4.8 Small swing check valves only in upward-flowing vertical lines at accessible locations.

5. Valve Types
   5.1 Bypass valves shall be gate valves if the design pressure drop is less than 50 psig. Bypass valves shall be globe valves if the design pressure drop is more than 50 psig. Bypass valves shall be gate valves in size 6" and larger and globes in 4" and smaller as a general rule.
   5.2 Manifolds are to be considered special when valves other than gates or globes are required or when weld-end valves are specified.
   5.3 Butterfly control valves do not require manifolds where they bypass unless noted otherwise on the P&ID.
   5.4 Block valves are usually gate valves unless globe valves are required for the reasons stated in 5.1.
6. Reducers and Swages
   6.1 Use concentric reducers and swages in vertical lines.
   6.2 Use eccentric reducers and swages in horizontal lines with this exception: for lines 2" and smaller, use concentric swaged nipples when complete drainage is not essential. Place reducers or swages (when an option exists) in locations which make the piping the largest size over the longest possible length of the run.
7. Clearances
   7.1 Allow 12" minimum clearance space between the bottom of the control valve and actual grade or platform.
   7.2 Provide 12" minimum clearance space between the top of the actuator and any obstruction or fixture directly above the diaphragm case. Caution–Determine true control valve height. Include all actuator accessories, bonnets, top positioners, etc.
   7.3 Make 6" minimum clearance between handrails and vertical lines or flanges. Provide 4" minimum clearance between diaphragm and vertical lines or flanges. Increase clearances by the insulation thickness where the vertical lines or flanges are insulated.
8. Precautions
   8.1 Arrange manifolds to provide flexibility for removal of control valve when ring or small tongue and groove joints are used.
   8.2 Do not use reducing tees where pressure drop is sufficient.
   8.3 Make certain of flow direction through all types of control valves.
   8.4 Avoid the use of screwed unions when seal welding is needed; if unions are necessary, use flanges.
   8.5 For blind flange requirements, consult engineer.
   8.6 Manifold design and location shall be reviewed with the engineers when the following conditions are involved: Control valve pressure is 100 psig or more, flashing exists at or near the control valve, and for erosive services such as decoking and slurry.
   8.7 Pay particular attention to the orientations, clearances, and access space required for handwheels, positioners, or other special accessories and control valves.
9. General
   9.1 Types of manifolds shall be kept to a minimum for any one project.
   9.2 Provide a drain valve between upstream block valve and control valve.
   9.3 Automatic pump start control valves do not usually require a manifold bypass or double block valves.
   9.4 1½" and smaller screwed socket welded manifolds need only overall width dimensions.
   9.5 Control valves shall be installed with stems vertically upward unless instrument engineer approves other location.
   9.6 Manifolds are sometimes in the horizontal plane, not always in vertical plane.
10. Dimensioning Practices
   10.1 Dimension the vertical location of all valves.
   10.2 Dimension the horizontal location of flanged and weld end valves.
   10.3 Screwed or socket welded manifolds need only overall width dimension in multiples of 3".

Figure 10-6 shows an excellent example of a control valve station with accompanying valves, reducers, fittings, piping, insulation, and instrumentation. This photograph actually shows two separate control stations: the large-diameter control valve in the center and a smaller-diameter control valve on the right side. Both are diaphragm-operated with bypasses and appropriate instrumentation. Notice how both are provided with independent supports and drip legs.

Instrumentation

Instrumentation is covered in detail in Chapter 6 and also in Chapters 11, 12, and 13. In this section, instrument locations and explanations of level gauges, and level controllers supplement the material provided in other chapters.

Instrumentation is shown on the P&ID and is located, designed, and called for by the instrument engineer. The pipe drafter will come into contact with the P&ID and other drawings which have been supplied by the instrumentation group. In general, the instrumentation, location, and installation drawings are provided by the instrument group and include the dimensions necessary for their construction. Included on the instrument drawing is the type of instrument, its operating connection, whether it is electrical or pneumatic, and all vendor information which may be necessary for its placement.

In general, all instrumentation drawings are completed outside the piping group, so the piping drafter need only have an understanding of the function, design capabilities, and location of the instruments. He or she need not be concerned with drawing the details. Instrumentation drawings show exactly where the instruments should be located, how they should be connected, and also the location of raceways, control panels, and other information the piping drafter needs to provide sufficient clearance around the instruments and to locate them for operation and reading. Figure 10-6 shows some control station instrumentation that will be required. Note the pneumatic connections to the diaphragm valves along with a variety of pressure gauges.

Although instrumentation is one of the last things to be added to the piping system, it is necessary to plan in the early stages for the location of gauge connections and other instrumentation; they must not become an afterthought. Provide sufficient space, clearance, and adequate location points. Instrument location lists are usually made in the instrument group and provided to the pipe drafting group before the piping system is designed.

The piping drafter uses the instrumentation drawings and location lists to represent the instruments, location, and configuration in relationship to piping, vessels, and other equipment. Only large instruments are drawn on the piping drawing–level glasses, level controllers, and the like. Small pressure gauges, temperature wells, and so forth may be shown on the piping drawing, but often they are not. Models are always tagged so that all instrument locations are identified. One of the main purposes of showing instrumentation connections and locations on piping drawings is so that connections, flanges, orifice places, and so on can be installed at the construction stage directly from the piping drawings. In most cases, instrumentation, design, callouts, and functions are in accordance with ISA-S5.1 (see Chapter 6).

The instrumentation group provides the piping and design group with instrumentation details and location lists. These drawings give the general arrangement, location, configuration, instrument function, instrument number, control panel hookup, piping associated with the instruments or electrical hookups, and any vendor dimensions which may be necessary for the piping designer—especially where the size of the instrument is extremely important. Level glass and level controllers, for example, are located near aisleways for easy reading. It is also necessary to know the exact location of instruments to establish platforms for reading, maintenance, and operation.

Instrument piping consists of control piping for either pneumatic or hydraulic apparatus. In many cases, instrument piping is heat-traced because of infrequent use or cold climate. Instruments are connected to equipment vessels and piping with elbowlets, threadolets, nippolets, swaged nipples, couplings, and other standard connections. These fittings are usually provided with screwed connections for small-diameter piping. Flanged nozzles are also used especially for bridles and level glass connections to vessels. The type of instrument connection depends on temperature, pressure, and the material to be monitored; thus high-pressure steam and low-pressure oil may require different connections for the same instrument. Of primary importance in the design and location of instruments for piping systems is ensuring that they can be easily read and operated from grade level or from adequate platforms. It is common procedure to locate the instruments near valving stations whenever possible.

Level Instruments The liquid level gauge enables the operator to make a visual reading of the liquid level inside a vessel (Fig. 10-7). The level glass is also referred to as the gauge glass. This instrument is not used with any controlling device, such as a control valve; its only function is to provide a quick way of checking the liquid level within a vessel.

Many different types of gauges are available, though they all provide the same service. Several lengths are manufactured and in most cases two or more glasses can be staggered to increase the length of the viewing area. This depends upon the service required and the vessel's internal configuration. By combining gauge glasses in this way it is possible to cover the complete float range required by the process.

Figure 10-7 shows a typical level glass gauge with its connection to either a horizontal or vertical vessel. Level gauges can be installed in tandem and triplet, depending on the viewing range which must be monitored. In Fig. 10-7, a single level gauge is shown with isolating valves and flange connections to the two vessels. In general, the viewing range for a single level gauge is 5 ft (1.5 m). For vessels where the viewing area must be greater, the use of tandem or triple level glass units is suggested; gauges should be overlapped so that they monitor all portions of the viewing area. Specifications for the installation, location, and design of level glass gauges must be taken from standard company procedures and vendor information.

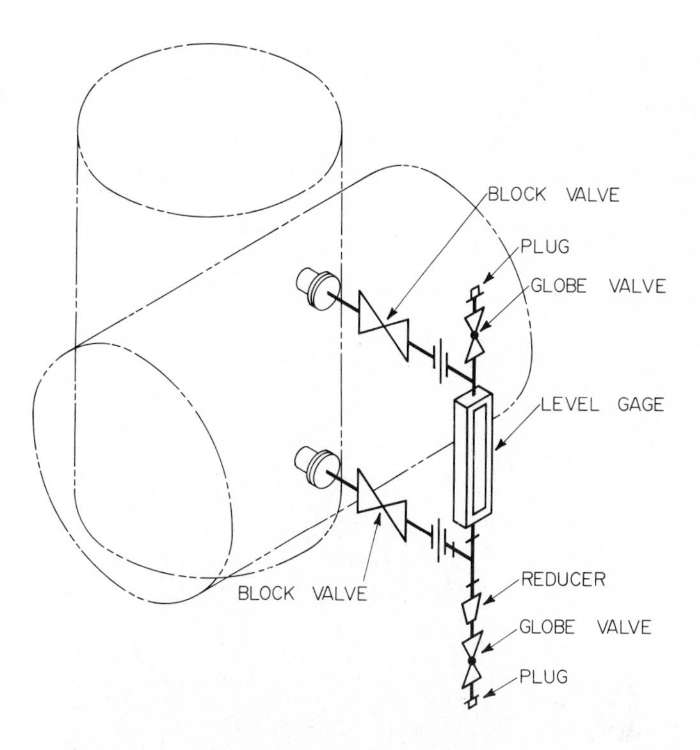

FIG. 10-7 Typical level glass connection for a vertical or horizontal vessel.

A level gauge can also be used to aid in adjusting or verifying a liquid level controller. A liquid level controller is used to maintain a predetermined liquid level in a vessel by transmitting a pneumatic or electrical signal to a valve which regulates the flow in or out of a vessel. Connected to a recorder, the controller may be used to record the liquid level periodically or constantly.

The level gauge and level controller work in conjunction with each other. The normal liquid level of the vessel, the center of the float range for the controller, and the center of the visible glass range for the level glass (gauge) are on a common centerline as shown in Figure 10-8.

In a pneumatically operated system, the air supply is introduced at the instrument (level controller). The level controller measures the level of liquid in the vessel and sends

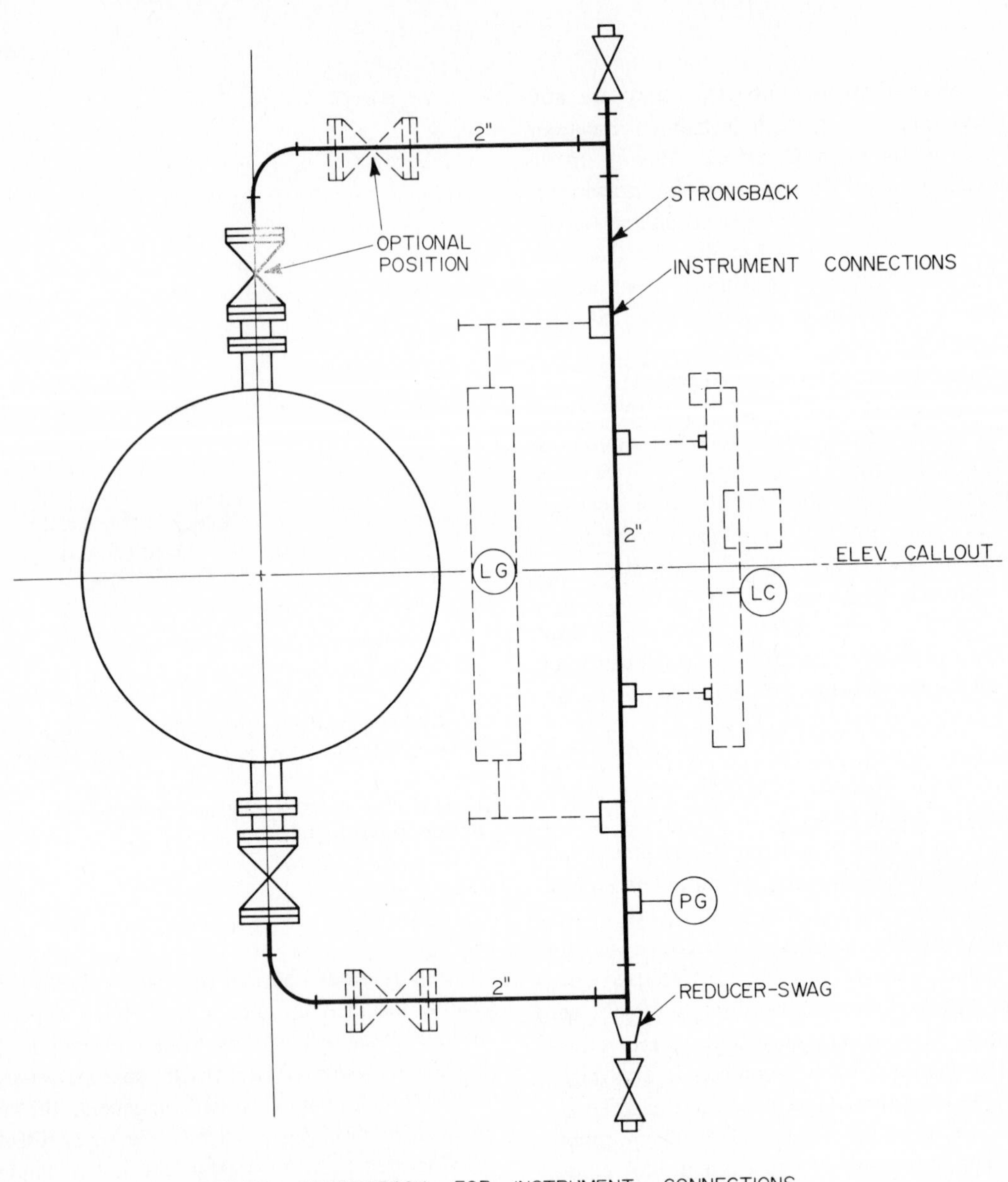

FIG. 10-8 Bridle and strongback for instrument connections to a horizontal vessel, including level glass, pressure gauge, and level control instruments.

a pneumatic signal to a board or to a locally mounted level indicator and to the control valve. The signal controls the opening and closing of the diaphragm control valve, thereby maintaining the desired liquid level in the vessel.

The level indicator is used to indicate high or low levels in the vessel. The level switch mounted on the outside of the vessel sends an electric signal to a light and/or horn in the control building. The level switch is used to shut off the warning mechanism.

Bridles Figure 10-8 shows a bridle hookup to a horizontal vessel. A bridle is a common line used for manifolding multiple instruments to one pair of connections on a vessel. The vertical piping where the level gauge or level controller is situated is called the *strongback* or *standpipe.* On this vertical piping, it is possible to locate various connections, such as pressure and temperature gauges, level gauges, and a level controller. Level controllers monitor the level of material in a vessel, keeping it at a predetermined level. In general, the bridle connection is designed so that the strongback is a larger-diameter pipe than the connecting horizontal members. In Fig. 10-8, the isolating valves are shown in a vertical position with an optional position in the horizontal run. The horizontal position is preferred because it allows the system to self-drain. Draining valves and plugs are also provided at both ends of the strongback. The level controller shown in hidden lines in Fig. 10-8 indicates the normal location for this instrument.

There are many other gauges and instruments which are covered in more detail in Chapter 6 and in Chapters 11 through 13 for petrochemical, conventional, and nuclear piping. Level alarms, pressure gauges, level controllers, temperature indicators—all are common instruments found on piping systems.

ELECTRICAL REQUIREMENTS

Electrical drawings are provided by the electrical group in a procedure similar to that of the instrumentation group. All electrical designs and requirements are determined by the electrical engineer separately from the piping group, though the piping design and drafting group must have access to the electrical drawings and requirements to show the location of conduit runs, control boxes, lighting fixtures, switch gear, cables, and circuits. The electrical requirements must be known in the beginning stages of the project whether the model method or the drafting method is used. All electrical needs are covered by specifications and codes laid down by different associations and government agencies.

One of the primary functions of the electrical group is to provide the electrical requirements, locations, and drawings (including substations). In many cases, the electrical equipment must be indicated to the structural engineers because electrical systems will be established underground and also because much concrete and foundation design depends on the type of electrical equipment which must be protected. In general, the electrical drawings are similar to the flow diagrams; they are drawn schematically and are not dimensioned. If major conduit runs, trays, and large pieces of electrical equipment must be located, these are dimensioned on the drawing. The size of conduit runs and electrical trays must be known so that the piping designer can leave sufficient room around pipe racks and other areas of concern, especially where interference between electrical, piping, and other systems are a factor.

Interference is always a problem in the design and implementation of a nuclear or a fossil fuel power plant where the total system is confined to a structure. In these cases, the electrical needs are enormous and must be critically planned in the beginning stages of the project to provide sufficient clearance for installation, operation, maintenance, and repair and so that pipe supports, pipe runs, HVAC, and other systems have clearance. Equipment location drawings, plot plans, and foundation plans are necessary for the electrical group to establish the location of electrical equipment, conduit runs, and so forth in relation to piping systems.

Industrial plants are divided into areas depending on the hazard presented by electrical equipment. Hazardous areas are sections of a plant where there is a possibility of fire from electrical sources. In general, process petrochemical plants have pneumatic and hydraulic equipment because electric arcs may cause ignition of vapors or flammable substances. Hazardous areas are typed according to the operation and level of hazard. The following comments are from the AEMS model manual:

In any plant design, the provision of electrical services becomes an integrated system analogous to a piping system. Power distribution begins at the transformer station and power is run to the plant areas through a series of breakers, motor control centers, starters, distribution panels, and switches to the point of end use.

While the diameter of any particular run of conduit may be small, the total number of runs and the associated equipment require significant amounts of space.

Because construction forces need light and power, electrical is one of the first trades to enter a building. Unless space is reserved early in the design stage for electrical distribution, and the craftsmen are restricted to this space, conflicts and interferences are inevitable and significant amounts of electrical gear may have to be relocated.

The location of switching, breakers, and starters is important to operating safety. Plant forces should be consulted for optimum locations and for lock-out arrangements. Access and laydown space must be provided for servicing and maintenance.

When electrical and instrument conduits are run on the same bridges or supports as process and service piping, the conduits must be protected from fluid leaks.

Electrical circuits are usually laid out on schematic diagrams. Plan, elevation, and section drawings are not usually made. Conduit is installed in the field from the schematics and conduit schedules. A few companies are beginning to use isometrics similar to piping isometrics for conduit except that a lesser degree of accuracy is required.

Lighting is usually installed on a regular grid from plan drawings. Here again, interferences are common and relocation is then necessary. A preferred method is to make lighting layouts on a model for evaluation and agreement by other design disciplines and by operating personnel.

HEATING, VENTILATION, AND AIR CONDITIONING

The following information on heating, ventilation, and air conditioning has been extracted from AEMS sources:

HEATING, VENTILATION, AND AIR CONDITIONING

Heating, ventilation, and air conditioning (HVAC) is a significant part of any industrial design. Since the equipment and the ducts needed to handle air and other gases are large and bulky, the HVAC systems are major space consumers in a plant layout. Therefore, it is important to consider these items early in the design stage so that the most economical systems and system arrangements can be employed.

Air intake and exhaust locations, equipment locations, duct runs, and piping runs should be developed in concert with the total plant.

Heating is the simplest of the three factors. If the building is air conditioned, heating is part of the air conditioning system. If the building is not air conditioned, heat will be supplied by steam, electric, or direct-fired systems either throughout the plant or on a local basis. In these cases the distribution systems and the heat exchanger units (radiators, space heaters, infrared units, etc.) must be located in appropriate spaces.

In the case of ventilation, the designer is interested in the characteristics of both the exhaust system and the supply system. Exhaust ventilation is often provided to remove some hazardous or noxious material from the immediate work areas. If so, the designer must provide an exhaust stream of sufficient volume and velocity to capture gases and/or particulate matter and carry them to a point of collection or disposal. To achieve air balance, an adequate air supply must be maintained by infiltration or by a separate supply (makeup) air system.

The trend in most modern designs and construction is to air conditioning. A number of reasons influence this trend. An obvious reason is operator comfort and improved operator efficiency. Another reason is that some electronic devices do not function reliably when variations in temperature or humidity occur. Modern plants operate on electronic instruments and a growing number of them are controlled by in-line computers. Additionally, processes such as textile production, papermaking, printing, etc., are sensitive to variations of humidity and temperature. Other process conditions such as the need to protect a product from contamination by airborne impurities may demand that the plant be air conditioned.

A complete HVAC system becomes a major design effort. Facilities must be installed to introduce air to the building, clear it, dehumidify it, reheat and rehumidify it to specified conditions, and distribute it throughout the building in controlled amounts and at controlled velocities. In like manner, the exhaust system must pick up the design quantity of air and carry it to the discharge point or recirculate it to the conditioning unit. The two systems must be coordinated or "balanced" so that their performances are equivalent. In recirculating systems, some amount of fresh air is usually added continuously and this stream also must be included in the balancing. The instrumentation needed to control a system can become a very complex design problem by itself.

An air handling system is a fluid flow system analogous to a piping system for liquids and the same design factors are considered. Air is moved by relatively small pressure differences provided by mechanical devices (fans and blowers). Turbulent flow is not desirable in duct systems because of pressure losses. Ducts are designed and sized for laminar flow and turning vanes are often used when changes of direction occur.

Supply air treatment is usually performed in some type of unit cabinet where filtration, washing, and tempering take place. These cabinets are usually quite large [and] have extensive service piping requirements and complex internals. Air supply units require regular maintenance and inspection. Easy access and adequate space must be provided for these servicing operations, especially for filter changing.

As the trend to energy conservation continues, other components will be introduced in HVAC systems such as waste heat recovery and solar heating devices. Their presence will introduce new design parameters.

Ventilation requirements have become quite stringent under the regulations issued by the Occupational Health and Safety Administration (OSHA). Where dust, heat, or vapors become a problem, localized exhaust is usually provided. Complex hoods may be required and modeling is the best way to communicate concepts between engineering disciplines, hygienists, constructors, and users. Also, provisions in environmental legislation, such as the Clean Air Act, are beginning to affect exhaust systems to a major extent. Exhaust infiltration, stack gas scrubbers, and other effluent treatment processes are now becoming commonplace and contribute to the complexity of design and the requirements for space.

A special area of HVAC design is the "clean room" used in electronic assembly, space work, and similar operations. These units are usually purchased as self-contained modules but they do require service connections and coordination with the main facility.

Controlled atmosphere facilities are another design area. Usually, but not always, these facilities use a recirculating inert atmosphere. Regenerating systems are normally purchased package units but they require gas handling supply and exhaust ducting and significant service piping. Internals of the system are complex and require frequent maintenance and servicing.

Vacuum systems and vessel vent systems are also a part of HVAC. In these cases, the necessity of operating at pressures lower than atmospheric requires the use of relatively large diameter pipe systems. If the volume of building air exhausted by these systems is significant, allowances must be made in the balancing schedule.

HVAC and dust collection systems can often be located so as to use "unwanted" floor space or building cubage. A good method is to develop an idealized process layout, add major auxiliary equipment and electrical gear, and lastly add the HVAC and dust equipment. Then the relative cost of pipe runs, electrical cable runs, and duct runs should be compared and equipment locations adjusted accordingly. Typically, no adjustments will be needed as the first try of the above sequence will usually be the least expensive and most efficient.

STRUCTURAL DESIGN

Structural and architectural drawings are required for all piping systems, especially for industrial installations which require buildings for housing the process (nuclear power plants, conventional power plants, petrochemical installations) and for warehouses, office buildings, cafeterias, control houses, and the like. The architectural and structural necessities of a project must be determined in the early stages, especially where the piping system, equipment, and components are confined to the structure. Therefore structural supports, location of major structural elements, and adequate installation, maintenance, and operation room must all be taken into account.

The structural design must be considered in the very early phases of the engineering design stage of the project. One of the first decisions is to determine whether the project is to be done with a model, drawings, or both. The design and configuration of a structural unit is based on the processes, configuration of equipment, piping, HVAC, electrical needs, and so forth, including locations for pipe supports and structural embedments. The structural group works closely with the piping and design group, and in some cases these groups are combined so there is no misunderstanding between these two essential functions. Structural drawings must show all framing, structural elements, connections, foundations, concrete, and flooring. In general, the dimensions for stairways, platforms, structural elements, and columns are all provided on the structural drawing.

"Structural framing in an industrial plant construction is usually of steel or reinforced concrete, or a combination of both. The chemical and petrochemical plants so common today are almost always outdoor plants supported by steel structures."

The piping drafter must be familiar with the terms associated with structural drawings and design. Many of these are covered later in this chapter in the pipe rack section and the stairway and platform section. The student should understand structural drawings thoroughly; there should be no mistakes regarding the design, placement, and support of the piping system and equipment. The Industrial Glossary in the appendix defines many of the essential terms—for example, column, truss, strut, anchor bolt, rebar, grout, and bent.

One of the main uses for structural drawings is in the pipe support group, where it is necessary to locate and attach pipe supports to structural elements such as structural steel, concrete, flooring, and foundations. All connections, interferences, and locations for pipe supports must be determined from the structural drawings. In general, most of the outlines for structural elements are shown on piping drawings. In the case of nuclear reactors, the piping and structural design are completely interwoven because of the configuration of the building and piping and their confinement to a limited space. In most cases, the pipe drafter will show structural elements in single line on the piping drawing including the outline of the foundation, flooring, and concrete needs (walls, columns, truss locations).

Steel structures are composed of vertical members called columns and horizontal members called beams. Columns are usually H sections whose flanges are the same width as the depth of the web. Beams are usually I sections or channel sections. Spandrel beams are the beams connecting the columns at the periphery of the building. They may have loads applied by the exterior walls of the building, as well as loads applied by the building's content. Primary beams are the beams that connect the columns inside the buildings. Secondary beams are heavier members used inside the building to support floors, process equipment, and service machinery. Tertiary beams are lighter members used to support equipment or to provide support for catwalks, platforms, ladders, etc. Other steel members used to supply wind breaking, wall anchors, framing for doors and walls, etc., are called miscellaneous steel.

Figure 10-9 is a column orientation drawing:

A row of columns and its connecting steel beams is referred to as a bent. When a building is erected, the columns are oriented so that the flanges are parallel with the long dimension of the building. When additions are made to an existing building, the flanges are oriented in the same direction as the existing steel.

Figure 10-10 is an example of a typical bent showing structural elements from a side or section view:

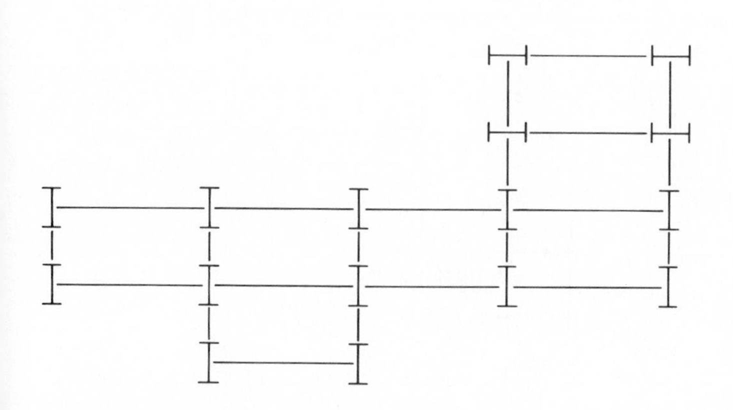

FIG. 10-9 Column orientation. (Courtesy American Engineering Model Society)

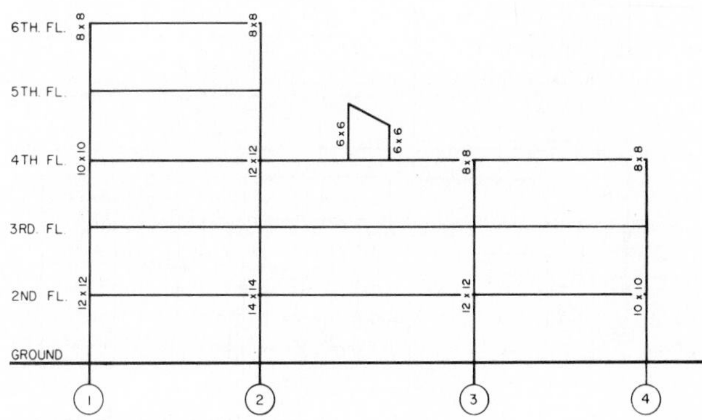

FIG. 10-10 Typical bent. (Courtesy American Engineering Model Society)

Note the difference between columns 3 and 4, and columns 1 and 2. They appear to be supporting the same floors, but they are different sizes. The internal columns, 2 and 3, support loads from the two bays of steel, whereas the external columns, 1 and 4, support loads from one bay only.

A change in size of a column usually occurs after two elevations because of the increasing structural load and because of the practical limit imposed on the length of members by shipping requirements. The column break usually comes about 2′ above the floor line to simplify calculation connections and erection problems.

LAYOUT AND ARRANGEMENT

Before the pipe designer can lay out a piping system or start the construction of a model, it is necessary to have a flow diagram—Fig. 10-11 shows a wood pulp processing plant and Fig. 10-12 shows a nuclear project. These two diagrams are only process diagrams; the P&ID is also needed for the startup of a project. Flow diagrams are covered in great detail in Chapter 6. Besides the P&ID, all manufacturer's drawings or vendor prints must be accessible to the piping drafter and designer. The information will establish dimensions and configurations for equipment such as pumps, compressors, heat exchangers, vessels, valves, towers and standard items. These drawings are essential for the accurate, economical placement and layout of an industrial plant. Plot size and off-plot buildings, roadways, control building, and existing pipeway (pipe racks, if any) must also be available to the designer. The model maker and piping designer must know the configuration and dimensions of the equipment to be used on a project.

For preliminary models the model maker will construct the general configuration of equipment from plastic or wood. In some cases, simple rectangular blocks are used to represent equipment.

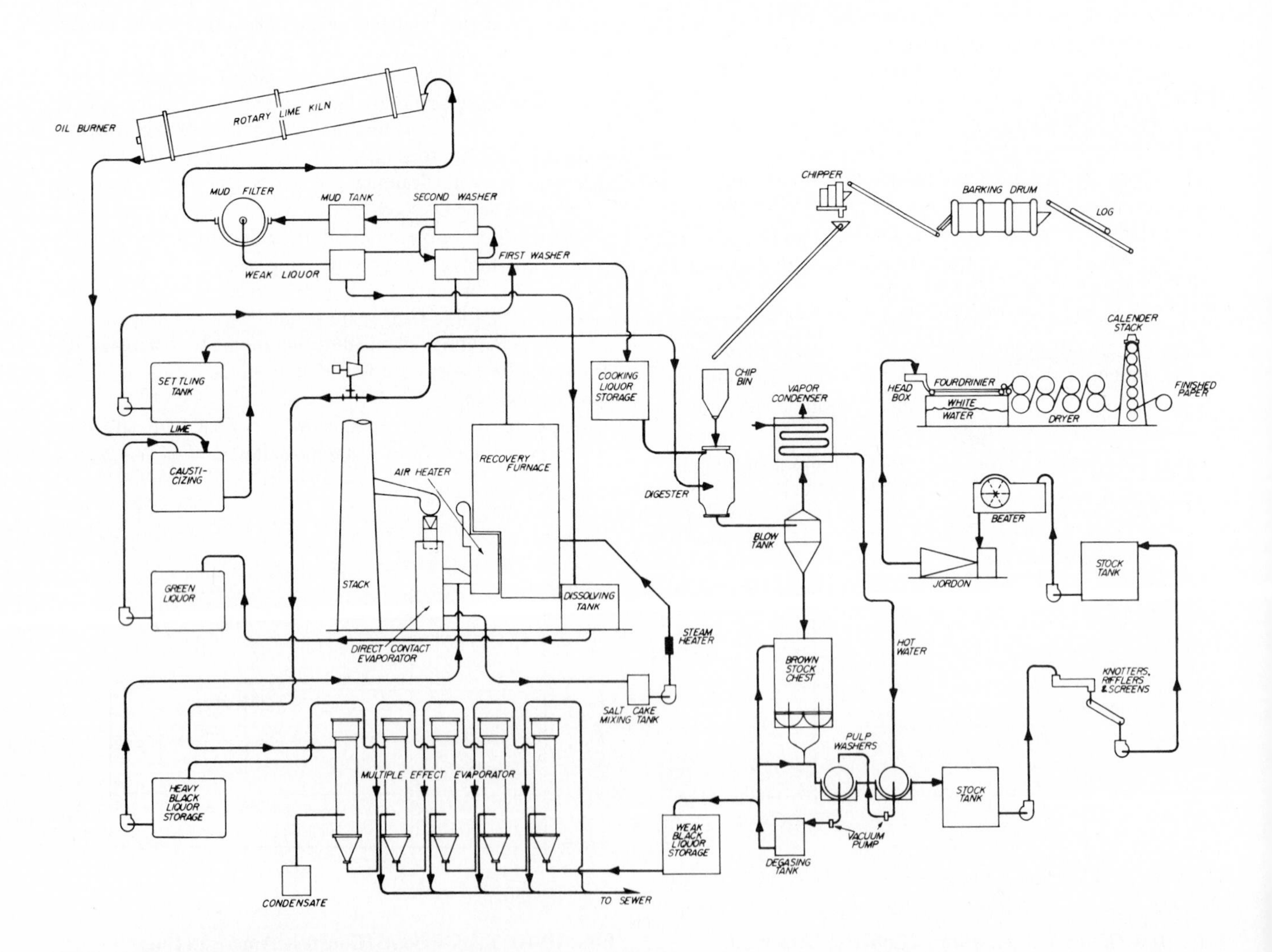

FIG. 10-11 Flow diagram for a pulpwood process unit.

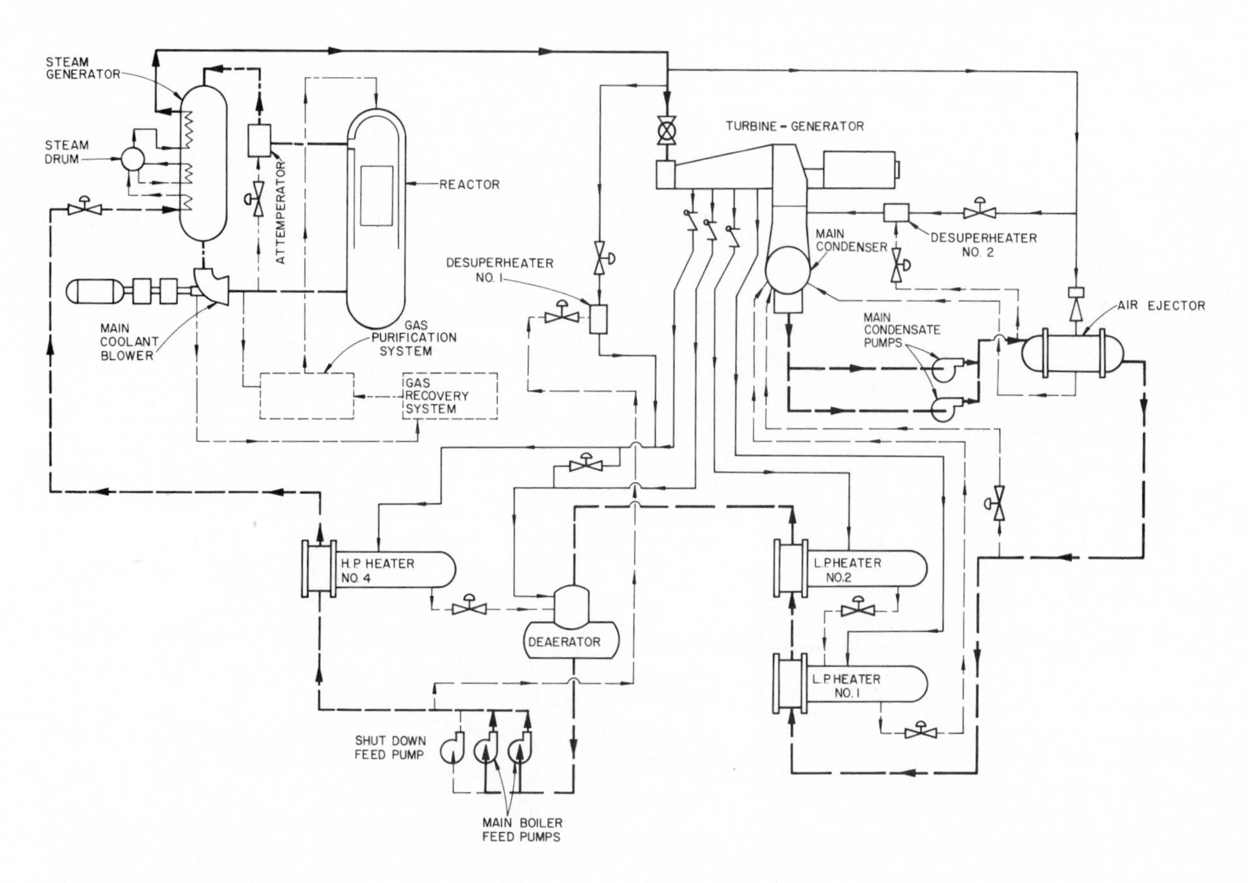

FIG. 10-12 Flow diagram for a nuclear project.

Arrangement of Piping

One of the most important duties of the piping designer is the layout and arrangement of piping runs in an industrial plant. To arrive at the most economically feasible and functional design, the piping designer must have access to the flow diagram, vendor's prints, possible equipment connection points, the plot plan, and vessel drawings. One of the most important considerations will be which lines are alloy, large diameter or in other words expensive. An attempt is made to run the most expensive pipe first, alloy piping, then large diameter carbon steel lines. In this way it is possible to keep these runs as short and economical as possible.

One of the first procedures after equipment location is to sketch on the plot plan all the major pipe requirements taken from the P&ID. Another method is to sketch on the plot plan the necessary pipelines and then transfer these to the model. Nozzles and vessel connections are not considered at this point in the design procedure. By determining the total piping needs in plan view, it is possible to establish pipeways and pipe rack locations, control valve stations and meter runs. For outdoor facilities such as petrochemical installations, most of the horizontal piping is run along major pipeways on pipe racks or, in some cases, on sleepers. After the major piping has been determined and located in the plan view, it is then possible to decide upon connections to equipment. Chapter 11 goes into much greater detail, covering requirements for layout and design of petrochemical installations; Chapter 12 discusses fossil fuel power generation piping; Chapter 13 is devoted to nuclear piping. These chapters should be consulted for specific types of piping systems, arrangements, and design. Chapter 5 lists many specifications for the drawing of elevation, plan, and section views when using the drawing method of piping design.

The main job of the pipe designer is to lay out equipment and piping, along with coordinating the piping group with subfields such as the structural, electrical, and HVAC groups to establish an economical, feasible piping installation which meets all design requirements. The piping designer also coordinates stress calculations and determines the selection and location of piping components such as valves and pipe supports.

When laying out a piping system, the designer must take into account variables of design requirements, the economics of the system, the stresses under which the system will be placed, accessibility for maintenance, repair, and installation, and the support of the system. The following guidelines apply to design and layout of industrial piping:

1. Arrange all piping that must be serviced frequently so that it can be readily dismantled without interruption to the rest of the system and without unnecessary removal of pieces of equipment and parallel piping.
2. The economics of the length of pipe, number of fittings, and types of valves must be taken into account in the early stages.

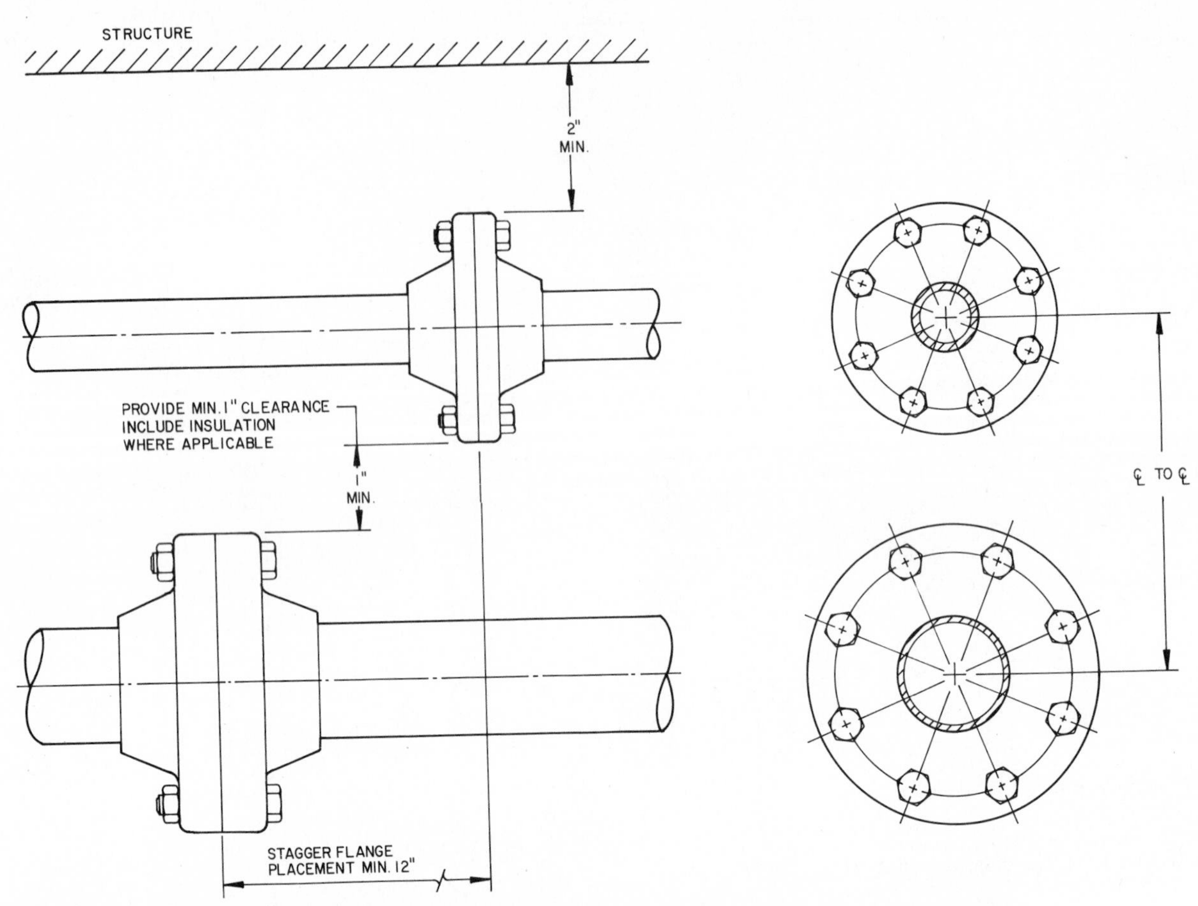

FIG. 10-13 Pipe spacing. Clearance must be provided between flanges on pipes running parallel to each other as on racks. Allow 2 in. between an outer flange and the nearest structure. Dimensions for insulated piping and flanges should be taken from the outside of the insulation. When designing parallel lines and establishing spacing for minimum thermal movement of hot lines, one assumes that all of the piping is flanged.

3. Care must be taken to provide access to all areas which must be serviced and maintained—*not* arranging piping and components with insufficient room for the operators and maintenance crew to function.
4. Mechanical and thermal shock, vibration, and other stresses must be determined in the early stages of the project in order to provide for sufficient change in direction, pipe loops, anchors, and expansion joints.
5. Process requirements must be consulted when arranging, connecting, and ordering piping and components. This is the job of the piping designer, not the drafter.
6. All piping and equipment should be arranged to meet maximum safety requirements for the area.
7. Arrange piping so that change of direction corresponds with change of elevation, though care must be taken not to create pockets in vertical runs.
8. Whenever possible, group similar equipment together to simplify piping.
9. Designers should take into account all specifications which determine materials, pipe thickness, schedule number, and insulation requirements.
10. Piping systems should be constructed with standard catalog items whenever possible in order to avoid expensive fabrication costs.
11. Piping should be arranged above grade and elevated on racks or supported from structural members in buildings. In some cases, large-diameter pipes such as for cooling water, and sewage may be located underground in trenches or 12 in. to 18 in. (30 cm to 45 cm) above grade on sleepers, because the size and weight of the line would make it impossible to support it on pipe racks or at higher elevations.
12. The piping arrangement should eliminate pocketing by sloping the lines toward the nearest piece of equipment or providing a slight general slope throughout the whole system.
13. The piping system should be constructed with a maximum of connections, such as flange joints, wherever it may be necessary to disassemble the system.
14. Piping should not be placed above equipment such as economizers, pumps, and compressors or above platforms. Whenever possible, keep piping to the side of equipment underneath in front or rear, not above.
15. Maintain the required clearance between piping and equipment. Check the specifications in Chapters 11, 12, and 13 for suggested minimum and maximum clearances between piping, vessels, equipment, platforms, and stairways.
16. Parallel piping should be arranged with a minimum of 1 in. (2.5 cm) clearance between the outer edge of the flanges. In Fig. 10-13, a minimum of 2 in. (5 cm)

is required between the outer edge of the flange and the nearest structure, and a minimum of 1 in. (2.54 cm) clearance is provided between flanges. All piping should be considered flanged in the design stage so that these clearances will always be maintained.

17. Stagger flanges as in Fig. 10-13 for easy maintenance, repair, replacement, and access to bolts.
18. The center-to-center distance between piping should be established according to the flange diameter and insulation requirements, not just the diameter of the pipeline.
19. Provide sufficient clearance around all piping for access to equipment and components.
20. Pipe spacing and clearances should be established from specification charts for different services.

The student should consult the specifications in Chapter 5 on the layout and drafting procedures for piping drawings in general. For specifications concerning subfields, consult the chapters dealing with nuclear power, conventional power, and petrochemical installations.

Arrangement of Equipment

Figures 10-14 and 10-15 give an idea of the dimensions that are required for equipment arrangement drawings. The preliminary model often takes the place of a location drawing. The large projects of the power generation unit in Chapter 12 and the nuclear power plant containment building in Chapter 13 are also equipment location drawings, only major vessels, equipment, and large piping are shown.

Figure 10-15 shows a turbine building where all the dimensions are set up on coordinate lines between structural elements. In many ways, equipment location drawings look more like architectural or structural drawings than piping drawings. Notice, in Fig. 10-15, the systematic layout and location of equipment, neatly arranged with adequate pull space provided for the turbines and other equipment.

Though the specifications for the piping arrangement and layout have already been discussed, in reality the layout and equipment arrangement of industrial plants is the first step in the design process. Depending on the type of industrial plant—nuclear, conventional, petrochemical, food processing, water treatment—the specifications and design requirements and layout will differ. It is therefore impossible to outline a definite set of horizontal and vertical spacing dimensions for equipment, though Chapters 11, 12, and 13 do provide many specifications for spacing and layout. Most equipment is located above grade level, though care should be taken not to elevate equipment in a way that will hamper maintenance, repair, and operation procedures. Many elevations are determined by the process itself and the amount of static head to be developed by the vessel. Using the plot plan, the piping designer can sketch the general arrangement of equipment. Using the model method, styrofoam or wood blocks representing the outer dimensions of the equipment can be used to try out various equipment locations on the plot plan or within the building structure. Remember that vertical and horizontal spacing of equipment is determined primarily by the type of process. **When laying out a project, the equipment should be positioned in the same sequence as the flow from equipment to equipment shown on the P&ID.**

Types of Equipment

Various types of equipment for piping systems—such as pumps, compressors, vessels, and exchangers—are common to all piping subfields. Other pieces of equipment—such as turbines, condensers, reactors, steam generators, boilers, and fractionating towers—are used primarily for one type of piping system, such as a nuclear power plant, water treatment plant, steam generation plant, or petrochemical installation. Therefore this section covers only those pieces of equipment and vessels found on a majority of piping systems. Chapters 11, 12, and 13 discuss the major subfields of industrial piping and should be consulted for the location and arrangement of relevant equipment and piping.

Mechanical equipment includes pumps, motors, and compressors, which are found on most industrial piping systems. Many of the specifications for pumps and compressors are also discussed in Chapter 11.

Compressors Compressors provide sufficient pressure for gases and vapors. In most piping installations, compressors are used primarily for the creation of highly pressurized air for different parts of the plant, such as instrument air, utility air, and various pneumatic systems such as those for pneumatic valves. Compressors can develop pressures as high as 20,000 psig, though their usual working range is somewhat lower. There are many types of compressors with which the designer should be familiar. They range from reciprocating compressors to various rotary compressors. The compressor needed for a particular installation will be determined from the process requirements by the engineer. Figure 10-16 shows a compressor unit installed indoors, which is usually the case. Filters, separators, receivers, silencers, and coolers are pieces of equipment that make up compressor units. As can be seen in Fig. 10-16, the compressor units are elevated above grade level and are provided with an enclosed area to keep the unit dry. This photograph also shows a maze of piping manifolds for the distribution of compressed air. The designer should consult the requirements and specifications for the arrangement of compressors, compressor piping, and compressor housing. In some cases, compressors can be installed without a compressor house, but these requirements should be examined by the process engineer. In general, compressors should be installed indoors on concrete foundations as in Fig. 10-16. Sufficient room around the compressor unit should be provided for maintenance and operation, including access to valving and

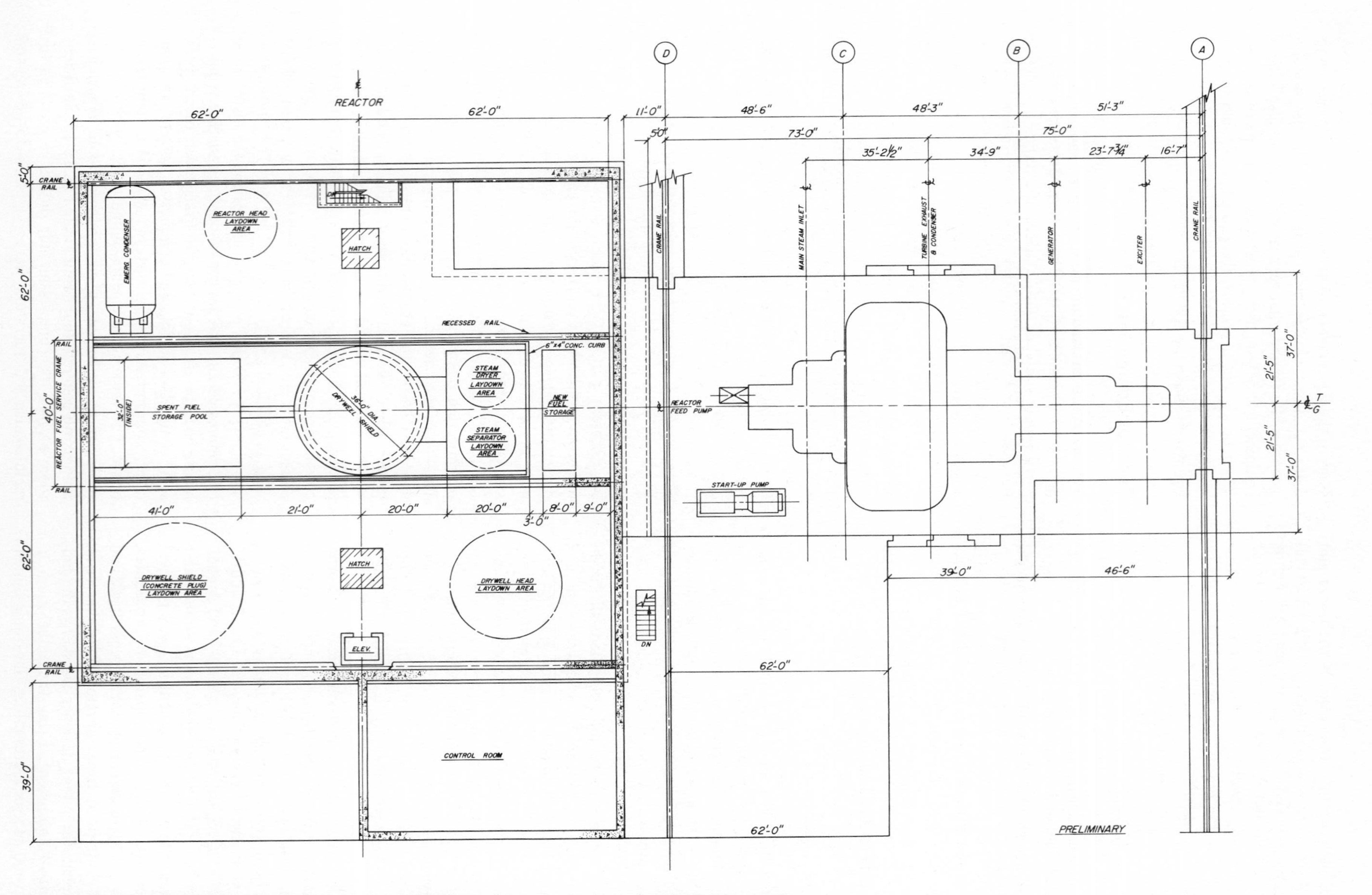

FIG. 10-14 Equipment location drawing for a BWR nuclear power plant showing major equipment and foundations.

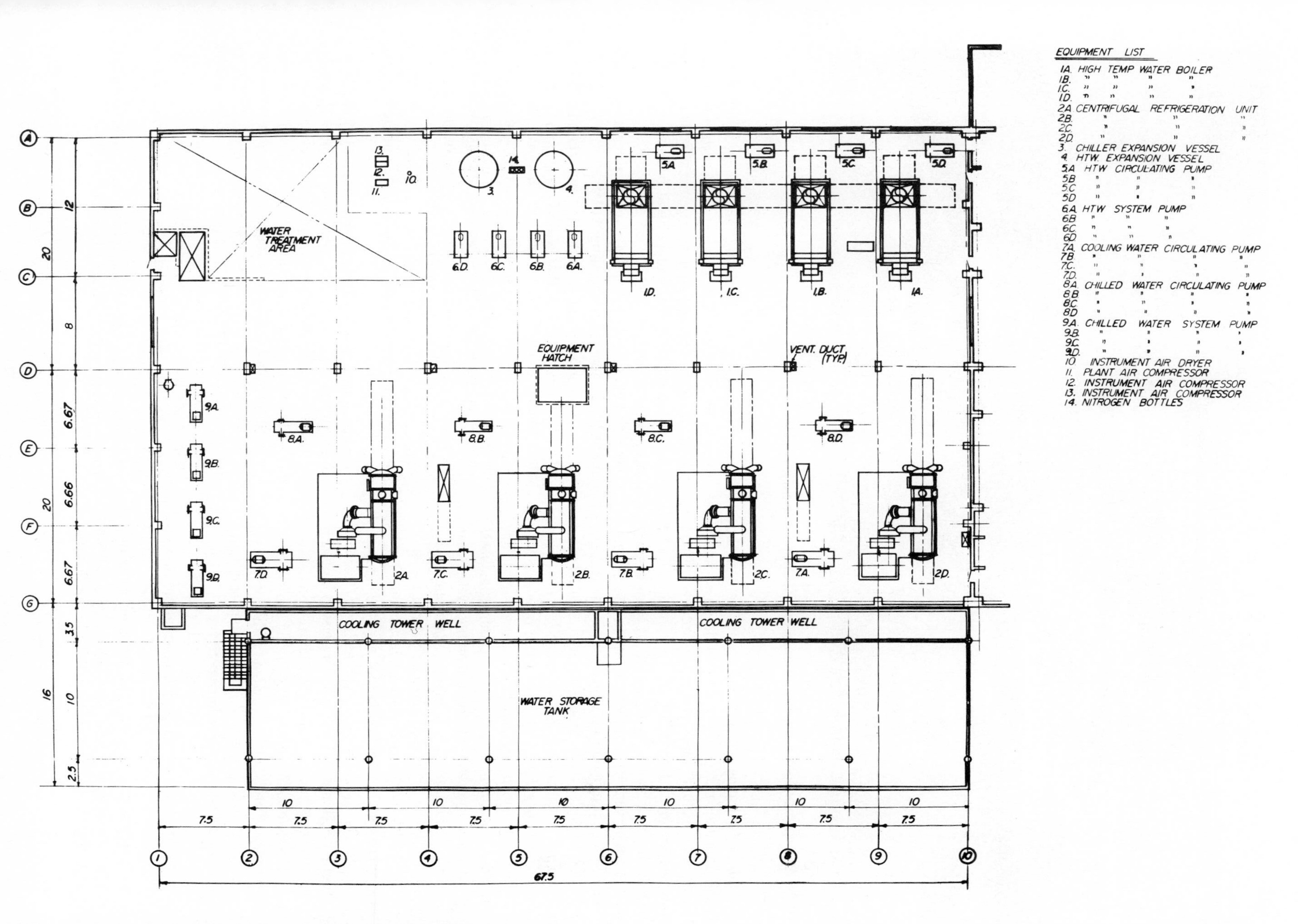

FIG. 10-15 Equipment placement drawing for turbine installation.

FIG. 10-16 Piping manifold for a compressor unit. (Courtesy Stockham Valves and Fittings)

piping. Compressor piping arrangements should be designed so that the unit does not create unwanted liquid pockets—the design must include vents and drains at areas where pocketing may happen. Depending on the substance to be compressed, the piping and general layout of the unit must take into account various hazardous gases in the system. Avoid unnecessary configurations which may cause leakage. Moreover, the building itself should be well ventilated so that gases do not accumulate.

Compressed air lines should be vented and provided with traps at low points which may collect unwanted liquid. The compressor installation must take into account the possibility of vibration transmitted to the piping system and accompanying fittings and components. Manifold valving and piping should be supported separately from the compressor unit. Use long-radius elbows whenever possible to avoid radical changes in direction in and near the compressor unit, especially the inlet and discharge lines. Overhead piping should be eliminated for all compressor units for ease of maintenance, repair, and replacement. Strainers are also required in suction lines, especially during startup operation. Two major problems associated with compressor units are unwanted foreign material in the system and liquid retention. Both can cause serious damage to the unit.

Pumps Pumps are used on all piping installations for increasing the pressure of the line material. Figure 10-17 shows a fuel oil pumping, heating, and straining set with a rotary pump; note the variety of instrumentation, valving, and supports. Figure 10-18 shows a typical pump installation at a refinery. Many specifications for pumps are also called out in Chapter 11, which should be consulted. Figure 10-19 is an example of an airport fueling system with associated pumping stations and storage vessels. Notice that most of the piping is run underground because of the hazards associated with jet fuel.

The following description of various aspects of pumps and piping systems is provided by AEMS:

A reciprocating pump is a piston and cylinder with a valving arrangement such that withdrawing the piston from the cylinder produces a vacuum on the inlet. When the piston is pushed into the cylinder, pressure is applied on the discharge.

These pumps have a higher suction lift than centrifugal pumps, and with proper design are capable of developing a higher pressure at the terminal point.

The jet pump (ejectors, injectors) is a device which utilizes the energy of a fluid under pressure for pumping a second fluid.

These pumps are particularly useful in pumping water from ditches, ship bilges, etc. Sediment in these fluids can be passed through the jet with a minimum of damage. Jets are useful in putting a vacuum on a closed tank and they are used on some boilers as feed pumps. Because they have no moving parts and no packing glands, they are often used to transfer radioactive fluids. The pumping fluids are usually steam, water under pressure, or compressed air.

There are many other types of pumps such as the airlift pump, the pulsometer, the gas displacement pump, and the diaphragm pump.

All of these pumps, and variations, have a single purpose—that of producing fluid flow against resistance or pressure.

Generally resistance and pressure are encountered in the same system and are therefore both reduced to a common

FIG. 10-17 Coen "AA" 20-gpm, 200-psi fuel oil pumping, heating, and straining set: one steam pump, one rotary pump, two Coen series B, multifilm, cleanable fuel oil heaters. The student should be able to identify a variety of valves, fittings, instruments, and supports. (Courtesy Coen Company, Burlingame, CA)

unit expressed in feet of water total head. Total head and the needed capacity, expressed in gallons per minute, are the factors used in determining the size and type pump to be used.

Total head is the total net head in feet of water against which the pump must operate.

$$\text{Head in feet} = \frac{\text{Pressure in lb/in.}^2 \times 2.31}{\text{specific gravity}}$$

Total head equals either total discharge head plus total suction lift or total discharge head minus total suction head.

Total discharge head is the static head existing when the pump is not in service plus pipe friction. Pipe friction is the resistance offered by pipe fittings, valves, etc., and is usually expressed in feet of head of water per hundred feet of pipe. For instance, a 2″ gate valve fully open is equivalent to 1.2 feet of straight pipe, a 2″ globe valve fully open is equivalent to 55 feet of straight pipe, and a 2″ long radius elbow is equivalent to 3.5 feet of straight pipe. Tables of values of friction are available for all sizes and kinds of pipe, valves, and fittings.

When a pump operates with a suction lift, creating a partial vacuum, liquid enters the pump because of atmospheric pressure on the surface of the liquid at the source of supply. In theory, the maximum suction lift with cold water at sea level is about 32 feet. The maximum suction lift generally practical for the average centrifugal pump is about 15 feet and for the average reciprocating pump it is 20 feet. Total suction lift is the sum of the static lift plus pipe friction when the surface of the liquid to be moved is below the pump.

Total suction head is the static suction head minus the pipe friction when the surface of the liquid to be pumped is above the pump.

Two values are required to determine the size of the pump needed: the amount of fluid to be moved and the pipe size.

The capacity can be determined by a number of methods that depend on the design conditions. Tables are available that give velocity and total loss in head through any size pipe for any rate of flow. A pipe size can be selected from the table that gives a fairly high velocity with as little loss in head as possible. Selection is largely a matter of judgment and experience. The pump, the distance to be pumped, the type of fluid to be pumped, must also be considered in this choice.

FIG. 10-18 Pumping installation showing the pump foundation, piping, valving, fittings, and supports.

When the pipe size has been selected the total loss in head can be calculated. Manufacturers' catalogs can then be used to select a suitable pump.

A few things to remember in running pipe to and from pumps are:

1. Never trap the suction.
2. Have a shut-off valve on both suction and discharge lines.
3. Have a strainer in the suction side.
4. Have a control (either hand or automatic) on the discharge.

Figure 10-18 shows that the pump itself is elevated above grade approximately 8 to 12 in. (20 cm to 30 cm), which is normal practice. The top of grout to grade distance should always be a minimum of 8 in. (20 cm) and usually a maximum of 12 in. (30 cm). Figure 10-18 also shows aisleways for general access around the pump, usually providing a minimum of 3 ft (91 cm) clearance for maintenance and operation. Though this installation does not follow the accepted practice of leaving the top of the pump free of piping, it is best to eliminate as much piping from above the pump area as possible. Aisleways or operating passages such as that shown to the right of the pumping installation are usually provided at the drive end of the pump so that mechanical equipment can be used to install and replace the pump station.

Chapter 11 covers many of the specifications for the location of pumps, and these should be consulted when doing the projects provided in the text. There are also a variety of photographs in Chapter 11 which show pumping stations and installations.

In multiple pumping installations, it is good practice to line up the foundations of pumps parallel to one another for an accurate, pleasing, and functional arrangement of equipment and piping. One general specification for the orientation of pumps is that the distance between the discharge centerline and pipe rack centerline is to be kept at

FIG. 10-19 Airport fueling system with storage tanks and pumping installations. Due to the highly flammable nature of airplane fuel, much of the piping for this installation is underground, except in the vicinity of the pumping unit. (Courtesy Stockham Valves and Fittings)

24 in. (60 cm) whenever the pumping unit is to serve a pipe rack area. Pumps should be located close to the tanks and vessels from which they take suction to reduce the piping and fitting requirements. The suction line should be slightly sloped toward the pump to eliminate pocketing of the line material.

Heat Exchangers and Economizers There are many different types of heat exchange units, and Chapter 11 covers them in great detail, including specifications for their location, arrangement, and design. The shell and tube exchanger is found throughout the petrochemical and power generation field. According to the AEMS:

A shell and tube exchanger has a rigid tube sheet supporting the tubes inside the shell. This type is designed for single passes of each fluid, usually in countercurrent flow. It has been modified with baffles or partition plates to increase the number of passes and raise the amount of heat transfer.

A variation of the shell and tube exchanger has a floating head. The sheet joins the tube rigidly, but the assembly slides within the shell to permit expansion due to uneven heating. An external stuffing box is often used to expose the floating head and simplify tube cleaning without allowing leaks.

Another type of shell and tube exchanger has U-shaped tubes supported rigidly within the shell. Expansion is taken up in the U bends. This type should be used for clean, noncaking, noncorrosive fluids since U tubes must be cleaned chemically.

Heat exchangers allow the transfer of heat from one material to another without mixing streams. When possible, heat exchange units should be stacked to a maximum of 12 ft (3 m) where multiple units are required. Parallel economizers or heat exchangers should be located in a line which is a consistent distance from the centerline of the pipe rack from which they draw their line material. In all cases, the front of the shell must have sufficient room for the removal of the tube bundle for cleaning and servicing plus extra space for clearance. Exchangers which are arranged parallel to one another should have a minimum of 3 ft (91 cm) clearance between the shells or between the piping of one unit and the shell of the next (which usually is installed between the units). Exchangers should be located so that they can be easily installed, serviced, maintained, and removed from an operating aisleway, large passageway, or roadway. The exchanger unit is always elevated sufficiently to allow for the piping, fittings, and valving which may run beneath the unit. Usually, 2 ft 6 in. (76 cm) from top of foundation (pier) grout to grade is the minimum requirement, though it often runs higher than this, especially for piping which must be insulated or on units which have piping coming from underneath the shell. The concrete piers and foundations for the economizer units should be arranged a standard distance from the centerline of the pipe rack and in line whenever possible (given that the units themselves are the same length). The distance between the centerline of the pipe rack and the exchanger head (elliptical head) should be oriented a consistent dimension, such as 8′. Multiple exchangers on the same unit can be lined up in this manner.

The flanged (bolted) end will always be facing away from the rack to allow sufficient pull space for the internal coil or tubes.

Vessels Both horizontal and vertical vessels are used on all types of piping installations, and Chapter 11 details most of the specifications and design requirements for both. Chapters 12 and 13 also cover vessels associated with power generation equipment. Many other types of equipment are common to all piping facilities: control buildings, motors, valve stations, condensers. Explanations and specifications for these units are found in Chapters 11, 12, and 13, which deal with specific piping installations and associated equipment.

DESIGNING A TYPICAL INSTALLATION

This section deals primarily with the initial steps which must be taken to design and lay out a typical industrial installation. Figures 10-20 through 10-25 show the general procedures for the creation of a piping installation, including the drawing of a *site plan* (Fig. 10-20), its breakdown into a feasible *plot plan* (Fig. 10-21), and its subsequent division into workable units. Figure 10-22 shows the *piping index drawing*, and the *foundation plan* is shown in Fig. 10-23. Figure 10-24 and Fig. 10-25 show the pipe rack arrangement for the same project.

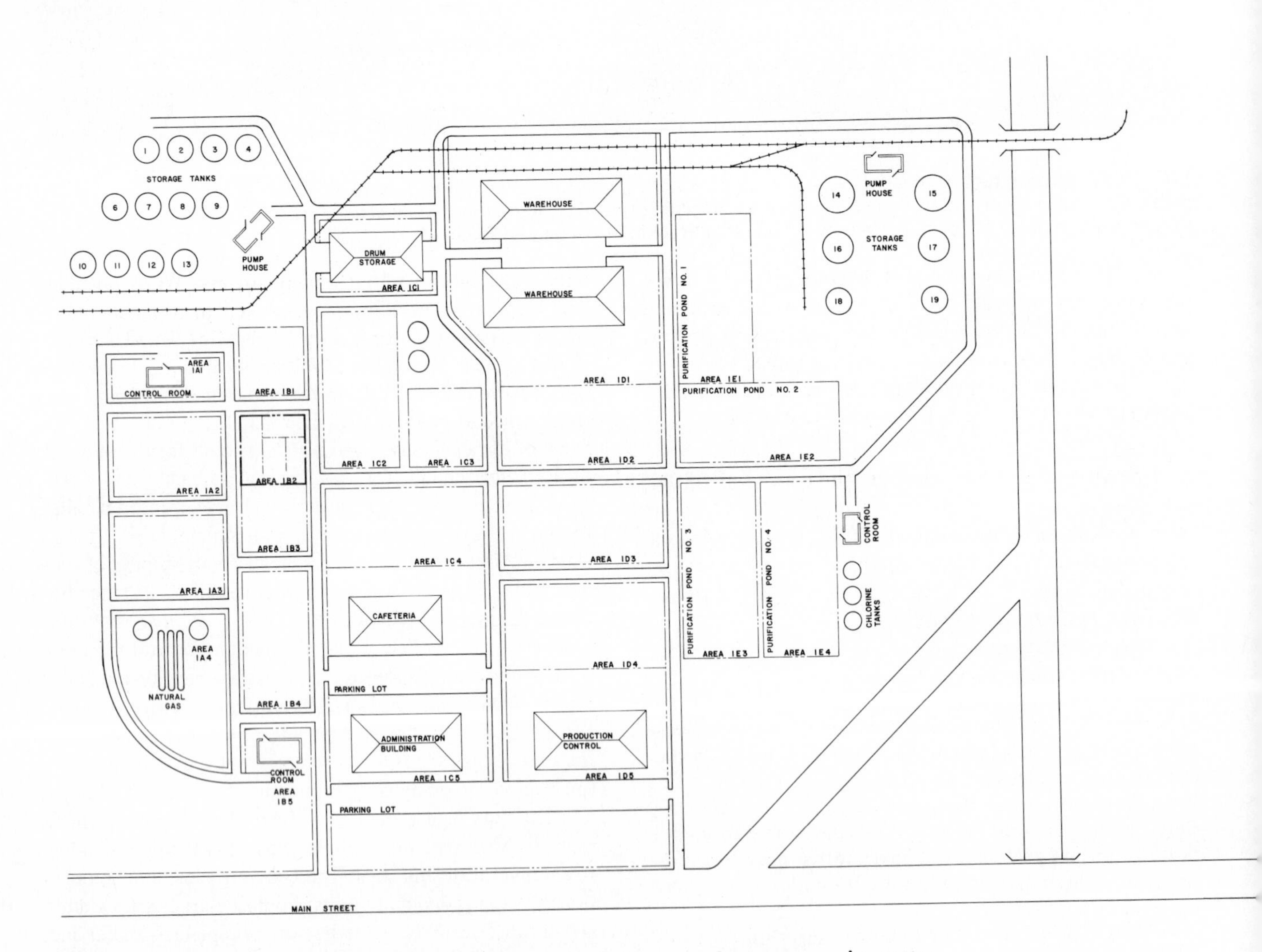

FIG. 10-20 Site plan for a petrochemical installation, including areas for process units, water supply, roadways, lanes, storage facilities, railroad loading, warehouses, and administration buildings.

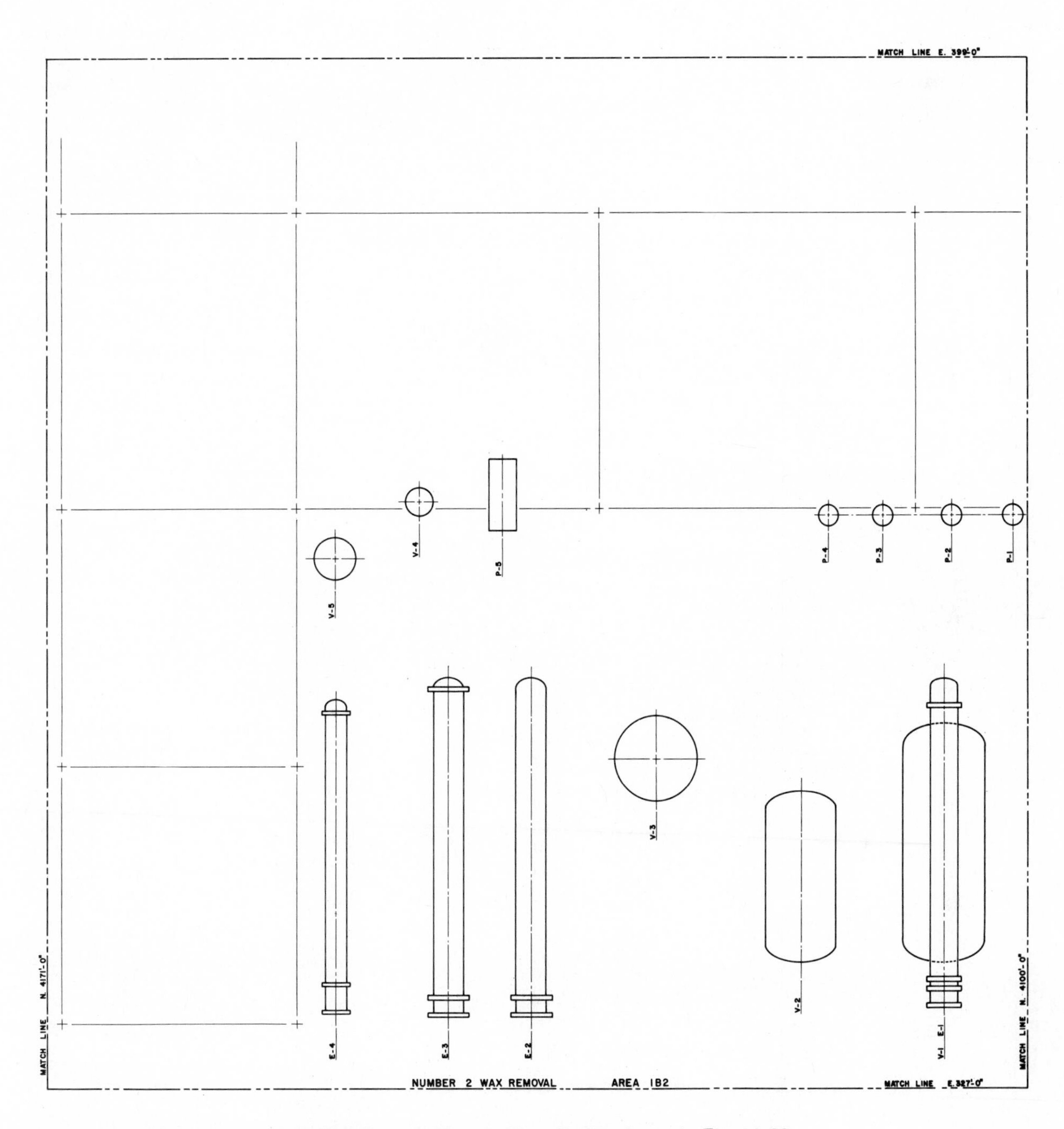

FIG. 10-21 Plot plan of area 1B2 of the petrochemical installation shown in Fig. 10-20.

The Site Plan

From the flow diagram and process requirements for a project, the pipe designer can begin to lay out in detail a portion of a plant, section, or unit. Figure 10-20 shows a site plan which takes into account all the necessary buildings, piping arrangements, pipeways, storage area, warehouses, shipping facilities, roadways, and personnel needs such as cafeterias and personnel buildings. The overall site plan shown here is used to determine the logical arrangement of a petrochemical installation. This site plan allows for the location and arrangement of all buildings according to specifications, safety requirements, and operational details. It is the first step in the construction and layout of a

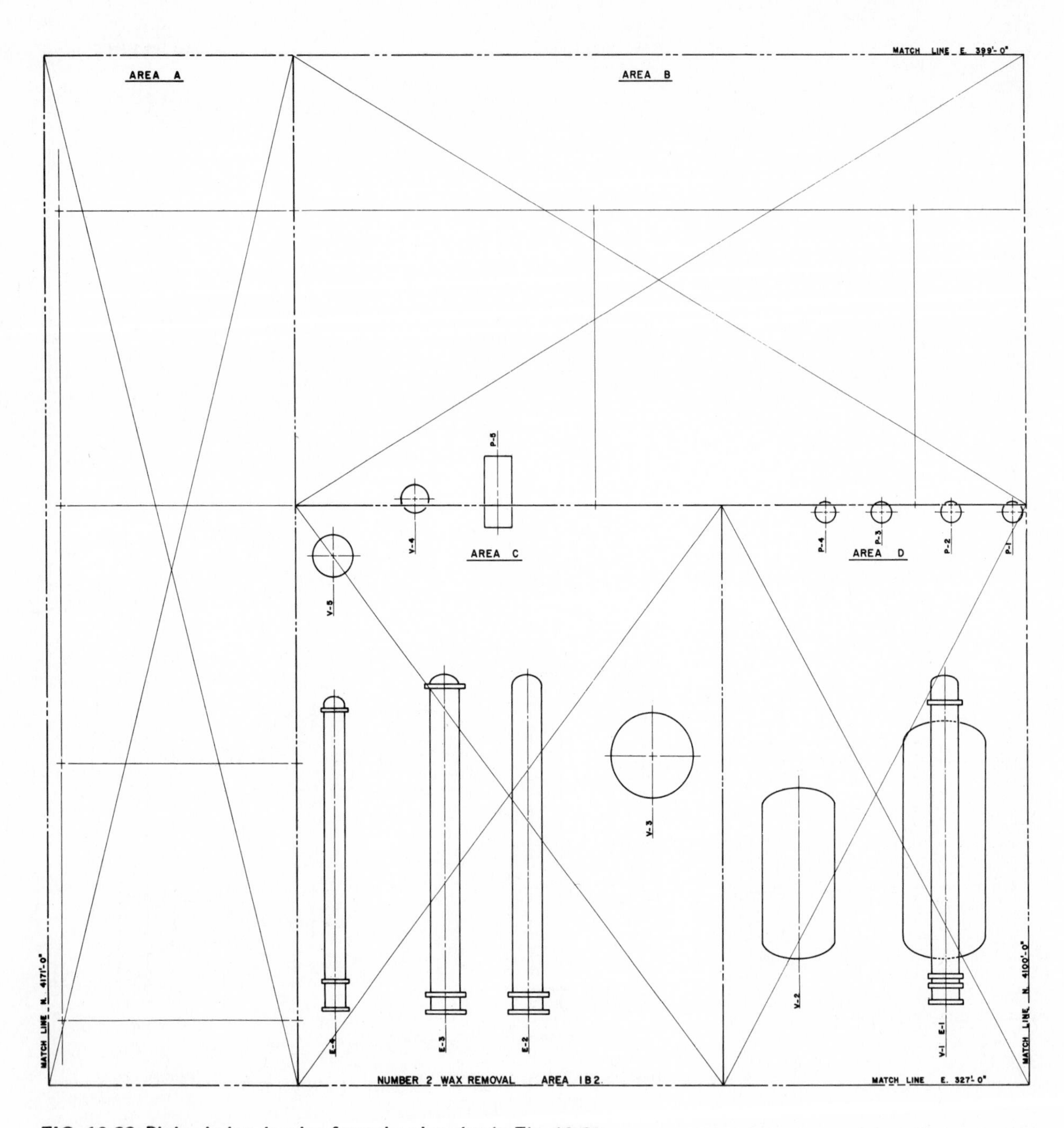

FIG. 10-22 Piping index drawing from the plot plan in Fig. 10-21.

large-scale refinery and is usually drawn or modeled to a small scale.

In many cases, the pipe drafter and designer will come in contact with the construction of an additional unit to an existing facility. In these cases, the original site plan for the whole unit will also be necessary in order to establish the general limits from which to create the plot plan for the new unit. The design and layout of a new unit or complete refinery must take into account many elements, including operation, specifications, design requirements, and construction sequences. This site plan, sometimes referred to as a *master plot plan,* shows the general terrain of the entire

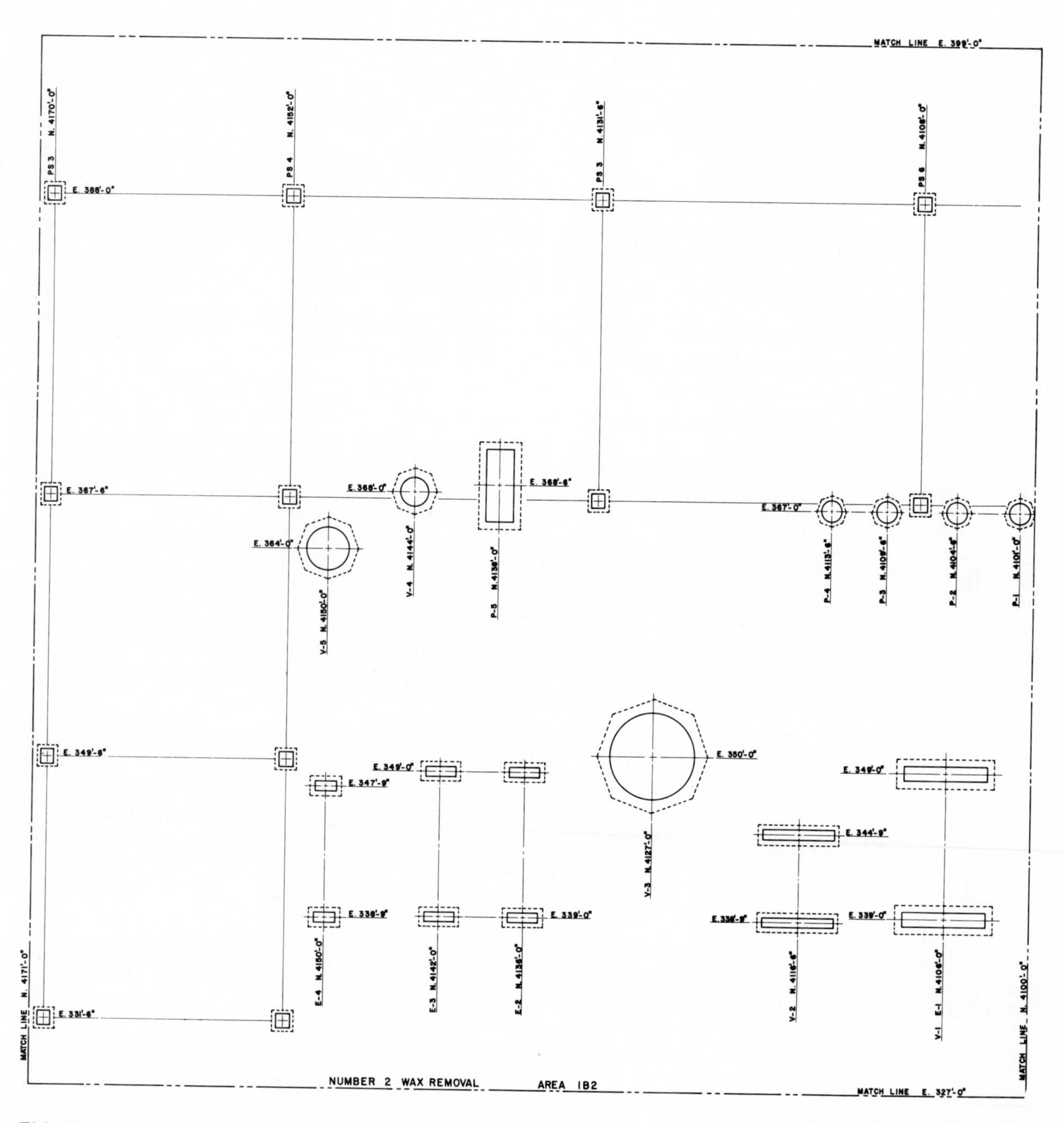

FIG. 10-23 Foundation plan for the project in Fig. 10-21.

facility, all existing roadways, railroad units, shipping facilities, and in some cases even the contours of the site if it is not on flat ground. This site plan must also show all the additional elements for a facility, including storage space, first aid stations, shops, offices, research and development laboratories, warehouses, and shipping stations.

In some projects, the facility is simply one large building or unit as in Fig. 10-14, which is an equipment location drawing for a nuclear project. As can be seen from this drawing, all the design and operation requirements are confined to a single structure, which includes areas for equipment, turbines, and reactor core.

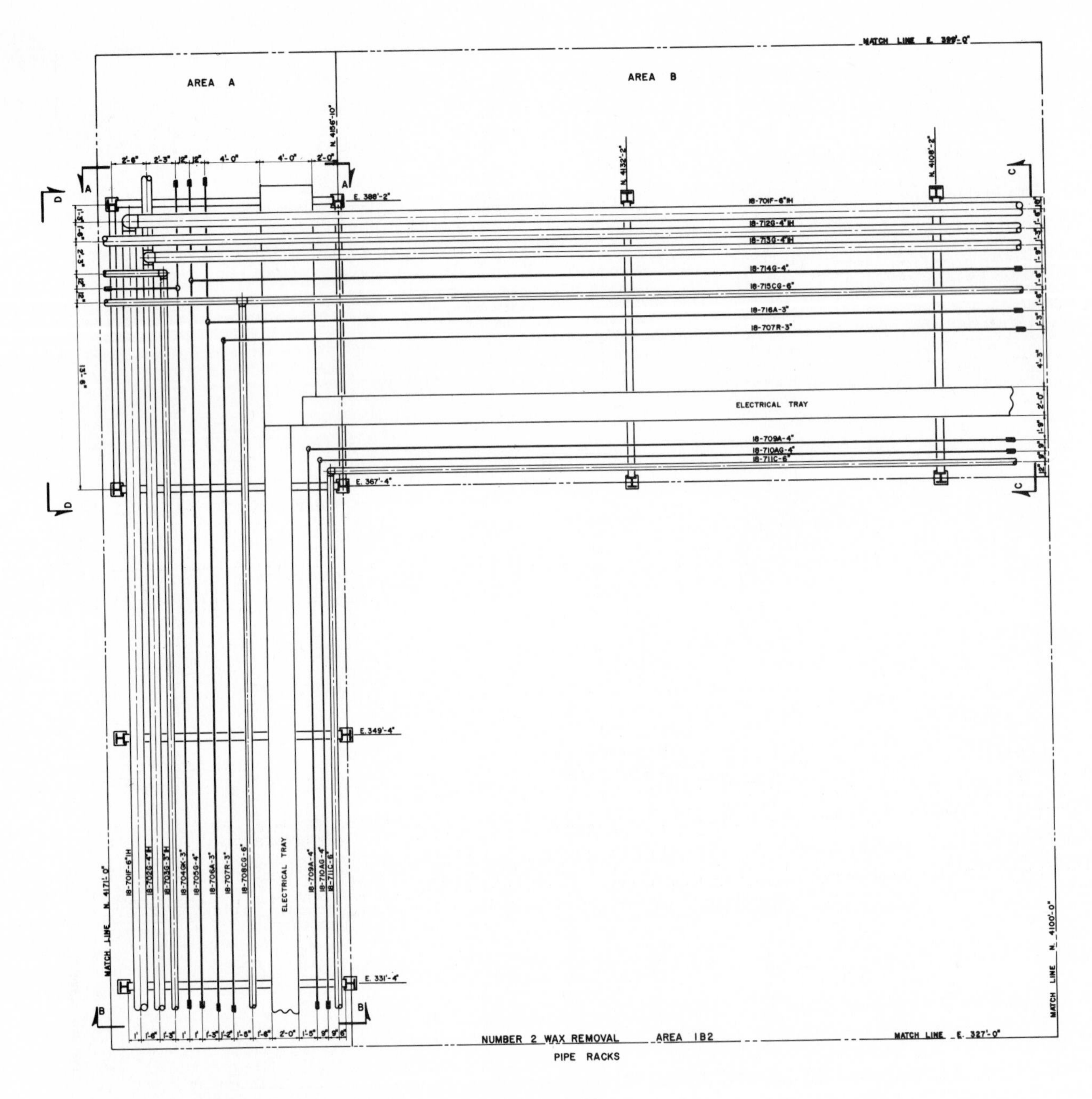

FIG. 10-24 Pipe rack from the project shown in Fig. 10-21.

When a model is used, the site plan corresponds to the site model. Simple blocks of styrofoam or wood can represent major structures, storage tanks, buildings, and so forth and can be easily rearranged. When the drawing method is used, the existing physical traits must be shown on the site plan and then the area must be divided into logical working units, depending on the facility and its design needs. In Fig. 10-20, the warehouse and shipping units are established close to the railroad lines whereas the storage tanks are located well away from the general processing areas. The cafeteria, administration building, and production control buildings are situated close to the main street, which is an existing roadway providing easy access for visitors and employees. Other roadways provide access to the general process area. Equipment is located on the site plan in order to determine the size of the facility and land use requirements. The equipment location is only estimated. Later it will be tied down to definite dimensions during the plot plan stage.

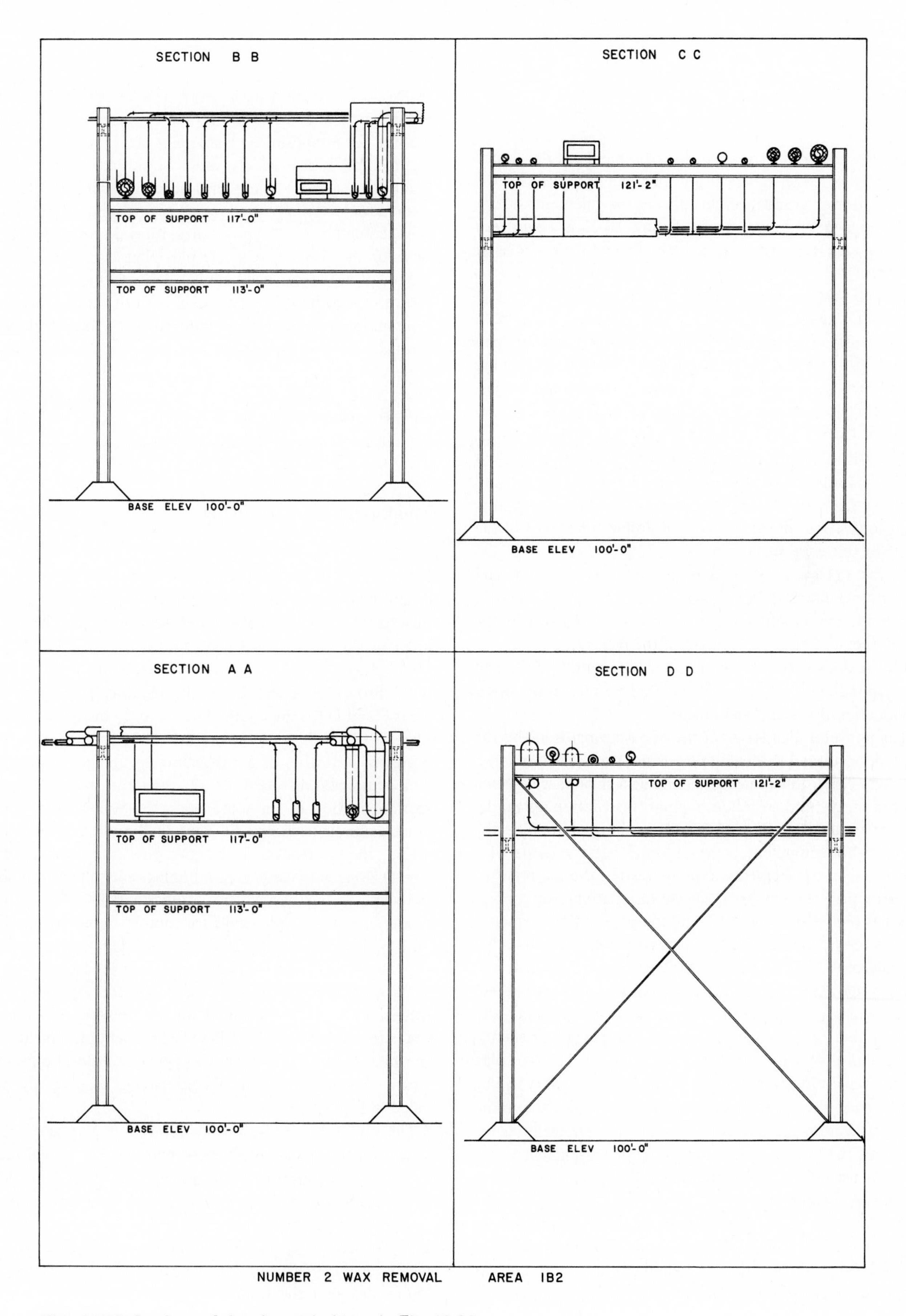

FIG. 10-25 Sections of the pipe rack shown in Fig. 10-25.

The Unit Plot Plan

After the site plan or master plot plan has been developed (or the site model has been finished and approved by the various design groups), each unit can be taken separately and designed in greater detail. Figure 10-21 is a *unit plot plan* of area 1B2 from Fig. 10-20 (the site plan or master plot plan). On this drawing, area 1B2 has been drawn to a sufficient scale to allow for the placement of major equipment. Plot plans show all the major pieces of equipment and identify them as in the Number 2 Wax Removal Area of Fig. 10-21. The major match lines and coordinates are established by north and south, east and west match lines, from which major equipment and pipe rack centerlines are located. Units E1, E2, E3, and E4 are heat exchange units; V2 is a horizontal drum; V3 is a vertical vessel, as are V4 and V5. Units P1, P2, P4, and P5 are pumping stations, and their centerlines are located on this drawing. The match lines look either north or east and define the extent of the unit. In this case, match lines east 327′ 0″ and 399′ 0″ define the extent of the unit in an east-west direction and match lines north 4100′ 0″ and 4171′ 0″ define the north-south boundaries. These may be referred to as **battery limits**, when they determine the extent of the facility.

Match lines determine limits that have mating units and corresponding match lines. Battery limits establish the outer boundaries of the unit and facility.

The unit plot plan is one of the first drawings done by the pipe designer and drafter. Piping is not shown at this stage. In some cases, this drawing is done at a smaller scale (such as 1 in. = 10 ft) than a typical piping installation unless the unit is small, as in Fig. 10-21. Many things determine the location of equipment, pipeways, and battery limits of a unit—design requirements, type of vessels, the materials of construction for pipelines, valving installations, amount of parallel piping runs required, and the general flow sequence between equipment. The layout, arrangement, and location of different vessels, such as heat exchangers, drums, and pumps, are discussed in Chapter 11 and should be consulted. The spacing for equipment and pipeways is determined by the number of units and the amount of piping components, valve stations, and so forth that are required. The plot plan or equipment location drawing of the nuclear facility in Fig. 10-14 must also consider these variables, though much more care is taken in the design and layout of these facilities because of the limited space. For petrochemical installations, there is usually far more land and open space available for piping and equipment.

At this stage of the plot plan the designer must know the equipment, processes, and design needs in order to establish the location and elevation of equipment and subunits. The flow sequence, process requirements, codes, specifications for equipment—all must be taken into account by the piping designer and drafter. Construction requirements, maintenance, and operation needs must also be considered at this point. The model method offers the easiest and most visible method for establishing a viable plot plan, though the drafting method is still widely used. Structural aspects of the plant must also be taken into account, such as extent of foundation footings, pier requirements, and all structural steel, columns, and fireproof structures that may be necessary for the project. A viable plot plan can only be established by an experienced piping designer and drafter. The novice cannot be expected to know all the design specifications and requirements for a unit—these are only established through many years on the job, working closely with experienced personnel. The piping drafter should study the many plot plans and plant arrangements available at the company where he or she is employed; company practices may differ from established procedures elsewhere.

The Piping Index Drawing

Figure 10-22 shows the piping index drawing for the project established in Fig. 10-21. In this drawing, the plot plan is divided into feasible working units which can then be drawn at a larger scale, such as 3/8 in. = 1 ft or 1/2 in. = 1 ft. Figure 10-22 shows the plot plan divided at coordinates or match lines into areas A and B for the pipeway pipe racks and areas C and D for the equipment areas. By breaking the plot plan into smaller indexed units, the piping drafter can represent a portion of a unit, showing piping, components, and equipment in great detail using the plan, elevation, section method or, in some cases, the model method. Related equipment should be kept in the same area—in Fig. 10-22, the horizontal drums or vessels are established in area D along with associated pumping units; the economizer units and vertical vessels are located in area C. Areas A and B are allocated to pipeways for multilevel pipe racks (Fig. 10-24), thus separating the pipe rack pipeway area from process areas C and D. This project has been used only to show general procedures. In actual practice the pipe rack (pipeway) is always included on the drawing of a unit in order to establish take-off lines, equipment placement, and the like. Figure 10-24 shows a typical multilevel pipe rack which changes direction; accompanying sections are provided in Fig. 10-25.

The piping index drawing and the plot plan are usually drawn on the same size sheet of paper and to similar scale, as is the **foundation location drawing** shown in Fig. 10-23. *This foundation drawing establishes, by coordinates, piers, foundations, footings, pedestals, and so forth required for a unit.* Notice in Fig. 10-23 how the foundations are lined up whenever possible to provide a neat and logical installation. (Note also that the foundation locations on this drawing, for the economizer units in area C, are not really adequate for the length of the vessels.) By establishing the size of the foundations and footing, interference problems and structural requirements can be determined in the earlier stages of the project. The hidden portions of the foundations shown in Fig. 10-23 indicate the extent of the buried footing.

Thus a logical sequence of development proceeds from the site plan to the plot plan to the piping index drawing to the foundation plan and subsequently to each indexed area, as shown in Fig. 10-24 of pipe rack. This series of drawings is an excellent project for the beginner. It will familiarize the student with the general procedure for layout and arrangement of typical petrochemical installation. When drawing this project, the student should attempt to follow design and layout specifications for equipment spacing and placement. Note that these drawings now violate some standard practices. Figure 10-26 is an example of a pipe rack constructed out of pipe.

Paving, Grading, and Drainage

The paving, grading, and drainage layout is completed after the plot plan, the unit index, and the foundation plan have been established. Figure 10-27 is an example of a paving, grading, and drainage drawing. The drainage systems to be used should be designated:

1. Storm water
2. Storm and oily water
3. Corrosive service
4. Sanitary service
5. Special services such as acid drainage.

The piping drafter will need to know the size and type of pipe to be used and the minimum and maximum depth of drainage lines for physical protection. ***The high point of finish grade*** (HPFG) is provided by the engineer along with any other special requirements that are entailed by the process or unit in question. In general, off-plot areas are controlled by storm runoff only and are not usually integrated into the paving, grading, and drainage drawing, which is defined by battery limits.

To lay out the paving, grading, and drainage drawing, the drafter must have a copy of the plot plan, site plan, and information supplied by the piping designer, including the location of process drains, valve boxes, and underground process or cooling water lines. The drafter must also find out from the electrical group the location of underground conduit banks and pull boxes.

These drawings should be completed with an on-plot minimum scale of 1 in. = 10 ft and for off-plot areas a minimum of 1 in. = 50 ft. The orientation and layout of the paving, grading, and drainage drawing must correspond to that of the plot plan. The same datum lines are used on both the paving, grading, and drainage drawings and the plot plan, as are the plant datum HPFG elevations and all other major callouts such as battery limits, match lines, and equipment centerlines. A drainage site plan should be established; when several plans are required to cover a large area, a key map may be called for. On the paving, grading, and drainage drawing, all roads, plot limits, building outlines, foundations, pull boxes, stanchion columns, and other physical characteristics which would impair the surface flow or interfere with drain line routing must be shown.

The first step in laying out the drainage area is to block out areas to be drained. It is general practice not to exceed 2500 sq ft per area. The drainage areas should be as nearly square as possible. One-quarter inch per foot should be the minimum slope for all drainage areas; one-half inch per foot is the maximum. The high point of finish grade should be set at the edge of drainage areas and the edge of pipeways, and the pavement should be sloped up from the high point of finish grade to the center of the pipeway unless there is a reason for not following this procedure. The following specifications for the layout of drainage areas are meant primarily for petrochemical installations:

1. Establish drain pavement areas to prevent hydrocarbon or acid spills from entering storm water and drainage systems.
2. Isolate all corrosive areas.
3. Set the elevation of concrete slabs to allow for the thickness of erosion-resistant pavement.
4. Set the elevation of the drain in corrosion areas so that the top of the drain conforms with final elevations of corrosion material used for paving.
5. Slope banks for roads, ditches, cuts, and fills according to engineer's specifications.
6. Slope road sections not more than 1/4 to 3/8 in. per foot on the road and 1 in. per foot on the shoulder.
7. Establish roads wide enough when rough graded to receive the finish paving layer.
8. Use vertical curves for abrupt grade changes on perimeter roads.
9. Use contours to show grade when the area to be graded is asphaltic concrete or when required by engineer specifications. Contour should not be used for concrete surfaces.
10. Draw each fifth contour heavier than the others.
11. For drainage lines, construction materials must be established by the engineer according to the corrosion-resistant nature of the material and the drainage to be handled. For corrosive service, polyvinyl chloride pipe, and other special plastic pipes may be required.
12. Use steel pipe for drain lines if soil conditions are unstable or if waste escaping into the ground might destroy the concrete foundation or release poisonous gases into the atmosphere.
13. Run main drain lines parallel or at right angles at all base lines established on the plot plan.
14. Locate oily water and storm water systems adjacent to each other whenever possible.
15. Specify the type of protective coating for all underground steel lines.

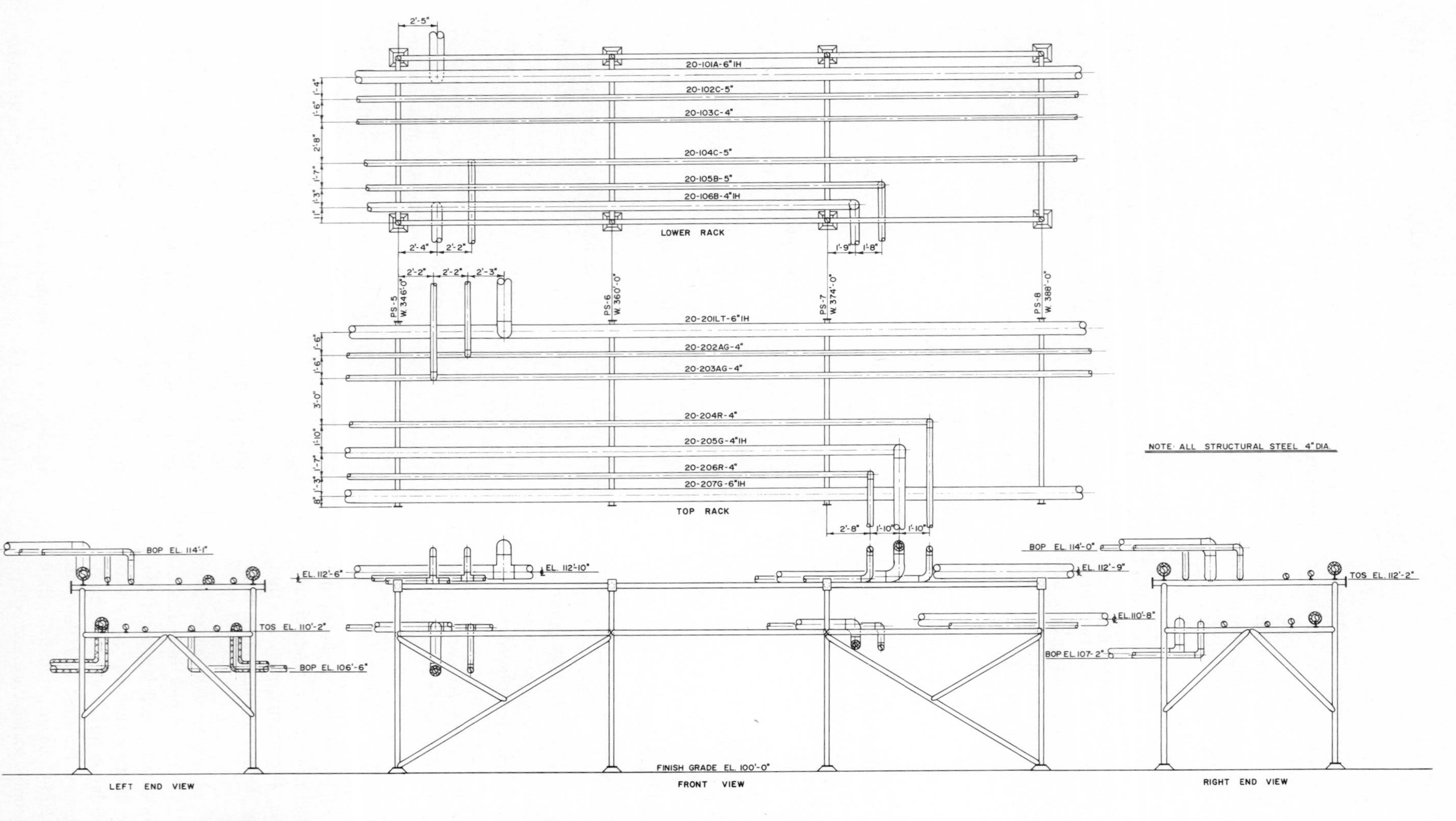

FIG. 10-26 Pipe rack constructed solely of structural piping. All bents, struts, columns, and bracing are fabricated from pipe.

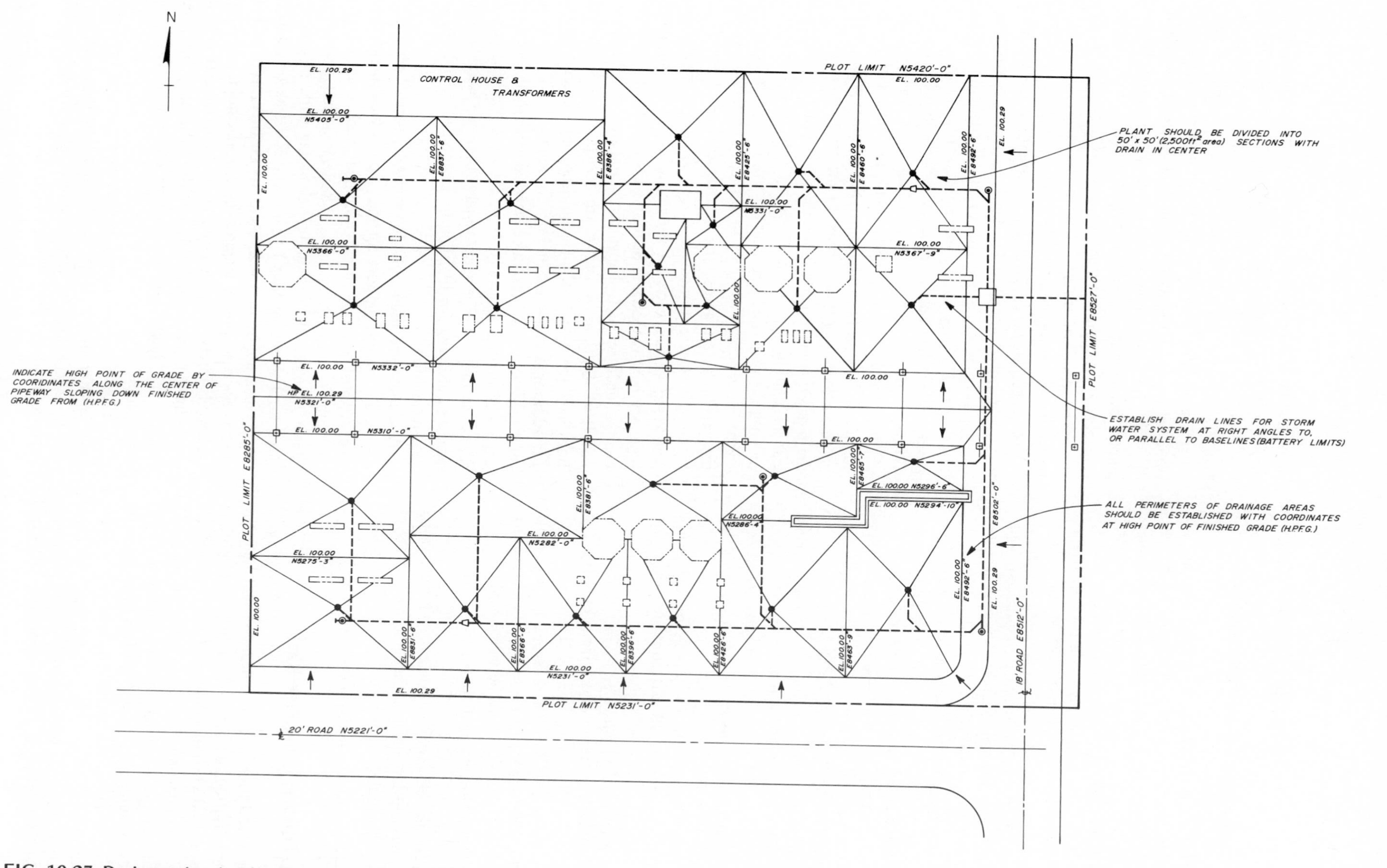

FIG. 10-27 Drainage drawing for a petrochemical installation.

16. Note all details on the drawing for steel pipe drainage systems.
17. Establish all drain lines and gas fields below the frost line.
18. For dimensions and elevations, locate by coordinates using base lines, battery limits and match lines which have been established on the plot plan—including those for centerlines of roadways, plot limits, centerlines of all drain lines, centers of manholes, area drains, process drains, catch basins, high points of all drainage areas, and other construction features that must be shown on the drawing, including curbs and separators. *Note:* Coordinates locating foundations and other features on the plot plan need not be shown unless absolutely necessary.
19. Give all elevations in feet and hundreds.
20. Give elevation of grade at high points, low points, and any other grade points that must be defined.
21. Invert elevations at all drain lines at high points, change in slopes, out points to entrance manholes, separators, or similar items.
22. Give elevation to grade at area drain, top of process drain, top of pull boxes, and top of curves.
23. Locate all information for catch basins, manholes, vents, seals, and process drains on the paving, grading, and drainage drawing.
24. Locate process drain hubs for pumps on the pump centerline and 1 ft 0 in. (30 cm) from the base of the foundation.
25. Raise the process drain hub 2 in. (5 cm) above finish grade to prevent surface runoff from entering the system.
26. Install manholes at points where several laterals join a main or for directional changes in large-diameter runs. Do not exceed 400 ft (121 m) between manholes.
27. Use standard manholes on all drawings.

Figure 10-27 should be carefully studied and used as an example to design and lay out a paving, grading, and drainage system for the project shown in Figs. 10-20 through 10-25.

Yard Piping, Pipeways, and Pipe Racks

The following description and specifications will help the prospective pipe drafter and designer establish yard piping, major pipeways, and corresponding pipe racks needed to support parallel piping runs. The photograph on page 327 is an excellent example of an adequately supported, neatly arranged pipeway and pipe rack system. This pipe rack runs down the center of a petrochemical unit, and it is possible to pick out the precast fireproofed concrete pipe rack bents on which most of the pipes are supported. Halfway between each concrete bent is an additional structural beam for strengthening the system. Pipeways must be established for all petrochemical and other process units to arrange horizontal pipe runs neatly and economically, grouping them in a logical sequence for easy takeoff to local equipment and other piping units. Figure 10-28 shows a pipe rack as seen from below. In this case, the structural elements are made of steel. This is a double-tier pipe rack. Many of the photographs throughout the book, especially in Chapter 11, show typical pipe rack, pipeway, and yard piping runs. Chapter 11 also gives specifications and descriptions of petrochemical pipe runs and corresponding pipe racks.

Pipe racks can be constructed of structural steel or precast, fireproofed concrete elements (page 327). Another method is to use pipe for the structural elements. Figure 10-26 shows a pipe rack constructed solely of structural piping. Piping has been used even for cross-bracing in this drawing. All petrochemical installations, nuclear power plants, fossil fuel power plants, and other industrial piping complexes have a multitude of parallel pipelines which must be channeled along logical runs. In the figure on page 327, the pipeway runs directly down the middle of a petrochemical installation with various refining units and equipment on both the right- and left-hand side of the pipe rack.

Pipe racks consist of bents or U-shaped structures where the pipelines rest across the horizontal support member. Figure 10-24 is a typical pipe rack drawing showing the plan view; section views are shown in Fig. 10-25. A typical pipe rack bent is shown in Section CC—in this case, the pipe bent is made up of U-shaped structural steel and has only one level. Sections A and B are also pipe bents, though both have the option of a second horizontal cross-member for additional pipelines, thereby forming a double-deck or multiple-level pipe rack. The vertical structural elements shown in sections A through D are referred to as *stanchions*. The area under a pipe rack usually forms a general service space or aisleway for installation, maintenance, and operational functions.

Pipe racks are used only for multiple, parallel pipe runs where a common support for many pipes is essential. They also support other pieces of equipment, such as cable trays, instrument lines, utility lines, electrical conduit, and other items necessary for an industrial piping unit. In general, structural shapes such as I beams and H beams are used for pipe racks unless the rack needs to be fireproofed. When this is the case, prefabricated concrete stanchions and bents are used. Fireproofing is essential in the hazardous areas of an industrial plant where fire could actually melt the structural steel shapes, causing further destruction during an emergency.

In general, all but large-diameter pipes are run on a pipe rack. Water mains, large sewer lines, cooling water, and other service lines of large-diameter pipe are usually laid underground and are seldom found on the pipe rack unless

FIG. 10-28 View of the underside of a pipe rack at a petrochemical installation.

it is important to run these pipes overhead for easy access to the process areas.

By grouping pipelines in banks or racks, it is possible to support multiple runs on similar cross-members. These adjacent piping runs are designed to withstand the total weight load of the pipelines and contents. They also provide supplemental support for lighting fixtures, control junction boxes, cable trays, and electrical lines, as well as serving other purposes. Elevating the pipelines permits access around equipment and process areas which would generally be cluttered with piping at lower elevations. Takeoff lines from the pipe rack, as shown on page 327, always drop or extend vertically from the major header (pipeline), crossing the pipe rack at 90 degrees to the piece of equipment to which it connects.

For large amounts of piping of various diameters, types, and materials in a typical industrial plant, the logical, economical layout of pipeways and pipe racks is essential. The designer and drafter must take into account a multitude of variables and specifications when laying out banks or pipe rack areas. After the plot plan has been established by battery limits, it is possible to divide it into a number of areas. In Fig. 10-24, the only two areas shown in detail are areas A and B, the pipeway and pipe rack areas. Various sections of this plan view of the pipe rack are needed to show all piping and takeoff lines. The designer must review the piping requirements for a unit (shown on the flow diagram) in order to lay out pipeway and pipe rack areas efficiently. The total number of lines to be supported determines the size of the pipe rack, whether it must be a single or double-level rack, and the spacing of stanchions and pipes. Always allow for the installation of electrical trays, utility lines, and conduit when allocating space and deciding on the limits of the pipe rack.

About 20 to 25 percent leeway should be given to excess space for expansion and future piping needs. It is essential that all major pipe racks be strutted; Fig. 10-28 shows the cross-bracing (struts) for a typical double-level pipe rack bent. Struts are also provided between bents in most cases. In Fig. 10-26, all the pipe bents and cross-bracing are constructed from pipe. The left-end view shows the two struts used to strengthen the pipe bent. The front view shows the cross-bracing (struts) necessary to provide sufficient structural stability for the pipe rack.

Figure 10-29 defines the placement of pipelines on a typical pipe rack. Chapter 11 discusses pipe racks also and should be consulted for petrochemical piping. In general, the designer and drafter should follow the recommended

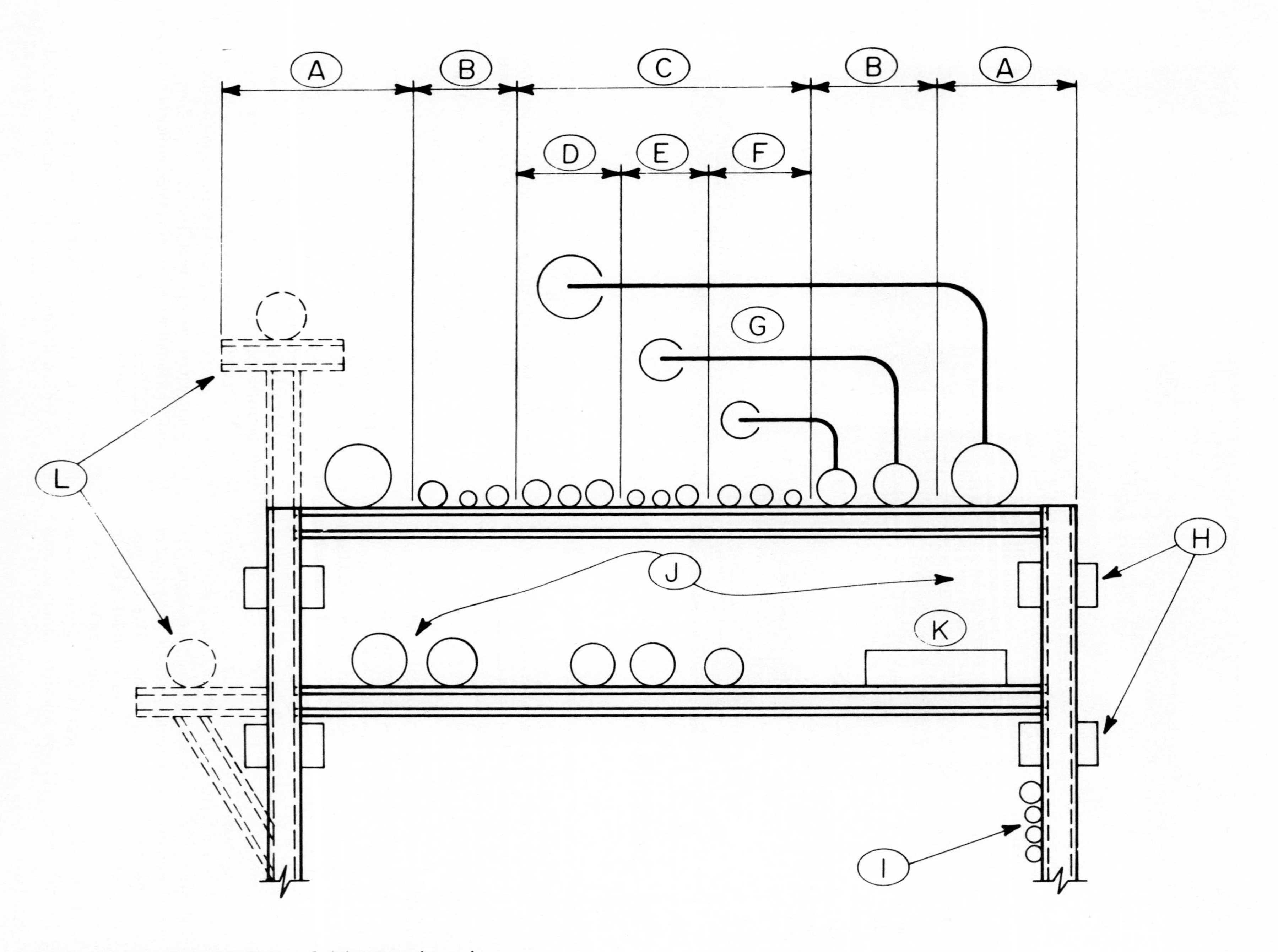

FIG. 10-29 Suggested arrangement of piping on a pipe rack.

placement of the different pipelines shown in Fig. 10-29. Large-diameter pipe and heavy lines should be placed to the outside of the pipe rack. These include large-diameter water lines, blowdown lines, and process lines shown in area A. In area B, it is customary to put medium-diameter process lines. Area C (the middle of the pipe rack) is usually relegated to utility pipelines, which are usually of smaller diameter and lighter. By placing the heavy lines in area A, it is possible to put most of the load on the outside of the pipe bent where the vertical structural elements will take up the support. Area D should be kept for fuel oil and chemical services, Area E for the lightest, smallest-diameter lines, such as gas and air, and Area F for steam, condensate, and the like. Areas D and F can be reversed under normal circumstances. By grouping utility lines and service lines in the middle of the rack, it is easier to establish takeoff lines, which generally run at 90 degrees to the pipe rack in both directions. An attempt should be made to place process lines on the side of the rack from which they connect to equipment.

When using a double-deck pipe rack, area J is normally allocated to process lines. Placing the larger-diameter and heaviest lines to the outside reduces the need for large, expensive, structural elements for cross-bracing the pipe bent by placing most of the load on the outside edges on the vertical members (stanchions). Utility and service lines should be allocated to the top cross-member (the upper level) of double-level pipe racks. Planning for a typical pipe rack must take into account many variables—especially the side of the rack from which a line must be taken to connect to equipment.

Area H on Fig. 10-29 is usually allocated to light utilities, instrument lines, and instrument and conduit runs. Area I can be used for small utilities or conduit. Area K on both sides of the lower rack is sometimes used for electrical trays or multiple conduit runs as in Figs. 10-24 and 10-25. Area L (both sides of the rack) can provide rack extension space.

Figure 10-26 (the right and left end views of the pipe rack) shows the use of shoes to elevate insulated lines above the cross-member in order to provide sufficient space between the insulation and the horizontal member of the pipe bent. This is an essential practice, especially where the pipeline must be heavily insulated around the shoe. As can be seen in this drawing, some lines have the insulation cut away so that the pipe can rest on the cross-member. This is done for medium-temperature lines only. Lines which need not be insulated may lie directly on the cross-members. Figure 10-26 also shows the use of vertical takeoff lines. Seldom should a takeoff line project from the pipeline on the rack without changing elevation first. It is standard practice to orient the takeoff line vertically whether above or below the mainline or header, as shown in Figs. 10-26 and 10-25. These sections show the general practice followed in industry where pipelines coming from the top level extend vertically above the main header line and then move horizontally at 90 degrees to the pipe rack. Figure 10-24 (area A) shows a double-level pipe rack. When the pipe rack changes direction and moves toward area B, the elevations of both the upper and lower part of the pipe bent change. This change can be seen in section BB, where the takeoff lines extend vertically above the main header line and move toward the right at a higher elevation and from there pick up the single-level pipe rack shown in section CC. **When pipelines change direction, whether they are on pipe racks or not, they must change elevation.** In section B, the top line is at 117 ft 0 in.; in section CC, it has moved to 121 ft 2 in. at the top of the support.

The following guidelines establish specifications for spacing and clearance, insulation, placement of lines, and other information which may be required:

1. **Placement of pipelines on pipe racks**
   (a) Allocate the outer edges of pipe racks, whether double-level or single-level, to heavy, large-diameter pipelines.
   (b) Allocate the center of pipe racks to light, small-diameter lines for utility services.
   (c) Place process lines on both sides of utility lines. If process lines go to equipment on the right, they should be on the right side of the rack; if process lines go to equipment on the left, they should be on the left.
   (d) When a large number of pipes are to be established in one pipeway, a multiple-level pipe rack should be considered. Utilities and service lines go on the top level of the rack and process lines on the lower level. If it is necessary to take process lines to higher elevations, the process lines should be located on the top level.
   (e) Establish all piping loops for expansion over the top of the pipe rack. Never extend loops out horizontally from pipe rack edges.
   (f) Pipelines should not be run over stanchions for single-line racks in case future piping needs require the construction of another level.
   (g) All small-diameter pipes should be supported by trays, grating, or other constant supports in order to eliminate sag; or they should be supported locally by placing them next to larger-diameter pipes.
   (h) Establish instrument lines, conduit lines or boxes, and trays on the outside of the pipe rack as shown in area H of Fig. 10-29, unless the electrical requirement is great enough to warrant its placement directly on the horizontal member (Figs. 10-24 and 10-25).

(i) Attempt to limit the diameter of piping on a rack to 18 in. (45 cm); place larger-diameter pipelines on sleepers or underground.

(j) All high-temperature insulated lines must be supported on shoes (and in some cases slide arrangements) to provide for thermal movement and elevate the pipeline and insulation above the rack.

(k) Medium-temperature insulated lines can have insulation cut away so that the pipeline lies directly on the rack. Care should be taken to cut away sufficient insulation so that surrounding insulation is not crushed during thermal movement.

(l) When using concrete pipe bents, it is important to provide steel plating for the pipeline to rest on. Pipelines must never rest directly on concrete.

(m) When takeoff lines are required from the pipeline on a pipe rack, they should be extended 90 degrees to the main line either above or below, thereby *always changing direction and elevation.* (In some cases 45 degree takeoff lines are used.)

(n) Low-temperature noninsulated lines should lie directly on the cross-member.

(o) Always design pipe racks with single levels whenever possible because of the additional expense of a double-deck pipe rack.

(p) All lines which require expansion loops for thermal stress should be grouped on a similar side of the pipe rack—never toward the outside of the rack—thus eliminating overhang caused by the expansion loop.

2. **Design and arrangement of pipe rack elements**

(a) Pipe rack bents and distances between bents should be established on a modular or even basis—for example, 20 ft (6 m) between bents.

(b) In general, 20 (16 m) to 25 ft (7.6 m) between bents is a common modular dimension for pipe racks.

(c) Allow 25 percent additional space for expansion and future piping needs.

(d) Multilevel racks should be a minimum of 3 ft (.9 m) between elevations; whenever possible, 4 ft (1.2 m) should be used as a standard elevation change.

(e) To determine the piperack width distances between stanchions should be established according to the total number of pipelines needed to keep the pipe rack at a single level whenever possible.

(f) Distances between stanchions (columns) should be set up on a modular basis with a maximum of 25 ft (7.6 m). This can be expanded to 30 ft (9.1 m) for pipe racks where air coolers are installed on the rack. Distances between stanchions should be a minimum of 8 ft (2.4 m), and this distance should be increased by 2 ft (60 cm) increments when establishing the width of a pipe rack.

(g) Main pipe racks should always be provided with supplemental structural elements (struts) for additional strength.

(h) Provide supplemental supports for all small-diameter piping (depending on distance between bents).

(i) The minimum clearance below the pipe rack should be established according to the equipment needed and the size of the aisleway required. The elevation is determined by whether or not the pipe rack must cross a roadway.

EXAMPLE: Specifications

1. 17′ 6″ (5.3 m) major roadways
2. 15′ (4.5 m) minor roadways
3. 12′ (3.6 m) pipe rack aisle (beneath rack)
4. 7′ 6″ (2.2 m) aisleways
5. 7′ 2″ (2.1 m) stairways
6. 6′ 6″ (2 m) passageways

(j) Pipe rack elevations should be established to slope the pipe lines slightly in order to avoid pocketing and accumulation of line material.

(k) Takeoff line elevations must be established in accordance with the level of equipment to which they will be connected. It is important to slant these lines toward equipment to avoid pocketing.

(l) Control valves, junction boxes, and utility stations should be located so they are easily operated and serviced from beneath the pipe rack. Stanchions are used to support these items.

(m) Avoid multilevel pipe racks whenever possible because of additional cost.

(n) Pipe rack members, such as cross-bracing and struts, should be consistent in size to reduce costs and eliminate problems in erection. This includes stringers, struts, cross-members, horizontal supports, and columns and stanchions. The pipe rack should be designed so that these elements are the same for each type—columns the same size (stanchions), stringers the same size, struts the same size.

(o) The vertical distance between levels in a pipe rack should be established according to the design requirements. In general, 3 ft (.9 m) is the minimum when changing directions on a pipe rack and 6 ft (1.8 m) between levels on the same bent whenever possible to provide sufficient clearance, though it is general practice to limit this distance to 4 ft (1.2 m) because of cost.

(p) Most of the area beneath the pipe rack should be kept free for aisleway space for mobile equipment, service, maintenance, and operation purposes.

These specifications are only guidelines for the piping drafter and designer who wants to establish an efficient and economical pipeway and pipe rack. Many variables must be taken into account, and much of the piping will be determined by the available space and whether the lines must be in a building or in a yard. The following excerpt from the American Engineering Model Society provides more information on the design and layout of piping systems:

Piping Systems

Before discussing the variables in piping systems, it is necessary to discuss equipment arrangements. There are two general classes of arrangements. In one the unit processes are contained in a structure. The other places the unit processes adjacent to a pipe bridge that serves as a connecting link. The first mentioned arrangement is commonly found in chemical processes while the second is more often used in petroleum processes.

Many of the details to be discussed under this heading will apply to both arrangements. However, they apply to a much greater degree to the process structure.

Most process buildings are long, narrow buildings in order to obtain maximum flexibility for future changes. For instance, if process changes or additions are required, they may be made adjacent to the unit process to which they will be connected. If the unit process equipment involved was in the center of a square structure, this would not be possible.

The elevation of the structure is often determined by gravity heads necessary for flow.

Since the structure is long and narrow, such items as ducts, header pipes, and electrical conduit must be run the length of the structure. The equipment arrangement should provide a maximum of space to carry the items mentioned above the entire length of the structure. In other words, the equipment should take up the minimum profile. Wherever possible, smaller equipment should be hidden in the shadow of the larger equipment pieces.

In outdoor plants, the process lines are usually carried on a supporting bridge. This is known as the yard or in-line arrangement. The pipe bridge in this case must carry the ducts, pipe headers, and electrical conduit. Its size, therefore, must be adequate to carry all of these items plus space for additions due to process changes and future additions.

To determine the bridge requirements, the piping between the unit processes is sketched freehand on a plot plan. The service piping is then added in a like manner. After all of the piping has been roughed in, freehand spacing section sketches are made of the most congested areas. Areas must be reserved for other services such as electrical conduit, instrument channels, and catwalks. If more than one level of piping is required, the service piping, electrical conduit, and instrument channels should be placed on the top level to avoid damage caused by process leaks.

When selecting the elevation for each pipe, thought must be given to eliminating pockets. All of the lines should be self-draining if possible. The high point should be located so that vent valves are accessible.

There are two possible methods for increasing the available bridge space. The simplest method is to add another level as the vertical space above the bridge is generally clear. The second method is to increase the width of the bridge. Increased width may require a relocation of the unit processes. From the designer's point of view, the simple method of adding another level may seem to be the best answer. However, from a structural viewpoint, the cost of additional height may far exceed the cost of the additional width. This is especially true if the service piping on the top level requires anchor points. The cost of revising the plot plan caused by increasing the width of the bridge will be small in comparison to the structural savings in most cases. However, should the designer fail to recognize the necessity of the increased width at an early stage of the design, the cost of the revision will be greatly increased. Therefore, this check should be made before any piping design is done.

After due consideration has been given to the equipment arrangement, the piping system design begins. As the piping system becomes more complicated, more elaborate designs must be included to assure a successful design. One design in use is the lane system. In this system, lanes or elevations are reserved for piping in one direction while other lanes are reserved for piping at 90° to the first mentioned piping. The object of the lane system is to prevent interferences between pipes crossing one another.

The lane system may eliminate many of the interference problems. In fact, it has been adopted as a rule by some engineering companies. However, the use of the lane system introduces additional costs and greater friction losses in the piping.

Common pipe hangers may be used to support piping that is grouped together. Considerable savings can be made in erection as the support for a group of pipes may be only slightly more expensive than a support for an individual pipe. Further, the problem of scaffolding for erection is simplified. If piling or footing is required, any additional supports will be very expensive. The pipe designer must be careful not to carry grouping too far. In some cases, the costs of the additional hangers and pipe required to reach the bridge may more than offset any savings obtained by grouping.

Particular attention must be given to the support problem if the pipe is fabricated from plastic, glass, or is glass-lined. In general, the unsupported span for such materials is much less than for carbon steel pipe. Glass piping is very susceptible to breakage and is usually backed up by wood for protection.

The branches from headers should be as short as possible consistent with good overall design. The designer should think carefully about the phrase "as short as possible." This does not mean the header should be run as close as possible to the branch designation in all cases. If steam or other hot material is piped through the header, the header should be located far enough away from the stiffest branch destination to provide sufficient flexibility to keep the stress due to thermal expansion within allowable limits. A pipe containing a flow sensing orifice should be located far enough from the branch destination to provide the required straight run of pipe.

There are two methods of handling vertical piping passing through a floor. The first is to use individual sleeves for each pipe. The second is an internal or external pipe chase. The internal chases are curbed slots in the concrete floor. The slots should be placed in the shadows if possible. The designer must locate the points where vertical piping will be passing through floors so that suitable openings can be installed before the floor is poured. A common practice is to allow a floor sleeve to project 4 to 6 inches above the surface of the concrete. This prevents liquids from draining to the floor below. The designer must be careful to see that the sleeves do not interfere with other fittings when a connecting horizontal run is close to the floor. The piping must clear the curbs as well as the floor.

Usually, the most practical approach is to combine the methods. Sometimes the use of sleeves will be unavoidable for process reasons. By using chases where it is practical to group the vertical piping, and sleeves where it is not, the best and most economical result can be achieved.

Foundation and Concrete Drawings

Foundation and concrete drawings are necessary aspects of all piping system layout and design. The general foundation drawing includes coordinates, boundary limits, battery limits, dimensions, and the location, size, and extent of rebar and reinforcement steel. The piping drafter may be called upon to draw or to work closely with the structural group, including the engineer and designers, which determines the calculations, extent, location, and procedures for establishing foundations, drains, sewers, grading, and other aspects.

All underground equipment, piping, and components must be selected prior to the location and determination of foundations. Often electrical conduit, sewers, and water lines are placed underground. These must be located and designed along with the appropriate drains and penetrations before the foundation is established. Underground networks are among the first drawings completed after the plot plan. When the model method is used, all underground lines must be indicated with appropriately color-coded tape. Sometimes the basement and underground piping are indicated by modeling which shows all the components below a clear Plexiglas base.

Concrete drawings are similar to foundation drawings, though they are primarily concerned with the top of grout for foundations, pipe stanchions, and other column and vessel foundations, including octagonal or rectangular foundations and piers for elevated vessels. The piping designer and drafter are sometimes called upon to complete these drawings, though in general the structural department does much of this work, at least in the design and preliminary stages.

Foundations Figure 10-30 shows typical foundations, pipe stanchions, and concrete stanchions. Detail A shows the top

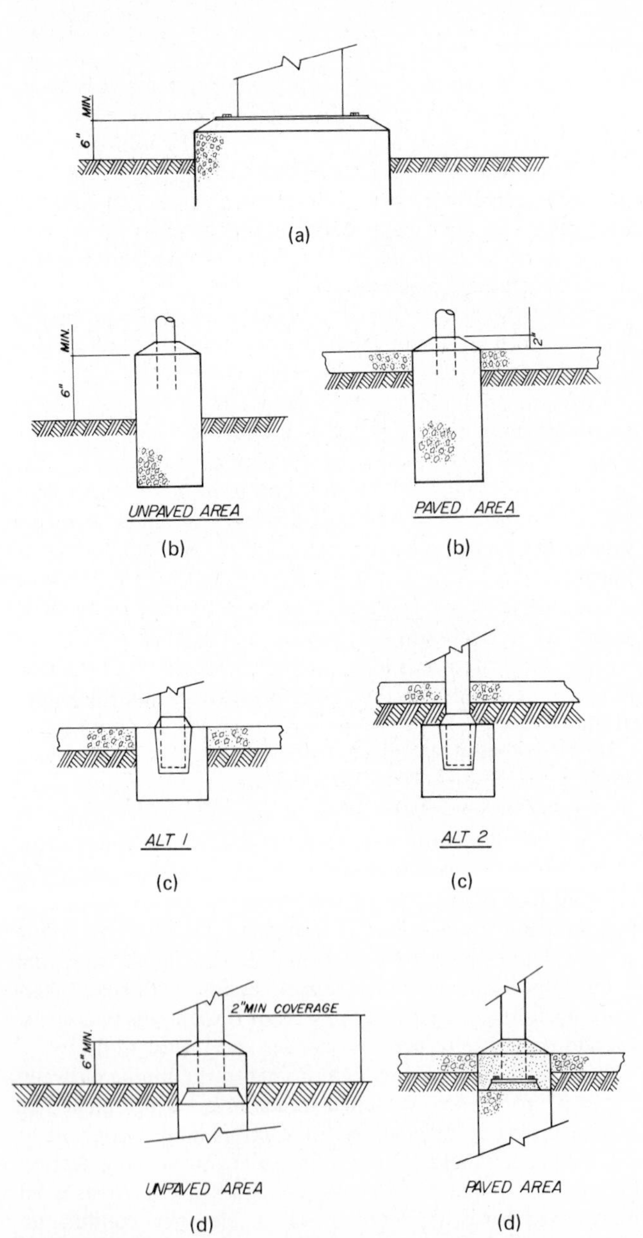

FIG. 10-30 Foundation, design, and drawing procedures.

of grout for the foundation usually used for mechanical equipment, machinery, or steel structures. Top of grout should be at a minimum of 6 in. (15 cm) above grade for steel structures where access is restricted or the pier is much larger than the base plate. Detail B shows steel pipe stanchions with embedded foundations placed with the top of the foundation a minimum of 2 in. (5 cm) above paving and paved areas and a minimum of 6 in. (15 cm) above grade in unpaved areas. In off-plot paved areas where access is not a problem, the top of the foundation may be placed 6 in. (15 cm) above paving. Detail C shows a precast concrete stanchion socket-type foundation. The grout should be placed flush with the top of the foundation or below the finished grade whenever possible. On pile-supported foundations where it is anticipated that paving will settle, it is important to place the top of the foundation far enough below paving to prevent the foundation from hanging up (see Detail D). Foundations for stanchions that have steel base plates or steel structures where access is restricted or piers are much larger than the base plate must, in paved areas, be placed with the top of the foundation below paving protected with a layer of concrete. In unpaved areas, steel must be protected to a minimum of 6 in. (15 cm) above grade. These four details show the common variations of equipment foundations and are useful for the projects in this book. These procedures for the establishment of foundations should be adhered to whenever possible.

Vertical Vessel Foundations Figure 10-31 shows a typical octagonal foundation for vertical vessels. Octagonal foundations should be established for all vertical vessels equipped with a skirt. For vessels designed with legs, square or rectangular foundations should be used. The design calculations for footings and pedestal for vertical and horizontal vessels are usually done by the foundation and structural designer.

The foundation designer must take into account many different calculations and design aspects when establishing the pedestal, footing, and foundation concrete requirements for vessels with skirts or with legs. The following information must be determined:

1. Weight of the footing
2. Weight of the vessel empty
3. Weight of the vessel under test
4. Weight of the vessel under operating conditions
5. Area of the footing
6. Area of the steel reinforcement
7. Computed soil pressure
8. Allowable soil pressure
9. Inside diameter of base
10. Depth of footing in inches

For vessels with legs, it is also essential to determine the area footing and dimension the bolt circle, which is indicated on the vessel or column drawing. For vessels or columns with skirts, octagonal foundations are used as in Fig. 10-31; here too the bolt circle must be established from the diameter of the vessel column drawing. For vessels with skirts or legs, the top of grout should be a minimum of 6 in. (15 cm) above grade for the foundation; this elevation is usually set by the pipe drafter. The minimum below grade should be 12 in. (30 cm) when the soil pressure is 1000 psf or greater, below the frost line when frost conditions are a factor, or established through calculations by the foundation engineer when soil conditions are poor. Many other considerations must be taken into account, such as problems with seismic disturbances. When steel is used to reinforce the foundation, the top of steel should be 2 in. (5 cm) below the top of the foundation. The bottom of steel should be 3 in. (7.6 cm) above the bottom of the foundation, and the reinforcements should be run in both directions in both

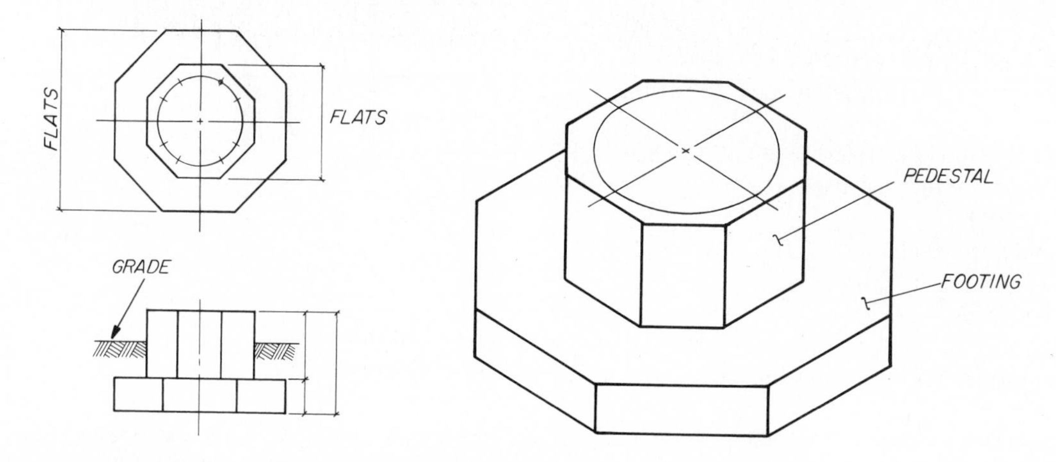

FIG. 10-31 Typical foundation for vertical vessel.

top and bottom. Where anchor bolts are used to control the depth of concrete, it is important to maintain 3 in. (7.6 cm) clearance from the bottom of the foundation to the anchor bolt.

Stairways and Platforms

Figures 10-32 through 10-36 are typical detail drawings for plan and elevation views of stairways, stairwells, and passageways and platforms showing headroom and clearances. These details should be drawn at 3/8 in. = 1 ft.

Plan views should always be shown and necessary elevation views should be established to show the stairwells, stairways, ladders, and bracing in the best possible view. The plan view should be oriented to designate the column centerline in the same way as it is shown on the overall plot plan. All dimensions should be referenced to these columns. It is general practice to draw combined steel and concrete structures with steel on a separate drawing. It is important to show the phantom outline of all concrete portions of a structure on a steel drawing, though it is not usually necessary to indicate the phantom lines of steel on concrete structures. One of the most important aspects of positioning stairways, platforms, and passageways is to ensure that piping and equipment will not interfere with structural members. All platforms and structural elements and elevations are established from piping study drawings. The following specifications apply to stairways, landings, platforms, passageways, and headroom clearances for these items.

1. A minimum horizontal walking width of 2 ft 6 in. is required.
2. Clearance on platforms from ladder to back handrail should be a minimum of 2 ft 6 in.
3. Clearance from manway to handrail should be a minimum of 3 ft 0 in.
4. Headroom clearance should always be a minimum of 6 ft 6 in.
5. Stairs should have 9¾-in. treads and 7¾-in. risers.
6. All elevations set by the piping group must agree with the riser requirement for stairways.
7. Elevation of platforms established on piping drawings should be adjusted so that the height requirement for risers can be applied.
8. Stairways should be a minimum of 2 ft 6 in. wide.
9. Landing areas at top or bottom of stair runs must be at least 2 ft 6 in. wide.
10. Handrail clearances should be a minimum of 2 in. from surrounding obstructions, including concrete walls or structural elements.
11. Headroom clearance on stairways should be 7 ft 2 in. (7 ft optional).
12. All structural framing members in platforms, including brackets, must be established after the space requirements for stairways have been resolved.
13. Concrete beams should be covered with steel floor plates when used as walking surfaces.

Figure 10-32 shows a plan view of a set of stairways and landings including clearances and dimensions; Fig. 10-33 shows the elevation view. The top drawing in this detail shows the landing and the first riser up or down. When bolted connections are used at top and bottom of stair stringers, the space required for the stairway must be equal to the length of run plus the landing space. The lower drawing shows the landing first riser up, first riser down, and tread; stair stringers and landing framing are welded together as one unit. The space for the stairway is the length of run plus one tread plus the landing space, as shown in this drawing. Figure 10-34 shows the plan and elevation view of the single stairway. Care must be taken to establish the minimum 7 ft 2 in. clearance, which is the 6 ft 6 in. headroom requirement plus one riser for stairways. Figure 10-35 shows the headroom clearance on platforms and at passageways. The platform working area must be established before any framing or structural members are added in order to provide the minimum 6 ft 6 in. clearance for headroom. The passageway headroom requirement is the same, along with the 2 ft 6 in. passageway minimum requirement. Figure 10-36 shows the rise and run of stairway of 7¾-in. riser and 9¾-in. tread (38°29′ angle for the staircase). Notice the establishment of 6 ft 6 in. clearance at the top of the landing and a minimum of 7 ft 2 in. clearance for headroom on the stairway.

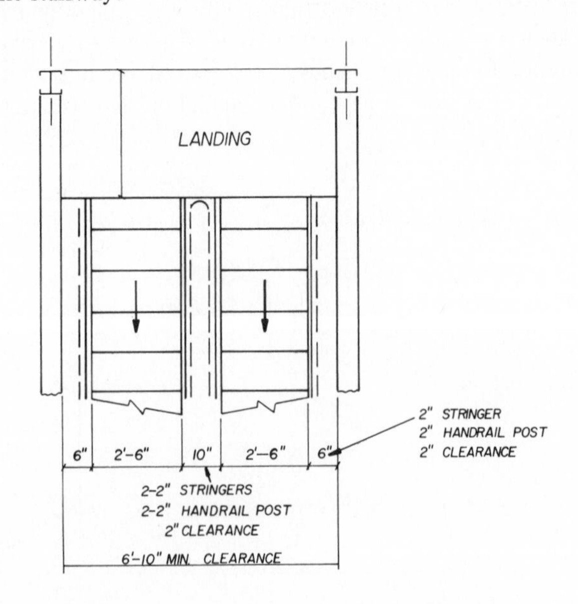

FIG. 10-32 Plan view of stairway and landing showing typical dimensioning and drawing procedures.

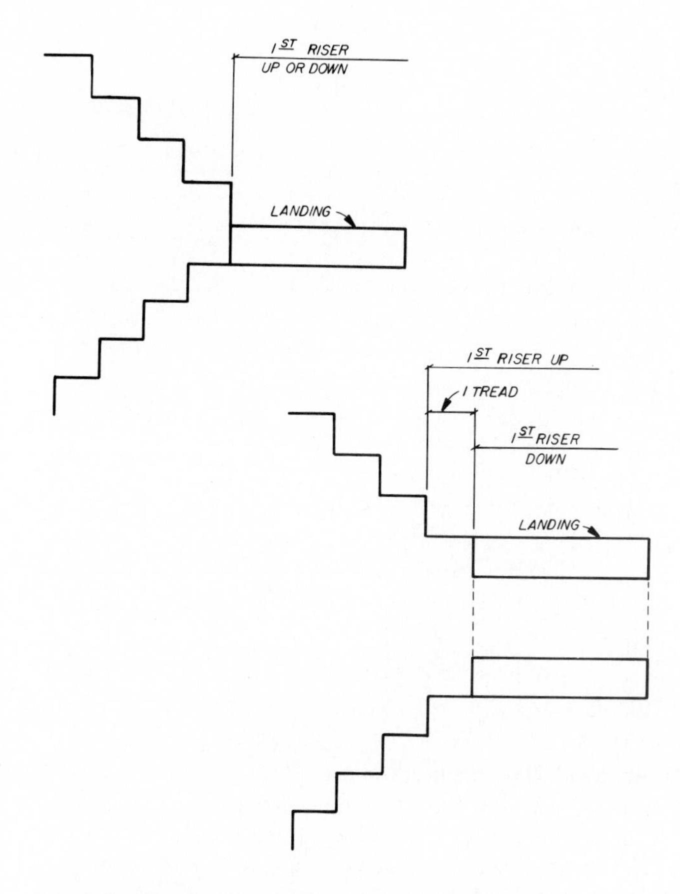

FIG. 10-33 Elevation view of stairways and landing.

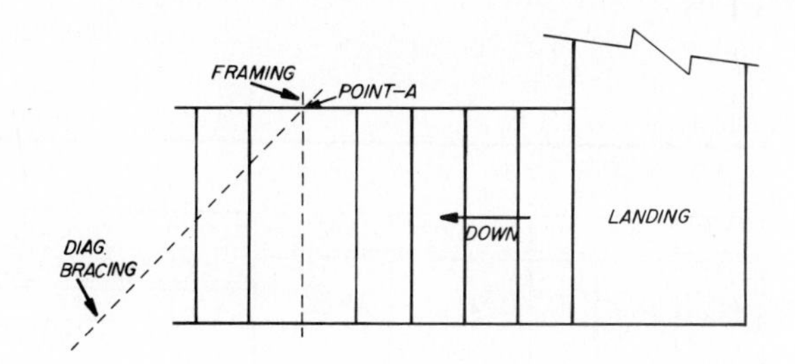

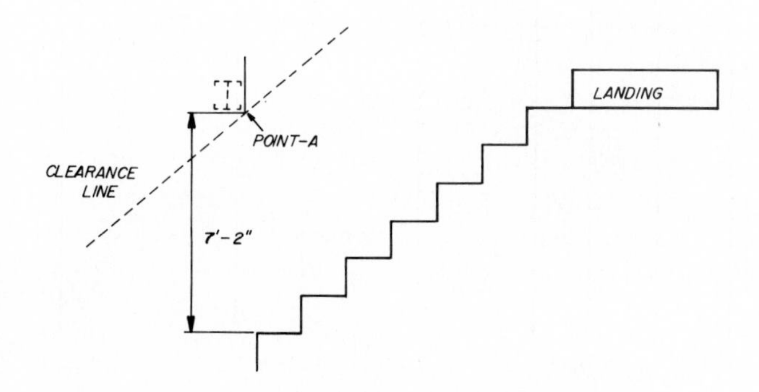

FIG. 10-34 Plan and elevation view showing clearance for a stairway.

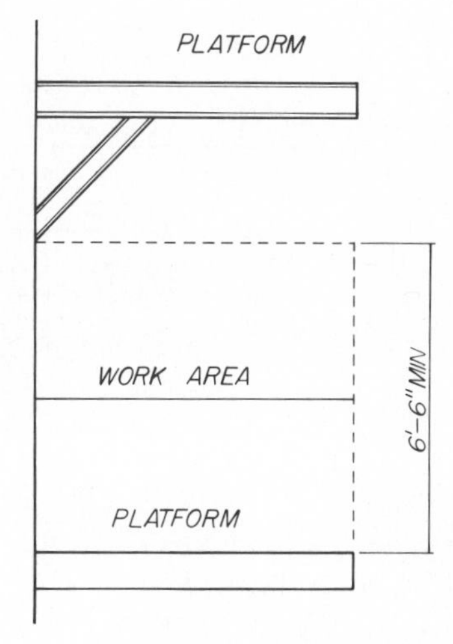

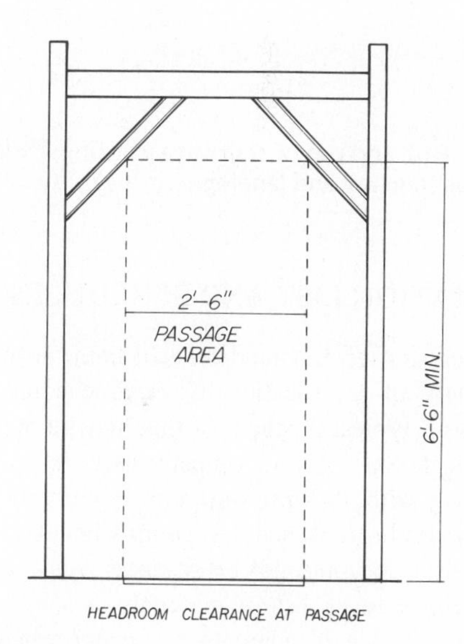

FIG. 10-35 Headroom clearance for platforms and passageways.

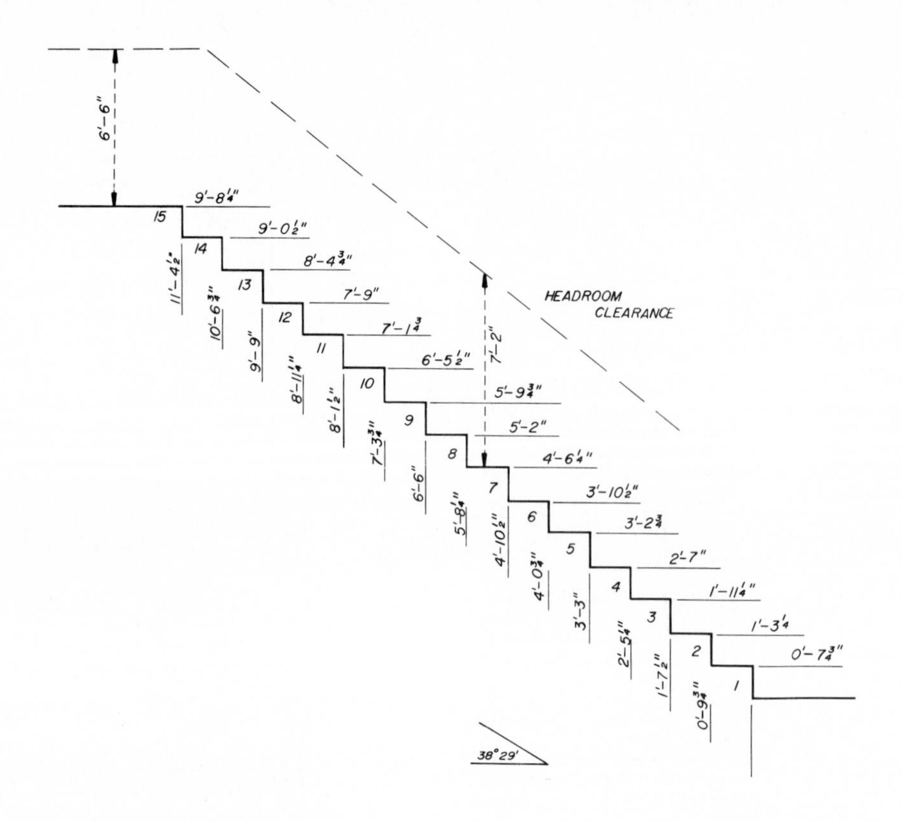

FIG. 10-36 Rise and run of stairways utilizing 7¾-in. riser and 9¾-in. tread. Note the headroom clearance for stairway and landing.

SPECIFICATION LIST AND SCHEDULES

Every piping drafter and modeler will come into contact with a variety of specification lists and schedules. Figure 10-37 shows a typical pipeline list that may be required for a pipe project. Since every company uses its own list sequence along with its own variation of plumbing fixture schedules, water heater schedules, pump schedules, and heat exchanger lists, it would not be advantageous to set down strict guidelines for a list of materials or schedules. The novice must be able to adapt to the many schedules and lists required on the job. In many cases, schedules for drains, plumbing fixtures, and other equipment will be needed. One of the basic lists the student will encounter is the *pipeline list,* which usually indicates the drawing the pipeline is shown on, the line number, line size, schedule, material, design requirements such as pressure and temperature, and the origin and termination points on the relevant drawing. In some cases, the flow diagram number is also given. Level gauge specification sheets, relief valve specification sheets, level instrument specification sheets, heat exchanger specification lists—all are important forms with which the piping drafter will come in contact. Most of these are completed by the process or instrument engineers and give essential information concerning a particular instrument, pipeline, or piece of equipment. Sooner or later, every piping drafter will have to understand, compile, and work with pipeline lists, pipe specialty lists, and other specification lists and sheets.

DRW#	LINE#	SIZE	SCHEDULE	MATERIAL	DESIGN REQUIREMENTS		NOTES
					PSI	TEMPERATURE	ORIG-TEF
101-5	0641 IW	6"	40	STEEL	320	500°F	BATT-SI

FIG. 10-37 Pipeline list.

QUIZ

1. What is the primary function of a piping system? Name three basic design considerations.
2. What is a slurry? Name two different types.
3. How does flashing happen in a piping system and how is it possible to reduce its likelihood?
4. What is the primary function of a control station?
5. What is meant by hazardous fluids?
6. Name two maintenance considerations which must be taken into account in the design of piping systems.
7. What are the primary causes of vibration and how can they be eliminated?
8. Describe three different ways of dealing with thermal movement in piping systems.
9. What is an expansion loop and how is it used?
10. Name three methods used for retention of piping temperature. Describe each method and explain when it should be used.
11. What is stream tracing?
12. Name three types of drains.
13. Why is it necessary to remove condensate from steam lines?
14. What are control valve stations and why are they so important to piping systems?
15. What type of instruments can be found in Fig. 10-6?
16. What are the primary functions of instruments and piping systems?
17. How does instrument piping differ from other piping in a typical system?
18. What are the normal methods for instrument connections?
19. What is a level glass and what is its function?
20. What are level controllers used for?
21. Name three types of lists used and prepared by the piping designer.
22. What is the *run* of a stairway?
23. What is the minimum clearance for headroom for a passageway?
24. What is a tread?
25. Define landing.
26. What does the top of grout mean for foundations?
27. What is a precast concrete stanchion?
28. Where should utility lines be located on a pipe rack?
29. What is the common distance between bents in a pipe rack arrangement?
30. What is the normal distance between levels of a pipe rack?
31. What does yard piping mean?
32. Insulated pipelines on pipe racks are always ________ ________ from a structural member.
33. What dimensions must be established from the plot plan in order to lay out the paving, grading, and drainage system for a piping installation?
34. What is a drawing index and how is it used?
35. What is the function of a site plan?
36. What is the difference between the site plan and unit plot plan?
37. What is electrical conduit?
38. Name three considerations for electrical design in piping systems.
39. What are columns, bents, and beams?
40. Name four important considerations in the original arrangement and layout of piping systems.
41. What are the considerations for establishing elevations of structures?
42. What is a pipe sleeve used for?
43. Name three types of pumps and compressors and give some basic information on each.
44. What are heat exchange units used for?

PROBLEMS

1. Redraw two of the control manifolds from Fig. 10-5 approximately three times the size in the book. Establish a scale for the drawing, labeling, detailing, and dimensioning of both control manifolds using specifications given in Fig. 10-4.
2. Redraw Fig. 10-8 approximately two times the book scale. Establish an appropriate scale; dimension; and call out all aspects of the detail, showing the level glass, pressure gauge, and level controller connections. nections.
3. Redraw Fig. 10-12.
4. Redraw Fig. 10-11.
5. Redraw Fig. 10-15 of the equipment location for the turbine house. (This drawing is in metrics.)
6. Redraw Fig. 10-14.
7. Figures 10-20 through 10-25 are all portions of the same project. Construct the site plan first and then the plot plan, trace the necessary coordinates and equipment to establish and draw the foundation plan and piping index drawings from the plot plan. Also prepare a paving, drainage, and grading drawing.
8. Redraw Fig. 10-26 of the pipe rack.
9. Redraw the drainage plan shown in Fig. 10-27.
10. Using specifications set down for paving, grading, and drainage, prepare a logical drainage drawing for the project in Problem 7.

Petrochemical refinery installation. (Courtesy Babcock & Wilcox, Tubular Products Div.)

11 PROCESS PETROCHEMICAL PIPING

This chapter is primarily an introduction to the petrochemical industry. It deals with petrochemical processes, equipment, and products—especially with oil refineries and all systems, facilities, processes, and products associated with the refining of oil, from the transportation of crude oil to the blending and shipping of the final products.

Because of the energy shortage and the demand for a comprehensive energy policy, oil and its associated products have become a critical industry. The number of jobs in this area for drafters, designers, and engineers has multiplied rapidly and will continue to do so for quite some time. The diminishing supplies of natural gas and the many uses of petrochemical products will make this area of employment extremely important for the rest of this century, as demand for oil and its products becomes greater than the supply.

Process petrochemical plants or refineries involve piping systems and associated equipment which process crude oil as the primary component in the production of a variety of end products and by-products used by society and industry. Crude oil is to be considered the primary *feed* or *charge* that is used throughout the refinery in the production of many products, all of which are based on hydrocarbon compounds. Figure 11-1 shows the general sequence of operations in petroleum refining.

The most important products obtained from these processes are different forms of gases, oils, and subproducts. *Light ends,* the form of hydrocarbon with a lower boiling temperature than other hydrocarbon compounds, include various pure and mixed hydrocarbon products: fuel gas, ethane, propane, butane, and liquid petroleum gas. The heavier compounds can be grouped together under oils and include diesel fuel, fuel oil, lubricating oil, gasoline, and kerosene. The heaviest hydrocarbon compounds associated with the petrochemical refining process are asphalt, grease,

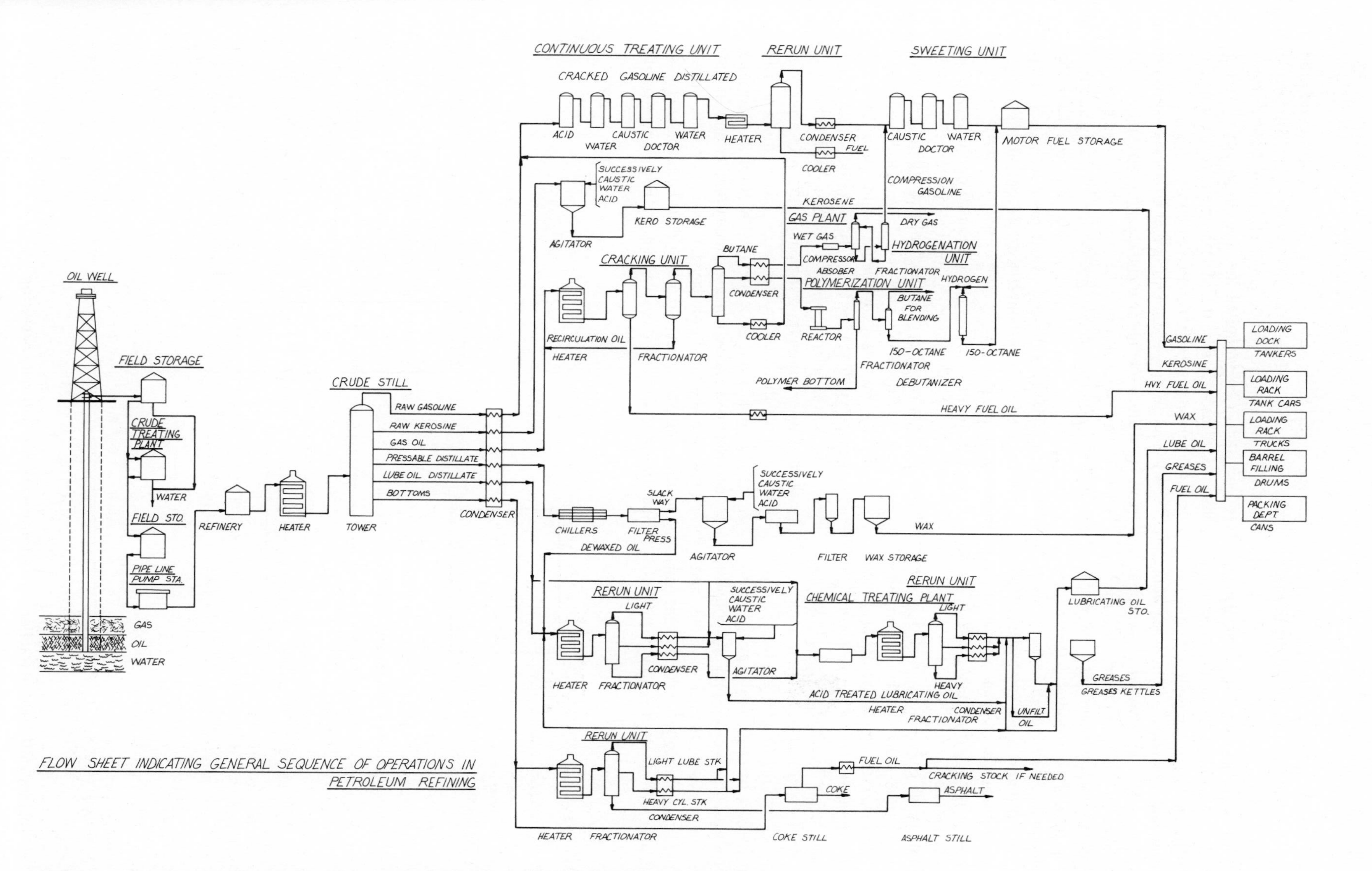

FIG. 11-1 Flow sheet indicating general sequence of operations in petroleum refining.

wax, and products that contain hydrocarbons as a primary ingredient, such as synthetic rubber, chemicals, medicines, drugs, and plastics.

The refining of crude oil and petroleum involves an assortment of processes, but the main one is the simple separation of hydrocarbons into light and heavy compounds by the use of fractional distillation and cracking. All hydrocarbons have a molecular structure containing hydrogen and carbon. Since crude oil ranges in consistency from very thin to almost solid, it is necessary to refine or fractionate the *stock*, or feed, in order to produce the required materials for society's needs. Gasoline and some of the lighter end products are far more in demand than the heavier hydrocarbon products. A variety of processes have been developed to separate the lighter and heavier compounds, such as the fractionization tower diagrammed in Fig. 11-2.

Hydrocarbon compounds which are composed of more hydrogen than carbon are lighter compounds and form the lighter end products. Unfortunately, most of the products that can be obtained by simple fractionation or distillation (separation of lower and higher hydrogen compounds) are heavy end products. Since fractional distillation does not meet the great demand for the lighter end products, many techniques have been created to *crack* the heavier hydrocarbons in order to produce light hydrocarbon compounds. Catalytic cracking, hydrogenation, thermal cracking, and several reactions are used to produce more of the light end products from the original crude stock. It is now possible to process crude oil in such a way as to create light end products in far more abundance than ever before. Because of this ability to create the desired quantity of light or heavy end products (depending on the needs of society), it is possible to adjust the amount of end products according to seasonal demand or other economic factors.

Besides the primary end products that are available through fractional distillation, thermal and catalytic cracking, hydrogenation, and other processes, an oil refinery also produces certain by-products, depending on the plant in question. Light gases that can be used in the refinery, petroleum coke, acid sludge, and many waste gases are just a few of the lower-quality by-products that are obtained by petrochemical refining.

Throughout this chapter, keep in mind that hydrocarbon compounds are the primary feed in all the processes going on in the various types of equipment and piping systems. A typical petrochemical complex or refinery consists of a maze of piping for moving hydrocarbons and for utilities and backup systems. There are many types of processing units associated with petrochemical refining: chemical plants, gasoline plants, ammonia plants, and others. In a majority of cases, more than one process takes place at the same site. A typical petrochemical refinery with all of its associated process sections may cover many square miles and include not only the plant and processing units, but also the storage facilities and vast array of piping that are associated with the transportation and movement of both raw material

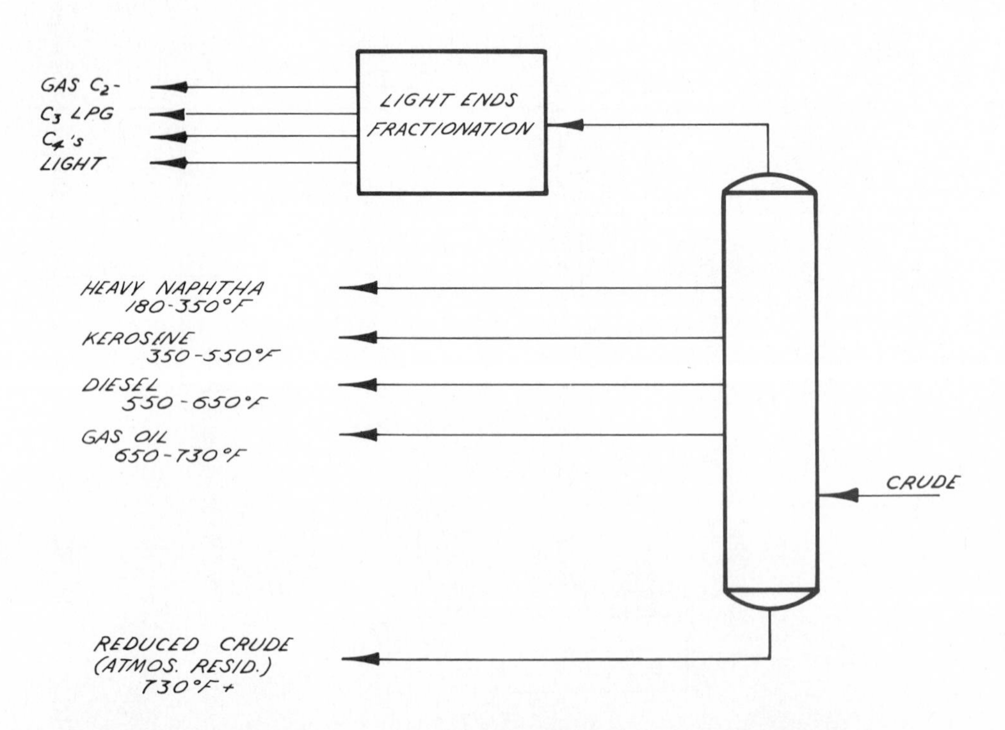

FIG. 11-2 Typical crude topping unit, used to fractionate hydrocarbons into naturally occurring light and heavy compounds.

and end products. By reviewing the many photographs of petrochemical installations throughout this book, students can familiarize themselves with the variations of complexes and piping systems. Figure 11-3 shows a typical petrochemical processing unit.

Petrochemical refineries are constructed to process numerous types and consistencies of hydrocarbon products in their crude form. The nature of the material determines the processing units. Installations that have access to natural gas along with crude oils from the oil-exporting nations of the world have a great range of plants and processing units at the same site—all adapted to process hydrocarbons into the many products associated with petrochemical refining.

Some types of crude oil have a higher sulfur content than others and may be in a semisolid state. Therefore equipment and transportation piping must meet the requirements of the raw material. Refineries that must process high-sulfur-content crude oil construct and design process units that can separate the sulfur from the oil. In this type of refinery, sulfur becomes a by-product. The thicker, almost solid types of crude oil require extensive heating before they can be pumped and moved through pipelines from the tanker or other shipping equipment to the refinery site. It is necessary to heat-trace most of these pipelines.

FIG. 11-3 View of a petrochemical complex showing a distillation column in the background with associated piping, platforms, and valves. In the foreground, a heat exchanger with a bolted blind-flange end enclosure rests on a concrete pier and is attached to welded piping. Notice the tangental flanged connections to heat exchange units on the right-hand side of the shell. (Courtesy Babcock & Wilcox, Tubular Products Div.)

A typical petrochemical installation is composed of more than the piping and equipment necessary for the production of gas and oil. Among the additional facilities that may be present at a refinery site or at their own separate installations are chemical plants which, in general, take hydrocarbon compounds that have been refined or semirefined and then mix, blend, change, and convert the molecular structure of the hydrocarbons along with numerous additives to produce the wide variety of chemicals that are available to both industry and society at large—medicines, plastics, drugs, fertilizers, insecticides, food additives. Another facility associated with petrochemical installations is the gasoline plant, which uses natural gas or liquid natural gas (instead of the heavier hydrocarbons associated with crude oil) to produce gasoline, propane, butane, methane, and olefins. Motor oil, asphalt, wax, grease, lubricating oil, and diesel fuel—the products of the heavier hydrocarbon compounds, which have a larger quantity of carbon atoms—are not manufactured in a gasoline plant.

The use of gasoline plants will become more of a problem in the future as natural gas supplies diminish, however. Although the discovery of oil is usually accompanied by a quantity of natural gas, it is often impossible to transfer the natural gas to a processing area because of the inaccessibility of the oil field, as in the case of the Alaskan fields. Natural gas is also found in areas with no substantial oil deposits (such as coal fields). Natural gas, though used sometimes in gasoline plants, is primarily a heating fuel for housing and buildings, especially in areas where it is abundant.

In the early stages of the petrochemical and oil industry, natural gas was considered almost a nuisance that made it harder and more dangerous to drill and pipe oil from the oil fields. Now natural gas is considered to be almost as important as oil, depending on the area of the world where it is located.

Ammonia plants and processing units may also be associated with the petrochemical refinery site or located separately. A typical petrochemical installation also has separate plants for the production of waxes, asphalt, tar, and many other products.

SAFETY AND HAZARDS

Apart from the diminishing supplies of natural gas and crude oil, another concern has been the environmental factors that have become exceedingly important in the last ten years, including the finding and mining of oil in environmentally sensitive areas such as the North Sea, Alaska Slope, and on the continental shelves offshore. These concerns, along with the hazards that have become readily apparent by the enormous oil spills which have taken place, have raised many questions over the use of oil and petrochemical products in such quantities that require excessive exploitation of the earth's environment and threaten the very quality of life itself. Many of the piping codes, standards, and requirements pertaining to the production, shipping, movement, and creation of petrochemical products have become necessary in order to monitor industry's drive to meet the demand of the modern industrial states. Many of the guidelines do not concern the piping drafter, though the petrochemical engineer and designer will, of necessity, be well versed in the creation, design, and construction of environmentally safe piping systems and their associated equipment and processing units.

The petrochemical refinery presents a wide range of other hazards and environmental problems as well. The typical installation requires piping and equipment that is able to transfer and process many types and quantities of products at much higher temperatures, pressures, and toxic levels than are normally associated with other piping systems. The following list describes the hazards and safety requirements commonly associated with petrochemical piping systems:

1. *Fire hazards:* Fire hazards include all the situations that may produce combustion in a petrochemical installation. Since all petrochemical products burn to some degree and their vapors produce combustible material, it is necessary to design systems that eliminate one of the elements necessary for combustion. The production of petrochemicals in an oil refinery is always accompanied by the presence of hydrocarbons and heat; therefore it is important to eliminate oxygen or air from the processes to keep the fire hazard to a minimum. Fires can also start by combining various chemicals and hydrocarbon compounds. In general, the problem of fire is far more prevalent in the production of the lighter hydrocarbon products—butane, ethane, methane, propane, LPG, fuel gas—because these compounds vaporize more easily than the heavier compounds such as crude oil, asphalt, tar, and oil. Moreover, the lighter hydrocarbon products present problems in the piping system and equipment in the form of leaks at flanges, valves, equipment connections, and nozzles. The heavier hydrocarbon compounds are much easier and safer to transport, although gasket materials reacting with hydrocarbon compounds have in some cases produced serious leakage problems.
2. *Disposal of waste:* Waste gases and waste products that may accumulate in drains or need to be discharged into fireboxes or flare stacks present difficulties along with the environmental problems associated with their disposal in a liquid state—not only hydrocarbon compounds but other toxic chemicals, gases, and acids associated with petrochemical products. All must be safely dealt with so they do not contaminate the surrounding environment. Since all chemical and hydrocarbon wastes must be broken down, a good portion of refinery design is devoted to this problem. Flare stacks burn off different waste gases, and furnace fireboxes are used in some cases to burn gases or solids that may present fire hazards.
3. *Pressure and temperature:* All petrochemical piping codes require the safe design and construction of processing equipment to ensure that pressure, temperature, and chemical reactions do not cause leakage or destructive problems that may destroy whole sections of a refinery. The use of steam in conjunction with hot oil creates expansion problems which may harm piping or equipment by causing excessive pressure surges and steam hammer.

These are just a few of many hazards that are prevalent in all petrochemical installations. History has shown that one of the most critical areas of a petrochemical complex is the storage area—it is there that most fires, explosions, and damage to property and life have taken place. Transportation of the end product, such as methane, ethane, propane, and natural gas, is especially hazardous also. Leaks and broken vessels have caused much damage to equipment used in transportation. In most cases, the hazards have been railroad accidents, explosions, and fires associated with moving the end product—hazards that are harder to control than those at the refinery.

EQUIPMENT FUNCTIONS

A wide variety of equipment is needed for the manufacturing operations associated with crude oil. These items must be specified and located early in the design sequence. The following list describes the basic functions of process equipment discussed in the ensuing paragraphs:*

1. Transportation and movement of line material
2. Storage of both raw and finished products
3. Heat generation for temperature requirements and auxiliary uses
4. Heat exchange units that transfer heat from one material to another
5. Chemical change or reaction for the production of various products
6. Combining or mixing of various products
7. Alteration of the state of the original material (including its particle size and consistency) and initial purification
8. Separation of the materials used for petrochemical processing

***All quoted passages in this chapter are courtesy of the American Engineering Model Society.**

Many of the different pieces of basic equipment that will be described are used primarily for one type of operation, although some are used for multiple operations and therefore span more than one category.

Transportation

Most of the information given here is essential to the petrochemical engineer and designer; the drafter must have at least a limited understanding of the equipment associated with the transportation of raw hydrocarbon compounds. A variety of equipment and piping is necessary—including all transportation piping for the movement of crude oil from the oil field to the processing units, railroad tank cars, trucks, supertankers, oceangoing freighters, and all pipeline components such as pumps and valves. Transportation of crude oil starts at its source—the drilling rig and oil field—and includes transferring the crude oil from the field to ships or vehicles, from the refineries to equipment, and to and from storage tanks. Piping transports vapors, liquids, gases, and in many cases solid or nearly solid mixtures.

The size and thickness of pipe varies dramatically, depending on whether the raw material is to be moved to a storage tank from the oil field or between different processing units. In general, piping systems associated with petrochemical products, transportation, and movement between equipment involve 1-in. to 48-in. (2.5 cm to 121 cm) diameter piping. The pressures, temperatures, chemical compositions, and reactions that take place in these transportation and movement systems determine the thickness and materials required for the pipeline. In general, welded piping systems are used more often than any other type, though it is necessary to flange many of the connections for valves, nozzles, and components in order to facilitate installation and maintenance.

FIG. 11-4 Oil pipeline being laid in the North African desert. A 40-in. block valve is being welded to the piping.

FIG. 11-5 A 48-in.-diameter oil pipeline being lowered into a trench in the Alaska wilderness. Far more equipment and a much higher level of technology are necessary to establish cross-country pipelines in adverse climates.

From Oil Well to Refinery In the oil fields, the raw material is usually stored, though in some cases it is immediately pumped through many miles of production flow lines such as those shown in Figs. 11-4 and 11-5. Natural gas and crude oil are the two products that are usually pumped from the original site. A series of storage tanks, tank farms, and pumping stations are required to move these raw materials. The location and environment determine much of the equipment and technology needed to transport the different hydrocarbon raw materials. In oil fields that are located in warmer climates or semidesert areas, such as the southwestern United States or the Mideast, as in Fig. 11-4, the level of technology needed for the design and construction of oil transportation lines is relatively simple in comparison to those shown in Fig. 11-5, where a large-diameter pipeline is being laid in a colder climate. In this case, the pipe must be heavily insulated, and the equipment and technology needed to construct a pipeline are far more complicated. Drilling and oil production from offshore rigs also present problems not associated with land-based oil wells. wells.

After the crude oil or natural gas has been piped through the transportation lines and pumping stations, it is either moved to a terminal or directed to a refinery. Figure 11-6

FIG. 11-6 Oil tanker being loaded in an Alaskan bay. The pipeline in Fig. 11-5 transports crude oil to an ice-free bay where tankers and supertankers load and carry the oil to refineries in the United States.

shows a tanker loading raw crude oil from the Alaska Slope which has been pumped hundreds of miles to this loading station. The tanker will transport the oil to one of many terminals in the United States for refineries that will do the processing. Designers of oil transportation systems such as that shown in Figs. 11-4, 11-5, and 11-6 and the associated terminals must take into account a vast array of physical conditions, such as seismic activity, tides, thermal movement, environmental protection, and the movement, construction, and laying of pipe over areas that are relatively inaccessible such as mountains and rivers. When oil production occurs in flat regions with warm climates, fewer problems are encountered between oil well and transportation terminal. The greatest production problem for the Mideast has been the movement of oil from storage facilities to refining facilities in other parts of the world. Oceangoing tankers and supertankers have been used to transport crude oil and have encountered many difficulties in spills, collisions, and other disasters on the high seas, and in the unloading and loading operations. For the movement of oil from regions like the Alaska Slope through mountains, rivers, and environmentally sensitive areas, the pipeline must be designed to withstand seismic disturbances. Moreover, the oil must be kept at an elevated temperature so it will not freeze—usually by providing heavy insulation and heating facilities at the source and at heating stations along the length of the pipeline. Special codes, requirements, and standards have been established to cover this system, since the piping code that applies to oil transportation piping, ASA B31.4, is limited to piping with metal temperatures under 200°F.

After the oil has been moved through pipelines to the tankers, it is shipped to the nations that do the majority of processing and consume most of the oil, such as the United States. Figure 11-7 shows an oil tanker docked at a refinery and unloading its crude oil cargo. Figure 11-8 shows the piping pier that connects the unloading and loading terminal to the tank farm and storage area. Many of the lines used for the movement of crude oil need to be steam or electrically traced to keep the oil in a liquid state. For high-sulfur-content oil and other oils that are almost solid in form, this arrangement is essential. The storage facilities shown in Fig. 11-8 are adjacent to a large petrochemical refinery with a number of different processing units.

The refinery is the end of the line for the raw material. At the refinery the transportation of the raw and semi-refined hydrocarbons to and from processing units—and the movement of finished products to the refinery terminals or shipping stations—also require piping and associated equipment.

As readily available supplies of natural gas and oil diminish, the necessity of piping systems to convey crude oil and natural gas from inaccessible areas, such as offshore drilling rigs in the North Sea, becomes far more important. The days of simple transportation lines over flat land in warm climates are really over.

FIG. 11-7 Oil tanker unloading crude oil at refinery.

FIG. 11-8 Pumped from tankers via long piers, oil is transported through heated pipelines to storage tank farms like this one. Because oil comes in a variety of viscosities and is not always in a highly fluid state, piping must be steam or electrically traced to keep the crude oil flowing.

Codes and Standards Oil transportation piping is covered under Section 4 of the Oil Transportation Piping Code, ASA B31.4 1959, a standard that applies to design, materials, fabrication, inspection, and testing of pipeline transportation systems dealing with liquid petroleum products. This code covers liquid petroleum piping used to convey crude oil, finished products, LPG, and so on between terminals, tanks, refineries, gasoline plants, and delivery (consumer) points. All piping and associated components such as gaskets, flanges, joining methods, fittings, valves, relief mechanisms, dampeners, and regulators are covered under this code. However it does not cover crude oil production tanks and the transportation lines from the oil wells to the storage facility. Moreover, it does not cover instrumentation and auxiliary piping, including pressure vessels, fabrication methods, and auxiliary components. ASA B31.3 covers all areas associated with the fabrication, erection, construction, assembly, and components required for oil refinery piping systems. This Oil Refinery Code is more inclusive than the one for oil transportation piping. It is important to consult both when determining which is applicable for a certain design situation. Other ASME codes cover the various pieces of equipment, storage tanks, and vessels not included in the Oil Refinery Code. Oil refinery piping is covered in more detail in the section on process and auxiliary equipment.

Oil Transportation Piping Code B31.4 covers a variety of areas that should be familiar to the petrochemical designer, engineer, or drafter. The general guidelines follow:

1. The Oil Transportation Code covers piping between -20°F and 250°F.
2. External loading such as seismic activity, thermal expansion, vibration, wind, environmental considerations such as ice and snow, and the total weight of the piping including the components and the material content must be considered in the design of any oil transportation system.
3. Any externally caused damage that may be encountered in the construction, implementation, and use of the transportation system must be considered—vibration caused by equipment or surrounding environment, overall weight, thermal forces, and environmental problems such as places where pipe must be buried or elevated over rivers, ridges, roadways, unstable ground, self-supported spans.
4. Erosion and corrosion must be taken into account if the soil condition or surrounding environment pose problems.
5. Design and construction of oil transportation and piping must meet rigid standards and a variety of protective procedures and designs may be necessary—including heavy insulation of piping in cold areas, steam tracing, special thermal expansion joints, wall thickness increasing for heavy stress areas or highly corrosive situations, and concrete encasing to protect the piping in high-stress areas.

Storage

Petrochemical products, both the raw form and the finished product, are stored in vessels, tanks, and enclosures which are usually cylindrical in shape. Vertical cylinders are used for raw products (natural gas and crude oil) and spheres for special high-temperature/high-pressure or volatile substances. *Tank farms* are areas that contain only large-diameter tanks for the storage of petrochemical products or raw crude oil. Figure 11-8 shows a tank farm used exclusively for the storage of crude oil and natural gas. Many storage tanks are of extremely large diameter and can store great quantities of the raw hydrocarbon material before processing. Tank farms and storage areas for refined petrochemical products are situated well away from processing areas of the refinery.

Heat Generation

The generation of heat is an extremely important aspect of all petrochemical processing plants. Figure 11-9 shows a fired heater with accompanying stack. Fired heaters, boilers, stills, and other heat generation equipment are an integral part of the chemical processing procedure, most of which entails the heating of the charge stock to a high temperature in order to fractionate or produce the required end products. Figure 11-10 shows an ammonia plant; at the right side of the photograph is a large heater unit. Stills are also considered heat generation equipment.

FIG. 11-9 Reformer and LPG distillation plant with a variety of heaters and towers. The building in the foreground is a fired heater—a large furnace with multiple burners.

Heat Transfer

Heat transfer equipment includes economizers, condensers, and heat exchangers. Heat exchangers are primarily unfired vessels which transfer heat from one material to another. Condensers also function as heat transfer units, since the condensing process transfers much of the latent heat of a vapor to another material in order to condense.

Figure 11-11 shows a whole section and series of heat exchange units; Fig. 11-12 shows a heat exchange economizer. Figure 9-18 is a detailed fabrication drawing of a heat exchange unit showing the external and internal parts. Most heat exchangers such as those in Fig. 9-18 are *shell and tube exchangers* consisting of a tube bundle (in this case copper) encased in a large-diameter pipe or shell. By conveying fluids at different temperatures through the shell and the coil arrangement, it is possible to exchange heat between the two materials without contaminating either one by mixing. There are many photographs and drawings throughout the text showing heat exchangers and other heat transfer equipment.

FIG. 11-10 Ammonia plant for production of NH_3, which is used to manufacture fertilizers and chemicals.

FIG. 11-11 A petrochemical plant (hydrogen reformer). Charge heaters are located on the left; the hydrogen reactor and stabilizer column are on the right.

FIG. 11-12 Close-up view of vessels and columns, showing heavily insulated pipelines with aluminum shielding. The vessels in this petrochemical unit are also insulated.

Chemical Change

Chemical reaction takes place primarily in *reactors*—pieces of equipment that house the raw material (or process material) and a catalyst. Reactors are used to facilitate the chemical reaction which changes the molecular structure of the stock. Chemical reactions are created in a variety of reactors such as those shown in Fig. 11-10 of an ammonia plant, which are vertical vessels with domed heads; these reactors include such equipment as autoclaves and furnaces to provide the conditions for the molecular change. Since the facilitation of chemical reaction requires the addition or subtraction of heat, many of the pieces of equipment associated with chemical change deal with great thermal stress. Therefore the design of these vessels and devices must take into account not only chemical but also thermal reactions to the materials of construction and high pressure levels in most cases.

Combining and Mixing

The mixing of materials may include the combination of liquid, solid, and gas and all variations that are possible. Mixing is also possible in many different heating and cooling units. Besides the use of heating and cooling units to mix line contents, special equipment has been developed: proportioning pumps and valves, mixers, blenders, and agitators.

Alteration

Equipment to change the consistency or particle size of materials is utilized primarily in food processing units and sometimes chemical plants for the production of drugs and medicines. Its functions include the creation of both larger and smaller particles for different materials by means of crushers, rollers, hammer mills, and agglomerators. This type of equipment is also used for the reduction of particle size for coal units and the production of oil from slate and oil shale, though it is not commonly associated with petrochemical installations in general.

Separation

Equipment for the separation of charge stock and other materials used in petrochemical processing is of primary importance. No chemical change occurs in this type of equipment. The primary function of separation equipment is to create lighter products from the heavier hydrocarbon compounds in crude feedstock. This type of equipment includes distillation and fractionation columns and towers, cyclones, defoamers, dryers, evaporators, filters, flotation tanks, scrubbers, settling tanks, centrifuges, and strippers. The most important are the distillation and fractionation columns, which enable the separation by temperature of lighter and heavier products (Fig. 11-2). Distillation does not chemically change the product, but only separates the components. Separation equipment also works with solid, liquid, and gaseous materials, slurries, and combinations. Many of these pieces of equipment are multifunctional—the distillation column involves a still with heat generation capabilities. Other equipment separates materials by movement or settling.

PROCESS AND AUXILIARY EQUIPMENT

This section deals with the description, function, and use of process and auxiliary equipment commonly found in a petrochemical installation.

Pipe

Piping, though not generally considered equipment, can be looked upon as a closed container or system. The function, use, and selection of pipe and components are critical elements in the total design of a petrochemical complex. The designer must consider all the variables present in the selection of piping materials, sizes, and related components. The properties of the pipe material, the effects of corrosion and erosion on the material from the line substance, resistance to environmentally caused corrosion and erosion, flammability, resistance to seismic and thermal shock, the efficient support of the piping, the selection and joining of piping components—all are questions that need to be answered by the design engineer or petrochemical engineer. The design engineer is responsible for the selection of piping and materials depending on all the environmental, economic, and design needs of the system in question.

Materials The selection of materials is based on many considerations:

1. Type of fluid conveyed
2. Operating temperature and pressure
3. Amount of fluid or gases to be conveyed
4. Effects of corrosion and erosion
5. Thermal and seismic shock
6. Flammability
7. Weldability and joinability of the pipe and related components

As has been stated, the material selected for the piping must be determined by the process engineer or design engineer. Materials require testing for the conditions under which they will convey liquids or gases. Several types of steel—carbon and low-alloy steels, ferritic chromium stainless steels, austenitic chromium nickel stainless steels, and others—require impact tests before being certified for use in certain petrochemical installations. Other testing is also required for piping components and pipe depending on the design requirements. See Chapter 1 for more detailed information on the materials that are available for piping.

Though steel pipe is most prevalent in a petrochemical installation, other kinds are also used, including copper, aluminum, copper alloys, numerous types of iron, plastic, glass, rubber, carbon, and ceramic materials.

Material Limitations Steel has limitations that need to be considered when specifying it as the primary material for pipe used at a refinery. For piping systems in operation in excess of 750°F for long periods of time, the possibility of graphitization of the carbon must be taken into account—especially for many types of carbon-based steel such as plain carbon steel, carbon manganese alloy, carbon silicon steel, and plain nickel alloy. Other carbon steels such as carbon molybdenum, manganese chromium vanadium, chromium vanadium, and carbon molybdenum vanadium are also affected by graphitization when used in piping systems that operate at 850°F over extended periods of time. Any extremely hazardous area of a petrochemical installation which involves high-temperature, high-pressure, and chemical change must be considered when one is selecting the piping material.

There are also many limitations on the use of cast iron. For flammable, hydrocarbon, high-pressure/high-temperature service, cast iron must generally be limited to temperatures below 300°F and operating pressures of 400 psi maximum. For both underground service and aboveground service, piping is limited to 150 psi if it is near process equipment. The use of malleable iron is restricted to a maximum of 650°F for pressurized parts; when highly flammable materials are conveyed, the design restrictions are 350°F and 400 psi.

Cast, wrought, malleable, and nodular iron are usually restricted to nonflammable, nontoxic service when used for refinery piping or components. Other materials such as glass, plastics, rubber, and ceramics are used in various situations depending on the design capabilities of the material in question, and they should be studied carefully before being used in any service in a petrochemical plant.

Copper alloys are restricted to nonflammable service because of low melting temperatures. Copper, when used, is usually found in the economizer or heat exchange units for the internal coils, though the design temperature for this type of heat exchanger must be taken into account when specifying copper (Fig. 9-18).

Both new and used piping can be specified for use in a petrochemical installation and, depending on the materials, they must meet certain specifications set down by the ASTM and other standards societies. External/internal linings and coatings of many types used on steel pipe are limited in their ability to withstand corrosion and erosion; pipe is not to be considered strengthened by a coating or lining.

Joining Pipe and Components Butt welding is the primary method used to join petrochemical piping systems. The piping code restricts many joints, depending on the service. For example, threaded joints are restricted to piping 4 in. (10 cm) and smaller, depending on pressure, temperature, and the material to be conveyed. Many other joints–including sleeve and coupled joints, bell and spigot, socket weld, and flanged joints–are used throughout the refinery for certain services, usually low-temperature/low-pressure, noncaustic material services, though many flanges and flange joints are necessary throughout the petrochemical complex. Chapter 2 gives detailed explanation of the use of welded joints for piping systems; Chapter 3 discusses methods of joining pipe.

Changing Direction As can be seen in the photographs in this chapter, piping systems for petrochemical refineries require a considerable assortment of change-of-direction runs and fittings. The codes governing refinery piping and fabrication of components need to be consulted when determining the change-of-direction and joining methods to be used for the construction and fabrication of components. See Chapter 9 for a detailed explanation of piping fabrications and bending methods.

The typical refinery requires both shop-constructed bends and standard elbows that can be catalog-ordered. The bending procedure, whether cold or hot, must be taken into account to avoid undue distortions or thickness variations that may affect the safety and operation of the system. Cold bending is restricted to bends which do not distort the diameter more than 2½ percent, and there are other restrictions on the size of the cold bend that can be used for certain services, though the modern shop has within its means new procedures that make cold bending quite compatible with code restrictions. Miter bends also have certain restrictions, and it is important to consult the piping code. Figure 11-3 shows a miter bend for vapor line connected to the top of the column in the background.

Flanges and associated flange faces must correspond to ASA B16.5 and MSS SP-44. A detailed explanation of flanges and flange facings is provided in Chapter 3. In general, all the different types of flanges–including weld neck, slip on, lapped, companion, reducing, blind, and screwed–are used in a refinery, depending on their size and pressure/temperature rating. Many end closures for equipment have blind flanges. Figure 11-3 shows a heat exchange unit with a bolted blind flange for an end enclosure; this arrangement can also be seen in Figs. 11-12 and 11-13. Besides blind flanges, spherical, conical, and ellipsoidal heads are used for vessel ends.

Gaskets also present problems when used on flanging in petrochemical installations because of the high temperatures and pressures and caustic materials that are conveyed through the piping system. They must be able to handle the high pressure and temperature and not react chemically with the line substance. Ring-type joints, male and female, tongue and groove, and raised face flanges, all with accompanying gasketing, are found on pipe connections depending on design needs. Chapter 3 discusses gasketing and bolting for flanged joints.

Apart from the pipe material, size is determined according to temperature, pressure, and flow requirements by the petrochemical engineer and design engineer. Both pressure and temperature affect the thickness required for the piping run. The petroleum refinery code ASA B31.3 deals with the

FIG. 11-13 Chemical plant. Notice the heat exchangers and horizontal vessels on the left-hand side and on the far right. The installation in the center is composed of desalter drums and columns in the background.

majority of design situations encountered, so that piping and size can be selected for safe operation of a process petrochemical installation. For the calculation of the minimum wall thickness, the formula is

$$t = M\left(\frac{P \times D}{2s} + C\right)$$

For this calculation, t equals the minimum accepted wall thickness for the pipe or fitting and includes the standard tolerance for manufacturing of pipe (12.5 percent). The other variables are

P = psig (pressure per square inch gauge, the internal design service pressure requirement plus 10 percent)

s = unit tensile stress or allowable stress in psi (determined by the operating temperature)

M = manufacturer's tolerance (in the case of steel pipe, usually 1.125)

D = OD of pipe (in inches)

C = corrosion allowance

The corrosion allowance depends entirely on material and manufacturer's suggested corrosion allowance for materials and design situations. In general, 0.125 in. can be used for carbon steel pipes in petrochemical installations, though this depends entirely on type of fluid, temperature, pressure, and flow requirements. Besides this formula, the student should consult Chapter 1 for the calculation of pipe thicknesses and Chapter 12 for formulas involving steam piping.

Condensers*

Condensers are heat exchange units used to condense vapor by cooling. A majority of the condensers used in a petrochemical installation are vertically or horizontally mounted shell and tube exchangers that have a large nozzle for the entry of the vapor to reduce pressure loss during the operation. According to the AEMS:

"The condenser is a piece of equipment that is used to reduce the temperature of the overhead vapor to a point where it will condense to a liquid. The vapor is piped to a chamber where it is cooled by a circulating cooling medium. Sometimes another liquid steam that must subsequently be heated is used as a cooling medium."

Condensers are also discussed in Chapters 10, 12, and 13.

Heat Exchangers

Heat exchangers are primarily horizontal vessels which enable one fluid or material to exchange its heat with another without coming in contact with it. A wide variety of sizes, shapes, and types are available, including the typical shell and tube type shown in Figs. 9-18 and 11-12. An air cooler type uses air to cool the process material as it passes over the coiled tubes that spiral within the vessel. Double-pipe exchangers, the simplest form of heat exchange units, have two different sizes of pipes, one inside the other, allowing the heat from one material to pass through to the material in another pipe. The two fluids or gases do not mix in any of these units.

Heat Transfer

"There are three mechanisms of heat transfer: conduction, convection, and radiation. Conduction is the primary method used in industrial heat exchange equipment. Conduction is the transfer of heat through a body, or bodies, in physical contact with each other without appreciable motion of the particles of the bodies. In heat exchangers, this is usually done between two liquids separated by a metal plate or tube. The effectiveness of conducted heat transfer depends on the temperature differential between the fluids and the heat conductivity of the separating material."

Thermal Conduction

"An exchanger is used to transfer heat from one fluid to another, either for heating or cooling. The rate of heat transfer depends on the amount of surface exposed, and the velocity and pressure of fluid passing through the exchanger, the turbulence of the fluids, triple wall effect, frequency of passes through the exchanger by each fluid, and the material of construction."

When designing and fabricating units, the following specifications must be considered:

1. All heat exchangers require pressure, safety, and relieving valves or other mechanisms to protect against internal failure.
2. Design exchangers so that the hotter material travels through the tube in order to limit heat loss through the exchanger vessel.
3. The more caustic, corrosive, and chemically active material should be run through the tubing instead of the body of the exchanger vessel.

*All quoted passages in this chapter are courtesy of the American Engineering Model Society.

4. When using a heat exchanger for cooling, the cooling fluid should be run through the tube instead of the vessel body.
5. Design exchangers to permit a countercurrent flow pattern between the two materials in order to achieve the best heat transfer.
6. Heat exchangers using steam must be designed to permit the gathering and extraction of condensate.
7. Insulate the heat exchanger shell for economy. Insulation also reduces condensate formation.
8. For heat exchangers using steam as the heating material, it is good economics to design the exchanger to pass the material to be heated through the coil and the steam through the shell of the exchanger. This practice allows for easier condensate extraction.

Reboilers Reboilers are a type of heat exchanger wherein the liquid contained in the bottom of the tower is continually boiled. The heating medium may be steam or hot process liquid which is run through the coil (tube) side of the reboiler while the crude stock is circulated in the shell side as shown in Fig. 11-14. The reboiling process provides the heat necessary for distillation of the stock.

The kettle type reboiler, shown in Fig. 11-15, has an enlarged portion of shell, which provides a large surface area where vapor can escape from the reheated charge stock.

A thermosyphon reboiler (vertical type) is attached directly to the tower, usually on the side opposite from the pipe rack. The natural movement of the heated stream circulates the crude stock between the reboiler and the tower, keeping the stream at a predetermined elevated temperature.

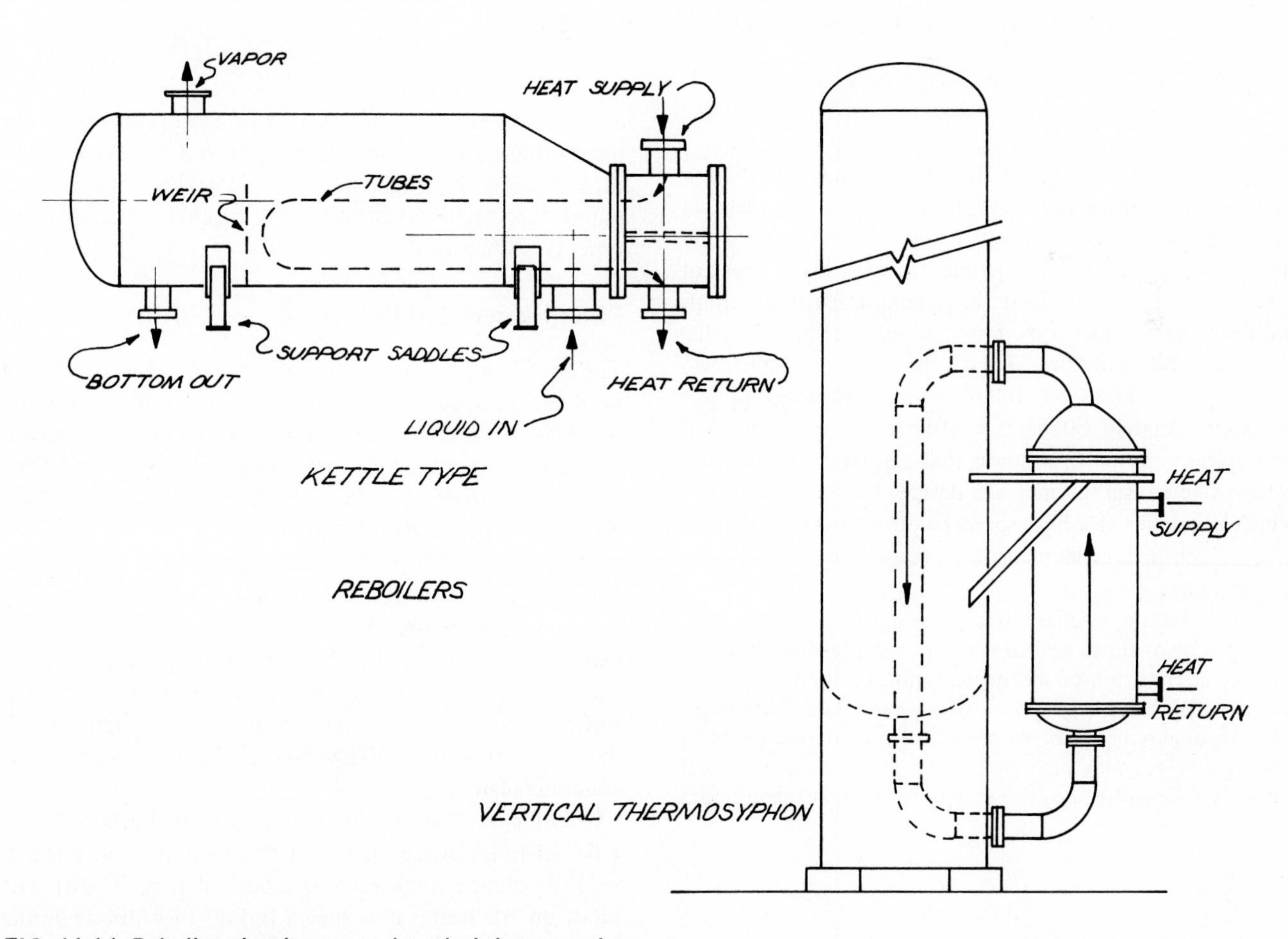

FIG. 11-14 Reboilers, kettle type and vertical thermosyphon type.

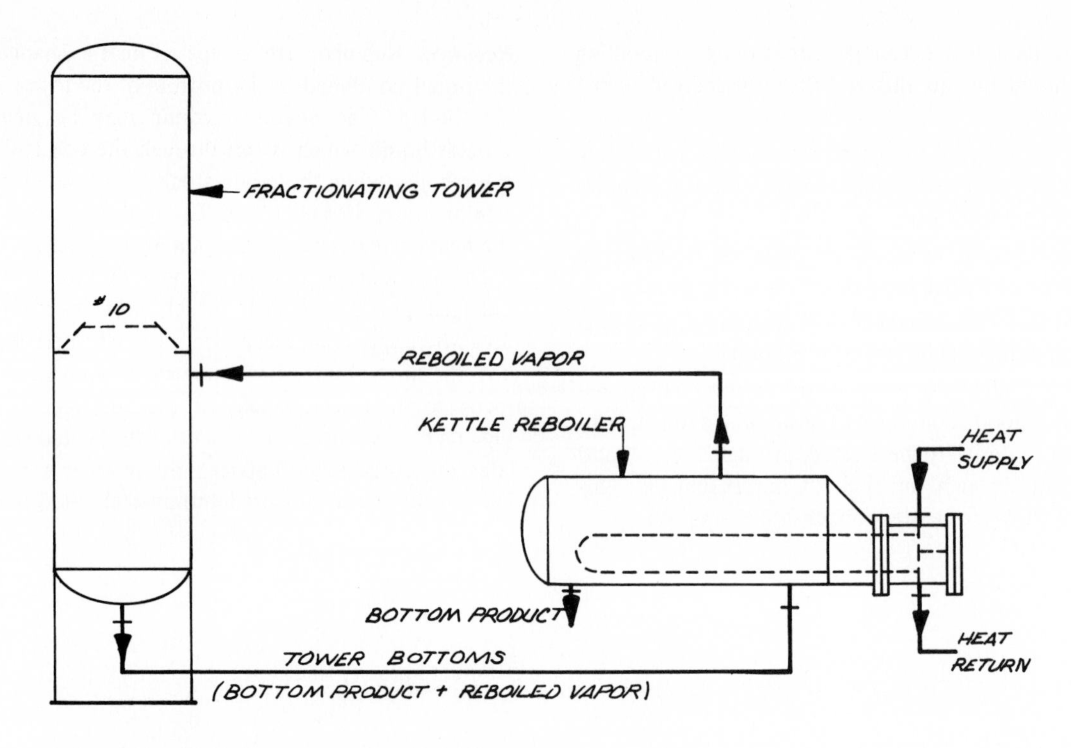

FIG. 11-15 Kettle reboiler to fractionating tower flow pattern.

Pumps

Pumps are used throughout the petrochemical installation to increase pressure and move the fluid from one piece of equipment to another. Pumps increase the pressure of the line in order to circulate the line material. Many types of pump and sizes are available. In general, centrifugal pumps are the most common type found in petrochemical installations and oil refineries. They can be divided into two basic types: turbines and volute pumps. Centrifugal pumps are approximately 80 percent efficient in operation and also come with multiple stages that can be connected in a series. Centrifugal pumps are designed with an impeller which is rotated at a high speed to produce the centrifugal force which in turn increases the pressure on an outlet side of the pump.

"By definition, a pump is a mechanical device for producing flow of fluids against resistance or pressure either by the communication of mechanical motion or by direct application of the stored power of another fluid. Pumps are used throughout a plant to move fluids from one point to another, overcoming pipe friction resistance and/or vertical lift resistance and/or positive pressures found at the terminal point."

Compressors

Compressors increase the pressure of gases or vapors whereas pumps work with liquids. Many instruments in the petrochemical installation function with air operators, as do many of the remote-controlled valves that can be operated through either hydraulic or air cylinders. Compressors, like pumps, are available in both centrifugal and reciprocating types (see Chapter 10).

Fired Heaters and Boilers

The fired heater is one of the basic pieces of equipment in a petrochemical complex; without it, most processes could not take place. Fired heaters are used to heat the charges that are processed in the petrochemical installation. They are really furnaces in which the heating process can take place in the most efficient manner (Fig. 11-16). The fired heater in Fig. 11-9 is of the horizontal type which has a rectangular shape with a series of burners. The stack, which is a part of every fired heater, is nothing more than a pipe extending well above the plant to carry off the waste gases and fumes from the heater. Fired heaters can also be of the vertical type, which are cylindrical in form. Figure 11-10 shows a series of fired heaters on the right side of the ammonia plant.

Whether vertical or horizontal, the fired heater has a series of tubes inside the heater box to convey the line material or charge stock such as crude oil (Fig. 11-16). The tubes on the heater unit shown in Fig. 11-9 run along the bottom of the heater and on the sides. The typical box-type rectangular, horizontal heater is used for large, heavy-duty

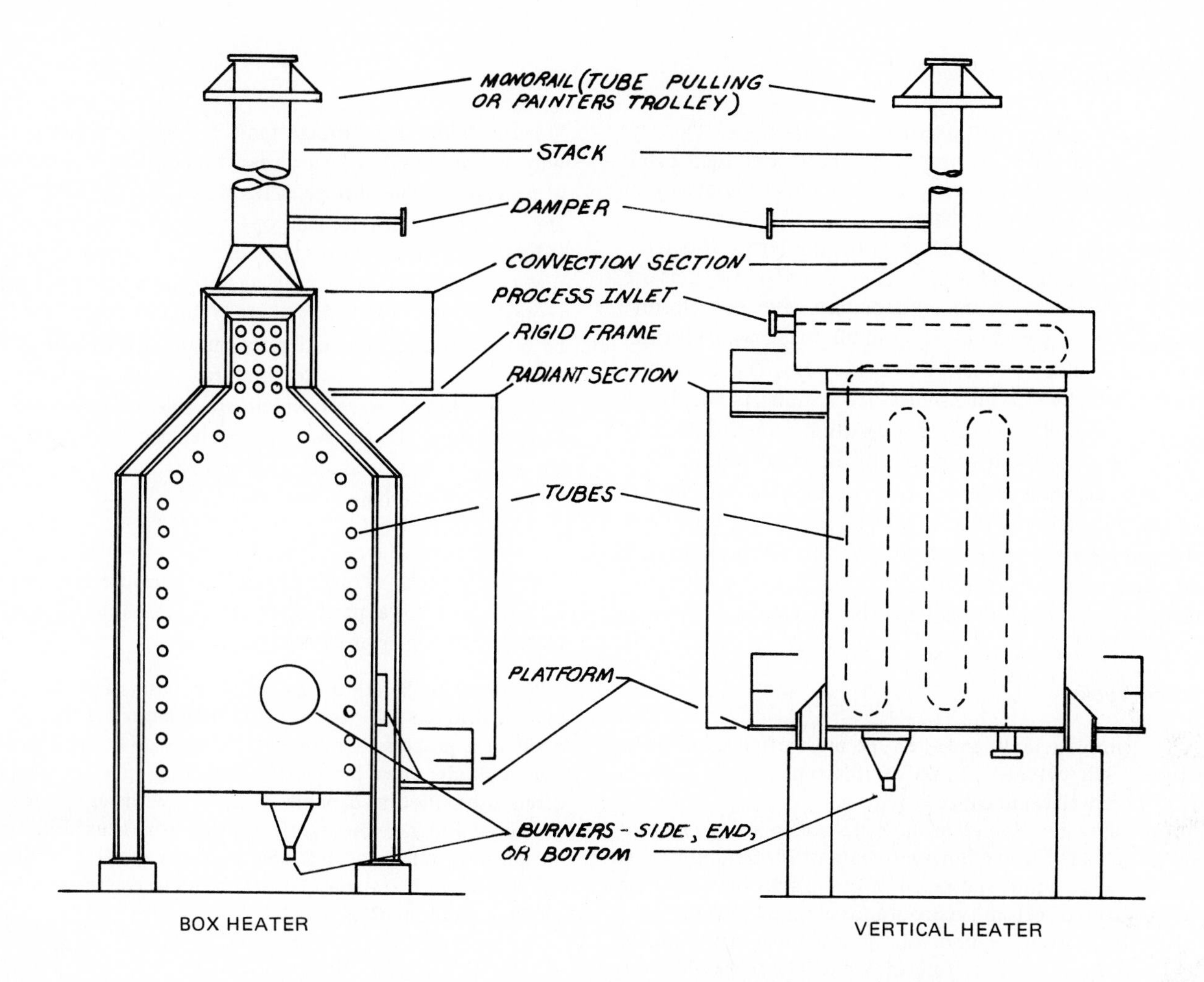

FIG. 11-16 Fired heaters: the vertical heater is a cylindrical steel fire box lined with bricks for insulation. Vertical "U" tubes suspended around the inside walls carry the process charge to be heated. The box type heater is a rectangular brick-lined steel structure with the process "U" tubes on a horizontal plane inside the walls. This type can be modified to produce superheated steam.

service in comparison to the small vertical fired heater, which is used primarily for special light service. Both vertical and horizontal heaters are made up of radiant and convection areas. Tubes receive heat directly from the burner in the radiant section, and the convection section receives heat from the tubes. Vertical heaters are basically large-diameter pipes elevated from ground level to permit access to the bottom. On both types, a stack with a damper for exhaust control is self-supporting. Ladders and platforming are provided for control and maintenance.

Boilers, which are similar to fired heaters, are discussed in Chapters 10, 12, and 13. Boilers are used for the production of steam for plant services, utilities, or process equipment. Some fired heaters can also be used to generate steam in the convection section. Both fired heaters and boilers are heat exchange units which have a firebox or combustion chamber to generate heat which is then transferred from the burning substance or fuel to the line material, such as water or crude oil. Fired heaters use gas or oil for fuel.

Tanks and Vessels*

This section considers only tanks, containers, and drums, which are used primarily as storage facilities, and horizontal vessels such as drums and pressure tanks. We will not discuss towers, columns, or stills, though they are referred to as vessels also.

Vessels and tanks can be divided into two separate categories: vertical and horizontal. According to the AEMS:

"Tanks of one kind or another are the basic process vessels. They are used for the storage of raw materials, intermediate, and finished products. Suitably modified, they are used as dissolvers, precipitators, reactors, fermenters, storage chambers, etc.

*Most of this section is based on information provided by the American Engineering Model Society.

Tanks are constructed in many sizes and shapes. The two most common forms are vertical and horizontal right cylinders with suitable end closures. Pressure tanks are fabricated as spheres or portions of spheres."

Students should be acquainted with the terms associated with tanks, vessels, and drums. The sides of the vessels are usually referred to as the *shell* or *wall.* The end closure, which can be a large blind flange or elliptical, semielliptical, or conical in shape, is referred to as the *head*. The horizontal vessel is usually set on saddles–steel supports which are cast or anchored to concrete supporting columns. *Nozzles* permit the attachment of piping to the vessel with a minimum of leakage. Nozzles are covered in detail later in this chapter. Insulation is another integral part of the vessel and may completely encase it as in Fig. 11-12. The following descriptions of the types and functions of vessels are by no means inclusive of all the varieties that may be encountered.

Vertical Tanks

"The simplest tank, and the type most often used, is the vertical right cylinder. Tanks of this type can be built to virtually any size and capacity required.

Tanks used for storage in the open are covered. The roof may be self-supporting; however, the roof is often supported by a central column and radial rafters. Tanks built at grade are of the flat bottom type; outdoor elevated tanks may be either flat bottom, supported by a suitable structure, or they may be suspended from the sides. The suspended bottom may be hemispherical, hemiellipsoidal, conispherical, or conical. The outdoor flat bottom tank, built at grade, may rest on a foundation of sand, or both sand and a curbing of concrete slightly larger than the tank diameter. Indoor storage tanks are generally flat bottom. Pressurized tanks usually have dished heads."

Horizontal Tanks

"Horizontal tanks are usually cylindrical in shape and have closed ends. Head shapes are the same as those used on vertical tanks. They are usually supported from below by saddles fastened to the foundation or floor. Occasionally, they are supported by slings hung from an overhead structure. Normally, the length of the horizontal tank should not be more than 5 times its diameter."

Horizontal tanks include overhead accumulators or reflex accumulators which are used to gather the products cut from the top of a fractionating column. One kind of vessel is a drum with a set of baffles as internals. In this drum, a mixture of gas and liquid passes through a chamber that separates the two. The gas-liquid mixture collides with the baffles and the impact frees the gases, allowing them to gather in the upper part of the chamber to be drawn off. All heavier substances are then accumulated and drawn from the bottom of the drum.

Storage Tanks The storage tank is a vessel which is used only for holding raw or processed materials–it is not used in processing. The storage tanks shown in Fig. 11-8, which are cylindrical with conical tops, are used to hold crude oil. (Chapter 12 shows two types of storage tanks that are a normal part of a fossil fuel steam generation plant.) The typical storage tank is anywhere from 150 to 200 ft (45 m to 60 m) in diameter and rests on a concrete foundation. Other tanks are also used for storage, such as spheres and thick horizontal vessels which are able to withstand the high pressures and temperatures which accompany the storage of light end products such as LNG, ammonia, butane, propane, and methane.

Pressure Vessels Pressure vessels or pressure tanks are an integral part of a petrochemical installation and can be used for storage or process of materials.

"Pressure tanks are used to store volatile materials. They are constructed to withstand the maximum pressure developed in a process. A number of shapes other than the upright cylinder have been developed. Full spherical shape is often used that is capable of safely withstanding higher pressures. For gasoline, the Hoton-spheroid is used. Ellipsoidal pressure tanks are available in large capacities. Cylindrical shapes with outward dished heads serve for storage at medium and low pressures."

Multisphere Tanks

"For storage gases and volumes at pressures that would require prohibitive thickness in a single sphere, a multisphere tank that resembles a honeycomb can be used."

Multicylinder Tanks

"A multicylinder tank consists of cylindrical shell segments and internal diaphragms, and with heads which may be partial multispheres."

Construction Materials

1. *Steel tanks:* "The great majority of tanks are made of steel because steel tanks are strong, comparatively light, relatively indestructible, and have an advantage of cost. The modern method of construction is by welding. The sides are welded to the base plate, which extends somewhat beyond the diameter of the tank, and the weld is made on both inner and outer surfaces. The curved sections for the cylinder are similarly welded together. Stainless steel tanks are also used extensively for process purposes."
2. *Wooden tanks:* "Wood is the construction material of choice for tanks in some of the chemical and alloy industries, because it is not attacked by many solutions, such as dilute hydrochloric acid, lactic acid, acetic acid, and salt solutions that do attack steel."
3. *Lined tanks:* "In some process industries, steel tanks coated on the inside with a material resistant to the process are used to obtain cost advantages. These linings include rubber, various plastics, lead, glass, and masonry tile or refractories."

Tank Covers and Vents

"Closed tanks are usually vented. If they contain a volatile liquid, they must be vented by a two-way valve, so that the vapors which form during the heat of the day may escape, preventing a rise in pressure, and so air may enter during the night, when the reduction in pressure develops.

There are other ways to avoid the losses resulting in this diurnal fluctuation in temperature. For the storage of volatile products, steel tanks with pontoon roofs are often used. Pontoon roof fits the tank with a small clearing, and has enough buoyance to float on the surface of the liquid; it rises and falls with it. At no time is there any vapor space which might undergo expansion or contraction. Another means of avoiding losses from temperature fluctuations is the use of a "graver" with an expansion roof, in which the roof is made in the form of a bell fitting in a liquid seal."

Columns and Towers

Columns and towers are in fact versions of vessels, though they are considered separately in this section. This group of equipment is the most essential aspect of the refinery system. Figures 11-3, 11-9, 11-10, 11-11, 11-12, and 11-17 all show variations of columns, towers, distillation units, and reactors. In general, towers and columns are divided into three separate functions: distillation, fractionation, and chemical operations. *Stills* are defined as cylindrical closed chambers where the application of heat to the charge stock changes from a liquid to a vapor, with provision for recondensation of the vapor in another vessel. This can be done with or without chemical decomposition of the charge stock. By conveying the vapor through a cooling chamber, it is condensed (liquefied) and channeled to another part of the refinery for further processing.

Columns and towers are stills which increase the degree of separation that can be obtained during the distillation of crude oil. The tower is really nothing more than an elongated still. Fractionation towers are used for light end products which require precise work and can produce the required substance according to rigid distillation specifications, such as those that are necessary for the production of propane, gasoline, and naphtha. *Bubble towers* separate by partial condensation and divide vapors from the still into liquid portions. Bubble towers do not have the extent of operations associated with the fractionation tower. Fractionation columns (towers) have the ability to take cuts at various levels, enabling the removal of many different distillants at the same time (Fig. 11-2). Towers are large cylindrical vessels that may have baffles or plates (trays spaced inside the shell to promote mixing of the downward flow of liquid with the upward flow of the vapors.

FIG. 11-17 Reformer and LPG distillation unit. This view of the petrochemical complex is the opposite of that shown in Fig. 11-9. In the background the stack rises from the top of the burner shed, which is shown close-up in Fig. 11-9. Note the piping racks, which are usually multilevel in a chemical plant. This plant is laid out with good access to all pieces of equipment and components.

Circular platforms, an integral part of a column or tower, are secured to the sides of the vessel, affording the service and operational personnel access to important portions of the tower. Manways are large-diameter nozzles with hinged blind flanges providing access to the interior. Columns are supported on *skirts*—large diameter structures welded to the bottom of the column which elevate it above grade and permit access to the bottom. The skirt is connected to the tower and rests on an octagonal pedestal (concrete foundation) which distributes the tower's weight evenly.

PROCESSES*

Distillation

According to the AEMS:

"Distillation is a separation process based on the relative volatility of the materials to be separated and on a change in phase in part of the original mixture. The simplest example is the vaporization by heating of a volatile component of a liquid mixture, leaving a relatively nonvolatile component as a liquid residue. Distillation may be used for the purpose of obtaining a vaporizable portion (overhead), a valuable nonvolatile portion (bottoms or residue), or both. In alcohol distillation, the volatile alcohol is the valuable product, and the residues are discarded. In petroleum refining, crude oil is distilled into overhead, side draw cuts, and bottoms, all of which are valuable. When distillation equipment is used primarily for the distillation of crude oil to remove light fractions exclusively, this process is referred to as topping.

Materials from the bottoms of the tower or distillation unit are the heavier hydrocarbon compounds which do not vaporize, such as asphalt.

The type of processes employed for distillation are:

Simple distillation in which the liquid is heated until a vapor is formed, the vapor is drawn off, condensed, and collected. This condensed overhead can be collected in one fraction or in successive fractions.

Multiple distillation can effect further separation by successively redistilling the desired component. Extensive multiple distillation would be inefficient and would indicate the need for fractionization.

Fractional distillation is a countercurrent, multiple distillation in which rising vapor makes many contacts with descending liquid. The vapor [is] mixed in the more volatile component as it rises and the liquid [becomes] enriched in the less volatile component.

Steam distillation is any of the previous procedures in which open steam is admitted to the still."

During steam distillation, vaporization of the volatile constituents can be effected at lower temperatures by introducing open steam directly into the stock or charge.

Distillation Equipment

"Distillation equipment can be classified broadly into equipment for batch or for continuous operation. The physical conditions needed for distillation vary over a wide range. Temperatures may range from liquid nitrogen (low pressure gas separation) to 700–800°F for some separations in petroleum refining. Pressures vary from hard vacuum to 1000 psi. Throughput may range from a few gallons to 50,000 gph."

Batch Distallation Equipment

"The simplest piece of distillation equipment is the batch still. In operation, a batch or charge is fed to a pot, heat is applied, and the vapor is condensed and collected in a receiver."

A pot still for low-pressure operation is usually a horizontal cylinder of welded construction. The body is fitted with some means of heating such as direct fire, a steam jacket, an internal coil, or a sparger for open steam. For pressure operation, the wall thickness of the still must be increased.

"Sometimes, due to the chemistry involved, it is necessary to carry on the distillation process under a vacuum, i.e., when the product would decompose at the temperature required at atmospheric pressure or above. The receiver is usually a simple tank with a pump."

When steam is the only source of heat in a still, tower, or pot still, whether coming from open or closed coils, the process is referred to as *steam refining.* These stills operate at low temperatures for the production of gasoline and naphtha. When open steam coils are utilized, the process is termed *steam distillation.* The primary significance of steam refining lies in the production of gasoline and naphtha. In this process, the product's color and odor are extremely important and can be carefully controlled.

Fractionation

Fractionation is defined by the AEMS as "a method of multiple distillation for the separation of one or more volatile compounds." Thermal cracking is one of the primary functions of a fractionation column. Thermal cracking requires the decomposition, rearrangement, or combination of hydrocarbon molecules without the use of a catalyst. The chemical structure does not change, only the structure and size of the hydrocarbon molecules. Pressure, temperature, time, and feed are the primary variables that must be controlled to achieve the desired product. When high-temperature cracking of hydrocarbon molecules is done in the presence of steam, the process is a form of thermal cracking referred to as *steam cracking.* Fractionation or multiple distillation can also be done by successive batch distillations.

"For economic reasons, equipment has been developed, known as [the] fractionating column or tower. The essential characteristics of the fractionating tower are an upward flow of vapor with a downward flow of liquid in a chamber which will provide many contacts between the liquid and vapor, i.e., the liquid is spread over as much surface as possible to provide this contact. Also, there must be a return of

***All quoted material in this section is provided by the American Engineering Model Society.**

a portion of the condensed overhead as reflux to the top of the column to provide contact to the top of the vapor and rich product. The return of reflux to the top is a necessary feature of a fractionization column, and the higher ratio of reflux to product, the greater degree of separation. The degree of purification of the fraction increases with an increase in the proportion of the overhead vapor that is condensed and returned to the top of the tower as reflux, and with the number of contacts of vapor and liquid from the top to the bottom of column (number of plates)."

Rising vapors in the tower make contact with the liquid material as it passes downward through a series of trays. In this way the heavier components of the vapor are condensed and the lighter hydrocarbons in the downward line of liquid material can be vaporized. This process is referred to as *refluxing*. Any vapors that are cut from the top of the column and can be condensed to form a volatile liquid are channeled to a condenser. Liquefied products are also cut from the various trays on the fractionation tower, and bottoms are channeled into holding tanks as are the other materials stored in different vessels, depending on their pressure, temperature, and liquid or gaseous state (Fig. 11-18).

Batch Fractionation

"When the fractionization operation is conducted batchwise with the vapors rising from a still pot, it is called rectification. Equipment for this operation is essentially the same for that of the batch still with a fractionating tower inserted between the pot and the condenser, and with a provision made for a return of the reflux liquid to the top of the tower.

Batch fractionating is used for the distillation of many different mixtures and for making sharp separations between several overhead fractions."

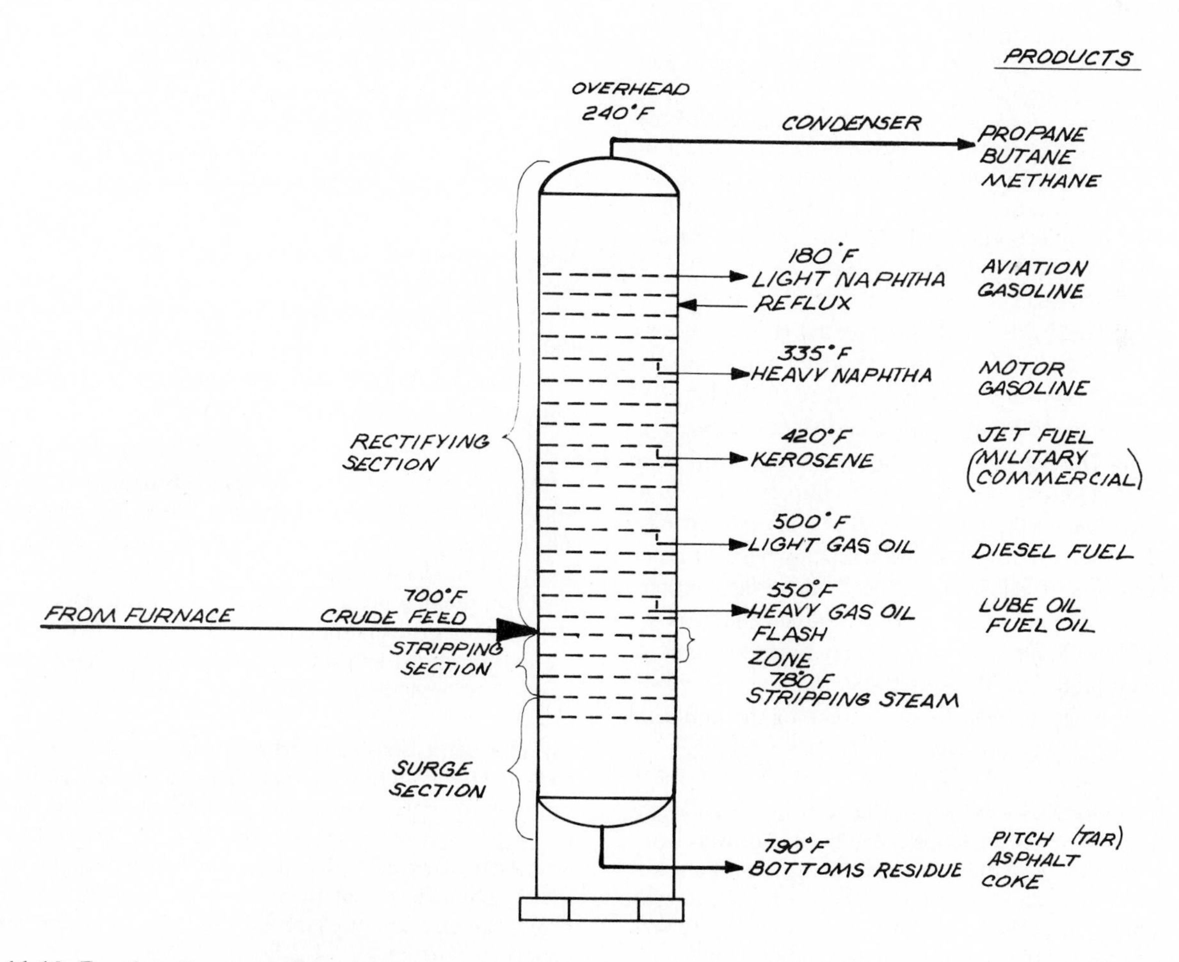

FIG. 11-18 Fractionating tower flow chart.

Continuous Fractionation

"Continuous fractionization is the most widely used of all distillation processes. The characteristics of continuous fractionization are continuous feed at the half-height point, or at the top of a fractionating column, and the continuous withdrawal of residues and the product. For this continuous fractionization, a source of heat is required at the base of the column. This is usually open steam, steam coil, or reboiler which is similar to a pot still except that it is operated at a constant level by the continuous addition of feed stock."

The feed (charge) is heated at the base of the column for continuous fractionation. In another column or tower operation, the charge stock temperature is elevated by a heat exchanger, furnace, or other equipment, such as a fired heater.

"When the liquid feed enters at the top of the column, the objective is to remove a volatile component from the residue, and the operation is called stripping. In a straight stripping operation, no reflux is returned to the top of the column. Here again, though, satisfactory separation is a function of a number of contacts between the vapor and liquid feed.

When the column feed is at or near the half-height of the column, the action above the entry point is termed rectification, and the action below the feed entry is stripping. Continuous fractionating is used for processes involving large volumes of uniform material and fractions with reasonably good separation are required."

Sections of the fractionating column are given names as shown in Fig. 11-18. Flash Zone: The feed stock initially separates into liquid and vapor in this area. Stripping Section: In this area, additional light materials are released as vapor (stripped) from the heavier liquid portion of the feed stock by the addition of steam (high temperature). Surge Section: Area where the liquid is held at a reserve capacity within the vessel for it and additional equipment to function properly. Rectifying Section: This is where the countercurrent flow of the reflux and vapor reaches an equilibrium. This means the liquid head on each tray is just enough to allow the vapor to rise through the liquid, and the vapor velocity is just fast enough to allow some of the liquid to drop through the same opening in the tray.

"Many fractions or good separation can be obtained by using columns in series, each one fed with the bottoms from the preceding one. Side draws, where the vapor is of a peculiar characteristic, can also be taken and used as is, or sent to a stripper. [In general, continuous fractionating towers operate under a steady, uniform charge material and a uniform temperature.]

Distillation columns are made in many sizes. Vessel wall construction can either be continuous, as is the case of a high pressure tower, or can be made of flanged sections that are stacked one on top of the other. The vessels are usually equipped with dished heads on both ends and a skirt is welded to the bottom end which has provisions to take anchor bolts from the foundation. [This type of column and tower is shown in many of the photographs in this chapter.]

Tray or plate type columns are fitted with trays at different intervals along the column. One type of tray is a sieve type which is a perforated plate enclosed by weirs with downcomer facilities."

The fractionation column uses trays to slow down and break up the rising oil vapors and to collect the liquid which condenses as the vapors cool below their boiling points. This counter-current flow of vapors going up and liquids coming down is controlled by the tray design. Two tray types are shown in Fig. 11-19. The type of tray used depends on the commodity being processed.

A desired fraction is removed from the vessel by means of an outlet pipe (usually a flanged nozzle) at a predetermined level. It is then piped to another section of the refinery for further processing. Above each tray in the column, vapors are continually condensing and falling onto the tray. As each tray overflows, the hot liquid falls down the downcomer to the tray below. Since the rising vapor is always hotter than the liquid at any given tray, the liquid picks up heat from the vapor which keeps it at its boiling point.

The *tray* collects the downcoming liquid and at the same time allows vapors to pass from the bottom of the tower to the top, constantly mixing the two as the liquid passes over the trays and the vapor rises and meets the liquid as it falls. Bubble-capped trays are a form of tray used for processing; cap size, distribution, and the number of caps employed influence the internal design of the column.

"A packed column is sometimes used to improve the operation of some types of distillation. These columns are usually less than 20" (50 cm) in diameter, because large diameters lead to channeling in the packing. Materials commonly used for packing are coke, gravel, lathe turnings, chain, and special shapes of ceramics or tiles.

Provisions must be made for access manholes for cleaning, inspection, and repair throughout columns, towers, and vessels.

Temperature sensing instruments are installed in thermal wells in all distillation columns. To function properly, these thermowells must extend into the liquid on the tray or into the vapor space above the tray. Usually two instrument nozzles for thermal wells are installed at each tray level of a tower. The location of the nozzles and other items must be accessible and are used to establish the level and size of operating platforms.

A common method for installing platforms on process vessels is to use bent channels attached to the column for primary members. The maximum width of the platform of

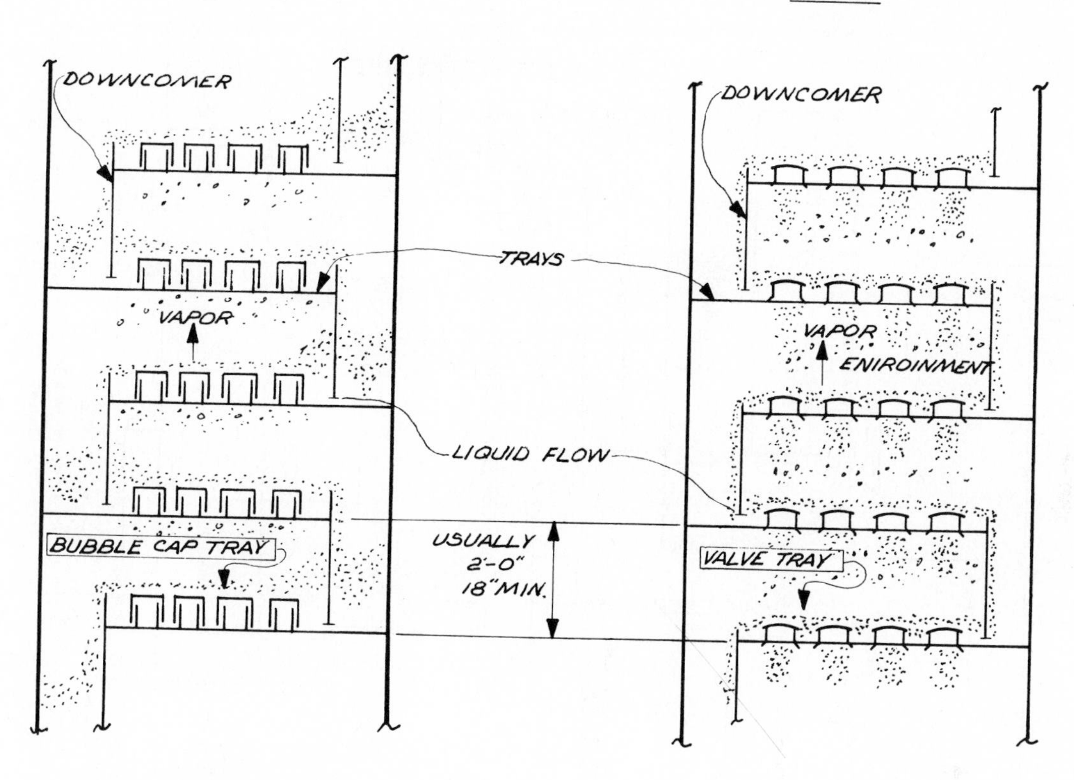

FIG. 11-19 Tower tray types.

this type is approximately 4′ 3″ (1.3 m) and the usual width is approximately 3′ (91 cm). The horizontal member closest to the column shell is placed about 6″ (15 cm) away from the side of the column to allow space for insulating the column. All vertical service piping is usually installed in this space.

For safety reasons, ladders are mounted on the vessel shell a minimum of 8″ (20 cm) away from any obstacle. For this reason, small pipe should not be installed behind the ladder. Landings are provided for ladders, often at 15′ (4.5 m) intervals.

Larger vertical piping such as feed, reflux, and drawoff lines are supported by standard structural clips. Usually such pipes are too close to the tower to elbow directly in; common practice is to use a pair of 45° elbows and then a 90° to enter the nozzle [in some cases, 45° mitered elbows]. This horizontal run into the nozzle provides a location for a valve. Adequate platform room must be provided under the valve for operation and maintenance.

The skirt of the column is usually surrounded by 2″ (5 cm) of fireproofing material. Piping must be kept clear of this area as well as from other access openings in the skirt."

Skirts are also provided with ventilation holes so vapors cannot collect in the chamber below the vessel. The accurate, systematic, and economical location and construction of the various processing units are of prime importance to the piping designer or drafter. Figure 11-13 shows a very orderly, efficiently designed series of columns, heat exchange units, and holding vessels. Figure 11-20 is a water-quench blowdown unit for a petrochemical facility.

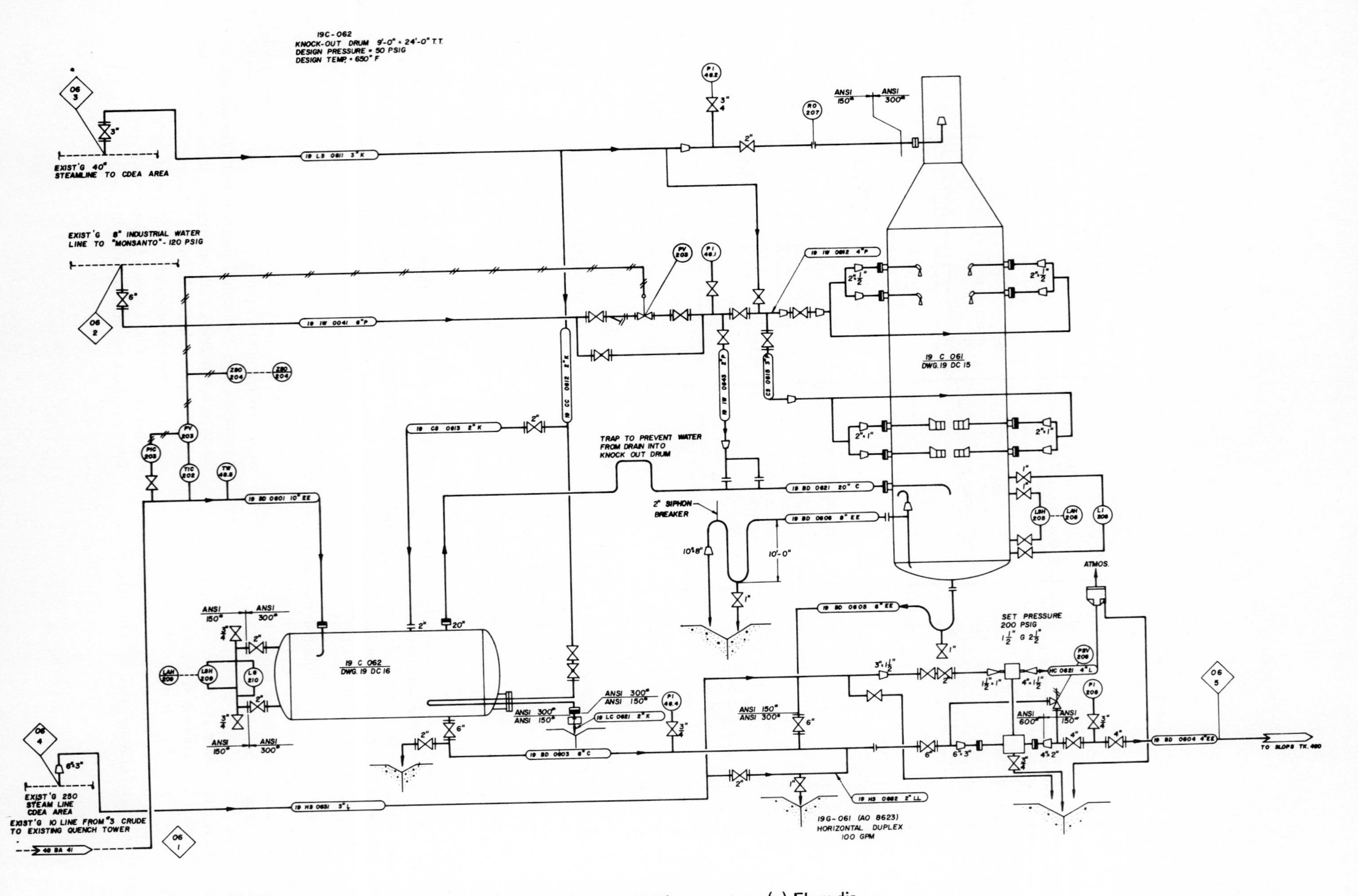

FIG. 11-20 Complete set of nine drawings for a water-quench blowdown system. (a) Flow diagram of water-quench blowdown.

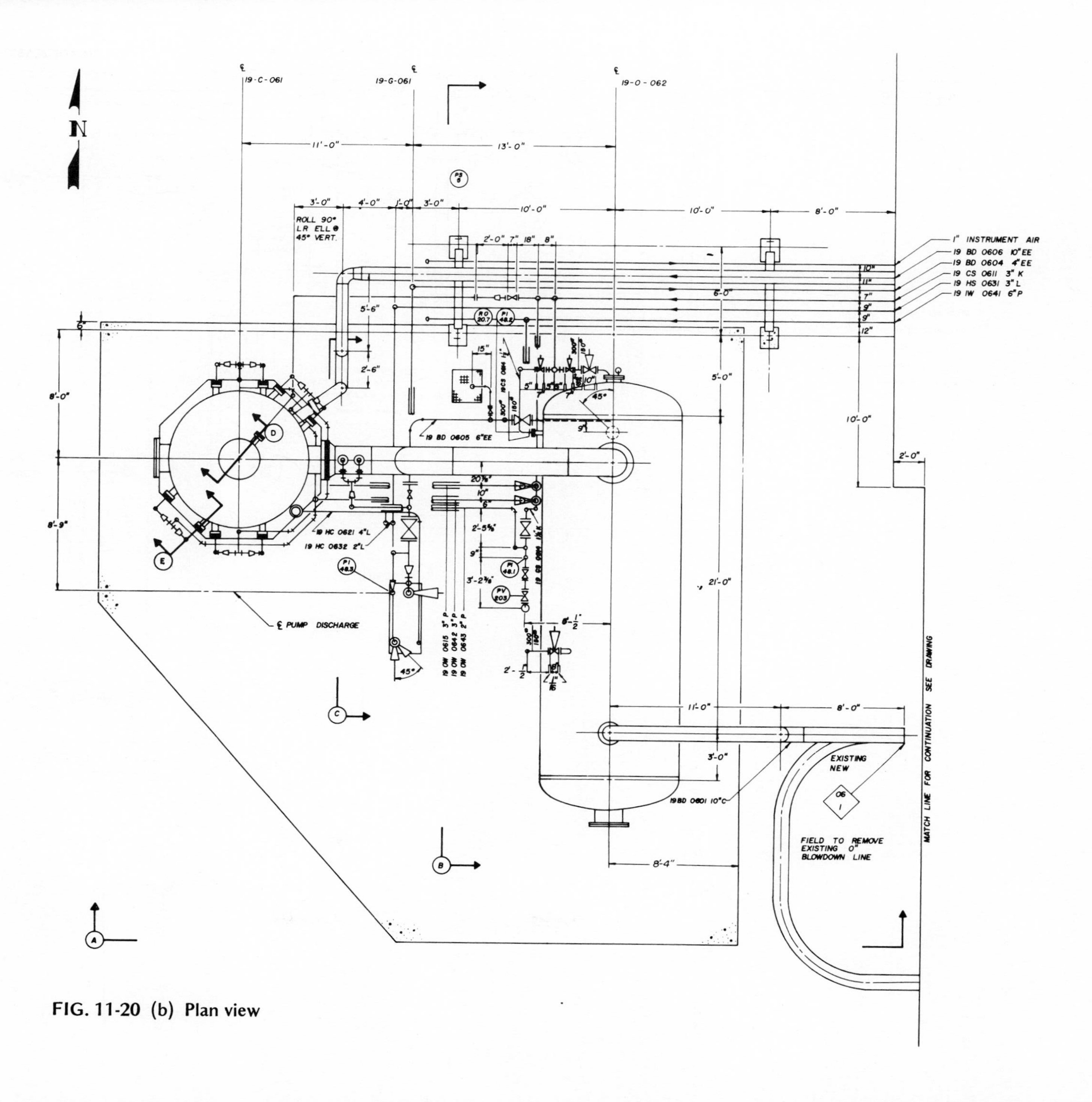

FIG. 11-20 (b) Plan view

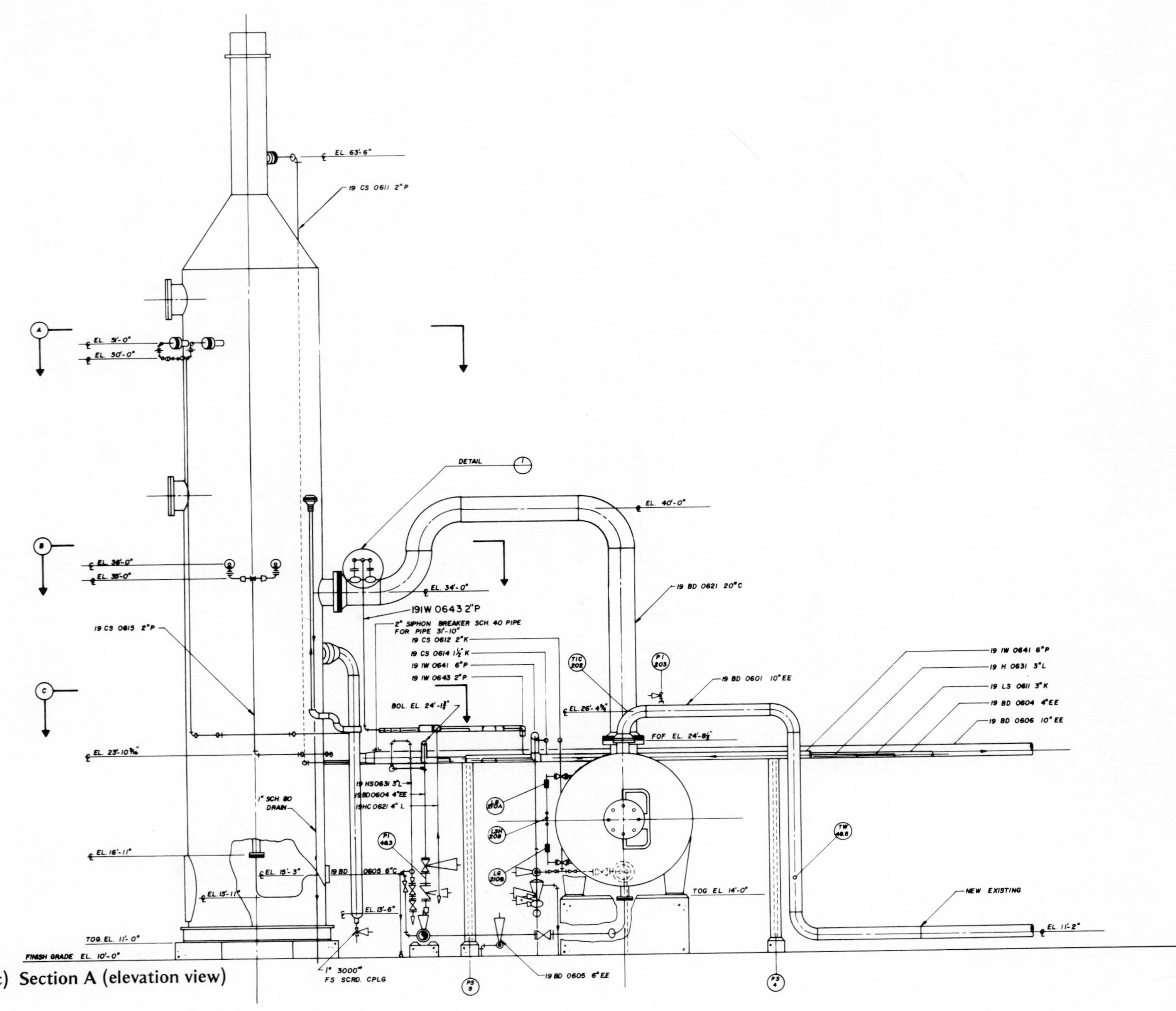

FIG. 11-20 (c) Section A (elevation view)

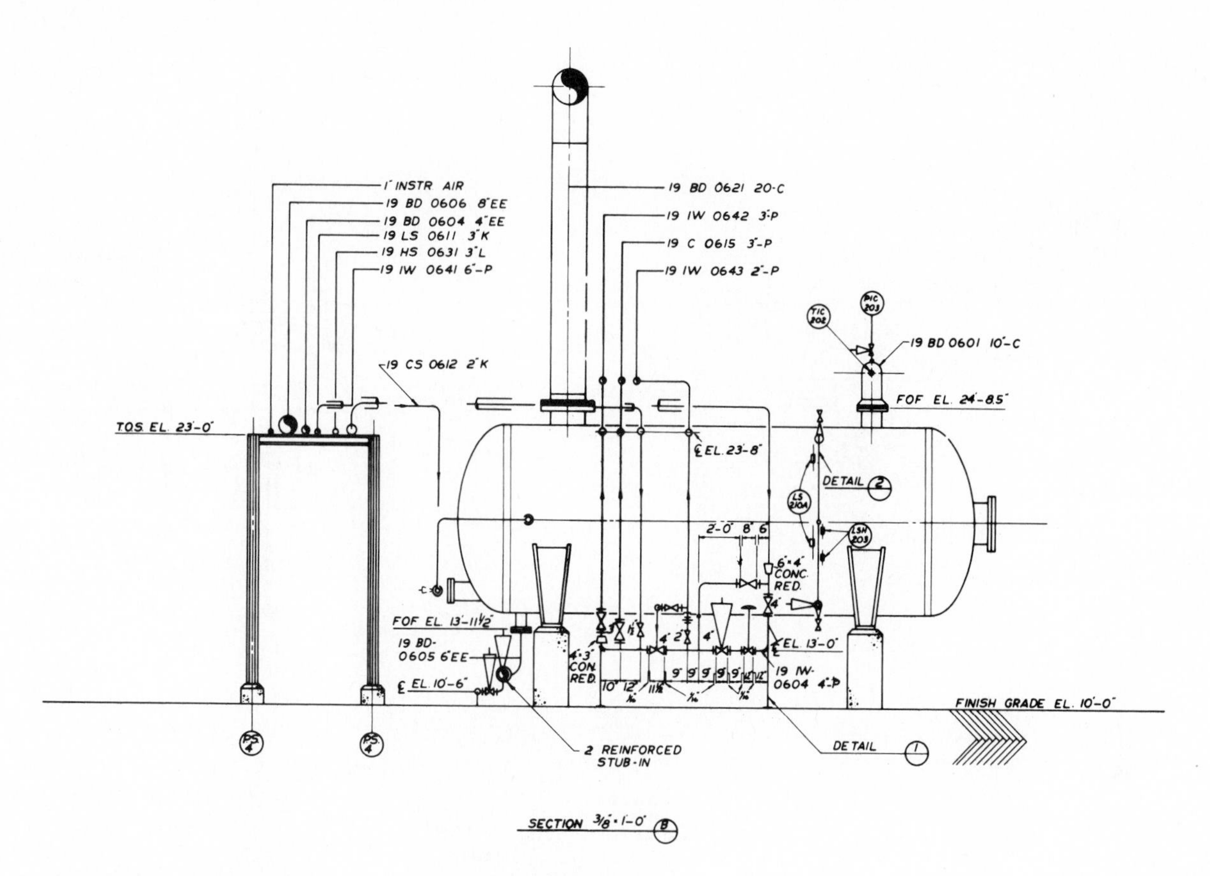

FIG. 11-20 (d) Section B

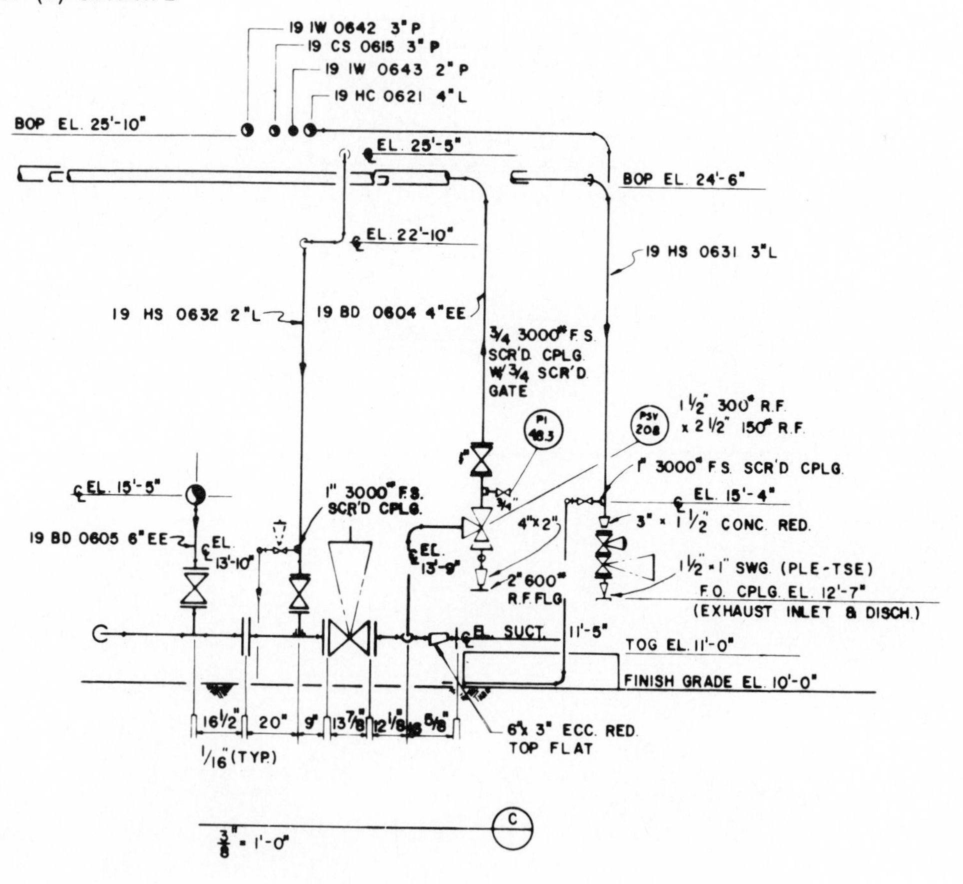

FIG. 11-20 (e) Section C from plan view

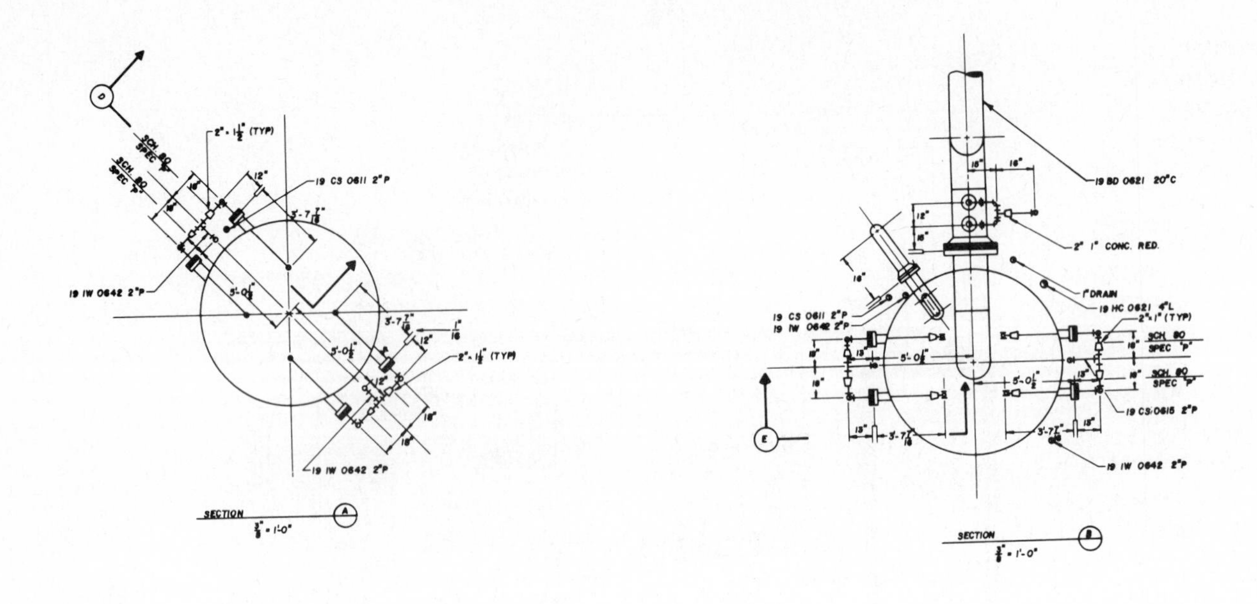

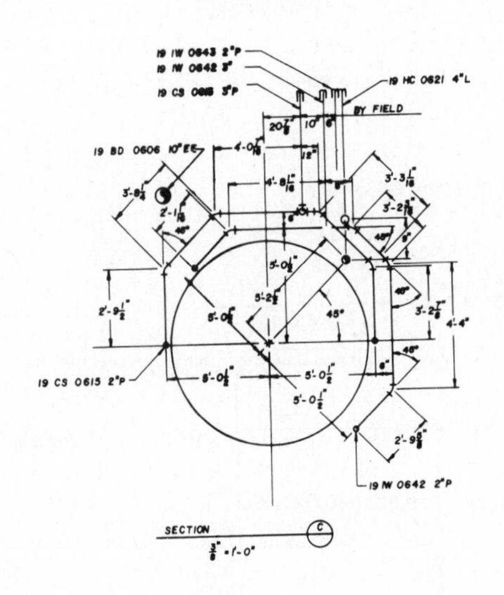

FIG. 11-20 (f) Quench tower sections

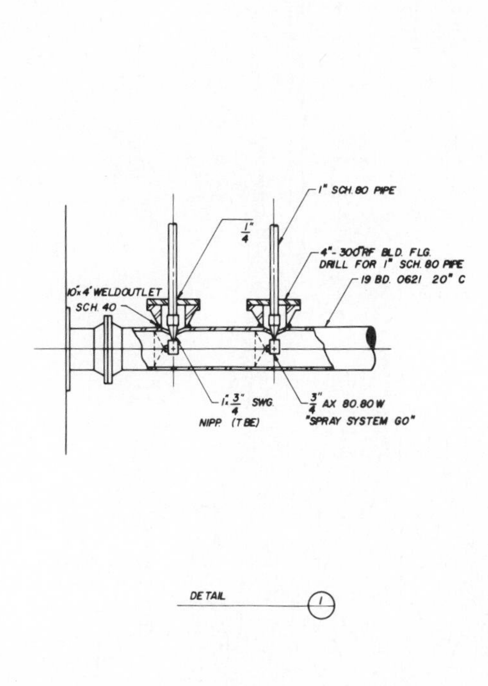

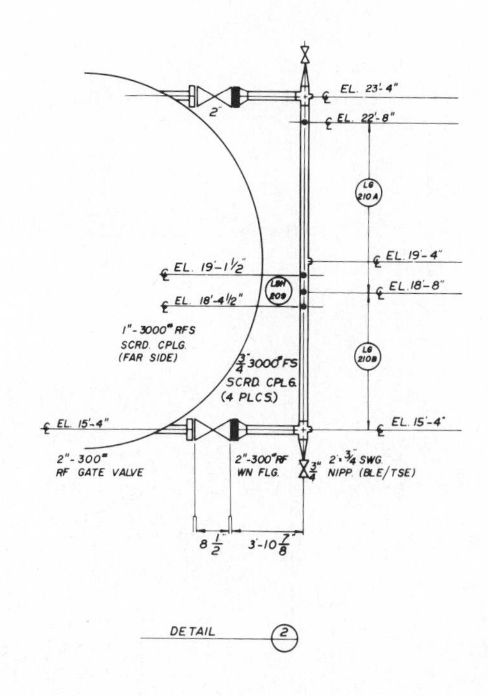

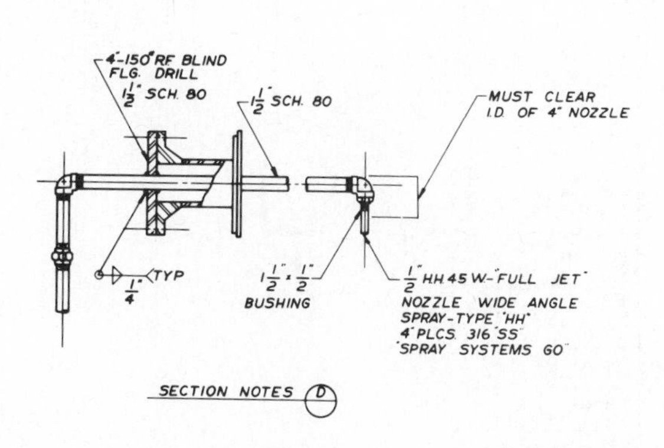

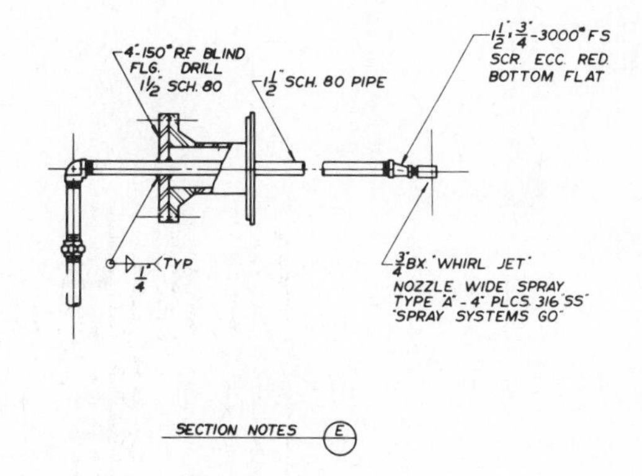

FIG. 11-20 (g) Details 1 and 2 and Section notes D and E

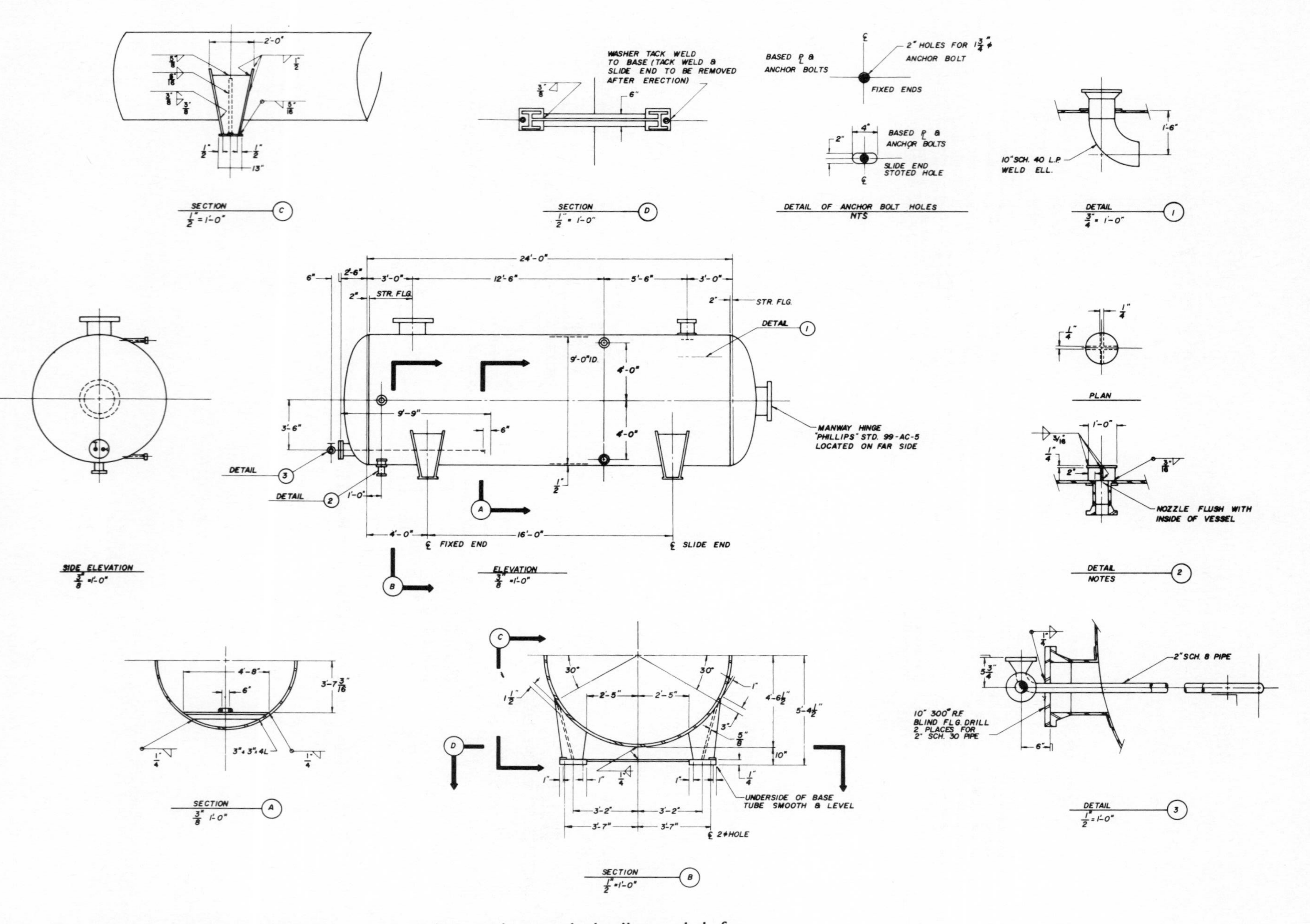

FIG. 11-20 (h) Knock-down drum with accompanying sections and details needed for fabrication

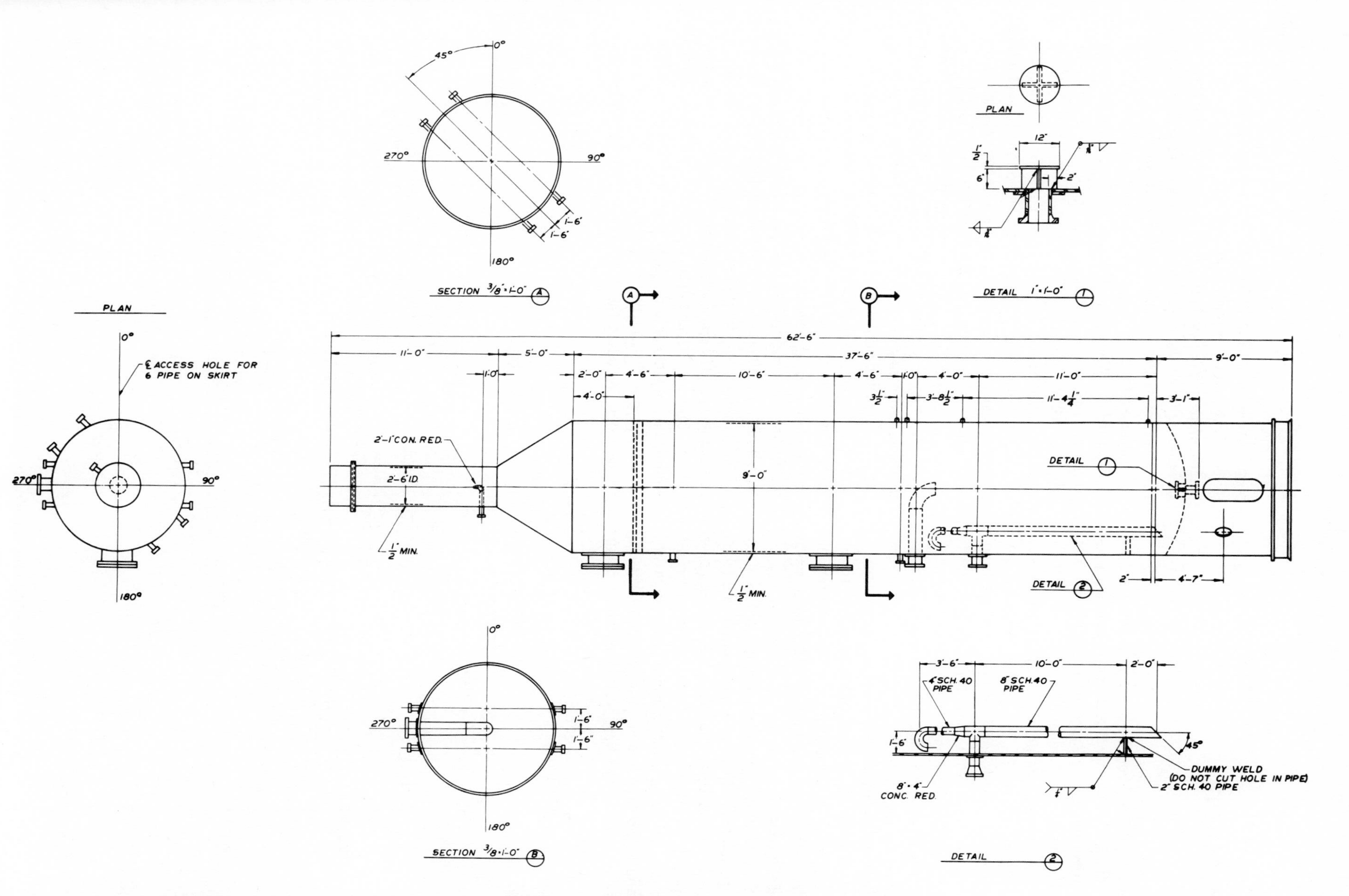

FIG. 11-20 (i) Fabrication drawing with accompanying sections and details for the quench tower

PIPING LAYOUT AND ARRANGEMENT

Many of the steps described here are also discussed in Chapter 10. The following specifications and design procedures are primarily for petrochemical complexes. The petrochemical designer and lead drafter will have at their disposal the flow diagram and piping specifications for the project in question, all vendor equipment drawings, any special instrumentation needs and locations, and vessel specifications in most cases. With these specifications determined by the petrochemical engineer, chemical engineer, and designer, it is possible to start the preliminary arrangement and layout of the plant. All company design groups at this time have either made their needs known or a meeting is called to allow the HVAC, instrumentation, structural, piping, and electrical group designers or lead drafters an opportunity to contribute to the most efficient use of space and economical layout of the plant.

Chapter 15 explains the general procedure at this stage when using preliminary models. A preliminary model is basically an equivalent step to the sketching and design of the overall plot plan. This is one of the most important stages of the general design effort because of the necessity to plan carefully the layout and arrangement of vessels, piping, roadways, racks, and the location of other major pieces of equipment in order to avoid future interferences or lack of operating or maintenance space.

The master plan or preliminary model includes the general positioning of all the major equipment and pipeways. When the drawing method is used, it is either roughly sketched or crayoned in on a large boundaried representation of the prospective plant area. The overall plan, which is sometimes referred to as a master plot plan, covers an area which is usually too large for adequate representation on paper in the detailed design stages. Therefore it is broken into major sections or blocked-out areas, and battery limits are established for the extent of each section along major match lines or coordinates. This practice is similar to the procedure followed in the sectioning of the model into workable units along logical points such as piping racks and between processing units. From this original preliminary model or master plot plan, the equipment can be placed according to the specifications for the processing unit.

The foundation drawing for the vessels, the drainage drawing, and the piping drawing index all are steps that follow the finished site and master plot plan. Establishing the battery limits, and major sectioning along match lines, is the primary function of the piping drawing index, as explained in Chapter 10. The master plot plan or preliminary model for the typical petrochemical installation includes all the equipment and major units that form the design requirements. In Fig. 11-21, the major equipment and primary connections, including the direction of flow, are shown for

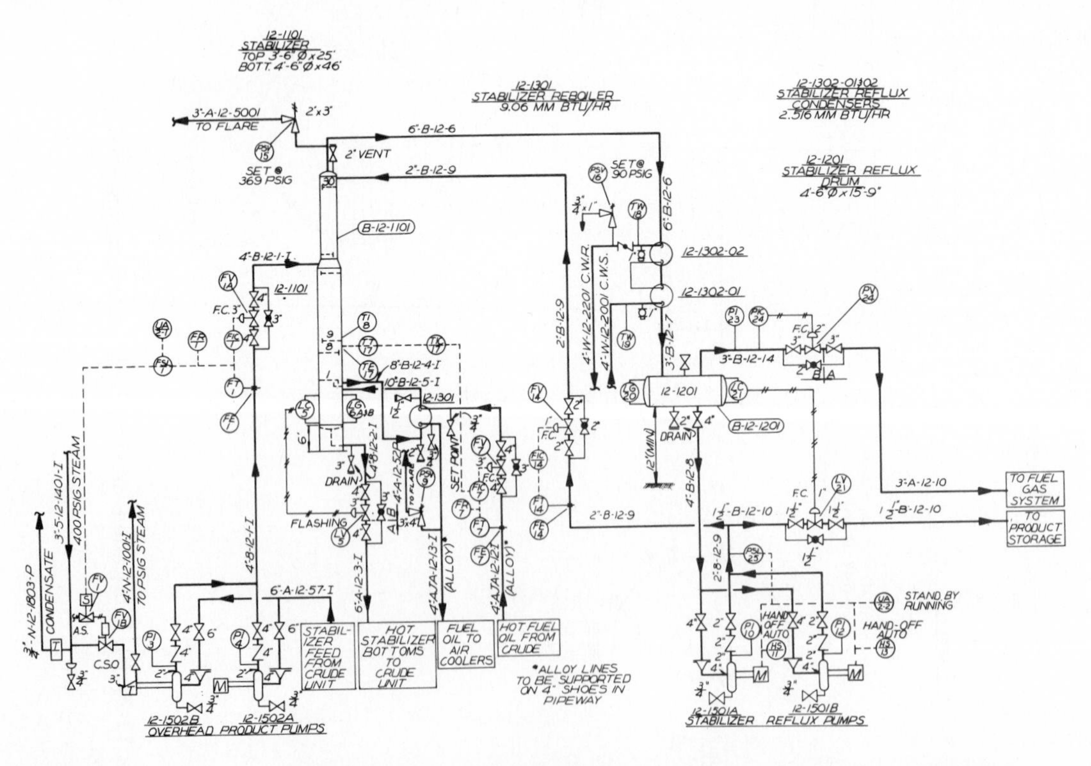

FIG. 11-21 Process flow diagram of a petrochemical unit showing typical refinery components

the petrochemical unit. By closely reviewing this process diagram, one can see many types of equipment: distillation column, exchangers, instruments, holding tanks, control stations, and a variety of pumps and major valves, filters, and large trap devices. Working with a similar flow diagram and P&ID, the petrochemical designer or layout drafter can, through the use of experience and a list of specifications and codes, complete the master plot plan, site plan, and preliminary model.

Figure 11-11 shows a petrochemical complex under construction with several pieces of equipment and piping runs. In the background, it is possible to see numerous buildings that must also be taken into account in the preliminary site plan stage–including the location of offices, workshops, laboratories, machine shops, warehouses, administration buildings, cafeterias, first aid stations, and even union meeting halls. In many instances, the construction of a new unit to an existing plant will only entail the addition of another portion of the refinery; offices, shops, warehouses, and so forth will already exist.

In the case of new industrial plants such as those built in foreign countries, the overall site plot plan will necessarily include all these facilities. By dividing the site plan and master plot plan into logical blocks with established battery limits, it is possible to work more efficiently with the smaller section. On the site plan, all separate units need to be labeled and described according to function and purpose, depending on the type of processing unit in that index area. The location of processing units in relationship to each other takes into account the requirements of the process and also safety hazards–such as positioning the refinery flare system too close to process units, which may be hazardous because of sparks, flames, gases, and vapors.

The location of major equipment that needs to be serviced, and nonprocessing structures such as the administration building and cafeteria, must afford easy access from major roadways. The design should limit the amount of movement and travel in volatile areas of the complex. Processing units should have easy access to secondary roads, but roadways and equipment should be positioned so they do not require the excessive passage of personnel and deliveries where petrochemical processing units are positioned. In other words, it is poor practice to establish the cafeteria adjacent to a processing unit–especially one that removes the sulfur from crude oil.

If the plant is being constructed from the ground up with no previous buildings at the site, all this must be taken into account, along with access to railroad and shipping facilities, future highway and road needs in the area, and the general physical contour of the plant. Units shown in Figs. 11-9, 11-11, and 11-17 show the construction of a plant on flat ground and in somewhat uncongested areas, although the unit in Fig. 11-11 has been established in the center of an existing plant near a city. All drainage and sewage lines must be determined by the time the site plan or preliminary model is completed to allow the construction crew early access to the design requirements for foundations, footings, water, sewage mains, and the like.

After the overall plot plan has been established and the battery limits and piping index have been determined, it is then possible to do the final arrangement of equipment and pipe racks. The layout designer or lead drafter determines the most advantageous location and selects the components to make up the processing unit in a particular area. Location of major piping runs can now be determined. The rough elevation of equipment, establishment of major piping between equipment, and the location of mechanical equipment, such as pumps and compressors, can also be done at this time, taking into account all the various safety, maintenance, and access necessities for the modern petrochemical complex. Equipment locations need to be established, along with foundation and footing requirements. Major pipe rack runs are established, as is the elevation of the pipe rack and whether it will be a single-tier or multiple-tier design. If the design is to be coordinated with existing equipment, piping runs, and roadways, the existing routing for pipeways and roadways must be adhered to.

Figure 11-22 shows the underside of a well-spaced, uncluttered design which enables the plant to operate in the most efficient manner by providing maintenance and operation areas in and around equipment. This is of primary concern to the piping designer. Figure 11-23 shows a parallel pump installation along with the layout and arrangement of piping and mechanical equipment which allows for replacement and maintenance of the pumps and surrounding components.

After the master plot plan has been completed, the piping designer or lead drafter can then do the necessary routing of the pipe and determine the location of the smaller equipment and piping components. At this time it is also necessary to have access to the P&ID vendor prints, electrical needs, vessel drawings, pipe specifications, instrument needs and locations, and a variety of other information. The designer sketches the pipe on the master plot plan first and then indicates the minor piping needs directly on the unit plot plan. Figures 11-20 and 11-33 show typical processing units that have been broken out of the master plot plan to provide a larger scale and better clarity for the drawing and design procedure. Chapter 10 describes the design sequence of piping layout; Chapter 15 discusses the model sequence.

Using the specifications for equipment and piping needs, the drafter and designer then lay out the plant. All the drawings are oriented to north-south and east-west coordinates. All major equipment centerlines, pipe rack centerlines, column centerlines, and other equipment centerlines are located from battery limits as shown in Fig. 11-20. Major piping may also be established according to this method by the use of match lines and coordinate dimen-

FIG. 11-22 View of a petrochemical complex showing a series of control valve stations and instrumentation. Notice the large horizontal drum supported on the piers to the right.

FIG. 11-23 Two pumping stations at a refinery. Notice the dummy leg supports for the elbow on the inlet port of the pumping unit. Piping systems in and around equipment that vibrates must be well supported and sometimes rigidly connected to surrounding structural members and foundation.

sions. The following specifications apply to the layout and arrangement of a petrochemical complex:

1. Locate all major equipment and pipe racks according to north-south and east-west coordinates.
2. Locate equipment according to specifications determined by process, function, safety, existing equipment, existing pipe runs, and maintenance and operation access needs.
3. Show the flow direction on all piping.
4. Use elevation designations and coordinates to dimension all major equipment elevation and section views instead of using dimensions.
5. Locate pipe racks, pipe runs, and all yard piping in horizontal runs where possible and shown to scale on the drawing.
6. Arrange piping in plan view first and then in elevation.
7. Locate all major piping connections between equipment and pipe racks in plan and elevation.
8. Locate all instrumentation.
9. Label all pipelines according to the pipeline code, showing line number, insulation specifications, pipe size, and pipe function.
10. Dimension minor pipe runs and smaller equipment from major equipment. (See Chapter 5 for specifications for dimensioning.)
11. Prepare drawings to scale—including equipment, line sizes (when the double-line method is used), foundation supports, racks, flange diameters, valve lengths.
12. Draw valve handwheel orientations and valve flange-to-flange dimensions to scale. Handwheels should be shown in open position to establish maximum open position and minimum clearance.
13. Follow the specified line thickness requirements when drawing petrochemical installations as described in Chapter 5.

Piping arrangement includes all safety, maintenance, and economically efficient design requirements for running the pipeline. For the establishment of piping arrangement on the final drawings, it is necessary to take into account all specifications and design needs. This general piping arrangement for process plants is not covered by the piping code since it does not concern itself with the design and material necessities of the pipe or pipe contents. The following ***general piping*** specifications apply:

1. Allow adequate space for future and auxiliary equipment such as electrical trays, electrical conduit, instrumentation lines and HVAC.
2. Refer constantly to the engineering flow diagram (P&ID) for the required pipe routing flow sequence.
3. Allow expansion space of 20 to 25 percent for all pipe ways.
4. Arrange pipe racks that route the pipe headers to equipment and processing units first.
5. Establish all piping on overhead racks where possible.
6. Provide minimum 7 ft 2 in. (2.1 m) clearance (headroom) in maintenance areas, walkways, and platforms.
7. Locate piping at grade only in areas where it is absolutely necessary for equipment connections.
8. Remember that utilities such as drainage, water supply, and pumpout lines are not usually run in overhead lines; major sewer lines are usually located below grade.
9. Arrange piping in efficient, logically spaced banks, piping runs, and horizontal groups whenever possible.
10. Allow for adequate space between equipment and between major process units for the future establishment of piping racks or pipe sleepers.
11. Arrange all pipelines that are at the same elevation so that the pipe bottoms fall in the same plane—except for insulated lines which are supported on saddles, shoes, and slide arrangements, in which case the pipe bottom will be slightly higher than the plane of noninsulated lines.
12. Keep piping which lies within structures and buildings at a minimum of 7 ft 6 in. (2.2 m) above grade level, platforms, and walkways.
13. Establish piping over major plant roadways at a minimum of 22 ft (6.7 m) in elevation in general. For secondary roadways and general access areas, piping should be a minimum of 15 ft (4.5 m) above grade.
14. Place piping on sleepers a minimum of 12 in. (30 cm) above grade.
15. Remember that horizontal clearance between piping and equipment, major structural elements, and so on should be a minimum of 12 in. (30 cm) in most cases.
16. Arrange horizontal pipe runs to allow 1 in. (2.54 cm) between flanges, even when lines are not flanged (see Chapter 10).
17. Stagger the flange connections to allow for adequate servicing and maintenance (see Chapter 10).
18. Provide adequate horizontal spacing between insulated pipes for service and maintenance.
19. Always change elevation when changing the direction of a pipe run.
20. Establish pipeline dimensions according to east-west and north-south lines.
21. Make elevation changes a minimum of 18 in. (45 cm) (usually larger) to permit access to the surrounding area.
22. Remember that lines below grade, underground lines, sewer lines, and so on should change elevation when changing direction (except for major sewer lines, drainage lines, and large-diameter piping).
23. Limit the pipeline to commonly stocked sizes.
24. Keep overhead pipelines 1 in. (2.54 cm) diameter and larger.
25. Remember that instrument lines are usually of smaller diameter and should be arranged in groups to provide either continuous or frequent supporting.
26. Make all utility connections for water, steam, and air 3/4 in. (1.9 cm) diameter minimum.
27. Make all sample connections, drain, and vent connections 3/4 in. (1.9 cm) minimum.
28. Eliminate pockets in all pipelines.
29. Slope pipelines, especially steam lines, to eliminate pocketing of condensate.
30. Be sure that pipelines which carry caustic, acid, freezable materials, and condensate are provided with mechanisms (sloping, vents, drains, steam traps) to eliminate the excess material from the line after use.
31. Eliminate pocketing especially in areas where valves and piping components may cause pipe sag.
32. Vent high points in piping to eliminate gas collection.
33. Eliminate temperature and safety problems at the design stage.
34. Do not run gas and oil piping in areas which might present hazards (such as through the control room, washrooms, switch rooms).
35. Run high-temperature/high-pressure lines to avoid close contact with electrical conduits, electrical trays, electrical connections, and motor-operated equipment (such as motor-operated valves) which may cause sparks.
36. Do not discharge steam condensate or other hot materials into ground drains that are located near electrical runs.
37. Group hot and insulated lines together where possible. Cold lines should be kept separate to eliminate corrosion, condensate, and vapor problems.
38. Locate pipelines carrying caustic materials, chemicals, acids, and the like away from equipment or pipelines which have elevated temperatures.
39. Consider all construction items and sequences in the design stage of the plant.
40. Remember that cranes, pulleys and moving equipment should have sufficient operating space during the construction and maintenance phases of the project.

41. Consider all operation and maintenance problems in the design and location of equipment and piping.
42. Provide adequate work space and maintenance space for equipment such as compressors and pumps.
43. Provide for replacement space in the general equipment location design–pull space for heat exchange units, spare pump space, and so on.
44. Place piping in trenches only where absolutely necessary, and be sure to provide adequate work space around piping components.

This list is by no means comprehensive, but it should give the student a general idea of the considerations that are necessary for piping in a petrochemical installation.

EQUIPMENT LAYOUT AND ARRANGEMENT

This section deals with layout and arrangement procedures for equipment. Arrangement of equipment must take into account both elevation and horizontal spacing. Consideration must be given to maintenance, operation, storage, and pull space. All safety conditions must be met, especially for hazardous equipment which may require various items and devices to be separated according to standard distances.

Both horizontal and elevation spacing is extremely important with regard to equipment. It has been general procedure, for processing units, to establish much of the equipment above grade to provide access around units and underneath pipe runs and other equipment. Examples of this practice are shown throughout the book in different photographs. In Fig. 11-11 almost every piece of equipment in the plant is elevated from grade. Though this is an accepted procedure, the excessive elevation of equipment can be costly, and the pros and cons must be taken into account in the beginning stages of the project.

It is up to the piping designer to establish equipment elevations for most of the smaller equipment and in some cases the larger items as well. All operation, safety, and maintenance requirements must be considered when establishing elevation spacing between equipment and horizontal spacing. In many cases, equipment can be placed at grade. Some mechanical equipment such as pumps and compressors is elevated only slightly from grade. As shown in Fig. 11-23, the placement of the pump directly at grade will inhibit certain operation and maintenance procedures. Therefore even mechanical equipment is slightly elevated (usually 12 in. (30 cm) minimum). In most cases, the operating and maintenance requirements must be known for the piece of equipment before determining the elevation.

As has been noted, the horizontal spacing between equipment as in Fig. 11-23 is a major aspect of the layout of structures and equipment for the processing unit. The following specifications apply to the layout and design of a process petrochemical unit:

1. Operation, maintenance, and safety considerations are paramount in the spacing of process equipment.
2. The minimum of 3 ft (91 cm) working space clearance should be provided in and around all equipment.
3. Motorized devices such as material handling equipment and carts must be considered in the determination of aisle sizes and access ways.
4. Repair and replacement areas must be provided next to equipment that needs frequent or periodic servicing or replacement, such as pull space for economizers and backup pumps for pumping stations.
5. Piping in the general area of equipment which needs frequent servicing, maintenance, or replacement should be established to permit easy access to the equipment in question without disturbing the piping configurations.
6. Access ways, roadways, major throughways, and aisles in processing units should be a minimum of 20 ft (6 m) wide.
7. Access around warehouses, shipping units, railway facilities, and the like must be considered when establishing equipment and piping.
8. Elevated items must conform to specifications for different pieces of process equipment. These specifications are usually standardized.
9. Elevations for hazardous equipment should always be established by engineers.
10. Establish all equipment relative to the grade or other equipment in the area. The centerline of an economizer should be located from grade, for example, but the dimension to the nozzle connection is sometimes established from the economizer (exchanger) itself.
11. Elevate all equipment that needs to develop sufficient static head in order to meet flow requirements for the unit.
12. Establish platforms at adequate elevations in relationship to the equipment needing attention.
13. Establish equipment elevations for mechanical devices, such as pumps and compressors, at a sufficient level above grade to provide ease of maintenance and operation.
14. All special equipment should be located in accordance with the installation, operating, and maintenance requirements supplied by the manufacturer.
15. Provide a minimum of 6 in. (15 cm) clearance between the grade and the bottom of any pipe which runs into equipment.
16. Provide a minimum of 6 in. (15 cm) clearance between the bottom of insulation and grade for all piping–especially piping connected to vessels.
17. Provide a minimum of 6 in. (15 cm) between valves, valve bodies, and flanges and grade for components located at grade–especially for components located beneath vessels and equipment.
18. Spacing of flammable equipment is determined by hazardous spacing requirements and codes.
19. Depending on the service, function, and material of the equipment and its associated piping, horizontal spacing is determined by the process designer according to pressure, flammability, and other considerations.
20. Locate hazardous units a minimum of 100 ft (30 m) from loading facilities, shipping facilities, railroads, public roadways, cooling towers, blowdown stacks, furnaces, fired heaters, boilers, high-pressure/high-temperature units, and storage tanks.
21. Equipment
    (a) Heat Exchangers and Economizers
        (1) Locate heat exchangers and economizers adjacent to a roadway, aisleway, or other major access area, and line up head of exchangers from roadway.
        (2) When necessary, stack exchangers that are in series up to a maximum of 12 ft (3.6 m).
        (3) For multiple exchangers, a gantry crane or monorail system should be considered.
        (4) Provide a minimum of 3 ft (91 cm) horizontal clearance between exchangers arranged in a row.
        (5) Exchangers should have access to major pipe racks.
        (6) Arrange exchangers so that pull space is oriented away from the pipe rack. Exchanger pull space should not be located underneath a rack.
        (7) Rear head of exchanger should be established a minimum of 8 ft (2.4 m) from the centerline of the pipe rack column.

(8) Exchangers should be arranged to permit easy access to piping and components attached to the exchanger.
(9) All exchangers should be elevated a minimum of 2 ft 6 in. (76 cm) above grade to the bottom of the exchanger shell to allow adequate space for piping below the exchangers, valves, and change-of-direction fittings.
(10) When necessary, exchangers should be elevated on platforms well above surrounding equipment. A minimum of 5 ft (1.5 m) platform access space should be established around elevated exchangers.
(11) All exchanger units should be designed to permit removal of the tube bundle for cleaning and repair.
(12) In general, exchangers should be arranged to allow access to piping, valves, instruments, and maintenance on one side. When exchangers are arranged in a series, piping for adjacent exchangers should be kept toward the same aisleway, providing a minimum of 2 ft 6 in. (76 cm) clearance between piping and each unit.
(13) Where possible, stack economizers that are in a series to minimize piping and valves.
(14) Some exchanger units should be slightly sloped to provide adequate drainage.
(15) Exchanger units should be mounted above horizontal drums or vessels, if possible, to use space efficiently and reduce piping requirements.
(16) Provide space for the tube bundle at ground level, especially for elevated exchanger units.
(17) Exchanger pull space should not interfere with major roadways, aisleways, or other general access runs.
(18) The plan view of the piping installation should designate the area necessary for pull space for the heat exchanger unit and tube bundle.
(19) If the unit is insulated, a minimum of 2 ft 6 in. (76 cm) should be provided between the insulation on the heat exchanger and grade.
(20) When possible, the front of the foundation of exchangers and vessels arranged in a series or parallel should be lined up.
(21) Locate the concrete footings, piers, and pedestals a minimum of 13 ft 6 in. (4.1 m) from the rack column centerline when possible.
(22) When exchanger units are similar lengths, the foundations, footings, and piers along with corresponding saddles should be lined up.
(23) The tube bundle removal space is determined by manufacturer's recommendations.
(24) Allow a minimum of 1½ times the exchanger length for removal space.
(25) When spacing exchangers that are parallel to one another, run the piping between the units to provide a minimum of 2 ft 6 in. (76 cm) clearance between the piping and shell. Keep a minimum of 3 ft (91 cm) between shells where no piping is run.
(26) Establish the dimensions and configuration of the exchanger (including supports, foundations, footings, and saddle orientation) before arranging piping, valves, and instrumentation.
(27) All insulation thicknesses should be specified on the heat exchanger shell unit, including insulation for piping.
(28) An operating space of 2 ft 6 in. (76 cm) to 3 ft (91 cm) is required between all units plus extra operating space wherever valves and handwheels may cause interference or inaccessibility.

(b) Compressors, Motors, and Pumps
(1) Locate pumps under pipe racks and major runs as in Fig. 11-23 (see Chapter 10).
(2) The centerline of the discharge side of the pump should be approximately 2 ft from the rack column centerline.
(3) Establish pump suction lines in a way that eliminates vapor pockets.
(4) Locate pump lines to avoid pocketing (including liquid pockets).
(5) Provide adequate maintenance space in and around pumps and compressors. A minimum of 3 ft (91 cm) should be established between parallel pumps or pumps and other equipment or piping.
(6) Establish an operating and maintenance aisle at the driver end of the pump.
(7) The operating aisle should be a minimum of 5 ft (1.5 m) wide.
(8) Arrange pump piping to provide for a minimum length of suction line.
(9) Orient the pump in relation to the suction side of the vessel.
(10) Arrange pumps parallel and in series wherever possible on the same side of the rack.
(11) Establish pumps in parallel rows so that drainage and other utilities are systematically oriented.
(12) The centerline of the discharge side of the pump should be the same for all pumps arranged in a row.
(13) Arrange pumps in areas that have access space for mechanical movement for easy repair, maintenance, and replacement.
(14) Provide adequate storage and replacement pump space near operating units.
(15) Install compressors inside buildings.
(16) Provide access areas for servicing and maintenance of cylinders.
(17) The minimum distance between compressors is usually specified at 8 ft (2.4 m).
(18) Motors and other mechanical equipment should have adequate clearance for maintenance and operation, which is determined by experienced personnel or the manufacturer's suggested space requirements.

(c) Fired Heaters
(1) Heaters, furnaces, and boilers should be located a minimum of 50 ft (15 m) from combustible equipment.
(2) Tube pulling space should be provided for all heaters.
(3) Establish heaters and boilers to have maximum access from roadways for fire equipment, maintenance, operation, and repair.
(4) Establish a minimum of 25 ft (7.6 m) between heater, furnace, boiler, and reactors to eliminate expensive piping. Figure 11-10 shows the heater unit in relationship to ammonia plant equipment including various reactors.
(5) Snuffing steam manifold, manual fuel shutoff valves, and other safety and control valve equipment should be located a minimum of 50 ft (15 m) from the fired heater.
(6) Provide adequate headroom under vertical heaters for access to burners for maintenance and other purposes.
(7) Vertical heaters should be a maximum of 7 ft (2.1 m) above grade, which does not include pipe extensions from the heater floor.

(d) Vertical Vessels: Towers, Columns
(1) Provide an octagonal foundation footing for optimum load requirements (see Chapter 10).
(2) Most vertical vessels require relief valve connections, as can be seen in most of the figures in this chapter.
(3) All vertical piping on the sides of columns should be established outside platforms wherever possible.
(4) Eliminate interference between piping, ladders, and other structures by establishing them well outside these areas.
(5) Pass only necessary piping through platforms and flooring, and provide toe boards and other protection equipment including protective insulation where necessary.
(6) Support all high-elevation piping in columns by structural attachments and supports to the column itself, not to the platform.
(7) Do not run piping over platform areas.
(8) Arrange manways, ladders, and cages to one side of the column only. This frees whole areas of the column or tower of structural items and permits straight runs of piping as in Figs. 11-11 and 11-13.

(9) When arranging piping from columns, pipe from the highest elevation should be laid out first along with the larger lines.
(10) One area along the full length of the column should be clear of all piping and equipment to allow easy access for hoisting equipment to the top of the column. Figure 11-11 shows a small platform at the top of the column in the foreground which is used for a pulley and hoisting arrangement (davit).
(11) In general, space towers a minimum of three to four diameters apart.
(12) Provide a minimum of 10 ft (3 m) between the edge of the tower shell and the centerline of the pipe rack, using the largest-diameter vertical vessel (14 ft (4.2 m) to 16 ft (4.8 m) from centerline of tower to pipe rack is sufficient).
(13) Locate all vertical equipment along a common centerline where possible. This arrangement depends on the type of unit and cannot be taken as a hard and fast rule. Figure 11-12 shows three columns arranged along the same centerline. These three columns and towers have a definite functional relationship to each other; the columns shown in Figs. 11-13 and 11-17 do not have this relationship.
(14) Establish all instrumentation, ladders, and valving installations on the rack side of the column where possible.
(15) Establish control valves between the column and pipe rack where possible.

(e) Horizontal Vessels: Drums, Tanks, Pressure Vessels
(1) Inlet and outlet nozzles and associated piping should be established at opposite ends of the vessel.
(2) Inlet nozzles should be provided on top of the vessel; outlet nozzles and piping go on the bottom under normal operating circumstances.
(3) Provide adequate platforms and cages for vessels that are above grade.
(4) Cages should be provided for all ladders to platforms 20 ft (6 m) above grade.
(5) The platforms and manways should be established for easy access to the internals of the horizontal vessel. Figure 11-13 shows an end manway for the two parallel vessels and side access manholes onto a platform (such as those shown on the drum in the foreground of Fig. 11-11).
(6) Establish the foundation a minimum of 13 ft 6 in. (4.1 m) from the pipe rack centerline as is the case with other horizontal vessels such as economizers.
(7) Arrange vessels to fall in rows as in Fig. 11-11, which shows a series of horizontal drums and vessels arranged in line with the round concrete pedestals on the same centerline for both drums and exchanger units.
(8) Provide adequate clearance between vessels arranged in series. A minimum of 4 ft (1.2 m) between vessel shells is required. (Specification for exchangers is different.)
(9) Locate piping to and from vessels so that it does not interfere with inspection, painting, maintenance, and operation space between the vessel shell and other equipment.
(10) All process and design elevation requirements for the establishment of vessels must be maintained. Figure 11-22 shows a horizontal drum supported on a concrete pedestal and steel saddle arrangement. Notice the access space underneath and around the vessel.
(11) If vessels do not need servicing, they can be arranged closer together as in Fig. 11-13, where the two desalter units are considerably closer than normal. Notice the rectangular pedestals and steel saddles which are welded directly to the vessel shell. Note also the location of the platform along the entire length of both vessels, providing access to the valving, instrumentation, and piping. The photographs in this chapter showing vessels and vessel arrangements give an excellent idea of the design and layout possibilities.

(f) Control Buildings
(1) Locate control buildings where they provide easy access to the total installation.
(2) Control buildings should be arranged in a nonhazardous portion of the plant.
(3) Materials for control buildings and their specifications should be considered in the early stages of the project.
(4) Adequate space for personnel facilities should be considered in the design and layout of the control building area.

The preceding guidelines are by no means an exhaustive list of specifications for the arrangement of equipment for process petrochemical installations, but they should serve as a general guide for the beginning drafter. Always consult vendor drawings, manufacturer's specifications, piping codes, and standards when designing a unit.

PIPING SPECIFICATIONS

The following specifications deal exclusively with piping in and around processing equipment, piping between equipment, and rack and yard piping. They cover heat exchangers, pumps, compressors, fired heaters, vessels, columns and towers, nozzles and connections, valves, and instruments.

Exchangers

The flow diagram shows the use and function of a particular exchanger unit including temperature and pressure, material handled, flow direction, and flow rate. One standard procedure is to pass the heated material into the bottom of the exchanger unit and out the top. The cooling stream of fluid usually passes in the opposite direction, entering at the top of the exchanger and being removed at the bottom. Exchangers are usually coded according to the Tubular Exchanger Manufacturers Association (TEMA) method, which designates the type of exchanger by using a letter coding. Exchanger data sheets and lists provided by the vessel group give the piping drafter or designer all the information concerning the TEMA coding, dimensional specifications and tolerances, the materials of construction, and the design process that may be necessary for the construction and selection of materials and fabrication methods. Drawings showing nozzle orientations and dimensions are provided for exchanger units. See Fig. 11-24 for an example of dimensioning and drawing procedures concerning exchangers. The following specifications apply to exchanger piping:

1. Design all piping to and from exchanger to provide sufficient lifting and removal space and to eliminate the dismantling of pipelines and valving areas during the cleaning, removal, maintenance, or pulling of the tubes.
2. Establish adequate bypass piping around exchanger units that are subject to corrosion, units that must be cleaned or main-

tained during the process operation, units that may need to be shut down during operation, units that need temperature control provisions, and in other special circumstances. If the exchanger is in a series, or a group of exchangers is designed to operate as one unit, the bypass line and accompanying valving should be provided for the entire series.

3. Design exchanger piping to be supported from grade or from a low-elevation pipe rack as in Fig. 11-3. In general, all piping and valves which may add loads to the nozzles should be supported independently from grade or from the pipe rack. This practice will eliminate disassembly of the piping and components during maintenance.
4. In general, piping should not run directly over the exchanger unit, though in some cases this is done if sufficient removal area is provided in the horizontal spaces next to the exchanger unit (as in Fig. 11-3). Figure 11-12 shows a typical exchanger series where the process material is channeled from one unit to the next.
5. Design piping in and around exchanger units to provide sufficient space between piping and the exchanger shell (usually 2 ft 6 in. (76 cm) minimum).
6. Lay out the piping for exchanger units so that the heated material or steam passes through the bottom of the exchanger unit and out the top (through the tubes). The stream that needs to be cooled should pass through the top of the unit (shell side) and out the bottom unless the equipment has been laid out to facilitate another type of process.
7. Provide a sufficient number of connections and flanged joints to facilitate the removal of any unit or section of piping without disturbing the rest of the area.
8. All piping and valving should be located on one side of the exchanger unit as in Fig. 11-3 so that operation and maintenance can be dealt with on one side of the exchanger unit only.
9. All connections should be provided with adequate bolting space and sufficient clearance for the easy removal of the unit or components such as valves.
10. Valves should be positioned on piping that runs parallel to the unit, not on vertical pipe runs.
11. Figure 11-24, which shows the dimensioning and drawing procedures for drums and exchangers.
    (a) All equipment nozzle dimensions are to the contact face and centerline of the flange. Do not show additional descriptive data such as "face of flange" on drawings. Show gasket thicknesses only when dimension lines extend past the contact face of flanges.
    (b) Place a distinct dot at the intersection of contact face and nozzle centerline to indicate location of the work point.
    (c) Show standard (largest number of similar kind) nozzle rating and face type.
    (d) Make short riser lines perpendicular to the equipment centerline. Make long riser lines vertical (plumb).
    (e) Location dimensions are generally shown on plan view.
    (f) Dimension to a match line. Dimension to the building structural column centerline when equipment is located in a building.
    (g) Do not slope drums or exchangers unless project specifications call for it. Though Fig. 11-24 refers to the dimensioning for sloping drums and exchangers, the dimensioning procedure can be used for nonsloping drums and exchangers.

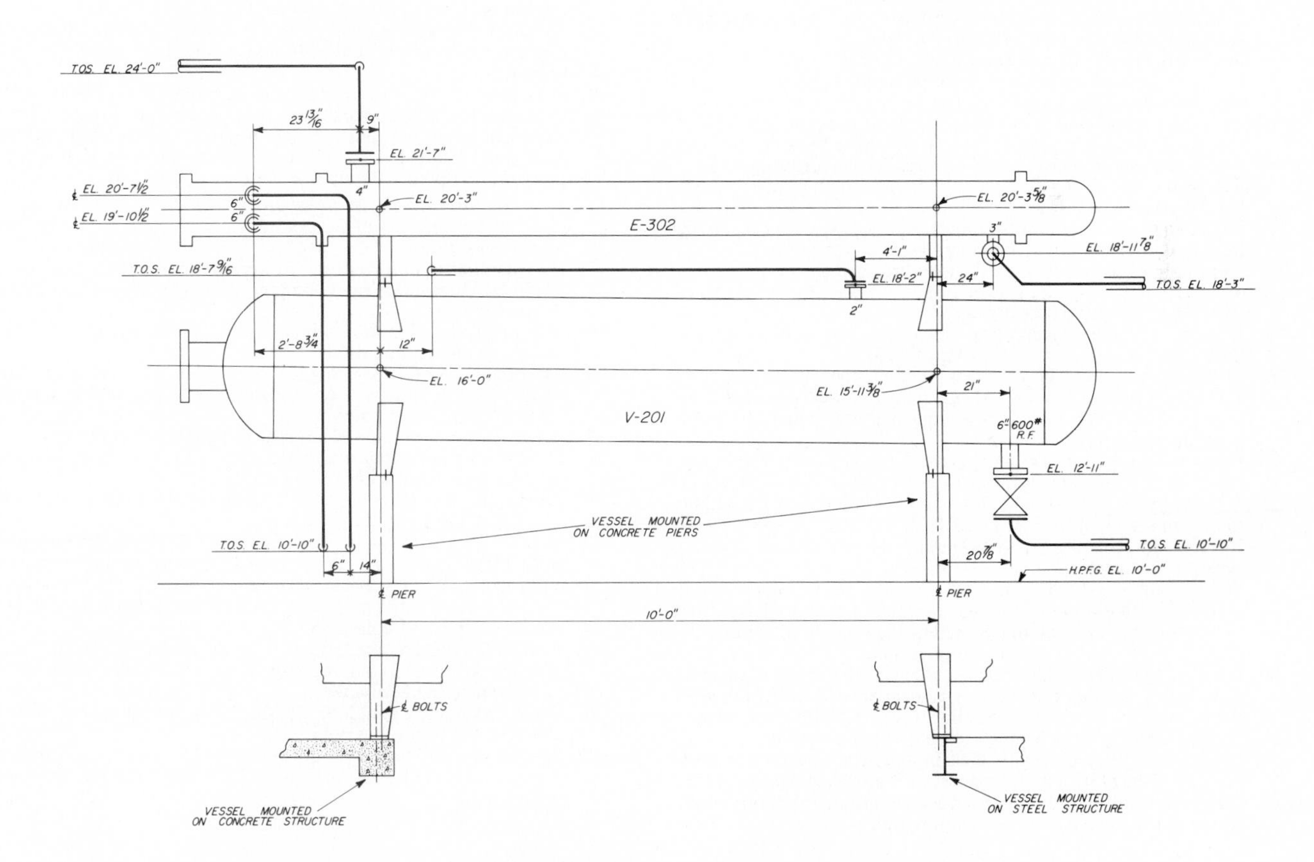

FIG. 11-24 Drafting procedures for drawing and dimensioning drums and exchangers (both sloping and nonsloping vessels).

Pumps

The following specifications concern the layout, arrangement, and fitting location for the piping of pumps. Chapter 10 should also be consulted.

1. Whenever possible, design piping so that it does not run directly over the pumping unit.
2. Provide sufficient maintenance and access space in and around pumps as in Fig. 11-23.
3. The motor end of pumps should be kept free of piping.
4. To reduce the total amount of piping, locate pumps as close as possible to the vessel from which they take suction.
5. Provide sufficient support for all piping and components so that no loading is transferred to the pump nozzles. In Fig. 11-23, dummy supports are provided for the pump lines to eliminate external loading on the pump connections.
6. Locate all valve handwheels so that they project toward the major maintenance side of the pump installation and away from the pump itself.
7. The suction and discharge lines should usually be independently anchored, depending on the type of pump and the design.
8. All auxiliary piping for cooling, oil, and so forth should be designed to facilitate the easy removal and maintenance of the pump.
9. To reduce the pressure drop from the pump suction line, the piping length should be kept to a minimum and changes of direction should be limited.
10. To avoid gas pocketing, eccentric reducers should be installed with sloping side down, on suction line connection.
11. Arrange piping so that flow turbulence is kept to a minimum, especially in the area of pump suction.
12. The use of short-radius elbows and bends, tees, and crosses and branch connections should be kept to a minimum around the suction inlet of the pump except where necessary.
13. As in Fig. 11-23, elbows should be installed so that they turn in the vertical plane. This practice eliminates problems in flow distribution.
14. To reduce turbulence at the inlet nozzle of the pump, sufficient length of straight piping should be provided between the suction nozzle and the last change of direction in the piping.
15. Design suction piping to eliminate all gas pockets where possible. When gas pockets are unavoidable, they should be vented.
16. In general, use long-radius elbows or bends for changes of direction on the suction and discharge side of the pump.
17. Suction lines should be slightly sloped toward the pump.
18. Suction lines using large-diameter pipes should be provided with strainers where necessary.
19. In some circumstances, valves should be avoided in suction lines (though this is not true in Fig. 11-23).
20. Check valves should be provided between the discharge nozzle and block valve as in Fig. 11-23. This practice will stop flow reversal and prevent backup of line material when the pump is not in operation.
21. Reciprocating pumps should be provided with relief valving between the pump and gate valve.
22. A block (gate) valve should always be provided on the discharge side of the pump as in Fig. 11-23.
23. Rotary and centrifugal pumps require the installation of check valves in the discharge line.
24. Provide shock absorbers or flexible connections where quick-closing or remote control valves are installed on discharge lines to prevent shock and undue stress loading on the pumping unit.
25. For line materials with a high incidence of foreign substances and during startup operation, strainers should be installed in the suction line of the pump downstream from the gate valve.
26. Provide warmup lines for high-temperature operations.
27. Bypass lines should be provided in and around pumping units in case of shutdown, maintenance, or removal of equipment. In many cases, this will require parallel piping to pump units where one is used as only a backup pump.

Compressors

Compressor piping requires extra protection in its design and arrangement because of certain hazards which accompany compressor units. Specifications provided by the manufacturer should be closely adhered to for operating and maintenance applications. Stress problems and fatigue failures are common for piping associated with compressors. Therefore stress and vibration analysis may be necessary before the piping and compressor unit are designed. Thermal and pressure conditions should also be analyzed for flexibility depending on the temperature and operation design requirements.

1. For reciprocating pumps, overhead piping should be avoided to eliminate vibration problems.
2. All compressor piping should be supported from grade or structural elements. No external loading should be put on nozzles or connections.
3. Piping should be designed to eliminate overhanging weight which may accompany the installation of heavy-duty valves. All valves and components should be supported adequately from grade or structural elements.
4. Design requirements should be consulted concerning the anchoring and supporting of pipe at nozzles. In general, discharge lines which are connected to the surge drum should not be anchored because of the thermal stress that usually accompanies this design.
5. All fittings should be butt-welded whenever possible except for flange connections provided for breakdown of the unit.
6. Shock-absorbing equipment, bracing, and the like should be provided for severe service.
7. As with pump piping, the use of tees, short-radius elbows, and other drastic change-of-direction fittings should be reduced to a minimum. Use long-radius elbows and bends instead.
8. Avoid the use of short-radius elbows and drastic change-of-direction fittings at both discharge and inlet sides of compressors.
9. Knockout drums with accompanying piping should be installed upstream for compressors which may have condensate problems.
10. Keep suction lines between compressor connections and the knockout drum free of pockets by sloping toward the compressor and making the lines as short as possible.
11. It is sometimes necessary to provide electric or steam tracing and insulation for wet-gas compressors to prevent the buildup of condensation.
12. Provide drains and venting to remove condensate and other liquid buildup where pocketing may occur in the suction line of the compressor.
13. In general, strainers should be installed in suction lines wherever necessary, depending on the service.
14. All piping should be designed to prevent pocketing and low points in the line where condensate buildup could severely damage compressor equipment.
15. Safety valves should be located near compressor buildings.
16. Provide gate or block valves on both the inlet and discharge side of compressors, depending on the service.
17. Provide motor-operated remote control valves for all compressors which are housed in buildings.
18. Check valves are usually provided on the discharge line for both centrifugal and rotary compressors, depending on design requirements. Where check valves are called for, they should be located close to the compressor connection to limit the distance between valve and compressor.
19. With centrifugal compressors, it is sometimes necessary to locate check valves on the discharge lines to prevent surging which may damage the equipment.
20. Surge chambers should be provided for the discharge side of compressors to protect equipment.
21. Pipe fittings and valves should be located to eliminate vibration, surge, and pocketing problems which may damage the compressor unit.

Fired Heaters

The following list of general piping specifications and valves applies to fired heaters. The manufacturer's suggested piping and specifications should always be consulted before designing and arranging piping in and around fired heaters. Figure 11-9 is an example of a fired heater.

1. For all units that require decoking, dropout spools should be provided. For mechanical decoking, flanged fittings should be provided. For frequent decoking units, permanent steam and air connections should be installed on the unit to reduce the amount of downtime for decoking and also eliminate the breaking down of process piping in and around the unit.
2. Blowdown valves should be located a sufficient distance from the heater unit to enable manual operation in all situations.
3. Install blowdown valves on transfer outlet and charge inlet lines below the level of the lowest coil unit.
4. Blowdown valves should be specified as a minimum requirement of 100 percent the diameter of the maximum coil size.
5. Install steam-out valves on all charge outlet and transfer lines (in conjunction with the blowdown system where necessary).
6. Steam-out valves should be operable a safe distance from the heater unit.
7. Fuel gas lines should be provided with condensate removal units, drains, and other connections to prevent buildup of line material, pocketing, and condensate. The only way to reduce this problem is to slant all pipelines toward the heater unit.
8. Install remote control, automatic fuel regulator valves upstream from the distribution header that may be required for the fuel supply lines.
9. All fuel oil lines should be designed to carry at least two times the requirement for the heater unit.
10. Provide block valves and shutoff valves between fuel line, firing line, and headers.
11. To regulate pressure at the burner installation, a pressure-reducing valve should be installed near the heater on the inlet line.
12. Provide steam traps where necessary on lines in which condensate may form (see also Chapter 10).

Vessels and Storage Tanks

This section covers specifications for vessel piping including pressure vessels and storage tanks. Vertical vessels are covered in the column and tower specification list in the next section. In general, the piping drafter and designer must consult all the specifications laid down by the ASME boiler and pressure vessel codes concerning piping to and from pressurized vessels. This code provides the design and fabrication specifications for piping, components, fittings, and valves in and around pressure vessels, including relief valves.

1. In the majority of cases, inlet lines are located at the top of the vessel and outlet lines at the bottom unless the design requirements or specifications say otherwise.
2. All valving installations, instruments, and sections that may be frequently dismantled must be accessible from platforms.
3. Design piping in and around vessels to leave one side of the vessel free of piping for general access to the vessel shell.
4. Since most vessels are elevated a considerable distance above grade, piping must be supported on racks running above the vessel or on one side of the vessel shell only.
5. Locate all nozzles, manways and other service points for convenient access from platforms.
6. Block valves should be provided at nozzles for safety; relief valves should be provided on reboiler and vapor lines, transfer lines, and steam draw-off lines.
7. For piping associated with storage tanks, it may be necessary to provide separate suction and discharge piping and connections, depending on the design.
8. All valves associated with storage tank piping should be constructed of steel to maximize fire resistance.
9. Provide expansion bends for all storage tank piping which may be subject to thermal or settlement stress.
10. Provide drainage systems, including drains, draw-off systems, and discharge lines, between storage tank and sewer.
11. Piping used for transfer to tank contents, including pump manifolds, should be designed to absorb any thermal or seismic shock.
12. Limit the length of pipeline between tanks and pumps.
13. Single suction lines should be used whenever necessary for tankage piping unless the design requirements call for a manifold pump.
14. Suction lines along with piping to tankage should be kept a minimum length.
15. Lines should be provided with draining, venting, and purging equipment and devices where necessary.
16. Provide an adequate number of flange connections between pumps, tankage, and other equipment to allow for frequent disassembly.
17. Piping in and around tank farms and other storage areas of volatile material should be run underground.
18. When piping extends aboveground in storage areas, it should be fireproofed.

Columns and Towers

Towers represent the most complex piping needs because of the high-temperature, high-pressure, and elevation problems encountered in the design and layout. If the vessel is pressurized, the ASME code should be consulted.

1. All piping connected to the vessel should be conveyed directly downward from the nozzle, using a 90-degree elbow or other change-of-direction fitting, as shown in Figs. 11-10 and 11-11. In Fig. 11-3, this procedure has been somewhat altered in order to run the vertical piping from the tower in parallel banks.
2. To establish a logical sequence of downcomer lines from the tower, the piping drafter and designer should consider the piping at the highest elevation first.
3. Arrange all tower piping, nozzles, connections, manways, instruments, and valves to be readily accessible from platforms for easy installation, maintenance, and operation of the equipment.
4. The drafter must consult the P&ID which specifies the column piping, including connections, blinds, valves, and instrumentation.
5. The detail drawing of the column should be made available to the piping designer. Figure 11-20(i) is an example of a fabrication drawing for the quench tower.
6. Piping drafters and designers require the following information in order to lay out, arrange, and design piping in and around columns:
   (a) P&ID
   (b) Plot plan
   (c) Fabrication drawing or sketch of column
   (d) Drawings or specifications of the internal aspects of the column
   (e) Line designation list
   (f) All code restrictions concerning plant operation, platform height, ladder lengths, cage requirements, and so forth.
   (g) Nozzle elevations and orientation (if important).
7. Arrange vertical piping on the column to run outside the platform area. Keep a minimum of 12 in. (30 cm) between piping and the column shell 16 in. (50 cm) if insulated.
8. Piping should be arranged so that one whole area of the column is free of platforms, piping, and other components in order to facilitate the lowering or lifting of equipment by the davit. Downcoming pipe runs from the top of the tower to the bottom are taken off horizontally, leaving a minimum of 7 ft 6 in. (2.2 m) clearance between grade and the bottom of pipe.
9. Trays should be spaced to permit the logical location of nozzles and manholes.
10. Vent holes should be provided in the skirt of the column along with necessary piping drains, siphon vents, and the like. In some cases, a bottoms pumpout line may be necessary (see Figs. 11-20(c), 11-20(d), and 11-25.
11. Piping should be arranged to provide adequate access to manways leading to the internal parts of the tower.
12. Arrange valving on column piping to eliminate pocketing.
13. Provide platforms to all valving on the column.
14. It is usually necessary to establish a nozzle, short pipe run, and relief valve on the top of columns, depending on the design requirements. Many of the photographs in this and other chapters show this installation.
15. Relief valves should be provided at the highest elevation of a line which requires relieving.

16. When possible, install valves from all lines coming from inside the skirt (underneath the tower) so that the tower is operable from outside the vessel.
17. Instrumentation and instrument piping–including connections for temperature, pressure, and level glasses–must be located according to process requirements.
18. All instrumentation should be accessible for easy maintenance and operation.
19. In general, manways and connection locations should be standardized for a particular tower.
20. Attempt to arrange nozzles and manways on one side of the tower only.
21. Because of process complexity and tower size, it is sometimes necessary to establish multiple mini-platforms as in Fig. 11-3; in this case, they should be arranged along one side of the column only. These small platforms, used between and below the larger circular platforms shown on this tower, allow access to valves, level gauges, and instrumentation. (Manways must have access from the larger platforms.)
22. Figure 11-9 shows the use of a variety of octagonal platforms. At the lower levels, many of these platforms are connected to provide solid grouping.
23. Most towers, as in Fig. 11-11 and elsewhere in this book, are provided with platforms at the highest elevation to afford access to the piping, nozzles, valving, and manways associated with that level of the column. The layout depends on the column and its use.

Nozzles and Vessel Connections

All the vessels, tanks, columns, and towers covered in this and other chapters require a number of different nozzles and connections, depending on the vessel and its use. The connecting pipe size and other factors determine the size, type, orientation, and arrangement of the connection. Nozzles are used for all pipe connections to vessels. Figure 11-3 shows a heat exchange unit with five separate nozzles for connecting the pipe to the vessel shell. The nozzles are pieces

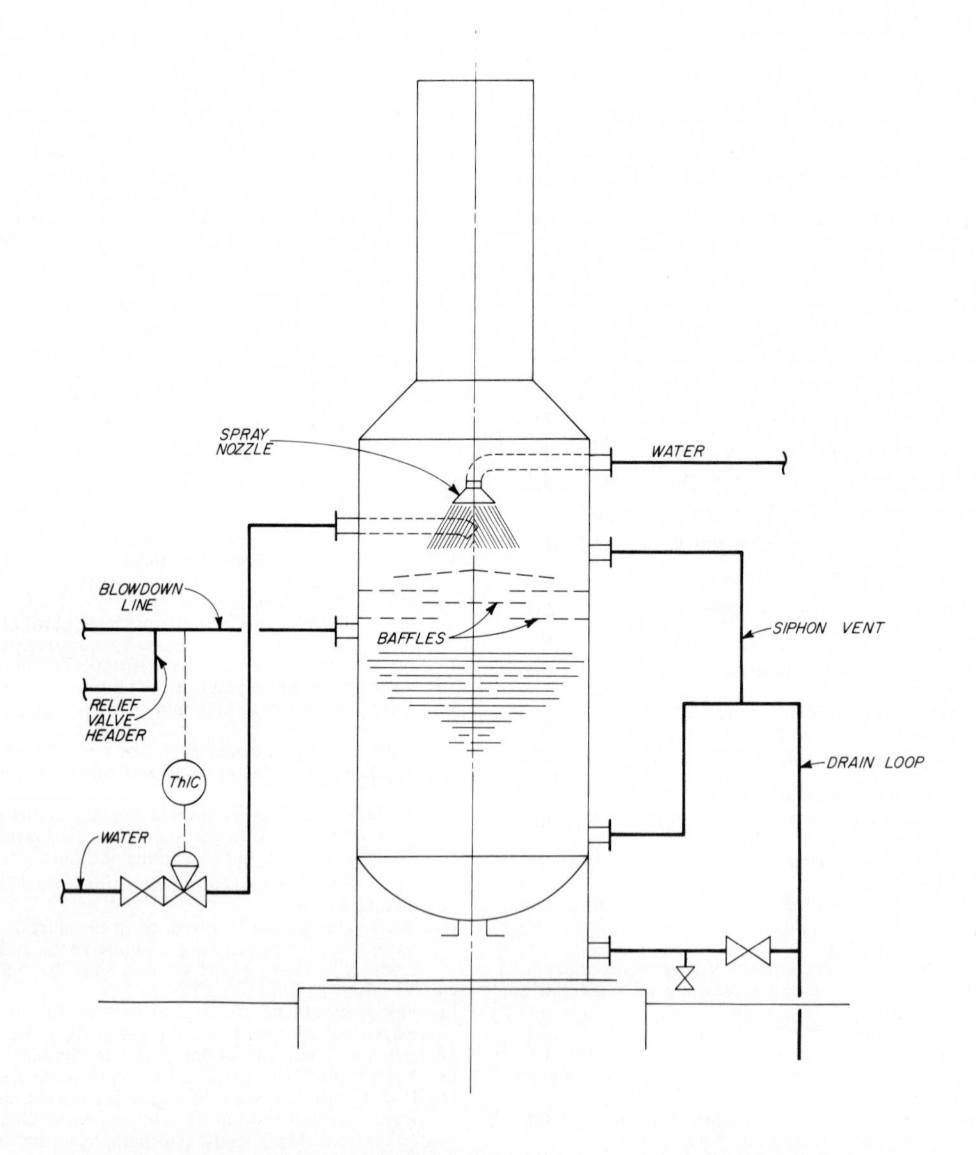

FIG. 11-25 Simplified flow diagram for a water-quench blowdown vessel.

of pipe (stub-ins) that are welded to the vessel shell; one end has a welded companion flange. In some cases, long-neck flanges are welded directly to the vessel.

Flanges connect with similar flanges welded to the pipe as in Fig. 11-3. Figure 11-12 shows another use of welded nozzles and flange connections. In this case, the vessel is insulated, as is the inlet and outlet piping. These vessels and piping are used for critical service—note the amount of insulation and multiple bolting of the vessel connections and vessel head. For pressure vessels such as these, the ASME boiler and pressure code must be consulted for the connection and nozzles to be used. In many cases, reinforced connections are called for as in Fig. 11-12. Figure 11-3 shows a pressure vessel with piping and connections for less critical service, this requiring no reinforced nozzles.

In some cases, the nozzle outlets are extruded. This extrusion is sometimes used for pressure vessel nozzles, though it is somewhat limited in use because of the cost. One advantage of extruded nozzles is the consistent flow pattern which can be obtained.

Other connections and nozzles are also available, depending on the pressure and temperature range and the service needs. Sockolets, threadolets, and weldolets are used for smaller lines. Instrument connections such as those for level glasses are usually simple. Pressure vessels and nonpressure vessels require many different connections for a variety of design circumstances—inlet, outlet, drains, and manway connections and different instrument connections such as level gauges, pressure gauges, and temperature gauges. All these connections must be made according to the appropriate piping codes and standards. Moreover, the drafter and designer should consult the company standard for common practices.

All connections to vessels must take into account a variety of design and service needs, such as flexibility, and anchoring certain connections. In general, vessel connections are considered anchors for piping, though most piping is designed and supported to eliminate external loading to the vessel shell which may cause excessive stress at the connection point during thermal or mechanical movement. When locating valves near nozzles and connections, it is important to take into account the extra weight imposed by the addition of the valve to the pipeline. In most cases, the valve and piping must be supported independently of the vessel and accompanying nozzle connection.

Figures 11-20(h) and 11-20(i) (and Fig. 9-14 in Chapter 9) are typical detail drawings of vessels and towers. In all these drawings, the nozzle size, type, and orientation have been given along with accompanying dimensions for the fabrication of the vessel. Nozzles and connections are always considered an integral part of the vessel drawing. In general, the engineer determines the orientation, location and elevation of nozzles and connections to all vessels. The piping drafter then takes the engineer's sketch and makes the complete fabrication drawing, showing all the locations, sizes, and dimensions. This drawing then goes to the piping fabricator (Chapter 9). Figure 11-24 shows the detailing of a drum and exchanger. Most of the dimensioning suggestions in this figure are for sloping drums and exchangers; note that the nozzle locations orientations, and elevations have been determined.

The elevations required for certain connections are critical items, especially when the vessels are situated close to grade as in Fig. 11-12. The distance between the flanged connection and grade becomes extremely important for installation, maintenance, and servicing. In general, most nozzles are oriented along the centerline of the vessel. This is not always the case, though. In Fig. 11-3, the two nozzle connections on the right side of the heat exchange unit enter and leave the vessel shell horizontally near the top and bottom of the exchanger (instead of the normal vertical inlet and outlet connections shown in Fig. 11-12).

Valves

The selection of all valves is usually done by the process engineer, though the piping drafter and designer should be well acquainted with the different valves that are commercially available. Chapter 4 covers valves in greater detail; Chapter 10 discusses valves and control valve stations.

The location and orientation of valves and handwheels must be determined according to the design and process needs. Figure 11-26 shows a valve installation; the handwheel orientation for some valves is vertical, for others it is horizontal, and some are at a 45-degree angle. All these locations and orientations depend on accessibility, piping arrangements, and the ability to support this part of the system. Access to valves is extremely important—especially in and around elevated vessels which require that valves be reached from platforms and ladders.

Valves, because of the added weight, make the piping system considerably heavier than the normal piping run. This weight must be taken into account when supporting the system and locating the valves. Moreover, the selection of valves must be determined according to pressure, temperature, and process requirements. The capacity of the valve body to handle loading in line stress should be taken into account. The following specifications will help determine the location and arrangement of valves:

1. The location of valves is usually determined by the piping designer or process engineer according to the process requirements shown on the P&ID flow diagram.
2. Locate all valves so they may be properly serviced, installed, repaired, and removed from the line.
3. Where possible, locate valves to be operable from grade or provide permanent platforms for valve installations at higher elevations.
4. Valve handwheels which are more than 6 ft (1.8 m) above grade or above platform elevation should be equipped with extension stems or chain operators unless the valve is 1 in. (2.5 cm) or smaller.

5. Valves that are located below grade and cannot be operated efficiently should be provided with extension stems for the handwheels.
6. Locate valves with extension stems and chain operators so that they do not interfere with access to other equipment and will not be damaged during maintenance and operation.
7. Locate valves whenever possible near, but not in, general access areas, such as aisles and walkways parallel to platforms.
8. Do not locate valves on pipe racks unless absolutely necessary.
9. Whenever possible, locate valving in horizontal runs only. Valves in vertical lines tend to trap excess line material when closed and cannot be drained completely.
10. Locate valves in vertical runs only where complete drainage is possible during normal operation of the system.
11. All important valving should be located in accessible situations and grouped where possible into valve stations.
12. Do not locate valve handwheels too close to piping around the valve; ensure that operators will have sufficient hand room.
13. Locate all safety control valves and emergency valving well away from hazardous equipment so they will be operable during an emergency.
14. Attempt to locate control valves and control valve stations a standard distance from the pipe rack stanchion, and establish a standard handwheel orientation from the station. (Example 2 ft. (60 cm) from pipe rack centerline.)
15. Locate control valves near pipe rack stanchions and away from equipment whenever possible. This practice permits grouping and aligning of control valve stations near aisles and walkways and allow for easier operation, maintenance, and repair.
16. Relief valve piping should be located in accordance with API, RP 520, and ASME boiler and pressure vessel codes, depending on the design requirements.
    (a) Locate pressure relief valves so that inlet piping is short and creates no pockets.
    (b) When pressure relief valves are used on compressors and pumps, it may be advantageous to install the valve a sufficient distance from the equipment to eliminate vibration and turbulence caused by the compressor or pumping equipment.
    (c) Pressure relief valves require the safe removal of discharged line material. In petrochemical plants, blowdown is usually disposed of into a closed system or channeled to a flare stack. Very seldom is discharging into the atmosphere possible with current environmental codes.
    (d) Provide blowdown stacks and blowdown drums for systems which require the separation of vapors and liquids.

FIG. 11-26 Piping manifold for alumina treaters. This plant has been in operation for a long time. Notice the wear, corrosion, and overall dirtiness in comparison to the plants shown in Figs. 11-9 and 11-11.

Instruments

Instrument piping includes all piping used to connect instruments to process equipment and to other piping. It includes level glass installations, level alarms, pressure gauge installations, level controllers, all hydraulic or air-operated equipment, and all control piping for remote control equipment. The drafter and designer should consult the relevant specifications and needs for a particular instrument and its associated piping.

The conditions under which the instrument piping must operate are an extremely important consideration in determining the type of piping and connections to be used. All external conditions such as temperature, external loading, and weather must be taken into account. Moreover, the internal pressure and temperature of the primary piping and equipment must be considered because of thermal and mechanical stress. The mechanical strength of the instrument connection must also be taken into account.

Most instrument piping uses 1 to 3/4-in. (2.5 cm to 1.9 cm) couplings, depending on the service. These connections are susceptible to external and internal shock because of their size. Since instrument piping is used infrequently, it is subject to freezing and other problems associated with infrequent use of a system containing fluids. It is sometimes necessary to steam trace or electrically trace instrument piping lines to ensure efficient operation of the instruments under all conditions. (Chapters 6 and 10 cover instrumentation and instrument piping including level glasses, level gauges, controllers, and pressure and temperature gauge connections.)

UTILITIES FOR REFINERIES

Utilities form an integral part of the total refinery piping system and rank only second to process piping in importance. Utilities include flare, blowdown, fuel, water, gas, electricity, air, steam, and condensate systems, along with waste disposal utilities which are usually considered part of the drainage system. (A general description of utilities is provided in Chapter 6.)

Flare System

A utility of primary importance is the flare system. This system requires its own set of piping for the disposal of vapors and gases to the flare stack, which is usually located away from process areas because of its combustible nature. Flare system piping must be designed so that the pipelines and flare headers are self-draining—condensate or liquid pockets must not form in the lines. The primary function of the flare system is to burn waste gases from relief valves, blowdown from process equipment, and vapors from other process and fuel lines so that they do not cause environmental pollution. Most refineries can be seen from great distances at night because of the many different flare stacks protruding into the sky.

Location of the flare system's pipelines, headers, and pipe runs should be given primary consideration in the early layout of the pipelines on the pipe rack. They should be located on the outer edge of the lower pipe rack and arranged so that the headers are on one side of the rack only. All flare lines, headers, and so on must be provided with drains and valving to draw off condensate or liquid buildup. In areas where temperatures reach freezing, consideration should be given to steam tracing or electrical tracing of the line along with appropriate insulation to prevent condensate or liquid from freezing in the line. Materials for flare system components and piping are determined by the service and the medium conveyed.

Fuel Utilities

All petrochemical installations use some form of fuel for combustion in order to create heat for the process. The fired heater shown in Fig. 11-9 is designed to burn both liquid and gaseous fuels. In many cases, the refinery actually burns some of the products it creates. Refineries in the past used to operate on natural gas, but with diminishing supplies refineries are now using their own end products for the combustible material needed for process heat. It is often possible to use hydrocarbon waste products or channel waste gases to fire furnaces, fired heaters, and boilers. Fuel oils are also used. Each fuel used for the refinery process has its own separate piping system and pumping unit to convey the gas or liquid from the storage area to the burner unit. These piping systems must be designed to limit the amount of pocketing and are also provided with traps and strainers to prevent foreign particles from entering heater or boiler burners.

In cases where liquid fuels are used, it is necessary to provide insulation and steam tracing, depending on the type and consistency of the fuel. For gas lines, condensate and steam traps are located at logical pocketing or low points in the pipelines. In general, cast iron and steel piping are used for the refinery fuel utility lines. Steel piping is used primarily for aboveground service because of its better fire protection safety record; cast iron is used for underground service.

Water Utilities

Water utilities are divided into high-pressure and low-pressure systems. High-pressure water utilities are used for firefighting; low-pressure services are primarily for drinking water, for cooling water, and, generally in a separate system, for sanitation. The different water services can sometimes be located underground in permanent lines to reduce piping congestion and also to reduce the possibility of adverse temperature effects on the water system, which can cause freezing or external loading due to snow and high winds.

In general, water piping in and around the process unit is conveyed on a pipe rack according to pipe rack specifications for the location. Steam and electrical tracing is also provided along with sufficient insulation to prevent freezing in cold climates, especially for pipelines which are infrequently used (for cleaning services, for instance). A low flow rate increases the possibility of freezing.

Cooling water systems are usually separate from the firefighting and sanitary service water and are used primarily for such items as the water-quench blowdown system shown in Fig. 11-20 and for cleaning equipment like heat exchangers. Cooling water is also used for the heat transfer for condensers and boilers and as makeup water for steam generation. In some cases, water is passed through a cooling tower or other cooling unit. An attempt is made to reuse the utility cooling water as much as possible by means of cooling towers and ponds, adding makeup water to the system as necessary. All water lines should be provided with sufficient drains and should be sloped to prevent pocketing when the water is not being used or when the system is shut down for maintenance or repair. In general, low-pressure water utilities are used for cooling water in processing hydrocarbon products.

Often separate water systems are interconnected so that cooling water or sanitation water can be directed for firefighting in case of emergency. This use requires the addition of emergency pump services so this water can be efficiently moved through the system. Strainers should also be provided in water pipelines depending on the service. In general, water used for cooling shell and tube coolers, condensers, and heat exchange units should pass through a strainer when recycled to eliminate foreign material from the lines.

Disposal of water requires separate drainage systems. Water used for process purposes obviously cannot be disposed of directly into the same sewers and drains in which sanitation and firefighting water is channeled. Most process cooling water must be treated before being discharged into local water sources. This procedure requires a separate system of piping for its disposal, though water used for condensers and heat exchangers can usually be disposed of in the sanitary water drainage system.

Air Utilities

Air services are divided into two separate systems: instrument air and utility air (primarily for air-driven motors and for cleaning equipment and pipelines). Air piping systems usually consist of small-diameter pipes and tubing. An integral part of the air piping system is the air compressor. In general, process plants use steel piping and fittings; in some cases galvanized steel, cast iron, and malleable iron fittings are also used.

The instrument air system is usually designed to be completely independent from the utility air system and equipment, though many of the specifications for the utility air system also apply for instrument air. Various filtering mechanisms, discharge connections, and drainage devices should be provided on all instrument lines, depending on the service. It is also good practice to design the instrument air system with local mini-compressors instead of a single instrument air system for the entire plant. Backup compressors should be installed in case of power failures and other problems. In general, service outlets for air should be kept to a minimum of 1 in. (2.5 cm) diameter. Outlet connections should be situated at various points in the process plant to minimize the length of hose connections to portable equipment.

Steam Utilities

Steam generation is a normal operation in most refinery systems, depending on the process. In general, fired heaters and boilers are an integral part of the equipment in a typical refinery. Steam is used for distillation and fractionating columns as a source of heat; sometimes it is channeled directly into the process material to produce a specific reaction with hydrocarbon compounds. Most refineries have more than one steam system as well as a separate steam exhausting system, all operated under a wide range of pressures and temperatures which must be taken into account when specifying valves, fittings, and pipes.

Steam piping systems must be designed to eliminate pocketing and the accumulation of condensate. One of the designer's primary concerns should be to create a system which will distribute clean, dry steam throughout the system. Numerous steam traps, drains, and other devices such as knockout drums and steam separators are provided to eliminate problems in the steam lines. *In general, steam branch lines should be designed to extract steam from the top of the main header instead of from below to eliminate condensate problems in this portion of the system.* Steam should be exhausted without the freezing of condensate and without burning operators and maintenance crews–in general, steam is exhausted into an appropriate sewer drain. Another problem with steam utilities is the thermal movement usually associated with high-temperature/high-pressure piping systems. Steam piping is explained in more detail in Chapters 12 and 13. Expansion loops, bends, and spring hanger arrangements are a necessary part of the steam utility piping system. In some cases, expansion joints are provided to limit the amount of thermal shock and stress on joints.

Condensate systems are another integral part of the steam utility service piping. Condensate must be extracted from all steam lines in order to provide dry, clean steam for an assortment of services. If condensate is reused in the steam condenser-condensate cycle, the system is said to be *partially closed.* Makeup water is added only when necessary. Condensate removal systems require that the condensate be channeled into an oil-free sewer or drainage system, not into the sanitary drains and sewers.

Steam traps form an essential part of the condensate removal system and must be provided for all steam lines, both exhaust and line systems. Steam traps should be located as near as possible to major pieces of equipment or major pipelines, though they must be serviceable during the operation of the plant. Steam traps should be located in conjunction with turbines, separators, heat units, and other major pieces of equipment along the steam line; they should be designed with discharge piping of sufficient size to allow for the maximum discharge that may occur. If the discharge line is open to the atmosphere, it must not cause any injury to surrounding equipment or personnel. Many types of steam traps are available and there are many specifications regarding the design of the relevant piping and valving. (See also Chapter 4 on valves.)

Waste Disposal

Waste disposal is one of the principal utilities of the process petrochemical plant. Wastes in the form of liquids, vapors, gases, and solids are usually contaminated with varying levels of hydrocarbons and pose a critical environmental problem. Therefore their disposal and treatment constitute a serious matter for consideration in the earliest stages of the design and layout of the process plant. Water treatment and hydrocarbon treatment areas take up considerable space in most refineries. Involved is the collecting, storing, and discharging of wastes, most of which must be treated before being released into the ground or a local water source. The design and layout of the sewer and drainage system become an essential step for process engineers after the plot plan has been established. The drainage, venting, and sewer systems are all part of the waste disposal utility (see Chapter 10). Waste disposal is usually divided into separate systems, depending on the waste—drainage and sewage systems are provided for each type of waste. In general, any water or waste containing hydrocarbons, such as process water which comes in contact with oil, must be channeled into the oil waste system. Most petrochemical installations have separate waste disposal sewer systems for oil-containing wastes, for general sanitary and runoff water drainage, and for waste waters which contain very low levels of oil contamination.

After the plot plan has been established, it is customary to do the drainage plan and, after that, the sewer system and waste disposal piping system. Any oil-bearing water must be handled separately from the sanitary disposal system and be channeled to a separator where the oil is removed from the water. Because drainage from the surface areas around the plant also contains a certain amount of hydrocarbons, these must be channeled into appropriate sewers. Surface drainage is usually designed as a separate system from the other waste disposal systems. If this is not possible because of the overall size of the plant, surface drainage containing oil can sometimes be channeled into the same sewer system as the waste process water.

All uncontaminated waste water from condensers, boiler blowdown, and so forth can be collected and channeled to the appropriate discharge area without treatment. In many cases, local codes require that even this water pass through oil separators. Sewage is usually hooked up to the municipal system when this is available. Water from drinking outlets, shower facilities, and toilets, as well as runoff water which is not contaminated with oil, is usually channeled into the local discharge system. Waste disposal systems also convey acids and caustic materials that must be collected and treated separately from the other waste systems. Treatment ponds and basins are supplied for caustic and acidic discharges and for the treatment of other oil-bearing waste.

Drainage and waste disposal systems are designed to be gravity-operated so that chemical water, and hydrocarbon-contaminated streams can be easily channeled to the appropriate sewer via drains and piping systems. The *flare system* is a waste disposal system that is used for flammable mixtures and vapors which must be channeled separately from normal sewage, though it is impossible to eliminate all flammable substances from sewers. Much of the waste disposal system is therefore oriented toward the removal of hydrocarbon-contaminated water. The design and arrangement of waste disposal systems and sewage systems are primarily the job of the process engineer and designer. It is they who must take the environmental and code specifications into account when arranging, locating, and designing the system. The details are beyond the scope of this book.

PIPE SUPPORT SYSTEMS

Pipe supports for process plants are similar to those used for nuclear and conventional power generation systems. Chapter 8 covers pipe supports in detail. This section provides some general specifications for the design, location, and arrangement of pipe support systems for process installations.

The area provided to install, run, and support pipe in a petrochemical installation is usually much greater than that provided in a building for nuclear or conventional power generation. In process plants, it is possible to establish pipeways, pipe runs, and pipe racks to carry multiple levels of piping for different services and utilities. This arrangement greatly simplifies the support system. Figure 11-17 is an excellent example of the use of piping racks to support multiple piping runs. Another factor in the design and construction of petrochemical installations is that most piping will be exposed to the elements. Therefore the design and construction of supporting devices for petrochemical and process piping systems must take into account all the external and internal conditions to which the piping system is subjected. External effects on piping systems include loading introduced by wind, snow, and seismic activity. Other considerations are pressure and temperature variations, thermal and mechanical shock, and thermal expansion and contraction. Because of the area available for most piping installations at a petrochemical plant, these negative factors can be greatly reduced by providing sufficient expansion space and adequate support by pipe racks.

All piping systems must be designed to support the system's total deadweight. Weight calculations must take into account the weight of liquid or water which is sometimes used for hydrostatic testing as well as the weight of pipe and contents during service. The various piping codes must be consulted concerning the design and layout of pipe support devices.

Joint leakage, vibration problems, thermal expansion and contraction, flexibility versus anchoring and restraint devices, interference problems between supports, piping, and equipment, adequate insulation for pipe runs and its effect upon the pipe support system, mechanical movement which may damage pipe joints or equipment connections, stresses which may develop in the piping—all must be taken into account in the design stage. Moreover, adequate compensation must be made for normal expansion and contraction of the piping system during operation. Petrochemical piping deals with a variety of high-temperature/high-pressure processes, and these must be considered in the design of the system. Piping sag and pockets must be eliminated where possible on piping racks. In some cases, this is done by providing a slight slope by placing the pipe on shoes which rest on the pipe rack structural elements. The design of a pipe support system should take into account the following:

1. When possible, pipe supports or structural elements on a pipe rack should be designed to hold equal weights for the same length of pipe.
2. Spring hangers should be used for pipelines which undergo thermal movement.
3. Place supports near changes of direction in the pipe run.
4. Arrange small-diameter pipelines in groups to be supported by trays or to facilitate the close spacing of supports.
5. Where small pipelines are routed individually or in small groups, they should be run near large pipes to be supported more easily.
6. Provide supports in areas where nozzles, connections, and other joints may require frequent dismantling so that the pipe is supported during downtime.
7. Channel pipelines that undergo thermal expansion so they do not interfere with surrounding equipment and other portions of the piping system.

Flexibility

The flexibility of a piping system is of extreme importance when the pressure and temperature range causes thermal movement. Thermal movement can result in the failure of portions of the piping system—including pipe joints, anchors, and nozzle connections. If thermal movement is not calculated and the pipe support system is not sufficiently flexible for contraction and expansion, the piping may develop leaks and put undue stress on portions of the system. Because of the processes carried out in a typical refinery, these failures could damage surrounding equipment and endanger operators and maintenance crews.

Most connections to vessels and nozzles are provided with anchors. Anchoring the pipe near the vessel connections ensures that the movement will not adversely affect the connection at the vessel. Expansion of the piping and fittings must be taken into account in other portions of the system. All pipelines must be able to withstand a certain amount of movement between anchors and connections.

Piping guides, restraints, and slide supports must be provided so that movement is guided along established lanes and does not interfere with parallel pipe runs and local equipment. Hot lines require movement regulation for vertical changes and should be provided with shoes when the pipeline is insulated. The pipe must move freely on the horizontal structural support without adversely affecting the insulation. Uninsulated piping lines are usually supported with direct contact to the supporting structure. Figure 11-17 shows uninsulated pipelines on a structural steel pipe rack. Pipelines are not anchored or secured to the rack but are allowed to float within prescribed limits to accommodate thermal movement. Shoes and sliding supports must be sufficiently long to allow maximum movement along the structural support so that the thermal movement of the pipe will not cause the pipe shoe to come off the horizontal structural support member (bent).

The pipe analysis group usually does a complex set of calculations to determine the expansion and flexibility of the piping system in question. Most pipelines that are buried are seldom in need of formal thermal calculations for flexibility and expansion unless the temperature range of the piping is sufficiently high to warrant it. It is general practice to run most hydrocarbon-carrying pipelines above grade, providing pipelines with the ability to slide on shoes or directly on structural supports. It may be necessary to provide expansion loops or a series of bends and offsets to absorb thermal movement. Expansion joints and couplings are seldom used for petrochemical lines, but in some cases they are called for.

Expansion loops are a primary means of dealing with thermal movement of a pipeline. Whenever possible, pipes which will experience thermal expansion should be grouped together; the largest, hottest line goes on the outside of the rack and the cooler lines go on the inside, providing a series of expansion loops which take up the thermal strain. These loops must be adequately supported. In Fig. 11-17, the change of direction in the pipeline can serve the same purpose as expansion loops and bends. When designing pipelines with expansion loops, an attempt should be made to locate the horizontal loops so that they extend over the piping, not out from the pipe rack, and are independently supported from the rack.

Changing the direction of a pipeline allows the system to become more flexible if the point of change is left unrestricted. U bends are used in some situations to provide flexibility for thermal expansion (Fig. 11-20). When pipelines must be anchored fairly close together, making it impossible for change-of-direction bends, loops, or offsets, it may be necessary to provide expansion joints. Expansion joints should be limited to situations where space limitations make their use imperative or where a pressure drop in the pipeline would be undesirable. It is important to consult the manufacturer's suggested use of expansion joints for a particular process.

Vibration must also be considered when designing piping systems which are connected to mechanical equipment such as reciprocating compressors and pumps. Adverse stress at the joint connections for such piping systems must be taken into account to reduce the possibility of failure.

Yard Piping Supports

Chapter 10 deals in more detail with pipe racks and should be consulted when designing a pipe support system. Pipe racks (or *banks*) are always used for multiple runs of horizontal pipe for refinery systems, as shown in Fig. 11-17 and throughout this book. It is important to fireproof the pipe rack for severe and flammable services. Yard piping should be neatly arranged to provide sufficient access to equipment. In some instances, piping is run on concrete sleepers only 12 to 24 in. (30 cm to 60 cm) above grade level. This practice should be avoided in most circumstances because of the problem of inaccessibility of surrounding areas; moreover, ramps must be constructed to go over such sleepers. Figure 11-8 shows pipelines which are run at grade; in this case, the practice is called for because the piping is on a pier which need not be elevated.

It is the job of the piping drafter and designer to establish sufficient supports, anchors, stops, guides, and shoes for the system in question. Running the pipeline along pipe racks on multiple levels allows for the most efficient support. Figure 11-22 shows a petrochemical installation where the pipe rack has been constructed of precast concrete, rendering it fireproof. Note that the pipelines are set directly on the pipe rack. Moreover, small-diameter instrumentation and utility lines are attached to the vertical members, providing an efficient system which can be easily installed, operated, and maintained.

In general, pipe racks should be designed to accommodate as many pipelines on one level as possible to eliminate the cost of providing extra levels to the rack. Water lines, drains, sewers, and large utilities should not be installed on pipe racks unless absolutely necessary or unless the size of the pipeline will not adversely affect the pipe rack. Cooling water lines, because of their large diameter, are usually channeled underground. Pipelines should be run on the pipe rack so that the heavier utilities are closer to the stanchions, not in the middle of the rack. Large-diameter process lines and large-diameter piping for blowdown and in some cases cooling water should be kept to the outside 20 percent of each side of the pipe rack. The center of the rack should be open for small-diameter pipelines, which are usually utility lines for fuel oil, gas, chemicals, air, instrumentation, steam, and condensate. Where multiple levels are used, utilities should be kept to the top rack; the lower rack should be saved primarily for process and large-diameter pipelines unless changes of direction require that the large-diameter or process lines be taken off the top lines, to higher elevation equipment.

Locating utility lines in the center of the pipe rack makes it easier to take connections running to either side of the piping rack. In all situations, pipelines should either drop or extend in a vertical direction when connecting to a pipeline on a pipe rack. Extension lines or headers should never be taken off horizontally from a pipe rack. The proper procedure can be seen in Fig. 11-17, where all the change-of-direction lines, takeoff lines, and headers accompany a change in elevation. (Figure 11-3 also shows this procedure.) The change of direction also provides flexibility in the system.

Another factor to consider when arranging pipelines on pipe racks is the necessity of locating the line closest to the side from which the takeoff line will connect to equipment. Therefore process pipe lines should be located on the side of the rack where equipment to which they will be connected is positioned.

Insulated pipelines should be provided with shoes which are welded to the pipe and allow free movement along the structural element which supports them. Uninsulated lines can be placed directly on the steel support members. For low and moderate-temperature lines, insulation can be cut away at the point of contact between the pipeline and the structural support member, though care must be taken to provide a sufficiently large cutaway portion lest the insulation be smashed against the structural element during thermal movement.

Instrument lines can be located, as in Fig. 11-22, around the vertical legs of the pipe rack, either on the inside or the outside, depending on the service. Sliding supports, which are provided as guides for shoes, should be designed for minimal resistance by friction to thermal movement. In general, steel is used for these services. Plates of steel are provided for pipelines which lie directly on concrete pipe racks, so that the pipe and the rack are not damaged during mechanical or thermal movement. It is important to design the pipe system so that the pipeline never comes in contact with concrete racks or other concrete such as on sleepers. Contact between piping and concrete causes greater corrosion problems than metal-to-metal contact.

STRESS ANALYSIS

Stress analysis is needed for petrochemical piping systems whenever the pressure, temperature, thermal movement, or external loading can damage surrounding piping and components. Weight and external loading are usually not so much of a problem as thermal expansion. In general, petrochemical piping systems are designed to use a sufficient amount of 90-degree bends and elbows for change of direction. In Fig. 11-13, the piping components connected to the major tower are provided with a multitude of 90-degree bends and loops, to reduce thermal stress. Stress analysis for flexibility and expansion must be done for piping configurations for which designers and engineers do not have adequate information and for special configurations. High-temperature/high-pressure lines for severe service are usually required to have an accurate computer stress analysis. Stress analysis must also be done for pipelines and portions of the system that must be securely anchored because of piping codes and standards. Most pipe analysis is done by the stress group.

CIRCULAR PLATFORMS

The following guidelines and specifications will enable users of the book to familiarize themselves with the procedures for design, layout, and detailing of circular platforms for towers, columns, and vertical vessels. These specifications and drawings can be used for Problems 3 and 4 at the end of the chapter, which requires the establishment of two platforms at appropriate elevations. The general procedure for the design and location of platforms around vertical vessels starts with the piping designer or lead drafter preparing a sketch or outline drawing of the platform showing the desired location of ladders, appropriate elevation for platforms, type and location of handrails, pipe that protrudes through the platform, and the location of pipe supports, if this is part of the platform. For this procedure the designer must have access to the plot plan and standard handrail, ladder, guard, and platform drawings, which are available from the company. Vendor drawings of standard items and accompanying dimensions and specifications are also necessary. The drafter will need this sketch of the platform design along with the appropriate reference material—including the loading data for design and special loads, the column or vessel drawing, the column foundation drawing, and the standard drawings that show company procedures for design and layout of circular platforms. The following specifications are only general:

1. All platform drawings should be completed in 3/8 in. = 1 ft scale.
2. Platforms should be numbered consecutively beginning with the lowest.
3. Ladders should be numbered consecutively beginning with number 1 at grade.
4. Plans should be drawn using the conventional single-line technique for all structural shapes.
5. Use heavy dashed lines to represent brackets and light lines to represent handrails.
6. Double-line drafting procedures should be used for special brackets or details which need clarification.
7. All platforms should be titled and include information establishing the elevation at the top of platform framing member (such as elevation 120 ft 6 in.).
8. Drafters should lay out and size platform framing members.
9. A complete list of materials and specifications must be given for each platform drawing.
10. In general, platform decking should be 1/4-in. (.6 cm) plate unless grating is required.
11. Toe boards should be specified for all pipe penetrations through the platform.
12. Platforms next to each other, though supported from separate columns or vessels, should be connected where possible.
13. Platforms should afford access to vessel shell penetrations, manways, nozzles, and instruments such as level glasses and controllers and provide for the operation, maintenance, and servicing of valves, nozzles, and pressure relief valves.
14. Platforms should be established at the top of the column in most circumstances.
15. Restrict the length of ladders between platforms or landings to 30 ft (9 m) depending on local codes and specifications. Provide intermediate platforms such as those shown in Fig. 11-3 for access to minor equipment and penetrations.
16. Design all platforms, ladders, handrails, and cages to comply with OSHA requirements and local specifications, using standard details provided by the company.
17. Platform drawings should show all the elevations and dimensions necessary for the construction, ordering, and specification of the platform, ladder, cage, and handrail assemblies.
18. Loads should be established on each platform.
19. Show location of manways and nozzles on platform drawings.
20. Vessel, shell, and radius of shell must be shown on platform drawings, which should be taken from vessel drawings.
21. Minimum walking space should be established at 2 ft 6 in. (76 cm) for platform size.
22. Where bracket spacing becomes greater than 6 ft (1.8 m), extra midway support brackets should be added.
23. When the ladder extends down to grade, 8 in. (20 cm) clearance is required at the foundation and the vessel base support ring.

24. Location of toe boards, location of handrail post, detailing and dimensioning of cage bar extension, handrails, loop guards, safety guards, and ladder cage must all be shown on the platform drawing.

Figure 11-27(a) shows a typical platform and ladder detail for one column. This drawing shows platforms at three separate elevations with different configurations depending on the nozzles and penetrations at each elevation. Whenever possible, all platforms for a tower should be kept on a separate sheet. The right-hand corner of the drawing should be reserved for platform 1, the lower left-hand column for platform 2, the upper left-hand corner for platform 3, and so on. All sections and blowups should be confined to this drawing.

Figure 11-27(b) starts the general procedure for laying out and detailing platforms for columns and vertical vessels:

1. Get the outside radius of the vessel shell from the vessel drawing.
2. Show the location of the manhole.
3. Locate the platform framing member 3 in. clear of the shell or 1 in. clear of insulation, whichever distance is greater.
4. Round out dimension A to the next higher 1/2 in.

For Fig. 11-27(c), the procedure is this:

1. Set dimensions R and B and angle θ from the piping study of the platform, maintaining 3 ft minimum clearance from the face of the manhole to the edge of the platform framing member.
2. The minimum unobstructed walking space is 2 ft 6 in. for the remainder of the platform.

For Fig. 11-27(d), the procedure is this:

1. Set bracket spacing (angle θ) based on radius R and the platform framing channel size (from Fig. 11-27b). Check for interference with piping and appurtenances.
2. Select the bracket from standard drawings and show it on the plan.
3. Check for headroom clearance (minimum of 6 ft 6 in.).
4. Add framing members at bracket locations.
5. If bracket spacing cannot be maintained due to piping or other interferences and C becomes greater than 6 ft 0 in., add extra framing members between brackets.
6. Locate the ladder centerline 1 ft 3 in. clear of platform framing.
7. If the ladder is located on the centerline of the vessel, this 1 ft 3 in. dimension will establish dimension B on Fig. 11-27(c).
8. When the ladder extends down to grade, 8 in. clearance is required at the foundation and the vessel base support ring.

For Fig. 11-27(e), the procedure is this:

1. Add handrails, loop guard, safety guards, and ladder cage according to standard drawings. *Note:* Use the standard dimensions provided in these sheets for the projects in the problem section of this chapter.
2. Show the extension of cage bars to handrail where applicable.
3. Use an X to show location of handrail posts.
4. Label and dimension the preceding items only when necessary to define special clearance or access requirements.
5. Show the location of the toe board only at the platform edge near the vessel and at holes in the floor plate.

Figure 11-27(f) is a finished drawing of platform 1 at elevation 110 ft 6 in. It is a typical example of the circular platform drawing which is required for each elevation where a platform is established on a tower or vertical vessel. All required dimensions, framing members, sizes, and codes should appear on this drawing.

Figure 11-27(g) shows a set of specifications and dimensions for the adequate design of the framing bracket. Brackets that exceed the conditions specified here must be specially designed, and calculations for loading and so forth must be taken into account. This detail is adequate for a minimum of 1500 lb superimposed dead plus live load. The bracket with knee brace is a typical method used for reinforcing platforms.

The student should scrutinize the photographs in the book showing the design and location of platforms, ladders, and cages. Figure 11-11 shows a number of platforms at various elevations, including interconnecting ladders and cages. Figure 11-13 is another good example of the platforms and accompanying devices necessary for vertical vessels; note the rectangular top platform, a common practice when the platform extends above the tower. Figure 11-10 shows a variety of ladders, cages, and platforms for process equipment and vertical vessels. Notice the rectangular platform at the top of the vessel to the far left (next to the large double-relief valve). Figures 11-9 and 11-17, which are different views of the same unit, show the use of octagonal platforms, which are similar to circular platforms in many ways. A variety of platforms, cages, and ladders are also shown for horizontal equipment, as in Figs. 11-10 and 11-13. Figures 11-28 through 11-32 show a model of Fig. 11-20. These photographs are of the completed model assembly; as can be seen, two platforms have been established on this model. Students doing Problem 3 at the end of the chapter—requiring the drawing of the complete set of plan, elevation, section, and details in Fig. 11-20—should incorporate into the project, the design, detailing, and modeling of circular platforms at appropriate elevations.

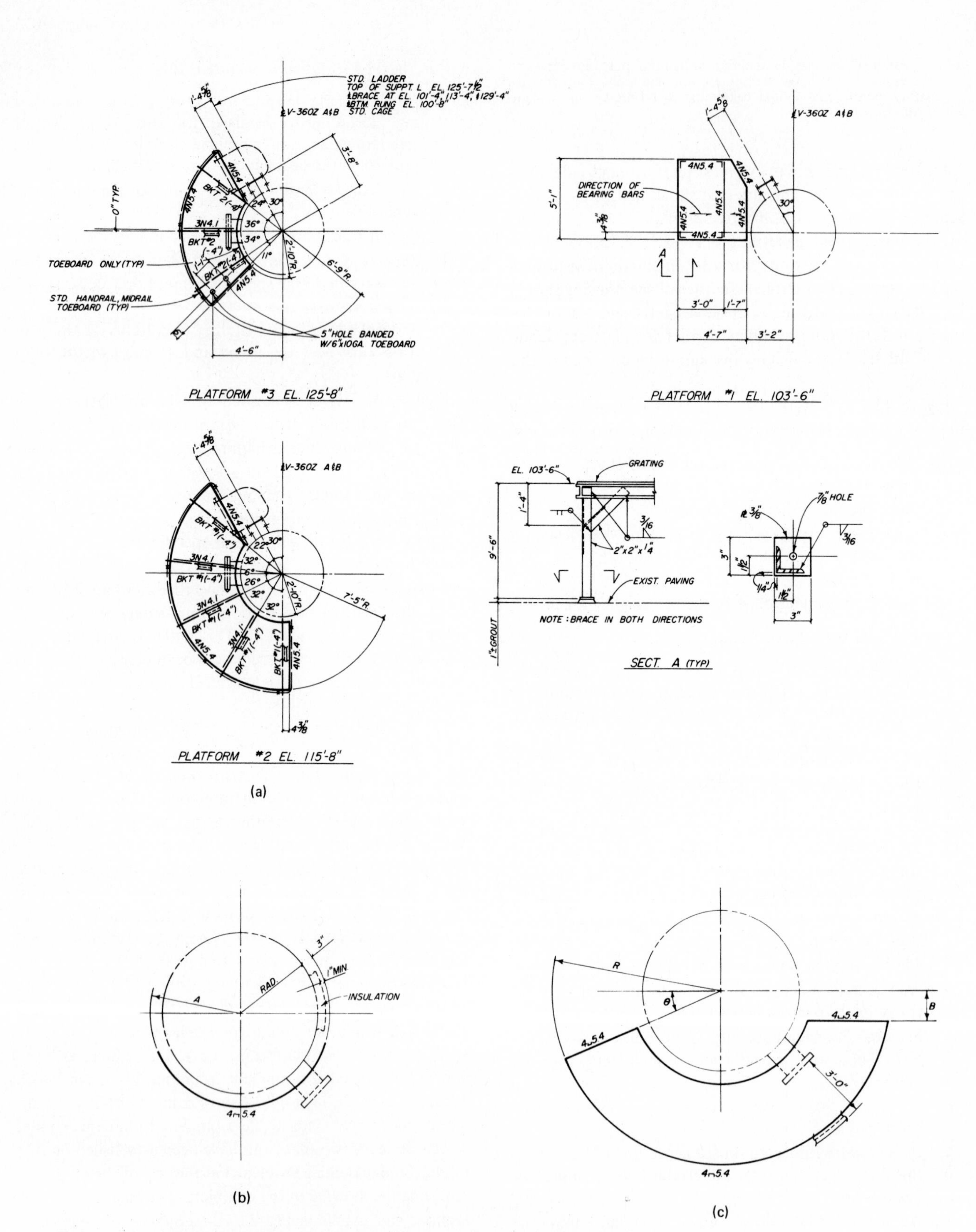

FIG. 11-27 Drafting and design of circular platforms used for vertical vessels.

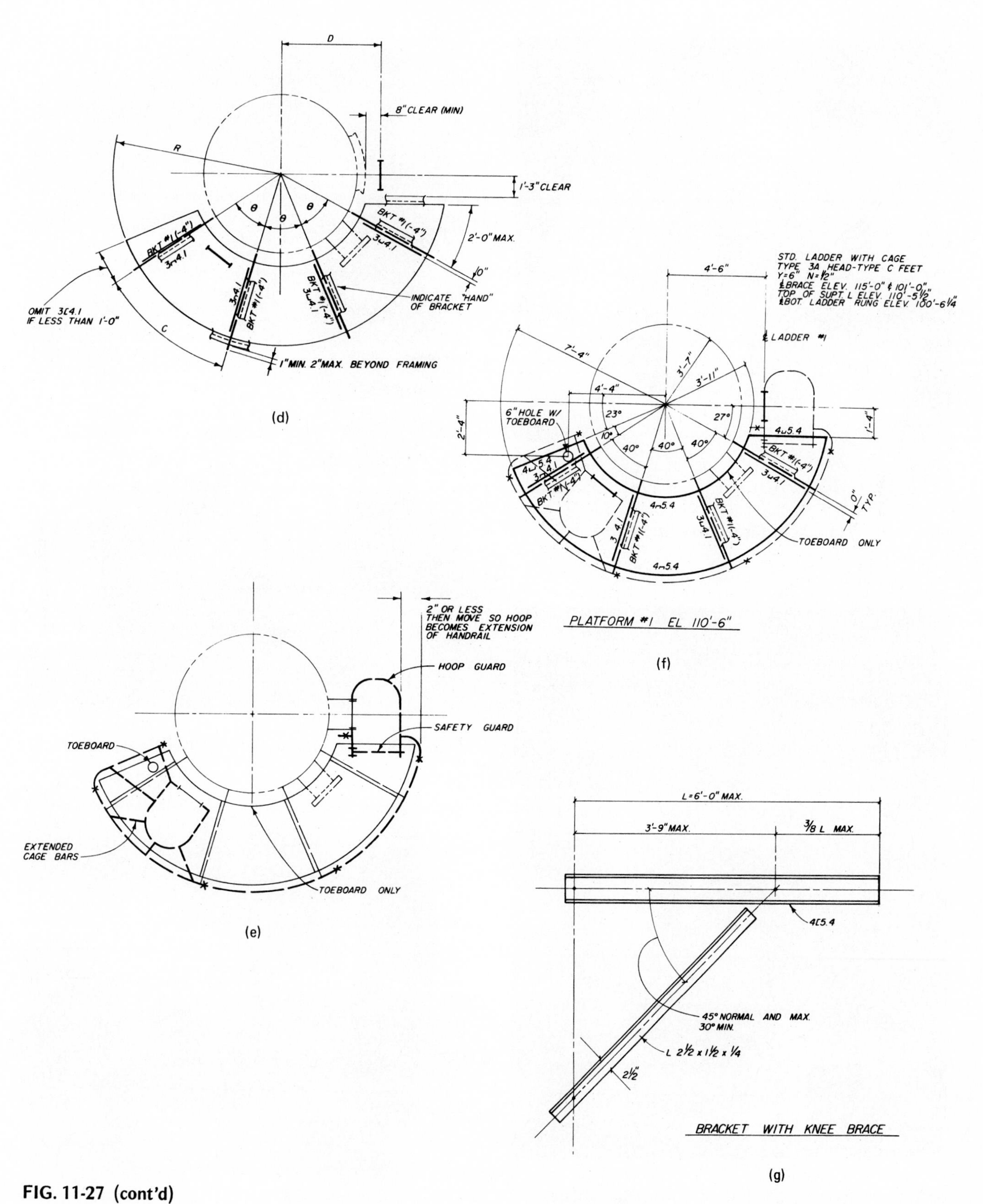

FIG. 11-27 (cont'd)

FIG. 11-28 Plan view of the refinery flare system.

FIG. 11-29 Rear elevation view of the refinery flare system.

FIG. 11-30 Front view of the refinery flare system.

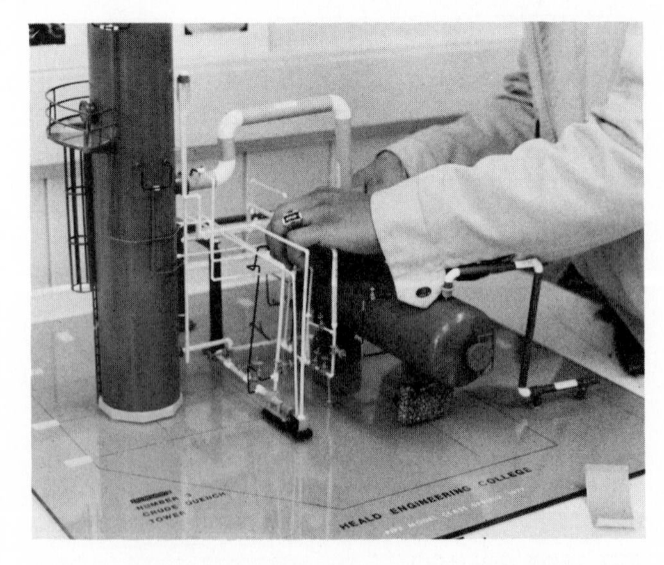

FIG. 11-31 Student tagging the model.

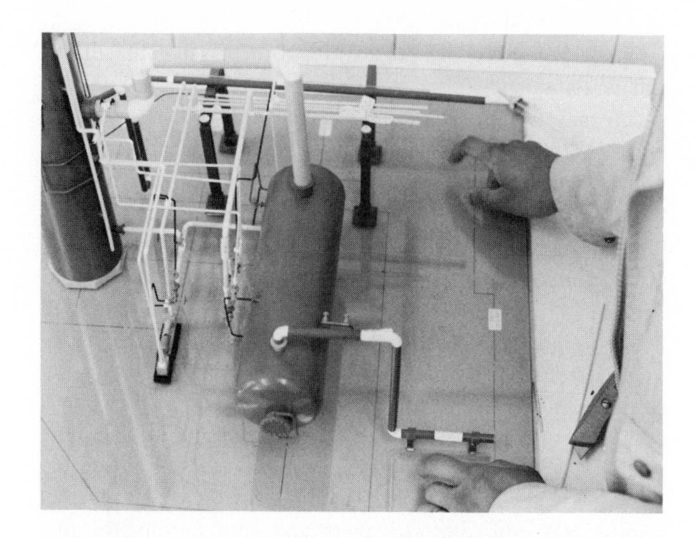

FIG. 11-32 This crude oil, water-quench blowdown model, constructed from the drawings in Fig. 11-20, provides an excellent project for students. By drawing the views provided in Fig. 11-20 and then constructing a model of the project, students are able to see how a typical petrochemical unit is created in industry.

WATER-QUENCH BLOWDOWN SYSTEM

The water-quench blowdown system provides for the sudden cooling of hot material discharged from another processing unit, such as a cracking coil. The cooling is done by injecting cooling water directly into the hot oil (Fig. 11-25).

Figure 11-20 is a set of plan, elevation, section, and detail views of this small unit. Figure 11-25 shows a simplified flow diagram of a water-quench blowdown for the disposal of hot liquids. This vertical vessel is equipped with a variety of connections, baffles, and piping along with a tall outlet stack which provides for vapor emissions. The hot hydrocarbon material is piped into the blowdown stack above the liquid level. The water flow is actuated by the entrance of the hot liquid into the vessel; additional water from sprays is introduced above the blowdown line. Steam (in some variations) is piped to the top of the stack, as in Fig. 11-20, providing a spray which helps disperse the vapors. The inside of the vessel is equipped with a series of baffles which speeds the cooling process. Cooling water is used for this service as it is mixed directly with the hydrocarbon material. A system of drains and siphons channels the cooled water into the sewer system.

A blowdown drum usually accompanies this type of unit, as in Fig. 11-20. The drum is used to accumulate liquid hydrocarbons that have a low vaporization level. The hydrocarbon mixture is channeled into the horizontal drum, which contains a series of baffles (as in Fig. 11-20) to separate the first stage of hydrocarbons and vapors.

The project shown in Figs. 11-20(a) through 11-20(i) is a typical petrochemical process unit. This series of drawings, including the flow diagram, plan, elevation, and section views, and the accompanying details of the drum and stack, are excellent examples of the drawings that might be required of a piping drafter. In constructing this project, the student should use 3/8 in. = 1 ft scale and 24 x 36 in. paper. The guidelines set down in Chapter 5 for line weights, lettering, and dimensioning should be strictly adhered to when completing this project. Refer to the first four chapters of the book for explanations on valves, fittings, and piping.

The general procedure for the construction of such a unit is as follows. The engineering flow diagram will have already been sketched by the process engineer. This drawing will then be laid out by the piping drafter in a completed form, as in Fig. 11-20(a), according to specifications provided in Chapter 6. Figure 11-20(a) should be the first drawing completed in this set. The P&ID covers all the essential equipment, piping, components, valving, and instrumentation.

As the student completes this project, it will become clear that certain aspects of this design—line routing, equipment location—could be improved. Moreover, certain aspects of the flow diagram (Fig. 11-20a) are not included on the set of drawings: the level indicator which should be attached to the water-quench blowdown stack and a number of drains, valves, and instrumentation. By paying strict attention to all the information on Fig. 11-20(a) and comparing it to the drawings for the plan, elevation, and section used, it is possible to add the different aspects of this project onto the drawings, locating drains, instruments, piping, valves, and so on.

Another way of completing this project is to use the model method directly from the drawings provided in the book. Using the drawings provided, it should be possible to order the necessary components, fabricate the base and structural elements, and construct a complete model of this unit.

When using the drawing method, the student should now start to lay out the plan view of the process unit as in Fig. 11-20(b). Note that these drawings are not dimensioned according to normal piping procedures: A mechanical system is used for dimensioning the plan, elevation, and details of this project. Follow the procedures laid down in Chapter 5 for the dimensioning of piping systems. (As a matter of company procedure, it would be necessary to have access to the foundation location plan, plot plan, drainage, paving, and grade plan, vessel fabrication requirements, P&ID, and other information essential for the layout and construction of this project.)

Start by establishing the match line for the continuation of the project, which is shown on the far right-hand side of the drawing. Using the dimensions provided on this drawing, set up a centerline of the horizontal vessel and the stack, establishing the centerlines and foundation location dimensions first. The next piece of equipment to locate is the pump foundation in plan and the drain. From these centerlines, the piers and foundations can be laid out. After major coordinates and dimensions have been established, the pipe rack centerlines and large-diameter piping should be established. It is now possible to construct the outline of the octagonal stack foundation, the diameter of the stack, and the outline of the knockout drum. (Note that in some cases the dimensions must be scaled.) The pipe rack pipelines can then be laid out.

At this point, the only thing that should be missing is the areas which show up better in sections B, C, D, and E. So before putting any dimensions onto the plan view or establishing any further pipelines, valving, or instrumentation, begin the elevation view shown in Fig. 11-20(c) by establishing the finish grade elevation at 10 ft 0 in. and locating the major equipment centerlines, including the stack, vessel, and pipe rack columns. This view (Fig. 11-20c) gives a much clearer idea of the general piping arrangement for the pump area and the level glass of the drum; it also shows the extent of the foundation and top of grout for both the stack and the horizontal vessel. Complete all of Fig. 11-20(c) except for the piping between the two large pieces of equipment.

At this point, begin section B shown in Fig. 11-20(d). The dimensions needed to establish the pipe supports and the horizontal vessel are provided primarily from the plan view shown in Fig. 11-20(b). This view gives a much clearer idea of the piping installation and control valve station.

After Fig. 11-20(d) is completed, go back to Fig. 11-20(b) and establish much of the piping in and around the vessel. The section view shown in Fig. 11-20(f) should be completed before attempting to lay out the piping in and around the tower. The steam and water spray pipes are a maze of 1-in. and 2-in. piping; section C (Fig. 11-20e) clarifies the piping much better than other views.

Figure 11-20(f) shows section E and details 1 and 2. Complete this drawing after the main elevation and section views are finished.

Figure 11-20(e) shows section C from the plan view. This drawing should be completed before the pump piping area on Fig. 11-20(b) is laid out to provide adequate dimensional locations of the various pipelines which connect to the pump station.

Lastly, complete the fabrication drawings for the drum and water quench blowdown stack. Any necessary details which may not have been established on sectional views in Figs. 11-20(b), (c), and (d) should be established on these drawings—including instrumentation, notes, pipeline numbers, and dimensioning. As part of this project, design and locate the platforms, ladders, and cages that may be required for access to the stack; use Fig. 11-27 as an example.

When constructing a model of this project, the first step in the layout of the equipment is to establish a base of adequate size. Match lines and foundation limits should be scribed on the base with centerline tape. Then the major equipment centerlines for the stack, drum, pump, and pipe racks should be established with centerline tape. Next, using the drawings provided, fabricate the vessels, assemble the pump, and construct the piers and foundations for the stack, pump, and knockout drum. After major equipment and foundations have been fabricated, they should be accurately located on the model base. At this point, it is possible to start establishing the pipe rack lines and the large-diameter pipelines, such as 19BD062 20" C and 19BD0601 10" EE. Follow all the details laid out on the drawings provided, using established procedures for the construction of models (Chapter 15).

Whether the student uses the model or the drawing method for this project, the specifications and dimensions established in this chapter and Chapter 10 for various types of equipment should be consulted frequently. These specifications will help in the design and location of the control valve station in Fig. 11-20(d). Equipment location and pump specifications are discussed in this chapter and Chapter 10.

Do not assume that all the procedures, design, and layouts in this book are absolutely correct or represent the best possible solution insofar as economy and placement are concerned. The talented piping designer and modeler should be able to take specifications and use them to their utmost—perfecting existing systems and drawings such as those provided in Fig. 11-20 and also establishing new facilities straight from the P&ID drawings. The background information provided in this book should enable the student to establish, arrange, draw, and construct a model of this project without much trouble.

The instructor may also wish to have the student draw individual isometric spools of the major pipelines—such as line 19BD0606 10" EE, which has a confusing configuration. Isometrics of 19BD0621 20" C and many of the other lines will also help the student to understand the extent of drawings necessary for a piping project.

Figure 11-33(a) through 11-33(m) shows another piping unit that makes an excellent student project for drawing and modeling.

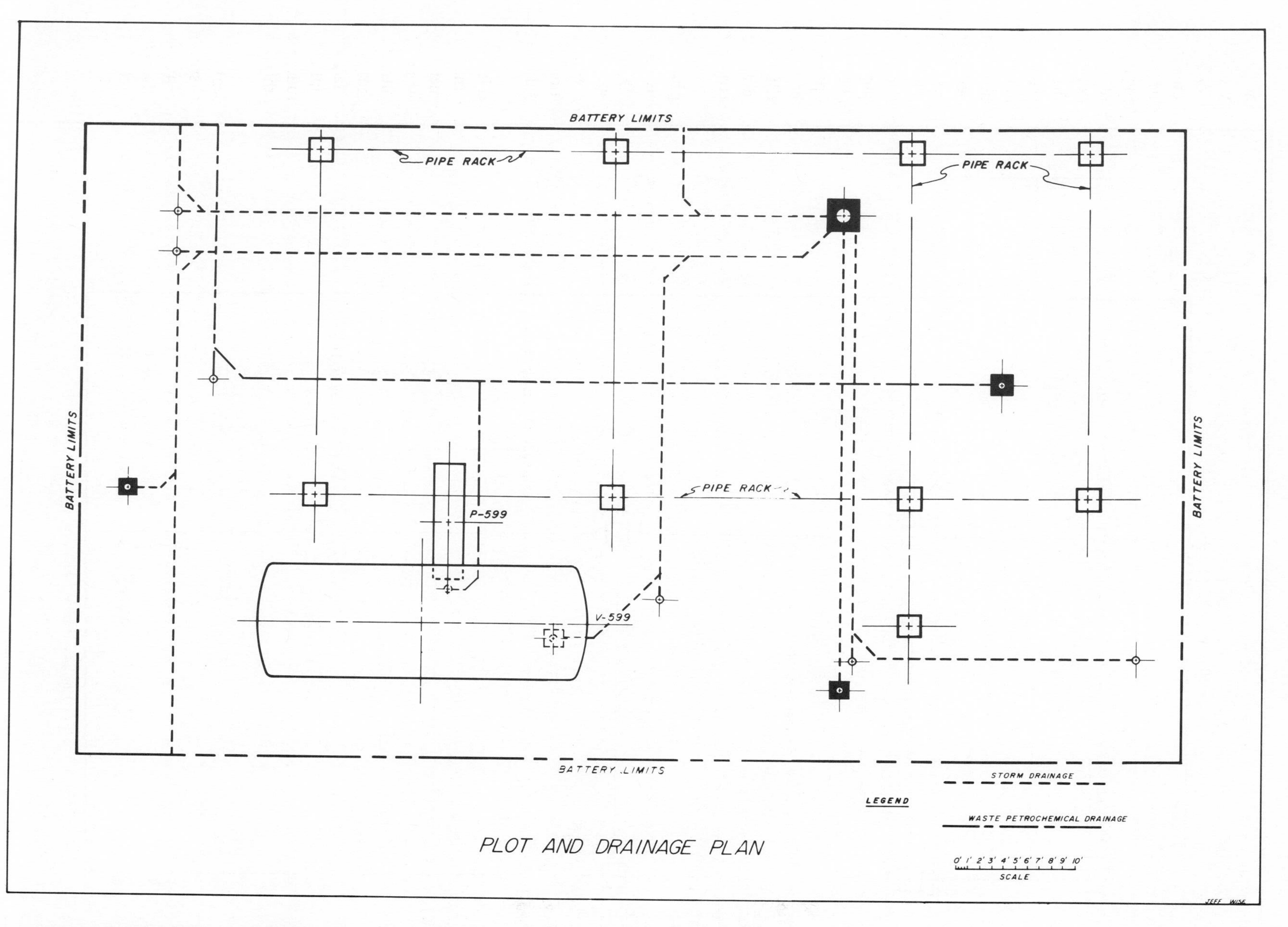

FIG. 11-33 This project is a series of drawings for a fin-fan air-cooler mounted on a pipe rack and includes a major valve station.

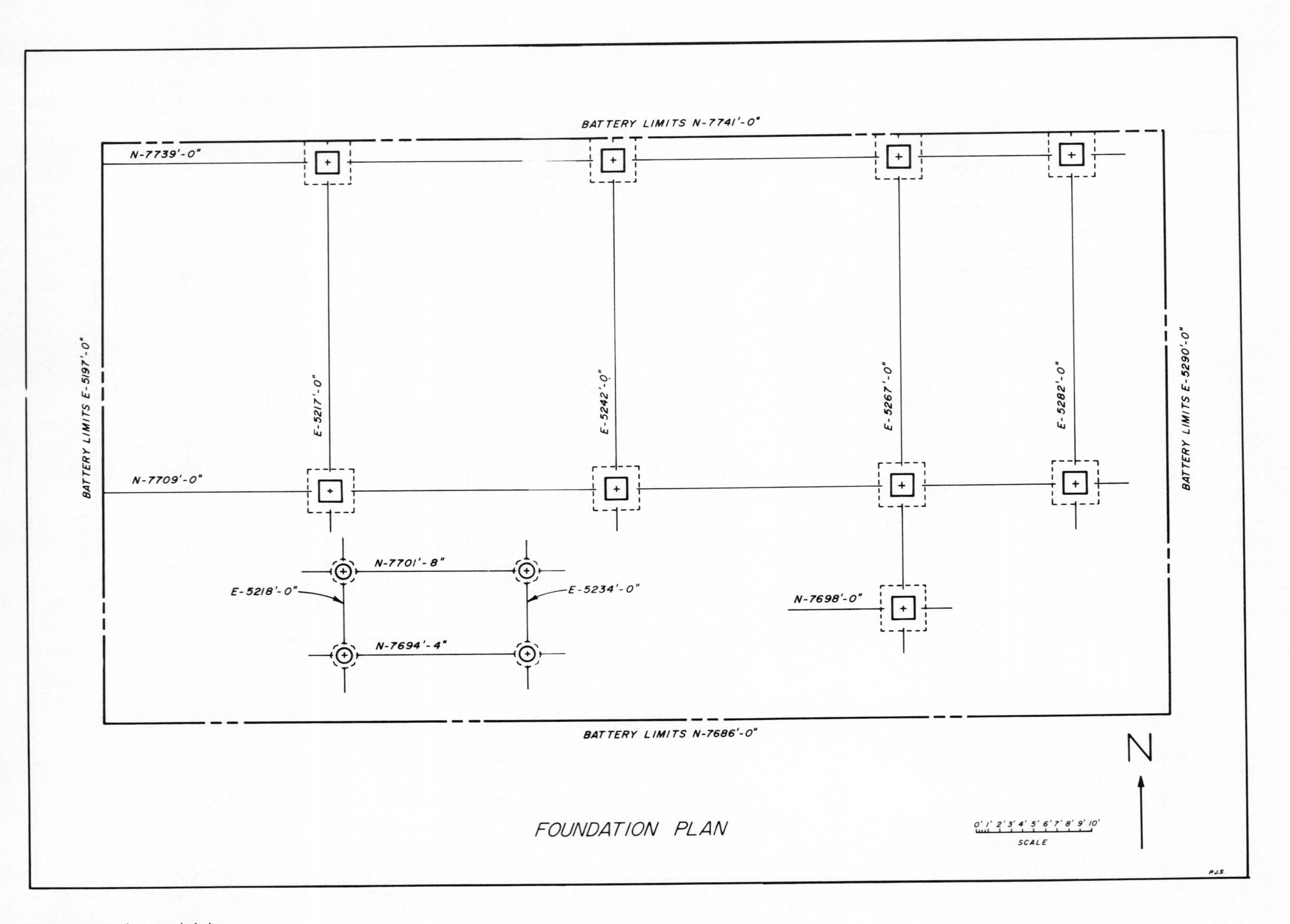

FIG. 11-33 (cont'd).(b)

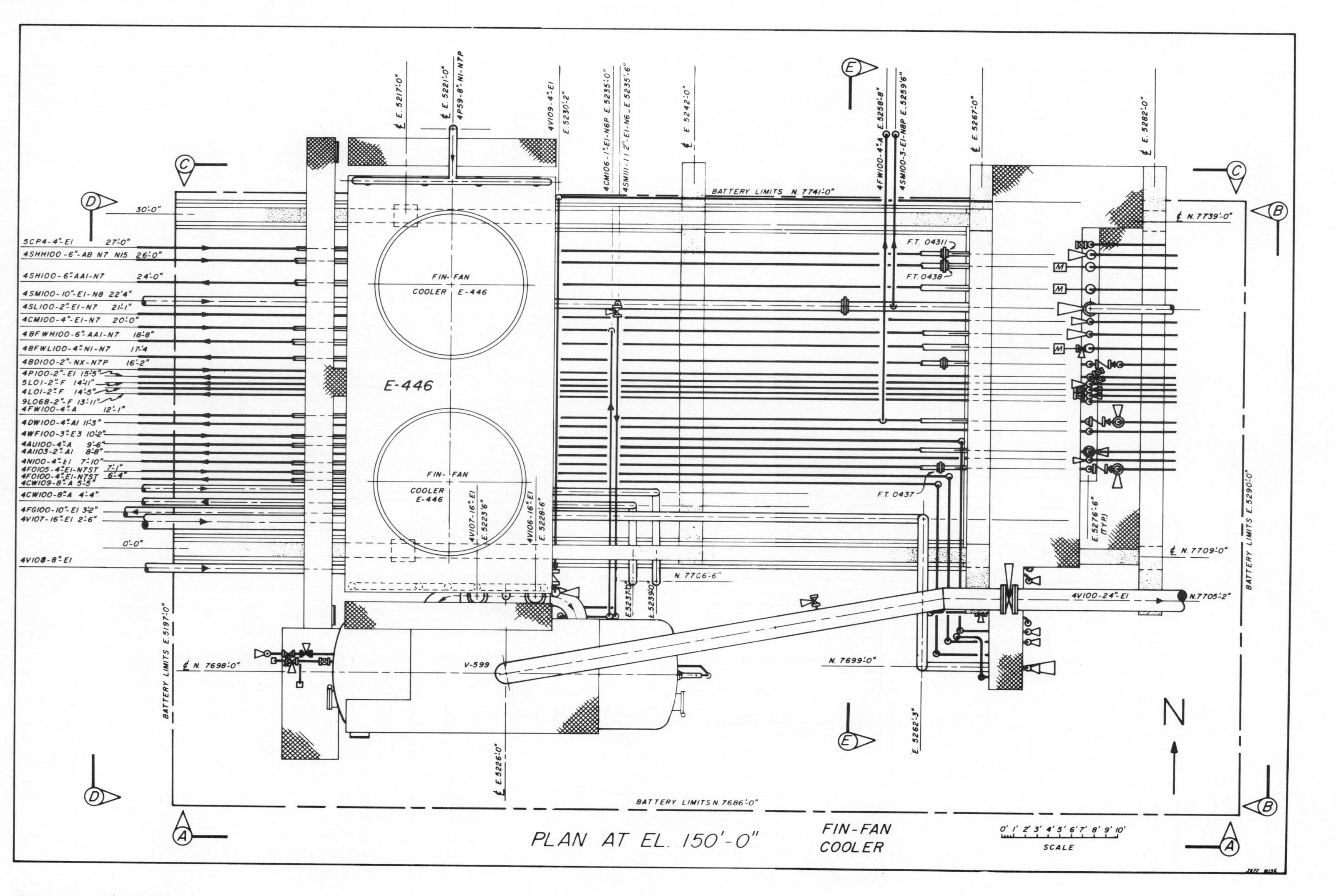

FIG. 11-33 (cont'd) (c)

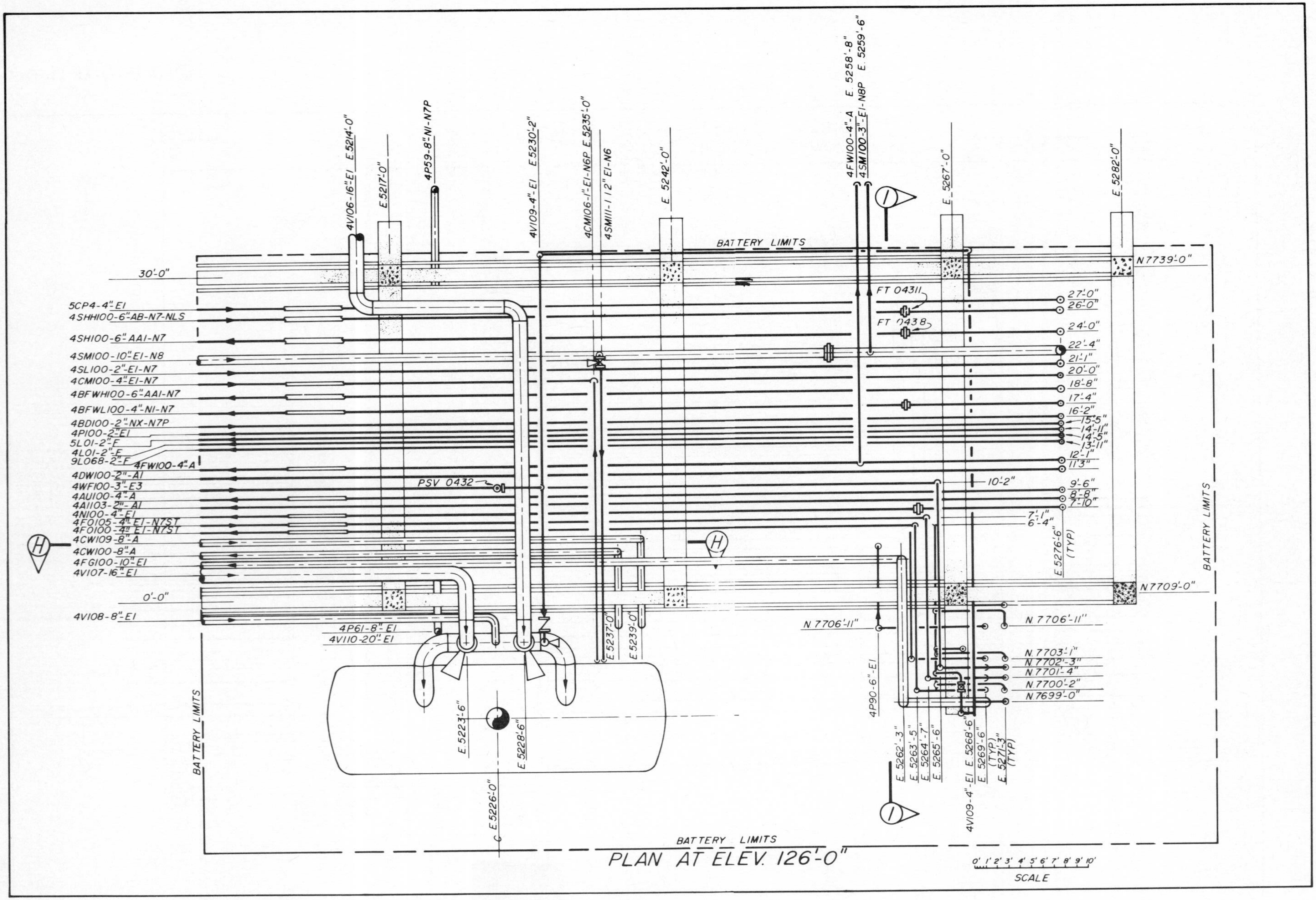

30'-0"
5CP4-4"-E1
4SHH100-6"-AB-N7-NLS
4SH100-6"-AA1-N7
4SM100-10"-E1-N8
4SL100-2"-E1-N7
4CM100-4"-E1-N7
4BFWH100-6"-AA1-N7
4BFWL100-4"-N1-N7
4BD100-2"-NX-N7P
4P100-2"-E1
5L01-2"-F
4L01-2"-F
9L068-2"-F
4FW100-4"-A
4DW100-2"-A1
4WF100-3"-E3
4AU100-4"-A
4A1103-2"-A1
4N100-4"-E1
4F0105-4"-E1-N7ST
4F0100-4"-E1-N7ST
4CW109-8"-A
4CW100-8"-A
4FG100-10"-E1
4V107-16"-E1
0'-0"
4V108-8"-E1

4V106-16"-E1 E 5214'-0"
E 5217'-0"
4P59-8"-N1-N7P
4V109-4"-E1 E 5230'-2"
4CM106-1"-E1-N6P E 5235'-0"
4SM111-1 1/2"-E1-N6
E 5242'-0"
4FW100-4"-A E 5258'-8"
4SM100-3"-E1-N8P E 5259'-6"
E 5267'-0"
E 5282'-0"

BATTERY LIMITS
N7739'-0"
FT 04311
FT 0438
27'-0"
26'-0"
24'-0"
22'-4"
21'-1"
20'-0"
18'-8"
17'-4"
16'-2"
15'-5"
14'-11"
14'-5"
13'-11"
12'-1"
11'-3"
10'-2"
9'-6"
8'-8"
7'-10"
7'-1"
6'-4"
E 5276'-6" (TYP)
PSV 0432
N7709'-0"
N 7706'-11"
N 7706'-11"
N 7703'-1"
N 7702'-3"
N 7701'-4"
N 7700'-2"
N 7699'-0"
4P61-8"-E1
4V110-20"-E1
E 5237'-0"
E 5239'-0"
4P90-6"-E1
E 5223'-6"
E 5228'-6"
E 5226'-0"
E 5262'-3"
E 5263'-5"
E 5264'-7"
E 5265'-6"
4V109-4"-E1 E 5268'-6"
E 5269'-6" (TYP)
E 5271'-3" (TYP)
BATTERY LIMITS

PLAN AT ELEV. 126'-0"

0' 1' 2' 3' 4' 5' 6' 7' 8' 9' 10'
SCALE

FIG. 11-33 (cont'd) (d)

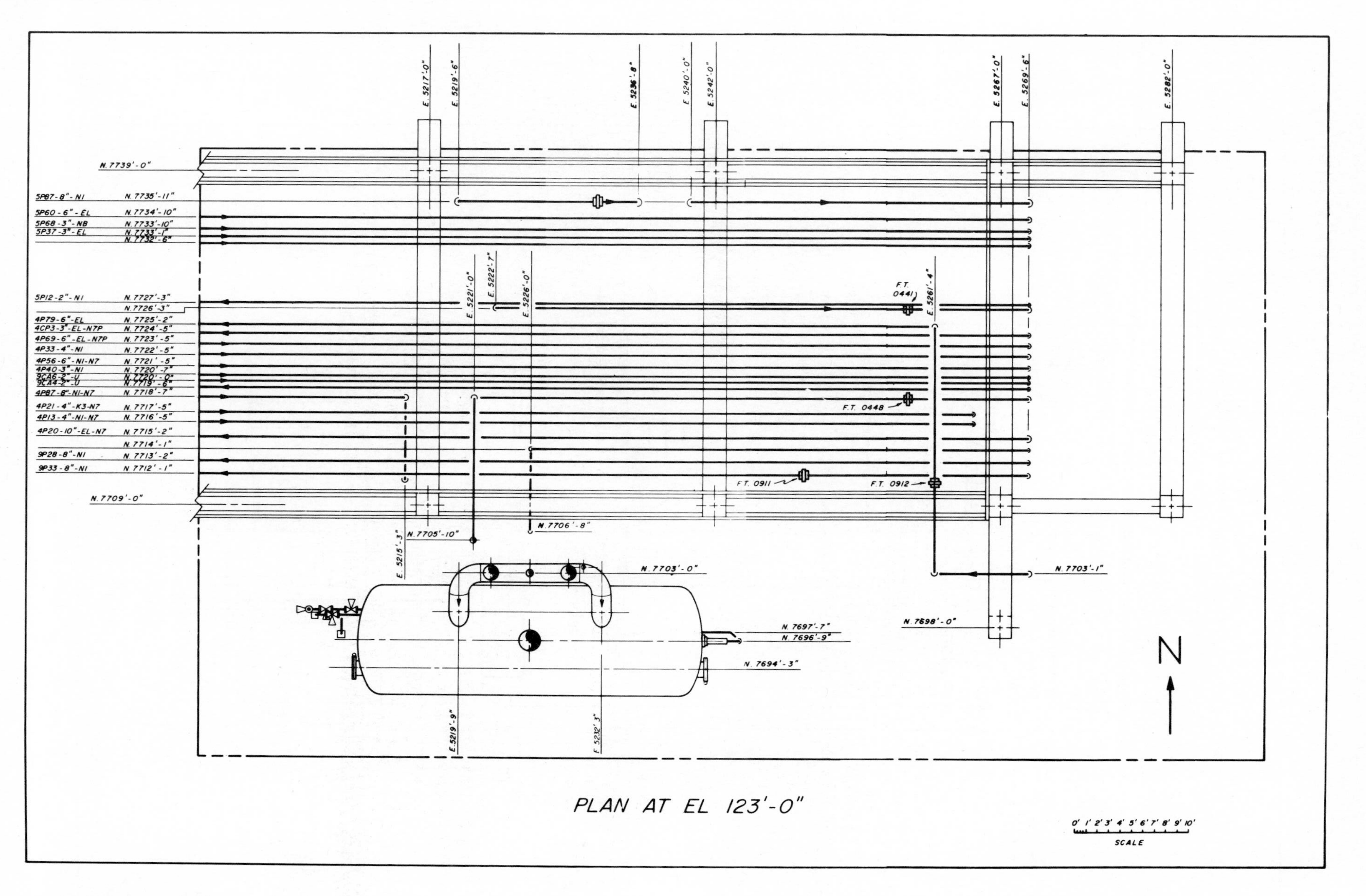

FIG. 11-33 (cont'd) (e)

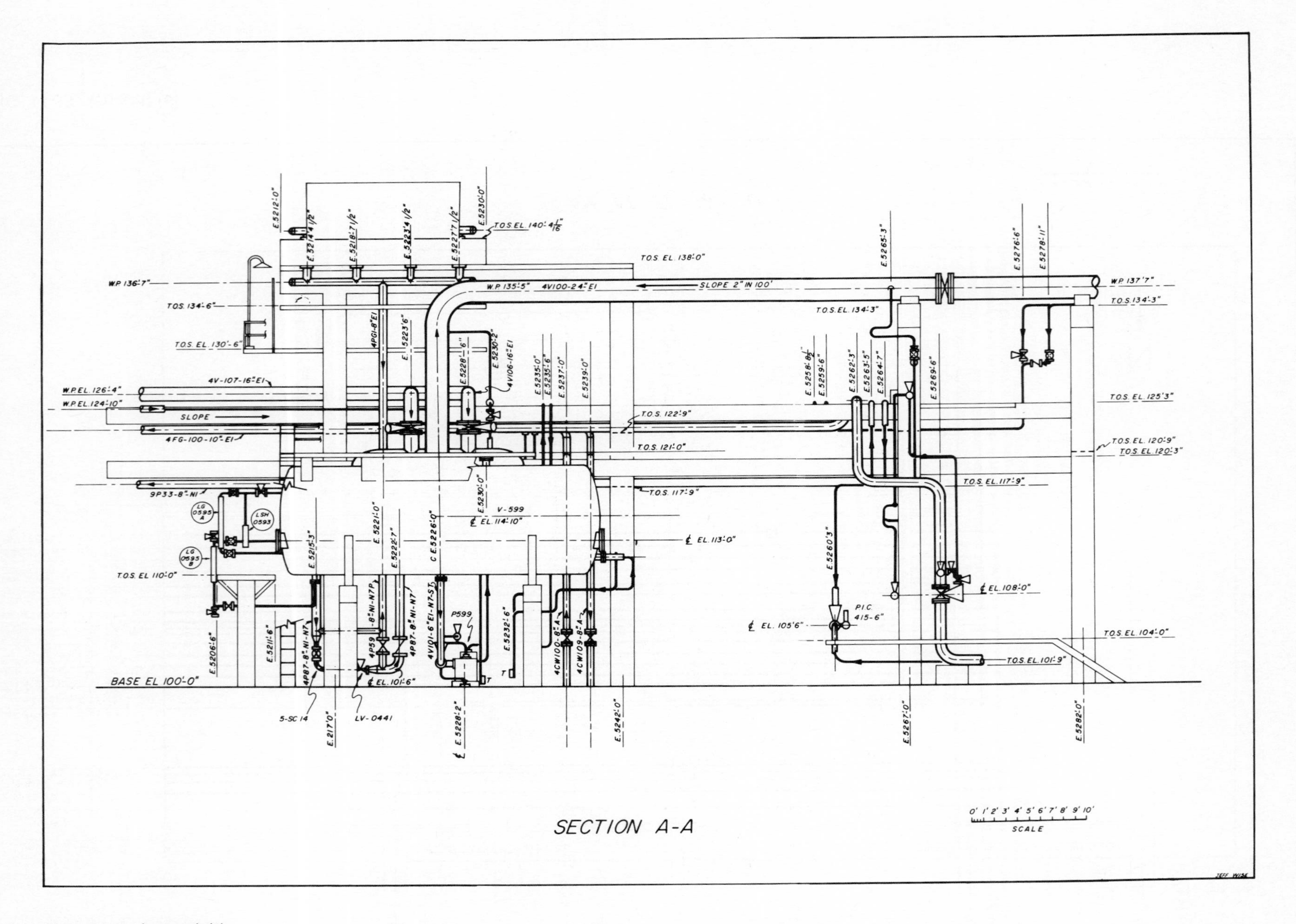

FIG. 11-33 (cont'd) (f)

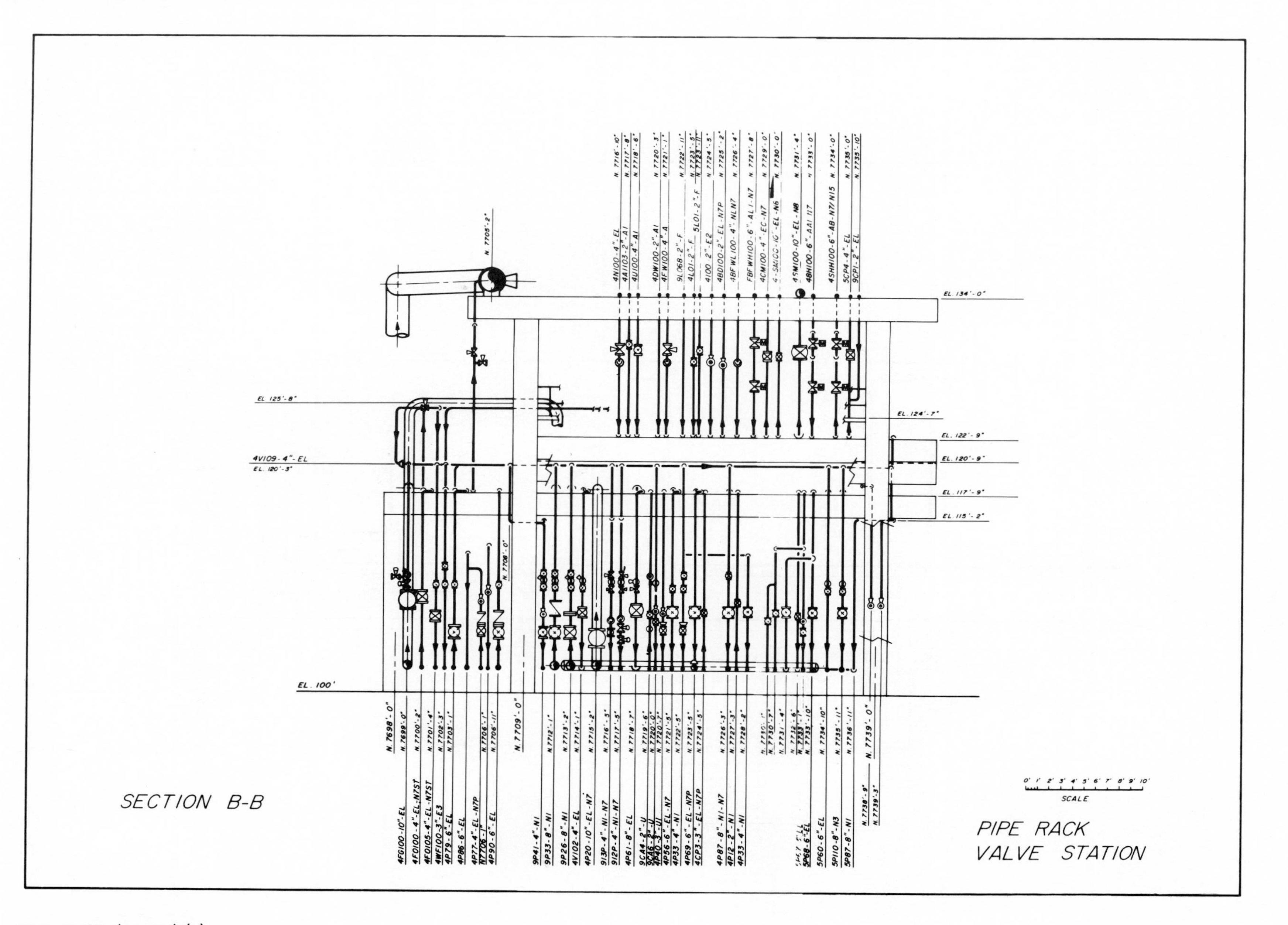

FIG. 11-33 (cont'd) (g)

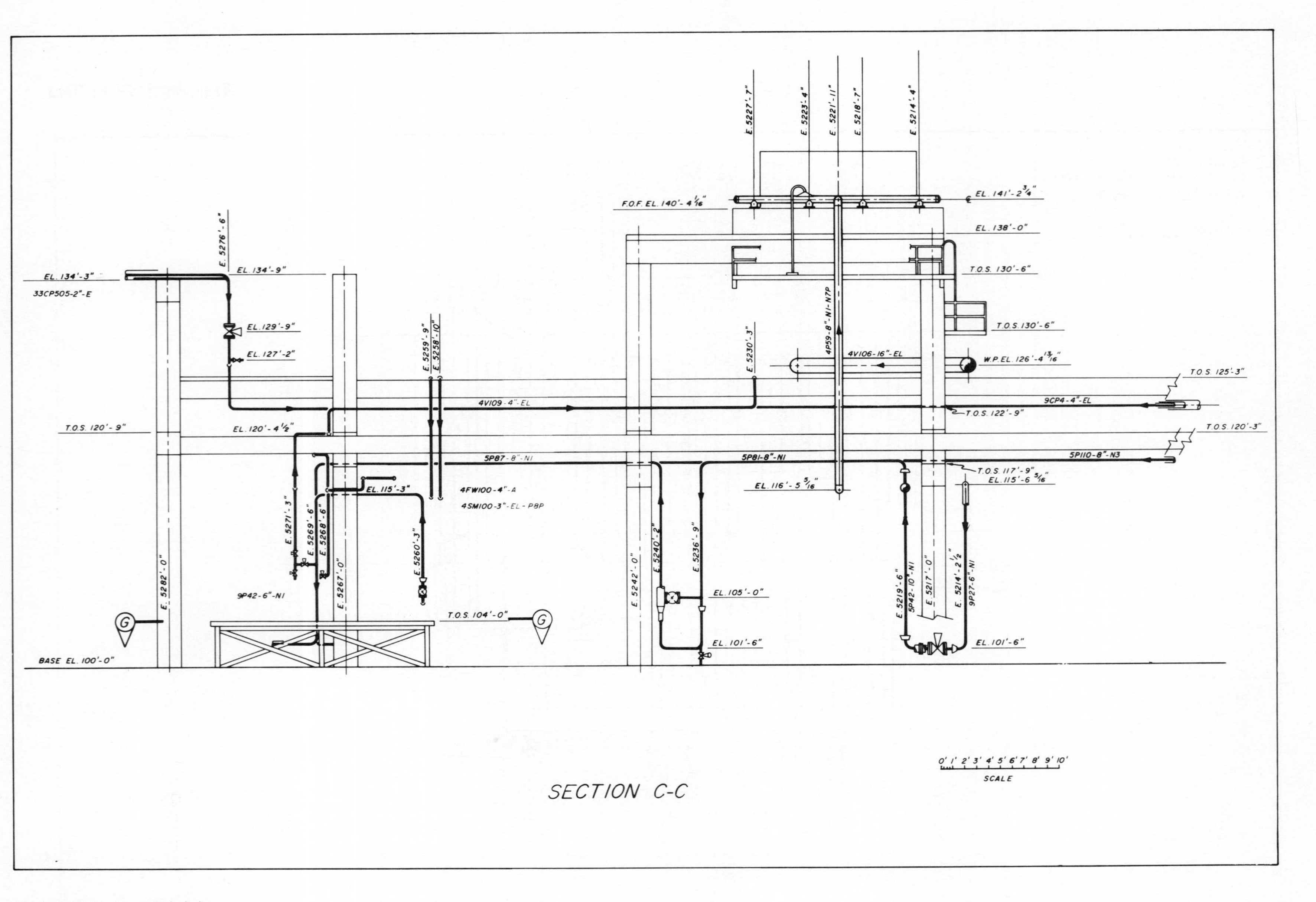

FIG. 11-33 (cont'd) (h)

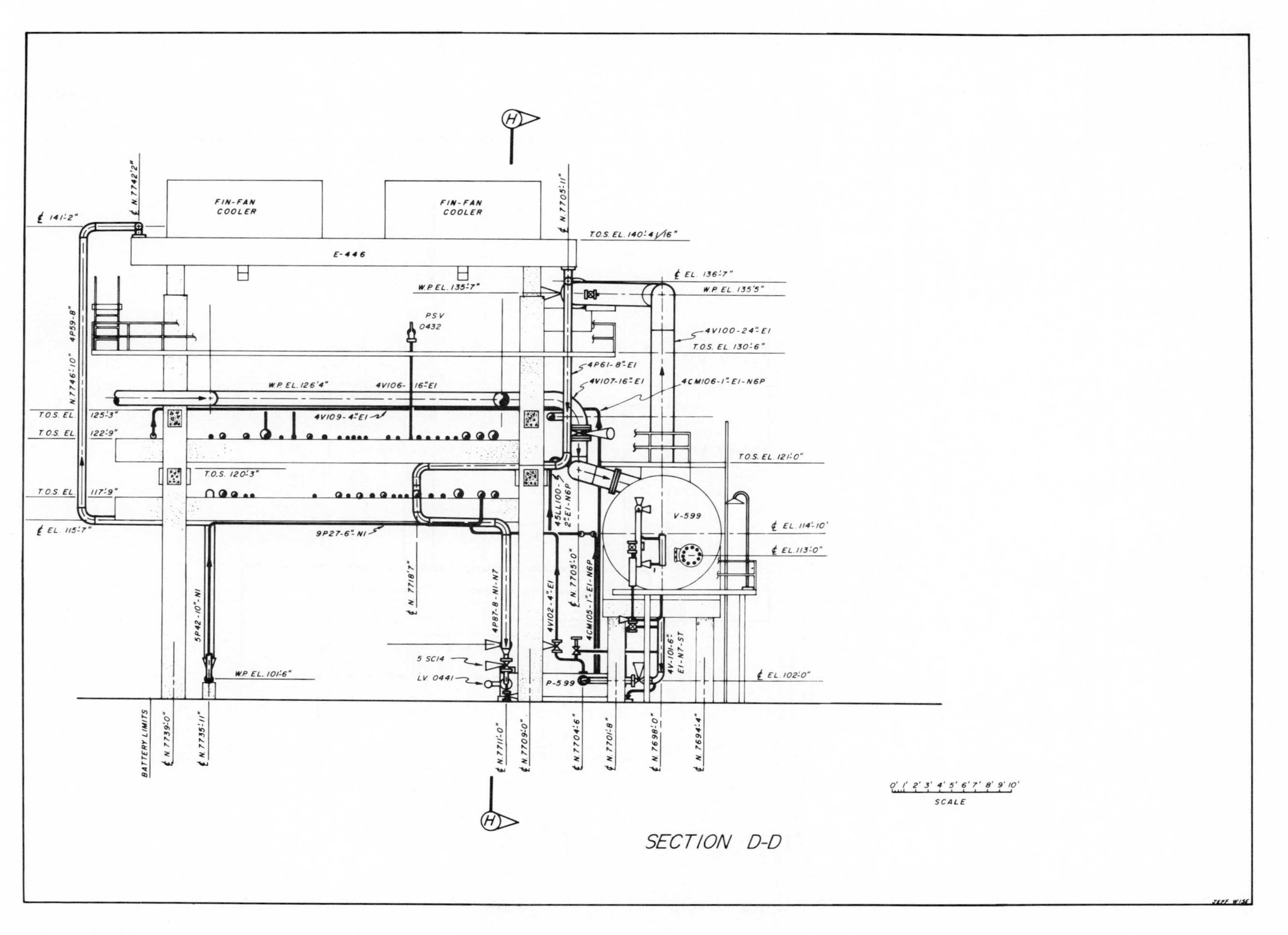

FIG. 11-33 (cont'd) (i)

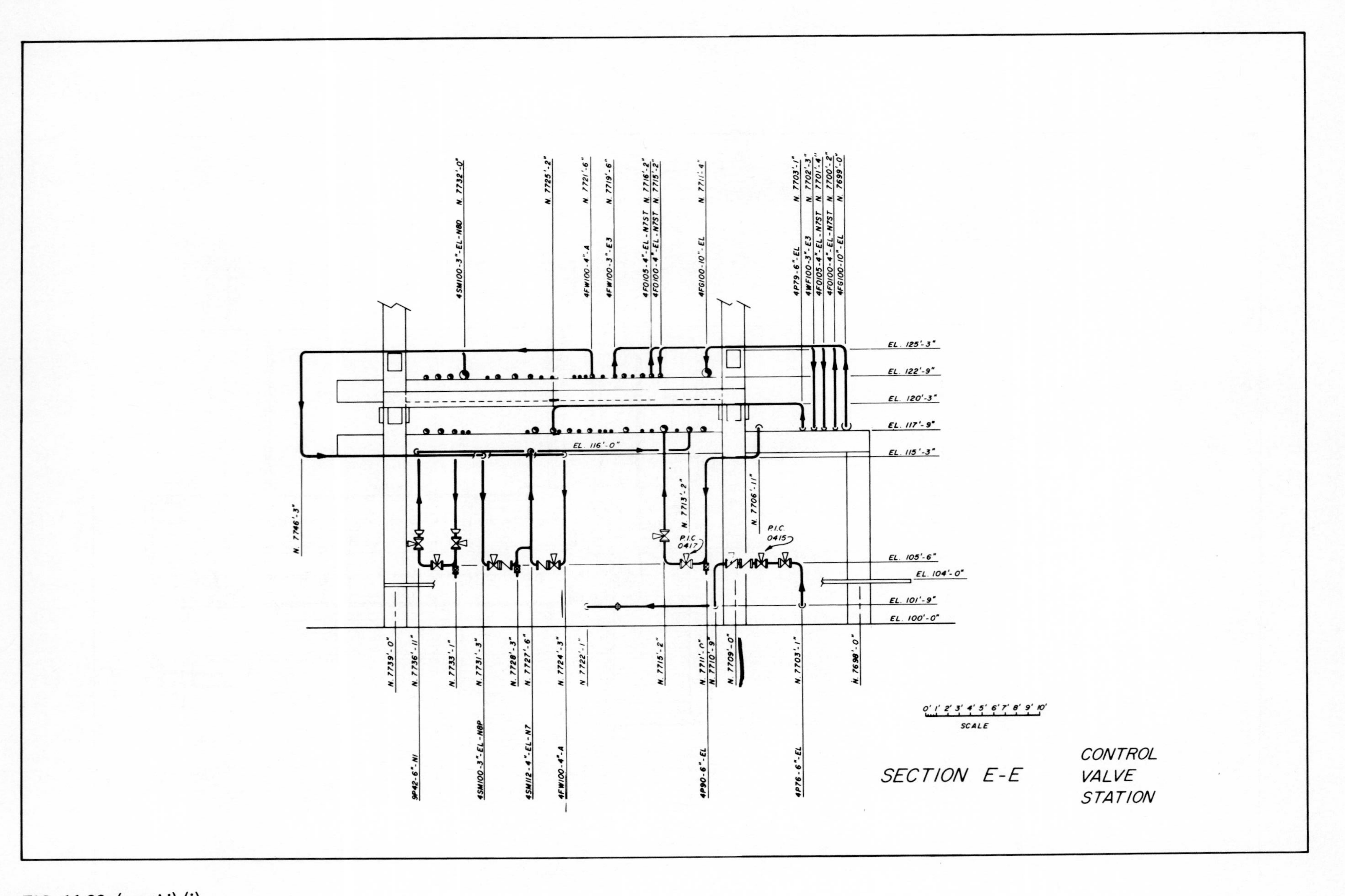

FIG. 11-33 (cont'd) (j)

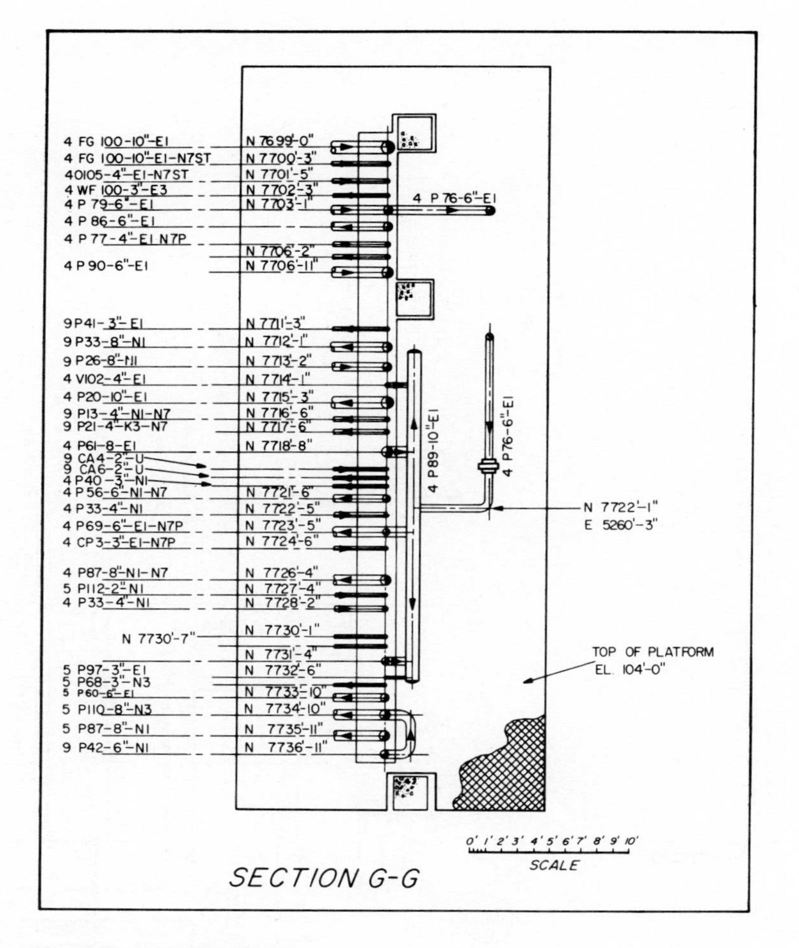

FIG. 11-33 (cont'd) (k)

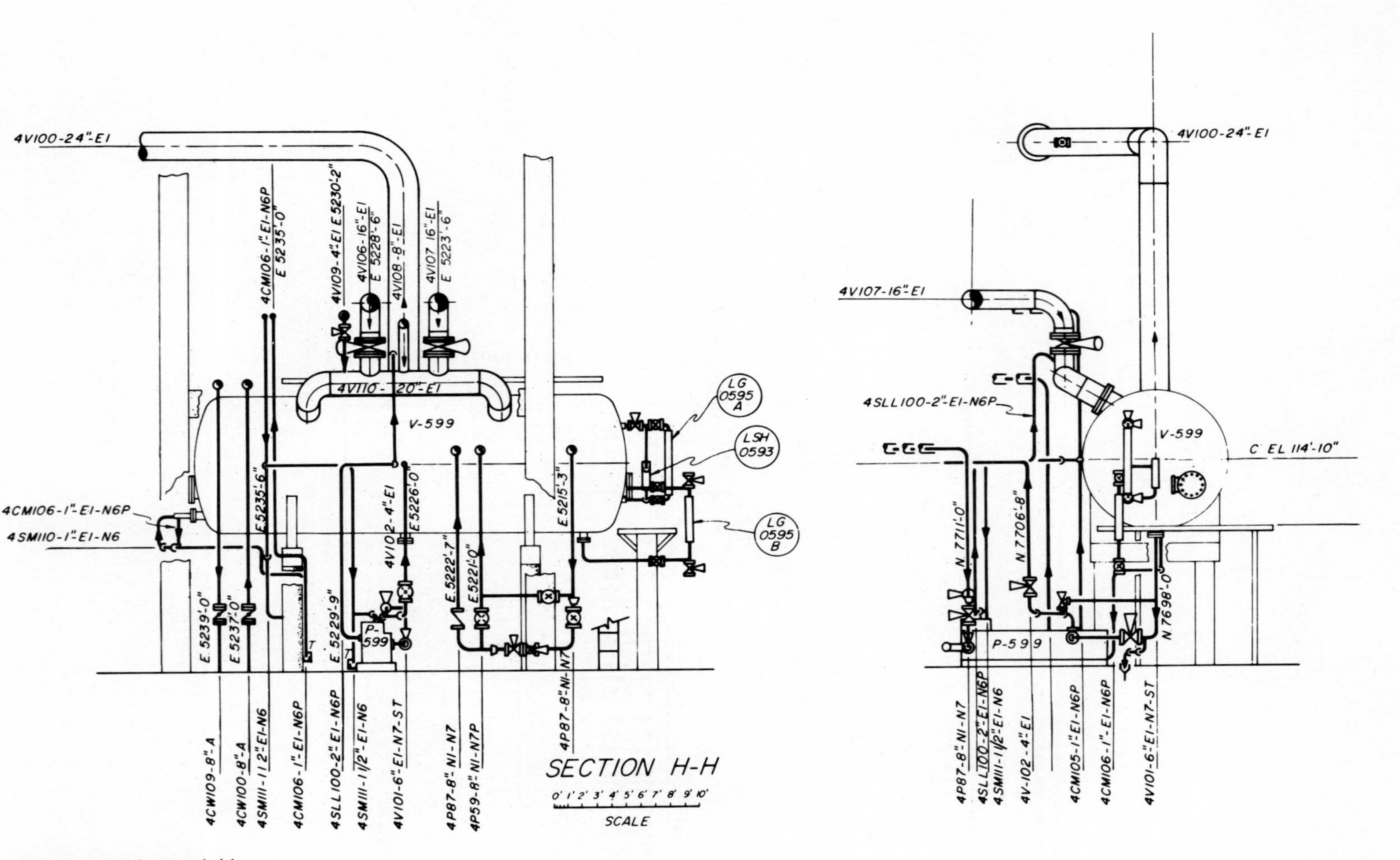

FIG. 11-33 (cont'd) (I)

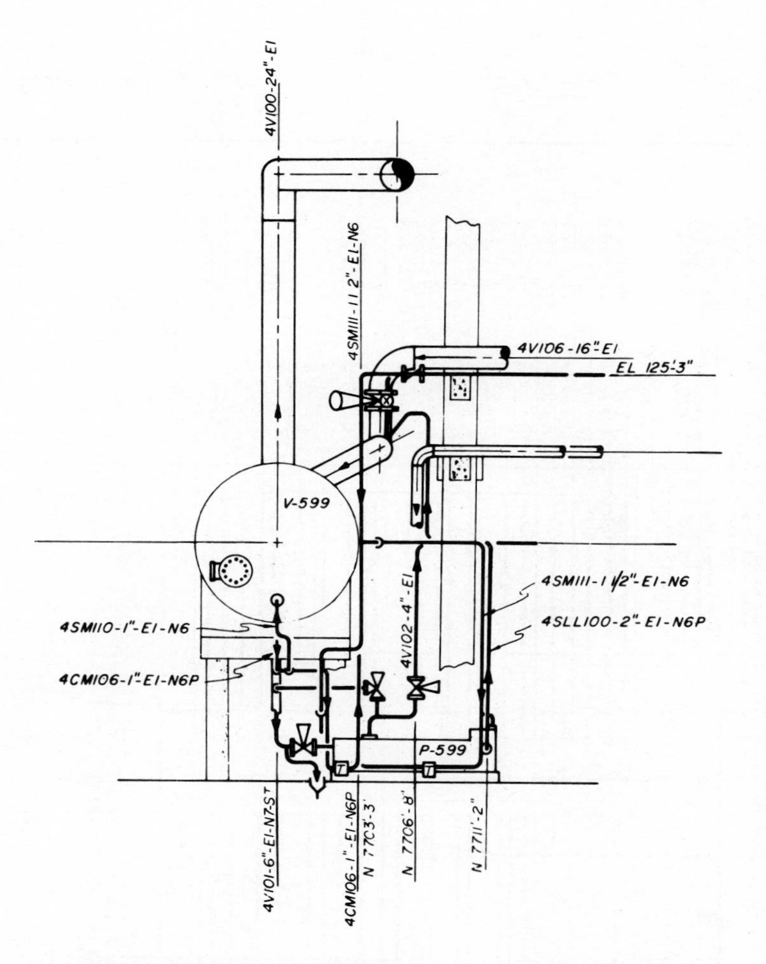

Left side view of Fig. 11-33(l) above.

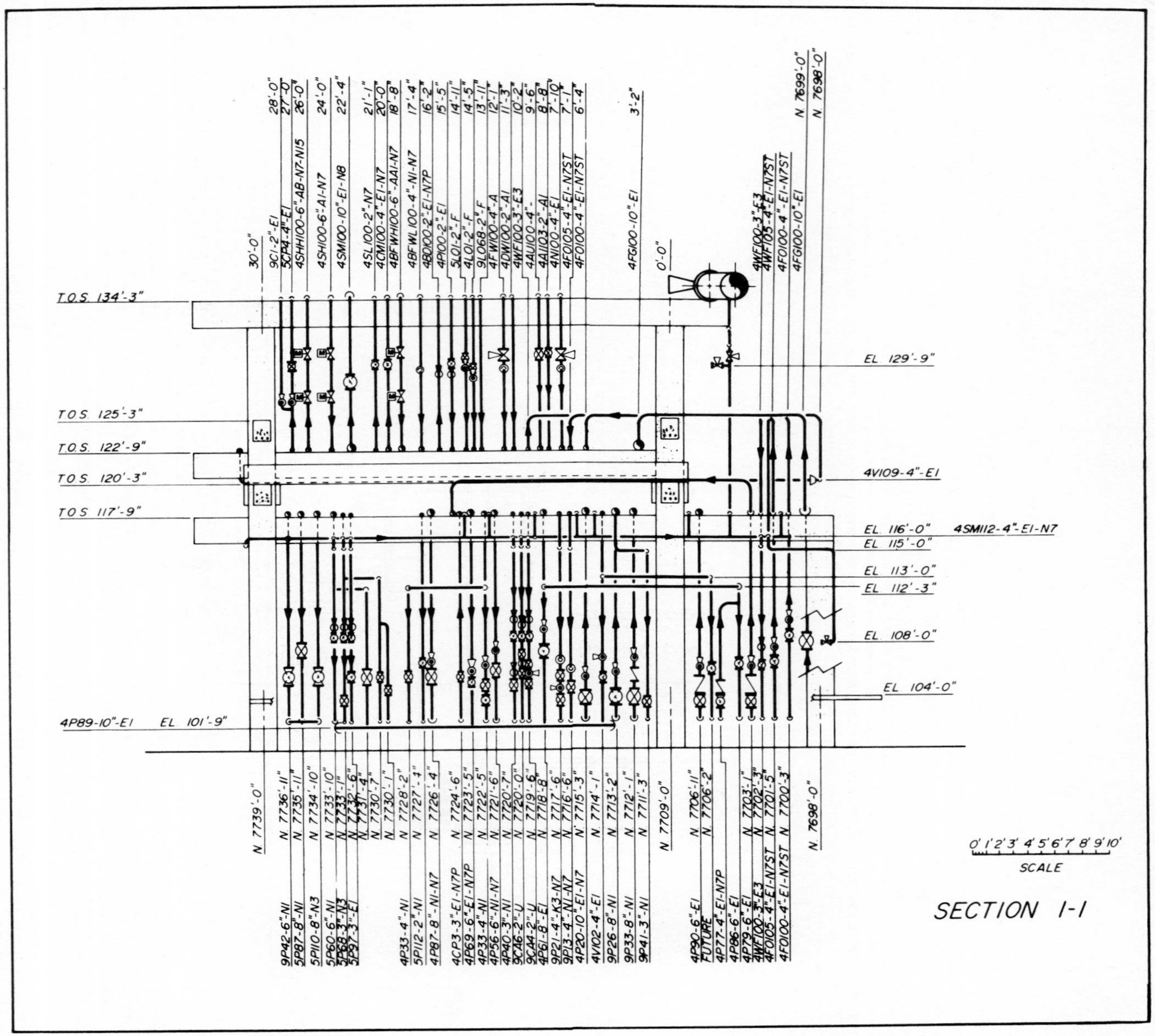
T.O.S. 134'-3"
T.O.S. 125'-3"
T.O.S. 122'-9"
T.O.S. 120'-3"
T.O.S. 117'-9"
EL 129'-9"
4V109-4"-E1
EL 116'-0"
4SM112-4"-E1-N7
EL 115'-0"
EL 113'-0"
EL 112'-3"
EL 108'-0"
EL 104'-0"
4P89-10"-E1
EL 101'-9"
0'-0"
30'-0"
3'-2"
4FG100-10"-E1
N 7739'-0"
N 7709'-0"
N 7698'-0"
N 7699'-0"
0' 1' 2' 3' 4' 5' 6' 7' 8' 9' 10'
SCALE
SECTION 1-1

FIG. 11-33 (cont'd) (m)

QUIZ

1. Crude oil before processing is also referred to as ______ ______ or ______ stock.
2. Why is it necessary to use fractionation to produce enough gasoline?
3. Name three uses for heat in a petrochemical processing plant.
4. Name two vessels that are typically found in a refinery and explain their uses.
5. Why is it necessary to provide a flare system at a petrochemical plant?
6. Vessels or spheres that utilize a catalyst are referred to as ______ .
7. When storage tanks are grouped together, this is termed a ______ .
8. Name three uses for insulation in a typical petrochemical complex.
9. What are platforms used for on vessels?
10. How can water hammer damage a piping system?
11. What is grout?
12. Name two uses for cooling water in a petrochemical installation.
13. What is a blowdown system?
14. A drum or a vessel that is constructed with the use of baffles through which petrochemical products are made to pass is sometimes referred to as a ______ ______ .
15. Vertical vessels are sometimes called columns and ______ .
16. Name two cracking processes used for the refining of petrochemicals.
17. Name three kinds of tanks and their uses.
18. What is distillation and how is it used in petrochemical processing?
19. Name four products usually associated with petrochemical refineries.
20. What problems are normally associated with supporting petrochemical refinery piping?
21. Whenever possible, supports should be placed ______ ______ in areas where the piping system changes direction.
22. Name three areas where pipe supports should be located in order to facilitate installation, maintenance, and operation of a petrochemical piping system.
23. When is it necessary to anchor or guide oil refinery piping?
24. How are sliding supports, shoes, or slide assemblies used for long runs of piping?
25. Why must oil refinery piping be flexible?
26. Most petrochemical projects are drawn to what scale?
27. What scale is usually used for modeling petrochemical complexes?
28. What is a penetration?
29. Name the various elements associated with circular or rectangular platforms that are necessary for columns, vertical vessels, or horizontal vessels.
30. What is a blowdown drum and how is it used?
31. Name four safety conditions that must be considered when designing petrochemical installations.
32. What is a control valve station?
33. Name four systems commonly found in petrochemical installations.
34. What codes cover pressure vessels?
35. Name three uses for water piping in a refinery.
36. Name three materials usually associated with piping for petrochemical installations.
37. Name five specifications for the location and design of pumping stations.
38. Why are manholes necessary on columns?
39. What is the normal distance associated with ladders between landings?
40. What is a plot plan?

PROBLEMS

1. Using the scale 3/8 in. = 1 ft, redraw the fabrication drawing in Fig. 11-20(h) of the knockout drum.
2. Using the scale 3/8 in. = 1 ft, redraw the fabrication detail of the column in Fig. 11-20(i).
3. Using the drawings provided in Fig. 11-20, prepare a complete set of plan, elevation, and section views of the refinery flare system, including the column and drum fabrication drawings. Note that the drawings are all dimensioned according to normal mechanical drafting procedures instead of pipe drafting and dimensioning specifications. Follow the specifications and procedures set down in Chapter 5 when redoing this project. (Much of this project can be redesigned according to specifications in Chapters 10 and 11.)
4. As a challenging individual project, complete a model of the project in Fig. 11-20.
5. Redo the project in Fig. 11-33 (3/8 in. = 1 ft scale).
6. Complete a model of Fig. 11-33 (same scale as drawings).

Fossil fuel power plant.

12 CONVENTIONAL POWER GENERATION PIPING

Chapters 12 and 13 deal with power generation using fossil fuels and nuclear energy. These chapters do not focus on design, but introduce to the student two important areas of piping in the world today. Power generation is a large field in itself. Companies that design and build power plants employ piping drafters for a vast array of positions. A piping drafter in a power company or construction firm is called upon to prepare many drawings—including the P&ID, the design of the power plant itself, the layout of the equipment, piping isometrics, and pipe support drafting and design.

Since there is great concern throughout the world for enough power to satisfy economic needs, new forms of power generation are coming into being. Fossil fuel and hydroelectric power have been the two main sources of electricity in modern industrial states and developing nations. As the year 2000 approaches, it becomes apparent that fossil fuels are reaching the end of their usefulness because of their pollution, the extreme amount of energy required to find, locate, and produce them, their rapidly rising cost, and the fast diminishing level of known reserves.

Coal is about the only fossil fuel left in great abundance. For a short period of time coal will be utilized as it was at the turn of the last century, but eventually it will become obsolete because of the environmental pollution. Conventional power plants using fossil fuels are basically a thing of the past. New combustion equipment for clean burning is being produced, it is true, but at increased cost, and even then only to extend the time of transition to solar power.

This chapter discusses conventional power plants with their diverse types of equipment. The equipment in the conventional power plant is similar to that of a nuclear installation. Both entail the generation of steam to produce electricity through a turbine and generator. The only difference lies in the means by which the steam is created, so much of this chapter also applies to the nuclear generation of electricity. This chapter provides a set of drawings to show elevations and equipment locations of a typical single-turbine power plant using conventional fossil fuels. These plans are in no way a complete set of all the detailed piping situations in the plant; they only represent the general layout of major piping and equipment at various elevations and cross-sections.

The fossil fuel power plant in this chapter and the nuclear containment drawings in Chapter 13 are good examples of modern power plants and, on the job, these installations would be designed with the use of a model.

FOSSIL FUEL POWER PLANTS

The uses of steam in industry are numerous. They include heating, food processing, oil refining, and several other chemical processes. Most of the design characteristics of steam piping are similar, depending of course on their relative pressure and temperature. In power plants, steam piping is used to transfer the energy created by the heating of water and production of steam to a turbine which, in turn, produces electricity by turning a generator (Fig. 12-1). In small industrial plants it is possible to generate steam at a higher pressure and install a turbine generator for the creation of in-plant electrical energy besides using the steam for production processes within the building.

The use of fossil fuel steam generation has a long and varied history. The first central station for power production was completed in 1881 in Philadelphia, where four Babcock & Wilcox boilers were installed to produce electricity for the Brush Electric Light Co. By the early 1900s, steam pressure rose considerably in the larger central stations being constructed throughout the nation, and by 1914 pressures of 150 psi at 500°F (260°C) were considered normal. During the years 1924 and 1925 steam pressure for central stations increased once again with major changes in the production and use of steam throughout this period. Larger boilers produced higher rates of output because of the ability to pulverize coal as fuel. This development created the need for the water-cooled furnace.

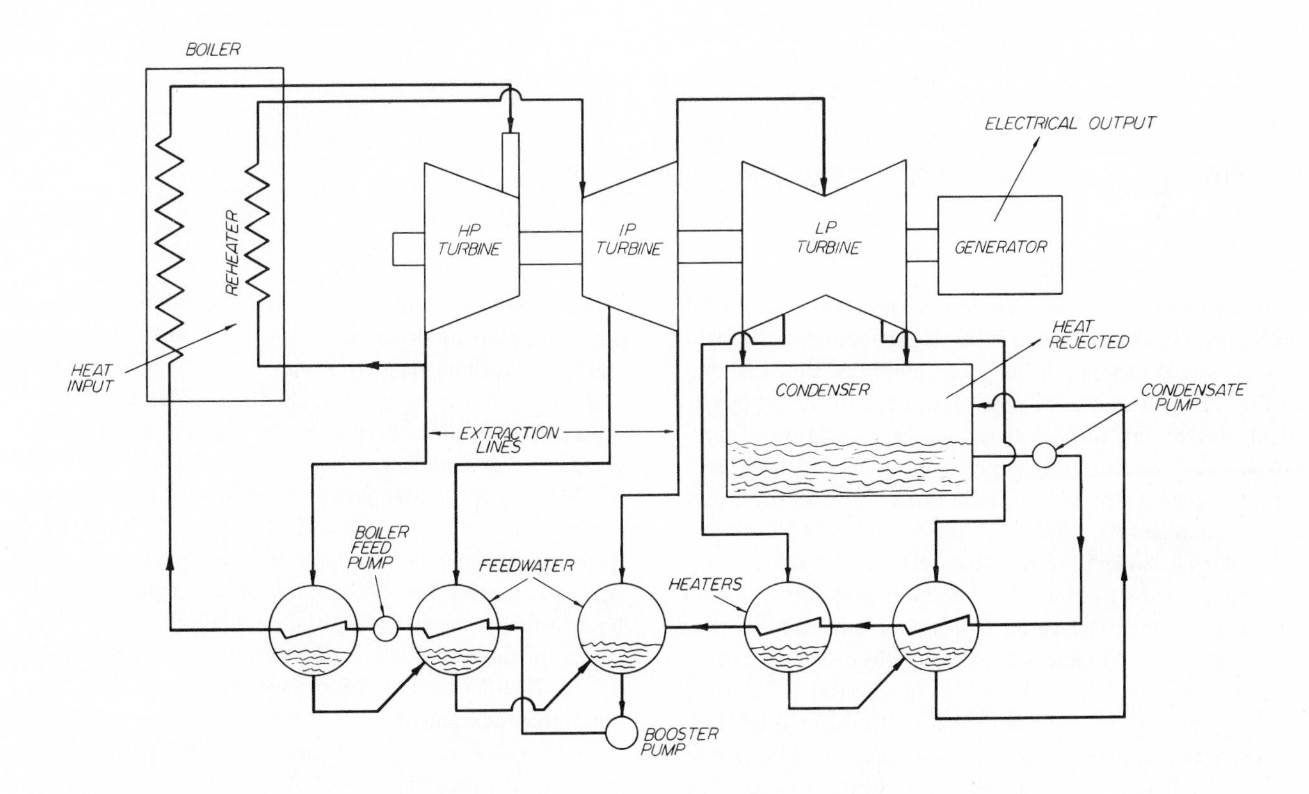

FIG. 12-1 Flow diagram showing the cycle for the generation of electric power by fossil fuels.

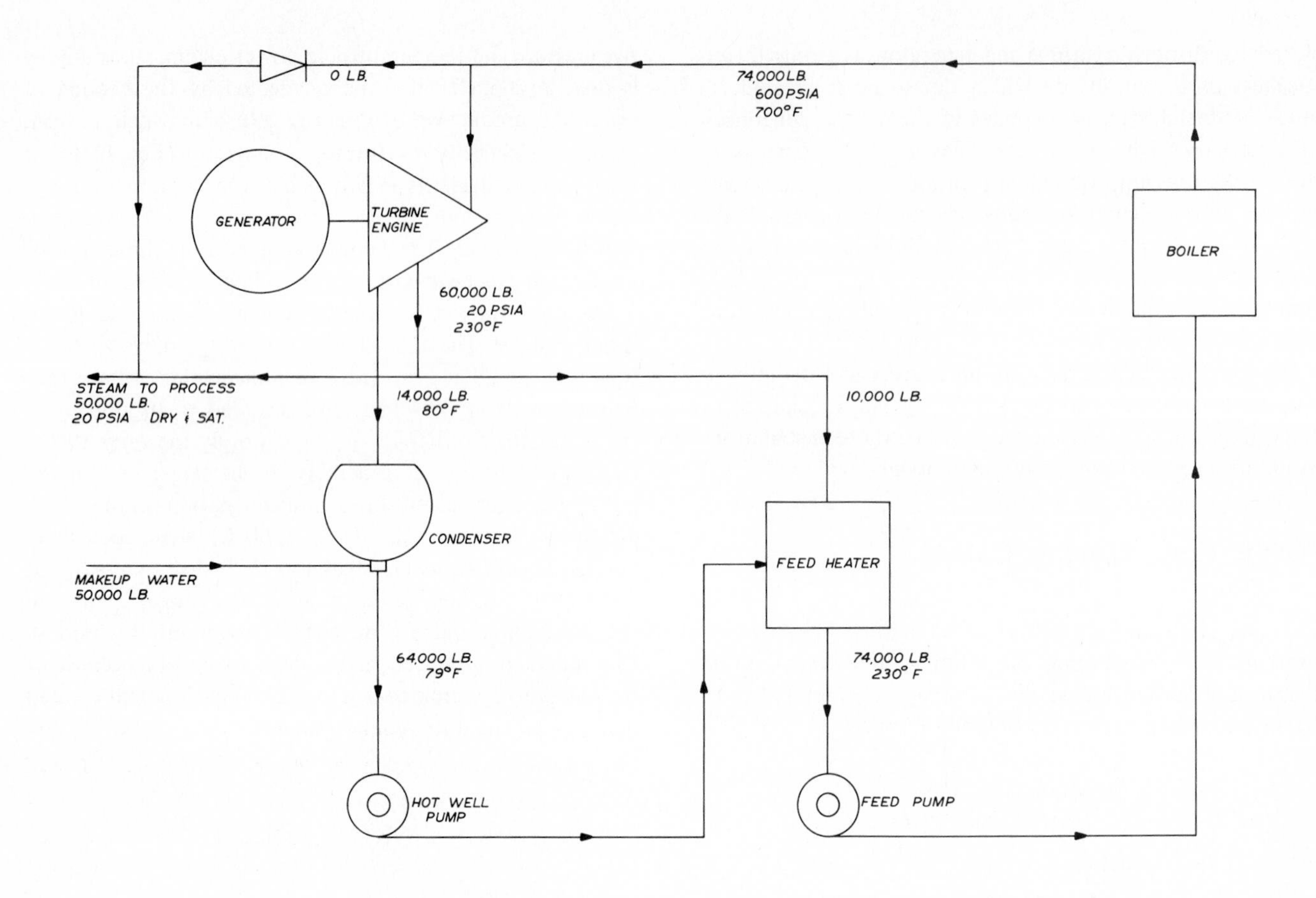

FIG. 12-2 Diagram of generation of electric power and process steam.

At the present time, temperatures of over 1000°F (537.7°C) with pressures from 2000 to 5000 psig are common throughout the power generation field. The simplified flow diagram in Fig. 12-1 shows the process by which a fossil fuel power plant creates and utilizes steam in the generation of electricity. The feedwater is pumped into the preheater area where it goes to the boiler which turns it into steam. From there it is piped to the high-pressure turbine and back out through the boiler through the reheater into the intermediate-pressure turbine. From there it goes into the low-pressure turbine. The turbine turns the generator to produce electricity. At many points along this cycle, extraction lines remove condensate from the steam after it has been used by the turbine and feeds it back into the original feedwater. Steam exhausted by the low-pressure turbine is sent to the condenser and converted to feedwater to begin the process again. In this way, the system loses very little of its original water and the process can be cycled over and over again with little waste.

The boiler itself can be fueled by coal, oil, natural gas, or a combination of other fossil fuels. In the modern power plant, the turbine receives steam from the boiler which is created by the burning of fossil fuel. Water from a local source is sometimes preheated by the flue gases that are saved for recycling from the power plant stack. Gases from the burning of the fuel pass over the boiler tubes and superheater tubes and then run out of the boiler to the economizer, where the heat is used to prewarm the feedwater before it enters the boiler. The steam generator or the boiler usually has either natural circulation or forced circulation. In the latter case, water is moved through the pipes by use of boiler feed pumps or booster pumps.

Another permanent item in the power plant is the crane used to move the turbine and generator for installation, replacement, periodic inspection, or repair. In the drawings of the power plant (Fig. 12-3), the crane can be seen in the sectional view along with the control cabinet and main hook assembly.

The modern power plant requires very few operating personnel. Most of the plant is on automatic controls for everything from boiler water level to the fuel and air supply. A good percentage of the valves and equipment are motor-driven or cylinder-operated and controlled from a central instrumentation room where the operation of the plant is monitored by a technician.

Design Options

The design of the plant is extremely important, and all the variables in the locality must be taken into account. The designer and the construction company must review all the external considerations including:

1. Local climate
2. Amount of available space
3. Type of fuel or fuels to be used
4. Amount of electricity to be generated
5. Equipment available
6. Total arrangement of the plant
7. Plant location
8. Optimal use of operating space
9. Appearance and ease of maintenance

All these items must be incorporated into the design in coordination with local authorities and power companies. Building and operating costs must also be taken into account. Because of the enormous complexity involved in building and designing a modern power plant, the power company often hires a large, experienced construction company for the work. In some cases the power company has its own design and engineering departments that are experienced in the complete construction and design of a power plant from start to finish.

Design Effects

Among the many variables to consider in the construction of a modern power plant, it is important to take into account, in the beginning stages, the *operation variables* such as convenience of controls and the adequate layout of all the various aspects of piping, instrumentation, and equipment. Difficult fuel access to the site, insufficient shop space, unsafe controls or location of controls, inadequate storage and housing facilities, along with mismatched power capacities of turbines, generators, and condensers, can influence the eventual operating cost and the ability of the power company to produce adequate and inexpensive electricity. The completeness of specifications of the units and construction items becomes increasingly important in the final cost and operating convenience of the plant. Underspecification will leave manufacturers guessing or allow them to cut corners which will prove expensive to the power company later—in on-site equipment modification or even total replacement. Thus the construction firm must have thoroughly experienced designers and drafters.

Models

Models (see Chapter 15) are used throughout the power generation industry for checking equipment and piping locations. It is most economical to use the model to design the plant itself. There are two basic types of models: design and check models.

Models show the adequacy of space for maintenance and operation and also the general appearance. Most power companies now use models in their initial stages even though the cost may appear prohibitive. In the long run the usefulness of models in eliminating costly errors far outweighs the initial sum invested. By using a model, all conflicts between piping, steel, electrical conduits, and so forth can be spotted long before construction in the field, therefore eliminating expensive changes on the actual plant. Models also result in far better looking and better arranged plants. Models are used together by departments in the construction company to help the structural, mechanical, electrical, and piping groups coordinate their efforts and eliminate interferences. It is much easier to change the run of a line or move pipe hangers on the model than it is to discover the interference after the plant is well into the construction stage.

Check models are created from the piping layouts and diagrams provided by the design group and drafters. Especially in areas where the piping is congested, a model is extremely useful.

Many companies have resisted the use of models because of their cost; moreover, some drafters and designers are reluctant to use a new tool. In many cases drafters feel that drawings are sufficient for detecting interferences and designing the total system. Many hard-core designers and drafters are convinced of a model's usefulness only after the company has instituted a program and its advantages become obvious.

The installation shown in Fig. 12-3 will make an excellent group model project. Modeling enables students to convert their drawings and layouts of a system into the concrete experience only a three-dimensional object can provide. Whether they are used for designing or checking, models will continue to become more and more necessary in the power generation area. And the drafter who has experience with them will be much more valuable to employers than one who does not.

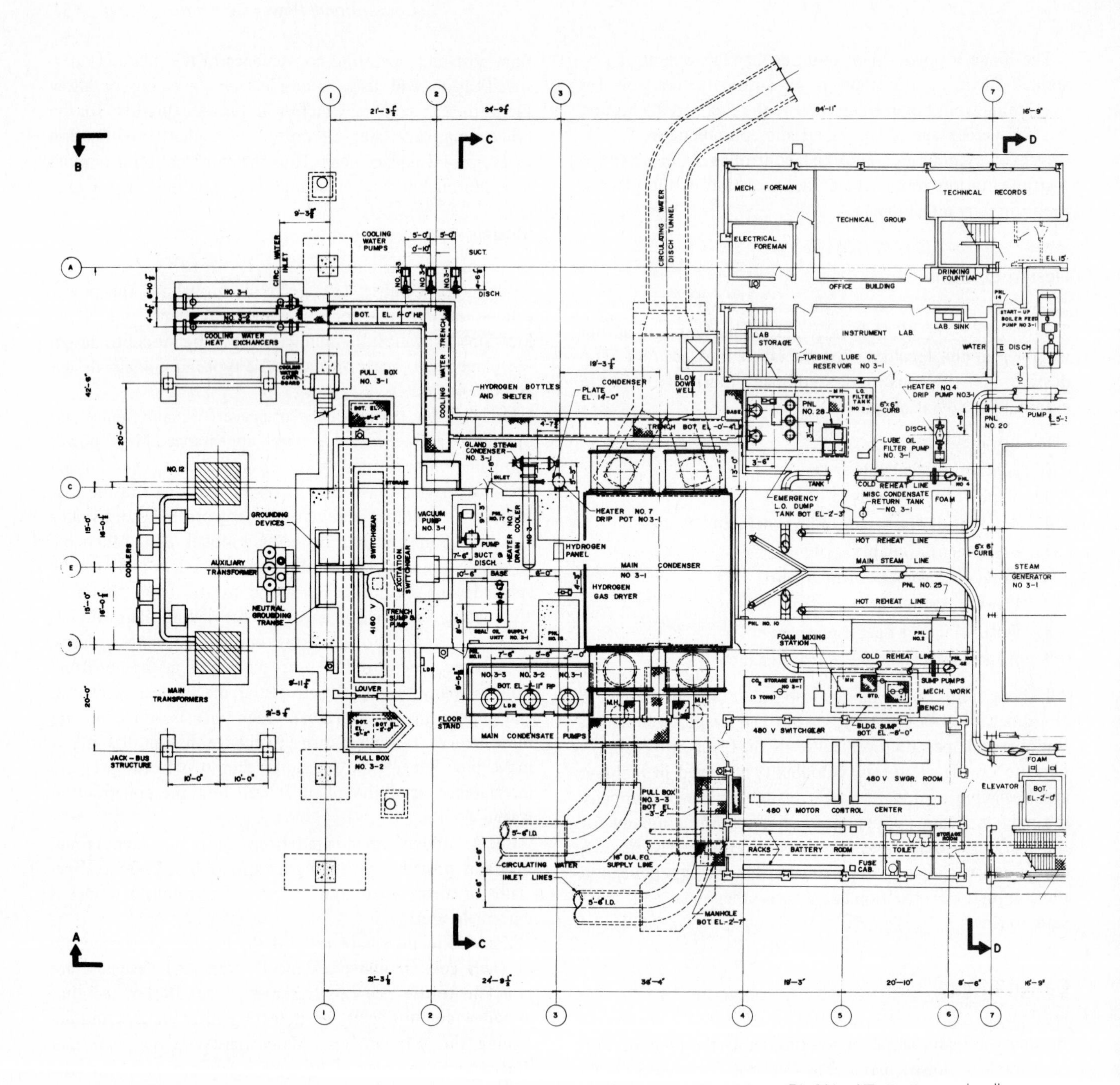

FIG. 12-3 Set of drawings for equipment placement and major lines for a single-turbine, single-boiler, fossil fuel electric generation power plant.

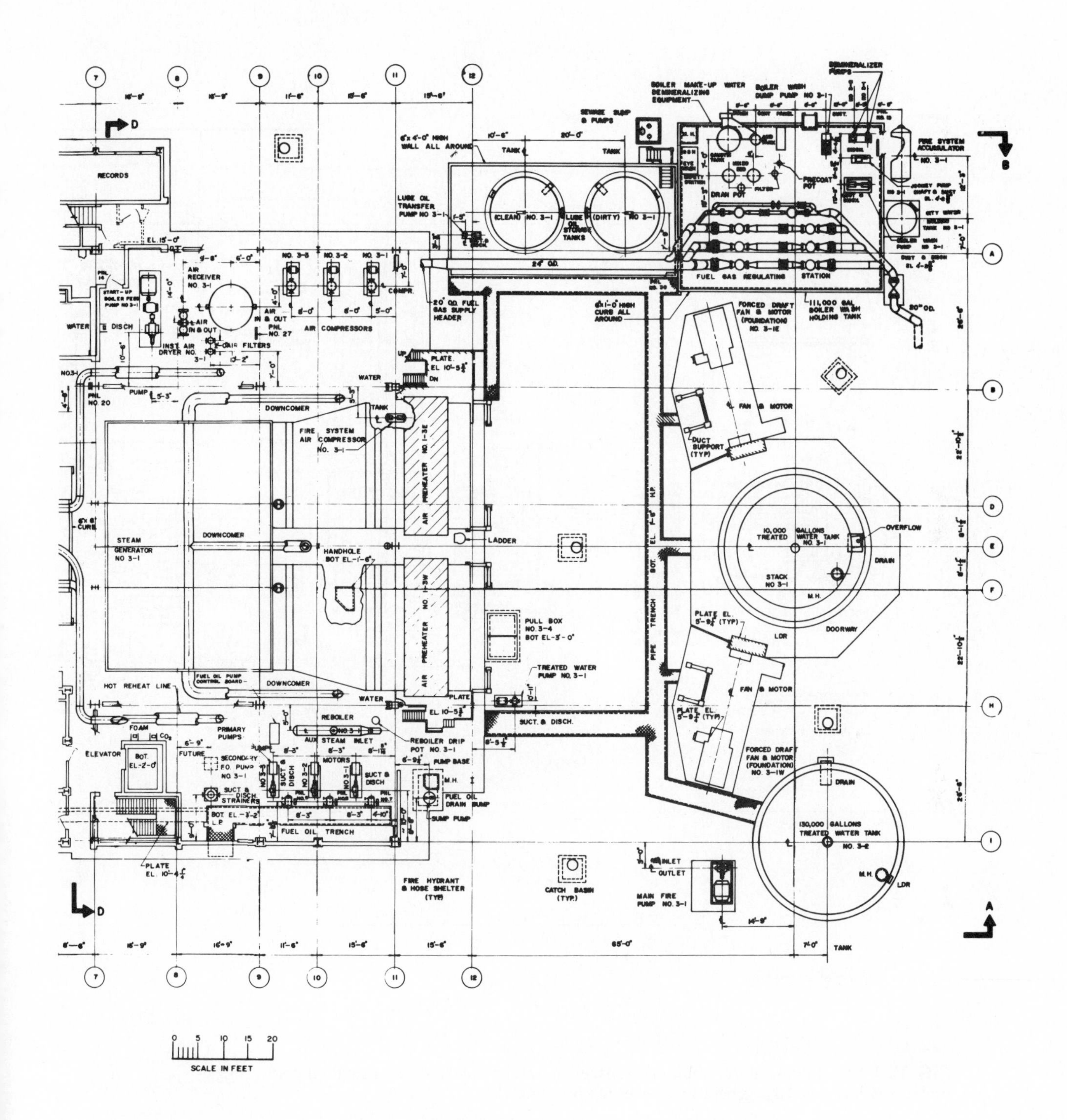
RECORDS
STEAM GENERATOR NO 3-1
DOWNCOMER
AIR COMPRESSORS
AIR RECEIVER NO. 3-1
FIRE SYSTEM AIR COMPRESSOR NO. 3-1
AIR PREHEATER NO. 1-3E
AIR PREHEATER NO. 1-3W
HANDHOLE BOT EL.-1'-6"
LADDER
PULL BOX NO. 3-4 BOT EL-3'-0"
TREATED WATER PUMP NO. 3-1
HOT REHEAT LINE
PRIMARY PUMPS
REBOILER
REBOILER DRIP POT NO. 3-1
AUX STEAM INLET
SECONDARY F.O. PUMP NO. 3-1
SUCT & DISCH STRAINERS
FUEL OIL TRENCH
FUEL OIL DRAIN SUMP
SUMP PUMP
ELEVATOR
FIRE HYDRANT & HOSE SHELTER (TYP)
CATCH BASIN (TYP)
MAIN FIRE PUMP NO. 3-1
LUBE OIL TRANSFER PUMP NO 3-1
LUBE OIL STORAGE TANKS
20" O.D. FUEL GAS SUPPLY HEADER
FUEL GAS REGULATING STATION
BOILER MAKE-UP WATER DEMINERALIZING EQUIPMENT
DEMINERALIZER PUMPS
FIRE SYSTEM ACCUMULATOR NO. 3-1
111,000 GAL BOILER WASH HOLDING TANK
FORCED DRAFT FAN & MOTOR (FOUNDATION) NO. 3-1E
FORCED DRAFT FAN & MOTOR (FOUNDATION) NO. 3-1W
FAN & MOTOR
DUCT SUPPORT (TYP)
PIPE TRENCH
10,000 GALLONS TREATED WATER TANK NO. 3-1
STACK NO. 3-1
OVERFLOW
DRAIN
DOORWAY
130,000 GALLONS TREATED WATER TANK NO. 3-2
SCALE IN FEET

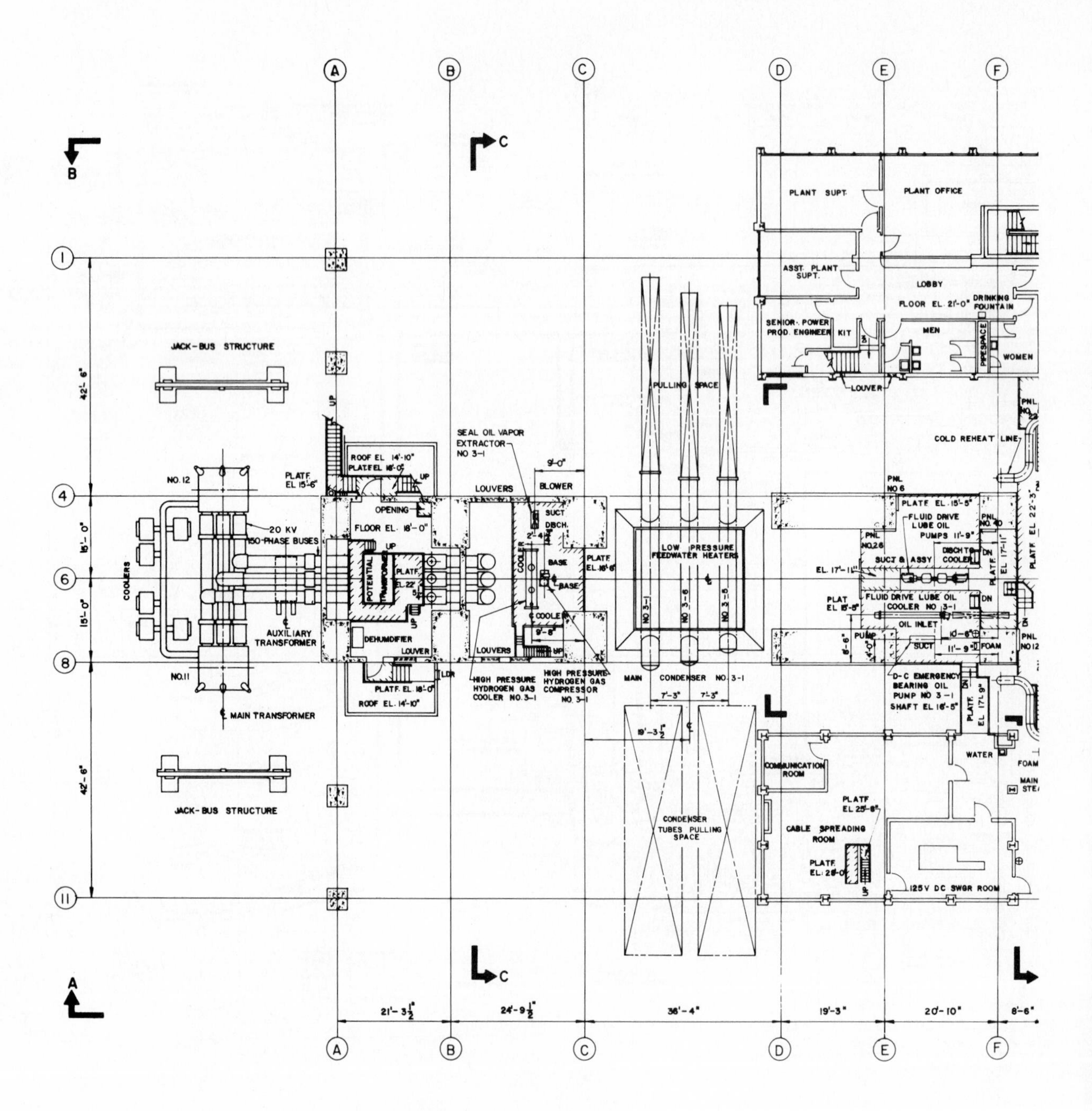

FIG. 12-3 Set of drawings for equipment placement and major lines for a single-turbine, single-boiler, fossil fuel electric generation power plant. (cont'd)

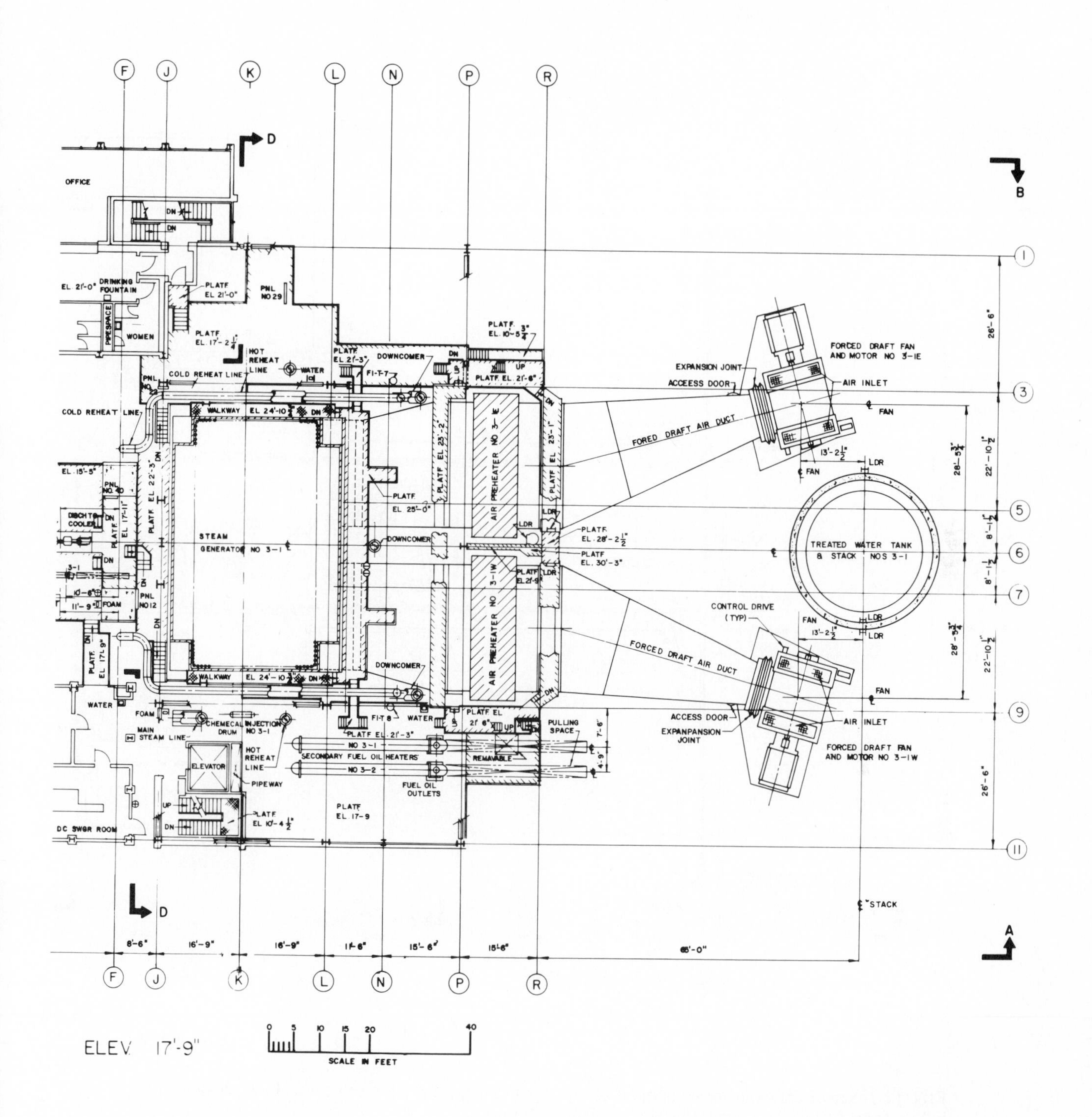

F
J
K
L
N
P
R
D
B
OFFICE
DN
EL 21'-0"
DRINKING FOUNTAIN
PLATF EL 21'-0"
PNL NO 29
PIPESPACE
WOMEN
PLATF EL 17'-2 1/2"
PLATF EL 10-5 3/4"
FORCED DRAFT FAN AND MOTOR NO 3-1E
EXPANSION JOINT
ACCESS DOOR
AIR INLET
FAN
HOT REHEAT LINE
WATER
PLATF EL 21'-3"
DOWNCOMER
FI-T-7
UP
PLATF EL 21'-6"
COLD REHEAT LINE
PNL NO 22
WALKWAY EL 24'-10
COLD REHEAT LINE
FORED DRAFT AIR DUCT
13'-2 1/2"
LDR
EL 15'-5"
PNL NO 40
PLATF EL 22'-3"
PLATF EL 25'-0"
PLATF EL 23'-2"
AIR PREHEATER NO 3-E
PLATF EL 23'-1"
DISCH TO COOLER
PLATF EL 17'-11"
STEAM GENERATOR NO 3-1
DOWNCOMER
PLATF EL 28'-2 1/2"
PLATF EL 30'-3"
TREATED WATER TANK & STACK NOS 3-1
3-1
10'-6"
11'-9"
FOAM
PNL NO 12
PLATF EL 21'-9"
AIR PREHEATER NO 3-1W
CONTROL DRIVE (TYP)
PLATF EL 17'-9"
DOWNCOMER
WALKWAY EL 24'-10
FORCED DRAFT AIR DUCT
WATER
FOAM
MAIN STEAM LINE
CHEMECAL INJECTION DRUM NO 3-1
FI-T 8
WATER
PLATF EL 21'-3"
PLATF EL 21' 6"
PULLING SPACE
ACCESS DOOR
EXPANPANSION JOINT
AIR INLET
FORCED DRAFT FAN AND MOTOR NO 3-1W
HOT REHEAT LINE
NO 3-1
SECONDARY FUEL OIL HEATERS
NO 3-2
REMAVABLE
ELEVATOR
PIPEWAY
FUEL OIL OUTLETS
7'-6"
4'-9"
PLATF EL 10'-4 1/2"
PLATF EL 17-9
DC SWGR ROOM
26'-6"
28'-5 3/4"
22'-10 1/2"
8'-1 1/2"
8'-1 1/2"
28'-5 3/4"
22'-10 1/2"
26'-6"
1
3
5
6
7
9
11
℄ STACK
A
8'-6"
16'-9"
16'-9"
11'-6"
15'-6"
15'-6"
65'-0"
ELEV. 17'-9"
0 5 10 15 20 40
SCALE IN FEET

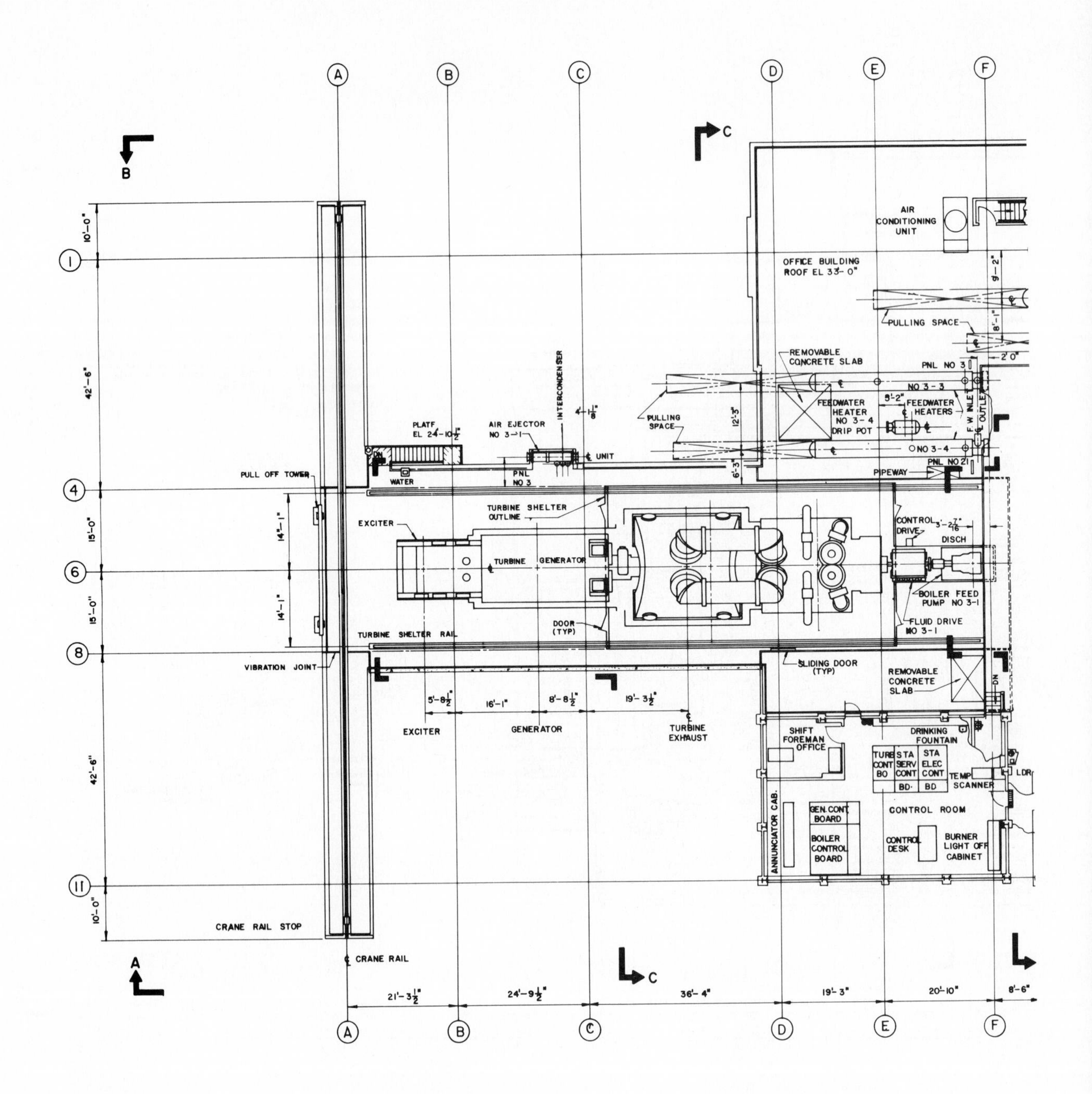

PLAN AT ELEV. 33'-0"

FIG. 12-3 Set of drawings for equipment placement and major lines for a single-turbine, single-boiler, fossil fuel electric generation power plant. (cont'd)

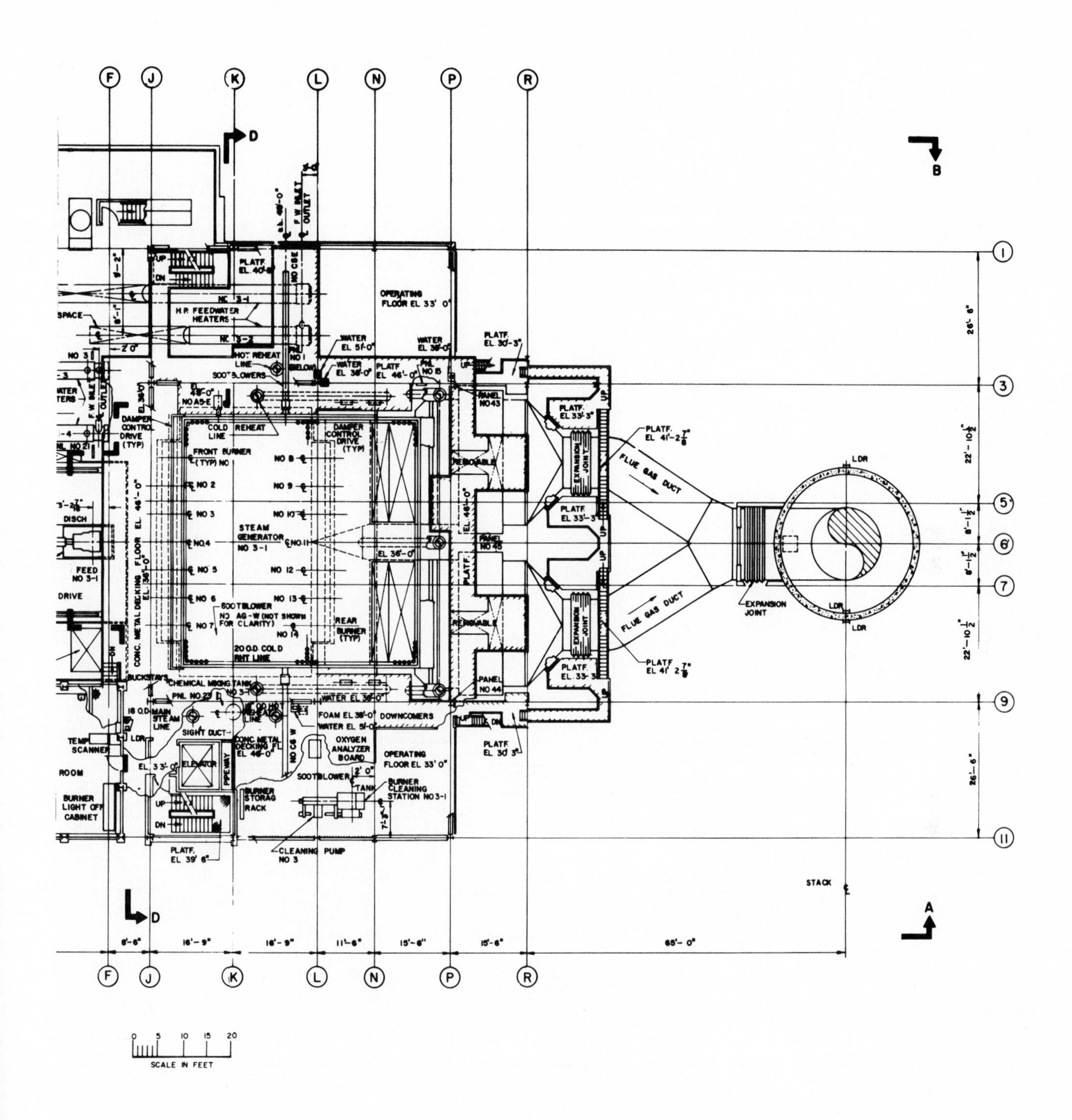

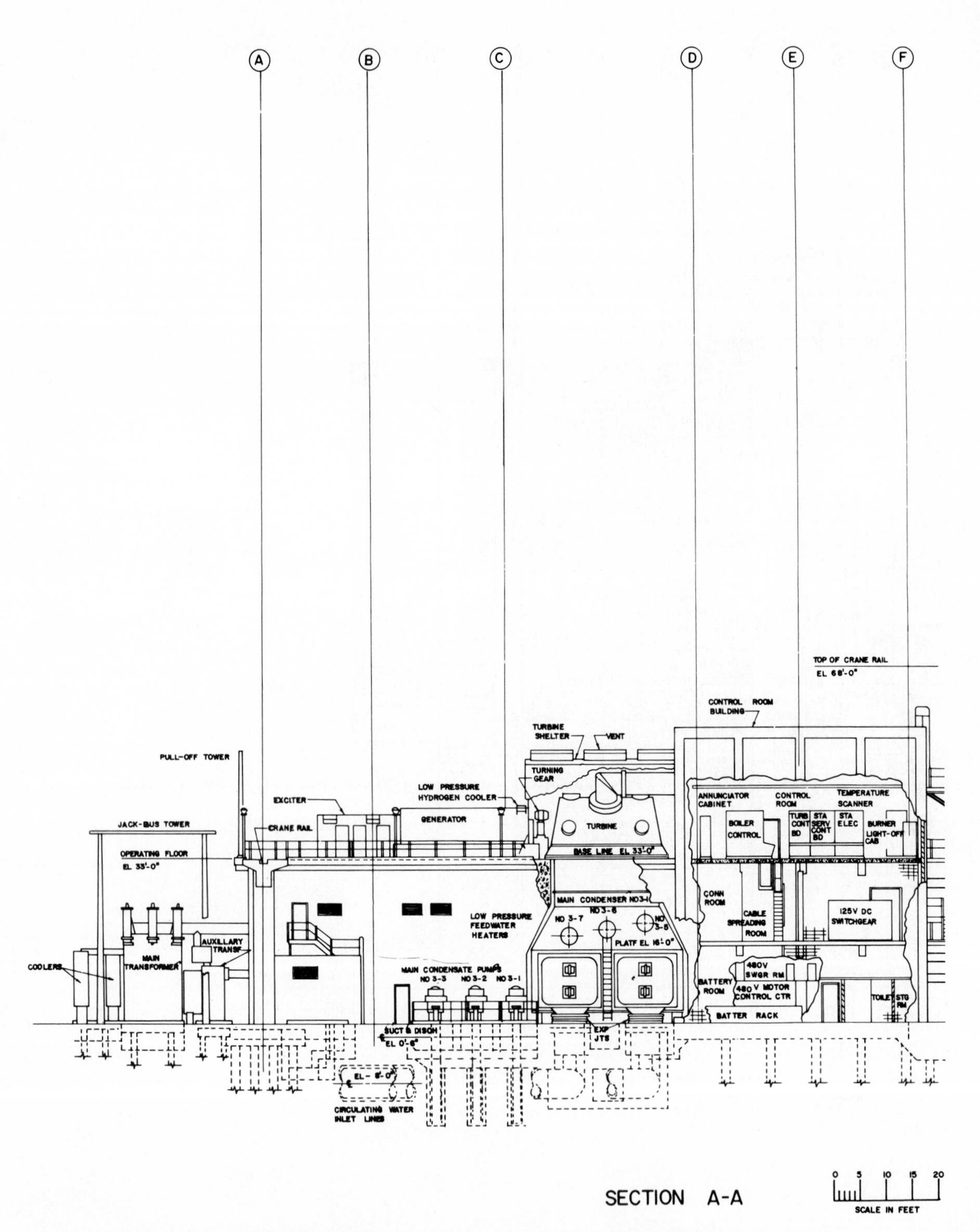

FIG. 12-3 Set of drawings for equipment placement and major lines for a single-turbine, single-boiler, fossil fuel electric generation power plant. (cont'd)

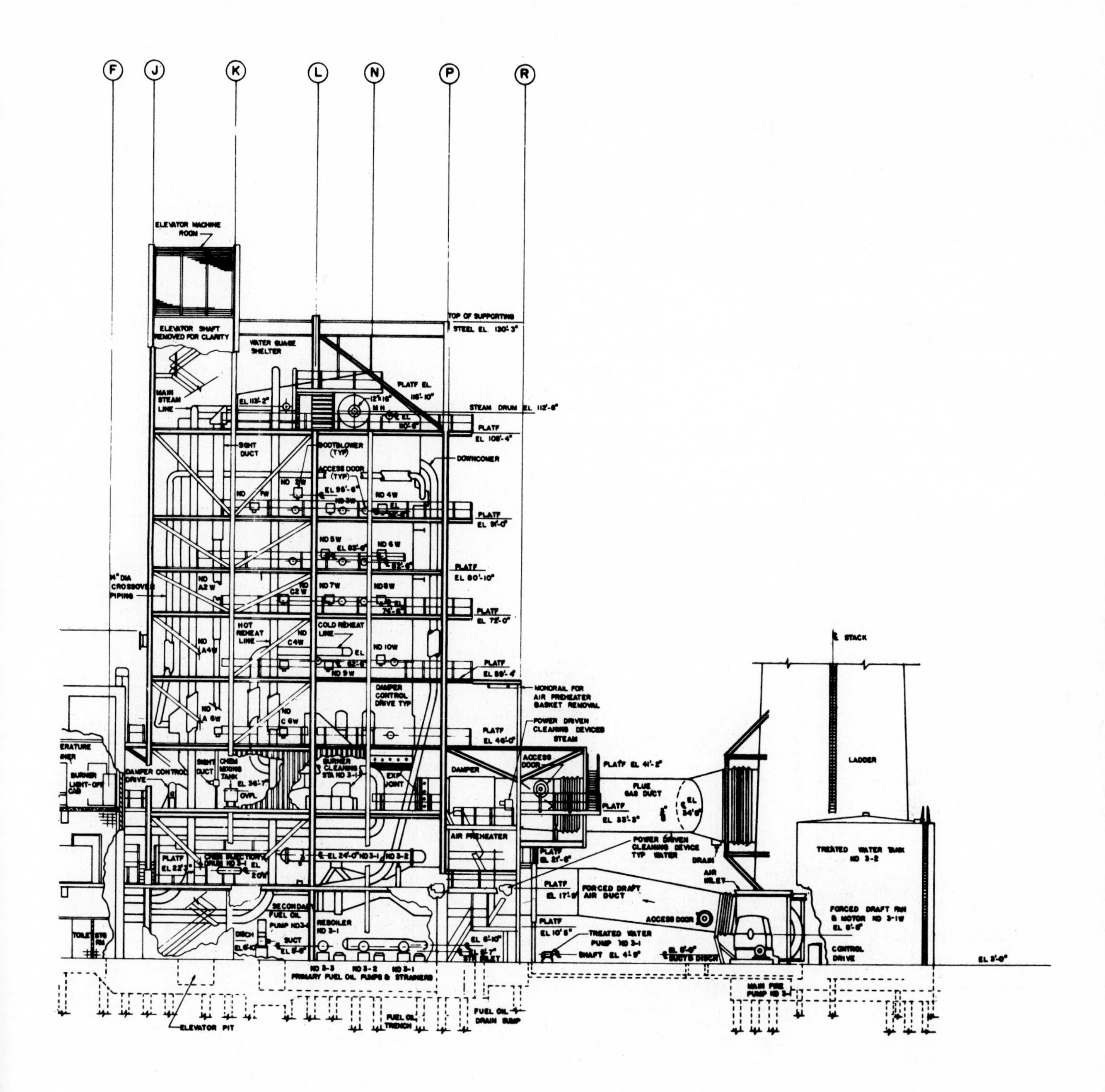

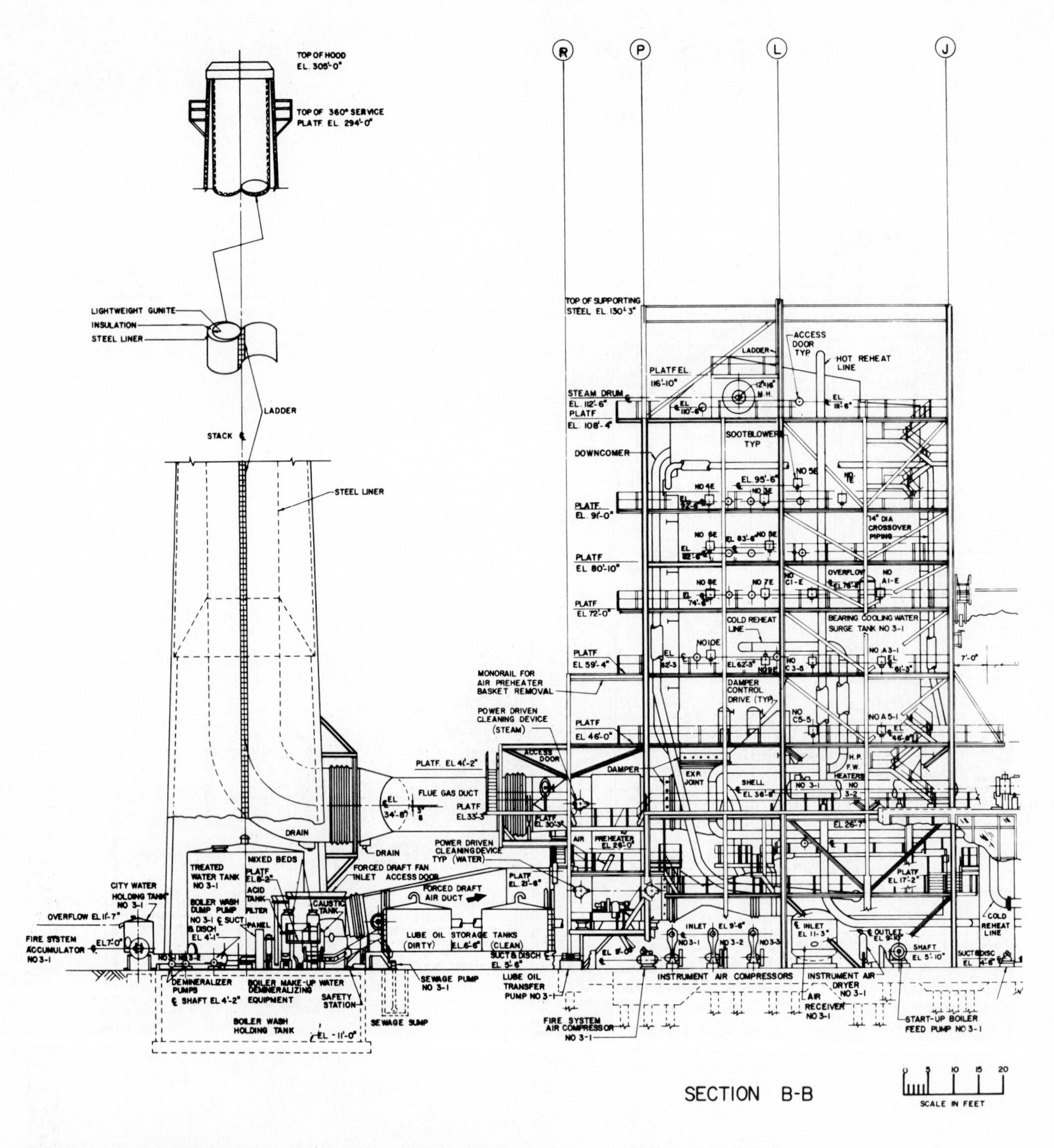

FIG. 12-3 Set of drawings for equipment placement and major lines for a single-turbine, single-boiler, fossil fuel electric generation power plant. (cont'd)

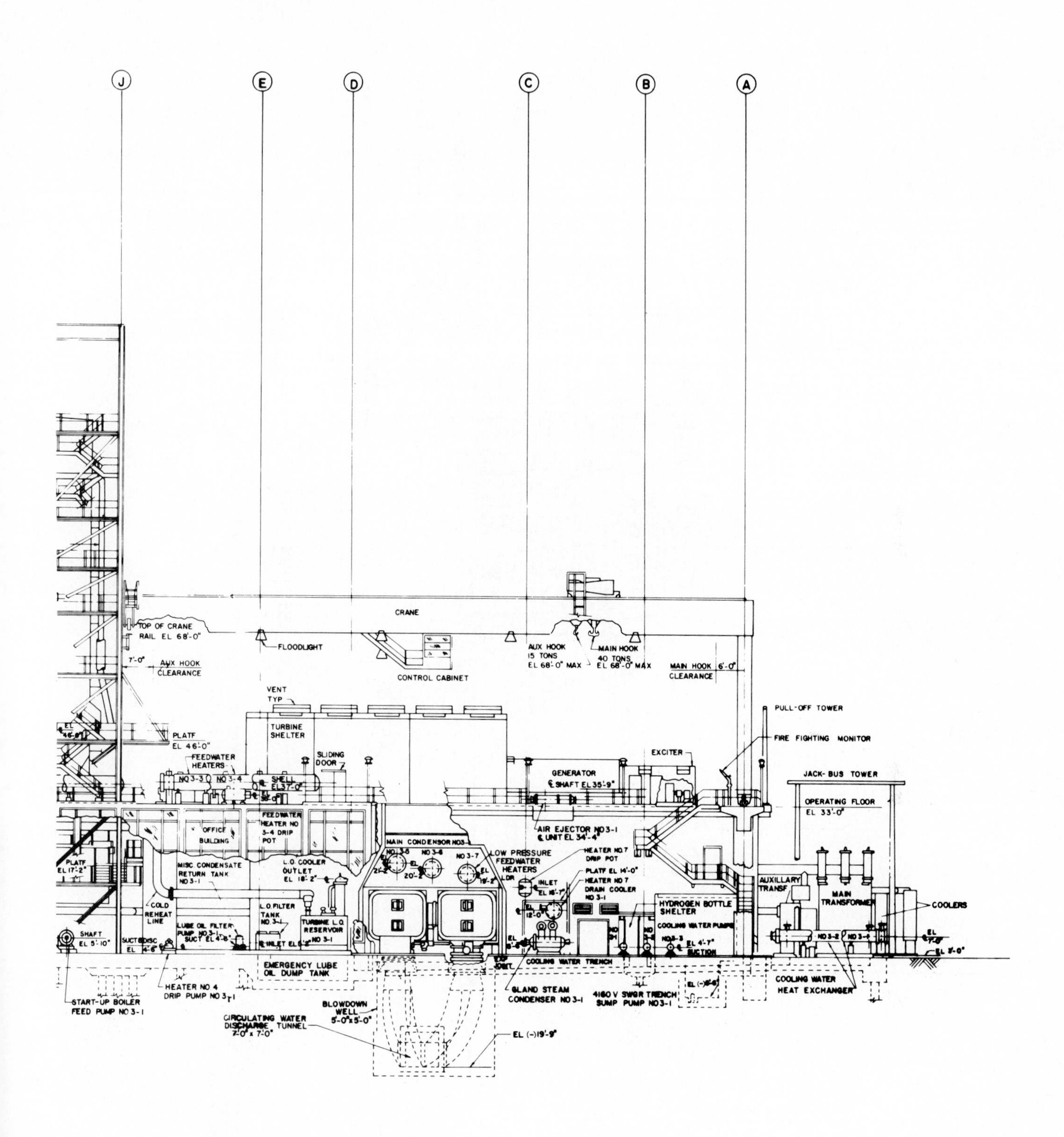
J
E
D
C
B
A
CRANE
TOP OF CRANE RAIL EL 68'-0"
FLOODLIGHT
AUX HOOK 15 TONS EL 68'-0" MAX
MAIN HOOK 40 TONS EL 68'-0" MAX
MAIN HOOK CLEARANCE 6'-0"
AUX HOOK CLEARANCE 7'-0"
CONTROL CABINET
VENT TYP
TURBINE SHELTER
PLATF EL 46'-0"
FEEDWATER HEATERS
SLIDING DOOR
PULL-OFF TOWER
FIRE FIGHTING MONITOR
EXCITER
GENERATOR
JACK-BUS TOWER
OPERATING FLOOR EL 33'-0"
AIR EJECTOR NO 3-1
OFFICE BUILDING
MAIN CONDENSOR
LOW PRESSURE FEEDWATER HEATERS
MISC CONDENSATE RETURN TANK NO 3-1
L.O. COOLER OUTLET
COLD REHEAT LINE
L.O FILTER TANK
TURBINE L.O. RESERVOIR NO 3-1
HYDROGEN BOTTLE SHELTER
COOLING WATER PUMPS
AUXILLARY TRANSF
MAIN TRANSFORMER
COOLERS
EMERGENCY LUBE OIL DUMP TANK
COOLING WATER TRENCH
GLAND STEAM CONDENSER NO 3-1
4160 V SWGR TRENCH SUMP PUMP NO 3-1
COOLING WATER HEAT EXCHANGER
HEATER NO 4 DRIP PUMP NO 3-1
START-UP BOILER FEED PUMP NO 3-1
BLOWDOWN WELL 5'-0"x5'-0"
CIRCULATING WATER DISCHARGE TUNNEL 7'-0" x 7'-0"
EL (-)19'-9"

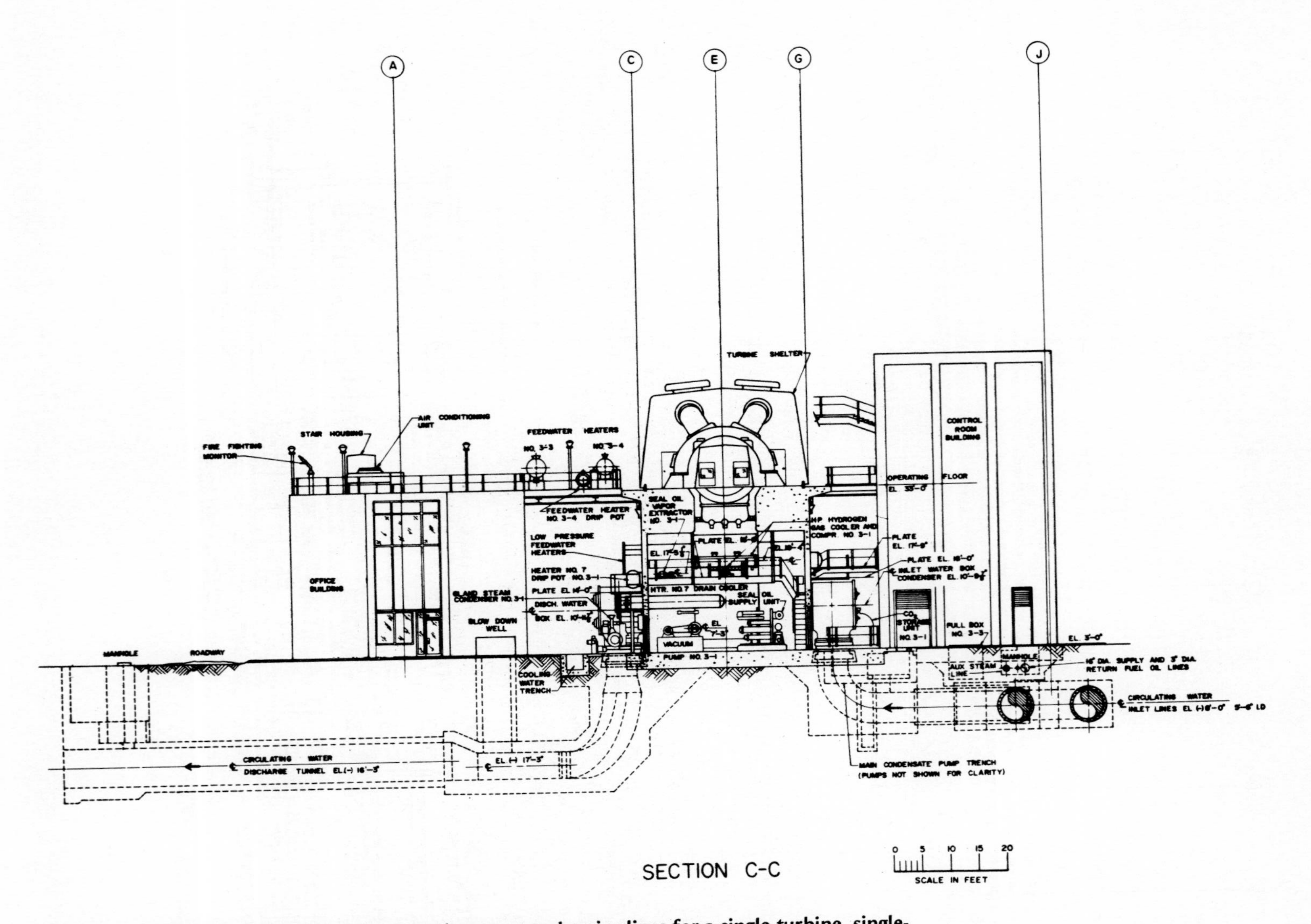

FIG. 12-3 Set of drawings for equipment placement and major lines for a single-turbine, single-boiler, fossil fuel electric generation power plant. (cont'd)

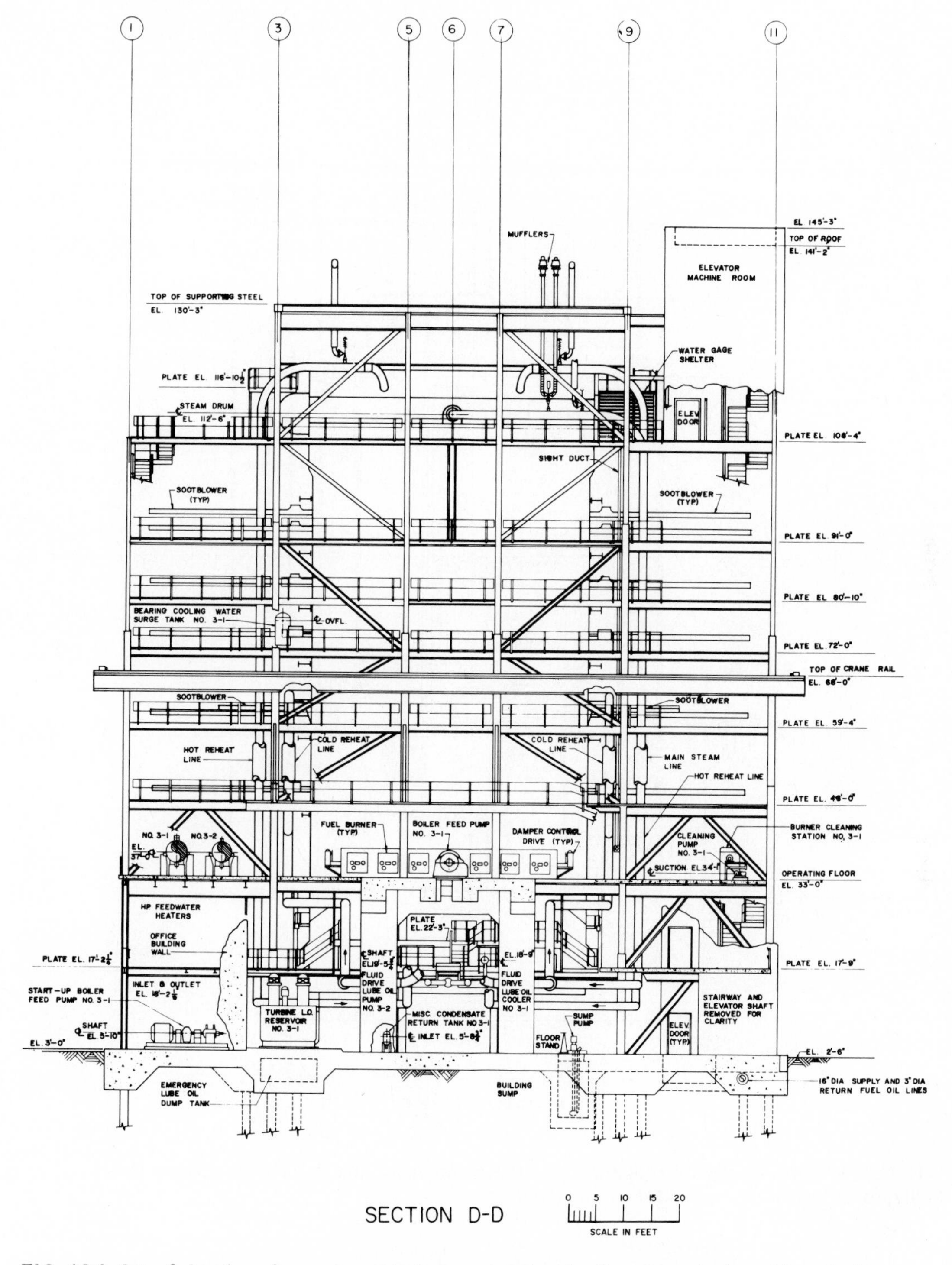

FIG. 12-3 Set of drawings for equipment placement and major lines for a single-turbine, single-boiler, fossil fuel electric generation power plant. (cont'd)

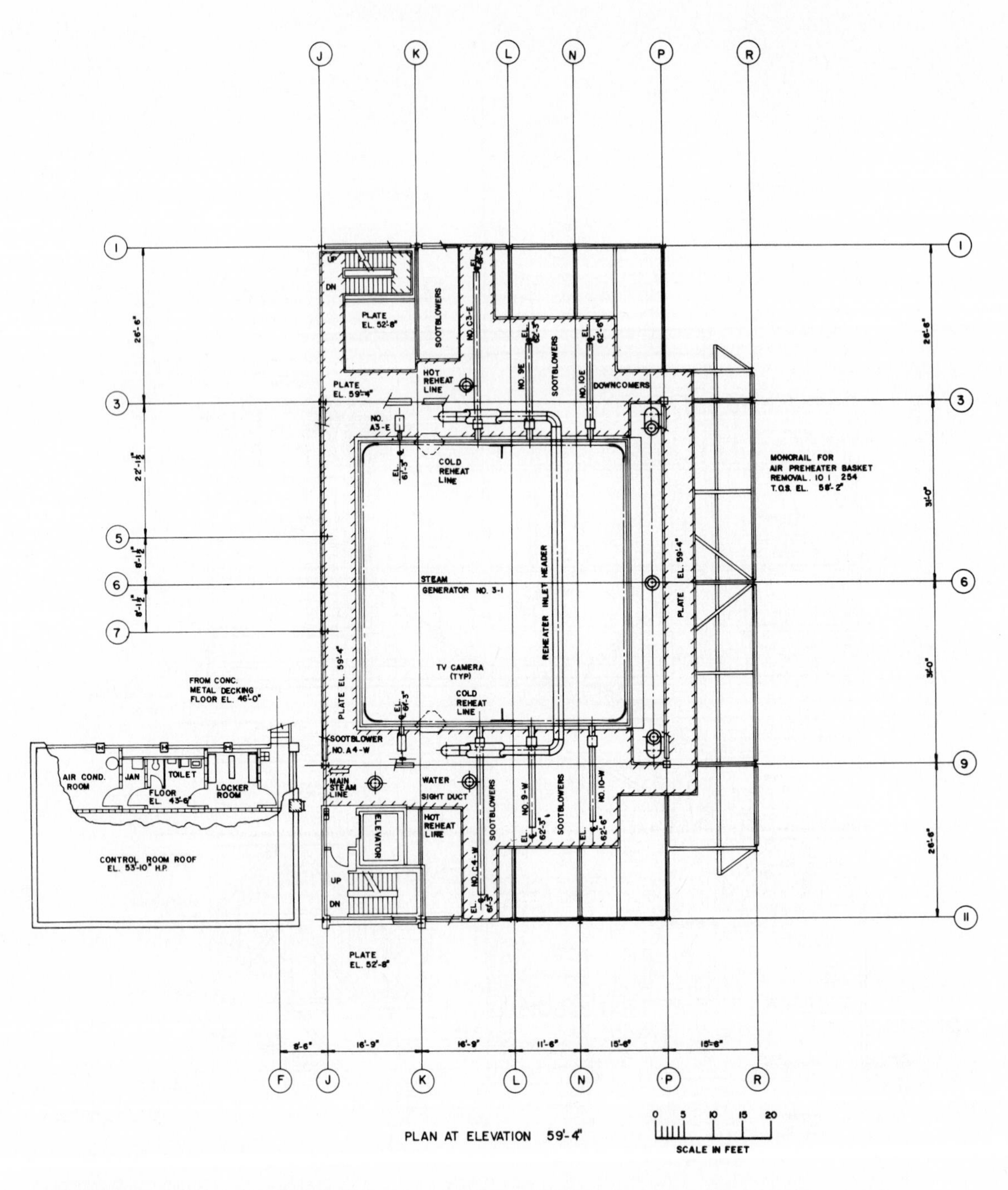

FIG. 12-3 Set of drawings for equipment placement and major lines for a single-turbine, single-boiler, fossil fuel electric generation power plant. (cont'd)

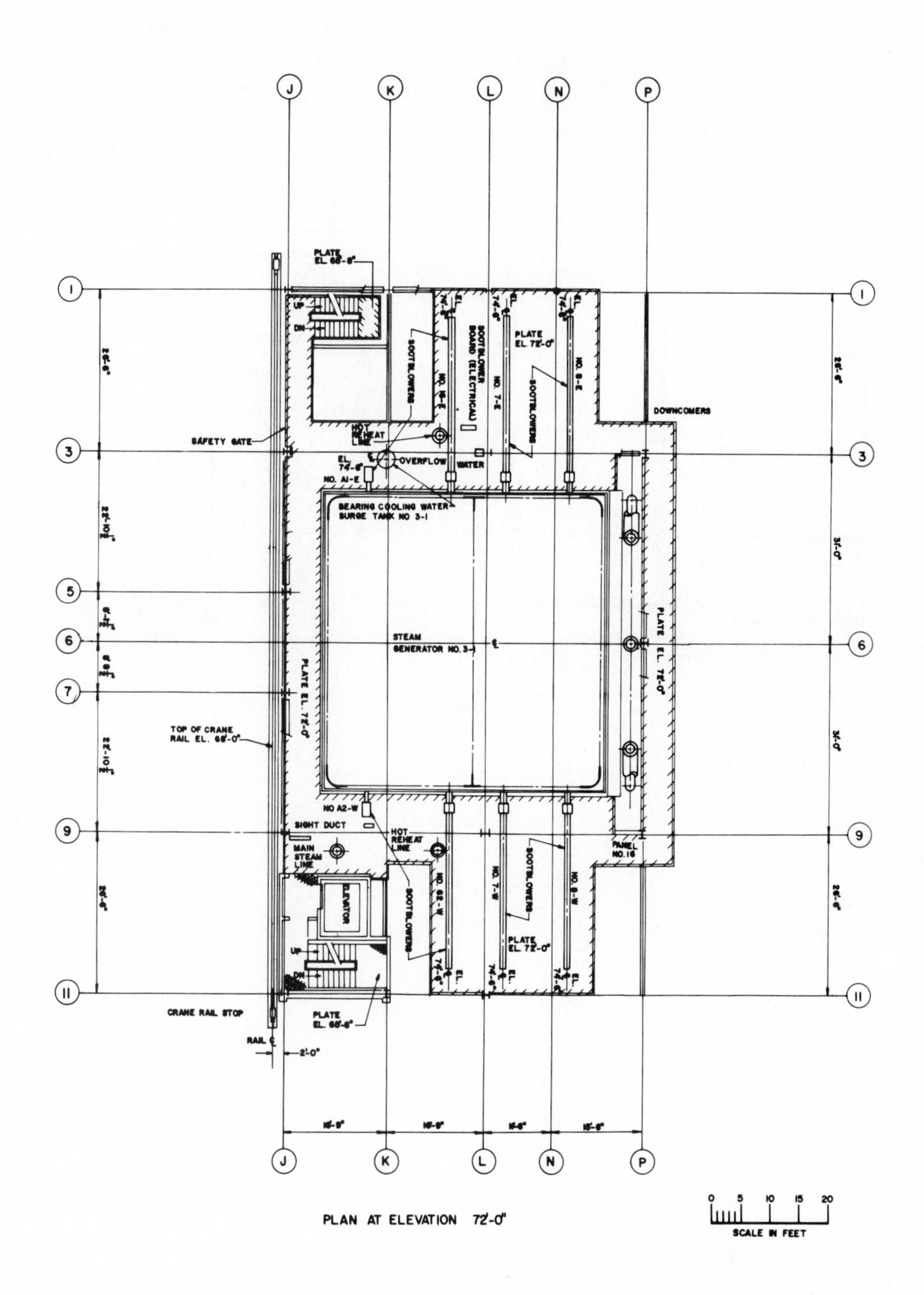

PLATE EL. 68'-6"
SAFETY GATE
HOT REHEAT LINE
OVERFLOW
WATER
SOOTBLOWERS
SOOTBLOWER BOARD (ELECTRICAL)
PLATE EL. 72'-0"
DOWNCOMERS
BEARING COOLING WATER SURGE TANK NO 3-1
STEAM GENERATOR NO. 3-1
PLATE EL. 72'-0"
TOP OF CRANE RAIL EL. 68'-0"
SIGHT DUCT
MAIN STEAM LINE
ELEVATOR
PANEL NO. 16
CRANE RAIL STOP
PLATE EL. 68'-6"
PLAN AT ELEVATION 72'-0"
SCALE IN FEET

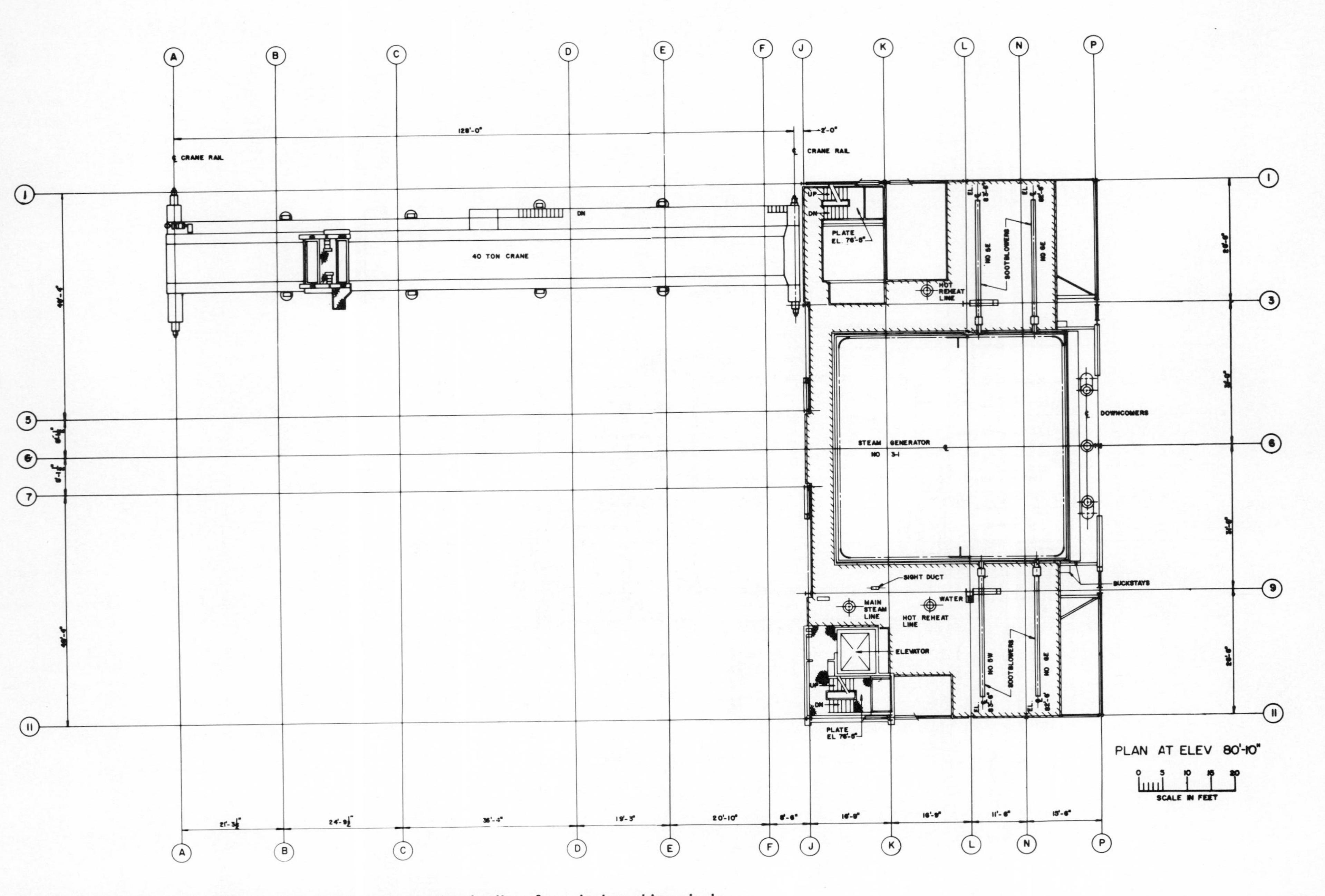

FIG. 12-3 Set of drawings for equipment placement and major lines for a single-turbine, single-boiler, fossil fuel electric generation power plant. (cont'd)

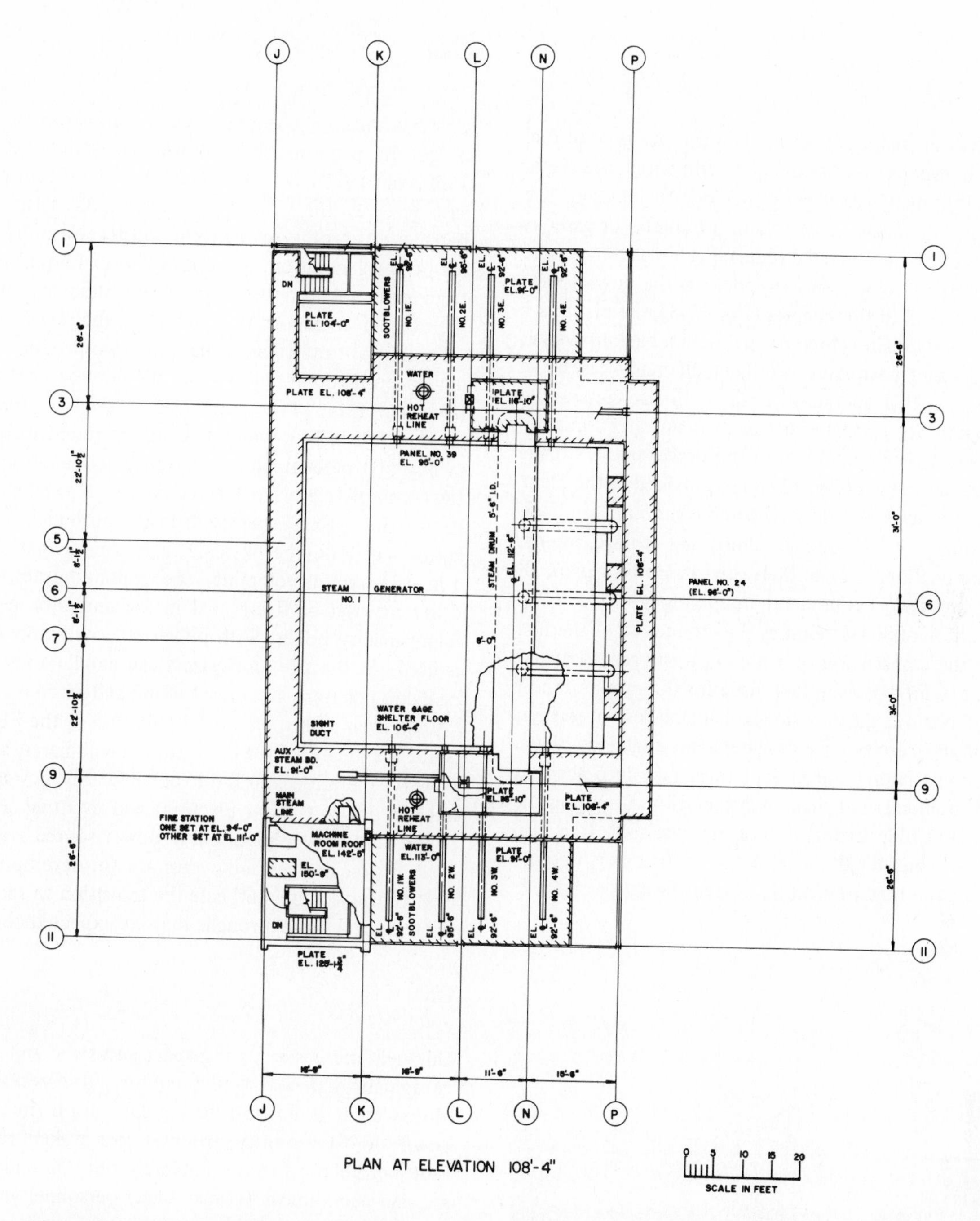

FIG. 12-3 Set of drawings for equipment placement and major lines for a single-turbine, single-boiler, fossil fuel electric generation power plant. (cont'd)

FUEL

All conventional steam power plants burn one type of fuel or another in order to release the heat with which the steam is generated. In many cases, more than one fuel is used. The fuel may be liquid (oil or petroleum products) or gaseous (natural gas or butane) or solid (coal). Each plant may store and utilize all the major fuel supplies. In the power plant drawings provided in this chapter (Fig. 12-3), the plant uses natural gas as its main source of fuel with a backup system for oil and other petroleum products. In many cases, the power company that operates the plant will also control the local gas supply or purchase the gas from another company and receive it through pipelines. In this power plant, storage facilities for natural gas (Fig. 12-4) are provided at the plant site. As might be expected, an enormous amount of piping is used to transport fuel supplies. Unloading, storage, piping, and heating facilities for the fuels used in a power plant all require a great deal of equipment and piping.

All generation of power requires the production of thermal energy by the combustion of a fuel with oxygen. Coal and oil, at present, are the main fuels used for the production of electricity. Natural gas has slowly become less important because of its scarcity. The designers and engineers of the company constructing the project must take into account economic, political, and social considerations affecting the various combustible materials that are available. It is also important to consider the local users or customers that are being served and how to meet their needs best.

FIG. 12-4 The holding tank in the background is a natural gas storage facility. As gas is used, the large water-filled cover keeps adequate pressure on the fuel to empty the tank completely.

Coal

Coal has once again become a major natural resource. Though it lost its prominence in power generation between the 1950s and 1970s, it is again becoming significant because of vast reserves throughout the world. As natural gas and petroleum become scarce, coal will receive more attention for power generation. Pulverized coal and liquid fuels made from coal are also becoming more important because of their ease in transportation and use in the power plant.

Large coal-fired steam generators that produce electricity have been made available since the development of pulverized coal firing. Pulverized coal is used in *fluidized bed combustion* (FBC), which has demonstrated increased efficiency (10 percent above conventional coal-burning facilities), complete ash separation, and low-tar fuel gases regardless of the type of coal used. In recent models, the pulverization need not be extreme; a maximum particle size of 1 in. (25 mm) is tolerable. The cyclone burner is the only other method of firing coal in the amounts needed for a high-capacity boiler. With pulverized coal, more equipment is used. At the site, pulverizers can handle a vast variety of coal compositions and create liquid and gaseous fuels.

Coal, even with the reduced pollution of the FBC process, is not the ideal fuel for the future. It will merely supplement power generation until the possible safe development of nuclear power (fusion reactors) and eventual use of solar power—the only replenishable power source available. Nuclear fusion devices show promise for development in the next century, and could ease the transition to solar power if the technical breakthroughs that are expected come to pass.

Oil

Oil fuels are easier to transport and store and require less equipment than coal, which requires pulverizers and stokers. Moreover, it is possible to regulate an oil fire and to shut down almost instantly, whereas coal stokers require more time before the fire is completely out. Oil and natural gas are also economical because fewer personnel are necessary. Petroleum is the major source of hydrocarbons available. The United States at one time was the largest producer of oil, though between 1950 and 1980 it has become increasingly an oil importer, making the price of electricity less favorable compared to coal and other fuel sources. Light fuel oils are the easiest to store and transport because they require no heating; the heavier fuel oils require tracer lines and heating coils to keep them sufficiently fluid. Storage tanks for oil (Fig. 12-5) are another aspect of the total design of the power plant that a drafter will encounter. They require many piping and valve installations.

FIG. 12-5 Fuel oil tank enclosed by a dike in case of bursting or fire. This facility provides oil for a modern conventional power plant.

Natural Gas

Natural gas was at one time considered almost inexhaustible since all areas of the world containing petroleum also contain natural gas and many areas that do not have petroleum also have natural gas. In recent times, however, it has become increasingly apparent that natural gas supplies are limited too. Natural gas supply lines are in many cases owned by the power company. The lines need to be well drained, and a sufficient number of traps must be provided to ensure that the accumulation of condensate will not interfere with gas flow or cause other problems because of impurities. All supply lines must be adequately sized for maximum gas flow and minimum pressure drop.

A promising form of solar power which may, in the future, fill our natural gas distribution lines is called the *hydrogen economy.* It may prove possible to convert seawater into hydrogen, oxygen, and minerals by means of the natural solar heating of the sea itself. In that case, hydrogen could substitute for natural gas, for it burns absolutely clean with no combustion product except water. Chimneys and stacks would be unnecessary. One drawback, however, is that hydrogen is corrosive to natural gas piping.

Oil Shale

Oil shale is a sedimentary rock which contains organic matter and will yield oil. Oil shale is widely distributed throughout the world. Because of its production expense, it has not been used as fuel except in Scandinavia and Scotland, though eventually it will become exceedingly important due to diminishing supplies of oil. Upon being heated, its products can be used in power generation just like oil.

Butane and Propane

Butane and propane can also be burned in a modern power plant. They are burned as gases in the boiler and heaters but stored as liquids under pressure. Various cooling and heating equipment must be provided when using this fuel supply.

Wood

Wood is another source of fuel that is replenishable, though it is becoming apparent that the forests of the world are diminishing. The burning of wood for steam generation is usually confined to parts of the world where other fuel products are not available. Wood waste products are used in some parts of the world to produce alcohol as fuel; even wood itself in the form of bark and slabs is often reduced to a convenient size for handling and burning in the power plant. Wood should not be considered a viable alternative to other forms of fuel unless nothing else is available.

EQUIPMENT

The equipment in a modern power plant is extremely sophisticated. The following paragraphs describe the basic components of every plant: steam turbines, generators, superheaters, economizers, heat exchangers, water systems, and cooling towers and ponds.

Steam Turbines

One of the most important pieces of equipment in the power plant is the steam turbine. Steam turbines have been used since 100 B.C.E. to produce mechanical energy. In 1884, Sir Charles Parsons developed a reaction turbine. His patent was eventually bought by the Westinghouse Electrical Corporation, which produced some of the first commercial turbines for power generation. During the same period, C. G. Curtis developed the impulse turbine, which was then produced by the General Electric Company. Steam is expanded through nozzles and, in the case of impulse turbines, impinges on a series of blades or buckets. The rotating blades are attached to a shaft which turns a generator which produces electricity. The force on the blades is created by the pressure drop of steam as it passes through the nozzle. Impulse turbines involve almost no pressure drop; in a reaction turbine the pressure drop is great. As shown in Fig. 12-1, the turbine permits steam to be extracted at different pressures in order to utilize the pressure drop throughout the turbine cycle. Therefore high-pressure, intermediate-pressure, and low-pressure steam can be used in the generation of electricity.

Generators

The generator is the most important piece of equipment in the power plant. Without it, electricity could not be produced. In many ways all other equipment is only supplemental in operation. Modern generators usually have a rotating field and stationary armature coils; when the magnetic field is interrupted by close loops of wire, electric current is produced.

Steam Generators (Boilers)

Boilers comprise one of the most important aspects of the power plant because this is where the steam is generated unless geothermal energy is available. In simplistic terms power plants make steam in a closed vessel (boiler) by adding water and heating it until a sufficient temperature is obtained to create steam.

In the power plant drawings (Fig. 12-3), the boiler is referred to as a *steam generator*. The steam generator includes the furnace (Fig. 12-6), the boiler, the preheater, and all fuel-burning equipment. The steam generator can be defined as an apparatus for producing and recovering heat and then transferring it to a working medium. Energy in the form of heat is produced in the steam generator by the combustion of fuel (Fig. 12-7) and is transferred to water, which converts it to steam for the production of electricity through the turbine and generator. The gases and heat created by the combustion of fuel are also used in superheaters, air preheaters, and economizers. Many large power plants are single-unit installations with one boiler and one turbine or, in some cases, tandem or multiple turbines on a single shaft. Figure 12-3 shows one large steam generator and a large aircraft turbine. In this way the power plant can make maximum use of minimum piping.

The efficiency of a steam generator is of extreme importance. With increasing fuel costs and decreasing supplies it becomes imperative to use the fuel completely. Modern advances include the use of high feedwater temperatures from air preheaters and cleaner heat transfer surfaces. Up to a point, increasing the maximum pressure and temperature increases the efficiency.

Boilers are usually classified by whether the tubes are filled with water or hot gas. Thus there are *water-tube* boilers and *fire-tube* boilers; the water-tube type is more common. Because of the value of high pressures and temperatures, boiler design has tended toward the implementation of extremely tall steam generators as in Fig. 12-3. Tall boilers help in the circulation of water (which is poor at high steam pressures) and provide large heat-absorbing capacities. The boiler heating surface is the part of the steam generator which includes the drums or tubes or shells of the boiler circulatory system which have hot gases on one side and water (or water and steam) on the other.

Boilers can be natural-circulation or forced-circulation types. In a natural circulation unit, heat from the hot gases flows through the walls of the boiler tubes, is transmitted to the water, and produces steam. Because of modern requirements for high temperatures and pressures in boilers, however, it is becoming increasingly important to have forced circulation so that water circulation in all parts of the steam generator is adequate.

Besides water-tube and fire-tube boilers, *shell boilers* are available—vessels that contain water. A portion of the shell is exposed to heat or the water itself is heated by electrodes. This type has a very low capacity. The use of tubes, whether fire tubes or water tubes, increases the heat transfer area between the hot gases and the water, but having water rather than fire in the tubes takes the system pressure off the container. The water-tube boiler is much more efficient than the shell type. Fire-tube boilers are not well suited for

FIG. 12-6 Steam generation burner assembly. Note the instruments above the burner. (Courtesy Coen Company, Burlingame, Ca.)

FIG. 12-7 Four Coen B-17 combination gas and oil burning registers firing an integral furnace boiler on induced draft to 35,000 lbs steam output. (Courtesy Coen Company, Burlingame, Ca.)

superheat and therefore cannot take advantage of superheated steam.

In Fig. 12-3, the steam generator has twin furnaces and two rows of burners. The steam generator in this drawing is suspended from large top-supported hangers which are shown in Fig. 12-8. When designing a boiler support it is important to take into account the maximum amount of expansion caused by changes in temperature in the boiler unit. Top-supported units like those in the drawings must be able to swing at proper angles and accommodate severe stresses and movements caused by thermal expansion. In general, many factors must be taken into account for adequate boiler design: steam flow pressure and temperature, feedwater temperature and conditions, amount of electrical energy to be generated, fuel or combination of fuels to be utilized, and efficiency and economy of operation.

FIG. 12-8 Boiler hangers for suspending the steam generator from the steel beams above. In the background is a large spring hanger which helps support an insulated steam line.

Superheaters

The use of superheated steam to drive high-speed turbines is essential to prevent condensation due to loss of energy in the steam. This is especially true of reaction turbines. The degree of superheat above the boiling point at the given pressure tells how much energy can be expended by the steam before condensation begins. In most high-temperature installations the superheater provides a constant high level of superheat that is maintained by various controls. Superheaters can also be drainable or nondrainable, depending on the installation.

Superheaters are boiler accessories and fall into two main categories: the *convection* type and the *radiant* type. The convection type has an increasing temperature as the boiler rating increases. It is the least expensive and simplest superheater and is used frequently. The radiant type has a decreasing curve: As the boiler rating increases, the temperature curve decreases.

Economizers

Economizers improve the overall efficiency of the power plant by utilizing the heat from the combustion gases to preheat the feedwater before it enters the boiler. The gases are therefore cooled and better economy is obtained. By heating the feedwater before it enters the boiler, the boiler does not experience the extreme temperature strains caused when the incoming feedwater is too cold. Economizers can be steaming or nonsteaming. Where water is heated enough to cause it to boil, special arrangements are made to allow the water to enter the drum so that the steam released does not cause priming or foaming. In the nonsteaming type, which is prevalent, the economizer does not heat the incoming feedwater to a boiling level. Though there is an advantage in producing part of the steam in the economizer, the economizer must be limited to 20 percent of total steam production. Where much new feedwater (makeup water) is required, it is impractical to use steam from the economizer.

Heat Exchangers

In the older steam power plants the used steam was exhausted to the atmosphere and thus much heat was lost to the system. The atmosphere in this case acted as the condenser in the cycle. By recovering the exhausted steam, the heat exchanger or condenser reduces the amount of makeup water that is necessary for the system from 100 percent to less than 5 percent. (Some makeup is always lost to the system through blowdown, leaks, contaminated steam, or steam traps.) Heat from the condenser and from the exhausted steam can also be used by adding feedwater heaters that raise the temperature of the condensed steam before it enters the boiler.

Two methods are used to condense steam. One is to mix the vapor with a liquid which increases the temperature of the liquid; the other is to transfer the latent energy of the vapor through a surface to another fluid (as in a feedwater heater) at a lower temperature. Both condensers and feedwater heaters operate on these principles. In a modern power plant a large amount of the expense goes into the installation of pumps, circulating water facilities, piping, and condensers so that the steam can be expanded and used as completely and efficiently as possible.

Condensers are used to improve the power plant's cycle efficiency. It is possible to recover over 90 percent of the original water used in the system and recycle it without the purifying methods which are usually necessary for the original feedwater because of salt content, mineral content, or pollution in the local water supply.

Water Supply and Treatment Systems

Water is used throughout the power plant in areas that include:

1. Production of steam
2. Condenser cooling
3. Boiler makeup
4. Fire protection
5. Sanitary and drinking use
6. Process cooling

Water supplies are transported from primary sources such as lakes, wells, rivers, or oceans to the plant by a series of piping. The characteristics of the water supply are always important and influence the placement of the plant site and the equipment used. Changes in water temperature and mineral/salt content are critical when considering placement of the power plant. Water treatment, transportation, and storage are basic aspects of the total power plant installation. Where there is no adequate supply of water, other means must be used to produce the water for the power plant. Where there is no large body of water to use for condensing, cooling towers or spray ponds must be installed—which adds another element to the total cost and site requirements of the plant.

Cooling Towers and Ponds

To compensate for the lack of a large supply of cooling water and condensing water, a power plant must use cooling towers and ponds. Small streams and wells can be used when makeup water and cooling water requirements are not great. Many types of cooling towers are available to mix or force air to mingle with water in order to cool it. Water is usually piped to the top of the tower and allowed to fall by gravity through trays and baffles or troughs with the air moving horizontally through them. There are three basic types of cooling towers: the *induced-draft* tower, the *forced-draft* tower, and the *natural-draft* tower (Fig. 12-9).

With the natural-draft tower, there is a need for a larger volume per unit of cooling capacity because of variable wind velocity and direction. In induced-draft towers, which are at present one of the most popular types in large installations, huge fans are used with vertical shafts to pull air up from the sides of the tower and create a draft that passes across the water as it flows down. In forced-draft cooling towers, fans on the horizontal-shaft project air into the falling water through baffles that are pointed upward.

FIG. 12-9 Natural-draft cooling towers.

STEAM PIPING PLACEMENT

Convenience in operating and maintaining the piping equipment is extremely important for overall efficiency of the plant. The following guidelines apply to design conveniences that may be encountered in a power plant:

1. Handwheels on valves should not be more than 6 ft (1.83 m) above the ground, unless they can be chainwheel operated or a platform walkway ladder is provided.
2. Valves below the operating floor should be equipped with extensions or other means of access.
3. Any valve that is not readily available should be remote-actuated.
4. When drawing valve positions, indicate stems and handwheels on piping drawings so their location is not left to the construction crew.
5. Adequate clearance should be left for the maintenance, dismantling, or replacement of all valves and equipment.
6. In screwed piping it is important to use sufficient unions to allow for easy disassembly of pipelines for maintenance and replacement of valves and fittings.
7. Overhead piping should be easily accessible for alterations and proper maintenance. If at all possible, construct piping lines so that they remain accessible from established platforms and walkways.
8. Clearances should be allowed between all piping and existing equipment or electrical installations.
tions.
9. When running horizontal piping lines, keep a minimum of 6 ft 6 in. (1.98 m) vertical clearance where possible to provide for personnel walkways and equipment passage.
10. Establish enough clearance around piping so that equipment may be installed, repaired, dismantled, or removed without interfering with the piping runs.

11. The use of a model is of extreme importance in checking design and determining clearances and accessibility of piping installations.
12. Piping should be arranged and marked so that the purpose and flow direction of each valve and line are obvious.
13. In most cases, it is possible to color code piping lines and service lines and stencil the name of the service at intervals along the pipe.
14. Where various piping systems come into contact with each other, design the system from a comprehensive standpoint rather than as separate systems.
15. Keep the system as symmetrical and uniform in appearance as possible.
16. The piping drafter should have available all the drawings of equipment location, steel structure, mechanical equipment, and passageways in order to design a total system without interferences.
17. Piping that travels in similar directions should be kept on the same plane.
18. Small piping should be taken into account in the original design of the major piping system to avoid shoddy installations after the main pipes are installed.
19. Avoid using 45-degree turns in piping runs.
20. Pipes should run parallel to the axis of the building wherever possible.
21. Where space permits and pressure drop is to be minimal, bends are preferable to elbows even though elbows are superior to bends in appearance and from an economic standpoint.
22. A pipe gallery should be provided at a suitable elevation within the boiler house.
23. Establish adequate subbasement pipe trenches and piping runs which may be needed in a lower structure of the power plant.
24. Arrange all repetitious equipment in groups with identical piping runs and installations where possible.
25. Lanes should be provided for piping to approach equipment.
26. Adequate piping areas must be provided for the main steam boiler feed and atmospheric exhaust pipes leading to and from the boiler system.

Sizing Steam Piping

The following review of methods for sizing steam pipelines will give the reader experience in calculating pipe sizes and velocities for steam uses. (They are not to be substituted for manufacturers' charts for suggested product uses for particular services.) It is important to be able to deliver the required quantities of steam at a specified pressure to specific parts of the system. The size of the piping must be adequate for the condition but not oversized. Large pipes are expensive and require more installation whereas a pipe too small produces excessive pressure drop and high velocities and does not deliver the required quantity of steam at the required pressure. It is impossible to set down definite rules for sizing steam piping; experience is the best guide. Variation of inside diameter and condition of pipe both influence the calculations for pipe sizing. The demands of the installation must be taken into account along with its performance and economy when determining pipe size for steam piping.

The following formulas will enable the designer and drafter to find the necessary pipe size for a service. The following variables are involved:

d = inside pipe diameter
S = superheat (in degrees Fahrenheit)
p = absolute pressure (in psi)
W = steam flow (in pounds per minute
V = velocity

Pipe sizes for saturated steam:

$$d = \sqrt{80{,}000W} \over pV$$

Thomas formula for sizing well-insulated steam lines:

$$d = 4.5(1 + 0.00025S)\frac{W}{p}$$

EXAMPLES

1. Find the correct pipe size and schedule number for a saturated steam line which has a velocity of 5000 fpm (maximum) at 250 psi and requires a flow rate of 1000 lb per hour.

$$\frac{1000 \text{ lb per hour}}{60 \text{ minutes}} = 16.67 \text{ lb per minute}$$

$$d = \frac{80{,}000(16.67)}{(250)(5000)} = 1.02 \text{ (inside diameter)}$$

Now refer to Chapter 1 (Table 1-2) for pipe sizes and schedule numbers for steel pipe. Under 1-in. pipe, inside diameter equals 1.049 for standard (schedule 40) pipe.
Answer: 1-in. standard pipe (schedule 40)

2. Using the Thomas formula, find the pipe size and schedule number for a steam service line requiring 200 lb per minute steam flow at 80°F superheat and 300 psig (using steel pipe).

$$d = 4.5(1 + 0.00025S)\frac{W}{p}$$

$$= 4.5(1 + 0.00025 \times 80)\frac{200}{300}$$

$$= 4.5(1.02)(0.81)$$

$$= 3.7179 \text{ (inside diameter)}$$

Answer (from Chapter 1, table 1-2): 4-in. pipe (schedule 80)

Determining Pipe Sizes by Velocity

Before laying out piping systems, it is important to determine all pipe sizes that will allow for reasonable velocity, taking into account friction losses. As far as economics is concerned, keep the velocity high without exceeding the maximum to avoid excessive operating losses through pressure drops. Maximum allowable velocity corresponds to the permissible pressure drop from the point of consumption to the point of supply. The chart on steam design velocity (Table 12-1) for the flow of fluids in pipes is used for the determination of reasonable velocity drops in saturated and superheated steam lines. The higher limits are used for larger pipes and the lower limits for smaller ones. In extreme services, use calculated velocities by computing the actual pressure drop. Steam with a high moisture content has a greater erosive action on seats and valves and exposed parts and will also reduce the velocity. Another general rule is to use the velocity of 1000 to 1200 feet per minute (fpm) per inch of inside pipe diameter—1200 fpm for inside diameters over 12 in. and 1000 fpm for inside diameters below 12 in.

EXAMPLE: Check the pipe size for giving a reasonable velocity by comparing recommended steam velocity to the calculated velocity. From Example 2 above:

$$V = \frac{80{,}000(1 + 0.0017S)\,W}{pd^2}$$

$$= \frac{80{,}000(1.02)200}{300(3.717)^2}$$

$$= \frac{16{,}320{,}000}{4144.8267} = 3937.438 \text{ fpm}$$

Using the calculation formula for recommended velocities for pipes under 12 in. in diameter:

$$V = 1000 \text{ fpm per inch of pipe diameter}$$

$$1000 \times 3.71 = 3710 \text{ fpm}$$

Answer: This velocity—3937 to 3710 fpm—is considered adequate.

TABLE 12-1 SATURATED STEAM DESIGN VELOCITIES FOR FLOW OF FLUIDS IN PIPE

Pressure (psig)	*Use*	*Velocity (fpm)*
0-15	Low pressures (heating	4000-6000
50-150	Medium pressures	6000-10,000
200+	High pressures (turbines, boilers for power plants)	10,000-20,000

POWER PLANT DESIGN

When laying out a typical power plant, it is important to consider the appearance and symmetry of the entire plant, the convenience of operation, and the mechanical functioning of the system as a whole. Figure 12-3 shows an adequate use of these three principles: The plant is laid out as symmetrically as possible and with easy access to all functioning parts for maintenance and operating personnel. This plant is also extremely compact, providing optimal use of space throughout.

Cost, of course, is always important in the design of the system, but it should not take precedence over ease of operation and the safe, efficient functioning of equipment. Most power plants include a great deal of insulation and other economically advantageous support equipment which take up space that cannot be neglected. High-quality products must be used throughout if the plant is to have the typical design life of 25 years minimum.

The designer and drafter should keep in mind the following power plant design essentials:

1. Refined design in the total layout of the piping system
2. Proper circuits for fluid flow
3. Proper circuits for bypasses, drains, and valves
4. Accurate shapes on drawings for boiler, turbine, condenser, and auxiliary equipment
5. Adequate P&ID drawings from which to design the total system
6. Complete shop fabrication drawings
7. Spool drawings detailing all lines in isometric throughout the plant
8. Detailed pipe support and placement drawings
9. Adequate standards for flanges, valves, and fittings (ASA standards and ASTM specifications for high-temperature service)
10. Specifying the welding option for securing equipment to a pipeline
11. Use of welded joints, especially in high-pressure/high-temperature steam situations, to minimize leaks
12. Minimal use of flanges with screw fittings
13. Use of ASA B31.1 (section 1), which covers installation of power piping systems and steam generating plants
14. Use of ASME boiler construction codes
15. Use of acceptable wall thicknesses for pipes and fittings that conform to both codes

Selection of Valves

It is important to consider the operating pressure and temperature under which valves will operate. (Chapter 4 discusses the background and particulars on valves and special problems of installation.) Economics will be involved in the selection when different types are equally useful.

Where pressure drops through the valves must be kept to a minimum, gate valves are preferred; where close regulation is necessary, globe valves are preferred although they have a high pressure drop through the valve body. Globe valves are used in power plants for turbine and engine throttles, bypasses around traps, hand-feed regulators, and boilers. Where possible, a gate valve should precede the globe valve used for throttling. Check valves are used in the following situations:

1. Where a trap discharges into a common header
2. Where various lines are joined together to discharge into a common header
3. In feed lines close to a boiler to prevent water or steam going back from the boiler
4. For protection of feedlines in case of rupture or failure

Welded ends are preferred in steam power plants because they are best for high-temperature/high-pressure uses. Flange connections can be used at points where valves must be removed from the line frequently. Safety and relief valves are used on all boiler or pressure vessels. It is advantageous to use more than one valve for safety valves, and set one on lower pressure than the rest to serve as a warning. Guidelines for the use of safety valves are as follows:

1. Safety valves should be the direct spring-loaded type.
2. No valve should be installed between a vessel and a safety valve.
3. A safety valve escape vent pipe should be full sized.
4. A safety valve should be set to prevent no more than a 10 percent rise in pressure in the vessel above the maximum allowable working pressure.
5. The manufacturer's recommendations for use of equipment and valves should always be consulted before installing or calling for optional equipment.
6. Safety valves must not be regarded as a means of controlling steam pressure, but only as a final resort. They do not, and must not, replace control instrumentation.

Pipe Thimbles

Pipe thimbles are sleeves made from pipe that are sufficiently strong to protect the pipeline where it passes through structural parts of the building; they must extend a minimum of 4 in. (10 cm) on either side of a wall or floor. Consult the manufacturer's schedule for the thimble sizes required for different nominal pipe diameters. Piping thimbles are generally used where there are concrete and masonry walls or floors through which the pipes must pass.

Types of Piping

Reheat steam piping usually requires special stainless steel or chrome-molybdenum steel because of the high-pressure/high-temperature conditions. Reheat steam piping is usually larger than the main steam lines. It is possible to follow the hot reheat line on Fig. 12-3, specifically on the 3 ft (0.91 m) elevation view.

Turbine bypass piping is also a supercritical service and is usually sized for 30 percent of the total flow.

Condensate piping is the suction and discharge piping of a hotwell trap, extractor, or other condensate pipe which operates between the condenser and the closed heater from which the feedwater to the boiler feed pumps is taken. In Fig. 12-3, the elevation view of the main condenser shows the low-pressure feedwater heaters passing through the condenser below the turbine.

Blowoff piping must conform to the ASME construction code. Blowdown tanks are provided where necessary for the blowdown from the boiler.

Instruments

Instruments are used throughout the modern power plant to ensure safety of operation and of personnel. Instruments give operating information and record temperature, flow, voltage, current, level, and pressure throughout the system. The majority of the valves and instruments are board-mounted in the control room to provide the operator with all the information needed to control energy production at the plant safely. Instruments for the lines recording temperature and pressure and flow may be located in the control room, giving the operator a permanent record of the line under observation.

Temperature Instruments Temperature instruments are among the essential measurement devices in the modern power plant because of the high temperatures involved in the production of electricity. Most mechanical equipment is designed for the Fahrenheit scale; most electrical equipment is designed for the Celsius scale. The various temperature recording (Fig. 12-10) and measurement devices in-

FIG. 12-10 Temperature gauge and recording device. Note the small pressure gauge for the instrument line.

clude remote indicating thermometers, glass stem thermometers, bimetallic thermometers, and thermometer wells, along with corresponding recording devices. Instruments must be adequately placed by the design engineer to facilitate safety throughout the power plant.

Pressure Gauges Pressure gauges and pressure recording instruments are also extremely important throughout the steam power plant. Again, because of the high pressures encountered in power generation, many types are provided throughout the plant to ensure adequate recording and observation of the pressure level.

Instrument Piping All fittings and the majority of piping for instruments should be suitable for the maximum temperature of the line in which the instruments are established, though on superheated lines in steam piping it is not necessary to provide the instrument piping with the full temperature-withstanding ability. Suggestions for instrument piping include:

1. Use a minimum of 1/2 in. (1.27 cm) nominal size for high pressures.
2. Use schedule 160 source nipples 3/4 in. (1.91 cm) with 1/2-in. (1.27-cm) valves counterbored to receive 3/4-in. (1.91-cm) pipe at the inlet and 1/2-in. (1.27-cm) pipe at the outlet.
3. Use copper tubing where possible because of installation and bending ease.
4. Use thin-walled steel tubing with flared connections or stainless steel pipe for temperatures above 400°F (204°C).
5. Use copper tubing and brass flared tube connectors for services under 400°F (204°C) and pressures of 200 to 1200 psi.
6. Locate all gauges of thermometers and flow meters where they are readily accessible and can easily be read.

Oil Piping

Oil piping is used throughout the power generation plant where lubrication or low-pressure fuel piping is needed. These situations are usually below 125-psig pressures, though in some fuel or pump discharge piping, 1000-psig pressures are encountered. In many cases screwed or flanged fittings are used in piping. Valves 2 in. (5.08 cm) and smaller are usually brass. Socket-welded joints are used throughout oil lines because of the ease of maintenance and installation.

Compressed Air Piping

Compressed air systems are used throughout the power plant for operation of remote control devices at pressures of 100 to 125 psig.

QUIZ

1. Name three uses of steam in industry.
2. What is a condenser's function in a power plant?
3. Name five external influences on a power plant design.
4. What is a design model and how is it utilized?
5. How does location affect a power plant's design?
6. Name five uses of a check model for a modern power plant.
7. What are the three most common fossil fuels for electrical power generation?
8. What are the positive and negative aspects of the use of coal for power generation?
9. Describe the working of a steam turbine.
10. What are the two types of turbines?
11. What is the function of a generator?
12. What type of boiler passes hot gases through its tubes?
13. Describe a shell-type boiler.
14. What is the function of a steam generator?
15. How does feedwater temperature affect the boiler?
16. What is the primary function of an economizer?
17. Why are cooling towers used in some power plants and not in others?
18. What are the three types of cooling towers?
19. When locating valves in a power plant, what are the main determining factors?
20. What is the minimum clearance for horizontal piping runs?
21. What is the function of chain wheels on a valve?
22. Define gpm.
23. What unit of measurement is used for calling out steam quantities?
24. What is the formula for determining pipe size for a saturated steam line?
25. What does the *velocity* of a pipeline mean?
26. What type of valve should be used when a large pressure drop through the valve must be avoided?
27. What is the function of a pipe thimble?
28. What is condensate piping?
29. Name some common gauges used throughout a power plant.
30. How is oil piping used in a power plant?

PROBLEMS

1. As a group or individual project, prepare a complete set of drawings for the power plant in Fig. 12-3. Scale the text drawing where necessary. Use a scale that is large enough to provide clarity, such as 1/8", 3/32", or 3/16" to 1'.
2. As a group project, prepare a complete list of parts needed to build a model of the drawings in Problem 1. Construct a model using as many ordered parts as

possible; fabricate out of wood and plastic sheet the portions of the plant that cannot be purchased. Build at the same scale as the drawings.

3. Draw one elevation and one section of the power plant in Fig. 12-3.
4. Using the Thomas formula, calculate the following pipe sizes:

	Steam Flow (lb per hr)	*Superheat (°F)*	*psig*	*Pipe Material*
(a)	12,000	100	350	Steel
(b)	18,000	75	250	Steel
(c)	10,000	105	274	Stainless steel
(d)	15,000	55	400	Steel
(e)	135,570	80	300	Stainless steel

5. Calculate the pipe size and schedule number for the following saturated steam lines:

	Velocity (fpm)	*psi*	*Steam Flow (lb per hr)*	*Pipe Material*
(a)	11,000	650	12,000	Steel
(b)	6000	325	1800	Stainless steel
(c)	13,000	450	10,000	Steel
(d)	4000	250	950	Steel
(e)	3332	150	1500	Stainless steel

Nuclear power plant with single reactor and containment building. Note the outdoor crane assembly used to dismantle the turbine.

13 NUCLEAR POWER GENERATION PIPING

Nuclear *fission* power is regarded by most specialists as a transitional source of energy. Fossil fuels are rapidly being depleted, and we have not yet made the technological breakthroughs necessary for large-scale solar and nuclear *fusion* power. Although a great deal of energy is produced by a small amount of fissionable material, these fuels, uranium and plutonium, are quite rare and the supply is definitely limited. In fact, all plutonium is produced from uranium, in nuclear plants, so it is mainly uranium deposits which concern us, although we can produce some fissile uranium from natural thorium.

Nuclear fission (splitting the nuclei of uranium or plutonium atoms) is a source of heat, as are fossil fuels and solar energy. As such, it supplies the energy for steam-driven turbines to produce electricity. That is, apart from the nuclear reactor, a nuclear power plant is very much like a conventional power plant (see Chapter 12). No special turbines or generators are needed. The major new considerations are not so much concerned with the generation of power, but with the fact that a tremendous release of nuclear radiation accompanies the power generation—which requires the isolation of the reactor itself and much remote instrumentation, along with heavy and bulky shields. Because the fuel material is not consumed entirely, used fuel must be chemically reprocessed to go into the new assembly and the radioactive waste must somehow be stored for hundred of years until the radioactivity subsides. All this processing must be done by remote control. Since it involves handling acids and radioactive liquids, much special piping is needed apart from the power plant piping itself. We will see that the problems associated with radioactivity impose special demands on the choice of piping materials.

BACKGROUND

Since the beginning of the twentieth century, it has been known that a tremendous amount of energy is contained in the nucleus of an atom. But until 1939, when fission was discovered, it was not considered possible to induce an atomic nucleus to release its energy, at least not on earth. It was only known that the incredible energy given off by the sun must be some sort of nuclear process. In that year, however, the fact that the nucleus of a uranium atom could be split by hitting it with a neutron was discovered in Germany and knowledge of the process was smuggled to the United States with the last scientists who escaped. Soon it was realized that a bomb could be made, but only by means of prodigious effort. It was of great consequence to world history that the scientists who remained in Germany did not believe the bomb could be made—and were able to convince their government to that effect.

The essential problem was that the splitting of the nucleus occurs only with the relatively rare isotope or uranium called uranium 235 (U-235). This rare variety, however, was mixed with the abundant isotope U-238 and could not be separated by chemical means. This fact concerning the bomb is important to realize, for it means that a nuclear reactor cannot become a nuclear bomb. For a bomb, one must use pure U-235 or, as we now know, plutonium 239. The natural mix of uranium 235 and uranium 238 is about 0.7 percent U-235 and 99.3 percent U-238. A nuclear reactor can operate using a mix enriched to 2 or 3 percent U-235. This slight purification of U-235 is much less expensive and makes reactor fuel more available. It also has important consequences for the design of the reactor. An explanation of how it is possible for the reactor to run on such low enrichment fuel will acquaint us with some of the terms encountered in nuclear technology.

In the fission process, the uranium nucleus splits into two *fission fragments* of roughly equal weight. These fragments are themselves nuclei of elements of about half the weight of uranium. The U-235 atoms that are used do not always form the same products as fission fragments, but they are all highly radioactive as a rule. These are the waste products of the process whose safe disposal is such a vexing problem. When the used fuel rods are removed from the reactor for reprocessing, these elements are dissolved chemically and stored in liquid form. Already there are millions of gallons of highly radioactive or "hot" nuclear waste stored in steel drums awaiting the discovery of a safe disposal system, which at the time of this writing is estimated to be at least fourteen years off. Meanwhile, the drums slowly corrode.

In addition to the fission products, several neutrons are released, usually two or three. Some are sprayed off in the splitting and are called *prompt neutrons.* A small percentage are given off by the fission fragments shortly after the fission and are called *delayed neutrons.* The fact that some are delayed is of extreme importance for the control of the process in a reactor.

The fission process takes many forms, but we can use one of them to illustrate how the energy is released:

U-235 + neutron → krypton 94 + barium 139 + 3 neutrons

Now the weight or mass of each of these is measurable in atomic mass units (amu):

U-235 =	235.117 amu	Ba-139 =	138.952 amu
Neutron =	1.009 amu	Kr-94 =	93.937 amu
	236.126 amu	3 neutrons =	3.027 amu
			235.916 amu

The original amount of matter is greater than that which is present after fission. The difference is 236.126 - 235.916 = 0.210 amu, which is the amount converted to energy by Einstein's law: $E = mC^2$. Seen as a percentage of the original, it is 0.210/236.126 = 0.000889 or a bit less than one-tenth of 1 percent of the matter converted to energy. If 10 kilograms (kg) of U-235 were to fission completely, however, about 9 grams (g) of mass would be converted to energy. Calculations using $E = mC^2$ show that this 9 g of mass, converted to energy, would yield 2.25 megawatts of power for 100,000 hours:

$$\begin{aligned} E &= mC^2 \\ &= 9\,(9 \times 10^{20}) = 8.1 \times 10^{21} \\ &= 8.1 \times 10^{14} \text{ ergs} \\ &= 2.25 \times 10^{11} \text{ watthours or } 2.25 \times 10^{5} \text{ joules} \end{aligned}$$

In the example just given, the fission fragments are barium 139 and krypton 94, both of which are highly radioactive. Other fission reactors yield different fragments. Thus fission of 10 kg of U-235 would leave us with 9.991 kg of nuclear waste, along with the quantity of chemicals needed to remove them from spent fuel rods.

The fact that neutrons cause U-235 to fission in the first place, and are also a product of the process, makes possible the *chain reaction* in which one of the product neutrons can hit another U-235 nucleus and cause another fission to occur. If each fission causes another fission, the chain reaction is sustained at a *critical reaction rate.* If each fission is responsible for more than one fission, the reaction is called *supercritical.* If the U-235 atoms are far apart, the product neutrons have only a small chance of hitting another U-235 atom and the reaction will fall off. The fact that pure U-235 is used in a bomb means the atoms are all close together and ensures that the neutrons will hit other U-235 atoms—and since they are traveling fast, they will do so very quickly. In less than a millionth of a second, most of the energy is released and the bomb explodes. But in the reactor, the U-235 atoms are far apart. How then can we get the neutrons to hit other U-235 nuclei? The answer lies in the use of a *moderator.*

It turns out that a U-235 nucleus is much more likely to absorb a slow neutron than a fast neutron. If we can slow the neutrons down, we can therefore compensate for the fact that the atoms are far apart. The moderator moderates the speed of the neutrons—that is, it slows them down—by means of a series of collisions with light elements. The first reactor used blocks of pure carbon for this purpose. Water is an excellent moderator, especially *heavy water.* Other materials will be mentioned later. If the atoms of the moderator are too heavy, it takes more collisions to slow a neutron down, and it escapes before slowing sufficiently to be captured by U-235.

The other effect of moderating materials is to reflect the neutrons back toward the fuel through these same collisions. Thus the moderator has a dual effect: making the neutrons more acceptable to the fuel by slowing them down and keeping them near the fuel via reflection. These two effects make it possible to have a sustained critical reaction even though the U-235 atoms are far apart. But the process is so time-consuming that a reactor could never release all its energy in an instant: The reactor, that is, cannot become a nuclear bomb. Though it can release enough energy to melt the fuel assembly or cause steam pressure to explode the reactor vessel, the containment vessels are supposedly designed to contain the most violent possible release of nuclear energy.

Thus far, we have a picture of a reactor which consists of fuel rods with some moderating material in between. In addition, we must consider the control rods, which are made of a neutron-absorbing material such as cadmium, boron, or hafnium. These can be inserted into the area occupied by the moderator to reduce the number of neutrons available for the chain reaction. In typical operation, they are adjusted by servomechanisms so that the reaction rate remains constant at the desired level. The last major component of the system is the coolant, which removes the heat energy released by the uranium and transfers it to the boilers to make steam for power generation. This completes the list of basic components: fuel, moderator, control rods, and coolant.

TYPES OF REACTORS

The materials used for reactor vessels are extremely important because of the high rate of neutron bombardment (high *neutron flux*) in the vicinity of the reactor chamber. Materials which absorb neutrons easily will quickly become radioactive. Even with materials that do not readily absorb neutrons, the radioactivity of the entire system slowly grows so that an entire facility must be abandoned after about 30 years of operation. For the same reason, most reactors transfer heat from the reactor core in two stages so that the radioactive coolant will not contaminate the steam turbines. In this type of reactor—the *pressurized water reactor* (PWR)—the common coolant is purified water under high pressure. Impurities in water are easily made radioactive and so must be carefully removed. High pressure is employed so that the water will not boil in the reactor core but will still be hot enough to boil water to produce sufficient steam pressure for the turbine. In one typical plant, the reactor coolant water is heated to over 600°F (315°C) and produces steam at about 500°F (260°C) and nearly 800 lb/in.^{3}. A disadvantage of the PWR is that the water in the primary coolant must be pumped, and there is also power loss because heat escapes from the additional piping needed for the loop.

A second type of reactor—the *boiling water reactor* (BWR)—has a different set of advantages and disadvantages. It has increased thermal efficiency because there is no primary loop, but radioactivity in the coolant can get to the turbines, which typically require much more maintenance than the primary coolant pumps. Moreover, service personnel must be much more aware of radioactivity. If the system is flushed and one waits a short time, however, the exposure is reasonably small. Another disadvantage is the greater danger of rapid boiling of the water at the surface of the fuel elements. If a steam film forms at the surface, the ability of heat to get to the coolant is vastly reduced and the fuel element will melt.

A fact which is usually an advantage for water-cooled and water-moderated reactors is the *negative temperature coefficient* of reactivity. If the reactor gets too hot, water expands, moderation is reduced, and the reaction rate goes down. Conversely, if the temperature drops, the water contracts, moderation is increased, and the reaction rate goes up. Using water, therefore, adds stability to the process as long as there is no way for a surge of cold water to enter the operating reactor.

From the earliest days of reactors, some have been cooled by liquid sodium or sodium potassium (NaK) and moderated by solid graphite (carbon). The *sodium graphite reactor* (SGR) has three great advantages. First, neither the coolant nor the moderator is pressurized, so a steam explosion of the vessel is impossible. Even a leak can easily be contained without an expensive containment vessel. Second, the boiling temperature of the coolant is over 1600°F (870°C), so high-pressure steam and superheated steam can be produced in the boiler and a more efficient turbine can be utilized. Along with this is the advantage of high heat-transfer capability of the liquid metal. Third, an electrical induction pump can be used for the metal—a pump which has no moving parts and is thus practically free of maintenance. This feature is important because of the radioactivity in the primary loop. A lesser advantage is the lower corrosion rate of the reactor vessel, but this is not ordinarily a limiting factor in the life of the reactor if care is exercised in fabrication. Disadvantages include a larger amount of power needed to pump the NaK. Moreover, since the coolant will solidify in the system during a down period in the operations, it must be remelted at up to 208°F (98°C) before it will flow again. Finally, this coolant has only minimal moderating effect and can produce a chemical explosion if a leak occurs in the boiler.

A number of modern reactors use helium gas as a coolant. Besides its high heat transfer efficiency, helium has no tendency to become radioactive under nuclear bombardment and can be used at higher temperatures than water systems. It also has some moderating ability, though at the low densities of gases it must be aided by solid (graphite) moderators. Continuous gas flow must be assured to prevent melting of the core element. Modern computerized control systems with extremely rapid reaction times can supposedly respond to unusual conditions.

Liquid still has the advantage of greater core meltdown protection, and some benzene derivatives now being used as coolant moderator materials. Reactor temperature must be lower than with water, but pressure and corrosion are greatly reduced, purification is easier, and neutron-induced radioactivity is lower than with water.

All the reactors described above are moderated reactors, also known as *thermal reactors* because they use slow neutrons whose average speed is determined by the operating temperature in the core. The bomb, on the other hand, is an example of a *fast reactor*—one that uses the fast neutrons as they emerge from the fission process. As noted above, the fissionable atoms must be close together and the core must be small. It is possible to produce and control small fast reactors which do not use moderators. Since the fuel must be nearly pure uranium 235 or plutonium 239, it is extremely expensive, but there is an advantage which outweighs all the considerations of fuel cost: It can be used for "breeding" plutonium 239 fuel from otherwise useless U-238.

If U-238 absorbs a neutron, it becomes U-239, which is highly radioactive. If U-239 is stored for a short time, the radioactive transformation will turn it into plutonium 239 (Pu-239). Theoretically, one could make a critical core of pure U-235 surrounded by a large amount of pure U-238. For each neutron which produces fission in U-235, on the average 2.25 neutrons are produced. If exactly one of these causes another fission, then 1.5 (again, one the average) could escape into the U-238 and one would have (if all these entered U-239 nuclei) 1.5 times as much fuel at the end of the process as one had at the beginning. This is not possible in a thermal reactor because too many neutrons are absorbed in the moderator, coolant, and structure. In actual practice, we cannot achieve a 50 percent increase in available fuel. But we can, nonetheless, produce more fuel than is used—which will in the long run result in the use of all the earth's uranium for fuel, whereas 99 percent of it would otherwise be wasted.

Fast reactors can also be used for power production. They generate heat, but the coolant must be liquid metal, for any other substance would have a moderating effect. The removal of large amounts of heat from a small core also requires the high heat-transfer efficiency of liquid sodium or NaK. Note that in thermal reactors with fuel of 2 to 3 percent U-235 and 97 to 98 percent U-238, some fuel is produced in the fuel rods in normal operation, which helps to extend core life. Still, the conversion efficiency is low. Much less than one atom of fuel is produced for each fission.

STRUCTURAL AND PIPING MATERIALS

It should be evident that both materials and system design in the nuclear field are subject to an entirely new set of considerations. Many metals commonly used, such as steel alloys, absorb neutrons and become radioactive. Fortunately, carbon and chromium, so important for steel and stainless steel, are relatively harmless in this respect. Aluminum is also useful, and zirconium is much used as a cladding material for the fuel element. Additionally, the neutron bombardment of most metals increases both their strength and their brittleness over a period of time. These effects are quite considerable and must be taken into account where there might be mechanical shock.
account where there might be mechanical shock.

A final consideration in the case of pressurized pipes and vessels is chloride stress corrosion, which is greatly accelerated if oxygen is present and inner flow surfaces are not perfectly smooth. In the design and fabrication of the reactor vessel and primary piping, only the highest-quality materials and work must be specified; otherwise materials will soon be radioactive and difficult to service. Any shutdown will require considerable surface decontamination and just plain waiting for the radioactivity in the metal to subside.

Safety and Radioactivity

Reactor shielding for personnel against radiation is a twofold consideration: One must be protected from both neutrons and gamma rays. Light materials stop neutrons but not gamma rays. Heavy materials stop gamma rays but not neutrons. It is desirable to place the neutron shielding close to the reactor itself, for it will also reflect neutrons back into the vessel. Any good moderator is a good neutron shield or reflector as well. Often polyethylene can be used—or, if neutron-absorbing capability is also desired, polyethylene impregnated with boron. Boron is a very effective neutron absorber, and boron steel can be used for control rods in the core. Outside the reflector shield is placed some heavy absorber for gamma rays. If space is not a consideration, thick concrete may be used; otherwise, lead is the most effective shield. The system must be designed so that there are no straight-line paths for radiation from the reactor where steam lines and control devices penetrate the shield.

Nuclear Piping

Nuclear piping is any piping whose content is radioactive. In general, this content is the coolant that runs through the core of the reactor and the material used in processing the spent fuel element and disposing of radioactive waste. Standards for nuclear piping are published by the nuclear piping committee (B31) of the ASA, whose latest publication should be consulted.

Experts in the field of nuclear piping usually take one of two approaches to the problem of safety and reliability: build additional safety factors into the piping itself or use standard specifications and surround the installation with a huge containment vessel. Since the latter alternative is much more visible and demonstrates safety concerns, the containment vessel is often used. Nonetheless, the cost of a breakdown in the primary loop necessitates that additional safety factors be incorporated there as well.

The following table gives a good idea of the range of temperatures and pressures encountered in the primary loop (nuclear piping) of different reactors:

	Primary Loop	
Reactor Type	*Maximum Temperature (°F)*	*Maximum Pressure (psig)*
PWR	635	2000
BWR	550	1000
SGR	930	0
Gas-cooled	1380	350
Fast	900	0
Organic-liquid-cooled	575	120

FUSION IN THE FUTURE

Nuclear fusion is the process by which our sun generates its energy. Solar power is, therefore, fusion power, but it is quite another thing to make the process work under conditions on earth. Temperatures of many millions of degrees are needed to make it go, and, of course, no container made of known materials can stand such temperatures. The technology for producing energy from fusion is that of confining the fuel in a "magnetic bottle." The physics of this phenomenon is beyond the scope of this book, but the process can be described to some extent. The fuel is hydrogen, which is ionized into a plasma by stripping the electrons from the atoms at a high temperature. The plasma is formed into a flowing ring of plasma in a magnetic field. Then the field is suddenly intensified, which compresses the plasma and simultaneously increases its temperature to the millions of degrees required. Hydrogen nuclei (protons) collide at high energy and stick together, or fuse, releasing energy. The tremendous release of fusion energy expands the plasma and pushes back against the confining magnetic field, which, in turn, sends a new surge of electric current through the windings of the magnet.

Thus fusion energy is converted directly into electricity without the use of steam. It seems likely that fusion power units will be small and even considered disposable, though some large designs are supercooled. Only in the supercooled units would piping be a major design factor in any usual sense for the fusion reactor. Nevertheless, a demonstration of the fusion process of energy release helps to clarify the operation of the process and to explain why fusion is clean in the sense that radioactive waste is not a problem. Consider the weight of the fuel and also the product. The process as shown is simplified in that four protons must be fused together in stages. Again weights are given in atomic mass units (amu):

4 protons	→ helium 4
1 proton	= 1.008 amu
x 4	
4 protons	= 4.032 amu
He-4	= 4.005 amu
Total mass at start	= 4.032 amu
Total mass after fusion	= 4.005 amu
Mass converted to energy	= 0.027 amu
	or 0.67% of original mass

In the case of fission, only 0.089 percent of the original mass is converted to energy, so this process has a much greater energy yield per pound of fuel.

The only end product of this process is helium 4, which is not radioactive. The intermediate product, hydrogen 2 (deuterium), is also stable. Since there is a yield of gamma radiation in the process, the reactor must be shielded. But gamma radiation does not cause secondary radioactivity as does neutron radiation, so the process is quite clean in comparison to fission reaction.

NUCLEAR POWER PLANTS AND FACILITIES

Nuclear reactors are similar to conventional power plants that burn fossil fuel in that they both generate steam to turn the turbines that drive their generators. The nuclear reactor takes the place of the boiler and burners of the conventional power plant. Uranium or plutonium fuels produce the heat that generates steam.

In this sense, the nuclear power plant is much more efficient and cleaner than plants in which fossil fuels must be continuously transported and utilized in burner and boiler

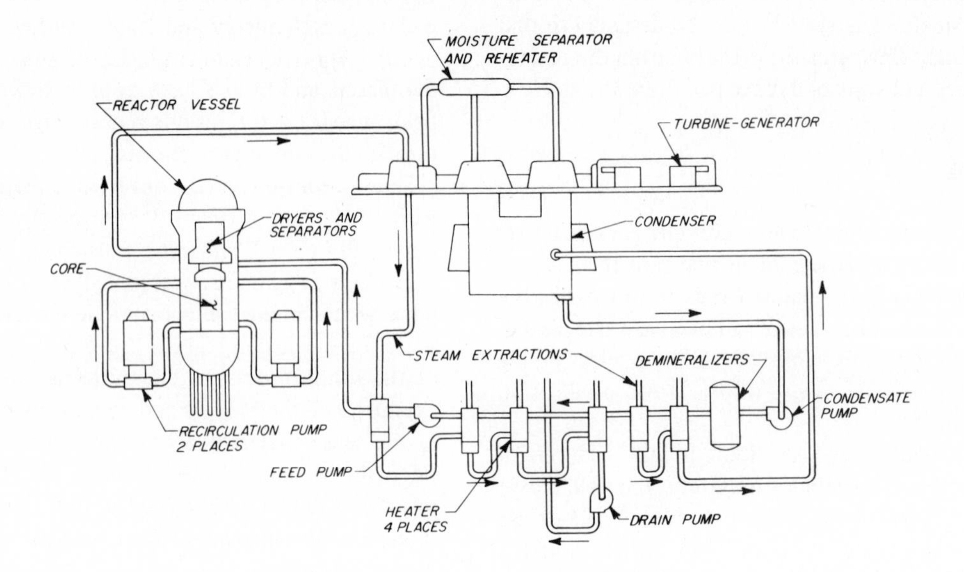

FIG. 13-1 BWR power plant schematic.

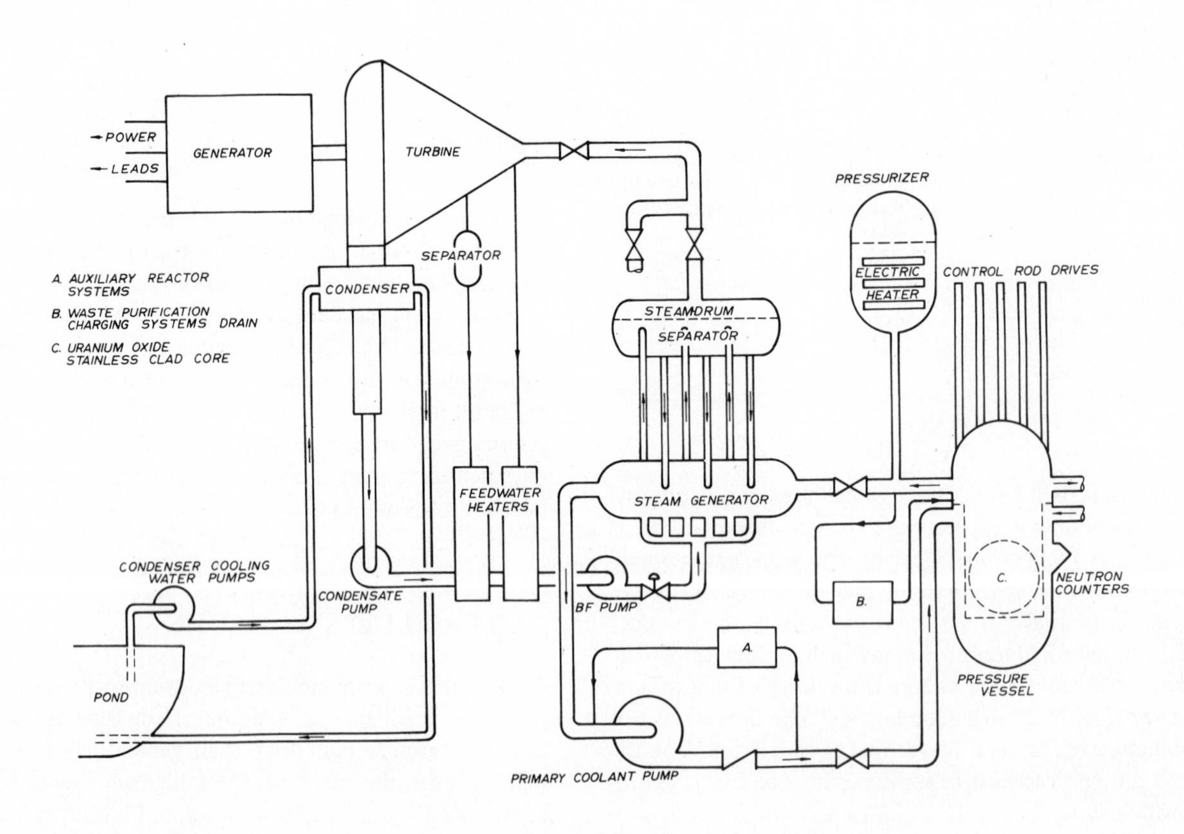

FIG. 13-2 PWR nuclear power plant schematic flow diagram showing auxiliary reactor systems, waste purification, charging systems, and uranium oxide stainless clad core for the nuclear pressure vessel.

equipment. Storage of used fuel, which is highly radioactive, has become a severe problem with the increased use and production of nuclear power plants.
become a severe problem with the increased use and production of nuclear power plants.

In the nuclear power plant, fission is used to heat the reactor coolant, which conveys the heat to other areas of the nuclear reactor to generate steam. This chapter deals primarily with reactors which utilize water as a coolant. We have already mentioned the boiling water reactor (BWR) and the pressurized water reactor (PWR). The BWR system is shown schematically in the flow diagram of Fig. 13-1; the PWR system is shown in the flow diagram of Fig. 13-2.

In Fig. 13-1, it is possible to trace the manner in which the nuclear fuel in the reactor vessel produces heat and thus results in the boiling of the reactor coolant, which in this case is water. The boiling water reactor creates steam directly in the reactor and from there it is piped to the turbine. In this flow diagram, it enters the high-pressure turbine to the left and goes onto the low-pressure turbine and from there to the condenser. The high-pressure and low-pressure turbines are arranged in line and turn the same generator. The spent steam is channeled into the condenser (or from steam extractions into the feed lines) and is then channeled back to the reactor vessel. Though this flow diagram depicts a closed system, the reactor system in fact requires a quantity of makeup water in cases where steam loss, leakage, or other factors are involved. Moreover, the recirculation pump and cooling system in the lower portion of the reactor must have cool water available at all times, especially in cases of emergency where a fast cooldown is necessary. What characterizes the boiling water reactor primarily is the production of steam directly in the reactor vessel and the use of that steam to turn the turbine. In this way, radioactivity contaminates a large portion of the total system.

Figure 13-2 shows a flow diagram for a pressurized water reactor. In the PWR, the reactor coolant is highly pressurized and therefore does not normally reach boiling point. The reactor area contains the coolant, which passes through the steam generator or heat exchangers. The heat from the coolant is transferred through the heat exchange unit into a separate system, which is at a lower pressure than that of the reactor water. In this way, heat in the steam drum or separator creates steam which turns the turbine. Because two separate loops are used, radioactivity from the reactor coolant is never transferred directly to the turbine area.

As has been stated, the production of steam using nuclear fuel versus fossil fuel is the only major difference between the two types of power generation. Once the steam is created, the systems are virtually the same. Thus the processes concerned with steam-driven generators, condensers, and water transfer including feedwater makeup are the same for both nuclear and fossil fuel power plants.

Figures 13-3 and 13-4 show a modern nuclear power plant which utilizes two or sometimes more reactors—an arrangement which provides advantages in economics, licensing,

FIG. 13-3 Nuclear power plant with tandem reactors and containment areas.

FIG. 13-4 Ocean view of a nuclear reactor power plant.

FIG. 13-5 Aerial view of nuclear power plant. Note the cooling pond in the background, the single-turbine generator unit with crane assembly in the foreground, and the nuclear containment area next to the operations buildings.

and the efficient use of land. Figure 13-5 shows a modern nuclear power plant with only one reactor area. Most of the equipment remains uncovered since the power plant is in a warm climate. Figure 13-4 shows a completely enclosed turbine and equipment building in a PWR system. The separation of the turbine and other systems from the nuclear reactor and containment building is a fairly common design for both the PWR and BWR systems.

In Fig. 13-5, it can be seen that the nuclear containment structure is separate from the administrative and control buildings and that the turbine generator units are exposed to the environment. A normal part of the nuclear power station is the transformer yard with the electrical hookups necessary to channel the generated electricity to areas where it is used. When power plants are situated next to large bodies of water, as in Figs. 13-4 and 13-5, it is unnecessary to provide a cooling tower or pond. These structures are much more prevalent in inland areas where the required amount of cool feedwater makeup and reactor coolant water is not generally available.

Another typical component of a nuclear power plant is the special area required for fuel handling and processing. The containment building and the reactor building are one and the same and may include a variety of subsystems for the production of electricity. The reactor containment buildings are constructed from steel and concrete. Figure 13-6 shows a nuclear reactor containment structure in its construction. The BWR system's containment area permits limited access during operation since radioactivity is spread through the system, whereas the PWR system allows more maintenance and operational procedures during the electricity generation because the systems are separated.
rate.

In Fig. 13-5, the auxiliary buildings are separate from the turbine area, though this may not always be the case. In Fig. 13-3, the turbine building is the main structure with two identical right-hand and left-hand reactor areas. A variety of subbuildings can also be seen in Fig. 13-3. These buildings house an assortment of auxiliary systems, which may include evaporators, analysis systems, safety systems, waste systems, demineralizers, storage facilities, and in some cases the administration offices, though it is common practice to construct a separate administration building. A great amount of space must be provided for labs, a machine shop, offices, a conference room, and even a model room for training and maintenance purposes. Special facilities and sometimes old buildings are relegated to the storage and handling of nuclear fuel. Extreme caution is taken in transferring and handling radioactive materials.

The turbine building shown in Figs. 13-7 and 13-8 suggests the immensity of the nuclear reactor's turbine system. The turbine building contains many systems, including the generator, condensers, feedwater systems, condensate systems, feed pump, moisture separators, demineralizers, and related equipment. Some equipment is locally controlled as

FIG. 13-6 Placement of rebar during construction of the containment building for a nuclear power plant. Notice the various pipe embedments in the walls.

in the turbine building shown in Fig. 13-7, and much of the related equipment in the turbine building is monitored from the control station. A centralized control room is a common feature of all nuclear power plants, which allows the operators to monitor the total system from one small area. Figure 13-9 shows a modern power plant control room. All instrumentation—meters, temperature gauges, flow indicators, and the like—is connected to the control room, thereby providing immediate monitoring of the nuclear power plant. Thus it is possible to make quick changes in the operation of the reactor system or to employ the backup and protection systems that are necessary for the nuclear power plant.

Figure 13-10 shows the reactor vessel for the nuclear power plant depicted in Fig. 13-3. The reactor vessel in this photograph is being lowered into place during the construction stage. The reactor, of course, is the central aspect of the nuclear power plant. It is here that the heat is created which in turn produces steam to turn the turbines and generate electricity.

There are some differences between pressure vessels in the PWR and BWR systems, though both serve the identical function of producing heat within the safe confines of

FIG. 13-7 Tandem turbine generator units in a nuclear power plant turbine building. Notice the control stations at the center of the building.

FIG. 13-8 Turbine operation floor. Notice the various piping configurations, vessels, and instrumentation.

FIG. 13-9 Control room for a modern power generation plant.

FIG. 13-10 Reactor vessel hanging from a polar crane. This vessel is being installed in the containment area.

the reactor vessel. The BWR reactor vessel is less thick and has a larger capacity in the rounded top for steam production. In this system, we recall, steam is produced directly in the reactor and then channeled to the turbine area. A multitude of outlets are provided in the BWR reactor, including those for steam outlet to the turbine, feedwater and coolant inlets, safety system inlets, and rod penetrations at the bottom of the vessel. The PWR pressure vessel (Fig. 13-10) is smaller and has a considerably thicker wall because it is under higher pressure than the BWR vessel. The PWR system contains no moisture separator and provides for a wider coolant outlet above the core region and an inlet below the core. Besides these inlets and outlets, there are safety system cooling water and backup system inlets. The control rod penetrations for the PWR system are at the top of the reactor vessel.

Figure 13-11 is a P&ID of the reactor area for a BWR system. The corresponding drawing for equipment placement is Fig. 13-12. Figure 13-13 shows the pressure vessel being lowered into position during the construction stages of a nuclear power plant. A variety of internal arrangements are necessary to provide for temperature changes and the control of heat production by the radioactive fuel in the core area by means of control rods and other systems.

Both the BWR and PWR vessels require leakproof vessels that are pressure and temperature resistant. One of the most important influences in the reactor area is, of course, the effect of high radioactivity on the vessel. Both types of reactor vessels weigh hundreds of tons. The BWR reactor is larger and bulkier than the PWR system. These reactor vessels are really large pipes that require extreme caution and monitoring for defects in the production stage. They are usually constructed of carbon steel with stainless steel clad interiors. Depending on the reactor type and the operating pressure and temperature, the reactor vessel can be anywhere from 5 in. (12.7 cm) to 15 in. (38 cm) thick. As has been stated, a variety of penetrations through the reactor vessel are necessary for reactor coolant, feedwater, and safety/backup systems, and these nozzles must be fabricated with extreme care to ensure safe operation of the system.

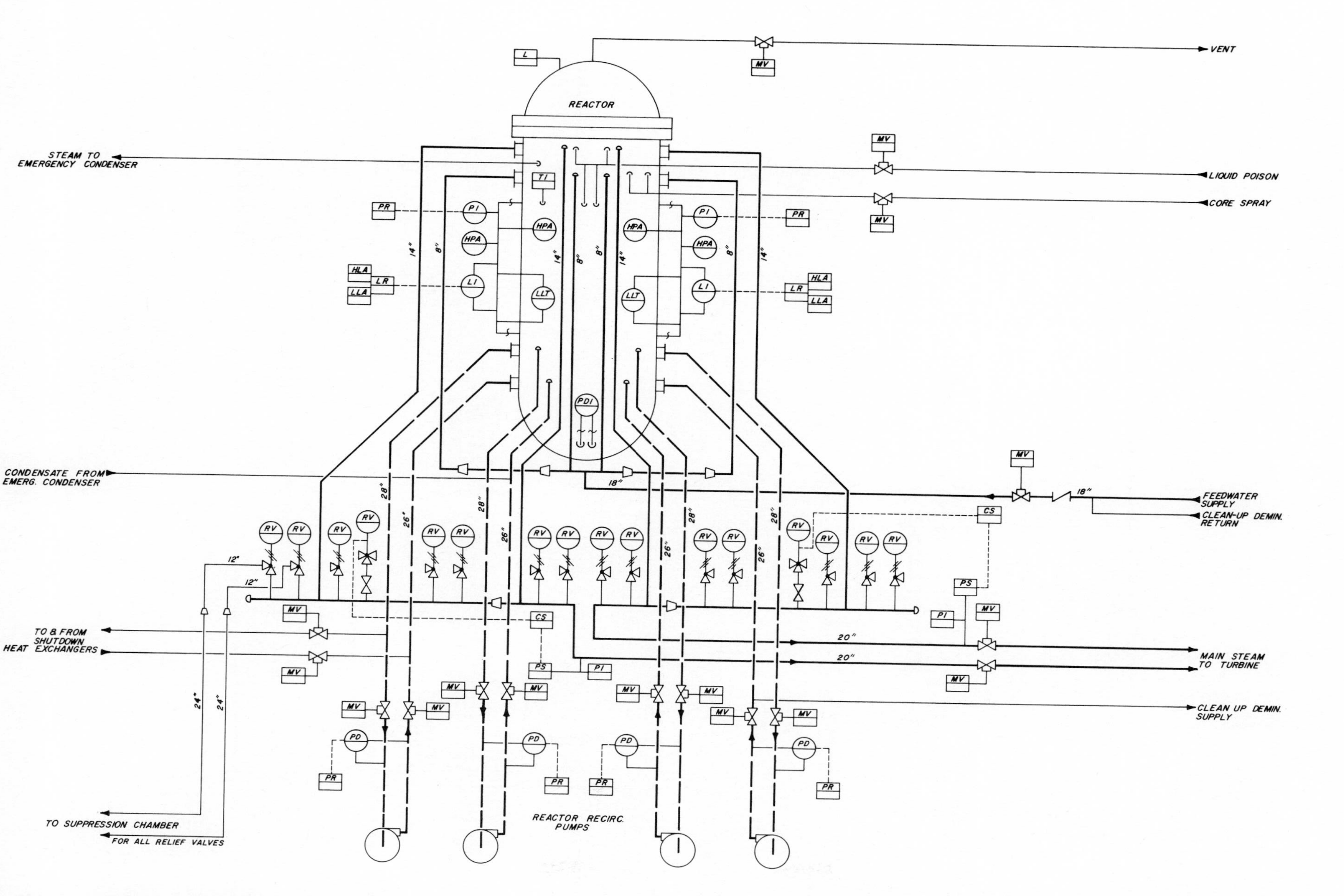

FIG. 13-11 Nuclear reactor P&ID.

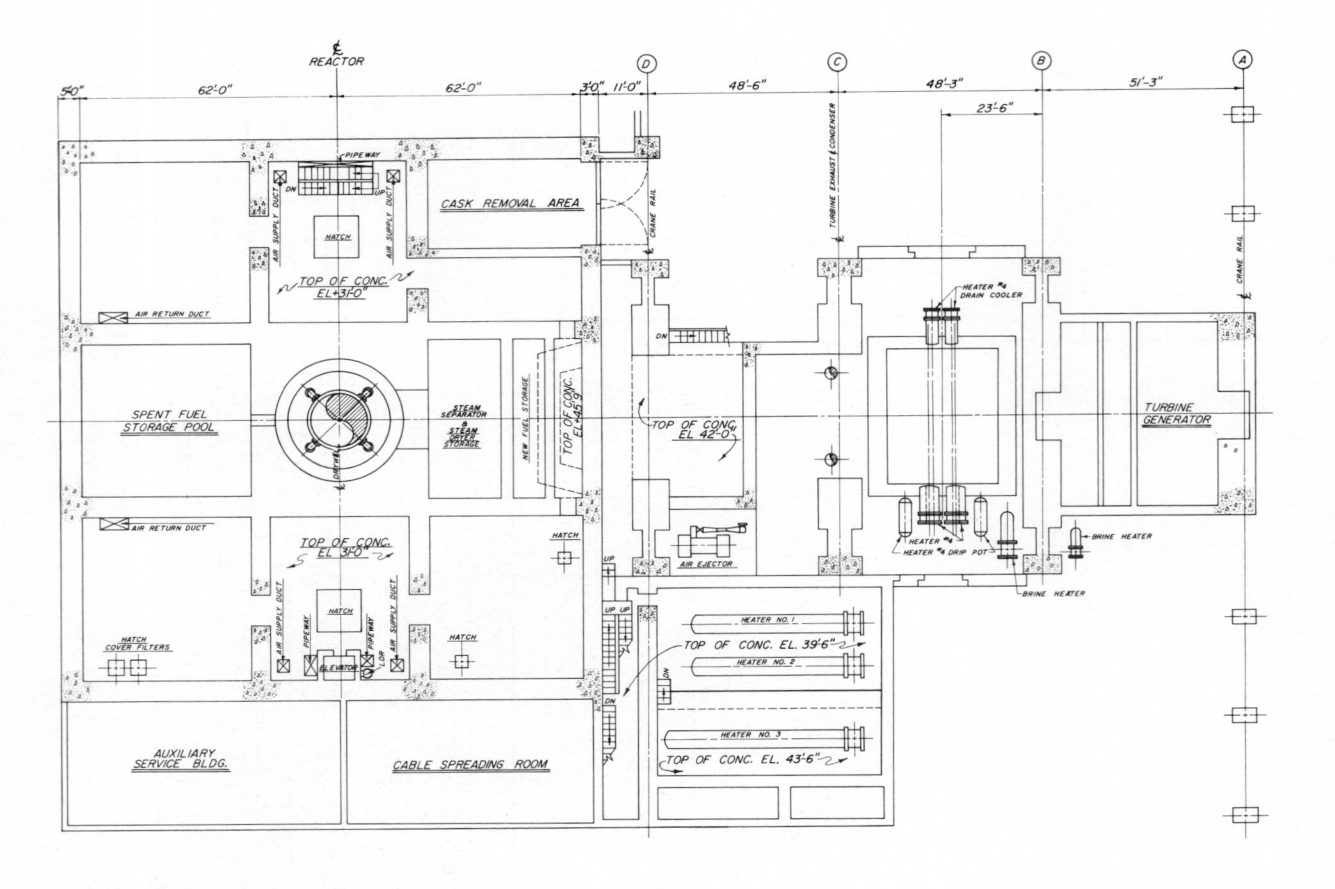

FIG. 13-12 Equipment placement drawing for a typical nuclear power plant.

FIG. 13-13 Reactor containment vessel being lowered into place in a nuclear reactor building.

Reactor Auxiliary Systems

All power plants contain major and minor and backup systems, such as the condensate system, feedwater system, and coolant system. All must be considered major areas of concern for the piping drafter working on a nuclear power plant. Numerous subsystems—cooling water, ventilation, fire protection, lubrication—provide support for major equipment. Auxiliary systems, which include all the above, are necessary for the adequate maintenance of a nuclear power plant. This section deals specifically with reactor auxiliary systems, which support the maintenance and operation of the reactor area.

These auxiliary reactor systems are for the most part unique to nuclear power plants in comparison to fossil fuel plants, though some elements, such as cooling and ventilation, are similar. Some auxiliary systems, for both PWR and BWR nuclear reactors, help maintain water levels in the coolant system, demineralization systems, and reactor coolant purification. Purification helps reduce the chemical effects on reactor materials; demineralization reduces the radioactivity, most of which resides in the minerals that water usually contains.

A chemical system is needed for corrosion control, and feedwater makeup systems are also necessary for the typical nuclear power plant. The chemical and volume control systems in the reactor area monitor the amount and quality of water that enters the reactor coolant area. Also monitored are the pressure and temperature of the subsystems. Many different systems are interlocked to provide diverse capabilities in case of reactor system malfunction.

The control rod drive system is another auxiliary system that is part of the reactor area. As has been stated, it provides the ability to move the control rods which regulate the nuclear reaction rate and the rate of power generation. The control rod drive system differs considerably for the two types of reactors.

Steam Systems In the BWR reactor, the water which is used as a coolant is boiled in the reactor vessel and the steam produced is then channeled through moisture separators, valves by piping systems to the turbine. In the PWR reactor, the steam generators are large tube-and-shell heat exchangers. The reactor coolant, which is at a high temperature and pressure, heats the feedwater in the shell, thereby transferring reactor heat to the feedwater, which in turn boils and becomes steam. This steam then goes through the accompanying moisture separators and enters the turbine as dry steam; the spent steam returns to the condenser and back through the feedwater system to the steam generator. The turbine area, which is the same for both fossil fuel and nuclear power generation, is described in Chapter 12.

After the steam has transferred its energy to the turbine, it is returned to a liquid state called *condensate.* The flow diagrams in the beginning of the chapter depict the movement of steam from reactor to turbine and into the condenser areas. *Condensers are heat exchange units which remove excess heat from the spent steam and convert this low-pressure steam back into water or condensate.* The heat removed from the steam in condensing it is transferred to a circulating water system, which is tied into a cool-water source such as the ocean or cooling towers. The condensate is channeled into the feedwater system and eventually goes through the cycle again. A large maze of piping with accompanying pumps and valves is necessary for these systems to operate adequately under high temperatures and pressure.

The feedwater system is usually preheated to eliminate thermal shock and the surge of reactivity that would accompany the introduction of cool water into the reactor vessel area. The feedwater or makeup water is heated by the steam condensed in the condenser. Many designs are available for steam generators, turbine generators, and condensers, including the use of different turbines and generating units. Turbines and condensers are described in Chapter 12. Figures 13-7, 13-8, 13-14, and 13-15 all show a variety of turbine generator units or flow diagrams, which should help the student become acquainted with the piping and equipment for a nuclear power plant.

A typical steam supply system for a BWR nuclear power plant includes the reactor and its auxiliary systems, including the reactor service systems and feedwater. All pressure vessels and the piping in the drywell must meet the requirements provided in the ASME boiler and pressure vessel code. Piping outside the drywell must comply with the ASME code for pressure piping. Stainless steel is used for all surfaces which are in contact with hot reactor water during operation of the nuclear reactor cycle because of the radioactivity that is transferred from the nuclear core and reactor vessel through the coolant. The main steam lines and feedwater system are carbon steel.

The Turbine

The conventional turbine systems described in Chapter 12 are similar to those in a nuclear power plant. Turbines in the modern nuclear power plant are usually combinations of various arrangements. The *single-flow turbine,* which is the most simple arrangement for a turbine, is shown in Fig. 13-2. As steam enters the single-flow turbine, it is at high temperature and pressure and makes a single pass over the turbine blades, expending most of its energy; from

FIG. 13-14 Turbine generator train. This view shows the turbine generating units inside the turbine building of a typical nuclear power plant.

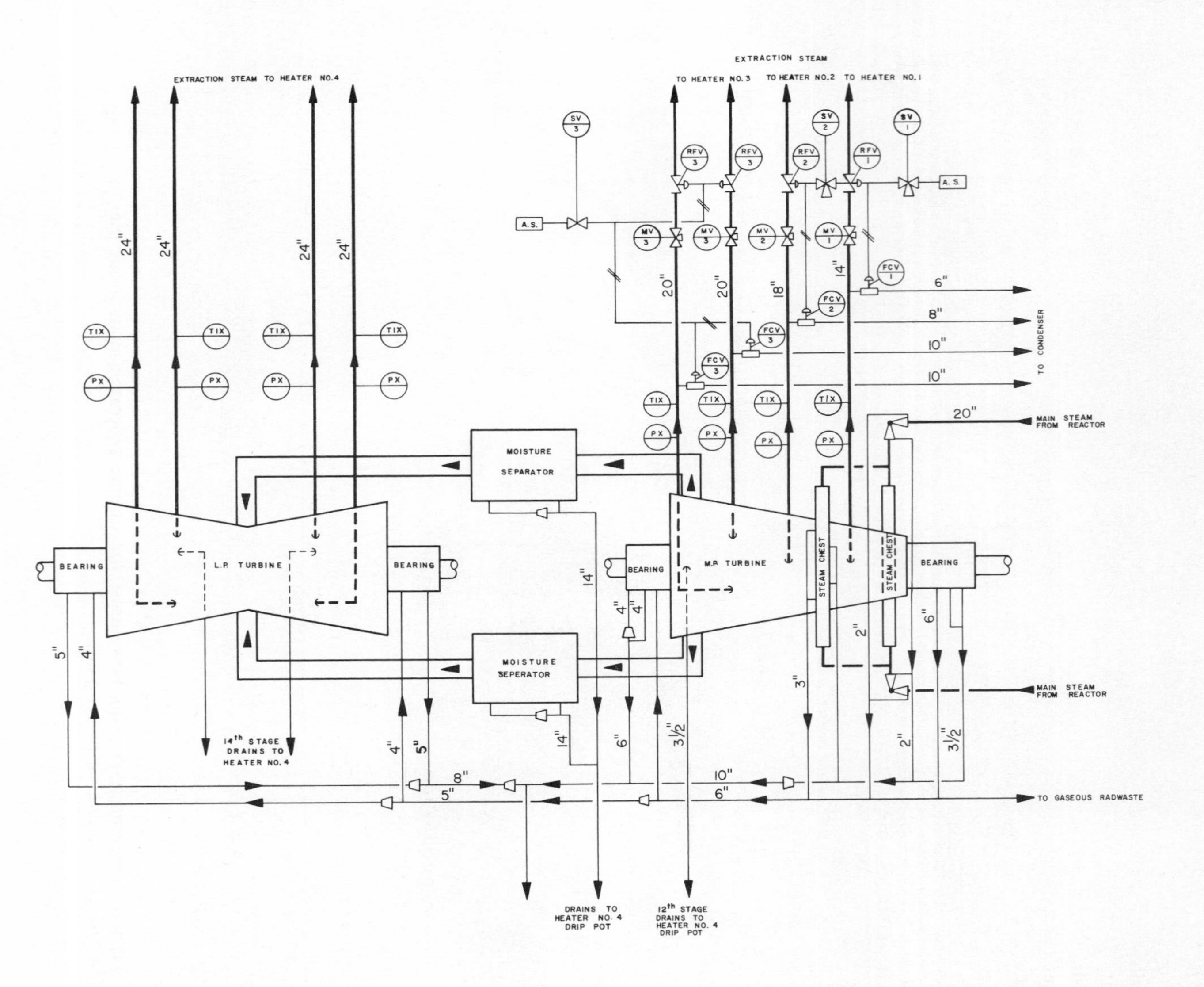

FIG. 13-15 Turbine P&ID.

there, the spent steam is channeled into the condenser. *Compound turbines* are also used in some power plants. The compound unit has both a high-pressure unit and a low-pressure unit, usually arranged in tandem. In the tandem arrangement, the turbines are connected to a single shaft, thereby driving the same generator. The *double-flow turbine* is another variation, though this equipment is usually limited to low-pressure situations. The double-flow turbine has a steam inlet in the center of the turbine, and the steam is split into two flows. Another turbine arrangement is similar to that shown in Fig. 13-1–which might be called a triple-tandem, double-flow, double-reheat, triple-extraction turbine. In this arrangement, the inlet steam passes through the high-pressure turbine, through a reheater into an intermediate-pressure turbine, and through another reheater into a double-flow, low-pressure turbine. All the turbines are arranged so that they operate the same shaft, thereby driving the same generator. This arrangement allows for the most efficient use of the steam.

The steam flow in a normal turbine is usually classified in three ways: straight flow, reheat flow, and extraction flow. *Straight-flow turbines* use the steam directly from the inlet and exhaust it into the condenser. This arrangement does not use all the steam efficiently, however, unless the turbine is extremely large. *Reheat-flow turbines* exhaust the steam into a reheater which increases the temperature of the high-pressure turbine outlet steam and channels it into an intermediate-pressure or low-pressure turbine. In *extraction-flow turbines,* the steam is used not only for turning the turbine but also for other portions of the industrial plant, such as heat exchangers and feedwater heaters. Most turbine arrangements and steam flow categories are not used exclusively. Combinations of turbines and types of steam flow are used to maximize the efficient use of steam. The only difference between steam generation in a PWR and a BWR plant is that the steam for the turbine comes from a steam generator in the pressurized water reactor and from the nuclear reactor vessel in the boiling water reactor. After the steam has been generated, it is used in the same way to turn the turbine and therefore drives the generator in both situations. Figure 13-15 is the P&ID for the BWR power plant showing the turbine with accompanying piping, valves, and instrumentation.

Condenser and Condensate System

The main condenser is designed to perform various functions, including condensing steam exhausted from the turbine to obtain the desired vacuum, deaerating the condensate, heater drains, and other returns, serving as a heat sink for excess reactor steam dumped by the bypass system, and holding up condensate in the hotwell for decay of short-lived radioactivity.

The condensate system is used in most plants to take the condensate from the main condenser hotwell and channel it back into either the steam generators or directly into the reactor vessel. The condensate passes through equipment that increases its heat level to avoid thermal shock. Pumps move the condensate from the condenser back to the steam generation area at low pressure. Air is also removed from the condensate before it is returned to the reactor or steam generator. As can be seen, the condensate steam system is almost a closed loop. Usually only a small amount of make-up water is needed to compensate for leakage and other minor situations. Therefore the condensate and feedwater system are connected at this point. Since the water used in this loop is purified, demineralized, and extremely expensive to maintain, the condensate system must be totally free of impurities and leaks.

Cooling Water

Cooling water is pumped through the reactor cooling water heat exchangers, through the turbine water heat exchanger, and into separate cooling water systems. Circulating water is pumped through the tube side of the heat exchangers by the auxiliary circulating water pumps and discharged to the main discharge tunnel for a BWR system. In a PWR system, the reactor coolant system consists of closed loops which circulate the reactor coolant between the steam generator and the nuclear vessel. This coolant system is connected to the reactor vessels, steam generators, and the pressurizer by means of coolant piping and is moved with coolant pumps. In general, two or sometimes more reactor coolant loops that are closed are connected in series or parallel, depending on the number of steam generating units.

Figure 13-16 shows a PWR power plant that has four steam generators per nuclear vessel: a total of eight steam generators. In the PWR system, the coolant loops must remain at a desired pressure to prevent boiling–hence the use of the pressurizer. All reactor coolant piping for the PWR system is constructed of either pure stainless steel or stainless clad carbon steel pipe. In general, pipe ranges between 2 ft (60 cm) and 3 ft (91 cm) ID; wall thickness depends on the design pressure, though the average is 3 in. (7.6 cm). The all-welded joint connection system is used, and pipe is totally insulated to prevent heat loss from the pressurized system.

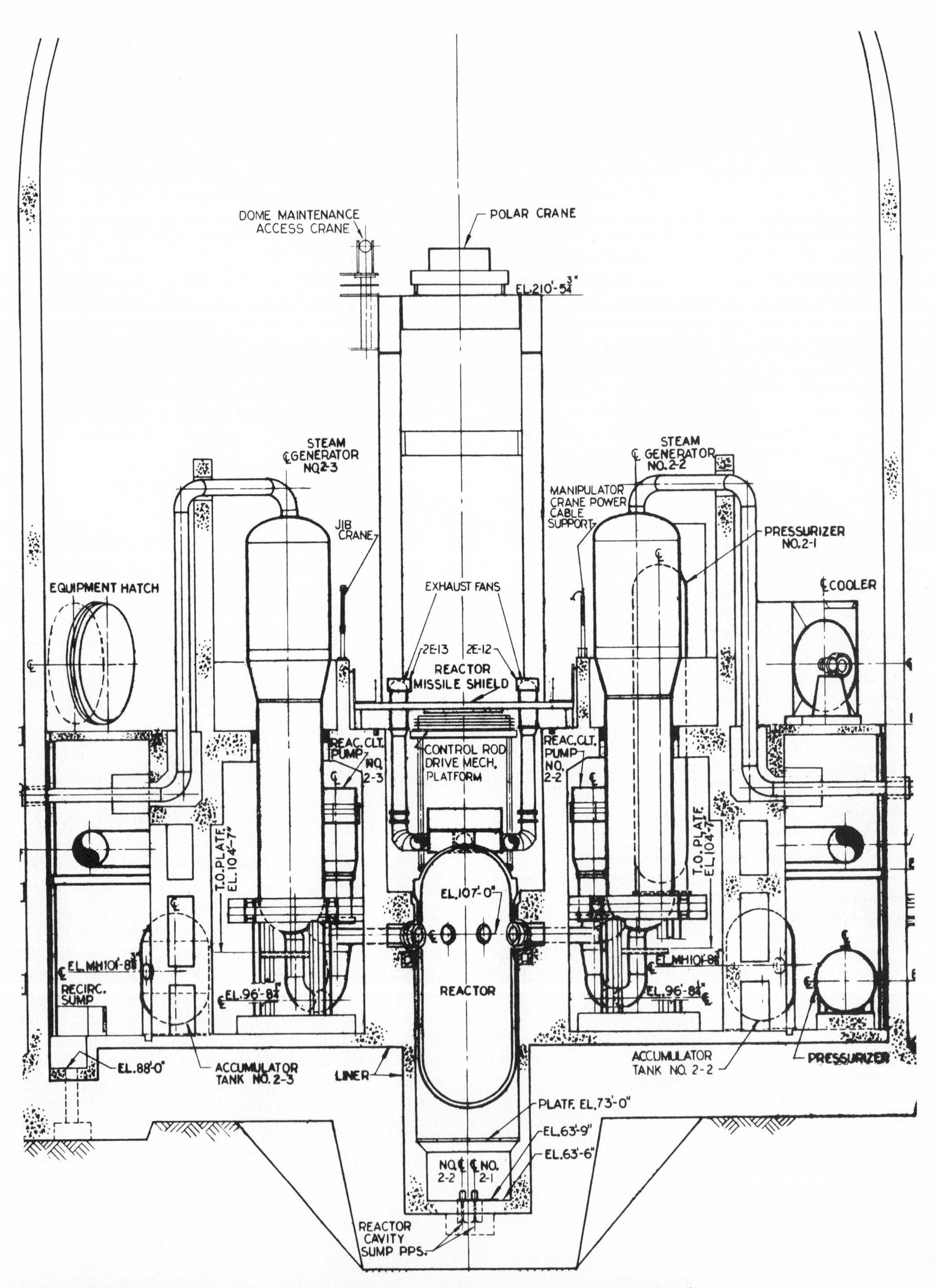

FIG. 13-16 Set of drawings for a typical PWR nuclear reactor containment building showing the various types of equipment and major piping for a nuclear reactor core and steam generators. A polar gantry crane is also shown. (a) Front elevation

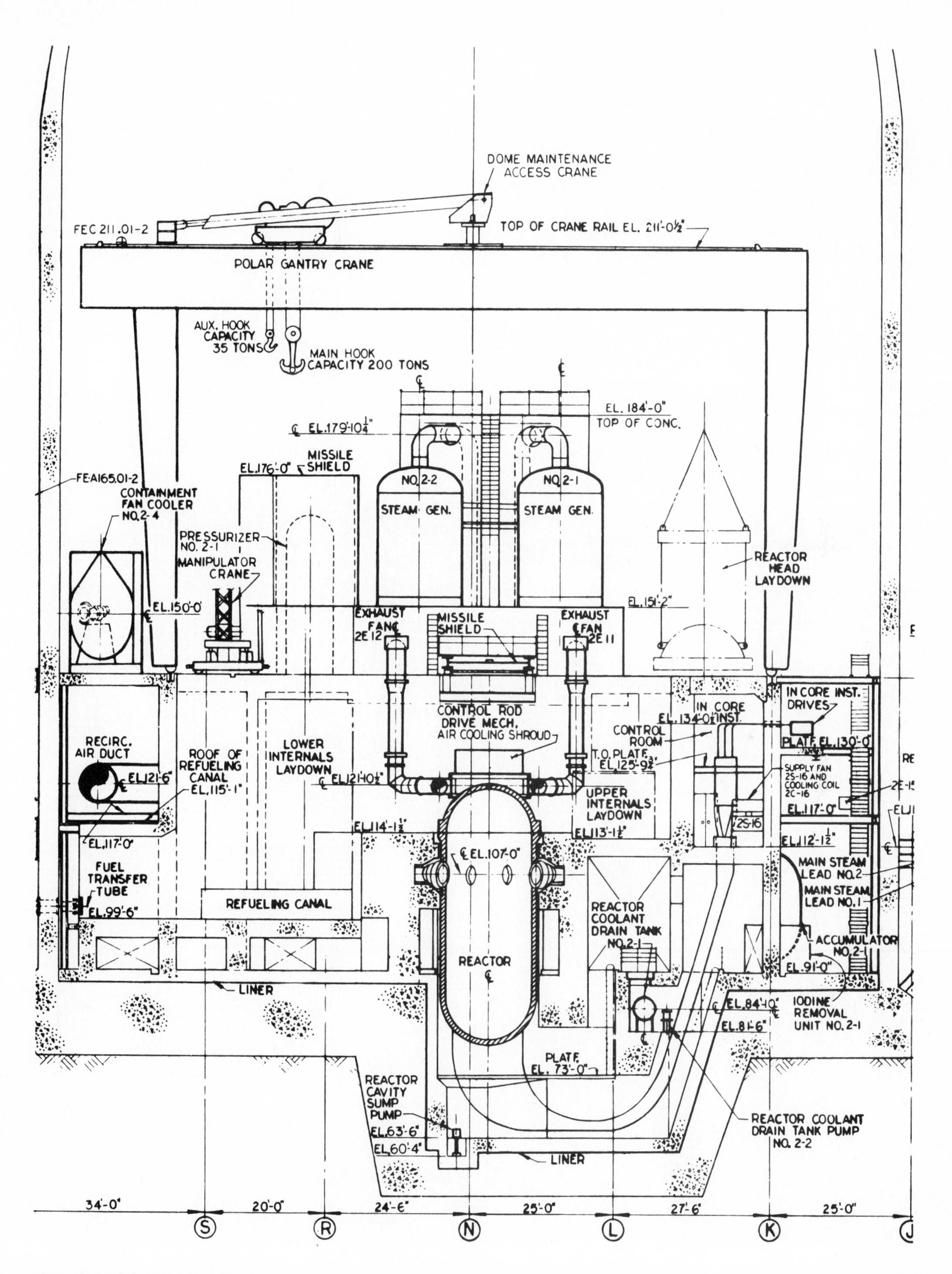

FIG. 13-16 (b) Side elevation

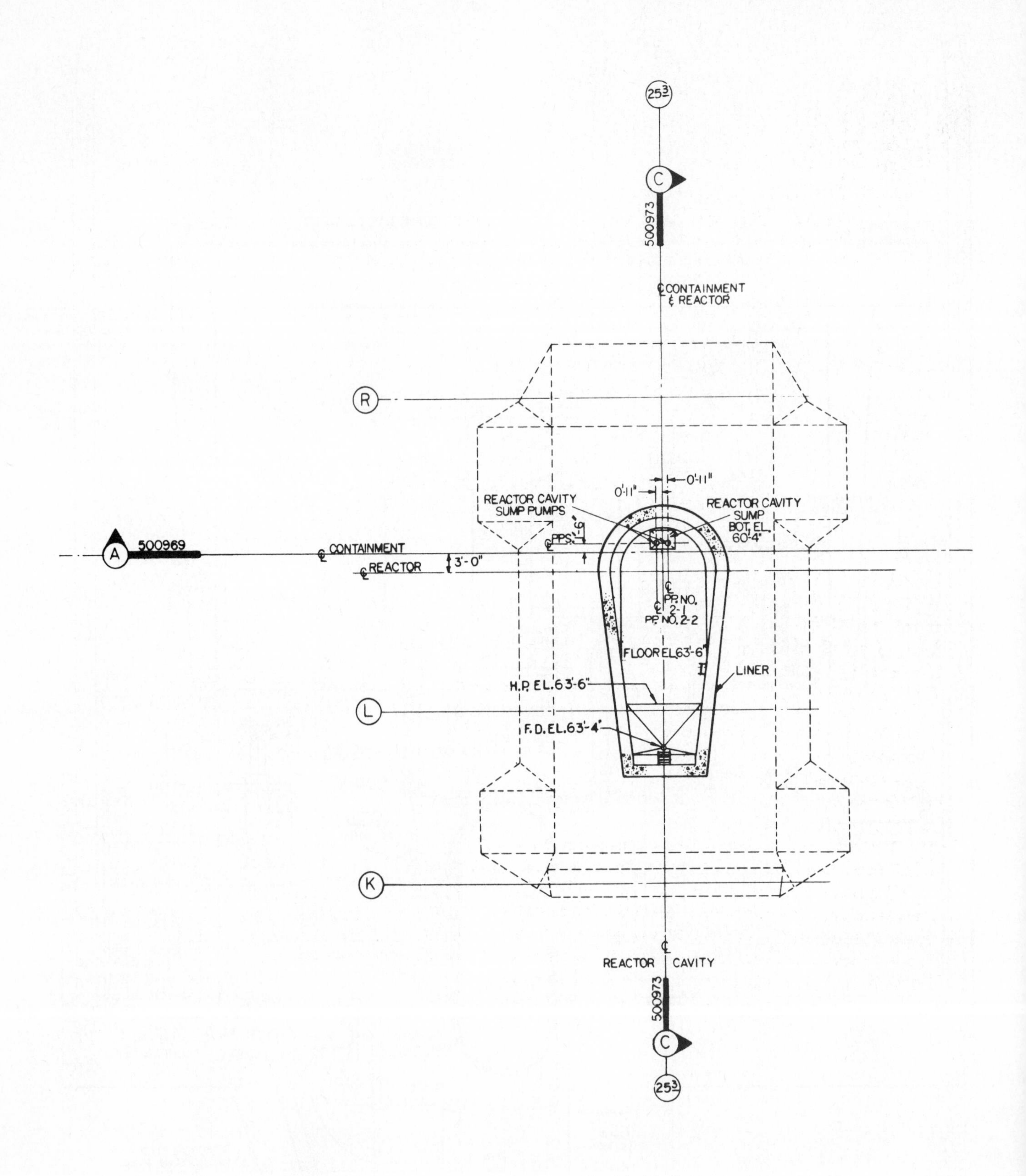

FIG. 13-16 (c) (cont'd)

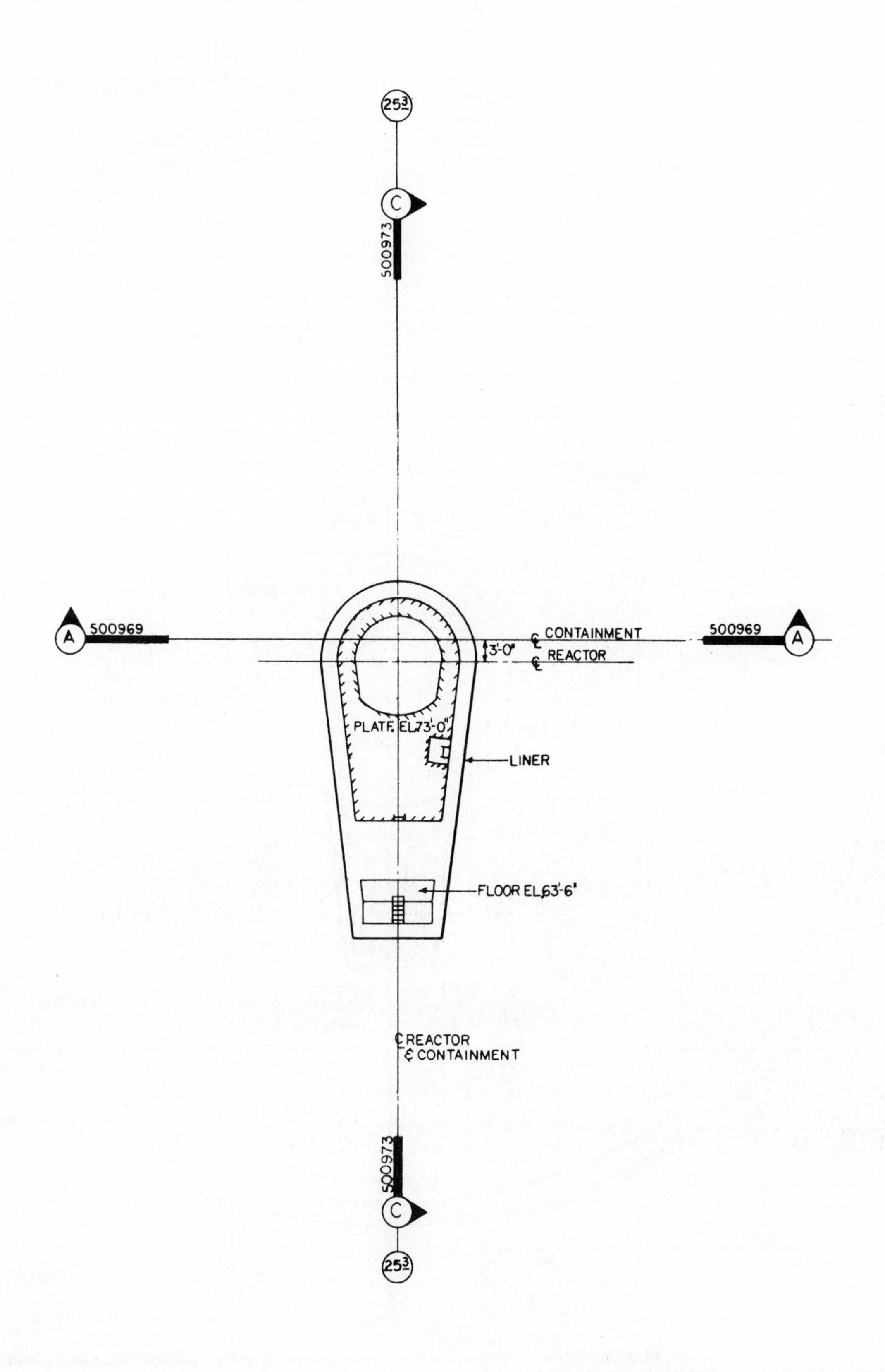

ELEVATION 73'-0"

FIG. 13-16 (d) (cont'd)

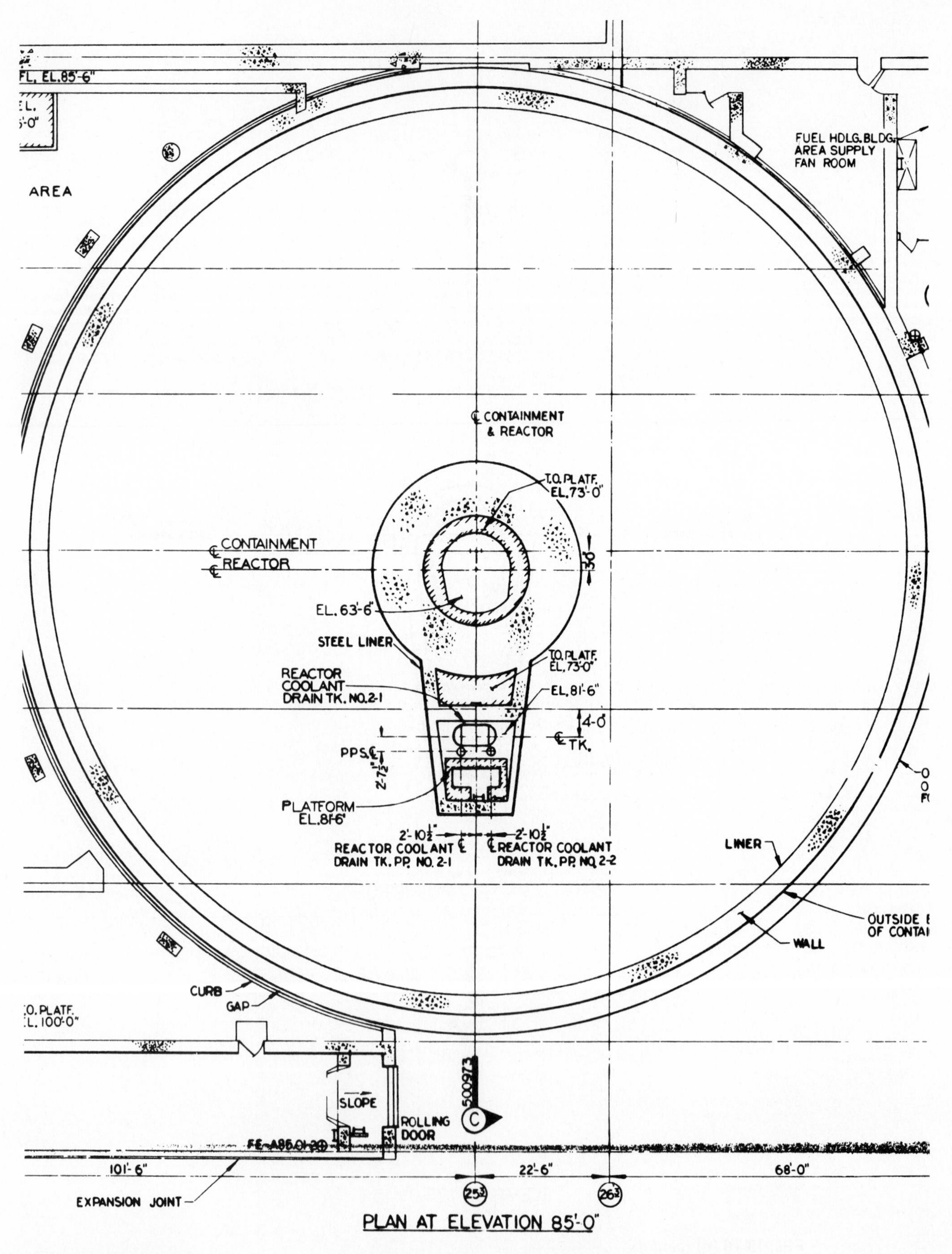

FIG. 13-16 (e) (cont'd)

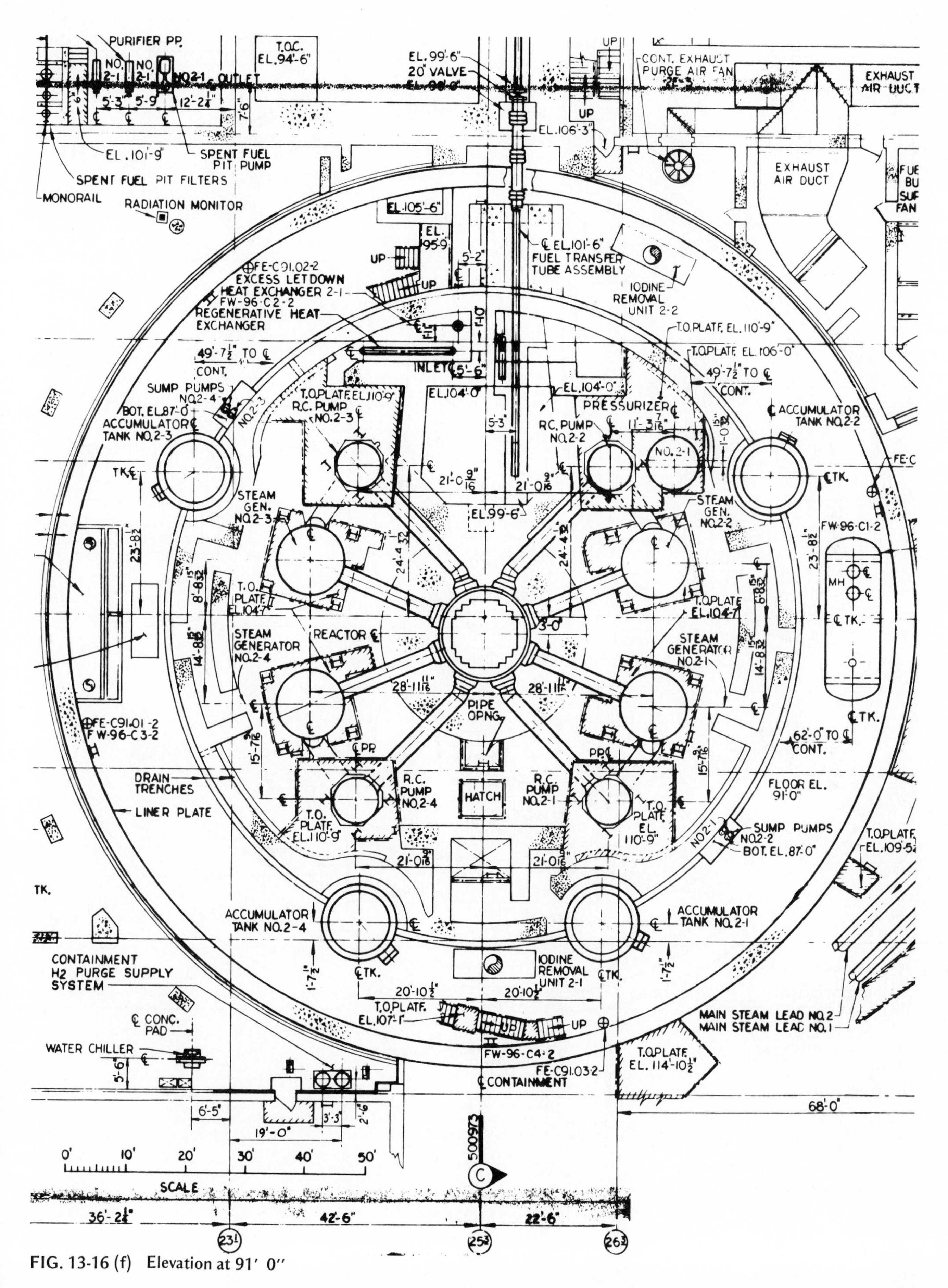

FIG. 13-16 (f) Elevation at 91′ 0″

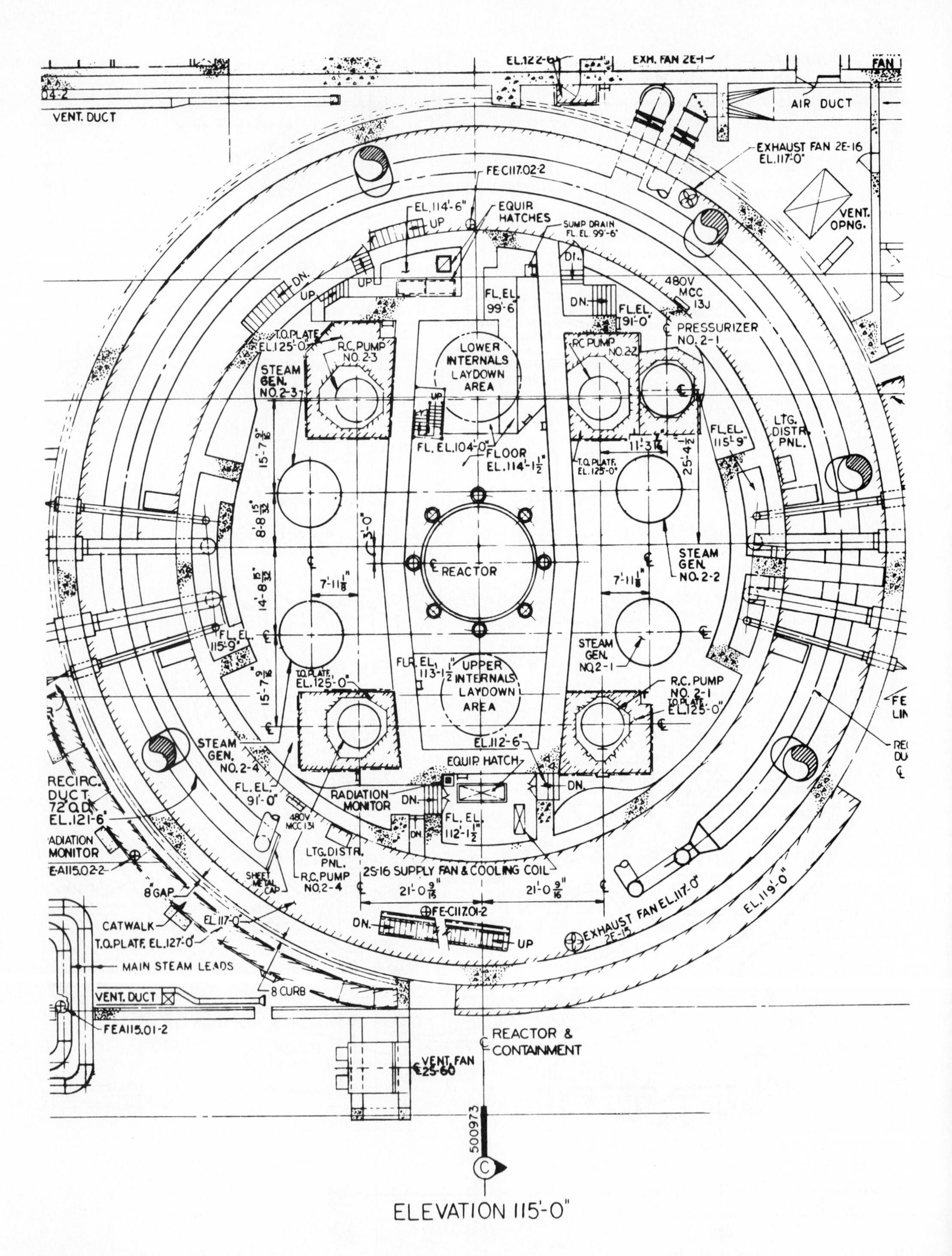

FIG. 13-16 (g) (cont'd)

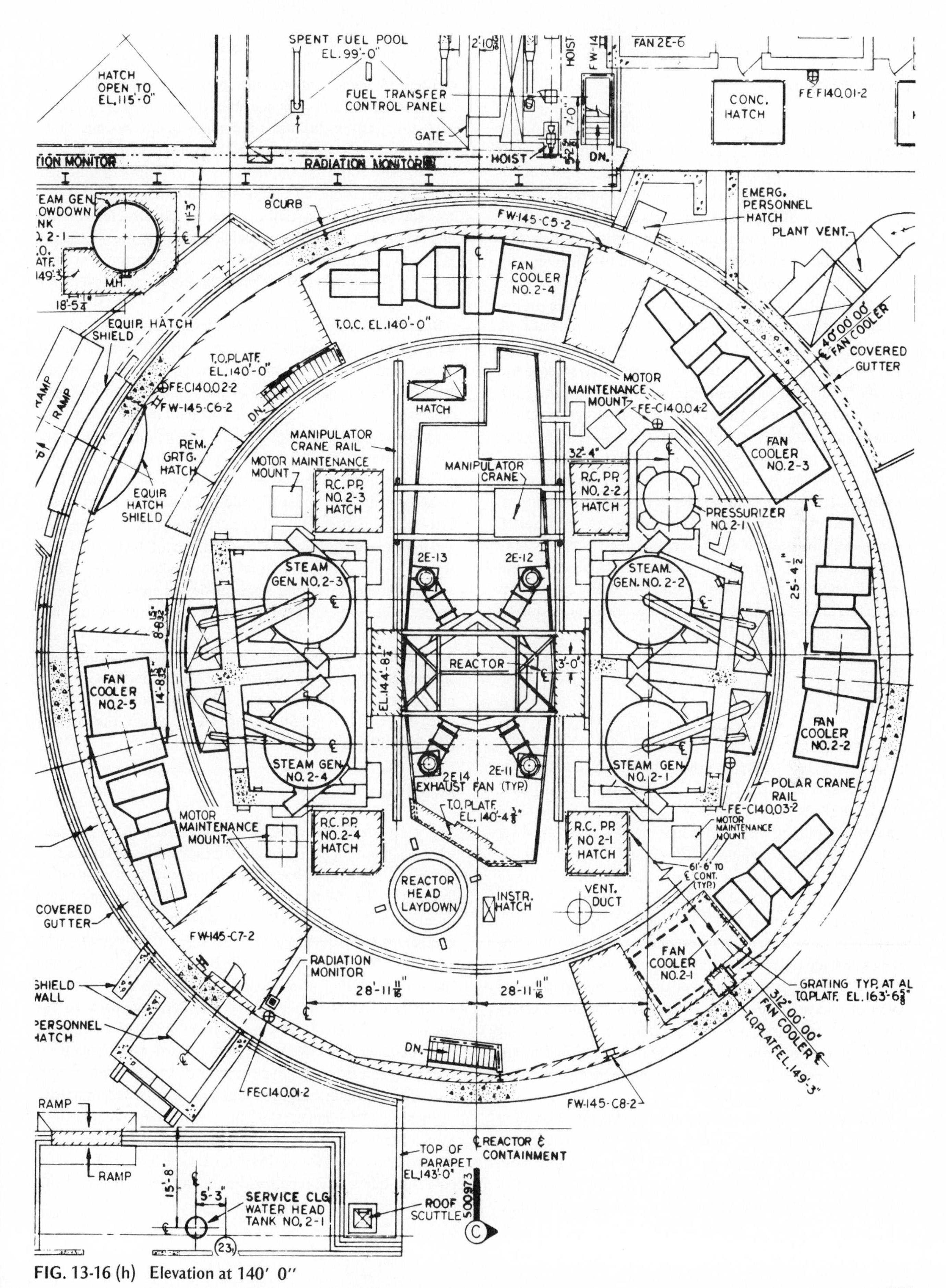

FIG. 13-16 (h) Elevation at 140′ 0″

Instrumentation

The instrumentation found in all portions of the nuclear power plant constitutes a total system which remotely operates and monitors pressure/temperature, fluid flow, liquid level, and mechanical control-rod position. This system includes a variety of detectors, sensors, amplifiers, displays, controllers, and other instrumentation units. A large portion of a modern nuclear power plant must be remotely controlled so that temperature, pressure, liquid level, and so on are monitored and kept within safe limits. The proper of backup instrumentation and safety systems must also be provided. The PWR plant has process instrumentation systems—including all of the flow and radioactivity sensors and controllers—that automatically control numerous units and equipment, including electrical output, reactor temperature, coolant temperature, pressure, pressurized water level, and water level in the steam generator. The BWR process instrumentation includes similar controls that are automatically monitored from the control room for electrical output, reactor pressure, and water level in the reactor vessel. Nuclear instrumentation for the BWR plant includes core detectors, radioactivity sensors, and a variety of alarms and meters located in the control room. The PWR system has similar instrumentation, but there are differences due to the location of steam generation.

NUCLEAR PIPING

Piping in the nuclear power plant is usually divided into conventional and nuclear (radioactive) piping. Nuclear piping contains and circulates radioactive substances; the amount of nuclear piping necessary depends on the type of power plant. Conventional piping is all piping used for service backup systems, feedwater systems, and the like, none of which contains any radioactive material.

As can be seen from Fig. 13-16 and the flow diagrams, the nuclear power plant is a vast maze of interconnecting and parallel piping systems—all of which concern the piping drafter. Over the years, a number of codes, standards, and regulations have been implemented by standards associations and government agencies to control nuclear piping because of its potential hazards. Even conventional piping in a nuclear power plant falls under more stringent guidelines than in a fossil fuel plant.

Certainly the nuclear power plant presents hazardous situations, especially the piping systems that carry radioactive waste, fluids, liquids, and steam. To ensure safe, cost-effective operation of the nuclear power plant, stringent design codes, recommendations, and standards are implemented. All nuclear piping with its associated valves, pumps, and transfer lines is designed with the utmost care to eliminate any possible leakage or damage to the system. Since nuclear piping systems undergo high levels of erosion, corrosion, radioactive embedment, and high-temperature/high-pressure operation, all piping and equipment that come in contact with radioactivity must be constructed of the highest-quality materials available under the most stringent fabrication guidelines that are possible with modern technology. The entire system must be completely leakproof, especially the nuclear piping.

Because of potentially severe thermal shock and seismic disturbances, all aspects of nuclear piping are highly engineered and closely monitored by government agencies. Nuclear reactor equipment, such as boilers, turbines, and condensers, is under more stringent requirements than conventional power plant equipment. The nuclear piping code for pressure piping, ASA B31.1, covers the type of nuclear piping designed to contain a radioactive medium which, if leaked to the atmosphere, could endanger operators or the surrounding population. All nuclear piping must be designed and constructed according to section 1 of B31.1-1 (1955) and all newer revisions of this code. Nuclear piping and related equipment demand rigorous design and fabrication in order to construct a piping system that is utterly reliable. It is not enough to overdesign aspects of the system in the hope that it will meet safety requirements. Even subsystems and auxiliary systems, which are not defined as nuclear piping, must be designed and fabricated with the utmost care—especially if their failure could affect the nuclear piping in the plant.

PWR Plants

The pressurized water reactor power plant requires somewhat different nuclear piping than the boiling water reactor. Since the PWR plant has a higher pressure than that established for the BWR system, it is not uncommon to find primary loops in the PWR plant capable of operating under conditions of +2000 psig and 600°F (315°C). Figure 13-16 shows a PWR power plant with four steam generators and the accompanying primary loops from the reactor to the steam generator, through the condenser, and back to the reactor area. The primary piping loops in this plant contain pure demineralized water, which is used as a coolant to transfer heat from the nuclear reactor to the steam generator. These primary piping loops are constructed primarily of high-quality stainless steel pipe; fittings are created from forged stainless steel pipe; valves are of cast stainless steel. All stainless steel used in this plant is type 304. ASTM A376 covers stainless steel pipe for the primary loop; ASTM A182 covers stainless steel fittings.

BWR Plants

All the characteristics of the BWR plant have been thoroughly described in the beginning of the chapter. BWR plants have special nuclear piping needs. Because of the contamination of the primary piping lines which connect the reactor to the turbine, radioactive steam and water are transferred from the core to the turbine area. This primary piping is constructed of materials similar to those used in PWR plants—namely, stainless steel type 304. In the BWR system, steam piping carries radioactive steam which has been generated directly in the nuclear vessel. In general, the radioactivity level of the turbine generator area is sufficiently low to permit access to the turbine for maintenance.

Piping Design

The design of nuclear piping must take into account stringent requirements to alleviate operational hazards and maintenance shutdowns. Costs for repair, maintenance, and replacement must be considered when ordering and specifying equipment for the nuclear power plant. All welding of pipelines and layout of pumps, equipment, and valves must be designed to permit easy access to the system and also deal with the radioactive problems that may be encountered. Though the temperature and pressure in nuclear power plants are somewhat below the levels for conventional plants, radioactivity and other considerations make the nuclear plant far more expensive to design and construct.

One of the most important aspects of nuclear piping is the new stringent codes and regulations governing design because of the possibility of seismic activity. Implementation of the power plant is sometimes delayed for 5 to 10 years while whole areas of the plant are redesigned and specially supported with seismic-quality pipe supports. Shock suppressors, snubbers, sway brackets, restraints—all have become extremely important in the total design effort of a nuclear power plant because of seismic activity and the postulated consequences.

High velocities in the coolant and recirculation loops, which are constructed primarily of corrosion-resistant stainless steel, demand the attention of all piping designers. Thermal expansion also causes a range of problems in a nuclear power plant not found in conventional power plants. Thermal and seismic shock studies must be done on all piping runs and fabrication in order to ensure maximum safety under adverse conditions. Flexibility under a wide variety of forces must be taken into account when running pipelines and placing pipe supports. Structural loadings on hanger supports, anchors, braces, restraints, and snubbers have to take into account the worst possible conditions that may occur in the power plant. Welded connections and bolted connections must be designed far more conservatively than those in a conventional power plant. Any welding used for the fabrication of the piping system must be thoroughly tested. High-quality fabrication and forming methods must be used for the construction. Containment and distribution of thermal movement and deadweight are two more important factors in the design of a nuclear piping system.

All nuclear piping designers and engineers must take into account the effects of radioactivity on the surrounding materials and equipment—radiation usually hardens them and makes them brittle. A high degree of standardization has been forthcoming because of the need to monitor the effects of radiation on construction materials. Nuclear radiation can have a considerable influence on corrosion also. Some water in the nuclear piping system carries sufficient radiation to affect equipment and piping far from the reactor core. Radioactive contamination and the development of radioactive deposits on any portion of the piping system must be dealt with in the design stages.

Pressure and temperature also need to be considered when designing the nuclear piping system, though the requirements are sometimes less stringent than those for the fossil fuel plant. Nuclear piping systems, including all pipe, valves, and pumps, must be monitored through a variety of safety standards, quality control checking, and testing methods that depend on the area and the design requirements for operation of the power plant. Temperatures involved in heat transfer can also create problems—especially in areas adjacent to the reactor core pressure vessel or during heat transfer in the steam generators in the PWR system. In many cases, the primary piping and all piping and components close to the nuclear reactor require greater thickness and more stringent and expensive materials for construction because of the susceptibility to thermal stress and radiation with accompanying corrosion and erosion. Materials are chosen for characteristics which will allow cost-effective implementation of the piping system with the highest possible degree of safety.

All fabrication methods, including welding and forming, must take into account the effects of radiation on fabrication processes and materials. Nuclear radiation has been noted to affect both the structural materials and the fabrication materials, creating problems with impact resistance and overall ductility. In general, carbon and alloy steels (including stainless steels) suffer these problems much more when used in the structural material and other materials used in nuclear power piping. Reactor radiation increases their hardness and strength, but they become less ductile and more brittle.

Pipe Forces An assortment of internal and external forces make nuclear piping far more critical than conventional piping. Most of the calculations for forces on nuclear piping are done with a computer–calculations for fluid flow, thermal expansion, seismic disturbance loading, pipeline flexibility, and pipe support locations are just a few.

Nuclear piping in the BWR system connects the nuclear vessel to the turbine area and from there it runs to the condenser and back to the vessel again. In the PWR system, it moves the nuclear heated fluid from the reactor vessel to the steam generator and back to the nuclear vessel. These primary piping loops create a number of problems for the piping designer because of the fixed connections to the vessels by way of the nozzles, which are, in reality, anchors. The piping connected to these welded nozzles on the nuclear reactor turbine and steam generator must be able to withstand great forces–such as the total weight of the piping system (the deadweight) which must be carried by pipe support mechanisms. This deadweight includes all the fittings and components that are part of the pipe run, including valves, meters, and other special equipment. Because of the large diameters and wall thickness of the pipe, the accompanying equipment is usually of greater weight than on a conventional pipeline, especially for items such as nuclear valves with their motorized or hydraulic actuators. Deadweight thus becomes the first major consideration for piping design.

Pressure and temperature are two other major design problems that accompany every piping system, though in the case of nuclear power plants the force is on the nuclear piping loops, which take into account the internal pressure at which the pipeline will be functioning and all the basic loads that act upon the pipe in a longitudinal and diametral or circumferential manner. Included in the calculations for pressure forces are special design requirements for the use of piping bends and fittings and other connections, which in most cases must be reinforced to handle fluctuation in pressure, water hammer, and steam hammer.

Thermal forces on nuclear piping can cause expansion that must be dealt with in the design phase of the system. In most cases, restraints, guides, spring hangers, and braces are used to create a flexible system which can withstand thermal expansion. Thermal expansion must be judged along with seismic forces, depending on the power plant's location. Seismic loads and forces often are dealt with by securing the system to a structural member and eliminating as much movement as possible for the piping system. But since thermal expansion would destroy such a restrained piping configuration, both forces, thermal and seismic, must be dealt with in conjunction. One must provide restraints and anchors at important intersections or connections to eliminate nonthermal movement of the pipeline while relieving the forces caused by thermal expansion at the connections, such as nozzles where the piping enters the vessels. The other portions of the piping system must be loosely supported with pipe support configurations which allow for thermal expansion while eliminating the dangers of seismic forces. Spring hangers, supports, and snubbers help eliminate the sudden violent movement of the piping system which usually accompanies seismic disturbances. Snubbers, sway braces, and sway struts allow for thermal movement while maintaining rigid restraint to deal with seismic movement. Thus the nuclear piping system must be rigid enough to withstand seismic disturbances but flexible enough to accommodate thermal expansion.

Other problems must also be considered. Pipe rupture and pipe whip, for example, may damage the surrounding piping by spraying fluid. A variety of vibrational disturbances from accompanying equipment or from the operation of the system itself must be taken into account. All these problems must be dealt with through computer analysis and made available to the nuclear pipe designer.

The pipe designer must also take into account the problem of interferences–not only the obvious interferences between piping and equipment but also the not so obvious interference with code inspection work and maintenance, both of which play a significant part in the functioning of a modern nuclear power plant. Periodic inspection of nuclear piping lines and the removal of insulation covering and storage during inspection are major considerations for the designer. Moreover, room must be provided for alternative and backup systems and replacement of equipment such as pumps, valves, motors, and compressors.

Seismic Movement All nuclear reactors in their planning stages must deal with the problems of seismic effects. An environmental impact statement must establish beyond any doubt that postulated earthquake activity will not have a detrimental effect on the nuclear piping system or any portion of the plant which has radioactivity that could harm operators or the population.

As has been said, snubbers and other pipe hanger equipment such as sway braces, sway struts, and shock arrestors are used to prevent sudden movement in the nuclear piping system which could break a pipeline or damage equipment such as valving and pumps. Anchoring the piping system and using restraints and guides helps eliminate the effects of seismic disturbances, though this procedure also creates enormous stress on the system during thermal expansion. Chapter 8 deals with the pipe support variations that are utilized with nuclear piping.

Each nuclear piping run that begins and ends at an obvious break point, such as a piece of equipment or nozzle or vessel, is set up on coordinates and programmed into a computer to provide a mathematical model of the prospective piping system. This model takes into account all the various forces and loads which accompany a typical nuclear piping spool. All movement in *X, Y, Z* directions is analyzed by the computer. The computer establishes selected portions or points along the pipe run where movement should be restricted, some in an *X, Y* direction and others in only one direction, such as the *Z* direction. In the case of restraints or anchors, all three directions, *X, Y,* and *Z,* are specified. These points, though ideally fixed, cannot always be used because of surrounding equipment, pipelines, or interference problems with HVAC or electrical runs. A certain amount of leeway is built into this system providing for movement of the suggested pipe support, snubber, anchor, or restraint between 6 (15 cm) and 12 in. (30 cm) on either side of the suggested point of support. These pipe support mechanisms must be connected to surrounding steel, concrete, or other structural items near the point of support.

Temperature Effects Thermal expansion requires a high degree of flexibility to avoid damaging, severing, or breaking pipelines—especially in connections to major equipment such as pumps, compressors, vessels, and turbines. The ability of equipment such as turbines, pumps, compressors, valves, and vessel connections to absorb thermal expansion movements and loads through their connections is an extremely important consideration for nuclear piping designers. Where fittings such as elbows, laterals, tees, and bends are prevalent, thermal expansion tends to create a multitude of ments and loads through their connections is an extremely important consideration for nuclear piping designers. Where fittings such as elbows, laterals, tees, and bends are prevalent, thermal expansion tends to create a multitude of intensified problems. All nuclear piping and pressure vessel piping and boiler codes must be followed when calculating the design requirements based on thermal expansion plus a well-considered safety factor.

Steam and Water Hammer The sudden operation of equipment in nuclear piping lines—such as remotely controlled pumps and valves—can damage the piping system by water hammer. The system must be designed to eliminate any accumulations of water or condensate which may harm the system. Steam hammer must also be taken into account where major steam lines connect to the nuclear vessel, the turbine, or steam generator.

Piping Fabrication

Nuclear piping is fabricated with most of the techniques similar to conventional piping, though standards and safety factors—and therefore the expense—are considerably greater. Because of the great number of standards, codes, and specifications for nuclear fabrication, many shop fabricators have turned down nuclear piping assignments. Moreover, the lack of profit in comparison to other piping work has made it difficult to interest reliable shop fabricators in nuclear piping work. Many nuclear projects are held up for long periods of time because of this, especially the critical areas of nuclear piping loops. It is generally agreed that shop fabrication of nuclear piping is more advantageous than field construction because of the testing and specifications and codes that must be dealt with for every portion of the nuclear piping system. Most 3-in. (7.62 cm) and larger nuclear piping is fabricated in the shop. Drawings, testing, and fabrication techniques are similar to those described in Chapter 9—including bending, welding, extruding, and flanging. When fabrications are of considerable size, they must be broken down into units that can be worked with by the fabricator and also the shipper.

Connections, Fittings, and Joints A variety of conventional change-of-direction and connection devices are employed throughout the nuclear power piping field. The use of bending is suggested wherever this procedure can be accomplished economically. The long-radius elbow can be substituted for bending if necessary. Both bending and the long-radius elbow provide a smooth, low-friction, less turbulent transition between piping directions. All the typical fittings described in Chapter 3 are used for nuclear piping, though reliance on tees, laterals, reducing tees, and reducing laterals should be kept to a minimum for critical piping unless these pieces of equipment are reinforced.

End preparation for welded piping systems is the same as described in Chapter 2. Various bevel preparations which correspond to the boiler and pressure vessel code, with accompanying specifications for nuclear piping procedures, are available. All aspects of welding come under the jurisdiction of the ASME boiler and pressure vessel codes for nuclear piping. Butt welds are the most prevalent. The use of socket welding for pipe joints in the nuclear field is somewhat limited because of corrosion and erosion problems. When socket-welded fittings are used for small-diameter pipe, they must be made up with extreme care to provide tight alignment of the piping runs—there must be no gaps between the ends where foreign materials and radioactivity can cause problems. The advantages of using welding for piping systems are well defined in Chapter 2 and apply here for nuclear piping systems as well as in other areas.

Other kinds of joints are also used in nuclear piping, depending on their application, especially in areas where the system must be broken down, repaired, maintained, or removed from the line frequently. In these cases, flanged joints are more commonly accepted. For steam lines, flanged joints are made up with extreme care to ensure leakproof service, usually with the use of RTJ connections. In general, flanged joints in nuclear piping loops or accompanying systems are more stringently designed and require considerably more bolting than normal flanged joints in other services. Weld-type flanges and RTJ gaskets present problems since radiation may affect seal and gasket materials.

Insulation

Nuclear insulation is one of the main areas for the nuclear piping designer's consideration. Insulation for a typical nuclear power plant requires more rigid specifications, codes, and standards than for other services because of the radiation, erosion, corrosion, heat, pressure, and seismic problems that may be encountered. The nuclear vessel area—with its nuclear piping loops and large-OD piping throughout the containment area connecting the nuclear vessel with the steam generator, turbines, and condenser—requires high-quality insulation in order to protect surrounding equipment and instrumentation and restrict the radiation given off from the piping. A variety of inspection procedures must accompany all nuclear piping, especially in the containment area and piping directly associated with nuclear processes. Thus all piping must be accessible to the inspector; moreover, insulation is never permanently bonded and is usually constructed to be easily removed, stacked, stored, and maintained during inspection. Insulation must also be able to withstand unusual problems in the containment area from the use of spraydown, chemical solutions, and water associated with certain power plants. Block insulation, metal insulation, and blanket insulation have been used with relative success by nuclear piping designers. Most protective insulation must be clad with stainless steel or aluminum sheets, especially where reflective insulation is necessary. Reflective insulation is usually composed of layers of thin, crumpled metal providing air spaces which help insulate the piping.

Valves, pipe supports, and other equipment which is part of the piping system or line require special attention to reduce the total amount of heat loss through the equipment. In some cases, valves and supports must be insulated also. Permanent insulation is used in all areas of the nuclear plant where inspection is unnecessary. Calcium silicate insulation, fiberglass, and mineral wool are some of the more commonly found types of insulation for permanent installation.

Pipe Supports

Chapter 8 covers in detail the design, drafting, and location of pipe supports for both conventional and nuclear piping. The common pipe supports in the nuclear field are rod hangers, struts, variable-spring hangers, constant-spring hangers, shock suppressors, sliding supports, anchors, and restraints, all of which vary in the design characteristics that are necessary for a particular job.

As stated in Chapter 8, pipe supports can be classified into seismic category 1 and seismic category 2. Category 1 includes pipe supports that are designed for seismic disturbances; category 2 contains conventional pipe supports which are not used where seismic problems are prevalent. Pipe supports can also be classified as hanger-critical or hanger-noncritical. Hanger-critical supports are associated with nuclear piping systems or piping systems subject to seismic activity. Therefore they fall into seismic category 1 and have temperature limits equal to or higher than 300°F (149°C). Any piping system, especially one associated with nuclear piping, which needs special consideration should be classified as hanger-critical. All other systems are considered hanger-noncritical.

The pipe support must be classified by the designer on the detail drawing of the pipe support in question. Since the design, fabrication, erection, and inspection of pipe supports are important considerations, classification must be done with care. All the piping systems which have been described—feedwater, condensate, turbine, nuclear vessel, main steam, radwaste—require consideration for adequate pipe support design and location. Moreover, a vast number of auxiliary systems are very important to the nuclear power plant. All these must be accurately located and supported. Because of the high level of radioactivity generated in a nuclear power plant through the fission process, it is important to ensure the safety of operating personnel and surrounding communities and environment. Therefore backup systems are provided, such as core spray, heat removal, and a variety of cooling backup systems and flushdown systems.

The typical nuclear plant may utilize 100,000 ft (3048 m) of piping 3 in. (7.62 cm) and larger per generating unit. The average number of pipe supports for such a unit can be between 10,000 and 30,000, many of which are for surrounding equipment, pipelines, electrical trays, HVAC, and so on. To avoid congested areas, the pipe support designer's work is stringently regulated and standardized wherever possible, providing for the safest possible system that can be designed. Many companies have had to redo pipe support systems completely, adding large numbers of snubbers, braces, struts, spring hangers, and other equipment to allow for seismic disturbances without disrupting the nuclear power plant operation or causing safety problems. One such plant has been under construction for 12 years without being licensed for operation. This added expense has pushed the total cost of the power plant to over $1 billion, and still the nuclear licensing has not been attained. Therefore, as one can see, the accurate, safe design of a nuclear power plant is extremely important, especially for pipe supports and hangers.

To review, one of the pipe designer's main tasks is to ensure the adequate support of the system and eliminate the negative effects of vibration, thermal expansion, seismic activity, or shock from internal movement of the line material and breakage in parallel lines which may affect surrounding equipment and connections. Therefore all piping designers must take into account the variables of insulation, containment penetrations, restraints, supports, and shock arrestors. Both variable-support and constant-support hangers are used in nuclear piping, depending on the design load, the thermal expansion, and the amount of travel that may be necessary to allow the system to adjust to thermal and seismic problems. Smaller-OD piping throughout the nuclear plant can be restrained, anchored, and supported much more easily than large piping loops, especially primary loops concerned with the movement of radioactive water and steam. Most of the hangers, supports, guides, and restraints described in Chapter 8 apply to the nuclear piping field; a variety of guides and restraints that are constructed of structural steel are shown in pipe support details in that chapter. Figure 13-17 shows a typical nuclear power plant pipe support detail for a snubber assembly on a 3-in. (7.6 cm) pipe insulated with 2½ in. (5.7 cm) of insulation. Figure 13-18 shows the pipe support detail for a variable-spring trapeze assembly.

Whenever possible, it is important to attach the pipe support system to large structural members—I beams, columns, concrete containment areas, flooring—whether it be pipe supports, snubbers, restraints, guides, or other alignment mechanisms. Pipe penetration through the containment wall or between elevations in the containment area is also important. Problems in expansion and external shock affect the design of the penetration area.

When noncritical piping penetrates walls or the containment building or floors, it is sometimes possible to anchor the pipe directly to the structural item, especially in low-pressure/low-temperature situations where thermal expansion is not a consideration. This provides not only for support but anchors the system with a rigid connection which will have a positive effect if seismic activity is prevalent in the area. Where large piping and critical piping must penetrate containment walls, floors, elevation levels, and other walls (especially in primary steam lines), nozzles are provided at the penetration entrance with a sleeve of larger OD/ID piping to run the piping line through the containment wall or structural item.

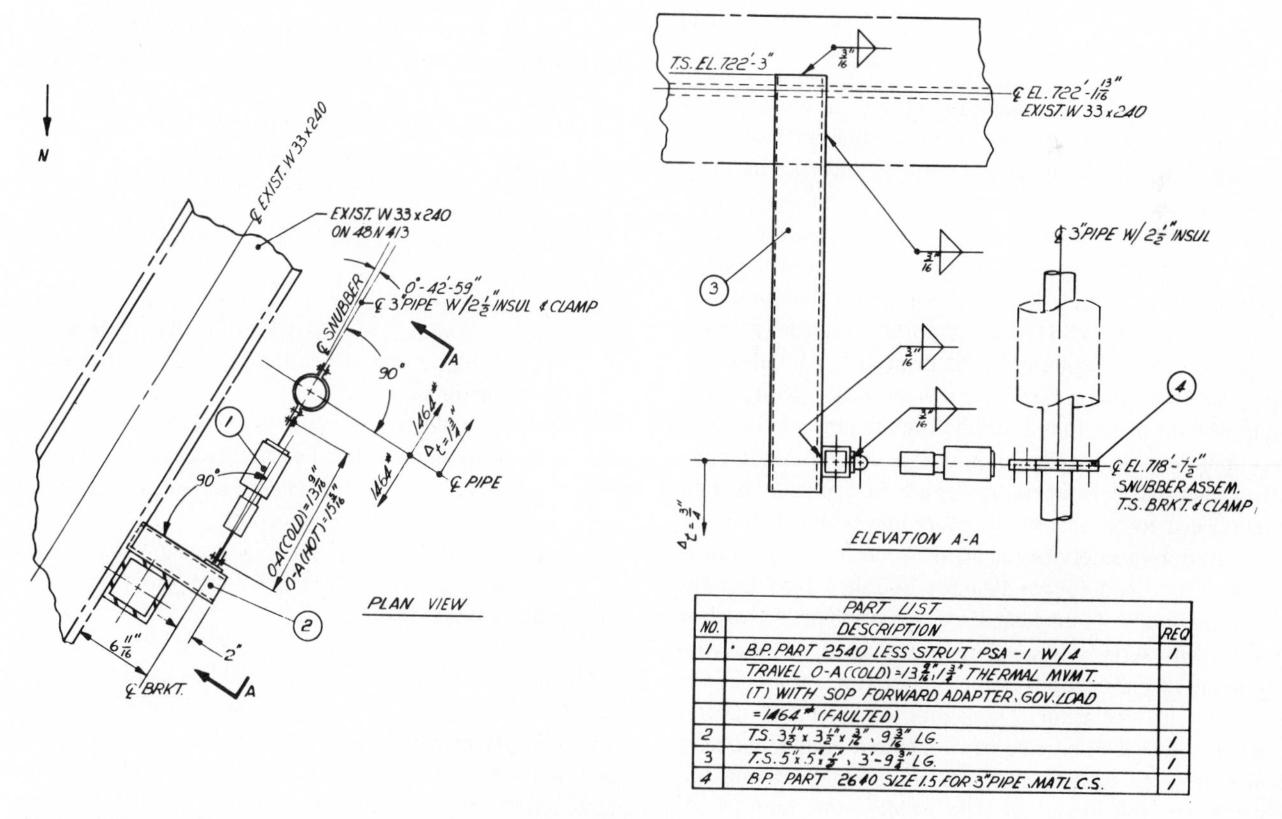

PART LIST		
NO.	DESCRIPTION	REQ
1	• B.P. PART 2540 LESS STRUT PSA -1 W/4	1
	TRAVEL 0-A (COLD) = 13 9/16", 1 3/4" THERMAL MVMT.	
	(T) WITH SOP FORWARD ADAPTER, GOV. LOAD	
	= 1464# (FAULTED)	
2	T.S. 3½" x 3½" x 3/16", 9 3/16" LG.	1
3	T.S. 5" x 5" x ½", 3'-9 3/4" LG.	1
4	B.P. PART 2640 SIZE 1.5 FOR 3" PIPE, MATL C.S.	1

FIG. 13-17 Typical nuclear power plant pipe snubber assembly.

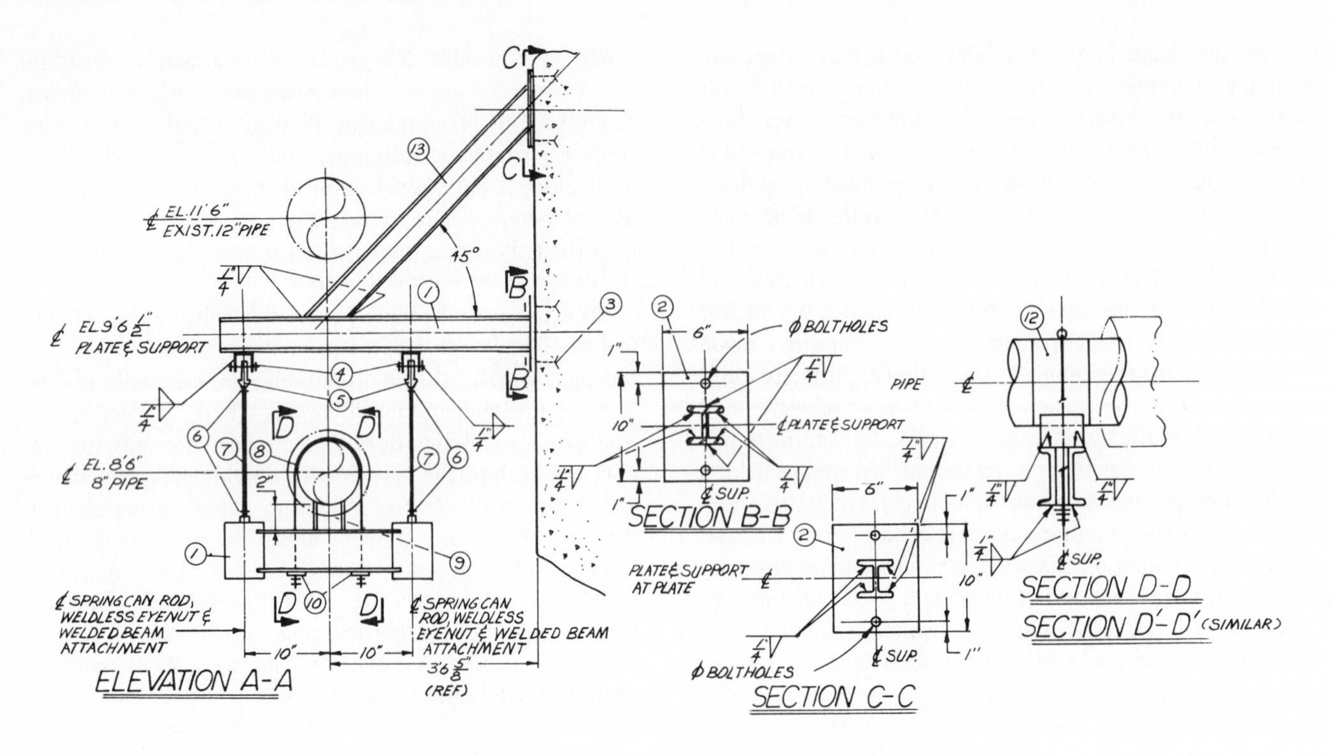

FIG. 13-18 Variable-spring trapeze pipe hanger for 8-in.-diameter pipe installed in a nuclear power plant. (See Chapter 8 to determine the items shown on this detail.)

Testing

Nuclear piping materials, design, and equipment are governed by many testing and quality control requirements. Nondestructive testing is used to determine the acceptability of the materials and fabricated products or shop fabrication configurations used in the nuclear power plant.
rication configurations used in the nuclear power plant.
Much of this examination is done prior to installation, but in-service inspection (ISI) is also important, especially where insulation covers the items to be inspected. All defects in pipe materials or accompanying components, fittings, and equipment must be found well before the power plant goes into operation. In the case of in-service inspection after the power plant has been operating for a period of time, certain defects need to be uncovered before they seriously hamper the operation or safety of the plant. Welds, materials, thicknesses, service conditions—all must be taken into account by the inspector. Testing the various aspects of the nuclear power plant system is standard procedure from the very beginning of fabrication to the completion of the power plant and its operation. The piping designer and drafter must ensure that inspecting personnel and equipment have sufficient space to determine the acceptability of the system's design and operation. The ASME code, section 11, covers in-service inspection requirements and must be adhered to by all nuclear piping designers. Design suggestions for ensuring adequate access for inspection include:

1. Provide adequate storage space for removable insulation.
2. Provide easy access to all circumferential welds by establishing these joints and connections away from major equipment, structural items, and penetration areas.
3. Establish pipe support locations and connections for snubbers, guides, supports, and the like in areas which do not hinder the inspection process (especially ultrasonic examination).
4. Provide adequate examination space in and around all piping areas and related equipment.
5. Design critical piping where radioactivity may be a problem to allow for inspection.
6. Design and monitor welding procedures to provide for clean, smooth, efficient welded joints which allow for adequate testing.

The ASA code for pressure piping (B31.1) and other codes which cover nuclear services must be consulted to provide a piping system which meets the requirements for certification and licensing. Pipe lengths, fittings, connections, welds, radioactive loops, vessel connections, nozzles, weld-neck flanges, valve connections—all require a variety of examination and testing procedures, depending on the design requirements for the portion of the plant in question.

Nuclear piping systems are tested for leaks by using such procedures as hydrostatic testing, pneumatic testing, soap bubble testing, and gas testing of the piping and its fittings and components. Minimum requirements for leakage in nuclear piping systems are covered by standards and codes for each test and must be consulted during the design process. Radiographic examination is also used in the nuclear power industry, especially for pressurized vessels and high-temperature piping loops.

MODELING

Besides offering obvious design advantages, the model also helps in nuclear licensing procedures and requirements. Since the model provides three-dimensional visibility, it can be established that redundant systems (backups) and safety features (mechanical, equipment, electrical) all meet with licensing codes and design requirements. The use of the model in nuclear piping design has become almost universally accepted by power construction firms because of the ease with which piping can be routed to avoid compromise or damage to safety equipment in the event of piping breaks and pipe whips. The location of seismic supports, restraints, and other pipe supports can be optimized with the model. Models allow the designer to plan for the removal and replacement of equipment. With the model, designers can develop detailed equipment locations, removal procedures, plant operation sequences, and construction sequences. Moreover, the model can be used later on for training operators and maintenance crews.

Nuclear power plant models, like all piping models, use color coding to highlight safety separations and fire protection provisions, access areas for installation, maintenance, and removal of equipment, and access for in-service inspection. Because of the vast amount of piping and associated components, fittings, and intricate configurations, especially around the reactor area, the model permits a much easier three-dimensional visualization than the multitude of drawings associated with the typical nuclear power plant with its thousands of overlays, elevations, sections, plans, and developments. Because of the size of the nuclear power plant, models are usually modular in form—that is, they can be sectioned at obvious break points, providing for easy assembly, easy access to the interior of the nuclear plant, and easy change in design and rerouting of piping. The areas to be modeled are carefully studied at a variety of elevations and then split horizontally and vertically to increase visibility and access to the modeled plant. This procedure also provides for the installation and design of piping hangers and related equipment in the latter stages of the design process. Material takeoff and isometric fabrication details are both facilitated by use of a model.

The typical nuclear power plant model may take up one whole floor of an engineering firm's building and cost up to $2 million in time, materials, and construction. Nevertheless the advantages, as detailed in Chapter 15, far outweigh the cost of the process. To be effective for the nuclear power industry, models must be begun in the early stages of the project to eliminate redundant, complicated drawings from a variety of fields (electrical, HVAC, isometric) and orthographic projection pipe design procedures. By using the model from the beginning stages all the way to the construction of the plant, it is possible to maximize its effectiveness. See Chapter 15 for a much more detailed explanation of the modeling process.

Figure 13-16 shows several elevations and sections of a nuclear containment area with the major equipment placement and major routing of large-diameter pipe. These drawings provide an excellent opportunity to become familiar with a nuclear power plant—especially in the containment area, which is the major difference between nuclear and conventional power plants. Careful review of illustrations in Chapter 12 and 13, particularly the P&ID drawings, will show the similarities in piping structures and the differences in the power generation areas.

Figure 13-16 can be used as a group project or advanced individual project. It is also possible to construct a model of this containment area. The liner or containment building need not be constructed, though a circular area should be described along the base of the model to show where the containment wall would be. It is possible to use thin acrylic sheets and, by using a jig and fixture arrangement, bend plastic sheeting into the desired configuration, usually using double wall construction. The fixture will provide adequate restraint during the gluing process. One problem with this method is that large holes or cutaways (dodges) need to be cut out of the containment wall to show equipment and pipe routing. Plastic sheeting eliminates this problem. However, it is probably more economical to omit the containment building and simply provide 1-in. (2.54 cm) or 2-in. (5 cm) containment walls around the reactor, symbolizing the extent of the containment building but not constructing it in total. The dome on top of the containment area should be eliminated in either case.

Figure 13-16 is a PWR system. In this power plant, four steam generators must also be fabricated along with the reactor core. Great care must be taken when attempting to construct a model from these drawings. It may be necessary to reroute piping or place equipment approximately. Figure 13-19 is a photograph of the model containment area of the power plant shown in Fig. 13-16. This photograph suggests the possibilities for this model project.

FIG. 13-19 Model of containment area for a nuclear power plant. This model is from the same project as Figs. 13-4, 13-10, and 13-16. Note the piping and placement of major vessels.

QUIZ

1. What is nuclear fission?
2. In a nuclear reactor, what causes fission to occur?
3. What is a chain reaction?
4. What is the major content of radioactive waste?
5. Where does the energy come from in the fission process?
6. What fraction of natural uranium is fissionable as it stands?
7. What is the source of fissionable plutonium?
8. What is a moderator and why is it used in a reactor?
9. What special demands does a nuclear reactor make on the choice of piping materials?
10. List the major kinds of nuclear reactors according to the coolant and moderators used and the major advantages of each.
11. How does a fast reactor differ from a thermal reactor?
12. What is the *breeding* of fuel? What is the ideal efficiency of the process?

13. What effect does neutron bombardment have upon metal?
14. Why must the inner surfaces be perfectly smooth in the primary loop of a pressurized water reactor?
15. What two kinds of radiation must be shielded against in a reactor? What materials are good for each?
16. What is nuclear fusion?
17. How are models used in the design of a nuclear power plant?
18. Name some uses for the model after the design and construction of the nuclear power plant.
19. What loads must piping hangers compensate for in a nuclear power plant?
20. What are snubbers used for in a nuclear power plant?
21. Name three design requirements for a typical nuclear pipe support.
22. What engineering societies influence the standardization of nuclear power piping materials and design?
23. What is the ASME boiler and pressure vessel code and what does it do?
24. What is nondestructive testing?
25. Name two types of testing used for nuclear piping and explain how they are done.
26. What is corrosion? How does it affect nuclear piping?
27. What are the effects of radioactivity and corrosion on nuclear installations?
28. What is the effect of nuclear radiation on structural metals?
29. What type of welding is used for piping joints in nuclear piping?
30. When are flanged joints used for nuclear piping?
31. What materials are used for recirculation piping?
32. What effect does erosion have on nuclear piping materials?
33. What pipe fittings are usually used in a nuclear reactor?
34. What effect does thermal shock have on tees in feedwater piping for a nuclear power plant?
35. What types of pipe fabrication are needed for a typical nuclear power plant piping system?
36. What is thermal expansion and how does it affect piping in a nuclear power plant?
37. Name three design requirements and problems encountered in nuclear piping.
38. What forces prevail on a typical nuclear pipeline?
39. How does seismic activity affect the design of nuclear power piping?
40. Name six basic systems found in a typical nuclear power plant.

PROBLEMS

1. Redo Fig. 13-1 using a Leroy set and ink. Use appropriate pen sizes and line thicknesses.
appropriate pen sizes and line thicknesses, representing pipelines darker and thicker than equipment.
2. Redo Fig. 13-2.
3. Draw in ink the reactor P&ID shown in Fig. 13-11. Use a large sheet of paper in order to accommodate all lines, equipment, and lettering.
4. Redo the turbine P&ID shown in Fig. 13-15. Use a large sheet of paper in order to accommodate all lines, equipment, and lettering.
5. On an appropriate size of paper, draw to scale the equipment placement drawing of the nuclear power plant (Fig. 13-12).
6. Draw one front-section view of the nuclear containment building in Fig. 13-16 and one plan and elevation of the same project.
7. Redo all the drawings provided in Fig. 13-16 of the nuclear power plant containment building as a class project or group project. Drawings should be done approximately two times the size in the text or to an appropriate scale. Most of the dimensions must be scaled from the drawings provided.
8. As a class project, build a model of the nuclear power plant containment building in Fig. 13-16. *Note:* This project should be done after the drawings have been completed by a group or individual so that there is sufficient background and knowledge of the project.
9. Using Fig. 13-17, redraw the pipe support detail. Reposition elevation AA so that it appears as an auxiliary view of the plan view.
10. Redraw Fig. 13-18. Attempt to reposition the section views in a more logical manner. Since there is no bill of materials for this figure, prepare a complete list of materials by relying on the information provided in Chapter 8.

Church of the Valley in San Ramon, California—one of the first solar-heated churches in the country. Designed by Haridson, Komatsu, Associates of San Francisco. The design for this complex of religious buildings incorporates a comprehensive system of energy conservation in heating, cooling, lighting, and ventilation. (Courtesy Haridson, Komatsu, Associates, San Francisco)

14 SOLAR PIPING

Solar energy has been with us since the beginning of time. In fact, it was the first source of energy ever used. The earliest use of the sun's energy was for heating and cooling living spaces. Solar heating and cooling are based on simple principles: In the summer, the sun appears high in the sky and its rays fall almost vertically to earth; in the winter, the sun appears lower on the horizon and its rays are slanted and less intense. This basic behavior of the sun's path has been exploited for millennia by different cultures all over the world.

SOLAR ENERGY—PAST AND PRESENT

Dwellings have been constructed throughout history utilizing the sun's energy to heat and cool the structure. In many areas, solar energy was the only source of heating. Cultures as widely separated as those of the Middle East and the American Indian have relied on solar movement for heating and cooling their houses. The Pueblo structures of the southwestern United States and many of the adobelike structures of the Middle East are very similar in materials and orientation to solar movement. The Greeks designed entire cities on an east-west axis so that the long sides of structures were exposed to the south. The Romans designed many of their structures with south-facing windows which provided not only sunlight but heat.

Many solar devices were produced during the Renaissance, though few were actually put to serious use. In the latter part of the eighteenth century, solar furnaces capable of melting copper, iron, and other metals were developed. These furnaces were used throughout the Middle East and Europe; some produced temperatures close to 3000°F (1648°C). These early attempts to harness the sun's energy laid the groundwork for present-day development.

In the early 1900s, the concept of the flat-plate collector was developed. This efficient device intercepts, collects, and transfers solar radiation over a horizontal surface. (Earlier solar devices focused the sun's rays to a single point or line where the energy would then be collected and transferred.) Flat-plate collectors (Fig. 14-1) put solar power for domestic uses within reach of the common person, being less expensive, simpler, easier to construct out of available materials, and also more effective or overcast days. Because coal, natural gas, and electricity were expensive at the turn of the century, solar power began to make headway in the southern portions of the country—especially Florida, where 50,000 to 60,000 solar units were in use, and southern California. In both places, sunlight is plentiful and weather conditions are adverse for only a minimal portion of the year. As fossil fuels became increasingly cheaper, however, solar energy lost out. By the 1950s, there were virtually no solar devices in use.

FIG. 14-1 Solar water heating system for an 18-unit apartment building in Oakland, Ca. by Sunworks Collectors, designed and installed by The Solar Center, San Francisco. This retrofit provides approximately 60 percent of the building's annual hot water energy load. (Courtesy The Solar Center, San Francisco)

Now, as history does an about-face, fossil fuels, with their increasing scarcity, expense, and environmental danger, are losing favor and solar advocates are having their finest hour. With new research and development perfecting many of the already efficient devices, the next age, instead of being referred to as the nuclear age, may well be termed the Solar Age. Solar energy is being developed much more rapidly abroad because of the scarcity of fossil fuels and their effect on the environment in other countries. Japan, Israel, Australia, France, the Soviet Union, and India all have been active in the development and utilization of solar energy. Humankind is thus returning to the only viable, efficient, inexpensive, renewable, and constant source of energy that has ever existed—the sun.

THE SOLAR SOLUTION

Solar power represents one solution to the energy shortage that has plagued the world in recent years. In Chapters 11, 12, and 13 we examined conventional energy sources and the nuclear alternative. At present, these forms of energy are of paramount importance to the world at large. Solar power represents a possible alternative which is not as intensive as other forms of power generation and can be designed to fit local domestic energy needs at a favorable cost-effective rate when compared to nuclear and conventional power plants. Although solar power at present cannot generate the large quantities of electricity demanded by an industrial nation, the systems that are presently available can diminish our dependence on fossil fuels and nuclear energy.

Solar power cannot be compared to nuclear, petrochemical, and fossil fuel power as far as the size and quantity of components and piping. In general, solar systems deal primarily with piping 2 or 3 in. (5.08 cm - 7.62 cm) in diameter and screwed or soldered fittings and valves. Nevertheless, solar power involves many of the basic items found in other systems—such as piping or tubing in its many different materials (copper, aluminum, steel); fittings, flanges, and various connections (welded, soldered, screwed); valves ranging from control to simple gate, globe, and check valves; and instrumentation, including pressure and temperature gauges. Pumps are also of primary importance to many solar energy systems, as are vessels and tanks for heat exchange units and heat storage components. Thus many of the components of solar energy systems are the same as those discussed elsewhere in the book for heavy industry. Since temperature and pressure in a typical solar-powered unit are low, the solar energy system is classified primarily as a domestic and commercial plumbing system, not a heavy industrial piping system. The future, however, will bring far greater use of solar power and more sophisticated systems.

At present, solar energy systems are used primarily in houses, apartments, and some commercial buildings for water heating and space heating and cooling. The flow diagrams in this chapter show similarities to piping systems presented elsewhere in the book, though the constituent parts are much simpler and cheaper. The student will use much of the information provided in the early chapters of the book, especially Chapters 1 through 4, when studying this solar chapter, since the components are virtually the same for all piping systems. Though their level of technology, cost, and type of components may vary, their basic design and function are the same.

Because solar energy systems have large components, such as collectors and storage tanks, they are highly architectural in nature. In fact, architectural firms generally purchase ready-made solar energy components and incorporate them into the design of their project. Therefore much of the system drawing of solar energy units is done by architectural firms, not industrial construction firms. Many manufacturers of components for solar units design, lay out, and prepare the mechanical drawings required for the industrial production of the component. Knowledge of mechanical and piping drafting is essential for these purposes. In the future, many jobs will be created by the solar industry, not only for drafters but in other areas of construction and manufacturing as well.

Solar energy represents one of the most promising means for heating and cooling domestic and commercial establishments—mainly because the basic technology has existed since the turn of the century. Present energy shortages are stimulating governmental and private research, development, and demonstration of solar energy projects. The acceptance of solar power will bring many benefits, not only in the reduction of our dependence on fossil fuel and nuclear power, but also in the decrease of pollution to atmosphere, ground, and water, the by-product of the technological-industrial state.

Solar power is available to virtually everyone on earth, though in the warmer climates its intensity and efficiency are greatly increased. No solar energy system operates at 100 percent efficiency. Though 50 to 60 percent efficiency is considered good at present, the future will bring a much higher return with increased research and development. Solar energy is fast becoming cost-effective in comparison to fossil fuels, which have increased in price dramatically in the last few years. Because of the many emerging industrial nations throughout the world, the demand for fossil fuels is not likely to decrease. Solar power can provide an increasingly effective alternative in many areas—such as home heating, cooling, and limited electrical generation—

thereby freeing fossil fuels for areas where technology cannot readily change from its present course.

TYPES OF SYSTEMS

Solar energy flows through the earth's region of the solar system at about 1 kilowatt per square meter, if that square meter is perpendicular to the sun's rays. The atmosphere of the earth filters out most of the infrared and ultraviolet light, and we are lucky that it does. The ability to stop infrared light stabilizes the temperature of the atmosphere at the surface of the earth, and the ability to stop ultraviolet light protects us from genetic disturbance. Still, most of the sun's inflowing energy is in the visible range and the amount lost to us is quite small.

The greatest losses of solar energy are due to atmospheric conditions and the *angle of incidence* of the sun's rays striking the atmosphere. On a hazy day, when shadows are not sharp, one has already lost over half the solar energy available through a clear atmosphere. A cloudy day is useless for solar energy collection. With a clear atmosphere and the sun directly overhead, the atmospheric loss of solar energy is about 14 percent. This loss rises to 28 percent when the sun is 30 degrees above the horizon at noon in midwinter. Thus year-round solar power operation would be most effective in the southern parts of the country—especially in the Southwest, where the atmosphere is clear and drier than elsewhere.

Apart from atmospheric losses, solar power is further diluted because sunlight arrives at an angle from the vertical. If the sunlight is 30 degrees from the horizon (60 degrees from the vertical), half of its intensity is lost over a given area of flat earth surface. Even though solar power collectors are usually tilted to receive the sun's rays directly perpendicular to the collector, the total amount of energy received, say, per square mile of the earth's surface is affected. Nevertheless, with reasonably direct sunlight and a reasonably clear atmosphere, several hundred watts of power strike each square meter. A collector with 550 m on a side can therefore collect 100 megawatts—enough power for a fairly large city if it could be converted to electricity at 100 percent efficiency. Typical conversion efficiencies are low, however, either by direct photocell conversion or by using the heat to power a steam cycle. Both methods may be drastically improved within a few years, but the first major use of solar power will be simply for heating water or other line contents either for domestic heating or for preliminary heating supplemented by heat from other sources.

This brings us to the basic division of types of solar energy used for heating. One type depends upon the degree of sunlight concentration at the collector; the other depends upon the means of heat transfer in the system.

Focused and Unfocused Systems

A simple lens can produce a spot of intense heat sufficient to set fire to wood. The same can be done with mirrors, particularly parabolic mirrors which are aimed at the sun and therefore must be continually changed in orientation to follow the sun across the sky. The category of *focused systems* includes simple solar ovens and solar furnaces, the largest of which can produce a temperature of 3000°F (1600°C). These furnaces produce a spot of heat for special tasks, but since they are not generally used to heat piped liquids, they are of little concern to us here. It is important to note, however, that high temperatures are available only with extreme focusing.

The next degree of focusing is obtained with a channel mirror, which can be pictured as a piece of pipe cut in half lengthwise and silvered to make a mirror on the inside (see Fig. 14-2). Down the center of the channel runs a small black pipe. A more expensive type, but with greater concentration, has a parabolic cross-section; this design must also be rotated so that the axis of the parabola follows the sun. Ideally, both these designs are also turned horizontally to face the sun's rays directly. In such an arrangement, sufficient temperatures can be produced to make steam directly in the central pipe. This saturated steam must be further heated to drive a turbine. Devices like these can be made to cover large areas and produce great amounts of steam for low-pressure turbines. The channel, whether it has a circular or a parabolic cross-section, can be made of polished sheet metal with suitable support from below. If it is made in great lengths, the central steam pipe must be supported at intervals. Efficiency can be increased if the central pipe is painted black (to absorb heat) on the side toward the reflector channel and silver (to reduce heat loss to the atmosphere by radiation) on the side toward the sun.

There are *unfocused systems* also. The simplest solar collector is the flat panel, which has no appreciable focus of its own but works by a variation of the *greenhouse effect* (see Fig. 14-3). The key to this effect is a clear glass (or special plastic) layer through which sunlight passes easily, but which blocks the flow of heat from materials warmed by the sunlight. The materials are usually painted flat black to absorb more sunlight. The flat-plate collector cannot produce steam, but it can heat water which can be used to stabilize the temperature of a building. The building can be further heated by other means if necessary. For stabilizing building temperature, the hot water is stored in large tanks for use at night while the storing of heat during the day prevents the building from becoming too warm. Such a unit is best conceived as aiding the conventional heating/cooling unit, for there are times when sunlight will not be available.

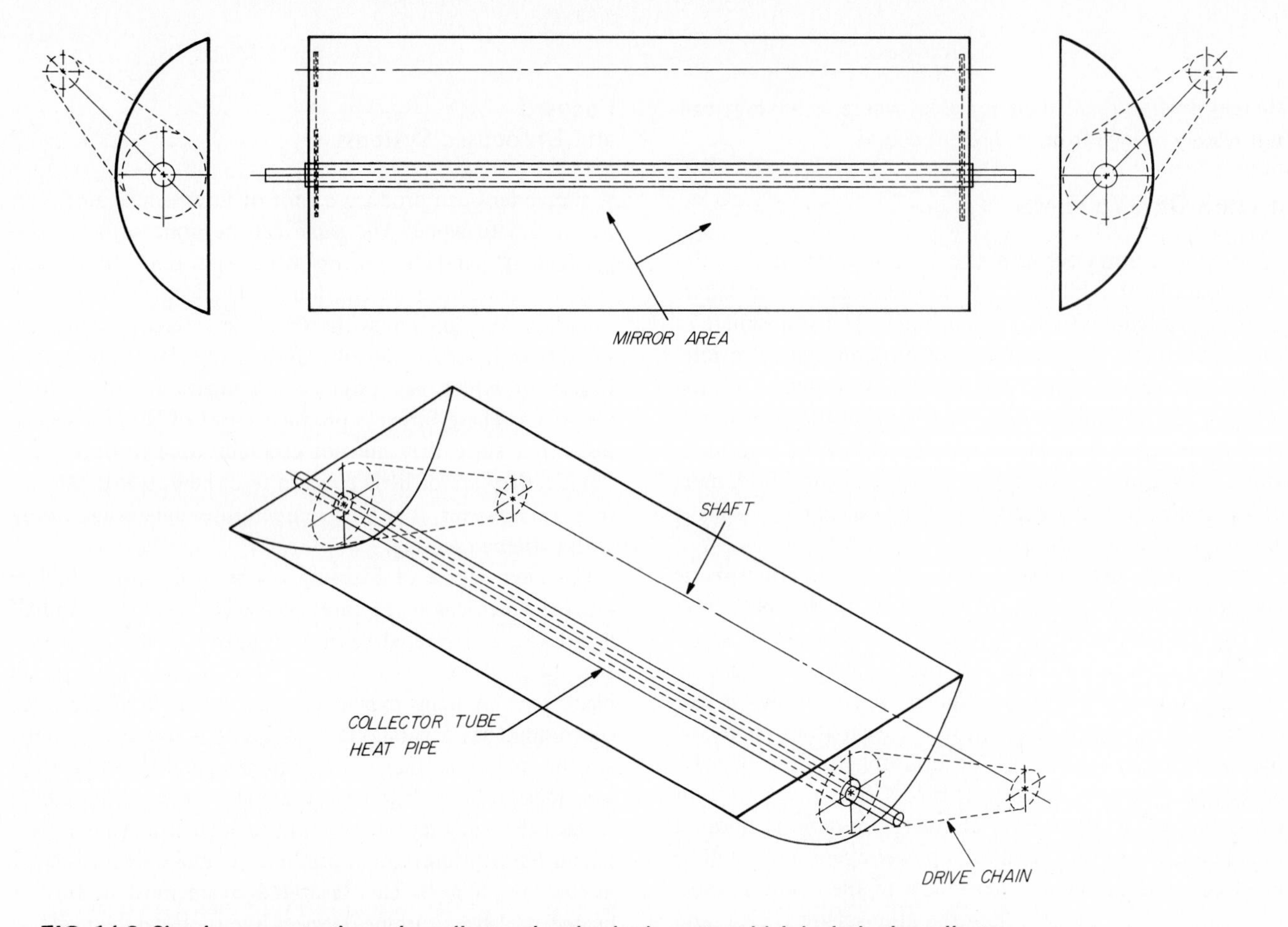

FIG. 14-2 Simple concentrating solar collector showing basic parts, which include the collector tube (pipe), the concentrating mirror, and a series of pulleys and drives. Concentrating solar collectors focus the rays of the sun on the collector tube to heat the line fluid (along an axis), which is then piped to the heat exchange unit.

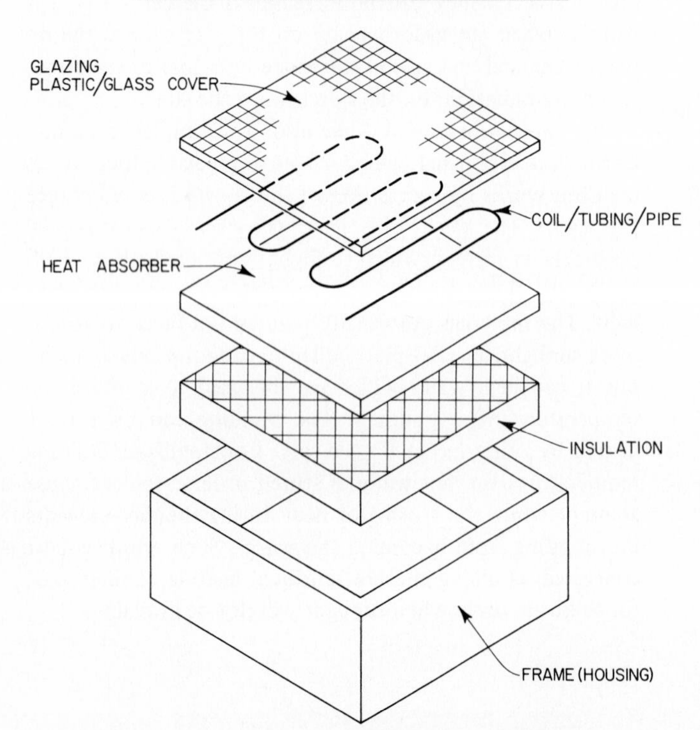

FIG. 14-3 Exploded view of a flat-plate solar collector—the basic component for all collectors or solar panels. The unit catches the rays of the sun and uses the generated heat to increase the temperature of water or air, which passes through the tubing. Usually a number of solar collectors are required to provide sufficient energy, and these are grouped together in a series or separately in a bank, or array, positioned on the roof or at ground level.

The cover plate, of plastic or glass, helps trap the sun's heat. After the heat is transferred to the tubing, the water is heated and then carried to other portions of the system for use. The heat absorber and tubing can also be incorporated in one unit. A heat absorber of metal or plastic with a black surface assures minimum heat loss by radiation. It is one of the main elements of the solar collector panel. Mineral wool or foam insulation helps to prevent heat from escaping through the sides and back of the collector panel. In a flat-plate panel, the collector does not concentrate the sunlight. (One that does so is shown in Fig. 14-2.)

Active and Passive Systems

A solar energy system is said to be *passive* if the water flows by means of convection—the natural tendency of hot water to rise as it expands. The hot water rises to a storage area *above* the collectors and can then be run to where it is needed. This system will not work for temperature stabilization, for the hot water will not flow downward to a space which is to be warmed. Almost all solar systems are therefore *active*, which means that hot water is pumped from one location to another. A passive system uses the natural orientation of the building: large south-facing windows, large overhangs to prevent the summer sun from heating the building, natural architecture designed to utilize environmental temperature changes. An active system uses piping, components, pumps, and collectors.

Large systems are much more economical than small ones. Since large volumes of hot water (or any other medium) lose heat much more slowly than small ones, heat should be stored in the largest feasible tanks as spherical as possible. Probably the cylinder, whose height is equal to its diameter, is the most cost-effective shape. Moreover, the technological breakthrough that is expected will probably be made in large facilities, for large systems will be competitive sooner. At present, solar power equipment is still too expensive in relation to other energy sources and sunshine is too unreliable in most areas for solar power to be the sole energy source for industrial processes. However, for various reasons most solar manufacturing has been oriented toward home heating, cooling, and other domestic services. There are a great many solar energy designs using collectors as their primary device in conjunction with passive features such as solar-oriented windows and natural architecture.

The prime consideration in all solar power systems is the ability to collect direct, reflected, or diffused radiation from the sun. *Direct radiation* is undeflected, parallel, solar rays reaching the building or collector with maximum intensity. *Diffused radiation* is multidirectional, is not as intense, and has been deflected and redirected by atmospheric conditions such as clouds, moisture, and pollution. *Reflected radiation* has been reflected by objects, including the earth, walls, and buildings. All these sources are important to the solar designer, though direct radiation is of primary concern to those interested in solar power systems incorporating collectors and piping. Diffused and reflected radiation, though helping to stabilize and increase the temperature surrounding the reflector and buildings, does not provide sufficient heat whereas direct radiation can heat solar collectors and furnaces.

The amount of solar radiation that is available in an area depends on the orientation of the sun, the time of day, the time of year, and atmospheric conditions, including the type of moisture and pollution content. Clear skies provide the maximum solar radiation in any of its three forms—especially direct solar radiation, the primary ingredient in all solar power systems. As has been stated, the southern portions of the continent have far more direct solar radiation than the northern parts. In the winter, much of the solar radiation is diminished because of the sun's angle of incidence and atmospheric conditions, along with the higher temperatures needed to operate the solar equipment because of ambient temperatures. Reflected and diffused radiation are important during clear periods in the north at wintertime because they help increase the ambient temperature which in turn helps the efficiency of the solar energy system. Another contribution to the heating of a structure is made by air movement. The surrounding temperature of a structure, or the ambient temperature, is affected by the weather, especially the wind. One of the primary concerns for the solar architect is to design systems using natural foliage, ground, and structures to provide windbreaks, thereby increasing the amount of heat derived from convection or lowered air movement in the area of the structure. Utilizing the earth's own stored heat can also help, considering the fact that the temperature becomes considerably warmer 6 ft (1.8 m) below the surface because of retained heat.

The main consideration for the piping architectural drafter is the collection of solar radiation to provide heat for a solar energy device, such as a collector or photovoltaic solar cell. In this chapter, we are primarily concerned with the development of solar collectors and piping systems that convey, distribute, and store heat trapped by the solar collector in the liquid medium or air medium. Solar collectors and solar architectural designs must take into account the ability of a system to collect the available solar radiation, to store the heat at night and in cloudy weather, and to distribute heat or hot water efficiently throughout the area required. The architecture itself must be able to retain heat by use of special glasses, insulation, and environmental design. All the various components available for piping systems—pumps, fans, heat exchangers, vessels, storage devices, valves, control systems—may be used in active systems. Active systems also incorporate aspects of passive systems which will help maximize efficiency, such as architectural considerations, insulation, and the heat absorption and retention qualities of material. The ability of a system to collect, circulate, distribute, and store heat is determined not only by site, structure, and environment, but also by the quality and type of system designed and installed.

Another consideration for the solar architect, designer, and installer is the system's ability to cool or at least prevent overheating of the structure during warm weather. Many environmental and architectural devices can be employed for cooling, such as shading, special ventilation ducts, and fans. In designs like the one shown on page 526, large overhangs shade the windows in summer when the sun is at a higher angle of incidence. Moreover, one may simply draw off the collected heat from roof collectors and store it for cooler times.

Heliochemical, Heliothermal, and Helioelectrical Systems

Three basic types of solar systems are being explored today: heliochemical, heliothermal, and helioelectrical. (The word *helios* is Greek for sun.) All provide variations on a theme for the utilization of the sun's energy in meeting today's energy needs. *Helioelectrical* designs use solar cells to collect solar radiation and convert it directly to electrical current. These devices, which have powered many of the space exploration vehicles, are at present extremely expensive and hard to manufacture, since they have to be assembled primarily by hand. Silicon solar cells and similar inventions may eventually be perfected to the degree that direct conversion of sunlight to electricity becomes cost-effective in comparison to fossil fuels. An example of a *heliochemical* process is photosynthesis, where sunlight causes carbon dioxide, water, and nutrients from the soil to nourish plants and also release oxygen, which is necessary not only for all life but also for normal burning of fuels. All fossil fuels are the end products of photosynthesis and can be used because photosynthesis continues to provide oxygen. Helioelectrical devices are outside the concern of this book, as are heliochemical ones.

The *heliothermal* process is our primary concern here. All solar collector devices fall into this category. In general, heliothermal energy or heat is developed by the heat absorber surface, which is usually black. Black surfaces can absorb more heat and transfer it to a working liquid. By pumping the hot liquid, we can redistribute the heat throughout an area. All the collector devices shown in this chapter are heliothermal. The flat-plate collector in Fig. 14-3 is a perfect example of a heliothermal device with a cover sheet, heat absorber area, coil or tubing, insulation, and housing designed to trap solar radiation and absorb solar heat, usually in the range of 90°F to 300°F (32.2°C to 148.8°C). This heat is transferred to the working fluid, which is circulated throughout the structure or piped to a collection tank for later distribution. Heliothermal devices can also be used to generate electricity by the production of steam if the sunlight is focused to create higher temperatures.

The temperature difference between different depths of ocean currents is also a solar phenomenon that may be harnessed eventually to drive a heat engine, and therefore a generator. Wave machines are also considered solar devices, as are wind mills and turbines, which derive their power from solar-driven wind.

One of the primary problems involved in the use of solar energy on a large scale—especially for anything larger than a single dwelling or apartment unit—is financing. The highly concentrated capital of energy industries presently operating in the United States will make it very hard to convert to individually operated or community-owned solar power systems. The energy conglomerates, which have a virtual monopoly on all energy production in the United States, will find it very hard to accept the loss of money to individually owned power sources, especially ones which will eventually pay for themselves and be nearly free after the initial investment has been recovered. Solar energy offers advantages with respect to many forms of power because users can control their own destiny in the utilization of power to meet their needs. If solar power is to become a viable energy source in the world, subsidies such as those given to the nuclear and conventional power producers must also be given to the inventors, designers, and producers of solar equipment. At present such industries as the power industry account for many of the positions available to the piping drafter. In the future, with the spread of solar power, we may see greatly expanded demands for locally produced energy systems—and solar power piping—which will supplement and eventually supplant the need for nuclear and fossil fuel power.

SOLAR HEATING AND COOLING

Passive systems are direct systems, since they use the sun's energy directly without intervening components and piping. Indirect systems are synonymous with active systems, which require some sort of mechanical energy to move the heated substance from the collector to the storage area and then circulate the heated fluid. An exception to this classification would be a solar hot water heater for washing which may not require a pump and is therefore passive (but still indirect). At the present time it is necessary to provide backup systems—which can be anything from the traditional oil-fired furnace to the pot-bellied stove, depending on the heating needs, the climate, and atmospheric conditions. In general, 60 to 95 percent of domestic and also many commercial establishments can be heated directly from solar radiation by means of both active and passive or indirect and direct systems in combination.

Active solar systems, which concern us most in this chapter, must provide not only the collection of solar energy, but storage, circulation, and controlling mechanisms. Collection involves the absorption of heat by a variety of solar collectors and devices—in other words, heliothermal mechanisms, both flat-plate and concentrated types. Storage methods depend on the amount of heat required over a period of time that includes cold weather, darkness, and atmospheric distrubances such as rain and pollution. Vessels, rocks, and bins can be integrated into storage systems. Circulation systems are extremely important since they move the heated medium throughout the system from the collector to the storage area and then to the areas which require heating or hot water.

A variety of controls are required for the active solar heating system, including pumps, temperature measuring devices, thermostats, temperature and pressure gauges, and remote-operated valves. Component systems are made up of storage, collection, and distribution components tied together by circulation devices and overseen by a system of controls. Controls or instrumentation are similar to those found in other piping systems. Thermostats are nothing more than temperature controllers, and other controls must be provided to move, redirect, and shut down the system when necessary. Pneumatic controls are used in elaborate systems. Temperature elements, pressure gauges, remote-operated valves, control valves—all are found in solar energy systems. The flow diagrams in this chapter show some of the possibilities.

This chapter deals primarily with liquid systems in which a liquid is heated from a collector and circulated throughout the system from storage to distribution points. Liquid systems can be used for heating swimming pools, heating hot water for domestic purposes, and simply heating. It has been found that solar energy systems utilizing air instead of liquid are somewhat more efficient and cheaper in small structures. The larger the structure, the more efficient the liquid.

One use of solar power in the future will become more pronounced as development is triggered by the rising cost of fossil fuels and electricity: the use of solar power for cooling. Nocturnal cooling is one possibility. By exposing the solar collector to nighttime temperatures and circulating the working fluids, the house can be cooled before the sun increases the temperature of the collector. One of the more promising (though less developed and therefore expensive) variations of solar cooling is the use of absorption refrigeration units. These units, in conjunction with heat pumps, make it possible to draw off heat from the air in a building. An absorption chiller uses the principle of refrigeration to exchange heat, and in the process the absorption liquid becomes chilled. This chilled substance can be used as a refrigerant through a complicated redistribution of the cooled fluid. Solar collectors supply the heat for this refrigeration process. Absorption chillers are very complicated and expensive. All chilling devices at present convert heat energy into mechanical energy by altering the state of circulating fluid from a liquid to a gas.

SITE PLANNING

One of the most important aspects of solar architecture and solar design, both domestic and commercial, is the site planning that must be done before the project can begin. Solar design factors, which must be considered not only at the job site but in the general area, include vegetation, topography, climate, geology, orientation of the property to the sun, and surrounding structures (Fig. 14-4). Every site for solar power is unique because of its orientation to the sun and other variables in the environment. Early consideration of these items will yield cost-effective placement of the equipment and efficient design.

The primary factor in all solar architecture is the location of the building. The geographic conditions of the area must be taken into account—especially the wind flow patterns, the general topography, the seasonal path of the sun with regard to the site (Fig. 14-4), and the temperature differential under normal and adverse wind conditions. The topographic data for the site must include the degree of incline and the presence of slopes, hills, or mountains in the area.

South-facing slopes provide for maximum solar exposure under almost all conditions (in the northern hemisphere); north-facing slopes provide the least possibility for the use of solar equipment. East-facing properties are exposed primarily in the morning hours; west-facing properties get the afternoon sun. All this must be taken into account when determining the feasibility, location, and orientation of solar equipment and the solar dwelling. Geological studies may also be required to determine the type of earth and rock on a particular surface. Certain soil types present engineering limitations and also determine the vegetation that may be supported at a site. The existing vegetation and the possibility of establishing new vegetation which would help solar heating and cooling devices (by acting, for example, as windbreaks) must be taken into consideration, as should also the location and size of vegetation which may interfere with solar collection devices.

Climate is of primary importance. The hours of sunlight per year, monthly, daily, and seasonal changes—all have a definite effect on the feasibility of a solar structure. The amount of winter versus summer sun, the wind currents in the general area, and the weather to which the equipment will be exposed (rain, snow, sleet, hail) are determining factors also. Fog and pollution may cause as many problems as natural weather variations, especially in areas such as southern California and the coastal zones. The area stretching from central California to Florida is the primary solar region in the United States. In the mountainous and northern portions of the country, long, bitter winters severely limit the total solar capacity. Even under these circumstances, an overall conversion of the United States to solar power could easily provide from 60 to 75 percent of the domestic heating, cooling, and other energy uses in the Sun Belt and 30 to 40 percent in most of the rest of the country.

Church of the Valley in San Ramon, California shown in Fig. 14-4 and on page 526 is an excellent example of site planning and the use of both active and passive solar energy design.

This building project, with day-care facilities, is intended to serve the needs of community groups seven days a week, as well as the needs of the congregation on Sunday. A pioneer in the area of energy conservation, it is intended to serve as a model for the construction activity taking place in surrounding communities.

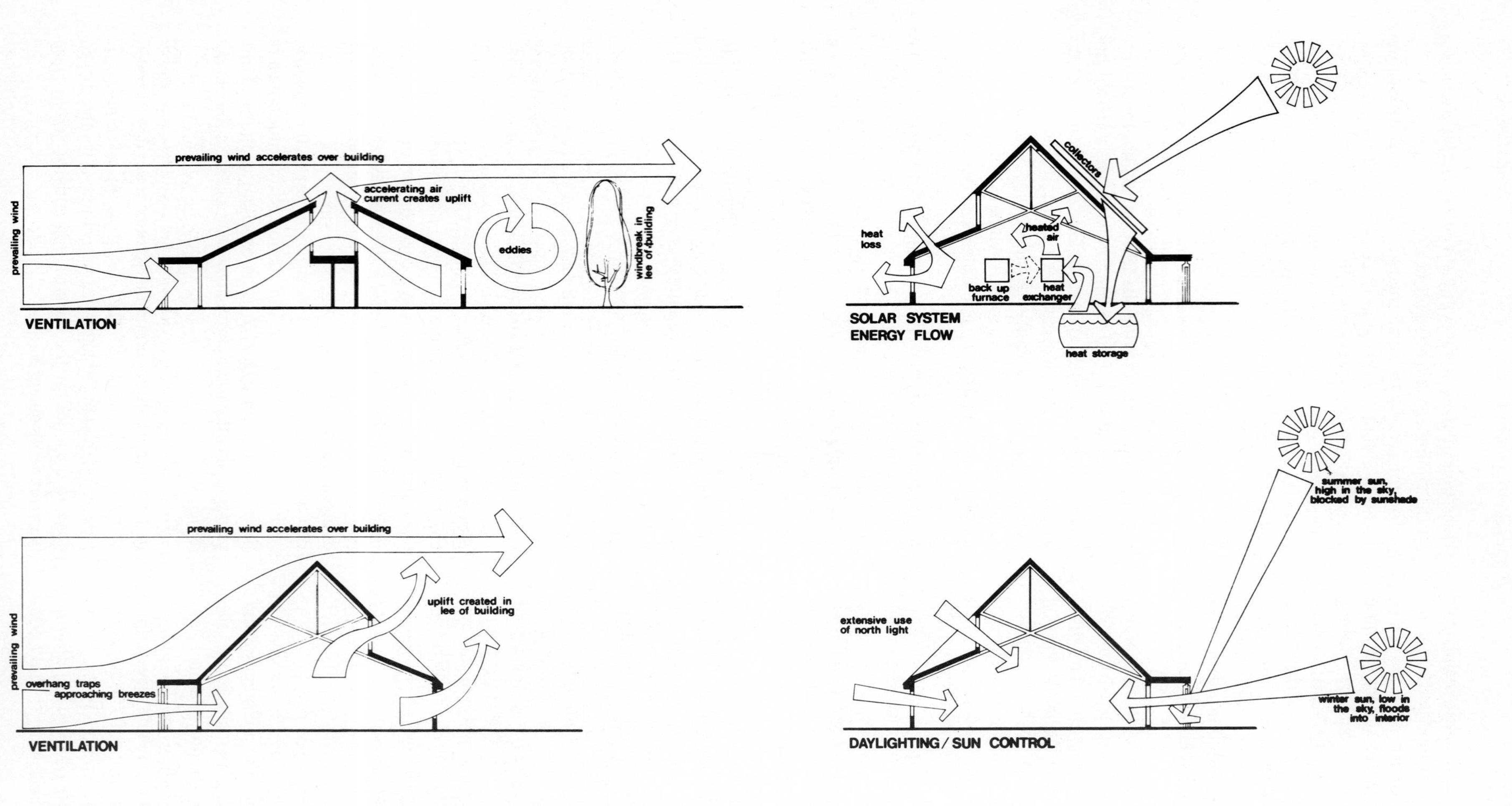

ENERGY SYSTEMS: NATURAL AND MECHANICAL

FIG. 14-4 Schematic representation of both natural and mechanical energy systems for the Church of the Valley shown on page 516. (Courtesy Haridson, Komatsu, Associates, San Francisco)

The design of the building grew out of a careful study of energy use, including lighting, cooling, ventilating, and heating. Heat will be provided by a hydronic solar energy system comprising 1500 sq ft of copper solar collector, two underground tanks to store heat, and a sophisticated electronic control system. Gas-fired furnaces will provide backup heat during extended periods of cloudy weather. Cooling will be provided entirely by natural means. The buildings have been oriented to avoid the hot summer sun while admitting most of the desired winter sun. Buildings have been placed to channel the prevailing summer breezes to the spaces which need cooling the most; overhangs and clerestory windows will accelerate the breeze to the buildings' interiors. The control system which governs the heating system in winter will operate during the summer as well; in the early morning hours fans will circulate cool air through the building interior, storing cold in the structure's mass, to enable it to stay cool naturally during the day. The most intensively used spaces are lit almost entirely by natural light.

The complex of buildings has been sited to form a natural part of the surrounding landscape of rolling hills and meadows. The buildings are grouped around a courtyard surrounded by trees; a stream flows through the middle.

COLLECTORS

The heart of all active solar power systems is the collector. This device, as we have seen, collects solar radiation and absorbs it as heat in a working fluid which is then transferred to different parts of the dwelling for different uses. Solar collectors are commercially available and are designed for economic, long-lasting, customized use for various solar energy systems. Along with the collection of heat, of primary importance is its storage and distribution. The collector must be able to absorb sufficient heat for practical purposes. Commercial solar collectors are designed as total systems or individual units which can be altered according to the dwelling, climate, and site.

The two most common uses of solar energy are for heating water and for heating and cooling a structure. Conventional boilers, furnaces, and electric heaters have a far greater ability to create high temperatures than the typical solar heating system, which operates with a maximum temperature range of 90°F to 200°F (32.2°C to 93.3°C). Therefore the system must be designed to use this lower temperature range efficiently. The storage, distribution, and collector devices suitable for a particular climate and desired power needs must all be taken into account in the design stages of the solar energy system. The ability to redirect heat, to store it, and to shut off the system both automatically and manually requires that the solar power system must be built with a series of gauges, thermostats, temperature devices, and automatically controlled valves.

Flat-Plate Collectors

Of primary concern in this chapter is the nonfocusing collector, or flat-plate collector, where the absorber surface is equal to the area upon which sunlight falls. (In the concentrating or focusing collector, the absorbing surface is smaller than the aperture for incident radiation.) In some cases, these two types are combined by means of a semiconcentrating collector in the form of a half shell which concentrates the sun's energy on a long black plate. Nevertheless, most research, development, and distribution have been confined to flat-plate collectors. Certainly this type is more cost-effective at the present stage of development and is the one which most solar energy enthusiasts in the architectural design field are specifying. Thermal energy is captured by the solar collector and transferred to a working gas or liquid designed as an integral part of the collector.

Figure 14-3 shows an exploded view of the flat-plate solar collector. The primary ingredients of flat-plate collectors are transparent sheets or glazing which help to reduce air current cooling and also trap radiation within the confines of the collector. Many types of glass and plastic have been designed for use as a cover plate. The common design needs are a high transmittance of short-wave (visible) solar radiation coupled with a high absorption of long-wave (infrared) radiation. The cover sheet, and therefore the collector surface, is usually designed so that it is perpendicular to the sun's rays—that is, the angle of incidence is as close to 90 degrees as possible. When the angle of incidence is less than 30 degrees, the loss of radiation by reflection can be greater than the amount of radiation collected by the solar device.

The next layer of the solar collector is the absorber section, which is usually designed for the insertion of tubing or pipes which carry the working fluid or air. Absorber materials are designed for high solar absorption and low emissivity to prevent reradiating the solar energy. High thermal conductivity is another essential property of the absorber. To meet both needs, absorber surfaces are coated with dark paint increasing the ability to absorb solar radiation. Special coatings have also been developed to minimize emissivity losses and increase the amount of solar absorption available.

Flat-plate collectors designed for domestic or commercial use are available from commercial manufacturers. These collectors are easier to fabricate, install, and maintain than other designs. Moreover, they can be designed directly into a building at ground or roof level and provided with brackets so their angle of incidence can be adjusted for climatic and seasonal changes. Flat-plate collectors are also designed to collect diffused as well as direct solar radiation and can attain temperatures of 250°F to 300°F (121.1°C to 148.8°C). They can be designed in parallel, in series, or individually, with components to provide space heating, cooling, and water heating for domestic purposes. The cover sheet and absorption plate, tubing, insulation, and framing shown in Fig. 14-3 reveal the simple nature of this component.

Flat-plate collectors can be air types or liquid types. The air type heats air as it passes through openings in the absorber portion of the collector; the air is then blown through ducting to the areas of use or into a heat storage bin. Liquid-type collectors, which are more prevalent, heat a working fluid such as antifreeze, oil, or water as it passes through the absorber section conveyed by tubing; the fluid is then pumped to a storage area, to a heat exchanger, or directly to areas of usage. Both types of collectors are designed to be controlled automatically, depending on the temperature needs of the structure and the time of day. For example, at night the collector system loop will not be activated; only the storage-to-usage circulation loop is used. During the sunny portion of the day, when the collector is actually absorbing heat, its collector loop will be activated and may work by itself or in conjunction with the circulation loop. The collector operates until the temperature of the fluid drawn from the collector drops below that necessary to maintain the temperature in the storage area.

The absorber area is one of the principal aspects of the flat-plate design. The type and color of the absorber dictate its heat absorption capacity based on the material, consistency, texture, and coloring. Black has been used primarily because it absorbs radiation, but it also increases emissivity, which is the reradiation of heat back to the outside. *Emissivity* can be offset by the type of coating and texture on the metal absorber. In no existing flat-plate collector system can the amount of solar radiation entering the device be equal to the amount that is absorbed. A loss of 20 to 60 percent, which varies according to type and climate, is common. A variety of selective surfaces have also been designed; chemical and electrolytic processes and the use of color provide low reflective qualities and decrease emissivity in comparison to the black absorber types.

As the typical flat-plate collector reaches maximum temperature, the air between the cover sheet and the absorber area becomes hot enough to conduct heat back through the cover sheet, thereby creating another source of heat loss that must be dealt with. Cover sheets (also called collector covers or glazing) are designed to reduce heat loss caused by wind cooling by providing an insulating space between the absorber and outside air. Some collector devices do not need cover sheets if climatic conditions are right. Many different plastics have been tried on collectors without much success except for a few exotic and expensive plastics which have properties similar to glass. Fiberglass and thermal-pane plastics have been used with great success because they are unaffected by conditions which distort other plastics at high temperatures.

In many cases, the use of multiple cover sheets is required in order to produce sufficient insulation from wind, provide strength for adverse climatic conditions, increase solar absorption, and reduce the loss of radiation. Multiple levels do, however, decrease the amount of solar heat that can be collected. Flat-plate collectors are designed to be stationary and operate in the realm of 100°F to 200°F (37.7°C to 93.3°C) with a fluid-flow system. Many materials, such as copper, steel, and aluminum, have been utilized for collectors. Collector materials are discussed in greater detail at the end of the chapter.

One of the primary advantages of flat-plate collectors, apart from their easy assembly and installation, is that they can be designed as an integral part of the structure. By designing collectors as part of the roof or walls, it is possible to minimize the distance between the collector and heat storage area—and thus the amount of circulation piping and ducting.

Air-Type Collectors Flat-plate air-type collectors are easier to maintain and are especially useful in areas where freezing and corrosion are associated with liquid collectors. Air can be used directly to heat the area or rock storage bin, avoiding heat transfer units such as economizers that must be used for liquid collector systems. The use of air-type collectors is limited in most cases to space heating; using air to heat water for domestic purposes is inefficient. Moreover, fans must usually be designed into the system, which require the use of electrical power.

Liquid-Type Collectors Piped water collectors and water-trickling collectors are two common types in use at the present time. The water-trickling collector, which was designed and patented by Harry Thomason, utilizes corrugated roofing panels covered with a glass cover sheet. Pipe manifolds distribute the water along the upper edges of the collectors, which are set at a specific angle to provide the necessary angle of incidence and also hold the flow. By establishing a very thin layer of water over the entire surface of the collector, the water absorbs and carries away much of the heat. It is then collected at the bottom of the system where it is carried to a heat storage area or used directly for heating. A variation on this theme is to enclose the water between two layers of metal or other material to inhibit condensation and provide a layer of water of specified thickness. This system is therefore more of a closed system than the original.

The large water surface and collector area necessary for this system make it more expensive than other units. Furthermore, condensation and reradiation and absorption problems plague this design. The large area required for heat absorption also creates freezing problems. Nevertheless, liquid collectors are currently popular because of their ease in manufacture and assembly away from the site.

In the typical system the water, with antifreeze or oil, is pumped through tubing which has been designed to be a portion of, or embedded into, the absorber area. Copper has been used most successfully for this type of system. Aluminum, because of corrosion problems, is somewhat restricted. Steel in its galvanized form has also been designed into some systems because it is less expensive than other materials, but it does not provide the heat absorption and retention properties of copper. All water collector systems require the extensive use of piping, fitting, and valves to transfer the working fluid to the collector through manifolding which can either be built into the collector or designed as part of the structure itself. The collector-to-storage loop is usually separate from the storage-to-distribution loop.

Liquid systems are sometimes designed to provide only domestic hot water for swimming pool, laundry, bathroom, and kitchen. These applications alone reduce dependence on other forms of energy. By reducing the hot water produced by other sources, energy requirements can be reduced 15 to 25 percent for a typical family. This is why solar energy pays for itself in comparison to the high cost of using conventional energy sources to heat hot water for domestic purposes.

It takes far less electrical power to pump the working fluid in a typical liquid system than it takes to move warm air through ducting in a typical air-type collector. The liquid system is the only logical system to use for retrofits on existing structures and for large structures such as commercial buildings, because piping is less expensive than ducting (which also takes up considerably more space). Storage facilities are also smaller for liquid systems in comparison to the hot-air storage bins associated with air-type collector systems. In general, the piping used for flat-plate collectors is small-diameter aluminum, copper, or galvanized steel tubing. Domestic plumbing piping, fittings, and valves can be used. Connections can be flared, socket welded, soldered, or screwed. For amateurs constructing their own solar energy system, screwed fittings provide the easiest method for installation, maintenance, and repair.

Focusing Collectors

Research and development of focusing collectors is receiving a great deal of attention because of their ability to produce much higher temperatures than the typical flat-plate collector. By directing all the available incident sunlight to a specified focal point or axis where the working fluid is concentrated, the focusing collector with either a paraboloid or cylindrical cross-section produces temperatures between 300°F and 1000°F (148°C to 537°C). The problem with focusing collectors is that they must be oriented constantly to the sun and are therefore composed of separate units for focusing, for piping the circulation medium, and for insulating the entire piping system. A highly reflective material must be used on the focusing unit such as aluminized mylar or some sort of mirror device. For low-temperature systems, aluminum or polished metal might be sufficient.

A typical focusing collector with a focal axis is shown in Fig. 14-2. Systems like this one have a black or heat-absorption coating on the focal access piping which enables it to absorb sufficient energy to achieve high temperatures and in some cases produce saturated steam. All piping to and from the unit must be heavily insulated to reduce heat loss. Cover sheets for protection against wind, dust, and other damage are also necessary for this type of system. Of primary importance in this type of collector is the ability to produce higher temperatures than flat-plate collectors. Therefore the standards for the piping system, valves, and instrumentation must reflect this higher temperature/pressure need.

A major drawback of focusing collectors is their automatic rotational mechanisms for continually tracking the sun, which are extremely costly in comparison to flat-plate collector systems. The use of concentrating collectors is primarily relegated to extremely clear, low-pollution, high-sun areas such as the desert or Sun Belt where precipitation is minimal. The use of concentrating collectors to run solar cooling systems is being explored, but they will require further research and development and also the introduction of a more economical unit.

Linear Concentrating Collectors

The linear concentrating collector is a channel-shaped device which enables solar radiation to be focused from the curved reflector onto a pipe at the axis. It is less expensive than the circular or dish-type concentrating collector, which focuses onto a single point instead of an axis. Both produce higher temperatures than flat-plate collectors, and circular concentrating collectors can reach very high temperatures indeed. Linear concentrating devices provide high temperatures with few mechanical devices and thus are economically feasible for large installations. Nevertheless, the flatplate collector is the most economically feasible unit available today for small projects and is therefore of primary concern to the solar architect and drafter.

CIRCULATION SYSTEMS

Two basic methods are used for circulating the working fluid in a solar energy system. One is based on gravity (the *thermosiphon effect*); the other uses mechanical devices such as pumps.

The Thermosiphon Effect

Natural circulation—the thermosiphon effect—is created by differences in density in the working fluid. Since cold fluid is considerably more dense than hot fluid, the warm fluid in a system tends to rise while the cold fluid sinks to the bottom. A naturally circulating collector-to-storage loop can be built into the system if it is possible to place the storage chamber higher than the collector unit. As the working fluid is heated in the collector area, the warm fluid naturally rises into the storage area.

When rooftop collectors are installed, this system of natural convection cannot be used because the storage area is considerably lower than the collector. One way to deal with this problem is to position the collector at ground level and the storage chamber several feet above. The inlet and outlet pipes of the storage vessel are also positioned to take advantage of the stratification of temperature in the storage tank, where cold water sinks to the bottom. Thus the cold water connection is located on the bottom of the tank so that cold water can be transferred to the collector, where it is heated and redeposited into the storage tank above the cold water outlet. This natural circulation is sustained because of the differing densities in the cold downcomer line and the warm collector riser pipeline.

The circulation loop between the distribution points and storage chamber can also use the thermosiphon effect. By locating the storage tank below the heating space, it is possible to draw off the warmer fluids as they naturally rise through the heat distribution area. It is almost impossible to use the thermosiphon method for both the collector storage loop and the heat distribution storage loop, however, because of the elevation differences necessary to construct such a unit. It could be done for a hilltop building with collectors located on the hillside below the building.

Many problems are associated with this system that must be carefully taken into account when designing and installing components. Flow reversal must be prevented by check valves. Trapping devices and draining drip legs should be provided to extract the trapped air and drain the system completely when needed. Friction effects in various lines must also be considered.

Forced-circulation systems require electricity to power pumps or blowers, depending on whether the system uses air or liquid for the distribution of heat. Wherever possible, the partial use of the thermosiphon effect should be considered. To use the thermosiphon effect in the collector-to-storage loop requires installation of the storage vessel a minimum of 2 ft (60.9 cm) above the collector. This distance is sufficient to prevent water from circulating in the wrong direction. The thermosiphon system does not require pumps and can also reduce the number of controls necessary. The thermosiphon system is well suited for heating water for domestic purposes because the storage area can also serve as the heat exchange unit for a coil which carries the domestic water; the two fluids never mix. Moreover, the amount of piping and fittings is greatly reduced when the thermosiphon system is used for a domestic water heater only.

In areas where the sun is fairly consistent, a 4 × 8 ft collector will produce approximately 45 gal of hot water a day, varying from 160°F (71°C) in summer to 120°F (48°C) in winter. The piping will need to be insulated, as will the tank, and it should also be shielded with a sheet jacket. By connecting solar collectors in parallel it is possible to generate more hot water than a single solar unit will provide. Designing the system so that the same collector banks are aligned in series provides preheating for the incoming water. By the time it reaches the second and third collector bank, it is at a very high temperature. The quality and type of collector may have to be varied in such a system.

Circulation Loops

There are two basic circulation loops in the typical solar energy system: the *closed loop* between the collector and heat exchanger and the *open loop* from the heat exchanger to the point of use such as kitchen or bathroom (and sometimes back for reheating). A second closed loop could also provide circulation between the heat exchange unit (or heat storage tank) and the distribution points for heating the living space.

The primary loop between the collector and the storage unit transfers the energy derived from solar radiation and absorbed in the collector working fluid to the storage area, which is usually a heavily insulated vessel or tank. The ambient temperature must be taken into account in order to minimize corrosion in this loop. Moreover, most of this loop is heavily insulated to reduce heat loss. The secondary loop (storage to distribution points) must be designed to distribute stored thermal heat to the living space efficiently. In general, this loop requires more piping because of the need to spread heat over a greater area.

Many of the established practices for distributing heat buildings also apply to solar power units. The circulation fluid is important. Whether it will be air or a liquid is the first consideration. Since we are dealing with active solar units, primarily piping and components, the liquid (or *hydronic*) system is far more important here.

There are quite a few hydronic heat-collecting systems available on the market, and these prefabricated items make the design and installation of solar energy structures much easier than self-constructed variations. One of the chief differences between air and liquid systems is the total amount of space that is required. Liquid systems with their piping require far less space in comparison to the amount of air ducting necessary for hot air solar systems. But hydronic systems require more maintenance to keep them leakproof and prevent damage to the system through freezing and corrosion.

In cold climates, freezing of the circulation medium must be taken into account for all piping loops and may require extensive insulation and the use of antifreeze. Damage due to freezing can be considerable, especially in portions of the loop that are noncirculating during cloudy weather or at night. It may even be necessary to drain the collector on extremely cold days and nights. The formation of snow and ice on the collector should also be avoided if possible. Since the continual circulation of fluids during cold days and nights will actually cause the building to cool down, this is not a good way to keep the system from freezing.

Antifreeze is widely used in solar power systems in cold climates. One problem associated with antifreeze when differing metallic compositions such as steel, copper, and aluminum are utilized is electrolytic corrosion. Since antifreeze undergoes serious decomposition in normal service, various acids are produced which increase the corrosion. Thus solar heating systems must be provided with adequate drainage connections to change and maintain antifreeze in the circulation loops to prevent acid buildup. By turning the storage tank into a heat exchange unit, it is possible to separate the antifreeze/liquid circulation loop from the heat distribution loop (Fig. 14-5). This separation prevents many of the problems commonly associated with hydronic systems with differing metallic compositions. In some cases, an expansion tank will be required because the expansion of heated liquids must be taken into account.

Piping for Hydronic Systems

In general, the piping systems associated with solar collectors are small-diameter (1/4 to 3 in. OD) (.63 cm to 7.6 cm) copper or aluminum tubing. Copper, the most expensive, is still the most popular because of its overall heat conduction and corrosion-resistant properties. Copper tubing is available in varying wall thicknesses; type K, L, and M. Tees, elbows, crosses, and other fittings are available for the typical plumbing piping system. It is also possible to use small-diameter galvanized steel piping based on the schedule numbers discussed in Chapter 1. The use of plastic piping (PVC, ABS, RS, or CPVC) has become popular because it is inexpensive, though the temperature range is much lower than for metallic materials—a maximum of 180°F (82°C) with a pressure rating of 100 psi. The pressure developed in a typical flat-plate hydronic energy solar system is usually a maximum of 20 to 25 psi. By reducing the pressure requirement, the temperature can be extended beyond the normal 180°F (82°C) limit. Manufacturer's suggested use of piping with standard solar collector systems should be consulted when plastic piping is considered.

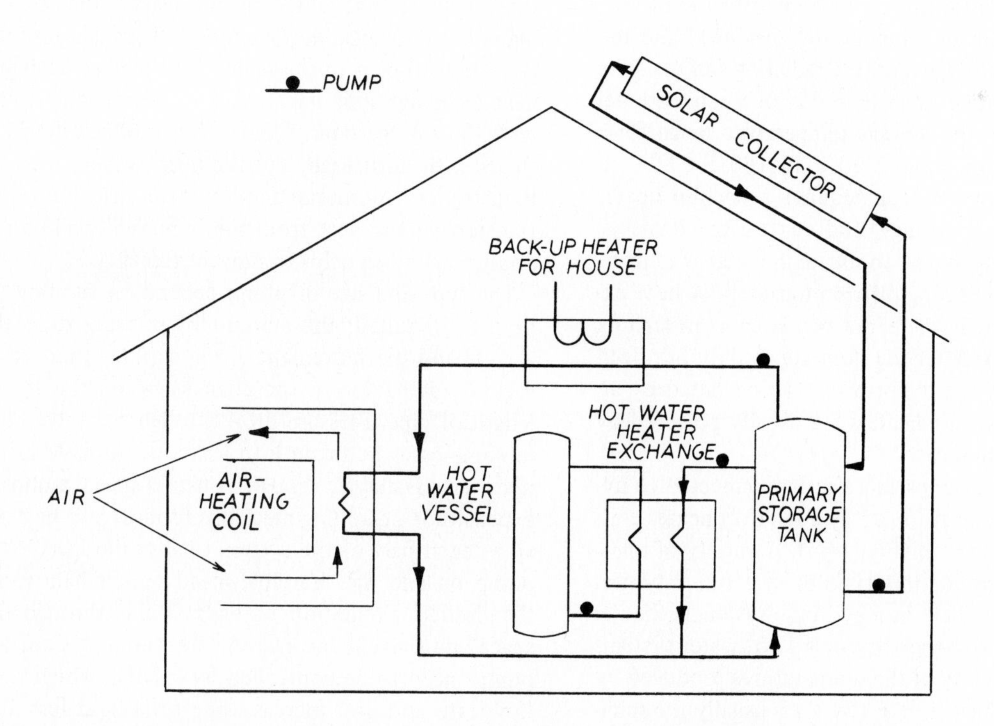

FIG. 14-5 Typical solar heating and cooling schematic for a home. Note the flow of fluid from the solar collector to the primary storage tank through a closed loop that goes from the backup heater for the house to the air heating coil and back into the primary storage tank. A second loop will go from the primary storage tank into the hot water heat exchange vessel, where the heat from the primary storage tank will be conveyed to the hot water vessel pipe loop.

Various other quantities must also be calculated in order to produce adequate flow. The gallons per minute flow requirement must be calculated along with the total amount of heating fluid required in the collector area. This calculation will also determine the dimensions of the circulation system (such as pipe size) and the head pressure developed by the size and height of the storage tank. The longer the piping in the pipe loop, the greater the loss in pressure. Therefore the psi requirement through pumps to compensate for pressure drop or head loss must be calculated. Flow sizes and flow rates for various types of piping must be taken into account, as must valves and change-of-direction fittings such as elbows and tees. In general, these mathematical calculations are simple in comparison to the sophisticated analyses required for nuclear, petrochemical, and other industrial uses of piping.

Distribution of Heat

The circulation loop between the heat exchanger or storage area and the point of heat usage is usually referred to as the *distribution loop.* Kitchen, washing machine, bathroom—the distribution loop must be designed to provide for all the domestic water and heating needs. Distribution loops are generally separated because the space heating loop is usually a closed loop containing antifreeze or some other additive.

Because of the range of temperatures produced and the heat losses from the storage unit, the radiating surface for a typical solar unit must be larger than that of a conventional heating system. Since the average temperature range for a solar power distribution loop is 90°F to 180°F (32°C to 82°C), the domestic water loop requires a backup heater with additional piping, fittings, and control mechanisms, including self-actuating valves, to maintain a higher temperature than the space heating distribution loop. A heat exchanger is also required where the two streams never mix (Figs. 14-6 and 14-7). After the domestic supply has been preheated in this way, its temperature can be raised by an auxiliary heater. Auxiliary heaters are usually powered by electricity, oil, or natural gas.

Warm-air distribution loops can also be connected to hydronic systems where air is forced by a fan over heater coils from the heat storage vessel (Fig. 14-6). A variety of ducts must also be provided for this type of system. A typical method of distributing heat in a hydronic system is the use of baseboard radiators, though baseboard hot-water systems do not operate efficiently at the temperatures produced by a solar power unit. Since 130°F (54°C) is usually the minimum temperature which will be effective for this type of distribution system, it must be redesigned for solar power units—unless high-performance collectors with accompanying heat pump systems can provide sufficiently high temperatures. Retrofit systems (solar power units installed in existing structures) can use baseboard radiators for preheating fluids before the temperature is increased by conventional methods. Figure 14-1 shows a typical retrofit on an apartment building.

Radiant floor or radiant ceiling configurations can also be used to distribute heat. A grid of small-diameter pipes embedded in the structure (usually concrete) circulates the low-temperature working fluid (approximately 120°F) (48°C), which is pumped through pipes so that the heat is absorbed directly in the structure. This heating system can work very well for a majority of structures, though it must be implemented in the design and construction stages and requires more time to heat the area and cools off more slowly as well. The amount of piping and other components required for this system is also greater.

Pumps

In all active liquid solar power systems, pumps move the working fluid from the storage tank to the collector and back and also between the storage tank and distribution points. Temperature control systems are used in conjunction with pumps to move the water through the collector loop only when the water in the collector is hotter than that in the storage tank—which is usually the case only during periods of sunshine. One way to establish this movement is to construct a dual system with a heat exchanger; the heat exchange unit has a backup heater and is combined with the storage tank. The collector loop contains working fluids with antifreeze, and the heat exchanger isolates this loop from the domestic and hot water distribution loops by transferring the heat from the immersed heating coil. This arrangement also helps to prevent freezing.

The type and size of pump depend on the flow rate and pressure required, the amount of pressure drop that may accompany the movement of the working fluid through the various piping loops, the change-of-direction fittings, and length of pipe. Pressure drop through the valve body must in some cases be taken into account. Variable-speed pump units are available, though constant-speed pumps are less expensive. Controlling mechanisms must also be designed as an integral part of the system to alter the speed or turn the pump on and off. Variable-speed pumps help to maintain the desired circulation of warm fluids through the living space and prevent overheating the building. Constant-speed pumps need to be controlled by valving, which retards the flow rate and also increases the total head loss in the system. Electrical energy for this type of pump is also wasted because the speed cannot be altered to reduce the amount of flow. Since pumps for solar units are small, seldom exceeding 5 gpm, pressure drop is minimal. For larger units, of course, larger pumps may be required, especially for apartment units and commercial buildings. In these applications, variable-speed pumps are economical.

Heat Pumps

Basically the *heat pump* is a form of refrigeration unit which is operated in reverse. A refrigerant is circulated in a closed loop by a mechanical compressor and the hot end of the condenser is kept under constant pressure. The cold end in the evaporator section is at a reduced pressure. By mechanically increasing the temperature and pressure of the refrigerant, it will condense, giving off heat. It will rapidly evaporate and drop in temperature, taking up heat as it passes through a pressure relief valve.

Thus heat pumps use compression to raise the temperature of working media. If solar collectors are used to preheat water, a heat pump will then raise the temperature of the fluid to a level suitable for a conventional space heating system, such as the radiant or baseboard type. Heat pumps are more energy efficient than other types of heating and cooling equipment. Used with solar collectors to supply low-temperature working fluid, the heat pump can be connected to warm-air, hot-water baseboard collectors or other distribution units. The heat pump is used even in nonsolar units. The use of solar collectors in conjunction with heat pumps is becoming increasingly popular.

STORAGE UNITS

The storage unit for a typical liquid solar power system is a low-pressure vessel, heavily insulated, which can hold the necessary volume of heated fluid from the collector over a period of time. In some cases, the storage tank and the exchanger coil are connected. In general, though, the heat transfer fluid which circulates between the collector and the storage tank is a separate stream from that of a heat exchanger coil, which receives fluid from the storage tank and passes it on to another loop for domestic hot water or heating.

Tanks holding 500 to 1000 gal are required and can be installed in the basement or buried. Storage units can be makeshift tanks or large-diameter concrete piping for sophisticated vessels, depending on the economics and type of system involved. To reduce the volume of storage, solar architecture must be integrated in the design stages to situate the building efficiently and provide sufficient insulation and special doors and windows. The pressure requirements for such a vessel are extremely small in comparison to industrial vessels. As temperatures rise in the system, of course, the pressure requirement will also increase. The size of the container depends on the amount of hot water required to heat a certain area or provide sufficient domestic heating services. The storage area in cubic feet and the desired temperature must be calculated to arrive at the proper vessel size and capacity.

Hot-water storage tanks can be fashioned from existing water heating vessels, though usually these are used only as backup heaters in conjunction with new storage vessels which are much larger (and also require far more insulation to maintain the required temperatures). A typical storage vessel would be a steel tank approximately 4 ft (1.2 m) in diameter and 11 ft (3.3 m) long. Concrete storage tanks or septic tanks can be used instead of steel tanks because they are less expensive and easier to install, though they are much heavier and sometimes harder to procure. Since it is good policy to locate the storage tank where it can be readily serviced, some septic tank units cannot be used. Moreover, the thermal insulation required for underground installation may create problems in installation and sealing it from moisture in the ground. Connecting piping, valving, and pumping may also be complicated by underground construction. A minimum of 6-in. (15.2 cm) fiberglass low-thermal-conductivity insulation is required for most storage units, regardless of the type of tank.

CONSTRUCTION MATERIALS

Copper, aluminum, and galvanized steel have been utilized for solar collectors. All possess positive and negative features. Copper's corrosion resistance makes it ideal for solar collection, though it is expensive. Aluminum, less expensive than copper, is lightweight and more versatile as a material of construction. Steel, on the other hand, has greater strength and is cheaper than aluminum and copper. Both aluminum and steel do have corrosion problems when used with liquid systems and antifreeze. Absorber plates can be constructed of steel, copper, aluminum, or a combination of these materials. Thermal performance, corrosion resistance, mechanical strength, cost, coating compatibility—all must be taken into account when designing solar collectors using any of these three materials. The thermal performance of steel may equal that of copper and aluminum, depending on the design of the absorber, though the thermal conductivity of steel is only 10 percent that of aluminum and 5 percent that of copper.

The *energy* cost of materials must also be taken into account when considering mass production of units. Aluminum used as an absorber material has a much higher production energy cost than copper, and carbon steel is less expensive than both. Therefore the amount of absorber materials necessary for a collector must be considered. Combining two metals in a solar unit will also create problems, especially where aluminum components are used with copper or steel loops. By inserting short portions of plastic or carbon steel piping at strategic points, it is possible to separate the materials sufficiently.

Steel

Steel offers one of the most reliable and longest-lasting materials for collector construction, and galvanized steel provides sufficient corrosion resistance. The resistance of materials to the operating environment must be considered when one is choosing between copper, aluminum, and steel. Various types of antifreeze and inhibitors must be added to the working fluid to prevent corrosion of the system. In closed-loop systems with inhibitors, the use of copper in absorbers requires the least maintenance; steel demands considerably more attention and care. The material, fabrication, and construction cost is also a major factor in absorber plate construction. Steel, in comparison to the other metals, is considerably cheaper.

Copper

Copper provides the best conductivity and greatest resistance to corrosion. It is also easier to fabricate by using a variety of soldering connections and is easily adapted to solar units because of its wide use in the plumbing trade. Moreover, copper and brass are available in a wide variety of fittings and valves. However, copper is much more expensive than other materials. Since its thermal conductivity is much greater than that of other materials, copper can collect the same amount of heat as other materials of thicker gauges. Corrosion resistance is extremely important in all solar collection and circulation systems. When mixed with other materials such as steel and aluminum, copper tends to coat and protect itself, though the adjoining metals are subject to galvanic corrosion. By designing the system to be all copper, it is therefore possible to eliminate corrosion. Some copper flat-plate collectors have been in use for a half century with only minimum damage caused by corrosion and other problems. Another way of eliminating corrosion and erosion is to reduce the total hot water velocity, keeping it a maximum of 5 fps and usually less, especially where small-diameter tubing is used.

Corrosion of copper becomes a problem primarily where antifreeze is added to the system, so inhibitors may be necessary. The fabrication of copper units is much simpler than for other systems. Electroplating can be used on copper surfaces, as can other types of electrical treatment, making copper an even more desirable material for solar units.

Aluminum

Aluminum is lightweight, easy to fabricate, and offers excellent heat transfer abilities. Moreover, it has an excellent strength-to-weight ratio, making it possible to use smaller amounts of material to produce a rugged and stable design. Aluminum can be joined through many methods including welding, brazing, and soldering. Adhesive and screwed fittings can also be used, though they are not common. common.

Corrosion presents one of the most frequently mentioned problems for aluminum—especially when it is used in conjunction with other materials such as copper. It is almost impossible to eliminate copper from a system, however, because a majority of fittings, valves, and pumps have copper parts. Aluminum is highly resistant to atmospheric corrosion compared to other materials, especially steel. The problems arise when it is used primarily for liquid heat transfer in a hydronic system. Therefore all working fluids for an aluminum solar unit must contain additives, chemicals, and other substances, such as antifreeze, which may cause corrosion as they decompose. Inhibitors must be added to prevent aluminum corrosion in a majority of cases. Although water is not usually corrosive to aluminum, impurities in the local water system can cause problems. Because aluminum reacts directly with oxygen, forming a thin oxide film, it has been used in a wide variety of industrial situations where this film helps protect the material over a period of time, making it seem almost indestructible from a corrosion standpoint. Again the designer must be cautioned against designing a solar piping system or collector unit where copper and aluminum are directly connected or where copper and carbon steel are connected in the piping loops. Insulator sections must be installed between dissimilar materials to avoid destruction of the aluminum.

Plastics and Other Materials

Many other materials have been used for collectors and piping loops for solar energy systems—including glass, stainless steel, iron, and plastic pipe. When glass pipe is used for solar absorption, it is usually painted black. This system is somewhat expensive, however, and very sensitive to freezing temperatures. Plastic piping has also been tried, usually polyvinyl chloride (PVC), though problems of decomposition have limited its use for solar units. Stainless steel piping has also been tried, but it is extremely expensive in comparison to other materials. Iron pipe for solar water heaters has also been developed in Japan and other countries.

THE PROMISE OF SOLAR POWER

Solar power devices using liquids require extensive plumbing, piping, and devices usually associated with other industrial piping systems, though their pressure and temperature range is lower and flow, storage and diameter requirements are also considerably less. Solar power at present is confined primarily to the architectural construction trades except for the manufacture of specific units, which may involve piping drafting to supplement the architectural design. The figures throughout the chapter represent only a few variations of solar devices. The flow diagrams should provide sufficient explanation for units described in the text. Figure 14-5 shows the solar collector-to-storage-tank loop connected directly to the space heat distribution; it is connected indirectly through a heat exchanger for the domestic hot water needs. Figures 14-6 and 14-7 are variations of solar collector loops. Figure 14-8 shows a typical flow diagram for heating a pool. Figure 14-9 shows a system of flat-plate solar collectors and their construction. This system is a retrofit unit—note the piping and valves connecting the solar panels.

Solar power is only in its infancy. Eventually, it will be developed economically for commercial and industrial purposes. It is a field which offers much opportunity for those willing to explore the job possibilities. Knowledge of piping systems, drafting, and design can be a starting point for those seeking employment in this extremely important field, upon which much of our future as a nation and world community will depend. The development of higher temperature and pressure systems—along with direct conversion of sunlight into electricity or through traditional turbine generator units—will become possible with advances in technology.

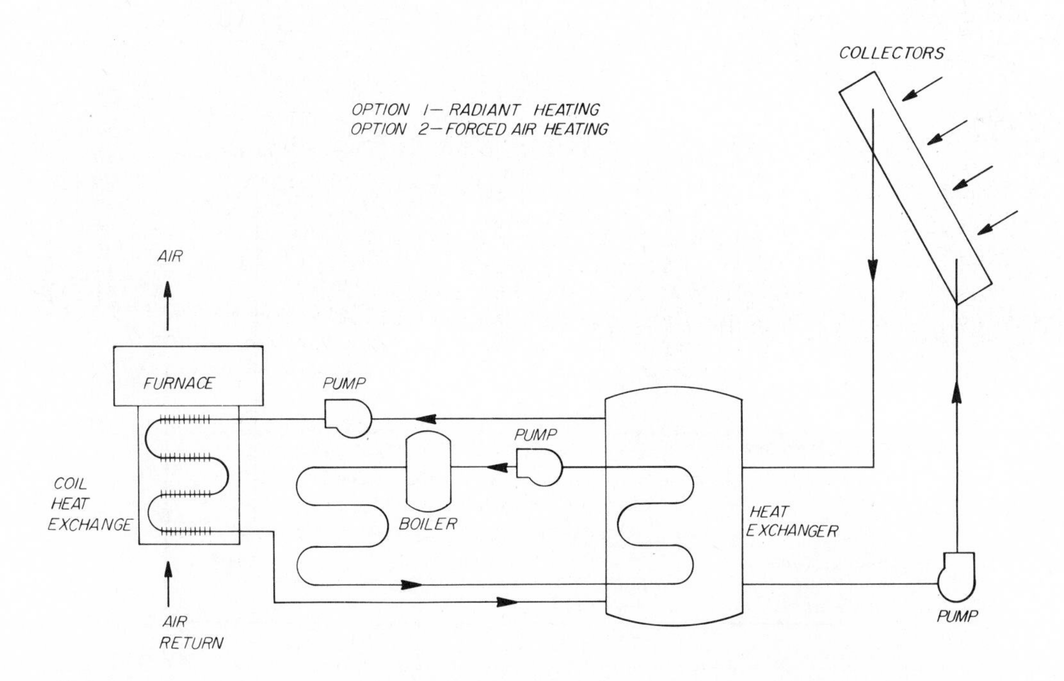

FIG. 14-6 This schematic representation of a solar energy system shows the combination of available solar heating systems, including backup furnace, boiler, pumps, filters, domestic water heaters, heat exchangers, storage tanks, solar collector, all of which enable the system to provide hot water for domestic use, hot air for space heating purposes, and heated water for a swimming pool.

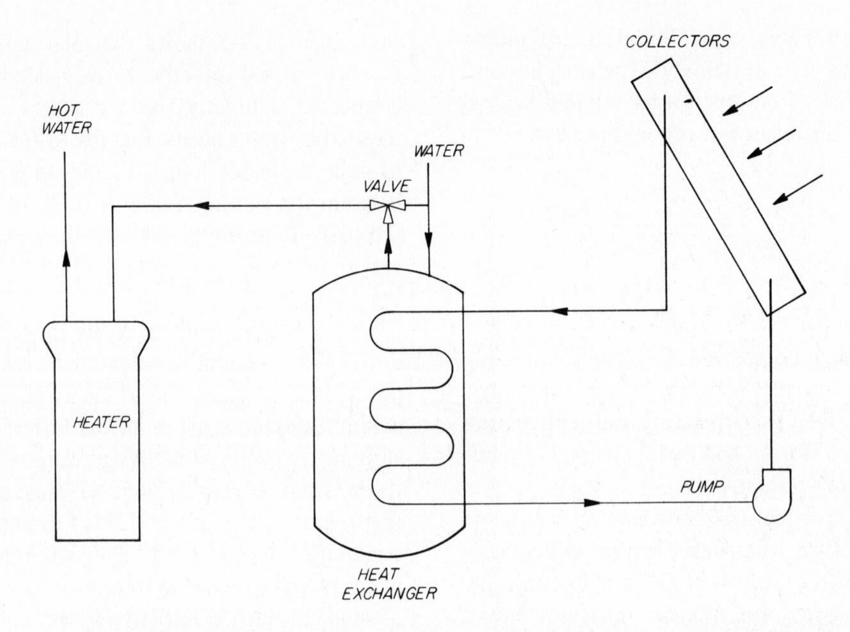

FIG. 14-7 This schematic shows a typical solar energy system used to provide domestic hot water. The solar collector traps heat from the sun in a closed water system, which is circulated by means of a pump. Utilizing a heat exchange unit, the pipe loop from the solar collector comes in contact with water from the water heater. The two water streams never actually contact each other.

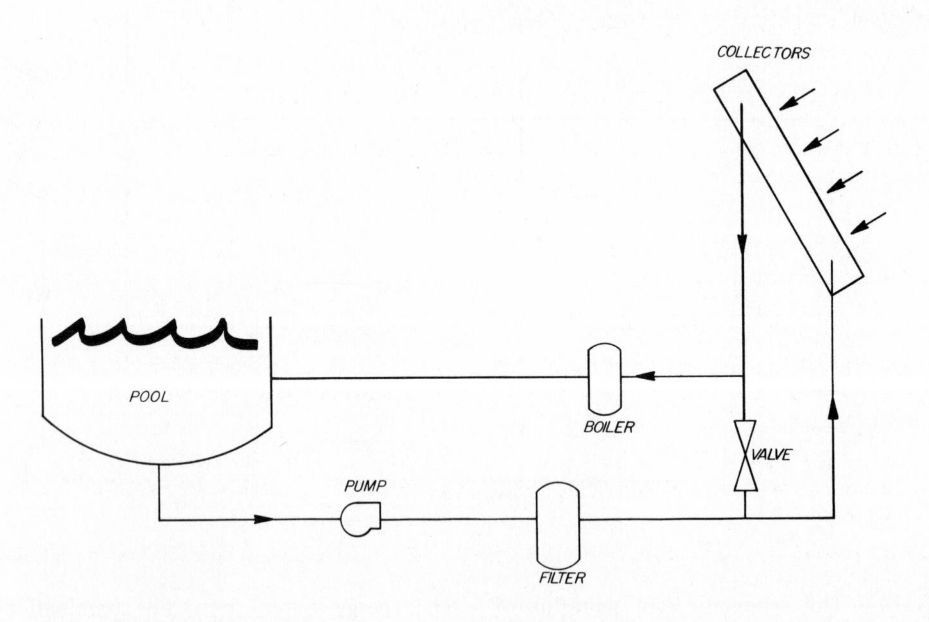

FIG. 14-8 Typical schematic for a solar-powered heating system for a pool. It works in much the same way as all hot water systems, using a collector and piping loop between the pool and a backup boiler.

FIG. 14-9 Solar water heating system for a 50-unit apartment building in Oakland, Ca. Collectors supplied by Western Solar Development, Inc.; design and installation by The Solar Center, San Francisco. This retrofit provides approximately 60 percent of the building's annual hot water energy load. (Courtesy The Solar Center, San Francisco, CA)

QUIZ

1. Describe how a flat-plate solar collector works.
2. Name three benefits from the use of solar energy systems.
3. What are the primary uses of solar energy at the present time?
4. Name the major components for a typical solar heating and cooling system.
5. Describe the difference between a passive and an active solar energy system.
6. What is a solar collector and what is its primary function?
7. How can solar energy help us meet today's energy needs?
8. What are the primary reasons for the trend toward solar energy?
9. What circulation fluids are generally used?
10. What piping systems are necessary for solar-powered units?
11. Can solar energy be used to cool as well as heat?
12. What drawings are required for a solar-powered unit for a building?
13. What are flow diagrams and P&IDs used for in solar work?
14. What are the physical and environmental requirements for utilizing solar energy?
15. What is the difference between a concentrator and a flat-plate collector?

16. How can solar collectors be used to create steam?
17. What are the main functions of a solar energy system?
18. How is solar energy stored in a typical system?
19. Name two ways of distributing heat in a solar energy system.
20. What is the difference between direct, diffused, and reflected radiation?
21. Describe heating by convection and by conduction.
22. What are the primary methods used to distribute solar energy throughout a house?
23. Define *heliochemical, helioelectrical,* and *heliothermal.* How do they relate to solar power?
24. What are the primary domestic uses for solar power?
25. What is the difference between direct and indirect solar systems?
26. Why are auxiliary heating systems usually required?
27. What are controls used for on a solar energy system?
28. What is meant by an *integral* system?
29. How can solar power be used to heat pools?
30. Why are pumps necessary for solar energy systems?
31. What vessels are required for a typical solar energy system?
32. Describe the possibilities for employment in the solar industry compared to the nuclear industry.
33. Give a short historical review of solar power in the world and in the United States.
34. Why has solar energy been neglected up until this time?
35. How does a solar furnace work? What are the maximum temperatures obtainable through this method?
36. What are the primary geographical considerations for site selection for solar power units?
37. What are the basic parts of an active solar energy system?
38. What are the normal temperature ranges for domestic solar power units?
39. Describe the nonfocusing collector and explain how it works.
40. What substances are usually conveyed in collector tubes?
41. What materials are used for solar collectors? Give the pros and cons for each type.
42. Why is glazing used for the cover sheet in a typical panel?
43. What pipes can be used for solar energy systems?
44. Describe the air-type collector.
45. What is a water-trickling collector and how does it work?
46. What is a liquid-type collector?
47. Describe the difference between a linear and a circular concentrating collector.
48. What does it mean to have parallel systems?
49. What is the difference between natural and forced circulation?

PROBLEMS

1. From a set of architectural drawings for an individual home or building, design and draw the plumbing, piping and components for a complete solar space and water heating project.
2. Draw a flow diagram for Problem 1.
3. Design a solar water heating system for a pool (in isometric).
4. Redraw Fig. 14-2, 14-5, 14-6, 14-7, or 14-8.
5. Draw an isometric view of a typical solar energy pipe loop and components for domestic space and hot water heating.

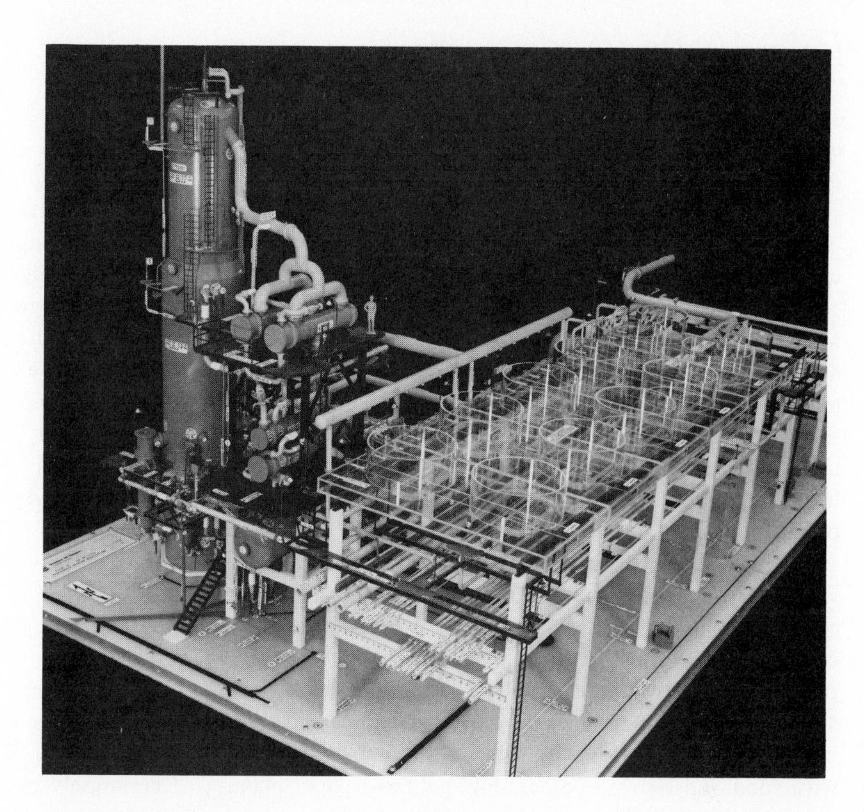

LSR splitter facility for petrochemical complex.

15 INDUSTRIAL MODEL BUILDING

MODELS

Models should be used in every piping drafting course to introduce prospective drafters to a field with which they will have much contact.* Even the smallest drafting room has sufficient space for the construction of a model by an individual student or group.

Though some of this chapter deals with the wide variety of tools that are useful in a model room, these tools are not absolutely necessary. Any drafting class can set up an inexpensive model area for less than $200 with an assortment tools for the construction of many of the piping drafting projects in this text. Only the larger, more complicated models need more sophisticated equipment.

*** This chapter has been written with the cooperation of Engineering Model Associates, Inc., and the American Engineering Model Society.**

THE USE OF MODELS

Models are used throughout the industrial piping field as scale representations of construction projects. A model shows a piping installation (including valves, fittings, and other components), structural aspects of the complex, equipment, instrumentation, architectural aspects, pipe supports, and other features. Models clearly reveal the basic design of a piping installation, and the advantages of the model as a tool for design and checking will become apparent as one reads further.

Because models are three-dimensional representations of an actual piping installation completed to an accurate scale, they eliminate many of the problems encountered in the use of piping drawings. Models provide an opportunity to design piping systems which are suited to the situation and permit the designer and checker to examine critically all the aspects of the design and construction for the project in question, thereby eliminating problems in the material procurement, construction, and installation phases of the job.

Models are not justified in every piping system design. Nevertheless their cost, which may run high, pays for itself when the complexity and intricacy of a large industrial complex might cause problems if only drawings are used. Models enable the transfer of engineering and construction data, transcending language differences and eliminating the majority of interferences. Some of the smaller projects in this book will not, of course, represent the complexity that might be encountered in industry, but they will nevertheless provide the student with valuable experience in the use of modeling materials and tools. The projects in Chapters 11, 12, and 13 are excellent examples of piping installations which may be made considerably more efficient by the use of a model. Models, therefore, are truly a design tool that can eliminate unnecessary problems, bad design, inefficient planning, and other expensive, time-consuming situations. They are used throughout the process, petrochemical, nuclear, and conventional power generation industries, along with food and beverage processing, pulp and paper manufacture, pharmaceutical processes, and other fields.

Traditionally, drawings have been used to convey design information to the construction crew, but in this age of complex piping systems, the exclusive use of drawings is diminishing rapidly because of the special skills necessary to read and interpret the complex, often overdetailed drawings. Drawings are only for a two-dimensional medium, and the advantages of the third dimension will become apparent after the construction of just one model (Fig. 15-1).

Models were first used for checking a design that was already completed by use of the traditional plan, elevation, and section method. The model enabled the designer to deal with many of the errors commonly associated with the use of orthographic projection in pipe design. Everything from water pollution control plants to bridges have been modeled, enabling the designers to convey, before construction, all the necessary information to the purchasers of the system. The benefits of models depend upon the complexity of the project, whether or not the model is used in the design stages, the degree to which modeling can replace drafting, and the accuracy to which it will be built.

In general, 3/8 in. = 1 ft to 3/4 in. = 1 ft are the most commonly used scales; 3/8 in. = 1 ft is used throughout the petrochemical industry and 1/2 in. = 1 ft is more common

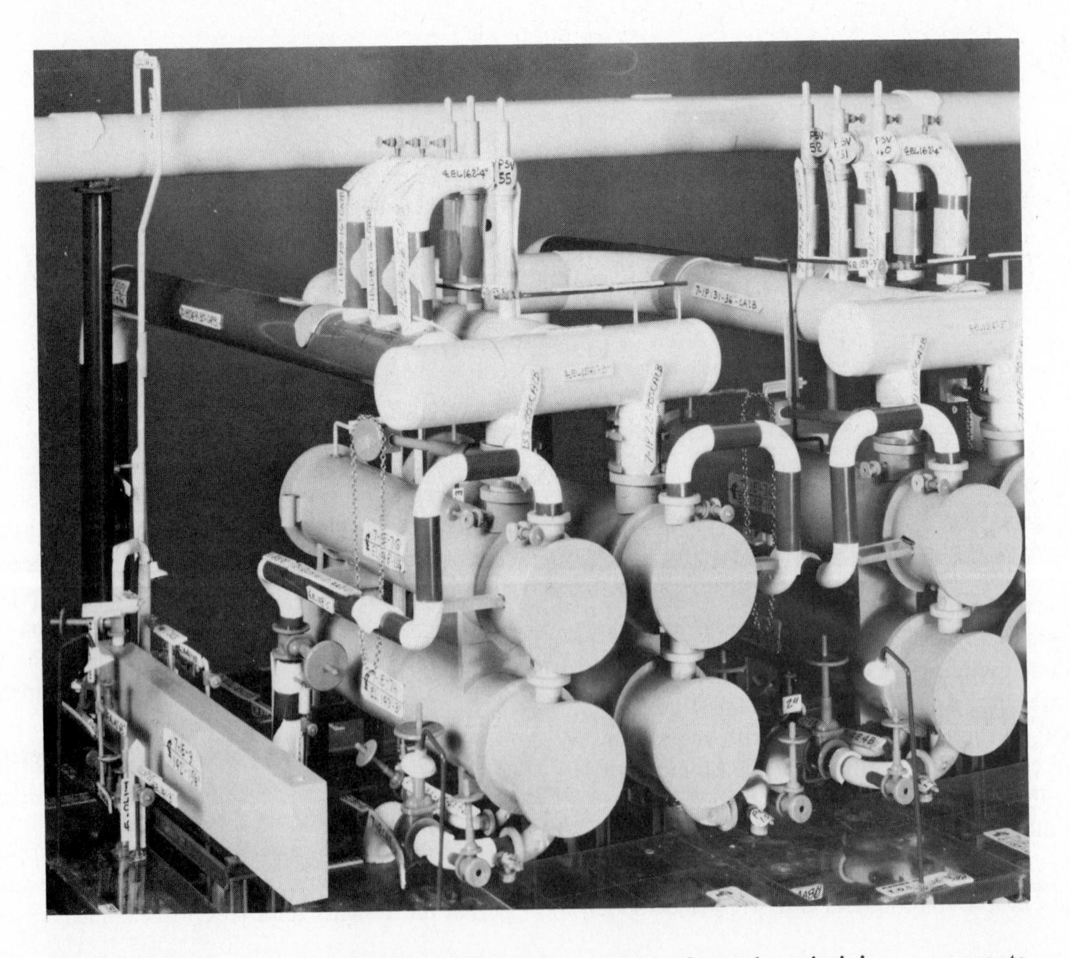

FIG. 15-1 Close-up showing model section with a variety of vessels and piping components. Notice the small chain-operated handwheel on the model.

in the power generation field. Larger scales are sometimes used for subsections in problematic areas (Fig. 15-2), and in some cases full-scale models of extremely important areas may be constructed. Industrial models include the use of model tables, model bases and frames, floors, and walls. There are representations of all structural elements, piping instrumentation and components, electrical or conduit trays, HVAC, and a variety of other equipment involved in the project, such as fossil fuel steam generators, turbines, economizers, and feedwater pumps. Each industry has its own variations of equipment and components that will make up its industrial model. Anyone involved in drawing piping systems is likely to encounter models sooner or later. In some cases, companies turn to the experienced designer and drafter to run the pipe and place other equipment directly onto the model from their drawings. This has been the case in the petrochemical industry, especially where design changes to existing installations have been made directly in the drafting room on models by the drafters. Model building offers a number of excellent job opportunities throughout the world.

Models are accurate, three-dimensional representations that offer advantages unavailable in two dimensions. Every complex piping installation has a maze of piping and struc-

FIG. 15-2 Large model of a power plant. Notice the small (scaled) figure of the man on top of the complex.

tural, mechanical, and electrical equipment. When using drawings, it may be impossible to spot interferences because of the mass of drawings, views, and sections that are necessary for the use of orthographic projection in piping systems. Two dimensions can show only a cross-section. Even with isometrics, one cannot represent all layers on the same drawing. And if there are several drawings, the points of correspondence must be kept in mind three-dimensionally. Add to this the fact that several departments —structural, electrical, pipe support—will have systems interpenetrating in the same space, and the number of correlations becomes enormous.

The drafters and designers of each department have their own set of standards that are shown on their drawings, which may not be always included on the piping system drawing. Therefore interferences and misconceptions in design can occur much more readily when using only drawings. These problems do not arise with the model since all systems are accurately placed to scale as they will appear.

When working in three-dimensions, the designer is able to visualize many of the design sequences and operations that are necessary for the project, which may not be possible when using a large quantity and variety of drawings. In some cases, the construction company will actually use the model to design the entire system without recording the design on drawings. Models provide their greatest benefits when used throughout the design stage as a working tool, though the beginning model, or preliminary model, may look nothing like the final designed model. Many stages may be needed in between to provide the designer an accurate three-dimensional means of dealing with any difficulties that may be encountered. Besides being a tool for interdepartmental communication, the model offers a three-dimensional view for recording on film the construction sequence for the system in question.

Where only the model is used, it becomes the only channel of communication between the designers and the field construction crew. Along with photographs, the design model then becomes an extremely valuable tool that is treated with great respect—as much as four years of design work could be lost if the model were destroyed. When recently touring a large power plant construction company, students were asked not to touch any portion of the model because it was the *only* record of the nuclear project being constructed. This shows the seriousness with which some companies have taken their model building. They have created an essential design service parallel to the drafting fields within their company. Shipped to the construction site, the model gives the construction crew access to an actual scale representation of what must be constructed. It can also be used to prepare construction and manpower schedules and to help determine the sequence and priorities of installation. The field crew uses the model constantly during the construction period and again for training personnel at startup time.

In general, the engineer develops the basic plant arrangement with the preliminary model, enabling the designers to prepare job specifications and maintain design uniformity throughout the project and helping to improve supervision in the design stage. In many cases, the model will eliminate a number of piping, plan, elevation, and section drawings. Moreover, the HVAC, instrumentation, and electrical groups may also find that the model eliminates formerly necessary drawings. During the construction stage, a great amount of time will be saved because there are no drawings to be interpreted. Models can also help to determine whether sufficient space has been provided for routing small piping and electrical components.

After the project has been constructed, the model shows itself an excellent tool for training maintenance personnel and plant operators. The economic advantages of models vary throughout the engineering fields that use them. Some companies have claimed a variety of benefits—such as the ability to tie down subcontractors with firm bids on components, the savings in total engineering and construction time (as much as 20 percent), the elimination of 10 to 40 percent of the drafting requirements, and the saving of large amounts of money by reducing piping footage and fittings.

Though the petrochemical field has found that using models in the design stage may cost about the same as orthographically projected engineering drawings, models provide more accurate, quicker designs and correspondingly lower engineering costs. The elimination of some drawings has facilitated more efficient designs, since drawings are very difficult to understand when they involve intricate systems.

PRESENT AND FUTURE PROCEDURES

One of the new procedures in modeling has been the use of photography. Though the process is still in its infancy, most companies that use complex, industrial piping models also use photography to exhibit a variety of views similar to those afforded by multiple drawings (Fig. 15-3). Color photographs have been found most useful because the model itself is constructed of color-coded materials. Photographs of the model assist in project orientation and can establish a detailed sequence of construction and pipe installation schedules. Moreover, photographs can be sent to the field prior to the completion of the model so that portions of the project can be started before the model reaches the job site. A flat-field camera can sometimes be used to photograph models, thereby permitting enlargement to drawing size.

Another new procedure in the modeling field is the use of computer-drawn isometrics instead of hand-drawn ones. The model is completely labeled to locate all components, elevations, pipe runs, and so forth. This information is then fed into a computer which will construct an isometric view of any line in question. Material specifications and bills of

FIG. 15-3 Model showing vessels and other equipment along with structural elements. This model has not reached the piping stage; the equipment is merely laid out. The two boxlike pieces of equipment at the bottom of the photograph represent pumps.

materials can all be prepared through the use of a computer in this way. Computer isometrics, though they have their disadvantages, are being developed into a viable tool for large industrial complexes. In smaller projects, the model after tagging is sent directly to the pipe fabrication unit, which then works directly from the model in sketching isometric or orthographic spools.

All these techniques are being developed throughout the industry. The American Engineering Model Society (AEMS) has pioneered in the coordination that is necessary to enable the model building field to be standardized with the cooperation of industry and the model component companies. Standardization will bring many advantages throughout the piping industry. The purpose of the AEMS is to disseminate codes and information. Every model building course should have access to AEMS publications, which provide a wealth of information concerning the model building field, including standards, components, tools, safety, and design.

Most of the components used in construction models have been standardized by manufacturing companies like Engineering Model Associates, Inc. (EMA). Thus valves, fittings, pipes, and other components are interchangeable with a high degree of accuracy. EMA provides a wealth of research and development activities available to the piping industry and has developed 6000 different components for piping models. Both the AEMS and EMA provide numerous publications, newsletters, and standards that should be part of every model building course.

ADVANTAGES OF MODELS

The following advantages have been documented by users of models throughout the piping industry. The model:

1. Provides essential three-dimensional visualization (Fig. 15-4)
2. Enables design groups, contractors, and clients, who may speak different languages, to communicate through the medium of the model
3. Serves as an extremely adaptable medium for the engineer and piping designer
4. Provides for quicker decision making and approval, thereby eliminating time-consuming evaluation of drawings
5. Eliminates costly drafting procedures
6. Provides an excellent system for checking engineering design
7. Can be used for checking construction
8. Can be used for contractor bidding
9. Provides for excellent scheduling of construction work
10. Presents a perfect example of an as-built record
11. Offers a selection of operator and maintenance training opportunities
12. Can be used for nontechnical graphic display for public relations
13. Can be used as a colorful, attractive sales tool
14. Can be used for modifications before construction
15. Allows for rearrangement of components and design without costly drafting changes
16. Allows for more accurate planning in the preliminary stages of the project concerning civil engineering and land use
17. Identifies pipeline uses and clearances by means of color coding
18. Shows access to components, valves, and instrumentation
19. Can be used for insurance examination and painting, insulation, and other contractors
20. Reduces the total time of construction
21. Can be used for construction planning, including the sequence of building
22. Provides for jurisdiction decisions
23. Cuts engineering and drafting time
24. Eliminates costly errors in the checking stage that may be encountered with drawings

FIG. 15-4 Close-up showing a variety of columns and piping installations. Notice the amount of tagging required for industrial piping models, including instrumentation. Note also the variety of piping and instrumentation components: temperature instruments, pressure differential indicators, pressure indicators, level glasses, and level control instruments.

25. Allows for the construction of alternative arrangements for piping, plot plans, and installations

26. Eliminates almost completely all major interferences in a complex piping installation

Models as Communication Tools

The area in which models provide the greatest asset to the design process for piping installations is that of communication—between design groups, between the company and outside vendors and planners, between engineers, designers, and construction personnel, and between the purchaser of the project and the engineering firm. Compared to drawings, models offer a much more reliable source of communication. Because of their two-dimensional aspect and their complexity, drawings present communication problems among all the groups in the design engineering firm. Models, on the other hand, when constructed to professional standards, facilitate communication by showing clearly the several aspects of the design project in a three-dimensional form. Because many different groups in an engineering firm must make use of the same drawing and design data, the model is an excellent means whereby all these subfields—electrical, piping, instrumentation, civil, pipe supports—can come together in common understanding of the project as a whole, working out the common problems of interference and priority that are part of every complex industrial plant. Models offer these groups the opportunity to rearrange, to redesign, and to solve problems at the design stage—not in the field during construction where such problems can lead to expensive and time-consuming changes.

As a means of communication, the model offers an opportunity to create a design that incorporates better operating techniques, alternative design options, and more efficient, less complicated overall installations. After they are used in the construction stage, models can also be utilized as a communication tool in writing operating manuals and training operators and maintenance crews. Moreover, long-range design possibilities can be tested on the model before they are implemented in the plant.

In the beginning stages, all models offer communication benefits that preliminary drawings do not. Above all, nontechnical personnel such as government agencies, business leaders, financial contributors, and the public at large can view the prospective project. In the earlier stages, models reduce the total engineering and development process because preliminary errors may be entirely eliminated.

The time required to study complicated drawings may delay large numbers of designers and construction workers. Models, on the other hand, can be considered instant communication. Those involved in the project can ignore aspects that do not concern them and focus on their specialty while still maintaining an overall perspective of the surrounding area. This, of course, increases the ability of engineers and designers to make quick modifications which then can be communicated to the various subgroups of the company. Process and instrument engineers, electrical engineers, HVAC engineers—all can be assured that their part of the project is functioning properly and that standards and specifications are being met.

Models also create a visual sequence of construction procedures, thus eliminating the trial and error method that is sometimes associated with the use of complicated, orthographically projected piping drawings. The construction crew can plan for their work and establish the sequence of construction well in advance of the actual startup time. In conjunction with essential drawings such as the process flow diagram, models offer an excellent means of checking systems. Design considerations, such as accessibility, economy, structure, pipe runs, placement of instruments and equipment, and maintenance space and requirements, can all be visualized and checked for adequate location directly on the model.

Models as Design Tools

The use of models in design involves two definite stages. The first stage is the preliminary equipment arrangement or plot plan model; the second is the final piping design model. Both stages are necessary for the most efficient use of modeling. Preliminary equipment layout models offer engineers, designers, and others an excellent opportunity to review the status of the project. Design concepts can be developed and examined for feasibility before drawings are even begun. This preliminary stage permits rearrangement and optional equipment placement. Though preliminary models are usually block models with only simple shapes and structural elements, they do offer an excellent three-dimensional overview. The preliminary model is never intended to resemble the completed project. It allows engineers, designers, and drafters who will later use the design model to participate in the original layout and arrangement of the project. The final piping model (Fig. 15-5, Fig. 15-6)

FIG. 15-5 Large piping project model. Notice the white walls that are cut away to show the interior of the plant.

FIG. 15-6 Close-up of a control station on a petrochemical model.

is an active design tool that is built simultaneously with, and sometimes in advance of, many of the other subareas of the project.

In the early days of model building, a check model was constructed after the fact—that is, after the drawings and other items of design were completed. This, of course, reduced the effectiveness of the model, though it did help eliminate costly interference errors. Check modeling is limited to checking, operator training, and construction planning.

To realize its full value, the model should be started as soon as the process flow diagrams and other necessary drawings have been completed. As the design model moves through its different stages of completion, the piping designer will gradually run the pipe, do the small-scale piping, and place components, pipe supports, and valves on the model. The design model offers the ability to revise the proposed location of equipment—moving pipe or a piece of equipment to allow more access, relocating a nozzle, repositioning a valve.

MODELING PROGRAMS

The implementation of a model program will, of course, be expensive in the beginning stages because of the purchase of equipment and the construction necessary to establish the model shop. The following paragraphs discuss program costs, the training involved, and the model shop. Specific tools are discussed later in the chapter.

Program Costs

Any school that wishes to create a model shop will do well to consult EMA or AEMS (Fig. 15-7). As has been stated, it is possible to do several types of projects with a minimum of equipment. The projects in Chapter 11 can be done without power tools, though it is obviously advantageous to have basic power equipment available. Costs in setting up the model shop depend on the amount of equipment and sophistication necessary for the shop to be efficient. Power tools, hand tools, furniture, model components, supplies, training—all add to the initial investment.

Regarding the costs for various models, it has been found that preliminary models, which are usually made of block materials and structural elements that can be reused, are rather inexpensive compared to the final design model with all its minute detailing of instrumentation, piping runs, and equipment. The materials used in the preliminary model, such as polyurethane foam and wood, keep costs to a minimum. The basic costs of major components such as tables, bases, large equipment, and structures are the expenses which are usually attributed to the design model at the model shop before it is piped up by the designer (usually in view of the drafting and design department).

The design group completes the smaller components—such as valves, fittings, pipe, platforms, stairs, ladders, bracing, pipe supports, and small equipment. Therefore most of the costs involved at the design stage are associated with labor and small standardized components. It is a good idea when doing any of the projects in this book to keep a record of the expenses incurred. In 1978, problem 4 in Chapter 11 cost $50 to $60, depending on the base material. Plexiglas, of course, is more expensive than wood or fiberboard. The projects in Chapters 12 and 13 will run considerably more and should be regarded as group projects. It is suggested that projects in the book be drawn first to familiarize the student with the design problems that may be encountered in modeling the project. This will eliminate cost errors and increase understanding. EMA offers a school model building kit shown in Fig. 15-8 which is an excellent basic educational model project. See *Industrial Model Building*, Prentice-Hall Inc., 1981.

FIG. 15-7 *EMA Reporters:* **model designer newspaper available to interested groups. (Courtesy Engineering Model Associates, Inc.)**

FIG. 15-8 EMA design manual kit project available to schools. (Courtesy Engineering Model Associates, Inc.)

Training Programs

Training differs according to the educational level of piping designers and model makers. Both must have sufficient background in piping and piping design to understand all that is involved in the modeling procedure. The model technicians spend the majority of their time in the model shop constructing bases, tables, vessels, and large structural items. Depending on the background and abilities of the model technician, the general procedure is for the modeler to advance into the piping design group. The essential background for a model technician includes:

1. The ability to use power tools and machinery for the construction of items made of a wide variety of materials, including plastics and wood.
2. Knowledge of the basics of piping, the ability to read piping and other engineering drawings, and the ability to sketch.
3. The ability to fabricate the several items necessary for the construction of a model, such as bases, major equipment, structural items, and vessels. (Pipe designers must have abilities similar to those of the model technician, but they must also have intricate knowledge of piping systems and other components, an ability to read, understand, and draw flow diagrams, isometrics, and other drawings or sketches, and knowledge of piping specifications, codes, and design characteristics that belong exclusively to subfields of piping industrial complexes.)
4. Background in the capabilities and functions of equipment, vessels, and so on.
5. Manual dexterity and a general ability to do intricate designing and pipe installations on a model.

Mastery of the information given throughout this book will provide a firm basis for the student who wishes to become a piping designer.

In the past, model builders and technicians have been trained on the job by companies which use models. Now a number of technical and vocational schools have model building programs which offer an excellent opportunity for training model technicians. Though nothing will take the place of on-the-job training, companies tend to regard model technicians as a group separate from the pipe design group, which may cause problems in some cases because the technician will not have the varied background of the person trained in the drafting and design of piping systems. Other companies use the model building program for training drafters and model technicians who will eventually become pipe designers. This scheme enables the worker to understand all the subfields and areas of concern involving models and industrial piping projects. A good background in mathematics, science, and engineering (including geometry, trigonometry, mechanical engineering, and electrical engineering) helps the prospective model technician or designer.

School programs should create a worker who has the ability to progress from model technician to piping designer. This ability is only brought about by education in a program that gives the student hands-on experience in model making in conjunction with drafting and design. All piping designers and model technicians should have a basic knowledge of materials, power tool use, hand tool use, model construction techniques, safety, model piping, and drafting and design. The piping drafter may be called upon to use the model in a variety of ways—such as tagging for identification, taking isometrics directly from a model, or utilizing the model for the placement and design of pipe supports. The more experience piping drafters have in the use of models, the greater their understanding and capabilities will be in the field.

This chapter covers modeling with regard to pipe design and model technician assignments. The book in general acquaints the student with drawings and piping projects that will be encountered in the field. Ninety percent of the drawings and problems in the text have been drawn from industrial sources. Thus the student should gain the background necessary to enter the piping and drafting field and be able to construct a model project. The model technician must learn to fabricate bases, tables, and large equipment and thus can be trained only with proper equipment in a model shop. Large items such as Plexiglas bases can be precut at the place of purchase. The projects in the book are sufficiently simple so that a great deal of fabrication will not be necessary (except in Chapters 12 and 13).

The main emphasis is on the construction of a clear, concise, accurate model that represents the original design project in its totality. This does not mean that any project that is drawn must be modeled exactly as the drawing states. The student will find that the projects in this book offer ample opportunity for redesigning and relocating pipelines, equipment, valves, and nozzles. This is an excellent opportunity for students to use their skill in determining alternative positions of equipment and pipes. All models should be at a sufficient scale to enable the student to understand fully the relationship of drafting to model making and the need to be accurate and concise. Any redesign of projects must follow specifications laid down in the preceding chapters.

The Model Shop

If a school cannot set up a separate model shop, a small area in the drafting room can be used, though it is advantageous to have access to more space. Space requirements depend on the emphasis the drafting program wishes to put on modeling and the sophistication of equipment. Industrial model shops can occupy entire floors of buildings that stretch out over a city block; other model shops may occupy only 1100 ft^2, depending on the number of model makers. Storage space must be provided for equipment, tools, students' supplies, and, in some cases, students' models and equipment. For a class of 10 drafters to use a model shop at the same time, it would be necessary to provide a minimum of 2000 ft^2, though it must be stressed that small individual models or large group models can be constructed directly in the drafting room as class or individual projects. All piping drafting classes should attempt the construction of at least one project.

DRAWING AND THE MODEL

Though this chapter emphasizes the use of models in piping design, by no means is it suggesting that a model will totally eliminate the use of drawings. The majority of the design disciplines in engineering firms, such as civil, electrical, HVAC, and structural, are still required for various projects. The model helps to reduce the number of drawings necessary, but it does not eliminate drafting entirely. Even with the use of a model in the design stage, fabrication isometrics, spool drawings, pipe support drawings, and the like still play an essential role in the total design procedure. Though the use of the orthographic, plan, section, and elevation method can be limited, in many cases isometrics and detail views of the project are still essential. Orthographic drawings then become the exception not the rule.

Isometrics can be sketched by the model designer or piping designer and then be redrawn by the drafter, though in some cases the drafter works directly from the model, sketching and drawing the pipe fabrication isometrics necessary for construction of the project. Drawings are still much more suitable for minor installations and changes to existing projects. The more complicated and larger the project, the more advantageous the use of models in comparison to drafting.

When the model has reached a certain stage, isometrics can be taken of the various lines, and modeling and drawing can be alternated. When the model is tagged and labeled, it is easier to recognize and construct isometrics.

In some cases, drawings are done from a model as an after-the-fact drafting procedure for recording elements of the project. Thus drafters with less experience can work on the drawings. With orthographic, plan, elevation, and section drawings in the design stage, however, a high level of competence and engineering experience on various systems is an absolute prerequisite. The drafter can depict the work on the background of the model, making sections, elevations, and isometrics, working in conjunction with the model making procedure. No matter how sophisticated the modern techniques of piping design become with the use of models and computers, drafters will always find a ready market for their services.

TYPES OF MODELS

The following paragraphs describe the kinds of models that may be encountered in industry. Industrial piping models are only one of the many kinds employed. Engineering models are used in every field from defense department prototypes for missiles to architectural models of airports. The piping design model concept is less than a quarter century old and is still developing. Models are also used for a variety of purposes already mentioned—engineering models for a design, architectural models for display and sales, prototype models for public acceptance, and models used for product development.

Architectural Models

Architectural models, one of the first models developed in industry, include a variety of study and design models, presentation models of finished projects, sales information models, experimental models, and check models. In many cases, architectural models end up being purely for show and appearance of the total facility. They are also used for environmental impact reports and may show a variety of topographic and geographic features of the area surrounding the intended structure, features which are not usually part of the engineering model. Thus architectural models offer the opportunity to view the prospective project in relationship to the surrounding area and construction site. Visual aspects such as orientation to sun and surrounding structures play an important part in the use of architectural models.

Architectural models help eliminate problems concerning the use of the facility, such as local zoning, restrictions on space, utilities, transportation facilities, government services such as fire and police, and the geographical relationship of the structure to the surrounding ecosystem. Architectural models are seldom used as design tools in the sense of piping industrial models.

Study Models

Study models are usually extremely small-scale planning tools used by government or industrial groups to gain a general perspective of the project. Models may include a wide range of existing and proposed facilities in relationship to the area as a whole. Study models are usually constructed of materials that are simple to use and inexpensive.

Presentation Models

Presentation models are similar to study models, but they involve the presentation of a proposed project. Therefore the model must be complete with detail, showing the design and environmental aspects of the project. Presentation models are used throughout government and the private sector to gain approval for projects from local communities. Design elements and overall aspects are shown directly and faithfully on the presentation model so that those concerned with the implementation of the project can make a thorough and honest evaluation.

Product Models

Product models are often used in mechanical engineering to help design various aspects of machinery or other mechanical devices. In many cases, these models are made to a larger scale than the eventual project.

Topographical Models

Topographical models are used throughout government and private industry to show geological configurations for the mapping, grade, elevation, and terrain of intended project property. They are primarily used to study the terrain and determine the most effective design use for projects such as highways and for overall city and county planning. In general, the topographical model is made to an extremely small scale because of the area it encompasses.

Prototype Models

Prototype models are basically similar to product models, but they are working simulations of the eventual product. In the case of automobiles, the company may build an actual prototype of a future car to gain knowledge of its aerodynamic engineering aspects and also to gain an estimation of public acceptance and sales. Some prototype models are just simplified mockups of the eventual product.

Piping Models

Piping models, the primary concern of this chapter, deal mainly with petrochemical, refinery, food processing, power generation, and manufacturing facilities. Piping models are made up of piping systems and related equipment.

Plot Plan Models Plot plan models, also referred to as site plan models, are usually constructed on an extremely small scale. They depict the major plant area in relationship to the entire site—all environmental characteristics, both natural and artificial, including adjacent structural items, roadways, railroad connections, pipe alleys and racks, existing equipment, and buildings. Since the plot plan model offers the ability to create alternative arrangements with regard to the proposed new facility, the designer can take an overall view of the surrounding site in its relationship to rivers, lakes, and other important features. Plot plan models cannot be made as accurately as models constructed to a larger scale, but they do simulate the available space and design options that are available. Plot plan models also permit engineering groups to discuss the options available within the prescribed physical limits. The plot plan may include topographical configurations, thus making its use more effective. Plot plan models are often used as display models to show the government and public agencies the proposed facility's configuration.

Plant Layout Models Plant layout models are the equivalent of plot plan models. They deal primarily with existing or proposed buildings where the use of space is limited and determined by an architectural structure. Plant layout models have a variety of block-type items to represent the equipment involved in the project. They often are used in architectural areas to divide office space and indicate other floor space options.

Preliminary Models Preliminary models are also referred to as plot plan models or block-type models. They are used to facilitate quick changes concerning plant alternatives, equipment arrangements, or plot plan situations. To review the preliminary model, all the engineering and design groups in a company meet to determine the best possible arrangement of the project in question. This model, therefore, takes on extreme significance for the establishment of important design decisions, including plot plan arrangement, expenditures for engineering subgroups, and operating and maintenance problems. The preliminary model is usually constructed of blocks made from styrofoam or wood that are only temporarily placed on the base of the model, enabling a variety of arrangements to be made quickly. Structural components are also used on the preliminary model and are usually clipped together for easy disassembly. All preliminary models must permit quick changes to be made by moving structural elements and identifiable blocks representing equipment variations.

Design Models The design model can be broken into two steps: the basic design model and the final design model. The basic model is assembled by the model technician in a modeling room with all the equipment, such as bases, tables, and other structural elements, needed to construct the model. Platforms, equipment, vessels, bases, structural steel, concrete and walls, flooring—all are completed on the basic model. Figure 15-2 shows an example of the basic design model; in this case, a few of the larger piping runs have already been established on the model. Figure 15-3 shows another design model in its basic stage; a variety of vessels, structural and concrete elements, and other equipment is in place, but no piping has yet been attached to the model.

At the piping design stage the basic model is sent to the drafting and design area and the piping designers and drafters run pipe connecting the equipment and establish the instrumentation, electrical, HVAC, and pipe support features. It is from this final design model that the drafting groups work, including the pipe support division and isometric group. A design model, when completed, includes all the elements of the future plant—equipment, vessels, structures, piping, piping components (valves, fittings, meters), instrumentation, electrical, HVAC, lighting. The final design model is the project in miniature.

Check Models Check models in many instances contain all the elements of the design model, but the timing of construction differs. Check models are made after all the drawings and engineering sketches and design have taken place. They are used only to locate interferences and problems that cannot be readily seen on the orthographic drawings. Therefore check modeling is an expensive, after-the-fact procedure that does not provide for the most efficient use of modeling.

Most of the modeling projects in this book should be constructed as check models, however. In other words, the drawings should be done first and the model built from the drawings. This procedure is advocated so that the student becomes familiar with the *drafting* procedures in piping design. In Chapters 12 and 13, it is also possible to have a group construct the model directly from the prints provided

DESIGN MODEL CONSTRUCTION

The piping design and engineering departments coordinate the overall construction of the design model. Therefore all of the subgroups—electrical, HVAC, and instrumentation—must deal directly with the model coordinator. This arrangement helps to eliminate problems in the use of space on the model that may be allocated for different equipment. The design model, after the basic modeling stage, is usually located in the drafting department. If this practice is not feasible, areas for drawing, drafting boards, and so forth can be provided in the model room.

The first step in the design of an industrial plant is the construction of the preliminary or block-type model where simplified blocks and cylinders are used to represent equipment variations and design possibilities. After an acceptable plant layout has been established and reviewed by all of the engineering disciplines, the arrangement is approved. At this point certain drawings are taken from the model, such as the plot plan drawing. With the use of the model, the drawings and layout of the projected plant can be done quickly by less experienced personnel. Any plan, section, and elevation drawings that may be needed can be taken directly from the preliminary model. The plan layout drawings, when used, become a working document of the prospective plant and can be distributed to the subgroups in the engineering department for detailing and location of equipment such as HVAC, ducting, electrical trays, ladders, platforms, electrical components, instrumentation, and valves.

The model allows for a constant state of review and approval by the various design groups. Electrical conduit runs and ductwork can be added to the model in the earlier stages, along with any small equipment that is not manufactured and is available in the model department. Simple freehand sketching can be used to offer design options for routing pipe and locating in-line valves and instruments. It may take two or three changes to obtain the best possible pipe routing. Any minor steel, ductwork, conduit runs, or equipment can also be added at this time. After the piping designer has established all the major piping runs and connections for the required process, the model is made accessible to the process and instrumentation engineers for review.

A continuous overview is maintained by all the engineering groups so that interferences and design problems can be worked out at the earliest possible stage. This feature, of course, is not available when using orthographically projected drawings. The design and construction progress can also be documented much more easily on the model than on the drawings. At this stage, the model offers complete and correct plant design possibilities. From this model, using the basic coordinates and locations for pipes and equipment, drafters and detailers are able to produce isometric views of each line and provide all the fabrication dimensions necessary.

When using models in the design stage, accuracy and the establishment of allowable tolerances are primary concerns. The design group checker must review the completed design model and check the dimensions for accuracy. At this time, it is also possible to redesign any areas in the plant that present interference or process problems.

If the overall model is of a large scale and the plant is of sufficient size, models must be constructed in modules so that portions can be disassembled or unfastened to permit the designer and construction crew to view all the areas. A typical nuclear power plant may have upward of 30 sections that can be pulled apart and viewed separately, ranging in size and on bases of 2 x 4 ft or 2 x 5 ft. This breakaway sectioning is essential in the modeling of large complexes.

Determination of the design model scale is extremely important. The preliminary model or site model scale is usually 1:200 or 1/4 in. = 1 ft. Topographical or architectural models are constructed as small as 1 in. = 50 ft. In general, the final design model is constructed from 1/4 in. = 1 ft to 3/4 in. = 1 ft, depending on the complexity and size of the intended plant. In some cases, 1½ in. or 3 in. to a foot are also used. Some engineering firms use full scale for modeling problem areas, such as access areas to instrumentation on nuclear reactors or modeling the control room for placement of instrumentation and operating controls. Among the factors that determine scale are the complexity or detail that is necessary, the relationship of this scale to available model components, the amount of space available for model construction, and the economics of model construction at a particular scale.

In most cases, 1/2 in. = 1 ft and 3/8 in. = 1 ft are the scales most common in the petrochemical and power generation fields. Petrochemical plants, refineries, and process plants offer more space than nuclear or conventional power plants, mainly because the power plant must be confined within a building whereas a petrochemical installation is seldom confined to one structure and can therefore be spread out over a wider area.

When determining scale, it is also important to consider the cost factor in the construction of bases and items that cannot be ordered. Fabrication of equipment, vessels, and bases that cannot be catalog ordered may increase the cost of the project considerably. The amount and size of piping to be shown and constructed on the model will also influence the scale. In problem areas, piping down to 1 in. must be constructed on the model. Though this practice prevents problems, including overcomplication of certain areas to be modeled, the basic procedure is to model all piping 2 in. OD and above unless smaller piping of extremely critical areas is necessary. Since the smallest pipe available from model component companies is 1/16 in. in diameter, this will determine its use as a minimum-OD pipe size. Larger scales may be necessary where small piping must be constructed on the model. Another problem in determining the scale is the model's intricacy. If the scale is too small, it may be impossible to construct certain areas and problems will arise in taking isometrics from the model. Large-scale models can increase the cost of the total engineering production and require the use of fictitious supports.

Most of the models that can be constructed from this book should be scaled at 3/8 in. = 1 ft where possible. In Chapter 12, that scale would present size and cost problems so the power plant should be modeled at a maximum of 3/32 in. = 1 ft. The same is true of the nuclear reactor in Chapter 13. The smaller projects in the book can be modeled to a larger scale to provide more detail and increase their value as teaching aids.

TOOLS

Model building tools are divided into two categories: those required in the model shop and those required in the design room. The school that wishes to set up a small model program in conjunction with the piping drafting class need only concern itself with tools used in the design room. Tool requirements can also be categorized as hand tools versus power tools.

Hand Tools

Hand tools for the design room include clamps, vises, saws, picks, razors, putty knives, scribers, steel rules, hemostats, pliers, hammers, combination squares, toolboxes, leather knives, Exacto knives, hacksaws, small twist drills, solvent applicators, marking pens, pencils, straightedges, scale tapes, protractors, dividers, circle templates, screwdrivers, files, tubing cutters, razor cutters, lightweight hammers, utility knives, pen vises, drill sets, tweezers, needlenose pliers, cutting pliers, wire cutters, ignition pliers, and V blocks.

Among the larger tools for common use in the design room are bending boards, vises, V blocks, extension cords, brushes, clamps, a variety of triangles, framing squares, combination squares, sets of screwdrivers, hammers, scissors, a variety of file configurations, glue dispensers, calipers, dividers, protractors, utility rules, large metal tapes and scales, hand drills, extension drills, and adjustable wrenches.

The hand tools in the model shop include most of the items for the piping designer, but also involve larger squares, framing squares, metal rules and tapes, wrenches, and hammers. Other equipment is not usually present in the design area but is necessary for the model builder: hand saws, large drafting triangles, tap and die sets, hand drills, needle-nose pliers, linesman's pliers, paintbrushes, bench vises, portable vises, large clamps, awls, knives, allen wrenches, box end wrenches, crescent wrenches, and screwdrivers of all types. This list, of course, can be expanded greatly. It must be noted that not all the tools listed for the design and model areas are necessary. It is possible to construct a model in a drafting room with very few tools indeed. Increasing the number of tools available will benefit the accuracy of the model, however.

Power Tools

The designer sander and designer table saw are two pieces of equipment that should be included in the design area and perhaps in the model shop. These high-quality, portable machines provide for accurate construction. Design area machines include the designer saw, a small disk sander, and a hand-held drill or small drill press. These power tools are all that is really necessary for the piping designer or pipe drafting student. The designer saw can cut a variety of angles and shapes on wood, plastics, and other materials. The sander is used for shaping, squaring, and eliminating rough edges on components. The hand drill or small drill press, in conjunction with a V block, can provide quick drilling of holes.

Model shop power tools include tilting arbor or table saws, precision saws, floor model drill press, lathe, belt and disk sander, joiner, portable router, portable hand drills, vacuum, spray paint equipment, radial saws, and band saws. All these pieces of equipment make the construction of a model more efficient and cost-effective. The disk sander is similar to the designer's sander, but it is usually larger and is used for sanding different shapes. The belt sander is used primarily for rectangular objects. The joiner is used for the construction of straight edges on sheet or board stock. The band saw is used to cut odd shapes. The radial saw is used for large items that must be cut to a specified length, such as model bases. The drill press is used for drum sanding and drilling and is essential for constructing models made of plastic. The lathe is used for turning a number of nonstandard items and for drilling items that cannot be done on other machines. It is also used for shaping wood or plastic.

Purchasing all these machines will run up the original cost of the model shop. In fact, schools may find it prohibitive to construct a model shop with all the equipment that has been mentioned. The minimum equipment for a model program includes only the design area hand tools and power tools, which can be purchased for approximately $300.

Safety Precautions

Because model building entails the use of power tools and hand tools that are extremely hazardous if operating procedures are not adhered to, it is important to institute a safety program in conjunction with every model building class. The Occupational Safety and Health Administration standards should be closely followed. Everyone involved in a model building program should review the safety and maintenance rules and conditions necessary to keep the model shop a safe warkplace. The most serious area is in the use of power tools, including saws, sanders, drills, lathes, and other equipment that can cause serious bodily injury to the operator and those in the surrounding area if not properly maintained and operated.

Hand tools, including knives and saws, hand drills, razor cutters, and scribers, all present safety problems if the worker is not properly trained in their use. The materials used in the model shop, including wood, plastic, metal, and standard component items, can present serious health hazards if handled improperly. Sharp edges on plastic and wood, metal slivers, edges on component parts that have been machined—all must be taken into consideration when instituting a set of safety rules.

Another area involving safety precautions is in the use of solvents, cements, and other flammable materials, which can cause serious environmental hazards if improperly used and maintained. Sloppiness in the model room, buildup of dust, plastic chips, paint supplies, and rags, are all fire hazards. Electrical equipment must be properly maintained and precautions taken in the use of extension cords, electrical outlets, and on and off switches. Air contamination from solvents and power tools must also be considered. Proper ventilation of the working area is essential.

MODEL COMPONENTS

Standard model components are available through Engineering Model Associates, Inc. The model industry has been standardized in many areas, therefore eliminating the necessity of creating components from scratch in the model room. The petrochemical models, process piping models, and power generation models used in industry have contributed to the research for standard model components. Over 6000 standard items are available through Engineering Model Associates, Inc. (Figs. 15-9 through 15-13); their catalogs are available upon request.

The use of standard parts makes the assembly of models much easier, especially in schools, where it may not be possible to fabricate model parts in the drafting or model room. Standard model parts can, in most instances, supply 80 to 95 percent of the components for models or piping installation projects. The model projects in Chapter 11 can be constructed entirely from standard components with only slight modifications. The large model projects in Chapters 12 and 13, on the other hand, require many of the components to be fabricated from wood or plastic.

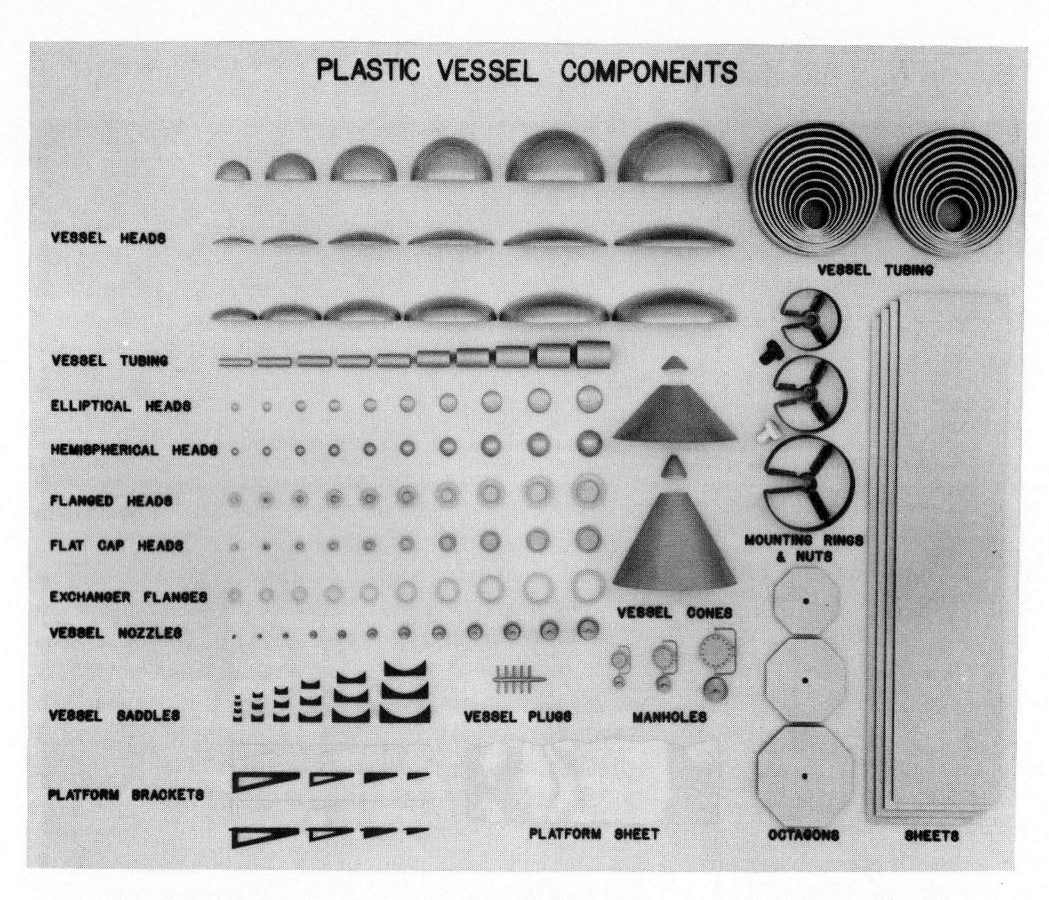

FIG. 15-9 Plastic vessel components. (Courtesy Engineering Model Associates, Inc.)

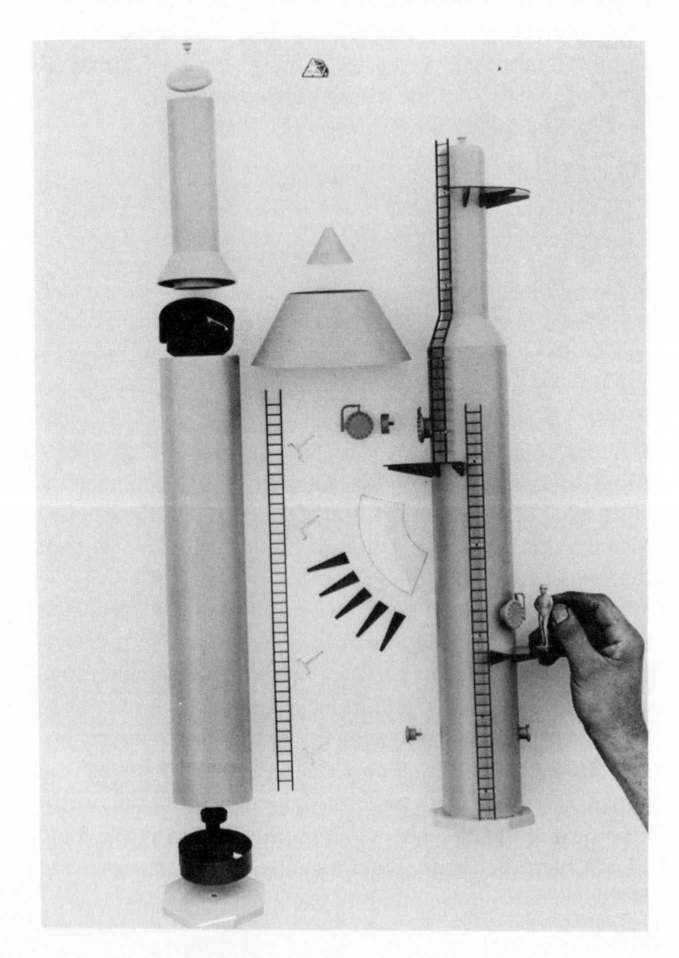

FIG. 15-10 Vessel and equipment assemblies. (Courtesy Engineering Model Associates, Inc.)

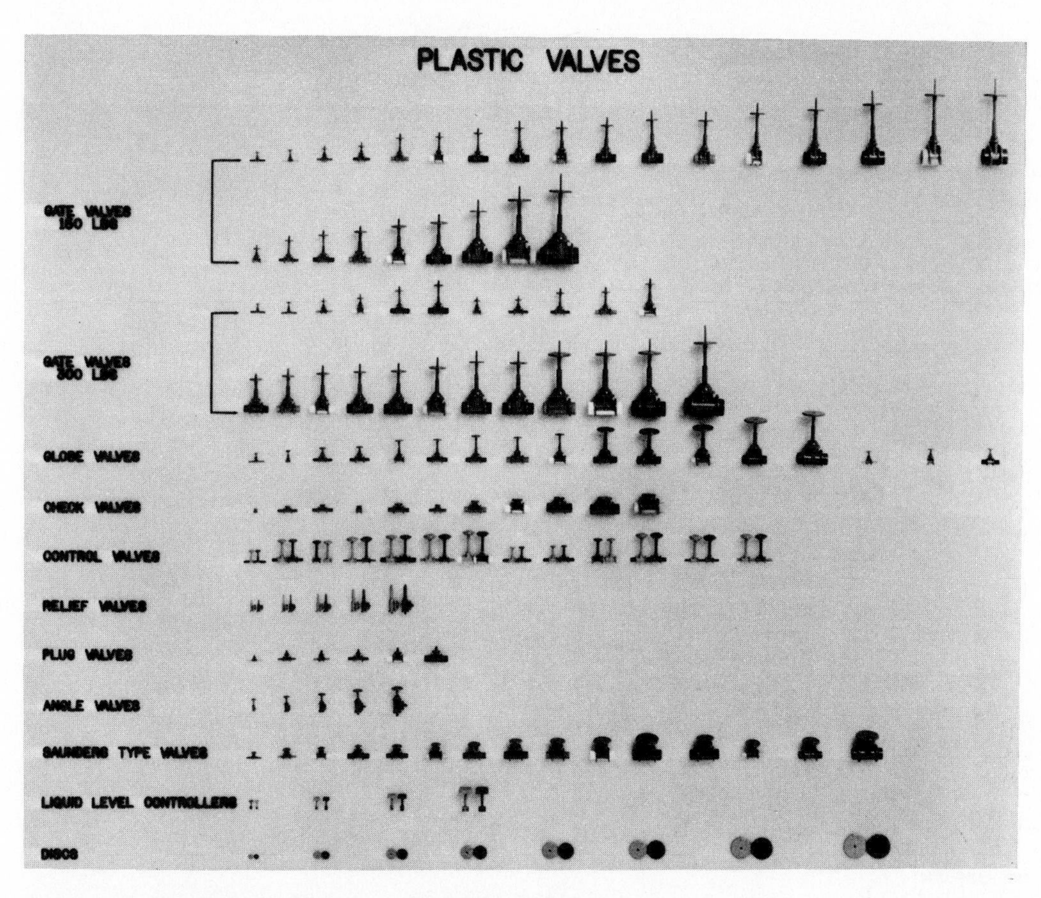

FIG. 15-11 Plastic valves. (Courtesy Engineering Model Associates, Inc.)

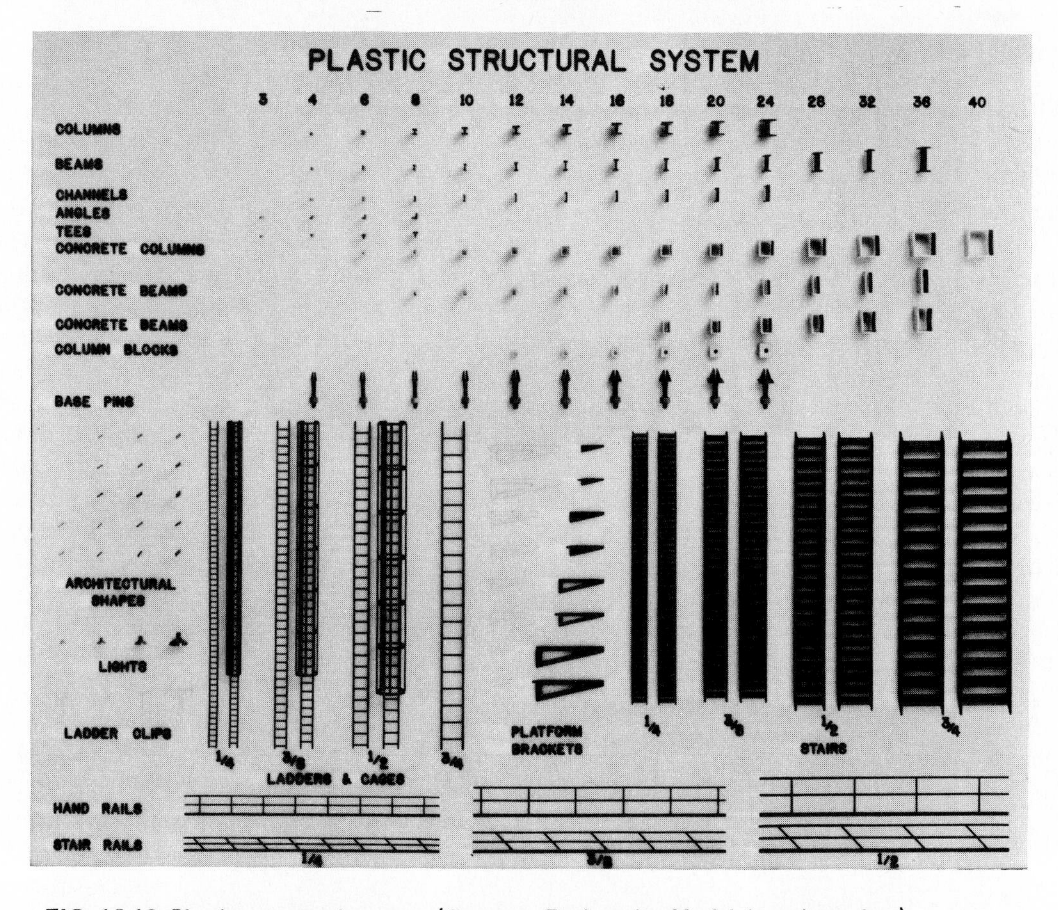

FIG. 15-12 Plastic structural system. (Courtesy Engineering Model Associates, Inc.)

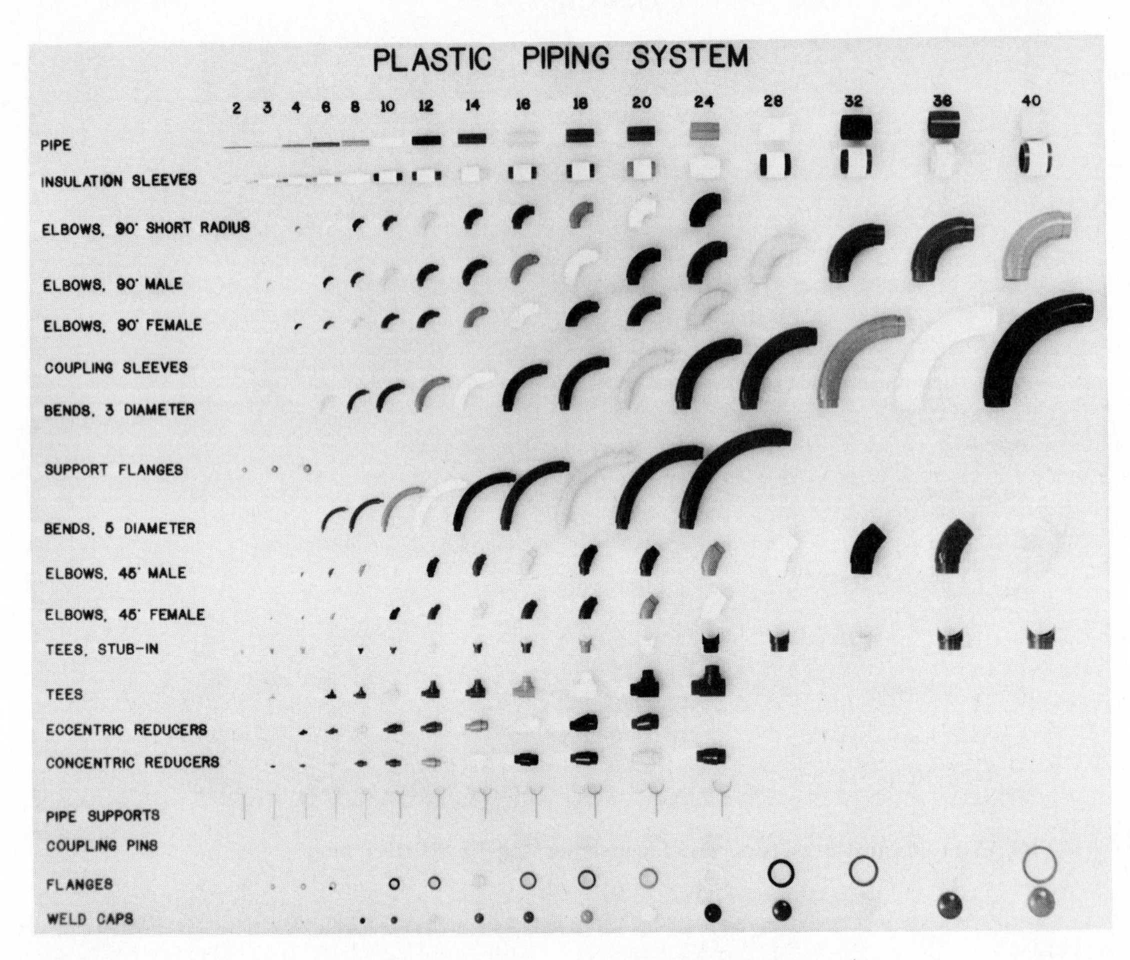

FIG. 15-13 Plastic piping system. (Courtesy Engineering Model Associates, Inc.)

Materials and Supplies

Besides the standard components, a wide selection of materials are also used in the construction of models: plastics, wood, metal, solvent, cement, paint. The model maker and piping designer must know how to use these materials in conjunction with each other, utilizing the proper bonding procedure for the different plastics, woods, or metals in the model.

Standard components which consist of molded plastics can be easily cemented together with solvent to form the configurations for the piping system. Whenever possible, these plastics should be used in place of expensive metals or woods which may not hold up as well. Use of standard components has greatly reduced the need to fabricate parts from heavy plastics, wood, or metal. Familiarity with materials and their most suitable application is a must for the piping designer and model builder.

The model that can be constructed from the drawings in Chapter 12 includes a number of fabricated items for the steam generator and ducting. Besides balsa, the modeler can also use poplar and pine in the construction of model parts. Plywood is often used as a base with plastic sheets over the top to provide a more accurate working surface. Total model cost will vary drastically according to the materials used. Fot the sake of economy, it is possible to use fiberboard or plywood for the model base and not cover it with expensive acrylic sheets, though this detracts from the appearance of the model. The model that can be built from project 4 in Chapter 11 may vary as much as $10 to $20, depending on the type of base that is used, which can be glass, plywood, fiberboard, metal, or plastic sheets.

Labor usually accounts for 80 to 90 percent of the modeling cost. Materials usually only run 10 to 20 percent of the total modeling construction, though for schools labor is not a problem like cost of materials. In some instances, schools can afford to keep a large inventory, though this practice is usually not possible because of the expense and the storage facility necessary to handle bulky items. If the school cannot afford to have standard component items stocked, the student should do a material takeoff from the book or while preparing the drawings for the model. In this way a majority of the components will be available when the drawings are done—instead of waiting for the order to be filled after the drawings have been completed. This arrangement, of course, depends on the school.

Color Coding

Standard components come in a selection of colors. Almost all are available in yellow, blue, orange, pink, green, white, gray, red, light blue, brown, and black. All models are constructed with a color code in mind. Though the code may differ from company to company, there are certain standard procedures. There are three different ways of color coding a model: by service, by materials, and by process and service.

When grouping by service, it is suggested that the following scheme be used:

1. Process: yellow
2. Utilities: blue (or broken down into steam and condensate: orange; water: blue; air: white; gases: black or brown; fire protection service: red)
3. Existing piping: white
4. Valves: gray
5. Insulation: white (or broken down into hot insulation: white sleeve; cold insulation: purple sleeve; steam-traced insulation: light blue sleeve; electrical-traced insulation: electrical green sleeve)
6. All electrical items (including trays, electrical equipment and conduit): green
7. HVAC: buff (or preferably light or dark blue)
8. Roads: brown
9. Plot plan or limits (including battery limits): black tape
10. Building walls: black tape
11. Trenches: black outline
12. Underground piping: black centerline tape
13. Underground electrical: electrical green tape
14. Underground instrumentation: instrumentation pink tape

When color coding by materials, the following scheme should be used:

1. Carbon steel: yellow
2. Alloy: red
3. Special materials: orange
4. Electrical: green
5. Galvanized: buff or blue

The following general list can also be used:

1. Structural steel: gray
2. Existing structures: white
3. Future structures: gray with speckles
4. Concrete: gray
5. Equipment: gray
6. Existing equipment: white
7. Future equipment: valve gray with speckles
8. Electrical: green
9. Instrumentation: pink
10. HVAC: light or dark blue
11. Fictitious ports: transparent (clear)
12. Safety equipment: red

When grouping by process and service, the following scheme is suggested:

1. Process: yellow
2. Utilities: blue
3. Special service: orange
4. Electrical: green
5. HVAC: buff or blue

It is hoped that any school or group using this book in the classroom will provide the student with the opportunity to fabricate at least one model in a group or individually. This hands-on experience is essential if the student is to understand piping systems drafting and design from a three-dimensional perspective, rather than just the two dimensions provided for in drafting procedures. For an in-depth coverage of modeling see the author's text Industrial Model Building, which covers all phases of design modeling. (Prentice-Hall, Inc., Englewood Cliffs, N.J., 1981).

QUIZ

1. When is a model called for in the production of a piping project?
2. Define an industrial model.
3. How can the use of a model save construction time?
4. How is a model used as a communication tool?
5. Name four advantages to the use of a model.
6. What are five common types of models?
7. How is a check model used?
8. What is the name of the model society?
9. Why are models color coded? Name three advantages.
10. How is a plot plan model used?
11. A design model is used for many reasons. Name four and explain.
12. How can a model help reduce interferences?
13. Do models always reduce the design cost? Why?
14. How can the use of a design model influence the type, amount, and production of drawings on a project?
15. How are models used after the project is finished?

PROBLEMS

1. Construct a model of the oil refinery addition in Chapter 11 (Problem 4). After completing a set of orthographically projected drawings and details of the refinery section at 3/8 in. = 1 ft scale, order and construct a complete model at the same scale (Fig. 11-20).
2. As a class or group project, prepare a complete set of drawings of the fossil fuel power generation plant in Chapter 12 (Problem 1) at a scale of 1/8 in. = 1 ft or 3/32 in. = 1 ft. Make a model of the power plant in the same scale (Fig. 12-3). A good percentage of this model will need to be fabricated.
3. As a class project, prepare a complete set of views and details of Fig. 13-16. Choose an appropriate scale based on the desired size of the model. (This model could be a single project for one student using a very small scale.) Complete a large-scale model of the nuclear containment area as a group project.
4. Construct a model of Problem 5 in Chapter 11. 3/8 in. = 1 ft scale (Fig. 11-33).
5. Many of the drawings throughout the text will make excellent model projects, particularly those in Chapters 5, 7, 8, and 9. The instructor should judge the desirability of each project before assigning it as a model project. Many of the smaller systems in Chapters 5 and 7 would make excellent beginning-level model projects.
6. Take an isometric drawing from an existing model and do a spool drawing, and material takeoff.
7. Take a series of photographs of a model that has been made in class. The instructor will review the photographs for clarity, view, camera angle, and other features.

APPENDIX A: Standards

NATIONAL BUREAU OF STANDARDS: SUPERINTENDENT OF DOCUMENTS, U.S. GOVERNMENT PRINTING OFFICE, WASHINGTON D.C. 20402

AIA	American Insurance Association
ANSI	American National Standards Institute
API	American Petroleum Institute
ASME	American Society of Mechanical Engineers
ASTM	American Society for Testing and Materials
AWS	American Welding Society
AWWA	American Waterworks Association
ISA	Instrument Society of America
MSS	Manufacturers' Standardization Society of the Valve and Fittings Industry
PFI	Pipe Fabrication Institute
ANSI	American National Standards Institute, 1430 Broadway, New York, NY 10018
ASME	The American Society of Mechanical Engineers, 345 East 47th Street, New York, NY 10017
ASTM	American Society for Testing Materials, 1916 Race Street, Philadelphia, PA 19103
MSS	Manufacturers Standardization Society of the Valve and Fitting Industry, 1815 North Fort Myer Drive, Arlington, VA 22209

PIPE FLANGES, FITTINGS, GASKETS & VALVES

B16.1	(1975)	Cast Iron Pipe Flanges and Flanged Fittings-Class 25, 125, 250 and 800.
B16.3	(1977)	Malleable-Iron Threaded Fittings, Class 150 and 300.
B16.4	(1977)	Cast Iron Threaded Fitting, Class 125 and 250.
B16.5	(1977)	Steel Pipe Flanges and Flanged Fittings (Including Ratings for Class 150, 300, 400, 600, 900, 1500 and 2500).
B16.9	(1971)	Factory-Made Wrought Steel Buttwelding Fittings.
B16.10	(1973)	Face-to-Face and End-to-End Dimensions of Ferrous Valves.
B16.11	(1973)	Forged Steel Fittings, Socket-Welding and Threaded.
B16.12	(1977)	Cast-Iron Threaded Drainage Fittings.
B16.14	(1977)	Ferrous Pipe Plugs, Bushings and Locknuts with Pipe Threads.
B16.15	(1971)	Cast Bronze Threaded Fittings, 125 and 250 lb.
B16.18	(1972)	Cast Bronze Solder Joint Pressure Fittings.
B16.20	(1973)	Ring-Joint Gaskets and Grooves for Steel Pipe Flanges.

B16.21 (1962) Nonmetallic Gaskets for Pipe Flanges.

B16.22 (1973) Wrought Copper and Bronze Solder-Joint Pressure Fittings.

B16.32 (1976) Cast Copper Alloy Solder Joint Drainage Fittings - DWV.

B16.24 (1971) Bronze Flanges and Flanged Fittings, 150 and 300 lb.

B16.25 (1972) Buttwelding Ends.

B16.28 (1964) (R1972) Wrought Steel Buttwelding Short Radius Elbows and Returns.

B16.29 (1973) Wrought Copper and Wrought Copper Alloy Solder-Joint Drainage Fittings.

B16.31 (1971) Non-Ferrous Pipe Flanges, 150, 300, 400, 600, 900, 1500 and 2500 lb.

B16.32 (1973) Cast Bronze Solder-Joint Fittings for Solvent Drainage Systems.

B16.33 (1973) Small Manually Operated Metallic Gas Valves in Gas Distribution Systems whose Maximum Allowable Operating Pressure Does Not Exceed 60 PSIG or 125 PSIG.

B16.34 (1977) Steel Valves (Flanged and Buttwelding End).

B16.36 (1975) Steel Orifice Flanges, Class 300, 600, 900, 1500 and 2500.

B16.39 (1977) Malleable Iron Threaded Pipe Unions, Class 150, 250, and 300.

PIPING

Code for Pressure Piping, B31 Interpretations:

B31 Guide, Corrosion Control for ANSI B31.1 Power Piping Systems.

B31.1 (1977) Edition, Power Piping (With Addenda up to the 1980 Edition).

B31.2 (1968) Fuel Gas Piping.

B31.3 (1976) Chemical Plant and Petroleum Refinery Piping.

B31.4 (1974) Liquid Petroleum Transportation Piping Systems.

B31.4a (1977) Referenced Standards Addendum to B31.4-1974, Liquid Petroleum Transportation Piping Systems.

B31.5 (1974) Refrigeration Piping.

B31.7 (1969) Nuclear Power Piping.

B31.8 (1975) Gas Transmission and Distribution Piping Systems.

ASME Guide for Gas Transmission and Distribution Piping Systems -

B32.1 (1952) (R1977) Preferred Thickness for Uncoated, Thin, Flat Metals. (Under 0.250 in.)

B32.2 (1969) (R1974) Preferred Diameters for Round Wire - 0.500 inches and under.

B32.3 (1977) Preferred Metric Sizes for Flat Metal Products.

B32.4 (1977) Preferred Metric Sized for Round, Square and Hexagon Metal Products.

B32.5 (1977) Preferred Metric Sizes for Tubular Metal Products Other Than Pipe.

B32.6 (1977) Preferred Metric Equivalents of Inch Sizes for Tubular Metal Products Other Than Pipe.

B36.10 (1975) Welded and Seamless Wrought Steel Pipe.

B36.19 (1976) Stainless Steel Pipe.

DRAFTING PRACTICES, CHARTS AND ILLUSTRATIONS

Y14 Report, Digital Representation of Physical Object Shapes.

Y14 Report, Number 2-Guideline for Documenting of Computer Systems Used in Computer-Aided Preparation of Product Definition Data-User instructions.

Y14 Report, Number 3-Guideline for Documenting of Computer Systems in Computer-Aided Preparation of Product Definition Data-Design Requirements.

Y14.1 (1975) Drawing Sheet Size and Format.

Y14.2 (1973) Line Conventions and Lettering.

Y14.3 (1975) Multi and Sectional View Drawings.

Y14.4 (1957) Pictorial Drawing.

ES STANDARDS AVAILABLE FROM PIPE FABRICATION INSTITUTE

Standards

ES1 End Preparation and Machined Booking Rings for Butt Welds (1974)

*ES2 Method of Dimensioning Piping Assemblies (1974)

ES3 Fabricating Tolerances (1974)

ES4 Hydrostatic Testing of Fabricated Piping (1975)

ES5 Cleaning of Fabricated Piping (1975)

ES6 Recommended Practice for Heat Treatment of Pipe Bends and Formed Components of Carbon and Ferritic Alloy Steels (1977)

ES7 Minimum Length and Spacing for Welded Nozzles (1962) (Reaffirmed 1975)

ES11 Permanent Marking on Piping Materials (1975)

ES16 Access Holes and Plugs for Radiographic Inspection of Pipe Welds (1974)

ES18 Ultrasonic Examination of Tubular Products (1975)

ES19 Preheat and Postheat Treatment of Welds (1972) (Reaffirmed 1975)

ES20 Wall Thickness Measurement of Pipe Bends by Ultrasonic Examination (1975)

ES21 Manual Gas Tungsten-Arc Root Pass Welding End Preparation and Dimensional Joint and Fit Up Tolerances (1974)

*ES22 Recommended Practice for Color Coding of Piping Materials (1969) (Reaffirmed 1975)

ES24 Pipe Bending Tolerances-Minimum Bending Radii-Minimum Tangents (1975)

ES25 Random Radiography of Pressure Retaining Girth Butt Welds (1973)

ES26 Welded Load Bearing Attachments to Pressure Retaining Piping Materials (1976)

ES27 Visual Examination - The Purpose, Meaning and Limitation of the Term (1974)

ES28 Recommended Practice for Welding of Transition Joints between Dissimilar Steel Combinations (1976)

LIST OF MSS STANDARD PRACTICES

SP-6-1974	Standard Finishes for Contact Faces of Pipe Flanges and Connecting-End Flanges of Valves and Fittings
SP-9-1970	Spot Facing for Bronze, Iron and Steel Flanges
SP-25-1964	Standard Marking System for Valves, Fittings, Flanges and Unions
SP-42-1959	150 lb. Corrosion Resistant Cast Flanged Valves
SP-43-1971	Wrought Stainless Steel Butt-Welding Fittings
SP-44-1975	Steel Pipe Line Flanges
SP-45-1971	Bypass and Drain Connection Standard (formerly SP-5 and SP-28)
SP-51-1957 (R 1965)	150 lb. Corrosion Resistant Cast Flanges and Flanged Fittings
SP-53-1971	Quality Standard for Steel Castings - Dry Partical Magnetic Inspection Method
SP-54-1971	Quality Standard for Steel Castings - Radiographic Inspection Method
SP-55-1971 (R 1975)	Quality Standard for Steel Castings - Visual Method
*SP-58-1975	Pipe Hangers and Supports - Materials, Design
SP-60-1969	Connecting Flange Joint Between Tapping Sleeves and Tapping Valves
SP-61-1961	Hydrostatic Testing of Steel Valves
SP-65-1962 (R 1968)	High Pressure Chemical Industry Flanges and Threaded Stubs for Use with Lens Gaskets
SP-67-1976	Butterfly Valves
SP-69-1976	Pipe Hangers and Supports - Selection and Application
SP-70-1970	Cast Iron Gate Valves, Flanged and Threaded Ends
SP-71-1970	Cast Iron Swing Check Valves, Flanged and Threaded Ends
SP-71-1970	Ball Valves with Flanged or Butt-Welding Ends for General Service
SP-73-1970	Silver Brazing Joints for Wrought and Cast Solder Joint Fittings
SP-75-1973	Specification for High Test Wrought Welding Fittings
SP-76-1970	Malleable Iron Threaded Pipe Unions - 150, 250, and 300 lb.
SP-77-1971	Guidelines for Pipe Support Contractual Relationships
SP-78-1972	Cast Iron Plug Valves
SP-79-1974	Socket-Welding Reducer Inserts
SP-80-1974	Bronze Gate, Globe, Angle and Check Valves
SP-81-1975	Stainless Steel, Bonnetless, Flanged, Wafer, Knife Gate Valves
SP-82-1976	Valve Pressure Testing Methods

APPENDIX B: Tables, Charts and Components

GUIDE TO SELECTION OF CORROSION RESISTANT MATERIALS FOR INDUSTRIAL VALVES

Key: E = Excellent; G = Good; F = Fair; P = Unsatisfactory; .. = Information Lacking

	Body and Bonnet Material						Seat and Disc Material					Stem Material
Fluid Handled	S-1 T-1 Valve Bronze	Cast Iron	Ni-Resist Type 2	Cast Carbon Steel	Monel Metal	N-3 SA Metal	NT-7 High Tin Nickel Alloy	NT-4 Nickel Alloy	Stainless Steel (13% Chrome)	Stainless Steel (18-8 Mo)	Brin-Alloy 600 B.H.N. LQ 600	Alloy 6 R.S.B. "Stemalloy"
Formaldehyde	G	..	..	..	G	G	G	G	G	G	G	G
Formic Acid	G	P	..	..	G	G	G	F	F	E	G	P
Freon: Dry	E	F	G	F	E	E	E	E	E	E	E	E
Wet	G	F	G	F	G	G	G	G	F	G	G	F
Fuel Oil	G	G	E	G	G	G	G	G	E	E	E	G
Gasoline: Refined	E	E	E	E	E	E	E	E	E	E	E	E
Sour	..	..	G	..	..	..	..	..	G	E	..	..
Glucose	E	E	E	E	E	E	E	E	E	E	E	E
Glue	E	E	E	E	E	E	E	E	E	E	E	E
Glycerine	E	E	E	E	E	E	E	E	E	E	E	E
Hydrofluoric Acid	F	P	F-P	P	E	F	F	F	P	P	G	P
Hydrochloric Acid	P	P	P	P	P	P	P	P	P	P	..	P
Hydrogen Gas (Cold)	E	E	E	E	E	E	E	E	E	E	E	E
Hydrogen Peroxide U.S.P.	P	G	F	..	G	F	..	P	F	F	G	..
Hydrogen Sulfide: Dry	P	G	E	G	P	P	P	P	E	E	P	P
Wet	P	P	G	P	P	P	P	P	F	F	P	P
Lacquers	G	G	..	G	E	E	E	E	E	E	E	G
Linseed Oil	G	F	..	F	E	E	E	E	E	E	E	G
Liquors, Paper Pulp Mill:												
Black Liquor	P	F	G	G	G	G	..	G	F	E	G	F
Green Liquor	P	F	G	G	E	F	..	G	G	E	E	F
White Liquor	P	F	G	G	E	F	..	F	G	E	E	F
Magnesium Chloride	G	..	P	..	G	G	G	G	..	..	G	F
Magnesium Hydroxide	F	G	E	G	E	E	E	E	G	E	E	F
Mercury (80° F.)	P	G	G	G	F	F	F	F	E	E	E	P
Methyl Chloride	G	G	G	G	G	G	G	G	G	G	G	G
Milk	..	..	P	..	E	G	G	G	G	E	E	..
Mine Water: Coal	P	P	G	P	F	P-F	P	P	P	G	P	P
Copper	P-F	P	G	..	P	F	..	..	..	G	..	..
Molasses	G	F	F-G	F	G	G	G	G	G	E	E	G
Natural Gas	G	G	E	G	E	E	E	E	E	E	E	G
Nitric Acid	P	P	P	P	P	P	P	P	P	G	P	P
Oleic Acid	F	P	G	P	G	..	E	F	P	E	G	F
Oxalic Acid 5%	F	F	..	P	F	..	P	P	F	E	F	..
Oxygen	E	G	E	G	E	E	E	E	E	E	E	E
Phenol: Crude	F	F	F-G	P	E	E	..	G	P	E	E	..
U.D.P.	E	E	E	E	E	E	..	E	E	E	E	..
Petroleum Oils: Crude	..	F	G	F	F	..	..	..	G	E	F	..
Refined	E	E	E	E	E	E	E	E	E	E	E	E
Phosphoric Acid: 75%	G	P	P	P	G	..	..	..	..	E	G	..
55%	P	P	P	P	P	..	..	..	..	F	P	P
Potassium Hydroxide	P	G	E	G	E	F	G	F	G	E	E	P
Pyroligneous Acid	G	G	G	..	F	G	E	..	G	E	G	..
Propane Gas	E	E	E	E	E	E	E	E	E	E	E	E
Sodium Bicarbonate 10%	G	F	G	F	G	E	G	G	E	E	G	F
Sodium Carbonate	G	E	E	E	E	E	E	E	E	E	E	E
Sodium Chloride	G	P	G	P	E	E	E	G	P	F	E	F
Sodium Hydroxide	P	G	E	G	E	E	E	F	G	E	E	P
Sodium Hypochlorite	P-F	P	P	P	P-F	P-F	P	P	P	F	G	P
Sodium Nitrate	P	G	G	G	G	G	G	G	E	E	G	E
Sodium Silicate	P	G	E	G	E	G	E	G	E	E	E	P
Sodium Sulfate (Pure)	F	..	F	..	G	G	G	G	..	..	G	F
Sodium Sulfide	P	F	F	F	F	F	F	P	G	G	F	P
Sodium Sulfite	G	F	..	..	G	F	..	G	G	G	G	..
Sulfuric Acid: 78% or less	F	P	P	P	E	E	G	G	P	P	G	F
78% to 90%	P	G	G	G	G	F	G	F	F	F	G	P
90% to 95%	P	G	G	G	P-F	F	F	F	G	E	F	P
Fuming	P	G	P	G	P	P	P	P	E	E	P	P
Sulfurous Acid	G	P	P	P	P	P	P	P	P	G	P	F
Tannic Acid	G	..	F	..	G	..	..	..	..	G	G	..
Toluol	E	E	E	E	E	E	E	E	E	E	E	E
Turpentine	E	..	G	..	E	E	E	E	E	E	E	..
Vinegar	F	P	F	P	G	F	F	F	F	E	G	F
Zinc Chloride	P	P	F	P	P	P	..	P	P	P	P	..

FLOW OF FLUIDS

In the field of hydraulics an important engineering problem is the calculation of the energy loss caused by the resistance of fluids to movement through pipe, valves, fittings, or other enclosed channels. Many empirical equations have been developed to determine the energy loss for specific fluids and flow conditions. Most of these empirical equations are based on the following equations:

$$h = f\,L\,\frac{v^x}{d^y}$$

The friction factor (f), the exponential coefficient (x) for the velocity, and the exponential coefficient (y) for the diameter are based on experimental data. These empirical equations give very good results when they are used for fluids and flow conditions within the limits for which they were developed. However, large errors may result if an empirical equation is used for conditions beyond the limits for which it was developed.

The Pipe-friction Equation was developed by the use of fundamental hydraulic principles and dimensional analysis to determine the energy loss for the steady flow of incompressible fluids through circular cross section pipe. The names Fanning, Darcy and Weisbach have been associated with variations of the following Pipe-friction Equation:

$$h = f\,\frac{L}{D}\,\frac{v^2}{2g}$$

The friction factor (f) was determined experimentally, and it was considered to be constant for a limited range of conditions. The Pipe-friction Equation gives good results when it is used within the limits for which the friction factor was determined. Investigations have shown that the friction factor is actually a function of the pipe diameter, roughness of the pipe, average flow velocity, fiuid density, and absolute viscosity of the fluid. These variables, except for the roughness of the pipe, can be expressed as a dimensionless ratio known as the Reynolds Number as shown by the following equation:

$$R = \frac{D\,v\,\rho}{\mu}$$

No simple mathematical equations have been developed to show the relationship between friction factor and Reynolds Number for all flow conditions. Curves developed by R. S. Pigott and Emery Kemler to show the relationship between Reynolds Number and friction factor are generally used to determine the friction factor.

When the Reynolds Number is less than 1200, the flow is considered viscous or laminar, and the friction factor is independent of the roughness of the pipe. Therefore, the friction factor (f) is proportional to the Reynolds Number (R) as shown by the following equation:

$$f = \frac{64}{R}$$

When the Reynolds Number is between 1200 and 2500, the flow can be either laminar, turbulent, or a combination of both. This region is called the critical zone. Since the friction factor depends on the condition of flow, it is impossible to accurately determine the friction factor in this region. However, it is general practice to consider the flow as being turbulent in the critical zone. When the Reynolds number is general practice to consider the flow as being turbulent in the critical zone. When the Reynolds number is greater than 2500, the flow is considered turbulent and the friction factor depends not only on the Reynolds Number, but also on the size and roughness of the pipe. The friction factor is usually obtained from the curves developed by R. S. Pigott & Emery Kemler.

The Pipe-friction Equation has been modified to include the use of Reynolds Number for obtaining the friction factor as shown by the following equation.

$$h = \text{function of } \frac{Dv\rho}{\mu}\,\frac{L}{D}\,\frac{v^2}{2g}$$

The use of the modified equation is generally called the Rational Method. Although these equations were developed for steady flow of incompressible fluids, they can be used for steady flow of compressible fluids, providing the change in density of the fluid does not exceed 10% of the fluid condition. There are many variations of the Pipe-friction Equation and two equations frequently used are as follows as shown at top of left hand column, next page.

FLOW OF FLUIDS (Continued)

For Viscous Flow: $\Delta p = .000668 \dfrac{\mu L v}{d^2}$

For Turbulent Flow: $\Delta p = .00129\, f \dfrac{L \rho v^2}{d}$

Δp = pressure loss in psi
L = length of pipe in ft.
v = average velocity of fluid in pipe in ft. per sec.
d = inside dia. of pipe in inches
f = friction factor
K = Flow Resistance Coefficient
h = head loss in ft. of fluid
μ = absolute viscosity lbs. per ft.-sec.
ρ = density in lbs. per cu. ft.
D = inside diameter of pipe in ft.
$R = \dfrac{D v \rho}{\mu}$ (Reynolds Number)
g = acceleration of gravity (32.2 ft. per sec. per sec.)

Investigations have shown that the energy loss for steady flow through valves and fittings can be expressed by the following Valve and Fittings Equation:

$$h = K \frac{v^2}{2g}$$

The Flow Resistance Coefficient (K) is related to the velocity of the fluid in nominal size pipe connected to the valve or fitting, and it remains relatively constant for a specific type of valve or fitting.

It is generally accepted that the energy loss through valves and fittings is affected by the same variables that affect the energy loss for flow through straight pipe. This fact makes it possible to express the flow resistance of valves and fittings as being equivalent to a length of straight pipe. By combining the Pipe-friction Equation with the Valve and Fitting Equation, the Flow Resistance Coefficient can be expressed by the following equation.

$$K = f \frac{L}{D}$$

By using an average friction factor for turbulent flow, this equation can be used to determine the approximate length of pipe that will be equivalent to any valve or fitting for which the Flow Resistance Coefficient (K) has been established. The equivalent length of straight pipe can be used to determine the approximate pressure drop for the required flow conditions by using equations or tables developed for determining the pressure drop for flow through straight pipe.

FLOW RESISTANCES FACTORS FOR VALVES AND FITTINGS

		GATE VALVE	ANGLE VALVE	60°-Y VALVE	GLOBE VALVE	45° ELBOW	90° ELBOW	TEE THRU RUN	TEE THRU BRANCH	180° CLOSE RETURN
COEFFICIENT		.2	5	6	10	.4	.9	.5	1.8	2
RATIO		7	180	216	360	14	32	18	64	72
SCHEDULE 40 PIPE										
Final Pipe Size Inches	Inside Diameter inches	L—EQUIVALENT LENGTH OF SCHEDULE 40 PIPE IN FEET								
½	.622	.4	10	12	20	.8	1.8	1.0	3.6	4.0
¾	.824	.5	13	15	25	1.0	2.3	1.3	4.5	5.0
1	1.049	.6	15	18	30	1.2	2.7	1.5	5.4	6.0
1¼	1.380	.8	20	24	40	1.6	3.6	2.0	7.2	8.0
1½	1.610	1.0	25	30	50	2.0	4.5	2.5	9.0	10
2	2.067	1.2	30	36	60	2.4	5.4	3.0	11	12
2½	2.469	1.5	38	45	75	3.0	6.8	3.8	14	15
3	3.068	2.0	50	60	100	4.0	9.0	5.0	18	20
4	4.026	2.4	60	72	120	4.8	11	6.0	22	24
5	5.047	3.0	75	90	150	6.0	14	7.5	28	30
6	6.065	3.6	90	108	180	7.2	16	9.0	32	36
8	7.981	4.8	120	144	240	9.6	22	12	44	48
10	10.02	6.0	150	180	300	12	27	15	54	60
12	11.94	7.2	180	216	360	14	32	18	64	72
14	13.13	8.0	200	240	400	16	36	20	72	80
16	15.00	9.0	225	270	450	18	40	23	80	90
18	16.88	10	250	300	500	20	45	25	90	100
20	18.81	11	280	336	560	22	50	28	100	110
24	22.63	14	340	408	680	27	60	34	120	140

Note: Flow Resistance Factors are average values based on data for flow of water through valves and fittings.

"KING-CLIP" GATE VALVES

IRON BODY BRONZE MOUNTED OR ALL-IRON

Screwed ¼ to 2 in., 150 LB. S.P. 450°F. — 225 LB. W.O.G.

Screwed 2½ to 4 in., 125 LB., S.P. 450°F. — 175 LB. W.O.G.

Flanged 1 to 4 in., 125 LB. S.P. 450°F. — 175 LB. W.O.G.

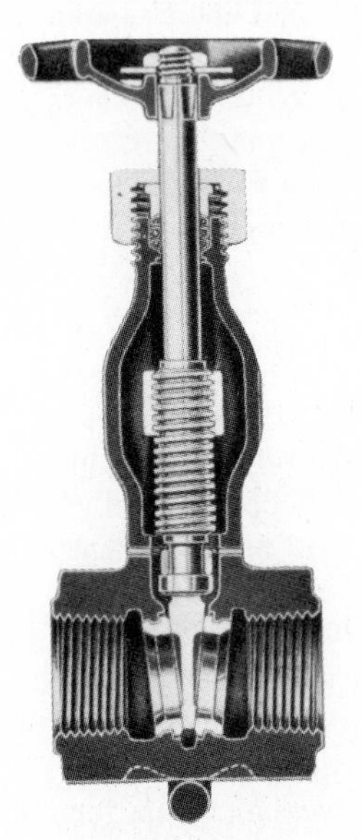

Bronze Mounted "Stemalloy" Stem Fig. 1640

Flanged I.B.B.M., Fig. 1641 All-Iron, Fig. 1645

Screwed OS&Y I.B.B.M., Fig. 1681 All-Iron, Fig 1683

Flanged, OS&Y I.B.B.M., Fig. 1682 All-Iron, Fig. 1684

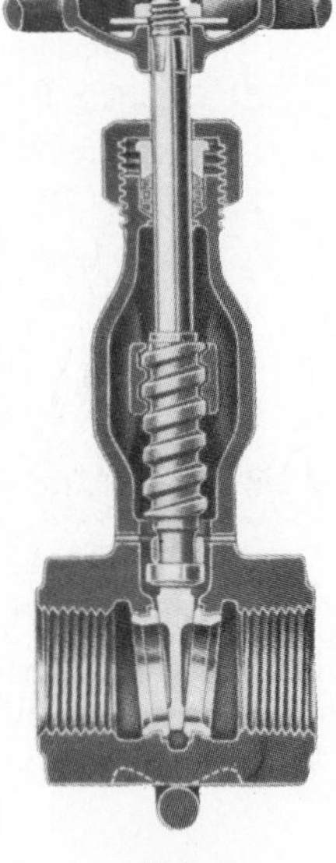

All-Iron Steel Stem Fig. 1644

The "King-clip" Gate Valve is the modern version of the clip valve developed and patented by Lunkenheimer years ago. It is rugged, exceptionally rigid, and can be readily disassembled by removing the two U-bolt nuts. It is built to do rough jobs where lighter valves fail.

Bronze Mounted valves are for general service on steam, water, gas, air, oil, and gasoline lines. Seat rings, discs, and bonnet bushings are bronze.

Stems are "Stemalloy," a bronze alloy highly resistant to stem thread wear.

Drain Channels, of ample size, provide for bonnet drainage—preventing damage to that part by freezing of trapped water.

All-Iron Valves handle solutions which attack bronze but not iron. Seats are integral in 2½ inch and larger sizes, tubular steel in smaller sizes. Discs are iron in larger sizes, forged steel in 1½ inch and smaller sizes.

Bodies and Bonnets are of close grained Lunkenheimer Cast Iron (A.S.T.M. A-126), more corrosion resistant than the average cast iron. See page 45.

Stems are steel which are chemically treated with phosphate to inhibit rust, with coarse loose fittings threads—to prevent seizing due to corrosive action.

Drain Channels in bonnet are for the purpose of clearing space above stem threads of media when draining pipe lines.

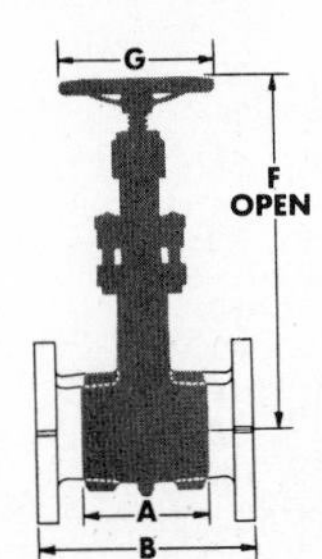

DIMENSIONS, IN INCHES

Size (inches)	¼	⅜	½	¾	1	1¼	1½	2	2½	3	4
A	1⅞	2 1/16	2¼	2⅝	2 15/16	3 5/16	3 7/16	4 1/16	4⅝	4 15/16	6 13/16
B	—	—	—	—	3 3/16	3½	6	7	7½	8	9
E	4⅞	4⅞	5⅝	6 13/16	7 15/16	8 15/16	10 5/16	12¾	15¼	17⅞	24 1/16
F	—	—	6¾	7 11/16	8 11/16	9½	11	13½	15 15/16	18⅝	—
G	2½	2½	2½	3	3½	4⅛	4⅝	5⅛	5½	7	8

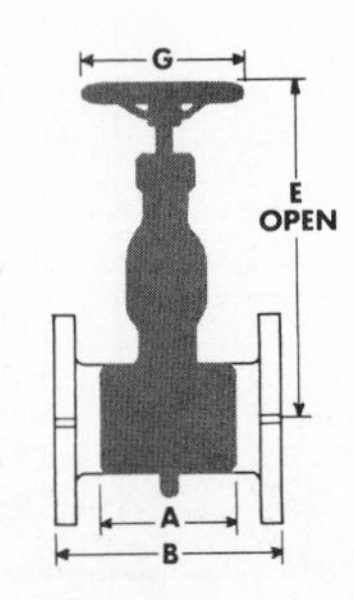

250 LB. IRON BODY GATE VALVES

BRONZE MOUNTED

2 to 12 inch . . . 250 LB. S.P. 450°F. — 500 LB. W.O.G.

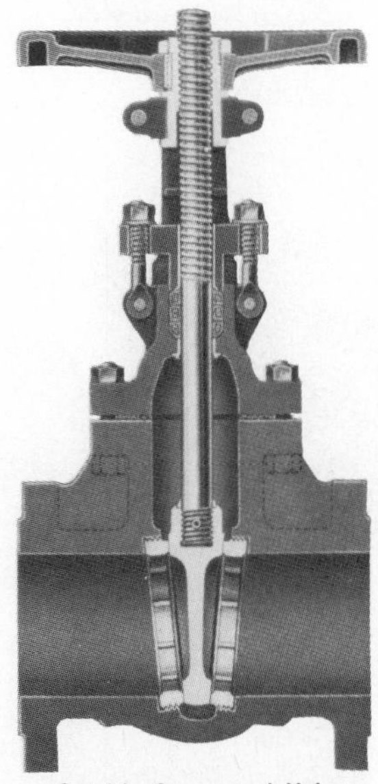

Outside Screw and Yoke
I.B.B.M.
Fig. 1436, Flanged

I.B.B.M. With By-Pass
Fig. 1438, Flanged

I.B.B.M.
Fig. 1435, Screwed

Lunkenheimer 250 lb. Iron Body Gate Valves in the O.S.&Y. design give dependable performance on high pressure steam, water, gas, and oil lines.

Bodies and Bonnets of rugged construction are of Lunkenheimer Cast Iron (A.S.T.M. A-126).

Seat Rings are bronze, heavily constructed and rectangular in section, renewable if necessary. Bottom faces seat against solid walls of body.

Discs are Cast Iron discs with bronze facing rings.

Stems are Manganese Bronze with Acme Standard threads.

Guide Channels in disc are accurately machined and guide ribs in body are precision as cast, assuring longer life.

Stuffing Boxes are deep. Valves repackable under pressure.

By-Passes and Drains — Bodies of 6 inch and larger sizes have bosses for by-pass and drain connections.

Flanges — Dimensions, drilling, and facing conform to American Cast Iron Flange Standard, Class 250 ANSI (B16.2-1960) shown on page 302. Flanges save 1/16 inch raised face. Valves will be shipped with flanges drilled unless ordered otherwise.

Flanged valves conform to American Standard for Face to Face Dimensions of Ferrous Flanged Valves ANSI (B16.10-1973) for 250 lb. Cast Iron Wedge Gate Valves.

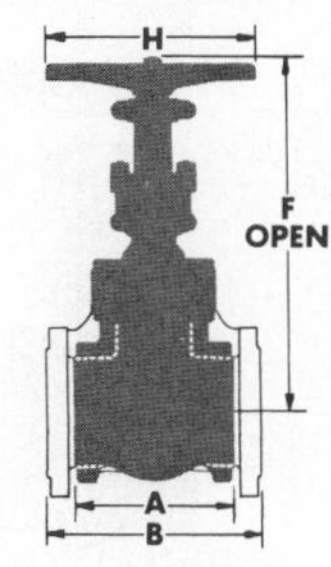

DIMENSIONS, IN INCHES AND WEIGHTS, IN POUNDS

Size inches)	2	2½	3	4	5	6	8	10	12
A	6	6⅝	7¼	8 9/16	—	—	—	—	—
B	8½	9½	11⅛	12	15	15⅞	16½	18	19¾
F	15 3/16	17 11/16	19 3/16	23½	27 11/16	32¼	39 5/16	47½	55⅜
H	8	9	9	12	14	16	18	20	22
Fig. 1435 Wts.	45.0	63.0	81.0	139.0	—	—	—	—	—
Fig. 1436 Wts.	57.0	80.0	103.0	174.0	262.0	359.0	558.0	827.0	1163.0
Fig. 1438 Wts.	—	—	—	—	—	367.0	565.0	831.0	1168.0

FORGED STEEL GATE VALVES

COMPACT STANDARD OR FULL PORT

150, 300, 600, 800 LB. S.P.

Inside Screw
800 lb. S.P.
Fig. 1011W, Screwed
Fig. 1017W, Socket Weld Ends

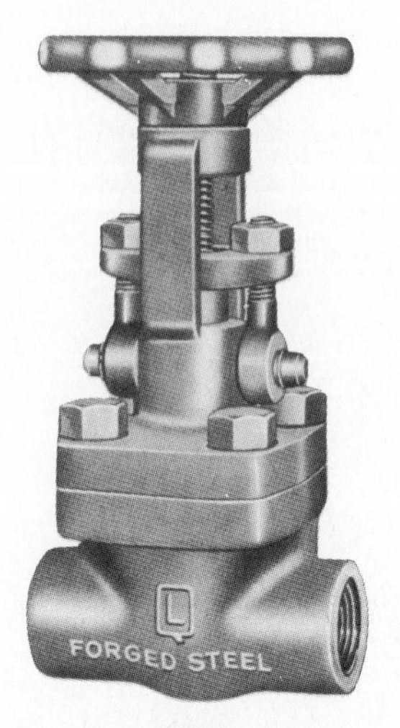

OS&Y
800 lb. S.P.
Fig. 1111W, Screwed
Fig. 1117W, Socket Weld Ends
Fig. 1311W, Screwed
Fig. 1317W, Socket Weld Ends

OS&Y
Flanged
Fig. 1100W, 150 lb. S.P.
Fig. 1105W, 300 lb. S.P.
600 lb. S.P.
Fig. 1110W

Lunkenheimer forged steel bolted bonnet gate valves have been designed to withstand the most severe service conditions. They are available with inside screw, rising stem, or outside screw and yoke with bolted bonnets.

Bodies & Bonnets—Forged Carbon Steel (A.S.T.M. A105 Grade 11).

Yoke Nuts—Corrosion resistant alloy steel having over 2000° F. melting point.

Packing—Braided asbestos jacket, reinforced with Inconel wire over plastic core. Inhibitor added to prevent stem corrosion.

Gland Bolts and Nuts—Stainless steel swing-away eyebolts with stainless steel adjusting nuts.

Gaskets—18 & 8 Stainless Steel spiral wound with asbestos filler.

Seat Rings—Rolled in-place to insure seal. Stellite applied to 13 Chromium stainless steel base for standard W & U trim.

Surface Finish — Ferrous-Phosphate rust resisting finish inside and out.

End Flanges—Drilled and spot faced in accordance with applicable requirements of ANSI B16.5 . . . 150 lb. and 300 lb. Series have 1/16″ raised face, 600 lb. Series has 1/4″ raised face—all raised faces have concentric serrated finish.

Face to Face—In accordance with applicable requirements of ANSI B16.10.

DIMENSIONS, IN INCHES AND WEIGHTS, IN POUNDS

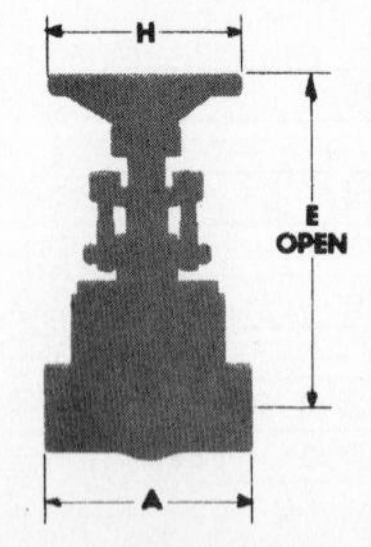

Size	1/4	3/8	1/2	3/4	1	1 1/4	1 1/2	2
A	3	3	3	3 1/2	4 1/4	4 1/2	4 15/16	5 5/8
E	5 11/16	5 11/16	5 11/16	6 13/16	7 3/8	8 3/8	10 5/8	11 29/32
F	6 9/32	6 9/32	6 9/32	7 3/16	8 5/16	9 17/32	11 13/32	12 15/16
H	3	3	3	3 1/2	4	4	5	6
J	3	3	3	3 1/2	4 1/4	4 1/4	5	6
Fig. 1011W, Wts.	3.0	3.0	3.0	4.5	7.5	9.7	13.7	21.2
Fig. 1117W, Wts.	3.2	3.2	3.0	4.5	7.4	9.3	14.0	22.0

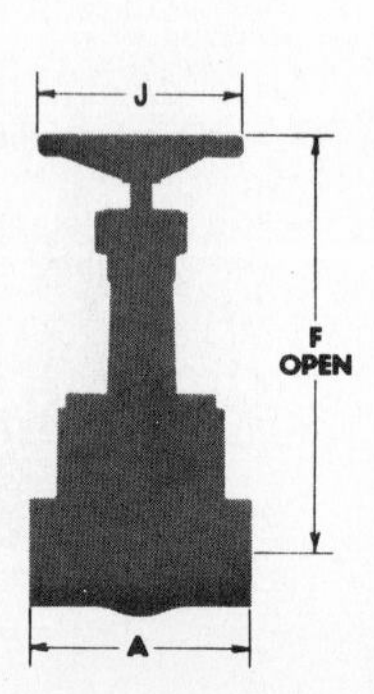

300 LB. BRONZE "RENEWO" GLOBE VALVE
350 LB. BRONZE "RENEWO" ANGLE VALVE

Globe Screwed: 300 LB. S.P. 550°F. — 600 LB. W.O.G.
Angle Screwed: 350 LB. S.P. 550°F. — 1000 LB. W.O.G.

RENEWABLE REGULAR AND PLUG TYPE SEAT AND DISC

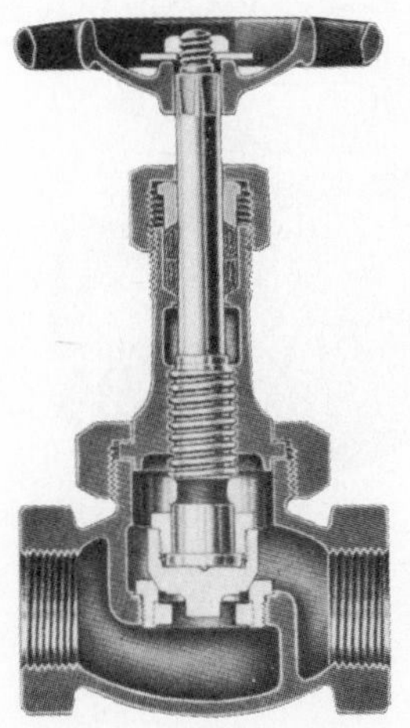

Globe
Regular Type Seat and Disc
"NT4" Nickel Alloy
Union Bonnet
Fig. 16, Screwed

Globe
Plug Type Seat and Disc
500 Brinell Stainless Steel
Union Bonnet
Fig. 16-PS, Screwed

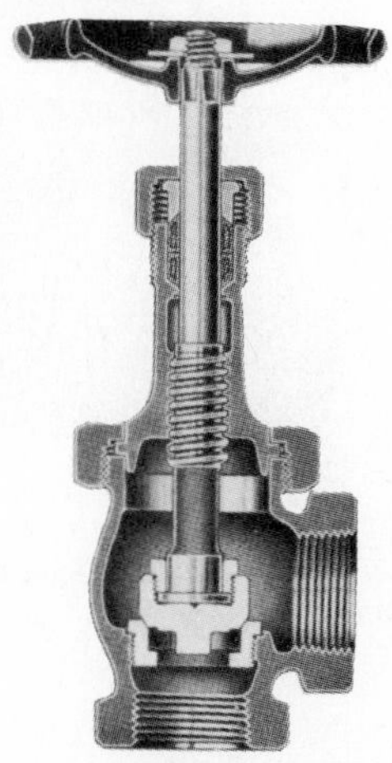

Angle
Regular Type Seat and Disc
Fig. 17, Screwed
Plug Type Seat and Disc
Fig. 17-PS, Screwed

Lunkenheimer 300 and 350 lb. "Renewo" globe and angle valves offer two distinct types of Seat-Disc combination.

Plug Type 500 Brinell Stainless Steel Seats and Discs—recommended for service demanding high resistance to destructive action on seat bearings. Valves are recommended for throttling, drain, drip, water column blow down and other severe services. Stainless Steel, heat treated to extreme hardness affords maximum resistance to severe erosive service.

Regular Type "NT4" Seats and Discs—Economical for numerous general service installations, including steam, water, oil and gas lines. For use where full flow is desired. "NT4" Nickel Alloy seating material is hard and resists wear.

Bodies and Bonnets are of Lunkenheimer S-1 Steam Bronze (A.S.T.M. B61).

Renewability—As the name "Renewo" implies, all parts including the seats and discs are renewable and interchangeable.

Handwheels are malleable iron, "Non Slip" design.

Stems of "Stemalloy" provide maximum resistance to thread wear, corrosion, and embrittlement.

Repacking is easily done while valve is wide open under pressure. Stuffing boxes are deep for extra packing. Hexagon head gland provides wrench hold to loosen gland without prying.

Wrenches for removing and inserting seat rings are shown on page 201

DIMENSIONS, IN INCHES AND WEIGHTS, IN POUNDS

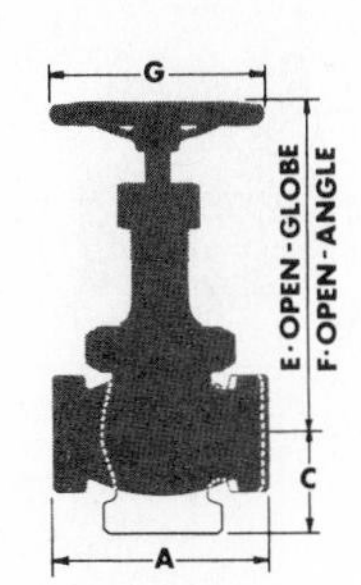

Size	¼	⅜	½	¾	1	1¼	1½	2	2½	3
A	2 5/16	2½	2⅞	3 17/32	4⅛	4 23/32	5¼	6⅜	7⅜	8⅜
C	1 3/32	1 5/32	1⅜	1⅝	1 29/32	2 7/32	2 7/16	3	3 19/32	4⅛
E	4⅝	4⅝	5 3/16	6⅛	6⅞	7 13/16	8⅜	9½	11⅛	12 11/16
F	4⅝	4⅝	5 3/16	6⅛	6⅞	7 13/16	8⅜	9½	11⅛	12 11/16
G	2½	2½	3	3½	4⅛	4⅝	5⅛	5½	8	10
Fig. 16 Wt.	1.2	1.3	2.0	3.4	5.2	7.3	10.0	17.0	30.0	43.0
Fig. 16 PS Wt.	1.2	1.4	2.0	3.4	5.2	7.3	10.0	17.0	30.0	43.0
Fig. 17 Wt.	1.2	1.3	1.9	3.2	5.0	7.0	9.2	15.0	28.0	—
Fig. 17 PS Wt.	1.2	1.3	1.9	3.2	4.8	7.2	9.3	16.0	26.0	—

150. LB. CAST STEEL SWING CHECK VALVES

BOLTED CAP

For Pressure-Temperature Ratings, See Page 210

Fig. 1572, Flanged Ends

Fig. 1573, Butt-Weld Ends

Lunkenheimer 150 lb. Steel Swing Check Valves are exceptionally sturdy and have many quality features usually found only in valves with higher pressure-temperature ratings.

See page 224 for complete description.

Design—The body and cap are made with heavy metal sections to assure safe, dependable service. Streamlined flow areas reduce turbulence and pressure drop. The valve may be used in either a horizontal line, or in a vertical line to prevent downward flow. For vertical line installations, a stop-lug in the body prevents cocking of the disc in the open position.

Materials—The carbon steel used in these valves has been selected to provide good casting properties, mechanical strength and safe weldability.

Castings are made by the electric furnace process using modern foundry equipment.

Body-Cap Connections are male-female joints with profile corrugated gaskets. Bolting consists of through-studs with a nut on each end.

Disc Assembly—The disc is fastened securely to the disc carrier by *a locknut.* Wear is eliminated on the disc as it is designed to swivel free.

Seat Rings are rectangular bottom seating construction to resist distortion. This also assures a tight seat ring joint. Seat rings are available in various trims.

DIMENSIONS, IN INCHES AND WEIGHTS, IN POUNDS

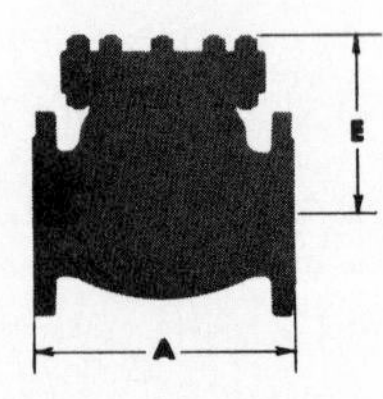

Size	1½	2	2½	3	3½	4	5	6	8	10	12
A	6½	8	8½	9½	10½	11½	13	14	19½	24½	27½
E	4⅝	5 11/16	6⅛	6⅝	7⅛	7½	8 11/16	9⅜	11⅛	13 11/16	15¼
Fig. 1572, Wts.	20	38	44	63	70	92	120	162	304	793	—
Fig. 1573, Wts.	16	30	35	51	56	74	96	130	240	632	—

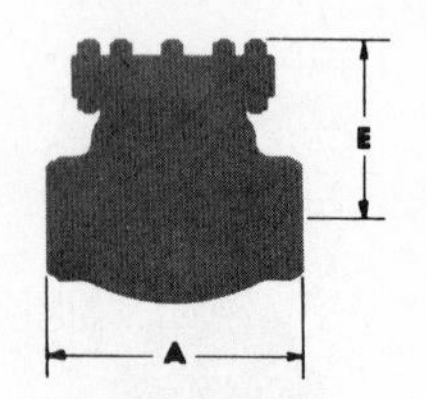

125 LB. IRON BODY SWING CHECK VALVES

BRONZE MOUNTED OR ALL-IRON

BOLTED CAP

125 LB. S.P. 450°F. — 200 LB. W.O.G.

Swing Check
I.B.B.M.
Fig. 1790, Flanged

Swing Check
Screwed
Fig. 1789, I.B.B.M.
Fig. 1791, All-Iron

Swing Check
All-Iron
Fig. 1792, Flanged

The design of Lunkenheimer Iron-Body Swing-Check Valves provides straightway flow with minimum pressure loss and permits free action of the disc. Valves may be installed in either horizontal or vertical lines.

Bronze-Mounted Valves are recommended for use with oil, steam, water, air, gas, and other fluids that do not attack bronze or iron.

Discs and Disc Assembly Parts, Seat Rings and Side Plugs are bronze. Discs in all sizes are solid bronze. (A.S.T.M. B-62).

All-Iron Valves are recommended for use on fluids which attack bronze but not iron.

Discs, Disc Carriers, and Seat Rings in All-Iron valves are cast iron. Side plugs and hinge pin are steel.

Bodies and Caps of both valves are cast iron (A.S.T.M. A126).

Two Renewable Side Plugs serve as bearings for the hinge pin. If long wear causes misalignment of parts and sluggish operation, original performance can be restored by replacing side plugs. If installation is close to a wall the hinge pin can be removed from either side.

Flanges—Dimensions, drilling and facing conform to the American Cast Iron Flange Standard, Class 125 (ANSI B16.1-1975). See page 302.

Flanged Valves conform to the American Standard for Face to Face Dimensions of Ferrous Flanged Valves (ANSI B16.10-1973), for 125 lb. Cast Iron Swing Check Valves.

DIMENSIONS, IN INCHES AND WEIGHTS, IN POUNDS

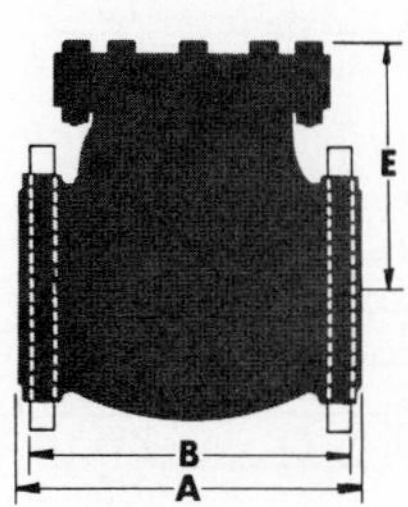

Size (inches)	2	2½	3	3½	4	5	6	8	10	12	14
A	8	8½	9½	10½	11½	13	14	—	—	—	—
B	8	8½	9½	10½	11½	13	14	19½	24½	27½	—
E	5³⁄₁₆	4⁵⁄₁₆	6¾	7½	8³⁄₁₆	8¹⁵⁄₁₆	9⅝	11⅞	13³⁄₁₆	15¹⁄₁₆	—
Fig. 1789 Wts.	24.0	36.0	48.0	—	81.0	—	149.0	—	—	—	—
Fig. 1790 Wts.	30.0	44.0	57.0	—	95.0	123.0	165.0	324.0	487.0	673.0	—
Fig. 1791 Wts.	24.5	35.0	48.0	—	81.0	—	—	—	—	—	—
Fig. 1792 Wts.	30.0	43.0	57.0	—	97.0	—	—	—	—	—	—

250 LB. IRON BODY "Y" BLOW-OFF VALVES

OUTSIDE SCREW AND YOKE—BOLTED BONNET

RENEWABLE SEAT AND DISC

250 LB. S.P. 450°—500 LB. W.O.G.

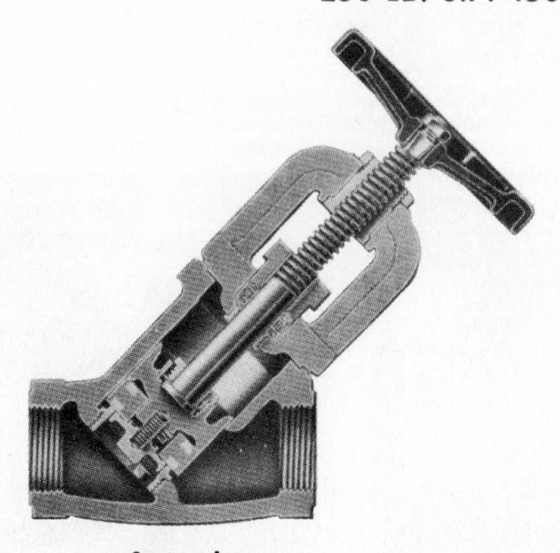

Screwed
Fig. 1351

Flanged
Fig. 1352

These iron body straightway or "Y" blow-off valves feature rugged construction and easy renewability of all parts subject to wear. Design conforms to the ASME Boiler Construction Code.

Bodies have large areas throughout. Position of seat opening permits practically unrestricted flow. Long cylindrical guides prevent accumulations of sediment in body necks and insure proper seating of the discs.

Non-Corrodible Bearings are provided for the stems by bronze yoke bushings, gland liners and stuffing box bushings. Stems are made of "Stemalloy," a highly corrosion-resistant long-wearing silicon bronze alloy.

Discs are reversible, and the alloy seat faces may be melted out and renewed when worn. An annular extension below the disc protects the seating surfaces by causing a washing action just before the valve is closed. Valves should be connected to receive pressure below the disc, as indicated by the flow direction arrow cast on the bodies.

Seat Rings are bronze with ample stock to allow for wear in the disc face alloy. Substantial lugs facilitate renewal, if necessary.

Body-Bonnet Joints are male and female with corrugated copper gaskets. They can be quickly disassembled for valve inspection or repairs by removing the two bonnet bolts.

Renewability—All parts of the valve are renewable.

Flanges have dimensions, drilling and facing conforming to the American Cast Iron Flange Standard Class 250 (ANSI B16.2-1960) shown on page 302. Flanges have 1/16-inch raised face. Valves will be shipped with flanges drilled unless ordered otherwise.

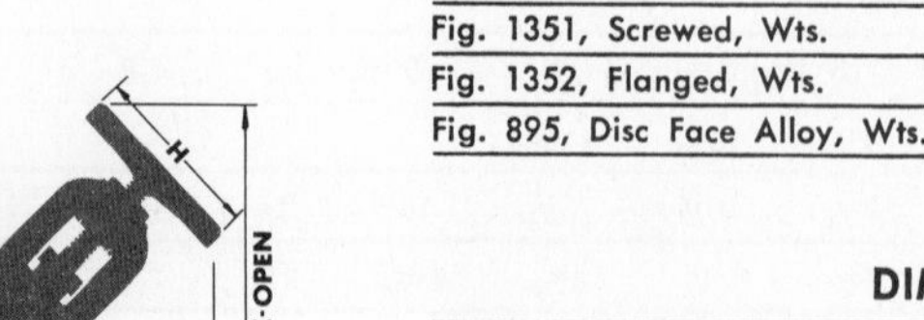

WEIGHTS, IN POUNDS

Size (Inches)	1½	2	2½
Fig. 1351, Screwed, Wts.	39.0	64.0	92.0
Fig. 1352, Flanged, Wts.	51.0	77.0	113.0
Fig. 895, Disc Face Alloy, Wts.	1.0	1.0	1.0

For Parts Identification, See Page 133

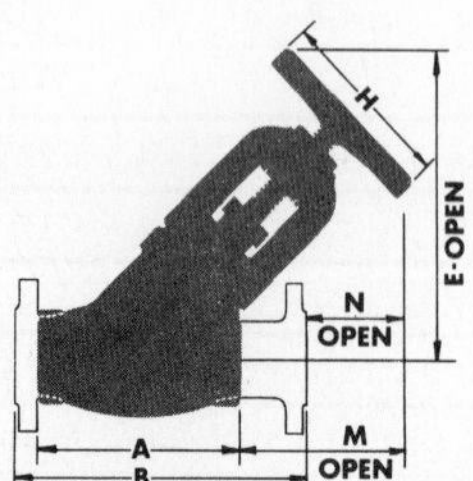

DIMENSIONS, IN INCHES

Size	1½	2	2½
A	7⅜	9⅛	10⅜
B	11½	12⅞	14¾
E	13 1/16	15 5/16	17 1/16
M	8⅝	9⅝	10⅝
N	5 5/16	6¾	7 1/16
H	7	8	9

150 LB. CAST STEEL GLOBE AND ANGLE VALVES

BOLTED BONNET

For Pressure-Temperature Ratings, See Page 210

Globe, Flanged Ends
Fig. 1532, Spherical Disc

Globe, Flanged Ends
Fig. 1542, Plug Disc

Angle, Flanged Ends
Fig. 1552, Spherical Disc

Globe, Butt-Weld Ends
Fig. 1543, Plug Disc

Angle, Butt-Weld Ends
Fig. 1563, Plug Disc

Angle, Flanged Ends
Fig. 1562, Plug Disc

Lunkenheimer 150 lb. Steel Globe and Angle Valves have many quality features usually found only in valves with higher-pressure temperature ratings. See page 218 for complete description.

Bonnet Joints are self centering male-female type connections. Bonnet bolting consists of through studs with nut on each end.

Discs are furnished in full-way design. In the 150 and 300 lb. pressure classes, a long swivel nut coupled with small clearances between the swivel nut and stem is utilized to accurately guide the disc to its seat. Trim "U" full-way valves are similarly guided.

Seats are shoulder seated and threaded.

Stem is heavy acme threads with liberal thread engagement with the yoke bushing, assuring long life of these parts.

Materials of bodies and bonnets are heavy shell-thickness to afford maximum safety. Long tapered fillets eliminate danger of casting defects at critical points.

Design. Available with either spherical or plug type discs. Adjustable to a wide variety of service requirements.

For material specifications, modifications and accessories, see page 218.

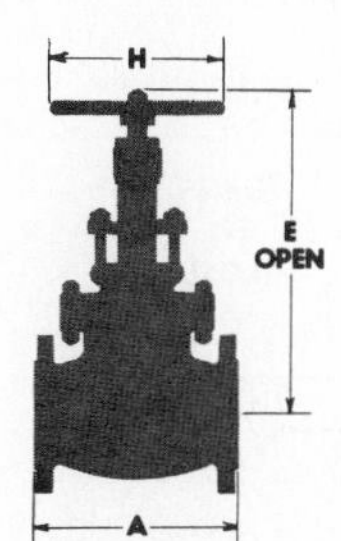

Note: Face-to-face dimension of flanged valves include 1/16-inch raised faces.
Drilling Templates and Flange Dimensions.................. page 304.
Butt Welding end valves same dimensions as flanged end.

DIMENSIONS, IN INCHES AND WEIGHTS, IN POUNDS

Size	2	2½	3	3½	4	5	6	8
A	8	8½	9½	10½	11½	14	16	19½
D	4	4½	4¾	5¼	5¾	7	8	9¾
E	13 13/16	14 9/16	16 23/32	17 13/32	20 1/32	22½	23 15/16	27 13/16
H	8	8	9	9	10	12	12	16
Fig. 1532, 1542, Wts.	48	62	83	98	120	169	210	352
Fig. 1543, Wts.	37	50	66	78	103	146	191	280
Fig. 1552, 1562, Wts.	46	61	77	97	120	168	210	338
Fig. 1563, Wts.	37	50	62	77	96	132	168	267

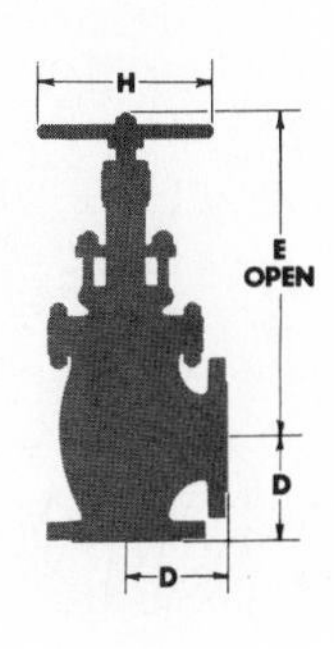

REGAL® BUTTERFLY VALVES–WAFER TYPE

150 LB. W.O.G.

LARGE SIZE — REPLACEABLE CROWN-SEAL® SEAT

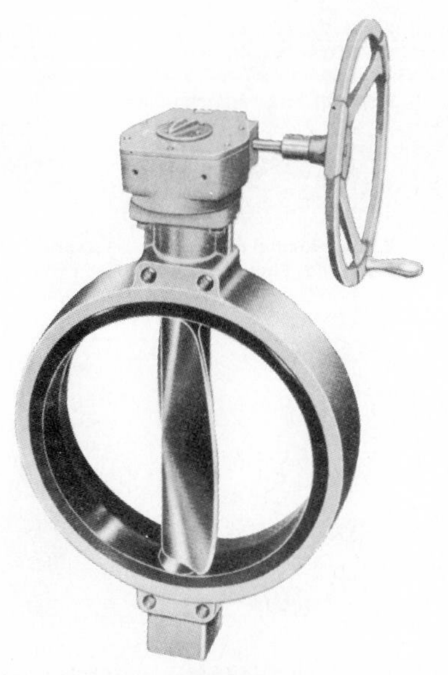

Fig. 4500-G
Manual Gear Actuator
Sizes 14 to 24 inch

Fig. 4500-G in sizes 14″ to 24″ is fully rated for service to 150 lb. W.O.G. This heavy duty line features a *replaceable,* rigidly reinforced resilient seat, self-lubricating stem bearings, steel stem, high strength cast iron body, ductile iron disc and heavy duty, weatherproof, manual worm gear actuator.

The large Regal Valves are companions to the 2-inch through 12-inch line illustrated and described on pages 272 and 273. These wafer type butterfly valves represent the most advanced design concept available. Complete field servicing, including seat and disc replacement contribute to the overall economical design inherent in the Regal valve family. All parts are precisely manufactured to assure complete interchangeability. Standard Buna-N seats may be replaced with other elastomers. Stem and discs may be interchanged with other than standard metals. These operations are readily accomplished in the field.

Lug holes are drilled and tapped to enable use of studs or bolts to position the valve in the flange assembly. Lug holes may be drilled and untapped for use of through bolts if specified. Tapped holes are standard. A correctly positioned wafer valve, held safely in place, contributes to ease and safety during installation. If adjacent equipment is to be serviced, a large Regal valve may be left bolted to a flange while routine maintenance on other system components is accomplishd. This convenient Regal design feature conserves time and manpower.

Manual worm gear actuator with indicator and removable handwheel assures ease of operation and infinite choice of disc position for fine throttling or positive shut-off.

The large Regal Valves are designed for ready use with variety of other actuators. See pages 276 and 289 for additional material specifications, construction features and typical actuated valve assemblies.

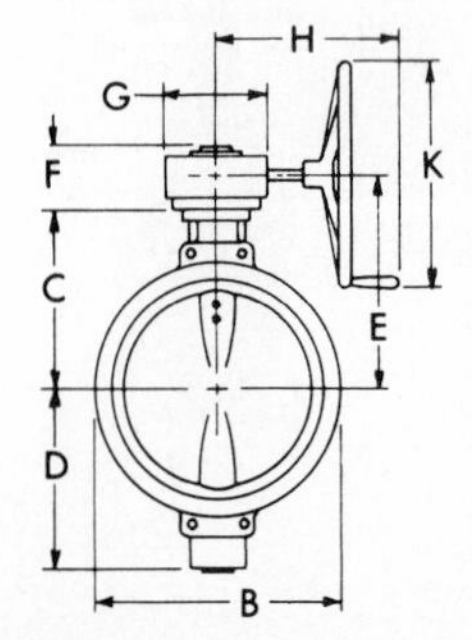

DIMENSIONS, IN INCHES AND WEIGHTS, IN POUNDS

Sizes Inches	A	B	C	D	E	F	G	H	J	K	L	Fig. 4500-G Wts.
14	3¾	17¾	12⅝	11⅞	15⅜	5¼	7½	13⅝	3½	12	13¾	180
16	4⅛	20¼	14	13½	16¾	5¼	7½	13⅝	3½	12	13¾	210
18	4½	21⅝	15	14¾	17¾	5¼	7½	13⅝	3½	12	13¾	250
20	5	23⅞	17¼	17⅛	19⅞	5¼	8⅜	15¼	4¼	18	17½	350
24	6 1/32	28¼	19¾	19⅝	22⅜	5¼	8⅜	15¼	4¼	18	17½	520

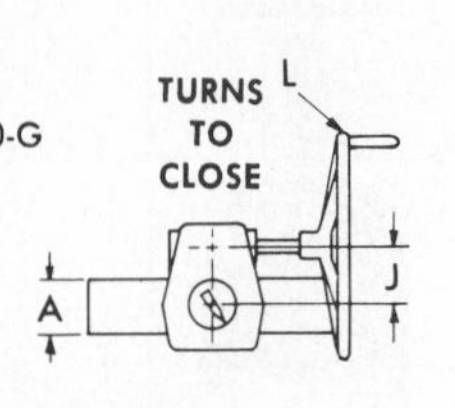

TOP VIEW

REGAL® BUTTERFLY VALVES

REPLACEABLE CROWN-SEAL® SEAT — LUG TYPE WAFER
150 LB. W.O.G. — 2 TO 12 INCHES

FIG. 4700, 4734, 4710

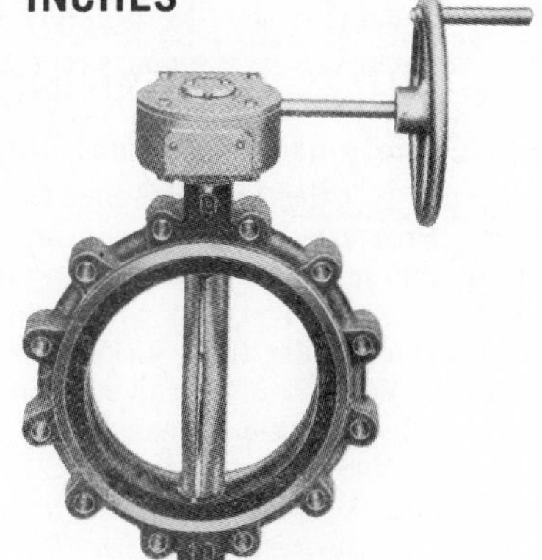

Fig. 4700-G, 4734-G, 4710-G
Gear Operated

In certain applications lug type construction is advantageous.

The lug type valve is self supporting when bolted to a flange. Since it is only necessary to bolt one flange at a time, the unit is easy to install. When in place, valve is automatically centered in the pipe run. The wafer can remain in place when either flange is removed.

Standard ANSI Class 125, Cast Iron or Class 150, Steel Flanges are used. Lug holes conform to ANSI flange drilling dimensions and have UNC threads. Unthreaded lug drilling is furnished if specified.

Alternative methods of bolting: 1. Full thread bolts or studs extending through both flanges and lugs. 2. Short studs or bolts attached through each flange and terminating within the threaded lugs.

Replaceable seat lug type valves are not recommended to hold pressure for terminal service with only one flange. A blind or companion flange should be bolted to the valve face away from the line connection.

Lug Type Regal Valve bodies are standard in ductile iron. There is the same choice of stem, disc, and seat materials as on the Wafer Type Regal Valve. For example, Fig. 4734 materials correspond to Fig. 4534 materials, except for bodies. Fig. 4510 and Fig. 4710 materials correspond except for bodies.

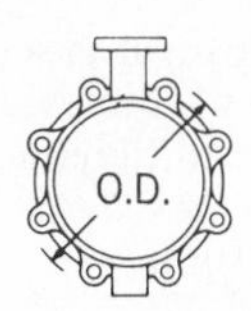

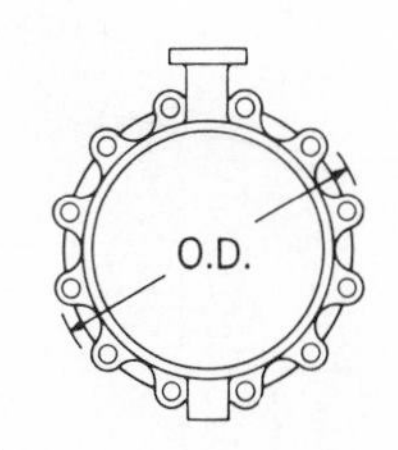

DIMENSIONS IN INCHES, WEIGHTS IN POUNDS

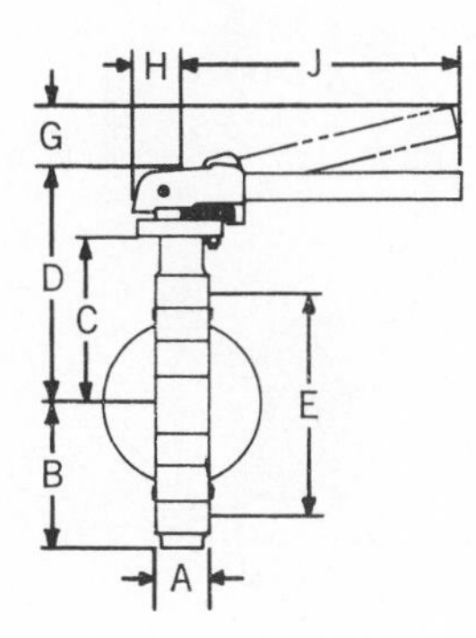

Valve Size	2	2 1/2	3	4	5	6	8	10	12
A	1 5/8	1 3/4	1 3/4	2	2 1/8	2 1/8	2 13/32	2 23/32	3
B	2 3/4	3 1/2	3 1/2	4 1/4	4 7/8	5 1/2	7	8 1/4	9 5/8
C	3 1/2	4 1/4	4 1/4	5 1/8	5 5/8	6 3/8	7 3/4	9	10 3/4
D	5 9/16	6 5/16	6 5/16	7 5/8	8 1/8	8 7/8	10 11/16	11 15/16	14 5/16
E (Face)	3 3/4	4 5/16	4 7/8	6 3/16	7 3/16	8 1/4	10 3/8	12 5/8	14 7/8
G	2	2	2	2 1/8	2 5/8	2 5/8	3 1/8	3 5/8	3 5/8
H	1 5/8	1 5/8	1 5/8	1 7/8	1 7/8	1 7/8	2 3/8	2 3/8	2 3/8
J	8	8	8	10	12	12	18	21	21
O.D.	5 7/8	6 5/8	7 1/8	8 3/4	9 7/8	10 7/8	12 7/8	15 7/8	18 5/8
Bolt Circle	4 3/4	5 1/2	6	7 1/2	8 1/2	9 1/2	11 3/4	14 1/4	17
No. of Holes	4	4	4	8	8	8	8	12	12
Bolt Size	5/8	5/8	5/8	5/8	3/4	3/4	3/4	7/8	7/8
Weights	8	10	10.5	20	24	28	46	66	100

BRONZE POP SAFETY AND RELIEF VALVES

The function of Pop Safety or Relief Valves is to open and protect the vessel on which they are installed, when the pressure within the vessel reaches or exceeds a safe limit. They are not intended for use as pressure-regulating devices.

RULES GOVERNING THE USE OF POP SAFETY VALVES

American Society of Mechanical Engineers

The requirements of the ASME Boiler Construction and Unfired Pressure Vessel Codes are generally accepted by the regulatory bodies of the various States as the rules governing the installation of Pop Safety Valves under their jurisdiction.

Lunkenheimer Pop Safety Valves comply with the ASME Rules for the Construction of Power Boilers, Locomotive Boilers, Miniature Boilers and Unfired Pressure Vessels. *They do not regularly comply with the requirements for low pressure steam heating boilers.*

Lunkenheimer Pop Safety Valves for Steam bear the ASME "V" Cloverleaf Symbol which signifies that these valves comply with the requirements of the ASME Code for Power Boilers. In addition, valves are stamped "NB" to indicate capacity certification by the National Board of Boiler and Pressure Vessel Inspectors.

Lunkenheimer Pop Safety Valves for Air and Gas bear the ASME "UV" Cloverleaf Symbol which signifies that these valves comply with the requirements of the ASME Code for Unfired Pressure Vessels. In addition, these valves are stamped "NB" to indicate capacity certification by the National Board of Boiler and Vessel Inspectors.

Bureau of Marine Inspection and Navigation

All Lunkenheimer Pop Safety and Relief Valves are made to comply with the requirements of the rules of this Bureau, and Lunkenheimer Pop Safety Valves are on the approved list. The bureau requires that approved safety valves be used on Main and Auxiliary Boilers, Super-heaters, Feed Water Heaters, and Reduced Pressure Pipe Lines not designed to withstand full boiler pressure.

Approved Pop Safety Valves must be of bronze or steel; cast-iron valves are prohibited. Valves 3 inches and larger must be made of steel. On superheated steam lines, all sizes must be made of steel. Valves 2½ inches and larger must have flanged ends and bolted bonnet. All valves must be fitted with lifting levers. Valves used on superheated steam must be of the outside spring type with spring protected by perforated cover plates.

American Petroleum Institute Field Boiler Code

Lunkenheimer Bronze Pop Safety Valves comply with the requirements of this code which in general duplicates the ASME Boiler Construction Code.

How does a Pop Safety Valve differ from a Relief Valve?

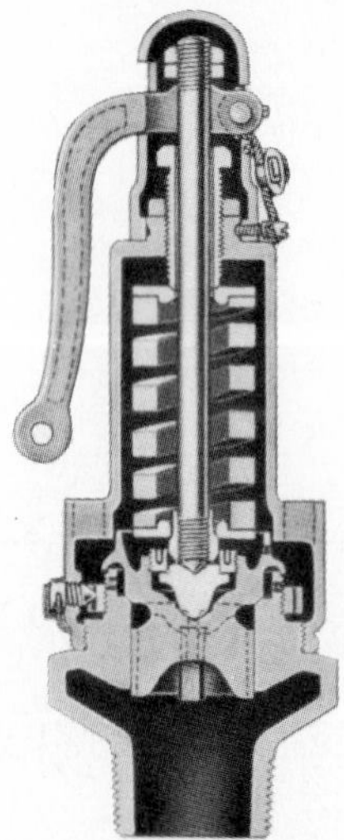

Pop Safety Valve Typical Construction Top Outlet

Pop safety valves are for use on gases, including steam.

Pop safety valves have an adjustable huddling chamber which regulates popping action and the amount of blowdown.

Lifting levers and drain holes on the discharge side of the valve to drain condensation are required by ASME Boiler Construction and Unfired Pressure Vessel Codes for all services excepting when release of noxious or inflammable gases may create a hazard.

For use on noxious or inflammable gases, pop safety valves with sealing caps and side outlets should be used. These valves do not have lifting levers or drain holes.

Pop safety valves function by "popping" wide open at the set pressure, remaining in that position until the pressure in the vessel has dropped the predetermined amount (blowback or blowdown). The valves will then snap shut instantly.

Relief valves are used on liquids.

Relief valves have a fixed lip in the disc to give increased lift as the pressure rises above the set pressure. They do not have a huddling chamber.

Relief valves start to open at the set pressure but require about 20 per cent over-pressure to open wide. As the pressure drops, they start to close and shut at approximately the set pressure.

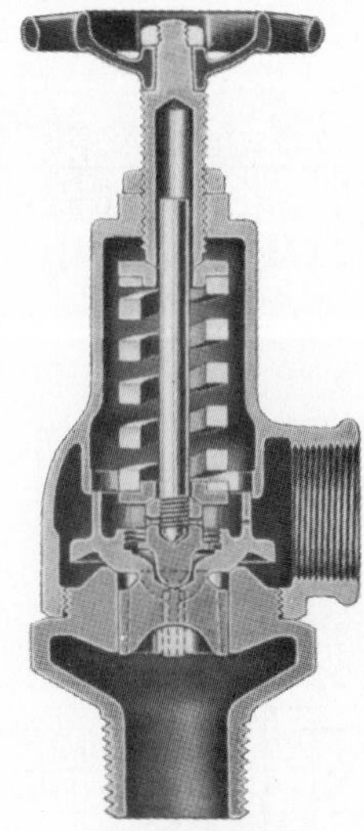

Relief Valve Typical Construction Side Outlet

BRONZE POP SAFETY VALVES FOR STEAM

A.S.M.E. STANDARD—NATIONAL BOARD CERTIFIED

Set at any single specified pressure from 1 lb. to 250 lbs., 450° F.
Orders should specify the Pressure Setting desired as well as the Figure Number.

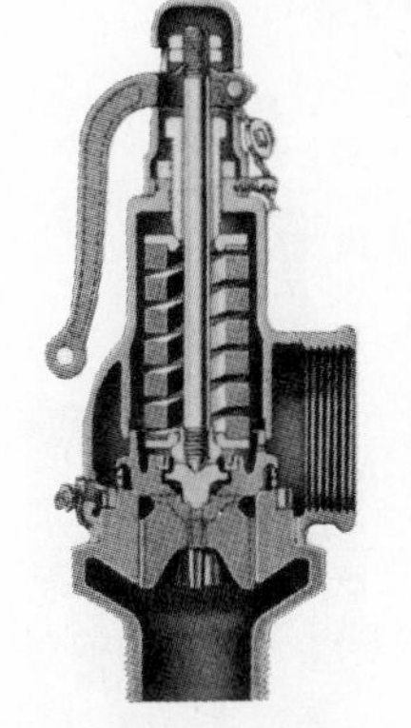

Side Outlet for Steam
Fig. 629, Male Inlet

The continued popularity of these bronze pop safety valves for steam testifies to their reliability and performance. The name plates indicate their conformance with the codes.

They are used on steam boilers, steam piping systems, steam provess vessels, receivers, etc.

Fig. 629 in ¾-to 3 inch sizes is approved by the Coast Guard for a maximum of 30 lbs. on Low Pressure Steam Heating Boilers. *Relieving Capacities* are on page 149. *Bells, Bases and Discs* are S1 steam bronze. *Springs* are cadmium plated steel. *Pressure Setting*—Valves are set at the factory at pressure specified, and sealed.

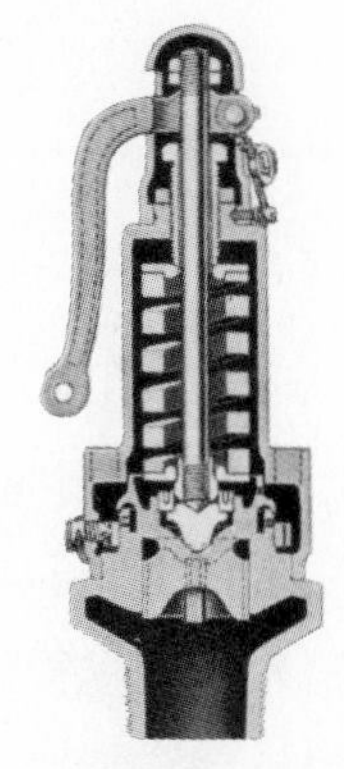

Top Outlet for Steam
Fig. 433, Male Inlet

Fig. 629

DIMENSIONS, IN INCHES AND WEIGHTS, IN POUNDS

Size of Inlet	3/8	1/2	3/4	1	1 1/4	1 1/2	2	2 1/2	3
Size of Outlet	1/2	3/4	1	1 1/4	1 1/2	2	2 1/2	3	3
Over-All Height	5 3/4	5 7/8	6 5/8	7 3/4	8 9/16	9 13/16	11 7/16	13 3/8	15 3/16
Center to Top of Bonnet	4	3 7/8	4 1/4	4 7/8	5 3/8	6 1/16	7	8 1/16	9 7/16
Center to End—Male Inlet	1 3/4	2	2 3/8	2 7/8	3 3/16	3 3/4	4 7/16	5 5/16	5 13/16
Center to End—Side Outlet	1 1/4	1 5/16	1 9/16	1 3/4	2	2 1/4	2 11/16	3 1/8	3 9/16
Fig. 629, Wts.	1.5	1.5	2.3	3.4	5.0	7.4	12.5	19.5	30.0
Fig. 433 Over-All Height	5 3/4	5 7/8	6 5/8	7 13/16	8 9/16	9 13/16	11 7/16	13 3/8	15 1/4
Diameter Bell at Base	1 13/16	1 13/16	2 1/8	2 7/16	2 13/16	3 5/16	4	4 11/16	5 9/16
Fig. 433, Wts.	1.3	1.3	2.0	2.8	4.1	6.1	10.0	16.0	26.0

For Parts Identification see page 143.

BRONZE POP SAFETY VALVES FOR STEAM

A.S.M.E. STANDARD—NATIONAL BOARD CERTIFIED

Set at any single specified pressure from 251 lbs. to 400 lbs., 450° F.

Orders should specify set pressure.

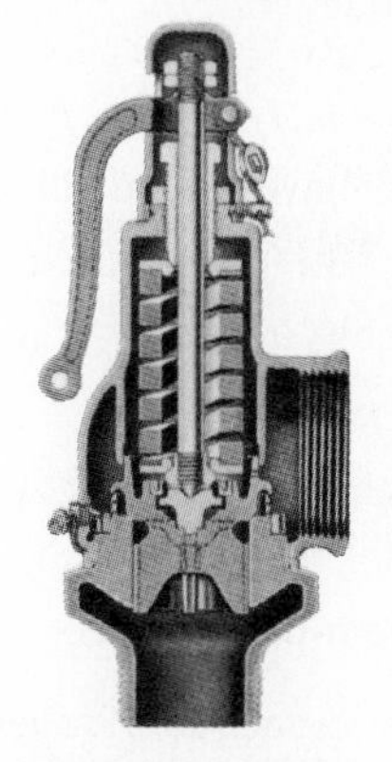

Side Outlet for Steam
Fig. 1483, Male Inlet

These valves are heavier than the valves shown above. Otherwise, they are identical in appearance, materials, and operation. The valves shown here carry a rating of 251 to 400 lbs. at 450° F. *Relieving Capacities* of these valves in pounds of steam per hour at stated gauge pressure are shown in Table 1 on page 148.

Bells and Bases are Lunkenheimer Steam Bronze S-1 (A.S.T.M. B61).

Fig. 1483

DIMENSIONS, IN INCHES AND WEIGHTS, IN POUNDS

Size of Inlet	1/2	3/4	1	1 1/4	1 1/2	2
Fig. 1483, Outlet Size	1	1 1/4	1 1/2	1 1/2	2 1/2	3
Over-All Height	6 11/16	7 11/16	8 7/16	8 9/16	11 7/16	13 5/16
Center to Top of Cap	4 3/8	4 13/16	5 5/16	5 5/16	7	8
Center to End—Male Inlet	2 5/16	2 7/8	3 1/8	3 1/4	4 7/16	5 5/16
Center to End—Female Outlet	1 9/16	1 3/4	2	2	2 11/16	3 1/8
Fig. 1483, Wts.	2.1	3.1	4.7	—	—	—

BRONZE RELIEF VALVES FOR LIQUIDS

Set at any single specified pressure from 1 lb. to 250 lbs. — 550°F.

Orders should specify the Pressure Setting desired.

Pressure Settings should be 20% above Working Pressure

These valves were originally designed for cylinder relief valves on reciprocating engines to relieve a slug of water which would ordinarily crack a cylinder head. Later applications developed for usage on water and other liquids, even steam. Although the valves do not have pop valve action they can be used on non-code installations.

They can be used on volatile and other liquids. They are used on the discharge side of the main pumps in gasoline bulk stations. They are frequently used on the discharge side of boiler feed pumps.

Bases and all working parts are of bronze, except the steel, cadmium-plated springs.

Seats are integral without huddling chamber.

Marine Service—Sizes 1/2 to 2 inches may be used on marine service.

Relief Valve
Fig. 658, Male Inlet

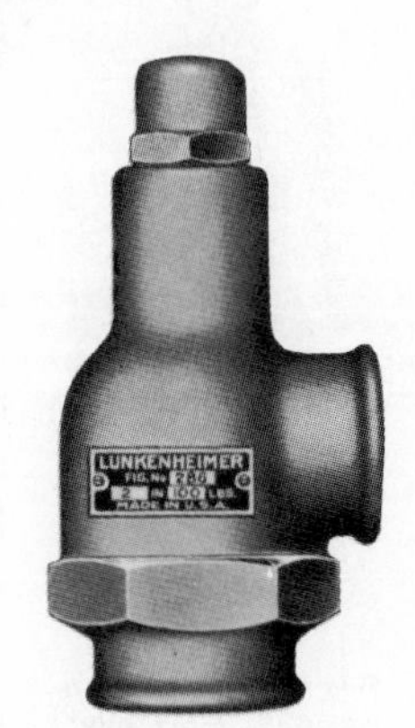

Relief Valve
Fig. 286, Female Inlet

DIMENSIONS, IN INCHES AND WEIGHTS, IN POUNDS

Size of Inlet	3/8	1/2	3/4	1	1 1/4	1 1/2	2	2 1/2	3
Size of Outlet	3/8	1/2	3/4	1	1 1/4	1 1/2	2	2 1/2	3
Over-All Height, Fig. 658	4 13/16	4 15/16	5 9/16	6 7/16	7 1/4	8 7/16	9 7/8	11 5/8	13 5/16
Over-All Height, Fig. 286	4 9/16	4 9/16	5 1/4	6	6 13/16	8	9 5/16	—	—
Center Line to Top of Cap	3 1/16	3 1/16	3 7/16	3 13/16	4 3/16	5	5 11/16	6 5/8	7 1/2
Center to End Inlet, Fig. 658	1 3/4	1 7/8	2 3/16	2 11/16	3 1/16	3 1/2	4 3/16	5	5 13/16
Center to End Inlet, Fig. 286	1 1/2	1 1/2	1 27/32	2 3/16	2 5/8	3	3 5/8	—	—
Center to End, Side Outlet	1 7/32	1 1/4	1 15/32	1 23/32	1 31/32	2 7/32	2 21/32	3 1/8	3 9/16
Fig. 658, Wts.	1.3	1.3	2.0	3.1	4.5	7.0	11.5	18.0	28.0
Fig. 286, Wts.	1.3	1.3	1.8	3.0	4.8	7.0	10.2	—	—

For Parts Identification, See Page 147.

BRONZE RELIEF VALVES FOR LIQUIDS

Set at any single specified pressure from 1 lb. to 250 lbs.—550°F.

Orders should specify the Pressure Setting desired.

Pressure Settings should be 20% above Working Pressure

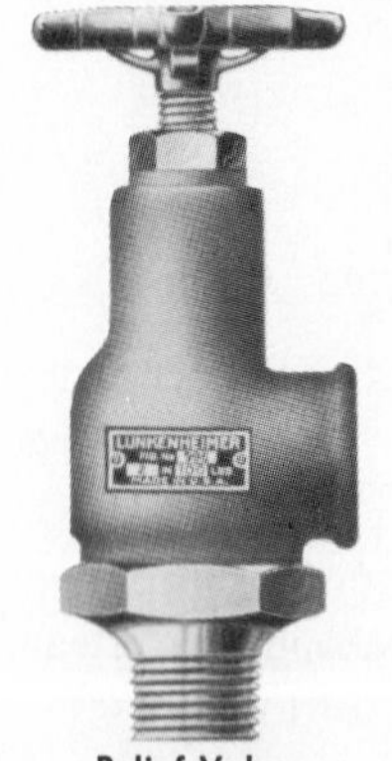

Relief Valve
With Regulating
Handwheel
Fig. 996, Male Inlet

Bronze relief valves for liquids with regulating handwheel are used on pumps, hydraulic presses, cylinders, pipe lines, etc. They are intended only to relieve excess pressures and are not designed to be used as regulating devices—to hold pressures at certain settings. When used on pipe lines, the valves should be at least one pipe size smaller.

Bells, Bases and Discs are of Lunkenheimer S-1 Bronze.

Seats are integral and have no huddling chamber.

Handwheels of malleable iron are "Non-Slip" design.

Pressure Setting should be 20% above normal working pressure.

Springs are of cadmium-plated steel.

Marine Service—Sizes 1/2 inch to 2 inches are suitable for marine service.

DIMENSIONS, IN INCHES AND WEIGHTS, IN POUNDS

Size of Inlet	3/8	1/2	3/4	1	1 1/4	1 1/2	2
Size of Outlet	3/8	1/2	3/4	1	1 1/4	1 1/2	2
Over-All Height, Fig. 996	5 5/8	5 3/4	6 5/8	7 3/8	8 5/16	9 1/2	10 7/8
Center to Top of Handwheel	3 7/8	3 7/8	4 7/32	4 23/32	5 1/4	6 1/32	6 11/16
Center to End, Inlet, Fig. 996	1 3/4	1 7/8	2 5/32	2 21/32	3 1/16	3 15/32	4 3/16
Center to End, Side Outlet	1 7/32	1 1/4	1 15/32	1 23/32	1 31/32	2 7/32	2 21/32
Fig. 996, Wts.	1.4	1.4	2.0	3.1	5.0	7.6	11.7

TABLE 3.30
DIMENSIONS OF STEEL-WELDING FITTINGS
(Based on ASA B16.9 except as noted)
All dimensions in inches

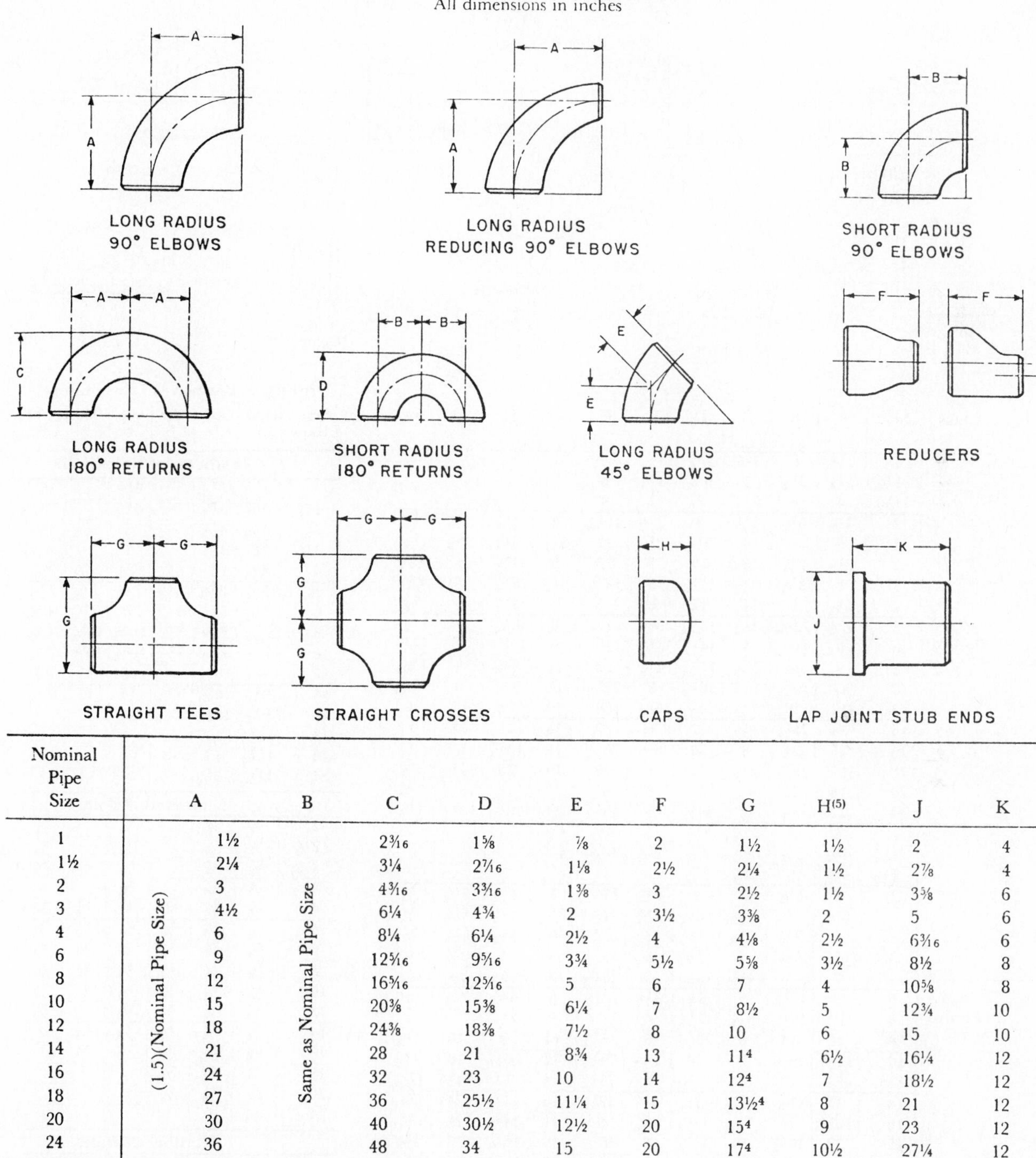

Nominal Pipe Size	A (1.5)(Nominal Pipe Size)	B	C	D	E	F	G	H[5]	J	K
1	1½	Same as Nominal Pipe Size	2 3/16	1⅝	⅞	2	1½	1½	2	4
1½	2¼		3¼	2 7/16	1⅛	2½	2¼	1½	2⅞	4
2	3		4 3/16	3 3/16	1⅜	3	2½	1½	3⅝	6
3	4½		6¼	4¾	2	3½	3⅜	2	5	6
4	6		8¼	6¼	2½	4	4⅛	2½	6 3/16	6
6	9		12 5/16	9 5/16	3¾	5½	5⅝	3½	8½	8
8	12		16 5/16	12 5/16	5	6	7	4	10⅝	8
10	15		20⅜	15⅜	6¼	7	8½	5	12¾	10
12	18		24⅜	18⅜	7½	8	10	6	15	10
14	21		28	21	8¾	13	11[4]	6½	16¼	12
16	24		32	23	10	14	12[4]	7	18½	12
18	27		36	25½	11¼	15	13½[4]	8	21	12
20	30		40	30½	12½	20	15[4]	9	23	12
24	36		48	34	15	20	17[4]	10½	27¼	12

[1] Dimensions for these fittings not an ASA standard but common commercial practice.
[2] For reducing tees, see ASA B16.9.
[3] For sizes larger than 24″, see MSS-SP-48.
[4] Center to end dimensions for outlet are not standardized in 14″ and larger. Dimensions given are in common use.
[5] For standard weight and extra strong. See ASA B16.9 for dimensions of other thicknesses.

Cast Steel Flanged Fittings

Dimensions, in Inches

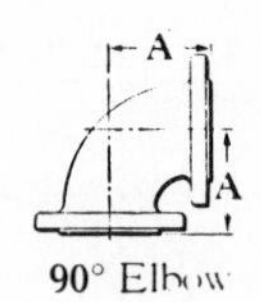

90° Elbow

Tee

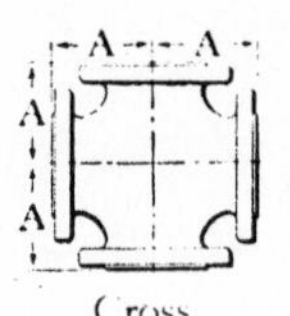

Cross

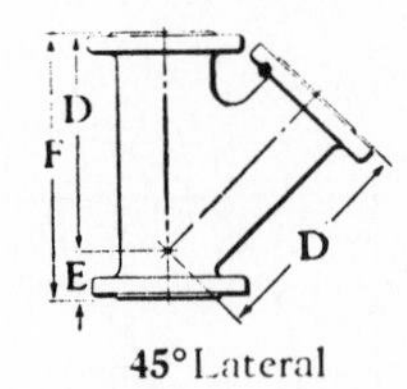

45° Lateral

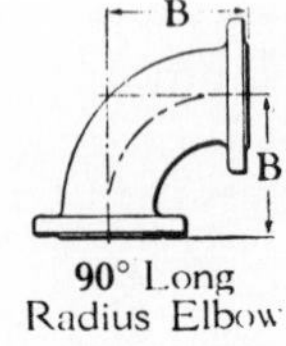

90° Long Radius Elbow

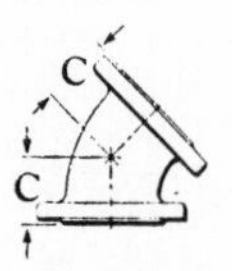

45° Elbow

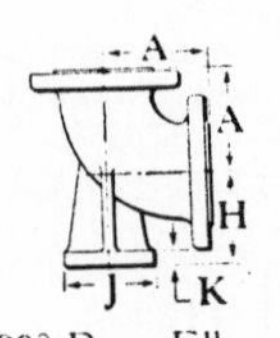

90° Base Elbow

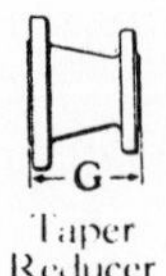

Taper Reducer

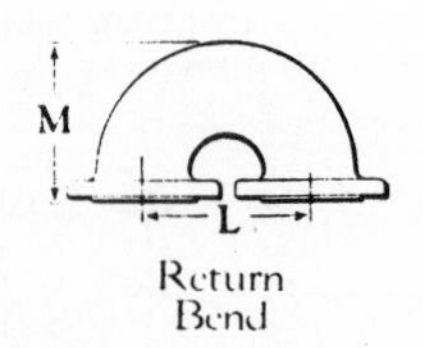

Return Bend

Class	Size	A	B	C	D	E	F	G	H	J	K
150 Pound	1	3½	5	1¾	5¾	1¾	7½				
	1¼	3¾	5½	2	6¼	1¾	8				
	1½	4	6	2¼	7	2	9				
	2	4½	6½	2½	8	2½	10½	5	4⅛	4⅝	½
	2½	5	7	3	9½	2½	12	5½	4½	4⅝	½
	3	5½	7¾	3	10	3	13	6	4⅞	5	9/16
	3½	6	8½	3½	11½	3	14½	6½	5¼	5	9/16
	4	6½	9	4	12	3	15	7	5½	6	⅝
	5	7½	10¼	4½	13½	3½	17	8	6¼	7	11/16
	6	8	11½	5	14½	3½	18	9	7	7	11/16
	8	9	14	5½	17½	4½	22	11	8⅜	9	15/16
	10	11	16½	6½	20½	5	25½	12	9¾	9	15/16
	12	12	19	7½	24½	5½	30	14	11¼	11	1
	14	14	21½	7½	27	6	33	16	12½	11	1
	16	15	24	8	30	6½	36½	18	13¾	11	1
	18	16½	26½	8½	32	7	39	19	15	13½	1⅛
	20	18	29	9½	35	8	43	20	16	13½	1⅛
	24	22	34	11	40½	9	49½	24	18½	13½	1⅛
300 Pound	1	4	5	2¼	6½	2	8½	4½			
	1¼	4¼	5½	2½	7¼	2¼	9½	4½			
	1½	4½	6	2¾	8½	2½	11	4½			
	2	5	6½	3	9	2½	11½	5	4½	5¼	¾
	2½	5½	7	3½	10½	2½	13	5½	4¾	5¼	¾
	3	6	7¾	3½	11	3	14	6	5¼	6⅛	13/16
	3½	6½	8½	4	12½	3	15½	6½	5⅝	6⅛	13/16
	4	7	9	4½	13½	3	16½	7	6	6½	⅞
	5	8	10¼	5	15	3½	18½	8	6¾	7½	1
	6	8½	11½	5½	17½	4	21½	9	7½	7½	1
	8	10	14	6	20½	5	25½	11	9	10	1¼
	10	11½	16½	7	24	5½	29½	12	10½	10	1¼
	12	13	19	8	27½	6	33½	14	12	12½	1 7/16
	14	15	21½	8½	31	6½	37½	16	13½	12½	1 7/16
	16	16½	24	9½	34½	7½	42	18	14¾	12½	1 7/16
	18	18	26½	10	37½	8	45½	19	16¼	15	1⅝
	20	19½	29	10½	40½	8½	49	20	17⅞	15	1⅝
	24	22½	34	12	47½	10	57½	24	20¾	17½	1⅞

Drilling of Base on 90° Base Elbow

Size Elbow	150 Lb.	300 Lb.	400 Lb.	600 Lb.	900 Lb.	1500 Lb.
	Diameter of Bolt Circle					
2	3½	3⅞		4½		5
2½	3½	3⅞		4½		5
3	3⅞	4½		5	5	5⅞
3½	3⅞	4½		5		
4	4¾	5	5	5⅞	5⅞	7⅞
5	5½	5⅞	5⅞	7⅞	7⅞	7⅞
6	5½	5⅞	5⅞	7⅞	7⅞	10⅝
8	7½	7⅞	7⅞	10⅝	10⅝	10⅝
10	7½	7⅞	7⅞	10⅝	10⅝	13
12	9½	10⅝	10⅝	13	13	13
14	9½	10⅝	10⅝	13	13	15¼
16	9½	10⅝	10⅝	13	13	
18	11¾	13	13			
20	11¾	13	13			
24	11¾	15¼	15¼			
	Diameter of Bolts					
2	½	⅝		¾		⅝
2½	½	⅝		¾		⅝
3	½	¾		⅝	⅝	¾
3½	½	¾		⅝		
4	⅝	⅝	⅝	¾	¾	¾
5	⅝	¾	¾	¾	¾	¾
6	⅝	¾	¾	¾	¾	¾
8	⅝	¾	¾	¾	¾	¾
10	⅝	¾	¾	¾	¾	⅞
12	¾	¾	¾	⅞	⅞	⅞
14	¾	¾	¾	⅞	⅞	1
16	¾	¾	¾	⅞	⅞	
18	¾	⅞	⅞			
20	¾	⅞	⅞			
24	¾	1	1			
	Number of Bolts					
All sizes	4	4	4	4	4	4

Return Bends

150 Pound												
Size	2	3			4							6
L	7	7½	8	9	9	10	11	12	13	14	18	12	15
M	6	7	7¼	7¾	8⅝	9⅛	9⅝	10⅛	10⅝	11⅛	13⅛	11¼	12¾

300 Pound																	
Size	2	3		4								6					8
L	7½	8½	12	10	11	11½	12	16½	17	17½	18	14	15	16½	17	18	17½
M	6⅝	8	9¾	9⅝	10⅛	10⅜	10⅝	12⅞	13⅛	13⅜	13⅝	13	13½	14¼	14½	15	16

DIMENSIONS OF CAST-IRON FLANGE FITTINGS

CLASS 125 LB. S.P. American National Standards Institute ANSI B16.1—1967

CLASS 250 LB. S.P. American National Standards Institute ANSI B16.2—1960

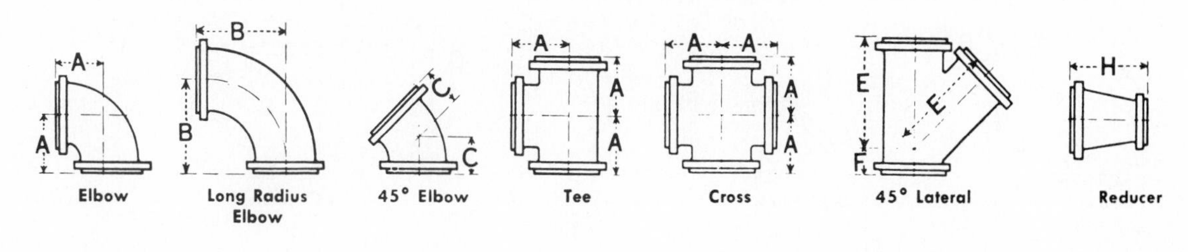

Nominal Pipe Size	Inside Diameter of Fitting	Wall Thickness	Center to Contact Surface, Metal Elbow, Tee and Cross [1,2,3] A	Center to Contact Surface, Long Radius Ell [1,2,3] B	Center to Contact Surface, 45° Ell [1,2,3] C	Long Center to Contact Surface, Lateral [1,2,3] E	Short Center to Contact Surface, Lateral [1,2,3] F	Contact Surface to Contact Surface, Reducer [1,2,3] H
CLASS 125 LB.								
1	1	5/16	3½	5	1¾	5¾	1¾	....
1¼	1¼	5/16	3¾	5½	2	6¼	1¾	....
1½	1½	5/16	4	6	2¼	7	2	....
2	2	5/16	4½	6½	2½	8	2½	5
2½	2½	5/16	5	7	3	9½	2½	5½
3	3	3/8	5½	7¾	3	10	3	6
3½	3½	7/16	6	8½	3½	11½	3	6½
4	4	½	6½	9	4	12	3	7
5	5	½	7½	10¼	4½	13½	3½	8
6	6	9/16	8	11½	5	14½	3½	9
8	8	5/8	9	14	5½	17½	4½	11
10	10	¾	11	16½	6½	20½	5	12
12	12	13/16	12	19	7½	24½	5½	14
14	14	7/8	14	21½	7½	27	6	16
16	16	1	15	24	8	30	6½	18
18	18	1 1/16	16½	26½	8½	32	7	19
20	20	1⅛	18	29	9½	35	8	20
24	24	1¼	22	34	11	40½	9	24
CLASS 250 LB.								
2	2	7/16	5	6½	3	9	2½	5
2½	2½	½	5½	7	3½	10½	2½	5½
3	3	9/16	6	7¾	3½	11	3	6
3½	3½	9/16	6½	8½	4	12½	3	6½
4	4	5/8	7	9	4½	13½	3	7
5	5	11/16	8	10¼	5	15	3½	8
6	6	¾	8½	11½	5½	17½	4	9
8	8	13/16	10	14	6	20½	5	11
10	10	15/16	11½	16½	7	24	5½	12
12	12	1	13	19	8	27½	6	14
14	13¼	1⅛	15	21½	8½	31	6½	16
16	15¼	1¼	16½	24	9½	34½	7½	18
18	17	1⅜	18	26½	10	37½	8	19
20	19	1½	19½	29	10½	40½	8½	20
24	23	1⅝	22½	34	12	47½	10	24

Notes: All dimensions are given in inches.

[1]All Class 125 cast-iron flanged fittings shall be plain faced; i.e., without projection or raised face.

All Class 250 cast-iron flanges and flanged fittings have a 1/16 in. raised face provided on the flange of each opening of these fittings and is included in (a) "thickness of flange," (b) "center to contact surface," and (c) "contact surface to contact surface dimenison."

[2]Where facings other than the 1/16 in. raised face are used (Class 250 only), the "center to contact surface" dimensions shall remain unchanged, and new "center to contact surface" dimensions shall be established to suit the facing used.

[3]Reducing fittings shall have the same "center to contact surface" dimensions as those of straight size fittings of the largest opening.

[4]Wall thickness at no point shall be less than 87½ per cent of the dimensions given in the table.

DIMENSIONS OF STEEL FLANGED FITTINGS

150-300 LB. S.P.

American National Standard Institute ANSI B16.5— 1973

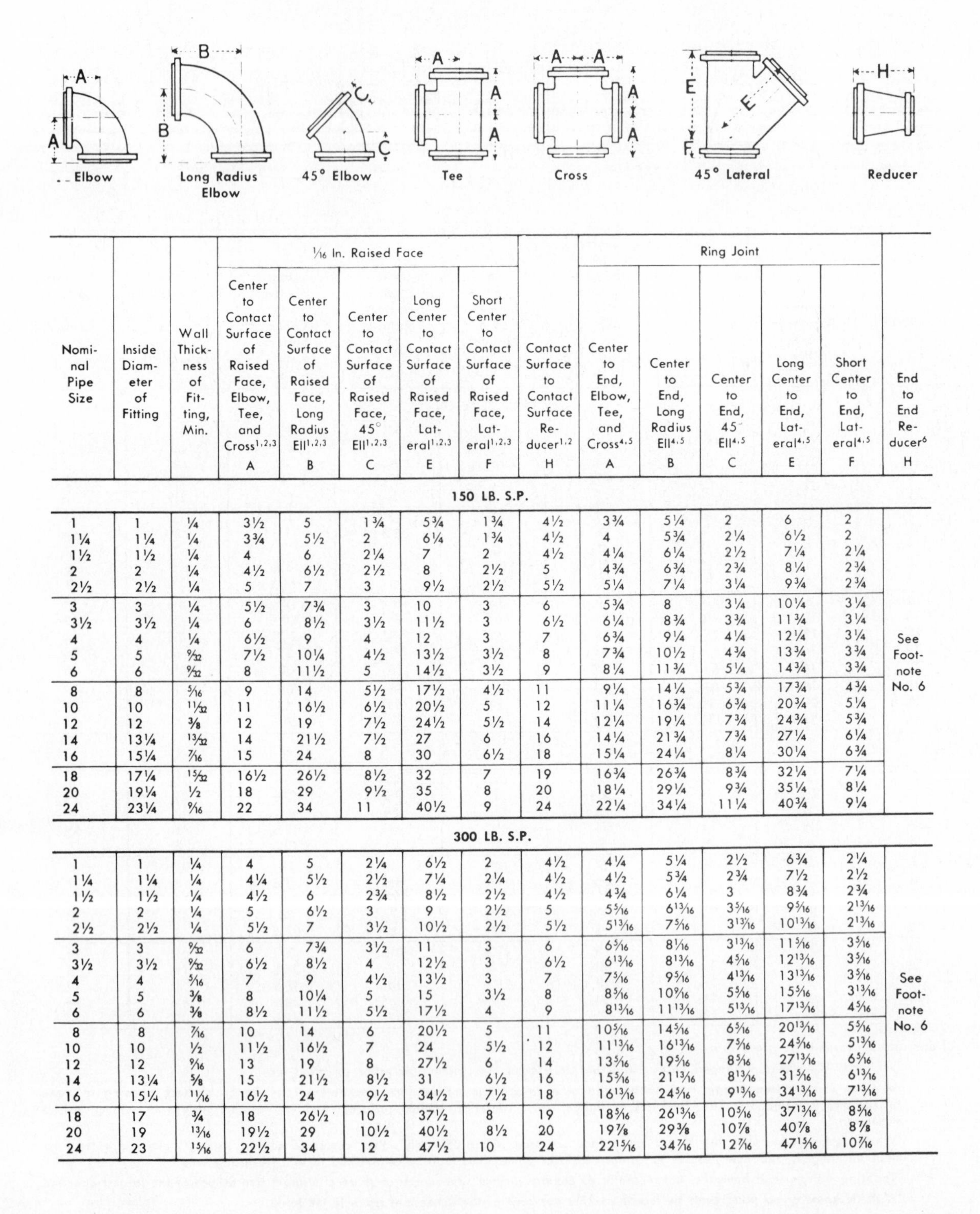

			1/16 In. Raised Face						Ring Joint					
Nominal Pipe Size	Inside Diameter of Fitting	Wall Thickness of Fitting, Min.	Center to Contact Surface of Raised Face, Elbow, Tee, and Cross[1,2,3] A	Center to Contact Surface of Raised Face, Long Radius Ell[1,2,3] B	Center to Contact Surface of Raised Face, 45° Ell[1,2,3] C	Long Center to Contact Surface of Raised Face, Lateral[1,2,3] E	Short Center to Contact Surface of Raised Face, Lateral[1,2,3] F	Contact Surface to Contact Surface Reducer[1,2] H	Center to End, Elbow, Tee, and Cross[4,5] A	Center to End, Long Radius Ell[4,5] B	Center to End, 45° Ell[4,5] C	Long Center to End, Lateral[4,5] E	Short Center to End, Lateral[4,5] F	End to End Reducer[6] H
150 LB. S.P.														
1	1	1/4	3 1/2	5	1 3/4	5 3/4	1 3/4	4 1/2	3 3/4	5 1/4	2	6	2	
1 1/4	1 1/4	1/4	3 3/4	5 1/2	2	6 1/4	1 3/4	4 1/2	4	5 3/4	2 1/4	6 1/2	2	
1 1/2	1 1/2	1/4	4	6	2 1/4	7	2	4 1/2	4 1/4	6 1/4	2 1/2	7 1/4	2 1/4	
2	2	1/4	4 1/2	6 1/2	2 1/2	8	2 1/2	5	4 3/4	6 3/4	2 3/4	8 1/4	2 3/4	
2 1/2	2 1/2	1/4	5	7	3	9 1/2	2 1/2	5 1/2	5 1/4	7 1/4	3 1/4	9 3/4	2 3/4	
3	3	1/4	5 1/2	7 3/4	3	10	3	6	5 3/4	8	3 1/4	10 1/4	3 1/4	
3 1/2	3 1/2	1/4	6	8 1/2	3 1/2	11 1/2	3	6 1/2	6 1/4	8 3/4	3 3/4	11 3/4	3 1/4	
4	4	1/4	6 1/2	9	4	12	3	7	6 3/4	9 1/4	4 1/4	12 1/4	3 1/4	See Footnote No. 6
5	5	9/32	7 1/2	10 1/4	4 1/2	13 1/2	3 1/2	8	7 3/4	10 1/2	4 3/4	13 3/4	3 3/4	
6	6	9/32	8	11 1/2	5	14 1/2	3 1/2	9	8 1/4	11 3/4	5 1/4	14 3/4	3 3/4	
8	8	5/16	9	14	5 1/2	17 1/2	4 1/2	11	9 1/4	14 1/4	5 3/4	17 3/4	4 3/4	
10	10	11/32	11	16 1/2	6 1/2	20 1/2	5	12	11 1/4	16 3/4	6 3/4	20 3/4	5 1/4	
12	12	3/8	12	19	7 1/2	24 1/2	5 1/2	14	12 1/4	19 1/4	7 3/4	24 3/4	5 3/4	
14	13 1/4	13/32	14	21 1/2	7 1/2	27	6	16	14 1/4	21 3/4	7 3/4	27 1/4	6 1/4	
16	15 1/4	7/16	15	24	8	30	6 1/2	18	15 1/4	24 1/4	8 1/4	30 1/4	6 3/4	
18	17 1/4	15/32	16 1/2	26 1/2	8 1/2	32	7	19	16 3/4	26 3/4	8 3/4	32 1/4	7 1/4	
20	19 1/4	1/2	18	29	9 1/2	35	8	20	18 1/4	29 1/4	9 3/4	35 1/4	8 1/4	
24	23 1/4	9/16	22	34	11	40 1/2	9	24	22 1/4	34 1/4	11 1/4	40 3/4	9 1/4	
300 LB. S.P.														
1	1	1/4	4	5	2 1/4	6 1/2	2	4 1/2	4 1/4	5 1/4	2 1/2	6 3/4	2 1/4	
1 1/4	1 1/4	1/4	4 1/4	5 1/2	2 1/2	7 1/4	2 1/4	4 1/2	4 1/2	5 3/4	2 3/4	7 1/2	2 1/2	
1 1/2	1 1/2	1/4	4 1/2	6	2 3/4	8 1/2	2 1/2	4 1/2	4 3/4	6 1/4	3	8 3/4	2 3/4	
2	2	1/4	5	6 1/2	3	9	2 1/2	5	5 5/16	6 13/16	3 5/16	9 5/16	2 13/16	
2 1/2	2 1/2	1/4	5 1/2	7	3 1/2	10 1/2	2 1/2	5 1/2	5 13/16	7 5/16	3 13/16	10 13/16	2 13/16	
3	3	9/32	6	7 3/4	3 1/2	11	3	6	6 5/16	8 1/16	3 13/16	11 5/16	3 5/16	
3 1/2	3 1/2	9/32	6 1/2	8 1/2	4	12 1/2	3	6 1/2	6 13/16	8 13/16	4 5/16	12 13/16	3 5/16	
4	4	5/16	7	9	4 1/2	13 1/2	3	7	7 5/16	9 5/16	4 13/16	13 13/16	3 5/16	See Footnote No. 6
5	5	3/8	8	10 1/4	5	15	3 1/2	8	8 5/16	10 9/16	5 5/16	15 5/16	3 13/16	
6	6	3/8	8 1/2	11 1/2	5 1/2	17 1/2	4	9	8 13/16	11 13/16	5 13/16	17 13/16	4 5/16	
8	8	7/16	10	14	6	20 1/2	5	11	10 5/16	14 5/16	6 5/16	20 13/16	5 5/16	
10	10	1/2	11 1/2	16 1/2	7	24	5 1/2	12	11 13/16	16 13/16	7 5/16	24 5/16	5 13/16	
12	12	9/16	13	19	8	27 1/2	6	14	13 5/16	19 5/16	8 5/16	27 13/16	6 5/16	
14	13 1/4	5/8	15	21 1/2	8 1/2	31	6 1/2	16	15 5/16	21 13/16	8 13/16	31 5/16	6 13/16	
16	15 1/4	11/16	16 1/2	24	9 1/2	34 1/2	7 1/2	18	16 13/16	24 5/16	9 13/16	34 13/16	7 13/16	
18	17	3/4	18	26 1/2	10	37 1/2	8	19	18 5/16	26 13/16	10 5/16	37 13/16	8 5/16	
20	19	13/16	19 1/2	29	10 1/2	40 1/2	8 1/2	20	19 7/8	29 3/8	10 7/8	40 7/8	8 7/8	
24	23	15/16	22 1/2	34	12	47 1/2	10	24	22 15/16	34 7/16	12 7/16	47 15/16	10 7/16	

Forged Steel Screwed Fittings

Dimensions, in Inches

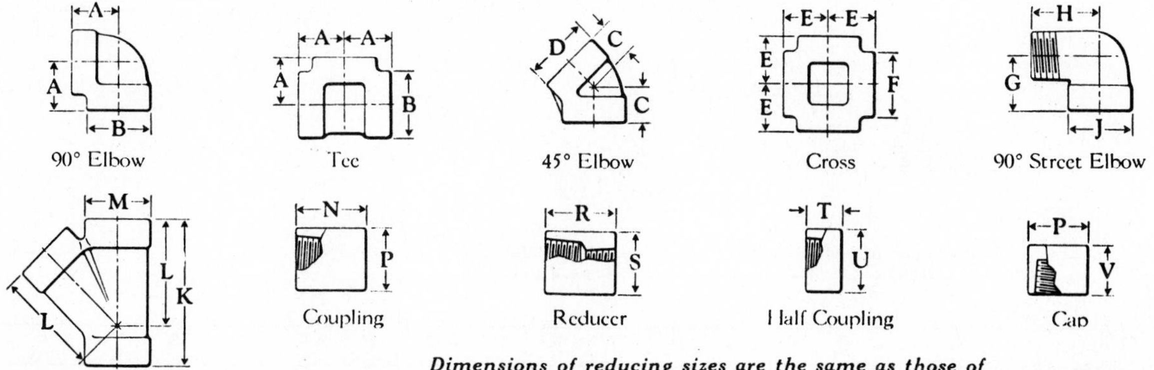

Dimensions of reducing sizes are the same as those of the straight size corresponding to the largest opening.

Size	A	B	C	D	E	F	G	H	J	K	L	M	N	P	R	S	T	U	V
2000-Pound W.O.G. Fittings																			
1/8	13/16	29/32	11/16	15/16	31/32	1													
1/4	13/16	29/32	3/4	1 1/32	31/32	1				2 5/16	1 5/8	13/16							
3/8	31/32	1 1/32	3/4	1 1/32	31/32	1				2 11/16	1 7/8	1							
1/2	1 1/8	1 5/16	7/8	1 5/16	1 1/8	1 5/16				3	2 1/8	1 5/16							
3/4	1 5/16	1 1/2	1	1 1/2	1 5/16	1 1/2				3 9/16	2 9/16	1 1/2							
1	1 1/2	1 13/16	1 1/8	1 13/16	1 1/2	1 13/16				4 1/8	3	1 13/16							
1 1/4	1 3/4	2 7/32	1 5/16	2 7/32	1 3/4	2 3/16				4 13/16	3 1/2	2 3/16							
1 1/2	2	2 15/32	1 7/16	2 15/32	2	2 7/16				5 3/8	3 15/16	2 7/16							
2	2 3/8	3	1 11/16	3	2 3/8	2 31/32				6 7/16	4 3/4	2 31/32							
2 1/2	3	3 5/8	2 1/16	4	3 1/4	4													
3	3 3/8	4 5/16	2 1/2	4 5/8	3 3/8	4 5/8													
4	4 3/16	5 3/4	3 1/8	5 3/4	4 3/16	5 3/4													
3000-Pound W.O.G. Fittings																			
1/8	13/16	29/32	5/8	7/8	31/32	1							1 1/8	19/32		3/4	5/8	3/4	1/2
1/4	31/32	1 1/32	3/4	1 1/32	31/32	1	7/8	1 1/4	1				1 3/8	3/4	1 3/8	3/4	11/16	3/4	11/16
3/8	1 1/8	1 5/16	7/8	1 5/16	1 1/8	1 5/16	1	1 1/2	1 1/4				1 5/8	15/16	1 1/2	7/8	3/4	7/8	3/4
1/2	1 5/16	1 1/2	1	1 1/2	1 5/16	1 1/2	1 1/4	1 5/8	1 1/2	3 9/16	2 9/16	1 1/2	1 7/8	1 1/8	1 7/8	1 1/8	15/16	1 1/8	1 1/4
3/4	1 1/2	1 13/16	1 1/8	1 13/16	1 1/2	1 13/16	1 3/8	1 7/8	1 3/4	4 1/8	3	1 13/16	2 1/8	1 3/8	2	1 3/8	1	1 3/8	1 1/4
1	1 3/4	2 7/32	1 5/16	2 7/32	1 3/4	2 3/16	1 3/4	2 1/4	2	4 13/16	3 1/2	2 3/16	2 3/8	1 3/4	2 3/8	1 3/4	1 3/16	1 3/4	1 1/2
1 1/4	2	2 15/32	1 7/16	2 15/32	2	2 7/16	2	2 5/8	2 7/16	5 3/8	3 15/16	2 7/16	2 7/8	2 1/4	2 5/8	2 1/4	1 5/16	2 1/4	1 5/8
1 1/2	2 3/8	3	1 11/16	3	2 3/8	2 31/32	2 1/8	2 13/16	2 3/4	6 7/16	4 3/4	2 31/32	2 7/8	2 1/2	3 1/8	2 1/2	1 9/16	2 1/2	1 5/8
2	2 1/2	3 9/16	1 23/32	3 5/16	2 1/2	3 5/16	2 1/2	3 5/16	3 5/16				3 3/8	3	3 3/8	3	1 11/16	3	2
2 1/2	3 3/8	4 3/8	2 1/16	4	3 1/4	4							3 5/8	3 5/8	3 5/8	3 5/8	1 13/16	3 5/8	2 3/8
3	3 3/4	4 3/4	2 1/2	4 5/8	3 3/8	4 5/8							4 1/4	4 1/4	4 1/4	4 1/4	2 1/8	4 1/4	2 9/16
3 1/2	4 1/2	6	3 1/8	5 3/4	4 3/16	5 3/4							4 1/2	4 3/4	4 1/2	4 3/4	2 1/4	4 3/4	2 5/8
4	4 1/2	6	3 1/8	5 3/4	4 3/16	5 3/4							4 3/4	5 1/2	4 3/4	5 1/2	2 3/8	5 1/2	2 11/16
5														6 1/2					3 1/16
6														8					3 3/16
6000-Pound W.O.G. Fittings																			
1/8	31/32	1 1/32	3/4	1 1/16	31/32	1							1 1/4	7/8			5/8	3/4	
1/4	1 1/8	1 5/16	7/8	1 5/16	1 1/8	1 5/16	1	1 1/2	1 1/4				1 3/8	1	1 3/8	1	11/16	1	
3/8	1 5/16	1 1/2	1	1 1/2	1 5/16	1 1/2	1 1/8	1 5/8	1 1/2	3 9/16	2 9/16	1 1/2	1 1/2	1 1/4	1 1/2	1 1/4	3/4	1 1/4	
1/2	1 1/2	1 13/16	1 1/8	1 13/16	1 1/2	1 13/16	1 3/8	1 7/8	1 3/4	4 1/8	3	1 13/16	1 7/8	1 1/2	1 7/8	1 1/2	15/16	1 1/2	
3/4	1 3/4	2 7/32	1 5/16	2 3/16	1 3/4	2 3/16	1 3/4	2 1/4	2	4 13/16	3 1/2	2 3/16	2	1 3/4	2	1 3/4	1	1 3/4	
1	2	2 15/32	1 11/32	2 7/16	2	2 7/16	2	2 5/8	2 7/16	5 3/8	3 15/16	2 7/16	2 3/8	2 1/4	2 3/8	2 1/4	1 3/16	2 1/4	
1 1/4	2 3/8	3	1 11/16	2 31/32	2 3/8	2 31/32	2 1/8	2 13/16	2 3/4	6 7/16	4 3/4	2 31/32	2 5/8	2 1/2	2 5/8	2 1/2	1 5/16	2 1/2	
1 1/2	2 1/2	3 9/16	1 23/32	3 5/16	2 1/2	3 5/16	2 1/2	3 5/16	3 5/16				3 1/8	3	3 1/8	3	1 9/16	3	
2	3 3/8	4 3/8	2 1/16	4	3 1/4	4							3 3/8	3 5/8	3 3/8	3 5/8	1 11/16	3 5/8	
2 1/2	3 3/4	4 3/4	2 1/2	4 5/8	3 3/8	4 5/8							3 5/8	4 1/4	3 5/8	4 1/4	1 13/16	4 1/4	
3	4 3/16	5 3/4	3 1/8	5 3/4	4 3/16	5 3/4							4 1/4	5	4 1/4	5	2 1/8	5	
3 1/2	4 1/2	6	3 1/8	5 3/4	4 3/16	5 3/4							4 1/2	5 3/4	4 1/2	5 3/4	2 1/4	5 3/4	
4													4 3/4	6 1/4	4 3/4	6 1/4	2 3/8	6 1/4	

Forged Steel Socket-Welding Fittings

Dimensions, in Inches

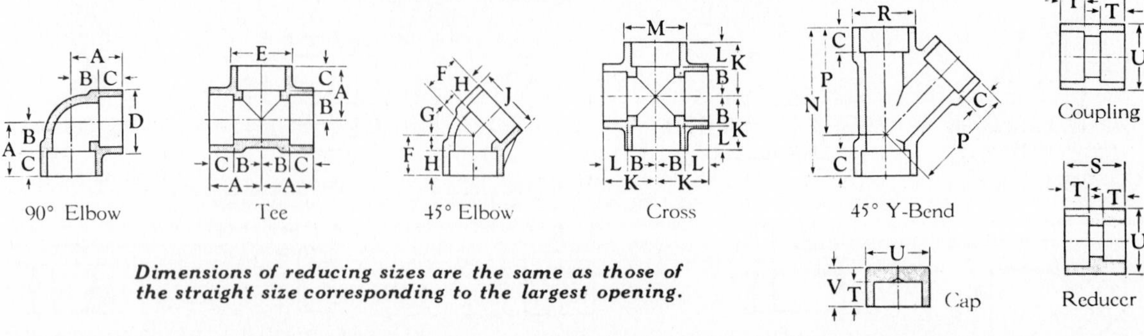

Dimensions of reducing sizes are the same as those of the straight size corresponding to the largest opening.

Size	A	B	C	D	E	F	G	H	J	K	L	M	N	P	R	S	T	U	V
2000-Pound W.O.G. Fittings, for use with Schedule 40 or Standard Pipe																			
1/8	13/16	7/16	3/8	15/16	7/8	11/16	5/16	3/8	15/16	31/32	17/32	1				1	3/8	3/4	5/8
1/4	13/16	7/16	3/8	29/32	29/32	3/4	5/16	7/16	1 1/32	31/32	17/32	1	2 5/16	1 5/8	13/16	1	3/8	3/4	5/8
3/8	31/32	17/32	7/16	1 1/32	1 1/32	3/4	5/16	7/16	1 1/32	31/32	7/16	1	2 11/16	1 7/8	1	1 1/8	7/16	1	11/16
1/2	1 1/8	5/8	1/2	1 5/16	1 5/16	7/8	7/16	7/16	1 5/16	1 1/8	1/2	1 5/16	3	2 1/8	1 1/4	1 3/8	1/2	1 1/4	3/4
3/4	1 5/16	3/4	9/16	1 1/2	1 1/2	1	1/2	1/2	1 1/2	1 5/16	9/16	1 1/2	3 9/16	2 9/16	1 1/2	1 1/2	9/16	1 1/2	13/16
1	1 1/2	7/8	5/8	1 13/16	1 13/16	1 1/8	9/16	9/16	1 13/16	1 1/2	5/8	1 13/16	4 1/8	3	1 13/16	1 3/4	5/8	1 3/4	1
1 1/4	1 3/4	1 1/16	11/16	2 7/32	2 7/32	1 5/16	11/16	5/8	2 7/32	1 3/4	11/16	2 3/16	4 13/16	3 1/2	2 3/16	1 7/8	11/16	2 1/4	1 1/16
1 1/2	2	1 1/4	3/4	2 15/32	2 15/32	1 7/16	13/16	5/8	2 15/32	2	3/4	2 7/16	5 3/8	3 15/16	2 7/16	2	3/4	2 1/2	1 3/16
2	2 3/8	1 1/2	7/8	3	3	1 11/16	1	11/16	3	2 3/8	7/8	2 31/32	6 7/16	4 3/4	2 31/32	2 1/2	7/8	3	1 3/8
2 1/2	3	1 5/8	1 3/8	3 5/8	3 5/8	2 1/16	1 1/8	15/16	4	3 1/4	1 5/8	4				2 1/2	7/8	3 5/8	1 1/2
3	3 3/8	2 1/4	1 1/8	4 5/16	4 5/16	2 1/2	1 1/4	1 1/4	4 5/8	3 3/8	1 1/8	4 5/8				2 3/4	1	4 1/8	1 5/8
4	4 3/16	2 5/8	1 9/16	5 3/4	5 3/4	3 1/8	1 5/8	1 1/2	5 3/4	4 3/16	1 9/16	5 3/4				3	1 1/8	5 1/4	1 7/8
3000-Pound W.O.G. Fittings, for use with Schedule 80 or Extra Strong Pipe																			
1/4	13/16	7/16	3/8	29/32	29/32	3/4	5/16	7/16	1 1/32	31/32	17/32	1	2 5/16	1 5/8	13/16	1	3/8	7/8	11/16
3/8	31/32	17/32	7/16	1 1/32	1 1/32	3/4	5/16	7/16	1 1/32	31/32	7/16	1	2 11/16	1 7/8	1	1 1/8	7/16	1	3/4
1/2	1 1/8	5/8	1/2	1 5/16	1 5/16	7/8	7/16	7/16	1 5/16	1 1/8	1/2	1 5/16	3	2 1/8	1 1/4	1 3/8	1/2	1 1/4	7/8
3/4	1 5/16	3/4	9/16	1 1/2	1 1/2	1	1/2	1/2	1 1/2	1 5/16	9/16	1 1/2	3 9/16	2 9/16	1 1/2	1 1/2	9/16	1 1/2	1
1	1 1/2	7/8	5/8	1 13/16	1 13/16	1 1/8	9/16	9/16	1 13/16	1 1/2	5/8	1 13/16	4 1/8	3	1 13/16	1 3/4	5/8	1 3/4	1 1/16
1 1/4	1 3/4	1 1/16	11/16	2 7/32	2 7/32	1 5/16	11/16	5/8	2 7/32	1 3/4	11/16	2 3/16	4 13/16	3 1/2	2 3/16	1 7/8	11/16	2 1/4	1 3/16
1 1/2	2	1 1/4	3/4	2 15/32	2 15/32	1 7/16	13/16	5/8	2 15/32	2	3/4	2 7/16	5 3/8	3 15/16	2 7/16	2	3/4	2 1/2	1 1/4
2	2 3/8	1 1/2	7/8	3	3	1 11/16	1	11/16	3	2 3/8	7/8	2 31/32	6 7/16	4 3/4	2 31/32	2 1/2	7/8	3	1 1/2
2 1/2	3	1 5/8	1 3/8	3 5/8	3 5/8	2 1/16	1 1/8	15/16	4	3 1/4	1 5/8	4				2 1/2	7/8	3 5/8	1 1/2
3	3 3/8	2 1/4	1 1/8	4 5/16	4 5/16	2 1/2	1 1/4	1 1/4	4 5/8	3 3/8	1 1/8	4 5/8				2 3/4	1	4 1/4	1 3/4
4	4 3/16	2 5/8	1 9/16	5 3/4	5 3/4	3 1/8	1 5/8	1 1/2	5 3/4	4 3/16	1 9/16	5 3/4				3	1 1/8	5 1/2	1 7/8
4000-Pound W.O.G. Fittings, for use with Schedule 160 Pipe																			
1/2	1 5/16	3/4	9/16	1 1/2	1 1/2	1	1/2	1/2	1 1/2	1 5/16	9/16	1 1/2	3 9/16	2 9/16	1 1/2	1 3/8	1/2	1 1/2	7/8
3/4	1 1/2	7/8	5/8	1 13/16	1 13/16	1 1/8	9/16	9/16	1 13/16	1 1/2	5/8	1 13/16	4 1/8	3	1 13/16	1 1/2	9/16	1 3/4	15/16
1	1 3/4	1 1/16	11/16	2 3/16	2 3/16	1 5/16	11/16	5/8	2 3/16	1 3/4	11/16	2 3/16	4 13/16	3 1/2	2 3/16	1 3/4	5/8	2 1/4	1 1/8
1 1/4	2	1 1/4	3/4	2 7/16	2 7/16	1 11/32	13/16	17/32	2 7/16	2	3/4	2 7/16	5 3/8	3 15/16	2 7/16	1 7/8	11/16	2 1/2	1 3/16
1 1/2	2 3/8	1 1/2	7/8	2 31/32	2 31/32	1 11/16	1	11/16	2 31/32	2 3/8	7/8	2 31/32	6 7/16	4 3/4	2 31/32	2	3/4	3	1 3/8
2	2 1/2	1 5/8	7/8	3 5/16	3 5/16	1 23/32	1 1/8	19/32	3 5/16	2 1/2	7/8	3 5/16				2 1/2	7/8	3 5/8	1 1/2
2 1/2	3 1/4	2 1/4	1	4	4	2 1/16	1 1/4	13/16	4	3 1/4	1	4				2 1/2	7/8	4 1/8	1 5/8
3	3 3/4	2 1/2	1 1/4	4 3/4	4 3/4	2 1/2	1 3/8	1 1/8	4 5/8	3 3/8	7/8	4 5/8				2 3/4	1	4 5/8	1 3/4
4	4 3/16	2 5/8	1 9/16	5 3/4	5 3/4	3 1/8	1 5/8	1 1/2	5 3/4	4 3/16	1 9/16	5 3/4				3	1 1/8	6	2
6000-Pound W.O.G. Fittings, for use with Double Extra Strong Pipe																			
3/8	1 1/8	17/32	19/32	1 5/16	1 5/16	7/8	3/8	1/2	1 5/16	1 1/8	19/32	1 5/16	3	2 1/8	1 1/4	1 1/8	7/16	1 5/16	15/16
1/2	1 5/16	5/8	11/16	1 1/2	1 1/2	1	3/8	5/8	1 1/2	1 5/16	11/16	1 1/2	3 9/16	2 9/16	1 1/2	1 3/8	1/2	1 1/2	1
3/4	1 1/2	3/4	3/4	1 13/16	1 13/16	1 1/8	7/16	11/16	1 13/16	1 1/2	3/4	1 13/16	4 1/8	3	1 13/16	1 1/2	9/16	1 3/4	1 1/16
1	1 3/4	7/8	7/8	2 3/16	2 3/16	1 5/16	1/2	13/16	2 3/16	1 3/4	7/8	2 3/16	4 13/16	3 1/2	2 3/16	1 3/4	5/8	2 1/4	1 1/4
1 1/4	2	1 1/16	15/16	2 7/16	2 7/16	1 11/32	5/8	23/32	2 7/16	2	15/16	2 7/16	5 3/8	3 15/16	2 7/16	1 7/8	11/16	2 1/2	1 5/16
1 1/2	2 3/8	1 1/4	1 1/8	2 31/32	2 31/32	1 11/16	19/32	1 3/32	2 31/32	2 3/8	1 1/8	2 31/32	6 7/16	4 3/4	2 31/32	2	3/4	3	1 3/8
2	2 1/2	1 1/2	1	3 5/16	3 5/16	1 23/32	7/8	27/32	3 5/16	2 1/2	1	3 5/16				2 1/2	7/8	3 5/8	1 5/8
2 1/2	3 1/4	1 3/4	1 1/2	4	4	2 1/16	1	1 1/16	4	3 1/4	1 1/2	4				2 1/2	7/8	4 1/4	1 5/8
3	3 3/4	2 1/8	1 5/8	4 3/4	4 3/4	2 1/2	1 1/4	1 1/4	4 5/8	3 3/8	1 1/4	4 5/8				2 3/4	1	5	1 7/8
4	4 1/2	2 5/8	1 7/8	6	6	3 1/8	1 5/8	1 1/2	5 3/4	4 3/16	1 9/16	5 3/4				3	1 1/8	6 1/4	2 1/8

150 lb. flanges

welding neck, fig. 1911

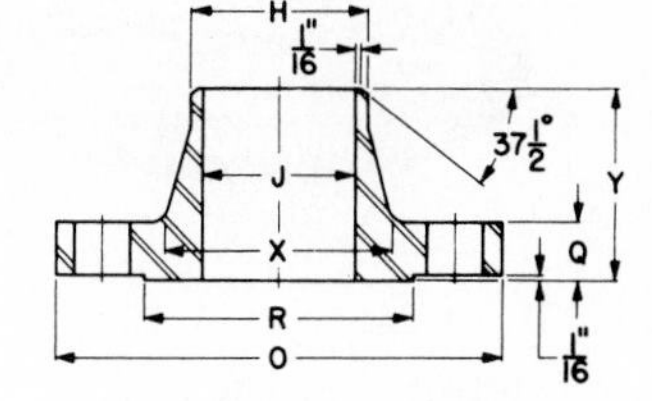

threaded, fig. 1931

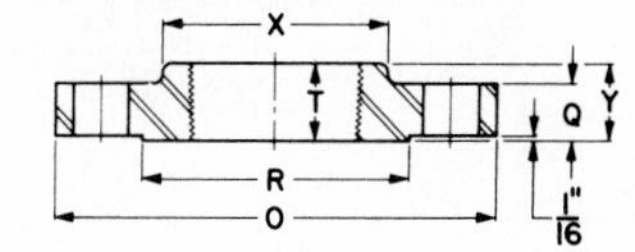

manufacture: ★150 lb flanges are carbon steel furnished to ASTM specifications A181 grade 1. ASTM A181 is the same as Boiler Construction Code Specification SA181. Flanges furnished faced, drilled, and spot faced or back faced. For standard flange facings, see page wff-85. For ring joint type details, see pages wff-86 to wff-88.

dimensions: All dimensions are in inches and in accordance with ANSI B16.5 where applicable. 22 inch is an interpolated dimension as given in MSS-SP-44. Sizes 26" to 42" have same flange drilling dimensions as 125 lb class cast iron flanges, ANSI B16.1. For dimensional tolerances see page wff-81.

pressure-temperature ratings: See page wff-89.

footnotes: ■Bolt holes are 1/8 inch larger than bolt diameter.

◆Flanges bored to dimensions shown, unless otherwise specified. Dimensions shown correspond to ANSI B36.10 inside diameter of standard wall pipe. The smallest bores to which forgings for these flanges may be machined are listed on page wff-82.

✦Thread lengths for 150 lb flanges are ANSI for Pipe Threads ANSI B2.1. Add depths or height of facing to thread length.

✣Lengths of alloy steel stud bolts include the thickness of two nuts, but do not include height of points.

✦Length shown does not include thickness of lap. For lapped to lapped, add thickness of both laps; for lapped to 1/16 inch raised face add one thickness of lap; for lapped to any other facing, add amount that such facing and one lap will cause the flanges to be separated.

nominal pipe size	outside diam. of flange O	thickness of flange Q (min.)	diam. of raised face R	diam. of hub at base X	no. of bolt holes	diam. of bolts■	diam. of bolt circle	length of bolts: stud✣ 1/16" raised face	length of bolts: stud✣ ring joint	length of bolts: machine 1/16" raised face	length thru hub: welding neck Y	length thru hub: slip-on, th'rd, socket Y	length thru hub: lap joint Y
1/2	3 1/2	7/16	1 3/8	1 3/16	4	1/2	2 3/8	2 1/4	......	1 3/4	1 7/8	5/8	5/8
3/4	3 7/8	1/2	1 11/16	1 1/2	4	1/2	2 3/4	2 1/4	......	2	2 1/16	5/8	5/8
1	4 1/4	9/16	2	1 15/16	4	1/2	3 1/8	2 1/2	3	2	2 3/16	11/16	11/16
1 1/4	4 5/8	5/8	2 1/2	2 5/16	4	1/2	3 1/2	2 1/2	3	2 1/4	2 1/4	13/16	13/16
1 1/2	5	11/16	2 7/8	2 9/16	4	1/2	3 7/8	2 3/4	3 1/4	2 1/4	2 7/16	7/8	7/8
2	6	3/4	3 5/8	3 7/16	4	5/8	4 3/4	3	3 1/2	2 3/4	2 1/2	1	1
2 1/2	7	7/8	4 1/8	3 9/16	4	5/8	5 1/2	3 1/4	3 3/4	3	2 3/4	1 1/8	1 1/8
3	7 1/2	15/16	5	4 1/4	4	5/8	6	3 1/2	4	3	2 3/4	1 3/16	1 3/16
3 1/2	8 1/2	15/16	5 1/2	4 13/16	8	5/8	7	3 1/2	4	3	2 13/16	1 1/4	1 1/4
4	9	15/16	6 3/16	5 5/16	8	5/8	7 1/2	3 1/2	4	3	3	1 5/16	1 5/16
5	10	15/16	7 5/16	6 7/16	8	3/4	8 1/2	3 3/4	4 1/4	3 1/4	3 1/2	1 7/16	1 7/16
6	11	1	8 1/2	7 9/16	8	3/4	9 1/2	3 3/4	4 1/4	3 1/4	3 1/2	1 9/16	1 9/16
8	13 1/2	1 1/8	10 5/8	9 11/16	8	3/4	11 3/4	4	4 1/2	3 1/2	4	1 3/4	1 3/4
10	16	1 3/16	12 3/4	12	12	7/8	14 1/4	4 1/2	5	3 3/4	4	1 15/16	1 15/16
12	19	1 1/4	15	14 3/8	12	7/8	17	4 1/2	5	4	4 1/2	2 3/16	2 3/16
14	21	1 3/8	16 1/4	15 3/4	12	1	18 3/4	5	5 1/2	4 1/4	5	2 1/4	3 1/8
16	23 1/2	1 7/16	18 1/2	18	16	1	21 1/4	5 1/4	5 3/4	4 1/2	5	2 1/2	3 7/16
18	25	1 9/16	21	19 7/8	16	1 1/8	22 3/4	5 3/4	6 1/4	4 3/4	5 1/2	2 11/16	3 13/16
20	27 1/2	1 11/16	23	22	20	1 1/8	25	6	6 1/2	5 1/4	5 11/16	2 7/8	4 1/16
22	29 1/2	1 13/16	25 1/4	24	20	1 1/4	27 1/4	6 1/2	7	5 1/2	5 7/8	3 1/8	4 1/4
24	32	1 7/8	27 1/4	26 1/8	20	1 1/4	29 1/2	6 3/4	7 1/4	5 3/4	6	3 1/4	4 3/8
26	34 1/4	2	29 1/4	28 1/2	24	1 1/4	31 3/4	7	......	6	5	3 3/8	......
28	36 1/2	2 1/16	31 1/4	30 3/4	28	1 1/4	34	7	......	6	5 1/16	3 7/16	......
30	38 3/4	2 1/8	33 3/4	32 3/4	28	1 1/4	36	7 1/4	......	6 1/4	5 1/8	3 1/2	......
32	41 3/4	2 1/4	35 3/4	35	28	1 1/2	38 1/2	8	......	6 3/4	5 1/4	3 5/8	......
34	43 3/4	2 5/16	37 3/4	37	32	1 1/2	40 1/2	8	......	7	5 5/16	3 11/16	......
36	46	2 3/8	40 1/4	39 1/4	32	1 1/2	42 3/4	8 1/4	......	7	5 3/8	3 3/4	......
42	53	2 5/8	47	46	36	1 1/2	49 1/2	8 3/4	......	7 1/2	5 3/8	4	......

lap joint, fig. 1901

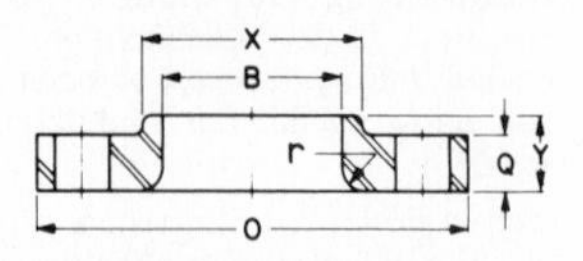

socket type, fig. 1968

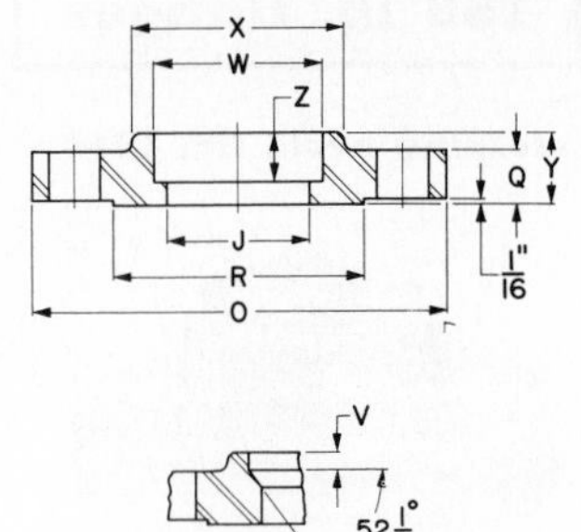

slip-on, fig. 1921

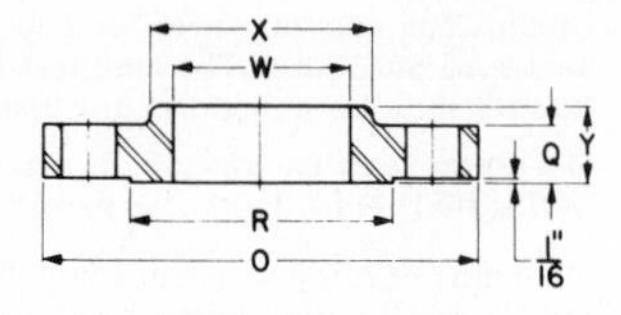

blind, fig. 1961

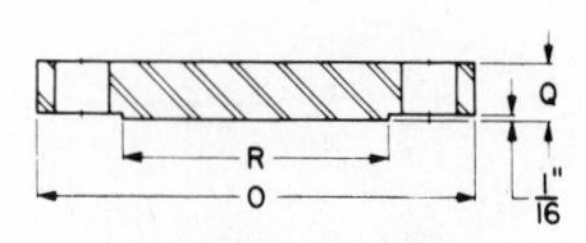

nominal pipe size	diameter of bore: lap joint B	slip-on, socket W	welding neck, socket J◆	minimum thread length T✦	diam. of hub at point of welding H	radius r	depth of socket Z	depth of socket V	weight (approx.) lb: welding neck	slip-on and threaded	lap joint	blind	socket
1/2	.90	.88	.62	5/8	.84	1/8	3/8	3/16	2	2	2	2	2
3/4	1.11	1.09	.82	5/8	1.05	1/8	7/16	1/8	2	2	2	2	2
1	1.38	1.36	1.05	11/16	1.32	1/8	1/2	1/8	2 1/2	2	2	2	2
1 1/4	1.72	1.70	1.38	13/16	1.66	3/16	9/16	3/16	2 1/2	3	3	3	3
1 1/2	1.97	1.95	1.61	7/8	1.90	1/4	5/8	3/16	4	3	3	3	3
2	2.46	2.44	2.07	1	2.38	5/16	11/16	1/4	6	5	5	4	5
2 1/2	2.97	2.94	2.47	1 1/8	2.88	5/16	3/4	1/4	10	7	7	7	7
3	3.60	3.57	3.07	1 3/16	3.50	3/8	13/16	1/4	11 1/4	8	8	9	8
3 1/2	4.10	4.07	3.55	1 1/4	4.00	3/8	7/8	5/16	12	11	11	13	11
4	4.60	4.57	4.03	1 5/16	4.50	7/16	15/16	3/8	15	13	13	17	13
5	5.69	5.66	5.05	1 7/16	5.56	7/16	15/16	1/2	19	15	15	20	15
6	6.75	6.72	6.07	1 9/16	6.63	1/2	1 1/16	9/16	24	19	19	26	19
8	8.75	8.72	7.98	1 3/4	8.63	1/2	1 1/4	5/8	39	30	30	45	30
10	10.92	10.88	10.02	1 15/16	10.75	1/2	1 5/16	3/4	52	43	43	70	43
12	12.92	12.88	12.00	2 3/16	12.75	1/2	1 9/16	15/16	80	64	64	110	64
14	14.18	14.14	13.25	2 1/4	14.00	1/2	1 5/8	7/8	120	85	99	131	85
16	16.19	16.16	15.25	2 1/2	16.00	1/2	1 3/4	1 1/16	127	93	128	170	93
18	18.20	18.18	17.25	2 11/16	18.00	1/2	1 15/16	1 1/8	140	120	146	209	120
20	20.25	20.20	19.25	2 7/8	20.00	1/2	2 1/8	1 3/16	170	155	185	272	155
22	22.25	22.22	21.25	3 1/8	22.00	1/2	2 3/8	...	224	159	245	333	185
24	24.25	24.25	23.25	3 1/4	24.00	1/2	2 1/2	1 3/8	260	210	260	411	210
26	...	26.25	as specified by purchaser	...	26.00	...	...	...	300	250	...	525	...
28	...	28.25		...	28.00	...	...	...	315	285	...	620	...
30	...	30.25		...	30.00	...	...	...	360	315	...	720	...
32	...	32.25		...	32.00	...	...	...	435	395	...	870	...
34	...	34.25		...	34.00	...	...	...	465	420	...	990	...
36	...	36.25		...	36.00	...	...	...	520	480	...	1125	...
42	...	42.25		...	42.00	...	...	...	750	680	...	1625	...

lap joint, fig. 1902

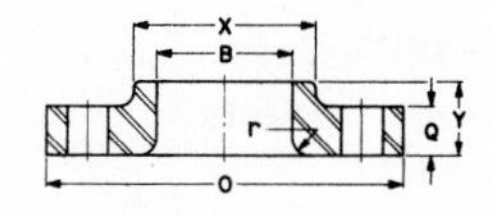

slip-on, fig. 1922

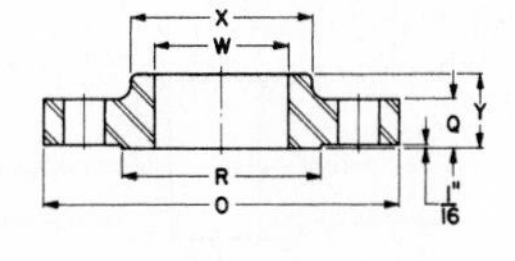

socket type, fig. 1969

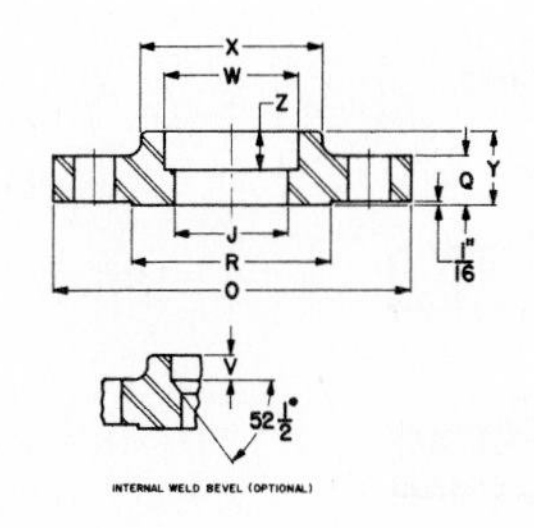

blind, fig. 1962

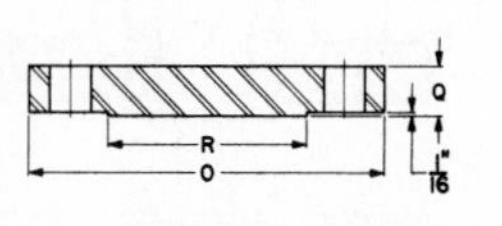

	diameter of bore									weight (approx.) lb			
nominal pipe size	threaded counter bore C	lap joint B	slip-on, socket W	welding neck, socket J◆	minimum thread length T✦	diam. of hub at point of welding H	radius r	depth of socket Z	depth of socket V	welding neck	slip-on, socket, th'rd	lap joint	blind
1/2	.93	.90	.88	.62	5/8	.84	1/8	3/8	5/16	2	3	3	2
3/4	1.14	1.11	1.09	.82	5/8	1.05	1/8	7/16	3/8	3	3	3	3
1	1.41	1.38	1.36	1.05	11/16	1.32	1/8	1/2	3/8	4	3	3	4
1 1/4	1.75	1.72	1.70	1.38	13/16	1.66	3/16	9/16	5/16	6	4	4	6
1 1/2	1.99	1.97	1.95	1.61	7/8	1.90	1/4	5/8	3/8	8	6	6	7
2	2.50	2.46	2.44	2.07	1 1/8	2.38	5/16	11/16	7/16	9	7	7	8
2 1/2	3.00	2.97	2.94	2.47	1 1/4	2.88	5/16	3/4	1/2	12	10	10	12
3	3.63	3.60	3.57	3.07	1 1/4	3.50	3/8	13/16	9/16	15	13	13	16
3 1/2	4.13	4.10	4.07	3.55	1 7/16	4.00	3/8	7/8	9/16	18	17	17	21
4	4.63	4.60	4.57	4.03	1 7/16	4.50	7/16	15/16	5/8	25	22	22	27
5	5.69	5.69	5.66	5.05	1 11/16	5.56	7/16	…	…	32	28	28	35
6	6.75	6.75	6.72	6.07	1 13/16	6.63	1/2	…	…	42	39	39	50
8	8.75	8.75	8.72	7.98	2	8.63	1/2	…	…	67	58	58	81
10	10.88	10.92	10.88	10.02	2 3/16	10.75	1/2	…	…	91	81	91	127
12	12.94	12.92	12.88	12.00	2 3/8	12.75	1/2	…	…	138	115	139	184
14	14.19	14.18	14.14	13.25	2 1/2	14.00	1/2	…	…	186	164	189	236
16	16.19	16.19	16.16	15.25	2 11/16	16.00	1/2	…	…	246	220	240	307
18	18.19	18.20	18.18	17.25	2 3/4	18.00	1/2	…	…	305	280	305	390
20	20.19	20.25	20.20	19.25	2 7/8	20.00	1/2	…	…	378	325	375	492
22	22.19	22.25	22.22	21.25	3 1/16	22.00	1/2	…	…	429	433	435	594
24	24.19	24.25	24.25	23.25	3 1/4	24.00	1/2	…	…	545	490	530	754
26	…	…	26.25	to be	…	26 1/4	…	…	…	670	570	…	1050
28	…	…	28.25	speci-	…	28 1/4	…	…	…	810	720	…	1275
30	…	…	30.25	fied	…	30 1/4	…	…	…	930	810	…	1500
32	…	…	32.25	by	…	32 1/4	…	…	…	1025	890	…	1775
33	…	…	34.25	pur-	…	34 5/16	…	…	…	1200	1075	…	2025
36	…	…	36.25	chaser	…	36 5/16	…	…	…	1300	1200	…	2275

FACING DIMENSIONS

FOR 150, 300, 400, 600, 900, 1500 AND 2500 LB. S.P. STEEL FLANGES

(Other than Ring Joints)

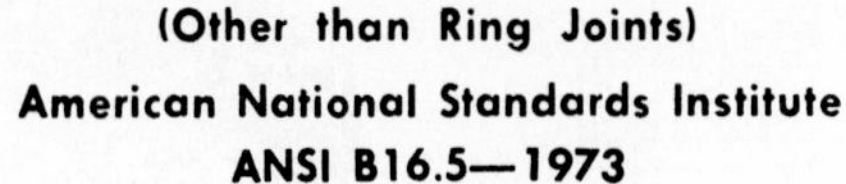

American National Standards Institute
ANSI B16.5—1973

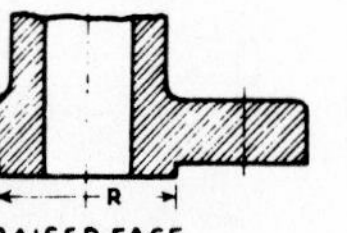

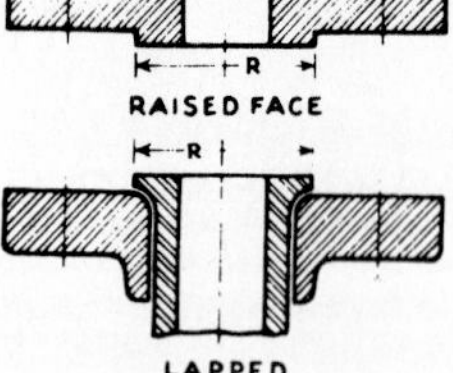

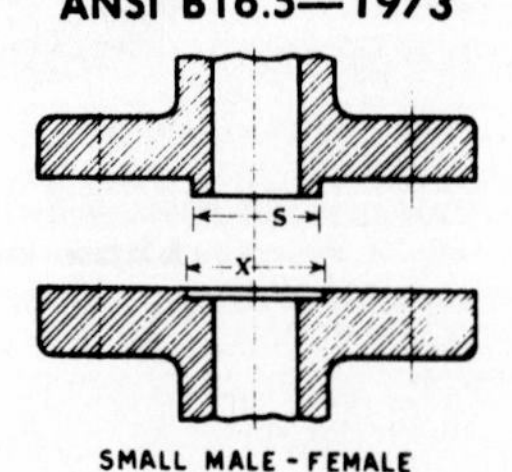

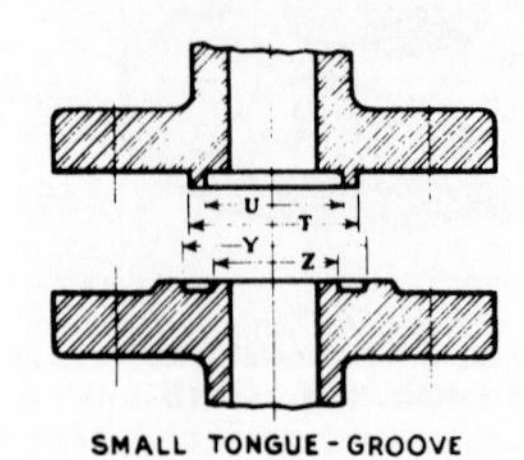

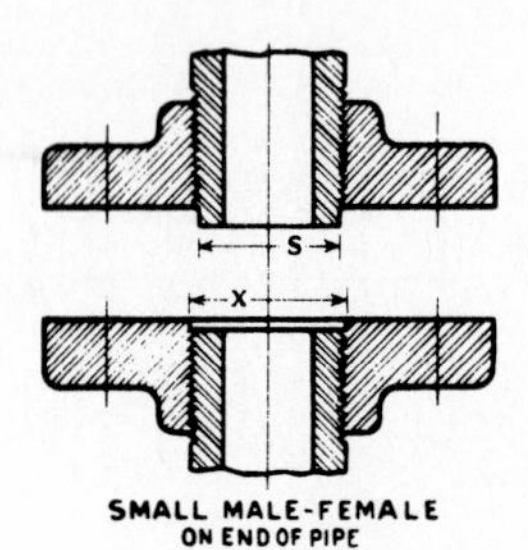

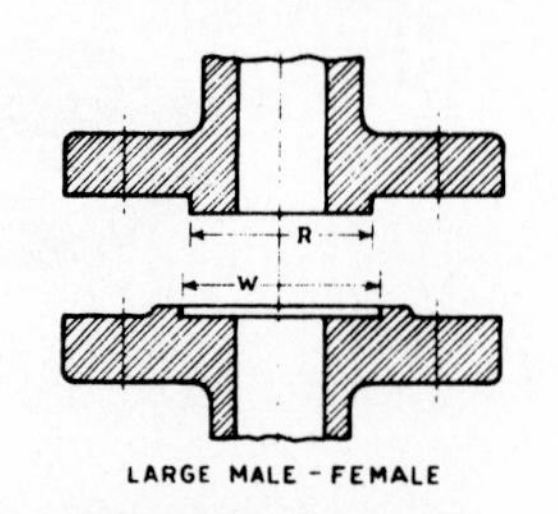

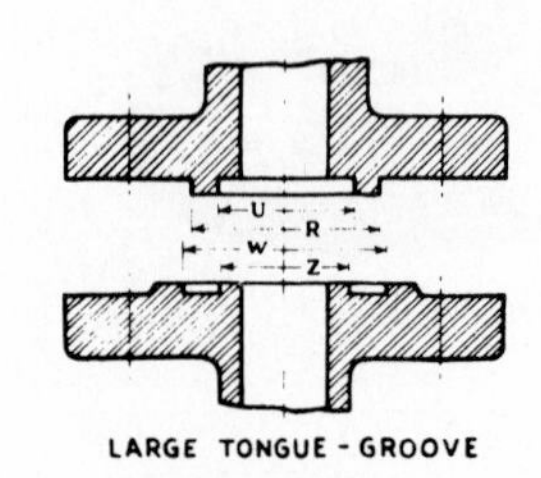

Nominal Pipe Size	Outside Diameter[3]			ID of Large and Small Tongue[3,5]	Outside Diameter[3]			ID of Large and Small Groove[3,5]	Height		Depth of Groove or Female
	Raised Face, Lapped, Large Male and Large Tongue[5]	Small Male[4,5]	Small Tongue[5]		Large Female and Large Groove[5]	Small Female[4,5]	Small Groove[5]		Raised Face 150 and 300 Lb. St'ds[1]	Raised Face, Large and Small Male and Tongue 400, 600, 900, 1500, and 2500 Lb. St'ds[2]	
	R	S	T	U	W	X	Y	Z			
1/2	1 3/8	23/32	1 3/8	1	1 7/16	25/32	1 7/16	15/16	1/16	1/4	3/16
3/4	1 11/16	15/16	1 11/16	1 5/16	1 3/4	1	1 3/4	1 1/4	1/16	1/4	3/16
1	2	1 3/16	1 7/8	1 1/2	2 1/16	1 1/4	1 15/16	1 7/16	1/16	1/4	3/16
1 1/4	2 1/2	1 1/2	2 1/4	1 7/8	2 9/16	1 9/16	2 5/16	1 13/16	1/16	1/4	3/16
1 1/2	2 7/8	1 3/4	2 1/2	2 1/8	2 15/16	1 13/16	2 9/16	2 1/16	1/16	1/4	3/16
2	3 5/8	2 1/4	3 1/4	2 7/8	3 11/16	2 5/16	3 5/16	2 13/16	1/16	1/4	3/16
2 1/2	4 1/8	2 11/16	3 3/4	3 3/8	4 3/16	2 3/4	3 13/16	3 5/16	1/16	1/4	3/16
3	5	3 5/16	4 5/8	4 1/4	5 1/16	3 3/8	4 11/16	4 3/16	1/16	1/4	3/16
3 1/2	5 1/2	3 13/16	5 1/8	4 3/4	5 9/16	3 7/8	5 3/16	4 11/16	1/16	1/4	3/16
4	6 3/16	4 5/16	5 11/16	5 3/16	6 1/4	4 3/8	5 3/4	5 1/8	1/16	1/4	3/16
5	7 5/16	5 3/8	6 13/16	6 5/16	7 3/8	5 7/16	6 7/8	6 1/4	1/16	1/4	3/16
6	8 1/2	6 3/8	8	7 1/2	8 9/16	6 7/16	8 1/16	7 7/16	1/16	1/4	3/16
8	10 5/8	8 3/8	10	9 3/8	10 11/16	8 7/16	10 1/16	9 5/16	1/16	1/4	3/16
10	12 3/4	10 1/2	12	11 1/4	12 13/16	10 9/16	12 1/16	11 3/16	1/16	1/4	3/16
12	15	12 1/2	14 1/4	13 1/2	15 1/16	12 9/16	14 5/16	13 7/16	1/16	1/4	3/16
14 OD	16 1/4	13 3/4	15 1/2	14 3/4	16 5/16	13 13/16	15 9/16	14 11/16	1/16	1/4	3/16
16 OD	18 1/2	15 3/4	17 5/8	16 3/4	18 9/16	15 13/16	17 11/16	16 11/16	1/16	1/4	3/16
18 OD	21	17 3/4	20 1/8	19 1/4	21 1/16	17 13/16	20 3/16	19 3/16	1/16	1/4	3/16
20 OD	23	19 3/4	22	21	23 1/16	19 13/16	22 1/16	20 15/16	1/16	1/4	3/16
24 OD	27 1/4	23 3/4	26 1/4	25 1/4	27 5/16	23 13/16	26 5/16	25 3/16	1/16	1/4	3/16

Notes: All dimensions given in inches.

[1]Regular facing for 150 and 300 lb. steel flanged fittings and companion flange standards is a 1/16 in. raised face included in the minimum flange thickness dimensions.

[2]Regular facing for 400, 600, 900, 1500 and 2500 lb. flange standards is a 1/4 in. raised face not included in minimum flange thickness dimensions.

[3]Tolerance of plus or minus 0.016 in. (1/64 in.) is allowed on the inside and outside diameters of all facings.

[4]For small male and female joints care should be taken in the use of these dimensions to insure that the inside diameter of fitting on pipe is small enough to permit sufficient bearing surface to prevent the crushing of the gasket. This applies particularly on lines where the joint is made on the end of the pipe. Inside diameter of fitting should match inside diameter of pipe as specified by purchaser. Several companion flanges for small male and female joints are furnished with plain face and are threaded with American Standard Locknut Thread (NPSL).

[5]Gaskets for male-female and tongue-groove joints shall cover the bottom of the recess with minimum clearances taking into account the tolerances prescribed in Note 3.

NUMBERS FOR RING-JOINT GASKETS AND GROOVES

American National Standards Institute ANSI B16.20 — 1973

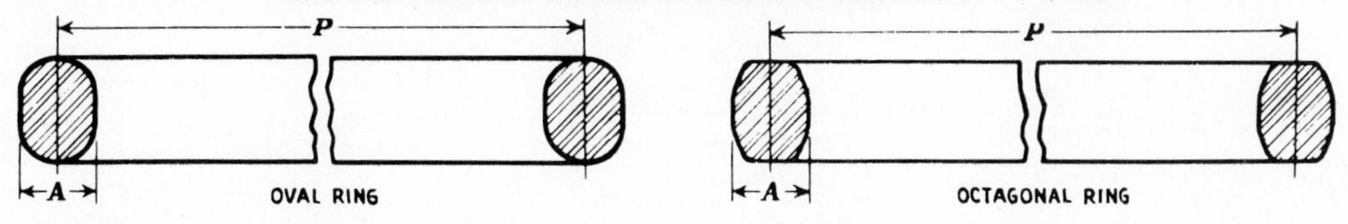

Number	Pitch Diameter P		Width of Ring A	Number	Pitch Diameter P		Width of Ring A	Number	Pitch Diameter P		Width of Ring A	Number	Pitch Diameter P		Width of Ring A
R 11	1 11/32	×	1/4	R 29	4 1/2	×	5/16	R 46	8 5/16	×	1/2	R 63	16 1/2	×	1
R 12	1 9/16	×	5/16	R 30	4 5/8	×	7/16	R 47	9	×	3/4	R 64	17 7/8	×	5/16
R 13	1 11/16	×	5/16	R 31	4 7/8	×	7/16	R 48	9 3/4	×	5/16	R 65	18 1/2	×	7/16
R 14	1 3/4	×	5/16	R 32	5	×	1/2	R 49	10 5/8	×	7/16	R 66	18 1/2	×	5/8
R 15	1 7/8	×	5/16	R 33	5 3/16	×	5/16	R 50	10 5/8	×	5/8	R 67	18 1/2	×	1 1/8
R 16	2	×	5/16	R 34	5 3/16	×	7/16	R 51	11	×	7/8	R 68	20 3/8	×	5/16
R 17	2 1/4	×	5/16	R 35	5 3/8	×	7/16	R 52	12	×	5/16	R 69	21	×	7/16
R 18	2 3/8	×	5/16	R 36	5 7/8	×	5/16	R 53	12 3/4	×	7/16	R 70	21	×	3/4
R 19	2 9/16	×	5/16	R 37	5 7/8	×	7/16	R 54	12 3/4	×	5/8	R 71	21	×	1 1/8
R 20	2 11/16	×	5/16	R 38	6 3/16	×	5/8	R 55	13 1/2	×	1 1/8	R 72	22	×	5/16
R 21	2 27/32	×	7/16	R 39	6 3/8	×	7/16	R 56	15	×	5/16	R 73	23	×	1/2
R 22	3 1/4	×	5/16	R 40	6 3/4	×	5/16	R 57	15	×	7/16	R 74	23	×	3/4
R 23	3 1/4	×	7/16	R 41	7 1/8	×	7/16	R 58	15	×	7/8	R 75	23	×	1 1/4
R 24	3 3/4	×	7/16	R 42	7 1/2	×	3/4	R 59	15 5/8	×	5/16	R 76	26 1/2	×	5/16
R 25	4	×	5/16	R 43	7 5/8	×	5/16	R 60	16	×	1 1/4	R 77	27 1/4	×	5/8
R 26	4	×	7/16	R 44	7 5/8	×	7/16	R 61	16 1/2	×	7/16	R 78	27 1/4	×	1
R 27	4 1/4	×	7/16	R 45	8 5/16	×	7/16	R 62	16 1/2	×	5/8	R 79	27 1/4	×	1 3/8
R 28	4 3/8	×	1/2												

Notes: The edge of each flange and the outside circumference of each ring shall carry corresponding identification marks: i.e., R 11, R 45, etc.

This standard shows only flat bottom grooves, because oval and octagon rings may be used. The former round bottom groove requires the use of an oval gasket.

All dimensions given in inches.

BRITISH STANDARD TAPER PIPE THREADS

Reproduced from British Standard No. 21—1957

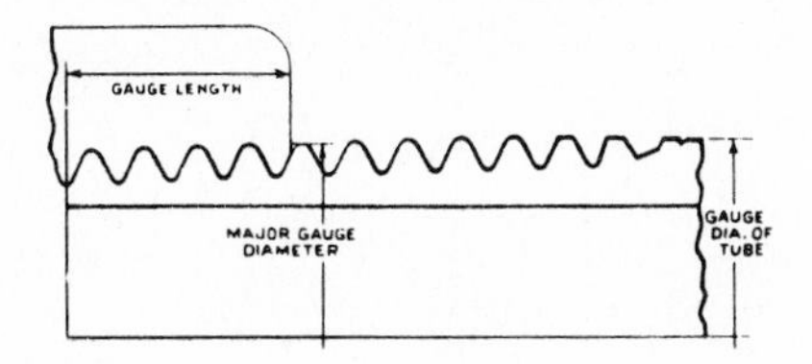

Nominal Size	Max. Outside Diameter	No. of Threads per Inch	Basic Gauge Diameters: Depth of Thread	Major	Effective Pitch Diameter	Minor	Gauge Length
1/8	0.412	28	0.0229	0.383	0.3601	0.3372	0.1563
1/4	0.550	19	0.0337	0.518	0.4843	0.4506	0.2367
3/8	0.688	19	0.0337	0.656	0.6223	0.5886	0.2500
1/2	0.859	14	0.0457	0.825	0.7793	0.7336	0.3214
3/4	1.075	14	0.0457	1.041	0.9953	0.9496	0.3750
1	1.351	11	0.0582	1.309	1.2508	1.1926	0.4091
1 1/4	1.692	11	0.0582	1.650	1.5918	1.5336	0.5000
1 1/2	1.924	11	0.0582	1.882	1.8238	1.7656	0.5000
2	2.403	11	0.0582	2.347	2.2888	2.2306	0.6250
2 1/2	3.021	11	0.0582	2.960	2.9018	2.8436	0.6875
3	3.526	11	0.0582	3.460	3.4018	3.3436	0.8125
3 1/2	4.021	11	0.0582	3.950	3.8918	3.8336	0.8750
4	4.526	11	0.0582	4.450	4.3918	4.3336	1.0000
5	5.536	11	0.0582	5.450	5.3918	5.3336	1.1250
6	6.541	11	0.0582	6.450	6.3918	6.3336	1.1250

Notes: Included thread angle 55°, formed at right angle to the pipe axis, and rounded equally top and bottom.

All dimensions given in inches.

ABBREVIATIONS

A	Anchor
AC	Air closes
ACCUM	Accumulator
AFD	Auxiliary feedwater
AI	All iron
AISC	American Institute of Steel Construction
Al	Aluminum
ALY	Alloy
ANSI	American National Standards Institute
AO	Air opens
API	American Petroleum Institute
APPROX	Approximate
ARCH	Architectural
ASA	American Standards Association
ASME	American Society of Mechanical Engineers
ASNT	American Society for Nondestructive Testing
ASSY	Assembly
ASTE	American Society of Testing Engineers
ASTM	American Society for Testing and Materials
AUT	Automatic vent trap
AUX	Auxiliary
AVG	Average
AWS	American Welding Society
AWWA	American Waterworks Association
AZ	Azimuth
B	Beveled
BB	Bolted bonnet
B&B	Bell and bell
BC	Bolt circle
BD	Blowdown
BE	Beveled ends
BF	Bottom flat
BF	Blind flange
BLDG	Building
BM	Beam
B/M (BM)	Bill of material
BOC	Bottom of concrete
BOP	Bottom of pipe
BOS (B/S)	Bottom of steel
BR	Bronze
BRS	Brass
B&S	Bell and spigot
BTM	Bottom
BUSH	Bushing
BW	Butt weld
BWG	Birmingham wire gage
C	Channel
°C	Degrees centigrade
CAS	Cast alloy steel

CBD	Continuous blowdown
C to C	Center to center
CCW	Component cooling water
CD	Closed drain
CFM	Cubic feet per minute
CHKD	Checked
CH-O	Chain-operated
CI	Cast iron
CL	Clearance
CL	Centerline
CO	Clean out
COL	Column
CONC	Concentric
COND	Condensate
CONN	Connection
CONSTR	Construction
CONT	Continuation
CORR	Corrosion
CPLG	Coupling
CR	Conductivity recorder
Cr	Chromium
Cr 13	Type 410 stainless steel
CS	Carbon steel
CS	Cast steel
CS	Cold spring
CSC	Car seal closed
CSO	Car seal opened
CTMT	Containment
CTR	Center
CTS	Containment spray
Cu	Copper
CVC	Chemical and volume control
CWP	Cold working pressure
DC	Density recorder
DC	Drain connection
DD	Double disk
DEG (°)	Degrees
DET	Detail
DF	Drain funnel
DI	Ductile iron
DIA	Diameter
DIM	Dimension
DISCH	Discharge
DPI	Differential pressure indicator
DWG	Drawing
DWN	Drawn
DXS	Double extra strength
E	East
EA	Each
EBD	Emergency blowdown valve
ECC	Eccentric
EF	Electric furnace
EFW	Electric fusion welded
EJMA	Expansion Joint Manufacturers Association
EL	Elevation
ELB	Elbowlet
ELEV	Elevation
ELL	Elbow
EMBED	Embedment
ENGR	Engineer
EP	Equipment piece
EQUIP	Equipment
ERW	Electric resistance welded
ESD	Emergency shutdown valve
EXCH	Exchanger
EXH	Exhaust
EXIST	Existing
FA	Flow alarm
FAB	Fabricate
FCV	Flow control valve
FD	Feedwater
F	Furnished by others
°F	Degrees Fahrenheit
FE	Flow element
FF	Full face
FF	Flat face
FFD	Flat, faced, and drilled
FI	Flow indicator
FIC	Flow indicating controller
FICV	Flow indicating control valve
FIG	Figure number
FL	Floor
FLD	Field
FLG (Flg.)	Flange
FmI	Displacement flowmeter
FO	Flow orifice
FOB	Flat on bottom
FPS	Feet per second
FR	Flow recorder
FRC	Flow recording controller
FRCV	Flow recorder control valve
FrRC	Flow ratio recording controller
FS	Far side
FS	Forged steel
FS	Flow switch
FSS	Forged stainless steel
FT (′)	Foot or feet
FTG	Fitting
FW	Field weld
GA	Gauge, gage
GAL	Gallon
GALV	Galvanized
GEN	General
GEN	Generator
GG	Gauge glass
GPM (gpm)	Gallons per minute
GR	Grade
GRD	Ground
GR JT	Ground joint
GRV	Groove
GU	Guide
HC	Hose connection
HC	Hydrocarbon
HC	Hand (manual) controller
HCV	Hand-operated control valve
HDR	Header
HEX	Hexagon
HF	Hard (stellite) face
HGR	Hanger
HIC	Hand-actuated pneumatic controller
HOR (HORIZ)	Horizontal
HR	Hanger (rod)
HS	Hanger (spring)
HVAC	Heating, ventilating, and air conditioning
HVY	Heavy
HYD	Hydraulic
I	Iron
IBBM	Iron body bronze mounted
IBD	Intermittent blowdown
ID	Inside diameter
IDD	Inside depth of dish
IN (″)	Inches
INS	Insulate
INT	Integral
INV	Invert (inside bottom of pipe)
IPS	Iron pipe size
ISO	Isometric
JCT	Junction
JT	Joint
L	Angle (structural 4-in. angle shape)
LA	Level alarm
LBS	Pounds
LC	Level controller
LC	Lock closed
LCR	Level control recorder
LCV	Level control valve
LG	Level glass
LI	Level indicator
LIC	Level indicating controller

Abbreviation	Meaning
LICV	Level indicating control valve
LLA	Liquid level alarm
LLC	Liquid level controller
LLI	Liquid level indicator
LLR	Liquid level recorder
LO	Lock opened
LOC	Location
LR	Long radius
LR	Level recorder
LRC	Level recording controller
LS	Level switch
LW	Lap weld
M	Miscellaneous shapes (steel)
M	Monel metal
MACH	Machine
MATL	Material
MAX	Maximum
MECH	Mechanical
M&F	Male and female
MFG	Manufacturing
MI	Malleable iron
MIN	Minimum
MISC	Miscellaneous
MR	Moisture recorder
MS	Main steam
MSS	Manufacturers Standardization Society (valve and fittings industry)
MW	Miter weld
N	North
NC	Normally closed
NEC	National Electrical Code
NEG (–)	Negative
NEMA	National Electrical Manufacturers Association
Ni	Nickel
NICU	Nickel copper alloy
NIP	Nipple
NO. (#)	Number
NO	Normally opened
NOM	Nominal
NOZ	Nozzle
NPS	National pipe size
NPT	National pipe thread
NRS	Nonrising stem
NS	Near side
NTS	Not to scale
NV	Needle valve
OD	Open drain
OD	Outside diameter
OH	Open hearth
OPP	Opposite
ORIG	Original
OS&Y	Outside screw and yoke
P	Personnel protection
PA	Pressure alarm
PC	Pressure controller
PCV	Pressure control valve
PdC	Pressure differential controller
PdI	Pressure differential indicator
PdRC	Pressure differential recording controller
PE	Plain end (not beveled)
PE	Pressure test connection
PERP	Perpendicular
PFI	Pipe Fabrication Institute
PI	Pressure indicator
PIC	Pressure controller
PICV	Pressure indicator control valve
P&ID	Piping and instrument diagram
PIM	Pressure indicating manometer
PL	Plate
PO	Pump out
POS (+)	Positive
PP	Pump
PR	Pair
PR	Pressure recording controller
PR	Pressure regulator
PRC	Pressure controller
PRESS	Pressure
PRI	Primary
PRV	Pressure-reducing valve
PS	Pipe support
PS	Pressure switch
PSD	Rupture disk
PSI (psi)	Pounds per square inch
PSIA	Pounds per square inch absolute
PSIG	Pounds per square inch gauge
PSV	Pressure safety valve
PT	Point
PVC	Polyvinyl chloride
QO	Quick opening
QTY	Quantity
QUAD	Quadrant
R	Radius
RB	Reactor building
RC	Reactor coolant
REAC	Reactor
RECD	Received
RED	Reducer
REF	Reference
REINF	Reinforce
REQ (REQ'D)	Required
REV	Revision
RF	Raised face
RFC	Ratio flow controller
RFI	Ratio flow indicator
RHR	Residual heat removal
RJ (RTJ)	Ring-type joint
R/L	Random length
RPM (rpm)	Revolutions per minute
RS	Rising stem
S	South
S	Steam pressure
SA	Sludge acid
SA	Sulfuric acid
SAE	Society of Automotive Engineers
SC	Sample connection
SCFH	Standard cubic feet per hour
SCFM	Standard cubic feet per minute
SCH	Schedule
SCR	Screwed ends
SCRD	Screwed
SD	Storm drain
SECT	Section
SF	Semifinished
SG	Sight glass
SGA	Special gravity alarm
SGC	Special gravity controller
SGI	Special gravity indicator
SGR	Special gravity recorder
SH (SHT)	Sheet
SI	Safety injection
SJ	Solder joints
SK	Sketch
SLOT	Slotted
SLV	Sleeve
SMLS	Seamless
SNUB	Snubber
SO	Slip-on
SO	Steam out
SOL	Sockolet
SP	Steam pressure
SPEC	Specification
SPG	Spring
SPI	Special
SQ	Square
SR	Short radius
SR	Speed recorder
SS	Stainless steel
STD	Standard
STIFF	Stiffener
STL	Steel
STM	Steam
STR (STRUCT)	Structure

SUCT	Suction
SUPT	Support
SW	Socket weld
SW	Socket welded ends
SWG	Swage
SWP	Standard working pressure
SYS	System
T	Steam trap
T	Threaded
TA	Temperature alarm
TAN	Tangent
TBE	Threaded both ends
TC	Temperature controller
TC	Test connection
T&C	Threaded and coupled
TCV	Temperature control valve
TdC	Temperature differential controller
TdI	Temperature differential indicator
TdR	Temperature differential recorder
TE	Threaded end
TECH	Technical
TEF	Teflon
TEMP	Temperature
TENS	Tension
T&G	Tongue and groove
THD	Threaded
THRU	Through
TI	Temperature indicator
TIC	Temperature controller
TICV	Temperature control valve
TOC	Top of concrete
TOE	Threaded on end
TOP	Top of pipe
TOS (T/S)	Top of steel
TP	Type
TR	Temperature recorder
TRC	Temperature recorder controller
TS	Temperature switch
TURB	Turbine
TV	Temperature valve
TW	Temperature well
TYP	Typical
UA	Unit alarm
VB	Vacuum breaker
VC	Vitrified clay
VERT	Vertical
VR	Viscosity recorder
W	West
W	Width
W/	With
WB	Welded bonnet
WC	Water column
WE	Weld end
WF	Welded flange
WF	Wide flange
WLD	Weld
WN	Weld neck
WOG	Water, oil, and gas pressure
WP	Working point
WP	Working pressure
WR	Weight recorder
WSP	Working steam pressure
WT	Weight
XH	Extra heavy
XS	Extra strong
XXH	Double extra heavy

GLOSSARY

ABSORBER Device used to absorb gases through liquids.

ABSORPTION The process by which radiation imparts energy to a material through which it passes.

ABSORPTION (SOLAR) The ability of a solar collector to soak up light and therefore capture the heat of the sun. Absorption (or absorptance) is measured as a percentage of the total radiation available.

ACCUMULATOR Container where fluids or gases are stored under pressure. The term also refers to a holding tank used for temporary storage.

ACTIVE SOLAR SYSTEM Any solar power system that uses solar collectors to absorb the sun's heat energy; a separate system to store excess heat for release during cooler periods; and a pumped circulation system (piping) to transfer heat from collector to storage area.

ACTUATOR Any device that will operate a valve by remote control (fluid motor, air cylinder, hydraulic cylinder, electric motor).

ADJUSTABILITY The ability of a pipe system and its support system to enable field installation and adequate functioning when the installation position differs from the design.

AFTERCOOLER Cooling unit for the reduction of heat in gas of a compressor; also a form of economizer or heat exchanger.

AIR, COMPRESSED Air having pressure greater than atmospheric pressure.

AIR, FREE Air subject only to atmospheric conditions and not contaminated.

AIR, INSTRUMENT Oil-free and moisture-free compressed air used to operate pneumatic control devices.

AIR, PLANT Compressed air used in industrial plants for air-operated tools. Not usually used for instruments or breathing air; in most cases, may not contact any process materials.

AIR, STANDARD Air at a pressure of 14.70 psia, a temperature of 68°F (20°C), and a relative humidity of 36 percent.

ALLEY, PIPE The main bank of pipe headers located inside the limits of a structure; known as a pipe bridge when outside the limits.

ANCHOR Rigid support that keeps the pipe from translation or rotation movement at one point along the piping system; also prevents transmission of forces and moments (thermal, shock, vibration) between both sides of the pipe.

ANCHOR BOLTS Hold-down points for equipment made from mechanical steel machine bolts, steel rods, or other devices and put in place before the concrete embedment is poured.

ANNULUS Doughnut-shaped duct or pipe.

ATTACHMENTS Simple mechanical devices or welded parts that allow a hanger or support to be connected to a structural shape or pipe.

AUTOCLAVE Vessel used to hold material, medium, or reactants under prescribed conditions including temperature, pressure, and movement.

BAFFLE Series of obstructions used in petrochemical vessels to guide or mix process materials.

BALLOON Circle with extension line used to label an item on a pipe support detail or isometric.

BASE PLATE Metal plate attached to equipment, vessels, or other components to provide a transition piece from equipment to foundation. Usually attached to the concrete foundation by anchor bolts embedded in concrete.

BASIC MODEL Model that includes all major structures; all equipment with nozzles and tagging is in place ready for the piping to be installed. Included in the basic model is a substructure or table with removable or folding legs.

BATTERY Group of similar reaction vessels or tanks.

BATTERY LIMIT Lines used on a plot plan to determine the outside limits of a unit. Usually established on the piping index drawing, plot plan, and site plan.

BAY OF STEEL Steel surrounding a space bounded by a floor and the four nearest columns.

BENDING BOARD Device used to bend plastic pipe for a model.

BENT Vertical plane of steel structure that extends along a column row in a building; consists of two columns with horizontal connecting members. In a pipe rack the bent is two columns and one or more horizontal connecting members which form a U shape.

BLENDER (MIXER) Equipment used for combining various stocks or charges in order to produce specified quantities of a material.

BLIND FLANGE Solid platelike fitting used to seal the end of a flanged-end pipeline (also known as dead end).

BLOWDOWN The discharged material from relief valves.

BLOWDOWN-BLOWBACK Difference between relieving and closing pressures in a relief or safety valve. Closing pressure is always less than the pressure at which the safety valve begins to open (relieve).

BLOWDOWN SYSTEM Discharge system which removes excess or relieved medium from vessels, relief valves, or other equipment commonly associated with a piping system.

BLOWDOWN TANK Vessel into which line material or contents of another vessel can be emptied immediately in times of emergency.

BOILER Vessel in which water is heated to generate steam under pressure.

BOILER, FIRE-TUBE Boiler system with heating tubes located within the shell. The use of tubes submerged in boiling water increases the area of contact between the water and the fire gases passing through the tubes.

BOILER, SHELL Vessel containing water with a portion of the shell exposed to a source of heat.

BOILER, WASTE HEAT Boiler that uses heat generated for another purpose which otherwise would be released to the atmosphere.

BOILER, WATER-TUBE The reverse of the fire-tube boiler. Hot gases are in contact with the outside surface of the tubes and the water is inside.

BUSHING Tapped or threaded fitting (male outside, female inside) used to reduce the size of the end opening of a fitting or valve.

BWR (BOILING WATER REACTOR) In the BWR, the heat produced in the nuclear fuels results in the boiling of the reactor coolant. In this design, the steam is created directly in the reactor and then used to turn a turbine.

BYPASS Pipe loop which provides partial or full flow of material around a piece of equipment or valve station.

BYPASS VALVE Small-diameter pipe connected from inlet to outlet side of valve with its own small valve to allow line contents to pass to the other side of disk or gate to equalize pressure for ease in operation.

CABLE TRAY Provides continuous support for electrical lines.

CAD (COMPUTER-AIDED DRAFTING) System by which a drawing can be made, dimensioned, and specified by the use of a computer.

CAGES Enclosures around ladders at a certain elevation to prevent personnel from falling.

CAPACITY The flow obtainable under given conditions of temperature, pressure, media, viscosity, and velocity.

CAP END The end enclosure of a vessel. Can be dished elliptical, or flanged and bolted.

CARBON STEEL Steel which has distinctive properties due mainly to its carbon content.

CATALYST Material which speeds up a chemical reaction without taking part in the reaction itself.

CATCH BASIN Device used to separate material from a waste stream.

CELSIUS Temperature scale where the boiling point of water is 100° and the freezing point of water is 0°.

CENTIGRADE See Celsius.

CENTIPOISE The unit of measurement of viscosity.

CENTRIFUGE Device used to separate elements of a mixture by centrifugal force.

CHARGE In a batch process, the original material which goes into a system for processing (usually crude stock).

CHECK MODEL Model fabricated after all the design work has been completed. It is simply a three-dimensional representation of the paper design.

CHEMICAL PLANT Plant that utilizes various hydrocarbon products from refineries to produce other products for consumer and industrial use.

CIRCULAR PLATFORM Steel grating or solid plate on a vessel or tower that provides an access area at different elevations to manways, valves, instruments, piping, and other areas.

CIRCULATION COMPONENTS Similar to distribution components, but usually taken to mean the specific fluid, piping, and ducting that is used.

CIRCULATION, FORCED The application of pressure to transfer fluid.

CLEARASITE Solvent-type cement for acrylic materials.

CODES Standards that provide industry with recognized specifications for convenience, safety, and uniform design.

COEFFICIENT OF EXPANSION The amount of thermal expansion or contraction of a material per degree of temperature change.

COLD BENDING Bending process for pipes with diameters (up to 42 in.) and walls (up to 3/4 in.). Wrinkling, excessive thinning, and ovality can be avoided with proper equipment and fixtures.

COLD LOAD Spring reading (a function of amount of compression) when pipe suspended from a spring hanger is at room temperature (see Hot Load). Cold load is higher than hot load if the thermal movement of the pipe is up and vice versa.

COLUMN Vertical vessel used for fractional distillation; also called tower or stanchion. Column also refers to vertical steel structural members for buildings, pipe racks, etc.

COLUMN LINE Straight row of steel columns represented on a model or drawing.

COMPANION FLANGE Any flange suited to connect with a flanged valve or fitting or with another flange to form a joint.

COMPONENTS, MODEL Standardized model parts and materials manufactured by various companies. Model components can be ordered from company catalogs.

COMPRESSION STRENGTH The ability of an object or material to resist being crushed.

COMPRESSOR Mechanical (piston and cylinder) device used to increase the pressure of air or gas.

COMPRESSOR, RECIPROCATING Device similar to the reciprocating pump. Pressurized gas or air is created by the use of cylinders and pistons.

COMPRESSOR, ROTARY Device used to compress air or gas by the rotation of valves.

COMPRESSOR, ROTARY LOBE Device used to produce pressure for gas and air by the use of lobed rotors.

CONCRETE DRAWINGS Drawing showing wall thickness, slab thickness, column location, concrete or block walls, floor and wall penetrations, openings and drains, embedments, channels, and in-place anchor bolts.

CONDENSATE Liquid resulting from condensation of vapor or gas in a line—especially condensed steam.

CONDENSATE UTILITY SYSTEM Separate piping system which collects the liquid formed from the condensing of steam.

CONDENSER Device used to condense vapor to liquid by cooling. Condensers are used to lower the turbine's exhaust pressure and also save water that may be lost to the system.

CONDENSER, BAROMETRIC Condenser that operates under a vacuum.

CONDENSER, DIRECT-CONTACT Equipment used to condense steam exhausted from a turbine. Can be the barometric or low-level jet type with parallel flow or counterflow design options.

CONDENSER, JET Condenses vapor and also removes noncondensable gases without additional equipment.

CONDENSER, SURFACE Large tubular device used mainly for condensing steam to even out the flow.

CONDENSING Process of vapor turning to liquid upon being cooled.

CONSTANT-LOAD HANGER Device that consists of a lever mechanism in a housing and a coil spring. The movement of piping, within certain limits, will not change the spring forces holding up the piping; thus no additional forces are introduced to the piping being supported.

CONTINUOUS BLOWDOWN The expulsion of water continuously from steam lines.

CONTROL BUILDING Centrally located structure that houses board-mounted instrumentation, control panels, and monitoring equipment necessary for the operation of the plant. Often provides laboratory, kitchen, locker room, work, and limited office space for control room workers.

CONTROL PANEL Board used for mounting instruments and controls for remote control operation of a process.

CONTROL STATION Pipe loop arrangement consisting of control valve, various fittings, and isolating (block) valves to regulate the flow, pressure, or quantity of material at a predetermined location.

CONDUCTION Transfer of heat or electrical energy through a body without displacement of the particles in the body.

CONDUCTIVITY The ease with which heat flows through a material. Copper is considered one of the better conductors of heat; insulating materials are extremely poor conductors.

CONDUIT Structural covering for electrical lines made from a variety of materials (plastics, pipe, steel, aluminum). Conduit provides the strength needed to run electrical lines throughout an industrial installation. Multiple conduits are usually placed on trays for running on pipe racks.

CONVECTION Transfer of heat between two surfaces—the cold one above, the hot one below—which are separated by a layer of air. The surfaces may be vertical, but the hot one cannot be above the cold one.

CONVECTOR Surface designed to transfer its heat by convection.

CONVENTIONAL PIPING Conventional piping for a nuclear power plant includes all auxiliary systems (including steam and service piping) which do not contain radioactive material.

COOLER Any piece of equipment or vessel which transfers heat from one material to another thereby cooling the process fluid.

COOLING WATER Any water used to remove heat from process material or equipment.

CORE Region of the nuclear reactor containing fissionable material.

CORROSION Chemical action or electrolysis causing the deterioration of materials.

CORROSION ALLOWANCE Extra metal added to the original calculated thickness to allow for corrosion.

COUPLING Female fitting used for joining two lengths of pipe.

COUPLING, DRESSER Coupling used for connecting plain-ended pipes.

CRACKING Continuous fractional distillation (see Fractionation) as applied to petroleum for the production of kerosene, gasoline, fuel oils, and the like.

CROSS Fitting with four openings.

CROSS, STRAIGHT Fitting in which all outlets are of the same diameter.

DAMPENING The reduction of noise, pulsation, or vibration.

DAVIT Small crane at the top of a column (tower) used to lift equipment.

DENSITY The ratio of mass to volume.

DESICCATOR Machine which applies vacuum, heat, or a chemical process to the process medium to remove unwanted liquid.

DESIGN MODEL Model which includes all equipment, vessels, piping, valves, fittings, electrical, instrumentation, and HVAC. The main purpose of a design model is to develop the ultimate piping arrangement with a minimum number of drawings. A design model is constructed in two stages: the basic model and the final model.

DESUPERHEATER Device which adds water to superheated steam to remove heat.

DEVELOPMENT, PIPE View on pipe fabrication drawing that shows the unfolding of pipe intersections in one plane. Developments are used for the construction of template drawings for pipe intersections.

DIAPHRAGM Takes the place of a disk in certain valves. Usually made of a rubberlike substance flexible enough to be stretched over the valve bore area.

DIKE Wall or enclosure around a tank or holding vessel to provide a backup system in case of ruptures. The dike must be able to hold the total contents of the enclosed vessel or tank.

DISCHARGE To relieve a line of excess pressure.

DISK (DISC) Part of valve which initiates, shuts off, or regulates flow.

DISK, RUPTURE Diaphragm placed between flanges on a vent line. The diaphragm has a specific breaking pressure and will fail before the safety limits of the system have been reached.

DISTILLATION Separation process based on the relative volatility of the materials to be separated and on a change in phase of the original mixture, which may be boiled off and recollected by condensation in another vessel. There are many processes including simple distillation, multiple distillation, fractional distillation, and steam distillation.

DISTILLATION EQUIPMENT, BATCH The batch still is the simplest piece of distillation equipment. During operation, a batch (or charge) is fed to a pot, heat is applied, and the vapor is condensed and collected from the receiver.

DISTILLATION, FRACTIONAL Countercurrent multiple distillation in which rising vapor makes contact with descending liquid flowing over baffles and trays. The vapor is enriched in the more volatile component as it rises and liquid is enriched in the less volatile components.

DISTILLATION, MULTIPLE Further separation by redistilling the desired component. Extensive multiple distillation is inefficient; fractionation is called for.

DISTILLATION, SIMPLE Liquid is heated until a vapor is formed; the vapor is then drawn off, condensed, and collected. This condensed overhead can be collected in one fraction or in a succession of fractions.

DISTILLATION, STEAM By introducing open steam directly into the charge, it is possible to vaporize the volatile constituents at a lower temperature.

DOUBLE EXTRA STRONG The schedule of wrought pipe weights in common use.

DOUBLE SPRING Tandem medium spring used for high deflection with large hot-cold travel variation.

DOWNCOMER Any line or section of equipment (tray) which provides for the downward movement of material.

DRAFT, FORCED Air flow through a combustion chamber when pressure is applied to the supply.

DRAFT, INDUCED Air flow produced in a combustion chamber when a vacuum is applied at the boiler flue.

DRAIN System which collects and redirects (through a

separate piping system) water runoff or unwanted substances which may be discharged in a process.

DRAINAGE DRAWING Drawing showing the necessary drainage lines, paving, and grading.

DRAINAGE FITTING Fitting used for plumbing drainage lines.

DRAINAGE SYSTEM All hubs, piping, fittings, connections, and holding tanks associated with the removal or relocation of unwanted materials that have been discharged from equipment; system includes runoff from rain or cleaning operations.

DRAIN HUB Funnel-shaped arrangement designed into flooring and concrete foundations for drainage. The drain hub is connected directly to the drain line, which removes the collected material.

DRAWING, DOUBLE-LINE Drawing using two outside lines and a centerline to represent a pipeline which is drawn to scale. All lines over 12 in. (30.84 cm) OD should be drawn in this manner.

DRAWING INDEX (MODEL INDEX) After the plot plan has been developed, the piping designer divides the plot plan along logical boundaries with coordinates defined by battery limits. The index systematically divides the plot plan into workable units to enable the designer and drafters to work on portions of the plan separately.

DRAWING, ORTHOGRAPHIC Drawing that reduces a three-dimensional object to a plane by projecting parallel lines at right angles.

DRAWING, SINGLE-LINE Drawing using only a single thick line to represent pipelines.

DRIP VALVE Valve installed on drip legs for the removal of material from pipelines.

DUMMY LEG Piece of pipe or rolled steel section which is welded to the pipe in order to support the line.

ECONOMIZER Heat exchanger used on a boiler by which exhaust gases are used to preheat the air entering the combustion chamber.

EJECTOR Type of pump which passes a pressurized medium such as steam or fluid through a venturi which has a branch at its narrowest end. The partial vacuum created in the branch moves the material.

ELASTIC LIMIT Limit beyond which material will not return to its original configuration after stress has been removed.

ELASTOMER Resilient material used for seals to prevent leakage and used as a coating to inhibit corrosion.

ELBOW Pipe fitting used to change direction.

ELBOW, LONG-RADIUS Elbow whose radius equals one and one-half pipe diameters.

ELBOW, SHORT-RADIUS Elbow whose radius equals one pipe diameter.

ELECTRICAL DRAWING Drawing which shows cable tray size and location, conduit and junction boxes, stacked trays, vertical trays, and tray supports.

ELEVATION, SECTION, AND PLAN METHOD Traditional method of orthographically projecting plan, elevation, and section views of a piping project.

ELL, STREET Elbow with male threads on one end and female threads on the other.

ENERGY The capacity to do work. Potential energy is the energy inherent in a mass because of its position with reference to other masses. Kinetic energy is the energy possessed by a mass because of its motion.

EROSION Process of wearing away metal by line flow due to velocity, impingement or presence of foreign particles.

EXCHANGER, HEAT Device used to transfer heat between two fluids.

EXPANSION JOINT Pressure-tight joint that permits the expansion or contraction of pipelines.

EXPANSION, THERMAL Expansion of a material caused by heating.

EXPLOSION PROOF Type of encasement used for electrical equipment to inhibit the spreading of fire to a hazardous area.

EXTRA STRONG Pipe sizes corresponding to schedule 80.

EXTRUDED NOZZLE Nozzle or outlet formed by pulling hemispherically or conically shaped dies through a circular hole from inside the pipe.

FABRICATION DRAWING Drawing used to represent pipeline configurations that must be made up in a shop or in the field.

FABRICATION ISOMETRIC Isometric drawing of pipe lines showing line designation, pipe size, elevations, pipe footage, elbow-to-elbow dimensions, weld numbers, piping configurations, valve types, pipe supports and symbols, and equipment locations and numbers.

FACE The finished contact surface of flanged-end piping or components (valves).

FACE-TO-FACE Dimensions from the face of the outlet port to the face of the inlet port of a valve or fitting.

FACING Part of valve body which connects to a companion flange.

FAHRENHEIT Temperature scale where the boiling point of water is 212° and the freezing point of water is 32°

FEED The primary medium in a process; the charge.

FEED, BOILER Water used to produce steam inside a boiler.

FEEDWATER Provides original and makeup water for the process.

FEMALE THREAD Internal thread in valves, fittings, and pipes for making screwed connections.

FIELD RUN Piping routes determined at the time of installation.

FILTER Device used for separating solids from fluids.

FILTER, ROTARY Continuous type of filter

FIN FAN Air cooler.

FIRED HEATER Any equipment which creates heat by internal combustion, such as a boiler or furnace.

FIRE PROTECTION SYSTEM System including water, CO_2, fog, and portable equipment.

FITTING Device used to assemble a piping system.

FLAG, MAP Small flag used to indicate interferences after a model has been well into construction.

FLAKER Device used in the production of chips, crystals, or flakes from a hot concentrated solution of melted crystals.

FLAME ARRESTOR Wire screen or parallel plates or tubes placed in lines to prevent the movement of flame upstream from a vessel.

FLANGE Rim on the end of a pipe, valve, or fitting for bolting to another piping element.

FLANGING The welding of standard or fabricated flanges to the ends of pipe sections or vessels.

FLARE STACK Tower which allows for the burning of waste gases or any other burnable substances which have been discharged.

FLASHING Sudden change from liquid to vapor until a state of equilibrium exists between the liquid and vapor phase. Occurs when a liquid is at or near its boiling point and the pressure is lowered.

FLIGHT Run of stairs.

FLOATING HEAD EXCHANGER Heat exchange unit, similar to the shell and tube exchanger, having U-shaped tubes running parallel with the shell, which are allowed to "float" or move during operation.

FLOTATION The process of separating the components of a mixture by suspending one in a froth that can be removed from the surface.

FLOW, CONCURRENT Two materials passing through a system in the same direction.

FLOW, COUNTERCURRENT Two materials passing through a system in opposite directions.

FLOW DIAGRAM Schematic diagram showing the elements and directions of a process. This diagram may indicate flow rate, temperatures, and pressures.

FLOW DISPLACEMENT METER Instrument placed directly in the line to indicate the total amount of flow.

FLOW ELEMENT Portable device, consisting of an orifice set with plugged tapped valves and a plate, used for testing flow. Vertical pipe sections are considered part of the measuring device.

FLOW INDICATING CONTROLLER Control valve and flow indicator used for controlling flow by connections to an orifice through a pneumatic transmitter.

FLOW INDICATOR Linear indicator or dial which shows the flow rate. The indicator can be of the differential type or a direct hookup.

FLOW, LAMINAR Flow in which each particle of the fluid is moving parallel to the walls of the pipe. The velocity of flow is greater at the center of the pipe.

FLOW MEDIUM See Fluid.

FLOWMETER Any device which indicates or records the quantity of liquid moving through a pipe.

FLOW RATE Volumetric flow per unit time (gpm, cfm).

FLOW RATIO RECORDING CONTROLLER Instrument that controls and records the flow ratio of the main line. An orifice must be within close range of the meter. Pneumatic recorders can be either board-mounted or locally mounted.

FLOW RECORDER Device which makes a permanent record of flow characteristics. It can be a board-mounted pneumatic type or a direct hookup differential variation.

FLOW RECORDING CONTROLLER Flow recorder which controls medium movement and is connected to an orifice through a pneumatic transmitter.

FLOW, TURBULENT Flow in which the velocity of each particle of the fluid changes constantly in magnitude and direction, though the fluid as a whole is moving in the direction parallel to the pipe.

FOREIGN MATTER Dirt, rust, scale, and the like in pipeline.

FORMING Method of pipe fabrication including bending, swaging, lapping, extruding, expanding, and belling. Most forming is done in the shop.

FOUNDATION PLAN Drawing based on the plot plan and equipment location drawings which establishes the extent of the foundations, footings, and concrete.

FRACTION In petrochemical processes, the products are often broken down into components of various weights (gases, light liquids, heavy liquids) called fractions.

FRACTIONATION Method of multidistillation for separation of volatile compounds in columns or towers. The essential characteristic of a fractionating tower is an upward flow of vapor with a downward flow of liquid in a chamber which will provide many contacts between liquid and vapor. Returning a portion of the condensed overhead as reflux to the top of the column provides contact between the vapor and the rich product—the higher the ratio of reflux to product, the greater degree of separation.

FRACTIONATION, BATCH Fractionation conducted batchwise with vapors rising from a still pot is called rectification. It is essentially the same as a batch still operation with a fractionating top inserted between the still top and the condenser; a provision is made for the return of reflux liquid to the top of the tower. Batch fractionation is used for making sharp separation between several overhead fractions and for distillation of a multitude of mixtures.

FRACTIONATION, CONTINUOUS In this type, the most common of all distillation processes, there is continuous feed at the half-height point, or at the top of the fractionating column, and the continuous withdrawal of residues and product. A source of heat is required at the base of the column—such as open steam, a steam coil, or reboiler (which is similar to a pot still but operated at a constant level by a continuous addition of feedstock).

FRICTION LOSS Pressure loss caused by friction between the pipe wall and the flowing liquid.

FUEL OIL Fuel necessary for the functioning of a petrochemical or power plant. In some cases, the refinery will use one of its own end products such as LP gas or oil to continue the production of other products.

FULL-BORE PORT Valve in which diameter of internal bore equals the inside diameter of pipe used on line.

FUSION, NUCLEAR The act of coalescing two or more atomic nuclei.

GALVANIC CORROSION Corrosion caused by the contact of different metals which are improperly isolated in the presence of a liquid that can conduct electricity.

GALVANIZING Process which covers the surface of iron or steel with a layer of zinc.

GAS State of matter distinguished from the solid and liquid states by low density and viscosity. A gas always fills its container entirely.

GAUGE LEVEL Device for measuring the level of a material in a vessel.

GLASS LINING Glass coating fused to a base material.

GLASS, SIGHT Glass plate or tube inserted in a vessel or a pipe wall for observation of internal conditions.

GRADE Ground elevation for a project. Usually set at 10 ft (3 m) elevation or 100 ft (30 m) elevation so that any portion of the installation that lies underground will have a positive elevation.

GRADE BEAM Beam used to support flooring at grade or ground level.

GRADIENT Slope.

GRAVITY SYSTEM Piping system that is under static head (pressure) only.

GROUT Thin, highly concentrated concrete covering between the foundation and equipment to hold the base plate; the top portion of a foundation for equipment (pumps, compressors).

GROUT LAYER Thin layer of cement which evens off the top of a foundation or other concrete form supporting the base or slide plate. Also a thin pad or foundation between the flooring and concrete foundation and a piece of equipment it will support. Structural embedments are utilized in the grout to permit bolting the base plate to the equipment.

GUIDE Type of hanger support that allows a pipe to move along its length but not sideways.

HAND HOLE Small hole in a vessel used for hand access for maintenance and adjustment.

HANGER Device which suspends piping.

HEAD Pressure in a fluid system expressed as equivalent feet of water. Also refers to the end closure of a vessel (horizontal or vertical, elliptical, spherical, or dish-shaped); in some cases large blind flanges may be bolted to the end of a vessel and used as its head.

HEAD, FRICTION The head required to overcome the friction between the fluid particles in motion and the interior surface of a conductor.

HEAD, STATIC The vertical distance from a given point in a fluid to the surface if the fluid is not moving. For fluids other than water, static head is given in equivalent feet of water.

HEAD, STATIC DISCHARGE Static head from the centerline of the pump to the free discharge surface.

HEAD, STATIC SUCTION The head from the surface of the supply source to the centerline of the pump.

HEAD, TOTAL STATIC The static head from the surface of the supply source to the free discharge surface.

HEAD, VELOCITY An equivalent head through which a liquid would have to fall to attain a given velocity.

HEADER Pipe to which two or more pipelines are joined to carry fluid from a common source to various points of use.

HEADROOM Physical space above a valve, pipe, platform, etc.

HEAT EXCHANGER Vessel and piping arrangement used to transfer heat from one material to another without mixing the two streams.

HEAT, LATENT Heat used to transform liquid to vapor or vapor to liquid.

HEAT PUMP Device used for heating or cooling by transferring heat via a mechanically driven thermodynamic process such as evaporation and condensation.

HEAT TRANSFER The transfer of heat from a relatively hot substance to a cooler one, usually in a heat exchanger.

HORTENSPHERE Spherical pressure storage tank.

HOT BENDING The bending of pipe to a predetermined radius after heating to a suitable temperature.

HOT LOAD Function of amount of compression when the pipe is at its hottest operation temperature (see Cold Load). Hot load is greater than cold load if pipe moves down when the system goes into operation. Sometimes the temperature change from cold to hot causes other supports to take up the load and the pipe rises locally so that its hot load is less than cold load.

HUB END End connection, caulked or leaded, used on valves, fittings, and pipe (mainly for water supply and sewage lines).

HUDDLING CHAMBER Area in a pipe safety valve which regulates popping action and the amount of blowdown.

HUMIDITY Water vapor contained in air expressed as a percentage of the maximum amount of water vapor the air could contain at the given temperature and pressure.

HVAC Heating, ventilating, and air conditioning.

HVAC DRAWING Drawing that shows ducting location and supports, openings, elevations to bottom of ducts, configuration, and ducting size.

HYDRATED Combined with water.

HYDRAULIC The production of motion by the use of liquids under pressure.

HYDROCARBON Any compound whose molecular structure is made up exclusively of molecules of hydrogen and carbon. Hydrocarbon compounds can be gas, liquid, or solid, depending on the temperature and structure. All petroleum products are derived from combinations of hydrocarbons.

IMPELLER Device that rotates in a centrifugal pump and produces the pumping action.

INCREASER Coupling that has a larger opening at one end (used to increase size of pipe opening). See also, Reducer.

INJECTOR Jet used for pumping.

INSTRUMENT Device that has measuring, recording, indicating, or controlling capabilities.

INSTRUMENT AIR Separate air system which operates instrumentation in a plant. Instrument air and utility air must be kept separate.

INSTRUMENT CHANNEL Channel-shaped tray that supports instrument lines; similar to electrical tray.

INSTRUMENTATION The application of industrial instruments to a process or manufacturing operation.

INSTRUMENTATION DRAWING Drawing which establishes points of connection, size, location, type, and other necessary information for the installation and operation of instruments.

INSULATION Material used to cover pipelines or vessels to maintain a constant temperature in the line; also used to prevent the transfer of heat between the atmosphere and the line fluid and to protect operators from burns.

INSULATION, ALUMINUM-ARMORED Insulation made of calcium silicate and asbestos fiber and covered with a weatherproof aluminum jacket; used for steam and process lines which operate at temperatures up to 1200°F (650°C).

INSULATION, ANTISWEAT Used to prevent cold water lines from sweating.

INSULATION, ASBESTOS-SPONGE FELTED Type of insulation applied to pipes, valves, and fittings with temperatures up to 700°F (370°C).

INSULATION, COLD Insulation applied to pipe, fittings, and valves in refrigeration services or to prevent heat penetration from local steam lines to the fluid in the insulated line.

INSULATION, FIBERGLASS A rigid, structurally strong insulation for temperatures up to 600°F (315°C).

INSULATION, HOT Insulation applied to pipe, fittings, and valves in steam services.

INSULATION, 85% MAGNESIA A durable, fireproof, molded insulation used for pipes at temperatures up to 600°F (315°C).

INSULATION RING Steel ring used to support insulation attached to the outside of a vessel.

INSULATION, ROCK CORK Mineral wool product bonded with a waterproof compound.

INSULATION, VEGETABLE CORK Compressed cork granules baked into a mold and used for low thermal conductivity and high moisture resistance.

INSULATION, WOOL FELT Commonly used for both hot and cold water lines. It has a temperature range of 40 to 212°F (5 to 100°C) and is made up of layers of wool felt with an inner waterproofed felt liner.

INTERCOOLER Device used for heat reduction of compressed gas; basically, a heat exchanger.

INTERFACE Common boundary between two materials or spaces.

JACKET, STEAM Shell around the outside of a pipe or vessel for steam heating.

JET PUMP (EJECTOR, INJECTOR) Device which uses the energy of another pressure for pumping a second fluid.

JOINT Point of connection between two piping elements.

KILOWATT One thousand watts of power (equal to about 1 1/3 horsepower).

KILOWATTHOUR The amount of energy equivalent to 1 kilowatt of power in use for one hour. One kilowatthour equals 3400 BTU.

KNOCKOUT DRUM Vessel used to separate a mixture of gas and liquid by means of a series of baffles. The impact of the material with the baffles frees the gas, thereby allowing it to rise while the heavier substances flow to the bottom of the vessel. Piping then removes the gases from the top of the unit and the liquid from the bottom.

LATENT HEAT Amount of heat absorbed or released in changing water from one phase to another.

LEVEL GLASS Reading device directly connected to a vessel from the low to high points of a level variation. The liquid in the vessel finds its own level and may be observed through glass in the instrument.

LEVEL INDICATING CONTROLLER Instrument that indicates a vessel's liquid level and regulates it by pneumatic signal to a control valve.

LEVEL INDICATOR Dial or linear indicator which shows the level of liquid in the vessel.

LEVEL RECORDER Instrument that makes a permanent record of liquid level in a vessel by pneumatic signal from a displacement-type transmitter on the vessel.

LEVEL RECORDING CONTROLLER Instrument which has the same type of transmitter as the level recorder with a pneumatic signal to a control valve as well as a recorder.

LINE Hose, pipe, or tube used for the transfer of fluids or gasses.

LINE PRESSURE Pressure in a line measured in pounds per square inch and rated in terms of SP or WOG.

LIQUID Substance capable of flowing or being poured. A liquid takes the shape of its container below its surface.

LOAD CELL Meter used to indicate weight.

LOCATION AREA OR LOCATION PLAN Reference drawing which appears on the pipe support detail giving coordinate information and location dimensions in two perpendicular directions from building columns, pipe centerline elevations, building area, and area number.

LUBRICANT Oil, grease, or plastic sealant used in some valves to facilitate operation and promote tight sealing. Pipe joint compound is a lubricant used when making up screwed joints to reduce friction and tearing of threads and to aid in sealing.

MAIN Primary piping; section of a piping system which contains the process fluid or a major service.

MAIN BAR Primary steel rebar used for reinforcement in concrete.

MAINTENANCE Cleaning, replacement, or repairing.

MAKEUP WATER New water that is added to a system when the process water has diminished through leakage, evaporation, or removal. Makeup water is used in petrochemical installations, nuclear reactors, and fossil fuel power plants.

MALE THREAD External thread on pipe, fittings, and valves for making screwed connections.

MALLEABLE FITTING Fitting made of malleable iron.

MALLEABLE IRON Cast iron which has been heat-treated in an oven to relieve its brittleness, improve its tensile strength, and enable it to be pounded to a given shape.

MATERIAL TAKEOFF FROM MODEL Using a model in compiling a complete list of materials, components, and standard catalog items to be ordered for the construction of an industrial piping project.

MILL, BALL Device used to pulverize solids.

MILL, HAMMER Device used to crush solids.

MILL LENGTH Usually 16 to 20 ft (4.80 to 6.00 m).

MILL, ROLL Device used to pulverize solids.

MANIFOLD Main line pipe with several branch connections; also referred to as a header.

MANOMETER Gauge used to measure differential pressure.

MANWAY Large nozzle and hinged flange on a vessel used to provide access to vessel internals.

MATERIAL Fluid, gas or solid substance.

MATERIAL BALANCE Series of calculations and descriptions of the process a certain material is to undergo. Usually completed by a petrochemical engineer, chemical engineer, or designer, the material balance shows the end product, chemical position, impurity level, and all important aspects of the material processed.

MATERIAL TAKEOFF Listing of materials, number of items, descriptions, quantities, and prices for various aspects of a project; usually made from the details, fabrication drawings, or models.

MEDIUM-SPRING HANGER Most common of the spring hangers; medium refers to its stiffness load abilities (medium-length spring).

MODEL DESIGNER A model designer is a person who is familiar with piping specifications, pipe fittings, and hardware, vessel, and equipment functions and has a knowledge of expansion, insulation, and support of piping. The designer should have sufficient experience to be able to design on the model and lay out and route pipe runs directly on the model from flowsheets and piping and instrument diagrams (P&IDs).

MODEL METHOD Piping design method that bypasses many of the traditional drawing procedures by building a scale model of the system in the design stage of a piping project.

MODEL TECHNICIAN Craftsman familiar with power tools and the various wood and plastic materials with which engineering models are built. The technician must have a knowledge of blueprint reading (including piping drawings) and be able to minimize model detail. The technician normally works in the model shop building tables, vessels, equipment, and structures for the various models. The model technician is capable of attaching or altering model pieces either in the shop or in the design area and is able to install piping on the model under the direction of a model or piping designer.

NIPPLE Short length of pipe, threaded (male) on both ends, used for joining piping elements.

NIPPLE, CLOSE Nipple which shows no unthreaded pipe between the threaded ends.

NONMETALLIC DISK (NMD) Composition disk which is usually made of asbestos or resins.

NORMALIZING Process in which an alloy or carbon steel is heated to a suitable temperature above the transformation range and is subsequently cooled in still air to room temperature.

NOTCH Small sharp edge in a weld area that may lead to weld failure.

NOZZLE Any piece of pipe (stub-in) which is welded to a vessel or piece of equipment and has a flanged end onto which the pipeline with a similar flanged connection can be bolted. It is also possible to use a long-neck welded flange as a nozzle. Nozzles provide for the attachment of a piping system to a vessel, column, or tower. Also: any device, fixed or adjustable, which controls a flow rate or discharge pattern from a pipe or line by means of special contour or size of an orifice.

NOZZLE ORIENTATION Nozzle location in a plan view.

NUCLEAR PIPING Piping for a nuclear power plant includes all steam and service piping which contain radioactive material.

NUCLEAR REACTOR Apparatus in which nuclear fission may be sustained through a self-supporting chain reaction. Components of a typical nuclear reactor include (1) fissionable material such as uranium or plutonium; (2) moderating material (unless it is a fast reactor); (3) reflector to conserve escaping neutrons; (4) heat removal system; and (5) measuring and controlling elements.

OCTAGON BASE Cement foundation associated with the support and footing area of columns (towers) or other vertical vessels.

OFF PLOT Area in the general vicinity of the project or on-plot area; battery limits determine what is on or off plot. Also refers to nonprocessing portion of plant, such as storage areas etc.

ON PLOT Area connected with a project and bounded by the battery limits. Any portion of the project or equipment which is on plot is drawn and constructed on the same set of drawings to provide a complete visual description of the system. Often the plot plan is divided up into areas to provide a smaller drawing of individual sections; all are considered to be on plot. Also refers exclusively to processing area of plant.

ON SITE In the field; at the construction site. Any portion of the project which must be designed, constructed, or fabricated at the construction site is referred to as on-site production or construction.

OPERATOR Any device, manual or remote, that activates a valve.

ORIFICE Precise hole in a metal plate which is placed in the line to produce differential pressures for flow measurement.

OVERHEAD ACCUMULATOR Horizontal holding tank containing overhead products which are produced in a fractionating column.

PADDLE Part of an agitator.

PEDESTAL Any concrete shape (pier), cylindrical, rectangular, etc., which supports a piece of equipment such as a vessel or pump.

PENETRATION DRAWING Drawing that shows sleeve schedule, type, size, location of pipe penetration, and all line designations.

P&ID Piping and instrumentation diagram.

PIPE Hollow cylinder used to carry fluids or gases.

PIPE BEND Directional change in pipeline obtained by bending the pipe. (Normal bend radius is five pipe diameters.)

PIPE BENDING Forming the pipe sections into predetermined radii by hot or cold bending procedures. Pipe sections are often bent to a five-diameter radius, though sharper radii of three pipe diameters are sometimes required. Bending to the radius of 6, 10 or 15 times the diameter of the pipe can be accomplished.

PIPE FABRICATION Production of sections, configurations, or assemblies of pipe and various pipe components such as valves and fittings. Fabrication includes a variety of processing: welding, forming, shaping, heating, cleaning, machining.

PIPE HANGER See Hanger.

PIPE INTERSECTION Pipe fabrication drawing that exhibits the necessary views and developments to construct pipe shapes where two or more pipes come together at a variety of angles.

PIPE LANE SYSTEM System where piping runs north to south at one elevation and east to west at another.

PIPELINE SCHEDULE List of all pipelines on a project giving line size, specifications, origin-termination, insulation, painting, and fabrication data, and other pertinent information.

PIPE RACK BENT Structure consisting of horizontal connecting members and two vertical columns (stanchions). The horizontal member is referred to as a strut when it connects two bents.

PIPE ROUTE Path of a pipe.

PIPE SCALE Hard flake-like material frequently found in new pipe caused by heating during fabrication.

PIPE STRAP Device used to hold lightweight pipe to wall or ceiling.

PIPE SUPPORT Devices which support piping.

PIPE SUPPORT DETAIL DRAWING Small detailed drawing showing design, location, and other information for the construction of a pipe support.

PIPE SUPPORT GROUP Special section which handles engineering, design, drafting, and field assistance for pipe supports.

PIPE SUPPORT SYSTEM Complete arrangement of pipe supports (hangers, anchors, guides, snubbers) that (1) hold the weight of the system with a minimum safety factor of 3; (2) permit thermal and seismic movement; (3) dampen vibration caused by mechanical equipment; and (4) maintain a safe stress limit.

PIPE THIMBLE Sleeve used where pipes pass through walls or floors or are embedded in concrete or masonry.

PIPE VOLUME Total interior volume, expressed in cubic measurement, which a cylindrical object can hold. The volume can be found by multiplying π times the inside radius squared by the length of pipe (area of cross-section times length).

PIPING General term used for piping systems or for pipe and fittings in pipelines.

PIPING INDEX List establishing logical, workable units for a process plant based on the master plot plan; also called drawing index.

PIPING, SERVICE Piping carrying cooling, heat, pressuring, and other media which will not be part of the final product.

PLASTOMER Type of plastic liner used in ball valves to ensure tight shutoff.

PLATE, CHECKER Steel plate flooring with nonslip embossing on its surface.

PLOT PLAN Plan which can be divided into various types, including master and unit plot plans. The master plot plan is the general plan which shows the overall location of equipment and structures, both existing and to be built; the unit plot plan is a smaller drawing showing only the equipment location for a single portion of a petrochemical or other unit. In general, neither type is dimensioned, though some companies do dimension the unit plot plan with coordinates.

PLUG Screwed fitting used for shutting off a tapped opening.

PNEUMATICS Engineering science pertaining to gases, especially where gas pressure is employed to do work.

POCKET Low point in a pipe where liquids can be trapped, or high point where vapors may become trapped.

PONTOON Floating roof used on storage tanks.

PORT, INLET Opening connected to the upstream side of a fluid system.

POSITIVE PRESSURE HEAD Driving force used to transfer fluids through pipe (expressed in feet of water or pounds per square inch).

POT STILL Horizontal cylinder of welded construction for low-pressure operation. The still body is fitted with some means of heating such as direct fire, jacket, or internal coil or a sparger for open steam. For pressure operation, the wall thickness of the still must be increased.

PRELIMINARY MODEL Simplified model used in the beginning stages of a project to experiment with design, placement, and other variables before the design is started. The main purpose of a preliminary model is to determine the ultimate equipment arrangement along with the location of major or critical lines.

PRESSURE, ABSOLUTE Total pressure measured from absolute zero.

PRESSURE ALARM Pressure switch attached by a coupling or flange to equipment or pipe; when pressure becomes excessive, the switch turns on an alarm.

PRESSURE, ATMOSPHERIC Pressure exerted in every direction upon a body by the atmosphere; equivalent to 14.7 psi at sea level (1 pascal).

PRESSURE, BACK Pressure on the upstream side of valve seats; or pressure surge in the downstream piping system.

PRESSURE CONTROLLER Instrument that regulates pressure by means of a control valve connected to a measurement point on the pipe or vessel.

PRESSURE DIFFERENTIAL Difference in pressure between any two points of a piping system.

PRESSURE DIFFERENTIAL CONTROLLER Instrument that controls pressure differential between two pipes or vessels.

PRESSURE DIFFERENTIAL INDICATOR Dial which indicates pressure differential between two pipes or vessels.

PRESSURE (DIFFERENTIAL) RECORDING CONTROLLER Instrument which controls pressure differential between two pipes or vessels by means of a valve.

PRESSURE, GAUGE Pressure in excess of atmospheric pressure.

PRESSURE INDICATING CONTROLLER Valve that has an indicator transmitter (or a remote mounted indicator) to control line or vessel pressure.

PRESSURE RECORDER Instrument that makes a permanent record of line or equipment pressure. The recorder can be local or board-mounted.

PRESSURE RECORDING CONTROLLER Instrument that is similar to a pressure recorder but has a control valve with a pneumatic signal hookup.

PRESSURE SWITCH A switch activated by pressure of the material.

PRESSURE VESSEL Any vessel (container) designed to withstand the internal pressure of gases or liquids.

PROCESS CHANGES Changes in the conditions of a process, such as flow, pressure, temperature.

PROCESS EQUIPMENT Any equipment which is not associated with utilities or services. Primary process, chemical change, or alteration of the material takes place within process equipment (including columns, reactors, and knockout drums). Various manufacturing operations are carried out in vessels referred to as process equipment. Materials are moved from one operation to another by mechanical devices that are also part of the process equipment.

PROCESS WATER Any water used in production or directly in the process line itself, such as a water-quench tank.

PROTECTION SADDLES Saddles used on insulated pipes that require free longitudinal movement.

PUMP Device which produces fluid flow against resistance or pressure.

PURGE To clean a line or piece of equipment in a piping system by flooding, washing out, or any method that removes foreign matter.

PWR (PRESSURIZED WATER REACTOR) The coolant in the PWR is kept at a high pressure. The reactor coolant heats up by flowing past the nuclear fuel. This heated coolant is then pumped into large exchangers or steam generators. The heat from the reactor coolant then heats a completely separate stream which also contains water. The second system is at a lower pressure and boils as it is heated, thus turning to steam which moves the turbine.

PYROMETER Device which measures high temperatures; usually associated with petrochemical processes.

QUENCHING Interjecting a cooler material into the discharge line, thereby lowering its temperature.

RADIATION Transfer of heat from one object to another by heat waves from the body of high temperature to the one of lower temperature without heating the air between the two objects.

RADIATION COOLANT Any cooling accomplished by radiation between fluids or solid surfaces.

RADIOACTIVITY Process by which certain nuclides, in a spontaneous disintegration in which energy is liberated, form new nuclides. This process is accompanied by the emission or radiation, such as alpha particles, beta particles, and gamma ray photons. Natural radioactivity is exhibited by more than 15 naturally occurring isotopes; artificial radioactivity can be induced in certain elements under controlled conditions.

REACTOR Vertical vessel or sphere in which a catalyst creates a chemical reaction that changes the molecular structure of the material being processed. Can be used in petrochemical refining to produce more of a specific end product.

REBAR Steel rod used for reinforcing concrete.

RECEIVER Simple tank with a pump.

RECIPROCATING PUMP Piston and cylinder device with a valving arrangement such that withdrawing the piston from the cylinder produces a vacuum on the inlet. When the piston is pushed into the cylinder, pressure is applied on the discharge.

REDUCER Coupling with a smaller opening at one end for reducing the size of the pipe opening (see Increaser).

REFINERY Petrochemical complex that utilizes crude oil in its original form as the primary process medium, converting it to a variety of products such as gasoline, tar, propane, fuel, oil, asphalt, and gas.

REFRIGERATION Reduction of temperature by means of mechanically driven thermodynamic processes in fluids.

REGULAR (STANDARD) Piping item regularly cataloged by a manufacturer.

REGULATOR Automatic valve which controls pressure or flow rate in a pipeline.

REINFORCING PAD Metal collar that is welded around a nozzle on a vessel in order to reinforce the opening area.

RESTRAINT Device used to restrain pipe from lateral and axial movement; there are one, two, and three-way types.

RETURN BEND U-shaped fitting used to reverse the direction of a pipe run.

REVISION Change in the existing design or drawing data.

RISER Any pipeline or piece of equipment which moves material upward.

ROD HANGER Type of hanger used extensively to carry vertical downward loads. It allows limited horizontal pipe movement in all directions.

RUN Main line of piping that has branch connections and side outlets.

SADDLE U-shaped piece of metal which provides support or reinforcement to insulated pipelines. It can also be used to establish a sloped run. Horizontal vessels or other pieces of equipment are supported on pedestals by means of thick steel saddles welded to the vessel which allow for the anchoring of equipment to the concrete pedestal. Saddles may also be used with guides.

SADDLE FLANGE Flange usually curved and riveted or welded to fit a boiler, tank, or other vessel and receive a threaded pipe. Also called a tank flange or boiler flange.

SAMPLE POINT Any area on a line, vessel, or piece of equipment where a sampling valve (or a branch line and accompanying equipment) is installed to enable the operator to sample the line material.

SARGOL Joint in which a lip is provided for welding to make the joint fluid-tight; mechanical strength is provided by bolted flanges.

SARLUN An improvement of the Sargol joint.

SCALE Mineral deposits in boilers and heat exchanger tubes.

SCALE, MODEL Same as a drawing scale. The scale determines the accurately reduced size of a model project, which is usually 3/8 in. = 1 ft (1:33 in metric) for petrochemical complexes and 1/2 in. = 1 ft (1:24 in metric) for power generation models.

SCALE PIPING SYSTEMS The full-scale piping system depicts all the pipes, elbows, valves, and other components in the model's full scale; the centerline piping system utilizes full-scale valving and components but not pipe runs. The centerline system uses 1/16-in. plastic-coated wire which can easily be formed by hand. Both systems indicate the true centerlines of all piping–automatically in the full-scale system and by the use of sizing sleeves in the centerline system.

SCHEDULE Measure of the relation of the wall thickness of pipe to the inside diameter.

SCHEDULE NUMBER Approximate value of the expression $1000 \times p/s$, where p is the service pressure and s is the allowable stress (in psi).

SCOPE MODEL One type of preliminary planning model. The purpose of the scope model is to evolve the most efficient process by designing the structures around the process, not the process around the structures. The scale for scope models should be no larger than 1/4 in. = 1 ft; common scales are 1/8 in. = 1 ft and 1 in. = 10 ft.

SCOPE SHEET Model specification sheet.

SCREWED END Pipe or fitting that is joined by threaded connections.

SCREWED FLANGE Flange that is attached to a pipe by a screwed connection.

SCRUBBER Device used for cleaning.

SEAMLESS Piercing and rolling of a solid billet or cupping a plate to form pipe.

SECTION, MODEL Vertical and horizontal splitting of the model into workable subunits to facilitate design, layout, and movement of the project.

SEISMIC Relating to vibration caused by an earthquake.

SEISMIC PIPE SUPPORT Pipe support hanger, restraint or snubber designed to withstand the piping load that may result from an earthquake.

SEPARATOR Device used to remove liquid from air compressors via a drain.

SERVICE FITTING A street tee or street ell having a male thread at one end.

SEWAGE SYSTEM Equipment or piping which conveys waste products, drainage, or other material to an appropriate treatment, storage, or disposal area.

SHELL Outer wall of a vessel, whether horizontal or vertical.

SHOE Metal piece attached (usually welded) to the underside of a pipe and resting on supporting steel. Used to reduce wear from sliding of lines subject to movement; also permits insulation to be applied to pipe, and can provide elevation from the support to allow for a light slope of the pipeline.

SHOP DRAWING Similar to a fabrication drawing but for shop fabrication only.

SHOP FABRICATION Fabrication completed in a shop as opposed to field construction. Shop fabrication has many advantages over field construction, offering a wider variety of production methods in a controlled environment.

SHORT SPRING Heavier spring than the medium type, giving 50 percent of the deflection a medium spring would provide under a given load.

SITE MODEL Model which provides a visual concept of the facility to be constructed in relationship to the surrounding area. After the site for a project has been selected, the site model can be started. In some cases, the site model may be used to evaluate sites. Site models require very little engineering information for the new project. Site models should include bridges, rail facilities, power right of ways, and utility hookups.

SKELP Plate prepared by forming and bending and welding into a pipe.

SKIRT Portion of a vertical vessel which extends from the vessel to the base and foundation.

SKIRT VENTS (SKIRT HOLES) Holes provided at the top of the skirt near the vessel bottom for ventilation.

SLEEPER Small low elevation concrete pier usually rectangular in shape and 12 to 24 in. in height used to support piping close to grade, especially large diameter piping.

SLIDE PLATE Metal plate upon which equipment is placed to separate it from the foundation or concrete; thus the equipment can alter its position depending on thermal movement, vibration, or bolting requirements.

SLIP-ON FLANGE Flange that slips onto a pipe and is welded in place.

SLURRY Flow medium which has solid material suspended in a liquid.

SNUBBER (SHOCK SUPPRESSOR, PIPE ARRESTOR) Device which absorbs shock forces which could damage the pipe system. Varieties are mechanical or hydraulic. Snubbers do not resist slow thermal movement.

SOCKET-WELDED FITTING OR VALVE Socket-end fitting or valve for low pressures and small diameters.

SOLID State of matter in which the relative motion of the molecules is restricted so that it has a definite shape and volume.

SOURCE NIPPLE Short length of heavy-walled pipe between high-pressure mains and the first valve of bypass, drain, or instrument connection.

SOUR WATER Any water which has an abnormal acid content.

SPECIAL Note stating that an item is different from the manufacturer's standard design.

SPECIFICATIONS List of design standards, codes, and procedures established by federal and state government vendors, and standards associations as guidelines.

SPECTACLE BLIND Double-flange unit that can be swiveled to either a blind position (closed) or an open position.

SPLITTER Device used to split the flow of liquid from a pipe into two streams.

SPOOL DRAWING Isometric or orthographic drawing used in industry to call out and dimension a pipeline unit for fabrication.

SPOOL SHEET Drawing in isometric or orthographic projection of a single pipeline section.

SPRAY POND Water-cooling device.

SPRING HANGER Support which allows variations in pipe position due to changes in temperature; often used for vertical lines.

SPRING SUPPORT (1) Variable load: spring hanger which responds only to the deadweight of a piping system and contents. This device allows both vertical and horizontal movement. Excellent for critical applications that need flexible support, this type of support has three variations based on travel distance. (2) Constant load: system of springs and lever arms used for high-temperature pipe installations where thermal movement may cause stress transfer of the system's load. The constant-load spring hanger balances the load and tension regardless of pipe movement (within certain established limits).

STANDARD Designation of cast iron flanges, fittings, and valves suitable for a maximum working steam pressure of 125 psig.

STANDARD WEIGHT Schedule of wrought iron pipe weights in common use.

STATIC PRESSURE Pressure of a nonflowing gas or liquid.

STEAM Vapor phase of water.

STEAM CRACKING Use of high temperature and steam in the cracking process of hydrocarbons.

STEAM, DRY Steam which is lacking suspended water particles.

STEAM REFINING Method using heat created from steam coils (open or closed) located at the bottom of a still. The process allows for low-temperature refining in the production of gasoline and naphtha. When open coils are used, the process is referred to as steam distillation because open steam is introduced directly into the charge. When closed coils are used, the process is termed steam refining because only the heat from the steam process is utilized. In steam refining, the primary concern is a product with a particular color and odor.

STEAM, SATURATED Steam in contact with the water from which it was generated; may be either wet or dry.

STEAM, SNUFFING Steam used to blanket fires.

STEAM, SUPERHEATED Steam heated above the temperature of the boiling point corresponding to the pressure.

STEAM TRACED Steam tubing that is wrapped around a pipe to keep line contents at a desired temperature.

STEAM, WET Steam carrying free water particles in suspension.

STEEL, FIREPROOFED Coating of special concrete placed on steel structures to prevent the steel from buckling from heat during a fire.

STEEL STRAP Simple support used for stationary and noninsulated pipelines in heating and common housing plumbing.

STEM Part of valve trim which moves the disk on and off the valve seat.

STILL Any closed vessel or chamber where heat can be introduced to a material in order to create vaporization. This process can take place with or without chemical decomposition, and the container or vessel is usually cylindrical. After the material has been vaporized, it is transferred to a cooling vessel or condenser where it can be returned to a liquid state and piped to the next process or storage facility.

STIRRUP U-shaped rebar used for structural strength in concrete foundations and other structures.

STRESS The load on an object, usually pipe or structural members, measured in pressure units (psi).

STRUCTURAL SHAPES Various steel items (I beams, T beams, channel beams, angle beams, pipe, plate and bar, stock).

STRUCTURAL STEEL DRAWING Drawing which shows column (stanchion) size and location, top of steel elevations, beam sizes, shapes, plate girders, platform drawings, grating, and concrete-enclosed steel.

STRUT Horizontal or angled structural member used primarily for longitudinal strength.

SUB HEADER Branch line from a large header or pipe.

SUPERHEAT Degree to which steam is at a higher temperature than its condensation point for the given pressure.

SUPPORT, FICTITIOUS Device used in models to support structure or piping that does not exist on the actual assembly.

SURGE Violent movement of fluid in a piping system caused by quick opening of valves or collection of condensate in pockets in a steam line; usually associated with steam or water hammer. Surges can cause carryover of the line material to vapor lines (puking).

SWAGING Process of reducing the ends of a pipe or tube section with rotating dies which are pressed intermittently or rotated against the pipe or tube end.

SWEATING Condensation of atmospheric water vapor onto pipes which are cooler than the surrounding air.

SWING ANGLE Rotation angle of the hanger rod caused by movement of pipe from cold to hot position.

SWING STRUT Pipe support component that can carry load in its axial direction while not restricting the pipe in any direction perpendicular to its axis within certain limits. A swing strut resists both tension and compression.

SYMBOL Graphic configuration that stands for a standard item such as a fitting, equipment, or valve.

TAGGING Procedure whereby all lines, coordinates, equipment, valves, and so forth are identified on the model by standard or special tags.

TAG, LINE Label used to identify pipes on a model.

TANK Basic process vessel used for the storage of raw materials, intermediates, and finished products. When modified, tanks can be used as dissolvers, precipitators, reactors, fermenters, and stills. Tanks are constructed in many sizes and shapes; the most common forms are vertical and horizontal right cylinders with end closures. Pressure tanks are fabricated as spheres or portions of spheres.

TANK CAR Railroad car (usually a vessel on wheels) used to transport raw or finished products in liquid, gas, or semisolid states.

TANKER Motorized vehicle for the transportation of petrochemical products on highways and roads. Can also refer to a large oceangoing ship; supertankers are considerably larger than the normal freighter.

TANK FARM Any area of a refinery or plant that is primarily used for the storage of liquid or gaseous products; usually a group of vessels or holding tanks.

TANK, HOLDING Any tank or vessel in which material is stored before it goes through processing.

TANK, HORIZONTAL Vessel, often cylindrical in shape, with closed ends. Head shapes are similar to those on vertical vessels or tanks. Usually supported below by saddles fastened to a foundation or floor or supported by slings hung from an overhead structure. Length of the horizontal tank should not be more than five times the diameter.

TANK, MULTICYLINDER Vessel with cylindrical shell segments and internal diaphragms with heads which may be partial multispheres.

TANK, MULTISPHERE Vessel used for storage of gas at pressures that would require prohibitive thickness in a single sphere.

TANK, PRESSURE Vessel used for storage of volatile materials. Can be constructed to withstand the maximum pressure developed in a process. Many shapes other than the upright cylinder are possible. A full spherical shape is sometimes used which is capable of withstanding higher pressures; for gasoline, the Horton spheroid is used. Ellipsoidal pressure tanks are available in large capacities. Cylindrical tanks with outward dished heads can serve for storage at medium and low pressures.

TANK, SETTLING Any vessel or tank which allows solids and liquids in the process material to separate gravitationally.

TANK, VERTICAL (VERTICAL VESSEL) Vertical right cylinder—the most common tank. Can be built to vertually any size and capacity required. Tanks used for storage in the open are covered. The roof may be self-supporting, but it is often supported by a central column and radial rafter. Tanks built at grade are of the flat-bottom type; outdoor elevated tanks may be flat-bottomed, supported by a suitable structure, or suspended from the sides. Suspended bottoms may be hemispherical, hemiellipsoidal, conispherical, or conical. Outdoor flat-bottom tanks at grade may rest on a foundation of sand or both sand and a curbing of concrete slightly larger than the tank diameter. Indoor storage tanks are usually flat-bottomed. Pressurized tanks may have dished heads.

TAP Tool used for forming female (internal) threads.

TEE Three-way fitting shaped like the letter *T*.

TEE JOINT Joint between two members located at right angles to each other forming a *T*.

TEMPERATURE, ABSOLUTE Temperature measured from absolute zero.

TEMPERATURE ALARM Temperature-sensitive device with an alarm which warns against rise or drop in the line or vessel temperature.

TEMPERATURE, AMBIENT Temperature of the medium surrounding an object; prevailing temperature in the immediate vicinity.

TEMPERATURE CONTROLLER Instrument which regulates pipeline or vessel temperature by a control valve which is actuated by pneumatic signals from a transmitter.

TEMPERATURE CONTROL VALVE Control valve that is regulated by temperature fluctuations.

TEMPERATURE DIFFERENTIAL CONTROLLER Temperature control valve combined with a temperature recording instrument for the controlled line or vessel which together control temperature differences between two vessels or pipelines.

TEMPERATURE ELEMENT Instrument that is thermocoupled without connections for measuring a line medium's temperature.

TEMPERATURE INDICATOR Device used for measuring temperature: (1) a locally mounted dial; (2) remote-mounted dial capillary tubes; or (3) electric thermocouple.

TEMPERATURE RECORDER Instrument which records a permanent continuous history of pipe or equipment temperature.

TEMPERATURE RECORDING CONTROLLER Device which records and regulates the temperature of a vessel or line by a pneumatic signal to control valve and recorder.

THERMOCOUPLE Temperature-sensing device consisting of two dissimilar metals joined together.

THREADER Tool used for cutting male threads on the end of a pipe (see Die).

THERMAL CRACKING Refining process through which hydrocarbon molecules are altered by the application of heat and sometimes with the aid of a catalyst. Time, feed, temperature, and pressure are important considerations. Thermal cracking involves the ability to decompose, rearrange, recombine, or create new molecular structures of hydrocarbon molecules. This process is essential in the production of gasoline because the normal by-products of crude oil contain only a small portion of gasoline. Most of the processes in a petrochemical complex are forms of thermal cracking.

THERMAL MOVEMENT Any expansion or contraction of the piping system due to a change in temperature associated with the line material or external conditions.

THERMAL SHOCK Shock caused by sudden rise or drop in temperature of the fluid in contact with a valve.

THERMAL STRESS Forces built up as anchored piping undergoes changes in temperature during operation. A rise in temperature causes expansion; as the pipe cools during downtime or between operations, the pipeline contracts.

THERMOSIPHON Use of natural circulation to distribute liquid without mechanical devices such as pumps. The system operates on the principle that warm substances rise (convection). The water heated therefore, is lighter and will flow up into the tank while cooler water on the bottom of the tank, being heavier, flows down into the collector and can be heated.

THOMAS FORMULA Formula for pipe sizing for insulated steamlines: $d = 4.5(1 + .00025S)\sqrt{W/P}$

THROTTLING Controlled variable restriction of flow rate usually accomplished by a regulator.

TIE Arrangement of one or more rods, bars, and the like to restrain movement of piping. Also refers to auxiliary rebar rod used to position and reinforce the main bar.

TOPPING Distillation of crude oil in order to remove light fractions.

TOWER Column or vertical vessel which increases the degree of separation that can be obtained during the fractionation and distillation of oil in a still. Towers used primarily for separation by fractionation are called for where accurate separation is absolutely necessary (as in the production of gasoline naphthas, which must meet rigid specifications). Towers used primarily for roughly dividing vapors from a still into different liquid portions (bubble towers) operate by partial condensation, not fractionation.

TOWER COOLING Device which removes heat from returned cooling water so that it may be reused.

TOWER, FORCED-DRAFT Cooling tower which forces air through the water being cooled by utilizing a horizontal-shaft fan on the side of the tower.

TOWER, INDUCED-DRAFT Cooling tower which draws air through the water being cooled by means of vertical fans.

TOWER, NATURAL-DRAFT Cooling tower which utilizes natural wind motions instead of induced-draft or forced-draft methods.

TRACING Using electrical or steam tubing to keep a pipeline or equipment at a constant predetermined temperature. Tracing methods usually require insulation of both tracer and pipeline.

TRANSMITTER Device which transmits instrument readings or pneumatic signals for recording and control.

TRAPEZE Pipe hanger arrangement formed by two suspended hanger rods and a crossbar which supports the pipe.

TRAY Device associated with vertical columns used in fractionation which provides a bafflelike series of protrusions in the tower where vapors and liquid mix.

TREPANNING Removal by destructive means of a small section of piping for evaluation of a weld and base-metal soundness.

TRIM Certain internal valve parts such as seat rings, disks, stems, and repacking seat bushings.

TRUSS Fabricated structure made of structural steel or piping that distributes weight from an overhead load such as that for a building or other structure.

TUBE BUNDLE Tubular (coil) part of a heat exchanger.

TUBING System of small-diameter, lightweight (usually copper, brass, or plastic) pipes where OD always equals tubing size.

TURBINE, DOUBLE-FLOW Turbine which splits steam in two directions before passing over the turbine blades.

TURBINE, SINGLE-FLOW Turbine which utilizes steam in a single path over the turbine blades.

TURBINE, STEAM Prime mover of electrical generators whose power is derived from high-velocity steam passing through nozzles and directed at the turbine's rotor blades.

TURBULENCE Nonparallel flow in a pipe; usually caused by obstructions, changes of direction, or other internal deviations from a straight run of pipe.

TURNBUCKLE Simple position-adjustment device for use on pipe supports.

UNION Three-piece flange-type screwed fitting used to join lengths of pipe to permit easy opening of a line.

UNION FITTING Fitting with a union at one or more ends.

UNIT FABRICATION Pipe components formed, welded, and produced by a variety of methods to create assemblies that can be shipped and installed in a complete section (unit) in the field.

UTILITY Any service necessary for the functioning of a plant: electricity, gas, water, fire protection, sewage disposal, steam, fuel, oil, instrument air, utility air, cooling water, drains, condensate, flare system, blowdown systems. In general, anything that is not directly associated with the primary processes in a plant can be considered a utility.

UTILITY AIR Air line used in cleanup operations, to drive motors, and other service functions.

UTILITY STEAM Steam that is generated and used for service in a plant.

VALVE Mechanical device used to interrupt or regulate flow in a piping system.

VALVE, FLOW-DIVIDING Valve which divides the flow from a single source into two or more branches.

VAPOR Gaseous state of any substance that is liquid or solid under ordinary conditions.

VAPOR LOCK Blocking of liquid flow by air or vapor trapped in a pipe.

VARIABLE-SPRING HANGER SUPPORT Spring support device (short, medium, or double spring). The load it exerts on the pipe depends on the amount of compression (loaded strength) in the spring (see Cold Load, Hot Load, Variability).

VARIABILITY Change in the load of a spring hanger during operation of the piping system.

$$\text{variability} = \frac{\text{cold load} - \text{hot load}}{\text{hot load}} \times 100\%$$

or

$$\text{variability} = \frac{\text{(required travel) (spring constant)}}{\text{hot load}} \times 100\%$$

VELOCITY Speed of flow in a pipeline (usually in feet per second).

VENDOR Firm or person that supplies material and equipment.

VENDOR PRINT Drawing provided by the manufacturing company for equipment used in piping and other systems, including pumps, instruments, economizers, heat exchangers, vessels, and also smaller items such as valves, bolts, and gaskets. Vendor prints show the information needed to draw and locate the equipment on the piping drawings. Also referred to as manufacturing drawings or vendor drawings.

VENTURI PORT (REDUCED PORT) Any valve in which the internal port is smaller than the pipeline bore.

VESSEL Any container used in conjunction with a piping system to hold, transform, or store the medium.

VISCOSITY Internal friction in a fluid; its "thickness" or resistance to flow.

VISCOSITY INDEX Measure of the viscosity-temperature characteristics of a fluid compared to other fluids.

VISCOUS MATERIAL Material having a high resistance to flow.

WATER HAMMER Shock or stress caused by liquid pressure waves that move through piping and come in contact with a change-in-direction fitting or closed piece of equipment. Can be caused by rapidly closing valves or (in the case of steam lines) condensate trapped in various portions of a valve body or in a line which is not sloped properly.

WEDGE Disk shape for certain types of gate valves.

WEIR PORT Particular shape of a valve body interior.

WELDED FITTING Forged or wrought steel elbow, tee, or similar piece for connection by welding to each other or to a pipe.

WELDED JOINT Union of two or more members produced by the application of a welding process.

WELDING END End of a fitting, pipe, or valve that is joined to other piping elements by welding.

WELDING-END VALVE Valves that are to be welded onto a pipeline or pipe assembly.

WELD-NECK FLANGE Flange with integral extended neck for welding to pipe.

WINTERIZING Insulation, jacketing, or heat tracing of piping systems including equipment and components.

WROUGHT PIPE Pipe worked as in the process of forming furnace-welded pipe from skelp; distinguished from cast pipe.

ZEOLITE Chemical used to demineralize water.

INDEX

W